Nonlinear Continuum Mechanics and Large Inelastic Deformations

SOLID MECHANICS AND ITS APPLICATIONS
Volume 174

---



*Aims and Scope of the Series*

The fundamental questions arising in mechanics are: *Why?*, *How?*, and *How much?* The aim of this series is to provide lucid accounts written by authoritative researchers giving vision and insight in answering these questions on the subject of mechanics as it relates to solids.

The scope of the series covers the entire spectrum of solid mechanics. Thus it includes the foundation of mechanics; variational formulations; computational mechanics; statics, kinematics and dynamics of rigid and elastic bodies: vibrations of solids and structures; dynamical systems and chaos; the theories of elasticity, plasticity and viscoelasticity; composite materials; rods, beams, shells and membranes; structural control and stability; soils, rocks and geomechanics; fracture; tribology; experimental mechanics; biomechanics and machine design.

The median level of presentation is the first year graduate student. Some texts are monographs defining the current state of the field; others are accessible to final year undergraduates; but essentially the emphasis is on readability and clarity.

For other titles published in this series, go to
www.springer.com/series/6557

Yuriy I. Dimitrienko

# Nonlinear Continuum Mechanics and Large Inelastic Deformations

Springer

Prof. Yuriy I. Dimitrienko
Bauman Moscow State Technical University
2nd Baumanskaya St. 5
105005 Moscow
Russia
dimit@serv.bmstu.ru

ISSN 0925-0042
ISBN 978-94-007-0033-8 e-ISBN 978-94-007-0034-5
DOI 10.1007/978-94-007-0034-5
Springer Dordrecht Heidelberg London New York

Library of Congress Control Number: 2010938719



*Cover design*: SPi Publisher Services

Printed on acid-free paper

Springer is part of Springer Science+Business Media (www.springer.com)

# Preface

Nonlinear continuum mechanics is the kernel of the general course 'Continuum Mechanics', which includes kinematics of continua, balance laws, general nonlinear theory of constitutive equations, relations at singular surfaces. Moreover, in the course of nonlinear continuum mechanics one also considers the theory of solids at finite (arbitrary) deformations. This arbitrariness of deformations makes the equations describing the behavior of continua extremely complex – nonlinear (so that sometimes the term '*strongly nonlinear*' is used), as the relationships contained in them cannot always be expressed in an explicit analytical way. If we drop the condition of the arbitrariness of continuum deformations and consider only infinitesimal deformations – usually the deformations till 1%, then the situation changes: the equations of continuum mechanics can be linearized. Hence for solving the applied problems one can exploit the wide range of analytical and numerical methods. However, many practical tasks demand the analysis not of the infinitesimal, but just the arbitrary (large) deformations of bodies, for example, such tasks include the rubber structure elements design (shock absorbers, gaskets, tires) for which the ultimate deformations can reach 100% and even higher. The various tasks of metal working under high pressure also belong to that class of problems, where large plastic deformations play a significant role, as well as the dynamical problems of barrier breakdown with a striker (aperture formation in the metal barrier while the breakdown is an example of large plastic deformations). Within this class of problems one can also find many problems of ground and rock mechanics, where there usually appears the need to consider large deformations, and modelling the processes in biological systems such as the functioning of human muscular tissue, and many others.

The theory of infinitesimal deformations of solids appeared in the XVII century in the works by Robert Hooke, who formulated one of the main assumptions of the theory: stresses are proportional to strains of bodies. Translating the assertion into mathematical language, this means that relations between stresses and displacements gradients of bodies are linear. Nowadays the theory of infinitesimal deformations is very deeply and thoroughly elaborated. On the different parts of this theory such as elasticity theory, plasticity theory, stability theory and many others there are many monographs and textbooks.

But the well-known Hooke's law does not hold for finite (or large) deformations: the basic relations between stresses and displacement gradients become 'strongly non-linear', and they cannot always be expressed analytically. The basis of finite deformations theory was laid in the XIX century by the eminent scientists A.L. Cauchy, J.L. Lagrange, L. Euler, G. Piola, A.J.C. Saint-Venant, G.R. Kirchhoff, and then developed by A.E.H. Love, G. Jaumann [28], M.A. Biot, F.D. Murnaghan [41] and other researchers. The works by M. Mooney and R.S. Rivlin written in the 1940s of the XX century contributed much to the formation of finite deformations theory as an independent part of continuum mechanics. The fundamental step was made in 1950–1960s of the XX century by the American mechanics school, first of all by B.D. Coleman [9], W. Noll and C. Truesdell [43, 54–56], who considered the nonlinear mechanics from the point of view of the formal mathematics. According to D. Hilbert, they introduced the axiomatics of nonlinear mechanics which structured the system of accumulated knowledge and made it possible to formulate the main directions of the further investigations in this theory. Together with R.S. Rivlin and A.J.M. Spencer [10, 13, 51] they elaborated the special mathematical apparatus for formulation of relationships, generalizing Hooke's law for finite deformations, namely the theory of nonlinear tensor functions. And also the tensor analysis widely used in continuum mechanics was considerably adapted to the problems of nonlinear mechanics. Equations of continuum mechanics got the invariant (i.e. independent of the choice of a reference system) form. The further development of this direction was made by A.C. Eringen, A.E. Green, W. Zerna, J.E. Adkins and others [1–7, 11, 12, 14–27, 29, 30, 32–35, 38–40, 42, 47–50, 52, 53, 57–60].

The role of Russian mechanics school in the development of contemporary nonlinear continuum mechanics principles is also quite substantial. In 1968 the first edition of the fundamental two-part textbook '*Continuum Mechanics*' by L.I. Sedov was published, which is still one of the most popular books on continuum mechanics in Russia. Outstanding results in the theory of finite elastic deformations were obtained by A.I. Lurie [36, 37], who wrote the principal monograph on the nonlinear theory of elasticity and systemized in it the problem classes of the theory of finite elastic deformations allowing for analytical solutions. Also the considerable step was made by K.F. Chernykh [8], who developed the theory of finite deformations for anisotropic media and elaborated the methods for solving the problems of nonlinear theory of shells and nonlinear theory of cracks. One can also mention the works by mechanics scientists: B.E. Pobedrya, V.I. Kondaurov, V.G. Karnauhov, A.A. Pozdeev, P.V. Trusov, Yu.I. Nyashin and many others who made considerable contributions to the theory of viscoelastic, elastoplastic and viscoplastic finite deformations.

This book is based on the lectures which the author has been giving for many years in Moscow Bauman State Technical University. The book has several fundamental traits:

1. It follows the mathematical style of course exposition, which assumes the usage of axioms, definitions, theorems and proofs.
2. It applies the tensor apparatus, mostly in the indexless form, as the latter combined with the special skills is very convenient in usage, and does not shade

the physical essence of the laws, and permits proceeding to any appropriate coordinate system.
3. It uses the divergence form of dynamic equations of deformation compatibility, that made it at last possible to write the complete system of balance laws of nonlinear mechanics in a single generalized form.
4. The theory of constitutive equations being the key part of nonlinear mechanics is for the first time exposed with the usage of all energetic couples of tensors, which were established by R. Hill [26] and K.F. Chernykh [8] and ordered by the author [12], and also with quasi-energetic couples of tensors found by the author [12].
5. To derive constitutive equations of nonlinear continuum mechanics, the author applied the theory of nonlinear tensor functions and tensor operators, elaborated by A.J.M. Spencer, R.S. Rivlin, J.L. Ericksen, V.V. Lokhin, Yu.I. Sirotin, B.E. Pobedrya, the author of this book and others.
6. The bases of theories of large elastic, viscoelastic and plastic deformations are explored from the uniform position.
7. The book uses a 'reader-friendly' style of material exposition, which can be characterized by the presence of quite detailed necessary mathematical calculations and proofs.

The axiomatic approach used in this book differs a bit from the analogous ones suggested by C. Truesdell [56] and other authors. The system of continuum mechanics axioms in the book is composed so as to minimize their total number, and give each axiom a clear physical interpretation. That is why the axioms by C. Truesdell connected with the logic relations between bodies are not included in the general list, the axioms on the bodies' mass are united into one axiom, the mass conservation law, and analogously the axioms on the existence of forces and inertial reference systems are united into one axiom, the momentum balance law. Though, the last axiom is split into the two parts: first the Sect. 3.2 considers the case of inertial reference systems, and then Sect. 4.10 deals with non-inertial ones. Unlike the axiomatics by C. Truesdell [56], in this book the axiom system includes so called principles of constitutive equations construction which play a fundamental role in the formation of a continuum mechanics equations system.

The axiomatic approach to the exploration of continuum mechanics possesses at least one merit – it permits the separation of all the values into two categories: primary and secondary. These are introduced axiomatically and consequently within the continuum mechanics there is no need to substantiate their appearance. The secondary category includes combinations of the first category's values. The axiomatic approach allows us also to distinguish from continuum mechanics statements between the definitions and corollaries of them (theorems); this is extremely useful for the initial acquaintance with the course.

To get acquainted with the specific apparatus of tensor analysis the reader is recommended to use the author's book '*Tensor Analysis and Nonlinear Tensor Functions*' [12], which uses the same main notations and definitions. All the references to the tensor analysis formulas in the text are addressed to the latter book.

This book covers the fundamental classical parts of nonlinear continuum mechanics: kinematics, balance laws, constitutive equations, relations at singular surfaces, the basics of theories of large elastic deformations, large viscoelastic deformations and large plastic deformations. Because of limits on space, important parts such as the theory of shells at large deformations, and the theory of media with phase transformations were not included in the book.

I would like to thank Professor B.E. Pobedrya (Moscow Lomonosov State University), Professor N.N. Smirnov (Moscow Lomonosov State University) and Professor V.S. Zarubin (Bauman Moscow State Technical University) for fruitful discussions and valuable advice on different problems in the book.

I am very grateful to Professor G.M.L. Gladwell of the University of Waterloo, Canada, who edited the book and improved the English text.

I also thank my wife, Dr. Irina D. Dimitrienko (Bauman Moscow State Technical University), who translated the book into English and prepared the camera-ready typescript.

I hope that the book proves to be useful for graduates and post-graduates of mathematical and natural-scientific departments of universities and for investigators, academic scientists and engineers working in solid mechanics, mechanical engineering, applied mathematics and physics. I hope that the book is of interest also for material science specialists developing advanced materials.

Russia *Yuriy Dimitrienko*

# Contents

# Chapter 1
# Introduction: Fundamental Axioms of Continuum Mechanics

Continuum mechanics, including nonlinear continuum mechanics, studies the behavior of *material bodies* or *continua*. We can mathematically define a *body* as follows: it is a set $\mathcal{B}$ consisting of elements $\mathcal{M}$ called *material points*. The concept of a material point in continuum mechanics is primary, i.e. axiomatic, as is the concept of a geometrical point in elementary geometry.

The mathematical description of a body $\mathcal{B}$ in continuum mechanics starts from the following definition.

**Definition 1.1.** A material body $\mathcal{B}$, for which there is a one-to-one correspondence $\widetilde{\mathcal{W}}$ between each material point $\mathcal{M} \in \mathcal{B}$ and its image in some metric space $\mathcal{X}$, i.e. $\widetilde{\mathcal{W}} : \mathcal{B} \longrightarrow \widetilde{\mathcal{W}}(\mathcal{B}) \subset \mathcal{X}$, or

$$\mathbf{a} = \widetilde{\mathcal{W}}(\mathcal{M}), \quad \mathcal{M} \in \mathcal{B}, \quad \mathbf{a} \in \mathcal{X}, \tag{1.1}$$

is called a *continuum*.

The set of all continua $\mathcal{B}$ is called the *universe* $\mathcal{U}$. The definition of one-to-one correspondence one can find, for example, in [31]. A *metric space* $\mathcal{X}$ is characterized by the presence of a distance function $l(\mathcal{M}, \mathcal{N})$, with the help of which one can measure the distance between any two points $\mathcal{M}$ and $\mathcal{N}$ of a body $\mathcal{B}$ [31]. The one-to-one correspondence between material points and points of a metric space $\mathcal{X}$ allows us to consider not the material body itself but only its image. Below, we draw no distinction between a material point and its image.

Definition 1.1 should be complemented with three fundamental axioms.

**Axiom 1 (on continuity).** *The image $\widetilde{\mathcal{W}}(\mathcal{B})$ of a body $\mathcal{B}$ is a continuous set (a continuum) in space $\mathcal{X}$.*

The concept of a *continuum* was considered in [31].

Axiom 1 introduces the main model of continuum mechanics, namely a continuous set, that is an idealization of real bodies consisting of discrete atoms and molecules. Physically, there is a limiting distance $l_{\min}$ such that for $l \leqslant l_{\min}$ the neighborhood of a material point $\mathcal{M} \in \mathcal{B}$ is empty. However, in a continuum (and in its image), due to the properties of a continuous set [31], any infinitesimal $\varepsilon$-neighborhood $U_\varepsilon(A)$ of a point $A \in \mathcal{W}(\mathcal{B}) \subset \mathcal{X}$ contains an infinite number

Y.I. Dimitrienko, *Nonlinear Continuum Mechanics and Large Inelastic Deformations*, Solid Mechanics and Its Applications 174, DOI 10.1007/978-94-007-0034-5_1,

of other points of this medium. In this sense, a continuum is only a model of a real body; and computed results in continuum mechanics referred to real bodies for distances between points $l \leqslant l_{\min}$ may prove to be incorrect. Nevertheless, Axiom 1 is of great importance: it allows us to apply the methods of tensor analysis in metric spaces.

**Axiom 2 (on an Euclidean space).** *As a metric space $\mathcal{X}$, in which continua are considered, we can choose a three-dimensional Euclidean metric space $\mathcal{E}_3^a$, i.e $\mathcal{X} = \mathcal{E}_3^a$.*

An *Euclidean point space* $\mathcal{E}_3^a$ (called also an affine Euclidean space) is a set of points $\mathcal{M}, \mathcal{N}, \ldots$, where there exists a mapping, uniquely assigning each ordered pair of points $\mathcal{M}, \mathcal{N}$ to an element (vector) $\mathbf{y}$ of Euclidean vector space $\mathcal{E}_3$ adjoined to $\mathcal{E}_3^a$ (it is also denoted by $\overrightarrow{\mathcal{M}\mathcal{N}} = \mathbf{y}$). In *Euclidean space* $\mathcal{E}_3$, there are operations of addition of elements (vectors) $\mathbf{x} + \mathbf{y}$, multiplication by a real number $\lambda\mathbf{y}$, $\lambda \in \mathbb{R}$ and *scalar product* of vectors [31], that can be given by a *metric matrix* $g_{ij}$ relative to some *basis* $\mathbf{e}_i$ in $\mathcal{E}_3$: $\mathbf{x} \cdot \mathbf{y} = g_{ij} x^i y^j$, where $x^i$ and $y^j$ are *coordinates of the vectors* with respect to the same basis: $\mathbf{x} = x^i \mathbf{e}_i$, $\mathbf{y} = y^i \mathbf{e}_i$. Moreover, in $\mathcal{E}_3$ we can introduce the vector product of vectors $\mathbf{a} = a^i \mathbf{e}_i$ and $\mathbf{b} = b^i \mathbf{e}_i$ as follows:

$$\mathbf{c} = \mathbf{a} \times \mathbf{b} = \sqrt{g}\, \epsilon_{ijk} a^i b^j \mathbf{e}^k = \frac{1}{\sqrt{g}}\, \epsilon^{ijk} a_i b_j \mathbf{e}_k. \tag{1.2}$$

Here $\epsilon_{ijk}$ and $\epsilon^{ijk}$ are the *Levi–Civita symbols* [12] (they are zero, if at least two of the indices $i, j, k$ are coincident; and they are equal to 1, if the indices $i, j, k$ form an even permutation, and are equal to $(-1)$ if the indices form an odd permutation), and $g = \det\,(g_{ij})$ is the determinant of the metric matrix.

Thus, Axiom 2 allows us to describe materials points of a continuum with the help of instruments of Euclidean point spaces: in the space $\mathcal{E}_3^a$ we can introduce a *rectangular Cartesian coordinate system* $O\bar{\mathbf{e}}_i$, being common for all continua and consisting of some point $O$ (the origin of the coordinate system) and an orthonormal basis $\bar{\mathbf{e}}_i$. In this system $O\bar{\mathbf{e}}_i$, every material point $\mathcal{M}$ is uniquely assigned to its *radius-vector* $\mathbf{x} = \overrightarrow{O\mathcal{M}}$. The distance $l(\mathcal{M}, \mathcal{N})$ between points $\mathcal{M}$ and $\mathcal{N}$ is measured by *the length of vector* $\overrightarrow{\mathcal{M}\mathcal{N}}$

$$l(\mathcal{M}, \mathcal{N}) = |\overrightarrow{\mathcal{M}\mathcal{N}}| = (\overrightarrow{\mathcal{M}\mathcal{N}} \cdot \overrightarrow{\mathcal{M}\mathcal{N}})^{1/2} = |\mathbf{y}| = (\mathbf{y} \cdot \mathbf{y})^{1/2}.$$

The length $l(\mathcal{M}, \mathcal{N})$ specifies a metric in the space $\mathcal{E}_3^a$; such a space is called a *metric space*. In a metric space we define the concept of a domain $V$, and in an Euclidean point space, the concepts of a plane and a straight line. In addition, in a metric space there are concepts of convergence of a point sequence, continuity and differentiability of functions etc.; Axiom 2 allows us to apply these concepts of mathematical analysis to continua.

Due to isomorphism (one-to-one correspondence) of Euclidean point spaces of the same dimension as a space $\mathcal{E}_3^a$, we can always consider the *space of elementary*

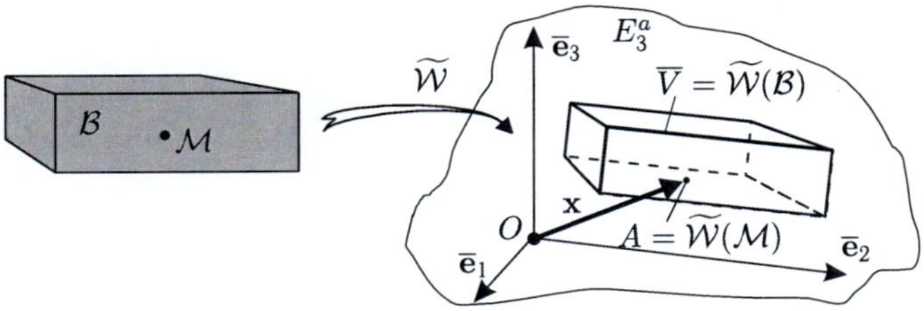


**Fig. 1.1** A real body $\mathcal{B}$ and its image $\widetilde{\mathcal{W}}(\mathcal{B})$ in $E_3^a$

*geometry* $E_3^a$, where points are usual geometric points, and vectors are directed straight-line segments in the space. The space $E_3^a$ allows us to show different objects of continuum mechanics geometrically.

*Example 1.1.* Figure 1.1 shows schematically a real material body $\mathcal{B}$ and its image $\widetilde{\mathcal{W}}(\mathcal{B})$ in the space $E_3^a$. □

Consider the pair $(\mathcal{M}, t)$, where $\mathcal{M} \in \mathcal{B}$, and $t \in \mathbb{R}_{+0}$ is some nonnegative real number. This pair is an element of the Cartesian product of the sets $\mathcal{B} \times \mathbb{R}_{+0}$.

**Axiom 3 (on the existence of absolute time).** *For every body $\mathcal{B}$ there exists a mapping*

$$\mathcal{W}: \quad \mathcal{B} \times \mathbb{R}_{+0} \longrightarrow \bar{V} \subset \mathcal{E}_3^a,$$

*in the form of a function*

$$A = \mathcal{W}(\mathcal{M}, t), \quad \mathcal{M} \in \mathcal{B}, \quad t \in \mathbb{R}_{+0}, \quad A \in \bar{V} \subset \mathcal{E}_3^a. \tag{1.3}$$

The parameter $t$ is called the *absolute time* (or simply *time*).

Notice that both Axioms 1 and 3 establish relations between points $\mathcal{M}$ and $A$. To avoid the ambiguity one should assume that the following *consistency condition* is satisfied: the mapping (1.1) coincides with (1.3) at some time $t = t_1$

$$\widetilde{\mathcal{W}}(\mathcal{M}) = \mathcal{W}(\mathcal{M}, t_1) \quad \forall \mathcal{M} \in \mathcal{B}. \tag{1.4}$$

Axiom 3 allows us to describe the *motion of a body* $\mathcal{B}$, which is defined as changes of the radius-vectors $\mathbf{x} = \overrightarrow{O\mathcal{M}}$ of material points $\mathcal{M}$ of the body in a coordinate system $O\bar{\mathbf{e}}_i$ being common for all $\mathcal{M} \in \mathcal{B}$ with time $t$.

For the same material point $\mathcal{M}$ at different times $t_1$ and $t_2$ we have, in general, distinct radius-vectors $\mathbf{x}_1$ and $\mathbf{x}_2$, respectively, in the system $O\bar{\mathbf{e}}_i$.

*Example 1.2.* Figure 1.2 shows the motion of a real body $\mathcal{B}$ and its images $\mathcal{W}(\mathcal{B}, t_1)$ and $\mathcal{W}(\mathcal{B}, t_2)$ in the space $E_3^a$ at times $t_1$ and $t_2$, respectively. □

The absolutism of time means that the time $t$ is independent of the radius-vector $\mathbf{x}$ of a point $\mathcal{M}$ in the coordinate system $O\bar{\mathbf{e}}_i$, i.e., in physical terms, the time varies

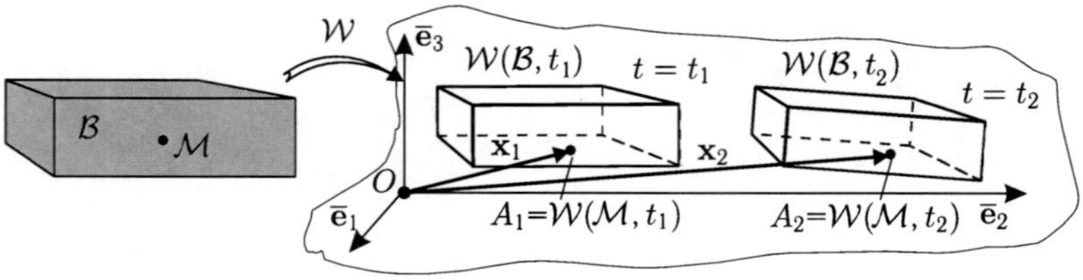


**Fig. 1.2** The motion of a body $\mathcal{B}$

in the same course for all material points $\mathcal{M}$. In a physical sense, this axiom remains valid only for the motion of bodies with speeds which are considerably smaller than the speed of light; otherwise, relativistic effects become essential, and Axiom 3 ceases to describe actual processes adequately. Relativistic effects are not considered in this work.

# Chapter 2
# Kinematics of Continua

## 2.1 Material and Spatial Descriptions of Continuum Motion

### *2.1.1 Lagrangian and Eulerian Coordinates: The Motion Law*

Let us consider a continuum $\mathcal{B}$. Due to Axiom 2, at time $t = 0$ there is a one-to-one correspondence between every material point $\mathcal{M} \in \mathcal{B}$ and its radius-vector $\overset{\circ}{\mathbf{x}} = \overrightarrow{O\mathcal{M}}$ in a Cartesian coordinate system $O\bar{\mathbf{e}}_i$. Denote Cartesian coordinates of the radius-vector by $\overset{\circ}{x}{}^i$ ($\overset{\circ}{\mathbf{x}} = \overset{\circ}{x}{}^i \bar{\mathbf{e}}_i$) and introduce curvilinear coordinates $X^i$ of the same material point $\mathcal{M}$ in the form of some differentiable one-to-one functions

$$\overset{\circ}{x}{}^i = \overset{\circ}{x}{}^i(X^k). \tag{2.1}$$

Since $\overset{\circ}{\mathbf{x}} = \overset{\circ}{x}{}^i \bar{\mathbf{e}}_i$, the relationship (2.1) takes the form

$$\overset{\circ}{\mathbf{x}} = \overset{\circ}{\mathbf{x}}(X^k). \tag{2.2}$$

Let us fix curvilinear coordinates of the point $\mathcal{M}$, and then material points of the continuum $\mathcal{B}$ are considered to be numbered by these coordinates $X^i$. For any motion of the continuum $\mathcal{B}$, coordinates $X^i$ of material points are considered to remain unchanged; they are said to be 'frozen' into the medium and move together with the continuum. Coordinates $X^i$ introduced in this way for a material point $\mathcal{M}$ are called *Lagrangian* (or *material*).

Due to Axiom 3, at every time $t$ there is a one-to-one correspondence between every point $\mathcal{M} \in \mathcal{B}$ with Lagrangian coordinates $X^i$ and its radius-vector $\mathbf{x} = \overrightarrow{O\mathcal{M}}$ with Cartesian coordinates $x^i$, where $\mathbf{x}$ and $x^i$ depend on $t$. This means that there is a connection between Lagrangian $X^i$, and the Cartesian $x^i$ coordinates of point $\mathcal{M}$ and time, i.e. there exist functions in the form (1.3)

$$x^i = x^i(X^k, t) \quad \forall X^k \in V_X. \tag{2.3}$$

Y.I. Dimitrienko, *Nonlinear Continuum Mechanics and Large Inelastic Deformations*, Solid Mechanics and Its Applications 174, DOI 10.1007/978-94-007-0034-5_2,

These functions determine a motion of the material point $\mathcal{M}$ in the Cartesian coordinate system $O\bar{\mathbf{e}}_i$ of space $\mathcal{E}_3^a$. The relationships (2.3) are said to be the *law of the motion of the continuum* $\mathcal{B}$.

Coordinates $x^i$ in (2.3) are called *Eulerian* (or *spatial*) coordinates of the material point $\mathcal{M}$.

Since $\mathbf{x} = x^i\bar{\mathbf{e}}_i$ and the coordinate system $O\bar{\mathbf{e}}_i$ is the same for all times $t$, the equivalent form of the motion law follows from (2.2):

$$\mathbf{x} = \mathbf{x}(X^k,\ t). \tag{2.4}$$

Since the consistency conditions (1.4) must be satisfied, from (2.2) and (2.4) we get the relationships

$$\mathbf{x}(X^k,\ 0) = \overset{\circ}{\mathbf{x}}(X^k), \quad x^i(X^k,\ 0) = \overset{\circ}{x}{}^i(X^k). \tag{2.5}$$

Here the initial time $t = 0$ is considered as the time $t_1$ in (1.4), because just at time $t = 0$ we introduced Lagrangian coordinates $X^i$ of point $\mathcal{M}$.

Unless otherwise stipulated, functions (2.3) are assumed to be regular in the domain $V_X \subset \mathbb{R}^3$ for all $t$, thus there exist the inverse functions

$$X^k = X^k(x^i,\ t) \quad \forall x^i \in V_x \subset \mathbb{R}^3.$$

The closed domain $\overset{\circ}{V} = \mathcal{W}(\mathcal{B},\ 0)$ in a fixed coordinate system $O\bar{\mathbf{e}}_i$, which is occupied by continuum $\mathcal{B}$ at the initial time $t = 0$, is called the *reference configuration* $\overset{\circ}{\mathcal{K}}$, and the domain $V = \mathcal{W}(\mathcal{B}, t)$ occupied by the same continuum $\mathcal{B}$ at the time $t > 0$ is called the *actual configuration* $\mathcal{K}$.

Figure 2.1 shows a geometric picture of the motion of a continuum from the reference configuration $\overset{\circ}{\mathcal{K}}$ to the actual one $\mathcal{K}$ at time $t$ in space $E_3^a$.

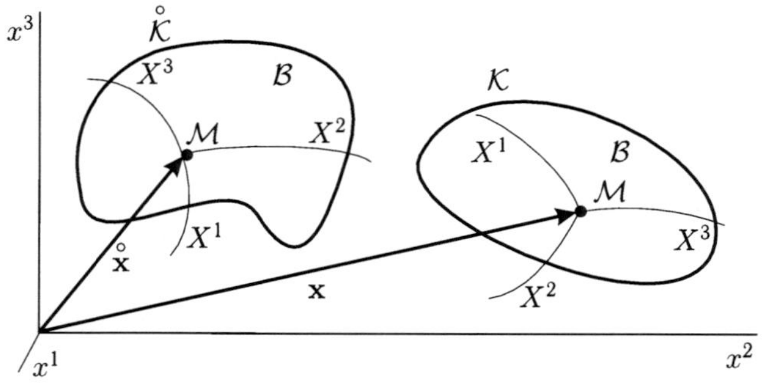


**Fig. 2.1** The motion of a continuum: positions of continuum $\mathcal{B}$ and material point $\mathcal{M}$ in reference and actual configurations

It should be noticed that if the continuum motion law (2.3) (or (2.4)) is known, then one of the main problems of continuum mechanics (to determine coordinates of all material points of the continuum at any time) will be resolved. However, in actual problems of continuum mechanics this law, as a rule, is unknown and must be found by solving some mathematical problems, whose statements are to be formulated. One of our objectives is to derive these statements.

*Example 2.1.* Let us consider a continuum $\mathcal{B}$, which at time $t = 0$ in the reference configuration $\mathring{\mathcal{K}}$ is a rectangular parallelepiped (a beam) with edge lengths $\mathring{h}_1$, $\mathring{h}_2$ and $\mathring{h}_3$, and in an actual configuration $\mathcal{K}$ at $t > 0$ the continuum is also a rectangular parallelepiped but with different edge lengths: $h_1$, $h_2$ and $h_3$. We assume that corresponding sides of both the parallelepipeds lie on parallel planes, and for one of the sides, which for example is situated on the plane $(x^2, x^3)$, points of diagonals' intersection in $\mathcal{K}$ and in $\mathring{\mathcal{K}}$ are coincident (Fig. 2.2). Then the motion law (2.3) for this continuum takes the form

$$x^\alpha = k_\alpha(t)\, X^\alpha, \qquad \alpha = 1,\ 2,\ 3, \tag{2.6}$$

i.e. coordinates $x^i$, $\mathring{x}^i = X^i$ of any material point $\mathcal{M}$ in $\mathring{\mathcal{K}}$ and $\mathcal{K}$ are proportional, and $k_\alpha(t) = h_\alpha(t)/\mathring{h}_\alpha$ is the proportion function. The motion law (2.6) is called the *beam extension law*. □

*Example 2.2.* In $\mathring{\mathcal{K}}$, let a continuum $\mathcal{B}$ be a rectangular parallelepiped oriented as shown in Fig. 2.3; its motion law (2.3) has the form

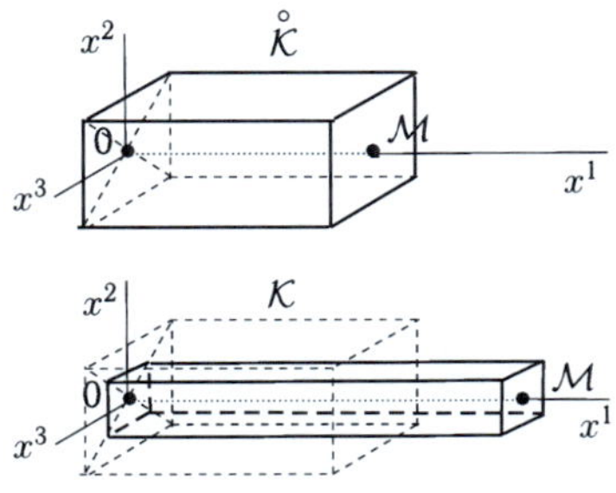


**Fig. 2.2** Extension of a beam

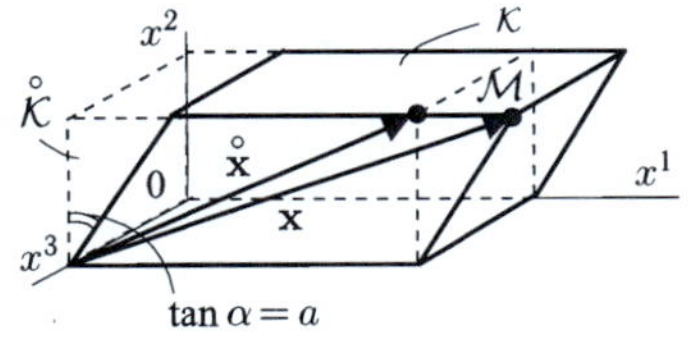


**Fig. 2.3** Simple shear of a beam

$$\begin{cases} x^1 = X^1 + a(t)X^2, \\ x^2 = X^2, \\ x^3 = X^3, \end{cases} \tag{2.7}$$

where $\overset{\circ}{x}{}^i = X^i$, $a(t)$ is a given function. In $\mathcal{K}$ this continuum $\mathcal{B}$ has become a parallelepiped, all cross-sections of which are planes orthogonal to the $Ox^3$ axis and are the same parallelograms.

This motion law is called *simple shear*; the tangent of the shear angle $\alpha$ is equal to $a$. □

*Example 2.3.* Consider a continuum $\mathcal{B}$, which in $\overset{\circ}{\mathcal{K}}$ is a rectangular parallelepiped (a beam) shown in Fig. 2.4; under the transformation from $\overset{\circ}{\mathcal{K}}$ to $\mathcal{K}$ this parallelepiped changes its dimensions without a change in its angles (as in Example 2.1) and rotates by an angle $\varphi(t)$ in the plane $Ox^1x^2$ around the point $O$ (Fig. 2.4). The motion law for the continuum is called the *rotation of a beam with extension*. In this case Eq. (2.1) have the form

$$x^i = F_0{}^i{}_j \overset{\circ}{x}{}^j, \quad \overset{\circ}{x}{}^j = X^j, \tag{2.8}$$

where the matrix $F_0{}^i{}_j$ is the product of two matrices, the rotation matrix $O_0$ and the stretch matrix $U_0$:

$$F_0{}^i{}_j = O_0{}^i{}_k U_0{}^k{}_j, \quad O_0{}^i{}_j = \begin{pmatrix} \cos\varphi & -\sin\varphi & 0 \\ \sin\varphi & \cos\varphi & 0 \\ 0 & 0 & 1 \end{pmatrix},$$

$$U_0{}^i{}_j = \begin{pmatrix} k_1 & 0 & 0 \\ 0 & k_2 & 0 \\ 0 & 0 & k_3 \end{pmatrix}, \quad F_0{}^i{}_j = \begin{pmatrix} k_1\cos\varphi & -k_2\sin\varphi & 0 \\ k_1\sin\varphi & k_2\cos\varphi & 0 \\ 0 & 0 & k_3 \end{pmatrix},$$

and $k_\alpha(t) = h_\alpha(t)/h_\alpha^0$ are the proportion functions characterizing the ratio of lengths of the beam edges in $\mathcal{K}$ and $\overset{\circ}{\mathcal{K}}$ (as in Example 2.1). □

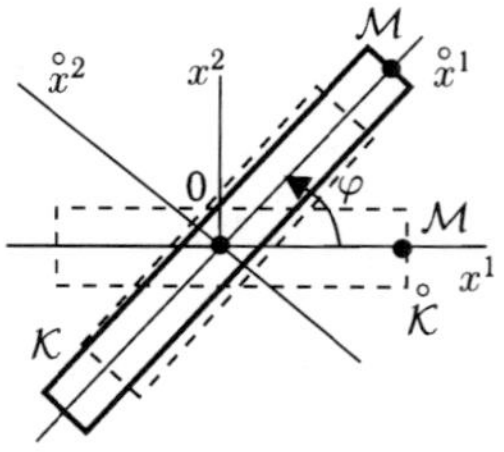


**Fig. 2.4** Rotation of a beam with extension

### 2.1.2 Material and Spatial Descriptions

In continuum mechanics, physical processes occurring in bodies are characterized by a certain set of *varying scalar fields* $\phi = \phi(\mathcal{M}, t)$, *vector fields* $\mathbf{a} = \mathbf{a}(\mathcal{M}, t)$, and *tensor fields* of the $n$th order ${}^n\boldsymbol{\Omega}(\mathcal{M}, t)$. We will consider tensors and tensor fields in detail in Sect. 2.1.4 (see also [12]).

Since in the Cartesian coordinate system $O\bar{\mathbf{e}}_i$ a material point $\mathcal{M}$ corresponds to both Lagrangian coordinates $X^i$ and Eulerian coordinates $x^i$, varying scalar and vector fields can be written as follows:

$$\phi(X^i,\ t) = \phi(X^i(x^j,\ t),\ t) = \widetilde{\phi}(x^j,\ t) = \widetilde{\phi}(\mathbf{x},\ t),$$
$$\mathbf{a}(X^i,\ t) = \widetilde{\mathbf{a}}(x^j,\ t) = \widetilde{\mathbf{a}}(\mathbf{x},\ t), \tag{2.9}$$

With the help of the motion law (2.3) (or (2.4)), we can pass from functions of Lagrangian coordinates to functions of Eulerian coordinates in formulae (2.9). In continuum mechanics the tilde $\widetilde{\ }$ is usually omitted (we will do this below).

For a fixed time $t$ in (2.9), we obtain stationary scalar and vector fields. If in (2.9) a material point $\mathcal{M}$ is fixed, and time $t$ changes within the interval $0 \leqslant t \leqslant t'$, then we get an ordinary scalar function $\phi = \phi(\mathcal{M},\ t)$ and vector function $\mathbf{a} = \mathbf{a}(\mathcal{M},\ t)$ depending on time.

According to relationships (2.9), there are two ways to describe different physical processes in continua.

In the *material (Lagrangian) description* of a continuum, all tensor fields describing physical processes are considered as functions of $X^i$ and $t$.

In the *spatial (Eulerian) description*, all tensor fields describing physical processes are functions of $x^i$ and $t$.

Both the descriptions are equivalent. It should be noted that for solids we more often use the material description, where it is convenient to fix coordinates $X^i$ of a material point $\mathcal{M}$ and to observe its motion at different times $t$. For gaseous and fluid continua, Eulerian description is more convenient; when an observer fixes a geometric point with coordinates $x^i$ and monitor the material points $\mathcal{M}$ passing through this point $x^i$ at different times $t$.

### 2.1.3 Local Bases in $\mathcal{K}$ and $\overset{\circ}{\mathcal{K}}$

Using the motion law (2.4) and relationship (2.1), at every material point $\mathcal{M}$ with coordinates $X^i$ in the actual and reference configurations we can introduce its *local basis vectors*:

$$\mathbf{r}_k = \frac{\partial \mathbf{x}}{\partial X^k} = \frac{\partial x^i}{\partial X^k}\bar{\mathbf{e}}_i = Q^i{}_k\bar{\mathbf{e}}_i, \quad \overset{\circ}{\mathbf{r}}_k = \frac{\partial \overset{\circ}{\mathbf{x}}}{\partial X^k} = \frac{\partial \overset{\circ}{x}{}^i}{\partial X^k}\bar{\mathbf{e}}_i = \overset{\circ}{Q}{}^i{}_k\bar{\mathbf{e}}_i, \tag{2.10}$$

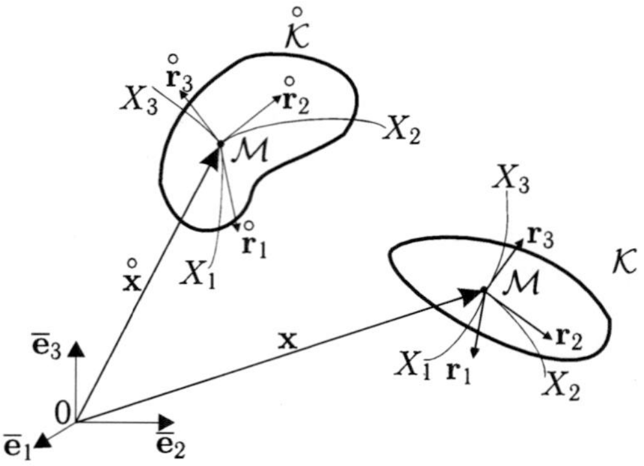


**Fig. 2.5** Local basis vectors in reference and actual configurations

where

$$Q^i{}_k = \partial x^i/\partial X^k, \quad \overset{\circ}{Q}{}^i{}_k = \partial \overset{\circ}{x}{}^i/\partial X^k,$$
$$P^i{}_k = \partial X^i/\partial x^k, \quad \overset{\circ}{P}{}^i{}_k = \partial X^i/\partial \overset{\circ}{x}{}^k \tag{2.11}$$

are *Jacobian matrices* and *inverse Jacobian matrices*.

Here and below all values referred to the configuration $\overset{\circ}{\mathcal{K}}$ will be denoted by superscript $^\circ$. As follows from the definition (2.11), local bases vectors $\mathbf{r}_k$ and $\overset{\circ}{\mathbf{r}}_k$ are directed tangentially to corresponding coordinate lines $X^k$ (Fig. 2.5).

In $\mathcal{K}$ and $\overset{\circ}{\mathcal{K}}$ introduce *metric matrices* $g_{kl}$, $\overset{\circ}{g}_{kl}$ and *inverse metric matrices* $g^{kl}$, $\overset{\circ}{g}{}^{kl}$ as follows:

$$g_{kl} = \mathbf{r}_k \cdot \mathbf{r}_l = Q^i{}_k Q^j{}_l \delta_{ij} = \frac{\partial x^i}{\partial X^k}\frac{\partial x^j}{\partial X^l}\delta_{ij},$$

$$\overset{\circ}{g}_{kl} = \overset{\circ}{\mathbf{r}}_k \cdot \overset{\circ}{\mathbf{r}}_l = \frac{\partial \overset{\circ}{x}{}^i}{\partial X^k}\frac{\partial \overset{\circ}{x}{}^j}{\partial X^l}\delta_{ij}, \; g^{kl} g_{lm} = \delta^k_m, \quad \overset{\circ}{g}{}^{kl}\overset{\circ}{g}_{lm} = \delta^k_m, \tag{2.12}$$

and also *vectors of reciprocal local bases*

$$\mathbf{r}^i = g^{im}\mathbf{r}_m, \quad \overset{\circ}{\mathbf{r}}{}^i = \overset{\circ}{g}{}^{im}\overset{\circ}{\mathbf{r}}_m, \tag{2.13}$$

which satisfy the reciprocity relations

$$\mathbf{r}_i \cdot \mathbf{r}^j = \delta_i{}^j, \quad \overset{\circ}{\mathbf{r}}_i \cdot \overset{\circ}{\mathbf{r}}{}^j = \delta_i{}^j, \tag{2.14a}$$

and also the following relations:

$$\mathbf{r}_n \times \mathbf{r}_m = \sqrt{g}\,\epsilon_{nmk}\mathbf{r}^k, \quad \overset{\circ}{\mathbf{r}}_n \times \overset{\circ}{\mathbf{r}}_m = \sqrt{\overset{\circ}{g}}\,\epsilon_{nmk}\,\overset{\circ}{\mathbf{r}}{}^k, \tag{2.14b}$$

With the help of the mixed multiplication, i.e. sequentially applying scalar and vector products to three different local bases vectors, we can determine the volumes $|V|$ and $|\overset{\circ}{V}|$ constructed by these vectors:

$$|\overset{\circ}{V}| = \overset{\circ}{\mathbf{r}}_1 \cdot (\overset{\circ}{\mathbf{r}}_2 \times \overset{\circ}{\mathbf{r}}_3) = \sqrt{\overset{\circ}{g}} = \sqrt{\det(\overset{\circ}{g}_{ij})} = \left|\partial \overset{\circ}{x}^k / \partial X^i\right|,$$
$$|V| = \mathbf{r}_1 \cdot (\mathbf{r}_2 \times \mathbf{r}_3) = \sqrt{g} = |\partial x^k / \partial X^i|. \tag{2.15}$$

It should be noted that although local bases $\mathbf{r}_i$ and $\overset{\circ}{\mathbf{r}}_i$ have been introduced in different configurations $\mathcal{K}$ and $\overset{\circ}{\mathcal{K}}$, they correspond to the same coordinates $X^i$ (if one consider the same point $\mathcal{M}$); therefore each of the bases can be carried as a rigid whole into the same point in $\mathcal{K}$ or in $\overset{\circ}{\mathcal{K}}$. Due to this, we can resolve any vector field $\mathbf{a}(\mathcal{M})$ for each of the bases $\mathbf{r}_i$, $\mathbf{r}^i$, $\overset{\circ}{\mathbf{r}}{}^i$ and $\overset{\circ}{\mathbf{r}}_i$:

$$\mathbf{a} = a^i \mathbf{r}_i = \overset{\circ}{a}{}^i \overset{\circ}{\mathbf{r}}_i = a_i \mathbf{r}^i = \overset{\circ}{a}_i \overset{\circ}{\mathbf{r}}{}^i. \tag{2.16}$$

If curvilinear coordinates $X^i$ are orthogonal, then the vectors $\overset{\circ}{\mathbf{r}}_i$ are orthogonal as well: $(\overset{\circ}{\mathbf{r}}_i \cdot \overset{\circ}{\mathbf{r}}_j = \delta_{ij})$, and matrices $\overset{\circ}{g}_{ij}$ and $\overset{\circ}{g}{}^{ij}$ are diagonal; hence we can introduce *Lamé's coefficients* $\overset{\circ}{H}_\alpha = \sqrt{\overset{\circ}{g}_{\alpha\alpha}}$ $(\alpha = 1, 2, 3)$ and the *physical orthonormal basis*

$$\widehat{\overset{\circ}{\mathbf{r}}}_\alpha = \overset{\circ}{\mathbf{r}}_\alpha / \overset{\circ}{H}_\alpha = \overset{\circ}{\mathbf{r}}{}^\alpha \overset{\circ}{H}_\alpha. \tag{2.17}$$

Components of a vector $\mathbf{a}$ with respect to this basis are called *physical*:

$$\mathbf{a} = \widehat{\overset{\circ}{a}}{}^i \, \widehat{\overset{\circ}{\mathbf{r}}}_i. \tag{2.18}$$

The actual basis $\mathbf{r}_i$ is in general not orthogonal even if the basis $\overset{\circ}{\mathbf{r}}_i$ is orthogonal; therefore we cannot introduce the corresponding physical basis in $\mathcal{K}$. One can introduce a physical basis in $\mathcal{K}$ not with the help of $\mathbf{r}_i$, but with the help of another special basis (see Sect. 2.1.7).

### 2.1.4 Tensors and Tensor Fields in Continuum Mechanics

For different local bases $\mathbf{r}_i$, $\overset{\circ}{\mathbf{r}}_i$, $\mathbf{r}^i$, $\overset{\circ}{\mathbf{r}}{}^i$ or $\bar{\mathbf{e}}_i$ at every point $\mathcal{M}$, and with the help of formulae given in the work [12] we can introduce different *dyadic (tensor) bases*: $\mathbf{r}_i \otimes \mathbf{r}_j$, $\overset{\circ}{\mathbf{r}}_i \otimes \overset{\circ}{\mathbf{r}}_j$, $\overset{\circ}{\mathbf{r}}_i \otimes \mathbf{r}^j$, $\mathbf{r}_i \otimes \overset{\circ}{\mathbf{r}}_j$, $\overset{\circ}{\mathbf{r}}{}^i \otimes \mathbf{r}_j$ etc., which are equivalence classes of vector sets consisting of $2 \cdot 3 = 6$ vectors (for example, $\mathbf{r}_1 \otimes \mathbf{r}_j = [\mathbf{r}_1 \mathbf{r}_j \mathbf{r}_2 \mathbf{0} \mathbf{r}_3 \mathbf{0}]$, where [ ] is the notation of an equivalence class), and $\otimes$ is the sign of *tensor product*. A field of *second-order tensor* $\mathbf{T}(\mathcal{M})$ can be represented as a linear combination of dyadic basis elements:

$$\mathbf{T} = T^{ij}\mathbf{r}_i \otimes \mathbf{r}_j = \overset{\circ}{T}{}^{ij}\overset{\circ}{\mathbf{r}}_i \otimes \overset{\circ}{\mathbf{r}}_j = \bar{T}^{ij}\bar{\mathbf{e}}_i \otimes \bar{\mathbf{e}}_j = \overset{\circ}{T}_{ij}\overset{\circ}{\mathbf{r}}{}^i \otimes \overset{\circ}{\mathbf{r}}{}^j. \tag{2.19}$$

During the passage from one basis to another, tensor components $T^{ij}$ are transformed by the *tensor law*:

$$\bar{T}^{ij} = P^i_{\ k} P^j_{\ l}\, T^{kl} = \overset{\circ}{P}{}^i_{\ k} \overset{\circ}{P}{}^j_{\ l}\, \overset{\circ}{T}{}^{kl}. \tag{2.20}$$

Metric matrices $g^{im}$, $g_{im}$, $\overset{\circ}{g}{}^{im}$ and $\overset{\circ}{g}_{im}$ are components of the *unit (metric) tensor* $\mathbf{E}$ with respect to different bases:

$$\mathbf{E} = g_{im}\mathbf{r}^i \otimes \mathbf{r}^m = \overset{\circ}{g}_{im}\overset{\circ}{\mathbf{r}}{}^i \otimes \overset{\circ}{\mathbf{r}}{}^m = g^{im}\mathbf{r}_i \otimes \mathbf{r}_m = \overset{\circ}{g}{}^{im}\overset{\circ}{\mathbf{r}}_i \otimes \overset{\circ}{\mathbf{r}}_m. \tag{2.21}$$

For a second-order tensor $\mathbf{T}$, in continuum mechanics one often uses the transpose tensor $\mathbf{T}^{\mathrm{T}} = T^{ij}\mathbf{r}_j \otimes \mathbf{r}_i$ and the inverse tensor $\mathbf{T}^{-1}$, where $\mathbf{T}^{-1} \cdot \mathbf{T} = \mathbf{E}$. The inverse tensor exists only for a nonsingular tensor (when $\det \mathbf{T} \neq 0$). The determinant of a tensor is defined by the determinant of its mixed components matrix: $\det \mathbf{T} = \det T^i_{\ j}$.

Besides second-order tensors, in continuum mechanics one sometimes uses tensors of higher orders [12]. To introduce the tensors, we define *polyadic bases* by induction: $\mathbf{r}_{i_1} \otimes \ldots \otimes \mathbf{r}_{i_n}$; the bases are equivalence classes of vector sets consisting of $n \cdot 3 = 3n$ vectors. A field of $n$th order tensor ${}^n\boldsymbol{\Omega}(\mathcal{M})$ can be represented by a linear combination of polyadic basis elements:

$${}^n\boldsymbol{\Omega} = \Omega^{i_1\ldots i_n}\mathbf{r}_{i_1} \otimes \ldots \otimes \mathbf{r}_{i_n} = \overset{\circ}{\Omega}{}^{i_1\ldots i_n}\overset{\circ}{\mathbf{r}}_{i_1} \otimes \ldots \otimes \overset{\circ}{\mathbf{r}}_{i_n},$$

where $\Omega^{i_1\ldots i_n}$ and $\overset{\circ}{\Omega}{}^{i_1\ldots i_n}$ are components of the $n$th order tensor with respect to the corresponding polyadic basis.

For fourth-order tensors, analogs of the tensor $\mathbf{E}$ are the first, second and third unit tensors defined as follows:

$$\begin{gathered}\boldsymbol{\Delta}_{\mathrm{I}} = \mathbf{e}_i \otimes \mathbf{e}^i \otimes \mathbf{e}_k \otimes \mathbf{e}^k = \mathbf{E} \otimes \mathbf{E},\\ \boldsymbol{\Delta}_{\mathrm{II}} = \mathbf{e}_i \otimes \mathbf{e}_k \otimes \mathbf{e}^i \otimes \mathbf{e}^k, \quad \boldsymbol{\Delta}_{\mathrm{III}} = \mathbf{e}_i \otimes \mathbf{e}_k \otimes \mathbf{e}^k \otimes \mathbf{e}^i,\end{gathered} \tag{2.22}$$

and also the symmetric fourth-order unit tensor

$$\boldsymbol{\Delta} = \frac{1}{2}(\boldsymbol{\Delta}_{\mathrm{II}} + \boldsymbol{\Delta}_{\mathrm{III}}). \tag{2.22a}$$

$$\boldsymbol{\Delta} = \Delta^{ijkl}\mathbf{e}_i \otimes \mathbf{e}_j \otimes \mathbf{e}_k \otimes \mathbf{e}_l, \quad \Delta^{ijkl} = \frac{1}{2}(\delta^{ik}\delta^{jl} + \delta^{il}\delta^{jk}).$$

We can transpose fourth-order tensors as follows:

$$^4\boldsymbol{\Omega}^{(m_1m_2m_3m_4)} = \Omega^{i_1i_2i_3i_4}\mathbf{r}_{i_{m_1}} \otimes \mathbf{r}_{i_{m_2}} \otimes \mathbf{r}_{i_{m_3}} \otimes \mathbf{r}_{i_{m_4}},$$

where $(m_1m_2m_3m_4)$ is some substitution, for example, $^4\boldsymbol{\Omega}^{(4321)} = \Omega^{i_1i_2i_3i_4}\mathbf{r}_{i_4} \otimes \mathbf{r}_{i_3} \otimes \mathbf{r}_{i_2} \otimes \mathbf{r}_{i_1}$.

### 2.1.5 Covariant Derivatives in $\mathcal{K}$ and $\overset{\circ}{\mathcal{K}}$

Introduce the following *nabla-operators* in configurations $\mathcal{K}$ and $\overset{\circ}{\mathcal{K}}$, respectively:

$$\nabla = \mathbf{r}^k \otimes \frac{\partial}{\partial X^k}, \quad \overset{\circ}{\nabla} = \overset{\circ}{\mathbf{r}}{}^k \otimes \frac{\partial}{\partial X^k}. \tag{2.23}$$

Applying the nabla-operators to a vector field, we get the *gradients of a vector* in $\mathcal{K}$ and $\overset{\circ}{\mathcal{K}}$:

$$\nabla \otimes \mathbf{a} = \mathbf{r}^k \otimes \frac{\partial \mathbf{a}}{\partial X^k} = \nabla_k a_i \mathbf{r}^k \otimes \mathbf{r}^i,$$
$$\overset{\circ}{\nabla} \otimes \mathbf{a} = \overset{\circ}{\mathbf{r}}{}^k \otimes \frac{\partial \mathbf{a}}{\partial X^k} = \overset{\circ}{\nabla}_k \overset{\circ}{a}_i \overset{\circ}{\mathbf{r}}{}^k \otimes \overset{\circ}{\mathbf{r}}{}^i = \overset{\circ}{\nabla}{}^k \overset{\circ}{a}_i \overset{\circ}{\mathbf{r}}_k \otimes \overset{\circ}{\mathbf{r}}{}^i = \overset{\circ}{\nabla}{}^k \overset{\circ}{a}{}^i \overset{\circ}{\mathbf{r}}_k \otimes \overset{\circ}{\mathbf{r}}_i, \tag{2.24}$$

where we have denoted the following *covariant derivatives* in different *tensor bases* in configurations $\overset{\circ}{\mathcal{K}}$ and $\mathcal{K}$:

$$\overset{\circ}{\nabla}_k \overset{\circ}{a}_i = \frac{\partial \overset{\circ}{a}_i}{\partial X^k} - \overset{\circ}{\Gamma}{}^m_{ik} \overset{\circ}{a}_m, \quad \overset{\circ}{\nabla}_k \overset{\circ}{a}{}^i = \frac{\partial \overset{\circ}{a}{}^i}{\partial X^k} + \overset{\circ}{\Gamma}{}^i_{km} \overset{\circ}{a}{}^m$$
$$\nabla_k a_i = \frac{\partial a_i}{\partial X^k} - \Gamma^m_{ik} a_m, \quad \nabla_k a^i = \frac{\partial a^i}{\partial X^k} + \Gamma^i_{km} a^m. \tag{2.25}$$

Here $\Gamma^m_{ij}$ and $\overset{\circ}{\Gamma}{}^m_{ij}$ are the *Christoffel symbols* in configurations $\mathcal{K}$ and $\overset{\circ}{\mathcal{K}}$. For the Christoffel symbols the following relations (see [12]) hold:

$$\Gamma^m_{ij} = \frac{1}{2} g^{km} \left( \frac{\partial g_{kj}}{\partial X^i} + \frac{\partial g_{ki}}{\partial X^j} - \frac{\partial g_{ij}}{\partial X^k} \right),$$
$$\overset{\circ}{\Gamma}{}^m_{ij} = \frac{1}{2} \overset{\circ}{g}{}^{km} \left( \frac{\partial \overset{\circ}{g}_{kj}}{\partial X^i} + \frac{\partial \overset{\circ}{g}_{ki}}{\partial X^j} - \frac{\partial \overset{\circ}{g}_{ij}}{\partial X^k} \right). \tag{2.26}$$

*Contravariant derivatives* in $\mathcal{K}$ and $\overset{\circ}{\mathcal{K}}$ are introduced as follows:

$$\overset{\circ}{\nabla}{}^k \overset{\circ}{a}_i = \overset{\circ}{g}{}^{km} \overset{\circ}{\nabla}_m \overset{\circ}{a}_i, \quad \nabla^k a^i = g^{km} \nabla_m a^i. \tag{2.27}$$

The covariant derivatives (2.25) are components of the second-order tensors $\overset{\circ}{\nabla} \otimes \mathbf{a}$ and $\nabla \otimes \mathbf{a}$, therefore during the passage from the local basis $\mathbf{r}_i$ to another one they are transformed by the tensor law (2.20).

The nabla-operators $\overset{\circ}{\nabla}$ and $\nabla$ in $\overset{\circ}{\mathcal{K}}$ and $\mathcal{K}$ can be applied to a tensor field ${}^n\mathbf{\Omega}(X^i)$ of $n$th order:

$$\overset{\circ}{\nabla} \otimes {}^n\mathbf{\Omega} = \overset{\circ}{\mathbf{r}}{}^k \otimes \frac{\partial\, {}^n\mathbf{\Omega}}{\partial X^k} = \overset{\circ}{\nabla}_k \overset{\circ}{\Omega}{}^{i_1\ldots i_n} \overset{\circ}{\mathbf{r}}{}^k \otimes \overset{\circ}{\mathbf{r}}_{i_1} \otimes \ldots \otimes \overset{\circ}{\mathbf{r}}_{i_n},$$

$$\nabla \otimes {}^n\mathbf{\Omega} = \mathbf{r}^k \otimes \frac{\partial\, {}^n\mathbf{\Omega}}{\partial X^k} = \nabla_k \Omega^{i_1\ldots i_n} \mathbf{r}^k \otimes \mathbf{r}_{i_1} \otimes \ldots \otimes \mathbf{r}_{i_n}, \tag{2.28}$$

where $\nabla_k \Omega^{i_1\ldots i_n}$ and $\overset{\circ}{\nabla}_k \overset{\circ}{\Omega}{}^{i_1\ldots i_n}$ are the covariant derivatives in $\mathcal{K}$ and $\overset{\circ}{\mathcal{K}}$, respectively:

$$\overset{\circ}{\nabla}_k \overset{\circ}{\Omega}{}^{i_1\ldots i_n} = \frac{\partial}{\partial X^k} \overset{\circ}{\Omega}{}^{i_1\ldots i_n} + \sum_{s=1}^{n} \overset{\circ}{\Gamma}{}^{i_s}_{mk} \overset{\circ}{\Omega}{}^{i_1\ldots i_s = m \ldots i_n}. \tag{2.29}$$

In the same way we can define operations of scalar product of the nabla-operator in $\overset{\circ}{\mathcal{K}}$ (*the divergence of a tensor*):

$$\overset{\circ}{\nabla} \cdot {}^n\mathbf{\Omega} = \overset{\circ}{\mathbf{r}}_k \cdot \frac{\partial\, {}^n\mathbf{\Omega}}{\partial X^k} = \overset{\circ}{\nabla}_k \overset{\circ}{\Omega}{}^{k i_2\ldots i_n} \overset{\circ}{\mathbf{r}}_{i_2} \otimes \ldots \otimes \overset{\circ}{\mathbf{r}}_{i_n}, \tag{2.30}$$

and vector product of the nabla-operator in $\overset{\circ}{\mathcal{K}}$ (*the curl of a tensor*):

$$\overset{\circ}{\nabla} \times {}^n\mathbf{\Omega} = \overset{\circ}{\mathbf{r}}{}^k \times \frac{\partial\, {}^n\mathbf{\Omega}}{\partial X^k} = \frac{1}{\sqrt{\overset{\circ}{g}}} \epsilon^{ijk} \overset{\circ}{\nabla}_i \overset{\circ}{\Omega}_{j i_2 \ldots i_n} \overset{\circ}{\mathbf{r}}_k \otimes \overset{\circ}{\mathbf{r}}{}^{i_2} \otimes \ldots \otimes \overset{\circ}{\mathbf{r}}{}^{i_n}. \tag{2.31}$$

### 2.1.6 The Deformation Gradient

Consider how a local *neighborhood* of a point $\mathcal{M}$ is transformed during the passage from configuration $\overset{\circ}{\mathcal{K}}$ to $\mathcal{K}$. Take an arbitrary elementary radius-vector $d\overset{\circ}{\mathbf{x}}$ connecting in $\overset{\circ}{\mathcal{K}}$ two infinitesimally close points $\mathcal{M}$ and $\mathcal{M}'$ (Fig. 2.6). In configuration $\mathcal{K}$, these material points $\mathcal{M}$ and $\mathcal{M}'$ are connected by the elementary radius-vector $d\mathbf{x}$.

The vectors $d\overset{\circ}{\mathbf{x}}$ and $d\mathbf{x}$ can always be resolved for local bases:

$$d\mathbf{x}(X^k) = \frac{\partial \mathbf{x}}{\partial X^k}\, dX^k = \mathbf{r}_k\, dX^k, \quad d\overset{\circ}{\mathbf{x}}(X^k) = \frac{\partial \overset{\circ}{\mathbf{x}}}{\partial X^k}\, dX^k = \overset{\circ}{\mathbf{r}}_k\, dX^k. \tag{2.32}$$

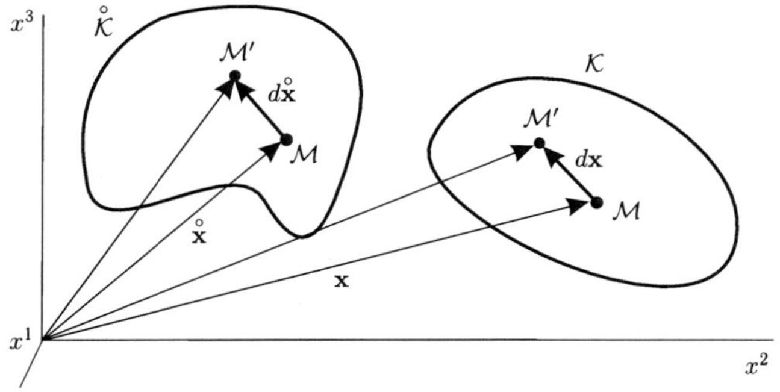


**Fig. 2.6** Transformation of an elementary radius-vector during the passage from the reference configuration to the actual one

On multiplying the first equation by $\mathbf{r}^m$ and the second – by $\overset{\circ}{\mathbf{r}}{}^m$, we get

$$\mathbf{r}^m \cdot d\mathbf{x} = \mathbf{r}^m \cdot \mathbf{r}_k \, dX^k = dX^m, \quad \overset{\circ}{\mathbf{r}}{}^m \cdot d\overset{\circ}{\mathbf{x}} = \overset{\circ}{\mathbf{r}}{}^m \cdot \overset{\circ}{\mathbf{r}}_k \, dX^k = dX^m. \tag{2.33}$$

Substitution of $dX^m$ (2.33) into the first equation of (2.32) yields $d\mathbf{x} = \mathbf{r}_k \otimes \overset{\circ}{\mathbf{r}}{}^k \cdot d\overset{\circ}{\mathbf{x}}$. Changing the order of the tensor and scalar products (that is permissible by the tensor analysis rules), we get the relation between $d\overset{\circ}{\mathbf{x}}$ and $d\mathbf{x}$:

$$d\mathbf{x} = \mathbf{F} \cdot d\overset{\circ}{\mathbf{x}}. \tag{2.34}$$

Here we have denoted the linear transformation tensor

$$\mathbf{F} = \mathbf{r}_k \otimes \overset{\circ}{\mathbf{r}}{}^k, \tag{2.35}$$

called the *deformation gradient*. As follows from (2.34), the deformation gradient connects elementary radius-vectors $d\overset{\circ}{\mathbf{x}}$ and $d\mathbf{x}$ of the same material point $\mathcal{M}$ in configurations $\mathcal{K}$ and $\overset{\circ}{\mathcal{K}}$.

Definition (2.19) allows us to give a geometric representation of the deformation gradient: if $\mathbf{r}_i$ are considered as the left vectors and $\overset{\circ}{\mathbf{r}}{}^i$ – as the right vectors, then, by formulae of Sect. 2.1.4 (see [12]), the tensor $\mathbf{F}$ takes the form

$$\mathbf{F} = \mathbf{r}_i \otimes \overset{\circ}{\mathbf{r}}{}^i = [\mathbf{r}_1 \overset{\circ}{\mathbf{r}}{}^1 \mathbf{r}_2 \overset{\circ}{\mathbf{r}}{}^2 \mathbf{r}_3 \overset{\circ}{\mathbf{r}}{}^3].$$

According to the geometric definition of a tensor (see Sect. 2.1.4), the tensor $\mathbf{F}$ can be represented as equivalence class of the ordered set of six vectors $\mathbf{r}_i, \overset{\circ}{\mathbf{r}}{}^i$ (Fig. 2.7).

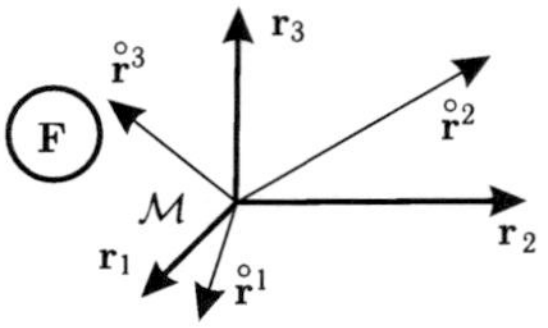


**Fig. 2.7** Geometric representation of the deformation gradient

Besides $\mathbf{F}$, in continuum mechanics one often uses the transpose tensor $\mathbf{F}^{\mathrm{T}}$, the inverse tensor $\mathbf{F}^{-1}$ and the inverse to the transpose tensor $\mathbf{F}^{-1\mathrm{T}}$:

$$\mathbf{F}^{\mathrm{T}} = \overset{\circ}{\mathbf{r}}^k \otimes \mathbf{r}_k = \overset{\circ}{\mathbf{r}}^k \otimes \frac{\partial \mathbf{x}}{\partial X^k} = \overset{\circ}{\nabla} \otimes \mathbf{x}, \quad \mathbf{F}^{-1} = \overset{\circ}{\mathbf{r}}_k \otimes \mathbf{r}^k,$$
$$\mathbf{F}^{-1\mathrm{T}} = (\overset{\circ}{\mathbf{r}}_k \otimes \mathbf{r}^k)^{\mathrm{T}} = \mathbf{r}^k \otimes \overset{\circ}{\mathbf{r}}_k = \mathbf{r}^k \otimes \frac{\partial \overset{\circ}{\mathbf{x}}}{\partial X^k} = \nabla \otimes \overset{\circ}{\mathbf{x}}. \tag{2.35a}$$

It follows from (2.35) that

$$\mathbf{F} \cdot \overset{\circ}{\mathbf{r}}_i = \mathbf{r}_k \otimes \overset{\circ}{\mathbf{r}}^k \cdot \mathbf{r}_i = \mathbf{r}_k \delta_i^k = \mathbf{r}_i . \tag{2.36}$$

i.e. the deformation gradient transforms local bases vectors of the same material point $\mathcal{M}$ from $\overset{\circ}{\mathcal{K}}$ to $\mathcal{K}$.

**Theorem 2.1.** *The transpose deformation gradient* $\mathbf{F}^{\mathrm{T}}$ *connects gradients of an arbitrary vector* $\mathbf{a}$ *in* $\overset{\circ}{\mathcal{K}}$ *and* $\mathcal{K}$*:*

$$\overset{\circ}{\nabla} \otimes \mathbf{a} = \mathbf{F}^{\mathrm{T}} \cdot \nabla \otimes \mathbf{a}, \quad \nabla \otimes \mathbf{a} = \mathbf{F}^{-1\mathrm{T}} \cdot \overset{\circ}{\nabla} \otimes \mathbf{a}. \tag{2.37}$$

▼ To derive formulae (2.37), we apply the definitions (2.24) and (2.35):

$$\nabla \otimes \mathbf{a} = \mathbf{r}^i \otimes \frac{\partial \mathbf{a}}{\partial X^i} = \mathbf{r}^j \delta^i_j \otimes \frac{\partial \mathbf{a}}{\partial X^i} = \mathbf{r}^j \otimes \overset{\circ}{\mathbf{r}}_j \cdot \overset{\circ}{\mathbf{r}}^i \otimes \frac{\partial \mathbf{a}}{\partial X^i} = \mathbf{F}^{-1\mathrm{T}} \cdot \overset{\circ}{\nabla} \otimes \mathbf{a}. \ \blacktriangle \tag{2.38}$$

## *2.1.7 Curvilinear Spatial Coordinates*

Notice that the choice of Cartesian basis $O\bar{\mathbf{e}}_i$ as a fixed (immovable) system in the spatial (Eulerian) description of the continuum motion is not a necessary condition. For some problems of continuum mechanics it is convenient to consider a *moving system* $O'\bar{\mathbf{e}}'_i$ with the origin at a moving point $O'$ ($\mathbf{x}_0 = \overrightarrow{OO'}$) and a moving orthonormal basis $\bar{\mathbf{e}}'_i$ (Fig. 2.8), which is connected to $\bar{\mathbf{e}}_i$ by the orthogonal tensor $\mathbf{Q}$:

$$\bar{\mathbf{e}}'_i = \mathbf{Q} \cdot \bar{\mathbf{e}}_i . \tag{2.39}$$

In this case, instead of Cartesian coordinates $x^i$ of a point $\mathcal{M}$ in the basis $\bar{\mathbf{e}}_i$ one consider its Cartesian coordinates $\widetilde{x}^i$ in basis $\bar{\mathbf{e}}'_i$:

$$\widetilde{\mathbf{x}} = \mathbf{x} - \mathbf{x}_0 = \widetilde{x}^i \bar{\mathbf{e}}'_i . \tag{2.40}$$

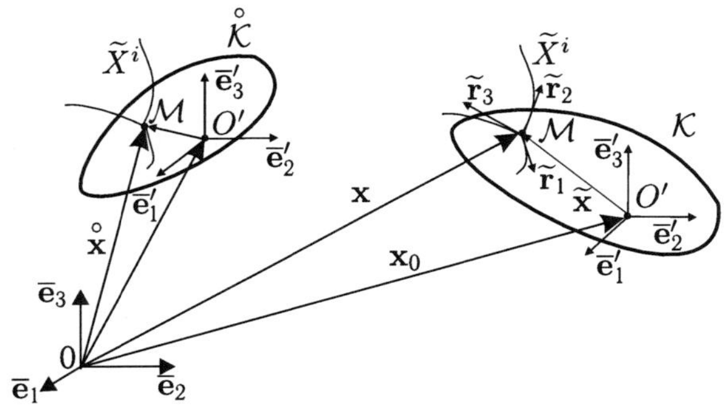


**Fig. 2.8** Moving bases $\bar{\mathbf{e}}'_i$ and $\widetilde{\mathbf{r}}_i$ and curvilinear spatial coordinates $\widetilde{X}^i$ in moving system $O'\bar{\mathbf{e}}'_i$

Let $Q^i{}_j$ be components of the tensor $\mathbf{Q}$ with respect to the basis $\bar{\mathbf{e}}_i$:

$$\mathbf{Q} = Q^i{}_j \bar{\mathbf{e}}_i \otimes \bar{\mathbf{e}}^j, \tag{2.41}$$

then relation (2.39) takes the form

$$\bar{\mathbf{e}}'_i = Q^j{}_i \,\bar{\mathbf{e}}_j, \tag{2.42}$$

and coordinates $\widetilde{x}^i$ and $x^i$ are connected as follows:

$$\widetilde{\mathbf{x}} = \mathbf{x} - \mathbf{x}_0 = (x^i - x^i_0)\bar{\mathbf{e}}_i = \widetilde{x}^i \bar{\mathbf{e}}'_i = \widetilde{x}^i Q^j{}_i \,\bar{\mathbf{e}}_j,$$

$$x^i - x^i_0 = Q^i{}_j \widetilde{x}^j, \partial x^i / \partial \widetilde{x}^j = Q^i{}_j, \quad \partial \widetilde{x}^j / \partial x^i = P^j{}_i \,. \tag{2.43}$$

Instead of Cartesian coordinates $\widetilde{x}^i$, we can consider special curvilinear coordinates $\widetilde{X}^k$ with the origin at point $O'$:

$$\widetilde{x}^i = \widetilde{x}^i(\widetilde{X}^k), \tag{2.44}$$

which, due to (2.27), are connected to $x^i$ by the relations

$$x^i = x^i_0(t) + Q^i{}_j(t)\widetilde{x}^j(\widetilde{X}^k) \equiv x^i(\widetilde{X}^k, t) \quad \text{or} \quad \widetilde{X}^j = \widetilde{X}^j(x^i, t). \tag{2.45}$$

The dependence on $t$ in the relations is defined by functions $x^i_0(t)$ and $Q^i{}_j(t)$ (i.e. only by the motion of system $O'\bar{\mathbf{e}}'_i$), which are assumed to be known in continuum mechanics.

Coordinates $\widetilde{X}^i$ are no longer Lagrangian (material): at different times they correspond to different material points. However, it is often convenient to choose coordinates $\widetilde{X}^i$ coincident with $X^i$ in the reference configuration $\overset{\circ}{\mathcal{K}}$. In this case we have the relations

$$\overset{\circ}{x}{}^i(X^i) = x^i(X^j, 0) = x^i(\widetilde{X}^j, 0). \tag{2.46}$$

With the help of transformation (2.45) we can use the spatial description in coordinates $\widetilde{X}^i$ as well when consider the functions

$$\mathbf{a} = \mathbf{a}(x^i, t) = \widetilde{\mathbf{a}}(\widetilde{X}^i, t); \tag{2.47}$$

therefore coordinates $\widetilde{X}^i$ are called *curvilinear spatial coordinates*.

Introduce local vectors

$$\widetilde{\mathbf{r}}_i = \frac{\partial \mathbf{x}}{\partial \widetilde{X}^i} = \frac{\partial x^j}{\partial \widetilde{X}^i}\bar{\mathbf{e}}_j. \tag{2.48}$$

In particular, the basis $\bar{\mathbf{e}}'_i$ may be fixed (Fig. 2.9), then $\bar{\mathbf{e}}'_i = \bar{\mathbf{e}}_i$, $\widetilde{\mathbf{x}} = \mathbf{x}$, and curvilinear spatial coordinates $\widetilde{X}^j = \widetilde{X}^j(x^i)$ are independent of $t$; the basis $\widetilde{\mathbf{r}}_i$ is independent of $t$ as well, and from (2.46) and (2.48) it follows that the basis coincides with $\overset{\circ}{\mathbf{r}}_i$:

$$\widetilde{\mathbf{r}}_i = \frac{\partial x^j}{\partial \widetilde{X}^i}\bar{\mathbf{e}}_j = \frac{\partial \overset{\circ}{x}{}^j}{\partial X^i}\bar{\mathbf{e}}_j = \overset{\circ}{\mathbf{r}}_i. \tag{2.49}$$

When the basis $\bar{\mathbf{e}}'_i$ is moving, bases $\widetilde{\mathbf{r}}_i$ and $\overset{\circ}{\mathbf{r}}_i$ are no longer coincident.

The vectors $\widetilde{\mathbf{r}}_i$ are directed tangentially to the coordinate lines $\widetilde{X}^i$ and defined simultaneously with $\mathbf{r}_i$ at every point $\mathcal{M}$ at any time $t \geqslant 0$.

A change of vectors $\widetilde{\mathbf{r}}_i$ in time is defined only by the motion of basis $\bar{\mathbf{e}}'_i$, because from (2.42), (2.43), and (2.48) it follows that

$$\bar{\mathbf{e}}'_i = Q^j{}_i \frac{\partial \widetilde{X}^k}{\partial x^j}\widetilde{\mathbf{r}}_k = \frac{\partial x^j}{\partial \widetilde{x}^i}\frac{\partial \widetilde{X}^k}{\partial x^j}\widetilde{\mathbf{r}}_k = \left(\frac{\partial \widetilde{X}^k}{\partial \widetilde{x}^i}\right)\widetilde{\mathbf{r}}_k, \tag{2.50}$$

and the matrix $\widetilde{P}^k{}_i \equiv \partial \widetilde{X}^k / \partial \widetilde{x}^i$ is independent of $t$ according to (2.44).

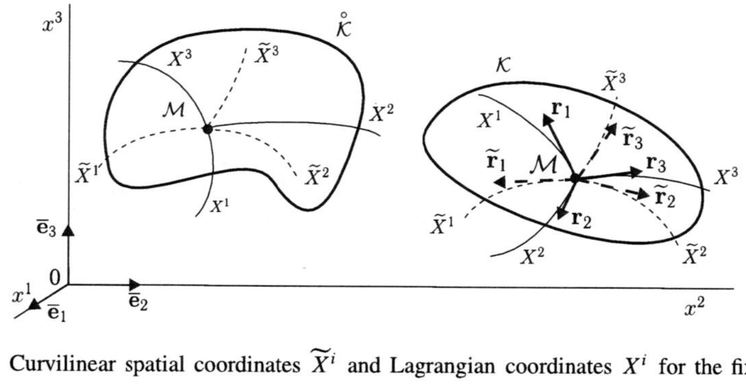


**Fig. 2.9** Curvilinear spatial coordinates $\widetilde{X}^i$ and Lagrangian coordinates $X^i$ for the fixed basis $\bar{\mathbf{e}}'_i = \bar{\mathbf{e}}_i$

The bases vectors $\mathbf{r}_i$ and $\widetilde{\mathbf{r}}_i$ are connected as follows:

$$\mathbf{r}_i = \frac{\partial \mathbf{x}}{\partial X^i} = \frac{\partial \mathbf{x}}{\partial \widetilde{X}^k}\frac{\partial \widetilde{X}^k}{\partial X^i} = \frac{\partial \widetilde{X}^k}{\partial X^i}\widetilde{\mathbf{r}}_k. \tag{2.51}$$

Just as in Sect. 2.1.2, we define the metric matrix $\widetilde{g}_{ij}$ and the inverse metric matrix $\widetilde{g}^{ij}$:

$$\widetilde{g}_{ij} = \widetilde{\mathbf{r}}_i \cdot \widetilde{\mathbf{r}}_j, \quad \widetilde{g}^{ij}\widetilde{g}_{jk} = \delta^i_k, \tag{2.52}$$

and the reciprocal basis vectors

$$\widetilde{\mathbf{r}}^i = \widetilde{g}^{ik}\widetilde{\mathbf{r}}_k = \frac{\partial \widetilde{X}^i}{\partial x^j}\bar{\mathbf{e}}^j = \frac{\partial \widetilde{X}^i}{\partial \widetilde{x}^k}\bar{\mathbf{e}}'^k. \tag{2.53}$$

According to formulae (2.51) and (2.52), we find the relation between matrices $g_{ij}$ and $\widetilde{g}_{kl}$:

$$g_{ij} = \mathbf{r}_i \cdot \mathbf{r}_j = \frac{\partial \widetilde{X}^k}{\partial X^i}\frac{\partial \widetilde{X}^l}{\partial X^j}\widetilde{\mathbf{r}}_k \cdot \widetilde{\mathbf{r}}_l = \frac{\partial \widetilde{X}^k}{\partial X^i}\frac{\partial \widetilde{X}^l}{\partial X^j}\widetilde{g}_{kl}. \tag{2.54}$$

The inverse matrix $g^{ij}$ is found from (2.54) by the rule of matrix product inversion (see Exercise 2.1.13):

$$g^{ij} = \frac{\partial X^i}{\partial \widetilde{X}^k}\frac{\partial X^j}{\partial \widetilde{X}^l}\widetilde{g}^{kl}. \tag{2.55}$$

From (2.51), (2.53) and (2.55) we can find the relation between vectors of reciprocal bases $\mathbf{r}^i$ and $\widetilde{\mathbf{r}}^i$:

$$\mathbf{r}^i = g^{ij}\mathbf{r}_j = \frac{\partial X^i}{\partial \widetilde{X}^k}\frac{\partial X^j}{\partial \widetilde{X}^l}\widetilde{g}^{kl}\frac{\partial \widetilde{X}^m}{\partial X^j}\widetilde{\mathbf{r}}_m = \frac{\partial X^i}{\partial \widetilde{X}^k}\widetilde{\mathbf{r}}^k. \tag{2.56}$$

Let there be a tensor ${}^n\boldsymbol{\Omega}$, then it can be resolved for the basis $\mathbf{r}_i$ and for the basis $\widetilde{\mathbf{r}}_i$:

$${}^n\boldsymbol{\Omega} = \Omega^{i_1\ldots i_n}\mathbf{r}_{i_1} \otimes \ldots \otimes \mathbf{r}_{i_n} = \widetilde{\Omega}^{i_1\ldots i_n}\widetilde{\mathbf{r}}_{i_1} \otimes \ldots \otimes \widetilde{\mathbf{r}}_{i_n}. \tag{2.57}$$

On substituting (2.51) into (2.57), we derive transformation formulae for tensor components during the passage from coordinates $X^i$ to $\widetilde{X}^i$:

$$\widetilde{\Omega}^{i_1\ldots i_n} = \Omega^{j_1\ldots j_n}\frac{\partial \widetilde{X}^{i_1}}{\partial X^{j_1}}\cdots\frac{\partial \widetilde{X}^{i_n}}{\partial X^{j_n}}. \tag{2.58}$$

Introduce the nabla-operator $\widetilde{\boldsymbol{\nabla}}$ of covariant differentiation in coordinates $\widetilde{X}^i$:

$$\widetilde{\boldsymbol{\nabla}} = \widetilde{\mathbf{r}}^i\frac{\partial}{\partial \widetilde{X}^i} \tag{2.59}$$

and contravariant derivatives of components $\widetilde{a}_i$ of a vector $\mathbf{a} = \widetilde{a}_i\widetilde{\mathbf{r}}^i$ in coordinates $\widetilde{X}^i$:

$$\widetilde{\nabla}_k\widetilde{a}_i = \frac{\partial\widetilde{a}_i}{\partial\widetilde{X}^k} - \widetilde{\Gamma}^m_{ik}\widetilde{a}_m, \quad \widetilde{\nabla}_k\widetilde{a}^i = \frac{\partial\widetilde{a}^i}{\partial\widetilde{X}^k} + \widetilde{\Gamma}^i_{km}\widetilde{a}^m. \tag{2.60}$$

The Christoffel symbols $\widetilde{\Gamma}^m_{ij}$ in coordinates $\widetilde{X}^i$ are connected to $\widetilde{g}_{ij}$ by the relations which are similar to (2.26).

**Theorem 2.2.** *The results of covariant differentiation in coordinates $\widetilde{X}^i$ and $X^i$ (in the configuration $\mathcal{K}$) are coincident:*

$$\nabla \otimes {}^n\boldsymbol{\Omega} = \widetilde{\nabla} \otimes {}^n\boldsymbol{\Omega}, \quad \nabla \cdot {}^n\boldsymbol{\Omega} = \widetilde{\nabla} \cdot {}^n\boldsymbol{\Omega}, \quad \nabla \times {}^n\boldsymbol{\Omega} = \widetilde{\nabla} \times {}^n\boldsymbol{\Omega}. \tag{2.61}$$

▼ Prove the first formula in (2.61). Due to (2.23), we have

$$\nabla \otimes {}^n\boldsymbol{\Omega} = \mathbf{r}^i \otimes \frac{\partial\, {}^n\boldsymbol{\Omega}}{\partial X^i} = \frac{\partial X^i}{\partial\widetilde{X}^k}\widetilde{\mathbf{r}}^k \otimes \frac{\partial\, {}^n\boldsymbol{\Omega}}{\partial\widetilde{X}^l}\frac{\partial\widetilde{X}^l}{\partial X^i} = \delta^l_k\mathbf{r}^k \otimes \frac{\partial\, {}^n\boldsymbol{\Omega}}{\partial\widetilde{X}^l} = \widetilde{\nabla} \otimes {}^n\boldsymbol{\Omega}. \tag{2.62}$$

The remaining two formulae in (2.61) can be derived in the same way (see Exercise 2.1.8). ▲

Going to components of a tensor ${}^n\boldsymbol{\Omega}$ with respect to bases $\mathbf{r}_i$ and $\widetilde{\mathbf{r}}_i$, from (2.58) we get the relation between the covariant derivatives:

$$\nabla_i\Omega^{j_1\dots j_n} = \widetilde{\nabla}_i\widetilde{\Omega}^{j_1\dots j_n}. \tag{2.63}$$

Determine the tensor $\mathbf{H}$ transforming coordinates $\widetilde{X}^i$ into $X^j$:

$$\mathbf{H} = \overset{\circ}{\mathbf{r}}_j \otimes \widetilde{\mathbf{r}}^j = \widetilde{H}^i{}_j\widetilde{\mathbf{r}}_i \otimes \widetilde{\mathbf{r}}^j = \bar{H}^i{}_j\bar{\mathbf{e}}_i \otimes \bar{\mathbf{e}}^j. \tag{2.64}$$

Then we get the relations (see Exercises 2.1.10 and 2.1.11):

$$\overset{\circ}{\mathbf{r}}_i = \mathbf{H} \cdot \widetilde{\mathbf{r}}_i = \widetilde{H}^j{}_i\widetilde{\mathbf{r}}_j, \quad \overset{\circ}{\mathbf{r}}{}^i = \mathbf{H}^{-1\mathrm{T}} \cdot \widetilde{\mathbf{r}}^i = (\widetilde{H}^i{}_j)^{-1}\widetilde{\mathbf{r}}^j. \tag{2.65}$$

The coordinates $\widetilde{X}^i$ are often chosen orthogonal, then the bases $\widetilde{\mathbf{r}}_i$ and $\widetilde{\mathbf{r}}^i$ are orthogonal as well, and the matrices $\widetilde{g}_{ij}$ and $\widetilde{g}^{ij}$ are diagonal; and we can introduce the physical (orthonormal) basis:

$$\widehat{\widetilde{\mathbf{r}}}_\alpha = \widetilde{\mathbf{r}}_\alpha / \widetilde{H}_\alpha, \tag{2.66}$$

where $\widetilde{H}_\alpha = \sqrt{\widetilde{g}_{\alpha\alpha}}$ are Lamé's coefficients, which are in general not coincident with the coefficients $\overset{\circ}{H}_\alpha = \sqrt{\overset{\circ}{g}_{\alpha\alpha}}$. Tensor components with respect to the basis $\widehat{\widetilde{\mathbf{r}}}_\alpha$ are called *physical*:

$$\mathbf{T} = \widehat{\widetilde{T}}{}^{ij}\widehat{\widetilde{\mathbf{r}}}_i \otimes \widehat{\widetilde{\mathbf{r}}}_j. \tag{2.67}$$

Relations between physical and covariant components of a tensor are determined by the known formulae (see [12]).

## Exercises for 2.1

**2.1.1.** With the help of formulae (2.10), (2.12), (2.13), and (2.17) show that if the motion law of a continuum describes extension of a beam (2.6) (see Example 2.1), then the local basis vectors $\overset{\circ}{\mathbf{r}}_i$ and the metric matrices have the forms

$$\overset{\circ}{\mathbf{r}}_i = \mathbf{e}_i, \quad \overset{\circ}{\mathbf{r}}{}^i = \mathbf{e}^i,$$

$$\mathbf{r}_\alpha = k_\alpha \bar{\mathbf{e}}_\alpha, \quad \mathbf{r}^\alpha = (1/k_\alpha)\bar{\mathbf{e}}^\alpha, \quad \alpha = 1, 2, 3,$$

$$\overset{\circ}{g}_{ij} = \delta_{ij}, \quad \overset{\circ}{g}{}^{ij} = \delta^{ij},$$

$$(g_{ij}) = \begin{pmatrix} k_1^2 & 0 & 0 \\ 0 & k_2^2 & 0 \\ 0 & 0 & k_3^2 \end{pmatrix}, \quad (g^{ij}) = \begin{pmatrix} k_1^{-2} & 0 & 0 \\ 0 & k_2^{-2} & 0 \\ 0 & 0 & k_3^{-2} \end{pmatrix},$$

i.e.

$$g_{\alpha\beta} = k_\alpha^2 \delta_{\alpha\beta}, \quad g^{\alpha\beta} = k_\alpha^{-2}\delta_{\alpha\beta}, \quad H_\alpha = \sqrt{g_{\alpha\alpha}} = k_\alpha, \quad \widehat{\mathbf{r}}_\alpha = \mathbf{e}_\alpha.$$

**2.1.2.** Show that if the motion law of a continuum describes a simple shear (see Example 2.2), then the local basis vectors and the metric matrices have the forms

$$\overset{\circ}{\mathbf{r}}_i = \mathbf{e}_i, \quad \overset{\circ}{\mathbf{r}}{}^i = \mathbf{e}^i, \quad \overset{\circ}{g}_{ij} = \delta_{ij}, \quad \overset{\circ}{g}{}^{ij} = \delta^{ij},$$

$$\mathbf{r}_1 = \bar{\mathbf{e}}_1, \quad \mathbf{r}_2 = a\bar{\mathbf{e}}_1 + \bar{\mathbf{e}}_2, \quad \mathbf{r}_3 = \bar{\mathbf{e}}_3,$$
$$\mathbf{r}^1 = \bar{\mathbf{e}}^1 - a\bar{\mathbf{e}}^2, \quad \mathbf{r}^2 = \bar{\mathbf{e}}^2, \quad \mathbf{r}^3 = \bar{\mathbf{e}}^3,$$

$$g_{ij} = \begin{pmatrix} 1 & a & 0 \\ a & 1+a^2 & 0 \\ 0 & 0 & 1 \end{pmatrix}, \quad g^{ij} = \begin{pmatrix} 1+a^2 & -a & 0 \\ -a & 1 & 0 \\ 0 & 0 & 1 \end{pmatrix}.$$

**2.1.3.** Show that if the motion law describes rotation of a beam with extension (see Example 2.3), then with introducing the rotation tensor $\mathbf{O}_0$ and the stretch tensor $\mathbf{U}_0$:

$$\mathbf{O}_0 = O_0{}^i{}_j \bar{\mathbf{e}}_i \otimes \bar{\mathbf{e}}^j, \quad \mathbf{U}_0 = \sum_{\alpha=1}^{3} k_\alpha \bar{\mathbf{e}}_\alpha \otimes \bar{\mathbf{e}}_\alpha,$$

we can rewrite the beam motion law in the tensor form

$$\mathbf{x} = \mathbf{F}_0 \cdot \overset{\circ}{\mathbf{x}}, \quad \mathbf{F}_0 = \mathbf{O}_0 \cdot \mathbf{U}_0.$$

Show that the local basis vectors and metric matrices for this problem have the forms

$$\mathbf{r}_i = F_0{}^k{}_i \bar{\mathbf{e}}_k, \quad \overset{\circ}{\mathbf{r}}_i = \bar{\mathbf{e}}_i,$$

$$g_{ij} = F_{0\ i}^{\ k} F_{0\ j}^{\ l} \delta_{kl} = \begin{pmatrix} k_1^2 \cos^2\varphi + k_2^2 \sin^2\varphi & (k_1^2 - k_2^2)\cos\varphi \sin\varphi & 0 \\ (k_1^2 - k_2^2)\cos\varphi\sin\varphi & k_1^2 \sin^2\varphi + k_2^2\cos^2\varphi & 0 \\ 0 & 0 & k_3^2 \end{pmatrix},$$

$$g = k_1 k_2 k_3,$$

$$g^{ij} = \begin{pmatrix} k_2^{-2}\sin^2\varphi + k_1^{-2}\cos^2\varphi & (k_1^{-2} - k_2^{-2})\cos\varphi\sin\varphi & 0 \\ (k_1^{-2} - k_2^{-2})\cos\varphi\sin\varphi & k_2^{-2}\cos^2\varphi + k_1^{-2}\sin^2\varphi & 0 \\ 0 & 0 & k_3^{-2} \end{pmatrix}.$$

**2.1.4.** Using the property (2.14) of reciprocal basis vectors, show that the following relations hold:

$$\overset{\circ}{\mathbf{r}}{}^i = \frac{\partial X^i}{\partial \overset{\circ}{x}{}^k}\bar{\mathbf{e}}^k, \quad \mathbf{r}^i = P^i_{\ k}\bar{\mathbf{e}}^k = \frac{\partial X^i}{\partial x^k}\bar{\mathbf{e}}^k.$$

**2.1.5.** Show that $\mathbf{F}$, $\mathbf{F}^{\mathrm{T}}$, $\mathbf{F}^{-1}$ and $\mathbf{F}^{-1\mathrm{T}}$ in the Cartesian coordinate system take the forms

$$\mathbf{F} = \frac{\partial x^m}{\partial \overset{\circ}{x}{}^i}\bar{\mathbf{e}}_m \otimes \bar{\mathbf{e}}^i, \quad \mathbf{F}^{\mathrm{T}} = \frac{\partial X^k}{\partial \overset{\circ}{x}{}^i}\bar{\mathbf{e}}^i \otimes \frac{\partial x^m}{\partial X^k}\bar{\mathbf{e}}_m = \frac{\partial x^m}{\partial \overset{\circ}{x}{}^i}\bar{\mathbf{e}}^i \otimes \bar{\mathbf{e}}_m,$$

$$\mathbf{F}^{-1} = \frac{\partial \overset{\circ}{x}{}^m}{\partial X^k}\bar{\mathbf{e}}_m \otimes \frac{\partial X^k}{\partial x^i}\bar{\mathbf{e}}^i = \frac{\partial \overset{\circ}{x}{}^m}{\partial x^i}\bar{\mathbf{e}}_m \otimes \mathbf{e}^i,$$

$$\mathbf{F}^{-1\mathrm{T}} = \frac{\partial X^k}{\partial x^i}\bar{\mathbf{e}}^i \otimes \frac{\partial \overset{\circ}{x}{}^m}{\partial X^k}\bar{\mathbf{e}}_m = \frac{\partial \overset{\circ}{x}{}^m}{\partial x^i}\bar{\mathbf{e}}^i \otimes \mathbf{e}_m.$$

**2.1.6.** Substituting (2.54), (2.52), and (2.55) into (2.12), derive formula (2.55).

**2.1.7.** Prove that

$$\mathbf{r}^i = \mathbf{F}^{-1\mathrm{T}} \cdot \overset{\circ}{\mathbf{r}}{}^i.$$

**2.1.8.** Derive the third formula of (2.61).

**2.1.9.** Prove that for any scalar function $\varphi(X^i)$ its gradients in $\overset{\circ}{\mathcal{K}}$ and $\mathcal{K}$ are connected by the relationship

$$\nabla\varphi = \mathbf{F}^{-1\mathrm{T}} \cdot \overset{\circ}{\nabla}\varphi.$$

**2.1.10.** Show that formulae (2.65) follow from (2.64).

**2.1.11.** Using (2.47), show that in formulae (2.64) the tensor $\mathbf{H}$ has the following components with respect to bases $\bar{\mathbf{e}}_i$ and $\widetilde{\mathbf{r}}_i$:

$$\bar{H}^i_{\ j} = \frac{\partial \overset{\circ}{x}{}^i}{\partial X^k}\frac{\partial \widetilde{X}^k}{\partial x^j}, \quad (\bar{H}^i_{\ j})^{-1} = \frac{\partial x^i}{\partial \widetilde{X}^k}\frac{\partial X^k}{\partial \overset{\circ}{x}{}^j}, \quad \widetilde{H}^i_{\ j} = \frac{\partial \overset{\circ}{x}{}^k}{\partial X^j}\frac{\partial \widetilde{X}^i}{\partial x^k}, \quad (\widetilde{H}^i_{\ j})^{-1} = \frac{\partial x^k}{\partial \widetilde{X}^j}\frac{\partial X^i}{\partial \overset{\circ}{x}{}^k}.$$

**2.1.12.** Introducing the notation $\overset{\circ}{F}{}^{ij}$ for components of the deformation gradient $\mathbf{F}$ with respect to basis $\overset{\circ}{\mathbf{r}}_i$: $\mathbf{F} = \overset{\circ}{F}{}^{ij}\overset{\circ}{\mathbf{r}}_i \otimes \overset{\circ}{\mathbf{r}}_j = \overset{\circ}{F}{}^{i}{}_{j}\overset{\circ}{\mathbf{r}}_i \otimes \overset{\circ}{\mathbf{r}}{}^j$, show that formula (2.36) yields

$$\mathbf{r}_j = \overset{\circ}{F}{}^{i}{}_{j}\overset{\circ}{\mathbf{r}}_i.$$

**2.1.13.** Show that the Levi-Civita symbols are connected by the relations

$$\epsilon_{ijk}\epsilon^{ijk} = 6, \quad \epsilon_{ijk}\epsilon^{ilm} = \delta^l_j\delta^m_k - \delta^l_k\delta^m_j, \qquad \epsilon_{ijk}\epsilon^{ijl} = 2\delta^l_i,$$

$$\sqrt{g}\,\epsilon_{ijk} = (1/\sqrt{g})\,\epsilon^{mnl}g_{mi}g_{nj}g_{lk}. \qquad \epsilon_{ijk}T^{jk} = 0$$

where $T^{jk}$ are components of an arbitrary symmetric tensor: $T^{jk} = T^{kj}$.

**2.1.14.** Using relations (2.14a), show that the local bases vectors are connected by the relations

$$\mathbf{r}_\alpha \times \mathbf{r}_\beta = \sqrt{g}\,\mathbf{r}^\gamma, \quad \overset{\circ}{\mathbf{r}}_\alpha \times \overset{\circ}{\mathbf{r}}_\beta = \sqrt{\overset{\circ}{g}}\,\overset{\circ}{\mathbf{r}}{}^\gamma, \quad \alpha \neq \beta \neq \gamma \neq \alpha.$$

**2.1.15.** Show that the unit fourth-order tensors $\mathbf{\Delta}_{\text{I}}$, $\mathbf{\Delta}_{\text{II}}$ and $\mathbf{\Delta}_{\text{III}}$ defined by formulae (2.22) have the following properties:

$$\mathbf{\Delta}_{\text{I}} \cdot\cdot\, \mathbf{T} = I_1(\mathbf{T})\mathbf{E}, \quad I_1(\mathbf{T}) = \mathbf{T} \cdot\cdot\, \mathbf{E}, \quad \mathbf{\Delta}_{\text{II}} \cdot\cdot\, \mathbf{T} = \mathbf{T}^{\text{T}},$$

$$\mathbf{\Delta}_{\text{III}} \cdot\cdot\, \mathbf{T} = \mathbf{T}, \quad \mathbf{\Delta} \cdot\cdot\, \mathbf{T} = \frac{1}{2}(\mathbf{T} + \mathbf{T}^{\text{T}}),$$

and

$$\mathbf{\Delta}_{\text{I}} \cdot\cdot\, {}^4\mathbf{\Omega} = \mathbf{E} \otimes \mathbf{E} \cdot\cdot\, {}^4\mathbf{\Omega}, \quad \mathbf{\Delta}_{\text{II}} \cdot\cdot\, {}^4\mathbf{\Omega} = \mathbf{\Omega}^{(2134)},$$

$$\mathbf{\Delta}_{\text{III}} \cdot\cdot\, {}^4\mathbf{\Omega} = {}^6\mathbf{\Omega}, \quad \mathbf{\Delta} \cdot\cdot\, {}^4\mathbf{\Omega} = \frac{1}{2}(\mathbf{\Omega}^{(2134)} + \mathbf{\Omega}),$$

for arbitrary second-order tensor $\mathbf{T}$ and fourth-order tensor ${}^4\mathbf{\Omega}$. As follows from these formulae, the tensor $\mathbf{\Delta}_{\text{III}}$ is the 'true' unit fourth-order tensor.

**2.1.16.** Show that components of the symmetric unit fourth-order tensor $\mathbf{\Delta}$ with respect to a tetradic basis have the form

$$\mathbf{\Delta} = \frac{1}{2}(\mathbf{e}_i \otimes \mathbf{e}^l \otimes \mathbf{e}^i \otimes \mathbf{e}_l + \mathbf{e}_i \otimes \mathbf{e}^l \otimes \mathbf{e}_l \otimes \mathbf{e}^i) = \Delta^{ijkl}\mathbf{e}_i \otimes \mathbf{e}_j \otimes \mathbf{e}_k \otimes \mathbf{e}_l.$$

$$\Delta^{ijkl} = \frac{1}{2}(\delta^{ik}\delta^{jl} + \delta^{il}\delta^{jk}).$$

**2.1.17.** Show that for any second-order tensor $\mathbf{T}$ and for any vector $\mathbf{a}$ the following formula of covariant differentiation hold:

$$\nabla \cdot (\mathbf{T} \cdot \mathbf{a}) = \mathbf{T} \cdot\cdot\, (\nabla \otimes \mathbf{a})^{\text{T}} + \mathbf{a} \cdot \nabla \cdot \mathbf{T}.$$

## 2.2 Deformation Tensors and Measures

### 2.2.1 Deformation Tensors

Besides **F**, important characteristics of the motion of a continuum are *deformation tensors*, which are introduced as follows:

$$\begin{aligned}
\mathbf{C} &= \frac{1}{2}\left(g_{ij} - \overset{\circ}{g}_{ij}\right)\overset{\circ}{\mathbf{r}}{}^i \otimes \overset{\circ}{\mathbf{r}}{}^j = \varepsilon_{ij}\overset{\circ}{\mathbf{r}}{}^i \otimes \overset{\circ}{\mathbf{r}}{}^j,\\
\mathbf{A} &= \frac{1}{2}\left(g_{ij} - \overset{\circ}{g}_{ij}\right)\mathbf{r}^i \otimes \mathbf{r}^j = \varepsilon_{ij}\mathbf{r}^i \otimes \mathbf{r}^j,\\
\mathbf{\Lambda} &= \frac{1}{2}\left(\overset{\circ}{g}{}^{ij} - g^{ij}\right)\overset{\circ}{\mathbf{r}}_i \otimes \overset{\circ}{\mathbf{r}}_j = \varepsilon^{ij}\overset{\circ}{\mathbf{r}}_i \otimes \overset{\circ}{\mathbf{r}}_j,\\
\mathbf{J} &= \frac{1}{2}\left(\overset{\circ}{g}{}^{ij} - g^{ij}\right)\mathbf{r}_i \otimes \mathbf{r}_j = \varepsilon^{ij}\mathbf{r}_i \otimes \mathbf{r}_j.
\end{aligned} \tag{2.68}$$

Here **C** is called the *right Cauchy–Green deformation tensor*, **A** – the *left Almansi deformation tensor*, **Λ** – the *right Almansi deformation tensor*, and **J** – the *left Cauchy–Green tensor*.

As follows from the definition of the tensors, covariant components of **C** and **A** are coincident, but they are defined with respect to different tensor bases. Components $\varepsilon_{ij}$ are called *covariant components of the deformation tensor*.

Contravariant components of the tensors **Λ** and **J** are also coincident and called *contravariant components* $\varepsilon^{ij}$ of the deformation tensor, but they are defined with respect to different tensor bases of the tensors **Λ** and **J**.

Notice that the deformation tensor components

$$\varepsilon_{ij} = \frac{1}{2}\left(g_{ij} - \overset{\circ}{g}_{ij}\right), \quad \varepsilon^{ij} = \frac{1}{2}\left(\overset{\circ}{g}{}^{ij} - g^{ij}\right), \tag{2.69}$$

have been defined independently of each other, therefore the formal rearrangement of indices is not permissible for these components, i.e.

$$\breve{\varepsilon}^{kl} = \varepsilon_{ij}g^{ik}g^{jl} \neq \varepsilon^{kl}, \quad \breve{\varepsilon}_{kl} = \varepsilon^{ij}g_{ik}g_{jl} \neq \varepsilon_{kl}. \tag{2.70}$$

Thus, when there is a need to obtain contravariant components from $\varepsilon_{ij}$ and covariant components from $\varepsilon^{ij}$, one should use the notation $\breve{\varepsilon}^{kl}$ and $\breve{\varepsilon}_{kl}$. We will also use the notation

$$\overset{\circ}{\varepsilon}_{kl} = \varepsilon^{ij}\overset{\circ}{g}_{ik}\overset{\circ}{g}_{jl}, \quad \overset{\circ}{\varepsilon}{}^{kl} = \varepsilon_{ij}\overset{\circ}{g}{}^{ik}\overset{\circ}{g}{}^{jl}. \tag{2.71}$$

**Theorem 2.3.** *The deformation tensors* **C**, **A**, **Λ** *and* **J** *are connected to the deformation gradient* **F** *as follows:*

$$\begin{aligned}
\mathbf{C} &= \tfrac{1}{2}(\mathbf{F}^{\mathrm{T}} \cdot \mathbf{F} - \mathbf{E}), & \mathbf{A} &= \tfrac{1}{2}(\mathbf{E} - \mathbf{F}^{-1\mathrm{T}} \cdot \mathbf{F}^{-1}),\\
\mathbf{\Lambda} &= \tfrac{1}{2}(\mathbf{E} - \mathbf{F}^{-1} \cdot \mathbf{F}^{-1\mathrm{T}}), & \mathbf{J} &= \tfrac{1}{2}(\mathbf{F} \cdot \mathbf{F}^{\mathrm{T}} - \mathbf{E}).
\end{aligned} \tag{2.72}$$

▼ Let us derive a relation between **C** and **F**. Having used the definitions of $g_{ij}$, $\overset{\circ}{g}_{ij}$ and **F**, we get

$$\mathbf{C} = \frac{1}{2}\left((\mathbf{r}_i \cdot \mathbf{r}_j)\overset{\circ}{\mathbf{r}}{}^i \otimes \overset{\circ}{\mathbf{r}}{}^j - \mathbf{E}\right) = \frac{1}{2}\left(\overset{\circ}{\mathbf{r}}{}^i \otimes \mathbf{r}_i \cdot \mathbf{r}_j \otimes \overset{\circ}{\mathbf{r}}{}^j - \mathbf{E}\right) = \frac{1}{2}(\mathbf{F}^{\mathrm{T}} \cdot \mathbf{F} - \mathbf{E}). \tag{2.73}$$

The remaining relations of (2.72) can be proved in the same way:

$$\mathbf{A} = \frac{1}{2}\left(\mathbf{E} - \mathbf{r}^i \otimes \overset{\circ}{\mathbf{r}}_i \cdot \overset{\circ}{\mathbf{r}}_j \otimes \mathbf{r}^j\right) = \frac{1}{2}\left(\mathbf{E} - \mathbf{F}^{-1\mathrm{T}} \cdot \mathbf{F}^{-1}\right),$$
$$\mathbf{\Lambda} = \frac{1}{2}\left(\mathbf{E} - \overset{\circ}{\mathbf{r}}_i \otimes \mathbf{r}^i \cdot \mathbf{r}^j \otimes \overset{\circ}{\mathbf{r}}_j\right) = \frac{1}{2}\left(\mathbf{E} - \mathbf{F}^{-1} \cdot \mathbf{F}^{-1\mathrm{T}}\right),$$
$$\mathbf{J} = \frac{1}{2}\left(\mathbf{r}_i \otimes \overset{\circ}{\mathbf{r}}{}^i \cdot \overset{\circ}{\mathbf{r}}{}^j \otimes \mathbf{r}_j - \mathbf{E}\right) = \frac{1}{2}\left(\mathbf{F} \cdot \mathbf{F}^{\mathrm{T}} - \mathbf{E}\right). \blacktriangle \tag{2.73a}$$

### 2.2.2 *Deformation Measures*

Besides the deformation tensors, we define *deformation measures*: the *right Cauchy–Green measure* **G** and the *left Almansi measure* **g**:

$$\mathbf{G} = g_{ij}\overset{\circ}{\mathbf{r}}{}^i \otimes \overset{\circ}{\mathbf{r}}{}^j = \mathbf{F}^{\mathrm{T}} \cdot \mathbf{F} = \mathbf{E} + 2\mathbf{C},$$
$$\mathbf{g} = \overset{\circ}{g}_{ij}\mathbf{r}^i \otimes \mathbf{r}^j = \mathbf{F}^{-1\mathrm{T}} \cdot \mathbf{F}^{-1} = \mathbf{E} - 2\mathbf{A}, \tag{2.74}$$

and also the *left Cauchy–Green measure* $\mathbf{g}^{-1}$ and the *right Almansi measure* $\mathbf{G}^{-1}$:

$$\mathbf{g}^{-1} = \overset{\circ}{g}{}^{ij}\mathbf{r}_i \otimes \mathbf{r}_j = \mathbf{F} \cdot \mathbf{F}^{\mathrm{T}} = \mathbf{E} + 2\mathbf{J},$$
$$\mathbf{G}^{-1} = g^{ij}\overset{\circ}{\mathbf{r}}_i \otimes \overset{\circ}{\mathbf{r}}_j = \mathbf{F}^{-1} \cdot \mathbf{F}^{-1\mathrm{T}} = \mathbf{E} - 2\mathbf{\Lambda}. \tag{2.75}$$

### 2.2.3 *Displacement Vector*

Introduce a *displacement vector* **u** of a point $\mathcal{M}$ from the reference configuration to the actual one as follows (Fig. 2.10):

$$\mathbf{u} = \mathbf{x} - \overset{\circ}{\mathbf{x}}. \tag{2.76}$$

**Theorem 2.4.** *The deformation tensors and the deformation gradient are connected to the displacement vector* **u** *by the relations*

$$\mathbf{F} = \mathbf{E} + (\overset{\circ}{\nabla} \otimes \mathbf{u})^{\mathrm{T}}, \quad \mathbf{F}^{-1} = \mathbf{E} - (\nabla \otimes \mathbf{u})^{\mathrm{T}},$$
$$\mathbf{F}^{\mathrm{T}} = \mathbf{E} + \overset{\circ}{\nabla} \otimes \mathbf{u}, \quad \mathbf{F}^{-1\mathrm{T}} = \mathbf{E} - \nabla \otimes \mathbf{u}, \tag{2.77}$$

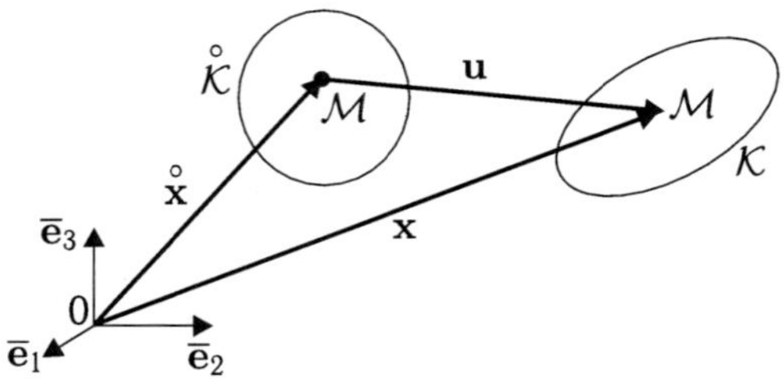


**Fig. 2.10** The displacement vector of a point $\mathcal{M}$ from the reference configuration to the actual one

*and also*

$$\mathbf{C} = \frac{1}{2}\left(\overset{\circ}{\nabla} \otimes \mathbf{u} + \overset{\circ}{\nabla} \otimes \mathbf{u}^{\mathrm{T}} + \overset{\circ}{\nabla} \otimes \mathbf{u} \cdot \overset{\circ}{\nabla} \otimes \mathbf{u}^{\mathrm{T}}\right),$$
$$\mathbf{A} = \frac{1}{2}\left(\nabla \otimes \mathbf{u} + \nabla \otimes \mathbf{u}^{\mathrm{T}} - \nabla \otimes \mathbf{u} \cdot \nabla \otimes \mathbf{u}^{\mathrm{T}}\right),$$
$$\mathbf{\Lambda} = \frac{1}{2}\left(\nabla \otimes \mathbf{u} + (\nabla \otimes \mathbf{u})^{\mathrm{T}} - \nabla \otimes \mathbf{u}^{\mathrm{T}} \cdot \nabla \otimes \mathbf{u}\right),$$
$$\mathbf{J} = \frac{1}{2}\left(\overset{\circ}{\nabla} \otimes \mathbf{u} + \overset{\circ}{\nabla} \otimes \mathbf{u}^{\mathrm{T}} + \overset{\circ}{\nabla} \otimes \mathbf{u}^{\mathrm{T}} \cdot \overset{\circ}{\nabla} \otimes \mathbf{u}\right). \tag{2.78}$$

▼ The definition (2.76) of the displacement vector and the properties (2.35) of the deformation gradient yield

$$\mathbf{F}^{\mathrm{T}} = \overset{\circ}{\nabla} \otimes \mathbf{x} = \overset{\circ}{\nabla} \otimes (\overset{\circ}{\mathbf{x}} + \mathbf{u})$$
$$= \overset{\circ}{\mathbf{r}}{}^{i} \otimes \frac{\partial \overset{\circ}{\mathbf{x}}}{\partial X^{i}} + \overset{\circ}{\nabla} \otimes \mathbf{u} = \overset{\circ}{\mathbf{r}}{}^{i} \otimes \overset{\circ}{\mathbf{r}}_{i} + \overset{\circ}{\nabla} \otimes \mathbf{u} = \mathbf{E} + \overset{\circ}{\nabla} \otimes \mathbf{u}. \tag{2.79}$$

Then the tensor $\mathbf{C}$ takes the form

$$\mathbf{C} = \frac{1}{2}\left((\mathbf{E} + \overset{\circ}{\nabla} \otimes \mathbf{u}) \cdot (\mathbf{E} + \overset{\circ}{\nabla} \otimes \mathbf{u}^{\mathrm{T}}) - \mathbf{E}\right)$$
$$= \frac{1}{2}\left(\overset{\circ}{\nabla} \otimes \mathbf{u} + \overset{\circ}{\nabla} \otimes \mathbf{u}^{\mathrm{T}} + \overset{\circ}{\nabla} \otimes \mathbf{u} \cdot \overset{\circ}{\nabla} \otimes \mathbf{u}^{\mathrm{T}}\right). \tag{2.80}$$

In a similar way, we can prove the remaining relations of the theorem. ▲

### *2.2.4 Relations Between Components of Deformation Tensors and Displacement Vector*

The displacement vector $\mathbf{u}$ can be resolved for both bases $\overset{\circ}{\mathbf{r}}_i$ and $\mathbf{r}_i$:

$$\mathbf{u} = \overset{\circ}{u}{}^{i} \overset{\circ}{\mathbf{r}}_{i} = u^{i} \mathbf{r}_{i}. \tag{2.81}$$

The derivative with respect to $X^i$ can be determined in both the bases as well:

$$\frac{\partial \mathbf{u}}{\partial X^i} = \overset{\circ}{\nabla}_i \overset{\circ}{u}^k \overset{\circ}{\mathbf{r}}_k = \nabla_i u^k \mathbf{r}_k. \tag{2.82}$$

Then the displacement vector gradients take the forms

$$\overset{\circ}{\boldsymbol{\nabla}} \otimes \mathbf{u} = \overset{\circ}{\mathbf{r}}{}^i \otimes \frac{\partial \mathbf{u}}{\partial X^i} = \overset{\circ}{\nabla}_i \overset{\circ}{u}^k \overset{\circ}{\mathbf{r}}{}^i \otimes \overset{\circ}{\mathbf{r}}_k = \overset{\circ}{\nabla}{}^i \overset{\circ}{u}^k \overset{\circ}{\mathbf{r}}_i \otimes \overset{\circ}{\mathbf{r}}_k, \tag{2.83}$$

$$\boldsymbol{\nabla} \otimes \mathbf{u} = \mathbf{r}^i \otimes \frac{\partial \mathbf{u}}{\partial X^i} = \nabla_i u^k \mathbf{r}^i \otimes \mathbf{r}_k = \nabla^i u^k \mathbf{r}_i \otimes \mathbf{r}_k. \tag{2.84}$$

Substitution of these expressions into (2.77) gives

$$\mathbf{F} = \left(\delta_i^k + \overset{\circ}{\nabla}_i \overset{\circ}{u}^k\right) \overset{\circ}{\mathbf{r}}_k \otimes \overset{\circ}{\mathbf{r}}{}^i = \overset{\circ}{F}{}^k_{\ i} \overset{\circ}{\mathbf{r}}_k \otimes \overset{\circ}{\mathbf{r}}{}^i. \tag{2.85}$$

Here we have introduced components of the deformation gradient in the reference configuration:

$$\overset{\circ}{F}{}^k_{\ i} = \delta_i^k + \overset{\circ}{\nabla}_i \overset{\circ}{u}^k. \tag{2.86}$$

The transpose gradient $\mathbf{F}^{\mathrm{T}}$ has the components

$$\mathbf{F}^{\mathrm{T}} = \overset{\circ}{F}{}^k_{\ i} \overset{\circ}{\mathbf{r}}{}^i \otimes \overset{\circ}{\mathbf{r}}_k = \overset{\circ}{F}{}_i^{\ k} \overset{\circ}{\mathbf{r}}_k \otimes \overset{\circ}{\mathbf{r}}{}^i, \tag{2.87}$$

$$\overset{\circ}{F}{}_i^{\ k} = \delta_i^k + \overset{\circ}{\nabla}{}^k \overset{\circ}{u}_i, \tag{2.88}$$

where $(\overset{\circ}{F}{}_i^{\ k})^{\mathrm{T}} = \overset{\circ}{F}{}^k_{\ i}$.

In a similar way, one can find the expression for the inverse gradient

$$\mathbf{F}^{-1} = \left(\delta_i^k - \nabla_i u^k\right) \mathbf{r}_k \otimes \mathbf{r}^i = \left(F^{-1}\right)^k_{\ i} \mathbf{r}_k \otimes \mathbf{r}^i \tag{2.89}$$

and for the inverse–transpose gradient

$$\mathbf{F}^{-1\mathrm{T}} = (F^{-1})^k_{\ i} \mathbf{r}^i \otimes \mathbf{r}_k = (F^{-1})_i^{\ k} \mathbf{r}_k \otimes \mathbf{r}^i, \tag{2.90}$$

where their components with respect to the actual configuration are expressed as follows:

$$(F^{-1})_i^{\ k} = \delta_i^k - \nabla^k u_i, \tag{2.91}$$

$$(F^{-1})^k_{\ i} = \delta_i^k - \nabla_i u^k. \tag{2.92}$$

Thus, we have proved the following theorem.

**Theorem 2.5.** *Components of the deformation gradients* $\mathbf{F}$, $\mathbf{F}^{\mathrm{T}}$, $\mathbf{F}^{-1}$ *and* $\mathbf{F}^{-1\mathrm{T}}$ *in local bases of configurations* $\mathring{\mathcal{K}}$ *and* $\mathcal{K}$ *are connected to components of the displacement vector* $\mathbf{u}$ *by relations* (2.86), (2.87), (2.91), *and* (2.92).

On substituting formulae (2.83) and (2.84) into (2.78) for $\mathbf{C}$ and $\mathbf{A}$ and comparing them with (2.68), we get

$$\varepsilon_{ij} = \frac{1}{2}\left(\mathring{\nabla}_i \mathring{u}_j + \mathring{\nabla}_j \mathring{u}_i + \mathring{\nabla}_i \mathring{u}^k \mathring{\nabla}_j \mathring{u}_k\right),$$
$$\varepsilon_{ij} = \frac{1}{2}\left(\nabla_i u_j + \nabla_j u_i - \nabla_i u^k \nabla_j u_k\right), \tag{2.93}$$

– The expressions for covariant components of the deformation tensor in terms of components of the displacement vector with respect to $\mathring{\mathcal{K}}$ and $\mathcal{K}$.

In a similar way, substituting (2.83) and (2.84) into (2.78) for $\boldsymbol{\Lambda}$ and $\mathbf{J}$, we obtain

$$\varepsilon^{ij} = \frac{1}{2}\left(\mathring{\nabla}^i \mathring{u}^j + \mathring{\nabla}^j \mathring{u}^i + \mathring{\nabla}^k \mathring{u}^i \mathring{\nabla}_k \mathring{u}^j\right),$$
$$\varepsilon^{ij} = \frac{1}{2}\left(\nabla^i u^j + \nabla^j u^i - \nabla^k u^i \nabla_k u^j\right) \tag{2.94}$$

– The relations between contravariant components of the deformation tensor and components of the displacement vector in $\mathring{\mathcal{K}}$ and $\mathcal{K}$.

Then with using relations (2.69), (2.93) and (2.94), we can find the connection between the metric matrices and displacement components:

$$g_{ij} = \mathring{g}_{ij} + \mathring{\nabla}_i \mathring{u}_j + \mathring{\nabla}_j \mathring{u}_i + \mathring{\nabla}_i \mathring{u}^k \mathring{\nabla}_j \mathring{u}_k = \mathring{g}_{ij} + \nabla_i u_j + \nabla_j u_i - \nabla_i u^k \nabla_j u_k, \tag{2.95}$$

$$g^{ij} = \mathring{g}^{ij} + \mathring{\nabla}^i \mathring{u}^j + \mathring{\nabla}^j \mathring{u}^i + \mathring{\nabla}^k \mathring{u}^i \mathring{\nabla}_k \mathring{u}^j = \mathring{g}^{ij} + \nabla^i u^j + \nabla^j u^i - \nabla^k u^i \nabla_k u^j. \tag{2.96}$$

Thus, we have proved the following theorem.

**Theorem 2.6.** *Components of the deformation tensor* $\varepsilon_{ij}$, $\varepsilon^{ij}$ *and metric matrices* $g_{ij}$, $g^{ij}$ *are connected to components of the displacement vector* $\mathbf{u}$ *by relations* (2.93)–(2.96).

### *2.2.5 Physical Meaning of Components of the Deformation Tensor*

Let us clarify now a physical meaning of components of the deformation tensor:

$$\varepsilon_{ij} = \frac{1}{2}\left(g_{ij} - \mathring{g}_{ij}\right) = \frac{1}{2}\left(\mathbf{r}_i \cdot \mathbf{r}_j - \mathring{\mathbf{r}}_i \cdot \mathring{\mathbf{r}}_j\right). \tag{2.97}$$

By the definition of the scalar product (see [12]), we have

$$\varepsilon_{\alpha\beta} = \frac{1}{2}\left(|\mathbf{r}_\alpha||\mathbf{r}_\beta|\cos\psi_{\alpha\beta} - |\overset{\circ}{\mathbf{r}}_\alpha||\overset{\circ}{\mathbf{r}}_\beta|\cos\overset{\circ}{\psi}_{\alpha\beta}\right), \tag{2.98}$$

where $\psi_{\alpha\beta}$ and $\overset{\circ}{\psi}_{\alpha\beta}$ are the angles between basis vectors $\mathbf{r}_\alpha$, $\mathbf{r}_\beta$ and $\overset{\circ}{\mathbf{r}}_\alpha$, $\overset{\circ}{\mathbf{r}}_\beta$ in $\mathcal{K}$ and $\overset{\circ}{\mathcal{K}}$, respectively.

Consider elementary radius-vectors $d\mathbf{x}$ and $d\overset{\circ}{\mathbf{x}}$ in configurations $\mathcal{K}$ and $\overset{\circ}{\mathcal{K}}$, and introduce their lengths $ds$ and $d\overset{\circ}{s}$, respectively:

$$ds^2 = d\mathbf{x}\cdot d\mathbf{x}, \qquad d\overset{\circ}{s}{}^2 = d\overset{\circ}{\mathbf{x}}\cdot d\overset{\circ}{\mathbf{x}}. \tag{2.99}$$

Since $d\overset{\circ}{\mathbf{x}}$ is arbitrary, we can choose it to be oriented along one of the basis vectors $\overset{\circ}{\mathbf{r}}_\alpha$. Then $d\mathbf{x}$ will be directed along the corresponding vector $\mathbf{r}_\alpha$ as well, because under this transformation $\overset{\circ}{\mathbf{r}}_\alpha$ becomes $\mathbf{r}_\alpha$ for the same material point $\mathcal{M}$ with Lagrangian coordinates $X^k$. In this case we have

$$|d\overset{\circ}{\mathbf{x}}| = d\overset{\circ}{s}_\alpha = \left|\frac{\partial\overset{\circ}{\mathbf{x}}}{\partial X^\alpha}\,dX^\alpha\right| = |\overset{\circ}{\mathbf{r}}_\alpha|\,dX^\alpha,$$

$$|d\mathbf{x}| = ds_\alpha = \left|\frac{\partial x}{\partial X^\alpha}\,dX^\alpha\right| = |\mathbf{r}_\alpha|\,dX^\alpha. \tag{2.100}$$

Hence

$$ds_\alpha/d\overset{\circ}{s}_\alpha = |\mathbf{r}_\alpha|/|\overset{\circ}{\mathbf{r}}_\alpha| = \delta_\alpha + 1, \tag{2.101}$$

where $\delta_\alpha$ is called the *relative elongation*. Formula (2.101) yields

$$|\mathbf{r}_\alpha| = |\overset{\circ}{\mathbf{r}}_\alpha|(1+\delta_\alpha). \tag{2.102}$$

On substituting this expression into (2.98), we get

$$\varepsilon_{\alpha\beta} = \frac{1}{2}|\overset{\circ}{\mathbf{r}}_\alpha||\overset{\circ}{\mathbf{r}}_\beta|\left((1+\delta_\alpha)(1+\delta_\beta)\cos\psi_{\alpha\beta} - \cos\overset{\circ}{\psi}_{\alpha\beta}\right). \tag{2.103}$$

Consider the case when $\alpha = \beta$, then $\psi_{\alpha\beta} = \overset{\circ}{\psi}_{\alpha\beta} = 0$ and

$$\varepsilon_{\alpha\alpha} = \frac{1}{2}|\overset{\circ}{\mathbf{r}}_\alpha|^2\left((1+\delta_\alpha)^2 - 1\right) = \frac{\overset{\circ}{g}_{\alpha\alpha}}{2}\left((1+\delta_\alpha)^2 - 1\right). \tag{2.104}$$

Let coordinates $X^i$ be coincident with Cartesian coordinates $x^i$, then $\overset{\circ}{g}_{\alpha\beta} = \delta_{\alpha\beta}$; and for infinitesimal values of the relative elongation, when $\delta_\alpha \ll 1$, we obtain

$$\varepsilon_{\alpha\alpha} \approx \delta_\alpha, \tag{2.105}$$

i.e. $\varepsilon_{\alpha\alpha}$ is coincident with the relative elongation.

In general, $\varepsilon_{\alpha\alpha}$ is a nonlinear function of corresponding elongations.

Consider $\alpha \neq \beta$ and assume that $X^i = x^i$, then $\overset{\circ}{\psi}_{\alpha\beta} = \pi/2$; and from (2.103) we get

$$\begin{aligned}\varepsilon_{\alpha\beta} &= \frac{1}{2}|\overset{\circ}{\mathbf{r}}_\alpha||\overset{\circ}{\mathbf{r}}_\beta|(1+\delta_\alpha)(1+\delta_\beta)\cos\psi_{\alpha\beta} \\ &= \frac{1}{2}\sqrt{\overset{\circ}{g}_{\alpha\alpha}}\sqrt{\overset{\circ}{g}_{\beta\beta}}(1+\delta_\alpha)(1+\delta_\beta)\sin\chi_{\alpha\beta} \\ &= \frac{1}{2}(1+\delta_\alpha)(1+\delta_\beta)\sin\chi_{\alpha\beta},\end{aligned} \tag{2.106}$$

where $\chi_{\alpha\beta} = \overset{\circ}{\psi}_{\alpha\beta} - \psi_{\alpha\beta} = (\pi/2) - \psi_{\alpha\beta}$ is the change of the angle between basis vectors $\mathbf{r}_\alpha$ and $\mathbf{r}_\beta$. For small relative elongations, when $\delta_\alpha \ll 1$, and small angles $\chi_{\alpha\beta} \ll 1$, from (2.106) we get

$$\varepsilon_{\alpha\beta} \approx \chi_{\alpha\beta}/2, \tag{2.107}$$

i.e. $\varepsilon_{\alpha\beta}$ is a half of the misalignment angle of the basis vectors.

### 2.2.6 Transformation of an Oriented Surface Element

In actual configuration $\mathcal{K}$ consider a smooth surface $\Sigma$, which contains two coordinate lines $X^\alpha$ and $X^\beta$.

Then we can introduce the normal $\mathbf{n}$ to the surface $\Sigma$ as follows:

$$\mathbf{n} = \frac{1}{\sqrt{\widetilde{g}}}\mathbf{r}_\alpha \times \mathbf{r}_\beta. \tag{2.108}$$

Here $\widetilde{g} = \det\,(\widetilde{g}_{\alpha\beta})$, and $\widetilde{g}_{\alpha\beta}$ is the two-dimensional matrix of the surface ($\alpha, \beta = 1, 2$):

$$\widetilde{g}_{\alpha\beta} = \mathbf{r}_\alpha \cdot \mathbf{r}_\beta \tag{2.109}$$

(it is not to be confused with the metric matrix $g_{ij} = \mathbf{r}_i \cdot \mathbf{r}_j$).

In configuration $\mathcal{K}$ consider a surface element $d\Sigma$ constructed on elementary radius-vectors $d\mathbf{x}_\alpha$, which are directed along local basis vectors, i.e. $d\mathbf{x}_\alpha = \mathbf{r}_\alpha dX^\alpha$ (Fig. 2.11). The value

$$d\Sigma = \sqrt{\widetilde{g}}\, dX^\alpha\, dX^\beta \tag{2.110}$$

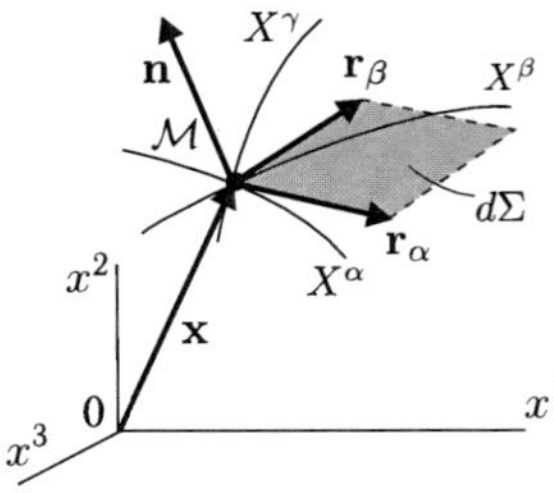


**Fig. 2.11** Introduction of oriented surface element $\mathbf{n}\,d\Sigma$

is called the *area of the surface element* $d\Sigma$ constructed on vectors $d\mathbf{x}_\alpha$ and $d\mathbf{x}_\beta$. Then formula (2.108) takes the form

$$\mathbf{n}\,d\Sigma = \mathbf{r}_\alpha\,dX^\alpha \times \mathbf{r}_\beta\,dX^\beta = d\mathbf{x}_\alpha \times d\mathbf{x}_\beta, \tag{2.111}$$

where $\mathbf{n}\,d\Sigma$ is called the *oriented surface element.*

Show that the normal $\mathbf{n}$ defined by formula (2.108) is a unit vector. According to the property (2.14b) of the vector product of basis vectors and the results of Exercise 2.1.13, we can rewrite Eq. (2.111) in the form

$$\begin{aligned}\mathbf{n}\,d\Sigma &= \mathbf{r}_\alpha \times \mathbf{r}_\beta\,dX^\alpha\,dX^\beta = \sqrt{g}\epsilon_{\alpha\beta\gamma}\mathbf{r}^\gamma\,dX^\alpha\,dX^\beta\\ &= (1/\sqrt{g})\epsilon^{ijk}g_{\alpha i}g_{\beta j}\mathbf{r}_k\,dX^\alpha\,dX^\beta\end{aligned} \tag{2.112}$$

(there is no summation over $\alpha,\beta$). Thus,

$$\begin{aligned}\mathbf{n}\,d\Sigma\cdot\mathbf{n}\,d\Sigma &= \sqrt{g}\,\epsilon_{\alpha\beta\gamma}\mathbf{r}^\gamma\cdot\frac{1}{\sqrt{g}}\,\epsilon^{ijk}g_{\alpha i}g_{\beta j}\mathbf{r}_k)(dX^\alpha\,dX^\beta)^2\\ &= (\epsilon_{\alpha\beta k}\epsilon^{ijk}g_{\alpha i}g_{\beta j})(dX^\alpha dX^\beta)^2 = (\delta^i_\alpha\delta^j_\beta - \delta^i_\beta\delta^j_\alpha)g_{\alpha i}g_{\beta j}(dX^\alpha dX^\beta)^2\\ &= (g_{\alpha\alpha}g_{\beta\beta} - g^2_{\alpha\beta})(dX^\alpha dX^\beta)^2 = \widetilde{g}(dX^\alpha dX^\beta)^2 = d\Sigma^2.\end{aligned} \tag{2.113}$$

Hence, $\mathbf{n}\cdot\mathbf{n} = 1$.

The surface element $d\Sigma$ in $\mathcal{K}$ corresponds to the surface element $d\overset{\circ}{\Sigma}$ in $\overset{\circ}{\mathcal{K}}$, which is constructed on elementary radius–vectors $d\overset{\circ}{\mathbf{x}}_\alpha$ and $d\overset{\circ}{\mathbf{x}}_\beta$:

$$\overset{\circ}{\mathbf{n}}\,d\overset{\circ}{\Sigma} = \overset{\circ}{\mathbf{r}}_\alpha dX^\alpha \times \overset{\circ}{\mathbf{r}}_\beta\,dX^\beta = \overset{\circ}{\mathbf{r}}_\alpha \times \overset{\circ}{\mathbf{r}}_\beta\,dX^\alpha\,dX^\beta. \tag{2.114}$$

Here $\overset{\circ}{\mathbf{n}}$ is the unit normal to $d\overset{\circ}{\Sigma}$.

Since $\mathbf{r}^\gamma = \mathbf{F}^{-1\mathrm{T}}\cdot\overset{\circ}{\mathbf{r}}{}^\gamma$, we get

$$\begin{aligned}\mathbf{n}\,d\Sigma &= \sqrt{g}\,\epsilon_{\alpha\beta\gamma}\mathbf{F}^{-1\mathrm{T}}\cdot\overset{\circ}{\mathbf{r}}{}^\gamma\,dX^\alpha\,dX^\beta\\ &= \sqrt{g/\overset{\circ}{g}}\,\mathbf{F}^{-1\mathrm{T}}\cdot\overset{\circ}{\mathbf{r}}_\alpha\times\overset{\circ}{\mathbf{r}}_\beta\,dX^\alpha\,dX^\beta\\ &= \sqrt{g/\overset{\circ}{g}}\,\mathbf{F}^{-1\mathrm{T}}\cdot\overset{\circ}{\mathbf{n}}\,d\overset{\circ}{\Sigma}.\end{aligned} \tag{2.115}$$

Thus, we have proved the following theorem.

**Theorem 2.7.** *The oriented surface elements* $\overset{\circ}{\mathbf{n}}\, d\overset{\circ}{\Sigma}$ *and* $\mathbf{n}\, d\Sigma$ *in* $\overset{\circ}{\mathcal{K}}$ *and* $\mathcal{K}$ *are connected by the relation*

$$\mathbf{n}\, d\Sigma = \sqrt{g/\overset{\circ}{g}}\, \overset{\circ}{\mathbf{n}} \cdot \mathbf{F}^{-1}\, d\overset{\circ}{\Sigma} = \sqrt{g/\overset{\circ}{g}}\, \mathbf{F}^{-1\mathrm{T}} \cdot \overset{\circ}{\mathbf{n}}\, d\overset{\circ}{\Sigma}. \tag{2.116}$$

With the help of the deformation measures we can derive formulae connecting the normals $\mathbf{n}$ and $\overset{\circ}{\mathbf{n}}$ to the surface element containing the same material points in $\mathcal{K}$ and $\overset{\circ}{\mathcal{K}}$.

Multiplying Eq. (2.116) by itself and taking the formula $\mathbf{n} \cdot \mathbf{n} = 1$ into account, we get

$$d\Sigma^2 = \frac{g}{\overset{\circ}{g}}(\overset{\circ}{\mathbf{n}} \cdot \mathbf{F}^{-1} \cdot \mathbf{F}^{-1\mathrm{T}} \cdot \overset{\circ}{\mathbf{n}})\, d\overset{\circ}{\Sigma}{}^2 = \frac{g}{\overset{\circ}{g}}(\overset{\circ}{\mathbf{n}} \cdot \mathbf{G}^{-1} \cdot \overset{\circ}{\mathbf{n}})\, d\overset{\circ}{\Sigma}{}^2. \tag{2.117}$$

Thus,

$$d\Sigma / d\overset{\circ}{\Sigma} = \sqrt{g/\overset{\circ}{g}}\, (\overset{\circ}{\mathbf{n}} \cdot \mathbf{G}^{-1} \cdot \overset{\circ}{\mathbf{n}})^{1/2}. \tag{2.118}$$

On the other hand, expressing $\overset{\circ}{\mathbf{n}}$ from (2.116) and then multiplying the obtained relation by itself, we obtain

$$d\overset{\circ}{\Sigma}{}^2 = \frac{\overset{\circ}{g}}{g}\, (\mathbf{n} \cdot \mathbf{F} \cdot \mathbf{F}^{\mathrm{T}} \cdot \mathbf{n})\, d\Sigma^2 = \frac{\overset{\circ}{g}}{g}(\mathbf{n} \cdot \mathbf{g}^{-1} \cdot \mathbf{n})\, d\Sigma^2, \tag{2.119}$$

Thus, we find that

$$d\overset{\circ}{\Sigma} / d\Sigma = \sqrt{\overset{\circ}{g}/g}\, (\mathbf{n} \cdot \mathbf{g}^{-1} \cdot \mathbf{n})^{1/2}. \tag{2.120}$$

On introducing the notation

$$\begin{aligned} \overset{\circ}{k} &= (\overset{\circ}{\mathbf{n}} \cdot \mathbf{G}^{-1} \cdot \overset{\circ}{\mathbf{n}})^{1/2} = (\overset{\circ}{\mathbf{n}} \cdot \mathbf{F}^{-1} \cdot \mathbf{F}^{-1\mathrm{T}} \cdot \overset{\circ}{\mathbf{n}})^{1/2}, \\ k &= (\mathbf{n} \cdot \mathbf{g}^{-1} \cdot \mathbf{n})^{1/2} = (\mathbf{n} \cdot \mathbf{F} \cdot \mathbf{F}^{\mathrm{T}} \cdot \mathbf{n})^{1/2}, \end{aligned} \tag{2.121}$$

relations (2.119) take the form

$$d\overset{\circ}{\Sigma} / d\Sigma = \sqrt{\overset{\circ}{g}/g}\, k = \sqrt{\overset{\circ}{g}/g}\, (1/\overset{\circ}{k}). \tag{2.122}$$

Thus, we get

$$k = 1/\overset{\circ}{k}. \tag{2.123}$$

Substitution of (2.119) and (2.120) into (2.116) gives the desired relations

$$\overset{\circ}{k}\mathbf{n} = \mathbf{F}^{-1\mathrm{T}} \cdot \overset{\circ}{\mathbf{n}}, \qquad k\overset{\circ}{\mathbf{n}} = \mathbf{F}^{\mathrm{T}} \cdot \mathbf{n}. \tag{2.124}$$

### 2.2.7 Representation of the Inverse Metric Matrix in terms of Components of the Deformation Tensor

Components $g_{ij}$ of the metric matrix are connected to components of the deformation tensor $\varepsilon_{ij}$ by relation (2.69). In continuum mechanics, one often needs to know the expression of the inverse metric matrix $g^{ij}$ in terms of $\varepsilon_{ij}$ (but not in terms of $\varepsilon^{ij}$). To derive this relation, we should use the connection between components of a matrix and its inverse (see [12]):

$$g^{ij} = \frac{1}{2g}\epsilon^{imn}\epsilon^{jkl}g_{mk}g_{nl}. \tag{2.125}$$

For $\overset{\circ}{g}{}^{ij}$, we have the similar formula

$$\overset{\circ}{g}{}^{ij} = \frac{1}{2\overset{\circ}{g}}\epsilon^{imn}\epsilon^{jkl}\overset{\circ}{g}_{mk}\overset{\circ}{g}_{nl}. \tag{2.126}$$

On substituting the relations (2.69) between $g_{mn}$, $\overset{\circ}{g}_{mn}$ and $\varepsilon_{mn}$ into (2.126), we get

$$\begin{aligned} g^{ij} &= \frac{1}{2g}\epsilon^{imn}\epsilon^{jkl}(\overset{\circ}{g}_{mk} + 2\varepsilon_{mk})(\overset{\circ}{g}_{nl} + 2\varepsilon_{nl}) \\ &= \frac{1}{2g}\epsilon^{imn}\epsilon^{jkl}(\overset{\circ}{g}_{mk}\overset{\circ}{g}_{nl} + 2\overset{\circ}{g}_{mk}\varepsilon_{nl} + 2\overset{\circ}{g}_{nl}\varepsilon_{mk} + 4\varepsilon_{mk}\varepsilon_{nl}). \end{aligned} \tag{2.127}$$

Removing the parentheses, modify four summands in (2.127) in the following way. The first summand with taking formula (2.126) into account gives the matrix $\overset{\circ}{g}{}^{ij}(\overset{\circ}{g}/g)$. To transform the second and the third summands, we should use the formulae

$$\left(1/\sqrt{\overset{\circ}{g}}\right)\epsilon^{jkl} = \sqrt{\overset{\circ}{g}}\,\epsilon_{tsp}\overset{\circ}{g}{}^{jt}\overset{\circ}{g}{}^{ks}\overset{\circ}{g}{}^{lp}. \tag{2.128}$$

$$\begin{aligned} (1/\overset{\circ}{g})\epsilon^{imn}\epsilon^{jkl}\overset{\circ}{g}_{mk} &= \epsilon^{imn}\epsilon_{tsp}\overset{\circ}{g}{}^{jt}\overset{\circ}{g}{}^{ks}\overset{\circ}{g}{}^{lp}\overset{\circ}{g}_{mk} = \epsilon^{imn}\epsilon_{tmp}\overset{\circ}{g}{}^{jt}\overset{\circ}{g}{}^{lp} \\ &= (\delta^i_t\delta^n_p - \delta^i_p\delta^n_t)\overset{\circ}{g}{}^{jt}\overset{\circ}{g}{}^{lp} = \overset{\circ}{g}{}^{ij}\overset{\circ}{g}{}^{nl} - \overset{\circ}{g}{}^{jn}\overset{\circ}{g}{}^{il}. \end{aligned} \tag{2.129}$$

$$(1/\overset{\circ}{g})\epsilon^{imn}\epsilon^{jkl}\overset{\circ}{g}_{mk}\varepsilon_{nl} = (\overset{\circ}{g}{}^{ij}\overset{\circ}{g}{}^{nl} - \overset{\circ}{g}{}^{jn}\overset{\circ}{g}{}^{il})\varepsilon_{nl}. \tag{2.130}$$

Formula (2.128) follows from the relation

$$\sqrt{g}\,\epsilon_{ijk} = (1/\sqrt{g})\,\epsilon^{mnl}g_{mi}g_{nj}g_{lk}$$

(see [12]), and relationship (2.129) has been obtained by using formula (2.128) and the properties of the Levi-Civita symbols (see Exercise 2.1.13).

On substituting formula (2.130) into (2.127), we get

$$g^{ij} = \frac{\mathring{g}}{g}\left(\mathring{g}^{ij} + 2\left(\mathring{g}^{ij}\mathring{g}^{nl} - 2\mathring{g}^{il}\mathring{g}^{jn}\right)\varepsilon_{nl} + \frac{2}{\mathring{g}}\epsilon^{imn}\epsilon^{jkl}\varepsilon_{mk}\varepsilon_{nl}\right). \quad (2.131)$$

Finally, we should express the determinant $g = \det(g_{ij})$ in terms of $\varepsilon_{ij}$. To do this, we multiply relation (2.131) by $g_{ij}$ and take formula (2.69) into account:

$$3g = \mathring{g}\left(\mathring{g}^{ij} + 2(\mathring{g}^{ij}\mathring{g}^{nl} - \mathring{g}^{il}\mathring{g}^{jn})\varepsilon_{nl} + \frac{2}{\mathring{g}}\epsilon^{imn}\epsilon^{jpl}\varepsilon_{mp}\varepsilon_{nl}\right)(\mathring{g}_{ij} + 2\varepsilon_{ij}). \quad (2.132)$$

Thus, we get

$$3g = \mathring{g}\Big(3 + 4\mathring{g}^{nl}\varepsilon_{nl} + (2/\mathring{g})\epsilon^{imn}\epsilon^{jpl}\mathring{g}_{ij}\varepsilon_{mp}\varepsilon_{nl} + 2\mathring{g}^{ij}\varepsilon_{ij} + 4(\mathring{g}^{ij}\mathring{g}^{nl} - \mathring{g}^{il}\mathring{g}^{jn})\varepsilon_{nl}\varepsilon_{ij} + (4/\mathring{g})\epsilon^{imn}\epsilon^{jpl}\varepsilon_{ij}\varepsilon_{mp}\varepsilon_{nl}\Big). \quad (2.133)$$

Modifying the third summand on the right-hand side by formula (2.130) and introducing the notation

$$I_{1\varepsilon} = \mathring{g}^{nl}\varepsilon_{nl}, \quad I_{2\varepsilon} = \frac{1}{2}\left(\mathring{g}^{ij}\mathring{g}^{nl} - \mathring{g}^{il}\mathring{g}^{jn}\right)\varepsilon_{ij}\varepsilon_{nl}, \quad I_{3\varepsilon} = \det(\varepsilon_{ij}\mathring{g}^{ik}), \quad (2.134)$$

from (2.133) we get the desired formula

$$g = \mathring{g}(1 + 2I_{1\varepsilon} + 4I_{2\varepsilon} + 8I_{3\varepsilon}). \quad (2.135)$$

Here we have taken account of formula (2.125) for the matrix determinant and also the relation $I_{3\varepsilon} = (1/\mathring{g})\det(\varepsilon_{ij})$. Thus, we have proved the following theorem.

**Theorem 2.8.** *The inverse metric matrix $g^{ij}$ is expressed in terms of components $\varepsilon_{ij}$ of the deformation tensor and $\mathring{g}^{ij}$ by formulae* (2.131) *and* (2.135).

Formulae (2.131) and (2.135) allow us to find the expression of contravariant components $\varepsilon^{ij}$ of the deformation tensor in terms of $\varepsilon_{ij}$. It follows from (2.69) that

$$\varepsilon^{ij} = (1/2)\left(\mathring{g}^{ij} - g^{ij}\right) = \mathring{g}^{ij}\frac{g - \mathring{g}}{2g} - \frac{\mathring{g}}{g}\left(\mathring{g}^{ij}\mathring{g}^{nl} - \mathring{g}^{il}\mathring{g}^{jn}\right)\varepsilon_{nl} - \frac{1}{g}\epsilon^{imn}\epsilon^{jkl}\varepsilon_{mk}\varepsilon_{nl}. \quad (2.136)$$

Substitution of formulae (2.93) or (2.94) into (2.131) and (2.136) gives the expression for components $\varepsilon^{ij}$ in terms of components of the displacement vector $\mathring{u}_i$ or $u_i$.

## Exercises for 2.2

**2.2.1.** Using the results of Exercise 2.1.1, show that the deformation gradient $\mathbf{F}$ and its inverse $\mathbf{F}^{-1}$ for the problem on a beam in tension (see Example 2.1) have the forms

$$\mathbf{F} = \mathbf{F}^{\mathrm{T}} = \sum_{\alpha=1}^{3} k_\alpha \bar{\mathbf{e}}_\alpha \otimes \bar{\mathbf{e}}_\alpha, \quad \mathbf{F}^{-1} = \mathbf{F}^{-1\mathrm{T}} = \sum_{\alpha=1}^{3} \frac{1}{k_\alpha} \bar{\mathbf{e}}_\alpha \otimes \bar{\mathbf{e}}_\alpha.$$

For this problem, the deformation tensors are determined by the formulae

$$\mathbf{C} = \boldsymbol{\Lambda} = \frac{1}{2}\sum_{\alpha=1}^{3}(k_\alpha^2 - 1)\bar{\mathbf{e}}_\alpha \otimes \bar{\mathbf{e}}_\alpha, \quad \mathbf{A} = \boldsymbol{\Lambda} = \frac{1}{2}\sum_{\alpha=1}^{3}(1 - k_\alpha^{-2})\bar{\mathbf{e}}_\alpha \otimes \bar{\mathbf{e}}_\alpha,$$

the deformation measures are determined as follows:

$$\mathbf{G} = \mathbf{g}^{-1} = \sum_{\alpha=1}^{3} k_\alpha^2 \bar{\mathbf{e}}_\alpha \otimes \bar{\mathbf{e}}_\alpha, \quad \mathbf{g} = \mathbf{G}^{-1} = \sum_{\alpha=1}^{3} k_\alpha^{-2} \bar{\mathbf{e}}_\alpha \otimes \bar{\mathbf{e}}_\alpha,$$

and components of the deformation tensor take the forms

$$\varepsilon_{\alpha\beta} = \frac{1}{2}(k_\alpha^2 - 1)\delta_{\alpha\beta}, \quad \varepsilon^{\alpha\beta} = \frac{1}{2}(1 - k_\alpha^{-2})\delta_{\alpha\beta}.$$

**2.2.2.** Using the results of Exercise 2.1.2, show that for the problem on a simple shear we have the following formulae for the deformation gradient:

$$\mathbf{F} = \overset{\circ}{F}{}^{ij} \bar{\mathbf{e}}_i \otimes \bar{\mathbf{e}}_j = \mathbf{E} + a\bar{\mathbf{e}}_1 \otimes \bar{\mathbf{e}}_2, \quad \mathbf{F}^{\mathrm{T}} = \mathbf{E} + a\bar{\mathbf{e}}_2 \otimes \bar{\mathbf{e}}_1,$$

$$\mathbf{F}^{-1} = \mathbf{E} - a\bar{\mathbf{e}}_1 \otimes \bar{\mathbf{e}}_2, \quad \mathbf{F}^{-1\mathrm{T}} = \mathbf{E} - a\bar{\mathbf{e}}_2 \otimes \bar{\mathbf{e}}_1,$$

i.e.

$$\overset{\circ}{F}{}^{ij} = \begin{pmatrix} 1 & a & 0 \\ 0 & 1 & 0 \\ 0 & 0 & 1 \end{pmatrix}, \quad \det \mathbf{F} = 1,$$

for the deformation tensors:

$$\mathbf{C} = (a/2)\mathbf{O}_3 + (a^2/2)\bar{\mathbf{e}}_2^2, \quad \mathbf{A} = (a/2)\mathbf{O}_3 - (a^2/2)\bar{\mathbf{e}}_2^2,$$

$$\boldsymbol{\Lambda} = (a/2)\mathbf{O}_3 - (a^2/2)\bar{\mathbf{e}}_1^2, \quad \mathbf{J} = (a/2)\mathbf{O}_3 + (a^2/2)\bar{\mathbf{e}}_1^2,$$

and for components of the deformation tensor:

$$(\varepsilon_{ij}) = \begin{pmatrix} 0 & a/2 & 0 \\ a/2 & a^2/2 & 0 \\ 0 & 0 & 0 \end{pmatrix}, \quad (\varepsilon^{ij}) = \begin{pmatrix} 0 & a/2 & 0 \\ a/2 & -a^2/2 & 0 \\ 0 & 0 & 0 \end{pmatrix}.$$

Here we have introduced the notation

$$\mathbf{O}_3 \equiv \bar{\mathbf{e}}_1 \otimes \bar{\mathbf{e}}_2 + \bar{\mathbf{e}}_2 \otimes \bar{\mathbf{e}}_1, \quad \bar{\mathbf{e}}_\alpha^2 \equiv \bar{\mathbf{e}}_\alpha \otimes \bar{\mathbf{e}}_\alpha, \quad \alpha = 1, 2, 3.$$

**2.2.3.** Using the formulae from Example 2.3 (see Sect. 2.1.1), show that for the problem on rotation of a beam with extension, the deformation gradient has the form

$$\mathbf{F} = F_0{}^i{}_j \bar{\mathbf{e}}_i \otimes \bar{\mathbf{e}}^j = \cos\varphi \sum_{\alpha=1}^{2} k_\alpha \bar{\mathbf{e}}_\alpha \otimes \bar{\mathbf{e}}_\alpha + k_3 \bar{\mathbf{e}}_3 \otimes \bar{\mathbf{e}}_3 + \sin\varphi k_2 (\bar{\mathbf{e}}_2 \otimes \bar{\mathbf{e}}_1 - \bar{\mathbf{e}}_1 \otimes \bar{\mathbf{e}}_2).$$

**2.2.4.** Using formulae (2.36), (2.85)–(2.92) and the results of Exercise 2.1.7, show that local basis vectors are connected to displacements by the relations

$$\mathbf{r}_i = \mathring{\mathbf{r}}_k (\delta_i^k + \mathring{\nabla}^k \mathring{u}_i), \quad \mathring{\mathbf{r}}_i = \mathbf{r}_k (\delta_i^k - \nabla_i u^k).$$

**2.2.5.** Using formulae (2.104) and (2.106), show that the physical components of the deformation tensor $\widehat{\mathring{\varepsilon}_{\alpha\beta}} = \varepsilon_{\alpha\beta} / \sqrt{\mathring{g}_{\alpha\alpha} \mathring{g}_{\beta\beta}}$ are connected to relative elongations $\delta_\alpha$ and angles $\chi_{\alpha\beta}$ by the relations

$$\widehat{\mathring{\varepsilon}_{\alpha\alpha}} = \frac{1}{2}((1 + \delta_\alpha)^2 - 1), \quad \widehat{\mathring{\varepsilon}_{\alpha\beta}} = \frac{1}{2}(1 + \delta_\alpha)(1 + \delta_\beta) \sin \chi_{\alpha\beta}.$$

**2.2.6.** Show that in the basis $\widetilde{\mathbf{r}}_i$ of curvilinear coordinate system $\widetilde{X}^i$ the expression of the tensor $\mathbf{F}^{-1}$ in terms of $\nabla \otimes \mathbf{u}$ can be rewritten in the form similar to (2.89)–(2.92):

$$\mathbf{u} = \widetilde{u}^k \widetilde{\mathbf{r}}_k, \quad \mathbf{F}^{-1} = (\widetilde{F}^{-1})^k{}_i \widetilde{\mathbf{r}}_k \otimes \widetilde{\mathbf{r}}^i, \quad (\widetilde{F}_{-1})^k_i = \delta_i^k - \widetilde{\nabla}_i \widetilde{u}^k.$$

**2.2.7.** Using formula (2.34), show that the following relationships hold:

$$|d\mathbf{x}|^2 = d\mathring{\mathbf{x}} \cdot \mathbf{G} \cdot d\mathring{\mathbf{x}}, \quad |d\mathring{\mathbf{x}}|^2 = d\mathbf{x} \cdot \mathbf{g} \cdot d\mathbf{x}.$$

## 2.3 Polar Decomposition

### 2.3.1 *Theorem on Polar Decomposition*

According to (2.36), the tensor $\mathbf{F}$ can be considered as a tensor of the linear transformation from the basis $\mathring{\mathbf{r}}_i$ to the basis $\mathbf{r}_i$. Since the vectors $\mathring{\mathbf{r}}_i$ and $\mathbf{r}_i$ are linearly independent, the tensor $\mathbf{F}$ is nonsingular. Then for this tensor the following theorem is valid.

**Theorem 2.9 (on the polar decomposition).** *Any nonsingular second-order tensor* $\mathbf{F}$ *can be represented as the scalar product of two second-order tensors:*

$$\mathbf{F} = \mathbf{O} \cdot \mathbf{U} \quad or \quad \mathbf{F} = \mathbf{V} \cdot \mathbf{O}. \tag{2.137}$$

*Here* $\mathbf{U}$ *and* $\mathbf{V}$ *are the symmetric and positive-definite tensors,* $\mathbf{O}$ *is the orthogonal tensor, and each of the decompositions* (2.137) *is unique.*

▼ Prove the existence of the decomposition (2.137) in the constructive way, i.e. we should construct the tensors **U**, **V** and **O**. To do this, consider the contractions of the tensor **F** with its transpose: $\mathbf{F}^{\mathrm{T}} \cdot \mathbf{F}$ and $\mathbf{F} \cdot \mathbf{F}^{\mathrm{T}}$. Both the tensors are symmetric, because

$$(\mathbf{F}^{\mathrm{T}} \cdot \mathbf{F})^{\mathrm{T}} = \mathbf{F}^{\mathrm{T}} \cdot (\mathbf{F}^{\mathrm{T}})^{\mathrm{T}} = \mathbf{F}^{\mathrm{T}} \cdot \mathbf{F} \text{ and } (\mathbf{F} \cdot \mathbf{F}^{\mathrm{T}})^{\mathrm{T}} = (\mathbf{F}^{\mathrm{T}})^{\mathrm{T}} \cdot \mathbf{F}^{\mathrm{T}} = \mathbf{F} \cdot \mathbf{F}^{\mathrm{T}}, \tag{2.138}$$

and positive-definite:

$$\mathbf{a} \cdot (\mathbf{F}^{\mathrm{T}} \cdot \mathbf{F}) \cdot \mathbf{a} = (\mathbf{a} \cdot \mathbf{F}^{\mathrm{T}}) \cdot (\mathbf{F} \cdot \mathbf{a}) = (\mathbf{F} \cdot \mathbf{a}) \cdot (\mathbf{F} \cdot \mathbf{a}) = \mathbf{b} \cdot \mathbf{b} = |\mathbf{b}|^2 > 0 \tag{2.139}$$

for any non-zero vector **a**, where $\mathbf{b} = \mathbf{F} \cdot \mathbf{a}$. Since any symmetric positive-definite tensor has three real positive eigenvalues [12], eigenvalues of tensors $\mathbf{F}^{\mathrm{T}} \cdot \mathbf{F}$ and $\mathbf{F} \cdot \mathbf{F}^{\mathrm{T}}$ can be denoted as $\overset{\circ}{\lambda}{}^2_\alpha$ and $\lambda^2_\alpha$. These tensors are diagonal in their eigenbases, i.e. they have the following forms:

$$\mathbf{F}^{\mathrm{T}} \cdot \mathbf{F} = \sum_{\alpha=1}^{3} \overset{\circ}{\lambda}{}^2_\alpha \overset{\circ}{\mathbf{p}}_\alpha \otimes \overset{\circ}{\mathbf{p}}_\alpha, \quad \mathbf{F} \cdot \mathbf{F}^{\mathrm{T}} = \sum_{\alpha=1}^{3} \lambda^2_\alpha \mathbf{p}_\alpha \otimes \mathbf{p}_\alpha. \tag{2.140}$$

Here $\overset{\circ}{\mathbf{p}}_\alpha$ are the eigenvectors of the tensor $\mathbf{F}^{\mathrm{T}} \cdot \mathbf{F}$ and $\mathbf{p}_\alpha$ – of the tensor $\mathbf{F} \cdot \mathbf{F}^{\mathrm{T}}$, which are real-valued and orthonormal:

$$\overset{\circ}{\mathbf{p}}_\alpha \cdot \overset{\circ}{\mathbf{p}}_\beta = \delta_{\alpha\beta}, \quad \mathbf{p}_\alpha \cdot \mathbf{p}_\beta = \delta_{\alpha\beta}. \tag{2.141}$$

The right-hand sides of (2.140) are the squares of certain tensors **U** and **V** defined as

$$\mathbf{U} = \sum_{\alpha=1}^{3} \overset{\circ}{\lambda}_\alpha \overset{\circ}{\mathbf{p}}_\alpha \otimes \overset{\circ}{\mathbf{p}}_\alpha, \ \overset{\circ}{\lambda}_\alpha > 0; \quad \mathbf{V} = \sum_{\alpha=1}^{3} \lambda_\alpha \mathbf{p}_\alpha \otimes \mathbf{p}_\alpha, \ \lambda_\alpha > 0. \tag{2.142}$$

Here signs at $\lambda_\alpha$ are always chosen positive.

In this case, the following relations are valid:

$$\mathbf{F}^{\mathrm{T}} \cdot \mathbf{F} = \mathbf{U}^2, \quad \mathbf{F} \cdot \mathbf{F}^{\mathrm{T}} = \mathbf{V}^2. \tag{2.143}$$

The constructed tensors **V** and **U** are symmetric due to formula (2.142) and positive-definite, because for any nonzero vector **a** we have

$$\mathbf{a}\cdot\mathbf{U}\cdot\mathbf{a}=\sum_{\alpha=1}^{3}\overset{\circ}{\lambda}_{\alpha}\mathbf{a}\cdot\overset{\circ}{\mathbf{p}}_{\alpha}\otimes\overset{\circ}{\mathbf{p}}_{\alpha}\cdot\mathbf{a}=\sum_{\alpha=1}^{3}\overset{\circ}{\lambda}_{\alpha}(\mathbf{a}\cdot\overset{\circ}{\mathbf{p}}_{\alpha})^{2}>0, \tag{2.144}$$

as $\overset{\circ}{\lambda}_{\alpha}>0$. In a similar way, we can prove that the tensor **V** is positive-definite.

Both the tensors **V** and **U** are nonsingular, because, under the conditions of the theorem, the tensor **F** is nonsingular. And from (2.143) we get

$$(\det\mathbf{U})^{2}=\det\mathbf{U}^{2}=\det(\mathbf{F}^{\mathrm{T}}\cdot\mathbf{F})=(\det\mathbf{F})^{2}\neq0. \tag{2.145}$$

Then there exist inverse tensors $\mathbf{U}^{-1}$ and $\mathbf{V}^{-1}$, with the help of which we can construct two more new tensors

$$\overset{\circ}{\mathbf{O}}=\mathbf{F}\cdot\mathbf{U}^{-1},\qquad \mathbf{O}=\mathbf{V}^{-1}\cdot\mathbf{F}, \tag{2.146}$$

which are orthogonal. Indeed,

$$\overset{\circ}{\mathbf{O}}{}^{\mathrm{T}}\cdot\overset{\circ}{\mathbf{O}}=(\mathbf{F}\cdot\mathbf{U}^{-1})^{\mathrm{T}}\cdot(\mathbf{F}\cdot\mathbf{U}^{-1})=\mathbf{U}^{-1}\cdot\mathbf{F}^{\mathrm{T}}\cdot\mathbf{F}\cdot\mathbf{U}^{-1}=\mathbf{U}^{-1}\cdot\mathbf{U}^{2}\cdot\mathbf{U}^{-1}=\mathbf{E}. \tag{2.147}$$

According to [12], this means that the tensor $\overset{\circ}{\mathbf{O}}$ is orthogonal. In a similar way, we can show that the tensor **O** is orthogonal as well.

Thus, we have really constructed the tensors **U** and $\overset{\circ}{\mathbf{O}}$, and also **V** and **O**, the product of which, due to (2.146), gives the tensor **F**:

$$\mathbf{F}=\overset{\circ}{\mathbf{O}}\cdot\mathbf{U}=\mathbf{V}\cdot\mathbf{O}. \tag{2.148}$$

Here **U** and **V** are symmetric, positive-definite tensors, **O** and $\overset{\circ}{\mathbf{O}}$ are orthogonal tensors.

Show that each of the decompositions (2.148) is unique. By contradiction, let there be one more resolution, for example

$$\mathbf{F}=\widetilde{\overset{\circ}{\mathbf{O}}}\cdot\widetilde{\mathbf{U}}. \tag{2.149}$$

But then

$$\mathbf{F}^{\mathrm{T}}\cdot\mathbf{F}=\widetilde{\mathbf{U}}^{2}=\mathbf{U}^{2}, \tag{2.150}$$

hence, $\widetilde{\mathbf{U}}=\mathbf{U}$, because the decomposition of the tensor $\mathbf{F}^{\mathrm{T}}\cdot\mathbf{F}$ for its eigenbasis is unique. Signs at $\overset{\circ}{\lambda}_{\alpha}$ and $\widetilde{\overset{\circ}{\lambda}}_{\alpha}$ are chosen positive by the condition. The coincidence of **U** and $\widetilde{\mathbf{U}}$ leads to the fact that $\widetilde{\overset{\circ}{\mathbf{O}}}$ and $\overset{\circ}{\mathbf{O}}$ are coincident as well, because

$$\widetilde{\overset{\circ}{\mathbf{O}}}=\mathbf{F}\cdot\widetilde{\mathbf{U}}^{-1}=\mathbf{F}\cdot\mathbf{U}^{-1}=\overset{\circ}{\mathbf{O}}. \tag{2.151}$$

This has proved uniqueness of the decomposition (2.148). We can verify uniqueness of the decomposition $\mathbf{F} = \mathbf{V} \cdot \mathbf{O}$ in a similar way.

Finally, we must show that the orthogonal tensors $\overset{\circ}{\mathbf{O}}$ and $\mathbf{O}$ are coincident, i.e. formula (2.137) follows from (2.148). To do this, we construct the tensor

$$\mathbf{F} \cdot \overset{\circ}{\mathbf{O}}^{\mathrm{T}} = \overset{\circ}{\mathbf{O}} \cdot \mathbf{U} \cdot \overset{\circ}{\mathbf{O}}^{\mathrm{T}}. \qquad (2.152)$$

Due to (2.148), this tensor satisfies the following relationship:

$$\overset{\circ}{\mathbf{O}} \cdot \mathbf{U} \cdot \overset{\circ}{\mathbf{O}}^{\mathrm{T}} = \mathbf{V} \cdot \mathbf{O} \cdot \overset{\circ}{\mathbf{O}}^{\mathrm{T}}. \qquad (2.153)$$

The tensor $\mathbf{O} \cdot \overset{\circ}{\mathbf{O}}^{\mathrm{T}}$ is orthogonal, because

$$(\mathbf{O} \cdot \overset{\circ}{\mathbf{O}}^{\mathrm{T}})^{\mathrm{T}} \cdot (\mathbf{O} \cdot \overset{\circ}{\mathbf{O}}^{\mathrm{T}}) = \overset{\circ}{\mathbf{O}} \cdot \mathbf{O}^{\mathrm{T}} \cdot \mathbf{O} \cdot \overset{\circ}{\mathbf{O}}^{\mathrm{T}} = \overset{\circ}{\mathbf{O}} \cdot \overset{\circ}{\mathbf{O}}^{\mathrm{T}} = \mathbf{E}. \qquad (2.154)$$

Then the relationship (2.153) can be considered as the polar decomposition of the tensor $\overset{\circ}{\mathbf{O}} \cdot \mathbf{U} \cdot \overset{\circ}{\mathbf{O}}^{\mathrm{T}}$. This tensor is symmetric, because

$$(\overset{\circ}{\mathbf{O}} \cdot \mathbf{U} \cdot \overset{\circ}{\mathbf{O}}^{\mathrm{T}})^{\mathrm{T}} = (\overset{\circ}{\mathbf{O}}^{\mathrm{T}})^{\mathrm{T}} \cdot (\overset{\circ}{\mathbf{O}} \cdot \mathbf{U})^{\mathrm{T}} = \overset{\circ}{\mathbf{O}} \cdot \mathbf{U} \cdot \overset{\circ}{\mathbf{O}}^{\mathrm{T}}. \qquad (2.155)$$

Then the formal equality

$$\overset{\circ}{\mathbf{O}} \cdot \mathbf{U} \cdot \overset{\circ}{\mathbf{O}}^{\mathrm{T}} = \overset{\circ}{\mathbf{O}} \cdot \mathbf{U} \cdot \overset{\circ}{\mathbf{O}}^{\mathrm{T}} \qquad (2.156)$$

is one more polar decomposition. However, as was shown above, the polar decomposition is unique; hence, the following relationships must be satisfied:

$$\mathbf{V} = \overset{\circ}{\mathbf{O}} \cdot \mathbf{U} \cdot \overset{\circ}{\mathbf{O}}^{\mathrm{T}} \quad \text{and} \quad \mathbf{O} \cdot \overset{\circ}{\mathbf{O}}^{\mathrm{T}} = \mathbf{E}. \qquad (2.157)$$

Thus, the orthogonal tensors $\mathbf{O}$ and $\overset{\circ}{\mathbf{O}}$ are coincident: $\mathbf{O} = \overset{\circ}{\mathbf{O}}$. ▲

The tensors $\mathbf{U}$ and $\mathbf{V}$ are called the *right and left stretch tensors*, respectively, and $\mathbf{O}$ is the *rotation tensor accompanying the deformation*.

The tensor $\mathbf{F}$ has nine independent components, the tensor $\mathbf{O}$ – three independent components, and each of the tensors $\mathbf{U}$ and $\mathbf{V}$ – six independent components.

*Remark 1.* Since the rotation tensor $\mathbf{O}$ is unique in the polar decomposition, from formula (2.157) we get that the stretch tensors $\mathbf{U}$ and $\mathbf{V}$ are connected by means of the tensor $\mathbf{O}$:

$$\mathbf{V} = \mathbf{O} \cdot \mathbf{U} \cdot \mathbf{O}^{\mathrm{T}}, \quad \mathbf{U} = \mathbf{O}^{\mathrm{T}} \cdot \mathbf{V} \cdot \mathbf{O}. \qquad (2.157a)$$

□

**Theorem 2.10.** *The Cauchy–Green and Almansi deformation tensors can be expressed in terms of the stretch tensors* **U** *and* **V** *as follows:*

$$\mathbf{C} = \frac{1}{2}(\mathbf{U}^2 - \mathbf{E}), \quad \mathbf{A} = \frac{1}{2}(\mathbf{E} - \mathbf{V}^{-2}),$$
$$\mathbf{\Lambda} = \frac{1}{2}(\mathbf{E} - \mathbf{U}^{-2}), \quad \mathbf{J} = \frac{1}{2}(\mathbf{V}^2 - \mathbf{E}). \tag{2.158}$$

▼ To see this, let us substitute the polar decomposition (2.137) into (2.72), and then we get the relationships (2.158). ▲

## 2.3.2 Eigenvalues and Eigenbases

**Theorem 2.11.** *Eigenvalues of the tensors* **U** *and* **V** *defined by* (2.142) *are coincident:*

$$\lambda_\alpha = \overset{\circ}{\lambda}_\alpha, \quad \alpha = 1, 2, 3, \tag{2.159}$$

*and eigenvectors* $\overset{\circ}{\mathbf{p}}_\alpha$ *and* $\mathbf{p}_\alpha$ *are connected by the rotation tensor* **O**:

$$\mathbf{p}_\alpha = \mathbf{O} \cdot \overset{\circ}{\mathbf{p}}_\alpha. \tag{2.159a}$$

▼ To prove the theorem, we use the definition (2.142) and the first formula of (2.157a):

$$\mathbf{V} = \sum_{\alpha=1}^{3} \lambda_\alpha \mathbf{p}_\alpha \otimes \mathbf{p}_\alpha = \mathbf{O} \cdot \mathbf{U} \cdot \mathbf{O}^{\mathrm{T}} = \sum_{\alpha=1}^{3} \overset{\circ}{\lambda}_\alpha \mathbf{O} \cdot \overset{\circ}{\mathbf{p}}_\alpha \otimes (\mathbf{O} \cdot \overset{\circ}{\mathbf{p}}_\alpha) = \sum_{\alpha=1}^{3} \overset{\circ}{\lambda}_\alpha \mathbf{p}'_\alpha \otimes \mathbf{p}'_\alpha,$$

where

$$\mathbf{p}'_\alpha = \mathbf{O} \cdot \overset{\circ}{\mathbf{p}}_\alpha.$$

According to the relationship, we have obtained two different eigenbases of the tensor **V** and two sets of eigenvalues, that is impossible. Therefore,

$$\mathbf{p}'_\alpha = \mathbf{O} \cdot \overset{\circ}{\mathbf{p}}_\alpha = \mathbf{p}_\alpha \text{ and } \lambda_\alpha = \overset{\circ}{\lambda}_\alpha,$$

as was to be proved. ▲

Due to (2.141), both the eigenbases are orthogonal. Therefore, reciprocal vectors of the eigenbases do not differ from $\mathbf{p}_\alpha$ and $\overset{\circ}{\mathbf{p}}_\alpha$:

$$\mathbf{p}_\alpha = \mathbf{p}^\alpha, \quad \overset{\circ}{\mathbf{p}}_\alpha = \overset{\circ}{\mathbf{p}}{}^\alpha. \tag{2.160}$$

The important problem for applications is to determine $\lambda_\alpha$, $\mathbf{p}_\alpha$ and $\overset{\circ}{\mathbf{p}}_\alpha$ by the given deformation gradient $\mathbf{F}$. To solve the problem, one should use the following method:

1. Construct the tensor $\mathbf{U}^2 = \mathbf{F}^{\mathrm{T}} \cdot \mathbf{F}$ (or $\mathbf{V}^2 = \mathbf{F} \cdot \mathbf{F}^{\mathrm{T}}$) and find its components in some basis being suitable for a considered problem; for example, in the Cartesian basis $\bar{\mathbf{e}}_i$:
$$\mathbf{U}^2 = (\bar{U}^2)^i{}_j \bar{\mathbf{e}}_i \otimes \bar{\mathbf{e}}^j \quad \text{and} \quad \mathbf{V}^2 = (\bar{V}^2)^i{}_j \bar{\mathbf{e}}_i \otimes \bar{\mathbf{e}}^j.$$
2. Find eigenvalues of the matrix $(\bar{U}^2)^i{}_j$ by solving the characteristic equation
$$\det(\mathbf{U}^2 - \lambda_\alpha^2 \mathbf{E}) = 0, \tag{2.161}$$
which in the basis $\bar{\mathbf{e}}_i$ takes the form
$$\det((U^2)^i{}_j - \lambda_\alpha^2 \delta^i_j) = 0. \tag{2.161a}$$
3. Find eigenvectors $\overset{\circ}{\mathbf{p}}_\alpha$ of the tensor $\mathbf{U}$ and eigenvectors $\mathbf{p}_\alpha$ of the tensor $\mathbf{V}$ from the equations
$$\mathbf{U}^2 \cdot \overset{\circ}{\mathbf{p}}_\alpha = \lambda_\alpha^2 \overset{\circ}{\mathbf{p}}_\alpha, \quad \mathbf{V}^2 \cdot \mathbf{p}_\alpha = \lambda_\alpha^2 \mathbf{p}_\alpha, \tag{2.162}$$
written, for example, in the basis $\bar{\mathbf{e}}_i$:
$$((\bar{U}^2)^i{}_j - \lambda_\alpha^2 \delta^i_j) \overset{\circ}{\widehat{Q}}{}^j{}_\alpha = 0, \quad ((\bar{V}^2)^i{}_j - \lambda_\alpha^2 \delta^i_j) \widehat{Q}^j{}_\alpha = 0, \tag{2.162a}$$
where $\widehat{Q}^j{}_\alpha$ and $\overset{\circ}{\widehat{Q}}{}^j{}_\alpha$ are Jacobian matrices of the eigenvectors:
$$\overset{\circ}{\mathbf{p}}_\alpha = \overset{\circ}{\widehat{Q}}{}^j{}_\alpha \bar{\mathbf{e}}_j, \quad \mathbf{p}_\alpha = \widehat{Q}^j{}_\alpha \bar{\mathbf{e}}_j. \tag{2.163}$$
To determine the matrices $\overset{\circ}{\widehat{Q}}{}^j{}_\alpha$ and $\widehat{Q}^j{}_\alpha$, one should consider only independent equations of the system (2.162a) and the normalization conditions (2.141):
$$|\mathbf{p}_\alpha| = 1, \quad |\overset{\circ}{\mathbf{p}}_\alpha| = 1, \tag{2.164}$$
which are equivalent to the quadratic equations
$$\overset{\circ}{\widehat{Q}}{}^i{}_\alpha \overset{\circ}{\widehat{Q}}{}^j{}_\alpha \delta_{ij} = 1, \quad \widehat{Q}^i{}_\alpha \widehat{Q}^j{}_\alpha \delta_{ij} = 1. \tag{2.164a}$$

4. Write the dyadic products (2.142) and find resolutions of the tensors **U** and **V** for the eigenbases; for example, for the Cartesian basis $\bar{\mathbf{e}}_i$:

$$\mathbf{U} = \sum_{\alpha=1}^{3} \lambda_\alpha \overset{\circ}{\widehat{Q}}{}^i{}_\alpha \overset{\circ}{\widehat{Q}}{}^j{}_\alpha \bar{\mathbf{e}}_i \otimes \bar{\mathbf{e}}_j, \quad \mathbf{V} = \sum_{\alpha=1}^{3} \lambda_\alpha \widehat{Q}^i{}_\alpha \widehat{Q}^j{}_\alpha \bar{\mathbf{e}}_i \otimes \bar{\mathbf{e}}_j.$$

Exercises 2.3.2–2.3.4 show examples of determination of the tensors **U** and **V**.

*Remark 2.* Notice that a solution of the quadratic equations (2.164a) may be not unique due to the choice of signs of matrix components $\overset{\circ}{\widehat{Q}}{}^i{}_\alpha$ and $\widehat{Q}^i{}_\alpha$, this ambiguity is resolved by applying one more additional condition, namely the condition of coincidence of the vectors $\overset{\circ}{\mathbf{p}}_\alpha$ and $\mathbf{p}_\alpha$ when $t \to 0_+$:

$$t \to 0_+ \quad \Rightarrow \quad \mathbf{p}_\alpha(t) = \overset{\circ}{\mathbf{p}}_\alpha(t), \quad \alpha = 1,\ 2,\ 3.$$

For the matrix $\overset{\circ}{\widehat{Q}}{}^i{}_\alpha$, the ambiguity of the sign choice remains. However, if there is a field of eigenvectors $\overset{\circ}{\mathbf{p}}_\alpha(\mathbf{x}, t)$, then this ambiguity may be retained only at one point $\mathbf{x}_0$ at one time, for example, $t = 0$; and for the remaining $\mathbf{x}$ and $t$, a sign at $\widehat{Q}^i{}_\alpha$ is chosen from the continuity condition of the vector field $\overset{\circ}{\mathbf{p}}_\alpha(\mathbf{x}, t)$ (for continuous motions). If the eigenvector field $\overset{\circ}{\mathbf{p}}_\alpha(\mathbf{x}_0, 0)$ contains the vectors $\bar{\mathbf{e}}_\alpha$, then the remaining ambiguity is resolved by the condition $\overset{\circ}{\mathbf{p}}_\alpha(\mathbf{x}_0, 0) = \bar{\mathbf{e}}_\alpha$.

The ambiguity of a solution of the system (2.162a), (2.164a) may also appear, if at some time $t_1$ at a point $\mathbf{x}$ the eigenvalues $\lambda_\alpha(t_1)$ prove to be triple. In this case, values of the matrices $\widehat{Q}^i{}_\alpha(t_1)$ and $\overset{\circ}{\widehat{Q}}{}^i{}_\alpha(t_1)$ are determined, as a rule, by passage to the limit:

$$\widehat{Q}^i{}_\alpha(t_1) = \lim_{t \to t_1} \widehat{Q}^i{}_\alpha(t), \quad \overset{\circ}{\widehat{Q}}{}^i{}_\alpha(t_1) = \lim_{t \to t_1} \overset{\circ}{\widehat{Q}}{}^i{}_\alpha(t), \quad \alpha = 1, 2, 3.$$

In the case of double eigenvalues $\lambda_\alpha$, these formulae are applied only to their corresponding matrix components $\widehat{Q}^i{}_\alpha$ and $\overset{\circ}{\widehat{Q}}{}^i{}_\alpha$. □

### 2.3.3 *Representation of the Deformation Tensors in Eigenbases*

**Theorem 2.12.** *In the tensor bases* $\mathbf{p}_\alpha \otimes \mathbf{p}_\beta$ *and* $\overset{\circ}{\mathbf{p}}_\alpha \otimes \overset{\circ}{\mathbf{p}}_\beta$, *the Cauchy–Green tensors* **C** *and* **J**, *the Almansi tensors* **A** *and* **Λ**, *and the deformation measures* **G**, $\mathbf{g}^{-1}$ *and* $\mathbf{G}^{-1}$, **g** *have the diagonal form:*

$$\mathbf{C} = \sum_{\alpha=1}^{3} \frac{1}{2}(\lambda_\alpha^2 - 1)\overset{\circ}{\mathbf{p}}_\alpha \otimes \overset{\circ}{\mathbf{p}}_\alpha, \quad \mathbf{\Lambda} = \sum_{\alpha=1}^{3} \frac{1}{2}(1 - \lambda_\alpha^{-2})\overset{\circ}{\mathbf{p}}_\alpha \otimes \overset{\circ}{\mathbf{p}}_\alpha,$$

$$\mathbf{A} = \sum_{\alpha=1}^{3} \frac{1}{2}(1 - \lambda_\alpha^{-2})\mathbf{p}_\alpha \otimes \mathbf{p}_\alpha, \quad \mathbf{J} = \sum_{\alpha=1}^{3} \frac{1}{2}(\lambda_\alpha^2 - 1)\mathbf{p}_\alpha \otimes \mathbf{p}_\alpha; \quad (2.165a)$$

and

$$\mathbf{G} = \sum_{\alpha=1}^{3} \lambda_\alpha^2 \overset{\circ}{\mathbf{p}}_\alpha \otimes \overset{\circ}{\mathbf{p}}_\alpha, \quad \mathbf{G}^{-1} = \sum_{\alpha=1}^{3} \lambda_\alpha^{-2} \overset{\circ}{\mathbf{p}}_\alpha \otimes \overset{\circ}{\mathbf{p}}_\alpha,$$

$$\mathbf{g}^{-1} = \sum_{\alpha=1}^{3} \lambda_\alpha^2 \mathbf{p}_\alpha \otimes \mathbf{p}_\alpha, \quad \mathbf{g} = \sum_{\alpha=1}^{3} \lambda_\alpha^{-2} \mathbf{p}_\alpha \otimes \mathbf{p}_\alpha. \quad (2.165b)$$

▼ On substituting formulae (2.142) into (2.158), we get (2.165a). Formulae (2.165b) follow from (2.165a) and (2.74), (2.75). ▲

Similarly to formulae (2.165), we can introduce new deformation tensors by determining their components with respect to the bases $\overset{\circ}{\mathbf{p}}_\alpha \otimes \overset{\circ}{\mathbf{p}}_\beta$ or $\mathbf{p}_\alpha \otimes \mathbf{p}_\beta$ as follows:

$$\overset{\circ}{\mathbf{M}} = \sum_{\alpha=1}^{3} f(\lambda_\alpha)\overset{\circ}{\mathbf{p}}_\alpha \otimes \overset{\circ}{\mathbf{p}}_\beta, \quad \mathbf{M} = \sum_{\alpha=1}^{3} f(\lambda_\alpha)\mathbf{p}_\alpha \otimes \mathbf{p}_\beta, \quad (2.166)$$

where $f(\lambda_\alpha)$ is a function of $\lambda_\alpha$. If $f(1) = 0$, then we get the deformation tensors; and if $f(1) = 1$, then we get the deformation measures.

Among the tensors (2.166), the *logarithmic deformation tensors and measures*

$$\overset{\circ}{\mathbf{H}} = \textstyle\sum_{\alpha=1}^{3} \lg \lambda_\alpha \overset{\circ}{\mathbf{p}}_\alpha \otimes \overset{\circ}{\mathbf{p}}_\beta, \quad \mathbf{H} = \sum_{\alpha=1}^{3} \lg \lambda_\alpha \mathbf{p}_\alpha \otimes \mathbf{p}_\beta,$$

$$\overset{\circ}{\mathbf{H}}_1 = \overset{\circ}{\mathbf{H}} + \mathbf{E}, \qquad \mathbf{H}_1 = \mathbf{H} + \mathbf{E}, \quad (2.167)$$

are the most widely known; they are called the *right* and *left Hencky tensors*, and also the *right* and *left Hencky measures*, respectively.

With the help of the eigenvectors $\mathbf{p}_\alpha$ and $\overset{\circ}{\mathbf{p}}_\alpha$ we can form the mixed dyads

$$\sum_{\alpha=1}^{3} \mathbf{p}_\alpha \otimes \overset{\circ}{\mathbf{p}}_\alpha = \sum_{\alpha=1}^{3} \mathbf{p}_\alpha \otimes \mathbf{p}_\alpha \cdot \mathbf{O} = \left(\sum_{\alpha=1}^{3} \mathbf{p}_\alpha \otimes \mathbf{p}^\alpha\right) \cdot \mathbf{O} = \mathbf{E} \cdot \mathbf{O}. \quad (2.168)$$

Here we have used the properties (2.159a) and (2.160), and the representation of the unit tensor $\mathbf{E}$ in an arbitrary mixed dyadic basis.

Thus, the rotation tensor $\mathbf{O}$ accompanying the deformation can be expressed in the eigenbasis as follows:

$$\mathbf{O} = \sum_{\alpha=1}^{3} \mathbf{p}_\alpha \otimes \overset{\circ}{\mathbf{p}}_\alpha = \mathbf{p}_i \otimes \overset{\circ}{\mathbf{p}}{}^i. \quad (2.169)$$

On substituting (2.169) and (2.142) into (2.137) and taking (2.141) into account, we get the following expression of the deformation gradient in the tensor eigenbasis:

$$\mathbf{F} = \mathbf{O} \cdot \mathbf{U} = \sum_{\alpha=1}^{3} \mathbf{p}_\alpha \otimes \overset{\circ}{\mathbf{p}}_\alpha \cdot \sum_{\beta=1}^{3} \overset{\circ}{\lambda}_\beta \overset{\circ}{\mathbf{p}}_\beta \otimes \overset{\circ}{\mathbf{p}}_\beta = \sum_{\alpha=1}^{3} \lambda_\alpha \mathbf{p}_\alpha \otimes \overset{\circ}{\mathbf{p}}_\alpha. \tag{2.170}$$

According to (2.170), the transpose $\mathbf{F}^{\mathrm{T}}$ and inverse $\mathbf{F}^{-1}$ gradients are expressed as follows:

$$\mathbf{F}^{\mathrm{T}} = \sum_{\alpha=1}^{3} \lambda_\alpha \overset{\circ}{\mathbf{p}}_\alpha \otimes \mathbf{p}_\alpha, \qquad \mathbf{F}^{-1} = \sum_{\alpha=1}^{3} \lambda_\alpha^{-1} \overset{\circ}{\mathbf{p}}_\alpha \otimes \mathbf{p}_\alpha. \tag{2.171}$$

### 2.3.4 Geometrical Meaning of Eigenvalues

Vectors of eigenbases $\overset{\circ}{\mathbf{p}}_\alpha$ and $\mathbf{p}_\alpha$ are connected by the transformation (2.159a). In $\overset{\circ}{\mathcal{K}}$ take elementary radius-vectors $d\overset{\circ}{\mathbf{x}}_\alpha$ oriented along the eigenbasis vectors $\overset{\circ}{\mathbf{p}}_\alpha$, then in $\mathcal{K}$ they correspond to radius-vectors $d\mathbf{x}_\alpha$:

$$d\overset{\circ}{\mathbf{x}}_\alpha = \overset{\circ}{\mathbf{p}}_\alpha |d\overset{\circ}{\mathbf{x}}_\alpha|, \qquad d\mathbf{x}_\alpha = \mathbf{F} \cdot d\overset{\circ}{\mathbf{x}}_\alpha. \tag{2.172}$$

Substitution of (2.170) into (2.172) yields

$$d\mathbf{x}_\alpha = \sum_{\beta=1}^{3} \lambda_\beta \mathbf{p}_\beta \otimes \overset{\circ}{\mathbf{p}}_\beta \cdot \overset{\circ}{\mathbf{p}}_\alpha |d\overset{\circ}{\mathbf{x}}_\alpha| = \lambda_\alpha |d\mathbf{x}_\alpha| \mathbf{p}_\alpha, \tag{2.173}$$

i.e. the elementary radius-vectors $d\mathbf{x}_\alpha$ in $\mathcal{K}$ will be also oriented along the corresponding eigenbasis vectors $\mathbf{p}_\alpha$.

Denote lengths of the vectors $d\overset{\circ}{\mathbf{x}}_\alpha$ and $d\mathbf{x}_\alpha$ by $d\overset{\circ}{s}_\alpha$ and $ds$, respectively, and derive relations between them:

$$\begin{aligned} ds_\alpha^2 &= d\mathbf{x}_\alpha \cdot d\mathbf{x}_\alpha = d\overset{\circ}{\mathbf{x}}_\alpha \cdot \mathbf{F}^{\mathrm{T}} \cdot \mathbf{F} \cdot d\overset{\circ}{\mathbf{x}}_\alpha = |d\overset{\circ}{\mathbf{x}}_\alpha|^2 \overset{\circ}{\mathbf{p}}_\alpha \cdot \mathbf{F}^{\mathrm{T}} \cdot \mathbf{F} \cdot \overset{\circ}{\mathbf{p}}_\alpha \\ &= d\overset{\circ}{s}{}_\alpha^2 \overset{\circ}{\mathbf{p}}_\alpha \cdot \mathbf{G} \cdot \overset{\circ}{\mathbf{p}}_\alpha = d\overset{\circ}{s}{}_\alpha^2 \lambda_\alpha^2. \end{aligned} \tag{2.174}$$

Here we have used Eqs. (2.165b) and (2.172). Formula (2.174) proves the following theorem.

**Theorem 2.13.** *Eigenvalues $\lambda_\alpha$ (principal stretches) are the elongation ratios for material fibres oriented along the principal (eigen-) directions:*

$$\lambda_\alpha = ds_\alpha / d\overset{\circ}{s}_\alpha. \tag{2.175}$$

### 2.3.5 Geometric Picture of Transformation of a Small Neighborhood of a Point of a Continuum

In $\overset{\circ}{\mathcal{K}}$, consider a small neighborhood of the material point $\mathcal{M}$ contained in a continuum; then every point $\mathcal{M}'$, connected to $\mathcal{M}$ by the elementary radius-vector $d\overset{\circ}{\mathbf{x}}$ (Fig. 2.12), will be connected to the same point $\mathcal{M}$ by radius-vector $d\mathbf{x}$ in $\mathcal{K}$. These radius-vectors are related as follows:

$$d\mathbf{x} = \mathbf{F} \cdot d\overset{\circ}{\mathbf{x}}. \tag{2.176}$$

The relation can be considered as the transformation of arbitrary radius-vector $d\overset{\circ}{\mathbf{x}}$ into $d\mathbf{x}$.

Rewrite the relation (2.176) in Cartesian coordinates:

$$dx^i = \bar{F}^i_m d\overset{\circ}{x}{}^m, \tag{2.177}$$

where $\bar{F}^i_m$ are components of the deformation gradient with respect to the Cartesian basis (see Exercise 2.1.5):

$$\bar{F}^i_m = (\partial x^i / \partial \overset{\circ}{x}{}^m), \tag{2.178}$$

which depend only on coordinates $\overset{\circ}{x}{}^m$ of the point $\mathcal{M}$, but they are independent of coordinates $d\overset{\circ}{x}{}^m$ of its neighboring points $\mathcal{M}'$. Therefore the transformation (2.177) is a linear transformation of coordinates $d\overset{\circ}{x}{}^m$ into $dx^i$, i.e. this is an affine transformation.

As follows from the general properties of affine transformations, straight lines and planes contained in a small neighborhood in $\overset{\circ}{\mathcal{K}}$ will be straight lines and planes in actual configuration $\mathcal{K}$. Parallel straight lines and planes are transformed into parallel straight lines and planes. Therefore if a small neighborhood in $\overset{\circ}{\mathcal{K}}$ is chosen

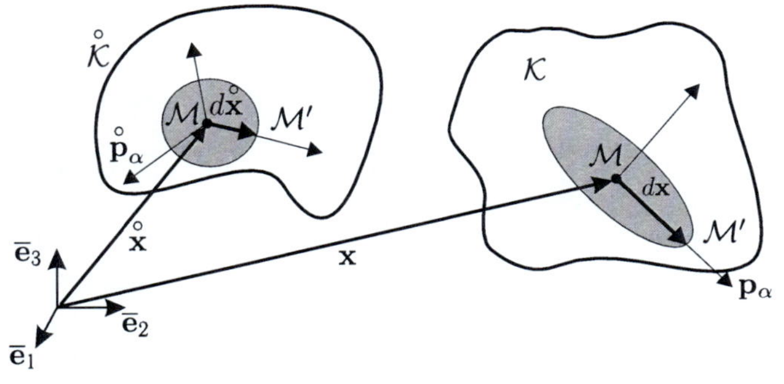


**Fig. 2.12** Transformation of a small neighborhood of the point contained in a continuum

to be a parallelogram, then in $\mathcal{K}$ the neighborhood will be a parallelogram as well (although angles between its edges, edge lengths and orientation of planes in space may change).

Since a second-order surface in $\overset{\circ}{\mathcal{K}}$ (and, in general, a surface specified by an algebraic expression of arbitrary $n$th order) is transformed into a surface of the same order in $\mathcal{K}$, a small spherical neighborhood in $\overset{\circ}{\mathcal{K}}$ is transformed into an ellipsoid in actual configuration $\mathcal{K}$ (Fig. 2.12).

As follows from formula (2.101), the ratio of lengths $ds_\alpha/d\overset{\circ}{s}_\alpha$ of an arbitrary vector (or of elementary radius-vector $d\mathbf{x}$ in $\mathcal{K}$ and $\overset{\circ}{\mathcal{K}}$) is independent of the initial length $d\overset{\circ}{s}_\alpha$ of the vector (because the relative elongation $\delta_\alpha$ is independent of $d\overset{\circ}{s}_\alpha$).

According to the polar decomposition (2.137), the transformation (2.176) from $\overset{\circ}{\mathcal{K}}$ to $\mathcal{K}$ can always be represented as the superposition of two transformations:

$$d\mathbf{x} = \mathbf{O}\cdot d\overset{\circ}{\mathbf{x}}{}', \quad d\overset{\circ}{\mathbf{x}}{}' = \mathbf{U}\cdot d\overset{\circ}{\mathbf{x}}, \tag{2.179}$$

realized with the help of the stretch tensor $\mathbf{U}$ and the rotation tensor $\mathbf{O}$, or

$$d\mathbf{x} = \mathbf{V}\cdot d\mathbf{x}', \quad d\mathbf{x}' = \mathbf{O}\cdot d\overset{\circ}{\mathbf{x}}. \tag{2.180}$$

The stretch tensor $\mathbf{U}$, which has three eigendirections $\overset{\circ}{\mathbf{p}}_\alpha$, transforms a small neighborhood of the point $\mathcal{M}$ with compressing or extending the neighborhood along these three directions $\overset{\circ}{\mathbf{p}}_\alpha$. The tensor $\mathbf{O}$ rotates the neighborhood deformed along $\overset{\circ}{\mathbf{p}}_\alpha$ as a rigid whole until the direction of $\overset{\circ}{\mathbf{p}}_\alpha$ becomes the direction of $\mathbf{p}_\alpha$. If one use the left stretch tensor $\mathbf{V}$, so rotation of axes $\overset{\circ}{\mathbf{p}}_\alpha$ in $\overset{\circ}{\mathcal{K}}$ till their coincidence with $\mathbf{p}_\alpha$ is first realized, and then compression or tension of the neighborhood occurs along the direction $\mathbf{p}_\alpha$. The result will be the same as for $\mathbf{U}$.

If a point $\mathcal{M}_\alpha$ is connected to $\mathcal{M}$ by radius-vector $d\overset{\circ}{\mathbf{x}}_\alpha$ oriented along the eigendirection $\overset{\circ}{\mathbf{p}}_\alpha$ (which is unknown before deformation), then in $\mathcal{K}$ the point $\mathcal{M}_\alpha$ will be connected to $\mathcal{M}$ by radius-vector $d\mathbf{x}_\alpha$ oriented along the corresponding eigendirection $\mathbf{p}_\alpha$.

If a small neighborhood of point $\mathcal{M}$ is chosen to be a sphere in $\overset{\circ}{\mathcal{K}}$ (see Fig. 2.12), then in $\mathcal{K}$ the sphere becomes an ellipsoid with principal axes oriented along the eigendirections $\mathbf{p}_\alpha$.

Thus, the transformation of a small neighborhood of every point $\mathcal{M}$ contained in a continuum under deformation can always be represented as a superposition of tension/compression along eigendirections and rotation of the neighborhood as a rigid whole, and also displacement as a rigid whole.

## Exercises for 2.3

**2.3.1.** Using the formula (2.157a), show that the following relations between **V** and **U** hold:

$$\mathbf{V}^m = \mathbf{O}\cdot\mathbf{U}^m\cdot\mathbf{O}^{\mathrm{T}}, \qquad \mathbf{U}^m = \mathbf{O}^{\mathrm{T}}\cdot\mathbf{V}^m\cdot\mathbf{O}$$

for all integer $m$ (positive and negative).

**2.3.2.** Using the results of Exercises 2.1.1 and 2.2.1, show that for the problem on tension of a beam, eigenvalues $\lambda_\alpha$ are

$$\lambda_\alpha = k_\alpha, \qquad \alpha = 1,2,3.$$

The stretch tensors **U** and **V** are coincident and have the form

$$\mathbf{U} = \mathbf{V} = \sum_{\alpha=1}^{3} k_\alpha \mathbf{e}_\alpha \otimes \mathbf{e}_\alpha,$$

and eigenvectors $\overset{\circ}{\mathbf{p}}_\alpha$ and $\mathbf{p}_\alpha$ coincide with $\mathbf{e}_\alpha$:

$$\overset{\circ}{\mathbf{p}}_\alpha = \mathbf{p}_\alpha = \mathbf{e}_\alpha, \qquad \alpha = 1,2,3.$$

The rotation tensor **O** for this problem is the unit one: $\mathbf{O} = \mathbf{E}$.

**2.3.3.** Using the results of Exercises 2.1.2, 2.2.2 and Remark 2, show that for the problem on simple shear (see Example 2.2 from Sect. 2.1.1), the tensors $\mathbf{U}^2$ and $\mathbf{V}^2$ are expressed as follows:

$$\mathbf{U}^2 = \mathbf{F}^{\mathrm{T}}\cdot\mathbf{F} = \mathbf{E} + a\mathbf{O}_3 + a^2\mathbf{e}_i^2 = (\bar{U}^2)^i{}_j \bar{\mathbf{e}}_i \otimes \bar{\mathbf{e}}^j,$$

$$\mathbf{V}^2 = \mathbf{E} + a\mathbf{O}_3 + a^2\mathbf{e}_i^2 = (\bar{V}^2)^i{}_j \bar{\mathbf{e}}_i \otimes \bar{\mathbf{e}}^j,$$

$$(\bar{U}^2)^i{}_j = \begin{pmatrix} 1 & a & 0 \\ a & 1+a^2 & 0 \\ 0 & 0 & 1 \end{pmatrix}, \quad (\bar{V}^2)^i{}_j = \begin{pmatrix} 1+a^2 & a & 0 \\ a & 1 & 0 \\ 0 & 0 & 1 \end{pmatrix},$$

eigenvalues $\lambda_\alpha$ are

$$\lambda_\alpha^2 = 1 + b_\alpha |a|, \quad \alpha = 1,2; \quad \lambda_3 = 1,$$

$$b_1 = \frac{a}{2} + \sqrt{1 + a^2/4}, \quad b_2 = \frac{a}{2} - \sqrt{1 + a^2/4},$$

eigenvectors $\overset{\circ}{\mathbf{p}}_\alpha$ and $\mathbf{p}_\alpha$ ($a > 0$) are

$$\overset{\circ}{\mathbf{p}}_\alpha = \frac{1}{\sqrt{1+b_\alpha^2}}(\bar{\mathbf{e}}_1 + b_\alpha \bar{\mathbf{e}}_2), \quad \overset{\circ}{\mathbf{p}}_3 = \bar{\mathbf{e}}_3,$$

$$\mathbf{p}_1 = \frac{1}{\sqrt{1+b_1^2}}(b_1\bar{\mathbf{e}}_1 + \bar{\mathbf{e}}_2), \quad \mathbf{p}_2 = -\frac{1}{\sqrt{1+b_2^2}}(b_2\bar{\mathbf{e}}_1 + \bar{\mathbf{e}}_2), \quad \mathbf{p}_3 = \bar{\mathbf{e}}_3,$$

the stretch tensors **U** and **V** are

$$\mathbf{U} = \bar{U}^i{}_j\bar{\mathbf{e}}_i \otimes \bar{\mathbf{e}}^j = U_0\bar{\mathbf{e}}_1^2 + U_1\mathbf{O}_3 + U_2\bar{\mathbf{e}}_2^2 + \bar{\mathbf{e}}_3^2,$$

$$\mathbf{V} = \bar{V}^i{}_j\bar{\mathbf{e}}_i \otimes \bar{\mathbf{e}}^j = U_2\bar{\mathbf{e}}_1^2 + U_1\mathbf{O}_3 + U_0\bar{\mathbf{e}}_2^2 + \bar{\mathbf{e}}_3^2,$$

$$\bar{U}^i{}_j = \begin{pmatrix} U_0 & U_1 & 0 \\ U_1 & U_2 & 0 \\ 0 & 0 & 1 \end{pmatrix}, \quad \bar{V}^i{}_j = \begin{pmatrix} U_2 & U_1 & 0 \\ U_1 & U_0 & 0 \\ 0 & 0 & 1 \end{pmatrix},$$

$$U_\beta = \frac{b_1^\beta\sqrt{1+b_1a}}{1+b_1^2} + \frac{b_2^\beta\sqrt{1+b_2a}}{1+b_2^2}, \quad \beta = 0, 1, 2,$$

and the rotation tensor **O** has the form

$$\mathbf{O} = \bar{O}^i{}_j\bar{\mathbf{e}}_i \otimes \bar{\mathbf{e}}^j = \cos\varphi(\bar{\mathbf{e}}_1^2 + \bar{\mathbf{e}}_2^2) + \sin\varphi(\bar{\mathbf{e}}_1 \otimes \bar{\mathbf{e}}_2 - \bar{\mathbf{e}}_2 \otimes \bar{\mathbf{e}}_1),$$

$$\bar{O}^i{}_j = \begin{pmatrix} \cos\varphi & \sin\varphi & 0 \\ -\sin\varphi & \cos\varphi & 0 \\ 0 & 0 & 1 \end{pmatrix},$$

$$\cos\varphi = \frac{b_1}{1+b_1^2} - \frac{b_2}{1+b_2^2}, \quad \sin\varphi = \frac{b_1^2}{1+b_1^2} - \frac{b_2^2}{1+b_2^2}.$$

Show that functions $b_1(a)$ and $b_2(a)$ satisfy the following relationships:

$$b_1 + b_2 = a, \quad b_1b_2 = -1, \quad b_1^2 + b_2^2 = 2 + a^2.$$

Show that at $a = 0$ for the considered problem the following equations really hold:

$$b_1 = 1, \quad b_2 = -1, \quad \lambda_1 = \lambda_2 = \lambda_3 = 1,$$

$$\overset{\circ}{\mathbf{p}}_1 = \mathbf{p}_1 = \frac{1}{\sqrt{2}}(\bar{\mathbf{e}}_1 + \bar{\mathbf{e}}_2), \quad \overset{\circ}{\mathbf{p}}_2 = \mathbf{p}_2 = \frac{1}{\sqrt{2}}(\bar{\mathbf{e}}_1 - \bar{\mathbf{e}}_2).$$

**2.3.4.** Using the results of Exercise 2.2.3, show that for the problem on rotation of a beam with tension (see Example 2.3 from Sect. 2.1.1), eigenvalues $\lambda_\alpha$ have the form

$$\lambda_\alpha = k_\alpha, \quad \alpha = 1, 2, 3,$$

and eigenvectors

$$\overset{\circ}{\mathbf{p}}_\alpha = \bar{\mathbf{e}}_\alpha, \quad \mathbf{p}_\alpha = \mathbf{O}_0 \cdot \bar{\mathbf{e}}_\alpha, \quad \alpha = 1, 2, 3.$$

Using formulae from Exercise 2.1.3 and data from Example 2.3, show that tensors **U**, **V**, **O**, and also **C**, **A**, **Λ** and **J** have the form

$$\mathbf{U} = \mathbf{U}_0 = \sum_{\alpha=1}^{3} k_\alpha \bar{\mathbf{e}}_\alpha \otimes \bar{\mathbf{e}}_\alpha, \quad \mathbf{O} = \mathbf{O}_0 = O_0{}^i{}_j \bar{\mathbf{e}}_i \otimes \bar{\mathbf{e}}^j,$$

$$\mathbf{V} = \mathbf{O}_0 \cdot \mathbf{U}_0 \cdot \mathbf{O}_0^{\mathrm{T}} = V_0 \bar{\mathbf{e}}_1^2 + V_1 \mathbf{O}_3 + V_2 \bar{\mathbf{e}}_2^2 + k_3 \bar{\mathbf{e}}_3^2 = V_0{}^i{}_j \bar{\mathbf{e}}_i \otimes \bar{\mathbf{e}}^j,$$

$$V_0{}^i{}_j = \begin{pmatrix} V_0 & V_1 & 0 \\ V_1 & V_2 & 0 \\ 0 & 0 & k_3 \end{pmatrix},$$

$$V_0 = k_1 \cos^2\varphi + k_2 \sin^2\varphi, \quad V_1 = (k_1 - k_2)\cos\varphi\,\sin\varphi,$$

$$V_2 = k_1 \sin^2\varphi + k_2 \cos^2\varphi,$$

$$\mathbf{C} = \frac{1}{2}(\mathbf{U}_0^2 - \mathbf{E}) = \sum_{\alpha=1}^{3} \frac{1}{2}(k_\alpha^2 - 1)\bar{\mathbf{e}}_\alpha \otimes \bar{\mathbf{e}}_\alpha,$$

$$\boldsymbol{\Lambda} = \frac{1}{2}(\mathbf{E} - \mathbf{U}_0^{-2}) = \sum_{\alpha=1}^{3} \frac{1}{2}(1 - k_\alpha^{-2})\bar{\mathbf{e}}_\alpha \otimes \bar{\mathbf{e}}_\alpha,$$

$$\mathbf{A} = \frac{1}{2}(\mathbf{E} - \mathbf{V}^{-2}) = \frac{1}{2}(\delta^{ij} - g^{ij})\bar{\mathbf{e}}_i \otimes \bar{\mathbf{e}}_j, \quad \mathbf{J} = \frac{1}{2}(\mathbf{V}^2 - \mathbf{E}) = \frac{1}{2}(g_{ij} - \delta_{ij})\bar{\mathbf{e}}^i \otimes \bar{\mathbf{e}}^j,$$

where metric matrices $g_{ij}$ and $g^{ij}$ are determined by formulae from Exercise 2.1.3.

We should take into consideration that the tensors **C** and **Λ** do not feel the beam rotation – they are coincident with the corresponding tensors for the problem on pure tension of the beam. Show that if we change the sequence of transformations (i.e. we first rotate and then extend the beam), then the tensors **A** and **J** do not feel the rotation.

## 2.4 Rate Characteristics of Continuum Motion

### *2.4.1 Velocity*

The *velocity* (vector) *of the motion* of a material point $\mathcal{M}$ with Lagrangian coordinates $X^i$ is determined as the partial derivative of the radius-vector $\mathbf{x}(X^i, t)$ with respect to time at fixed values of $X^i$:

$$\mathbf{v}(X^i, t) = \frac{\partial \mathbf{x}}{\partial t}(X^i, t)\Big|_{X^i}. \tag{2.181}$$

Velocity components $\bar{v}^i$ with respect to the basis $\bar{\mathbf{e}}_i$ have the form

$$\mathbf{v} = \bar{v}^i \bar{\mathbf{e}}_i = \frac{\partial x^i}{\partial t}\bar{\mathbf{e}}_i, \quad \bar{v}^i = \frac{\partial x^i}{\partial t}(X^j, t). \tag{2.182}$$

## 2.4.2 Total Derivative of a Tensor with Respect to Time

Any vector field $\mathbf{a}(\mathbf{x}, t)$ (and also scalar or tensor field) varying with time, which describes some physical process in a continuum, can be expressed in both Eulerian and Lagrangian descriptions with the help of the motion law (2.3):

$$\mathbf{a}(\mathbf{x}, t) = \mathbf{a}(\mathbf{x}(X^j, t), t). \tag{2.183}$$

Determine the derivative of the function with respect to time at fixed $X^i$ (i.e. for a fixed point $\mathcal{M}$):

$$\left.\frac{\partial \mathbf{a}}{\partial t}\right|_{X^i} = \left.\frac{\partial \mathbf{a}}{\partial t}\right|_{x^i} + \left.\frac{\partial \mathbf{a}}{\partial x^j}\frac{\partial x^j}{\partial t}\right|_{X^i}. \tag{2.184}$$

**Definition 2.1.** The partial derivative of a varying vector field $\mathbf{a}$ (2.183) with respect to time $t$ at fixed coordinates $X^i$ is called the *total derivative* of the function (2.183) *with respect to time*:

$$\dot{\mathbf{a}} \equiv \frac{d\mathbf{a}}{dt} = \left.\frac{\partial \mathbf{a}}{\partial t}\right|_{X^i}. \tag{2.185}$$

According to formulae (2.182), (2.11), and (2.23), the second summand on the right-hand side of (2.184) can be rewritten as follows:

$$\frac{\partial \mathbf{a}}{\partial x^j}\frac{\partial x^j}{\partial t} = \bar{v}^j P^k_j \frac{\partial \mathbf{a}}{\partial X^k} = \bar{v}^i \bar{\mathbf{e}}_i \cdot \bar{\mathbf{e}}^j P^k_j \otimes \frac{\partial \mathbf{a}}{\partial X^k} = \mathbf{v}\cdot\mathbf{r}^k \otimes \frac{\partial \mathbf{a}}{\partial X^k} = \mathbf{v}\cdot\nabla\otimes\mathbf{a}. \tag{2.186}$$

Then the relationship (2.184) yields

$$\frac{d\mathbf{a}}{dt} = \frac{\partial \mathbf{a}}{\partial t} + \mathbf{v}\cdot\nabla\otimes\mathbf{a}, \tag{2.187}$$

where we have introduced the notation for the partial derivative with respect to time which will be widely used below:

$$\frac{\partial \mathbf{a}}{\partial t} = \left.\frac{\partial \mathbf{a}}{\partial t}(x^i, t)\right|_{x^i}. \tag{2.188}$$

In formula (2.187) the vector $\mathbf{a}$ is considered as a function $\mathbf{a}(x^j, t)$. It is evident that if $\mathbf{a}$ is considered as a function of $(X^j, t)$, then from the definition (2.185) we get

$$\frac{d\mathbf{a}}{dt}(x^i, t) = \left.\frac{\partial}{\partial t}\mathbf{a}(X^j, t)\right|_{X^i}. \tag{2.189}$$

The total derivative $(d\mathbf{a}/dt)$ is also called the *material (substantial, individual) derivative* with respect to time, $(\partial\mathbf{a}/\partial t)$ in (2.187) is the *partial (local) derivative* with respect to time, and $\mathbf{v}\cdot\nabla\otimes\mathbf{a}$ is the *convective derivative*.

The material derivative $d\mathbf{a}/dt$ characterizes a change of the vector field $\mathbf{a}$ in a fixed material point $\mathcal{M}$, the local derivative determines a change of values of $\mathbf{a}$ in time at a fixed point $\mathbf{x}$ in space, and from (2.186) we get that the convective derivative characterizes a change of the field due to transfer of the material particle $\mathcal{M}$ from a point $\mathbf{x}$ to a point $\mathbf{x}+\mathbf{v}dt$ in space.

If we choose the vector $\mathbf{v}$ as $\mathbf{a}$, then the relationship between the displacement $\mathbf{u}$ and the velocity $\mathbf{v}$ vectors has the form

$$\mathbf{v} = \frac{d\mathbf{x}}{dt} = \frac{d\mathbf{u}}{dt} = \frac{\partial\mathbf{u}}{\partial t} + \mathbf{v}\cdot\nabla\otimes\mathbf{u}. \tag{2.190}$$

Similarly to formula (2.185), we can define the *total derivative* of the $n$th-order *tensor* ${}^n\mathbf{\Omega}$ with respect to time:

$${}^n\dot{\mathbf{\Omega}} = \frac{d}{dt}{}^n\mathbf{\Omega}(x^i,t) = \frac{\partial\,{}^n\mathbf{\Omega}}{\partial t}(X^i,t)\Big|_{X^i}. \tag{2.191}$$

**Theorem 2.14.** *The total derivative* (2.191) *of a varying tensor field* ${}^n\mathbf{\Omega}(x^i,t)$ *can be written as a sum of local and convective derivatives:*

$$\frac{d}{dt}{}^n\mathbf{\Omega} = \frac{\partial\,{}^n\mathbf{\Omega}}{\partial t} + \mathbf{v}\cdot\nabla\otimes{}^n\mathbf{\Omega}. \tag{2.192}$$

▼ Proof of the theorem is similar to the proof of the relationship (2.187). Details are left as Exercise 2.4.6 ▲

Let us consider now the question on components of the total derivative tensor.

**Theorem 2.15.** *Components of the total derivative tensor* ${}^n\dot{\mathbf{\Omega}}$ *are connected with the corresponding components of a tensor* ${}^n\mathbf{\Omega}$ *with respect to stationary bases* $\overset{\circ}{\mathbf{r}}_i$, $\bar{\mathbf{e}}_i$ *and* $\widetilde{\mathbf{r}}_i$ *and a moving basis* $\mathbf{r}_i$ *as follows:*

$$\frac{d}{dt}\overset{\circ}{\Omega}{}^{i_1\ldots i_n} = \frac{\partial}{\partial t}\overset{\circ}{\Omega}{}^{i_1\ldots i_n}(X^i,t)\Big|_{X^i}, \tag{2.193}$$

$$\frac{d}{dt}\bar{\Omega}^{i_1\ldots i_n} = \frac{\partial}{\partial t}\bar{\Omega}^{i_1\ldots i_n}(x^i,t) + \bar{v}^k\frac{\partial}{\partial x^k}\bar{\Omega}^{i_1\ldots i_n}(x^i,t), \tag{2.194}$$

$$\frac{d}{dt}\widetilde{\Omega}^{i_1\ldots i_n} = \frac{\partial}{\partial t}\widetilde{\Omega}^{i_1\ldots i_n}(\widetilde{X}^i,t) + \widetilde{v}^k\widetilde{\nabla}_k\widetilde{\Omega}^{i_1\ldots i_n}(\widetilde{X}^i,t), \tag{2.195}$$

$$\frac{d}{dt}\Omega^{i_1\ldots i_n} = \frac{\partial}{\partial t}\Omega^{i_1\ldots i_n}(X^i,t) + \sum_{\alpha=1}^{n}(\Omega^{i_1\ldots k\ldots i_n}\nabla_k v^{i_\alpha})(X^i,t), \tag{2.196}$$

where

$$
{}^n\dot{\mathbf{\Omega}} = \frac{d}{dt}\bar{\Omega}^{i_1\dots i_n}(x^i,t)\bar{\mathbf{e}}_{i_1}\otimes\dots\otimes\bar{\mathbf{e}}_{i_n} = \frac{d}{dt}\overset{\circ}{\Omega}{}^{i_1\dots i_n}(X^i,t)\overset{\circ}{\mathbf{r}}_{i_1}\otimes\dots\otimes\overset{\circ}{\mathbf{r}}_{i_n}
$$
$$
= \frac{d}{dt}\widetilde{\Omega}^{i_1\dots i_n}(\widetilde{X}^i,t)\widetilde{\mathbf{r}}_{i_1}\otimes\dots\otimes\widetilde{\mathbf{r}}_{i_n} = \frac{d}{dt}\Omega^{i_1\dots i_n}(X^i,t)\mathbf{r}_{i_1}\otimes\dots\otimes\mathbf{r}_{i_n}. \tag{2.197}
$$

In formula (2.196), the component $\Omega^{i_1\dots k\dots i_n}$ as the $\alpha$th superscript has index $k$ in place of $i_\alpha$.

▼ To prove the theorem, we resolve the tensor ${}^n\mathbf{\Omega}$ for different bases:

$$
{}^n\mathbf{\Omega} = \bar{\Omega}^{i_1\dots i_n}(x^i,t)\bar{\mathbf{e}}_{i_1}\otimes\dots\otimes\bar{\mathbf{e}}_{i_n} = \overset{\circ}{\Omega}{}^{i_1\dots i_n}(X^i,t)\overset{\circ}{\mathbf{r}}_{i_1}\otimes\dots\otimes\overset{\circ}{\mathbf{r}}_{i_n}
$$
$$
= \widetilde{\Omega}^{i_1\dots i_n}(\widetilde{X}^i,t)\widetilde{\mathbf{r}}_{i_1}\otimes\dots\otimes\widetilde{\mathbf{r}}_{i_n} = \Omega^{i_1\dots i_n}(X^i,t)\mathbf{r}_{i_1}\otimes\dots\otimes\mathbf{r}_{i_n}, \tag{2.198}
$$

and choose arguments of components of the tensor ${}^n\mathbf{\Omega}$ as in formula (2.198).

Then, substituting the resolution (2.198) for the basis $\overset{\circ}{\mathbf{r}}_i$ into the definition (2.191), we get the expression (2.193), because $d\overset{\circ}{\mathbf{r}}_i/dt = 0$.

On substituting the resolution (2.198) for the basis $\bar{\mathbf{e}}_i$ into the relationship (2.192), we obtain

$$
{}^n\dot{\mathbf{\Omega}} = \frac{\partial\bar{\Omega}^{i_1\dots i_n}}{\partial t}\bar{\mathbf{e}}_{i_1}\otimes\dots\otimes\bar{\mathbf{e}}_{i_n} + \bar{v}^k\bar{\mathbf{e}}_k\cdot\frac{\partial}{\partial x^m}\bar{\Omega}^{i_1\dots i_n}\bar{\mathbf{e}}^m\otimes\bar{\mathbf{e}}_{i_1}\otimes\dots\otimes\bar{\mathbf{e}}_{i_n}. \tag{2.198a}
$$

It is evident that formula (2.194) follows from (2.198a).

In a similar way, substituting the resolution (2.198) for the basis $\widetilde{\mathbf{r}}_i$ into the relation (2.192) and using the property (2.61) of nabla-operators $\nabla$ and $\widetilde{\nabla}$, and also the equation $\partial\widetilde{\mathbf{r}}_i/\partial t = 0$ (for the stationary basis $\widetilde{\mathbf{r}}_i$ see Sect. 2.1.7), we get

$$
{}^n\dot{\mathbf{\Omega}} = \frac{\partial\widetilde{\Omega}^{i_1\dots i_n}}{\partial t}\widetilde{\mathbf{r}}_{i_1}\otimes\dots\otimes\widetilde{\mathbf{r}}_{i_n} + \widetilde{v}^k\widetilde{\mathbf{r}}_k\cdot\widetilde{\nabla}_m\widetilde{\Omega}^{i_1\dots i_n}\widetilde{\mathbf{r}}^m\otimes\widetilde{\mathbf{r}}_{i_1}\otimes\dots\otimes\widetilde{\mathbf{r}}_{i_n}; \tag{2.199}
$$

and formula (2.195) follows from (2.199) at once.

Finally, substituting the resolution (2.198) for the moving basis $\mathbf{r}_i$ into the definition of the total derivative (2.192), we obtain

$$
{}^n\dot{\mathbf{\Omega}} = \left.\frac{\partial\Omega^{i_1\dots i_n}}{\partial t}\right|_{X^i}\mathbf{r}_{i_1}\otimes\dots\otimes\mathbf{r}_{i_n} + \sum_{\alpha=1}^{n}\Omega^{i_1\dots i_\alpha\dots i_n}\mathbf{r}_{i_1}\otimes\dots\left.\frac{\partial\mathbf{r}_{i_\alpha}}{\partial t}\right|_{X^i}\dots\otimes\mathbf{r}_{i_n}. \tag{2.200}
$$

Due to the definition (2.10) of local bases vectors and the definition (2.181) of the velocity, we have

$$
\left.\frac{\partial\mathbf{r}_{i_\alpha}}{\partial t}(X^j,t)\right|_{X^j} = \left.\frac{\partial^2\mathbf{x}}{\partial t\,\partial X^{i_\alpha}}\right|_{X^j} = \left.\frac{\partial\mathbf{v}}{\partial X^{i_\alpha}}\right|_{X^j} = \nabla_{i_\alpha}v^k\mathbf{r}_k. \tag{2.201}
$$

On substituting (2.201) into (2.200) and then collecting components at the same elements of the polyadic basis, we derive the formula (2.196). ▲

It should be noted that arguments of the resolutions (2.198) and of the derivatives of tensor components (2.193)–(2.196) have been chosen in the specific way.

### 2.4.3 *Differential of a Tensor*

**Definition 2.2.** For a tensor field ${}^n\boldsymbol{\Omega}(x^i, t)$, the following object

$$d\,{}^n\boldsymbol{\Omega} = \frac{d\,{}^n\boldsymbol{\Omega}}{dt}dt \tag{2.202}$$

is called the *differential of a tensor field* (or the differential of a tensor) ${}^n\boldsymbol{\Omega}(x^i, t)$.

According to formula (2.192) for the total derivative of a tensor with respect to time, we get that the differential of a tensor can be written in the form

$$d\,{}^n\boldsymbol{\Omega}(x^i, t) = \left(\frac{\partial\,{}^n\boldsymbol{\Omega}}{\partial t} + \mathbf{v}\cdot\nabla\otimes{}^n\boldsymbol{\Omega}\right)dt. \tag{2.203}$$

According to (2.190), the relation (2.203) takes the form

$$d\,{}^n\boldsymbol{\Omega} = \frac{\partial\,{}^n\boldsymbol{\Omega}}{\partial t}dt + d\mathbf{x}\cdot\nabla\otimes{}^n\boldsymbol{\Omega} \tag{2.204}$$

When a tensor field is *stationary* (i.e. $\partial\,{}^n\boldsymbol{\Omega}/\partial t = 0$), the differential of the tensor field has the form

$$\widehat{d}\,{}^n\boldsymbol{\Omega} = d\mathbf{x}\cdot\nabla\otimes{}^n\boldsymbol{\Omega}. \tag{2.205}$$

For stationary tensor fields $\widehat{d}\,{}^n\boldsymbol{\Omega} = d\,{}^n\boldsymbol{\Omega}$, but in general these differentials are not coincident.

According to Theorem 2.15, components of the tensor $d\,{}^n\boldsymbol{\Omega}$ with respect to the fixed basis $\overset{\circ}{\mathbf{r}}_i$ are written as follows:

$$d\,{}^n\boldsymbol{\Omega} = d\overset{\circ}{\Omega}{}^{j_1\dots j_n}\overset{\circ}{\mathbf{r}}_{j_1}\otimes\dots\otimes\overset{\circ}{\mathbf{r}}_{j_n}, \quad d\overset{\circ}{\Omega}{}^{j_1\dots j_n} = \frac{d\,\overset{\circ}{\Omega}{}^{j_1\dots j_n}}{dt}dt. \tag{2.206}$$

From (2.202) and (2.187) we get the following expression for the *differential of a vector*:

$$d\mathbf{a}(X^i, t) = \frac{d\mathbf{a}}{dt}dt = \left(\frac{\partial\mathbf{a}}{\partial t} + \mathbf{v}\cdot\nabla\otimes\mathbf{a}\right)dt, \tag{2.207}$$

and from (2.205) we have

$$d\widehat{\mathbf{a}} = (\mathbf{v}\cdot\nabla\otimes\mathbf{a})dt = (\nabla\otimes\mathbf{a})^{\mathrm{T}}\cdot d\mathbf{x}. \tag{2.208}$$

In particular, if $\mathbf{a} = \overset{\circ}{\mathbf{x}}$, then, by formulae (2.208) and (2.35a), we obtain

$$\widehat{d\overset{\circ}{\mathbf{x}}} = (\nabla \otimes \overset{\circ}{\mathbf{x}})^{\mathrm{T}} \cdot d\mathbf{x} = \mathbf{F}^{-1} \cdot d\mathbf{x}, \tag{2.209}$$

or

$$d\mathbf{x} = \mathbf{F} \cdot \widehat{d\overset{\circ}{\mathbf{x}}}. \tag{2.210}$$

On comparing formulae (2.210) with (2.34), we find that the elementary radius-vector $d\overset{\circ}{\mathbf{x}}$, introduced in Sect. 2.1 and connecting two infinitesimally close material points $\mathcal{M}$ and $\mathcal{M}'$, coincides with the vector $\widehat{d\overset{\circ}{\mathbf{x}}}$ in the notation (2.205).

## *2.4.4 Properties of Derivatives with Respect to Time*

Let us establish now important properties of partial and total derivatives of vector fields with respect to time.

**Theorem 2.16.** *The partial derivative of the vector product of basis vectors with respect to time has the form*

$$\frac{\partial}{\partial t}(\mathbf{r}_\alpha \times \mathbf{r}_\beta) = \frac{\partial \mathbf{r}_\alpha}{\partial t} \times \mathbf{r}_\beta + \mathbf{r}_\alpha \times \frac{\partial \mathbf{r}_\beta}{\partial t}. \tag{2.211}$$

▼ Determine the derivative of the vector product of two local basis vectors with respect to time:

$$\begin{aligned} \frac{\partial}{\partial t}(\mathbf{r}_\alpha \times \mathbf{r}_\beta) &= \frac{\partial}{\partial t}({Q^i}_\alpha \bar{\mathbf{e}}_i \times {Q^j}_\beta \bar{\mathbf{e}}_j) = \frac{\partial}{\partial t}({Q^i}_\alpha {Q^j}_\beta)\bar{\mathbf{e}}_i \times \bar{\mathbf{e}}_j \\ &= \frac{\partial {Q^i}_\alpha}{\partial t}\bar{\mathbf{e}}_i \times {Q^j}_\beta \bar{\mathbf{e}}_j + {Q^i}_\alpha \bar{\mathbf{e}}_i \times \frac{\partial {Q^j}_\beta}{\partial t}\bar{\mathbf{e}}_j. \end{aligned}$$

With use of relation (2.10) we really get (2.211). ▲

**Theorem 2.17.** *For arbitrary continuously differentiable vector fields* $\mathbf{a}(\mathbf{x}, t) = \bar{a}^i(x^k, t)\bar{\mathbf{e}}_i$ *and* $\mathbf{b}(\mathbf{x}, t) = \bar{b}^i(x^k, t)\bar{\mathbf{e}}_i$, *we have the formulae*

$$\frac{\partial}{\partial t}(\mathbf{a} \times \mathbf{b}) = \frac{\partial \mathbf{a}}{\partial t} \times \mathbf{b} + \mathbf{a} \times \frac{\partial \mathbf{b}}{\partial t}, \tag{2.212}$$

$$\frac{\partial}{\partial t}(\mathbf{a} \otimes \mathbf{b}) = \frac{\partial \mathbf{a}}{\partial t} \otimes \mathbf{b} + \mathbf{a} \otimes \frac{\partial \mathbf{b}}{\partial t}, \tag{2.213}$$

$$\frac{\partial}{\partial t}(\mathbf{a} \cdot \mathbf{b}) = \frac{\partial \mathbf{a}}{\partial t} \cdot \mathbf{b} + \mathbf{a} \cdot \frac{\partial \mathbf{b}}{\partial t}. \tag{2.214}$$

▼ A proof is similar to the proof of Theorem 2.16. ▲

**Theorem 2.18.** *The total derivatives of the vector and scalar products of two arbitrary vector fields* $\mathbf{a}(\mathbf{x},t)$ *and* $\mathbf{b}(\mathbf{x},t)$ *with respect to time have the forms*

$$\frac{d}{dt}(\mathbf{a}\times\mathbf{b}) = \frac{d\mathbf{a}}{dt}\times\mathbf{b} + \mathbf{a}\times\frac{d\mathbf{b}}{dt}, \tag{2.215}$$

$$\frac{d}{dt}(\mathbf{a}\cdot\mathbf{b}) = \frac{d\mathbf{a}}{dt}\cdot\mathbf{b} + \mathbf{a}\cdot\frac{d\mathbf{b}}{dt}. \tag{2.216}$$

▼ To prove formula (2.215), one should use the property of the total derivative (2.187):

$$\frac{d}{dt}(\mathbf{a}\times\mathbf{b}) = \frac{\partial}{\partial t}(\mathbf{a}\times\mathbf{b}) + \mathbf{v}\cdot\nabla\otimes(\mathbf{a}\times\mathbf{b}).$$

Modify the first summand by formula (2.212) and the second summand – by the formula $\nabla\otimes(\mathbf{a}\times\mathbf{b}) = (\nabla\otimes\mathbf{a})\times\mathbf{b} - (\nabla\otimes\mathbf{b})\times\mathbf{a}$ [12], then we get

$$\frac{d}{dt}(\mathbf{a}\times\mathbf{b}) = \frac{\partial\mathbf{a}}{\partial t}\times\mathbf{b} - \frac{\partial\mathbf{b}}{\partial t}\times\mathbf{a} + \mathbf{v}\cdot(\nabla\otimes\mathbf{a})\times\mathbf{b} - \mathbf{v}\cdot(\nabla\otimes\mathbf{b})\times\mathbf{a}.$$

Collecting the first summand with the third one and the second summand with the fourth one, and using the property (2.187) of the total derivative of a vector, we obtain

$$\frac{d}{dt}(\mathbf{a}\times\mathbf{b}) = \frac{d\mathbf{a}}{dt}\times\mathbf{b} - \frac{d\mathbf{b}}{dt}\times\mathbf{a} = \frac{d\mathbf{a}}{dt}\times\mathbf{b} + \mathbf{a}\times\frac{d\mathbf{b}}{dt}.$$

Formula (2.216) can be proved in a similar way. ▲

**Theorem 2.19.** *The total derivative of* $\sqrt{g}$ *with respect to time is connected to the divergence of the velocity* $\mathbf{v}$ *by*

$$\frac{d}{dt}\sqrt{g} = \sqrt{g}\,\nabla_i v^i = \sqrt{g}\,\nabla\cdot\mathbf{v}. \tag{2.217}$$

▼ Let us differentiate the second relation of (2.15) with taking formula (2.211) into account:

$$\begin{aligned}\frac{d}{dt}\sqrt{g} &= \frac{d}{dt}\mathbf{r}_1\cdot(\mathbf{r}_2\times\mathbf{r}_3)\\ &= \frac{\partial^2\mathbf{x}}{\partial t\partial X^1}\cdot(\mathbf{r}_2\times\mathbf{r}_3) + \mathbf{r}_1\cdot\left(\frac{\partial^2\mathbf{x}}{\partial t\partial X^2}\times\mathbf{r}_3\right) + \mathbf{r}_1\cdot\left(\mathbf{r}_2\times\frac{\partial^2\mathbf{x}}{\partial t\partial X^3}\right).\end{aligned} \tag{2.218}$$

Since

$$\frac{\partial^2\mathbf{x}}{\partial t\partial X^i} = \frac{\partial\mathbf{v}}{\partial X^i} = \nabla_i\mathbf{v} = \nabla_i v^j\mathbf{r}_j,$$

we get

$$\frac{d}{dt}\sqrt{g} = \nabla_1\mathbf{v}\sqrt{g}\cdot\mathbf{r}^1 + \mathbf{r}_1\cdot(\nabla_2\mathbf{v}\times\mathbf{r}_3) + \mathbf{r}_1\cdot(\mathbf{r}_2\times\nabla_3\mathbf{v}). \tag{2.219}$$

Here we have used the relations from Exercise 2.1.14.

According to the definition of the vector product (1.2), we obtain

$$\mathbf{r}_1\cdot\nabla_2\mathbf{v}\times\mathbf{r}_3 = \mathbf{r}_1\cdot\sqrt{g}\,\epsilon_{ijk}\nabla_2 v^i\delta_3^j\mathbf{r}^k = \sqrt{g}\,\epsilon_{i31}\nabla_2 v^i = \sqrt{g}\,\nabla_2 v^2. \tag{2.220}$$

On substituting (2.220) into (2.219), we really get formula (2.217). ▲

### 2.4.5 *The Velocity Gradient, the Deformation Rate Tensor and the Vorticity Tensor*

Consider elementary radius-vectors $d\overset{\circ}{\mathbf{x}}$ and $d\mathbf{x}$ connecting two infinitesimally close points $\mathcal{M}$ and $\mathcal{M}'$ in configurations $\overset{\circ}{\mathcal{K}}$ and $\mathcal{K}$, respectively. Determine the velocity of the point $\mathcal{M}'$ relative to the configuration connected to the point $\mathcal{M}$. To do this, determine the velocity differential $\widehat{d}\mathbf{v}$:

$$\widehat{d}\mathbf{v} = \frac{\partial}{\partial t}d\mathbf{x} = \frac{\partial^2\mathbf{x}}{\partial X^i\partial t}dX^i = \frac{\partial^2\mathbf{x}}{\partial X^i\partial t}\otimes\overset{\circ}{\mathbf{r}}{}^i\cdot d\overset{\circ}{\mathbf{x}} = \left(\overset{\circ}{\mathbf{r}}{}^i\otimes\frac{\partial\mathbf{v}}{\partial X^i}\right)^{\mathrm{T}}\cdot d\overset{\circ}{\mathbf{x}} = \left(\overset{\circ}{\nabla}\otimes\mathbf{v}\right)^{\mathrm{T}}\cdot d\overset{\circ}{\mathbf{x}}. \tag{2.221}$$

Here we have used the second equation of (2.33), the definition of the gradient (2.24) and formula (2.181). In a similar way, using the first equation of (2.33): $dX^i = \mathbf{r}^i\cdot d\mathbf{x}$, we get one more expression for the vector $\widehat{d}\mathbf{v}$:

$$\widehat{d}\mathbf{v} = (\nabla\otimes\mathbf{v})^{\mathrm{T}}\cdot d\mathbf{x}. \tag{2.222}$$

The second-order tensor $(\nabla\otimes\mathbf{v})^{\mathrm{T}}$ is called the *velocity gradient*, which connects the relative velocity $\widehat{d}\mathbf{v}$ of an elementary radius-vector $d\mathbf{x}$ to the vector $d\mathbf{x}$ itself:

$$\widehat{d}\mathbf{v} = \mathbf{L}\cdot d\mathbf{x}, \qquad \mathbf{L} = (\nabla\otimes\mathbf{v})^{\mathrm{T}}. \tag{2.223}$$

Just as any second-order tensor (see [12]), the tensor $\mathbf{L}$ can be represented by a sum of the symmetric tensor $\mathbf{D}$ and the skew-symmetric tensor $\mathbf{W}$:

$$\mathbf{L} = \mathbf{D} + \mathbf{W}. \tag{2.224}$$

The symmetric *deformation rate tensor* $\mathbf{D}$ is determined as follows:

$$\mathbf{D} = \frac{1}{2}(\nabla\otimes\mathbf{v} + \nabla\otimes\mathbf{v}^{\mathrm{T}}). \tag{2.225}$$

This tensor has six independent components.

The skew-symmetric *vorticity tensor* $\mathbf{W}$ is determined as follows:

$$\mathbf{W} = \frac{1}{2}(\nabla \otimes \mathbf{v}^{\mathrm{T}} - \nabla \otimes \mathbf{v}). \tag{2.226}$$

Since the tensor $\mathbf{W}$ is skew-symmetric and has three independent components, we can put the tensor $\mathbf{W}$ in correspondence with the *vorticity vector* $\boldsymbol{\omega}$ connected to the tensor (see [12]) as follows:

$$\boldsymbol{\omega} = \frac{1}{2}\mathbf{W} \cdot\cdot\, \boldsymbol{\varepsilon}, \qquad \mathbf{W} = \boldsymbol{\omega} \times \mathbf{E}. \tag{2.227}$$

where $\boldsymbol{\varepsilon}$ is the *Levi-Civita tensor*, which has the third order (see [12]). This tensor is determined as follows:

$$\boldsymbol{\varepsilon} = \frac{1}{\sqrt{g}}\epsilon^{ijk}\mathbf{r}_i \otimes \mathbf{r}_j \otimes \mathbf{r}_k. \tag{2.228}$$

On substituting (2.224)–(2.227) into (2.222), we prove the following theorem.

**Theorem 2.20 (Cauchy–Helmholtz).** *The velocity $\mathbf{v}(\mathcal{M}')$ of an arbitrary point $\mathcal{M}'$ in a neighborhood of the material point $\mathcal{M}$ consists of the translational motion velocity $\mathbf{v}(\mathcal{M})$ of the point $\mathcal{M}$, the velocity $\boldsymbol{\omega} \times d\mathbf{x}$ of rotation as a rigid whole and the deformation rate $\mathbf{D} \cdot d\mathbf{x}$, i.e.*

$$\widehat{d}\mathbf{v} = \boldsymbol{\omega} \times d\mathbf{x} + \mathbf{D} \cdot d\mathbf{x} \tag{2.229}$$

*or*

$$\mathbf{v}(\mathcal{M}') = \mathbf{v}(\mathcal{M}) + \boldsymbol{\omega} \times d\mathbf{x} + \mathbf{D} \cdot d\mathbf{x} + o\,(|d\mathbf{x}|). \tag{2.229a}$$

*Example 2.4.* Determine the tensor $\mathbf{L}$ for the problem on tension of a beam (see Example 2.1), substituting (2.182) into (2.223):

$$\mathbf{L}^{\mathrm{T}} = \bar{\mathbf{e}}^i \frac{\partial}{\partial X^i} \otimes \mathbf{v} = \bar{\mathbf{e}}^i \otimes \frac{\partial}{\partial X^i} \sum_{\alpha=1}^{3} \dot{k}_\alpha X^\alpha \bar{\mathbf{e}}_\alpha = \sum_{\alpha=1}^{3} \dot{k}_\alpha \bar{\mathbf{e}}_\alpha \otimes \bar{\mathbf{e}}_\alpha = \mathbf{L}.$$

Since the velocity gradient $\mathbf{L}$ in this case proves to be a symmetric tensor, from (2.225) and (2.226) it follows that

$$\mathbf{D} = \mathbf{L}, \qquad \mathbf{W} = \mathbf{0}.$$

Thus, in this case $\boldsymbol{\omega} = \mathbf{0}$. □

*Example 2.5.* Determine the tensor $\mathbf{L}$ for the problem on simple shear (see Example 2.2), substituting formula (2.182) into (2.223):

$$\mathbf{L}^{\mathrm{T}} = \bar{\mathbf{e}}^i \otimes \frac{\partial \mathbf{v}}{\partial X^i} = \frac{\partial \bar{v}^j}{\partial X^i}\bar{\mathbf{e}}^i \otimes \bar{\mathbf{e}}_j = \frac{\partial \bar{v}^1}{\partial X^2}\bar{\mathbf{e}}_2 \otimes \bar{\mathbf{e}}_1 = \dot{a}\bar{\mathbf{e}}_2 \otimes \bar{\mathbf{e}}_1.$$

According to formulae (2.225) and (2.226), we get

$$\mathbf{D} = (\dot{a}/2)(\bar{\mathbf{e}}_1 \otimes \bar{\mathbf{e}}_2 + \bar{\mathbf{e}}_2 \otimes \bar{\mathbf{e}}_1),$$

$$\mathbf{W} = (\dot{a}/2)(\bar{\mathbf{e}}_1 \otimes \bar{\mathbf{e}}_2 - \bar{\mathbf{e}}_2 \otimes \bar{\mathbf{e}}_1) = (\dot{a}/2)(\delta_1^i \delta_2^j - \delta_2^i \delta_1^j)\bar{\mathbf{e}}_i \otimes \bar{\mathbf{e}}_j.$$

Using formula (2.227), we determine the vorticity vector

$$\boldsymbol{\omega} = \frac{1}{2}\mathbf{W} \cdot\cdot\, \boldsymbol{\varepsilon} = \frac{\dot{a}}{4}(\delta_1^i \delta_2^j - \delta_2^i \delta_1^j)\epsilon_{jik}\bar{\mathbf{e}}^k = \frac{\dot{a}}{4}(\epsilon_{21k} - \epsilon_{12k})\bar{\mathbf{e}}^k = -\frac{\dot{a}}{2}\bar{\mathbf{e}}^3,$$

which is orthogonal to the shear plane. □

### 2.4.6 *Eigenvalues of the Deformation Rate Tensor*

Just as any symmetric tensor, the deformation rate tensor **D** has three orthonormal real-valued eigenvectors and three real positive eigenvalues (see [12]). Denote the eigenvectors by $\mathbf{q}_\alpha$ (these vectors, in general, are not coincident with $\mathbf{p}_\alpha$) and the eigenvalues – by $D_\alpha$. Then the tensor **D** can be resolved for its dyadic eigenbasis as follows:

$$\mathbf{D} = \sum_{\alpha=1}^{3} D_\alpha \mathbf{q}_\alpha \otimes \mathbf{q}_\alpha, \quad \mathbf{q}_\alpha \cdot \mathbf{q}_\beta = \delta_{\alpha\beta}. \tag{2.230}$$

Take in the actual configuration $\mathcal{K}$ an elementary radius-vector $d\mathbf{x}_\alpha$, connecting points $\mathcal{M}$ and $\mathcal{M}'$, so that the vector is oriented along the eigenvector $\mathbf{q}_\alpha$ of the tensor **D**; then, similarly to (2.172), we can write

$$d\mathbf{x}_\alpha = \mathbf{q}_\alpha |d\mathbf{x}_\alpha|, \quad |d\mathbf{x}_\alpha| = (d\mathbf{x}_\alpha \cdot d\mathbf{x}_\alpha)^{1/2}. \tag{2.231}$$

Apply the Cauchy–Helmholtz theorem (2.229) to the elementary radius-vector:

$$\widehat{d}\mathbf{v}_\alpha = \boldsymbol{\omega} \times d\mathbf{x}_\alpha + \mathbf{D} \cdot d\mathbf{x}_\alpha. \tag{2.232}$$

Multiplying the left and right sides of the equation by $d\mathbf{x}_\alpha$ and taking account of the property of the mixed derivative $d\mathbf{x}_\alpha \cdot (\boldsymbol{\omega} \times d\mathbf{x}_\alpha) = 0$, we get

$$d\mathbf{x}_\alpha \cdot \widehat{d}\mathbf{v}_\alpha = d\mathbf{x}_\alpha \cdot \mathbf{D} \cdot d\mathbf{x}_\alpha. \tag{2.233}$$

Substituting in place of **D** its expression (2.230) and in place of $d\mathbf{x}_\alpha$ their expressions (2.231), we obtain

$$d\mathbf{x}_\alpha \cdot \widehat{d}\mathbf{v}_\alpha = |d\mathbf{x}_\alpha|^2 \sum_{\beta=1}^{3} D_\beta \mathbf{q}_\alpha \cdot \mathbf{q}_\beta \otimes \mathbf{q}_\beta \cdot \mathbf{q}_\alpha = D_\alpha |\mathbf{q}_\alpha|^2. \tag{2.234}$$

Here we have used the property (2.230) of orthonormal vectors $\mathbf{q}_\alpha$.

Modify the scalar product on the left-hand side as follows:

$$d\mathbf{x}_\alpha \cdot \widehat{d}\mathbf{v}_\alpha = d\mathbf{x}_\alpha \cdot \frac{\partial}{\partial t} d\mathbf{x}_\alpha = \frac{1}{2}\frac{\partial}{\partial t}(d\mathbf{x}_\alpha \cdot d\mathbf{x}_\alpha) = |d\mathbf{x}_\alpha| \frac{\partial}{\partial t}|d\mathbf{x}_\alpha|. \tag{2.235}$$

On comparing (2.234) with (2.235), we obtain the following theorem.

**Theorem 2.21.** *Eigenvalues $D_\alpha$ of the deformation rate tensor $\mathbf{D}$ are the rates of relative elongations of elementary material fibres oriented along the eigenvectors $\mathbf{q}_\alpha$:*

$$D_\alpha = \frac{1}{|d\mathbf{x}_\alpha|}\frac{\partial}{\partial t}|d\mathbf{x}_\alpha|. \tag{2.236}$$

### 2.4.7 Resolution of the Vorticity Tensor for the Eigenbasis of the Deformation Rate Tensor

Modify the right-hand side of (2.232) as follows:

$$\widehat{d}\mathbf{v}_\alpha = \boldsymbol{\omega} \times d\mathbf{x}_\alpha + \mathbf{D} \cdot d\mathbf{x}_\alpha = (\boldsymbol{\omega} \times \mathbf{q}_\alpha + D_\alpha \mathbf{q}_\alpha)|d\mathbf{x}_\alpha|, \tag{2.237}$$

and the left-hand side of (2.232) with taking (2.236) into account:

$$\widehat{d}\mathbf{v}_\alpha = \frac{\partial}{\partial t} d\mathbf{x}_\alpha = \frac{\partial}{\partial t}(|d\mathbf{x}_\alpha|\mathbf{q}_\alpha) = \frac{\partial |d\mathbf{x}_\alpha|}{\partial t}\mathbf{q}_\alpha + |d\mathbf{x}_\alpha| \frac{\partial \mathbf{q}_\alpha}{\partial t} = |d\mathbf{x}_\alpha| \left( D_\alpha \mathbf{q}_\alpha + \frac{\partial \mathbf{q}_\alpha}{\partial t} \right). \tag{2.238}$$

On comparing (2.237) with (2.238), we get the following theorem.

**Theorem 2.22.** *The vorticity tensor $\mathbf{W}$ (or the vorticity vector $\boldsymbol{\omega}$) connects the rate of changing the eigenvectors $\mathbf{q}_\alpha$ to the vectors $\mathbf{q}_\alpha$ themselves:*

$$\dot{\mathbf{q}}_\alpha = \frac{\partial \mathbf{q}_\alpha}{\partial t} = \boldsymbol{\omega} \times \mathbf{q}_\alpha = \mathbf{W} \cdot \mathbf{q}_\alpha. \tag{2.239}$$

Using formula (2.239), we can resolve the tensor $\mathbf{W}$ for the eigenbasis $\mathbf{q}_\alpha$ of the deformation rate tensor as follows:

$$\mathbf{W} = \sum_{\alpha=1}^{3} \dot{\mathbf{q}}_\alpha \otimes \mathbf{q}_\alpha = \dot{\mathbf{q}}_i \otimes \mathbf{q}^i. \tag{2.240}$$

### 2.4.8 Geometric Picture of Infinitesimal Transformation of a Small Neighborhood of a Point

If in configuration $\mathcal{K}$ at time $t$ we consider an elementary radius-vector $d\mathbf{x}$ connecting two infinitesimally close material points $\mathcal{M}$ and $\mathcal{M}'$, then for infinitesimal time $dt$ the radius-vector is transformed into radius-vector $d\mathbf{x}'$ in configuration $\mathcal{K}(t + dt)$ (Fig. 2.13):

$$d\mathbf{x} = \mathbf{x}'(t) - \mathbf{x}(t), \qquad d\mathbf{x}' = \mathbf{x}'(t + dt) - \mathbf{x}(t + dt), \tag{2.241}$$

where $\mathbf{x}(t)$ and $\mathbf{x}'(t)$ are radius-vectors of the points $\mathcal{M}$ and $\mathcal{M}'$ in configuration $\mathcal{K}(t)$, respectively; and $\mathbf{x}(t + dt)$ and $\mathbf{x}'(t + dt)$ – in configuration $\mathcal{K}(t + dt)$. Displacements of points $\mathcal{M}$ and $\mathcal{M}'$ for infinitesimal time are defined by the velocity vectors $\mathbf{v}(\mathcal{M})$ and $\mathbf{v}(\mathcal{M}')$, respectively:

$$\mathbf{x}(t + dt) - \mathbf{x}(t) = \mathbf{v}(\mathcal{M})\, dt, \quad \mathbf{x}'(t + dt) - \mathbf{x}'(t) = \mathbf{v}(\mathcal{M}')\, dt. \tag{2.242}$$

Formulae (2.241), (2.242), and simple geometric relations (see Fig. 2.13) give

$$\mathbf{v}(\mathcal{M}')dt - \mathbf{v}(\mathcal{M})dt = d\mathbf{x}' - d\mathbf{x}. \tag{2.243}$$

On substituting (2.243) into (2.229a), we obtain the relation between elementary radius-vectors $d\mathbf{x}'$ and $d\mathbf{x}$:

$$d\mathbf{x}' = d\mathbf{x} + dt\boldsymbol{\omega} \times d\mathbf{x} + dt\mathbf{D} \cdot d\mathbf{x} + dt\, o(|d\mathbf{x}|). \tag{2.244}$$

The relation (2.244) can be considered as the transformation of coordinates $dx^i \to dx'^i$ in a small neighborhood of the point contained in a continuum. Since $dt\boldsymbol{\omega}$ and $dt\mathbf{D}$ are independent of $d\mathbf{x}$ and $d\mathbf{x}'$, so the transformation is

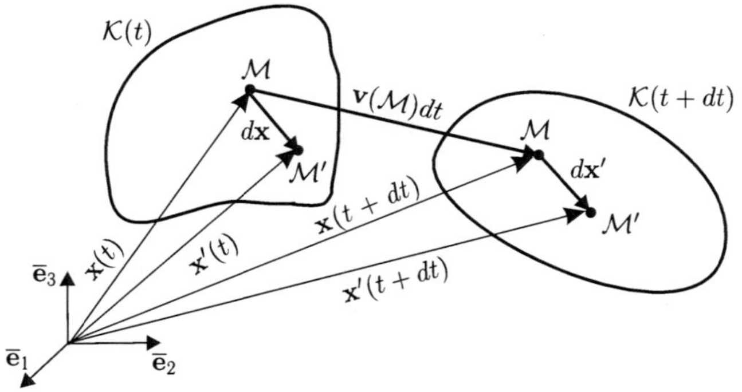


**Fig. 2.13** Infinitesimal transformation of an elementary radius-vector

linear, i.e. affine. The relation (2.244) can be represented as a superposition of two transformations up to an accuracy of $o(|d\mathbf{x}|)$:

$$d\mathbf{x}'' = \mathbf{A}_D \cdot d\mathbf{x}, \qquad \mathbf{A}_D = \mathbf{E} + dt\mathbf{D}, \tag{2.245}$$

$$d\mathbf{x}' = \mathbf{Q}_\omega \cdot d\mathbf{x}'', \qquad \mathbf{Q}_\omega = \mathbf{E} + dt\boldsymbol{\omega} \times \mathbf{E}. \tag{2.246}$$

The tensor $\mathbf{A}_D$ is symmetric and has three eigendirections, which are coincident with the eigendirections $\mathbf{q}_\alpha$ of the deformation rate tensor $\mathbf{D}$.

So just as the tensor $\mathbf{U}$, the tensor $\mathbf{A}_D$ transforms a small neighborhood of a point $\mathcal{M}$ by extending or compressing the neighborhood along the principal directions $\mathbf{q}_\alpha$. The material segments $|d\mathbf{x}''_\alpha|$ oriented along the eigendirections $\mathbf{q}_\alpha$ retain their orientation under the transformations (2.245), but their lengths vary as follows:

$$d\mathbf{x}_\alpha = |d\mathbf{x}_\alpha|\mathbf{q}_\alpha, \qquad d\mathbf{x}''_\alpha = (1 + D_\alpha dt|d\mathbf{x}_\alpha|)\mathbf{q}_\alpha = (1 + D_\alpha dt)d\mathbf{x}_\alpha.$$

The tensor $\mathbf{Q}_\omega$ (2.246) is orthogonal up to an accuracy of values $\sim (dt)^2$, because

$$\mathbf{Q}_\omega \cdot \mathbf{Q}_\omega^{\mathrm{T}} = (\mathbf{E} + dt\mathbf{W}) \cdot (\mathbf{E} + dt\mathbf{W}^{\mathrm{T}}) = \mathbf{E} - (dt)^2\mathbf{W}^2. \tag{2.247}$$

Here we have taken into account that the vorticity tensor $\mathbf{W}$ is skew-symmetric.

Thus, the transformation (2.246) determined by the tensor $\mathbf{Q}_\omega$ is rotation of the $\mathcal{M}$-point neighborhood as a rigid whole for infinitesimal time $dt$.

The vorticity vector $\boldsymbol{\omega}$ forming the tensor $\mathbf{Q}_\omega$ can be considered as instantaneous angular rate of rotation of the small neighborhood as a rigid whole, or as instantaneous angular rate of rotation of the eigentrihedron $\mathbf{q}_\alpha$ of the deformation rate tensor relative to the fixed basis $\bar{\mathbf{e}}_i$. This fact will be considered in detail in Sect. 2.5.7.

On uniting the properties of the transformations (2.245) and (2.246), we can make the following conclusion.

**Theorem 2.23.** *The infinitesimal transformation of a small neighborhood of the point contained in a continuum is a superposition of tension-compression of the neighborhood along the eigendirections $\mathbf{q}_\alpha$ and rotation of the axes $\mathbf{q}_\alpha$ as a rigid whole about the axis with the direction vector $\boldsymbol{\omega}$.*

Thus, we have the certain analogy between the eigendirections $\overset{\circ}{\mathbf{p}}_\alpha$ of the tensor $\mathbf{U}$ and the directions $\mathbf{q}_\alpha$ of the tensor $\mathbf{D}$: elementary material fibres oriented along $\overset{\circ}{\mathbf{p}}_\alpha$ and along $\mathbf{q}_\alpha$ remain mutually orthogonal and undergo only tension-compression. The axes $\overset{\circ}{\mathbf{p}}_\alpha$ remain mutually orthogonal under any finite transformations from $\overset{\circ}{\mathcal{K}}$ to $\mathcal{K}$, but $\mathbf{q}_\alpha$ – only under infinitesimal transformations from $\mathcal{K}(t)$ to $\mathcal{K}(t + dt)$.

### 2.4.9 Kinematic Meaning of the Vorticity Vector

Just as any orthogonal tensor, the orthogonal tensor $\mathbf{Q}_\omega$ of infinitesimal rotation from $\mathcal{K}(t)$ to $\mathcal{K}(t+dt)$ can be represented in the form (see [12])

$$\mathbf{Q}_\omega = \mathbf{E}\cos(d\varphi) + (1-\cos(d\varphi))\mathbf{e}\otimes\mathbf{e} - \mathbf{e}\times\mathbf{E}\sin(d\varphi), \tag{2.248}$$

where $d\varphi$ is the infinitesimal angle of rotation of the trihedron $\mathbf{q}_\alpha$ about the axis with the direction vector $\mathbf{e}$. Since values of $d\varphi$ are infinitesimal, we have

$$\mathbf{Q}_\omega = \mathbf{E} - \mathbf{e}\times\mathbf{E}d\varphi. \tag{2.249}$$

On comparing (2.249) with (2.246), we get

$$\boldsymbol{\omega} = -\frac{d\varphi}{dt}\mathbf{e}, \qquad |\boldsymbol{\omega}| = \frac{d\varphi}{dt}, \tag{2.250}$$

i.e. the vorticity vector $\boldsymbol{\omega}$ is really oriented along the instantaneous rotation axis $\mathbf{e}$, and the length $|\boldsymbol{\omega}|$ is equal to the instantaneous angular rate of rotation of the trihedron $\mathbf{q}_\alpha$ of the deformation rate tensor.

Let us consider now the question: relative to what system the vorticity vector $\boldsymbol{\omega}$ defines the rotation rate.

To answer the question, we introduce another orthogonal rotation tensor

$$\mathbf{O}_W = \mathbf{q}_i \otimes \bar{\mathbf{e}}_i, \tag{2.251}$$

which transforms the Cartesian trihedron $\bar{\mathbf{e}}_i$ as a rigid whole into the orthonormal trihedron $\mathbf{q}_i$:

$$\mathbf{q}_i = \mathbf{O}_W \cdot \bar{\mathbf{e}}_i. \tag{2.252}$$

The tensor $\mathbf{O}_W$ is a function of time $t$, because $\mathbf{q}_i = \mathbf{q}_i(t)$.

According to (2.240) and (2.252), the vorticity tensor $\mathbf{W}$ takes the form

$$\mathbf{W} = \dot{\mathbf{q}}_i \otimes \mathbf{q}_i = \dot{\mathbf{O}}_W \cdot \mathbf{O}_W^{\mathrm{T}}. \tag{2.253}$$

With the help of (2.253) we can represented the orthogonal tensor $\mathbf{Q}_\omega$ as follows:

$$\mathbf{Q}_\omega = \mathbf{E} + dt\mathbf{W} = \mathbf{E} + dt\dot{\mathbf{O}}_W \cdot \mathbf{O}_W^{\mathrm{T}}. \tag{2.254}$$

Thus, at each time $t$ two orthogonal tensors $\mathbf{O}_W$ and $\mathbf{Q}_\omega$ connect local neighborhoods of a point $\mathcal{M}$ in the reference $\overset{\circ}{\mathcal{K}}$ and actual configurations $\mathcal{K}(t+dt)$. If in $\mathcal{K}$ we consider an elementary radius-vector $d\mathbf{x}''$, then in $\overset{\circ}{\mathcal{K}}$ we find its corresponding radius-vector $d\overset{\circ}{\mathbf{x}}''$ obtained with the help of the rotation tensor $\mathbf{O}_W$, and in $\mathcal{K}(t+dt)$ – radius-vector $d\mathbf{x}'$:

$$d\mathbf{x}'' = \mathbf{O}_W \cdot d\overset{\circ}{\mathbf{x}}'', \quad d\mathbf{x}' = \mathbf{Q}_\omega \cdot d\mathbf{x}''.$$

A fixed observer connected to the Cartesian trihedron $\bar{\mathbf{e}}_i$ sees both the transformations: finite rotation for time $t$, which is described by the tensor $\mathbf{O}_W$, and instantaneous rotation of a local neighborhood for time $dt$, which is described by the infinitesimal rotation tensor $\mathbf{Q}_\omega$

Thus, the vorticity vector $\boldsymbol{\omega}$ is the vector of instantaneous angular rate of rotation of the trihedron $\mathbf{q}_\alpha$ relative to the trihedron $\bar{\mathbf{e}}_i$.

Comparing (2.246) with (2.254) (or (2.253) with (2.227)), we get

$$\dot{\mathbf{O}}_W \cdot \mathbf{O}_W^{\mathrm{T}} = \boldsymbol{\omega} \times \mathbf{E}. \tag{2.255}$$

### 2.4.10 Tensor of Angular Rate of Rotation (Spin)

In Sect. 2.3 we have introduced the tensor $\dot{\mathbf{O}}_W \cdot \mathbf{O}_W^{\mathrm{T}}$, where $\mathbf{O}_W$ is the orthogonal tensor of rotation. Such tensor $\boldsymbol{\Omega} = \dot{\mathbf{Q}} \cdot \mathbf{Q}^{\mathrm{T}}$ can be set up for any orthogonal tensor $\mathbf{Q}$ depending on time $t$.

The tensor $\dot{\mathbf{Q}} \cdot \mathbf{Q}^{\mathrm{T}}$ is skew-symmetric, because

$$\dot{\mathbf{Q}} \cdot \mathbf{Q}^{\mathrm{T}} + \mathbf{Q} \cdot \dot{\mathbf{Q}}^{\mathrm{T}} = (\mathbf{Q} \cdot \mathbf{Q}^{\mathrm{T}})^{\bullet} = (\mathbf{E})^{\bullet} = 0, \tag{2.256}$$

i.e.

$$\boldsymbol{\Omega}^{\mathrm{T}} = (\dot{\mathbf{Q}} \cdot \mathbf{Q}^{\mathrm{T}})^{\mathrm{T}} = \mathbf{Q} \cdot \dot{\mathbf{Q}}^{\mathrm{T}} = -\dot{\mathbf{Q}} \cdot \mathbf{Q}^{\mathrm{T}} = -\boldsymbol{\Omega}. \tag{2.257}$$

This tensor characterizes the angular rate of rotation of the orthonormal trihedron $\mathbf{h}_i$ formed with the help of $\mathbf{Q}$:

$$\mathbf{h}_i = \mathbf{Q} \cdot \bar{\mathbf{e}}_i, \tag{2.258}$$

relative to the Cartesian trihedron $\bar{\mathbf{e}}_i$.

Indeed, with the help of the tensor $\dot{\mathbf{Q}} \cdot \mathbf{Q}^{\mathrm{T}}$ we can form the tensor (2.254):

$$\mathbf{Q}_\omega = \mathbf{E} + dt\dot{\mathbf{Q}} \cdot \mathbf{Q}^{\mathrm{T}}, \tag{2.259}$$

which, according to (2.247), is the orthogonal tensor of infinitesimal rotation; and this tensor can be represented in the form (2.248) or (2.249):

$$\mathbf{Q}_\omega = \mathbf{E} - d\varphi \mathbf{e} \times \mathbf{E}, \tag{2.260}$$

where $d\varphi$ is the infinitesimal angle of rotation of the trihedron $\mathbf{h}_i$ about the axis with the direction vector $\mathbf{e}$. Comparing (2.259) with (2.260), we get the expression

$$\dot{\mathbf{Q}} \cdot \mathbf{Q}^{\mathrm{T}} = -\frac{d\varphi}{dt} \mathbf{e} \times \mathbf{E}, \tag{2.261}$$

which makes clear the sentence that the tensor $\dot{\mathbf{Q}}\cdot\mathbf{Q}^{\mathrm{T}}$ characterizes the instantaneous angular rate $d\varphi/dt$ of rotation of the trihedron $\mathbf{h}_i$ about the axis $\mathbf{e}$.

The tensor $\dot{\mathbf{Q}}\cdot\mathbf{Q}^{\mathrm{T}}$ is called the *tensor of angular rate of rotation* or the *spin*.

Expressing the tensor $\mathbf{Q}$ from (2.258) in terms of the bases $\mathbf{h}_i$ and $\bar{\mathbf{e}}_i$ ($\mathbf{h}_\alpha = \mathbf{h}^\alpha$, $\alpha = 1, 2, 3$, as the vectors are orthonormal):

$$\mathbf{Q} = \mathbf{h}^i \otimes \bar{\mathbf{e}}_i\,, \tag{2.262}$$

we get another representation of the spin:

$$\dot{\mathbf{Q}}\cdot\mathbf{Q}^{\mathrm{T}} = \dot{\mathbf{h}}_i \otimes \mathbf{h}^i\,. \tag{2.263}$$

Thus, we have proved the following theorem.

**Theorem 2.24.** *The spin connects the rates* $\dot{\mathbf{h}}_i$ *and the vectors* $\mathbf{h}_i$ *defined by formula* (2.258) *as follows:*

$$\dot{\mathbf{h}}_i = (\dot{\mathbf{Q}}\cdot\mathbf{Q}^{\mathrm{T}})\cdot\mathbf{h}_i\,. \tag{2.264}$$

Since the spin $\dot{\mathbf{Q}}\cdot\mathbf{Q}^{\mathrm{T}}$ is a skew-symmetric tensor, we can introduce the corresponding vorticity vector $\boldsymbol{\omega}_h$:

$$\dot{\mathbf{Q}}\cdot\mathbf{Q}^{\mathrm{T}} = \boldsymbol{\omega}_h \times \mathbf{E} \tag{2.265}$$

(from formulae (2.261) and (2.265) it follows that $\boldsymbol{\omega}_h = -(d\varphi/dt)\mathbf{e}$); then formula (2.264) takes the form

$$\dot{\mathbf{h}}_i = \boldsymbol{\omega}_h \times \mathbf{h}_i\,. \tag{2.266}$$

Resolving the vector $\boldsymbol{\omega}_h$ for the orthonormal basis $\mathbf{h}_i$: $\boldsymbol{\omega}_h = \omega^j_h\mathbf{h}_j$, we get one more representation of formula (2.264):

$$\dot{\mathbf{h}}_i = \omega^j_h\mathbf{h}_j \times \mathbf{h}_i = \epsilon_{jik}\omega^j_h\mathbf{h}^k\,. \tag{2.267}$$

This formula can also be rewritten in the form

$$\dot{\mathbf{h}}_\alpha = \omega^\beta_h\mathbf{h}_\gamma, \quad \alpha \neq \beta \neq \gamma \neq \alpha. \tag{2.268}$$

Taking different orthogonal tensors (or orthonormal bases) as $\mathbf{Q}$ (or $\mathbf{h}_i$), we obtain different spins:

(1) If we choose eigenvectors of the stretch tensor $\mathbf{U}$ as $\mathbf{h}_i$, i.e. $\mathbf{h}_i = \overset{\circ}{\mathbf{p}}_i$, then, according to (2.263), the corresponding spin $\boldsymbol{\Omega}_U$ takes the form

$$\boldsymbol{\Omega}_U = \dot{\mathbf{O}}_U\cdot\mathbf{O}^{\mathrm{T}}_U = \overset{\circ}{\mathbf{p}}{}^{\bullet}_i \otimes \overset{\circ}{\mathbf{p}}{}^i, \quad \mathbf{O}_U = \overset{\circ}{\mathbf{p}}{}^i \otimes \bar{\mathbf{e}}_i\,, \tag{2.269}$$

and formula (2.264) yields

$$\dot{\overset{\circ}{\mathbf{p}}}_i = \boldsymbol{\Omega}_U\cdot\overset{\circ}{\mathbf{p}}_i\,. \tag{2.270}$$

(2) If $\mathbf{h}_i = \mathbf{p}_i$, then the corresponding spin $\boldsymbol{\Omega}_V$ and the rotation tensor $\mathbf{O}_V$ have the forms

$$\boldsymbol{\Omega}_V = \dot{\mathbf{O}}_V \cdot \mathbf{O}_V^{\mathrm{T}} = \dot{\mathbf{p}}_i \otimes \mathbf{p}^i, \quad \mathbf{O}_V = \mathbf{p}_i \otimes \bar{\mathbf{e}}_i, \tag{2.271}$$

$$\dot{\mathbf{p}}_i = \boldsymbol{\Omega}_V \cdot \mathbf{p}_i. \tag{2.272}$$

(3) If $\mathbf{h}_i = \mathbf{q}_i$, then the corresponding spin $\boldsymbol{\Omega}_W$ coincides with the vorticity tensor $\mathbf{W}$ (see (2.253)):

$$\boldsymbol{\Omega}_W = \dot{\mathbf{O}}_W \cdot \mathbf{O}_W^{\mathrm{T}} = \dot{\mathbf{q}}_i \otimes \mathbf{q}_i = \mathbf{W}, \tag{2.273}$$

$$\dot{\mathbf{q}}_i = \boldsymbol{\Omega}_W \cdot \mathbf{q}_i. \tag{2.274}$$

(4) If we take the rotation tensor $\mathbf{O}$ accompanying deformation as $\mathbf{Q}$, then, as shown in (2.159a), the tensor $\mathbf{O}$ connects two moving bases $\mathbf{p}_i$ and $\overset{\circ}{\mathbf{p}}_i$:

$$\mathbf{p}_i = \mathbf{O} \cdot \overset{\circ}{\mathbf{p}}_i. \tag{2.275}$$

The tensor $\mathbf{O}$ can be expressed in terms of $\mathbf{O}_V$ and $\mathbf{O}_U$ as follows:

$$\mathbf{O} = \mathbf{p}_i \otimes \overset{\circ}{\mathbf{p}}_i = \mathbf{p}_i \otimes \bar{\mathbf{e}}_i \cdot \bar{\mathbf{e}}_j \otimes \overset{\circ}{\mathbf{p}}_j = \mathbf{O}_V \cdot \mathbf{O}_U^{\mathrm{T}}. \tag{2.276}$$

The corresponding spin $\boldsymbol{\Omega}$ has the form

$$\begin{aligned} \boldsymbol{\Omega} &= \dot{\mathbf{O}} \cdot \mathbf{O}^{\mathrm{T}} = (\dot{\mathbf{O}}_V \cdot \mathbf{O}_U^{\mathrm{T}} + \mathbf{O}_V \cdot \dot{\mathbf{O}}_U^{\mathrm{T}}) \cdot \mathbf{O}_U \cdot \mathbf{O}_V^{\mathrm{T}} \\ &= \dot{\mathbf{O}}_V \cdot \mathbf{O}_V^{\mathrm{T}} + \mathbf{O}_V \cdot \dot{\mathbf{O}}_U^{\mathrm{T}} \cdot \mathbf{O}_U \cdot \mathbf{O}_V^{\mathrm{T}} = \boldsymbol{\Omega}_V - \mathbf{O}_V \cdot \boldsymbol{\Omega}_U \cdot \mathbf{O}_V^{\mathrm{T}}. \end{aligned} \tag{2.277}$$

Unlike the cases (1)–(3), the spin tensor $\boldsymbol{\Omega}$ characterizes the angular rate of rotation of the trihedron $\mathbf{p}_i$ relative to the moving trihedron $\overset{\circ}{\mathbf{p}}_i$, but not relative to the trihedron $\bar{\mathbf{e}}_i$ being fixed.

Therefore, for the cases (1)–(3) the spins characterize the total angular rate, and for the case (4) – the relative rate.

### *2.4.11 Relationships Between Rates of Deformation Tensors and Velocity Gradients*

In continuum mechanics, one often needs the relations between rates of the deformation tensors (and also measures) and the velocity gradients $\mathbf{L} = (\boldsymbol{\nabla} \otimes \mathbf{v})^{\mathrm{T}}$ and

$$\overset{\circ}{\mathbf{L}} = \left(\overset{\circ}{\boldsymbol{\nabla}} \otimes \mathbf{v}\right)^{\mathrm{T}}. \tag{2.278}$$

Let us derive these relations.

**Theorem 2.25.** *The rates of varying the gradient* $\dot{\mathbf{F}}$ *and the inverse gradient* $(\mathbf{F}^{-1})^{\bullet}$ *are connected to* $\mathbf{L}$ *and* $\overset{\circ}{\mathbf{L}}$ *by the relations*

$$\dot{\mathbf{F}} = \mathbf{L} \cdot \mathbf{F}, \quad \dot{\mathbf{F}} = \overset{\circ}{\mathbf{L}},$$

$$(\mathbf{F}^{-1})^{\bullet} = -\mathbf{F}^{-1} \cdot \mathbf{L}, \quad (\mathbf{F}^{-1})^{\bullet} = -\mathbf{F}^{-1} \cdot \mathbf{F}^{-1\mathrm{T}} \cdot \overset{\circ}{\mathbf{L}}. \tag{2.279}$$

▼ Differentiating the relationships (2.35a) with respect to time $t$ and taking the definition of the velocity (2.181) into account, we get

$$\dot{\mathbf{F}}^{\mathrm{T}} = \overset{\circ}{\nabla} \otimes \mathbf{v} = \mathbf{F}^{\mathrm{T}} \cdot \nabla \otimes \mathbf{v}, \quad \dot{\mathbf{F}} = (\nabla \otimes \mathbf{v})^{\mathrm{T}} \cdot \mathbf{F}. \tag{2.279a}$$

According to the definitions of tensors $\mathbf{L}$ (2.223) and $\overset{\circ}{\mathbf{L}}$ (2.278), from (2.279a) we obtain formulae (2.279).

Differentiating the identity $(\mathbf{F} \cdot \mathbf{F}^{-1})^{\bullet} = \dot{\mathbf{E}} = 0$, we find that $\dot{\mathbf{F}} \cdot \mathbf{F}^{-1} = -\mathbf{F} \cdot (\mathbf{F}^{-1})^{\bullet}$; whence we get

$$(\mathbf{F}^{-1})^{\bullet} = -\mathbf{F}^{-1} \cdot \dot{\mathbf{F}} \cdot \mathbf{F}^{-1}. \tag{2.280}$$

On substituting the first two formulae of (2.279) into (2.280), we obtain

$$(\mathbf{F}^{-1})^{\bullet} = -\mathbf{F}^{-1} \cdot (\nabla \otimes \mathbf{v})^{\mathrm{T}}, \quad (\mathbf{F}^{-1\mathrm{T}})^{\bullet} = -(\nabla \otimes \mathbf{v}) \cdot \mathbf{F}^{-1\mathrm{T}}, \tag{2.281}$$

i.e. the third and the fourth relationships of (2.279) hold as well. ▲

According to formulae (2.225), (2.279a), and (2.281), we find that the rate of the deformation gradient is connected to the deformation rate tensor $\mathbf{D}$ by the relations

$$\mathbf{D} = \frac{1}{2}(\dot{\mathbf{F}} \cdot \mathbf{F}^{-1} + \mathbf{F}^{-1\mathrm{T}} \cdot \dot{\mathbf{F}}^{\mathrm{T}}), \quad \mathbf{D} = -\frac{1}{2}(\mathbf{F} \cdot \dot{\mathbf{F}}^{-1} + \dot{\mathbf{F}}^{-1\mathrm{T}} \cdot \mathbf{F}^{\mathrm{T}}). \tag{2.282}$$

Here and below we will use the notation $\dot{\mathbf{F}}^{-1} \equiv (\mathbf{F}^{-1})^{\bullet}$.

**Theorem 2.26.** *The rates of the deformation tensors* $\dot{\mathbf{C}}$, $\dot{\mathbf{A}}$, $\dot{\Lambda}$, $\dot{\mathbf{J}}$ *and deformation measures* $\dot{\mathbf{G}}$, $\dot{\mathbf{g}}$, $(\mathbf{G}^{-1})^{\bullet}$ *and* $(\mathbf{g}^{-1})^{\bullet}$ *are connected to the velocity gradients* $\mathbf{L}$ *and* $\overset{\circ}{\mathbf{L}}$ *by the relationships*

$$\begin{cases} \dot{\mathbf{C}} = \mathbf{F}^{\mathrm{T}} \cdot \mathbf{D} \cdot \mathbf{F}, & \dot{\mathbf{A}} = \mathbf{D} - \mathbf{L}^{\mathrm{T}} \cdot \mathbf{A} - \mathbf{A} \cdot \mathbf{L}, \\ \dot{\Lambda} = \mathbf{F}^{-1} \cdot \mathbf{D} \cdot \mathbf{F}^{-1\mathrm{T}}, & \dot{\mathbf{J}} = \mathbf{D} + \mathbf{L} \cdot \mathbf{J} + \mathbf{J} \cdot \mathbf{L}^{\mathrm{T}}, \end{cases} \tag{2.283}$$

*and*

$$\begin{cases} \dot{\mathbf{G}} = 2\mathbf{F}^{\mathrm{T}} \cdot \mathbf{D} \cdot \mathbf{F}, & \dot{\mathbf{g}} = -\mathbf{L}^{\mathrm{T}} \cdot \mathbf{g} - \mathbf{g} \cdot \mathbf{L}, \\ (\mathbf{G}^{-1})^{\bullet} = -2\mathbf{F}^{-1} \cdot \mathbf{D} \cdot \mathbf{F}^{-1\mathrm{T}}, & (\mathbf{g}^{-1})^{\bullet} = \mathbf{L} \cdot \mathbf{g}^{-1} + \mathbf{g}^{-1} \cdot \mathbf{L}^{\mathrm{T}}; \end{cases} \tag{2.284}$$

*and also*

$$\begin{cases} \dot{\mathbf{C}} = (1/2)\Big(\mathbf{F}^{\mathrm{T}} \cdot \overset{\circ}{\mathbf{L}} + \overset{\circ}{\mathbf{L}}{}^{\mathrm{T}} \cdot \mathbf{F}\Big), \quad \dot{\mathbf{J}} = (1/2)\Big(\overset{\circ}{\mathbf{L}} \cdot \mathbf{F}^{\mathrm{T}} + \mathbf{F}^{\mathrm{T}} \cdot \overset{\circ}{\mathbf{L}}\Big), \\ \dot{\mathbf{A}} = (1/2)\Big((\mathbf{E} - 2\mathbf{A}) \cdot \overset{\circ}{\mathbf{L}} \cdot \mathbf{F}^{-1} + \mathbf{F}^{-1\mathrm{T}} \cdot \overset{\circ}{\mathbf{L}}{}^{\mathrm{T}} \cdot (\mathbf{E} - 2\mathbf{A})\Big), \\ \dot{\boldsymbol{\Lambda}} = (1/2)\Big(\mathbf{F}^{-1} \cdot \overset{\circ}{\mathbf{L}} \cdot (\mathbf{E} - 2\boldsymbol{\Lambda}) + (\mathbf{E} - 2\boldsymbol{\Lambda}) \cdot \overset{\circ}{\mathbf{L}}{}^{\mathrm{T}} \cdot \mathbf{F}^{-1\mathrm{T}}\Big), \end{cases} \tag{2.285}$$

*and*

$$\begin{cases} \dot{\mathbf{G}} = \mathbf{F}^{\mathrm{T}} \cdot \overset{\circ}{\mathbf{L}} + \overset{\circ}{\mathbf{L}}{}^{\mathrm{T}} \cdot \mathbf{F}, \quad \dot{\mathbf{g}} = -\Big(\mathbf{g} \cdot \overset{\circ}{\mathbf{L}} \cdot \mathbf{F}^{-1} + \mathbf{F}^{-1\mathrm{T}} \cdot \overset{\circ}{\mathbf{L}}{}^{\mathrm{T}} \cdot \mathbf{g}\Big), \\ (\mathbf{G}^{-1})^{\bullet} = -\Big(\mathbf{F}^{-1} \cdot \overset{\circ}{\mathbf{L}} \cdot \mathbf{G}^{-1} + \mathbf{G}^{-1} \cdot \overset{\circ}{\mathbf{L}}{}^{\mathrm{T}} \cdot \mathbf{F}^{-1\mathrm{T}}\Big), \\ (\mathbf{g}^{-1})^{\bullet} = \overset{\circ}{\mathbf{L}} \cdot \mathbf{F}^{\mathrm{T}} + \mathbf{F}^{\mathrm{T}} \cdot \overset{\circ}{\mathbf{L}}. \end{cases} \tag{2.286}$$

▼ To prove formula (2.283), we must differentiate the relationships (2.72) with respect to $t$ and apply formulae (2.279):

$$\begin{aligned} \dot{\mathbf{C}} &= \frac{1}{2}(\dot{\mathbf{F}}^{\mathrm{T}} \cdot \mathbf{F} + \mathbf{F}^{\mathrm{T}} \cdot \dot{\mathbf{F}}) = \frac{1}{2}(\mathbf{F}^{\mathrm{T}} \cdot \mathbf{L}^{\mathrm{T}} \cdot \mathbf{F} + \mathbf{F}^{\mathrm{T}} \cdot \mathbf{L} \cdot \mathbf{F}) \\ &= \frac{1}{2}\mathbf{F}^{\mathrm{T}} \cdot (\mathbf{L}^{\mathrm{T}} + \mathbf{L}) \cdot \mathbf{F} = \mathbf{F}^{\mathrm{T}} \cdot \mathbf{D} \cdot \mathbf{F}, \end{aligned} \tag{2.287}$$

$$\begin{aligned} \dot{\mathbf{A}} &= \frac{1}{2}(\dot{\mathbf{F}}^{-1\mathrm{T}} \cdot \mathbf{F}^{-1} + \mathbf{F}^{-1\mathrm{T}} \cdot \dot{\mathbf{F}}^{-1}) \\ &= \frac{1}{2}(\mathbf{L}^{\mathrm{T}} \cdot \mathbf{F}^{-1\mathrm{T}} \cdot \mathbf{F}^{-1} + \mathbf{F}^{-1\mathrm{T}} \cdot \mathbf{F}^{-1} \cdot \mathbf{L}) \\ &= \frac{1}{2}(\mathbf{L}^{\mathrm{T}} \cdot (\mathbf{E} - 2\mathbf{A}) + (\mathbf{E} - 2\mathbf{A}) \cdot \mathbf{L}) = \mathbf{D} - \mathbf{L}^{\mathrm{T}} \cdot \mathbf{A} - \mathbf{A} \cdot \mathbf{L}, \end{aligned} \tag{2.288}$$

$$\begin{aligned} \dot{\boldsymbol{\Lambda}} &= -\frac{1}{2}(\dot{\mathbf{F}}^{-1} \cdot \mathbf{F}^{-1\mathrm{T}} + \mathbf{F}^{-1} \cdot \dot{\mathbf{F}}^{-1\mathrm{T}}) = \frac{1}{2}(\mathbf{F}^{-1} \cdot \mathbf{L} \cdot \mathbf{F}^{-1\mathrm{T}} + \mathbf{F}^{-1} \cdot \mathbf{L}^{\mathrm{T}} \cdot \mathbf{F}^{-1\mathrm{T}}) \\ &= \frac{1}{2}\mathbf{F}^{-1} \cdot (\mathbf{L} + \mathbf{L}^{\mathrm{T}}) \cdot \mathbf{F}^{-1\mathrm{T}} = \mathbf{F}^{-1} \cdot \mathbf{D} \cdot \mathbf{F}^{-1\mathrm{T}}, \end{aligned} \tag{2.289}$$

$$\dot{\mathbf{J}} = \frac{1}{2}(\dot{\mathbf{F}} \cdot \mathbf{F}^{\mathrm{T}} + \mathbf{F} \cdot \dot{\mathbf{F}}^{\mathrm{T}}) = \frac{1}{2}(\mathbf{L} \cdot \mathbf{F} \cdot \mathbf{F}^{\mathrm{T}} + \mathbf{F} \cdot \mathbf{F}^{\mathrm{T}} \cdot \mathbf{L}^{\mathrm{T}})$$
$$= \frac{1}{2}(\mathbf{L} \cdot (\mathbf{E} + 2\mathbf{J}) + (\mathbf{E} + 2\mathbf{J}) \cdot \mathbf{L}^{\mathrm{T}}) = \mathbf{D} + \mathbf{L} \cdot \mathbf{J} + \mathbf{J} \cdot \mathbf{L}^{\mathrm{T}}. \quad (2.290)$$

Formulae (2.284) follow from (2.283), if we have used the connections (2.74) and (2.75) between the deformation tensors and measures.

Formulae (2.285) follow from (2.287)–(2.290), if we have gone from $\mathbf{L}$ to $\overset{\circ}{\mathbf{L}}$:

$$\mathbf{L} = \overset{\circ}{\mathbf{L}} \cdot \mathbf{F}^{-1}. \quad (2.291)$$

Using the connections (2.74) and (2.75) between the deformation tensors and measures, from (2.285) we get formulae (2.286). ▲

Relations between the tensors $\dot{\mathbf{U}}$, $\dot{\mathbf{V}}$ and the velocity gradients $\mathbf{L}$ and $\overset{\circ}{\mathbf{L}}$ are more complicated. To derive them, we should use representations of $\mathbf{L}$ and $\overset{\circ}{\mathbf{L}}$ in terms of the eigenbasis vectors.

**Theorem 2.27.** *The following expressions for the velocity gradients hold:*

$$\overset{\circ}{\mathbf{L}}{}^{\mathrm{T}} = \overset{\circ}{\nabla} \otimes \mathbf{v} = \frac{\partial}{\partial t}\overset{\circ}{\nabla} \otimes \mathbf{x} = \sum_{\alpha=1}^{3} \dot{\lambda}_\alpha \overset{\circ}{\mathbf{p}}_\alpha \otimes \mathbf{p}_\alpha + \boldsymbol{\Omega}_U \cdot \mathbf{F}^{\mathrm{T}} - \mathbf{F}^{\mathrm{T}} \cdot \boldsymbol{\Omega}_V, \quad (2.292)$$

$$\mathbf{L} = (\nabla \otimes \mathbf{v})^{\mathrm{T}} = \sum_{\alpha,\beta=1}^{3} \left( \frac{\dot{\lambda}_\alpha}{\lambda_\alpha}\delta_{\alpha\beta} + \frac{\lambda_\alpha}{\lambda_\beta}\overset{\circ}{\Omega}_{U\beta\alpha} \right) \mathbf{p}_\alpha \otimes \mathbf{p}_\beta + \boldsymbol{\Omega}_V, \quad (2.293)$$

*where $\overset{\circ}{\Omega}_{U\alpha\beta}$ are components of the tensor $\boldsymbol{\Omega}_U$ with respect to the eigenbasis $\overset{\circ}{\mathbf{p}}_\alpha$:*

$$\boldsymbol{\Omega}_U = \overset{\circ}{\mathbf{p}}{}^{\bullet}_i \otimes \overset{\circ}{\mathbf{p}}{}^i = \sum_{\alpha=1}^{3} \overset{\circ}{\Omega}_{U\alpha\beta} \overset{\circ}{\mathbf{p}}_\alpha \otimes \overset{\circ}{\mathbf{p}}_\beta, \quad \overset{\circ}{\Omega}_{U\alpha\beta} = \overset{\circ}{\mathbf{p}}_\alpha \cdot \boldsymbol{\Omega}_U \cdot \overset{\circ}{\mathbf{p}}_\beta = \overset{\circ}{\mathbf{p}}{}^{\bullet}_\alpha \cdot \overset{\circ}{\mathbf{p}}_\beta. \quad (2.294)$$

▼ To prove formula (2.292), we should consider the first formula of (2.279a) and substitute into this formula the expression (2.171) for $\mathbf{F}$ in the eigenbasis:

$$\overset{\circ}{\mathbf{L}}{}^{\mathrm{T}} = \dot{\mathbf{F}}^{\mathrm{T}} = \sum_{\alpha=1}^{3} (\dot{\lambda}_\alpha \overset{\circ}{\mathbf{p}}_\alpha \otimes \mathbf{p}_\alpha + \lambda_\alpha(\overset{\circ}{\mathbf{p}}{}^{\bullet}_\alpha \otimes \mathbf{p}_\alpha + \overset{\circ}{\mathbf{p}}_\alpha \otimes \dot{\mathbf{p}}_\alpha)). \quad (2.295)$$

Using formulae (2.270) and (2.272), from (2.295) we derive the relationship (2.292).

To prove formula (2.293), we use formulae (2.292) and (2.291), having expressed $\mathbf{F}^{-1}$ in the form (2.171); then we get

$$\mathbf{L} = \overset{\circ}{\mathbf{L}} \cdot \mathbf{F}^{-1} = \left( \sum_{\alpha=1}^{3} \dot{\lambda}_\alpha \mathbf{p}_\alpha \otimes \overset{\circ}{\mathbf{p}}_\alpha + \lambda_\alpha (\mathbf{p}_\alpha \otimes \overset{\circ}{\mathbf{p}}{}^{\bullet}_\alpha + \dot{\mathbf{p}}_\alpha \otimes \overset{\circ}{\mathbf{p}}_\alpha) \right) \cdot \sum_{\beta=1}^{3} \lambda_\beta^{-1} \overset{\circ}{\mathbf{p}}_\beta \otimes \mathbf{p}_\beta$$

$$= \sum_{\alpha,\beta=1}^{3} \left( \frac{\dot{\lambda}_\alpha}{\lambda_\beta} \delta_{\alpha\beta} \mathbf{p}_\alpha \otimes \mathbf{p}_\alpha + \frac{\lambda_\alpha}{\lambda_\beta} (\overset{\circ}{\mathbf{p}}{}^{\bullet}_\alpha \cdot \overset{\circ}{\mathbf{p}}_\beta) \mathbf{p}_\alpha \otimes \mathbf{p}_\beta \right) + \sum_{\alpha=1}^{3} \dot{\mathbf{p}}_\alpha \otimes \mathbf{p}_\alpha. \quad (2.296)$$

Here we have taken into account that the vectors $\overset{\circ}{\mathbf{p}}_\alpha$ are orthonormal. Using formulae (2.270) and (2.272), from (2.296) we derive the formula (2.293). ▲

From formula (2.293) it follows that the deformation rate tensor $\mathbf{D}$ and the vorticity tensor can be represented in terms of the eigenbasis $\mathbf{p}_\alpha$ as follows:

$$\mathbf{D} = \sum_{\alpha,\beta=1}^{3} \left( \frac{\dot{\lambda}_\alpha}{\lambda_\alpha} \delta_{\alpha\beta} + \frac{1}{2} \left( \frac{\lambda_\alpha}{\lambda_\beta} - \frac{\lambda_\beta}{\lambda_\alpha} \right) \overset{\circ}{\Omega}_{U\beta\alpha} \right) \mathbf{p}_\alpha \otimes \mathbf{p}_\beta, \quad (2.297)$$

$$\mathbf{W} = \frac{1}{2} \sum_{\alpha,\beta=1}^{3} \left( \frac{\lambda_\alpha}{\lambda_\beta} + \frac{\lambda_\beta}{\lambda_\alpha} \right) \overset{\circ}{\Omega}_{U\beta\alpha} \mathbf{p}_\alpha \otimes \mathbf{p}_\beta + \boldsymbol{\Omega}_V. \quad (2.298)$$

Here we have taken into account that the tensors $\boldsymbol{\Omega}_U$ and $\boldsymbol{\Omega}_V$ are skew-symmetric.

Denote components of the tensor $\mathbf{D}$ with respect to the basis $\mathbf{p}_\alpha$ by $D_{\alpha\beta}$:

$$\mathbf{D} = \sum_{\alpha,\beta=1}^{3} D_{\alpha\beta} \mathbf{p}_\alpha \otimes \mathbf{p}_\beta, \quad D_{\alpha\beta} = \mathbf{p}_\alpha \cdot \mathbf{D} \cdot \mathbf{p}_\beta. \quad (2.299)$$

Then from (2.297) and (2.299) we get that diagonal components of the deformation rate tensor $D_{\alpha\alpha}$ with respect to the eigenbasis $\mathbf{p}_\alpha$ determine the relative rates of lengthening the material fibres oriented along the eigenvectors $\mathbf{p}_\alpha$ (compare with formula (2.236)):

$$D_{\alpha\alpha} = \dot{\lambda}_\alpha / \lambda_\alpha = d\dot{s}_\alpha / ds_\alpha, \quad \alpha = 1, 3; \quad (2.300)$$

and off-diagonal components $D_{\alpha\beta}$ are connected to $\overset{\circ}{\Omega}_{U\alpha\beta}$ by the relations

$$D_{\alpha\beta} = \frac{1}{2} \left( \frac{\lambda_\alpha^2 - \lambda_\beta^2}{\lambda_\alpha \lambda_\beta} \right) \overset{\circ}{\Omega}_{U\beta\alpha}, \quad \alpha \neq \beta. \quad (2.301)$$

From formulae (2.299) and (2.301) we can express the components $\overset{\circ}{\Omega}_{U\alpha\beta}$ in terms of components of the deformation rate tensor:

$$\overset{\circ}{\Omega}_{U\alpha\beta} = \frac{2\lambda_\alpha \lambda_\beta}{\lambda_\beta^2 - \lambda_\alpha^2} D_{\alpha\beta}, \quad \alpha \neq \beta; \quad \overset{\circ}{\Omega}_{U\alpha\alpha} = 0. \quad (2.302)$$

The diagonal components $\overset{\circ}{\Omega}_{U\alpha\alpha}$ are equal to zero, because the tensor $\boldsymbol{\Omega}_U$ is skew-symmetric.

On substituting the relationships (2.302) into (2.298), we find the expression for components $\overset{\circ}{\Omega}_{V\alpha\beta}$ of the tensor $\boldsymbol{\Omega}_V$ with respect to the basis $\mathbf{p}_\alpha$ in terms of the tensors $\mathbf{W}$ and $\mathbf{D}$ (and, hence, in terms of the velocity gradient $\mathbf{L}$):

$$\overset{\circ}{\Omega}_{V\alpha\beta} = \mathbf{p}_\alpha \cdot \boldsymbol{\Omega}_V \cdot \mathbf{p}_\alpha = \mathbf{p}_\alpha \cdot \mathbf{W} \cdot \mathbf{p}_\beta - \frac{\lambda_\alpha^2 + \lambda_\beta^2}{\lambda_\alpha^2 - \lambda_\beta^2} D_{\alpha\beta}, \quad \alpha \neq \beta, \tag{2.303}$$

where

$$\boldsymbol{\Omega}_V = \sum_{\alpha,\beta=1}^{3} \Omega_{V\alpha\beta} \mathbf{p}_\alpha \otimes \mathbf{p}_\beta. \tag{2.304}$$

*Remark.* The expressions (2.302) and (2.303) are valid only if the eigenvalues are not multiple: $\lambda_\alpha \neq \lambda_\beta$ ($\alpha \neq \beta$; $\alpha, \beta = 1, 2, 3$). If within the interval $[t_1, t_2]$ all three eigenvalues are coincident: $\lambda_\alpha = \lambda$ ($\alpha = 1, 2, 3$), then the stretch tensors are spherical: $\mathbf{U} = \lambda \overset{\circ}{\mathbf{p}}_i \otimes \overset{\circ}{\mathbf{p}}{}^i = \lambda \mathbf{E}$, $\mathbf{V} = \lambda \mathbf{E}$, and the eigenbases are not uniquely defined: as $\overset{\circ}{\mathbf{p}}_i$ and $\mathbf{p}_i$ we can take any orthonormal triple of vectors. In particular, one of the bases can be taken as fixed $\forall t \in [t_1, t_2]$, for example, $\overset{\circ}{\mathbf{p}}_i$ can be chosen as coincident with $\overset{\circ}{\mathbf{p}}_i(t_1)$; and the second basis $\mathbf{p}_i$ can depend on time $t$. In this case $\overset{\circ}{\mathbf{p}}{}_i^{\bullet} \equiv 0$ $\forall t \in [t_1, t_2]$, and from (2.294) and (2.298) it follows that within the considered time interval:

$$\overset{\circ}{\Omega}_{U\alpha\beta} = 0, \quad \boldsymbol{\Omega}_U = \mathbf{0}, \tag{2.305}$$

$$\boldsymbol{\Omega}_V = \mathbf{W}, \quad \Omega_{V\alpha\beta} = \mathbf{p}_\alpha \cdot \mathbf{W} \cdot \mathbf{p}_\beta, \quad \alpha, \beta = 1, 2, 3. \tag{2.306}$$

These relationships take the place of formulae (2.302), (2.303) in this case.

If within the time interval $[t_1, t_2]$ only two of three eigenvalues are coincident, for example, $\lambda_\alpha = \lambda_\beta$, then their corresponding eigenvectors $\overset{\circ}{\mathbf{p}}_\alpha$ and $\overset{\circ}{\mathbf{p}}_\beta$ are not uniquely defined as well: only their orthogonality to the vector $\overset{\circ}{\mathbf{p}}_\gamma$, corresponding to the third eigenvalue $\lambda_\gamma$, is given. Then we can extend the definition of $\overset{\circ}{\mathbf{p}}_\alpha$ and $\overset{\circ}{\mathbf{p}}_\beta$ so that $\overset{\circ}{\mathbf{p}}{}_\alpha^{\bullet} \cdot \mathbf{p}_\beta = 0$ $\forall t \in [t_1, t_2]$. In this case it follows from (2.294) that the only component $\overset{\circ}{\Omega}_{U\alpha\beta}$ vanishes, but $\overset{\circ}{\Omega}_{U\alpha\gamma} \neq 0$ and $\overset{\circ}{\Omega}_{U\beta\gamma} \neq 0$.

It follows from (2.298) that the component $\Omega_{V\alpha\beta}$ is determined by the formula

$$\Omega_{V\alpha\beta} = \mathbf{p}_\alpha \cdot \mathbf{W} \cdot \mathbf{p}_\beta, \quad \overset{\circ}{\Omega}_{U\alpha\beta} = 0. \tag{2.307}$$

The remaining components $\overset{\circ}{\Omega}_{U\alpha\gamma}$, $\overset{\circ}{\Omega}_{U\beta\gamma}$ and $\Omega_{V\alpha\gamma}$, $\Omega_{V\beta\gamma}$ are determined by formulae (2.302) and (2.303).

If the situation with multiple roots appears only at some time $t$, then the values of $\overset{\circ}{\Omega}_{U\alpha\beta}(t)$ and $\Omega_{V\alpha\beta}(t)$ can be determined by passing to the limit. □

Substituting formulae (2.294) and (2.298) into Eq. (2.277), and taking expressions (2.269) and (2.271) for $\mathbf{O}_U$ and $\mathbf{O}_V$ into consideration, we obtain the representation of the spin $\boldsymbol{\Omega}$ in the basis $\mathbf{p}_\alpha$:

$$\boldsymbol{\Omega} = \mathbf{W} - \frac{1}{2}\sum_{\alpha,\beta=1}^{3} \frac{\lambda_\alpha^2 + \lambda_\beta^2}{\lambda_\alpha \lambda_\beta} \overset{\circ}{\Omega}_{U\beta\alpha} \mathbf{p}_\alpha \otimes \mathbf{p}_\beta - \sum_{\alpha,\beta=1}^{3} \overset{\circ}{\Omega}_{U\alpha\beta} (\mathbf{p}_i \otimes \bar{\mathbf{e}}^i) \cdot \overset{\circ}{\mathbf{p}}_\alpha \otimes \overset{\circ}{\mathbf{p}}_\beta \cdot (\bar{\mathbf{e}}^j \otimes \mathbf{p}_j). \tag{2.308}$$

Introducing the notation for direction cosines

$$\overset{\circ}{l}_{\alpha\beta} = \overset{\circ}{\mathbf{p}}_\alpha \cdot \bar{\mathbf{e}}_\beta, \quad l_{\alpha\beta} = \mathbf{p}_\alpha \cdot \bar{\mathbf{e}}_\beta, \tag{2.309}$$

substituting (2.302) into (2.308) and collecting like terms, we obtain the following expression of the spin $\boldsymbol{\Omega}$ in terms of $\mathbf{W}$ and $\mathbf{D}$ (i.e. in terms of $\mathbf{L}$):

$$\boldsymbol{\Omega} = \mathbf{W} + \widetilde{\boldsymbol{\Omega}}, \tag{2.310}$$

$$\widetilde{\boldsymbol{\Omega}} = \sum_{\gamma,\rho=1}^{3} \widetilde{\Omega}_{\gamma\rho} \mathbf{p}_\gamma \otimes \mathbf{p}_\rho, \quad \widetilde{\Omega}_{\gamma\rho} = \sum_{\alpha,\beta=1}^{3} \widetilde{\Omega}_{\alpha\beta},$$

$$\widetilde{\Omega}_{\gamma\rho} = \sum_{\alpha,\beta=1}^{3} \frac{1}{\lambda_\beta^2 - \lambda_\alpha^2} ((\lambda_\alpha^2 + \lambda_\beta^2)\delta_{\alpha\gamma}\delta_{\beta\rho} - 2\lambda_\alpha\lambda_\beta l_{\alpha\gamma} l_{\beta\rho}) D_{\alpha\beta}.$$

**Theorem 2.28.** *Rates of the deformation measures* $\dot{\mathbf{U}}$, $(\mathbf{U}^{-1})^\bullet$ *and* $\dot{\mathbf{V}}$, $(\mathbf{V}^{-1})^\bullet$ *are connected to the velocity gradient* $\mathbf{L}$ *by the formulae*

$$\begin{aligned}
\dot{\mathbf{U}} &= \frac{1}{2}(\mathbf{F}^{\mathrm{T}} \cdot (\mathbf{D} + \widetilde{\boldsymbol{\Omega}}) \cdot \mathbf{O} + \mathbf{O}^{\mathrm{T}} \cdot (\mathbf{D} + \widetilde{\boldsymbol{\Omega}}^{\mathrm{T}}) \cdot \mathbf{F}), \\
(\mathbf{U}^{-1})^\bullet &= -\frac{1}{2}(\mathbf{F}^{-1} \cdot (\mathbf{D} - \widetilde{\boldsymbol{\Omega}}) \cdot \mathbf{O} + \mathbf{O}^{\mathrm{T}} \cdot (\mathbf{D} - \widetilde{\boldsymbol{\Omega}}^{\mathrm{T}}) \cdot \mathbf{F}^{-1\mathrm{T}}), \\
\dot{\mathbf{V}} &= \frac{1}{2}((\mathbf{L} + \boldsymbol{\Omega}) \cdot \mathbf{V} + \mathbf{V} \cdot (\mathbf{L}^{\mathrm{T}} + \boldsymbol{\Omega}^{\mathrm{T}})), \\
(\mathbf{V}^{-1})^\bullet &= \frac{1}{2}((\boldsymbol{\Omega} - \mathbf{L}^{\mathrm{T}}) \cdot \mathbf{V}^{-1} + \mathbf{V}^{-1} \cdot (\boldsymbol{\Omega}^{\mathrm{T}} - \mathbf{L})).
\end{aligned} \tag{2.311}$$

▼ Let us express the tensors $\mathbf{V}$ and $\mathbf{U}$ from the polar decomposition (2.137):

$$\mathbf{U} = \mathbf{O}^{\mathrm{T}} \cdot \mathbf{F}, \quad \mathbf{V} = \mathbf{F} \cdot \mathbf{O}^{\mathrm{T}}. \tag{2.312}$$

Since $\mathbf{U}$ and $\mathbf{V}$ are symmetric tensors, these expressions can be rewritten in the symmetrized form

$$\mathbf{U} = \frac{1}{2}(\mathbf{F}^{\mathrm{T}} \cdot \mathbf{O} + \mathbf{O}^{\mathrm{T}} \cdot \mathbf{F}), \quad \mathbf{V} = \frac{1}{2}(\mathbf{F} \cdot \mathbf{O}^{\mathrm{T}} + \mathbf{O} \cdot \mathbf{F}^{\mathrm{T}}). \tag{2.313}$$

Let us differentiate these relationships:

$$\dot{\mathbf{U}} = \frac{1}{2}(\dot{\mathbf{F}}^{\mathrm{T}} \cdot \mathbf{O} + \mathbf{F}^{\mathrm{T}} \cdot \dot{\mathbf{O}} + \dot{\mathbf{O}}^{\mathrm{T}} \cdot \mathbf{F} + \mathbf{O}^{\mathrm{T}} \cdot \dot{\mathbf{F}}),$$
$$\dot{\mathbf{V}} = \frac{1}{2}(\dot{\mathbf{F}} \cdot \mathbf{O}^{\mathrm{T}} + \mathbf{F} \cdot \dot{\mathbf{O}}^{\mathrm{T}} + \dot{\mathbf{O}} \cdot \mathbf{F}^{\mathrm{T}} + \mathbf{O} \cdot \dot{\mathbf{F}}^{\mathrm{T}}). \quad (2.314)$$

On substituting formulae (2.279) and the expression (2.277) for the spin $\boldsymbol{\Omega}$ into (2.314), we obtain

$$\dot{\mathbf{U}} = \frac{1}{2}(\mathbf{F}^{\mathrm{T}} \cdot \mathbf{L}^{\mathrm{T}} \cdot \mathbf{O} + \mathbf{F}^{\mathrm{T}} \cdot \dot{\mathbf{O}} \cdot \mathbf{O}^{\mathrm{T}} \cdot \mathbf{O} + \mathbf{O}^{\mathrm{T}} \cdot \mathbf{O} \cdot \dot{\mathbf{O}}^{\mathrm{T}} \cdot \mathbf{F} + \mathbf{O}^{\mathrm{T}} \cdot \mathbf{L} \cdot \mathbf{F})$$
$$= \frac{1}{2}(\mathbf{F}^{\mathrm{T}} \cdot (\mathbf{L}^{\mathrm{T}} + \boldsymbol{\Omega}) \cdot \mathbf{O} + \mathbf{O}^{\mathrm{T}} \cdot (\mathbf{L} + \boldsymbol{\Omega}^{\mathrm{T}}) \cdot \mathbf{F}), \quad (2.315)$$
$$\dot{\mathbf{V}} = \frac{1}{2}(\mathbf{L} \cdot \mathbf{F} \cdot \mathbf{O}^{\mathrm{T}} + \mathbf{F} \cdot \mathbf{O}^{\mathrm{T}} \cdot \mathbf{O} \cdot \dot{\mathbf{O}}^{\mathrm{T}} + \dot{\mathbf{O}} \cdot \mathbf{O}^{\mathrm{T}} \cdot \mathbf{O} \cdot \mathbf{F}^{\mathrm{T}} + \mathbf{O} \cdot \mathbf{F}^{\mathrm{T}} \cdot \mathbf{L}^{\mathrm{T}})$$
$$= \frac{1}{2}((\mathbf{L} + \boldsymbol{\Omega}) \cdot \mathbf{V} + \mathbf{V} \cdot (\mathbf{L}^{\mathrm{T}} + \boldsymbol{\Omega}^{\mathrm{T}})).$$

Taking formulae (2.310) and (2.224) into consideration, we find that

$$\mathbf{L}^{\mathrm{T}} + \boldsymbol{\Omega} = \mathbf{D} + \mathbf{W}^{\mathrm{T}} + \mathbf{W} + \widetilde{\boldsymbol{\Omega}} = \mathbf{D} + \widetilde{\boldsymbol{\Omega}}. \quad (2.316)$$

Substituting (2.316) into (2.315), we really get the first and the third formulae of (2.311).

The remaining two formulae in (2.311) can be proved in a similar way. ▲

In continuum mechanics the *deformation tensors* **B** and **Y** are applied, which have no explicit expression; they are defined by their derivatives and initial values:

$$\dot{\mathbf{B}} = \frac{1}{2}(\dot{\mathbf{U}} \cdot \mathbf{U}^{-1} + \mathbf{U}^{-1} \cdot \dot{\mathbf{U}}), \quad \mathbf{B}(0) = \mathbf{0}, \quad (2.317)$$

$$\dot{\mathbf{Y}} = \frac{1}{2}(\dot{\mathbf{V}} \cdot \mathbf{V}^{-1} + \mathbf{V}^{-1} \cdot \dot{\mathbf{V}}), \quad \mathbf{Y}(0) = \mathbf{0}. \quad (2.318)$$

After substitution of the expressions (2.312), formula (2.317) takes the form

$$\dot{\mathbf{B}} = \frac{1}{2}((\dot{\mathbf{O}}^{\mathrm{T}} \cdot \mathbf{F} + \mathbf{O}^{\mathrm{T}} \cdot \dot{\mathbf{F}}) \cdot \mathbf{U}^{-1} + \mathbf{U}^{-1} \cdot (\dot{\mathbf{F}}^{\mathrm{T}} \cdot \mathbf{O} + \mathbf{F}^{\mathrm{T}} \cdot \dot{\mathbf{O}}))$$
$$= \frac{1}{2}(\mathbf{O}^{\mathrm{T}} \cdot (\boldsymbol{\Omega} + \mathbf{L}) \cdot \mathbf{F} \cdot \mathbf{U}^{-1} + \mathbf{U}^{-1} \cdot \mathbf{F}^{\mathrm{T}} \cdot (\mathbf{L}^{\mathrm{T}} + \boldsymbol{\Omega}^{\mathrm{T}}) \cdot \mathbf{O})$$
$$= \frac{1}{2}(\mathbf{O}^{\mathrm{T}} \cdot (\boldsymbol{\Omega} + \mathbf{L}) \cdot \mathbf{O} + \mathbf{O}^{\mathrm{T}} \cdot (\mathbf{L}^{\mathrm{T}} + \boldsymbol{\Omega}^{\mathrm{T}}) \cdot \mathbf{O}), \quad (2.319)$$

and formula (2.318) is rewritten as follows:

$$\dot{\mathbf{Y}} = \frac{1}{2}((\dot{\mathbf{F}}\cdot\mathbf{O}^{\mathrm{T}} + \mathbf{F}\cdot\dot{\mathbf{O}}^{\mathrm{T}})\cdot\mathbf{V}^{-1} + \mathbf{V}^{-1}\cdot(\dot{\mathbf{O}}\cdot\mathbf{F}^{\mathrm{T}} + \mathbf{O}\cdot\dot{\mathbf{F}}^{\mathrm{T}}))$$
$$= \frac{1}{2}(\mathbf{L}\cdot\mathbf{F}\cdot\mathbf{O}^{\mathrm{T}}\cdot\mathbf{V}^{-1} + \mathbf{F}\cdot\mathbf{O}^{\mathrm{T}}\cdot\boldsymbol{\Omega}^{\mathrm{T}}\cdot\mathbf{V}^{-1} + \mathbf{V}^{-1}\cdot\boldsymbol{\Omega}\cdot\mathbf{O}\cdot\mathbf{F}^{\mathrm{T}} + \mathbf{V}^{-1}\cdot\mathbf{O}\cdot\mathbf{F}^{\mathrm{T}}\cdot\mathbf{L}^{\mathrm{T}})$$
$$= \frac{1}{2}(\mathbf{L} + \mathbf{V}\cdot\boldsymbol{\Omega}\cdot\mathbf{V}^{-1} + \mathbf{V}^{-1}\cdot\boldsymbol{\Omega}\cdot\mathbf{V} + \mathbf{L}^{\mathrm{T}}). \qquad (2.320)$$

Finally, we get the following expressions for $\dot{\mathbf{B}}$ and $\dot{\mathbf{Y}}$:

$$\dot{\mathbf{B}} = \mathbf{O}^{\mathrm{T}}\cdot\mathbf{D}\cdot\mathbf{O}, \qquad (2.321)$$

$$\dot{\mathbf{Y}} = \mathbf{D} + \tfrac{1}{2}(\mathbf{V}\cdot\boldsymbol{\Omega}^{\mathrm{T}}\cdot\mathbf{V}^{-1} + \mathbf{V}^{-1}\cdot\boldsymbol{\Omega}\cdot\mathbf{V}). \qquad (2.322)$$

### *2.4.12 Trajectory of a Material Point, Streamline and Vortex Line*

Having fixed coordinates $X^i$ of a material point $\mathcal{M}$ in the motion law (2.3), we get the parametric equation of a certain curve, where time $t$ is a parameter:

$$x^i = x^i(X^k, t), \qquad 0 \leqslant t \leqslant t'. \qquad (2.323)$$

The origin of the curve at $t = 0$ is a point with Cartesian coordinates $\overset{\circ}{x}{}^i(X^k)$ of the material point $\mathcal{M}$ in $\overset{\circ}{\mathcal{K}}$, and the end of the curve at $t = t'$ is a point with Cartesian coordinates $x^i(X^k, t')$ of the point $\mathcal{M}$ in $\mathcal{K}(t')$ (Fig. 2.14). The curve (2.323) is called the *trajectory of the point* $\mathcal{M}$ in the Cartesian coordinate system $O\bar{\mathbf{e}}_i$.

In the spatial description, the trajectory (2.323) at fixed $X^k$ is a solution of the kinematic equation (2.190):

$$dx^i/dt = \bar{v}^i(x^j, t), \qquad 0 < t \leqslant t' \qquad (2.324)$$

with the initial condition

$$t = 0: \quad x^i = \overset{\circ}{x}{}^i.$$

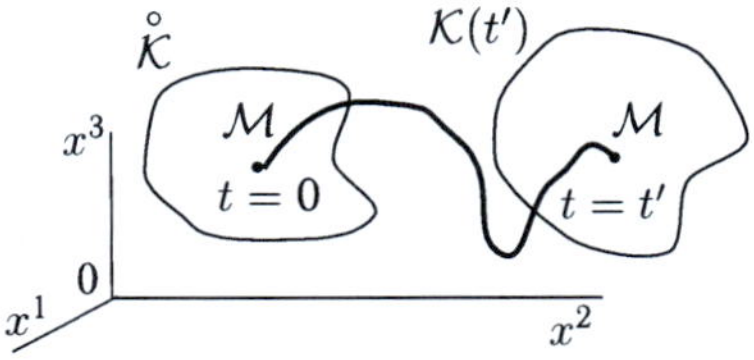


**Fig. 2.14** Trajectory of material point $\mathcal{M}$

Here $\bar{v}^i(x^j,t)$ are the velocity components with respect to the Cartesian basis $\bar{\mathbf{e}}_i$, which are assumed to be known.

Let a field of velocities $\mathbf{v}(x^j,t) = \bar{v}^i\bar{\mathbf{e}}_i$ be given. Fix a time $t$, and take a point $\mathcal{M}_1$ with Eulerian coordinates $x_1^i$ and Lagrangian coordinates $X_1^k$. Then a *streamline* passing through the point $\mathcal{M}_1$ is the curve

$$x^i = x^i(X^k,\tau), \quad \tau_1 \leqslant \tau \leqslant \tau_2, \tag{2.325}$$

which at its every point $x^i$ has a tangent being parallel to the velocity $\mathbf{v}(x^i,t)$ at the considered point and at the considered time. The equation of the streamline has the form

$$\begin{aligned} dx^i/d\tau = \bar{v}^i(x^j,t), \qquad \tau_1 < \tau \leqslant \tau_2, \\ \tau = \tau_1: \qquad x^i = x_1^i. \end{aligned} \tag{2.326}$$

Thus, the trajectory of a material point and the streamline are described, in general, by different equations, and so they are not coincident.

However, if the motion of a continuum is *steady-state* within time interval $t_1 \leqslant t \leqslant t'$, then in Eulerian description all the partial derivatives of all values, describing the motion, with respect to time vanish, in particular $\partial\mathbf{v}(x^i,t)/\partial t = 0$. So the trajectory equations (2.324) and the streamline equation (2.326) become coincident within the interval $t_1 \leqslant t \leqslant t'$, if they have at least one common point $\mathcal{M}_1$:

$$\begin{aligned} dx^i/d\tau = \bar{v}^i(x^j), \qquad t_1 = \tau_1 < \tau \leqslant \tau_2 = t', \\ \tau = \tau_1: \qquad x^i = x_1^i = x^i(X_1^k, t_1). \end{aligned} \tag{2.327}$$

In other words, in the steady-state motion a material point $\mathcal{M}$ moves along a streamline: at time $t = t_1$ its coordinates are the same as coordinates of point $\mathcal{M}_1$ at parameter value $\tau = \tau_1$, and at time $t = t_2$ they are the same as coordinates of point $\mathcal{M}_2$ at parameter value $\tau = \tau_2$ (Fig. 2.15).

Multiplying the Eq. (2.326) by the basis vectors $\bar{\mathbf{e}}_i$, we can rewrite the streamline equation in the vector form

$$\begin{aligned} d\mathbf{x} = \mathbf{v}(\mathbf{x},t)d\tau, \quad \tau_1 < \tau \leqslant \tau_2, \\ \mathbf{x} = \mathbf{x}_1, \quad \tau = \tau_1. \end{aligned} \tag{2.328}$$

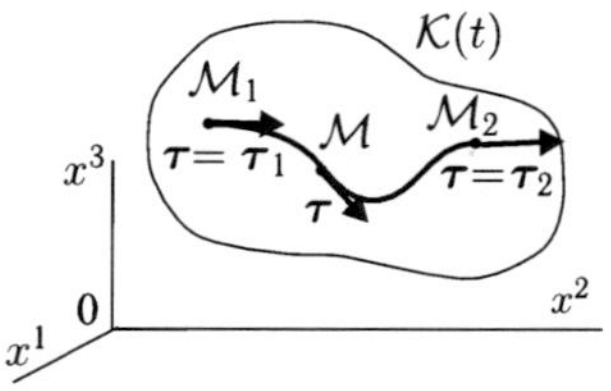


**Fig. 2.15** The streamline

Let us define now a *vortex line* passing through a point $\mathcal{M}_1$: this is a curve, which at its every point $x^i$ has a tangent being parallel to the vorticity vector $\boldsymbol{\omega}(x^j, t)$ at the considered point and at the fixed time $t$. The vortex line is described by the equation

$$\begin{aligned} d\mathbf{x} &= \boldsymbol{\omega}(\mathbf{x}, t)d\tau, \qquad \tau_1 < \tau \leqslant \tau_2, \\ \mathbf{x} &= \mathbf{x}_1, \qquad \tau = \tau_1 \end{aligned} \tag{2.329}$$

## 2.4.13 *Stream Tubes and Vortex Tubes*

Consider a curve $L$ in coordinates $x^i$ and draw a streamline through each point of the curve $L$. If $L$ is not a streamline itself, then we get a surface $\Sigma_v$, at each point of which the velocity $\mathbf{v}$ lies on the tangent plane to the surface. This surface is called the *stream surface*.

Let

$$f_v(x^i) = 0 \tag{2.330}$$

be the equation of the stream surface. Since the vector $\nabla f$ is normal to the surface $\Sigma_v$ [12], so it is orthogonal to the velocity $\mathbf{v}$, i.e. we have the relation

$$\mathbf{v} \cdot \nabla f_v = 0, \tag{2.331}$$

which is a partial differential equation for determination of the function $f(x^i)$ by the known velocity field $\mathbf{v}(x^i, t)$ at fixed $t$.

If a curve $L$ is closed, then the set of streamlines drawn through its points is called the *stream tube*.

Let a curve $L$ be not a vortex line. Drawing a vortex line through each point of the curve $L$, we obtain the *vortex surface* $\Sigma_\omega$, which is described by the equation $f_\omega(x^i) = 0$. This relation is a solution of the differential equation

$$\boldsymbol{\omega} \cdot \nabla f_\omega = 0. \tag{2.332}$$

If $L$ is a closed curve, then the surface $\Sigma_\omega$ is called the *vortex tube*.

## Exercises for 2.4

**2.4.1.** Show that the tensors $\mathbf{Q}_U$ and $\mathbf{Q}_V$ are orthogonal.

**2.4.2.** Using formulae (2.284) and (2.124), show that for the coefficient $\overset{\circ}{k}$ determined by formula (2.121) we have the following relationship:

$$\overset{\circ}{k}{}^{\bullet} = \overset{\circ}{k}(\mathbf{n} \cdot \mathbf{D} \cdot \mathbf{n}).$$

**2.4.3.** Using formulae (2.124), (2.279) and the result of Exercise 2.4.2, show that a rate of changing the normal **n** is determined as follows:

$$\dot{\mathbf{n}} = \gamma\mathbf{n} - \mathbf{n}\cdot\mathbf{L}, \qquad \gamma = \mathbf{n}\cdot\mathbf{D}\cdot\mathbf{n}.$$

**2.4.4.** Using the results of Exercise 2.4.2, show that for the coefficient $k$ determined by formula (2.121) we have the following equation:

$$\dot{k} = -k(\mathbf{n}\cdot\mathbf{D}\cdot\mathbf{n}).$$

**2.4.5.** Show that the transformations (2.245) of infinitesimal tension–compression (2.245) and infinitesimal rotation (2.246) are commutative up to terms of order $(dt)^2$:

$$\mathbf{Q}_\omega\cdot\mathbf{A}_D\cdot d\mathbf{x} = \mathbf{A}_\omega\cdot\mathbf{Q}_D\cdot d\mathbf{x},$$

while transformations of a small neighborhood determined by the tensors **O** and **U** or **O** and **V** are not commutative in general.

**2.4.6.** Prove Theorem 2.14.

**2.4.7.** Using the representation (2.142) for tensors **U** and **V** and formulae (2.270) and (2.272), show that rates of stretch tensors are expressed as follows:

$$\dot{\mathbf{U}} = \sum_{\alpha=1}^{3}\dot{\lambda}_\alpha\overset{\circ}{\mathbf{p}}_\alpha\otimes\overset{\circ}{\mathbf{p}}_\alpha + \boldsymbol{\Omega}_U\cdot\mathbf{U} - \mathbf{U}\cdot\boldsymbol{\Omega}_U, \quad \dot{\mathbf{V}} = \sum_{\alpha=1}^{3}\dot{\lambda}_\alpha\mathbf{p}_\alpha\otimes\mathbf{p}_\alpha + \boldsymbol{\Omega}_V\cdot\mathbf{V} - \mathbf{V}\cdot\boldsymbol{\Omega}_V.$$

**2.4.8.** Show that expressions for rates of the Hencky tensors (2.167) have the form

$$\overset{\circ}{\mathbf{H}}{}^{\bullet} = \sum_{\alpha=1}^{3}\frac{\dot{\lambda}_\alpha}{\lambda_\alpha}\overset{\circ}{\mathbf{p}}_\alpha\otimes\overset{\circ}{\mathbf{p}}_\alpha + \boldsymbol{\Omega}_U\cdot\overset{\circ}{\mathbf{H}} - \overset{\circ}{\mathbf{H}}\cdot\boldsymbol{\Omega}_U, \quad \mathbf{H}^{\bullet} = \sum_{\alpha=1}^{3}\frac{\dot{\lambda}_\alpha}{\lambda_\alpha}\mathbf{p}_\alpha\otimes\mathbf{p}_\alpha + \boldsymbol{\Omega}_V\cdot\mathbf{H} - \mathbf{H}\cdot\boldsymbol{\Omega}_V.$$

**2.4.9.** Using representations (2.294) and (2.304) for tensors $\boldsymbol{\Omega}_U$ and $\boldsymbol{\Omega}_V$, and also Eq. (2.300) and the result of Exercise 2.4.8, show that rates of the Hencky tensors can be expressed in the form

$$\overset{\circ}{\mathbf{H}}{}^{\bullet} = \sum_{\alpha,\beta=1}^{3} D_{\alpha\beta}\overset{\circ}{\mathbf{p}}_\alpha\otimes\overset{\circ}{\mathbf{p}}_\beta + \sum_{\substack{\alpha,\beta=1\\ \alpha\neq\beta}}^{3}\left(\frac{2\lambda_\alpha\lambda_\beta}{\lambda_\beta^2-\lambda_\alpha^2}\lg\frac{\lambda_\beta}{\lambda_\alpha} - 1\right)D_{\alpha\beta}\overset{\circ}{\mathbf{p}}_\alpha\otimes\overset{\circ}{\mathbf{p}}_\beta,$$

$$\dot{\mathbf{H}} = \mathbf{D} + \sum_{\substack{\alpha,\beta=1\\ \alpha\neq\beta}}^{3}\left(\frac{2\lambda_\alpha\lambda_\beta}{\lambda_\beta^2-\lambda_\alpha^2}\lg\frac{\lambda_\beta}{\lambda_\alpha} - 1\right)(\mathbf{p}_\alpha\cdot\mathbf{D}\cdot\mathbf{p}_\beta)\mathbf{p}_\alpha\otimes\mathbf{p}_\beta.$$

**2.4.10.** Show that expressions for $\overset{\circ}{\mathbf{H}}{}^{\bullet}$ and $\dot{\mathbf{H}}$ derived in Exercise 1.4.9 can be rewritten as follows:

$$\overset{\circ}{\mathbf{H}}{}^{\bullet} = {}^4\overset{\circ}{\mathbf{X}}_H \cdot\cdot\, \mathbf{D}, \quad \dot{\mathbf{H}} = {}^4\mathbf{X}_H \cdot\cdot \mathbf{D},$$

where the following fourth-order tensors are denoted:

$${}^4\overset{\circ}{\mathbf{X}}_H = X_{H\ ijkl}\overset{\circ}{\mathbf{p}}{}^i \otimes \overset{\circ}{\mathbf{p}}{}^j \otimes \mathbf{p}^k \otimes \mathbf{p}^l, \quad {}^4\mathbf{X}_H = X_{H\ ijkl}\mathbf{p}^i \otimes \mathbf{p}^j \otimes \mathbf{p}^k \otimes \mathbf{p}^l,$$

$$X_{H\ ijkl} = \begin{cases} \left(\dfrac{2\lambda_\alpha\lambda_\beta}{\lambda_\beta^2 - \lambda_\alpha^2}\lg\dfrac{\lambda_\beta}{\lambda_\alpha}\right)\Delta_{\alpha\beta kl}, & \alpha \neq \beta, \\ \Delta_{\alpha\beta kl}, & \alpha = \beta, \end{cases} \qquad \Delta_{\alpha\beta kl}=(1/2)(\delta_{\alpha k}\delta_{\beta l} + \delta_{\alpha l}\delta_{\beta k}).$$

**2.4.11.** Show that relations (2.294) and (2.302) for $\boldsymbol{\Omega}_U$, Eq. (2.303) and (2.304) for $\boldsymbol{\Omega}_V$ and Eq. (2.310) for $\boldsymbol{\Omega}$ can be rewritten as follows:

$$\boldsymbol{\Omega}_U = {}^4\boldsymbol{\Omega}_U \cdot\cdot\, \mathbf{D}, \quad \boldsymbol{\Omega}_V = {}^4\boldsymbol{\Omega}_V \cdot\cdot\, \mathbf{D} + \mathbf{W}, \quad \boldsymbol{\Omega} = {}^4\widetilde{\boldsymbol{\Omega}} \cdot\cdot\, \mathbf{D} + \mathbf{W},$$

where

$${}^4\boldsymbol{\Omega}_U = \Omega_{Uijkl}\overset{\circ}{\mathbf{p}}{}^i \otimes \overset{\circ}{\mathbf{p}}{}^j \otimes \mathbf{p}^k \otimes \mathbf{p}^l, \quad {}^4\boldsymbol{\Omega}_V = \Omega_{Vijkl}\mathbf{p}^i \otimes \mathbf{p}^j \otimes \mathbf{p}^k \otimes \mathbf{p}^l,$$

$${}^4\widetilde{\boldsymbol{\Omega}} = \widetilde{\Omega}_{ijkl}\mathbf{p}^i \otimes \mathbf{p}^j \otimes \mathbf{p}^k \otimes \mathbf{p}^l,$$

$$\Omega_{U\alpha\beta kl} = \begin{cases} \dfrac{2\lambda_\alpha\lambda_\beta}{\lambda_\beta^2 - \lambda_\alpha^2}\Delta_{\alpha\beta kl}, & \alpha \neq \beta, \\ 0, & \alpha = \beta, \end{cases} \qquad \Omega_{V\alpha\beta kl} = \begin{cases} \dfrac{\lambda_\alpha^2 + \lambda_\beta^2}{\lambda_\beta^2 - \lambda_\alpha^2}\Delta_{\alpha\beta kl}, & \alpha \neq \beta, \\ 0, & \alpha = \beta, \end{cases}$$

$$\widetilde{\Omega}_{\alpha\beta\gamma\rho} = \Omega_{V\alpha\beta\gamma\rho} - \frac{\lambda_\gamma\lambda_\rho}{\lambda_\rho^2 - \lambda_\gamma^2}(l_{\gamma\alpha}l_{\rho\beta} - l_{\rho\alpha}l_{\gamma\beta}), \quad \alpha \neq \beta, \ \gamma \neq \rho,$$

if $\alpha = \beta$ (or $\gamma = \rho$), then the first (or the second) summand vanishes.

**2.4.12.** Using the definitions (2.225) and (2.226) of the tensors **D** and **W**, and also the properties of unit tensors, show that **D**, **W** and **L** are connected by the formulae

$$\mathbf{D} = \boldsymbol{\Delta} \cdot\cdot\, \mathbf{L}, \quad \mathbf{W} = \widetilde{\boldsymbol{\Delta}} \cdot\cdot\, \mathbf{L}, \quad \widetilde{\boldsymbol{\Delta}} = (1/2)(\Delta_{\mathrm{III}} - \Delta_{\mathrm{II}}).$$

## 2.5 Co-rotational Derivatives

### *2.5.1 Definition of Co-rotational Derivatives*

Besides the total derivative $d\mathbf{a}/dt$ introduced in Sect. 2.4.1 and partial derivative of vectors and tensors with respect to time $\partial\mathbf{a}/\partial t$, so-called *co-rotational derivatives* are of great importance in continuum mechanics. They determine rates of changing tensors relative to some *moving basis* $\mathbf{h}_i$, i.e. the relative rates.

Let in a actual configuration $\mathcal{K}(t)$ there be some moving bases $\mathbf{h}_i$ or $\mathbf{h}^i$ and arbitrary varying scalar $\psi(X^i, t)$, vector $\mathbf{a}(X^i, t)$ and second-order tensor $\mathbf{T}(X^i, t)$ fields with the following components with respect to the bases:

$$\mathbf{a} = a^i \mathbf{h}_i = a_i \mathbf{h}^i, \tag{2.333}$$

$$\mathbf{T} = T^{ij} \mathbf{h}_i \otimes \mathbf{h}_j = T_{ij} \mathbf{h}^i \otimes \mathbf{h}^j = T^i{}_j \mathbf{h}_i \otimes \mathbf{h}^j = T_i{}^j \mathbf{h}^i \otimes \mathbf{h}_j. \tag{2.334}$$

Since any scalar function $\psi(X^i, t)$ is not connected to any basis (moving or fixed), it is evident that the co-rotational derivative of the function must be coincident with the total derivative with respect to time:

$$\psi^h = \dot{\psi}. \tag{2.335}$$

For a vector $\mathbf{a}$ and a tensor $\mathbf{T}$ we can introduce *co-rotational derivatives* $\mathbf{a}^h$ and $\mathbf{T}^h$ as vectors or tensors, components of which with respect to the same basis $\mathbf{h}_i$ coincide with rates of changing vector $\mathbf{a}$ and tensor $\mathbf{T}$ components, respectively:

$$\mathbf{a}^h = \frac{da^i}{dt} \mathbf{h}_i, \quad \mathbf{T}^h = \frac{d}{dt} T^{ij} \mathbf{h}_i \otimes \mathbf{h}_j. \tag{2.336}$$

If we consider the basis $\mathbf{h}^i$, then for the basis we can determine other co-rotational derivatives:

$$\mathbf{a}^H = \frac{da_i}{dt} \mathbf{h}^i, \tag{2.337}$$

$$\mathbf{T}^H = \frac{d}{dt} T_{ij} \mathbf{h}^i \otimes \mathbf{h}^j. \tag{2.338}$$

Thus, the co-rotational derivative $\mathbf{a}^h$ (or $\mathbf{T}^h$) determines the rate of varying a vector $\mathbf{a}$ (or a tensor $\mathbf{T}$) for an observer moving together with the basis $\mathbf{h}_i$. For the observer, the basis $\mathbf{h}_i$ is fixed, and hence in (2.336) the basis is not differentiated with respect to time. In a similar way, the derivatives $\mathbf{a}^H$ and $\mathbf{T}^H$ determine the rates of changing $\mathbf{a}$ and $\mathbf{T}$ for an observer moving together with the basis $\mathbf{h}^i$.

We can determine the co-rotational derivatives of a second-order tensor $\mathbf{T}$ in mixed moving dyadic bases $\mathbf{h}_i \otimes \mathbf{h}^j$ and $\mathbf{h}^i \otimes \mathbf{h}_j$, respectively:

$$\mathbf{T}^d = \frac{d}{dt} T^i{}_j \mathbf{h}_i \otimes \mathbf{h}^j, \quad \mathbf{T}^D = \frac{d}{dt} T_i{}^j \mathbf{h}^i \otimes \mathbf{h}_j. \tag{2.339}$$

Since vector components $a^i$ and $a_i$ can always be expressed in the form

$$a^i = \mathbf{a} \cdot \mathbf{h}^i, \quad a_i = \mathbf{a} \cdot \mathbf{h}_i, \tag{2.340}$$

and tensor components $T_{ij}$, $T^{ij}$, $T^{i}_{\ j}$ and $T_{i}^{\ j}$, with the help of the scalar product of (2.334) by $\mathbf{h}^i$ or $\mathbf{h}_j$, can be written as follows:

$$T^{ij} = \mathbf{h}^i\cdot\mathbf{T}\cdot\mathbf{h}^j, \quad T_{ij} = \mathbf{h}_i\cdot\mathbf{T}\cdot\mathbf{h}_j, \quad T^{i}_{\ j} = \mathbf{h}^i\cdot\mathbf{T}\cdot\mathbf{h}_j, \quad T_{i}^{\ j} = \mathbf{h}_i\cdot\mathbf{T}\cdot\mathbf{h}^j, \quad (2.341)$$

so rates of changing vector and tensor components in (2.336), (2.337), and (2.339) can be represented in the explicit form:

$$\frac{da^i}{dt} = \frac{d\mathbf{a}}{dt}\cdot\mathbf{h}^i + \mathbf{a}\cdot\frac{d\mathbf{h}^i}{dt}, \quad \frac{da_i}{dt} = \frac{d\mathbf{a}}{dt}\cdot\mathbf{h}_i + \mathbf{a}\cdot\frac{d\mathbf{h}_i}{dt}, \quad (2.342)$$

and also

$$\begin{aligned}
\frac{dT^{ij}}{dt} &= \mathbf{h}^i\cdot\frac{d\mathbf{T}}{dt}\cdot\mathbf{h}^j + \frac{d\mathbf{h}^i}{dt}\cdot\mathbf{T}\cdot\mathbf{h}^j + \mathbf{h}^i\cdot\mathbf{T}\cdot\frac{d\mathbf{h}^j}{dt},\\
\frac{dT_{ij}}{dt} &= \mathbf{h}_i\cdot\frac{d\mathbf{T}}{dt}\cdot\mathbf{h}_j + \frac{d\mathbf{h}_i}{dt}\cdot\mathbf{T}\cdot\mathbf{h}_j + \mathbf{h}_i\cdot\mathbf{T}\cdot\frac{d\mathbf{h}_j}{dt},\\
\frac{d}{dt}T^{i}_{\ j} &= \mathbf{h}^i\cdot\frac{d}{dt}\mathbf{T}\cdot\mathbf{h}_j + \frac{d\mathbf{h}^i}{dt}\cdot\mathbf{T}\cdot\mathbf{h}_j + \mathbf{h}^i\cdot\mathbf{T}\cdot\frac{d\mathbf{h}_j}{dt},\\
\frac{d}{dt}T_{i}^{\ j} &= \mathbf{h}_i\cdot\frac{d}{dt}\mathbf{T}\cdot\mathbf{h}^j + \frac{d\mathbf{h}_i}{dt}\cdot\mathbf{T}\cdot\mathbf{h}^j + \mathbf{h}_i\cdot\mathbf{T}\cdot\frac{d}{dt}\mathbf{h}^j. \quad (2.343)
\end{aligned}$$

Here total derivatives $d\mathbf{a}/dt$ and $d\mathbf{T}/dt$ are determined by the rules (2.187) and (2.192), respectively. Rates of changing basis vectors $d\mathbf{h}^i/dt$ and $d\mathbf{h}_i/dt$ are defined by the choice of basis $\mathbf{h}_i$ or $\mathbf{h}^j$.

Taking different bases as $\mathbf{h}_i$ and $\mathbf{h}^j$, we get different co-rotational derivatives. Let us consider the most widely used bases.

## 2.5.2 *The Oldroyd Derivative* ($\mathbf{h}_i = \mathbf{r}_i$)

If we choose the general local vector basis $\mathbf{r}_i$ as $\mathbf{h}_i$, then the derivative $\mathbf{a}^h = \mathbf{a}^{\mathrm{Ol}}$ (or $\mathbf{T}^h = \mathbf{T}^{\mathrm{Ol}}$) determines the rate of changing $\mathbf{a}$ (or $\mathbf{T}$) relative to the Lagrangian coordinate system $X^i$ moving together with the continuum. This derivative is called the *Oldroyd derivative* [44].

The derivative $d\mathbf{r}_i/dt$ is determined as follows:

$$\frac{d\mathbf{h}_i}{dt} = \frac{d\mathbf{r}_i}{dt} = \frac{\partial^2\mathbf{x}}{\partial t\partial X^i} = \frac{\partial\mathbf{v}}{\partial X^i} = \mathbf{r}_i\cdot\mathbf{r}^j\otimes\frac{\partial\mathbf{v}}{\partial X^j} = \mathbf{r}_i\cdot\nabla\otimes\mathbf{v} = (\nabla\otimes\mathbf{v})^{\mathrm{T}}\cdot\mathbf{r}_i. \quad (2.344)$$

In this case, as a basis $\mathbf{h}^i$ we consider the reciprocal local basis $\mathbf{r}^i$, the derivative of which $d\mathbf{r}^i/dt$ with respect to time has the form

$$\frac{d}{dt}(\mathbf{r}^i\otimes\mathbf{r}_i) = \frac{d}{dt}\mathbf{E} = 0, \quad (2.345)$$

or

$$\frac{d\mathbf{r}^i}{dt} \otimes \mathbf{r}_i = -\mathbf{r}^i \otimes \frac{d\mathbf{r}_i}{dt} = -\mathbf{r}^i \otimes (\nabla \otimes \mathbf{v})^{\mathrm{T}} \cdot \mathbf{r}_i. \tag{2.346}$$

Multiplying the equation by $\mathbf{r}^j$ from the right, we get

$$\frac{d\mathbf{h}^j}{dt} = \frac{d\mathbf{r}^j}{dt} = -\mathbf{r}^j \cdot (\nabla \otimes \mathbf{v})^{\mathrm{T}} = -(\nabla \otimes \mathbf{v}) \cdot \mathbf{r}^j. \tag{2.347}$$

On substituting the expressions (2.347) for the derivatives $d\mathbf{h}^i/dt$ and $d\mathbf{h}_j/dt$ into (2.342), we get the formula for the Oldroyd derivative in the basis $\mathbf{r}_i$:

$$\mathbf{a}^{\mathrm{Ol}} = \frac{da^i}{dt}\mathbf{r}_i = \frac{d\mathbf{a}}{dt} \cdot \mathbf{r}^i \otimes \mathbf{r}_i - \mathbf{a} \cdot (\nabla \otimes \mathbf{v}) \cdot \mathbf{r}^i \otimes \mathbf{r}_i = \frac{d}{dt}\mathbf{a} - \mathbf{a} \cdot \nabla \otimes \mathbf{v}, \tag{2.348}$$

$$\begin{aligned}\mathbf{T}^{\mathrm{Ol}} &= \frac{d\mathbf{T}^{ij}}{dt}\mathbf{r}_i \otimes \mathbf{r}_j = \mathbf{r}_i \otimes \frac{dT^{ij}}{dt}\mathbf{r}_j \\ &= \mathbf{r}_i \otimes \mathbf{r}^i \cdot \frac{d}{dt}\mathbf{T} \cdot \mathbf{r}^j \otimes \mathbf{r}_j - \mathbf{r}_i \otimes \mathbf{r}^i \cdot (\nabla \otimes \mathbf{v})^{\mathrm{T}} \cdot \mathbf{T} \cdot \mathbf{r}^j \otimes \mathbf{r}_j \\ &\quad -\mathbf{r}_i \otimes \mathbf{r}^i \cdot \mathbf{T} \cdot (\nabla \otimes \mathbf{v}) \cdot \mathbf{r}^j \otimes \mathbf{r}_j = \frac{d}{dt}\mathbf{T} - \mathbf{T} \cdot \nabla \otimes \mathbf{v} - (\nabla \otimes \mathbf{v})^{\mathrm{T}} \cdot \mathbf{T}. \end{aligned} \tag{2.349}$$

Here we have taken into account that $\mathbf{r}^i \otimes \mathbf{r}_i = \mathbf{E}$. Thus, we have proved the following theorem.

**Theorem 2.29.** *The Oldroyd derivative is related to the total derivative with respect to time as follows (for a vector* **a** *and for a tensor* **T**, *respectively):*

$$\mathbf{a}^{\mathrm{Ol}} = \dot{\mathbf{a}} - \mathbf{a} \cdot \nabla \otimes \mathbf{v}, \quad \mathbf{T}^{\mathrm{Ol}} = \dot{\mathbf{T}} - \mathbf{T} \cdot \nabla \otimes \mathbf{v} - (\nabla \otimes \mathbf{v})^{\mathrm{T}} \cdot \mathbf{T}. \tag{2.350}$$

### 2.5.3 The Cotter–Rivlin Derivative ($\mathbf{h}^i = \mathbf{r}^i$)

If we choose the reciprocal local basis $\mathbf{r}^i$ as a moving basis $\mathbf{h}^i$, then the derivative $\mathbf{a}^H$ (or $\mathbf{T}^H$) characterizes the rate of changing **a** (or **T**) relative to the basis $\mathbf{r}^i$ moving together with the Lagrangian coordinate system $X^i$. This derivative is called the *Cotter–Rivlin derivative* [10].

Because of formulae (2.342) and (2.347), we get the following theorem.

**Theorem 2.30.** *The Cotter–Rivlin derivative is related to the total derivative as follows (for a vector* **a** *and for a tensor* **T**, *respectively):*

$$\mathbf{a}^H \equiv \mathbf{a}^{\mathrm{CR}} = \frac{da_i}{dt}\mathbf{r}^i = \dot{\mathbf{a}} + (\nabla \otimes \mathbf{v}) \cdot \mathbf{a}, \tag{2.351}$$

$$\mathbf{T}^H \equiv \mathbf{T}^{\mathrm{CR}} = \frac{dT_{ij}}{dt}\mathbf{r}^i \otimes \mathbf{r}^j = \dot{\mathbf{T}} + \nabla \otimes \mathbf{v} \cdot \mathbf{T} + \mathbf{T} \cdot (\nabla \otimes \mathbf{v})^{\mathrm{T}}. \tag{2.352}$$

### 2.5.4 Mixed Co-rotational Derivatives

Since any vector **a** is defined by its components with respect to a vector basis, for example, in a moving basis $\mathbf{h}_i$ or $\mathbf{h}^j$, so for the vector in the moving bases we can determine only two co-rotational derivatives: by Oldroyd and by Cotter–Rivlin.

Any second-order tensor **T** is defined by its components with respect to a dyadic basis. Therefore, besides the Oldroyd and Cotter–Rivlin derivatives, which specify the rates of changes of a tensor **T** in moving dyadic bases $\mathbf{r}_i \otimes \mathbf{r}_j$ and $\mathbf{r}^i \otimes \mathbf{r}^j$, by formulae (2.339) we can determine two more derivatives in moving mixed dyadic bases:

$$\mathbf{T}^d = \frac{dT^i{}_j}{dt}\mathbf{r}_i \otimes \mathbf{r}^j, \quad \mathbf{T}^D = \frac{d}{dt}T_i{}^j\mathbf{r}^i \otimes \mathbf{r}_j. \tag{2.353}$$

On substituting the expressions (2.344) and (2.347) into (2.343), we get the following formulae for the rates of changing mixed components of the tensor **T**:

$$\frac{d}{dt}T^i{}_j = \mathbf{r}^i \cdot \dot{\mathbf{T}} \cdot \mathbf{r}_j - \mathbf{r}^i \cdot (\nabla \otimes \mathbf{v})^{\mathrm{T}} \cdot \mathbf{T} \cdot \mathbf{r}_j + \mathbf{r}^i \cdot \mathbf{T} \cdot (\nabla \otimes \mathbf{v})^{\mathrm{T}} \cdot \mathbf{r}_j,$$

$$\frac{d}{dt}T_i{}^j = \mathbf{r}_i \cdot \dot{\mathbf{T}} \cdot \mathbf{r}^j + \mathbf{r}_i \cdot (\nabla \otimes \mathbf{v}) \cdot \mathbf{T} \cdot \mathbf{r}^j - \mathbf{r}_i \cdot \mathbf{T} \cdot (\nabla \otimes \mathbf{v}) \cdot \mathbf{r}^j. \tag{2.354}$$

Having substituted (2.354) into (2.353), we get the following theorem.

**Theorem 2.31.** *The mixed derivatives* (2.353) *are connected to the total derivative by the relations*

$$\mathbf{T}^d = \dot{\mathbf{T}} - \mathbf{L} \cdot \mathbf{T} + \mathbf{T} \cdot \mathbf{L}, \quad \mathbf{T}^D = \dot{\mathbf{T}} + \mathbf{L}^{\mathrm{T}} \cdot \mathbf{T} - \mathbf{T} \cdot \mathbf{L}^{\mathrm{T}}. \tag{2.355}$$

The derivatives (2.353) are called the *left and right mixed co-rotational derivatives*, where $\mathbf{L} = (\nabla \otimes \mathbf{v})^{\mathrm{T}}$ is the velocity gradient (see (2.223)).

It should be noticed that, unlike other co-rotational derivatives considered in this paragraph, the mixed derivatives $\mathbf{T}^d$ and $\mathbf{T}^D$ do not form a symmetric tensor when they are applied to a symmetric tensor **T**. This fact explains a scarcer application of mixed derivatives in continuum mechanics.

### 2.5.5 The Derivative Relative to the Eigenbasis $\overset{\circ}{\mathbf{p}}_i$ of the Right Stretch Tensor

If we choose the eigenbasis $\overset{\circ}{\mathbf{p}}_i$ of the right stretch tensor **U** as a moving basis $\mathbf{h}_i$, then, since $\overset{\circ}{\mathbf{p}}_i$ are orthonormal, we get that $\mathbf{h}^i$ and $\mathbf{h}_j$ are coincident: $\mathbf{h}^\alpha = \mathbf{h}_\alpha$, $\alpha = 1, 2, 3$, and $|\mathbf{h}^i| = 1$. At every time, the moving coordinate system defined by the trihedron $\overset{\circ}{\mathbf{p}}_i$ executes an instantaneous rotation, which is characterized by the spin $\boldsymbol{\Omega}_U$ (2.269), and due to (2.270) we have

$$\frac{d\mathbf{h}_i}{dt} = \dot{\overset{\circ}{\mathbf{p}}}_i = \boldsymbol{\Omega}_U \cdot \overset{\circ}{\mathbf{p}}_i = \overset{\circ}{\mathbf{p}}_i \cdot \boldsymbol{\Omega}_U^{\mathrm{T}} = -\overset{\circ}{\mathbf{p}}_i \cdot \boldsymbol{\Omega}_U. \tag{2.356}$$

On substituting (2.356) into (2.343), we get

$$\mathbf{a}^h \equiv \mathbf{a}^U = \frac{da^i}{dt}\overset{\circ}{\mathbf{p}}_i = \dot{\mathbf{a}} \cdot \overset{\circ}{\mathbf{p}}{}^i \otimes \overset{\circ}{\mathbf{p}}_i + \mathbf{a} \cdot \boldsymbol{\Omega}_U \cdot \overset{\circ}{\mathbf{p}}{}^i \otimes \overset{\circ}{\mathbf{p}}_i = \dot{\mathbf{a}} + \mathbf{a} \cdot \boldsymbol{\Omega}_U, \tag{2.357}$$

$$\begin{aligned}\mathbf{T}^h \equiv \mathbf{T}^U &= \frac{T^{ij}}{dt}\overset{\circ}{\mathbf{p}}_i \otimes \overset{\circ}{\mathbf{p}}_j = \overset{\circ}{\mathbf{p}}_i \otimes \frac{dT^{ij}}{dt}\overset{\circ}{\mathbf{p}}_j \\ &= \overset{\circ}{\mathbf{p}}{}^i \otimes \overset{\circ}{\mathbf{p}}_i \cdot \dot{\mathbf{T}} \cdot \overset{\circ}{\mathbf{p}}{}^j \otimes \overset{\circ}{\mathbf{p}}_j - \overset{\circ}{\mathbf{p}}{}^i \otimes \overset{\circ}{\mathbf{p}}_i \cdot \boldsymbol{\Omega}_U \cdot \mathbf{T} \cdot \overset{\circ}{\mathbf{p}}{}^j \otimes \overset{\circ}{\mathbf{p}}_j \\ &\quad + \overset{\circ}{\mathbf{p}}{}^i \otimes \overset{\circ}{\mathbf{p}}_i \cdot \mathbf{T} \cdot \boldsymbol{\Omega}_U \cdot \overset{\circ}{\mathbf{p}}{}^j \otimes \overset{\circ}{\mathbf{p}}_j = \dot{\mathbf{T}} - \boldsymbol{\Omega}_U \cdot \mathbf{T} + \mathbf{T} \cdot \boldsymbol{\Omega}_U.\end{aligned} \tag{2.358}$$

The co-rotational derivative of a vector $\mathbf{a}$ (or a tensor $\mathbf{T}$) determined by (2.358) is called the *right derivative relative to the eigenbasis.*

Thus, we have proved the following theorem.

**Theorem 2.32.** *The right derivative relative to the eigenbasis is connected to the total derivative as follows (for a vector* $\mathbf{a}$ *and for a tensor* $\mathbf{T}$*, respectively):*

$$\mathbf{a}^U = \dot{\mathbf{a}} + \mathbf{a} \cdot \boldsymbol{\Omega}_U, \quad \mathbf{T}^U = \dot{\mathbf{T}} - \boldsymbol{\Omega}_U \cdot \mathbf{T} + \mathbf{T} \cdot \boldsymbol{\Omega}_U. \tag{2.359}$$

### 2.5.6 The Derivative in the Eigenbasis ($\mathbf{h}_i = \mathbf{p}_i$) of the Left Stretch Tensor

Take the eigenbasis $\mathbf{p}_i$ of the left stretch tensor $\mathbf{V}$ as a moving basis $\mathbf{h}_i$ and define the following co-rotational derivatives

$$\mathbf{a}^H \equiv \mathbf{a}^V = \frac{da^i}{dt}\mathbf{p}_i, \quad \mathbf{T}^H \equiv \mathbf{T}^V = \frac{dT^{ij}}{dt}\mathbf{p}_i \otimes \mathbf{p}_j, \tag{2.360}$$

called the *left derivatives in the eigenbasis.*

**Theorem 2.33.** *The left derivatives* (2.360) *in the eigenbasis are connected to the total derivative with respect to time by the following relations (for a vector* $\mathbf{a}$ *and for a tensor* $\mathbf{T}$*, respectively):*

$$\mathbf{a}^V = \dot{\mathbf{a}} - \boldsymbol{\Omega}_V \cdot \mathbf{a}, \quad \mathbf{T}^V = \dot{\mathbf{T}} - \boldsymbol{\Omega}_V \cdot \mathbf{T} + \mathbf{T} \cdot \boldsymbol{\Omega}_V. \tag{2.361}$$

▼ A proof follows from (2.337), (2.342), and (2.343), because from (2.272) we have

$$\frac{d\mathbf{h}_i}{dt} = \dot{\mathbf{p}}_i = \boldsymbol{\Omega}_V \cdot \mathbf{p}_i. \tag{2.362}$$

Since the bases $\overset{\circ}{\mathbf{p}}_i$ and $\mathbf{p}_i$ are orthonormal, all the co-rotational derivatives relative to the mixed dyadic bases $\overset{\circ}{\mathbf{p}}_i \otimes \overset{\circ}{\mathbf{p}}^i$, $\mathbf{p}^i \otimes \mathbf{p}_i$ coincide with $\mathbf{T}^U$ or $\mathbf{T}^V$, respectively. ▲

## 2.5.7 The Jaumann Derivative ($\mathbf{h}_i = \mathbf{q}_i$)

If we choose the eigenbasis of the deformation rate tensor as a moving basis $\mathbf{h}_i = \mathbf{q}_i$ (it should be noted that the basis $\mathbf{q}_i$ is also orthonormal and coincides with $\mathbf{q}^i$), then from (2.336) we get the *co-rotational Jaumann derivatives*:

$$\mathbf{a}^h \equiv \mathbf{a}^J = \frac{da^i}{dt}\mathbf{q}_i, \quad \mathbf{T}^h \equiv \mathbf{T}^J = \frac{dT^{ij}}{dt}\mathbf{q}_i \otimes \mathbf{q}_j. \tag{2.363}$$

**Theorem 2.34.** *The Jaumann derivatives* (2.363) *are connected to the total derivatives with respect to time by the relations (for a vector* $\mathbf{a}$ *and for a tensor* $\mathbf{T}$*, respectively)*

$$\mathbf{a}^J = \dot{\mathbf{a}} + \mathbf{a} \cdot \mathbf{W}, \tag{2.364}$$

$$\mathbf{T}^J = \dot{\mathbf{T}} - \mathbf{W} \cdot \mathbf{T} + \mathbf{T} \cdot \mathbf{W}. \tag{2.365}$$

▼ According to the relationship (2.274), we get

$$\frac{d\mathbf{h}_i}{dt} = \dot{\mathbf{q}}_i = \mathbf{W} \cdot \mathbf{q}_i,$$

therefore, due to formulae (2.337) and (2.342) we find

$$\mathbf{a}^J = \dot{\mathbf{a}} \cdot \mathbf{q}_i \otimes \mathbf{q}^i + \mathbf{a} \cdot \mathbf{W} \cdot \mathbf{q}_i \otimes \mathbf{q}^i = \dot{\mathbf{a}} + \mathbf{a} \cdot \mathbf{W}.$$

In a similar way, we can prove the relation (2.365). ▲

## 2.5.8 Co-rotational Derivatives in a Moving Orthonormal Basis

Let $\mathbf{h}_i$ be a moving orthonormal basis. In this case we denote co-rotational derivatives by the following way: $\mathbf{a}^h \equiv \mathbf{a}^Q$ and $\mathbf{T}^h \equiv \mathbf{T}^Q$. Due to orthonormalization of the basis $\mathbf{h}_i$, the total derivatives of $\mathbf{a}$ and $\mathbf{T}$ with taking account of (2.333), (2.334), and (2.336) can be written as follows:

$$\dot{\mathbf{a}} \equiv \frac{d\mathbf{a}}{dt} = \frac{da^i}{dt}\mathbf{h}_i + a^i\frac{d\mathbf{h}_i}{dt} = \mathbf{a}^Q + a^i\boldsymbol{\omega}_h \times \mathbf{h}_i, \tag{2.366}$$

$$\dot{\mathbf{T}} = \frac{dT^{ij}}{dt}\mathbf{h}_i \otimes \mathbf{h}_j + T^{ij}\Big(\frac{d\mathbf{h}_i}{dt} \otimes \mathbf{h}_j + \mathbf{h}_i \otimes \frac{d\mathbf{h}_j}{dt}\Big)$$
$$= \mathbf{T}^Q + T^{ij}\boldsymbol{\omega}_h \times \mathbf{h}_i \otimes \mathbf{h}_j - \mathbf{h}_i \otimes \mathbf{h}_j \times \boldsymbol{\omega}_h.$$

Here we have used formula (2.266) for derivative $\dot{\mathbf{h}}_i$ of the moving basis, where $\boldsymbol{\omega}_h$ is the *vorticity vector* giving a rotation of the basis $\mathbf{h}_i$ relative to the fixed basis $\bar{\mathbf{e}}_i$ (see (2.258) and (2.265)). With taking account of (2.333) and (2.334), formulae (2.366) can be written in the form

$$\mathbf{a}^Q = \dot{\mathbf{a}} - \boldsymbol{\omega}_h \times \mathbf{a}, \quad \mathbf{T}^Q = \dot{\mathbf{T}} - \boldsymbol{\omega}_h \times \mathbf{T} + \mathbf{T} \times \boldsymbol{\omega}_h. \tag{2.367}$$

It should be noticed that if $\mathbf{a} = \boldsymbol{\omega}_h$, then

$$\dot{\boldsymbol{\omega}}_h = \boldsymbol{\omega}_h^h, \tag{2.367a}$$

because $\boldsymbol{\omega}_h \times \boldsymbol{\omega}_h = \mathbf{0}$ due to properties of the vector product.

### 2.5.9 Spin Derivative

Take an arbitrary orthonormal basis $\bar{\mathbf{h}}_i$ at a point $\mathcal{M}$ of a continuum in $\mathcal{K}$. The trihedron must have the only property that at any time $t$ the basis $\bar{\mathbf{h}}_i$ rotates with the instantaneous angular rate, which is equal to the rotation rate of the trihedron $\mathbf{p}_i$ relative to the trihedron $\overset{\circ}{\mathbf{p}}_i$. As shown in Sect. 2.4.10, the instantaneous rotation of the trihedron is characterized by the spin tensor $\boldsymbol{\Omega} = \dot{\mathbf{O}} \cdot \mathbf{O}^{\mathrm{T}}$ determined by (2.277).

Then we can define the co-rotational derivative in the basis, which is called the *spin derivative* (of a vector $\mathbf{a}$ and of a tensor $\mathbf{T}$, respectively):

$$\mathbf{a}^h \equiv \mathbf{a}^S = \frac{da^i}{dt}\bar{\mathbf{h}}_i, \tag{2.368}$$

$$\mathbf{T}^h \equiv \mathbf{T}^S = \frac{dT^{ij}}{dt}\bar{\mathbf{h}}_i \otimes \bar{\mathbf{h}}_j. \tag{2.369}$$

**Theorem 2.35.** *The spin derivative is related to the total derivative with respect to time as follows (for a vector* $\mathbf{a}$ *and for a tensor* $\mathbf{T}$*, respectively):*

$$\mathbf{a}^S = \dot{\mathbf{a}} + \mathbf{a} \cdot \boldsymbol{\Omega}, \tag{2.370}$$

$$\mathbf{T}^S = \dot{\mathbf{T}} - \boldsymbol{\Omega} \cdot \mathbf{T} + \mathbf{T} \cdot \boldsymbol{\Omega}. \tag{2.371}$$

▼ A proof of Theorem 2.35 follows from (2.342), (2.343) and the relation

$$\dot{\bar{\mathbf{h}}}^i = \boldsymbol{\Omega} \cdot \bar{\mathbf{h}}^i, \tag{2.372}$$

which is a consequence of (2.264). The relation (2.371) follows from (2.343). ▲

## 2.5.10 Universal Form of the Co-rotational Derivatives

On comparing formulae (2.350), (2.351), (2.352), (2.359), (2.361), (2.365), (2.370), and (2.371), we can notice that all the representations of the co-rotational derivatives and also the total derivative with respect to time can be written in the universal form:

$$\begin{gathered} \mathbf{a}^h = \dot{\mathbf{a}} - \mathbf{Z}_h \cdot \mathbf{T}, \quad \mathbf{T}^h = \dot{\mathbf{T}} - \mathbf{Z}_h \cdot \mathbf{T} - \mathbf{T} \cdot \mathbf{Z}_h^{\mathrm{T}}, \\ h = \{\,\cdot,\ \mathrm{Ol},\ \mathrm{CR},\ U,\ V,\ J,\ S\,\}, \end{gathered} \tag{2.373}$$

where tensors $\mathbf{Z}_h$ have the following representation for different $h$:

$$\begin{gathered} \mathbf{Z}_h = \{\,\mathbf{0},\ \mathbf{L},\ -\mathbf{L}^{\mathrm{T}},\ \boldsymbol{\Omega}_U,\ \boldsymbol{\Omega}_V,\ \mathbf{W},\ \boldsymbol{\Omega}\,\}, \\ h = \{\,\cdot,\ \mathrm{Ol},\ \mathrm{CR},\ U,\ V,\ J,\ S\,\}. \end{gathered} \tag{2.374}$$

Since tensors $\boldsymbol{\Omega}_U$, $\boldsymbol{\Omega}_V$ and $\boldsymbol{\Omega}$ are linearly expressed in terms of $\mathbf{W}$ and $\mathbf{D}$ (see Exercise 2.4.11), so tensors $\mathbf{Z}_h$ can be written as linear functions of $\mathbf{W}$ and $\mathbf{D}$:

$$\mathbf{Z}_h = {}^4\mathbf{Z}_{Dh} \cdot\cdot\, \mathbf{D} + {}^4\mathbf{Z}_{Wh} \cdot\cdot\, \mathbf{W}. \tag{2.375}$$

Table 2.1 gives expressions for fourth-order tensors ${}^4\mathbf{Z}_{Dh}$ and ${}^4\mathbf{Z}_{Wh}$, where tensors ${}^4\boldsymbol{\Omega}_U$, ${}^4\boldsymbol{\Omega}_V$ and ${}^4\tilde{\boldsymbol{\Omega}}$ are defined in Exercise 2.4.11.

## 2.5.11 Relations Between Co-rotational Derivatives of Deformation Rate Tensors and Velocity Gradient

In Sect. 2.4.11 we have derived the relationships between rates of deformation tensors and velocity gradient $\mathbf{L}$. Similar connections also exist between co-rotational derivatives of the tensors and $\mathbf{L}$. Let us establish them.

**Table 2.1** Expressions of tensors ${}^4\mathbf{Z}_{Dh}$, ${}^4\mathbf{Z}_{Wh}$, and ${}^4\mathbf{E}_h$ for different co-rotational derivatives

| $h$ | $\cdot$ | Ol | CR | $U$ | $V$ | $J$ | $S$ |
|---|---|---|---|---|---|---|---|
| ${}^4\mathbf{Z}_{Dh}$ | $\mathbf{0}$ | $\boldsymbol{\Delta}_{\mathrm{III}}$ | $\boldsymbol{\Delta}_{-\mathrm{III}}$ | ${}^4\boldsymbol{\Omega}_U$ | ${}^4\boldsymbol{\Omega}_V$ | $\mathbf{0}$ | ${}^4\tilde{\boldsymbol{\Omega}}$ |
| ${}^4\mathbf{Z}_{Wh}$ | $\mathbf{0}$ | $\boldsymbol{\Delta}_{\mathrm{III}}$ | $\boldsymbol{\Delta}_{\mathrm{III}}$ | $\mathbf{0}$ | $\boldsymbol{\Delta}_{\mathrm{III}}$ | $\boldsymbol{\Delta}_{\mathrm{III}}$ | $\boldsymbol{\Delta}_{\mathrm{III}}$ |
| ${}^4\mathbf{E}_h$ | $\mathbf{0}$ | $-2\boldsymbol{\Delta}$ | $2\boldsymbol{\Delta}$ | $\mathbf{0}$ | $\mathbf{0}$ | $\mathbf{0}$ | $\mathbf{0}$ |

On substituting representations (2.283), (2.284), and (2.311) for rates $\dot{\mathbf{A}}$, $\dot{\mathbf{J}}$, $\dot{\mathbf{g}}$, $(\mathbf{g}^{-1})^{\bullet}$, $\dot{\mathbf{V}}$ and $(\mathbf{V}^{-1})^{\bullet}$ into formula (2.373), we get

$$\begin{aligned}
\mathbf{A}^h &= \mathbf{D} - (\mathbf{Z}_h + \mathbf{L}^{\mathrm{T}}) \cdot \mathbf{A} - \mathbf{A} \cdot (\mathbf{Z}_h^{\mathrm{T}} + \mathbf{L}),\\
\mathbf{J}^h &= \mathbf{D} - (\mathbf{Z}_h - \mathbf{L}) \cdot \mathbf{J} - \mathbf{J} \cdot (\mathbf{Z}_h^{\mathrm{T}} - \mathbf{L}^{\mathrm{T}}),\\
\mathbf{g}^h &= -(\mathbf{Z}_h + \mathbf{L}^{\mathrm{T}}) \cdot \mathbf{g} - \mathbf{g} \cdot (\mathbf{Z}_h^{\mathrm{T}} + \mathbf{L}),\\
(\mathbf{g}^{-1})^h &= -(\mathbf{Z}_h - \mathbf{L}) \cdot \mathbf{g}^{-1} - \mathbf{g}^{-1} \cdot (\mathbf{Z}_h^{\mathrm{T}} - \mathbf{L}^{\mathrm{T}}),\\
\mathbf{V}^h &= -\left(\mathbf{Z}_h - \frac{1}{2}(\mathbf{L} + \boldsymbol{\Omega})\right) \cdot \mathbf{V} - \mathbf{V} \cdot \left(\mathbf{Z}_h^{\mathrm{T}} - \frac{1}{2}(\mathbf{L}^{\mathrm{T}} + \boldsymbol{\Omega}^{\mathrm{T}})\right),\\
(\mathbf{V}^{-1})^h &= -\left(\mathbf{Z}_h - \frac{1}{2}(\boldsymbol{\Omega} - \mathbf{L}^{\mathrm{T}})\right) \cdot \mathbf{V}^{-1} - \mathbf{V}^{-1} \cdot \left(\mathbf{Z}_h^{\mathrm{T}} - \frac{1}{2}(\boldsymbol{\Omega}^{\mathrm{T}} - \mathbf{L})\right),\\
h &= \{\,\cdot,\ \mathrm{Ol},\ \mathrm{CR},\ U,\ V,\ J,\ S\,\}.
\end{aligned} \tag{2.376}$$

From these relationships we can find the following expressions:

$$\begin{aligned}
(\mathbf{V} - \mathbf{E})^h &= \mathbf{D} - {}^4\mathbf{E}_h \cdot\cdot\, \mathbf{D} - (\mathbf{Z}_h + \mathbf{Z}_h^{\mathrm{T}})\\
&\quad - \left(\mathbf{Z}_h - \frac{1}{2}(\boldsymbol{\Omega} + \mathbf{L})\right) \cdot (\mathbf{V} - \mathbf{E}) - (\mathbf{V} - \mathbf{E}) \cdot \left(\mathbf{Z}_h^{\mathrm{T}} - \frac{1}{2}(\boldsymbol{\Omega}^{\mathrm{T}} + \mathbf{L}^{\mathrm{T}})\right),
\end{aligned} \tag{2.377}$$

$$\begin{aligned}
(\mathbf{E} - \mathbf{V}^{-1})^h &= \mathbf{D} + {}^4\mathbf{E}_h \cdot\cdot\, \mathbf{D} + \mathbf{Z}_h + \mathbf{Z}_h^{\mathrm{T}}\\
&\quad - \left(\mathbf{Z}_h - \frac{1}{2}(\boldsymbol{\Omega} - \mathbf{L}^{\mathrm{T}})\right) \cdot (\mathbf{E} - \mathbf{V}^{-1}) - (\mathbf{E} - \mathbf{V}^{-1}) \cdot \left(\mathbf{Z}_h^{\mathrm{T}} - \frac{1}{2}(\boldsymbol{\Omega}^{\mathrm{T}} - \mathbf{L})\right).
\end{aligned}$$

Here we have denoted the co-rotational derivative of the metric tensor by

$$\mathbf{E}^h = {}^4\mathbf{E}_h \cdot\cdot\, \mathbf{D}. \tag{2.378}$$

The tensor ${}^4\mathbf{E}_h$ differs from zero-tensor only when $h = \{\mathrm{CR}, \mathrm{Ol}\}$ (see Exercise 2.5.3), its expressions are given in Table 2.1 (see Sect. 2.5.10).

Since tensors $\mathbf{Z}_h$ and $\boldsymbol{\Omega}$ are linearly expressed in terms of $\mathbf{W}$ and $\mathbf{D}$ (see formulae (2.375), (2.310), (2.302)–(2.304)), so on the right-hand sides of Eqs. (2.376) there are also linear functions of $\mathbf{W}$ and $\mathbf{D}$, their explicit expressions will be given in Sect. 4.2.22.

## Exercises for 2.5

**2.5.1.** Show that the mixed co-rotational derivatives, the left and the right co-rotational derivatives relative to the eigenbasis and also the Jaumann and spin derivatives satisfy the differentiation rules of scalar products:

$$(\mathbf{A}\cdot\mathbf{B})^h = \mathbf{A}^h\cdot\mathbf{B} + \mathbf{A}\cdot\mathbf{B}^h, \quad (\mathbf{a}\cdot\mathbf{A})^h = \mathbf{a}^h\cdot\mathbf{A} + \mathbf{a}\cdot\mathbf{A}^h,$$

$$(\psi\mathbf{A})^h = \psi^h\mathbf{A} + \psi\mathbf{A}^h, \quad h = \{\cdot, d, D, U, V, J, S\},$$

and the Oldroyd and Cotter–Rivlin derivatives do not satisfy this rule.

**2.5.2.** Show that for the co-rotational derivatives, the following rules of differentiation of scalar products of two vectors **a** and **b** and also of two tensors **T** and **B** remain valid:

$$(\mathbf{a}\cdot\mathbf{b})^h = (\mathbf{a}\cdot\mathbf{b})^{\bullet} = \mathbf{a}^h\cdot\mathbf{b} + \mathbf{a}\cdot\mathbf{b}^h, \qquad h = \{U, V, J, S\};$$

$$(\mathbf{T}\cdot\cdot\,\mathbf{B})^h = (\mathbf{T}\cdot\cdot\,\mathbf{B})^{\bullet} = \mathbf{T}^h\cdot\cdot\,\mathbf{B} + \mathbf{T}\cdot\cdot\,\mathbf{B}^h, \qquad h = \{d, D, U, V, J, S\}.$$

**2.5.3.** Show that the following co-rotational derivatives of the unit tensor **E** give the zero-tensor:

$$\mathbf{E}^h = 0, \quad h = \{\cdot, d, D, U, V, J, S\},$$

and the Oldroyd and Cotter–Rivlin derivatives of **E** are different from zero:

$$\mathbf{E}^{\mathrm{CR}} = 2\mathbf{D}, \quad \mathbf{E}^{\mathrm{Ol}} = -2\mathbf{D}.$$

**2.5.4.** Using formulae (2.283) and (2.352), show that the Cotter–Rivlin derivatives of the left Almansi deformation tensor **A** and of the left Almansi measure **g** have the form

$$\mathbf{A}^{\mathrm{CR}} = \mathbf{D}, \qquad \mathbf{g}^{\mathrm{CR}} = 0.$$

**2.5.5.** Using formulae (2.283) and (2.350), show that the Oldroyd derivatives of the right Cauchy–Green tensor **J** and of the right deformation measure $\mathbf{g}^{-1}$ have the form

$$\mathbf{J}^{\mathrm{Ol}} = \mathbf{D}, \qquad (\mathbf{g}^{-1})^{\mathrm{Ol}} = 0.$$

**2.5.6.** Using the expressions for the tensors **U** (2.142), **C** and $\mathbf{\Lambda}$ (2.165), and also **G** and $\mathbf{G}^{-1}$ (2.165), show that we can write the right derivative relative to the eigenbasis $\overset{\circ}{\mathbf{p}}_\alpha$ in the form

$$\mathbf{U}^U = \sum_{\alpha=1}^{3} \dot{\lambda}_\alpha \overset{\circ}{\mathbf{p}}_\alpha \otimes \overset{\circ}{\mathbf{p}}_\alpha, \quad \mathbf{C}^U = \sum_{\alpha=1}^{3} \lambda_\alpha\dot{\lambda}_\alpha \overset{\circ}{\mathbf{p}}_\alpha \otimes \overset{\circ}{\mathbf{p}}_\alpha,$$

$$\mathbf{\Lambda}^U = \sum_{\alpha=1}^{3} \frac{\dot{\lambda}_\alpha}{\lambda_\alpha^3} \overset{\circ}{\mathbf{p}}_\alpha \otimes \overset{\circ}{\mathbf{p}}_\alpha, \quad (\mathbf{U}^{-1})^U = \sum_{\alpha=1}^{3} \frac{\dot{\lambda}_\alpha}{\lambda_\alpha^2} \overset{\circ}{\mathbf{p}}_\alpha \otimes \overset{\circ}{\mathbf{p}}_\alpha,$$

$$\mathbf{G}^U = 2\mathbf{C}^U, \quad (\mathbf{G}^{-1})^U = -2\mathbf{\Lambda}^U.$$

**2.5.7.** Using the expressions for the tensors $\mathbf{V}$ (2.142), for $\mathbf{A}$ and $\mathbf{J}$ (2.165), and also for $\mathbf{g}$ and $\mathbf{g}^{-1}$ (2.165), show that we can write the left derivative relative to the eigenbasis $\mathbf{p}_\alpha$ in the form

$$\mathbf{V}^V = \sum_{\alpha=1}^{3} \dot{\lambda}_\alpha \mathbf{p}_\alpha \otimes \mathbf{p}_\alpha, \quad \mathbf{A}^V = \sum_{\alpha=1}^{3} \frac{\dot{\lambda}_\alpha}{\lambda_\alpha^3} \mathbf{p}_\alpha \otimes \mathbf{p}_\alpha,$$

$$\mathbf{J}^V = \sum_{\alpha=1}^{3} \lambda_\alpha \dot{\lambda}_\alpha \mathbf{p}_\alpha \otimes \mathbf{p}_\alpha, \quad \mathbf{g}^V = -2\mathbf{A}^V, \quad (\mathbf{g}^{-1})^V = 2\mathbf{J}^V.$$

**2.5.8.** Show that the Oldroyd and Jaumann derivatives of a second-order tensor $\mathbf{T}$ are connected by the relationship

$$\mathbf{T}^{\mathrm{Ol}} = \mathbf{T}^J - \mathbf{T} \cdot \mathbf{D} - \mathbf{D} \cdot \mathbf{T}.$$

**2.5.9.** Show that if for an arbitrary symmetric tensor $\mathbf{T}$ its co-rotational derivatives are equal to zero:

$$\mathbf{T}^h = 0, \qquad h = \{d, D, U, V, J, S\},$$

then the first invariant of the tensor: $I_1(\mathbf{T}) = \mathbf{T} \cdot\cdot\, \mathbf{E}$ has its stationary value, i.e.

$$\dot{I}_1(\mathbf{T}) = 0.$$

Show that for the co-rotational Oldroyd and Cotter–Rivlin derivatives this statement is not valid.

**2.5.10.** Using the results of Exercise 2.4.3, show that the co-rotational derivatives of the normal vector $\mathbf{n}$ satisfy the following relations:

$$\mathbf{n}^{\mathrm{CR}} = \gamma \mathbf{n}, \quad \gamma = \mathbf{n} \cdot \mathbf{D} \cdot \mathbf{n},$$

$$\mathbf{n}^{\mathrm{Ol}} = \gamma \mathbf{n} - \mathbf{n} \cdot \mathbf{L} - \mathbf{L} \cdot \mathbf{n}, \quad \mathbf{n}^{\mathrm{J}} = \gamma \mathbf{n} - \mathbf{n} \cdot \mathbf{D}.$$

**2.5.11.** Show that the following co-rotational derivatives of a symmetric tensor give a symmetric tensor:

$$\text{if } \mathbf{A} = \mathbf{A}^{\mathrm{T}}, \text{ then } (\mathbf{A}^h)^{\mathrm{T}} = \mathbf{A}^h, \;\; h = \{U,\ V,\ J,\ S,\ \mathrm{Ol}, \mathrm{CR}\},$$

and also a skew-symmetric tensor, if they are applied to a skew-symmetric tensor:

$$\text{if } \mathbf{B} = -\mathbf{B}^{\mathrm{T}}, \text{ then } (\mathbf{B}^h)^{\mathrm{T}} = -\mathbf{B}^h, \;\; h = \{U,\ V,\ J,\ S,\ \mathrm{Ol},\ \mathrm{CR}\}.$$

The mixed co-rotational derivatives $h = d, D$ have no such properties.

# Chapter 3
# Balance Laws

## 3.1 The Mass Conservation Law

### 3.1.1 Integral and Differential Forms

Let us consider *fundamental laws of continuum mechanics*, and complement the set of Axioms 1–3 (see Introduction) with new axioms.

**Axiom 4 (The mass conservation law).** *For every continuum $\mathcal{B}$, there exists a scalar function $M(\mathcal{B}, t)$ (i.e. the transformation $M : \mathcal{U} \times \mathbb{R}_{+0} \to \mathbb{R}_{+}$), called the* m a s s *of the continuum, which has the following properties:*

- *It is positive: $M > 0$,*
- *It is additive: $M(\mathcal{B}_1 + \mathcal{B}_2, t) = M(\mathcal{B}_1, t) + M(\mathcal{B}_2, t)$, $\forall \mathcal{B}_1$, $\mathcal{B}_2$, and $\forall t \geqslant 0$,*
- *It is invariant relative to any transformation of coordinates (2.1) and relative to any motion (2.3).*

Due to the third property, the mass in any actual configuration remains constant:

$$M(\mathcal{B}, t) = \text{const.} \tag{3.1}$$

*Remark.* The mass conservation law is valid only for a continuum containing the same material points for a considered time interval $[0, t]$. If a continuum $\mathcal{B}$ loses or acquires material points as time goes on (in this case phase transformations are said to occur in the continuum (see Sect. 5.1)), then the mass conservation law in the form (3.1) is no longer valid. The mass conservation law does not hold as well, if we exclude Axiom 3 and consider the motions with speeds close to the light speed; however, relativistic phenomena are not considered in classical continuum mechanics. □

The law (3.1) can be written in another form

$$dM/dt = 0. \tag{3.2}$$

Y.I. Dimitrienko, *Nonlinear Continuum Mechanics and Large Inelastic Deformations*, Solid Mechanics and Its Applications 174, DOI 10.1007/978-94-007-0034-5_3,

Since the mass is additive, $M$ may be expressed as follows:

$$M = \int_V dm, \tag{3.3}$$

where $dm$ is the mass of an elementary volume $dV$ involving a material point $\mathcal{M}$ belonging to the considered continuum $V$.

**Definition 3.1.** The ratio

$$\rho = dm/dV \tag{3.4}$$

is called the *density* of substance at the point $\mathcal{M}$.

Since the mass $M$ and volume $dV$ are positive, the mass $dm$ and density $\rho$ are always positive as well:

$$\rho > 0, \qquad dm > 0. \tag{3.4a}$$

On substituting (3.3) and (3.4) into (3.2), we can rewrite the mass conservation law (3.2) in the form

$$\frac{d}{dt}\int_V \rho dV = 0. \tag{3.5}$$

Having applied this relationship to an elementary volume, we get

$$\rho\, dV = \overset{\circ}{\rho}\, d\overset{\circ}{V} = \text{const}, \tag{3.6}$$

where $\rho$ and $\overset{\circ}{\rho}$ are the densities in configurations $\mathcal{K}$ and $\overset{\circ}{\mathcal{K}}$, respectively, and $d\overset{\circ}{V}$ is the elementary volume in the configuration $\overset{\circ}{\mathcal{K}}$.

The relationship (3.5) is called the *mass conservation law* in the integral form, and (3.6) – in the differential form.

### 3.1.2 The Continuity Equation in Lagrangian Variables

Consider in $\overset{\circ}{\mathcal{K}}$ an elementary volume $d\overset{\circ}{V}$ constructed on elementary radius-vectors oriented along the local basis vectors $d\overset{\circ}{\mathbf{x}}_\alpha = \overset{\circ}{\mathbf{r}}_\alpha dX^\alpha$ (see Sect. 2.2.6). This volume corresponds to a volume $dV$ constructed on vectors $\mathbf{r}_\alpha dX^\alpha$ in the actual configuration $\mathcal{K}$. The volumes $d\overset{\circ}{V}$ and $dV$ are determined by formulae (2.15):

$$\begin{aligned} d\overset{\circ}{V} &= \overset{\circ}{\mathbf{r}}_1 \cdot (\overset{\circ}{\mathbf{r}}_2 \times \overset{\circ}{\mathbf{r}}_3) dX^1 dX^2 dX^3 = \sqrt{\overset{\circ}{g}}\, dX^1 dX^2 dX^3 = \left|\frac{\partial \overset{\circ}{x}{}^k}{\partial X^i}\right| dX^1 dX^2 dX^3, \\ dV &= \mathbf{r}_1 \cdot (\mathbf{r}_2 \times \mathbf{r}_3) dX^1 dX^2 dX^3 = \sqrt{g}\, dX^1 dX^2 dX^3 = \left|\frac{\partial x^k}{\partial X^i}\right| dX^1 dX^2 dX^3. \end{aligned} \tag{3.7}$$

On substituting (3.7) into (3.6), we get the following theorem.

**Theorem 3.1.** *A change of the density in going from $\overset{\circ}{\mathcal{K}}$ to $\mathcal{K}$ can be determined by one of the following equations:*

$$\frac{\overset{\circ}{\rho}}{\rho} = \sqrt{\frac{g}{\overset{\circ}{g}}} = \frac{|\partial x^k/\partial X^i|}{|\partial \overset{\circ}{x}{}^j/\partial X^n|} = \left|\frac{\partial x^k}{\partial \overset{\circ}{x}{}^i}\right| = \det \mathbf{F}. \tag{3.8}$$

Unlike (3.6), these equations are called the *continuity equations in Lagrangian variables*.

The ratio of the elementary volumes in configurations $\mathcal{K}$ and $\overset{\circ}{\mathcal{K}}$, which follows from (3.7), is also widely used:

$$dV/d\overset{\circ}{V} = \sqrt{g/\overset{\circ}{g}}. \tag{3.9}$$

### *3.1.3 Differentiation of Integral over a Moving Volume*

Consider a vector field $\mathbf{a}(x^i, t)$, which is a continuously differentiable function of $x^i$ and $t$ in a domain $V(t)$, $\forall t \geqslant 0$, where the domain $V(t)$ contains the same material points (such a domain is usually called a *moving volume*).

Let us integrate the field $\mathbf{a}(x^i, t)$ over the domain $V(t)$ and determine the derivative of the integral: $\frac{d}{dt}\int_{V(t)} \mathbf{a}\, dV$.

To do this, we should change variables in the integral: $x^i \to \overset{\circ}{x}{}^i$, where $x^i \in V$, $\overset{\circ}{x}{}^i \in \overset{\circ}{V}$, and, according to (3.8), the Jacobian of the transformation is $|\partial x^k/\partial \overset{\circ}{x}{}^i| = \sqrt{g/\overset{\circ}{g}}$. Then we get

$$\frac{d}{dt}\int\limits_{V(t)} \mathbf{a}\, dV = \frac{d}{dt}\int\limits_{\overset{\circ}{V}} \sqrt{\frac{g}{\overset{\circ}{g}}}\mathbf{a}\, d\overset{\circ}{V}. \tag{3.10}$$

Since this change in variables means the transformation from the configuration $\mathcal{K}$ to $\overset{\circ}{\mathcal{K}}$, where the domain $\overset{\circ}{V}$ is time–independent, the derivative with respect to $t$ may be introduced under the integral sign:

$$\frac{d}{dt}\int\limits_{V(t)} \mathbf{a}\, dV = \int\limits_{\overset{\circ}{V}} \frac{d}{dt}\left(\frac{\sqrt{g}}{\sqrt{\overset{\circ}{g}}}\mathbf{a}\right) d\overset{\circ}{V} = \int\limits_{\overset{\circ}{V}} \frac{1}{\sqrt{\overset{\circ}{g}}}\left(\mathbf{a}\frac{d}{dt}\sqrt{g} + \sqrt{g}\frac{d\mathbf{a}}{dt}\right) d\overset{\circ}{V}. \tag{3.10a}$$

On substituting (2.187) and (2.217) into (3.10a), we get

$$\frac{d}{dt}\int\limits_{V(t)} \mathbf{a}\,dV = \int\limits_{\overset{\circ}{V}} \sqrt{g/\overset{\circ}{g}}\left(\mathbf{a}\nabla\cdot\mathbf{v} + \frac{\partial \mathbf{a}}{\partial t} + \mathbf{v}\cdot\nabla\otimes\mathbf{a}\right) d\overset{\circ}{V}$$

$$= \int\limits_{\overset{\circ}{V}} \sqrt{\frac{g}{\overset{\circ}{g}}}\left(\frac{\partial \mathbf{a}}{\partial t} + \nabla\cdot(\mathbf{v}\otimes\mathbf{a})\right) d\overset{\circ}{V} = \int\limits_{V}\left(\frac{\partial \mathbf{a}}{\partial t} + \nabla\cdot(\mathbf{v}\otimes\mathbf{a})\right) dV. \tag{3.11}$$

The last equality we have obtained by the reverse substitution $\overset{\circ}{x}{}^i \to x^i$.

Thus, we have proved following theorem.

**Theorem 3.2 (The Rule of Differentiation of the Integral over a Moving Volume).** *For any vector field* $\mathbf{a}(x^i, t)$ *specified in* $V(t)$ $\forall t \geqslant 0$ *and being a continuously differentiable function of* $x^i$ *and* $t$*, the following relationship holds:*

$$\frac{d}{dt}\int\limits_{V(t)} \mathbf{a}(x^i, t)\,dV = \int\limits_{V}\left(\frac{\partial \mathbf{a}}{\partial t} + \nabla\cdot\mathbf{v}\otimes\mathbf{a}\right) dV. \tag{3.12}$$

Taking in formula (3.12) $\varphi(x^i, t)\bar{\mathbf{e}}_\alpha$ as a vector field $\mathbf{a}(x^i, t)$, where $\bar{\mathbf{e}}_\alpha$ is one of the Cartesian basis vectors, and $\varphi(x^i, t)$ is a scalar field, and factoring $\bar{\mathbf{e}}_\alpha$ outside the integral sign on the left and right sides, we get the formula for differentiation of the integral of a scalar field:

$$\frac{d}{dt}\int\limits_{V(t)} \varphi(x^i, t)\,dV = \int\limits_{V}\left(\frac{\partial \varphi}{\partial t} + \nabla\cdot(\varphi\mathbf{v})\right) dV. \tag{3.13}$$

### *3.1.4 The Continuity Equation in Eulerian Variables*

Putting $\varphi = \rho$ in formula (3.13) and using the mass conservation law (3.5), we obtain

$$\int\limits_{V}\left(\frac{\partial \rho}{\partial t} + \nabla\cdot\rho\mathbf{v}\right) dV = 0 \tag{3.14}$$

– the *mass conservation law in Eulerian variables*. Unlike the law (3.5) which is formulated for a volume $V$ of a continuum containing at different times $t \geqslant 0$ the same material points, the law (3.14) holds for an arbitrary geometric domain $V$ of space $\mathcal{E}_3^a$, which at different times $t$ may involve different material points.

Since formula (3.14) holds true for an arbitrary domain $V \subset \mathcal{E}_3^a$, the integrand must vanish. So we have proved the following theorem.

**Theorem 3.3.** *If the function $\rho$, satisfying the mass conservation law (3.5), and the velocity* $\mathbf{v}$ *are continuously differentiable in $V(t)$ for all considered times $t \geqslant 0$, then at every point* $\mathbf{x}$ *of the domain $V \subset \mathcal{E}_3^a$ of a continuum the following relation holds $\forall t \geqslant 0$:*

$$\frac{\partial \rho}{\partial t} + \nabla \cdot \rho \mathbf{v} = 0. \quad (3.15)$$

Equation (3.15) is called the *continuity equation in Eulerian variables.*

According to the definition (2.187) of the total derivative with respect to time, Eq. (3.15) can be rewritten in the form

$$\frac{\partial \rho}{\partial t} + \mathbf{v} \cdot \nabla \rho + \rho \nabla \cdot \mathbf{v} = \frac{d\rho}{dt} + \rho \nabla \cdot \mathbf{v} = 0, \quad (3.16)$$

or

$$\frac{1}{\rho}\frac{d\rho}{dt} = -\nabla \cdot \mathbf{v}. \quad (3.17)$$

Due to (3.17), the continuity equation takes the form

$$\rho \frac{d}{dt}\left(\frac{1}{\rho}\right) = \nabla \cdot \mathbf{v}. \quad (3.18)$$

### *3.1.5 Determination of the Total Derivatives with respect to Time*

With the help of the continuity equation (3.15) we will prove several relationships which are frequently used in continuum mechanics.

**Theorem 3.4.** *If there is a tensor field ${}^n\boldsymbol{\Omega}(\mathbf{x}, t)$ of the nth order, which is continuously differentiable in a domain $V(t)$ $\forall t \geqslant 0$, then the following relations hold:*

$$\rho \frac{d}{dt}\left(\frac{{}^n\boldsymbol{\Omega}}{\rho}\right) = \frac{\partial}{\partial t}\,{}^n\boldsymbol{\Omega} + \nabla \cdot (\mathbf{v} \otimes {}^n\boldsymbol{\Omega}), \quad (3.19)$$

$$\rho \frac{d}{dt}({}^n\boldsymbol{\Omega}) = \frac{\partial}{\partial t}(\rho\, {}^n\boldsymbol{\Omega}) + \nabla \cdot (\rho \mathbf{v} \otimes {}^n\boldsymbol{\Omega}). \quad (3.20)$$

▼ According to the definition of the total derivative with respect to time (2.192), we have

$$\rho \frac{d}{dt}\left(\frac{{}^n\boldsymbol{\Omega}}{\rho}\right) = \rho \frac{\partial}{\partial t}\left(\frac{{}^n\boldsymbol{\Omega}}{\rho}\right) + \rho \mathbf{v} \cdot \nabla \otimes \left(\frac{{}^n\boldsymbol{\Omega}}{\rho}\right)$$

$$= \frac{\partial\, {}^n\boldsymbol{\Omega}}{\partial t} - \frac{\rho}{\rho^2}\frac{\partial \rho}{\partial t}\,{}^n\boldsymbol{\Omega} - \frac{\rho}{\rho^2}(\mathbf{v} \cdot \nabla \rho \otimes {}^n\boldsymbol{\Omega}) + \mathbf{v} \cdot \nabla \otimes {}^n\boldsymbol{\Omega}.$$

(3.21)

Collecting the second and the third summands and taking the following consequence of the continuity equation into account: $(\partial\rho/\partial t) + \mathbf{v}\cdot\nabla\rho = -\rho\nabla\cdot\mathbf{v}$, we finally get

$$\rho\frac{d}{dt}\left(\frac{{}^n\boldsymbol{\Omega}}{\rho}\right) = \frac{\partial\,{}^n\boldsymbol{\Omega}}{\partial t} + (\nabla\cdot\mathbf{v})\,{}^n\boldsymbol{\Omega} + \mathbf{v}\cdot\nabla\otimes\,{}^n\boldsymbol{\Omega} = \frac{\partial}{\partial t}\,{}^n\boldsymbol{\Omega} + \nabla\cdot(\mathbf{v}\otimes\,{}^n\boldsymbol{\Omega}). \quad (3.22)$$

To prove formula (3.20), it is sufficient to make the substitution ${}^n\boldsymbol{\Omega} \to \rho\,{}^n\boldsymbol{\Omega}$ in (3.19). ▲

### 3.1.6 The Gauss–Ostrogradskii Formulae

In continuum mechanics, the *Gauss–Ostrogradskii formulae* are frequently used for tensor fields ${}^k\boldsymbol{\Omega}(\mathbf{x})$ of arbitrary $k$th order in reference and actual configurations:

$$\int\limits_{\overset{\circ}{\Sigma}} \overset{\circ}{\mathbf{n}}\otimes{}^k\boldsymbol{\Omega}\,d\overset{\circ}{\Sigma} = \int\limits_{\overset{\circ}{V}} \overset{\circ}{\nabla}\otimes{}^k\boldsymbol{\Omega}\,d\overset{\circ}{V}, \qquad \int\limits_{\Sigma} \mathbf{n}\otimes{}^k\boldsymbol{\Omega}\,d\Sigma = \int\limits_{V} \nabla\otimes{}^k\boldsymbol{\Omega}\,dV. \quad (3.23)$$

$$\int\limits_{\overset{\circ}{\Sigma}} \overset{\circ}{\mathbf{n}}\cdot{}^k\boldsymbol{\Omega}\,d\overset{\circ}{\Sigma} = \int\limits_{\overset{\circ}{V}} \overset{\circ}{\nabla}\cdot{}^k\boldsymbol{\Omega}\,d\overset{\circ}{V}, \qquad \int\limits_{\Sigma} \mathbf{n}\cdot{}^k\boldsymbol{\Omega}\,d\Sigma = \int\limits_{V} \nabla\cdot{}^k\boldsymbol{\Omega}\,dV. \quad (3.24)$$

According to the second formula of (3.23), Eq. (3.12) takes the form

$$\frac{d}{dt}\int\limits_{V(t)} \mathbf{a}\,dV = \int\limits_{V} \frac{\partial\mathbf{a}}{\partial t}\,dV + \int\limits_{\Sigma} \mathbf{n}\cdot\mathbf{v}\otimes\mathbf{a}\,d\Sigma. \quad (3.25)$$

## Exercises for 3.1

**3.1.1.** Show that the continuity equation (3.8) can be written in the differential form:

$$\frac{d\rho}{dt} = -\frac{\rho}{\sqrt{g}}\frac{d\sqrt{g}}{dt}.$$

**3.1.2.** Using formulae (3.10), (3.19), and (3.20), show that the following integral relations (called *the rule of differentiation of the integral of a tensor field over a moving volume*) hold:

$$\frac{d}{dt}\int\limits_{V(t)} {}^n\boldsymbol{\Omega}\,dV = \int\limits_{V(t)} \rho\frac{d}{dt}\left(\frac{{}^n\boldsymbol{\Omega}}{\rho}\right)dV = \int\limits_{V}\left(\frac{\partial\,{}^n\boldsymbol{\Omega}}{\partial t} + \nabla\cdot(\mathbf{v}\otimes{}^n\boldsymbol{\Omega})\right)dV$$

$$= \int\limits_{V} \frac{\partial\,{}^n\boldsymbol{\Omega}}{\partial t}\,dV + \int\limits_{\Sigma} \mathbf{n}\cdot(\mathbf{v}\otimes\,{}^n\boldsymbol{\Omega})\,d\Sigma,$$

$$\frac{d}{dt}\int\limits_{V(t)} \rho\, {}^n\boldsymbol{\Omega}\, dV = \int\limits_{V(t)} \rho \frac{d\, {}^n\boldsymbol{\Omega}}{dt}\, dV = \int\limits_{V} \left(\frac{\partial}{\partial t}(\rho\, {}^n\boldsymbol{\Omega}) + \nabla \cdot (\rho \mathbf{v} \otimes {}^n\boldsymbol{\Omega})\right) dV$$

$$= \int\limits_{V} \frac{\partial}{\partial t}(\rho\, {}^n\boldsymbol{\Omega})\, dV + \int\limits_{\Sigma} \mathbf{n} \cdot (\rho \mathbf{v} \otimes\, {}^n\boldsymbol{\Omega})\, d\Sigma.$$

**3.1.3.** Using formulae (3.19), show that for an arbitrary vector field $\mathbf{a}(X^k, t)$ the following relation holds:

$$\rho \frac{d}{dt}\left(\frac{\mathbf{a}}{\rho}\right) = \frac{d}{dt}\mathbf{a} + \mathbf{a}(\nabla \cdot \mathbf{v}).$$

**3.1.4.** Similarly to formula (3.10), prove that for a tensor field ${}^k\boldsymbol{\Omega}(\mathbf{x})$ of arbitrary $k$th order the following formula of change of variables is valid:

$$\int\limits_{V} {}^k\boldsymbol{\Omega} \rho\, dV = \int\limits_{\overset{\circ}{V}} {}^k\boldsymbol{\Omega} \overset{\circ}{\rho}\, d\overset{\circ}{V}.$$

## 3.2 The Momentum Balance Law and the Stress Tensor

### *3.2.1 The Momentum Balance Law*

**Definition 3.2.** The vector

$$\mathbf{I} = \int\limits_{V} \mathbf{v}\, dm = \int\limits_{V} \rho \mathbf{v}\, dV, \tag{3.26}$$

is called the *momentum of a continuum.*

*Remark 1.* As in the mass conservation law (3.5), a domain $V(t)$ contains the same material points for all considered time interval. □

*Remark 2.* It should be noted that since the velocity $\mathbf{v} = d\mathbf{x}/dt$ of a material point $\mathcal{M}$, due to (2.181), has been introduced in a certain coordinate system $O\bar{\mathbf{e}}_i$, where $\mathbf{x}(\mathcal{M})$ is the radius-vector of the point $\mathcal{M}$, the momentum vector $\mathbf{I}$ is connected to this coordinate system. If we have chosen another coordinate system $O'\mathbf{e}'_i$, then, applying formula (3.26), we obtain another momentum vector $\mathbf{I}' = \int_V \rho \mathbf{v}'\, dV$, where $\mathbf{v}' = d\mathbf{x}'/dt$, and $\mathbf{x}'(\mathcal{M})$ is the radius-vector of the same material point $\mathcal{M}$ but in the system $O'\mathbf{e}'_i$. □

This remark is of great importance for understanding the following axiom.

**Axiom 5 (The momentum balance law).** *For any two continua $\mathcal{B}$ and $\mathcal{B}_1$ at any time $t$, there exists a vector function $\boldsymbol{\mathcal{F}} = \boldsymbol{\mathcal{F}}(\mathcal{B}, \mathcal{B}_1, t)$ (i.e. the transformation $\boldsymbol{\mathcal{F}}$ : $\mathcal{U} \times \mathcal{U} \times \mathbb{R}_{+0} \to \mathcal{E}_3^a$), which is possible to be zero-valued (i.e. $\boldsymbol{\mathcal{F}} = \mathbf{0}$). This vector function is called the force of interaction of the bodies $\mathcal{B}$ and $\mathcal{B}_1$, and has the following properties:*

- *It is additive:*

$$\boldsymbol{\mathcal{F}}(\mathcal{B}' + \mathcal{B}'',\ \mathcal{B}_1,\ t) = \boldsymbol{\mathcal{F}}(\mathcal{B}',\ \mathcal{B}_1,\ t) + \boldsymbol{\mathcal{F}}(\mathcal{B}'',\ \mathcal{B}_1,\ t),$$

$$\boldsymbol{\mathcal{F}}(\mathcal{B},\ \mathcal{B}_1' + \mathcal{B}_1'',\ t) = \boldsymbol{\mathcal{F}}(\mathcal{B},\ \mathcal{B}_1',\ t) + \boldsymbol{\mathcal{F}}(\mathcal{B},\ \mathcal{B}_1'',\ t),$$

  *where $\mathcal{B} = \mathcal{B}' \cup \mathcal{B}''$, $\mathcal{B}_1 = \mathcal{B}_1' \cup \mathcal{B}_1''$,*
- *The rate of changing the momentum $\mathbf{I}$ of a continuum $\mathcal{B}$ at any time $t$ is equal to the vector $\boldsymbol{\mathcal{F}}(\mathcal{B}, t) = \boldsymbol{\mathcal{F}}(\mathcal{B}, \mathcal{B}^e, t)$ which is the summarized vector of external forces, acting onto the body $\mathcal{B}$ (where $\mathcal{B}^e = \mathcal{U} \setminus \mathcal{B}$ is the outside of the body $\mathcal{B}$):*

$$d\mathbf{I}/dt = \boldsymbol{\mathcal{F}}. \tag{3.27}$$

The relationship (3.27) is called the *momentum balance law*.

*Remark 3.* Since the vector $\mathbf{I}$ has been introduced in a certain coordinate system $O\bar{\mathbf{e}}_i$, the momentum balance law (3.27) is written for this coordinate system $O\bar{\mathbf{e}}_i$. In another coordinate system $O'\mathbf{e}_i'$ the vector $d\mathbf{I}'/dt$ may be different from $d\mathbf{I}/dt$, and the law (3.27) also changes its form. The momentum balance law for such systems will be given in Sect. 4.10.11. The coordinate systems, in which the momentum balance law has the form (3.27), are called *inertial* (the system $O\bar{\mathbf{e}}_i$ is also inertial). The existence of inertial coordinate systems is declared by the first Newton law, and Eq. (3.27) expresses the second Newton law. In continuum mechanics both these laws are introduced by Axiom 5 together with Axioms 1–3 on the existence of the special coordinate system $O\bar{\mathbf{e}}_i$.

In more detail, noninertial coordinate systems will be considered in Sect. 4.9. □

As a body $\mathcal{B}$, consider the elementary volume $dV$ of a continuum involving a point $\mathcal{M}$. Then, by Axiom 5, the summarized vector of external forces denoted by $d\boldsymbol{\mathcal{F}}$ acts on $dV$. Here the following two cases are possible:

(1) $\mathcal{M}$ is an interior point of the domain $V$.
(2) $\mathcal{M}$ is on the surface $\Sigma$ of the domain $V$ (in this case the intersection of the surface $\Sigma$ with closure of the domain $d\overline{V}$ is denoted by $d\Sigma$).

**Definition 3.3.** The vector

$$\mathbf{f} = \frac{d\boldsymbol{\mathcal{F}}}{dm} = \frac{d\boldsymbol{\mathcal{F}}}{\rho\, dV} \tag{3.28}$$

is called the *specific mass force* and the vector

$$\mathbf{s} = d\mathcal{F}/d\Sigma \tag{3.29}$$

is called the *specific surface force*. Here $dm$ is the mass of an elementary volume $\mathbf{dV}$ of a continuum.

Since the vector $\mathcal{F}$ is additive, we have the following relations for the whole continuum occupying the volume $V$ in $\mathcal{K}$:

$$\mathcal{F} = \mathcal{F}_m + \mathcal{F}_\Sigma, \tag{3.30}$$

$$\mathcal{F}_m = \int\limits_V \mathbf{f}\,dm = \int\limits_V \rho\mathbf{f}\,dV, \quad \mathcal{F}_\Sigma = \int\limits_\Sigma \mathbf{s}\,d\Sigma. \tag{3.31}$$

The vector $\mathcal{F}_m$ is called the *summarized vector of external mass forces* acting on the considered continuum, and $\mathcal{F}_\Sigma$ is called the *summarized vector of external surface forces*.

Due to (3.30), the momentum balance law (3.27) takes the *integral form*:

$$\frac{d}{dt}\int\limits_V \rho\mathbf{v}\,dV = \int\limits_V \rho\mathbf{f}\,dV + \int\limits_\Sigma \mathbf{s}\,d\Sigma. \tag{3.32}$$

### 3.2.2 External and Internal Forces

Mass $\mathbf{f}$ and surface $\mathbf{s}$ forces are *external* forces for a continuum $V$, if they are induced by objects not belonging to the considered continuum $V$ (i.e. by external objects).

External mass forces may be of the following types: (1) the gravity force $\mathbf{f} = g_\Sigma\bar{\mathbf{e}}$, where $g_\Sigma$ is the acceleration in a free fall at a surface of a planet (for the Earth $g_\Sigma \approx 9.8\ \mathrm{kg/m\cdot s^2}$), and $\bar{\mathbf{e}}$ is the normal vector to the planet surface; (2) external forces of inertia caused by the motion of a body relative to a moving coordinate system (see Sect. 4.11); (3) electromagnetic forces.

External surface forces are forces of interaction of two continua contacting with each other, for example, at impact of two solid bodies.

Besides external forces, there are *internal forces* considered in continuum mechanics.

Modify Eq. (3.32). According to the rule (3.12) and the continuity equation (3.15), we have

$$\frac{d}{dt}\int\limits_V \rho\mathbf{v}\,dV = \int\limits_V \left(\frac{\partial}{\partial t}\rho\mathbf{v} + \nabla\cdot(\rho\mathbf{v}\otimes\mathbf{v})\right)dV$$
$$= \int\limits_V \left(\frac{\partial\rho}{\partial t}\mathbf{v} + \rho\frac{\partial\mathbf{v}}{\partial t} + (\nabla\cdot\rho\mathbf{v})\mathbf{v} + \rho\mathbf{v}\cdot\nabla\otimes\mathbf{v}\right)dV = \int\limits_V \rho\frac{d\mathbf{v}}{dt}\,dV. \tag{3.33}$$

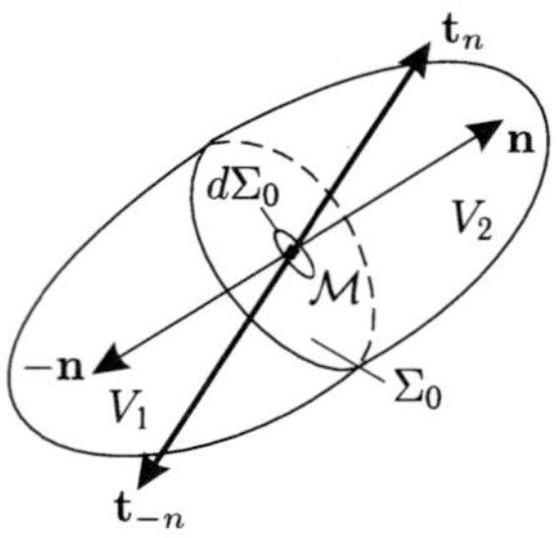


**Fig. 3.1** Internal surface forces on the surface $\Sigma_0$

Then Eq. (3.32) takes the form

$$\Omega = \int\limits_V \rho\left(\mathbf{f} - \frac{d\mathbf{v}}{dt}\right) dV + \int\limits_\Sigma \mathbf{s}\, d\Sigma = 0. \tag{3.34}$$

Hence, the *acceleration* $(d\mathbf{v}/dt)$ is the specific mass force but it is an *internal force* caused by inertia effects. Therefore, it is also called the *internal inertia force*.

Consider an example of *internal surface forces*.

Take an arbitrary continuum volume $V$ and divide the volume by a surface $\Sigma_0$ into two parts $V_1$ and $V_2$ (Fig. 3.1). Let $\mathbf{n}$ be a normal vector to $\Sigma_0$ at a point $\mathcal{M} \in \Sigma_0$, and the vector be directed outwards from $V_1$. Then each of volumes $V_1$ and $V_2$ may be considered as a separate continuum undergoing the action of the external forces $\mathcal{F}_1$ and $\mathcal{F}_2$.

If we consider the surface element $d\Sigma \in \Sigma_0$, which contains the point $\mathcal{M}$, then for the domain $V_1$, according to (3.29), the surface element $d\Sigma$ undergoes the action of the surface force $d\mathcal{F}_1$ (or the specific surface force $\mathbf{s}_1$), and for the domain $V_2$ the same surface element $d\Sigma$ undergoes the action of the surface force $d\mathcal{F}_2$ (or the specific surface force $\mathbf{s}_2$). Denote these specific surface forces as follows:

$$\mathbf{t}_n = \mathbf{s}_1 = d\mathcal{F}_1/d\Sigma \quad \text{and} \quad \mathbf{t}_{-n} = \mathbf{s}_2 = d\mathcal{F}_2/d\Sigma. \tag{3.35}$$

The vectors $\mathbf{t}_n$ and $\mathbf{t}_{-n}$ are called the *stress vectors*. They are *specific internal surface forces* relative to the whole volume $V$ of the continuum (because they are defined for interior points $\mathcal{M}$ of the continuum).

### 3.2.3 *Cauchy's Theorems on Properties of the Stress Vector*

The split of forces into external and internal ones is relative: the same forces may be internal or external with respect to different subdomains of a considered continuum.

Let us write the momentum balance equation (3.34) for the whole volume $V$ and for its separate parts $V_1$ and $V_2$:

$$\int_{V_1} \rho\left(\mathbf{f}_1 - \frac{d\mathbf{v}}{dt}\right) dV + \int_{\Sigma_1} \mathbf{s}_1 \, d\Sigma + \int_{\Sigma_0} \mathbf{t}_n \, d\Sigma = 0, \tag{3.36}$$

$$\int_{V_2} \rho\left(\mathbf{f}_2 - \frac{d\mathbf{v}}{dt}\right) dV + \int_{\Sigma_2} \mathbf{s}_2 \, d\Sigma + \int_{\Sigma_0} \mathbf{t}_{-n} \, d\Sigma = 0, \tag{3.37}$$

where $\mathbf{f}_i$ and $\mathbf{s}_i$ are the forces acting in the volumes $V_i$ and on their surfaces $\Sigma_i$, i.e.

$$\mathbf{f} = \mathbf{f}_i \text{ in } V_i \quad \text{and} \quad \mathbf{s} = \mathbf{s}_i \text{ on } \Sigma_i .$$

Since all the functions $\mathbf{s}_i$ and $\mathbf{f}_i$ are continuous, so having subtracted Eqs. (3.36) and (3.37) from (3.34), we get

$$\int_{\Sigma_0} (\mathbf{t}_n + \mathbf{t}_{-n}) \, d\Sigma = 0. \tag{3.38}$$

Since the surface $\Sigma_0$ is arbitrary, we get $\mathbf{t}_n + \mathbf{t}_{-n} = 0$. Thus, we have proved the following theorem.

**Theorem 3.5 (The first Cauchy theorem – on continuity of the stress vector).** *For the same point $\mathcal{M}$, which is an interior point of a volume $V$, the stress vectors defined relative to the surface elements $\mathbf{n}d\Sigma_0$ and $-(\mathbf{n}d\Sigma_0)$ differ only in sign:*

$$\mathbf{t}_n = -\mathbf{t}_{-n}, \tag{3.39}$$

*i.e. the field $\mathbf{t}_n(\mathbf{x})$ is continuous in the volume $V$.*

*Remark.* The result (3.39) is a consequence of the assumption that there are no discontinuities of the functions, which was made at the derivation of Eq. (3.34). When the functions are not continuous (for example, for shock waves in a continuum), Eq. (3.39) does not hold. □

Let us consider one more important property of the stress vector. At any point $\mathcal{M}$ we may construct an elementary volume $dV$ in the form of a tetrahedron (Fig. 3.2),

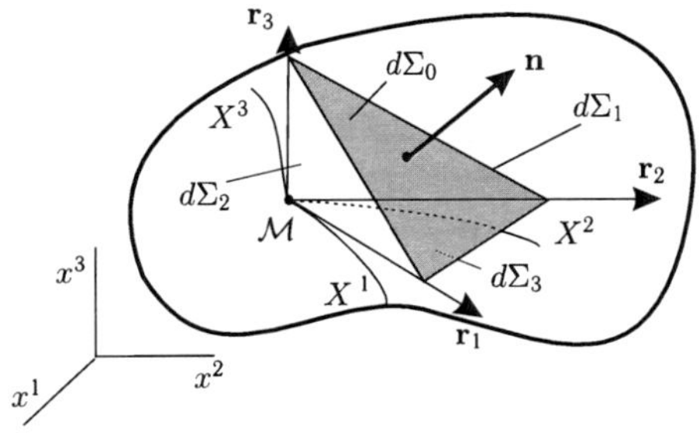


**Fig. 3.2** The properties of the internal stresses

whose edges are oriented along vectors $dx_\alpha = \mathbf{r}_\alpha dX^\alpha$. Let $d\Sigma_\alpha$ be areas of three sides lying on the coordinate planes, and $d\Sigma_0$ be the area of the inclined side of the tetrahedron.

The exterior normal vector to the surface element $d\Sigma_0$ is denoted by $\mathbf{n}$. On the surfaces $d\Sigma_\alpha$, the normal vectors are defined as $-(\mathbf{r}^\alpha/|\mathbf{r}^\alpha|)$, because the reciprocal basis vectors $\mathbf{r}^\alpha$ are orthogonal to the planes $d\Sigma_\alpha$.

Areas of the sides $d\Sigma_0$ and $d\Sigma_\alpha$ are connected by the relation

$$\mathbf{n}\, d\Sigma_0 - \sum_{\alpha=1}^{3} \frac{\mathbf{r}^\alpha}{|\mathbf{r}^\alpha|}\, d\Sigma_\alpha = 0. \tag{3.40}$$

This relation follows, for example, from the equation

$$\int_\Sigma \mathbf{n}\, d\Sigma = \int_\Sigma \mathbf{n} \cdot \mathbf{E}\, d\Sigma = \int_V \nabla \cdot \mathbf{E}\, dV = 0, \tag{3.41}$$

applied to the tetrahedron, where $\Sigma$ and $V$ are the total area and the total volume of the tetrahedron, respectively. Here we have used the Gauss–Ostrogradskii formula (3.23).

The scalar product of (3.40) by $\mathbf{r}_\alpha$ yields

$$|\mathbf{r}^\alpha| \mathbf{n} \cdot \mathbf{r}_\alpha\, d\Sigma_0 = d\Sigma_\alpha. \tag{3.42}$$

Having applied Eq. (3.34) to the tetrahedron, we get

$$\mathbf{t}_n\, d\Sigma_0 + \sum_{\alpha=1}^{3} \mathbf{t}_{-\alpha}\, d\Sigma_\alpha + \rho \left( \mathbf{f} - \frac{d\mathbf{v}}{dt} \right) dV = 0, \tag{3.43}$$

where $\mathbf{t}_\alpha$ is the stress vector on the surface element $d\Sigma_\alpha$ with the normal $(\mathbf{r}^\alpha/|\mathbf{r}^\alpha|)$, and $\mathbf{t}^{-\alpha}$ – with the normal $(-\mathbf{r}^\alpha/|\mathbf{r}^\alpha|)$. Taking formulae (3.39) and (3.42) into account, we have

$$\mathbf{t}_n - \sum_{\alpha=1}^{3} \mathbf{n} \cdot \mathbf{r}_\alpha |\mathbf{r}^\alpha| \mathbf{t}_\alpha = -\rho \left( \mathbf{f} - \frac{d\mathbf{v}}{dt} \right) \frac{|dV|}{d\Sigma_0}. \tag{3.44}$$

Since $dV/d\Sigma_0$ is an infinitesimal value, we get the following theorem.

**Theorem 3.6 (The second Cauchy theorem).** *The stress vector* $\mathbf{t}_n$ *on an arbitrary surface element with a normal* $\mathbf{n}$ *is expressed in terms of stress vectors* $\mathbf{t}_\alpha$ *on three coordinate areas as follows:*

$$\mathbf{t}_n = \sum_{\alpha=1}^{3} \mathbf{n} \cdot \mathbf{r}_\alpha |\mathbf{r}^\alpha| \mathbf{t}_\alpha. \tag{3.45}$$

Theorems 3.5 and 3.6 have been proved by Cauchy in another form.
Since

$$|\mathbf{r}^\alpha| = (\mathbf{r}^\alpha \cdot \mathbf{r}^\alpha)^{1/2} = \sqrt{g^{\alpha\alpha}}, \tag{3.46}$$

Equation (3.45) takes the form

$$\mathbf{t}_n = \mathbf{n} \cdot \mathbf{T}, \tag{3.47}$$

where $\mathbf{T}$ is the second-order tensor called the *Cauchy stress tensor*:

$$\mathbf{T} = \sum_{\alpha=1}^{3} \mathbf{r}_\alpha \otimes \mathbf{t}^\alpha = \mathbf{r}_i \otimes \mathbf{t}^i, \tag{3.48}$$

$$\mathbf{t}^\alpha \equiv \mathbf{t}_\alpha \sqrt{g^{\alpha\alpha}}. \tag{3.49}$$

Thus, Theorem 3.6 may be enunciated in another way.

**Theorem 3.6a (The Cauchy theorem).** *For a continuous in $V \cup \Sigma$ stress vector field $\mathbf{s}(\mathbf{x}) = \mathbf{t}_n(\mathbf{x})$ satisfying Eq. (3.32), there always exists a tensor field $\mathbf{T}(\mathbf{x})$ satisfying the relation (3.47) in $V \cup \Sigma$.*

### *3.2.4 Generalized Cauchy's Theorem*

Theorem 3.6a may be stated for an arbitrary vector or even tensor field $\mathbf{\Phi}(\mathbf{x}, t)$ satisfying the equation, which is similar to (3.32).

**Theorem 3.6b.** *Let there be a continuously differentiable tensor field ${}^m\mathbf{A}(\mathbf{x}, t)$, $m \geqslant 0$, and a continuous tensor field ${}^m\mathbf{C}(\mathbf{x}, t)$ in a volume $V$, and there also be a continuous in $V \cup \Sigma$ vector field ${}^m\mathbf{B}_n(\mathbf{x}, t)$ depending on the choice of normal vector field $\mathbf{n}(\mathbf{x}, t)$, which satisfy the equation*

$$\frac{d}{dt}\int_{\widetilde{V}} \rho\, {}^m\mathbf{A}\, dV = \int_{\widetilde{V}} \rho\, {}^m\mathbf{C}\, dV + \int_{\widetilde{\Sigma}} {}^m\mathbf{B}_n\, d\Sigma \quad \forall \widetilde{V} \subset V, \tag{3.50}$$

*where $\widetilde{\Sigma}$ is the boundary of the domain $\widetilde{V}$, then the field ${}^m\mathbf{B}_n(\mathbf{x}, t)$ may depend on the field $\mathbf{n}(\mathbf{x}, t)$ only linearly, i.e. there exists a tensor field ${}^{m+1}\mathbf{B}(\mathbf{x}, t)$ of the $(m+1)$th order, such that in $V \cup \Sigma$ the following relation holds:*

$${}^m\mathbf{B}_n = \mathbf{n} \cdot {}^{m+1}\mathbf{B}. \tag{3.51}$$

▼ A proof of this theorem is similar to the proof of Theorem 3.6, where as the domain $\widetilde{V}$ we should take the tetrahedron, whose edges are oriented along the local basis vectors $\mathbf{r}_i$ (see Exercise 3.2.3). ▲

Equation (3.50) is called the *balance equation*.

### 3.2.5 The Cauchy and Piola–Kirchhoff Stress Tensors

The Cauchy stress tensor $\mathbf{T}$ defined by formulae (3.47)–(3.49), as a second-order tensor, can be resolved for any tensor basis, for example:

$$\mathbf{T} = T^{ij}\mathbf{r}_i \otimes \mathbf{r}_j = T_{ij}\mathbf{r}^i \otimes \mathbf{r}^j. \tag{3.52}$$

The definition (3.48) allows us to give a geometric representation of the Cauchy stress tensor. Indeed, if we take vectors $\mathbf{r}_i$ as the left vectors and consider $\mathbf{t}^i$ as the right vectors, then we can represent the tensor $\mathbf{T}$ in terms of equivalency classes (see Sect. 2.1.4 and [12]):

$$\mathbf{T} = \mathbf{r}_i \otimes \mathbf{t}^i = [\mathbf{r}_1\mathbf{t}^1\mathbf{r}_2\mathbf{t}^2\mathbf{r}_3\mathbf{t}^3]. \tag{3.53}$$

According to the geometric definition of a tensor (see [12]), the tensor $\mathbf{T}$ can be represented as the ordered set of six vectors $\mathbf{r}_i, \mathbf{t}^i$ with the common origin at a considered point $\mathcal{M}$ (Fig. 3.3), where the basis vectors $\mathbf{r}_i$ are defined.

The Cauchy stress tensor $\mathbf{T}$ is defined on a deformable surface element $d\Sigma$ in the actual configuration. We can determine the stress tensor on the corresponding nondeformed surface element $d\overset{\circ}{\Sigma}$ in the reference configuration as well. To do this, write out the relation (2.116) which connects oriented surface elements in $\mathcal{K}$ and $\overset{\circ}{\mathcal{K}}$:

$$\mathbf{n}\, d\Sigma = \sqrt{g/\overset{\circ}{g}}\, \overset{\circ}{\mathbf{n}} \cdot \mathbf{F}^{-1}\, d\overset{\circ}{\Sigma}_0, \tag{3.54}$$

and consider the stress vector $\mathbf{t}_n$ on the surface element $d\Sigma$:

$$\mathbf{t}_n\, d\Sigma = \mathbf{n} \cdot \mathbf{T}\, d\Sigma = \sqrt{g/\overset{\circ}{g}}\, \overset{\circ}{\mathbf{n}} \cdot \mathbf{F}^{-1} \cdot \mathbf{T}\, d\overset{\circ}{\Sigma} = \overset{\circ}{\mathbf{n}} \cdot \mathbf{P}\, d\overset{\circ}{\Sigma} = \overset{\circ}{\mathbf{t}}_n\, d\overset{\circ}{\Sigma}. \tag{3.55}$$

Here we have introduced the tensor

$$\mathbf{P} = \sqrt{g/\overset{\circ}{g}}\, \mathbf{F}^{-1} \cdot \mathbf{T}, \tag{3.56}$$

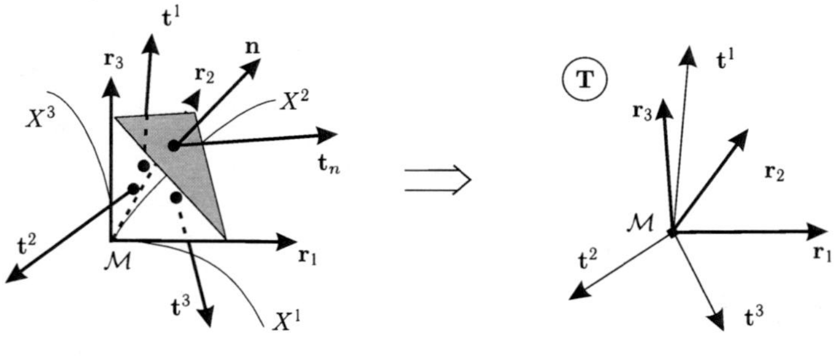


**Fig. 3.3** Geometric representation of the Cauchy stress tensor

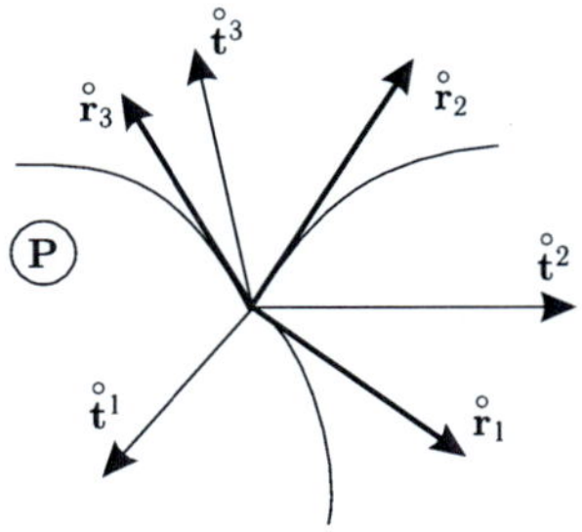


**Fig. 3.4** Geometric representation of the Piola–Kirchhoff stress tensor **P**

called the *first Piola–Kirchhoff stress tensor*, which is defined on the nondeformed surface element $d\overset{\circ}{\Sigma}$.

The vector $\overset{\circ}{\mathbf{t}}_n$ is called the *Piola–Kirchhoff stress vector*; this vector is connected to the tensor **P** by the Cauchy formula (3.47):

$$\overset{\circ}{\mathbf{t}}_n = \overset{\circ}{\mathbf{n}} \cdot \mathbf{P}. \tag{3.57}$$

With the help of the relations (2.35) and (3.48), the expression (3.56) can be represented in terms of equivalency classes (see Sect. 2.1.4)

$$\mathbf{P} = \overset{\circ}{\mathbf{r}}_i \otimes \overset{\circ}{\mathbf{t}}{}^i = [\overset{\circ}{\mathbf{r}}_1\overset{\circ}{\mathbf{t}}{}^1\overset{\circ}{\mathbf{r}}_2\overset{\circ}{\mathbf{t}}{}^2\overset{\circ}{\mathbf{r}}_3\overset{\circ}{\mathbf{t}}{}^3], \tag{3.58}$$

where the following vectors have been denoted

$$\overset{\circ}{\mathbf{t}}{}^\alpha = \mathbf{t}_\alpha \sqrt{g^{\alpha\alpha} g/\overset{\circ}{g}}. \tag{3.59}$$

With the help of this expression we can also give a geometric representation of the Piola–Kirchhoff stress tensor in the basis $\overset{\circ}{\mathbf{r}}_i$ of the reference configuration (Fig. 3.4).

### 3.2.6 Physical Meaning of Components of the Cauchy Stress Tensor

Formulae (3.53) and (3.58) give a geometric representation of the tensors **T** and **P**. Let us explain a physical meaning of components $\widehat{T}^{ij}$ of the Cauchy stress tensor with respect to an orthonormal basis.

Let there be a local basis $\mathbf{r}_i$ in $\mathcal{K}$, then, using the orthogonalization process, we can always construct an orthogonal basis $\widetilde{\mathbf{r}}_i$, and, normalizing the vectors $\widetilde{\mathbf{r}}_i$, we obtain the orthonormal basis $\widehat{\mathbf{r}}_i$ called the *physical basis*. In this basis the Cauchy stress tensor (3.52) has components $\widehat{T}^{ij}$, which are coincident with $\widehat{T}_{ij}$:

$$\mathbf{T} = \widehat{T}^{ij}\widehat{\mathbf{r}}_i \otimes \widehat{\mathbf{r}}_j. \tag{3.60}$$

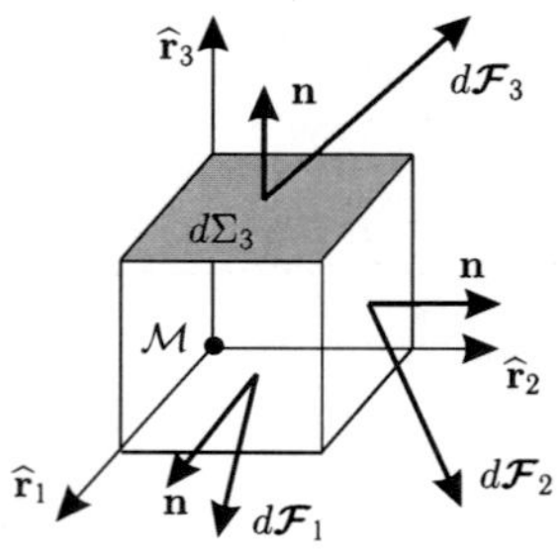


**Fig. 3.5** Physical meaning of components of the Cauchy stress tensor

Take an arbitrary material point $\mathcal{M} \in V$ and consider an elementary volume $dV \in V$ involving this point and having the form of a cube whose sides are orthogonal to the vectors $\widehat{\mathbf{r}}_\alpha$ (Fig. 3.5). Denote the area of the cube side orthogonal to $\widehat{\mathbf{r}}_\alpha$ ($\alpha = 1, 2, 3$) by $d\Sigma_\alpha$. Since we consider the elementary volume $dV$, the stress tensor $\mathbf{T}(\mathbf{x})$ is the same at every point $\mathbf{x} \in dV$ and $\mathbf{x} \in d\Sigma_\alpha$ ($\alpha = 1, 2, 3$).

Each of the sides $d\Sigma_\alpha$ undergoes the action of the surface force $d\mathcal{F}_\alpha$ connected to the stress vector $\mathbf{t}_n$ on this side by the relation (3.35):

$$\text{on } d\Sigma_\alpha : \quad \frac{d\mathcal{F}_\alpha}{d\Sigma_\alpha} = \mathbf{t}_n = \mathbf{n} \cdot \mathbf{T}. \tag{3.61}$$

Resolving the vectors $d\mathcal{F}_\alpha$ and $\mathbf{n}$ for the basis $\widehat{\mathbf{r}}_i$:

$$d\mathcal{F}_\alpha = d\mathcal{F}^i_\alpha\, \widehat{\mathbf{r}}_i, \qquad \mathbf{n} = \widehat{n}^i\, \widehat{\mathbf{r}}_i, \tag{3.62}$$

from (3.61) we get

$$\text{on } d\Sigma_\alpha : \quad \frac{d\mathcal{F}^i_\alpha}{d\Sigma_\alpha} = \widehat{n}_j \widehat{T}^{ji} = \widehat{T}^{\alpha i}, \qquad \alpha = 1, 2, 3, \tag{3.63}$$

because on $d\Sigma_\alpha$: $\widehat{n}_\alpha = 1, \widehat{n}_\beta = \widehat{n}_\gamma = 0, \alpha \neq \beta \neq \gamma \neq \alpha$.

Formula (3.63) allows us to explain the physical meaning of components $\widehat{T}^{ij}$ of the Cauchy stress tensor:

$$\text{on } d\Sigma_\alpha : \ \widehat{T}^{\alpha\alpha} = \frac{d\mathcal{F}^\alpha_\alpha}{d\Sigma_\alpha}, \ \widehat{T}^{\alpha\beta} = \frac{d\mathcal{F}^\beta_\alpha}{d\Sigma_\alpha}, \ \widehat{T}^{\alpha\gamma} = \frac{d\mathcal{F}^\gamma_\alpha}{d\Sigma_\alpha}, \ \alpha \neq \beta \neq \gamma \neq \alpha. \tag{3.64}$$

The *normal stress* $\widehat{T}^{\alpha\alpha}$ is the ratio of the corresponding normal component $d\mathcal{F}^\alpha_\alpha$ of the surface force $d\mathcal{F}_\alpha$, acting on the surface element $d\Sigma_\alpha$, to the area of this surface element; and the *tangential stresses* $\widehat{T}^{\alpha\beta}$, $\widehat{T}^{\alpha\gamma}$ are the ratios of tangential components $d\mathcal{F}^\beta_\alpha$, $d\mathcal{F}^\gamma_\alpha$ of the same force $d\mathcal{F}_\alpha$, acting on the surface element $d\Sigma_\alpha$, to the area of the same surface element.

*Remark 1.* Since the tensor $\mathbf{T}$ is the same in the whole cube $dV$, by Theorem 3.5, the forces $d\mathcal{F}_\alpha$ on the opposite sides of the cube differ only in sign. But the normals $\mathbf{n}$

on these sides differ also only in sign, therefore relations (3.61) on these sides are the same as well as the relations (3.64). □

*Remark 2.* Due to Remark 1, there are three different relations in formulae (3.64) on each of three sides $d\Sigma_\alpha$, i.e. we have nine different relations: for three normal stresses $\widehat{T}^{\alpha\alpha}$ and six tangential stresses $\widehat{T}^{\alpha\beta}$, $\alpha \neq \beta$. The pair tangential stresses $\widehat{T}^{\alpha\beta}$ and $\widehat{T}^{\beta\alpha}$, defined on different sides $d\Sigma_\alpha$ and $d\Sigma_\beta$, in general, are not coincident (Fig. 3.6).

Below, we will know that for many problems of continuum mechanics the stress tensor proves to be symmetric $\mathbf{T} = \mathbf{T}^{\mathrm{T}}$ (due to additional assumptions); it is the only case when pair tangential stresses coincide: $\widehat{T}^{\alpha\beta} = \widehat{T}^{\beta\alpha}$. □

Let us clarify now the physical meaning of components $\widehat{P}^{ij}$ of the Piola–Kirchhoff stress tensor $\mathbf{P}$ with respect to the orthonormal basis $\hat{\mathring{\mathbf{r}}}_i$ of the reference configuration:

$$\mathbf{P} = \widehat{P}^{ij}\hat{\mathring{\mathbf{r}}}_i \otimes \hat{\mathring{\mathbf{r}}}_j. \tag{3.65}$$

Take a material point $\mathcal{M}$ in $\mathring{\mathcal{K}}$ and consider an elementary volume $d\mathring{V}$ being a cube and containing this point. Let sides of the cube $d\mathring{\Sigma}_\alpha$ be orthogonal to the vectors $\hat{\mathring{\mathbf{r}}}_\alpha$ (Fig. 3.7).

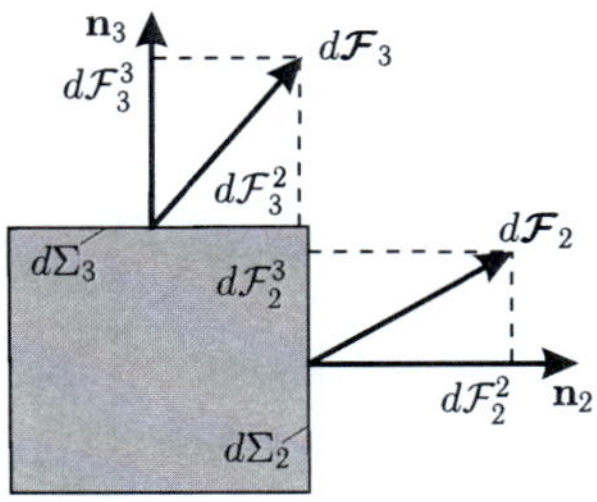


**Fig. 3.6** Difference of pair tangential stresses

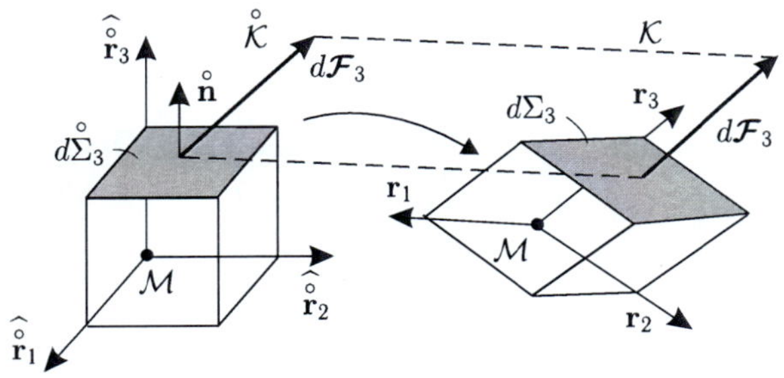


**Fig. 3.7** Physical meaning of components of the Piola–Kirchhoff stress tensor

In configuration $\mathcal{K}$ a distorted volume $dV$ corresponds to the cube $d\overset{\circ}{V}$. According to the geometric representation of transformation of a small neighborhood of a material point (see Sect. 2.3.5), the volume $dV$ is a parallelepiped, in general, with inclined sides $d\Sigma_\alpha$ remaining plane.

For elementary volumes $d\overset{\circ}{V}$ and $dV$ the stress tensors $\mathbf{T}(\mathbf{x})$ and $\mathbf{P}(\overset{\circ}{\mathbf{x}})$ are the same at every point $\mathbf{x} \in d\overline{V}$ and $\overset{\circ}{\mathbf{x}} \in d\overset{\circ}{\overline{V}}$.

The deformed (but plane) side $d\Sigma_\alpha$ undergoes the action of the surface force $d\mathcal{F}_\alpha$. Carry the force in the parallel way in $\overset{\circ}{\mathcal{K}}$ onto the corresponding side $d\overset{\circ}{\Sigma}_\alpha$, then the relations (3.55) on the side $d\overset{\circ}{\Sigma}_\alpha$ take the form

$$\text{on } d\overset{\circ}{\Sigma}_\alpha: \quad \frac{d\mathcal{F}_\alpha}{d\Sigma_\alpha}\left(\frac{d\Sigma_\alpha}{d\overset{\circ}{\Sigma}_\alpha}\right) = \mathbf{t}_n \left(\frac{d\Sigma_\alpha}{d\overset{\circ}{\Sigma}_\alpha}\right) = \overset{\circ}{\mathbf{n}} \cdot \mathbf{P}. \tag{3.66}$$

Resolving the vectors $d\mathcal{F}_\alpha$ and $\overset{\circ}{\mathbf{n}}$ for the basis $\widehat{\overset{\circ}{\mathbf{r}}}_i$:

$$d\mathcal{F}_\alpha = d\overset{\circ}{\mathcal{F}}{}^i_\alpha \widehat{\overset{\circ}{\mathbf{r}}}_i, \qquad \overset{\circ}{\mathbf{n}} = \widehat{\overset{\circ}{n}}{}^i\, \widehat{\overset{\circ}{\mathbf{r}}}_i,$$

from (3.66) we get

$$\text{on } d\overset{\circ}{\Sigma}_\alpha: \quad \frac{d\overset{\circ}{\mathcal{F}}{}^i_\alpha}{d\Sigma_\alpha} = \widehat{\overset{\circ}{n}}_i \widehat{P}^{ij} = \widehat{P}^{\alpha i}, \quad \alpha = 1, 2, 3; \tag{3.67}$$

because on $\overset{\circ}{\Sigma}_\alpha$: $\widehat{\overset{\circ}{n}}_\alpha = 1$, and $\widehat{\overset{\circ}{n}}_\beta = \widehat{\overset{\circ}{n}}_\gamma = 0$.

As follows from (3.67),

$$\text{on } d\overset{\circ}{\Sigma}_\alpha: \quad \widehat{P}^{\alpha\alpha} = \frac{d\overset{\circ}{\mathcal{F}}{}^\alpha_\alpha}{d\overset{\circ}{\Sigma}_\alpha}, \quad \widehat{P}^{\alpha\beta} = \frac{d\overset{\circ}{\mathcal{F}}{}^\beta_\alpha}{d\overset{\circ}{\Sigma}_\alpha}, \quad \widehat{P}^{\alpha\gamma} = \frac{d\overset{\circ}{\mathcal{F}}{}^\gamma_\alpha}{d\overset{\circ}{\Sigma}_\alpha}, \quad \alpha \neq \beta \neq \gamma \neq \alpha. \tag{3.68}$$

Thus, according to (3.68), the *normal stress* $\widehat{P}^{\alpha\alpha}$ is the ratio of normal component $d\overset{\circ}{\mathcal{F}}{}^\alpha_\alpha$ of the surface force $d\mathcal{F}_\alpha$, acting on deformed surface element $d\Sigma_\alpha$, to the area of corresponding undistorted surface element $d\overset{\circ}{\Sigma}_\alpha$. The *tangential stresses* $\widehat{P}^{\alpha\beta}$, $\widehat{P}^{\alpha\gamma}$ are the ratios of tangential components $d\overset{\circ}{\mathcal{F}}{}^\beta_\alpha$, $d\overset{\circ}{\mathcal{F}}{}^\gamma_\alpha$ of the same surface force $d\mathcal{F}_\alpha$, acting on the deformed surface element $d\Sigma_\alpha$, to the area of undistorted surface element $d\overset{\circ}{\Sigma}_\alpha$.

Remarks 1 and 2 for the tensor $\mathbf{P}$ are also valid with the exception of the fact that the tensor $\mathbf{P}$ remains nonsymmetric even if $\mathbf{T}$ is symmetric.

### 3.2.7 The Momentum Balance Equation in Spatial and Material Descriptions

Substituting the notation (3.35) for internal surface forces and Cauchy's formula (3.47) into (3.32), we get one more widely used form of the momentum balance law:

$$\frac{d}{dt}\int\limits_{V} \rho\mathbf{v}\, dV = \int\limits_{\Sigma} \mathbf{n}\cdot\mathbf{T}\, d\Sigma + \int\limits_{V} \rho\mathbf{f}\, dV. \tag{3.69}$$

Having substituted (3.35) and (3.47) into (3.34):

$$\int\limits_{V} \left(\rho\frac{d\mathbf{v}}{dt} - \rho\mathbf{f}\right) dV = \int\limits_{\Sigma} \mathbf{n}\cdot\mathbf{T}\, d\Sigma, \tag{3.70}$$

and then having transformed the surface integral into the volume one by the Gauss–Ostrogradskii formula (3.24), we obtain

$$\int\limits_{V} \left(\rho\frac{d\mathbf{v}}{dt} - \rho\mathbf{f} - \nabla\cdot\mathbf{T}\right) dV = 0. \tag{3.71}$$

Since the volume $V$ is arbitrary, the integrand must always vanish. Thus, we have proved the following theorem.

**Theorem 3.7.** *If the functions* $\mathbf{F}$, $\mathbf{v}$, $\mathbf{T}$ *and* $\mathbf{f}$, *which satisfy the momentum balance law (3.69) and depend on* $x^i, t$, *are continuously differentiable in* $V(t)$ *for all considered times* $t \geqslant 0$, *then at every point* $\mathcal{M} \in V(t)$ *the following momentum balance equation in Eulerian description (i.e. in* $\mathcal{K}$*) holds:*

$$\rho\frac{d\mathbf{v}}{dt} = \nabla\cdot\mathbf{T} + \rho\mathbf{f}. \tag{3.72}$$

According to the property (3.20) of the total derivative, we can rewrite the momentum balance equation in the *divergence form*:

$$\frac{\partial\rho\mathbf{v}}{\partial t} + \nabla\cdot\rho\mathbf{v}\otimes\mathbf{v} = \nabla\cdot\mathbf{T} + \rho\mathbf{f}. \tag{3.73}$$

Modify the momentum balance equation (3.70) with taking (3.55) into account:

$$\int\limits_{\overset{\circ}{V}} \overset{\circ}{\rho}\left(\frac{d\mathbf{v}}{dt} - \mathbf{f}\right) d\overset{\circ}{V} - \int\limits_{\overset{\circ}{\Sigma}} \overset{\circ}{\mathbf{n}}\cdot\mathbf{P}\, d\overset{\circ}{\Sigma} = 0. \tag{3.74}$$

With the help of the Gauss–Ostrogradskii theorem (3.24), we can rewrite this equation in the form

$$\int\limits_{\overset{\circ}{V}} \left( \overset{\circ}{\rho} \left( \frac{d\mathbf{v}}{dt} - \mathbf{f} \right) - \overset{\circ}{\nabla} \cdot \mathbf{P} \right) d\overset{\circ}{V} = 0. \tag{3.75}$$

Since the volume $\overset{\circ}{V}$ is arbitrary, we obtain the following theorem.

**Theorem 3.8.** *If the conditions of Theorem 3.7 are satisfied, then at every point* $\mathcal{M} \in \overset{\circ}{V}$ *for all considered times* $t \geqslant 0$ *the following momentum balance equation in Lagrangian description (i.e. in* $\overset{\circ}{\mathcal{K}}$*) holds:*

$$\overset{\circ}{\rho} \frac{d\mathbf{v}}{dt} = \overset{\circ}{\rho}\mathbf{f} + \overset{\circ}{\nabla} \cdot \mathbf{P}. \tag{3.76}$$

Since in Lagrangian description the total derivative with respect to time coincides with the partial one (see (2.189)), the divergence form of the momentum balance equation in $\overset{\circ}{\mathcal{K}}$ coincides with (3.76):

$$\overset{\circ}{\rho} \frac{\partial \mathbf{v}}{\partial t} = \overset{\circ}{\nabla} \cdot \mathbf{P} + \overset{\circ}{\rho}\mathbf{f}. \tag{3.77}$$

## Exercises for 3.2

**3.2.1.** Using formulae (2.35a) and (3.56), show that the Piola–Kirchhoff tensor $\mathbf{P}$ has the components $((\overset{\circ}{\rho}/\rho)T^{ij})$ with respect to the mixed dyadic basis $\overset{\circ}{\mathbf{r}}_i \otimes \mathbf{r}_j$:

$$\mathbf{P} = (\overset{\circ}{\rho}/\rho) T^{ij} \overset{\circ}{\mathbf{r}}_i \otimes \mathbf{r}_j.$$

**3.2.2.** Using the results of Exercises 3.2.1 and 2.1.12, show that components $\overset{\circ}{P}{}^{ij}$ of the tensor $\mathbf{P}$ with respect to the reference configuration basis can be written in the form

$$\mathbf{P} = \overset{\circ}{P}{}^{ij} \overset{\circ}{\mathbf{r}}_i \otimes \overset{\circ}{\mathbf{r}}_j, \quad \mathbf{F} = \overset{\circ}{F}{}^j{}_m \mathbf{r}_j \otimes \overset{\circ}{\mathbf{r}}{}^m, \quad \overset{\circ}{P}{}^{ij} = (\overset{\circ}{\rho}/\rho) T^{im} \overset{\circ}{F}{}^j{}_m.$$

**3.2.3.** Prove Theorem 3.6b by the same method as we used in the proof of Theorem 3.6.

## 3.3 The Angular Momentum Balance Law

### 3.3.1 The Integral Form

Consider the next fundamental law of continuum mechanics.

**Definition 3.4.** The vectors

$$\mathbf{k}' = \int_V \mathbf{x} \times \mathbf{v}\, dm = \int_V \mathbf{x} \times \rho \mathbf{v}\, dV,$$
$$\boldsymbol{\mu}'_m = \int_V \mathbf{x} \times \mathbf{f}\, dm = \int_V \mathbf{x} \times \rho \mathbf{f}\, dV, \tag{3.78}$$
$$\boldsymbol{\mu}'_\Sigma = \int_\Sigma \mathbf{x} \times \mathbf{t}_n\, d\Sigma, \quad \boldsymbol{\mu}' = \boldsymbol{\mu}'_m + \boldsymbol{\mu}'_\Sigma,$$

are called as follows: $\mathbf{k}'$ is the angular momentum vector of a continuum, $\boldsymbol{\mu}'_m$ is the vector of mass moments of a continuum, $\boldsymbol{\mu}'_\Sigma$ is the vector of surface moments of a continuum, $\boldsymbol{\mu}'$ is the vector of moments of a continuum.

**Axiom 6 (The angular momentum balance law).** *For any continuum $\mathcal{B}$ there are two additive vector-functions: $\mathbf{k}''(\mathcal{B}, t)$ called the own angular momentum vector and $\boldsymbol{\mu}''(\mathcal{B}, t)$ called the vector of own moments, and also the positive constant $\gamma_s > 0$ called the spin constant, such that $\forall t \geqslant 0$ the following equation holds:*

$$\frac{d\mathbf{k}}{dt} = \boldsymbol{\mu}, \tag{3.79}$$

*where $\mathbf{k}$ is the vector of the total angular momentum of the continuum, and $\boldsymbol{\mu}$ is the total moment vector:*

$$\mathbf{k} = \mathbf{k}' + \frac{1}{\gamma_s}\mathbf{k}'', \qquad \boldsymbol{\mu} = \boldsymbol{\mu}' + \boldsymbol{\mu}''.$$

*Remark.* Axiom 6 postulates the existence of two new vectors: $\mathbf{k}''$ and $\boldsymbol{\mu}''$, similarly to Axiom 5 introducing the force vector $\mathcal{F}$. Physical interpretation of the vectors $\mathbf{k}''$ and $\boldsymbol{\mu}''$ is more complicated than for the force $\mathcal{F}$ (for example, effects caused by the gravity force or the inertia force are intuitively clear and well known). The appearance of the vectors $\mathbf{k}''$ and $\boldsymbol{\mu}''$ can be caused by electromagnetic effects in continua, which have an ordered magnetic structure (for example, ferromagnetic materials). However, the vectors $\mathbf{k}''$ and $\boldsymbol{\mu}''$ can appear for some continua, where there are no electromagnetic effects (such continua are often called the *Cosserat continua*). It should be noted that although the spin constant $\gamma_s$ in the law (3.79) seems to be unnecessary (in place of $(1/\gamma_s)\mathbf{k}''$ we could introduce the vector $\mathbf{k}''$), its appearance is very essential; this will be explained in Sect. 3.4.6. □

Since the vectors $\mathbf{k}''$ and $\boldsymbol{\mu}''$ are additive, so, according to Sect. 3.2.1, we obtain the following classification.

**Definition 3.5.** The specific own angular momentum is the vector $\mathbf{k}_m$, the specific own mass moment is the vector $\mathbf{h}_m$, and the specific own surface moment is the vector $\mathbf{h}_\Sigma$, which are defined at a point $\mathcal{M}$ of a continuum as follows:

$$\mathbf{k}_m = \frac{d\mathbf{k}''}{dm}, \quad \mathbf{h}_m = \frac{d\boldsymbol{\mu}''}{dm}, \quad \mathbf{h}_\Sigma = \frac{d\boldsymbol{\mu}''}{d\Sigma}. \tag{3.80}$$

Since the vectors $\mathbf{k}''$ and $\boldsymbol{\mu}''$ are additive, for the whole considered volume of a continuum we have

$$\mathbf{k}'' = \int_V \rho \mathbf{k}_m \, dV, \qquad \boldsymbol{\mu}'' = \boldsymbol{\mu}''_m + \boldsymbol{\mu}''_\Sigma, \tag{3.81}$$

where the following notation has been introduced:

$$\boldsymbol{\mu}''_m = \int_V \rho \mathbf{h}_m \, dV, \quad \boldsymbol{\mu}''_\Sigma = \int_\Sigma \mathbf{h}_\Sigma \, d\Sigma, \tag{3.81a}$$

On substituting (3.78) and (3.81) into (3.79), the angular momentum balance law takes the integral form

$$\frac{d}{dt} \int_V \left( \mathbf{x} \times \rho \mathbf{v} + \frac{\rho}{\gamma_s} \mathbf{k}_m \right) dV = \int_V (\mathbf{x} \times \rho \mathbf{f} + \rho \mathbf{h}_m) \, dV + \int_\Sigma (\mathbf{x} \times \mathbf{t}_n + \mathbf{h}_\Sigma) \, d\Sigma. \tag{3.82}$$

### 3.3.2 Tensor of Moment Stresses

Equation (3.82) has the form of (3.50), if we introduce the following notation:

$$\mathbf{A} = \mathbf{x} \times \mathbf{v} + \mathbf{k}_m, \quad \mathbf{C} = \mathbf{x} \times \mathbf{f} + \mathbf{h}_m \quad \text{and} \quad \mathbf{B} = \mathbf{x} \times \mathbf{t}_n + \mathbf{h}_\Sigma.$$

Then to Eq. (3.82) we can apply Theorem 3.6b, which asserts the existence of a second-order tensor field $\widetilde{\mathbf{M}}(\mathbf{x}, t)$, defined in $V \cup \Sigma$, such that the following relation being an analog of (3.47) holds:

$$\mathbf{h}_\Sigma + \mathbf{x} \times \mathbf{t}_n = \mathbf{n} \cdot \widetilde{\mathbf{M}}. \tag{3.83}$$

With new notation

$$\mathbf{M} = \widetilde{\mathbf{M}} + \mathbf{T} \times \mathbf{x}, \tag{3.84}$$

this result can be written in another form.

**Theorem 3.9.** *For vector fields* $\mathbf{t}_n$ *and* $\mathbf{h}_\Sigma$ *being continuous in* $V \cup \Sigma$ *and satisfying Eqs. (3.32) and (3.82), there exists a second-order tensor field* $\mathbf{M}(\mathbf{x}, t)$ *such that in* $V \cup \Sigma$ *the following relation holds:*

$$\mathbf{h}_\Sigma = \mathbf{n} \cdot \mathbf{M}. \tag{3.85}$$

The tensor $\mathbf{M}$ is called the *tensor of moment stresses.*

### 3.3.3 Differential Form of the Angular Momentum Balance Law

At first, consider the left-hand side of (3.82) and pass from $V$ to $\overset{\circ}{V}$:

$$\begin{aligned}
&\frac{d}{dt}\int\limits_{\overset{\circ}{V}} \overset{\circ}{\rho}\Big(\mathbf{x}\times\mathbf{v} + \frac{1}{\gamma_s}\mathbf{k}_m\Big)\, d\overset{\circ}{V} \\
&= \frac{d}{dt}\int\limits_{\overset{\circ}{V}} \overset{\circ}{\rho}\Big(\mathbf{x}\times\mathbf{v} + \frac{1}{\gamma_s}\mathbf{k}_m\Big)\, d\overset{\circ}{V} = \int\limits_{\overset{\circ}{V}} \overset{\circ}{\rho}\Big(\frac{d}{dt}(\mathbf{x}\times\mathbf{v}) + \frac{1}{\gamma_s}\frac{d\mathbf{k}_m}{dt}\Big) d\overset{\circ}{V} \\
&= \int\limits_{V} \rho\Big(\mathbf{v}\times\mathbf{v} + \mathbf{x}\times\frac{d\mathbf{v}}{dt} + \frac{1}{\gamma_s}\frac{d\mathbf{k}_m}{dt}\Big) dV = \int\limits_{V} \rho\Big(\mathbf{x}\times\frac{d\mathbf{v}}{dt} + \frac{1}{\gamma_s}\frac{d\mathbf{k}_m}{dt}\Big) dV,
\end{aligned} \tag{3.86}$$

as $\mathbf{v}\times\mathbf{v} = 0$.

Modify the right-hand side of Eq. (3.82) by using the Gauss–Ostrogradskii theorem (3.24) and formula (3.47):

$$\int\limits_\Sigma \mathbf{x}\times\mathbf{t}_n\, d\Sigma = \int\limits_\Sigma \mathbf{x}\times(\mathbf{n}\cdot\mathbf{T}) d\Sigma = -\int\limits_\Sigma \mathbf{n}\cdot(\mathbf{T}\times\mathbf{x})\, d\Sigma = -\int\limits_V \nabla\cdot(\mathbf{T}\times\mathbf{x})\, dV. \tag{3.87}$$

The nabla-operator $\nabla$ in (3.87) is determined by the formula

$$\begin{aligned}
\nabla\cdot(\mathbf{T}\times\mathbf{x}) &= \mathbf{r}^i\cdot\frac{\partial}{\partial X^i}(\mathbf{T}\times\mathbf{x}) = (\nabla\cdot\mathbf{T})\times\mathbf{x} + \mathbf{r}^i\cdot\mathbf{T}\times\mathbf{r}_i \\
&= -\mathbf{x}\times(\nabla\cdot\mathbf{T}) + \boldsymbol{\varepsilon}\cdot\cdot\,\mathbf{T}.
\end{aligned} \tag{3.88}$$

Substitution of (3.88) into (3.87) yields

$$\int\limits_\Sigma \mathbf{x}\times\mathbf{t}_n\, d\Sigma = \int\limits_V \mathbf{x}\times\nabla\cdot\mathbf{T}\, dV - \int\limits_V \boldsymbol{\varepsilon}\cdot\cdot\,\mathbf{T}\, dV. \tag{3.89}$$

Having substituted (3.89) and (3.86) into (3.82), we get

$$\int\limits_V \rho \mathbf{x} \times \left( \frac{d\mathbf{v}}{dt} - \mathbf{f} - \frac{1}{\rho} \nabla \cdot \mathbf{T} \right) dV + \int\limits_V \frac{\rho}{\gamma_s} \frac{d\mathbf{k}_m}{dt}\, dV$$

$$= \int\limits_V \rho \mathbf{h}_m \, dV + \int\limits_V \nabla \cdot \mathbf{M} \, dV - \int\limits_V \boldsymbol{\varepsilon} \cdot\cdot\, \mathbf{T} \, dV. \tag{3.90}$$

Here we have used the definition of the tensor of moment stresses $\mathbf{M}$ (3.84).

Due to formula (3.72), Eq. (3.90) takes the final form

$$\int\limits_V \left( \frac{\rho}{\gamma_s} \frac{d\mathbf{k}_m}{dt} - \rho \mathbf{h}_m - \nabla \cdot \mathbf{M} + \boldsymbol{\varepsilon} \cdot\cdot\, \mathbf{T} \right) dV = 0. \tag{3.91}$$

Since the volume $V$ is arbitrary, we get the following theorem.

**Theorem 3.10.** *If the functions* $\mathbf{F}$, $\mathbf{v}$, $\mathbf{T}$, $\mathbf{f}$, *satisfying Eq. (3.69), and also the functions* $\mathbf{k}_m$, $\mathbf{h}_m$, $\mathbf{M}$, *satisfying Eq. (3.90), are continuously differentiable in* $V(t)$ *for all considered times* $t \geqslant 0$, *then at every point* $\mathcal{M} \in V(t)$ *the following angular momentum balance equation holds:*

$$\frac{\rho}{\gamma_s} \frac{d\mathbf{k}_m}{dt} = \rho \mathbf{h}_m + \nabla \cdot \mathbf{M} - \boldsymbol{\varepsilon} \cdot\cdot\, \mathbf{T}. \tag{3.92}$$

Thus, the angular momentum balance equation includes only one classical mechanical characteristic, namely the stress tensor $\mathbf{T}$, the remaining functions $\mathbf{k}_m$, $\mathbf{h}_m$ and $\mathbf{M}$, as it was mentioned above, either have a nonmechanical nature or are caused by nonclassical mechanical properties.

### *3.3.4 Nonpolar and Polar Continua*

For classical continua, for which there are not any nonmechanical values, we have

$$\mathbf{k}_m = 0, \quad \mathbf{h}_m = 0, \quad \mathbf{M} = 0. \tag{3.93}$$

Such continua are called *nonpolar*, and for them it follows from (3.93) and (3.92) that

$$\boldsymbol{\varepsilon} \cdot\cdot\, \mathbf{T} = -\sqrt{g} \sum_{\substack{\alpha=1 \\ \alpha\neq\beta\neq\gamma\neq\alpha}}^{3} (T^{\alpha\beta} - T^{\beta\alpha}) \mathbf{r}^\gamma = 0, \tag{3.94}$$

and hence

$$\mathbf{T} = \mathbf{T}^{\mathrm{T}}, \tag{3.95}$$

i.e. the Cauchy stress tensor for nonpolar continua is symmetric.

It should be noted that the corresponding Piola–Kirchhoff tensor is not symmetric even for nonpolar continua; that follows from its definition (3.56).

Thus, for nonpolar continua the angular momentum balance law reduces to the condition (3.95) that the Cauchy stress tensor $\mathbf{T}$ is symmetric.

A continuum, for which values of $\mathbf{k}_m$, $\mathbf{h}_m$ and $\mathbf{M}$ are different from identically zero:

$$\mathbf{k}_m \neq 0, \quad \mathbf{h}_m \neq 0, \quad \mathbf{M} \neq 0, \tag{3.96}$$

is called *polar*. For polar continua, the Cauchy stress tensor $\mathbf{T}$ is not symmetric.

### 3.3.5 The Angular Momentum Balance Equation in the Material Description

The integral form of the angular momentum balance equation (3.82) can be represented in the material description. To do this, we should modify the left-hand side of formula (3.82) according to (3.86):

$$\frac{d}{dt}\int_V \rho\left(\mathbf{x}\times\mathbf{v}+\frac{1}{\gamma_s}\mathbf{k}_m\right)dV = \frac{d}{dt}\int_{\overset{\circ}{V}} \overset{\circ}{\rho}\left(\mathbf{x}\times\mathbf{v}+\frac{1}{\gamma_s}\mathbf{k}_m\right)d\overset{\circ}{V}$$

$$= \int_{\overset{\circ}{V}} \overset{\circ}{\rho}\left(\mathbf{x}\times\frac{d\mathbf{v}}{dt}+\frac{1}{\gamma_s}\frac{d\mathbf{k}_m}{dt}\right)d\overset{\circ}{V} \tag{3.97}$$

and introduce *Lagrangian tensor of moment stresses* $\overset{\circ}{\mathbf{M}}$ in $\overset{\circ}{\mathcal{K}}$ similarly to the Piola–Kirchhoff stress tensor:

$$\mathbf{n}\cdot\mathbf{M}\,d\Sigma = \overset{\circ}{\mathbf{n}}\cdot\overset{\circ}{\mathbf{M}}\,d\overset{\circ}{\Sigma}, \tag{3.98}$$

and the specific own surface moment $\overset{\circ}{\mathbf{h}}_\Sigma$ in $\overset{\circ}{\mathcal{K}}$:

$$\overset{\circ}{\mathbf{h}}_\Sigma = \overset{\circ}{\mathbf{n}}\cdot\overset{\circ}{\mathbf{M}}. \tag{3.99}$$

Then the surface integral in (3.82) can be written in $\overset{\circ}{\mathcal{K}}$ as follows:

$$\int_\Sigma (\mathbf{x}\times\mathbf{t}_n+\mathbf{h}_\Sigma)\,d\Sigma = \int_\Sigma \mathbf{n}\cdot(-\mathbf{T}\times\mathbf{x}+\mathbf{M})\,d\Sigma = \int_{\overset{\circ}{\Sigma}} \overset{\circ}{\mathbf{n}}\cdot(-\mathbf{P}\times\mathbf{x}+\overset{\circ}{\mathbf{M}})\,d\overset{\circ}{\Sigma}, \tag{3.100}$$

and Eq. (3.82) takes the form

$$\frac{d}{dt}\int\limits_{\mathring{V}} \mathring{\rho}\left(\mathbf{x}\times\mathbf{v}+\frac{1}{\gamma_s}\mathbf{k}_m\right)\,d\mathring{V}$$

$$\int\limits_{\mathring{V}} \mathring{\rho}\,(\mathbf{x}\times\mathbf{f}+\mathbf{h}_m)\,d\mathring{V}+\int\limits_{\mathring{\Sigma}} \mathring{\mathbf{n}}\cdot(-\mathbf{P}\times\mathbf{x}+\mathring{\mathbf{M}})\,d\mathring{\Sigma}. \qquad (3.101)$$

## Exercises for 3.3

**3.3.1.** Show that from (3.101) we can get the angular momentum balance equation in the material description

$$\frac{\mathring{\rho}}{\gamma_s}\frac{d\mathbf{k}_m}{dt}=\mathring{\rho}\mathbf{h}_m+\mathring{\nabla}\cdot\mathring{\mathbf{M}}-\boldsymbol{\varepsilon}\cdot\cdot\,(\mathbf{F}\cdot\mathbf{P}).$$

**3.3.2.** Show that for nonpolar continua the angular momentum balance equation in the material description (see Exercise 3.3.1) takes the form

$$\boldsymbol{\varepsilon}\cdot\cdot\,(\mathbf{F}\cdot\mathbf{P})=0 \quad \text{or} \quad \mathbf{F}\cdot\mathbf{P}=\mathbf{P}^{\mathrm{T}}\cdot\mathbf{F}^{\mathrm{T}}$$

and coincides exactly with (3.96).

**3.3.3.** Using equation (3.98) and the formula of transformation of an oriented surface element, show that the tensors $\mathbf{M}$ and $\mathring{\mathbf{M}}$ are connected by the relation

$$\mathring{\mathbf{M}}=\sqrt{g/\mathring{g}}\,\mathbf{F}^{-1}\cdot\mathbf{M}.$$

## 3.4 The First Thermodynamic Law

### 3.4.1 *The Integral Form of the Energy Balance Law*

The mass, momentum and angular momentum balance laws describe the motion of a continuum. If we need thermal effects to be taken into account, then we must use *thermodynamic laws.*

Let us consider nonpolar continua, for which the relations (3.93) are satisfied.

**Definition 3.6.** The scalar function

$$K=\int\limits_{V}\frac{\mathbf{v}\cdot\mathbf{v}}{2}\,dm=\int\limits_{V}\rho\frac{|v|^2}{2}\,dV, \quad |v|^2=\mathbf{v}\cdot\mathbf{v} \qquad (3.102)$$

is called the *kinetic energy* of a continuum $V$.

**Definition 3.7.** The *power of external mass forces* $W_m$ and the *power of external surface forces* $W_\Sigma$ are the following scalar functions:

$$W_m = \int_V \mathbf{f}\cdot\mathbf{v}\,dm = \int_V \rho\mathbf{f}\cdot\mathbf{v}\,dV, \quad W_\Sigma = \int_\Sigma \mathbf{t}_n\cdot\mathbf{v}\,d\Sigma. \tag{3.103}$$

**Axiom 7 (The first thermodynamic law (or the energy balance law)).** *For any continuum $\mathcal{B}$ there are two scalar additive functions: $U(\mathcal{B},t)$ called the internal energy of the continuum and $Q(\mathcal{B},t)$ called the heating rate for the continuum such that $\forall t \geqslant 0$ the following equation is satisfied:*

$$\frac{dE}{dt} = W + Q, \tag{3.104}$$

*where $E$ is called the total energy of the continuum and consists of $U$ and $K$:*

$$E = U + K, \qquad W = W_m + W_\Sigma. \tag{3.105}$$

*Remark.* There are different statements of the thermodynamic law, and the above is called Truesdell's statement [54, 55]. The statement is convenient, because its form corresponds to Axioms 4–6. Moreover, this statement (unlike others) is really universal, i.e. it is independent of the type of a continuum. □

Since the functions $U$ and $Q$ are additive, similarly to Sect. 3.2.1 we can introduce the corresponding specific functions.

**Definition 3.8.** The specific internal energy is the function $e$, the specific heat flux from mass sources is the function $q_m$ and the specific heat flux from surface sources is the function $q_\Sigma$, which are defined at every point $\mathcal{M}$ of a continuum as follows:

$$e = \frac{dU}{dm}, \quad q_m = \frac{dQ}{dm}, \quad q_\Sigma = \frac{dQ}{d\Sigma}. \tag{3.106}$$

Since the functions $Q$ and $U$ are additive, for a whole continuum we have

$$Q = Q_m + Q_\Sigma,$$
$$Q_m = \int_V q_m\,dm = \int_V \rho q_m\,dV, \quad Q_\Sigma = \int_\Sigma q_\Sigma\,d\Sigma, \tag{3.107}$$
$$U = \int_V e\,dm = \int_V \rho e\,dV.$$

Substitution of Eqs. (3.102), (3.103), (3.105)–(3.107) into (3.104) yields the *energy balance law* in the integral form

$$\frac{d}{dt}\int\limits_V \rho\left(e+\frac{|v|^2}{2}\right)\,dV=\int\limits_V \rho(\mathbf{f}\cdot\mathbf{v}+q_m)\,dV+\int\limits_\Sigma (\mathbf{t}_n\cdot\mathbf{v}+q_\Sigma)\,d\Sigma. \quad (3.108)$$

Using on the left-hand side of (3.108) the differentiation rules for a volume integral (see Exercise 3.1.2):

$$\frac{d}{dt}\int\limits_V \rho\left(e+\frac{|v|^2}{2}\right)\,dV=\int\limits_V \rho\left(\frac{de}{dt}+\mathbf{v}\cdot\frac{d\mathbf{v}}{dt}\right)\,dV, \quad (3.109)$$

we obtain the following form of the energy balance law:

$$\int\limits_V \rho\left(-\frac{d}{dt}\left(e+\frac{1}{2}|v|^2\right)+\mathbf{f}\cdot\mathbf{v}+q_m\right)\,dV+\int\limits_\Sigma (\mathbf{t}_n\cdot\mathbf{v}+q_\Sigma)\,d\Sigma=0. \quad (3.110)$$

### *3.4.2 The Heat Flux Vector*

Equation (3.108) has the form (3.50), if in (3.50) we put

$$A=e+|v^2|/2,\quad C=\mathbf{f}\cdot\mathbf{v}+q_m\quad\text{and}\quad B=\mathbf{t}_n\cdot\mathbf{v}+q_\Sigma.$$

Then we can apply Theorem 3.6b to this equation; the theorem claims the existence of the vector $(-\mathbf{q})$ such that in $V\cup\Sigma$ the following relation holds:

$$q_\Sigma=-\mathbf{n}\cdot\mathbf{q}. \quad (3.111)$$

The vector $\mathbf{q}$ is called the *heat flux vector.*

Prove the analog of Theorem 3.6 allowing us to determine components of the vector $\mathbf{q}$.

As was made in Sect. 3.2.3, at a point $\mathcal{M}$ we construct an elementary volume $dV$ in the form of a tetrahedron, whose edges are oriented along vectors $\mathbf{r}_\alpha dX^\alpha$. Then for the volume all formulae (3.40)–(3.42) hold. Applying Eq. (3.110) to the tetrahedron, when there is no action of any mass or surface forces on the continuum (i.e. $\mathbf{t}_n=0$, $\mathbf{f}=0$), we get

$$q_\Sigma\,d\Sigma_0+\sum_{\alpha=1}^{3} q_\alpha\,d\Sigma_\alpha+\rho\left(q_m-\frac{d}{dt}\left(e+\frac{|v|^2}{2}\right)\right)\,dV=0. \quad (3.112)$$

Here $q_\alpha$ is the heat flux from surface sources on the surface element $d\Sigma_\alpha$ with the normal $(-\mathbf{r}^\alpha/|\mathbf{r}^\alpha|)$, and $q_\Sigma$ – on the surface element $d\Sigma_0$.

Then with taking (3.42) into account, we have

$$q_\Sigma + \sum_{\alpha=1}^{3} \mathbf{n}\cdot\mathbf{r}_\alpha|\mathbf{r}^\alpha|q_\alpha = -\rho\left(q_m - \frac{d}{dt}\left(e + \frac{|v|^2}{2}\right)\right)\frac{dV}{d\Sigma_0}. \quad (3.113)$$

Since the value $dV/d\Sigma_0$ is infinitesimal, we get the following theorem.

**Theorem 3.11.** *On every surface element with a normal* **n**, *the heat flux from surface sources* $q_\Sigma$ *is expressed in terms of heat fluxes on three coordinate surface elements:*

$$q_\Sigma = -\sum_{\alpha=1}^{3} \mathbf{n}\cdot\mathbf{r}_\alpha\sqrt{g^{\alpha\alpha}}\, q_\alpha. \quad (3.114)$$

Comparing (3.114) with (3.111), we find components of the heat flux vector:

$$\mathbf{q} = \sum_{\alpha=1}^{3} \mathbf{r}_\alpha\sqrt{g^{\alpha\alpha}}\, q_\alpha = \mathbf{r}_i q^i, \qquad q^\alpha = q_\alpha\sqrt{g^{\alpha\alpha}}. \quad (3.114a)$$

### 3.4.3 The Energy Balance Equation

Let us consider now formula (3.110) and transform the surface integrals by the Gauss–Ostrogradskii formula (3.24) with taking (3.111) into account as follows:

$$\begin{aligned}\int_\Sigma q_\Sigma d\Sigma &= -\int_\Sigma \mathbf{n}\cdot\mathbf{q}\, d\Sigma = -\int_V \nabla\cdot\mathbf{q}\, dV,\\ \int_\Sigma \mathbf{t}_n\cdot\mathbf{v}\, d\Sigma &= \int_\Sigma \mathbf{n}\cdot(\mathbf{T}\cdot\mathbf{v})\, d\Sigma = \int_V \nabla\cdot(\mathbf{T}\cdot\mathbf{v})\, dV.\end{aligned} \quad (3.115)$$

Substituting the expression (3.115) into (3.108), we obtain another integral form of the energy balance law:

$$\frac{d}{dt}\int_V \rho\left(e + \frac{|v|^2}{2}\right) dV = \int_V \rho(\mathbf{f}\cdot\mathbf{v} + q_m)\, dV + \int_\Sigma \mathbf{n}\cdot(\mathbf{T}\cdot\mathbf{v} - \mathbf{q})\, d\Sigma. \quad (3.116)$$

Substitution of formulae (3.115) into (3.110) yields

$$\int_V \left(\rho\left(\frac{de}{dt} + \frac{d|v|^2/2}{dt}\right) - \rho\mathbf{f}\cdot\mathbf{v} - \rho q_m + \nabla\cdot\mathbf{q} - \nabla\cdot(\mathbf{T}\cdot\mathbf{v})\right) dV = 0. \quad (3.117)$$

Since the volume $V$ is arbitrary, we get the following theorem.

**Theorem 3.12.** *If the functions* $\mathbf{F}$, $\mathbf{v}$, $\mathbf{T}$, $\mathbf{f}$, $q_m$, $\mathbf{q}$ *and* $e$, *satisfying Eq. (3.116), are continuously differentiable in* $V(t)$ *for all considered times* $t \geqslant 0$, *then at every point* $\mathcal{M} \in V(t)$ *the energy balance equation has the following differential form:*

$$\rho \frac{d}{dt}\left(e + \frac{|v|^2}{2}\right) = \rho \mathbf{f} \cdot \mathbf{v} + \rho q_m - \nabla \cdot \mathbf{q} + \nabla \cdot (\mathbf{T} \cdot \mathbf{v}). \quad (3.118)$$

According to formula (3.20), we can rewrite the energy balance equation in the *divergence form*

$$\frac{\partial \rho \epsilon}{\partial t} + \nabla \cdot (\rho \epsilon \mathbf{v}) = -\nabla \cdot \mathbf{q} + \nabla \cdot (\mathbf{T} \cdot \mathbf{v}) + \rho \mathbf{f} \cdot \mathbf{v} + \rho q_m, \quad \epsilon = e + \frac{|v|^2}{2}, \quad (3.119)$$

where $\epsilon$ is the *specific total energy* of the continuum.

### 3.4.4 Kinetic Energy and Heat Influx Equation

The scalar product of the momentum balance equation (3.72) by $\mathbf{v}$ yields

$$\rho \frac{d\mathbf{v}}{dt} \cdot \mathbf{v} = \rho \mathbf{f} \cdot \mathbf{v} + (\nabla \cdot \mathbf{T}) \cdot \mathbf{v}. \quad (3.120)$$

Using the property of covariant differentiation (see Exercise 2.1.17), we can reduce this equation to the form

$$\frac{1}{2}\rho \frac{d|v|^2}{dt} = \rho \mathbf{f} \cdot \mathbf{v} + \nabla \cdot (\mathbf{T} \cdot \mathbf{v}) - \mathbf{T} \cdot\cdot (\nabla \otimes \mathbf{v})^{\mathrm{T}}. \quad (3.121)$$

**Definition 3.9.** The scalar function

$$W_{(i)} = -\int_V \mathbf{T} \cdot\cdot (\nabla \otimes \mathbf{v})^{\mathrm{T}}\, dV \quad (3.122)$$

is called the *power of internal surface forces* of a continuum $V$.

Integrating Eq. (3.121) over a volume $V$ and using Definitions 3.6, 3.7, 3.9, and the rule of differentiation of an integral over a moving volume (see Exercise 3.1.2), we get the following theorem.

**Theorem 3.13 (The kinetic energy balance equation).** *A change of the kinetic energy of a continuum is equal to the summarized power of external and internal forces acting on the continuum:*

$$\frac{dK}{dt} = W + W_{(i)}. \quad (3.123)$$

Subtracting the kinetic energy equation (3.121) from the energy balance equation (3.118), we obtain the *differential heat influx equation*

$$\rho \frac{de}{dt} = \mathbf{T} \cdot\cdot (\nabla \otimes \mathbf{v})^{\mathrm{T}} + \rho q_m - \nabla \cdot \mathbf{q}. \tag{3.124}$$

This equation can be rewritten in the divergence form

$$\frac{\partial \rho e}{\partial t} + \nabla \cdot (\rho e \mathbf{v}) = -\nabla \cdot \mathbf{q} + \mathbf{T} \cdot\cdot (\nabla \otimes \mathbf{v})^{\mathrm{T}} + \rho q_m. \tag{3.125}$$

Subtracting Eq. (3.123) from (3.104), we get the *heat influx equation in the integral form*

$$\frac{d}{dt} U + W_{(i)} = Q. \tag{3.126}$$

### 3.4.5 The Energy Balance Equation in Lagrangian Description

Let us write now the energy balance equation in Lagrangian description. Use the equation in the integral form (3.108) and transform the surface integrals as follows:

$$\int_{\Sigma} (\mathbf{t}_n \cdot \mathbf{v} + q_{\Sigma})\, d\Sigma = \int_{\Sigma} \mathbf{n} \cdot (\mathbf{T} \cdot \mathbf{v} - \mathbf{q}_{\Sigma})\, d\Sigma$$

$$= \int_{\overset{\circ}{\Sigma}} \sqrt{g/\overset{\circ}{g}}\, \overset{\circ}{\mathbf{n}} \cdot (\mathbf{F}^{-1} \cdot \mathbf{T} \cdot \mathbf{v} - \mathbf{F}^{-1} \cdot \mathbf{q}_{\Sigma})\, d\overset{\circ}{\Sigma} = \int_{\overset{\circ}{\Sigma}} \overset{\circ}{\mathbf{n}} \cdot \left(\mathbf{P} \cdot \mathbf{v} - \overset{\circ}{\mathbf{q}}\right)\, d\Sigma. \tag{3.127}$$

Here we have used the definition (3.56) of the Piola–Kirchhoff tensor and introduced the *heat flux vector* in the reference configuration:

$$\overset{\circ}{\mathbf{q}} = \sqrt{g/\overset{\circ}{g}}\, \mathbf{F}^{-1} \cdot \mathbf{q}. \tag{3.128}$$

**Theorem 3.14.** *Heat flux vectors* $\mathbf{q}$ *and* $\overset{\circ}{\mathbf{q}}$ *are connected by the relation*

$$\overset{\circ}{\mathbf{q}} \cdot \overset{\circ}{\mathbf{n}}\, d\overset{\circ}{\Sigma} = \mathbf{q} \cdot \mathbf{n}\, d\Sigma. \tag{3.129}$$

▼ Indeed, multiplying (3.128) by $\overset{\circ}{\mathbf{n}}\, d\overset{\circ}{\Sigma}$ and using formula (2.116), we obtain

$$\overset{\circ}{\mathbf{q}} \cdot \overset{\circ}{\mathbf{n}}\, d\overset{\circ}{\Sigma} = \sqrt{g/\overset{\circ}{g}}\, \mathbf{q} \cdot \mathbf{F}^{-1\mathrm{T}} \cdot \overset{\circ}{\mathbf{n}}\, d\overset{\circ}{\Sigma} = \mathbf{q} \cdot \mathbf{n}\, d\Sigma. \quad \blacktriangle$$

Then Eq. (3.108) in the reference configuration takes the form

$$\int\limits_{\overset{\circ}{V}} \left( \overset{\circ}{\rho} \frac{d}{dt} \left( e + \frac{|v|^2}{2} \right) - \overset{\circ}{\rho} \mathbf{f} \cdot \mathbf{v} - \overset{\circ}{\rho} q_m - \overset{\circ}{\nabla} \cdot (\mathbf{P} \cdot \mathbf{v}) + \overset{\circ}{\nabla} \cdot \overset{\circ}{\mathbf{q}} \right) d\overset{\circ}{V} = 0. \quad (3.130)$$

Since the volume $\overset{\circ}{V}$ is arbitrary, the integrand vanishes. Thus, we have proved the following theorem.

**Theorem 3.15.** *Under the conditions of Theorem 3.12, at every point $\mathcal{M} \in \overset{\circ}{V}$ for all considered times $t \geqslant 0$ we have the energy balance equation in Lagrangian description*

$$\overset{\circ}{\rho} \frac{d}{dt} \left( e + \frac{|v|^2}{2} \right) = \overset{\circ}{\rho} \mathbf{f} \cdot \mathbf{v} + \overset{\circ}{\rho} q_m + \overset{\circ}{\nabla} \cdot (\mathbf{P} \cdot \mathbf{v}) - \overset{\circ}{\nabla} \cdot \overset{\circ}{\mathbf{q}}. \quad (3.131)$$

Similarly to formula (3.120), we can write the kinetic energy balance equation in the reference configuration

$$\overset{\circ}{\rho} \frac{d\mathbf{v}}{dt} \cdot \mathbf{v} = \overset{\circ}{\rho} \mathbf{f} \cdot \mathbf{v} + \mathbf{v} \cdot (\overset{\circ}{\nabla} \cdot \mathbf{P}), \quad (3.132)$$

with the help of which Eq. (3.131) takes the form

$$\overset{\circ}{\rho} e \frac{d}{dt} = \mathbf{P} \cdot\cdot (\overset{\circ}{\nabla} \otimes \mathbf{v})^{\mathrm{T}} + \overset{\circ}{\rho} q_m - \overset{\circ}{\nabla} \cdot \overset{\circ}{\mathbf{q}}. \quad (3.133)$$

This equation is called the *heat influx equation* in material (Lagrangian) description.

Since in Lagrangian description the total derivative coincides with the partial one, the divergence form of the heat influx equation does not differ from (3.133):

$$\overset{\circ}{\rho} \frac{\partial e}{\partial t} = -\overset{\circ}{\nabla} \cdot \overset{\circ}{\mathbf{q}} + \mathbf{P} \cdot\cdot (\overset{\circ}{\nabla} \otimes \mathbf{v})^{\mathrm{T}} + \overset{\circ}{\rho} q_m. \quad (3.134)$$

Integrate the heat influx equation (3.133) over the domain $\overset{\circ}{V}$ and take into account that the internal energy $U$ and the heating rate $Q$, according to (3.107), can also be represented in Lagrangian description:

$$U = \int\limits_V \rho e \, dV = \int\limits_{\overset{\circ}{V}} \overset{\circ}{\rho} e \, d\overset{\circ}{V}, \quad Q = Q_m + Q_\Sigma,$$
$$Q_m = \int\limits_V \rho q_m \, dV = \int\limits_{\overset{\circ}{V}} \overset{\circ}{\rho} q_m \, d\overset{\circ}{V}, \quad Q_\Sigma = \int\limits_\Sigma q_\Sigma \, d\Sigma = \int\limits_{\overset{\circ}{\Sigma}} \overset{\circ}{q}_\Sigma \, d\overset{\circ}{\Sigma}, \quad (3.135)$$

here we have used Eq. (3.129). As a result, from (3.133) we get the heat influx equation in the integral form (3.126) again, where the power of internal surface forces is defined by the following relations in Eulerian and Lagrangian descriptions, respectively:

$$W_{(i)} = -\int\limits_V \mathbf{T}\cdot\cdot(\nabla\otimes\mathbf{v})^{\mathrm{T}}\,dV = -\int\limits_{\overset{\circ}{V}} \mathbf{P}\cdot\cdot(\overset{\circ}{\nabla}\otimes\mathbf{v})^{\mathrm{T}}\,d\overset{\circ}{V}. \quad (3.136)$$

According to relation (3.55), we can represent the power of external forces (3.103) in Lagrangian description as follows:

$$W_m = \int\limits_V \rho\mathbf{f}\cdot\mathbf{v}\,dV = \int\limits_{\overset{\circ}{V}} \overset{\circ}{\rho}\mathbf{f}\cdot\mathbf{v}\,d\overset{\circ}{V},\quad W_\Sigma = \int\limits_\Sigma \mathbf{t}_n\cdot\mathbf{v}\,d\Sigma = \int\limits_{\overset{\circ}{\Sigma}} \overset{\circ}{\mathbf{t}}_n\cdot\mathbf{v}\,d\overset{\circ}{\Sigma}. \quad (3.137)$$

### 3.4.6 The Energy Balance Law for Polar Continua

Let us consider now polar continua, for which the relations (3.96) hold, i.e. the values of $\mathbf{k}_m$, $\mathbf{h}_m$ and $\mathbf{M}$ are different from identically zero. The energy balance law for polar continua differs from the corresponding law (3.104), (3.105) for nonpolar continua.

Similarly to (3.102), introduce the *kinetic energy of the own motion* of a continuum:

$$K_s = \int\limits_V \frac{|\mathbf{k}_m|^2}{2\gamma_s}\,dm = \int\limits_V \frac{\rho}{\gamma_s}\,|\mathbf{k}_m|^2\,dV,\quad |\mathbf{k}_m|^2 = \mathbf{k}_m\cdot\mathbf{k}_m. \quad (3.138)$$

Similarly to (3.103), introduce the *power of mass moments* $W_m^s$ and the *power of surface moments* $W_\Sigma^s$:

$$W_m^s = \int\limits_V \rho\mathbf{h}_m\cdot\mathbf{k}_m\,dV, \quad (3.139)$$

$$W_\Sigma^s = \int\limits_\Sigma \mathbf{h}_\Sigma\cdot\mathbf{k}_m\,d\Sigma = \int\limits_\Sigma \mathbf{n}\cdot\mathbf{M}\cdot\mathbf{k}_m\,d\Sigma. \quad (3.140)$$

The energy balance law for a polar continuum has the universal form (3.104), but the total energy $E$ and the summarized power $W$ of a continuum are defined not by (3.105) but by the formulae

$$E = U + K + K_s,\qquad W = W_m + W_m^s + W_\Sigma + W_\Sigma^s. \quad (3.141)$$

As a result, we obtain the following form of the *energy balance law for a polar continuum*:

$$\frac{d}{dt}(U + K + K_s) = W_m + W_m^s + W_\Sigma + W_\Sigma^s + Q. \tag{3.142}$$

From (3.142) with the help of (3.138)–(3.140) we get the integral form of the law:

$$\frac{d}{dt}\int_V \rho\Big(e + \frac{|v|^2}{2} + \frac{|\mathbf{k}_m|^2}{2\gamma_s}\Big)\,dV = \int_V \rho(\mathbf{f}\cdot\mathbf{v} + \mathbf{h}_m\cdot\mathbf{k}_m + q_m)\,dV + \int_\Sigma (\mathbf{t}_\Sigma\cdot\mathbf{v} + \mathbf{h}_\Sigma\cdot\mathbf{k}_m + q_\Sigma)\,d\Sigma. \tag{3.143}$$

Similarly to Sect. 3.4.2, the heat flux vector $\mathbf{q}$ for polar continua is introduced by formulae (3.111) and (3.114). Thus, in Eq. (3.143) we can transform the surface integral to the volume one, then we obtain the following *differential form* of the energy balance law for a polar continuum (the *energy equation*):

$$\rho\frac{d}{dt}\left(e + \frac{|v|^2}{2} + \frac{|\mathbf{k}_m|^2}{2\gamma_s}\right) = \rho\mathbf{f}\cdot\mathbf{v} + \rho\mathbf{h}_m\cdot\mathbf{k}_m + \rho q_m - \nabla\cdot\mathbf{q} + \nabla\cdot(\mathbf{T}\cdot\mathbf{v} + \mathbf{M}\cdot\mathbf{h}_m). \tag{3.144}$$

*Remark.* The energy equation (3.143) clarifies the meaning of introduction of the spin constant $\gamma_s$ in the angular momentum balance law (3.79): if in (3.79) or in (3.92) we eliminated the constant $\gamma_s$ by introducing the vector $\widetilde{\mathbf{k}}_m = \mathbf{k}_m/\gamma_s$ instead of $\mathbf{k}_m$, then the energy equation (3.144) would include the constant $\gamma_s$ all the same. This constant $\gamma_s$ plays a role of the energetic equivalent between the kinetic energy of a body and the kinetic energy of the own body motion caused by the own angular momentum for polar continua. □

According to relation (3.20), the energy equation (3.144) takes the divergence form

$$\frac{\partial}{\partial t}\rho\varepsilon + \nabla\cdot(\rho\mathbf{v}\varepsilon + \mathbf{q} - \mathbf{T}\cdot\mathbf{v} - \mathbf{M}\cdot\mathbf{k}_m) = \rho\mathbf{f}\cdot\mathbf{v} + \rho\mathbf{h}_m\cdot\mathbf{k}_m + \rho q_m,$$
$$\varepsilon = e + \frac{|v|^2}{2} + \frac{|\mathbf{k}_m|^2}{2\gamma_s}. \tag{3.145}$$

Let us formulate now an analog of the kinetic energy balance equation (3.123) for the kinetic energy of own motion $K_s$. Consider the angular momentum balance equation (3.92) and multiply the equation by $\mathbf{k}_m$:

$$\frac{\rho}{\gamma_s}\mathbf{k}_m\cdot\frac{d\mathbf{k}_m}{dt} = \rho\mathbf{h}_m\cdot\mathbf{k}_m + \mathbf{k}_m\cdot\nabla\cdot\mathbf{M} - \mathbf{k}_m\cdot{}^3\boldsymbol{\varepsilon}\cdot\cdot\,\mathbf{T}. \tag{3.146}$$

According to the property of the nabla-operator from Exercise 2.1.17, we can reduce Eq. (3.146) to the form

$$\frac{\rho}{\gamma_s}\frac{d|\mathbf{k}_m|^2}{dt} = \rho\mathbf{h}_m\cdot\mathbf{k}_m + \nabla\cdot(\mathbf{M}\cdot\mathbf{k}_m) - \mathbf{M}\cdot\cdot(\nabla\otimes\mathbf{k}_m)^{\mathrm{T}} - \mathbf{k}_m\cdot{}^3\boldsymbol{\varepsilon}\cdot\cdot\mathbf{T}. \quad (3.147)$$

Integrating this equation over $V$ and using the Gauss–Ostrogradskii formula and the differentiation rule of an integral over a moving volume (3.12), from (3.147) we get

$$\frac{dK_s}{dt} = W^s_m + W^s_\Sigma + W^s_{(i)} + W_s. \quad (3.148)$$

Here we have introduced the following notation:

$$W^s_{(i)} = -\int\limits_V \mathbf{M}\cdot\cdot(\nabla\otimes\mathbf{k}_m)^{\mathrm{T}}\,dV \quad (3.149)$$

– the *power of internal surface moments* and

$$W_s = -\int\limits_V \mathbf{k}_m\cdot{}^3\boldsymbol{\varepsilon}\cdot\cdot\mathbf{T}\,dV \quad (3.150)$$

– the *power of internal surface forces for the own body motion.*

Relation (3.148) is called the *kinetic energy balance equation for the own body motion.*

Subtracting Eqs. (3.123) and (3.148) from the energy balance law (3.142), we get the *integral heat influx equation for polar continua*:

$$\frac{dU}{dt} = Q - W_{(i)} - W^s_{(i)} - W_s. \quad (3.151)$$

Subtracting Eqs. (3.147) and (3.121) from the energy equation (3.144), we obtain the *differential heat influx equation for polar continua*:

$$\rho\frac{de}{dt} = \mathbf{T}\cdot\cdot(\nabla\otimes\mathbf{v})^{\mathrm{T}} + \mathbf{M}\cdot\cdot(\nabla\otimes\mathbf{k}_m)^{\mathrm{T}} + \mathbf{k}_m\cdot{}^3\boldsymbol{\varepsilon}\cdot\cdot\mathbf{T} - \nabla\cdot\mathbf{q} + \rho q_m. \quad (3.152)$$

If we assume $\mathbf{k}_m = 0$, $\mathbf{M} = 0$ and $\mathbf{h}_m = 0$, then the heat influx equation in the forms (3.151) and (3.152) coincides with the corresponding equations (3.124) and (3.126) for nonpolar continua.

According to the definition (3.97) of Lagrangian tensor of moment stresses $\overset{\circ}{\mathbf{M}}$, we can rewrite the *heat influx equation* (3.152) *in the material description*

$$\overset{\circ}{\rho}\frac{de}{dt} = \mathbf{P}\cdot\cdot(\overset{\circ}{\nabla}\otimes\mathbf{v})^{\mathrm{T}} + \overset{\circ}{\mathbf{M}}\cdot\cdot(\overset{\circ}{\nabla}\otimes\mathbf{k}_m)^{\mathrm{T}} + \mathbf{k}_m\cdot{}^3\boldsymbol{\varepsilon}\cdot\cdot(\mathbf{F}\cdot\mathbf{P}) - \overset{\circ}{\nabla}\cdot\overset{\circ}{\mathbf{q}} + \overset{\circ}{\rho}q_m. \quad (3.153)$$

This assertion should be proved as Exercise 3.4.1.

## Exercises for 3.4

**3.4.1.** Prove the relation (3.153).

# 3.5 The Second Thermodynamic Law

## 3.5.1 The Integral Form

To state the second thermodynamic law, we introduce a new local characteristic of a continuum which is called a *temperature*.

**Axiom 8 (on the existence of an absolute temperature).** *For every material point* $\mathcal{M}$ *of any continuum* $\mathcal{B}$ *for all* $t \geqslant 0$ *there is a scalar positive function*

$$\theta(X^i, t) = \theta(x^i, t) > 0, \tag{3.154}$$

*called an absolute temperature.*

The axiom on the existence of an absolute temperature is sometimes called the *zero thermodynamic law*.

It should be noted that we have introduced a temperature as local but not integral characteristic of a continuum (unlike mass $M$ or angular momentum). Although we can always introduce the integral temperature $\Theta$, for example, as $\Theta = \int_V \theta dm$, but this value does not appear in any fundamental laws of continuum mechanics.

However, there are integral characteristics containing a temperature, which are of great importance for formulating the second thermodynamic law.

**Definition 3.10.** The *entropy production by external mass sources* is the scalar function $\bar{Q}_m$ and the *entropy production by external surface sources* is the scalar function $\bar{Q}_\Sigma$, which are defined for a volume $V$ as follows:

$$\bar{Q}_m = \int_V \frac{q_m}{\theta}\, dm = \int_V \frac{\rho q_m}{\theta}\, dV, \quad \bar{Q}_\Sigma = \int_\Sigma \frac{q_\Sigma}{\theta}\, d\Sigma = -\int_\Sigma \frac{\mathbf{n}\cdot\mathbf{q}}{\theta}\, d\Sigma. \tag{3.155}$$

*Remark.* If by analogy with (3.106) we consider the entropy production by external mass sources $d\bar{Q}_m$, supplied to elementary volume $dV$ by mass $dm$, then from (3.155) and (3.106) we get

$$d\bar{Q}_m = q_m\, dm/\theta = dQ/\theta, \tag{3.155a}$$

i.e. $d\bar{Q}_m$ is the heat influx due to external mass sources divided by temperature $\theta$ of the volume $dV$ heated. Thus, $d\bar{Q}_m$ characterizes the efficiency of heating the volume $dV$, and $d\bar{Q}_\Sigma$ characterizes the efficiency of heating the surface element $d\Sigma$. □

**Axiom 9 (The second thermodynamic law).** *For any continuum $\mathcal{B}$ there are two scalar additive functions: $H(\mathcal{B}, t)$ called the entropy of the continuum and $\bar{Q}^*(\mathcal{B}, t)$ called the entropy production by internal sources, such that for all $t \geqslant 0$ the following equation holds:*

$$\frac{dH}{dt} = \bar{Q} + \bar{Q}^*, \tag{3.156}$$

*where we have denoted the summarized entropy production by external sources as follows:*

$$\bar{Q} = \bar{Q}_m + \bar{Q}_\Sigma, \tag{3.157}$$

*and a value of $\bar{Q}^*$ is always nonnegative (the Planck inequality):*

$$\bar{Q}^* \geqslant 0. \tag{3.158}$$

Since the functions $H$ and $\bar{Q}^*$ are additive, we can introduce the corresponding specific functions at every point $\mathcal{M} \in V$.

**Definition 3.11.** The *specific entropy* $\eta$ and the *specific internal entropy production* $q^*$ are defined for every point $\mathcal{M}$ of a continuum as follows:

$$\eta = \frac{dH}{dm}, \qquad q^* = \theta \frac{d\bar{Q}^*}{dm}. \tag{3.159}$$

As follows from the Planck inequality (3.158), and also the inequalities (3.154) and (3.4a), the function $q^*$ is always nonnegative:

$$q^* \geqslant 0. \tag{3.158a}$$

This inequality is also called the *Planck inequality*.

Since the functions $H$ and $\bar{Q}^*$ are additive for a whole continuum, we have

$$\bar{Q}^* = \int_V \frac{q^*}{\theta}\, dm = \int_V \frac{\rho q^*}{\theta}\, dV, \quad H = \int_V \eta\, dm = \int_V \rho\eta\, dV. \tag{3.160}$$

Two relations (3.156) and (3.158) are equivalent to the *Clausius inequality*

$$dH/dt \geqslant \bar{Q}. \tag{3.161}$$

On substituting (3.155) and (3.160) into (3.156), we obtain the *integral form of the second thermodynamic law*

$$\frac{d}{dt}\int_V \rho\eta\, dV = \int_V \frac{\rho(q_m + q^*)}{\theta}\, dV - \int_\Sigma \frac{\mathbf{n}\cdot\mathbf{q}}{\theta}\, d\Sigma. \tag{3.162}$$

### 3.5.2 Differential Form of the Second Thermodynamic Law

Transform the surface integral in (3.162) by the Gauss–Ostrogradskii formula (3.24):

$$\int_{\Sigma} \frac{\mathbf{n}\cdot\mathbf{q}}{\theta}\, d\Sigma = \int_{V} \nabla\cdot\left(\frac{\mathbf{q}}{\theta}\right)\, d\Sigma = \int_{V} \left(\frac{\nabla\cdot\mathbf{q}}{\theta} + \mathbf{q}\cdot\nabla\left(\frac{1}{\theta}\right)\right)\, dV. \tag{3.163}$$

Then Eq. (3.162) takes the form

$$\int_{V} \left(\rho\frac{d\eta}{dt} - \frac{\rho(q_m + q^*)}{\theta} + \nabla\cdot\left(\frac{\mathbf{q}}{\theta}\right)\right)\, dV = 0. \tag{3.164}$$

Since the volume $V$ is arbitrary, we get the following theorem.

**Theorem 3.16.** *If the functions $\eta$, $\mathbf{q}$, $q^*$, $q_m$ and $\theta$, satisfying Eq. (3.162), are continuously differentiable in $V(t)$ for all considered times $t \geqslant 0$, then at every point $\mathcal{M} \in V$ the second thermodynamic law has the following differential form (the entropy balance equation):*

$$\rho\frac{d\eta}{dt} = -\nabla\cdot\left(\frac{\mathbf{q}}{\theta}\right) + \rho\frac{q_m + q^*}{\theta}. \tag{3.165}$$

According to the property (3.20), we can rewrite Eq. (3.165) in the divergence form

$$\frac{\partial\rho\eta}{\partial t} + \nabla\cdot\left(\rho\mathbf{v}\eta + \frac{\mathbf{q}}{\theta}\right) = \rho\frac{q_m + q^*}{\theta}. \tag{3.165a}$$

According to the equation

$$\nabla\cdot\left(\frac{\mathbf{q}}{\theta}\right) = \frac{1}{\theta}\nabla\cdot\mathbf{q} - \frac{1}{\theta^2}\mathbf{q}\cdot\nabla\theta,$$

formulae (3.165) and (3.165a) can be rewritten in the form

$$\theta\rho\frac{d\eta}{dt} = -\nabla\cdot\mathbf{q} + \rho q_m + w^*, \tag{3.166}$$

or

$$\frac{\partial\rho\eta}{\partial t} + \nabla\cdot\rho\mathbf{v}\eta = -(1/\theta)\nabla\cdot\mathbf{q} + (1/\theta)(\rho q_m + w^*). \tag{3.166a}$$

Here we have introduced the function

$$w^* = \rho q^* + \frac{\mathbf{q}}{\theta}\cdot\nabla\theta, \tag{3.167}$$

called the *dissipation function.*

Besides the Planck inequality (3.158), there is one more important inequality in continuum mechanics.

**Axiom 10.** *At every point $\mathcal{M}$ of a continuum for all $t \geqslant 0$ the Fourier inequality holds:*

$$\mathbf{q} \cdot \nabla \theta \leqslant 0. \tag{3.168}$$

Rewrite (3.167) in the form

$$\rho q^* = w^* - \frac{\mathbf{q}}{\theta} \cdot \nabla \theta. \tag{3.169}$$

Equation (3.169) means that the specific internal entropy production is caused by two factors: dissipation (i.e. irreversible conversion of part of mechanical energy to heat energy) and nonuniform heating of a continuum (when there are irreversible processes of heat transfer from hotter parts of the continuum to its colder ones). The first cause corresponds to the function $w^*$, and the second one corresponds to the function

$$-\frac{1}{\theta} \mathbf{q} \cdot \nabla \theta \equiv w_q. \tag{3.170}$$

A mathematical expression for the dissipation function $w^*$ will be given in Chap. 4; however, it should be noted here that if $w^*$ is independent of the temperature gradient $\nabla \theta$ (in Chap. 4 we will show that this assumption always holds), then, considering the motion of a continuum from $\overset{\circ}{\mathcal{K}}$ to $\mathcal{K}$ only with uniform temperature field $\theta = \theta(t)$ ($\nabla \theta \equiv 0$ in $V$), from (3.169) and the Planck inequality we get the *dissipation inequality*

$$w^* \geqslant 0. \tag{3.171}$$

Since the function $w^*$ is independent of $\nabla \theta$, the inequality (3.171) holds under nonuniform heating as well.

### *3.5.3 The Second Thermodynamic Law in the Material Description*

In the material description, Eq. (3.162) takes the form

$$\int\limits_{\overset{\circ}{V}} \overset{\circ}{\rho} \frac{d\eta}{dt}\, d\overset{\circ}{V} = \int\limits_{\overset{\circ}{V}} \frac{\overset{\circ}{\rho}(q_m + q^*)}{\theta}\, d\overset{\circ}{V} - \int\limits_{\overset{\circ}{\Sigma}} \frac{\overset{\circ}{\mathbf{n}} \cdot \overset{\circ}{\mathbf{q}}}{\theta}\, d\overset{\circ}{\Sigma}, \tag{3.172}$$

then with the help of the usual manipulations we get the *second thermodynamic law in the material description*

$$\overset{\circ}{\rho} \frac{d\eta}{dt} = -\overset{\circ}{\nabla} \cdot \left( \frac{\overset{\circ}{\mathbf{q}}}{\theta} \right) + \overset{\circ}{\rho} \frac{q_m + q^*}{\theta}. \tag{3.173}$$

This equation can be rewritten in the form

$$\theta\overset{\circ}{\rho}\frac{d\eta}{dt} = -\overset{\circ}{\nabla}\cdot\overset{\circ}{\mathbf{q}} + \overset{\circ}{\rho}q_m + \overset{\circ}{w}^*, \tag{3.174}$$

where

$$\overset{\circ}{w}^* = \overset{\circ}{\rho}q^* + \frac{\overset{\circ}{\mathbf{q}}}{\theta}\cdot\overset{\circ}{\nabla}\theta \geqslant 0 \tag{3.175}$$

is the *dissipation function* in $\overset{\circ}{\mathcal{K}}$.

The Clausius inequality (3.161) in the material (Lagrangian) description retains its form, but entropy $H$ and entropy production $\bar{Q}$ by external sources, defined by (3.160), in Lagrangian description take the forms

$$H = \int_V \rho\eta\, dV = \int_{\overset{\circ}{V}} \overset{\circ}{\rho}\eta\, d\overset{\circ}{V}, \quad \bar{Q} = \bar{Q}_m + \bar{Q}_\Sigma,$$

$$\bar{Q}_m = \int_V \frac{\rho q_m}{\theta}\, dV = \int_{\overset{\circ}{V}} \frac{\overset{\circ}{\rho}q_m}{\theta}\, d\overset{\circ}{V}, \quad \bar{Q}_\Sigma = \int_\Sigma \frac{q_\Sigma}{\theta}\, d\Sigma = \int_{\overset{\circ}{\Sigma}} \frac{\overset{\circ}{q}_\Sigma}{\theta}\, d\overset{\circ}{\Sigma}, \tag{3.176}$$

$$\bar{Q}^* = \int_V \frac{\rho q^*}{\theta}\, dV = \int_{\overset{\circ}{V}} \frac{\overset{\circ}{\rho}q^*}{\theta}\, d\overset{\circ}{V}.$$

### 3.5.4 Heat Machines and Their Efficiency

The present-day *heat machines* are facilities, which convert heat into mechanical work (for example, internal-combustion engines, jet and turbo-jet engines, gas-turbine engines, nuclear engines etc.), or, on the contrary, convert mechanical work into heat (for example, some heating and cooling devices), or use both the conversions in turn (for example, diesel engines). The operation of all the heat machines is based on the thermodynamic laws.

Conventionally, the structure of a heat machine consists of a vessel with solid walls, which is filled with a working body of volume $V$ (it is usually gas or fluid).

Heat machines can be split into the three main types:

(1) Machines, for which the domain $V(t)$ is varying but material points of a working body are the same, i.e. $\overset{\circ}{V} = \text{const}$.
(2) Machines, for which $V$ remains unchanged with time but material points of a working body, occurring in $V$, are different.
(3) Machines, for which the domain $V$ is varying and material points occurring in $V$ are different.

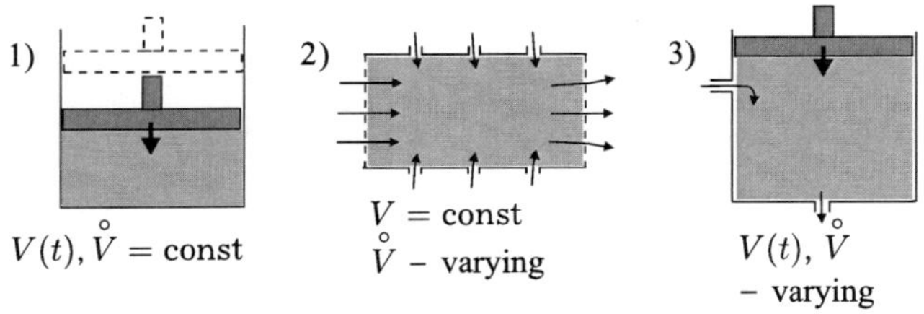


**Fig. 3.8** The scheme of three main types of heat machines

These three types of heat machines are shown in Fig. 3.8.

The first type involves cooling and heating facilities, and also machines operating on the Carnot cycle (see below). The second type includes air-jet and rocket engines etc. The third type involves gas-turbine, turbo-jet, internal-combustion engines etc.

Lagrangian description proves to be more convenient for computations of heat machines of the first type, and Eulerian description – for the ones of the second and third types.

Further, we will consider heat machines only of the first type; machines of the second and third types can be considered in a similar way (see Exercise 3.5.2).

All heat machines are characterized by their *efficiency*. To introduce the concepts of heat, mechanical work and the efficiency, we should return to the integral statements (3.126) and (3.161) of the thermodynamic laws.

Consider an arbitrary time interval $t_1 \leqslant t \leqslant t_2$ and integrate Eqs. (3.126) and (3.161) with respect to $t$ from $t_1$ to $t_2$; as a result, we obtain

$$\Delta U + A_{(i)} = C, \tag{3.177}$$

$$\Delta H \geqslant \bar{C}. \tag{3.178}$$

Here we have introduced the notation

$$\Delta U = U(t_2) - U(t_1), \quad \Delta H = H(t_2) - H(t_1); \tag{3.179}$$

$$\begin{aligned} A_{(i)} &= \int_{t_1}^{t_2} W_{(i)}(t)\,dt = -\int_{t_1}^{t_2}\int_{\overset{\circ}{V}} \mathbf{P} \cdot\cdot \overset{\circ}{\nabla} \otimes \mathbf{v}^{\mathrm{T}}\, d\overset{\circ}{V}\, dt, \\ A_{(e)} &= \int_{t_1}^{t_2} W(t)\,dt = \int_{t_1}^{t_2}\int_{\overset{\circ}{V}} \overset{\circ}{\rho}\mathbf{f}\cdot\mathbf{v}\, d\overset{\circ}{V}\, dt + \int_{t_1}^{t_2}\int_{\overset{\circ}{\Sigma}} \overset{\circ}{\mathbf{t}}_n \cdot \mathbf{v}\, d\overset{\circ}{\Sigma}\, dt; \end{aligned} \tag{3.180}$$

$$C = \int_{t_1}^{t_2} Q(t)\, dt = \int_{t_1}^{t_2} \int_{\overset{\circ}{V}} \overset{\circ}{\rho} q_m \, d\overset{\circ}{V}\, dt + \int_{t_1}^{t_2} \int_{\overset{\circ}{\Sigma}} \overset{\circ}{q}_\Sigma \, d\overset{\circ}{\Sigma}\, dt,$$

$$\bar{C} = \int_{t_1}^{t_2} \bar{Q}(t)\, dt = \int_{t_1}^{t_2} \int_{\overset{\circ}{V}} \frac{\overset{\circ}{\rho q_m}}{\theta} \, d\overset{\circ}{V}\, dt + \int_{t_1}^{t_2} \int_{\overset{\circ}{\Sigma}} \frac{\overset{\circ}{q}_\Sigma}{\theta} \, d\overset{\circ}{\Sigma}\, dt, \tag{3.181}$$

where we used Lagrangian representation of the integral values $U$, $H$, $W_{(i)}$, $Q$ and $\bar{Q}$ (see (3.135), (3.136), and (3.176)).

The value $\Delta U$ is called the *change of the internal energy* of a body $\mathcal{B}$, $\Delta H$ – the *change of the entropy* of a body $\mathcal{B}$, $A_{(i)}$ – the *mechanical work done by a body* $\mathcal{B}$ (or simply the *work*), $C$ – the *heat supplied to a body* (or simply the *heat*), $\bar{C}$ – the *entropy influx* to a body $\mathcal{B}$, $A_{(e)}$ – the *work done on a body* $\mathcal{B}$ (or the *work done by external forces*).

Let us use Lagrangian description in variables $X^i, t$, where $X^i \in \overset{\circ}{V}(t)$, $t \in [t_1, t_2]$, and divide the four-dimensional integration domain $\overset{\circ}{V} \times (t_1, t_2)$ of volume integrals into subdomains $\overset{\circ}{V}_{t\alpha}$ such that: (1) in each of these subdomains a sign of the scalar function $q_m$ remains unchanged, i.e. either $q_m \geqslant 0$, or $q_m < 0$, and (2) $\bigcup_\alpha \overset{\circ}{V}_{t\alpha} = \overset{\circ}{V} \times (t_1, t_2)$. In a similar way, divide the integration domain $\overset{\circ}{\Sigma} \times (t_1, t_2)$ of surface integrals into subdomains $\overset{\circ}{\Sigma}_{t\alpha}$ such that: (1) in each $\overset{\circ}{\Sigma}_{t\alpha}$ a sign of the function $\overset{\circ}{q}_\Sigma$ remains unchanged, i.e. either $\overset{\circ}{q}_\Sigma \geqslant 0$, or $\overset{\circ}{q}_\Sigma < 0$, and (2) $\bigcup_\alpha \overset{\circ}{\Sigma}_{t\alpha} = \overset{\circ}{\Sigma} \times (t_1, t_2)$. Then the heat $C$ supplied to the body can be split into two nonnegative parts:

$$C = C_+ - C_-, \tag{3.182}$$

$$C_\pm = \int_{t_1}^{t_2} \int_{\overset{\circ}{V}} \overset{\circ}{\rho} q_m^\pm \, d\overset{\circ}{V}\, dt + \int_{t_1}^{t_2} \int_{\overset{\circ}{\Sigma}} \overset{\circ}{q}{}_\Sigma^\pm \, d\overset{\circ}{\Sigma}\, dt \geqslant 0, \tag{3.183}$$

$$q_m^\pm = (|q_m| \pm q_m)/2 \geqslant 0, \quad \overset{\circ}{q}{}_\Sigma^\pm = (|\overset{\circ}{q}_\Sigma| \pm \overset{\circ}{q}_\Sigma)/2 \geqslant 0. \tag{3.184}$$

Here $C_+$ is the *heat absorbed by the body* (or supplied to the body), and $C_-$ is the *heat released by the body* (or withdrawn from the body).

Since the temperature $\theta$ of a body is nonnegative, we can also split the entropy influx $\bar{C}$ to the body into two nonnegative parts:

$$\bar{C} = \bar{C}_+ - \bar{C}_-, \tag{3.185}$$

$$\bar{C}_\pm = \int_{t_1}^{t_2} \int_{\overset{\circ}{V}} \frac{\overset{\circ}{\rho} q_m^\pm}{\theta} \, d\overset{\circ}{V}\, dt + \int_{t_1}^{t_2} \int_{\overset{\circ}{\Sigma}} \frac{\overset{\circ}{q}{}_\Sigma^\pm}{\theta} \, d\overset{\circ}{\Sigma}\, dt \geqslant 0. \tag{3.186}$$

Substituting (3.182) into (3.177), we can rewrite the heat influx equation in the form

$$C_- = C_+ - A_{(i)} - \Delta U. \tag{3.187}$$

Assume that $C_+ \neq 0$, then we can introduce the ratio of the mechanical work done by the body $\mathcal{B}$ to the heat supplied:

$$k_{\mathcal{B}} = A_{(i)}/C_+, \tag{3.188}$$

that is called the *efficiency* of the body $\mathcal{B}$ for the time interval $[t_1, t_2]$ considered.

According to the axiom (3.154), in each body $\mathcal{B}$ for any time interval $[t_1, t_2]$ the conditions $0 < \theta(X^i, t) < +\infty$ are satisfied; then we can introduce minimum and maximum values of the temperature of a body:

$$0 < \theta_{\min} \leqslant \theta(X^i, t) \leqslant \theta_{\max} < +\infty, \quad \forall X^i \in \overset{\circ}{\overline{V}}, \quad \forall t \in [t_1, t_2]. \tag{3.189}$$

From relations (3.189) we obtain

$$\begin{aligned} \overset{\circ}{\rho} q_m^+/\theta \geqslant \overset{\circ}{\rho} q_m^+/\theta_{\max} \quad &\text{and} \quad \overset{\circ}{\rho} q_m^-/\theta \leqslant \overset{\circ}{\rho} q_m^-/\theta_{\min}, \\ \overset{\circ}{q}{}_\Sigma^+/\theta \geqslant \overset{\circ}{q}{}_\Sigma^+/\theta_{\max} \quad &\text{and} \quad \overset{\circ}{q}{}_\Sigma^-/\theta \leqslant \overset{\circ}{q}{}_\Sigma^-/\theta_{\min}, \end{aligned} \tag{3.190}$$

because $q_m^+$, $\overset{\circ}{q}{}_\Sigma^+$, $q_m^-$, $\overset{\circ}{q}{}_\Sigma^-$, $\overset{\circ}{\rho}$ and $\theta$ are nonnegative.

Substituting (3.190) into the integrals (3.183), (3.186) and taking into account that the integrands are nonnegative, we find that

$$\bar{C}_+ \geqslant C_+/\theta_{\max}, \qquad \bar{C}_- \leqslant C_-/\theta_{\min}. \tag{3.191}$$

From formulae (3.178), (3.182), and (3.191) we get

$$\Delta H \geqslant \bar{C} = \bar{C}_+ - \bar{C}_- \geqslant \frac{C_+}{\theta_{\max}} - \frac{C_-}{\theta_{\min}}. \tag{3.192}$$

Substitution of the expression (3.187) into (3.192) yields

$$\Delta H \geqslant C_+ \left( \frac{1}{\theta_{\max}} - \frac{1}{\theta_{\min}} \right) + \frac{1}{\theta_{\min}} (A_{(i)} + \Delta U). \tag{3.193}$$

Dividing this relation by $C_+$ and multiplying by $\theta_{\min}$ (both the values are nonnegative), we obtain

$$\frac{A_{(i)}}{C_+} + \frac{\Delta U}{C_+} + \left( \frac{\theta_{\min}}{\theta_{\max}} - 1 \right) \leqslant \frac{\Delta H \theta_{\min}}{C_+}. \tag{3.194}$$

According to the definition of the efficiency $k_{\mathcal{B}}$ (3.188), we get the final *Truesdell estimate for the efficiency*:

$$k_{\mathcal{B}} \leqslant 1 - \frac{\theta_{\min}}{\theta_{\max}} - \frac{\Delta U - \theta_{\min} \Delta H}{C_+}. \tag{3.195}$$

Notice that since a sign of the third summand on the right-hand side of (3.195) is unknown, we cannot claim that in any process for each body $k_{\mathcal{B}} < 1$. This assertion, which is well known from the course of physics, holds for special processes of motion and heating of a body. Consider them in the next paragraph.

### 3.5.5 Adiabatic and Isothermal Processes. The Carnot Cycles

One can say that in a body $\mathcal{B}$ there occurs a *locally adiabatic* process in the subdomain $\overset{\circ}{V}_{t\alpha}$, if the heat influx due to mass sources into $\overset{\circ}{V}_{t\alpha}$ is zero:

$$q_m \equiv 0 \quad \forall (X^i, t) \in \overset{\circ}{V}_{t\alpha}. \tag{3.196}$$

If in some subdomain $\overset{\circ}{\Sigma}_{t\alpha}$ the heat influx due to surface sources is zero:

$$\overset{\circ}{q}_{\Sigma} \equiv 0 \quad \forall (X^i, t) \in \overset{\circ}{\Sigma}_{t\alpha}, \tag{3.197}$$

then one can say that a locally adiabatic process occurs in the subdomain $\overset{\circ}{\Sigma}_{t\alpha}$ of the body $\mathcal{B}$.

If in a body $\mathcal{B}$ within the time interval $[t_1, t_2]$ the following conditions are satisfied:

$$q_m \equiv 0 \ \forall X^i \in \overset{\circ}{V}, \quad \overset{\circ}{q}_{\Sigma} \equiv 0 \ \forall X^i \in \overset{\circ}{\Sigma}, \tag{3.198}$$

then one can say that there occurs an *adiabatic process in the restricted sense* in the body $\mathcal{B}$.

Finally, if in a whole body $\mathcal{B}$ within the time interval $[t_1, t_2]$ the following conditions are satisfied:

$$q_m \equiv 0 \ \forall X^i \in \overset{\circ}{V}, \quad \overset{\circ}{\mathbf{q}} \equiv 0 \ \forall X^i \in \overline{\overset{\circ}{V}}, \tag{3.199}$$

then one can say that there occurs an *adiabatic process* in the body $\mathcal{B}$ *in the broad sense* (or simply an *adiabatic process*).

Since (3.196) and (3.197) follow from the condition (3.199) (but not conversely), one can say that if in a body $\mathcal{B}$ an adiabatic process in the broad sense occurs, then in this body there occurs also an adiabatic process in the restricted sense (but not conversely).

One can say that in a body $\mathcal{B}$ there occurs a *locally isothermal* process in the subdomain $\overset{\circ}{V}_{t\alpha}$, if the temperature in this subdomain remains constant:

$$\theta(X^i, t) \equiv \text{const} \quad \forall (X^i, t) \in \overset{\circ}{V}_{t\alpha}. \tag{3.200}$$

If the temperature remains constant within a time interval $[t_1, t_2]$ in a whole body $\mathcal{B}$:

$$\theta(X^i, t) \equiv \text{const} \quad \forall X^i \in \overline{\overset{\circ}{V}}, \quad \forall t \in [t_1, t_2], \tag{3.201}$$

then one can say that in the body $\mathcal{B}$ there occurs an *isothermal process*.

One can say that in a body $\mathcal{B}$ there occurs a *thermodynamic cycle* within the time interval $[t_1, t_2]$, if the following conditions of periodicity are satisfied simultaneously:

$$\begin{gathered} \Delta U = U(t_2) - U(t_1) = 0, \quad \Delta H = H(t_2) - H(t_1) = 0, \\ C_+ \neq 0, \\ \theta(X^i, t_2) - \theta(X^i, t_1) \equiv 0 \quad \forall X^i \in \overset{\circ}{V}. \end{gathered} \tag{3.202}$$

Notice that not for every continuum we may ensure that the conditions (3.202) be satisfied, and, thus, not for every body a thermodynamic cycle can be realized. However, there are whole classes of bodies, for which the conditions (3.202) are satisfiable.

*Example 3.1.* Consider a continuum $\mathcal{B}$, where the specific internal energy $e$ and the specific entropy $\eta$ uniquely depend only on the density $\rho$ and temperature $\theta$:

$$e = e(\rho,\, \theta), \qquad \eta = \eta(\rho,\, \theta). \tag{3.203}$$

In Chap. 4 we will show that such a model of a continuum describes an ideal fluid or gas.

If relations (3.203) hold, then the conditions (3.202) can easily be satisfied by giving such external heat and mechanical actions that the fields of temperature $\theta(X^i, t)$ and density $\rho(X^i, t)$ in the body $\mathcal{B}$ be the same at $t = t_1$ and $t = t_2$: $\rho(X^i, t_1) = \rho(X^i, t_2)$, $\theta(X^i, t_1) = \theta(X^i, t_2)$. Indeed, in this case the following conditions are satisfied:

$$e(X^i,\, t_1) = e(\rho(X^i, t_1),\, \theta(X^i, t_1)) = e(\rho(X^i, t_2),\, \theta(X^i, t_2)) = e(X^i, t_2),$$

and hence

$$U(t_1) = \int_{\overset{\circ}{V}} \overset{\circ}{\rho} e(X^i, t_1)\, d\overset{\circ}{V} = \int_{\overset{\circ}{V}} \overset{\circ}{\rho} e(X^i, t_2)\, d\overset{\circ}{V} = U(t_2),$$

i.e. $\Delta U = 0$. The condition $\Delta H = 0$ can be satisfied in a similar way.

A thermodynamic cycle, occurring in an ideal fluid (gas), at every material point $X^i$ is assigned to some closed curve on the plane $(\rho, \theta)$ (Fig. 3.9). □

*Example 3.2.* Consider a continuum $\mathcal{B}$, where $e$ and $\eta$ are one-valued functions of temperature $\theta$ and the deformation gradient $\mathbf{F}$:

$$e = e(\mathbf{F},\, \theta), \qquad \eta = \eta(\mathbf{F},\, \theta). \tag{3.204}$$

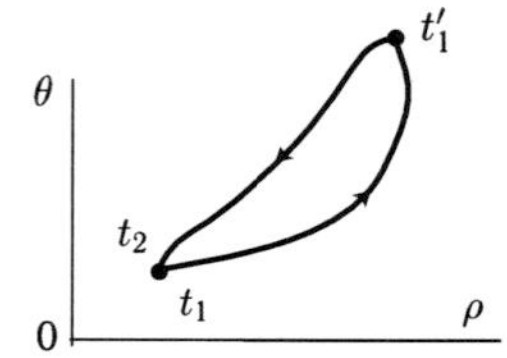


**Fig. 3.9** A thermodynamic cycle at the material point $X^i$ of an ideal fluid

Since density $\rho$ is a function of $\mathbf{F}$ ($\rho = \overset{\circ}{\rho}/\det\mathbf{F}$ due to (3.8)), the relations (3.203) are a particular case of (3.204). If dependences (3.204) cannot be reduced to (3.203), then they correspond to the model of an ideal solid (these models will be considered in detail in Chap. 8). For an ideal solid, one can readily ensure the conditions (3.202) to be satisfied by using relations (3.204). For this, it is sufficient that the following periodicity conditions for the cycle be satisfied:

$$\mathbf{F}(X^i, t_1) = \mathbf{F}(X^i, t_2), \quad \theta(X^i, t_1) = \theta(X^i, t_2) \quad \forall X^i \in \overset{\circ}{V}.$$

A thermodynamic cycle, occurring in an ideal solid, at every material point $X^i$ is also assigned to some closed curve in the generalized seven-dimensional space $(\bar{F}^i{}_j, t)$, where $\bar{F}^i{}_j$ are components of the tensor $\mathbf{F}$ with respect to the Cartesian basis $\bar{\mathbf{e}}_i$. □

One can say that a *generalized Carnot cycle* occurs in a body $\mathcal{B}$, if the following conditions are satisfied simultaneously:

(1) In the body $\mathcal{B}$ there occurs a thermodynamic cycle within $[t_1, t_2]$.
(2) The domains $\overset{\circ}{V} \times [t_1, t_2]$ and $\overset{\circ}{\Sigma} \times [t_1, t_2]$ of the body $\mathcal{B}$ consist only of the subdomains $\overset{\circ}{V}_{t\alpha}$ and $\overset{\circ}{\Sigma}_{t\alpha}$, where there occur locally adiabatic or locally isothermal processes.
(3) Locally isothermal processes may be only of the two types: either with the minimum temperature $\theta_{\min}$ and heat absorbtion:

$$\begin{aligned} \text{in } \overset{\circ}{V}_{t\alpha}: \quad & \theta(X^i, t) = \theta_{\min}, \quad q_m < 0, \\ \text{on } \overset{\circ}{\Sigma}_{t\alpha}: \quad & \theta(X^i, t) = \theta_{\min}, \quad \overset{\circ}{q}_\Sigma < 0, \end{aligned} \tag{3.205}$$

or with the maximum temperature $\theta_{\max}$ and heat release:

$$\begin{aligned} \text{in } \overset{\circ}{V}_{t\alpha}: \quad & \theta(X^i, t) = \theta_{\max}, \quad q_m > 0, \\ \text{on } \overset{\circ}{\Sigma}_{t\alpha}: \quad & \theta(X^i, t) = \theta_{\max}, \quad \overset{\circ}{q}_\Sigma > 0. \end{aligned} \tag{3.206}$$

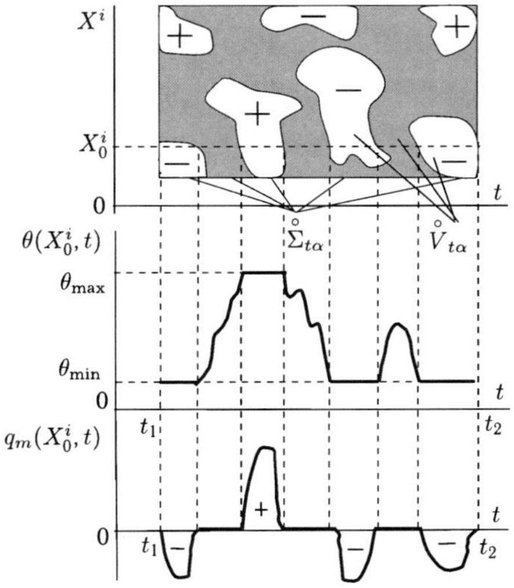


**Fig. 3.10** The typical picture of location of subdomains in the generalized Carnot cycle

The subdomains $\overset{\circ}{V}_{t\alpha}$ and $\overset{\circ}{\Sigma}_{t\alpha}$, where there occur locally isothermal processes, have no common points (and even their closures have no common points too) due to continuous differentiability of the temperature $\theta(X^i, t)$ within $\overset{\circ}{V} \times [t_1, t_2]$.

Figure 3.10 shows a typical location of the subdomains $\overset{\circ}{V}_{t\alpha}$ and $\overset{\circ}{\Sigma}_{t\alpha}$ in the generalized Carnot cycle.

One can say that in a body $\mathcal{B}$ there occurs an *uniform thermomechanical process* within the time interval $[t_1, t_2]$, if fields of its mechanical and thermodynamic values are uniform, i.e. they are independent of coordinates $X^i$ but may depend on time $t$:

$$\begin{aligned} \Omega(X^i, t) &= \Omega(t) \ \ \forall X^i \in \overset{\circ}{V}, \ \ \forall t \in [t_1, t_2], \\ \Omega &= \{\, \rho, \ \mathbf{q}, \ \mathbf{F}, \ \mathbf{T}, \ \mathbf{P}, \ q_m, \ e, \ \eta, \ \theta, \ q^* \,\}. \end{aligned} \tag{3.207}$$

(Notice that the velocity $\mathbf{v}$ and displacement $\mathbf{u}$ vectors of material points in an uniform process may depend on the coordinates $X^i$.)

The generalized Carnot cycle, which occurs in a body under the conditions of a uniform thermomechanical process, is called simply the *Carnot cycle*.

The Carnot cycle, according to the definitions given above, consists only of isothermal and adiabatic processes in a body $\mathcal{B}$, which alternate with each other; here within the time subinterval $\mathcal{T}_0 \in [t_1, t_2]$, when the process is adiabatic in the whole body, the rate of heating $Q$, according to (3.198) and (3.107), is zero; within the time subinterval $\mathcal{T}_- \in [t_1, t_2]$, when the temperature in the whole body reaches its minimum value $\theta_{\min}$, the rate of heating $Q$ is negative; and within the time subinterval $\mathcal{T}_+ \in [t_1, t_2]$, when the temperature in the whole body has its maximum value $\theta_{\max}$, the rate of heating $Q$ is positive:

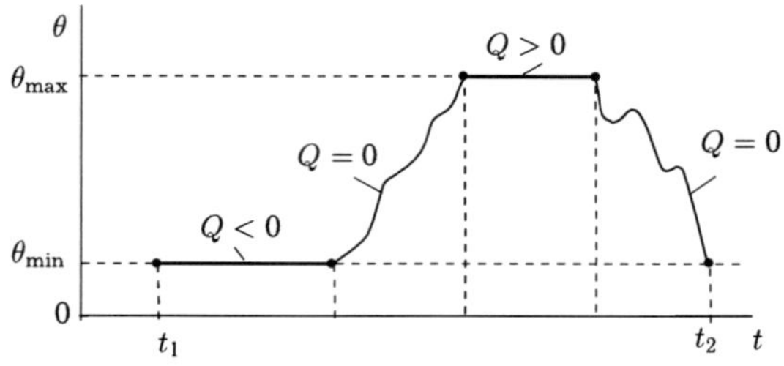


**Fig. 3.11** The elementary Carnot cycle

$$\begin{aligned} \mathcal{T}_0: &\quad Q(t) = 0, \\ \mathcal{T}_+: &\quad \theta = \theta_{\max}, \quad Q(t) > 0, \\ \mathcal{T}_-: &\quad \theta = \theta_{\min}, \quad Q(t) < 0. \end{aligned} \tag{3.208}$$

Figure 3.11 shows the *elementary Carnot cycle*. This cycle consists of four processes: two isothermal and two adiabatic ones. Other Carnot cycles for the same body $\mathcal{B}$ can differ from the elementary Carnot cycle only by the number of isothermal sections.

### 3.5.6 Truesdell's Theorem

**Theorem 3.16a (of Truesdell).**

*(1) If for the body $\mathcal{B}$ there exists a thermodynamic cycle with conditions (3.202) and given values $\theta_{\min}$ and $\theta_{\max}$ in this cycle, then the efficiency reaches its maximum possible value*

$$k_{\mathcal{B}\max} = 1 - \frac{\theta_{\min}}{\theta_{\max}} < 1, \tag{3.209}$$

*only in generalized Carnot cycles.*

*(2) Among all bodies $\mathcal{B}$, for which the thermodynamic cycle (3.202) is possible under the conditions of an uniform thermomechanical process (3.207), the efficiency reaches its maximum value (3.209) for such bodies, in the cycle of which there is no entropy production by internal sources:*

$$\bar{Q}^* \equiv 0 \quad \forall t \in [t_1, t_2]. \tag{3.210}$$

▼ (1) To prove the first assertion of Truesdell's Theorem we consider an arbitrary thermodynamic cycle; for this cycle, relationships (3.187) and (3.188) yield

$$A_{(i)} = C_+ - C_-, \tag{3.211}$$

and hence

$$k_{\mathcal{B}} = 1 - \frac{C_-}{C_+}. \tag{3.212}$$

As follows from (3.211) and (3.209), in order for the efficiency to reach its maximum value $k_{\mathcal{B}} = k_{\mathcal{B}\max}$ in the thermodynamic cycle, it is necessary and sufficient that the following condition be satisfied:

$$C_-/C_+ = \theta_{\min}/\theta_{\max}. \tag{3.213}$$

Since, according to (3.202), for any thermodynamic cycle $\Delta H = 0$, relation (3.192) with use of (3.213) takes the form

$$0 \geqslant \bar{C}_+ - \bar{C}_- \geqslant \frac{C_+}{\theta_{\max}} - \frac{C_-}{\theta_{\min}} = 0; \tag{3.214}$$

that is possible only if

$$\bar{C}_+ = \bar{C}_-. \tag{3.215}$$

Due to this relation, the system of two inequalities (3.191) becomes

$$\frac{C_+}{\theta_{\max}} \leqslant \bar{C}_+ = \bar{C}_- \leqslant \frac{C_-}{\theta_{\min}} = \frac{C_+}{\theta_{\max}}. \tag{3.216}$$

Since the left-hand and right-hand sides of (3.216) coincide, the inequality signs in this relation should be replaced by the equality sign. Thus, the following two equations must hold:

$$\frac{C_+}{\theta_{\max}} = \bar{C}_+, \quad \frac{C_-}{\theta_{\min}} = \bar{C}_-, \tag{3.217}$$

or with use of the notations (3.183) and (3.186)

$$\begin{aligned}
&\int_{t_1}^{t_2}\int_{\overset{\circ}{V}} \overset{\circ}{\rho} q_m^+ \left(\frac{1}{\theta} - \frac{1}{\theta_{\max}}\right) d\overset{\circ}{V}\,dt + \int_{t_1}^{t_2}\int_{\overset{\circ}{\Sigma}} \overset{\circ}{q}{}_\Sigma^+ \left(\frac{1}{\theta} - \frac{1}{\theta_{\max}}\right) d\overset{\circ}{\Sigma}\,dt = 0,\\
&\int_{t_1}^{t_2}\int_{\overset{\circ}{V}} \overset{\circ}{\rho} q_m^- \left(\frac{1}{\theta_{\min}} - \frac{1}{\theta}\right) d\overset{\circ}{V}\,dt + \int_{t_1}^{t_2}\int_{\overset{\circ}{\Sigma}} \overset{\circ}{q}{}_\Sigma^- \left(\frac{1}{\theta_{\min}} - \frac{1}{\theta}\right) d\overset{\circ}{\Sigma}\,dt = 0.
\end{aligned} \tag{3.218}$$

Since all integrands in these equations are non-negative, the sum of integrals is zero only if the integrands vanish; i.e. the following equations must hold:

$$\begin{aligned}
&q_m^+\left(\frac{1}{\theta} - \frac{1}{\theta_{\max}}\right) = 0, \quad \overset{\circ}{q}{}_\Sigma^+\left(\frac{1}{\theta} - \frac{1}{\theta_{\max}}\right) = 0,\\
&q_m^-\left(\frac{1}{\theta_{\min}} - \frac{1}{\theta}\right) = 0, \quad \overset{\circ}{q}{}_\Sigma^-\left(\frac{1}{\theta_{\min}} - \frac{1}{\theta}\right) = 0,\\
&\forall X^i \in \overline{V}, \quad \forall t \in [t_1, t_2].
\end{aligned} \tag{3.219}$$

These equations are satisfied if at least one of the cofactors vanishes, i.e. if in subdomains $\overset{\circ}{V}_{t\alpha}$ the following conditions are satisfied:

$$\text{in } \overset{\circ}{V}_{t\alpha}: \quad q_m > 0, \ \theta = \theta_{\max}, \ \text{ or } q_m < 0, \ \theta = \theta_{\min}, \ \text{ or } q_m = 0, \tag{3.220}$$

$$\text{at } \overset{\circ}{\Sigma}_{t\alpha}: \quad \overset{\circ}{q}_\Sigma > 0, \ \theta = \theta_{\max}, \ \text{ or } \overset{\circ}{q}_\Sigma < 0, \ \theta = \theta_{\min}, \ \text{ or } \overset{\circ}{q}_\Sigma = 0. \tag{3.221}$$

But, according to (3.205) and (3.206), this means that the thermodynamic cycle considered is a generalized Carnot cycle. Thus, we have shown that the requirement that the efficiency reach its maximum $k_\mathcal{B} = k_{\mathcal{B}\max}$ in a thermodynamic cycle leads to the requirement that this cycle be a Carnot cycle, and vice versa.

(2) To prove the second assertion of the theorem we use the fact established above that the efficiency reaches its maximum value at the generalized Carnot cycles; hence, under the conditions of uniform thermomechanical processes it is sufficient to prove only that the relation (3.210) holds for simple Carnot cycles.

Take into account that for a thermodynamic cycle $\Delta H = 0$, and represent $\Delta H$ as a sum of three terms:

$$0 = \Delta H = H(t_2) - H(t_1) = \int_{t_1}^{t_2} \dot{H}\, dt = \int_{\mathcal{T}_0} \dot{H}\, dt + \int_{\mathcal{T}_+} \dot{H}\, dt + \int_{\mathcal{T}_-} \dot{H}\, dt, \tag{3.222}$$

where the subsets $\mathcal{T}_0$, $\mathcal{T}_+$ and $\mathcal{T}_-$ are defined by the conditions (3.208) in an uniform thermomechanical process.

Within $\mathcal{T}_+$, according to (3.208), $\theta = \theta_{\max}$ and $Q > 0$, then due to the Clausius inequality (3.161), we have

$$\text{within } \mathcal{T}_+: \quad \dot{H} \geqslant \bar{Q} = \theta_{\max} Q > 0. \tag{3.223}$$

Assume that within some finite time interval being a subset $\widetilde{\mathcal{T}}_+$ of the set $\mathcal{T}_+$, the strict inequality in (3.223) holds: $\dot{H} > \theta_{\max} Q$. Then, integrating (3.223) over the whole $\mathcal{T}_+$, we obtain the strict inequality

$$\int_{\mathcal{T}_+} \dot{H}\, dt > \theta_{\max} \int_{\mathcal{T}_+} Q\, dt = \theta_{\max} C_+. \tag{3.224}$$

(The case, when the inequality $\dot{H} > \theta_{\max} Q$ is satisfied only at an isolated point, is excluded, because we assume that all the functions $H(t)$, $Q(t)$ and $\theta(t)$ are continuous.) Due to the Clausius inequality (3.161) and (3.208),

$$\int_{\mathcal{T}_-} \dot{H}\, dt \geqslant \int_{\mathcal{T}_-} \bar{Q}\, dt = \theta_{\min} \int_{\mathcal{T}_-} Q dt = -\theta_{\min} C_-. \tag{3.225}$$

(Here we have used the notations (3.183) and (3.184), which for uniform processes reduce to $C_- = -\int_{\mathcal{T}_-} Q\,dt = (1/2)\int_{t_1}^{t_2}(|Q| - Q)dt$.)

Addition of two inequalities (3.224) and (3.225) yields

$$\int\limits_{\mathcal{T}_+} \dot{H}\ dt + \int\limits_{\mathcal{T}_-} \dot{H}\ dt > \theta_{\max} C_+ - \theta_{\min} C_- = 0, \tag{3.226}$$

because we consider the Carnot cycle, for which the conditions (3.213) are satisfied. As follows from (3.222) and (3.226), the integral over $\mathcal{T}_0$ must be negative:

$$\int\limits_{\mathcal{T}_0} \dot{H}\ dt < 0. \tag{3.227}$$

However, due to the Clausius inequality (3.161), within $\mathcal{T}_0$ the conditions $\dot{H} \geqslant \bar{Q} = \theta Q = 0$ and $\int_{\mathcal{T}_0} \dot{H} dt \geqslant 0$ must be satisfied; therefore, the inequality (3.227) is impossible. Thus, the made assumption that within $\widetilde{\mathcal{T}}_+$: $\dot{H} > \theta_{\max} Q$ is not true, and from (3.223) it follows that within the whole $\mathcal{T}_+$ the following equality must hold:

$$\mathcal{T}_+ : \quad \dot{H} = \theta_{\max} Q. \tag{3.228}$$

In a similar way, one can prove that

$$\text{within } \mathcal{T}_- : \quad \dot{H} = \theta_{\min} Q. \tag{3.229}$$

Then, integrating (3.228) over $\mathcal{T}_+$ and (3.229) over $\mathcal{T}_-$ and substituting the results into (3.222), we obtain

$$0 = \Delta H = \int\limits_{\mathcal{T}_0} \dot{H}\ dt + \theta_{\max} C_+ - \theta_{\min} C_- = \int\limits_{\mathcal{T}_0} \dot{H}\ dt. \tag{3.230}$$

Here we have taken into account once more that the condition (3.213) is satisfied for Carnot's cycle.

As noted above, due to the Clausius inequality (3.161), within $\mathcal{T}_0$ we have $\dot{H} \geqslant 0$; then (3.230) yields

$$\mathcal{T}_0 : \quad \dot{H} = 0 = \theta Q. \tag{3.231}$$

Collecting the three relations (3.228), (3.229), (3.231), and taking (3.208) into account, we find that in the Carnot cycle with the maximum efficiency for the whole time interval considered the following relation holds:

$$\dot{H} = \theta Q = \bar{Q} \quad \forall t \in [t_1, t_2]. \tag{3.232}$$

This relationship, according to (3.156), means that there is no entropy production $\bar{Q}^*$ by internal sources, i.e. the condition (3.210) is really satisfied. ▲

*Remark 3.* Since the assertion (2) of Truesdell's theorem holds for uniform thermomechanical processes, for which, by (3.207), the gradient of a temperature field is zero in a whole body $\mathcal{B}$:

$$\overset{\circ}{\nabla}\theta \equiv 0 \quad \forall t \in [t_1, t_2], \quad \forall X^i \in \overset{\circ}{\overline{V}}; \tag{3.233}$$

so it follows from (3.175), (3.176), and (3.210) that the body $\mathcal{B}$, satisfying the assertion (2), must be *nondissipative*, i.e. in this body, at least, within the interval $[t_1, t_2]$ the dissipation function is zero:

$$w^* \equiv 0 \quad \forall t \in [t_1, t_2], \quad \forall X^i \in \overset{\circ}{\overline{V}}. \tag{3.234}$$

The processes, for which in a whole body the conditions (3.233) and (3.234) are satisfied simultaneously, i.e. the specific internal entropy production $q^*$ is identically zero, are often called *reversible*; and the processes, in which at least one of these conditions is violated, are called *irreversible*. □

## Exercises for 3.5

**3.5.1.** Using formula (3.128) and the result of Exercise 2.1.9, show that the dissipation functions $w^*$ and $\overset{\circ}{w}{}^*$, defined by (3.167) and (3.175), are connected by

$$w^* = (\rho/\overset{\circ}{\rho})\overset{\circ}{w}{}^*.$$

**3.5.2.** For heat machines of the second and third types, by analogy with (3.179)–(3.181) introduce the concepts of $\Delta H$, $A_{(i)}$, $C$ and $\bar{C}$ and prove that Truesdell's estimate (3.195) is valid.

Introduce the concepts of adiabatic and isothermal processes by analogy with the definition from Sect. 3.5.5 and also the concept of a thermodynamic cycle for $[t_1, t_2]$, having replaced the last relation in (3.202) by the two ones

$$V(x^i, t_1) = V(x^i, t_2), \quad \theta(x^i, t_1) - \theta(x^i, t_2) = 0 \quad \forall x^i \in V,$$

(which mean the coincidence of domains $V$ occupied by a body at times $t_1$ and $t_2$ and coincidence of temperature fields in the domains). Prove Truesdell's theorem for thermodynamic cycles occurring in heat machines of the second and third types.

## 3.6 Deformation Compatibility Equations

### 3.6.1 Compatibility Conditions

So we have completed the formulation of balance laws of continuum mechanics. However, the conditions of continuity of the motion of bodies, which have been used in Chap. 2, can also be rewritten in the form of some formal 'conservation law'. This formal law is of great importance for closing the equation system in continuum mechanics; thus, it will be consider in Sects. 3.6–3.8.

Formulate the requirement for continuity of a motion as follows: let it be known that in configuration $\overset{\circ}{\mathcal{K}}$ we can put every material point $\mathcal{M}$ of a continuum in one-to-one correspondence with its radius-vector $\overset{\circ}{\mathbf{x}}(X^k)$ in the Cartesian coordinate system $O\bar{\mathbf{e}}_i$; then we must define the conditions, under which there exists a one-valued function (the displacement vector $\mathbf{u}(X^k, t)$) connecting positions of the point $\mathcal{M}$ in $\overset{\circ}{\mathcal{K}}$ and $\mathcal{K}$. If in $\mathcal{K}$ some discontinuities appear (cracks, pores, etc.) (Fig. 3.12), then the motion is no longer continuous.

**Definition 3.12.** Necessary and sufficient conditions of the existence of a one-valued vector-function $\mathbf{u}(X^k, t)$ for a continuum $V$ are called *deformation compatibility conditions (equations)* for the continuum $V$.

If there is no one-valued vector-function $\mathbf{u}(X^k, t)$ for all $X^k \in \overset{\circ}{V}$, this means that in $\mathcal{K}$ we cannot introduce one-valued radius-vector $\mathbf{x}(X^k, t)$. Thus, the configuration $\mathcal{K}$ does not belong to Euclidean space $\mathcal{E}_3^a$. The converse statement holds as well. So we have proved the following theorem.

**Theorem 3.17.** *The deformation compatibility conditions for a continuum $V$ are equivalent to the condition that an actual configuration of the continuum belong to Euclidean space $\mathcal{E}_3^a$.*

*Remark.* At first sight it seems strange that an usual medium with cracks does not belong to an Euclidean space. But this continuum is not quite usual, because it *is*

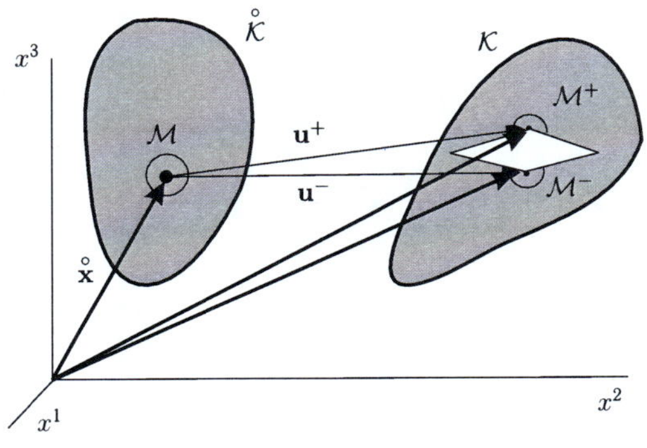


**Fig. 3.12** The example of violation of the compatibility conditions

*defined* in actual configuration $\mathcal{K}$ in such a way that in the reference configuration $\overset{\circ}{\mathcal{K}}$ this medium is assigned to a continuum already without discontinuities. If we defined a considered continuum with cracks in another way, namely as belonging to some new reference configuration $\overset{\circ}{\tilde{\mathcal{K}}}$, then each material point of the continuum would have its individual radius-vector and such a continuum would belong to Euclidean space $\mathcal{E}_3^a$. □

## 3.6.2 Integrability Condition for Differential Form

There are two types of compatibility equations: *dynamic* and *static*. In this paragraph we consider static compatibility equations. To derive the equations we should consider some differential form

$$\sum_{\alpha=1}^{3} A_\alpha \, dX^\alpha, \tag{3.235}$$

where $A_\alpha = A_\alpha(X^i)$ are smooth functions of variables $X^j$.

As known from the course of mathematical analysis, this form gives a *total differential* $dA$ if and only if the following integrability conditions hold

$$\frac{\partial A_\alpha}{\partial X^\beta} = \frac{\partial A_\beta}{\partial X^\alpha}, \quad \alpha, \beta = 1, 2, 3. \tag{3.236}$$

In this case we have the representation

$$dA = \sum_{\alpha=1}^{3} A_\alpha \, dX^\alpha. \tag{3.237}$$

This expression can be considered as an equation in differentials. The equation can be resolved for $A$ if and only if the conditions (3.236) are satisfied.

## 3.6.3 The First Form of Deformation Compatibility Conditions

Let $A$ be radius-vectors $\mathbf{x}$ and $\overset{\circ}{\mathbf{x}}$ of the same material point $\mathcal{M}$. For them the relation (3.237) is written as follows:

$$d\mathbf{x} = \mathbf{r}_i \, dX^i, \quad d\overset{\circ}{\mathbf{x}} = \overset{\circ}{\mathbf{r}}_i \, dX^i. \tag{3.238}$$

These relations can be considered from different points of view. On the one hand, if the motion law is known and in $\overset{\circ}{\mathcal{K}}$ there is a common Cartesian coordinate system, then the following relations hold

$$\overset{\circ}{\mathbf{x}} = \overset{\circ}{\mathbf{x}}(X^j), \quad \mathbf{x} = \mathbf{x}(X^j, t), \tag{3.239}$$

which are smooth functions of their arguments. Then Eqs. (3.238) are consequences of the relations (3.239). This approach has been used above.

On the other hand, if basis vectors $\mathbf{r}_i$ are known and radius-vector $\mathbf{x}$ is unknown (by the condition, the radius-vector $\overset{\circ}{\mathbf{x}}$ is always known, and it is a one-valued function), then, similarly to the expression (3.237), the first equation of (3.238) is an equation in differentials for radius-vector $\mathbf{x}$.

A solution of this equation always exists, because the integrability conditions (3.236) are satisfied. Indeed,

$$\frac{\partial \mathbf{r}_\alpha}{\partial X^\beta} = \frac{\partial \mathbf{x}}{\partial X^\beta \partial X^\alpha} = \frac{\partial \mathbf{x}}{\partial X^\alpha \partial X^\beta} = \frac{\partial \mathbf{r}_\beta}{\partial X^\alpha}. \tag{3.240}$$

Thus, if there exist local basis vectors at every point $X^i$ in $\mathcal{K}$, then there also exist radius-vectors of these points in $\mathcal{K}$.

If every material point $\mathcal{M}$ with coordinates $X^i$ in $\mathcal{K}$ may be uniquely assigned to its radius-vector $\mathbf{x}$, this means that there exist one-valued functions $\mathbf{u}(X^k, t) = \mathbf{x} - \overset{\circ}{\mathbf{x}}$ of displacements of the points from $\overset{\circ}{\mathcal{K}}$ to $\mathcal{K}$.

Conversely, if there exist one-valued displacements $\mathbf{u}(X^k, t)$ of material points from $\overset{\circ}{\mathcal{K}}$ into $\mathcal{K}$, then in $\mathcal{K}$ we can always introduce the radius-vector $\mathbf{x} = \overset{\circ}{\mathbf{x}} + \mathbf{u}$ and the local basis vectors $\mathbf{r}_i = \partial \mathbf{x}/\partial X^i$. Thus, we have proved the following theorem.

**Theorem 3.18.** *The deformation compatibility conditions for a continuum are satisfied if and only if in $\mathcal{K}$ there exist local basis vectors $\mathbf{r}_i$, i.e. functions having the following properties:*

- *they are linearly independent at every point $X^i$;*
- *they are one-valued and smooth $\forall X^i \in V$;*
- *they have a vector potential $\mathbf{x}$: $\mathbf{r}_i = \partial \mathbf{x}/\partial X^i$.*

### *3.6.4 The Second Form of Compatibility Conditions*

As $A$, we choose local bases vectors $\mathbf{r}_i$ and $\overset{\circ}{\mathbf{r}}_i$. For them, the relation (3.237) becomes

$$d\mathbf{r}_i = \frac{\partial \mathbf{r}_i}{\partial X^j}\, dX^j = (\Gamma_{ij}^m \mathbf{r}_m)\, dX^j, \quad d\overset{\circ}{\mathbf{r}}_i = \left(\overset{\circ}{\Gamma}{}_{ij}^m \overset{\circ}{\mathbf{r}}_m\right) dX^j. \tag{3.241}$$

If the motion law is known (the second equation in (3.239)) and the deformation compatibility conditions are satisfied, then relations (3.241) follow from the motion law (3.239). If the Christoffel symbols $\Gamma_{ij}^m$ are known, then (3.241) are equations for functions $\mathbf{r}_i$. The integrability conditions (3.236) for these equations have the form (the indices $j$ and $n$ change places)

$$\frac{\partial}{\partial X^n}(\Gamma_{ij}^m \mathbf{r}_m) = \frac{\partial}{\partial X^j}(\Gamma_{in}^m \mathbf{r}_m). \tag{3.242}$$

Removing the parentheses, we obtain

$$\begin{gathered} \frac{\partial \Gamma_{ij}^m}{\partial X^n} \mathbf{r}_m + \Gamma_{ij}^m \Gamma_{mn}^k \mathbf{r}_k = \frac{\partial \Gamma_{in}^m}{\partial X^j} \mathbf{r}_m + \Gamma_{in}^m \Gamma_{mj}^k \mathbf{r}_k, \\ \left( \frac{\partial \Gamma_{ij}^m}{\partial X^n} - \frac{\partial \Gamma_{in}^m}{\partial X^j} + \Gamma_{ij}^k \Gamma_{kn}^m - \Gamma_{in}^k \Gamma_{kj}^m \right) \mathbf{r}_m = 0. \end{gathered} \tag{3.243}$$

Since $\mathbf{r}_m$ are arbitrary, we have

$$R_{nji}^m \equiv \frac{\partial \Gamma_{ij}^m}{\partial X^n} - \frac{\partial \Gamma_{in}^m}{\partial X^j} + \Gamma_{ij}^k \Gamma_{kn}^m - \Gamma_{in}^k \Gamma_{kj}^m = 0, \tag{3.244}$$

where $R_{nji}^m$ is some notation for the present.

Using the condition (3.244), we can formulate the following theorem.

**Theorem 3.19.** *The deformation compatibility conditions are satisfied if and only if in $\mathcal{K}$ there exist Christoffel symbols $\Gamma_{ij}^m$ satisfying the integrability conditions (3.244).*

▼ Indeed, if there exist $\Gamma_{ij}^m$ satisfying the conditions (3.244), then the integrability conditions (3.242) remain valid. Hence the first form in (3.241) has a differential and there exist local basis vectors $\mathbf{r}_i$. Then, according to Theorem 3.18, the deformation compatibility conditions will be satisfied.

Conversely, if the compatibility conditions are satisfied, then, according to Theorem 3.18, there exist vector functions $\mathbf{r}_i$. Performing all manipulations (3.241)–(3.244), we verify that the conditions (3.244) hold. ▲

Notice that in $\overset{\circ}{\mathcal{K}}$ there always exist vectors $\overset{\circ}{\mathbf{x}}$ and $\overset{\circ}{\mathbf{r}}_i$. Hence we can perform similar manipulations for the second form in (3.241) too; as a result, we obtain the equations

$$\frac{\partial}{\partial X^n}\left(\overset{\circ}{\Gamma}{}_{ij}^m \overset{\circ}{\mathbf{r}}_m\right) = \frac{\partial}{\partial X^j}\left(\overset{\circ}{\Gamma}{}_{in}^m \overset{\circ}{\mathbf{r}}_m\right) \tag{3.245}$$

and

$$\overset{\circ}{R}{}_{nji}^m \equiv \frac{\partial \overset{\circ}{\Gamma}{}_{ij}^m}{\partial X^n} - \frac{\partial \overset{\circ}{\Gamma}{}_{in}^m}{\partial X^j} + \overset{\circ}{\Gamma}{}_{ij}^k \overset{\circ}{\Gamma}{}_{kn}^m - \overset{\circ}{\Gamma}{}_{in}^k \overset{\circ}{\Gamma}{}_{kj}^m = 0. \tag{3.246}$$

Values $R^m_{nji}$ and $\overset{\circ}{R}{}^m_{nji}$ are called components of the *Riemann–Christoffel* fourth-order tensors ${}^4\mathbf{R}$ and ${}^4\overset{\circ}{\mathbf{R}}$ in configurations $\overset{\circ}{\mathcal{K}}$ and $\mathcal{K}$, respectively. We will verify that these are really components of tensors in paragraph 3.6.6.

### 3.6.5 The Third Form of Compatibility Conditions

Introduce the *Christoffel symbols of the first kind* $\Gamma_{ijk}$ and $\overset{\circ}{\Gamma}_{ijk}$, which are connected to $\Gamma^m_{ij}$ and $\overset{\circ}{\Gamma}{}^m_{ij}$, called the *Christoffel symbols of the second kind*, by the relations (see [12])

$$\Gamma_{ijk} = g_{km}\Gamma^m_{ij}, \quad \overset{\circ}{\Gamma}_{ijk} = \overset{\circ}{g}_{km}\overset{\circ}{\Gamma}{}^m_{ij}. \tag{3.247}$$

The Christoffel symbols of the first kind and the metric matrix are connected by the formula

$$\Gamma_{ijk} = \frac{1}{2}\left(\frac{\partial g_{ik}}{\partial X^j} + \frac{\partial g_{kj}}{\partial X^i} - \frac{\partial g_{ij}}{\partial X^k}\right), \tag{3.248}$$

which follows from (2.26). Interchanging indices $i$ and $k$ in (3.248) and summing the result with $\Gamma_{ijk}$, we get

$$\Gamma_{ijk} + \Gamma_{kji} = \frac{\partial g_{ik}}{\partial X^j} + \frac{1}{2}\left(\frac{\partial g_{kj}}{\partial X^i} - \frac{\partial g_{ij}}{\partial X^k} + \frac{\partial g_{ij}}{\partial X^k} - \frac{\partial g_{kj}}{\partial X^i}\right) = \frac{\partial g_{ik}}{\partial X^j}, \tag{3.249}$$

Purely covariant components of the Riemann–Christoffel tensor are defined as follows:

$$R_{njik} = R^m_{nji}g_{mk} = g_{mk}\left(\frac{\partial(\Gamma_{ijl}g^{ml})}{\partial X^n} - \frac{\partial(\Gamma_{ikl}g^{ml})}{\partial X^j}\right) + g^{sl}(\Gamma_{ijl}\Gamma_{snk} - \Gamma_{inl}\Gamma_{sjk}). \tag{3.250}$$

According to the equation

$$g_{mk}\frac{\partial g^{ml}}{\partial X^n} = -g^{ml}\frac{\partial g_{mk}}{\partial X^n} = -g^{ml}(\Gamma_{mnk} + \Gamma_{knm}), \tag{3.251}$$

we can rewrite $R_{njik}$ in the form

$$\begin{aligned} R_{njik} &= \frac{\partial \Gamma_{ijk}}{\partial X^n} - \frac{\partial \Gamma_{ink}}{\partial X^j} - \Gamma_{ijl}(\Gamma_{mnk} + \Gamma_{knm})g^{ml} \\ &\quad + g^{ml}\Gamma_{inl}(\Gamma_{kjm} + \Gamma_{mjk}) + g^{sl}(\Gamma_{ijl}\Gamma_{snk} - \Gamma_{inl}\Gamma_{sjk}) \\ &= \frac{\partial \Gamma_{ijk}}{\partial X^n} - \frac{\partial \Gamma_{ink}}{\partial X^j} - g^{ml}(\Gamma_{inl}\Gamma_{kjm} - \Gamma_{ijl}\Gamma_{knm}). \end{aligned} \tag{3.252}$$

Substituting in place of derivatives of $\Gamma_{ijk}$ their expressions (3.248) in terms of the metric matrix components, we get

$$R_{njik} = \frac{1}{2}\left(\frac{\partial^2 g_{ik}}{\partial X^j \partial X^n} + \frac{\partial^2 g_{kj}}{\partial X^i \partial X^n} - \frac{\partial^2 g_{ij}}{\partial X^k \partial X^n} - \frac{\partial^2 g_{ik}}{\partial X^n \partial X^j} - \frac{\partial^2 g_{kn}}{\partial X^i \partial X^j} + \frac{\partial^2 g_{in}}{\partial X^k \partial X^j}\right) + g^{ml}(\Gamma_{inl}\Gamma_{kjm} - \Gamma_{ijl}\Gamma_{knm}),$$

or

$$R_{njik} = \frac{1}{2}\left(\frac{\partial^2 g_{kj}}{\partial X^i \partial X^n} - \frac{\partial^2 g_{ij}}{\partial X^k \partial X^n} + \frac{\partial^2 g_{in}}{\partial X^k \partial X^j} - \frac{\partial^2 g_{kn}}{\partial X^i \partial X^j}\right) + g^{ml}(\Gamma_{inl}\Gamma_{kjm} - \Gamma_{ijl}\Gamma_{knm}) = 0. \quad (3.253)$$

In the same way we can find the expression for $\overset{\circ}{R}_{njik}$ in terms of components of the metric matrix $\overset{\circ}{g}_{ij}$:

$$\overset{\circ}{R}_{njik} = \frac{1}{2}\left(\frac{\partial^2 \overset{\circ}{g}_{kj}}{\partial X^i \partial X^n} - \frac{\partial^2 \overset{\circ}{g}_{kn}}{\partial X^i \partial X^j} + \frac{\partial^2 \overset{\circ}{g}_{in}}{\partial X^k \partial X^j} - \frac{\partial^2 \overset{\circ}{g}_{ij}}{\partial X^k \partial X^n}\right) + \overset{\circ}{g}{}^{ml}\left(\overset{\circ}{\Gamma}_{inl}\overset{\circ}{\Gamma}_{kjm} - \overset{\circ}{\Gamma}_{ijl}\overset{\circ}{\Gamma}_{knm}\right) = 0. \quad (3.254)$$

**Theorem 3.19a.** *The deformation compatibility conditions are satisfied if and only if in $\mathcal{K}$ there exists a metric matrix $g_{ij}$ satisfying the Eqs. (3.253).*

▼ If the matrix $g_{ij}$ satisfying Eqs. (3.253) is known, then their equivalent relations (3.244) hold. As shown above, the relations (3.244) define a space to be Euclidean; and, according to Theorem 3.19, this means that the deformation compatibility conditions are satisfied too.

Conversely, if the deformation compatibility conditions are satisfied, then relations (3.244) hold and in Euclidean space the coefficients $\Gamma_{ij}^m$ have a potential being $g_{ij}$; thus, relations (3.248) hold too. Then their equivalent relations (3.253) remain valid as well. ▲

### 3.6.6 *Properties of Components of the Riemann–Christoffel Tensor*

We can check that functions $R_{njik}$ are symmetric in pairs of indices $n, j$ and $i, k$:

$$R_{njik} = R_{iknj}, \quad (3.255)$$

and also skew-symmetric in indices $n, j$ and $i, k$:

$$-R_{njik} = R_{njki}, \qquad R_{njik} = -R_{jnik}. \tag{3.256}$$

Thus, among 81 components $R_{njik}$ there are only six independent ones. They are usually chosen as follows:

$$R_{1212}, \quad R_{2323}, \quad R_{3131}, \quad R_{1223}, \quad R_{1231}, \quad R_{2331}, \tag{3.257}$$

the remaining components either are equal to zero or can be expressed in terms of components (3.257). Functions $\overset{\circ}{R}_{njik}$ have the same properties.

With the help of components $R_{njik}$ and $\overset{\circ}{R}_{njik}$ we can set up the following fourth-order tensors:

$${}^4\mathcal{R} = R_{njik}\mathbf{r}^n \otimes \mathbf{r}^j \otimes \mathbf{r}^i \otimes \mathbf{r}^k, \quad {}^4\overset{\circ}{\mathcal{R}} = \overset{\circ}{R}_{njik}\overset{\circ}{\mathbf{r}}{}^n \otimes \overset{\circ}{\mathbf{r}}{}^j \otimes \overset{\circ}{\mathbf{r}}{}^i \otimes \overset{\circ}{\mathbf{r}}{}^k, \tag{3.258}$$

which are called the *Riemann–Christoffel tensors* in $\mathcal{K}$ and $\overset{\circ}{\mathcal{K}}$.

We can verify that $R_{njik}$ and $\overset{\circ}{R}_{njik}$ are components of a tensor, for example, in the following way.

Consider an arbitrary vector $\mathbf{a} = a^k \mathbf{r}_k = \overset{\circ}{a}{}^k \overset{\circ}{\mathbf{r}}_k$ and determine its covariant derivative

$$\nabla_i a^k = \frac{\partial a^k}{\partial X^i} + \Gamma^k_{is} a^s \tag{3.259}$$

and the second covariant derivative

$$\begin{aligned}\nabla_j \nabla_i a^k &= \frac{\partial}{\partial X^j}(\nabla_i a^k) + \Gamma^k_{jm}\nabla_i a^m - \Gamma^m_{ji}\nabla_m a^k = \frac{\partial^2 a^k}{\partial X^j \partial X^i} + \frac{\partial \Gamma^k_{is}}{\partial X^j} a^s \\ &+ \Gamma^k_{is}\frac{\partial a^s}{\partial X^j} + \Gamma^k_{jm}\left(\frac{\partial a^m}{\partial X^i} + \Gamma^m_{is} a^s\right) - \Gamma^m_{ji}\left(\frac{\partial a^k}{\partial X^m} + \Gamma^k_{ms} a^s\right).\end{aligned} \tag{3.260}$$

Interchange indices $i$ and $j$ and write the difference:

$$\nabla_j \nabla_i a^k - \nabla_i \nabla_j a^k = \left(\frac{\partial \Gamma^k_{is}}{\partial X^j} - \frac{\partial \Gamma^k_{js}}{\partial X^i} + \Gamma^k_{jm}\Gamma^m_{is} - \Gamma^k_{im}\Gamma^m_{js}\right) a^s = R_{jis}{}^k a^s. \tag{3.261}$$

It follows from (3.261) that $R^k_{jis}$ are components of a tensor, because $a^s$ and the left-hand side of (3.261) are tensor components.

### 3.6.7 Interchange of the Second Covariant Derivatives

In Euclidean space $\mathcal{E}_3^a$, the following relations have been established:

$$\overset{\circ}{R}{}^{k}_{jis} = R^{k}_{jis} = 0, \tag{3.262}$$

(see formulae (3.253) and (3.254)); then from (3.261) it follows that the second covariant derivatives may be interchanged.

So we have proved the following theorem.

**Theorem 3.20.** *If the deformation compatibility conditions (3.253), and also (3.254) are satisfied, then the second covariant derivatives may be interchanged:*

$$\nabla_j \nabla_i a^k = \nabla_i \nabla_j a^k \quad \text{and} \quad \overset{\circ}{\nabla}_j \overset{\circ}{\nabla}_i \overset{\circ}{a}{}^k = \overset{\circ}{\nabla}_i \overset{\circ}{\nabla}_j \overset{\circ}{a}{}^k. \tag{3.263}$$

This differentiation rule remains valid for tensors of any order.

### 3.6.8 The Static Compatibility Equation

Just as in Sect. 3.6.5, we assume that in $\mathcal{K}$ there is a metric matrix $g_{ij}$ satisfying Eq. (3.253), and in $\overset{\circ}{\mathcal{K}}$ the continuity conditions (i.e. conditions (3.254)) are satisfied. Then in place of $g_{ij}$ we can consider components of the deformation tensor:

$$\varepsilon_{ij} = \frac{1}{2}(g_{ij} - \overset{\circ}{g}_{ij}) \tag{3.264}$$

and express the deformation compatibility conditions in terms of $\varepsilon_{ij}$.

Subtracting (3.254) from (3.253), we obtain

$$\begin{aligned} R_{njik} - \overset{\circ}{R}_{njik} = {} & \frac{\partial^2 \varepsilon_{kj}}{\partial X^i \partial X^n} - \frac{\partial^2 \varepsilon_{kn}}{\partial X^i \partial X^j} + \frac{\partial^2 \varepsilon_{in}}{\partial X^k \partial X^j} - \frac{\partial^2 \varepsilon_{ij}}{\partial X^k \partial X^n} \\ & + g^{ml}(\Gamma_{inl}\Gamma_{kjm} - \Gamma_{ijl}\Gamma_{knm}) - \overset{\circ}{g}{}^{ml}(\overset{\circ}{\Gamma}_{inl}\overset{\circ}{\Gamma}_{kjm} - \overset{\circ}{\Gamma}_{ijl}\overset{\circ}{\Gamma}_{knm}) = 0, \end{aligned} \tag{3.265}$$

where

$$\Gamma_{ijk} = \overset{\circ}{\Gamma}_{ijk} + \varepsilon_{ijk}, \quad \varepsilon_{ijk} = \frac{\partial \varepsilon_{ik}}{\partial X^j} + \frac{\partial \varepsilon_{jk}}{\partial X^i} - \frac{\partial \varepsilon_{ij}}{\partial X^k}. \tag{3.266}$$

The inverse metric matrix $g^{ml}$ can be expressed in terms of deformation tensor components according to formulae (2.131). Since all the functions $\overset{\circ}{\Gamma}_{inl}$ and $\overset{\circ}{g}{}^{ml}$ in the reference configuration are assumed to be known, the relations (3.265), with taking (3.266) and (2.131) into account, state the system of six scalar equations (because among components $R_{njik}$ there are only six independent ones) for six scalar functions $\varepsilon_{ij}$. These equations are called *static equations of compatibility* (or *deformation compatibility equations*).

Thus, we have proved the following theorem.

**Theorem 3.21.** *The deformation compatibility conditions for a continuum are satisfied if and only if in $\mathcal{K}$ the deformation tensor components $\varepsilon_{ij}$ satisfy Eqs. (3.265).*

Finally, let us formulate one more important theorem.

**Theorem 3.22.** *The deformation compatibility equations (3.265) have the solution*

$$\varepsilon_{ij} = \frac{1}{2}\left(\mathring{\nabla}_i \mathring{u}_j + \mathring{\nabla}_j \mathring{u}_i + \mathring{\nabla}_i \mathring{u}_k \mathring{\nabla}_j \mathring{u}_l \mathring{g}^{kl}\right). \tag{3.267}$$

Sometimes this result is said as follows: a solution of Eqs. (3.265) admits a potential, i.e. six functions $\varepsilon_{ij}$ are expressed in terms of covariant derivatives of three functions $\mathring{u}_i$.

▼ To prove the theorem we can replace the covariant derivatives by the partial ones:

$$\mathring{\nabla}_i \mathring{u}_j = \mathring{u}_{j/i} - \mathring{\Gamma}^m_{ij} \mathring{u}_m, \tag{3.268}$$

and, substituting the expressions (3.268) and (3.267) into Eq. (3.265), verify that this equation will be identically satisfied. However, this method is very clumsy. So let us prove the theorem in another way.

Let there be a solution of Eqs. (3.265) (functions $\varepsilon_{ij}$). According to Theorem 3.21, this means that the deformation compatibility equations are satisfied. Then, by Definition 3.12, there exists a one-valued vector-function $\mathbf{u}(X^k, t) = \mathbf{x} - \mathring{\mathbf{x}}$ being the displacement vector of material points. By components of this vector $\mathring{u}_i$, according to formulae (2.93), we can always find components $\widetilde{\varepsilon}_{ij}$, which coincide with $\varepsilon_{ij}$ (3.267). Indeed, if $\widetilde{\varepsilon}_{ij} \neq \varepsilon_{ij}$, then this means that there exist two distinct solutions of the deformation compatibility equations (3.265) and their corresponding, by Definition 3.12, different displacement vectors; that is impossible due to uniqueness of displacements $\mathbf{u}(X^k, t)$ of a continuum under the transformation from $\mathring{\mathcal{K}}$ to $\mathcal{K}$. Thus, $\widetilde{\varepsilon}_{ij} = \varepsilon_{ij}$, and formula (3.267) really holds. ▲

## 3.7 Dynamic Compatibility Equations

### *3.7.1 Dynamic Compatibility Equations in Lagrangian Description*

The deformation compatibility equations can be written in one more equivalent form, namely in terms of the velocity.

Consider Eq. (2.77) connecting $\mathbf{F}$ to $\overset{\circ}{\nabla} \otimes \mathbf{u}^{\mathrm{T}}$, and differentiate the equation with respect to $t$, taking the definition (2.181) of the velocity $\mathbf{v}$ into account:

$$\frac{d}{dt}\overset{\circ}{\nabla} \otimes \mathbf{u} = \overset{\circ}{\mathbf{r}}{}^{i} \otimes \frac{\partial^2 \mathbf{u}}{\partial X^i \partial t} = \overset{\circ}{\mathbf{r}}{}^{i} \otimes \frac{\partial \mathbf{v}}{\partial X^i} = \overset{\circ}{\nabla} \otimes \mathbf{v} = \frac{d\mathbf{F}^{\mathrm{T}}}{dt}. \tag{3.269}$$

Then we get the *dynamic compatibility equation in Lagrangian description*:

$$\frac{d\mathbf{F}^{\mathrm{T}}}{dt} = \overset{\circ}{\nabla} \otimes \mathbf{v}. \tag{3.270}$$

**Theorem 3.23.** *The deformation compatibility conditions are satisfied if and only if in $\mathcal{K}$ there is a tensor field $\mathbf{F}(X^i, t)$ satisfying the following conditions:*

- $\det \mathbf{F} \neq 0$ *at every point* $X^i$*;*
- $\mathbf{F}(X^i, 0) = \mathbf{E}$ *at* $t = 0$*;*
- *the field* $\mathbf{F}$ *has a vector potential, i.e. there exists a vector field* $\mathbf{v}$ *such that equation (3.270) holds* $\forall t > 0$ *and* $\forall X^i \in V$.

▼ Let the deformation compatibility conditions be satisfied, then, by Definition 3.12, there is a displacement vector $\mathbf{u}(X^i, t)$ with its gradient $\overset{\circ}{\nabla} \otimes \mathbf{u}$. On making the manipulations (3.269), we verify that Eq. (3.270) remains valid.

Prove that the converse assertion is valid. Let there be a vector-function $\mathbf{v}$ satisfying Eq. (3.270). Consider the function $\widetilde{\mathbf{u}}(X^i, t) = \int_0^t \mathbf{v}(X^i, \tau)\, d\tau$. This function satisfies the equation

$$\overset{\circ}{\nabla} \otimes \widetilde{\mathbf{u}} = \overset{\circ}{\nabla} \otimes \int_0^t \mathbf{v}\, d\tau = \int_0^t \overset{\circ}{\nabla} \otimes \mathbf{v}\, d\tau = \int_0^t \frac{d\mathbf{F}^{\mathrm{T}}}{d\tau}\, d\tau = \mathbf{F}^{\mathrm{T}} - \mathbf{E}. \tag{3.271}$$

Then with the help of this function we can construct the radius-vector $\widetilde{\mathbf{x}} = \overset{\circ}{\mathbf{x}} + \widetilde{\mathbf{u}}$. As a result, the tensor $\mathbf{F}$ takes the form

$$\mathbf{F}^{\mathrm{T}} = \mathbf{E} + \overset{\circ}{\nabla} \otimes \widetilde{\mathbf{u}} = \overset{\circ}{\mathbf{r}}{}^{i} \otimes \frac{\partial(\overset{\circ}{\mathbf{x}} + \widetilde{\mathbf{u}})}{\partial X^i} = \overset{\circ}{\mathbf{r}}{}^{i} \otimes \widetilde{\mathbf{r}}_i,$$

where $\widetilde{\mathbf{r}}_i = \partial\widetilde{\mathbf{x}}/\partial X^i$. But this means that the vector $\widetilde{\mathbf{u}}$ is the desired displacement vector $\mathbf{u}$, and $\mathbf{F}$ is the deformation gradient, because they satisfy all the kinematic relations: (2.35), (2.77), etc. Thus, the displacement vector really exists $\mathbf{u}$, and hence the deformation compatibility conditions are satisfied. ▲

### 3.7.2 Dynamic Compatibility Equations in Spatial Description

At first, let us prove the following auxiliary statement.

**Theorem 3.24.** *If the continuity equation (3.15) holds, then the deformation gradient satisfies the following relationship:*

$$\nabla \cdot (\rho \mathbf{F}) = 0. \tag{3.272}$$

▼ Represent the deformation gradient in the dyadic basis:

$$\rho \mathbf{F} = \rho F^{ij} \mathbf{r}_i \otimes \mathbf{r}_j. \tag{3.273}$$

Let us write the following formula for the divergence of a tensor (see [12]):

$$\nabla \cdot (\rho \mathbf{F}) = \frac{1}{\sqrt{g}} \frac{\partial}{\partial X^i} \left(\rho \sqrt{g}\, F^{ij} \mathbf{r}_j\right) = \frac{\overset{\circ}{\rho}}{\sqrt{g}} \frac{\partial}{\partial X^i} \left(\sqrt{\overset{\circ}{g}}\, \overset{\circ}{\mathbf{r}}{}^i\right). \tag{3.274}$$

Here we have used the continuity equation $\rho = \overset{\circ}{\rho}\sqrt{\overset{\circ}{g}/g}$, and also the evident relations

$$F^{ik}\mathbf{r}_k = F^{jk}\mathbf{r}^i \cdot \mathbf{r}_j \otimes \mathbf{r}_k = \mathbf{r}^i \cdot \mathbf{F} = (\mathbf{r}^i \cdot \mathbf{r}_k) \otimes \overset{\circ}{\mathbf{r}}{}^k = \overset{\circ}{\mathbf{r}}{}^i. \tag{3.275}$$

Differentiating (3.274) by parts, we get

$$\nabla \cdot (\rho \mathbf{F}) = \frac{\overset{\circ}{\rho}}{\sqrt{g}} \left( \frac{\partial \sqrt{\overset{\circ}{g}}}{\partial X^i} \overset{\circ}{\mathbf{r}}{}^i + \sqrt{\overset{\circ}{g}} \frac{\partial \overset{\circ}{\mathbf{r}}{}^i}{\partial X^i} \right) = \frac{\overset{\circ}{\rho}}{\sqrt{g}} \left( \sqrt{\overset{\circ}{g}}\, \overset{\circ}{\Gamma}{}^s_{is} \overset{\circ}{\mathbf{r}}{}^i - \sqrt{\overset{\circ}{g}}\, \overset{\circ}{\Gamma}{}^s_{is} \overset{\circ}{\mathbf{r}}{}^i \right) = 0. \tag{3.276}$$

Here we have used the properties of the Christoffel symbols (see [12]). ▲

Modify Eq. (3.270) with use of (2.37):

$$\frac{d\mathbf{F}^{\mathrm{T}}}{dt} = \mathbf{F}^{\mathrm{T}} \cdot \nabla \otimes \mathbf{v}, \tag{3.277}$$

and use the continuity equation (3.15) multiplied by $\mathbf{F}^{\mathrm{T}}$:

$$\frac{\partial \rho}{\partial t}\mathbf{F}^{\mathrm{T}} + \mathbf{F}^{\mathrm{T}} \nabla \cdot \rho \mathbf{v} = 0. \tag{3.278}$$

Multiply (3.277) by $\rho$ and use the definition (2.187) of the total derivative with respect to time:

$$\rho \frac{d\mathbf{F}^{\mathrm{T}}}{dt} = \rho \frac{\partial \mathbf{F}^{\mathrm{T}}}{\partial t} + \rho \mathbf{v} \cdot \nabla \otimes \mathbf{F}^{\mathrm{T}} = \rho \mathbf{F}^{\mathrm{T}} \cdot \nabla \otimes \mathbf{v}. \tag{3.279}$$

On summing Eqs. (3.278) and (3.279), we get

$$\frac{\partial \rho \mathbf{F}^{\mathrm{T}}}{\partial t} + \nabla \cdot (\rho \mathbf{v} \otimes \mathbf{F}^{\mathrm{T}}) - \rho \mathbf{F}^{\mathrm{T}} \cdot \nabla \otimes \mathbf{v} = 0. \tag{3.280}$$

Tensor multiplication of Eq. (3.272) by $-\mathbf{v}$ (i.e. $-\nabla \cdot (\rho \mathbf{F}) \otimes \mathbf{v} = 0$) and summation of the obtained expression with (3.280) give the *dynamic compatibility equation in the spatial description*:

$$\frac{\partial \rho \mathbf{F}^{\mathrm{T}}}{\partial t} + \nabla \cdot \left(\rho \mathbf{v} \otimes \mathbf{F}^{\mathrm{T}} - \rho \mathbf{F} \otimes \mathbf{v}\right) = 0. \tag{3.281}$$

In order for the deformation compatibility conditions be satisfied in $\mathcal{K}$, it is necessary and sufficient that Eq. (3.281), as well as (3.270), hold.

Transforming the first and the second summands in (3.281) by formula (3.20), we can rewrite the dynamic compatibility equation in the form

$$\rho \frac{d\mathbf{F}^{\mathrm{T}}}{dt} = \nabla \cdot (\rho \mathbf{F} \otimes \mathbf{v}). \tag{3.282}$$

## Exercises for 3.7

**3.7.1.** Show that the dynamic compatibility equation (3.270) can be written in the divergence form

$$d\mathbf{F}^{\mathrm{T}}/dt = \overset{\circ}{\nabla} \cdot (\mathbf{E} \otimes \mathbf{v}).$$

**3.7.2.** Using the relation $\mathbf{F}^{-1\mathrm{T}} \cdot \mathbf{F}^{-1} = \mathbf{E}$, show that the dynamic compatibility equation (3.270) can be rewritten as follows:

$$d\mathbf{F}^{-1\mathrm{T}}/dt = -(\nabla \otimes \mathbf{v}) \cdot \mathbf{F}^{-1\mathrm{T}}.$$

## 3.8 Compatibility Equations for Deformation Rates

Equations (3.270) and (3.281) express a condition of the existence of displacements $\mathbf{u}$, when a field of the deformation gradient $\mathbf{F}$ is given. The deformation compatibility conditions can also be formulated as conditions for a field of the deformation rate tensor $\mathbf{D}$.

**Theorem 3.25.** *If a continuum does not contain discontinuities in* $\overset{\circ}{\mathcal{K}}$ *and the symmetric deformation rate tensor* $\mathbf{D}$ *(2.225) is given in* $\mathcal{K}$*, then the continuum does not contain discontinuities in* $\mathcal{K}$ *if and only if the following compatibility conditions for deformation rates are satisfied:*

$$\mathrm{Ink}\,\mathbf{D} = 0. \tag{3.283}$$

Here the differential operator of incompatibility (see [12]) has been introduced as follows:

$$\mathrm{Ink}\,\mathbf{D} = \nabla \times (\nabla \times \mathbf{D}^{\mathrm{T}}) = (1/g)\epsilon^{ijk}\epsilon^{mnl}\nabla_i\nabla_m D_{jn}\mathbf{r}_k \otimes \mathbf{r}_l, \tag{3.284}$$

or in the component form

$$(\mathrm{Ink}\,\mathbf{D})^{kl} = (1/g)\epsilon^{ijk}\epsilon^{\alpha\beta l}(\nabla_i\nabla_\alpha D_{j\beta} - \nabla_i\nabla_\beta D_{j\alpha}), \;\; \alpha \neq \beta \neq l. \tag{3.285}$$

▼ Show the necessity of (3.283). If a continuum in $\mathcal{K}$ does not contain discontinuities, then there exist one-valued functions of coordinates: a radius-vector $\mathbf{x}$, a displacement vector $\mathbf{u}$ and a velocity $\mathbf{v}$. According to the definitions of the tensor $\mathbf{D}$ (2.225) and of the operator Ink (3.284), we get

$$\mathrm{Ink}\,\mathbf{D} = \frac{1}{2g}\epsilon^{ijk}\epsilon^{mnl}(\nabla_i\nabla_m\nabla_j v_n + \nabla_i\nabla_m\nabla_n v_j)\mathbf{r}_k \otimes \mathbf{r}_l = 0. \tag{3.286}$$

Here we have used the property (3.263) of interchange of covariant derivatives $\nabla_i\nabla_j$ and $\nabla_m\nabla_n$, and also the property of contraction of the Levi-Civita symbols with components of an arbitrary symmetric tensor (see Exercise 2.1.13).

Prove the sufficiency of the conditions (3.283). Let the symmetric tensor $\mathbf{D}$, satisfying the conditions (3.283), be given. Then we can write the following differential form similar to (3.235):

$$d\omega = (\nabla \times \mathbf{D}) \cdot d\mathbf{x} = (\nabla \times \mathbf{D} \cdot \mathbf{r}_i)\, dX^i, \tag{3.287}$$

which is a total differential if and only if the following integrability conditions of the type (3.236) are satisfied:

$$\frac{\partial}{\partial X^j}(\nabla \times \mathbf{D} \cdot \mathbf{r}_i) = \frac{\partial}{\partial X^i}(\nabla \times \mathbf{D} \cdot \mathbf{r}_j). \tag{3.288}$$

These conditions with use of the definition (2.31) of the tensor curl $(\nabla \times \mathbf{D})$ take the form

$$\nabla_j\left(\frac{1}{\sqrt{g}}\,\epsilon^{psk}\nabla_p D_{si}\right)\mathbf{r}_k = \nabla_i\left(\frac{1}{\sqrt{g}}\,\epsilon^{psk}\nabla_p D_{sj}\right)\mathbf{r}_k. \tag{3.289}$$

Due to Ricci's theorem (see [12]), the metric matrix $g_{ij}$ and $\sqrt{g}$ can be placed outside the covariant derivative sign; therefore the relation (3.289) is rewritten as follows:

$$\epsilon^{psk}(\nabla_p\nabla_j D_{si} - \nabla_p\nabla_i D_{sj}) = 0. \tag{3.290}$$

On comparing formulae (3.290) with (3.285), we obtain that condition (3.288) is equivalent to the condition

$$\nabla \times (\nabla \times \mathbf{D})^{\mathrm{T}} = 0 \tag{3.291}$$

(formula (3.283)). Thus, the conditions (3.288) are always satisfied, because the condition (3.291) is assumed to be satisfied and hence the integral form (3.287) is a total differential, and $\boldsymbol{\omega}$ is a one-valued function $\boldsymbol{\omega}(\mathbf{x})$. Then the total differential can be written in the form

$$d\boldsymbol{\omega} = \nabla \otimes \boldsymbol{\omega}^{\mathrm{T}} \cdot d\mathbf{x}. \tag{3.292}$$

It follows from (3.287) and (3.292) that the curl of the tensor $\mathbf{D}$ under the condition (3.291) can always be represented as the transpose gradient of some vector $\boldsymbol{\omega}$:

$$\nabla \times \mathbf{D} = \nabla \otimes \boldsymbol{\omega}^{\mathrm{T}}. \tag{3.293}$$

With the help of the vector $\boldsymbol{\omega}$ by formula (2.47) we can form the second-order tensor

$$\mathbf{W} = \boldsymbol{\omega} \times \mathbf{E} = \sqrt{g}\, \epsilon_{mik} \omega^m \mathbf{r}^i \otimes \mathbf{r}^k, \tag{3.294}$$

which is skew-symmetric, because $\epsilon_{mik}\omega^m = -\epsilon_{mki}\omega^m$.

With the help of $\mathbf{D}$ and $\mathbf{W}$ we can write one more differential form:

$$d\mathbf{v} = (\mathbf{D} + \mathbf{W}) \cdot d\mathbf{x}. \tag{3.295}$$

Similarly to the integrability conditions (3.291) for the form (3.287), the integrability conditions for the form (3.295) can be represented as follows:

$$\nabla \times (\mathbf{D} + \mathbf{W})^{\mathrm{T}} = 0. \tag{3.296}$$

But this condition is always satisfied, because

$$\begin{aligned} \nabla \times (\mathbf{D} + \mathbf{W})^{\mathrm{T}} &= \nabla \otimes \boldsymbol{\omega}^{\mathrm{T}} - \nabla \times \mathbf{W} = \nabla \otimes \boldsymbol{\omega}^{\mathrm{T}} - \nabla \times (\boldsymbol{\omega} \times \mathbf{E}) \\ &= \nabla \otimes \boldsymbol{\omega}^{\mathrm{T}} + (\nabla \cdot \boldsymbol{\omega})\mathbf{E} - \nabla \otimes \boldsymbol{\omega}^{\mathrm{T}} + \boldsymbol{\omega} \cdot \nabla \otimes \mathbf{E} - \boldsymbol{\omega} \otimes \nabla \cdot \mathbf{E} \\ &= \nabla \otimes \boldsymbol{\omega}^{\mathrm{T}} - \nabla \otimes \boldsymbol{\omega}^{\mathrm{T}} = 0. \end{aligned} \tag{3.297}$$

Here we have used formula (3.294) and the property of the operator $\nabla \times (\boldsymbol{\omega} \times \mathbf{E})$ (see [12]), and also taken into account that $\nabla \otimes \mathbf{E} = \mathbf{0}$, $\nabla \cdot \mathbf{E} = 0$ and $\nabla \cdot \boldsymbol{\omega} = 0$ due to

$$\begin{aligned} \nabla_m \omega^m &= \nabla \otimes \boldsymbol{\omega}^{\mathrm{T}} \cdot\cdot\, \mathbf{E} = \nabla \times \mathbf{D} \cdot\cdot\, \mathbf{E} \\ &= \frac{1}{\sqrt{g}} \epsilon^{ijk} \nabla_i D_{jl} \mathbf{r}_k \otimes \mathbf{r}^l \cdot\cdot\, \mathbf{r}_m \otimes \mathbf{r}^m = \frac{1}{\sqrt{g}} \epsilon^{ijk} \nabla_i D_{jk} = 0. \end{aligned} \tag{3.298}$$

Thus, the form (3.295) is a total differential, and there exists a vector $\mathbf{v}$, being a one-valued function of $\mathbf{x}$, hence by (2.222) we can represent its differential in the form

$$d\mathbf{v} = (\nabla \otimes \mathbf{v})^{\mathrm{T}} \cdot d\mathbf{x}. \tag{3.299}$$

On comparing (3.299) with (3.295), we get that $\nabla \otimes \mathbf{v}^{\mathrm{T}}$ can be represented as a sum of symmetric $\mathbf{D}$ and skew-symmetric $\mathbf{W}$ tensors:

$$\nabla \otimes \mathbf{v}^{\mathrm{T}} = \mathbf{D} + \mathbf{W}, \tag{3.300}$$

but this decomposition is unique, therefore by (2.225) and (2.226) we get

$$\mathbf{D} = (1/2)(\nabla \otimes \mathbf{v}^{\mathrm{T}} + \nabla \otimes \mathbf{v}), \tag{3.301}$$
$$\mathbf{W} = (1/2)(\nabla \otimes \mathbf{v} - \nabla \otimes \mathbf{v}^{\mathrm{T}}), \tag{3.302}$$

that proves the existence of the vector $\mathbf{v}$, satisfying Eq. (3.301) and hence being the velocity. ▲

As known from Sect. 2.4.5, the tensor $\mathbf{W}$, defined by (3.294) and (3.302), is called the *vorticity tensor*, and $\boldsymbol{\omega}$ – the *vorticity vector*.

## 3.9 The Complete System of Continuum Mechanics Laws

### *3.9.1 The Complete System in Eulerian Description*

Consider a nonpolar continuum.

The system of continuum mechanics laws in the spatial description (3.18), (3.72), (3.118), (3.165), and (3.282) can be rewritten in the generalized form

$$\rho \frac{d\bar{A}_\alpha}{dt} = \nabla \cdot \bar{B}_\alpha + \rho C_\alpha, \qquad \alpha = 1, \ldots, 6, \tag{3.303}$$

where the following generalized vectors are denoted:

$$\bar{A}_\alpha = \begin{pmatrix} 1/\rho \\ \mathbf{v} \\ e + |v|^2/2 \\ \eta \\ \mathbf{u} \\ \mathbf{F}^{\mathrm{T}} \end{pmatrix}, \quad \bar{B}_\alpha = \begin{pmatrix} \mathbf{v} \\ \mathbf{T} \\ \mathbf{T}\cdot\mathbf{v} - \mathbf{q} \\ -\mathbf{q}/\theta \\ \mathbf{0} \\ \rho\mathbf{F}\otimes\mathbf{v} \end{pmatrix}, \quad C_\alpha = \begin{pmatrix} 0 \\ \mathbf{f} \\ \mathbf{f}\cdot\mathbf{v} + q_m \\ (q_m + q^*)/\theta \\ \mathbf{v} \\ \mathbf{0} \end{pmatrix}. \tag{3.304}$$

Here index $\alpha = 1$ corresponds to the continuity equation,
$\alpha = 2$ – to the momentum balance equation,
$\alpha = 3$ – to the energy balance equation,
$\alpha = 4$ – to the entropy balance equation,
$\alpha = 5$ – to the kinematic equation,
$\alpha = 6$ – to the dynamic compatibility equation.

Equation (3.303) at $\alpha = 5$ has been obtained from (2.190) with the help of its multiplication by $\rho$:

$$\rho \frac{d\mathbf{u}}{dt} = \rho \mathbf{v}. \tag{3.305}$$

This relation is called the *kinematic equation*. The system (3.303) is called the universal system of continuum mechanics laws in total differentials.

This system can be reduced to the divergence form. To do this, we can collect the corresponding divergence forms of Eqs. (3.15), (3.73), (3.119), (3.165a), and (3.281), or can transform the left-hand side of the system (3.303) with the help of the continuity equation as follows:

$$\rho \frac{\partial \bar{A}_\alpha}{\partial t} + \rho \mathbf{v} \cdot \nabla \otimes \bar{A}_\alpha + \left(\frac{\partial \rho}{\partial t} + \nabla \cdot \rho \mathbf{v}\right) \bar{A}_\alpha = \frac{\partial \rho \bar{A}_\alpha}{\partial t} + \nabla \cdot \rho \mathbf{v} \otimes \bar{A}_\alpha,$$
$$\alpha = 2, \ldots, 6. \tag{3.306}$$

Then together with the continuity equation, the following complete system of continuum mechanics laws in the spatial description in the *divergence form* holds:

$$\frac{\partial \rho \, A_\alpha}{\partial t} + \nabla \cdot (\rho \mathbf{v} \otimes \, A_\alpha - \, B_\alpha) = \rho C_\alpha, \quad \alpha = 1, \ldots, 6, \tag{3.307}$$

where

$$A_\alpha = \begin{pmatrix} 1 \\ \mathbf{v} \\ e + |v|^2/2 \\ \eta \\ \mathbf{u} \\ \mathbf{F}^{\mathrm{T}} \end{pmatrix}, \quad B_\alpha = \begin{pmatrix} 0 \\ \mathbf{T} \\ \mathbf{T} \cdot \mathbf{v} - \mathbf{q} \\ -\mathbf{q}/\theta \\ \mathbf{0} \\ \rho \mathbf{F} \otimes \mathbf{v} \end{pmatrix}. \tag{3.308}$$

In particular, the kinematic equation (3.305) has the divergence form

$$\frac{\partial \rho \mathbf{u}}{\partial t} + \nabla \cdot \rho \mathbf{v} \otimes \mathbf{u} = \rho \mathbf{v}. \tag{3.309}$$

Thus, we have proved the following theorem.

**Theorem 3.26.** *The complete system of continuum mechanics laws in Eulerian description can be represented in the universal form in total differentials (3.303) and in the equivalent divergence form (3.307).*

### 3.9.2 The Complete System in Lagrangian Description

In Lagrangian description, the system of continuum mechanics laws (3.8), (3.76), (3.131), (3.173), (2.190), and (3.270) can also be written in the universal form

$$\overset{\circ}{\rho}\frac{d\overset{\circ}{A}_\alpha}{dt} = \overset{\circ}{\nabla}\cdot\overset{\circ}{B}_\alpha + \overset{\circ}{\rho}C_\alpha, \qquad \alpha = 1,\ldots,6, \tag{3.310}$$

where two more generalized vectors appear:

$$\overset{\circ}{A}_\alpha = \begin{pmatrix} (\rho/\overset{\circ}{\rho})\det\mathbf{F} \\ \mathbf{v} \\ e + |v^2|/2 \\ \eta \\ \mathbf{u} \\ \mathbf{F}^{\mathrm{T}} \end{pmatrix}, \qquad \overset{\circ}{B}_\alpha = \begin{pmatrix} 0 \\ \mathbf{P} \\ \mathbf{P}\cdot\mathbf{v} - \overset{\circ}{\mathbf{q}} \\ -\overset{\circ}{\mathbf{q}}/\theta \\ 0 \\ \overset{\circ}{\rho}\mathbf{E}\otimes\mathbf{v} \end{pmatrix}. \tag{3.311}$$

Since in a Lagrangian coordinate system the total derivative $d/dt$ coincides with the partial derivative $\partial/\partial t$, the equation set (3.310) already has the divergence form

*Remark.* It should be noted, that the first equation in the system (3.310) at $\alpha = 1$ has been obtained by differentiation of the continuity equation (3.8) with respect to Lagrangian variables. In this case we must complement this differential equation with the initial condition $\det\mathbf{F} = 1$. Then Eq. (3.310) at $\alpha = 1$ with this initial condition always has a solution, which is the relation (3.8): $(\rho/\overset{\circ}{\rho})\det\mathbf{F} = 1$. Hence, as function $\overset{\circ}{A}_1$ in (3.311) we can always use its actual value: $\overset{\circ}{A}_1 = 1$. This means that the generalized vectors $\overset{\circ}{A}_\alpha$ in (3.311) and $A_\alpha$ in (3.308) are coincident: $\overset{\circ}{A}_\alpha = A_\alpha$, $\alpha = 1,\ldots,6$. □

### 3.9.3 Integral Form of the System of Continuum Mechanics Laws

The system of continuum mechanics laws in the integral form (3.5), (3.69), (3.116), and (3.162) can be written in the universal form as well.

**Theorem 3.27.** *The system of continuum mechanics laws in the differential form (3.307) in the spatial description is equivalent to the integral form*

$$\frac{d}{dt}\int_V \rho A_\alpha\, dV = \int_\Sigma \mathbf{n}\cdot B_\alpha\, d\Sigma + \int_V \rho C_\alpha\, dV, \qquad \alpha = 1,\ldots,6. \tag{3.312}$$

▼ To prove the theorem for $\alpha = 1,\ldots,4$, it is sufficient to substitute the generalized vectors $A_\alpha$, $B_\alpha$ and $C_\alpha$ into (3.312) and to write Eq. (3.312) for each of $\alpha = 1,\ldots,4$. As a result, we obtain the relations (3.5), (3.69), (3.116), and (3.162) proved above.

For $\alpha = 5, 6$, we should integrate Eqs. (3.309) and (3.281) over $V$ and then apply the rule of differentiation of an integral over a moving volume (see Exercise 3.1.2) and the Gauss–Ostrogradskii theorem (3.24). ▲

In the same way we can prove the following theorem (see Exercise 3.9.2).

**Theorem 3.28.** *The system of continuum mechanics laws in the differential form in the material description (3.310) is equivalent to the integral form*

$$\frac{d}{dt}\int\limits_{\overset{\circ}{V}} \overset{\circ}{\rho}\overset{\circ}{A}_\alpha\, d\overset{\circ}{V} = \int\limits_{\overset{\circ}{\Sigma}} \overset{\circ}{\mathbf{n}}\cdot \overset{\circ}{B}_\alpha\, d\overset{\circ}{\Sigma} + \int\limits_{\overset{\circ}{V}} \overset{\circ}{\rho} C_\alpha\, d\overset{\circ}{V}, \quad \alpha = 1,\ldots,6. \tag{3.313}$$

In spite of the above-mentioned equivalence of differential and integral forms of the continuum mechanics laws, there is a considerable difference between them: in the integral form the angular momentum balance equation (3.82) even for nonpolar continua is not identically satisfied, while the corresponding equation in the differential form (see Sect. 3.3), reduced to symmetry of the stress tensor $\mathbf{T}$, has been excluded from the system (3.303). Thus, the integral form (3.312) must be complemented with the law (3.82); in other words, the index $\alpha$ ranges from 1 to 7 for the generalized vectors $A_\alpha$, $B_\alpha$ and $C_\alpha$ in (3.312):

$$A_\alpha = \begin{pmatrix} 1 \\ \mathbf{v} \\ e + |v|^2/2 \\ \eta \\ \mathbf{u} \\ \mathbf{F}^{\mathrm{T}} \\ \mathbf{x}\times\mathbf{v} \end{pmatrix}, \quad B_\alpha = \begin{pmatrix} 0 \\ \mathbf{T} \\ \mathbf{T}\cdot\mathbf{v}-\mathbf{q} \\ -\mathbf{q}/\theta \\ \mathbf{0} \\ \rho\mathbf{F}\otimes\mathbf{v} \\ -\mathbf{T}\times\mathbf{x} \end{pmatrix}, \quad C_\alpha = \begin{pmatrix} 0 \\ \mathbf{f} \\ \mathbf{f}\cdot\mathbf{v}+q_m \\ (q_m+q^*)/\theta \\ \mathbf{v} \\ \mathbf{0} \\ \mathbf{x}\times\mathbf{f} \end{pmatrix}. \tag{3.314}$$

The integral form (3.313) in the material description must consist of seven equations as well; and the generalized vectors $\overset{\circ}{A}_\alpha$ and $\overset{\circ}{B}_\alpha$ have the forms

$$\overset{\circ}{A}_\alpha = \begin{pmatrix} (\rho/\overset{\circ}{\rho})\det\mathbf{F} \\ \mathbf{v} \\ e + |v^2|/2 \\ \eta \\ \mathbf{u} \\ \mathbf{F}^{\mathrm{T}} \\ \mathbf{x}\times\mathbf{v} \end{pmatrix}, \quad \overset{\circ}{B}_\alpha = \begin{pmatrix} 0 \\ \mathbf{P} \\ \mathbf{P}\cdot\mathbf{v}-\overset{\circ}{\mathbf{q}} \\ -\overset{\circ}{\mathbf{q}}/\theta \\ 0 \\ \overset{\circ}{\rho}\mathbf{E}\otimes\mathbf{v} \\ -\mathbf{P}\times\mathbf{x} \end{pmatrix}. \tag{3.315}$$

## Exercises for 3.9

**3.9.1.** Prove Theorem 3.27.

**3.9.2.** Prove Theorem 3.28.

**3.9.3.** Consider an arbitrary varying domain $\tilde{V}(t) \subset V$, whose boundary points $\tilde{\Sigma}(t)$ move with velocity $\mathbf{c} = d\mathbf{x}_\Sigma/dt$, $\mathbf{x}_\Sigma \in \tilde{\Sigma}(t)$ (see Sect. 5.1.6), and show with use of formula (2.12) that in this case the complete system of continuum mechanics laws in the divergence form (3.307) yields one more integral formulation

$$\frac{d}{dt}\int\limits_{\tilde{V}} \rho A_\alpha \, dV + \int\limits_{\tilde{\Sigma}} \mathbf{n} \cdot (\rho(\mathbf{v} - \mathbf{c}) \otimes A_\alpha - B_\alpha) \, d\Sigma = \int\limits_{\tilde{V}} \rho C_\alpha \, dV. \qquad (3.316)$$

In the particular case when the domain $\tilde{V}$ is fixed, the last formula reduces to

$$\frac{d}{dt}\int\limits_{\tilde{V}} \rho A_\alpha \, dV + \int\limits_{\tilde{\Sigma}} \mathbf{n} \cdot (\rho\mathbf{v} \otimes A_\alpha - B_\alpha) \, d\Sigma = \int\limits_{\tilde{V}} \rho C_\alpha \, dV. \qquad (3.317)$$

# Chapter 4
# Constitutive Equations

## 4.1 Basic Principles for Derivation of Constitutive Equations

Consider nonpolar continua.

The equation system (3.307) consists of 18 scalar equations (each of the vector equations in (3.307) is equivalent to three scalar equations, and each of the tensor ones is equivalent to nine scalar equations), but involves 29 scalar unknowns:

$$\rho,\ \mathbf{v},\ \mathbf{u},\ \mathbf{T},\ e,\ \eta,\ \theta,\ \mathbf{q},\ \mathbf{F},\ q^*. \tag{4.1}$$

The functions $q_m$ and $\mathbf{f}$ are usually assumed to be known when electromagnetic effects are absent. Thus, the set (3.307) is incomplete.

To close the equation system (3.307) we need additional relations. These relations are called *constitutive equations*, because just they specify the type of a continuum (the universal balance laws do not differ types of continua, they are the same for all bodies). If constitutive equations are given in any way, then a continuum model is said to be specified.

Derivation of constitutive equations is based on some additional *principles*, i.e. on physical assumptions of a general character, which cannot be formulated in the form of partial differential equations. We will accept the following basic principles:

- the principle of thermodynamically consistent determinism
- the principle of local action
- the principle of equipresence
- the principle of material indifference
- the principle of material symmetry
- the Onsager principle

Besides them, other additional principles may be formulated for special models of a continuum.

Y.I. Dimitrienko, *Nonlinear Continuum Mechanics and Large Inelastic Deformations*, Solid Mechanics and Its Applications 174, DOI 10.1007/978-94-007-0034-5_4,

## 4.2 Energetic and Quasienergetic Couples of Tensors

### 4.2.1 Energetic Couples of Tensors

Before formulating the above-mentioned principles, let us give definitions of special stress and deformation tensors, which are of great importance for the theory of constitutive equations in continuum mechanics.

Consider the power of internal forces $W_{(i)}$ (3.122) and introduce the *stress power* $w_{(i)}$ and the *elementary work* $d'A$ *of stresses*:

$$w_{(i)} = \mathbf{T}\cdot\cdot(\nabla\otimes\mathbf{v})^{\mathrm{T}}, \quad W_{(i)} = -\int_V w_{(i)}\,dV, \quad d'A = w_{(i)}\,dt. \tag{4.2}$$

According to relation (2.37) between the gradients $\nabla\otimes\mathbf{v}$ and $\overset{\circ}{\nabla}\otimes\mathbf{v}$, and also the dynamic compatibility equation (3.270), we can find one more representation for $w_{(i)}$:

$$w_{(i)} = \mathbf{T}\cdot\cdot\,\nabla\otimes\mathbf{v}^{\mathrm{T}} = \mathbf{T}\cdot\cdot\,(\overset{\circ}{\nabla}\otimes\mathbf{v}^{\mathrm{T}}\cdot\mathbf{F}^{-1}) = (\mathbf{F}^{-1}\cdot\mathbf{T})\cdot\cdot\,\frac{d\mathbf{F}}{dt}. \tag{4.3}$$

Here we have used the formula of rearrangement of three tensors in the scalar product (see [12]):

$$\mathbf{A}\cdot\cdot\,(\mathbf{B}\cdot\mathbf{C}) = (\mathbf{C}\cdot\mathbf{A})\cdot\cdot\,\mathbf{B} = (\mathbf{A}\cdot\mathbf{B})\cdot\cdot\,\mathbf{C}. \tag{4.4}$$

The following properties of contraction of two arbitrary tensors (see [12]) will be frequently used below:

$$\mathbf{A}\cdot\cdot\,\mathbf{B} = \mathbf{A}^{\mathrm{T}}\cdot\cdot\,\mathbf{B}^{\mathrm{T}}, \tag{4.5}$$

$$\mathbf{A}\cdot\cdot\,\mathbf{B} = \frac{1}{2}(\mathbf{A}\cdot\cdot\,\mathbf{B} + \mathbf{A}^{\mathrm{T}}\cdot\cdot\,\mathbf{B}^{\mathrm{T}}), \tag{4.6}$$

$$\mathbf{A}\cdot\cdot\,\mathbf{B} = \mathbf{A}^{S}\cdot\cdot\,\mathbf{B}^{S} + \mathbf{A}^{K}\cdot\cdot\,\mathbf{B}^{K}. \tag{4.7}$$

Here $\mathbf{A}^S$, $\mathbf{B}^S$ are symmetric tensors and $\mathbf{A}^K$, $\mathbf{B}^K$ are skew-symmetric tensors:

$$\mathbf{A}^{S} = (\mathbf{A} + \mathbf{A}^{\mathrm{T}})/2, \qquad \mathbf{A}^{K} = (\mathbf{A} - \mathbf{A}^{\mathrm{T}})/2.$$

According to the definition (3.56) of the Piola–Kirchhoff tensor and formula (4.3), we obtain the following expression for $w_{(i)}$ in terms of the tensor $\mathbf{P}$:

$$w_{(i)} = (\rho/\overset{\circ}{\rho})\,\mathbf{P}\cdot\cdot\,\frac{d\mathbf{F}}{dt}. \tag{4.8}$$

The representation (4.8) of the stress power as the contraction of some stress tensor and the rate of some tensor describing deformation proves to be not unique. There are several different such representations. Since they are of great importance for derivation of constitutive equations in continuum mechanics, let us find the representations.

To derive the representations, we use the definition (4.2) and formula (4.7) with substitutions $\mathbf{A} \to \mathbf{T}$, $\mathbf{B} \to \mathbf{L} = \nabla \otimes \mathbf{v}^{\mathrm{T}}$; then we get

$$w_{(i)} = \mathbf{T}^S \cdot\cdot \mathbf{D} + \mathbf{T}^K \cdot\cdot \mathbf{W}, \tag{4.9}$$

where $\mathbf{D}$ is the deformation rate tensor (2.225), $\mathbf{W}$ is the vorticity tensor (2.226) introduced by the decomposition (2.224) of the velocity gradient $\mathbf{L}$; $\mathbf{T}^S$ and $\mathbf{T}^K$ are symmetric and skew-symmetric parts of the tensor $\mathbf{T}$:

$$\mathbf{T}^S = \frac{1}{2}(\mathbf{T} + \mathbf{T}^{\mathrm{T}}), \quad \mathbf{T}^K = \frac{1}{2}(\mathbf{T} - \mathbf{T}^{\mathrm{T}}). \tag{4.10}$$

For nonpolar continua, the tensor $\mathbf{T}$ is symmetric, and Eq. (4.9) takes the form

$$w_{(i)} = \mathbf{T} \cdot\cdot \mathbf{D}. \tag{4.11}$$

**Theorem 4.1.** *There exists a denumerable set of couples of symmetric tensors* $(\overset{(\mathrm{n})}{\mathbf{T}}, \overset{(\mathrm{n})}{\mathbf{C}})$, *with the help of which the stress power* $w_{(i)}$ *can be expressed in the form*

$$w_{(i)} = \overset{(\mathrm{n})}{\mathbf{T}} \cdot\cdot \frac{d}{dt}\overset{(\mathrm{n})}{\mathbf{C}} + \mathbf{T}^K \cdot\cdot \mathbf{W}, \quad n \in \mathbb{Z}. \tag{4.12}$$

*If the tensor* $\mathbf{T}$ *is symmetric, then the expression (4.12) reduces to the form*

$$w_{(i)} = \overset{(\mathrm{n})}{\mathbf{T}} \cdot\cdot \frac{d}{dt}\overset{(\mathrm{n})}{\mathbf{C}}, \quad n \in \mathbb{Z}. \tag{4.13}$$

Tensors $\overset{(\mathrm{n})}{\mathbf{T}}$ are called the *energetic tensors of stresses*, and $\overset{(\mathrm{n})}{\mathbf{C}}$ – the *energetic tensors of deformations.*

Here $\mathbb{Z}$ is the set of integers, and the index $n$ may be positive or negative integer, or zero. To emphasize the special status of the *energetic couples* we introduce the following notation: particular values of index $n$ will be denoted by Roman numerals: I, II, III, IV, V etc., or −I, −II, −III etc. The first five couples of energetic tensors are the most important, namely $(\overset{\mathrm{I}}{\mathbf{T}}, \overset{\mathrm{I}}{\mathbf{C}})$, $(\overset{\mathrm{II}}{\mathbf{T}}, \overset{\mathrm{II}}{\mathbf{C}})$, $(\overset{\mathrm{III}}{\mathbf{T}}, \overset{\mathrm{III}}{\mathbf{C}})$, $(\overset{\mathrm{IV}}{\mathbf{T}}, \overset{\mathrm{IV}}{\mathbf{C}})$ and $(\overset{\mathrm{V}}{\mathbf{T}}, \overset{\mathrm{V}}{\mathbf{C}})$; they will be called *principal*, because these couples are most often used in solving different problems, and these energetic tensors may be expressed explicitly in terms of the stress and deformation tensors introduced before (see Table 4.1).

For the symmetric tensor $\mathbf{T}$, the couples at $n$ = I, III, IV and V were derived by Hill [26], and the couple at $n$ = II was found by K.F. Chernykh [8]. The special

**Table 4.1** Principal energetic couples of tensors

| Number $n$ of couple | Energetic stress tensors $\overset{(n)}{\mathbf{T}}$ | Energetic deformation tensors $\overset{(n)}{\mathbf{C}}$ | Energetic deformation measures $\overset{(n)}{\mathbf{G}}$ |
|---|---|---|---|
| I | $\mathbf{F}^{\mathrm{T}}\cdot\mathbf{T}^S\cdot\mathbf{F}$ | $\mathbf{\Lambda}=\frac{1}{2}(\mathbf{E}-\mathbf{U}^{-2})$ | $-\frac{1}{2}\mathbf{G}^{-1}$ |
| II | $\frac{1}{2}(\mathbf{F}^{\mathrm{T}}\cdot\mathbf{T}^S\cdot\mathbf{O}+\mathbf{O}^{\mathrm{T}}\cdot\mathbf{T}^S\cdot\mathbf{F})$ | $\mathbf{E}-\mathbf{U}^{-1}$ | $-\mathbf{U}^{-1}$ |
| III | $\mathbf{O}^{\mathrm{T}}\cdot\mathbf{T}^S\cdot\mathbf{O}$ | $\mathbf{B}$ | $\overset{\mathrm{III}}{\mathbf{G}}$ |
| IV | $\frac{1}{2}(\mathbf{F}^{-1}\cdot\mathbf{T}^S\cdot\mathbf{O}+\mathbf{O}^{\mathrm{T}}\cdot\mathbf{T}^S\cdot\mathbf{F}^{-1\mathrm{T}})$ | $\mathbf{U}-\mathbf{E}$ | $\mathbf{U}$ |
| V | $\mathbf{F}^{-1}\cdot\mathbf{T}^S\cdot\mathbf{F}^{-1\mathrm{T}}$ | $\mathbf{C}=\frac{1}{2}(\mathbf{U}^2-\mathbf{E})$ | $\frac{1}{2}\mathbf{G}$ |

notation $\overset{(n)}{\mathbf{T}}$ and $\overset{(n)}{\mathbf{C}}$ and the method of ordering of these tensors, when each tensor $\overset{(n)}{\mathbf{C}}$ is expressed in terms of the corresponding power of the right stretch tensor $\mathbf{U}$, were suggested by the author [12].

Prove the theorem for each of the energetic couples separately.

### *4.2.2 The First Energetic Couple* $(\overset{\mathrm{I}}{\mathbf{T}}, \mathbf{\Lambda})$

Consider the expression $\mathbf{T}^S\cdot\cdot\,\mathbf{D}$ and substitute in place of the tensor $\mathbf{D}$ its representation (2.282) in terms of the deformation gradient rate $\dot{\mathbf{F}}$:

$$\mathbf{D}=\frac{1}{2}(\dot{\mathbf{F}}\cdot\mathbf{F}^{-1}+\mathbf{F}^{-1\mathrm{T}}\cdot\dot{\mathbf{F}}^{\mathrm{T}}). \tag{4.14}$$

Then

$$w_S^{(i)}=\mathbf{T}^S\cdot\cdot\mathbf{D}=\frac{1}{2}\mathbf{T}^S\cdot\cdot\,(\dot{\mathbf{F}}\cdot\mathbf{F}^{-1}+\mathbf{F}^{-1\mathrm{T}}\cdot\dot{\mathbf{F}}^{\mathrm{T}}). \tag{4.15}$$

Differentiation of the identities $\mathbf{F}\cdot\mathbf{F}^{-1}=\mathbf{E}$ and $\mathbf{F}^{\mathrm{T}}\cdot\mathbf{F}^{-1\mathrm{T}}=\mathbf{E}$ gives

$$\dot{\mathbf{F}}\cdot\mathbf{F}^{-1}=-\mathbf{F}\cdot\dot{\mathbf{F}}^{-1},\qquad \mathbf{F}^{-1\mathrm{T}}\cdot\dot{\mathbf{F}}^{\mathrm{T}}=-\dot{\mathbf{F}}^{-1\mathrm{T}}\cdot\mathbf{F}^{\mathrm{T}}. \tag{4.16}$$

On substituting these expressions into (4.15), we get

$$-w_S^{(i)}=\frac{1}{2}\mathbf{T}^S\cdot\cdot\,\mathbf{F}\cdot\dot{\mathbf{F}}^{-1}+\frac{1}{2}\mathbf{T}^S\cdot\cdot\,\dot{\mathbf{F}}^{-1\mathrm{T}}\cdot\mathbf{F}^{\mathrm{T}}. \tag{4.17}$$

Rewrite the tensors $\dot{\mathbf{F}}^{-1\mathrm{T}}$ and $\dot{\mathbf{F}}^{-1}$ as the scalar products of themselves by the metric tensor $\mathbf{E}$:

$$\dot{\mathbf{F}}^{-1\mathrm{T}}=\mathbf{E}\cdot\dot{\mathbf{F}}^{-1\mathrm{T}}=(\mathbf{F}\cdot\mathbf{F}^{-1})\cdot\dot{\mathbf{F}}^{-1\mathrm{T}},\quad \dot{\mathbf{F}}^{-1}=\dot{\mathbf{F}}^{-1}\cdot\mathbf{E}=\dot{\mathbf{F}}^{-1}\cdot(\mathbf{F}^{-1\mathrm{T}}\cdot\mathbf{F}^{\mathrm{T}}). \tag{4.18}$$

Substitution of these expressions into (4.17) yields

$$-w_S^{(i)} = \frac{1}{2}\mathbf{T}^S \cdot\cdot (\mathbf{F}\cdot\mathbf{F}^{-1}\cdot\dot{\mathbf{F}}^{-1\mathrm{T}})\cdot\mathbf{F}^{\mathrm{T}} + \frac{1}{2}\mathbf{T}^S \cdot\cdot (\mathbf{F}\cdot\dot{\mathbf{F}}^{-1}\cdot\mathbf{F}^{-1\mathrm{T}})\cdot\mathbf{F}^{\mathrm{T}}. \quad (4.19)$$

Applying the rule (4.4) of rearrangement of three arbitrary tensors in the scalar product, we obtain

$$\begin{aligned} -w_S^{(i)} &= \frac{1}{2}(\mathbf{F}^{\mathrm{T}}\cdot\mathbf{T}^S\cdot\mathbf{F})\cdot\cdot(\mathbf{F}^{-1}\cdot\dot{\mathbf{F}}^{-1\mathrm{T}}) + \frac{1}{2}(\mathbf{F}^{\mathrm{T}}\cdot\mathbf{T}^S\cdot\mathbf{F})\cdot\cdot(\dot{\mathbf{F}}^{-1}\cdot\mathbf{F}^{-1\mathrm{T}}) \\ &= \mathbf{F}^{\mathrm{T}}\cdot\mathbf{T}^S\cdot\mathbf{F}\cdot\cdot\frac{1}{2}\left(\mathbf{F}^{-1}\cdot\dot{\mathbf{F}}^{-1\mathrm{T}} + \dot{\mathbf{F}}^{-1}\cdot\mathbf{F}^{-1\mathrm{T}}\right) = -(\mathbf{F}^{\mathrm{T}}\cdot\mathbf{T}^S\cdot\mathbf{F})\cdot\cdot\dot{\mathbf{\Lambda}}. \end{aligned} \quad (4.20)$$

Here we have used the expression (2.72) for the right Almansi deformation tensor

$$\mathbf{\Lambda} = \frac{1}{2}(\mathbf{E} - \mathbf{F}^{-1}\cdot\mathbf{F}^{-1\mathrm{T}}) = \frac{1}{2}(\mathbf{E} - \mathbf{U}^{-2}), \quad (4.21)$$

and

$$-\dot{\mathbf{\Lambda}} = \frac{1}{2}\left(\dot{\mathbf{F}}^{-1}\cdot\mathbf{F}^{-1\mathrm{T}} + \mathbf{F}^{-1}\cdot\dot{\mathbf{F}}^{-1\mathrm{T}}\right). \quad (4.22)$$

Introducing the new tensor

$$\overset{\mathrm{I}}{\mathbf{T}} = \mathbf{F}^{\mathrm{T}}\cdot\mathbf{T}^S\cdot\mathbf{F}, \quad (4.23)$$

from (4.9) and (4.20) we obtain

$$w_{(i)} = \overset{\mathrm{I}}{\mathbf{T}}\cdot\cdot\dot{\mathbf{\Lambda}} + \mathbf{T}^K\cdot\cdot\mathbf{W}. \quad (4.24)$$

Thus, the representation (4.12) really exists, and the first energetic couple of tensors is $(\overset{\mathrm{I}}{\mathbf{T}}, \overset{\mathrm{I}}{\mathbf{C}}) = (\overset{\mathrm{I}}{\mathbf{T}}, \mathbf{\Lambda})$.

### 4.2.3 *The Fifth Energetic Couple* $(\overset{\mathrm{V}}{\mathbf{T}}, \mathbf{C})$

Introduce the following new tensor:

$$\overset{\mathrm{V}}{\mathbf{T}} = \mathbf{F}^{-1}\cdot\mathbf{T}^S\cdot\mathbf{F}^{-1\mathrm{T}} \quad (4.25)$$

and rewrite the expression (4.15) as follows:

$$
\begin{aligned}
w_S^{(i)} &= \frac{1}{2}\left(\mathbf{F}^{-1}\cdot\mathbf{T}^S\cdot\cdot\,\dot{\mathbf{F}}+\mathbf{T}^S\cdot\mathbf{F}^{-1\mathrm{T}}\cdot\cdot\,\dot{\mathbf{F}}^{\mathrm{T}}\right)\\
&= \frac{1}{2}(\mathbf{F}^{-1}\cdot\mathbf{T}^S\cdot\mathbf{F}^{-1\mathrm{T}}\cdot\cdot\,\mathbf{F}^{\mathrm{T}}\cdot\dot{\mathbf{F}}+\mathbf{F}^{-1}\cdot\mathbf{T}^S\cdot\mathbf{F}^{-1\mathrm{T}}\cdot\cdot\,\dot{\mathbf{F}}^{\mathrm{T}}\cdot\mathbf{F})\\
&= \overset{\mathrm{V}}{\mathbf{T}}\cdot\cdot\,\frac{1}{2}\left(\mathbf{F}^{\mathrm{T}}\cdot\dot{\mathbf{F}}+\dot{\mathbf{F}}^{\mathrm{T}}\cdot\mathbf{F}\right)=\overset{\mathrm{V}}{\mathbf{T}}\cdot\cdot\,\dot{\mathbf{C}}.
\end{aligned}
\tag{4.26}
$$

Here we have used the property (4.4) of the scalar product of tensors, have written the tensor $\mathbf{T}^S$ in the form

$$
\mathbf{T}^S=\mathbf{T}^S\cdot\mathbf{E}=\mathbf{T}^S\cdot\mathbf{F}^{-1\mathrm{T}}\cdot\mathbf{F}^{\mathrm{T}},\quad \mathbf{T}^S=\mathbf{E}\cdot\mathbf{T}^S=\mathbf{F}\cdot\mathbf{F}^{-1}\cdot\mathbf{T}^S, \tag{4.27}
$$

and have taken account of the expression for the derivative of the right Cauchy–Green deformation tensor $\mathbf{C}$ (2.287):

$$
\dot{\mathbf{C}}=\frac{1}{2}\left(\mathbf{F}^{\mathrm{T}}\cdot\dot{\mathbf{F}}+\dot{\mathbf{F}}^{\mathrm{T}}\cdot\mathbf{F}\right). \tag{4.28}
$$

Thus, there is one more energetic couple of tensors:

$$
(\overset{\mathrm{V}}{\mathbf{T}},\ \overset{\mathrm{V}}{\mathbf{C}})=(\overset{\mathrm{V}}{\mathbf{T}},\ \mathbf{C}).
$$

### 4.2.4 *The Fourth Energetic Couple* $(\overset{\mathrm{IV}}{\mathbf{T}},\ (\mathbf{U}-\mathbf{E}))$

Rewrite the expression (4.26), passing from $\overset{\mathrm{V}}{\mathbf{T}}$ to $\mathbf{T}^S$ by formula (4.25) and from $\mathbf{C}$ to $\mathbf{U}$ by (2.158). Then we get

$$
w_S^{(i)}=\overset{\mathrm{V}}{\mathbf{T}}\cdot\cdot\,\dot{\mathbf{C}}=\frac{1}{2}(\mathbf{F}^{-1}\cdot\mathbf{T}^S\cdot\mathbf{F}^{-1\mathrm{T}})\cdot\cdot\,(\mathbf{U}\cdot\dot{\mathbf{U}}+\dot{\mathbf{U}}\cdot\mathbf{U}). \tag{4.29}
$$

Use the rules (4.4) of rearrangement of three tensors in the scalar product:

$$
w_S^{(i)}=\frac{1}{2}\mathbf{F}^{-1}\cdot\mathbf{T}^S\cdot\mathbf{F}^{-1\mathrm{T}}\cdot\mathbf{U}\cdot\cdot\,\dot{\mathbf{U}}+\frac{1}{2}\mathbf{U}\cdot\mathbf{F}^{-1}\cdot\mathbf{T}^S\cdot\mathbf{F}^{-1\mathrm{T}}\cdot\cdot\,\dot{\mathbf{U}}. \tag{4.30}
$$

On taking the polar decomposition (2.137) into account:

$$
\mathbf{F}=\mathbf{O}\cdot\mathbf{U},\quad \mathbf{F}^{-1}=\mathbf{U}^{-1}\cdot\mathbf{O}^{\mathrm{T}},\quad \mathbf{F}^{-1\mathrm{T}}=\mathbf{O}\cdot\mathbf{U}^{-1}, \tag{4.31}
$$

we finally get

$$w_S^{(i)} = \frac{1}{2}\left(\mathbf{F}^{-1}\cdot\mathbf{T}^S\cdot\mathbf{O}+\mathbf{O}^{\mathrm{T}}\cdot\mathbf{T}^S\cdot\mathbf{F}^{-1\mathrm{T}}\right)\cdot\cdot\,\dot{\mathbf{U}} = \overset{\mathrm{IV}}{\mathbf{T}}\cdot\cdot\,\overset{\mathrm{IV}}{\mathbf{C}}{}^{\bullet}.$$

Here we have introduced the fourth energetic stress tensor $\overset{\mathrm{IV}}{\mathbf{T}}$:

$$\overset{\mathrm{IV}}{\mathbf{T}} = \frac{1}{2}\left(\mathbf{F}^{-1}\cdot\mathbf{T}^S\cdot\mathbf{O}+\mathbf{O}^{\mathrm{T}}\cdot\mathbf{T}^S\cdot\mathbf{F}^{-1\mathrm{T}}\right), \tag{4.32}$$

coupled for which is the tensor $\overset{\mathrm{IV}}{\mathbf{C}} = \mathbf{U}-\mathbf{E}$. We have taken into account that $\overset{\mathrm{IV}}{\mathbf{C}}{}^{\bullet} = \dot{\mathbf{U}}$, as $\dot{\mathbf{E}} = 0$.

### 4.2.5 *The Second Energetic Couple* $(\overset{\mathrm{II}}{\mathbf{T}},\ (\mathbf{E}-\mathbf{U}^{-1}))$

Rewrite the first energetic couple (4.24), replacing the tensor $\boldsymbol{\Lambda}$ by its expression (2.158): $\boldsymbol{\Lambda} = \frac{1}{2}(\mathbf{E}-\mathbf{U}^{-2})$. Then from (4.24) we get

$$\begin{aligned} w_S^{(i)} &= -\frac{1}{2}\mathbf{F}^{\mathrm{T}}\cdot\mathbf{T}^S\cdot\mathbf{F}\cdot\cdot\left(\mathbf{U}^{-1}\cdot\dot{\mathbf{U}}^{-1}+\dot{\mathbf{U}}^{-1}\cdot\mathbf{U}^{-1}\right)\\ &= -\frac{1}{2}\mathbf{F}^{\mathrm{T}}\cdot\mathbf{T}^S\cdot\mathbf{F}\cdot\mathbf{U}^{-1}\cdot\cdot\,\dot{\mathbf{U}}^{-1}-\frac{1}{2}\mathbf{U}^{-1}\cdot\mathbf{F}^{\mathrm{T}}\cdot\mathbf{T}\cdot\mathbf{F}\cdot\cdot\,\dot{\mathbf{U}}^{-1}. \end{aligned}$$

According to the property (4.31) of the polar decomposition, we obtain

$$w_S^{(i)} = \overset{\mathrm{II}}{\mathbf{T}}\cdot\cdot\,\overset{\mathrm{II}}{\mathbf{C}}{}^{\bullet}.$$

Here we have introduced the *second energetic stress tensor*

$$\overset{\mathrm{II}}{\mathbf{T}} = \frac{1}{2}\left(\mathbf{F}^{\mathrm{T}}\cdot\mathbf{T}^S\cdot\mathbf{O}+\mathbf{O}^{\mathrm{T}}\cdot\mathbf{T}^S\cdot\mathbf{F}\right), \tag{4.33}$$

coupled for which is the tensor $(\mathbf{E}-\mathbf{U}^{-1}) = \overset{\mathrm{II}}{\mathbf{C}}$, as $\overset{\mathrm{II}}{\mathbf{C}}{}^{\bullet} = (-\mathbf{U}^{-1})^{\bullet}$.

### 4.2.6 *The Third Energetic Couple* $(\overset{\mathrm{III}}{\mathbf{T}},\ \mathbf{B})$

We can derive one more energetic couple from Eq. (4.30), if in place of $\mathbf{F}^{-1}$ and $\mathbf{F}^{-1\mathrm{T}}$ substitute their polar decompositions (4.31):

$$w_S^{(i)} = \frac{1}{2}\mathbf{U}^{-1}\cdot\mathbf{O}^{\mathrm{T}}\cdot\mathbf{T}^S\cdot\mathbf{O}\cdot\mathbf{U}^{-1}\cdot\mathbf{U}\cdot\cdot\,\dot{\mathbf{U}}+\frac{1}{2}\mathbf{U}\cdot\mathbf{U}^{-1}\cdot\mathbf{O}^{\mathrm{T}}\cdot\mathbf{T}^S\cdot\mathbf{O}\cdot\mathbf{U}^{-1}\cdot\cdot\,\dot{\mathbf{U}}.$$

Changing the order in the scalar product, we get

$$w_S^{(i)} = \frac{1}{2}\mathbf{O}^{\mathrm{T}} \cdot \mathbf{T}^S \cdot \mathbf{O} \cdot\cdot \dot{\mathbf{U}} \cdot \mathbf{U}^{-1} + \frac{1}{2}\mathbf{O}^{\mathrm{T}} \cdot \mathbf{T}^S \cdot \mathbf{O} \cdot\cdot \mathbf{U}^{-1} \cdot \dot{\mathbf{U}}.$$

Hence, there exists the third energetic couple

$$w_S^{(i)} = \overset{\mathrm{III}}{\mathbf{T}} \cdot\cdot \dot{\mathbf{B}},$$

where $\overset{\mathrm{III}}{\mathbf{T}}$ is the *third energetic stress tensor*:

$$\overset{\mathrm{III}}{\mathbf{T}} = \mathbf{O}^{\mathrm{T}} \cdot \mathbf{T}^S \cdot \mathbf{O}, \tag{4.34}$$

and $\mathbf{B} = \overset{\mathrm{III}}{\mathbf{C}}$ is the *the third energetic deformation tensor* determined by its derivative and its initial value at $t = 0$ (see (2.317)).

### *4.2.7 General Representations for Energetic Tensors of Stresses and Deformations*

**Theorem 4.2.** *Each of the energetic deformation tensors* $\overset{(\mathrm{n})}{\mathbf{C}}$ *can be expressed in terms of the corresponding power of the right stretch tensor as follows:*

$$\overset{(\mathrm{n})}{\mathbf{C}} = \frac{1}{(n - \mathrm{III})}(\mathbf{U}^{n-\mathrm{III}} - \mathbf{E}), \quad \forall n \in \mathbb{Z},\ n \neq \mathrm{III}, \tag{4.35}$$

*and the energetic stress tensors* $\overset{(\mathrm{n})}{\mathbf{T}}$ *satisfy the equations*

$$\overset{\mathrm{IV}}{\mathbf{T}} = \frac{1}{n - \mathrm{III}} \sum_{k=0}^{n-\mathrm{IV}} \mathbf{U}^{n-k-\mathrm{IV}} \cdot \overset{(\mathrm{n})}{\mathbf{T}} \cdot \mathbf{U}^k, \quad \textit{if} \;\; n \geqslant \mathrm{IV}; \tag{4.35a}$$

$$\overset{\mathrm{II}}{\mathbf{T}} = \frac{1}{\mathrm{III} - n} \sum_{k=0}^{\mathrm{II}-n} \mathbf{U}^{n-\mathrm{II}+k} \cdot \overset{(\mathrm{n})}{\mathbf{T}} \cdot \mathbf{U}^{-k}, \quad \textit{if} \;\; n \leqslant \mathrm{II}. \tag{4.35b}$$

Here and below in such expressions at particular values of $n$ we must replace Roman numerals by corresponding Arabic ones and then perform arithmetic operations. For example, the relation (4.35) at $n = \mathrm{I}$ has the form

$$\overset{\mathrm{I}}{\mathbf{C}} = \frac{1}{1-3}(\mathbf{U}^{1-3} - \mathbf{E}) = \frac{1}{2}(\mathbf{E} - \mathbf{U}^{-2}).$$

▼ (1) A proof of formula (4.35) for $n$ = I, II, IV, and V is evident, if we write out the formula for these values of $n$ and compare the obtained expressions with the tensors $\overset{(n)}{\mathbf{C}}$ from Table 4.1.

(2) In order to show that the tensors $\overset{(n)}{\mathbf{C}}$ (when $n >$ V) with tensors $\overset{(n)}{\mathbf{T}}$ also constitute energetic couples of the form (4.12), we should determine the derivative of $\overset{(n)}{\mathbf{C}}$:

$$\overset{(n)}{\mathbf{C}}{}^{\bullet} = \frac{1}{n-\mathrm{III}}(\underbrace{\mathbf{U}\cdot\ldots\cdot\mathbf{U}}_{n-\mathrm{III}})^{\bullet} = \frac{1}{n-\mathrm{III}}(\dot{\mathbf{U}}\cdot\mathbf{U}^{n-\mathrm{IV}} + \mathbf{U}\cdot\dot{\mathbf{U}}\cdot\mathbf{U}^{n-\mathrm{V}} + \ldots$$

$$\ldots + \mathbf{U}^{n-\mathrm{V}}\cdot\dot{\mathbf{U}}\cdot\mathbf{U} + \mathbf{U}^{n-\mathrm{IV}}\cdot\dot{\mathbf{U}}) = \frac{1}{n-\mathrm{III}}\sum_{k=0}^{n-\mathrm{IV}}\mathbf{U}^{k}\cdot\dot{\mathbf{U}}\cdot\mathbf{U}^{n-\mathrm{IV}-k}.$$

So, if there exists energetic stress tensor $\overset{(n)}{\mathbf{T}}$, then it satisfies the equation

$$\overset{(n)}{\mathbf{T}}\cdot\cdot\overset{(n)}{\mathbf{C}}{}^{\bullet} = \frac{1}{n-\mathrm{III}}\overset{(n)}{\mathbf{T}}\cdot\cdot\sum_{k=0}^{n-\mathrm{IV}}\mathbf{U}^{k}\cdot\dot{\mathbf{U}}\cdot\mathbf{U}^{n-\mathrm{IV}-k}$$

$$= \left(\frac{1}{n-\mathrm{III}}\sum_{k=0}^{n-\mathrm{IV}}\mathbf{U}^{n-\mathrm{IV}-k}\cdot\overset{(n)}{\mathbf{T}}\cdot\mathbf{U}^{k}\right)\cdot\cdot\dot{\mathbf{U}}.$$

Since the tensor $(\mathbf{U}-\mathbf{E}) = \overset{\mathrm{IV}}{\mathbf{C}}$ is assigned to the fourth energetic stress tensor $\overset{\mathrm{IV}}{\mathbf{T}}$, the expression in parentheses should form the tensor $\overset{\mathrm{IV}}{\mathbf{T}}$. Thus, for all $\overset{(n)}{\mathbf{T}}$ (when $n >$ V) the formula (4.35a) really holds.

While $n$ = V, formula (4.35a) takes the form

$$\overset{\mathrm{IV}}{\mathbf{T}} = \frac{1}{2}(\mathbf{U}\cdot\overset{\mathrm{V}}{\mathbf{T}} + \overset{\mathrm{V}}{\mathbf{T}}\cdot\mathbf{U}).$$

Since the tensors $\overset{\mathrm{IV}}{\mathbf{T}}$ and $\overset{\mathrm{V}}{\mathbf{T}}$ have been introduced before according to Table 4.1, we should verify the last formula. Substitution of expressions (4.32) and (4.25) into this formula yields

$$\frac{1}{2}(\mathbf{F}^{-1}\cdot\mathbf{T}^{S}\cdot\mathbf{O} + \mathbf{O}^{\mathrm{T}}\cdot\mathbf{T}^{S}\cdot\mathbf{F}^{-1\mathrm{T}}) = \frac{1}{2}(\mathbf{U}\cdot\mathbf{F}^{-1}\cdot\mathbf{T}^{S}\cdot\mathbf{F}^{-1\mathrm{T}} + \mathbf{F}^{-1}\cdot\mathbf{T}^{S}\cdot\mathbf{F}^{-1\mathrm{T}}\cdot\mathbf{U}).$$

This formula is valid due to the polar decomposition (2.137) as $\mathbf{O}^{\mathrm{T}} = \mathbf{U}\cdot\mathbf{F}^{-1}$. Hence formula (4.35a) really holds when $n$ = V.

(3) To prove formula (4.35b) we determine the derivative of tensors $\overset{(n)}{\mathbf{C}}$ when $n < \mathrm{I}$:

$$\overset{(n)}{\mathbf{C}}{}^{\bullet} = \frac{1}{n-\mathrm{III}}(\underbrace{\mathbf{U}^{-1}\cdot\ldots\cdot\mathbf{U}^{-1}}_{\mathrm{III}-n})^{\bullet} = \frac{1}{n-\mathrm{III}}\sum_{k=0}^{\mathrm{II}-n}(\mathbf{U}^{-1})^{k}\cdot(\mathbf{U}^{-1})^{\bullet}\cdot(\mathbf{U}^{-1})^{\mathrm{II}-n-k}.$$

Then

$$\overset{(n)}{\mathbf{T}}\cdot\cdot\,\overset{(n)}{\mathbf{C}} = \left(\frac{1}{\mathrm{III}-n}\sum_{k=0}^{\mathrm{II}-n}\mathbf{U}^{n-\mathrm{II}+k}\cdot\overset{(n)}{\mathbf{T}}\cdot\mathbf{U}^{-k}\right)\cdot\cdot\,(-\mathbf{U}^{-1})^{\bullet}.$$

Since the tensor $(\mathbf{E}-\mathbf{U}^{-1}) = \overset{\mathrm{II}}{\mathbf{C}}$ is assigned to the second energetic stress tensor $\overset{\mathrm{II}}{\mathbf{T}}$, the expression in parentheses just forms the tensor $\overset{\mathrm{II}}{\mathbf{T}}$. Hence formula (4.35b) holds for all $n < \mathrm{I}$.

When $n = \mathrm{I}$, formula (4.35b) takes the form

$$\overset{\mathrm{II}}{\mathbf{T}} = \frac{1}{2}(\mathbf{U}^{-1}\cdot\overset{\mathrm{I}}{\mathbf{T}} + \overset{\mathrm{I}}{\mathbf{T}}\cdot\mathbf{U}^{-1}).$$

We can verify this formula by substituting expressions (4.33) and (4.23) into the last equation. ▲

A general relation between $\overset{(n)}{\mathbf{T}}$ and $\mathbf{T}$ is more complicated than the relation (4.35) between $\overset{(n)}{\mathbf{C}}$ and $\mathbf{U}$; however, the tensors $\overset{(n)}{\mathbf{T}}$ and $\mathbf{T}^{S}$ are evident to be connected by the linear relation

$$\overset{(n)}{\mathbf{T}} = {}^{4}\overset{(n)}{\mathbf{E}}{}^{-1}\cdot\cdot\,\mathbf{T}^{S}, \quad n \in \mathbb{Z}, \tag{4.36}$$

where ${}^{4}\overset{(n)}{\mathbf{E}}{}^{-1}$ are the *inverse tensors of energetic equivalence* (fourth-order tensors), whose expressions for the principal energetic couples are given in Table 4.2. These representations should be derived as Exercise 4.2.11. The notation $(\mathbf{F}^{\mathrm{T}}\otimes\mathbf{F})^{(1432)}$ for the transpose fourth-order tensor has been introduced in Sect. 2.1.4.

Expressions for the *tensors of energetic equivalence* ${}^{4}\overset{(n)}{\mathbf{E}}$ can be found by inverting the relations (4.36):

$$\mathbf{T}^{S} = {}^{4}\overset{(n)}{\mathbf{E}}\cdot\cdot\,\overset{(n)}{\mathbf{T}}, \quad n \in \mathbb{Z}. \tag{4.37}$$

**Table 4.2** Inverse tensors of energetic equivalence

| Number $n$ | ${}^{4}\overset{(n)}{\mathbf{E}}{}^{-1}$ |
|---|---|
| I | $(\mathbf{F}^{\mathrm{T}}\otimes\mathbf{F})^{(1432)}$ |
| II | $\frac{1}{2}(\mathbf{F}^{\mathrm{T}}\otimes\mathbf{O}+\mathbf{O}^{\mathrm{T}}\otimes\mathbf{F})^{(1432)}$ |
| III | $(\mathbf{O}^{\mathrm{T}}\otimes\mathbf{O})^{(1432)}$ |
| IV | $\frac{1}{2}(\mathbf{F}^{-1}\otimes\mathbf{O}+\mathbf{O}^{\mathrm{T}}\otimes\mathbf{F}^{-1\mathrm{T}})^{(1432)}$ |
| V | $(\mathbf{F}^{-1}\otimes\mathbf{F}^{-1\mathrm{T}})^{(1432)}$ |

Substituting the relationship (4.35a) into formula (4.37) where $n = \mathrm{IV}$, we obtain the expression for tensors ${}^4\overset{(\mathrm{n})}{\mathbf{E}}$ (when $n \geqslant \mathrm{V}$) in terms of the tensor ${}^4\overset{\mathrm{IV}}{\mathbf{E}}$:

$$ {}^4\overset{(\mathrm{n})}{\mathbf{E}} = \frac{1}{n-\mathrm{III}}\,{}^4\overset{\mathrm{IV}}{\mathbf{E}} \cdot\cdot \sum_{k=0}^{n-\mathrm{IV}} (\mathbf{U}^{n-k-\mathrm{IV}} \otimes \mathbf{U}^k)^{(1432)}, \quad n \geqslant \mathrm{V}. \tag{4.38a} $$

In a similar way, we can find the expression for tensors ${}^4\overset{(\mathrm{n})}{\mathbf{E}}$ when $n \leqslant \mathrm{I}$:

$$ {}^4\overset{(\mathrm{n})}{\mathbf{E}} = \frac{1}{\mathrm{III}-n}\,{}^4\overset{\mathrm{II}}{\mathbf{E}} \cdot\cdot \sum_{k=0}^{\mathrm{II}-n} (\mathbf{U}^{n-\mathrm{II}-k} \otimes \mathbf{U}^{-k})^{(1432)}, \quad n \leqslant \mathrm{I}. \tag{4.38b} $$

Components of all the tensors ${}^4\overset{(\mathrm{n})}{\mathbf{E}}$ with respect to the eigenbases $\mathbf{p}_i$ and $\overset{\circ}{\mathbf{p}}_i$ can be expressed in terms of only eigenvalues $\lambda_\alpha$.

**Theorem 4.3.** *The tensors of energetic equivalency* ${}^4\overset{(\mathrm{n})}{\mathbf{E}}$*, connecting the tensors* $\mathbf{T}^S$ *and* $\overset{(\mathrm{n})}{\mathbf{T}}$ *with the help of the linear relations (4.37), have the form*

$$ {}^4\overset{(\mathrm{n})}{\mathbf{E}} = \sum_{\alpha,\beta=1}^{3} \overset{(\mathrm{n})}{E}_{\alpha\beta}\, \mathbf{p}_\alpha \otimes \mathbf{p}_\beta \otimes \overset{\circ}{\mathbf{p}}_\beta \otimes \overset{\circ}{\mathbf{p}}_\alpha, \quad n \in \mathbb{Z}, \tag{4.38} $$

*where*

$$ \overset{\mathrm{I}}{E}_{\alpha\beta} = \frac{1}{\lambda_\alpha\lambda_\beta}, \quad \overset{\mathrm{II}}{E}_{\alpha\beta} = \frac{2}{\lambda_\alpha+\lambda_\beta}, \quad \overset{\mathrm{III}}{E}_{\alpha\beta} = 1, \quad \overset{\mathrm{IV}}{E}_{\alpha\beta} = \frac{2\lambda_\alpha\lambda_\beta}{\lambda_\alpha+\lambda_\beta}, \quad \overset{\mathrm{V}}{E}_{\alpha\beta} = \lambda_\alpha\lambda_\beta, $$

$$ \overset{(\mathrm{n})}{E}_{\alpha\beta} = \frac{2\lambda_\alpha\lambda_\beta}{(\lambda_\alpha+\lambda_\beta)(n-\mathrm{III})} \sum_{k=0}^{n-\mathrm{IV}} \lambda_\alpha^{n-k-\mathrm{IV}} \lambda_\beta^{k}, \quad n \geqslant \mathrm{V}, $$

$$ \overset{(\mathrm{n})}{E}_{\alpha\beta} = \frac{2}{\lambda_\alpha+\lambda_\beta} \sum_{k=0}^{\mathrm{II}-n} \lambda_\alpha^{n-\mathrm{II}+k} \lambda_\beta^{-k}, \quad n \leqslant \mathrm{I}. \tag{4.39} $$

▼ (a) Consider the case when $n = \mathrm{I}$, and represent the tensors $\mathbf{F}$, $\mathbf{O}$, $\mathbf{T}^S$ and $\overset{(\mathrm{n})}{\mathbf{T}}$ in the eigenbases $\overset{\circ}{\mathbf{p}}_i$ and $\mathbf{p}_i$ (see formulae (2.169)–(2.171)):

$$ \mathbf{F} = \sum_{\alpha=1}^{3} \lambda_\alpha \mathbf{p}_\alpha \otimes \overset{\circ}{\mathbf{p}}_\alpha, \quad \mathbf{F}^{\mathrm{T}} = \sum_{\alpha=1}^{3} \lambda_\alpha \overset{\circ}{\mathbf{p}}_\alpha \otimes \mathbf{p}_\alpha, \quad \mathbf{O} = \sum_{\alpha=1}^{3} \mathbf{p}_\alpha \otimes \overset{\circ}{\mathbf{p}}_\alpha, $$

$$ \mathbf{T}^S = \sum_{\alpha,\beta=1}^{3} T^{(p)}_{\alpha\beta} \mathbf{p}_\alpha \otimes \mathbf{p}_\beta, \quad \overset{(\mathrm{n})}{\mathbf{T}} = \sum_{\alpha,\beta=1}^{3} \overset{(\mathrm{n})\circ}{T}{}^{(p)}_{\alpha\beta} \overset{\circ}{\mathbf{p}}_\alpha \otimes \overset{\circ}{\mathbf{p}}_\beta, \tag{4.40} $$

where $T^{(p)}_{\alpha\beta}$ are components of the tensor $\mathbf{T}^S$ with respect to the basis $\mathbf{p}_\alpha$, and $\overset{(n)\circ}{T}{}^{(p)}_{\alpha\beta}$ are components of the tensor $\overset{(n)}{\mathbf{T}}$ with respect to the basis $\overset{\circ}{\mathbf{p}}_\alpha$.

On substituting these representations into formula (4.36) at $n = \mathrm{I}$, we get

$$\overset{\mathrm{I}}{\mathbf{T}} = \sum_{\alpha,\beta=1}^{3} \overset{\mathrm{I}\circ}{T}{}^{(p)}_{\alpha\beta} \overset{\circ}{\mathbf{p}}_\alpha \otimes \overset{\circ}{\mathbf{p}}_\beta = \mathbf{F}^{\mathrm{T}} \cdot \mathbf{T}^S \cdot \mathbf{F}$$

$$= \sum_{\alpha,\beta,\gamma,\omega=1}^{3} \lambda_\alpha \overset{\circ}{\mathbf{p}}_\alpha \otimes \mathbf{p}_\alpha \cdot T^{(p)}_{\gamma\omega} \mathbf{p}_\gamma \otimes \mathbf{p}_\omega \cdot \lambda_\beta \mathbf{p}_\beta \otimes \overset{\circ}{\mathbf{p}}_\beta = \sum_{\alpha,\beta=1}^{3} \lambda_\alpha \lambda_\beta T^{(p)}_{\alpha\beta} \overset{\circ}{\mathbf{p}}_\alpha \otimes \overset{\circ}{\mathbf{p}}_\beta .$$

Here we have taken into account that the basis $\mathbf{p}_\alpha$ is orthonormal (see (2.141)).

Hence,

$$\overset{\mathrm{I}\circ}{T}{}^{(p)}_{\alpha\beta} = T^{(p)}_{\alpha\beta} \lambda_\alpha \lambda_\beta .$$

Use once again the representation of the tensor $\mathbf{T}^S$ in the basis $\mathbf{p}_\alpha$ and take into account that

$$\overset{\mathrm{I}\circ}{T}{}^{(p)}_{\alpha\beta} = \overset{\circ}{\mathbf{p}}_\alpha \cdot \overset{\mathrm{I}}{\mathbf{T}} \cdot \overset{\circ}{\mathbf{p}}_\beta ,$$

then

$$\mathbf{T}^S = \sum_{\alpha,\beta=1}^{3} T^{(p)}_{\alpha\beta} \mathbf{p}_\alpha \otimes \mathbf{p}_\beta = \sum_{\alpha,\beta=1}^{3} \frac{1}{\lambda_\alpha \lambda_\beta} \mathbf{p}_\alpha \otimes \overset{\mathrm{I}\circ}{T}{}^{(p)}_{\alpha\beta} \otimes \mathbf{p}_\beta$$

$$= \sum_{\alpha,\beta=1}^{3} \frac{1}{\lambda_\alpha \lambda_\beta} \mathbf{p}_\alpha \otimes \overset{\circ}{\mathbf{p}}_\alpha \cdot \overset{\mathrm{I}}{\mathbf{T}} \cdot \overset{\circ}{\mathbf{p}}_\beta \otimes \mathbf{p}_\beta$$

$$= \sum_{\alpha,\beta=1}^{3} \frac{1}{\lambda_\alpha \lambda_\beta} (\mathbf{p}_\alpha \otimes \overset{\circ}{\mathbf{p}}_\alpha \otimes \overset{\circ}{\mathbf{p}}_\beta \otimes \mathbf{p}_\beta)^{(1432)} \cdot\cdot \overset{\mathrm{I}}{\mathbf{T}}$$

$$= \sum_{\alpha,\beta=1}^{3} \frac{1}{\lambda_\alpha \lambda_\beta} (\mathbf{p}_\alpha \otimes \mathbf{p}_\beta \otimes \overset{\circ}{\mathbf{p}}_\beta \otimes \overset{\circ}{\mathbf{p}}_\alpha) \cdot\cdot \overset{\mathrm{I}}{\mathbf{T}}.$$

Thus, the representation (4.37) with the tensor ${}^4\overset{\mathrm{I}}{\mathbf{E}}$ (4.38), (4.39) really holds at $n = \mathrm{I}$.

(b) Consider the case when $n = \mathrm{II}$. According to the representations (4.40), we can rewrite relation (4.36) for $n = \mathrm{II}$ as follows:

$$\overset{\text{II}}{\mathbf{T}} = \sum_{\alpha,\beta=1}^{3} \overset{\text{II}\circ}{T}{}^{(p)}_{\alpha\beta} \overset{\circ}{\mathbf{p}}_\alpha \otimes \overset{\circ}{\mathbf{p}}_\beta = \frac{1}{2}(\mathbf{F}^{\mathrm{T}} \cdot \mathbf{T} \cdot \mathbf{O} + \mathbf{O}^{\mathrm{T}} \cdot \mathbf{T} \cdot \mathbf{F})$$

$$= \frac{1}{2} \sum_{\alpha,\beta,\gamma,\omega=1}^{3} (\lambda_\alpha \overset{\circ}{\mathbf{p}}_\alpha \otimes \mathbf{p}_\alpha \cdot T^{(p)}_{\gamma\omega} \mathbf{p}_\gamma \otimes \mathbf{p}_\omega \cdot \mathbf{p}_\beta \otimes \overset{\circ}{\mathbf{p}}_\beta$$

$$+ \overset{\circ}{\mathbf{p}}_\alpha \otimes \mathbf{p}_\alpha \cdot T^{(p)}_{\gamma\omega} \mathbf{p}_\gamma \otimes \mathbf{p}_\omega \cdot \lambda_\beta \mathbf{p}_\beta \otimes \overset{\circ}{\mathbf{p}}_\beta) = \frac{1}{2} \sum_{\alpha,\beta=1}^{3} (\lambda_\alpha + \lambda_\beta) T^{(p)}_{\alpha\beta} \overset{\circ}{\mathbf{p}}_\alpha \otimes \overset{\circ}{\mathbf{p}}_\beta.$$

Thus, we get

$$\overset{\text{II}\circ}{T}{}^{(p)}_{\alpha\beta} = \frac{1}{2}(\lambda_\alpha + \lambda_\beta) T^{(p)}_{\alpha\beta}.$$

The further proof is analogous to the proof for $n = \text{I}$.

The theorem for $n = \text{III}$, IV and V can be proved in the same way (see Exercise 4.2.18).

(c) To prove the theorem for the cases when $n \geqslant \text{V}$ and $n \leqslant \text{I}$, one should use formulae (4.38), (4.39) at $n = \text{II}$ and IV, and also the expression (2.142) for the tensor $\mathbf{U}$ and substitute them into (4.38a) and (4.38b).

In proving the theorem, we have established that components $\overset{(\text{n})\circ}{T}{}^{(p)}_{\alpha\beta}$ of the tensor $\overset{(\text{n})}{\mathbf{T}}$ with respect to the basis $\overset{\circ}{\mathbf{p}}_\alpha$ and components $T^{(p)}_{\alpha\beta}$ of the tensor $\mathbf{T}$ with respect to the basis $\mathbf{p}_\alpha$ are connected by the relations

$$\overset{\text{I}\circ}{T}{}^{(p)}_{\alpha\beta} = \lambda_\alpha \lambda_\beta T^{(p)}_{\alpha\beta}, \quad \overset{\text{II}\circ}{T}{}^{(p)}_{\alpha\beta} = \frac{1}{2}(\lambda_\alpha + \lambda_\beta) T^{(p)}_{\alpha\beta},$$
$$\overset{\text{III}\circ}{T}{}^{(p)}_{\alpha\beta} = T^{(p)}_{\alpha\beta}, \quad \overset{\text{IV}\circ}{T}{}^{(p)}_{\alpha\beta} = \frac{1}{2}\Big(\frac{1}{\lambda_\alpha} + \frac{1}{\lambda_\beta}\Big) T^{(p)}_{\alpha\beta}, \quad \overset{\text{V}\circ}{T}{}^{(p)}_{\alpha\beta} = \frac{1}{\lambda_\alpha \lambda_\beta} T^{(p)}_{\alpha\beta}, \tag{4.41}$$

or

$$T^{(p)}_{\alpha\beta} = \overset{(\text{n})}{E}_{\alpha\beta} \overset{(\text{n})\circ}{T}{}^{(p)}_{\alpha\beta}, \quad \forall n \in \mathbb{Z}. \blacktriangle$$

### 4.2.8 Energetic Deformation Measures

Besides the energetic tensors $\overset{(\text{n})}{\mathbf{C}}$, introduce the *energetic deformation measures*

$$\overset{(\text{n})}{\mathbf{G}} = \frac{1}{n - \text{III}} \mathbf{U}^{n-\text{III}}, \quad \forall n \in \mathbb{Z}, \quad n \neq \text{III}, \tag{4.42}$$

which are the corresponding powers of the right stretch tensor $\mathbf{U}$.

One can readily establish the relation between $\overset{(n)}{\mathbf{G}}$ and $\overset{(n)}{\mathbf{C}}$:

$$\overset{(n)}{\mathbf{C}} = \overset{(n)}{\mathbf{G}} - \frac{1}{n - \mathrm{III}}\mathbf{E}. \tag{4.43}$$

Since the derivative of the metric tensor is zero, the derivative tensors $\overset{(n)}{\mathbf{G}}{}^{\bullet}$ and $\overset{(n)}{\mathbf{C}}{}^{\bullet}$ are coincident:

$$\overset{(n)}{\mathbf{C}}{}^{\bullet} = \overset{(n)}{\mathbf{G}}{}^{\bullet}. \tag{4.44}$$

Introduce the third energetic deformation measure

$$\overset{\mathrm{III}}{\mathbf{G}} = \mathbf{E} + \overset{\mathrm{III}}{\mathbf{C}}. \tag{4.43a}$$

Due to (2.317), the measure satisfies the equation

$$\frac{d\overset{\mathrm{III}}{\mathbf{G}}}{dt} = \frac{1}{2}\left(\frac{d\mathbf{U}}{dt}\cdot\mathbf{U}^{-1} + \mathbf{U}^{-1}\cdot\frac{d\mathbf{U}}{dt}\right), \quad \overset{\mathrm{III}}{\mathbf{G}}(0) = \mathbf{E}, \tag{4.43b}$$

and also satisfies the general equation (4.44).

Due to (4.44), the relation (4.12) for $w_{(i)}$ takes the form

$$w_{(i)} = \overset{(n)}{\mathbf{T}} \cdot\cdot \overset{(n)}{\mathbf{G}}{}^{\bullet} + \mathbf{T}^{K} \cdot\cdot \mathbf{W}, \qquad \forall n \in \mathbb{Z}. \tag{4.45}$$

If the tensor $\mathbf{T}$ is symmetric, then this formula reduces to the form

$$w_{(i)} = \overset{(n)}{\mathbf{T}} \cdot\cdot \frac{d}{dt}\overset{(n)}{\mathbf{G}}, \qquad \forall n \in \mathbb{Z}, \tag{4.46}$$

i.e. besides the couples of the energetic stress and deformation tensors $(\overset{(n)}{\mathbf{T}}, \overset{(n)}{\mathbf{C}})$, there are couples of the energetic stress tensors and deformation measures $(\overset{(n)}{\mathbf{T}}, \overset{(n)}{\mathbf{G}})$ (see Table 4.1).

It should be noted that besides the tensors $\overset{(n)}{\mathbf{G}}$ we can introduce a set of tensors of the form $\overset{(n)}{\mathbf{G}} + \mathbf{N}$, where $\mathbf{N}$ is an arbitrary constant-tensor: $\dot{\mathbf{N}} \equiv 0$, which also satisfy the relation (4.44). However, only the tensors $\overset{(n)}{\mathbf{G}}$, which were already used above, have some physical meaning:

$$\overset{\mathrm{I}}{\mathbf{G}} = -\frac{1}{2}\mathbf{G}^{-1}, \quad \overset{\mathrm{II}}{\mathbf{G}} = -\mathbf{U}^{-1}, \quad \overset{\mathrm{IV}}{\mathbf{G}} = \mathbf{U}, \quad \overset{\mathrm{V}}{\mathbf{G}} = \frac{1}{2}\mathbf{G}; \tag{4.47}$$

they are the right Cauchy–Green and Almansi deformation measures, and also the right stretch tensor and its inverse. All the tensors $\overset{(n)}{\mathbf{T}}$, $\overset{(n)}{\mathbf{C}}$ and $\overset{(n)}{\mathbf{G}}$ are symmetric.

The principal energetic deformation measures for $n = \mathrm{I}, \dots, \mathrm{V}$ are given in Table 4.1.

For the first time, formulae (4.35) and (4.42), and also the existence of energetic deformation measures have been derived in the systematized form by the author in [12].

### 4.2.9 *Relationships Between Principal Invariants of Energetic Deformation Measures and Tensors*

In continuum mechanics, three principal invariants of a second-order tensor (see [12]) are frequently used; they are determined as follows:

$$I_1(\overset{(n)}{\mathbf{C}}) = \mathbf{E} \cdot\cdot \overset{(n)}{\mathbf{C}}, \quad I_2(\overset{(n)}{\mathbf{C}}) = \frac{1}{2}(I_1^2(\overset{(n)}{\mathbf{C}}) - I_1(\overset{(n)}{\mathbf{C}}{}^2)), \quad I_3(\overset{(n)}{\mathbf{C}}) = \det(\overset{(n)}{\mathbf{C}}). \qquad (4.48)$$

The concept of an invariant will be considered in detail in Sect. 4.8.4.

Now, let us derive relations between principal invariants $I_\alpha(\overset{(n)}{\mathbf{C}})$ ($\alpha = 1, 2, 3$) of the energetic tensors $\overset{(n)}{\mathbf{C}}$ and principal invariants $I_\alpha(\overset{(n)}{\mathbf{G}})$ of the energetic deformation measures.

**Theorem 4.4.** *The principal invariants of the tensors* $\overset{(n)}{\mathbf{C}}$ *and* $\overset{(n)}{\mathbf{G}}$ *are connected by the relations*

$$\begin{aligned}
I_1(\overset{(n)}{\mathbf{C}}) &= I_1(\overset{(n)}{\mathbf{G}}) - \frac{3}{n - \mathrm{III}},\\
I_2(\overset{(n)}{\mathbf{C}}) &= I_2(\overset{(n)}{\mathbf{G}}) - \frac{2}{n - \mathrm{III}} I_1(\overset{(n)}{\mathbf{G}}) + \frac{3}{(n - \mathrm{III})^2},\\
I_3(\overset{(n)}{\mathbf{C}}) &= I_3(\overset{(n)}{\mathbf{G}}) - \frac{I_2(\overset{(n)}{\mathbf{G}})}{n - \mathrm{III}} - \frac{I_1(\overset{(n)}{\mathbf{G}})}{(n - \mathrm{III})^2} - \frac{1}{(n - \mathrm{III})^3},\\
I_1(\overset{(n)}{\mathbf{G}}) &= I_1(\overset{(n)}{\mathbf{C}}) + \frac{3}{n - \mathrm{III}},\\
I_2(\overset{(n)}{\mathbf{G}}) &= I_2(\overset{(n)}{\mathbf{C}}) + \frac{2}{n - \mathrm{III}} I_1(\overset{(n)}{\mathbf{C}}) + \frac{3}{(n - \mathrm{III})^2},\\
I_3(\overset{(n)}{\mathbf{G}}) &= I_3(\overset{(n)}{\mathbf{C}}) + \frac{I_2(\overset{(n)}{\mathbf{C}})}{n - \mathrm{III}} + \frac{I_1(\overset{(n)}{\mathbf{C}})}{(n - \mathrm{III})^2} + \frac{1}{(n - \mathrm{III})^3},\\
n &= \mathrm{I},\ \mathrm{II},\ \mathrm{IV},\ \mathrm{V}.
\end{aligned} \qquad (4.49)$$

▼ To prove the theorem, one should express the first, second and third powers of the tensors $\overset{(n)}{\mathbf{C}}$ in terms of powers of the tensors $\overset{(n)}{\mathbf{G}}$:

$$\overset{(n)}{\mathbf{C}} = \overset{(n)}{\mathbf{G}} - \frac{1}{n - \mathrm{III}}\mathbf{E},$$

$$\overset{(n)}{\mathbf{C}}{}^2 = \overset{(n)}{\mathbf{G}}{}^2 - \frac{2}{n - \mathrm{III}}\overset{(n)}{\mathbf{G}} + \frac{1}{(n - \mathrm{III})^2}\mathbf{E},$$

$$\overset{(n)}{\mathbf{C}}{}^3 = \overset{(n)}{\mathbf{G}}{}^3 - \frac{3}{n - \mathrm{III}}\overset{(n)}{\mathbf{G}}{}^2 + \frac{3}{(n - \mathrm{III})^2}\overset{(n)}{\mathbf{G}} - \frac{1}{(n - \mathrm{III})^3}\mathbf{E}.$$

Write out the following invariants of the tensors: $I_1(\overset{(n)}{\mathbf{G}})$, $I_1(\overset{(n)}{\mathbf{G}}{}^2)$ and $I_1(\overset{(n)}{\mathbf{G}}{}^3)$, and then, according to the formulae (see [12])

$$\begin{aligned} I_3(\mathbf{T}) = \det(\mathbf{T}) &= \frac{1}{6}\left(I_1^3(\mathbf{T}) - 3I_1(\mathbf{T})I_1(\mathbf{T}^2) + 2I_1(\mathbf{T}^3)\right), \\ I_3(\mathbf{T}) &= \frac{1}{3}\left(I_1(\mathbf{T}^3) - I_1^3(\mathbf{T}) + 3I_1(\mathbf{T})I_2(\mathbf{T})\right), \end{aligned} \quad (4.50)$$

we get the desired expressions for the first, second and third invariants of $\overset{(n)}{\mathbf{C}}$. The derivation should be performed in detail as Exercise 4.2.10. ▲

### 4.2.10 Quasienergetic Couples of Stress and Deformation Tensors

With the help of the left stretch tensor $\mathbf{V}$, we can introduce couples of the tensors $\overset{(n)}{\mathbf{S}}$ and $\overset{(n)}{\mathbf{A}}$ as well. However, in this case, the stress power $w_{(i)}$ depends, in addition, on the derivative of the rotation tensor $\mathbf{O}^{\mathrm{T}}$. Therefore, such couples $(\overset{(n)}{\mathbf{S}}, \overset{(n)}{\mathbf{A}})$ are called *quasienergetic*.

**Definition 4.1.** The tensors $\overset{(n)}{\mathbf{S}}$ and $\overset{(n)}{\mathbf{A}}$, defined by Table 4.3, are called the *principal quasienergetic stress and deformation tensors*, respectively.

**Theorem 4.5.** *The stress power $w_{(i)}$ (4.2) can always be expressed in terms of each of the quasienergetic tensor couples:*

$$w_{(i)} = \overset{(n)}{\mathbf{S}} \cdot\cdot \frac{d}{dt}\overset{(n)}{\mathbf{A}} + \overset{\circ}{\mathbf{S}} \cdot\cdot \frac{d}{dt}\mathbf{O}^{\mathrm{T}} + \mathbf{T}^K \cdot\cdot \mathbf{W}, \quad \forall n \in \mathbb{Z}. \quad (4.51)$$

**Table 4.3** Principal quasienergetic couples of tensors

| Number $n$ of couple | Quasienergetic stress tensors $\overset{(n)}{\mathbf{S}}$ | Quasienergetic deformation tensors $\overset{(n)}{\mathbf{A}}$ | Quasienergetic deformation measures $\overset{(n)}{\mathbf{g}}$ |
|---|---|---|---|
| I | $\mathbf{V}\cdot\mathbf{T}^S\cdot\mathbf{V}$ | $\mathbf{A}=\frac{1}{2}(\mathbf{E}-\mathbf{V}^{-2})$ | $-\frac{1}{2}\mathbf{g}^{-1}$ |
| II | $\frac{1}{2}(\mathbf{V}\cdot\mathbf{T}^S+\mathbf{T}^S\cdot\mathbf{V})$ | $\mathbf{E}-\mathbf{V}^{-1}$ | $-\mathbf{V}^{-1}$ |
| III | $\mathbf{T}^S$ | $\mathbf{Y}$ | $\overset{\text{III}}{\mathbf{g}}$ |
| IV | $\frac{1}{2}(\mathbf{V}^{-1}\cdot\mathbf{T}^S+\mathbf{T}^S\cdot\mathbf{V}^{-1})$ | $\mathbf{V}-\mathbf{E}$ | $\mathbf{V}$ |
| V | $\mathbf{V}^{-1}\cdot\mathbf{T}^S\cdot\mathbf{V}^{-1}$ | $\mathbf{J}=\frac{1}{2}(\mathbf{V}^2-\mathbf{E})$ | $\frac{1}{2}\mathbf{g}$ |

*If the tensor* $\mathbf{T}$ *is symmetric, then the expression for* $w_{(i)}$ *takes the form*

$$w_{(i)}=\overset{(n)}{\mathbf{S}}\cdot\cdot\frac{d}{dt}\overset{(n)}{\mathbf{A}}+\overset{\circ}{\mathbf{S}}\cdot\cdot\frac{d}{dt}\mathbf{O}^{\mathrm{T}},\qquad\forall n\in\mathbb{Z}.\tag{4.51a}$$

*Here* $\overset{(n)}{\mathbf{S}}$, $\overset{(n)}{\mathbf{A}}$ *are symmetric second-order quasienergetic tensors; and the tensor* $\overset{\circ}{\mathbf{S}}$ *is the same for all the couples, and it is called the rotation tensor of stresses:*

$$\overset{\circ}{\mathbf{S}}=\frac{1}{2}(\mathbf{V}\cdot\mathbf{T}^S\cdot\mathbf{V}^{-1}-\mathbf{V}^{-1}\cdot\mathbf{T}^S\cdot\mathbf{V})\cdot\mathbf{O}.\tag{4.52}$$

Let us prove the theorem for each of the principal quasienergetic couples separately.

### 4.2.11 The First Quasienergetic Couple $(\overset{\mathrm{I}}{\mathbf{S}},\mathbf{A})$

Consider the first energetic couple (4.24) and pass from the tensor $\mathbf{U}^{-2}$ to $\mathbf{V}^{-2}$ with the help of the relations $\mathbf{U}^{-2}=\mathbf{O}^{\mathrm{T}}\cdot\mathbf{V}^{-2}\cdot\mathbf{O}$ (see Exercise 2.3.1):

$$\begin{aligned}w_S^{(i)}&=\mathbf{T}^S\cdot\cdot\mathbf{D}=\overset{\mathrm{I}}{\mathbf{T}}\cdot\cdot\overset{\mathrm{I}}{\mathbf{\Lambda}}^{\bullet}=-\frac{1}{2}\mathbf{F}^{\mathrm{T}}\cdot\mathbf{T}^S\cdot\mathbf{F}\cdot\cdot\dot{\mathbf{U}}^{-2}\\&=-\frac{1}{2}\mathbf{F}^{\mathrm{T}}\cdot\mathbf{T}^S\cdot\mathbf{F}\cdot\cdot\left(\dot{\mathbf{O}}^{\mathrm{T}}\cdot\mathbf{V}^{-2}\cdot\mathbf{O}+\mathbf{O}^{\mathrm{T}}\cdot\dot{\mathbf{V}}^{-2}\cdot\mathbf{O}+\mathbf{O}^{\mathrm{T}}\cdot\mathbf{V}^{-2}\cdot\dot{\mathbf{O}}\right).\end{aligned}\tag{4.53}$$

According to the polar decomposition (2.137) and the rules (4.4) of rearrangement of tensors in the scalar product, we obtain

$$\begin{aligned}w_S^{(i)}=-\frac{1}{2}\Big(&\mathbf{V}^{-1}\cdot\mathbf{T}^S\cdot\mathbf{V}\cdot\cdot\mathbf{O}\cdot\dot{\mathbf{O}}^{\mathrm{T}}+\mathbf{V}\cdot\mathbf{T}^S\cdot\mathbf{V}\cdot\cdot\dot{\mathbf{V}}^{-2}\\&+\mathbf{V}\cdot\mathbf{T}^S\cdot\mathbf{V}^{-1}\cdot\cdot\dot{\mathbf{O}}\cdot\mathbf{O}^{\mathrm{T}}\Big).\end{aligned}\tag{4.54}$$

Taking into account that $\mathbf{O}\cdot\mathbf{O}^{\mathrm{T}} = \mathbf{E}$ and, hence,

$$\dot{\mathbf{O}}\cdot\mathbf{O}^{\mathrm{T}} = -\mathbf{O}\cdot\dot{\mathbf{O}}^{\mathrm{T}}, \tag{4.55}$$

we obtain the following final expression of the form (4.51):

$$w_S^{(i)} = \overset{\mathrm{I}}{\mathbf{S}}\cdot\cdot\left(-\frac{1}{2}\mathbf{V}^{-2}\right)^{\bullet} + \overset{\circ}{\mathbf{S}}\cdot\cdot\dot{\mathbf{O}}^{\mathrm{T}}. \tag{4.56}$$

Here the first quasienergetic tensors $\overset{\mathrm{I}}{\mathbf{S}}$, $\overset{\mathrm{I}}{\mathbf{A}}$ are defined as follows: $\overset{\mathrm{I}}{\mathbf{S}} = \mathbf{V}\cdot\mathbf{T}^S\cdot\mathbf{V}$ and $\overset{\mathrm{I}}{\mathbf{A}} = \mathbf{A} = \frac{1}{2}(\mathbf{E}-\mathbf{V}^{-2}) = \frac{1}{2}(\mathbf{E}-\mathbf{F}^{-1\mathrm{T}}\cdot\mathbf{F}^{-1})$, and the tensor $\overset{\circ}{\mathbf{S}}$ is defined by formula (4.52). Thus, we have proved the existence of the first quasienergetic couple $(\overset{\mathrm{I}}{\mathbf{S}}, \overset{\mathrm{I}}{\mathbf{A}}) = (\overset{\mathrm{I}}{\mathbf{S}}, \mathbf{A})$.

### 4.2.12 *The Second Quasienergetic Couple* $(\overset{\mathrm{II}}{\mathbf{S}}, (\mathbf{E}-\mathbf{V}^{-1}))$

Going from the derivative $\dot{\mathbf{V}}^{-2}$ to $\dot{\mathbf{V}}^{-1}$ in (4.55), we get

$$\begin{aligned} w_{(i)} &= \overset{\circ}{\mathbf{S}}\cdot\cdot\dot{\mathbf{O}}^{\mathrm{T}} - \frac{1}{2}\mathbf{V}\cdot\mathbf{T}^S\cdot\mathbf{V}\cdot\cdot\left(\mathbf{V}^{-1}\cdot\dot{\mathbf{V}}^{-1} + \dot{\mathbf{V}}^{-1}\cdot\mathbf{V}^{-1}\right) \\ &= \overset{\circ}{\mathbf{S}}\cdot\cdot\dot{\mathbf{O}}^{\mathrm{T}} - \frac{1}{2}(\mathbf{V}\cdot\mathbf{T}^S + \mathbf{T}^S\cdot\mathbf{V})\cdot\cdot\dot{\mathbf{V}}^{-1}. \end{aligned} \tag{4.57}$$

Here we have used the rule (4.4) again. As a result, we find the second quasienergetic couple:

$$w_{(i)} = \overset{\mathrm{II}}{\mathbf{S}}\cdot\cdot(\mathbf{E}-\mathbf{V}^{-1})^{\bullet} + \overset{\circ}{\mathbf{S}}\cdot\cdot\dot{\mathbf{O}}^{\mathrm{T}}, \tag{4.58}$$

where the second quasienergetic stress tensor $\overset{\mathrm{II}}{\mathbf{S}}$ has been introduced as follows:

$$\overset{\mathrm{II}}{\mathbf{S}} = \frac{1}{2}(\mathbf{V}\cdot\mathbf{T}^S + \mathbf{T}^S\cdot\mathbf{V}), \tag{4.59}$$

and the second quasienergetic deformation tensor is $\overset{\mathrm{II}}{\mathbf{A}} = (\mathbf{E}-\mathbf{V}^{-1})$.

### 4.2.13 *The Third Quasienergetic Couple* $(\mathbf{Y},\ \mathbf{T}^S)$

Consider the third energetic couple (see Sect. 4.2.6) and pass from the tensor $\dot{\mathbf{U}}$ to $\dot{\mathbf{V}}$:

$$w_S^{(i)} = \frac{1}{2}\mathbf{O}^{\mathrm{T}}\cdot\mathbf{T}^S\cdot\mathbf{O}\cdot\cdot(\dot{\mathbf{U}}\cdot\mathbf{U}^{-1} + \mathbf{U}^{-1}\cdot\dot{\mathbf{U}})$$

$$= \frac{1}{2}\mathbf{O}^{\mathrm{T}} \cdot \mathbf{T}^S \cdot \mathbf{O} \cdot\cdot (\dot{\mathbf{O}}^{\mathrm{T}} \cdot \mathbf{V} \cdot \mathbf{O} + \mathbf{O}^{\mathrm{T}} \cdot \dot{\mathbf{V}} \cdot \mathbf{O} + \mathbf{O}^{\mathrm{T}} \cdot \mathbf{V} \cdot \dot{\mathbf{O}}) \cdot \mathbf{O}^{\mathrm{T}} \cdot \mathbf{V}^{-1} \cdot \mathbf{O}$$
$$+ \frac{1}{2}\mathbf{O}^{\mathrm{T}} \cdot \mathbf{T}^S \cdot \mathbf{O} \cdot\cdot \mathbf{O}^{\mathrm{T}} \cdot \mathbf{V}^{-1} \cdot \mathbf{O}(\dot{\mathbf{O}}^{\mathrm{T}} \cdot \mathbf{V} \cdot \mathbf{O} + \mathbf{O}^{\mathrm{T}} \cdot \dot{\mathbf{V}} \cdot \mathbf{O} + \mathbf{O}^{\mathrm{T}} \cdot \mathbf{V} \cdot \dot{\mathbf{O}}).$$

Changing the order in the scalar product of tensors in each of the summands, we get

$$w_S^{(i)} = \frac{1}{2}(\mathbf{T}^S \cdot\cdot \mathbf{O} \cdot \dot{\mathbf{O}}^{\mathrm{T}} + \mathbf{V}^{-1} \cdot \mathbf{T}^S \cdot\cdot \dot{\mathbf{V}} + \mathbf{V}^{-1} \cdot \mathbf{T}^S \cdot \mathbf{V} \cdot\cdot \dot{\mathbf{O}} \cdot \mathbf{O}^{\mathrm{T}}$$
$$+ \mathbf{V} \cdot \mathbf{T}^S \cdot \mathbf{V}^{-1} \cdot\cdot \mathbf{O} \cdot \dot{\mathbf{O}}^{\mathrm{T}} + \mathbf{T}^S \cdot \mathbf{V}^{-1} \cdot\cdot \dot{\mathbf{V}} + \mathbf{T}^S \cdot\cdot \dot{\mathbf{O}} \cdot \mathbf{O}^{\mathrm{T}}). \quad (4.60)$$

According to the relation (4.55), we see that the first and the last terms are canceled in pair. Thus, Eq. (4.60) yields

$$w_S^{(i)} = \mathbf{T}^S \cdot\cdot \dot{\mathbf{Y}} + \overset{\circ}{\mathbf{S}} \cdot\cdot \dot{\mathbf{O}}^{\mathrm{T}}. \quad (4.61)$$

where we have introduced the new tensor $\mathbf{Y}$ similar to the tensor $\mathbf{B}$, which is defined by its derivative (see (2.318)):

$$\dot{\mathbf{Y}} = \frac{1}{2}(\dot{\mathbf{V}} \cdot \mathbf{V}^{-1} + \mathbf{V}^{-1} \cdot \dot{\mathbf{V}}), \quad \mathbf{Y}(0) = \mathbf{0}. \quad (4.62)$$

Thus, we have proved the existence of the third quasienergetic couple: $(\overset{\mathrm{III}}{\mathbf{A}}, \overset{\mathrm{III}}{\mathbf{S}}) = (\mathbf{Y}, \mathbf{T}^S)$.

### 4.2.14 *The Fourth Quasienergetic Couple* $(\overset{\mathrm{IV}}{\mathbf{S}}, (\mathbf{V} - \mathbf{E}))$

Due to (4.62), formula (4.61) takes the form

$$w_S^{(i)} = \frac{1}{2}(\mathbf{V}^{-1} \cdot \mathbf{T}^S + \mathbf{T}^S \cdot \mathbf{V}^{-1}) \cdot\cdot \dot{\mathbf{V}} + \overset{\circ}{\mathbf{S}} \cdot\cdot \dot{\mathbf{O}}^{\mathrm{T}}; \quad (4.63)$$

that proves the existence of the fourth quasienergetic couple:

$$w_S^{(i)} = \overset{\mathrm{IV}}{\mathbf{S}} \cdot\cdot \overset{\mathrm{IV}}{\mathbf{A}}{}^{\bullet} + \overset{\circ}{\mathbf{S}} \cdot\cdot \dot{\mathbf{O}}^{\mathrm{T}},$$

where the fourth quasienergetic tensors are defined as follows (see Table 4.3):

$$\overset{\mathrm{IV}}{\mathbf{S}} = \frac{1}{2}(\mathbf{V}^{-1} \cdot \mathbf{T}^S + \mathbf{T}^S \cdot \mathbf{V}^{-1}), \quad \overset{\mathrm{IV}}{\mathbf{A}} = \mathbf{V} - \mathbf{E}. \quad (4.63a)$$

### 4.2.15 The Fifth Quasienergetic Couple $(\overset{\mathrm{V}}{\mathbf{S}}, \mathbf{J})$

Rewrite the relation (4.63) as follows:

$$w_S^{(i)} = \frac{1}{2}(\mathbf{V}^{-1} \cdot \mathbf{T}^S \cdot\cdot \mathbf{V}^{-1} \cdot \mathbf{V} \cdot \dot{\mathbf{V}} + \mathbf{T}^S \cdot \mathbf{V}^{-1} \cdot\cdot \dot{\mathbf{V}} \cdot \mathbf{V} \cdot \mathbf{V}^{-1})$$
$$+ \overset{\circ}{\mathbf{S}} \cdot\cdot \dot{\mathbf{O}}^{\mathrm{T}} = \mathbf{V}^{-1} \cdot \mathbf{T}^S \cdot \mathbf{V}^{-1} \cdot\cdot \frac{1}{2}(\mathbf{V} \cdot \dot{\mathbf{V}} + \dot{\mathbf{V}} \cdot \mathbf{V}) + \overset{\circ}{\mathbf{S}} \cdot\cdot \dot{\mathbf{O}}^{\mathrm{T}}.$$

Hence, there exists the fifth quasienergetic couple:

$$w_S^{(i)} = \overset{\mathrm{V}}{\mathbf{S}} \cdot\cdot \dot{\mathbf{J}} + \overset{\circ}{\mathbf{S}} \cdot\cdot \dot{\mathbf{O}}^{\mathrm{T}},$$

where the fifth quasienergetic stress and deformation tensors are defined as follows:

$$\overset{\mathrm{V}}{\mathbf{S}} = \mathbf{V}^{-1} \cdot \mathbf{T}^S \cdot \mathbf{V}^{-1}, \quad \overset{\mathrm{V}}{\mathbf{A}} = \mathbf{J} = \frac{1}{2}(\mathbf{V}^2 - \mathbf{E}) = \frac{1}{2}(\mathbf{F} \cdot \mathbf{F}^{\mathrm{T}} - \mathbf{E}). \tag{4.64}$$

For the first time, the principal quasienergetic couples have been derived by the author in [12].

### 4.2.16 General Representation of Quasienergetic Tensors

**Theorem 4.6.** *Each of the quasienergetic deformation tensors* $\overset{(n)}{\mathbf{A}}$ *can be expressed in terms of the corresponding power of the left stretch tensor* $\mathbf{V}$ *as follows:*

$$\overset{(n)}{\mathbf{A}} = \frac{1}{(n - \mathrm{III})}(\mathbf{V}^{n-\mathrm{III}} - \mathbf{E}), \quad \forall n \in \mathbb{Z}, \quad n \neq \mathrm{III}, \tag{4.65}$$

*and quasienergetic stress tensors* $\overset{(n)}{\mathbf{S}}$ *satisfy the equations*

$$\overset{\mathrm{IV}}{\mathbf{S}} = \frac{1}{n - \mathrm{III}} \sum_{k=0}^{n-\mathrm{IV}} \mathbf{V}^{n-k-\mathrm{IV}} \cdot \overset{(n)}{\mathbf{S}} \cdot \mathbf{V}^k, \quad \text{if } n \geqslant \mathrm{IV}; \tag{4.65a}$$

$$\overset{\mathrm{II}}{\mathbf{S}} = \frac{1}{\mathrm{III} - n} \sum_{k=0}^{\mathrm{II}-n} \mathbf{V}^{n-\mathrm{II}+k} \cdot \overset{(n)}{\mathbf{S}} \cdot \mathbf{V}^{-k}, \quad \text{if } n \leqslant \mathrm{II}. \tag{4.65b}$$

▼ The theorem for $n$ =I, II, IV and V can be proved by immediate comparison of formula (4.65) with the tensors $\overset{(n)}{\mathbf{A}}$ from Table 4.3, and for $n >$ V and $n <$ I – by the same method as was used in the proof of Theorem 4.2. ▲

**Table 4.4** Inverse tensors of quasienergetic equivalency

| Number $n$ of couple | ${}^4\overset{(n)}{\mathbf{Q}}{}^{-1}$ |
|---|---|
| I | $(\mathbf{V}\otimes\mathbf{V})^{(1432)}$ |
| II | $\frac{1}{2}(\mathbf{V}\otimes\mathbf{E}+\mathbf{E}\otimes\mathbf{V})^{(1432)}$ |
| III | $\boldsymbol{\Delta}$ |
| IV | $\frac{1}{2}(\mathbf{V}^{-1}\otimes\mathbf{E}+\mathbf{E}\otimes\mathbf{V}^{-1})^{(1432)}$ |
| V | $(\mathbf{V}^{-1}\otimes\mathbf{V}^{-1})^{(1432)}$ |

The quasienergetic stress tensors $\overset{(n)}{\mathbf{S}}$ can be represented in the general form (see Exercise 4.2.17)

$$\overset{(n)}{\mathbf{S}} = {}^4\overset{(n)}{\mathbf{Q}}{}^{-1}\cdot\cdot\,\mathbf{T}^S, \tag{4.66}$$

where ${}^4\overset{(n)}{\mathbf{Q}}{}^{-1}$ are the *inverse tensors of quasienergetic equivalency*, whose expressions for $n=\mathrm{I},\ldots,\mathrm{V}$ are given in Table 4.4.

Expressions for the *tensors of quasienergetic equivalency* ${}^4\overset{(n)}{\mathbf{Q}}$ can be obtained by inverting the relations (4.66):

$$\mathbf{T}^S = {}^4\overset{(n)}{\mathbf{Q}}\cdot\cdot\,\overset{(n)}{\mathbf{S}}. \tag{4.67}$$

To invert them, we use the representations of the tensors in the eigenbasis $\mathbf{p}_i$ of the left stretch tensor $\mathbf{V}$, as it was made in Sect. 4.2.7.

**Theorem 4.7.** *The tensors of quasienergetic equivalency* ${}^4\overset{(n)}{\mathbf{Q}}$*, connecting the tensors* $\mathbf{T}^S$ *and* $\overset{(n)}{\mathbf{S}}$ *with the help of the linear relations (4.67), have the form*

$${}^4\overset{(n)}{\mathbf{Q}} = \sum_{\alpha,\beta=1}^{3}\overset{(n)}{E}_{\alpha\beta}\,\mathbf{p}_\alpha\otimes\mathbf{p}_\beta\otimes\mathbf{p}_\beta\otimes\mathbf{p}_\alpha,\quad n\in\mathbb{Z}, \tag{4.68}$$

*where* $\overset{(n)}{E}_{\alpha\beta}$ *are expressed by formulae (4.38), (4.39).*

▼ Consider only the case when $n=\mathrm{I}$; for the remaining cases a proof is analogous and should be performed as Exercise 4.2.19.

Expressing the tensors $\overset{(n)}{\mathbf{S}}$ in the eigenbasis $\mathbf{p}_i$:

$$\overset{(n)}{\mathbf{S}} = \sum_{\alpha,\beta=1}^{3}\overset{(n)}{S}{}^{(p)}_{\alpha\beta}\,\mathbf{p}_\alpha\otimes\mathbf{p}_\beta, \tag{4.69}$$

and using the representations (4.40) for $\mathbf{T}^S$ and (2.142) for $\mathbf{V}$, we can rewrite the relation (4.66) for $n = \mathrm{I}$ as follows:

$$\overset{\mathrm{I}}{\mathbf{S}} = \sum_{\alpha,\beta=1}^{3} \overset{\mathrm{I}}{S}{}^{(p)}_{\alpha\beta} \mathbf{p}_\alpha \otimes \mathbf{p}_\beta = \mathbf{V} \cdot \mathbf{T}^S \cdot \mathbf{V}$$

$$= \sum_{\alpha,\beta,\gamma,\omega=1}^{3} \lambda_\alpha \mathbf{p}_\alpha \otimes \mathbf{p}_\alpha \cdot T^{(p)}_{\gamma\omega} \mathbf{p}_\gamma \otimes \mathbf{p}_\omega \cdot \lambda_\beta \mathbf{p}_\beta \otimes \mathbf{p}_\beta = \sum_{\alpha,\beta=1}^{3} \lambda_\alpha \lambda_\beta T^{(p)}_{\alpha\beta} \mathbf{p}_\alpha \otimes \mathbf{p}_\beta,$$

i.e. $\overset{\mathrm{I}}{S}{}^{(p)}_{\alpha\beta} = \lambda_\alpha \lambda_\beta T^{(p)}_{\alpha\beta}$. Then

$$\mathbf{T}^S = \sum_{\alpha,\beta=1}^{3} T^{(p)}_{\alpha\beta} \mathbf{p}_\alpha \otimes \mathbf{p}_\beta = \sum_{\alpha,\beta=1}^{3} \frac{1}{\lambda_\alpha \lambda_\beta} \mathbf{p}_\alpha \otimes \overset{\mathrm{I}}{S}{}^{(p)}_{\alpha\beta} \otimes \mathbf{p}_\beta$$

$$= \sum_{\alpha,\beta=1}^{3} \frac{1}{\lambda_\alpha \lambda_\beta} \mathbf{p}_\alpha \otimes \mathbf{p}_\alpha \cdot \overset{\mathrm{I}}{\mathbf{S}} \cdot \mathbf{p}_\beta \otimes \mathbf{p}_\beta$$

$$= \sum_{\alpha,\beta=1}^{3} \frac{1}{\lambda_\alpha \lambda_\beta} (\mathbf{p}_\alpha \otimes \mathbf{p}_\beta \otimes \mathbf{p}_\beta \otimes \mathbf{p}_\alpha) \cdot\cdot \overset{\mathrm{I}}{\mathbf{S}}. \ \blacktriangle$$

*Remark.* It follows from the performed proof and formula (4.41) that components $\overset{\mathrm{I}}{S}{}^{(p)}_{\alpha\beta}$ and $\overset{\mathrm{I}\circ}{T}{}^{(p)}_{\alpha\beta}$ of the tensors $\overset{\mathrm{I}}{\mathbf{S}}$ and $\overset{\mathrm{I}}{\mathbf{T}}$ are coincident. The same result is valid for all the tensors $\overset{(\mathrm{n})}{\mathbf{S}}$ and $\overset{(\mathrm{n})}{\mathbf{T}}$ (see Exercises 4.2.18 and 4.2.19):

$$\overset{(\mathrm{n})}{S}{}^{(p)}_{\alpha\beta} = \overset{(\mathrm{n})\circ}{T}{}^{(p)}_{\alpha\beta}. \quad \square \tag{4.70}$$

### 4.2.17 *Quasienergetic Deformation Measures*

Similarly to the measures $\overset{(\mathrm{n})}{\mathbf{G}}$, let us introduce *quasienergetic deformation measures* $\overset{(\mathrm{n})}{\mathbf{g}}$:

$$\overset{(\mathrm{n})}{\mathbf{g}} = \frac{1}{n - \mathrm{III}} \mathbf{V}^{n-\mathrm{III}}, \quad n \in \mathbb{Z}, \quad n \neq \mathrm{III}, \tag{4.71}$$

being the corresponding powers of the left stretch tensor $\mathbf{V}$. The third quasienergetic deformation measure $\overset{\mathrm{III}}{\mathbf{g}}$ is introduced as follows:

$$\overset{\mathrm{III}}{\mathbf{g}} = \mathbf{E} + \overset{\mathrm{III}}{\mathbf{A}}. \tag{4.71a}$$

Due to (4.62), the measure satisfies the equation

$$\frac{d\overset{\mathrm{III}}{\mathbf{g}}}{dt} = \frac{1}{2}\left(\frac{d\mathbf{V}}{dt}\cdot\mathbf{V}^{-1} + \mathbf{V}^{-1}\cdot\frac{d\mathbf{V}}{dt}\right), \quad \overset{\mathrm{III}}{\mathbf{g}}(0) = \mathbf{E}. \tag{4.71b}$$

The tensors $\overset{(\mathrm{n})}{\mathbf{g}}$ and $\overset{(\mathrm{n})}{\mathbf{A}}$ are connected by the relation

$$\overset{(\mathrm{n})}{\mathbf{A}} = \overset{(\mathrm{n})}{\mathbf{g}} - \frac{1}{n-\mathrm{III}}\mathbf{E}, \tag{4.72}$$

(when $n = \mathrm{III}$ the multiplier $(n-\mathrm{III})$ is replaced by 1), therefore for all the couples

$$\overset{(\mathrm{n})}{\mathbf{A}}{}^{\bullet} = \overset{(\mathrm{n})}{\mathbf{g}}{}^{\bullet}. \tag{4.73}$$

According to this relation, the power (4.51) can be expressed in another form

$$w_{(i)} = \overset{(\mathrm{n})}{\mathbf{S}}\cdot\cdot\overset{(\mathrm{n})}{\mathbf{g}}{}^{\bullet} + \overset{\circ}{\mathbf{S}}\cdot\cdot\dot{\mathbf{O}}^{\mathrm{T}} + \mathbf{T}^{K}\cdot\cdot\mathbf{W}, \quad n\in\mathbb{Z}. \tag{4.74}$$

For the symmetric tensor $\mathbf{T}$, this expression reduces to

$$w_{(i)} = \overset{(\mathrm{n})}{\mathbf{S}}\cdot\cdot\overset{(\mathrm{n})}{\mathbf{g}}{}^{\bullet} + \overset{\circ}{\mathbf{S}}\cdot\cdot\dot{\mathbf{O}}^{\mathrm{T}}, \quad n\in\mathbb{Z}. \tag{4.74a}$$

The tensors $\overset{(\mathrm{n})}{\mathbf{g}}$ at $n = \mathrm{I, II, IV, V}$ are the left Cauchy–Green and Almansi deformation measures, and also the left stretch tensor and its inverse:

$$\overset{\mathrm{I}}{\mathbf{g}} = -\frac{1}{2}\mathbf{g}^{-1}, \quad \overset{\mathrm{II}}{\mathbf{g}} = -\mathbf{V}^{-1}, \quad \overset{\mathrm{IV}}{\mathbf{g}} = \mathbf{V}, \quad \overset{\mathrm{V}}{\mathbf{g}} = \frac{1}{2}\mathbf{g}. \tag{4.75}$$

All the tensors are symmetric.

### *4.2.18 Representation of Rotation Tensor of Stresses in Terms of Quasienergetic Couples of Tensors*

**Theorem 4.8.** *The rotation tensor of stresses* $\overset{\circ}{\mathbf{S}}$ *(4.52) can be represented in terms of the quasienergetic couples of the stress and deformation tensors as follows:*

$$\overset{\circ}{\mathbf{S}} = (\overset{(\mathrm{n})}{\mathbf{A}}\cdot\overset{(\mathrm{n})}{\mathbf{S}} - \overset{(\mathrm{n})}{\mathbf{S}}\cdot\overset{(\mathrm{n})}{\mathbf{A}})\cdot\mathbf{O}, \quad \forall n\in\mathbb{Z}, \quad n\neq\mathrm{III}. \tag{4.76}$$

▼ We first consider the principal quasienergetic couples when $n = \mathrm{I, II, IV, V}$. The definition of the tensor $\overset{\circ}{\mathbf{S}}$ (4.52)

$$\overset{\circ}{\mathbf{S}} = \frac{1}{2}(\mathbf{V}\cdot\mathbf{T}^S\cdot\mathbf{V}^{-1} - \mathbf{V}^{-1}\cdot\mathbf{T}^S\cdot\mathbf{V})\cdot\mathbf{O} \tag{4.77}$$

can be rewritten in the following four equivalent forms:

$$\begin{aligned}
\overset{\circ}{\mathbf{S}} &= \frac{1}{2}((\mathbf{V}\cdot\mathbf{T}^S\cdot\mathbf{V})\cdot\mathbf{V}^{-2} - \mathbf{V}^{-2}\cdot(\mathbf{V}\cdot\mathbf{T}^S\cdot\mathbf{V}))\cdot\mathbf{O},\\
\overset{\circ}{\mathbf{S}} &= \frac{1}{2}((\mathbf{V}\cdot\mathbf{T}^S + \mathbf{T}^S\cdot\mathbf{V})\cdot\mathbf{V}^{-1} - \mathbf{V}^{-1}\cdot(\mathbf{V}\cdot\mathbf{T}^S + \mathbf{T}^S\cdot\mathbf{V}))\cdot\mathbf{O},\\
\overset{\circ}{\mathbf{S}} &= \frac{1}{2}(\mathbf{V}\cdot(\mathbf{V}^{-1}\cdot\mathbf{T}^S + \mathbf{T}^S\cdot\mathbf{V}^{-1}) - (\mathbf{V}^{-1}\cdot\mathbf{T}^S + \mathbf{T}^S\cdot\mathbf{V}^{-1})\cdot\mathbf{V})\cdot\mathbf{O},\\
\overset{\circ}{\mathbf{S}} &= \frac{1}{2}(\mathbf{V}^2\cdot(\mathbf{V}^{-1}\cdot\mathbf{T}^S\cdot\mathbf{V}^{-1}) - (\mathbf{V}^{-1}\cdot\mathbf{T}^S\cdot\mathbf{V}^{-1})\cdot\mathbf{V}^2)\cdot\mathbf{O}.
\end{aligned} \tag{4.78}$$

Comparison of these relations with the definition of the quasienergetic tensors $\overset{(n)}{\mathbf{S}}$ and $\overset{(n)}{\mathbf{A}}$ (see Table 4.3) gives the formulae

$$\begin{aligned}
\overset{\circ}{\mathbf{S}} &= \frac{1}{2}(\overset{\mathrm{I}}{\mathbf{S}}\cdot(\mathbf{E} - 2\overset{\mathrm{I}}{\mathbf{A}}) - (\mathbf{E} - 2\overset{\mathrm{I}}{\mathbf{A}})\cdot\overset{\mathrm{I}}{\mathbf{S}})\cdot\mathbf{O},\\
\overset{\circ}{\mathbf{S}} &= \frac{1}{2}(\overset{\mathrm{II}}{\mathbf{S}}\cdot(\mathbf{E} - \overset{\mathrm{II}}{\mathbf{A}}) - (\mathbf{E} - \overset{\mathrm{II}}{\mathbf{A}})\cdot\overset{\mathrm{II}}{\mathbf{S}})\cdot\mathbf{O},\\
\overset{\circ}{\mathbf{S}} &= \frac{1}{2}((\overset{\mathrm{IV}}{\mathbf{A}} + \mathbf{E})\cdot\overset{\mathrm{IV}}{\mathbf{S}} - \overset{\mathrm{IV}}{\mathbf{S}}\cdot(\mathbf{E} + \overset{\mathrm{IV}}{\mathbf{A}}))\cdot\mathbf{O},\\
\overset{\circ}{\mathbf{S}} &= \frac{1}{2}((\mathbf{E} + 2\overset{\mathrm{V}}{\mathbf{A}})\cdot\overset{\mathrm{V}}{\mathbf{S}} - \overset{\mathrm{V}}{\mathbf{S}}\cdot(\mathbf{E} + 2\overset{\mathrm{V}}{\mathbf{A}}))\cdot\mathbf{O}.
\end{aligned} \tag{4.79}$$

Since the scalar product by the unit tensor is always commutative: $\overset{\mathrm{I}}{\mathbf{S}}\cdot\mathbf{E} = \mathbf{E}\cdot\overset{\mathrm{I}}{\mathbf{S}}$ etc.; so from (4.79) we obtain the desired relation (4.76).

Let us consider now the case when $n > \mathrm{V}$. Since for $n = \mathrm{IV}$ formula (4.76) has already been proved, we can use it and substitute relationship (4.65a) into this formula:

$$\begin{aligned}
\overset{\circ}{\mathbf{S}}\cdot\mathbf{O}^T &= \overset{\mathrm{IV}}{\mathbf{A}}\cdot\overset{\mathrm{IV}}{\mathbf{S}} - \overset{\mathrm{IV}}{\mathbf{S}}\cdot\overset{\mathrm{IV}}{\mathbf{A}} = \mathbf{V}\cdot\overset{\mathrm{IV}}{\mathbf{S}} - \overset{\mathrm{IV}}{\mathbf{S}}\cdot\mathbf{V}\\
&= \frac{1}{n-\mathrm{III}}\mathbf{V}\cdot(\mathbf{V}^{n-\mathrm{IV}}\cdot\overset{(n)}{\mathbf{S}} + \mathbf{V}^{n-\mathrm{V}}\cdot\overset{(n)}{\mathbf{S}}\cdot\mathbf{V} + \ldots + \mathbf{V}\cdot\overset{(n)}{\mathbf{S}}\cdot\mathbf{V}^{n-\mathrm{V}} + \overset{(n)}{\mathbf{S}}\cdot\mathbf{V}^{n-\mathrm{IV}})\\
&\quad - \frac{1}{n-\mathrm{III}}(\mathbf{V}^{n-\mathrm{IV}}\cdot\overset{(n)}{\mathbf{S}} + \mathbf{V}^{n-\mathrm{V}}\cdot\overset{(n)}{\mathbf{S}}\cdot\mathbf{V} + \ldots + \mathbf{V}\cdot\overset{(n)}{\mathbf{S}}\cdot\mathbf{V}^{n-\mathrm{V}} + \overset{(n)}{\mathbf{S}}\cdot\mathbf{V}^{n-\mathrm{IV}})\cdot\mathbf{V}.
\end{aligned}$$

If we remove parentheses in this formula, all the terms are canceled there except the first and last ones; therefore, we get

$$\overset{\circ}{\mathbf{S}} \cdot \mathbf{O}^T = \frac{1}{n - \text{III}}(\mathbf{V}^{n-\text{III}} \cdot \overset{(n)}{\mathbf{S}} - \overset{(n)}{\mathbf{S}} \cdot \mathbf{V}^{n-\text{III}}) = \overset{(n)}{\mathbf{A}} \cdot \overset{(n)}{\mathbf{S}} - \overset{(n)}{\mathbf{S}} \cdot \overset{(n)}{\mathbf{A}},$$

as $\mathbf{E} \cdot \overset{(n)}{\mathbf{S}} = \overset{(n)}{\mathbf{S}} \cdot \mathbf{E}$. Thus, formula (4.76) really holds for $n > \text{V}$.

The theorem for the case when $n < \text{I}$ can be proved in a similar way with the help of formula (4.65b). ▲

From this theorem we can get the following significant corollary, which will be used below.

**Theorem 4.9.** *The rotation tensor of stresses* $\overset{\circ}{\mathbf{S}}$ *is zero-tensor if and only if the tensors* $\overset{(n)}{\mathbf{S}}$ *and* $\overset{(n)}{\mathbf{A}}$ *commute with each other.*

### 4.2.19 Relations Between Density and Principal Invariants of Energetic and Quasienergetic Deformation Tensors

Relations between the density $\rho$ and principal invariants of the energetic and quasienergetic deformation tensors and measures will be frequently used below. Let us establish them.

**Theorem 4.10.** *For every continuum the ratio of densities* $\rho/\overset{\circ}{\rho}$ *is a one-valued function of the third principal invariant of the energetic and quasienergetic measures* $I_3(\overset{(n)}{\mathbf{G}})$ *and* $I_3(\overset{(n)}{\mathbf{g}})$, *and also a function of principal invariants of the energetic and quasienergetic deformation tensors* $I_\alpha(\overset{(n)}{\mathbf{C}})$ *and* $I_\alpha(\overset{(n)}{\mathbf{A}})$:

$$I_3(\overset{(n)}{\mathbf{G}}) = I_3(\overset{(n)}{\mathbf{g}}) = \frac{1}{(n - \text{III})^3}(\rho/\overset{\circ}{\rho})^{\text{III}-n}, \tag{4.80}$$

$$\rho/\overset{\circ}{\rho} = (1 + (n - \text{III})I_1(\overset{(n)}{\mathbf{C}}) + (n - \text{III})^2 I_2(\overset{(n)}{\mathbf{C}}) + (n - \text{III})^3 I_3(\overset{(n)}{\mathbf{C}}))^{1/(\text{III}-n)}, \tag{4.81}$$

$$\rho/\overset{\circ}{\rho} = (1 + (n - \text{III})I_1(\overset{(n)}{\mathbf{A}}) + (n - \text{III})^2 I_2(\overset{(n)}{\mathbf{A}}) + (n - \text{III})^3 I_3(\overset{(n)}{\mathbf{A}}))^{1/(\text{III}-n)}. \tag{4.82}$$

To prove formula (4.80), one should use the continuity equation (3.8) and the polar decomposition (2.137):

$$\overset{\circ}{\rho}/\rho = \det \mathbf{F} = \det (\mathbf{O} \cdot \mathbf{U}) = \det \mathbf{U} = \det \mathbf{V}. \tag{4.83}$$

Using the definition (4.42) of the measures $\overset{(n)}{\mathbf{G}}$ and formula (4.83), we verify that formula (4.80) for $I_3(\overset{(n)}{\mathbf{G}})$ is valid:

$$I_3(\overset{(n)}{\mathbf{G}}) = \det\left(\frac{1}{n-\mathrm{III}}\mathbf{U}^{n-\mathrm{III}}\right) = \frac{1}{(n-\mathrm{III})^3}(\det\mathbf{U})^{n-\mathrm{III}} = \frac{1}{(n-\mathrm{III})^3}(\rho/\overset{\circ}{\rho})^{\mathrm{III}-n}. \tag{4.84}$$

Similarly, according to the definition (4.75) of quasienergetic measures and formula (4.83), we can prove the remaining relations in formula (4.80):

$$I_3(\overset{(n)}{\mathbf{g}}) = \det\left(\frac{1}{n-\mathrm{III}}\mathbf{V}^{n-\mathrm{III}}\right) = \frac{1}{(n-\mathrm{III})^3}(\det\mathbf{V})^{n-\mathrm{III}} = \frac{1}{(n-\mathrm{III})^3}(\rho/\overset{\circ}{\rho})^{\mathrm{III}-n}. \tag{4.85}$$

On substituting the sixth formula of (4.50) into (4.80), we obtain the relations (4.81) between $(\rho/\overset{\circ}{\rho})$ and $I_\alpha(\overset{(n)}{\mathbf{C}})$. Since there are connections between the invariants $I_\alpha(\overset{(n)}{\mathbf{g}})$ and $I_\alpha(\overset{(n)}{\mathbf{A}})$, which are similar to (4.49) (see Exercise 4.2.8), so values of $(\rho/\overset{\circ}{\rho})$ and $I_\alpha(\overset{(n)}{\mathbf{A}})$ are also connected by relations (4.82) similar to (4.81). ▲

### *4.2.20 The Generalized Form of Representation of the Stress Power*

Consider only nonpolar continua, when the tensor $\mathbf{T}$ is symmetric; then

$$\mathbf{T}^S = \mathbf{T} \quad \text{and} \quad \mathbf{T}^K \equiv \mathbf{0}, \tag{4.86}$$

and, according to formulae (4.13), (4.46), (4.51a), and (4.74a), the stress power $w_{(i)}$ (4.2) can be expressed in the following equivalent forms:

$$w_{(i)} = \overset{(n)}{\mathbf{T}} \cdot\cdot \frac{d\overset{(n)}{\mathbf{C}}}{dt}, \quad n \in \mathbb{Z}, \tag{4.87}$$

$$w_{(i)} = \overset{(n)}{\mathbf{T}} \cdot\cdot \frac{d\overset{(n)}{\mathbf{G}}}{dt}, \quad n \in \mathbb{Z}, \tag{4.88}$$

$$w_{(i)} = \overset{(n)}{\mathbf{S}} \cdot\cdot \frac{d\overset{(n)}{\mathbf{A}}}{dt} + \overset{\circ}{\mathbf{S}} \cdot\cdot \frac{d\mathbf{O}^{\mathrm{T}}}{dt}, \quad n \in \mathbb{Z}, \tag{4.89}$$

$$w_{(i)} = \overset{(n)}{\mathbf{S}} \cdot\cdot \frac{d\overset{(n)}{\mathbf{g}}}{dt} + \overset{\circ}{\mathbf{S}} \cdot\cdot \frac{d\mathbf{O}^{\mathrm{T}}}{dt}, \quad n \in \mathbb{Z}. \tag{4.90}$$

The *stress power* $w_{(i)}$ is said to be written in the *form* $A_n$, $B_n$, $C_n$ or $D_n$, if we use expressions (4.87), (4.88), (4.89) or (4.90), respectively.

Introduce the *generalized energetic stress tensors* $\overset{(n)}{\mathbf{T}}_G$, the *generalized energetic deformation tensors* $\overset{(n)}{\mathbf{C}}_G$, the *generalized energetic deformation measures* $\overset{(n)}{\mathbf{G}}_G$ and the *generalized rotation tensor of stresses* $\mathbf{S}_G$ as follows:

$$\overset{(n)}{\mathbf{T}}_G = \begin{cases} \overset{(n)}{\mathbf{T}}, & \text{if } G = A, \\ \overset{(n)}{\mathbf{T}}, & \text{if } G = B, \\ \overset{(n)}{\mathbf{S}}, & \text{if } G = C, \\ \overset{(n)}{\mathbf{S}}, & \text{if } G = D, \end{cases} \quad \overset{(n)}{\mathbf{C}}_G = \begin{cases} \overset{(n)}{\mathbf{C}}, & \text{if } G = A, \\ \overset{(n)}{\mathbf{G}}, & \text{if } G = B, \\ \overset{(n)}{\mathbf{A}}, & \text{if } G = C, \\ \overset{(n)}{\mathbf{g}}, & \text{if } G = D, \end{cases}$$

$$\overset{(n)}{\mathbf{G}}_G = \begin{cases} \overset{(n)}{\mathbf{G}}, & \text{if } G = A, \\ \overset{(n)}{\mathbf{G}}, & \text{if } G = B, \\ \overset{(n)}{\mathbf{g}}, & \text{if } G = C, \\ \overset{(n)}{\mathbf{g}}, & \text{if } G = D, \end{cases} \quad \mathbf{S}_G = \begin{cases} \mathbf{0}, & \text{if } G = A, \\ \mathbf{0}, & \text{if } G = B, \\ \overset{\circ}{\mathbf{S}}, & \text{if } G = C, \\ \overset{\circ}{\mathbf{S}}, & \text{if } G = D, \end{cases} \tag{4.91}$$

where $G$ is the index taking on the letter values $G = A, B, C$, and $D$.

Then all the expressions (4.87)–(4.90) can be written in the single generalized form

$$w_{(i)} = \overset{(n)}{\mathbf{T}}_G \cdot\cdot \frac{d\overset{(n)}{\mathbf{C}}_G}{dt} + \mathbf{S}_G \cdot\cdot \frac{d\mathbf{O}^{\mathrm{T}}}{dt}, \quad n \in \mathbb{Z}, \ G = A, B, C, D. \tag{4.92}$$

### *4.2.21 Representation of Stress Power in Terms of Co-rotational Derivatives*

Two symmetric tensors $\mathbf{M}$ and $\mathbf{N}$ are called *coaxial*, if they have the same eigenbasis $\mathbf{p}_\alpha^{(M)}$ (but their eigenvalues $\lambda_\alpha^{(M)}$ and $\lambda_\alpha^{(N)}$ may in general be different):

$$\mathbf{M} = \sum_{\alpha=1}^{3} \lambda_\alpha^{(M)} \mathbf{p}_\alpha^{(M)} \otimes \mathbf{p}_\alpha^{(M)}, \quad \mathbf{N} = \sum_{\alpha=1}^{3} \lambda_\alpha^{(N)} \mathbf{p}_\alpha^{(M)} \otimes \mathbf{p}_\alpha^{(M)}.$$

Since the eigenbasis vectors $\mathbf{p}_\alpha^{(M)}$ are orthogonal with each other, the scalar product of coaxial tensors $\mathbf{M}$ and $\mathbf{N}$ is independent of order of the multipliers: $\mathbf{M}\cdot\mathbf{N} = \mathbf{N}\cdot\mathbf{M}$, i.e. the tensors $\mathbf{M}$ and $\mathbf{N}$ commutate with each other. The converse assertion holds as well: commuting symmetric tensors are coaxial (see Exercise 4.2.24).

**Theorem 4.11.** *Let the tensors* $\overset{(n)}{\mathbf{T}}$ *and* $\overset{(n)}{\mathbf{C}}$, *and also* $\overset{(n)}{\mathbf{S}}$ *and* $\overset{(n)}{\mathbf{A}}$ *be either coaxial in pairs or commuting in pairs for each n, then the stress power* $w_{(i)}$ *(4.92) can be represented in terms of the co-rotational derivatives as follows:*

$$w_{(i)} = \overset{(n)}{\mathbf{T}} \cdot\cdot \overset{(n)}{\mathbf{C}}{}^{h} = \overset{(n)}{\mathbf{T}} \cdot\cdot \overset{(n)}{\mathbf{G}}{}^{h} = \overset{(n)}{\mathbf{S}} \cdot\cdot \overset{(n)}{\mathbf{A}}{}^{h} = \overset{(n)}{\mathbf{S}} \cdot\cdot \overset{(n)}{\mathbf{g}}{}^{h}, \tag{4.93}$$

*where* $h = \{\cdot,\ U,\ V,\ J,\ S\}$.

▼ Notice that all the co-rotational derivatives, mentioned in the theorem, have the same structure (see (2.373)):

$$\overset{(n)}{\mathbf{C}}{}^{h} = \overset{(n)}{\mathbf{C}}{}^{\bullet} - \mathbf{Z}_h \cdot \overset{(n)}{\mathbf{C}} - \overset{(n)}{\mathbf{C}} \cdot \mathbf{Z}_h^{\mathrm{T}}, \tag{4.94}$$

where $\mathbf{Z}_h = \{\mathbf{0},\ \boldsymbol{\Omega}_U,\ \boldsymbol{\Omega}_V,\ \mathbf{W},\ \boldsymbol{\Omega}\}$ is a second-order tensor (see (2.374)). Then, due to the property (4.3) of contraction of three tensors, we obtain

$$\begin{aligned} \overset{(n)}{\mathbf{T}} \cdot\cdot \overset{(n)}{\mathbf{C}}{}^{h} &= \overset{(n)}{\mathbf{T}} \cdot\cdot \overset{(n)}{\mathbf{C}}{}^{\bullet} - \mathbf{T} \cdot\cdot \mathbf{Z}_h \cdot \overset{(n)}{\mathbf{C}} - \mathbf{T} \cdot\cdot \overset{(n)}{\mathbf{C}} \cdot \mathbf{Z}_h^{\mathrm{T}} \\ &= \overset{(n)}{\mathbf{T}} \cdot\cdot \overset{(n)}{\mathbf{C}}{}^{\bullet} - (\overset{(n)}{\mathbf{C}} \cdot \mathbf{T}) \cdot\cdot \mathbf{Z}_h - (\mathbf{T} \cdot \mathbf{C}) \cdot\cdot \mathbf{Z}_h^{\mathrm{T}}. \end{aligned} \tag{4.95}$$

Since the tensors $\overset{(n)}{\mathbf{C}}$ and $\overset{(n)}{\mathbf{T}}$ are either commuting or coaxial, so $\overset{(n)}{\mathbf{T}} \cdot \overset{(n)}{\mathbf{C}} = \overset{(n)}{\mathbf{C}} \cdot \overset{(n)}{\mathbf{T}}$, and the tensor $\mathbf{Z}_h$ is either skew-symmetric or zero-tensor. Then we find that

$$\overset{(n)}{\mathbf{T}} \cdot\cdot \overset{(n)}{\mathbf{C}}{}^{h} = \overset{(n)}{\mathbf{T}} \cdot\cdot \overset{(n)}{\mathbf{C}}{}^{\bullet} = w_{(i)}, \tag{4.96}$$

as was to be shown.

Using commutativity of the tensors $\overset{(n)}{\mathbf{A}}$ and $\overset{(n)}{\mathbf{S}}$ and Theorem 4.9, we find that $\overset{\circ}{\mathbf{S}} \equiv \mathbf{0}$ and $\overset{\circ}{\mathbf{S}} \cdot\cdot \dot{\mathbf{O}}^{\mathrm{T}} = 0$. Then, following the manipulations made above, we get

$$w_{(i)} = \overset{(n)}{\mathbf{S}} \cdot\cdot \overset{(n)}{\mathbf{A}}{}^{\bullet} + \overset{\circ}{\mathbf{S}} \cdot\cdot \dot{\mathbf{O}}^{\mathrm{T}} = \overset{(n)}{\mathbf{S}} \cdot\cdot \overset{(n)}{\mathbf{A}}{}^{h}. \ ▲ \tag{4.97}$$

### *4.2.22 Relations Between Rates of Energetic and Quasienergetic Tensors and Velocity Gradient*

In Sect. 2.5.11 we have derived the relations between the co-rotational derivatives (and the total derivative with respect to time) of the Almansi and Cauchy–Green deformation tensors and measures, on the one hand, and the velocity gradient $\mathbf{L}$, on the other hand. Using Eqs. (2.376) and (2.377), we can readily pass to the energetic and quasienergetic deformation tensors and measures and rewrite these relations in the generalized form

$$\overset{(n)}{\mathbf{C}}_G{}^h = {}^4\overset{(n)}{\mathbf{X}}_{Gh} \cdot\cdot \mathbf{D} + {}^4\overset{(n)}{\mathbf{Y}}_{Gh} \cdot\cdot \mathbf{W},$$
$$h = \{\cdot,\ \text{Ol},\ \text{CR},\ d,\ D,\ U,\ V,\ J,\ S\};\quad G = A,\ B,\ C,\ D;\quad n = \text{I}, \ldots, \text{V}, \tag{4.98}$$

where ${}^4\overset{(n)}{\mathbf{Y}}_{Gh}$ and ${}^4\overset{(n)}{\mathbf{X}}_{Gh}$ are the fourth-order tensors, their general expressions for $G = C, D$ have the forms

$$\begin{aligned}
{}^4\overset{(n)}{\mathbf{X}}_{Gh} &= {}^4\overset{(n)}{\mathbf{X}}{}^0_{Gh} - ({}^4\overset{(n)}{\mathbf{M}}{}^{(1342)}_h \cdot \overset{(n)}{\mathbf{C}}_G)^{(1423)} - \overset{(n)}{\mathbf{C}}_G \cdot {}^4\overset{(n)}{\mathbf{M}}{}^{(2134)}_h,\\
{}^4\overset{(n)}{\mathbf{Y}}_{Gh} &= {}^4\overset{(n)}{\mathbf{Y}}{}^0_{Gh} - ({}^4\overset{(n)}{\widetilde{\mathbf{M}}}{}^{(1342)}_h \cdot \overset{(n)}{\mathbf{C}}_G)^{(1423)} - \overset{(n)}{\mathbf{C}}_G \cdot {}^4\overset{(n)}{\widetilde{\mathbf{M}}}{}^{(2134)}_h,\\
G &= C,\ D, \quad n = \text{I}, \ldots, \text{V}.
\end{aligned} \tag{4.99}$$

The tensors ${}^4\overset{(n)}{\mathbf{X}}{}^0_{Gh}$ and ${}^4\overset{(n)}{\mathbf{M}}_h$ are given in Table 4.5.

Table 4.5 shows the following tensors: tensors ${}^4\mathbf{Z}_{Dh}$ and ${}^4\mathbf{Z}_{Wh}$ determined by formula (2.375) and Table 2.1, tensor ${}^4\widetilde{\mathbf{\Omega}}$ determined in Exercise 2.4.11, and tensor ${}^4\mathbf{E}_h$ determined by Table 2.1 (here $n = \text{I}, \ldots, \text{V}$).

The tensors ${}^4\overset{(n)}{\mathbf{X}}_{Dh}$ and ${}^4\overset{\text{III}}{\mathbf{Y}}$ have the forms

$$\begin{aligned}
{}^4\overset{(n)}{\mathbf{X}}_{Dh} = \mathbf{0}, \quad n = \text{I, II, IV, V};\quad {}^4\overset{\text{III}}{\mathbf{X}}_{Dh} &= {}^4\overset{\text{III}}{\mathbf{X}}_{Ch},\\
{}^4\overset{\text{III}}{\mathbf{Y}} &= \frac{1}{2}(\mathbf{V}^{-1} \otimes \mathbf{V} - \mathbf{V} \otimes \mathbf{V}^{-1}).
\end{aligned} \tag{4.100}$$

For the energetic deformation measures ($G = A, B$) we usually apply only the total derivatives with respect to time:

$${}^4\overset{(n)}{\mathbf{X}}_{Gh} = {}^4\overset{(n)}{\mathbf{X}}, \quad {}^4\overset{(n)}{\mathbf{Y}}_{Gh} = \mathbf{0}, \quad G = A, B. \tag{4.101}$$

**Table 4.5** Tensors included in formula (4.99) at different $n$

| $n$ | ${}^4\overset{(n)}{\mathbf{X}}{}^0_{Gh}$ | ${}^4\overset{(n)}{\mathbf{M}}_h$ | ${}^4\overset{(n)}{\widetilde{\mathbf{M}}}_h$ | ${}^4\overset{(n)}{\mathbf{Y}}{}^0_{Gh}$ |
|---|---|---|---|---|
| I | $\mathbf{\Delta}_{\text{III}}$ | ${}^4\mathbf{Z}_{Dh} + \mathbf{\Delta}_{\text{III}}$ | ${}^4\mathbf{Z}_{Wh} - \mathbf{\Delta}_{\text{III}}$ | $\mathbf{0}$ |
| II | $\mathbf{\Delta}_{\text{III}} - {}^4\mathbf{E}_h - ({}^4\mathbf{Z}_{Dh} + {}^4\mathbf{Z}^{(2134)}_{Dh})$ | ${}^4\mathbf{Z}_{Dh} - \frac{1}{2}({}^4\widetilde{\mathbf{\Omega}} + \mathbf{\Delta}_{\text{III}})$ | ${}^4\mathbf{Z}_{Wh} - \mathbf{\Delta}_{\text{III}}$ | $\mathbf{0}$ |
| III | $\mathbf{\Delta}_{\text{III}} + {}^4\overset{\text{III}}{\mathbf{Y}} \cdot\cdot {}^4\widetilde{\mathbf{\Omega}}$ | ${}^4\mathbf{Z}_{Dh}$ | ${}^4\mathbf{Z}_{Wh}$ | ${}^4\overset{\text{III}}{\mathbf{Y}}$ |
| IV | $\mathbf{\Delta}_{\text{III}} + {}^4\mathbf{E}_h + {}^4\mathbf{Z}_{Dh} + {}^4\mathbf{Z}^{(2134)}_{Dh}$ | ${}^4\mathbf{Z}_{Dh} - \frac{1}{2}({}^4\widetilde{\mathbf{\Omega}} - \mathbf{\Delta}_{\text{III}})$ | ${}^4\mathbf{Z}_{Wh} - \mathbf{\Delta}_{\text{III}}$ | $\mathbf{0}$ |
| V | $\mathbf{\Delta}_{\text{III}}$ | ${}^4\mathbf{Z}_{Dh} - \mathbf{\Delta}_{\text{III}}$ | ${}^4\mathbf{Z}_{Wh} - \mathbf{\Delta}_{\text{III}}$ | $\mathbf{0}$ |

Here the fourth-order tensors ${}^4\overset{(n)}{\mathbf{X}}$ are the same for deformation measures and tensors, they differ only for different $n$:

$$
\begin{aligned}
&{}^4\overset{\mathrm{V}}{\mathbf{X}} = (\mathbf{F}^{\mathrm{T}} \otimes \mathbf{F})^{(1432)}, \quad {}^4\overset{\mathrm{III}}{\mathbf{X}} = (\mathbf{O}^{\mathrm{T}} \otimes \mathbf{O})^{(1432)}, \quad {}^4\overset{\mathrm{I}}{\mathbf{X}} = (\mathbf{F}^{-1} \otimes \mathbf{F}^{-1\mathrm{T}})^{(1432)},\\
&{}^4\overset{\mathrm{IV}}{\mathbf{X}} = \frac{1}{2}(\mathbf{F}^{\mathrm{T}} \cdot (\boldsymbol{\Delta} + {}^4\widetilde{\boldsymbol{\Omega}}^{(1342)} \cdot \mathbf{O})^{(1423)} + \mathbf{O}^{\mathrm{T}} \cdot ((\boldsymbol{\Delta} + {}^4\widetilde{\boldsymbol{\Omega}}^{(2134)})^{(1342)} \cdot \mathbf{F})^{(1423)}),\\
&{}^4\overset{\mathrm{II}}{\mathbf{X}} = \frac{1}{2}(\mathbf{F}^{-1} \cdot (\boldsymbol{\Delta} - {}^4\widetilde{\boldsymbol{\Omega}}^{(1342)} \cdot \mathbf{O})^{(1423)} + \mathbf{O}^{\mathrm{T}} \cdot ((\boldsymbol{\Delta} - {}^4\widetilde{\boldsymbol{\Omega}}^{(2134)})^{(1342)} \cdot \mathbf{F}^{-1\mathrm{T}})^{(1423)}).
\end{aligned}
\tag{4.102}
$$

Formula (4.101) should be proved as Exercise 4.2.22.

Thus, the rates of the energetic deformation tensors and measures are independent of the vorticity tensor; they are linear tensor functions only of the deformation rate tensor:

$$
\overset{(n)}{\mathbf{C}}{}^{\bullet}_{G} = {}^4\overset{(n)}{\mathbf{X}} \cdot\cdot \mathbf{D}, \quad G = A,\ B. \tag{4.103}
$$

If the tensors **D** and **W** are expressed in terms of **L** (see Exercise 2.4.12):

$$
\mathbf{D} = \boldsymbol{\Delta} \cdot\cdot \mathbf{L}, \quad \mathbf{W} = \widetilde{\boldsymbol{\Delta}} \cdot\cdot \mathbf{L}, \quad \widetilde{\boldsymbol{\Delta}} = \frac{1}{2}(\boldsymbol{\Delta}_{\mathrm{III}} - \boldsymbol{\Delta}_{\mathrm{II}}), \tag{4.104}
$$

then relation (4.98) connects $\overset{(n)}{\mathbf{C}}{}^{h}_{G}$ to **L**:

$$
\begin{aligned}
&\overset{(n)}{\mathbf{C}}{}^{h}_{G} = {}^4\overset{(n)}{\mathbf{B}}_{Gh} \cdot\cdot \mathbf{L}, \quad G = A,\ B,\ C,\ D;\\
&h = \{\cdot,\ \mathrm{Ol},\ \mathrm{CR},\ U,\ V,\ J,\ S\}, \quad n = \mathrm{I}, \ldots, \mathrm{V},
\end{aligned}
\tag{4.105}
$$

where

$$
{}^4\overset{(n)}{\mathbf{B}}_{Gh} = {}^4\overset{(n)}{\mathbf{X}}_{Gh} \cdot\cdot \boldsymbol{\Delta} + {}^4\overset{(n)}{\mathbf{Y}}_{Gh} \cdot\cdot \widetilde{\boldsymbol{\Delta}}. \tag{4.106}
$$

## Exercises for 4.2

**4.2.1.** Let the symmetric Cauchy stress tensor **T** in an actual configuration have the following components (3.52): $\mathbf{T} = T^{ij}\mathbf{r}_i \otimes \mathbf{r}_j = T_{ij}\mathbf{r}^i \otimes \mathbf{r}^j$. Using the relations between $\overset{(n)}{\mathbf{T}}$ and **T** (see Table 4.1) and Eq. (2.36), show that the energetic tensors have the same components but with respect to other tensor bases:

$$
\begin{aligned}
&\overset{\mathrm{I}}{\mathbf{T}} = T_{ij}\overset{\circ}{\mathbf{r}}{}^i \otimes \overset{\circ}{\mathbf{r}}{}^j, \quad \overset{\mathrm{V}}{\mathbf{T}} = T^{ij}\overset{\circ}{\mathbf{r}}_i \otimes \overset{\circ}{\mathbf{r}}_j, \quad \overset{\mathrm{II}}{\mathbf{T}} = \frac{1}{2}T_{ij}(\overset{\circ}{\mathbf{r}}{}^i \otimes \widehat{\mathbf{r}}{}^j + \widehat{\mathbf{r}}{}^i \otimes \overset{\circ}{\mathbf{r}}{}^j),\\
&\overset{\mathrm{IV}}{\mathbf{T}} = \frac{1}{2}T^{ij}(\overset{\circ}{\mathbf{r}}_i \otimes \widehat{\mathbf{r}}_j + \widehat{\mathbf{r}}_i \otimes \overset{\circ}{\mathbf{r}}_j), \quad \overset{\mathrm{III}}{\mathbf{T}} = T_{ij}\widehat{\mathbf{r}}{}^i \otimes \widehat{\mathbf{r}}{}^j = T^{ij}\widehat{\mathbf{r}}_i \otimes \widehat{\mathbf{r}}_j,
\end{aligned}
$$

where

$$\widehat{\mathbf{r}}^i = \mathbf{O}^{\mathrm{T}} \cdot \mathbf{r}^i, \quad \widehat{\mathbf{r}}_i = \mathbf{O}^{\mathrm{T}} \cdot \mathbf{r}_i.$$

**4.2.2.** Show that the quasienergetic stress tensors $\overset{(\mathrm{n})}{\mathbf{S}}$ have the same components $T_{ij}$, $T^{ij}$ as the Cauchy tensor $\mathbf{T} = T^{ij}\mathbf{r}_i \otimes \mathbf{r}_j = T_{ij}\mathbf{r}^i \otimes \mathbf{r}^j$, but with respect to other tensor bases:

$$\overset{\mathrm{I}}{\mathbf{S}} = T_{ij}\check{\mathbf{r}}^i \otimes \check{\mathbf{r}}^j, \quad \overset{\mathrm{III}}{\mathbf{S}} = \mathbf{T} = T_{ij}\mathbf{r}^i \otimes \mathbf{r}^j, \quad \overset{\mathrm{II}}{\mathbf{S}} = \frac{1}{2}T_{ij}(\mathbf{r}^i \otimes \check{\mathbf{r}}^j + \check{\mathbf{r}}^i \otimes \mathbf{r}^j),$$

$$\overset{\mathrm{IV}}{\mathbf{S}} = \frac{1}{2}T^{ij}(\mathbf{r}_i \otimes \check{\mathbf{r}}_j + \check{\mathbf{r}}_i \otimes \mathbf{r}_j), \quad \overset{\mathrm{V}}{\mathbf{S}} = T^{ij}\check{\mathbf{r}}_i \otimes \check{\mathbf{r}}_j,$$

where

$$\check{\mathbf{r}}^i = \mathbf{O} \cdot \overset{\circ}{\mathbf{r}}{}^i, \quad \check{\mathbf{r}}_i = \mathbf{O} \cdot \overset{\circ}{\mathbf{r}}_i.$$

**4.2.3.** Using the fact that the components $\overset{(\mathrm{n})}{S}{}^{(p)}_{\alpha\beta}$ are determined with respect to the basis $\mathbf{p}_i$, and $\overset{(\mathrm{n})\circ}{T}{}^{(p)}_{\alpha\beta}$ – with respect to the basis $\overset{\circ}{\mathbf{p}}_i$, and taking formula (2.159a) into account, show that all the energetic and quasienergetic stress tensors are connected with the help of the rotation tensor $\mathbf{O}$ accompanying deformation as follows:

$$\overset{(\mathrm{n})}{\mathbf{S}} = \mathbf{O} \cdot \overset{(\mathrm{n})}{\mathbf{T}} \cdot \mathbf{O}^{\mathrm{T}}, \qquad n = \mathrm{I, II, III, IV, V}.$$

**4.2.4.** Show that all the energetic tensors $\overset{(\mathrm{n})}{\mathbf{C}}$ and $\overset{(\mathrm{n})}{\mathbf{G}}$ ($n = \mathrm{I, II, IV, V}$) have the following resolutions for the eigenbasis $\overset{\circ}{\mathbf{p}}_\alpha$:

$$\overset{(\mathrm{n})}{\mathbf{C}} = \frac{1}{n - \mathrm{III}} \sum_{\alpha=1}^{3} (\lambda_\alpha^{n-\mathrm{III}} - 1)\overset{\circ}{\mathbf{p}}_\alpha \otimes \overset{\circ}{\mathbf{p}}_\alpha, \quad \overset{(\mathrm{n})}{\mathbf{G}} = \frac{1}{n - \mathrm{III}} \sum_{\alpha=1}^{3} \lambda_\alpha^{n-\mathrm{III}}\overset{\circ}{\mathbf{p}}_\alpha \otimes \overset{\circ}{\mathbf{p}}_\alpha,$$

and all the quasienergetic tensors $\overset{(\mathrm{n})}{\mathbf{A}}$ and $\overset{(\mathrm{n})}{\mathbf{g}}$ ($n = \mathrm{I, II, IV, V}$) have the following resolutions for the basis $\mathbf{p}_\alpha$:

$$\overset{(\mathrm{n})}{\mathbf{A}} = \frac{1}{n - \mathrm{III}} \sum_{\alpha=1}^{3} (\lambda_\alpha^{n-\mathrm{III}} - 1)\mathbf{p}_\alpha \otimes \mathbf{p}_\alpha, \quad \overset{(\mathrm{n})}{\mathbf{g}} = \frac{1}{n - \mathrm{III}} \sum_{\alpha=1}^{3} \lambda_\alpha^{n-\mathrm{III}}\mathbf{p}_\alpha \otimes \mathbf{p}_\alpha.$$

**4.2.5.** Using formulae (2.36), (2.74), (2.75), (2.312), (4.47) and the results of Exercise 2.1.7, show that components of the energetic deformation measures $\overset{(\mathrm{n})}{\mathbf{G}}$ are the metric matrices $g_{ij}$ and $g^{ij}$ in different tensor bases:

$$\overset{\mathrm{I}}{\mathbf{G}} = -\frac{1}{2}\mathbf{G}^{-1} = -\frac{1}{2}g^{ij}\overset{\circ}{\mathbf{r}}_i \otimes \overset{\circ}{\mathbf{r}}_j, \quad \overset{\mathrm{II}}{\mathbf{G}} = -\frac{1}{2}g^{ij}(\overset{\circ}{\mathbf{r}}_i \otimes \widehat{\mathbf{r}}_j + \widehat{\mathbf{r}}_i \otimes \overset{\circ}{\mathbf{r}}_j),$$

$$\overset{\mathrm{IV}}{\mathbf{G}} = \frac{1}{2} g_{ij} (\overset{\circ}{\mathbf{r}}{}^i \otimes \widehat{\mathbf{r}}^j + \widehat{\mathbf{r}}^i \otimes \overset{\circ}{\mathbf{r}}{}^j), \quad \overset{\mathrm{V}}{\mathbf{G}} = \frac{1}{2}\mathbf{G} = \frac{1}{2} g_{ij} \overset{\circ}{\mathbf{r}}{}^i \otimes \overset{\circ}{\mathbf{r}}{}^j.$$

**4.2.6.** Show that components of the quasienergetic deformation measures $\overset{(\mathrm{n})}{\mathbf{g}}$ ($n =$ I, II, IV, V) are the metric matrices $g_{ij}$, $g^{ij}$ in different tensor bases:

$$\overset{\mathrm{I}}{\mathbf{g}} = -\frac{1}{2}\mathbf{g}^{-1} = -\frac{1}{2} g^{ij} \breve{\mathbf{r}}_i \otimes \breve{\mathbf{r}}_j, \quad \overset{\mathrm{II}}{\mathbf{g}} = -\frac{1}{2} g^{ij} (\breve{\mathbf{r}}_i \otimes \mathbf{r}_j + \mathbf{r}_i \otimes \breve{\mathbf{r}}_j),$$
$$\overset{\mathrm{IV}}{\mathbf{g}} = \frac{1}{2} g_{ij} (\breve{\mathbf{r}}^i \otimes \mathbf{r}^j + \mathbf{r}^i \otimes \breve{\mathbf{r}}^j), \quad \overset{\mathrm{V}}{\mathbf{g}} = \frac{1}{2}\mathbf{g} = \frac{1}{2} g_{ij} \breve{\mathbf{r}}^i \otimes \breve{\mathbf{r}}^j.$$

**4.2.7.** Show that the energetic tensors $\overset{\mathrm{I}}{\mathbf{A}}$ and $\overset{\mathrm{V}}{\mathbf{A}}$ have components $\varepsilon^{ij}$ and $\varepsilon_{ij}$ with respect to the following bases:

$$\overset{\mathrm{I}}{\mathbf{A}} = \mathbf{A} = \varepsilon^{ij} \breve{\mathbf{r}}_i \otimes \breve{\mathbf{r}}_j, \quad \overset{\mathrm{V}}{\mathbf{A}} = \mathbf{J} = \varepsilon_{ij} \breve{\mathbf{r}}^i \otimes \breve{\mathbf{r}}^j.$$

**4.2.8.** Using the results of Exercise 2.3.1, show that the tensors $\overset{(\mathrm{n})}{\mathbf{G}}$ and $\overset{(\mathrm{n})}{\mathbf{g}}$ are connected by the relation

$$\overset{(\mathrm{n})}{\mathbf{g}} = \mathbf{O} \cdot \overset{(\mathrm{n})}{\mathbf{G}} \cdot \mathbf{O}^{\mathrm{T}}, \quad n = \mathrm{I, \ II, \ IV, \ V};$$

and, using formulae (4.72) and (4.43), show that the tensors $\overset{(\mathrm{n})}{\mathbf{C}}$ and $\overset{(\mathrm{n})}{\mathbf{A}}$ are connected by the similar relation

$$\overset{(\mathrm{n})}{\mathbf{A}} = \mathbf{O} \cdot \overset{(\mathrm{n})}{\mathbf{C}} \cdot \mathbf{O}^{\mathrm{T}}, \quad n = \mathrm{I, \ II, \ IV, \ V}.$$

Using these relations between $\overset{(\mathrm{n})}{\mathbf{A}}$ and $\overset{(\mathrm{n})}{\mathbf{C}}$, and also between $\overset{(\mathrm{n})}{\mathbf{g}}$ and $\overset{(\mathrm{n})}{\mathbf{G}}$, show that the principal invariants of the energetic and quasienergetic deformation tensors and measures coincide:

$$I_\alpha(\overset{(\mathrm{n})}{\mathbf{A}}) = I_\alpha(\overset{(\mathrm{n})}{\mathbf{C}}), \quad I_\alpha(\overset{(\mathrm{n})}{\mathbf{g}}) = I_\alpha(\overset{(\mathrm{n})}{\mathbf{G}}), \quad \alpha = 1, 2, 3.$$

**4.2.9.** Show that the bases $\breve{\mathbf{r}}^i$, $\breve{\mathbf{r}}_i$ and $\widehat{\mathbf{r}}^j$, $\widehat{\mathbf{r}}_j$, introduced in Exercises 4.2.1 and 4.2.2, have the following properties:

$$\breve{\mathbf{r}}^i = \overset{\circ}{g}{}^{ij} \breve{\mathbf{r}}_j, \quad \widehat{\mathbf{r}}^i = g^{ij} \widehat{\mathbf{r}}_j, \quad \breve{\mathbf{r}}^i \cdot \breve{\mathbf{r}}^j = \overset{\circ}{g}{}^{ij}, \quad \breve{\mathbf{r}}_i \cdot \breve{\mathbf{r}}_j = \overset{\circ}{g}_{ij}, \quad \widehat{\mathbf{r}}^i \cdot \widehat{\mathbf{r}}^j = g^{ij}, \quad \widehat{\mathbf{r}}_i \cdot \widehat{\mathbf{r}}_j = g_{ij}.$$

**4.2.10.** Prove Theorem 4.4.

**4.2.11.** Prove that the energetic stress tensors $\overset{(\mathrm{n})}{\mathbf{T}}$ can always be expressed in the form (4.36).

**4.2.12.** Using the result of Exercise 2.5.6, show that the tensors $\overset{(n)}{\mathbf{C}}$ and $\overset{(n)}{\mathbf{C}}{}^U$, and also $\overset{(n)}{\mathbf{G}}$ and $\overset{(n)}{\mathbf{G}}{}^U$ are commutative in pairs:

$$\overset{(n)}{\mathbf{C}} \cdot \overset{(n)}{\mathbf{C}}{}^U = \overset{(n)}{\mathbf{C}}{}^U \cdot \overset{(n)}{\mathbf{C}}, \quad \overset{(n)}{\mathbf{G}} \cdot \overset{(n)}{\mathbf{G}}{}^U = \overset{(n)}{\mathbf{G}}{}^U \cdot \overset{(n)}{\mathbf{G}}, \quad n = \text{I, II, IV, V}.$$

Using the result of Exercise 2.5.7, show that the tensors $\overset{(n)}{\mathbf{A}}$ and $\overset{(n)}{\mathbf{A}}{}^V$, and also $\overset{(n)}{\mathbf{g}}$ and $\overset{(n)}{\mathbf{g}}{}^V$ are commutative in pairs:

$$\overset{(n)}{\mathbf{A}} \cdot \overset{(n)}{\mathbf{A}}{}^V = \overset{(n)}{\mathbf{A}}{}^V \cdot \overset{(n)}{\mathbf{A}}, \quad \overset{(n)}{\mathbf{g}} \cdot \overset{(n)}{\mathbf{g}}{}^V = \overset{(n)}{\mathbf{g}}{}^V \cdot \overset{(n)}{\mathbf{g}}.$$

**4.2.13.** Using the results of Exercises 2.1.1, 2.2.1, and 2.3.2, show that in the problem on tension of a beam (see Example 2.1) the tensors $\overset{(n)}{\mathbf{C}}$, $\overset{(n)}{\mathbf{G}}$, $\overset{(n)}{\mathbf{A}}$ and $\overset{(n)}{\mathbf{g}}$ have the forms

$$\overset{(n)}{\mathbf{C}} = \overset{(n)}{\mathbf{A}} = \frac{1}{n - \text{III}} \sum_{\alpha=1}^{3} (k_\alpha^{n-\text{III}} - 1)\bar{\mathbf{e}}_\alpha \otimes \bar{\mathbf{e}}_\alpha \equiv \sum_{\alpha=1}^{3} \overset{(n)}{C}_\alpha \bar{\mathbf{e}}_\alpha \otimes \bar{\mathbf{e}}_\alpha,$$

$$\overset{(n)}{\mathbf{G}} = \overset{(n)}{\mathbf{g}} = \frac{1}{n - \text{III}} \sum_{\alpha=1}^{3} k_\alpha^{n-\text{III}} \bar{\mathbf{e}}_\alpha \otimes \bar{\mathbf{e}}_\alpha, \quad n = \text{I, II, IV, V};$$

$$\overset{\text{III}}{\mathbf{C}} = \overset{\text{III}}{\mathbf{A}} = \sum_{\alpha=1}^{3} \lg k_\alpha \bar{\mathbf{e}}_\alpha \otimes \bar{\mathbf{e}}_\alpha, \quad \overset{\text{III}}{\mathbf{G}} = \overset{\text{III}}{\mathbf{g}} = \sum_{\alpha=1}^{3} (1 + \lg k_\alpha)\bar{\mathbf{e}}_\alpha \otimes \bar{\mathbf{e}}_\alpha.$$

Using formulae (4.80)–(4.82), show that for this problem the principal invariants $I_\alpha(\overset{(n)}{\mathbf{C}})$, $I_\alpha(\overset{(n)}{\mathbf{G}})$ and change in density $J = \rho/\overset{\circ}{\rho}$ have the forms

$$I_1(\overset{(n)}{\mathbf{C}}) = \frac{1}{n - \text{III}} \left( \sum_{\alpha=1}^{3} k_\alpha^{n-\text{III}} - 3 \right), \quad I_1(\overset{\text{III}}{\mathbf{C}}) = \sum_{\alpha=1}^{3} \lg k_\alpha, \quad J = \frac{\rho}{\overset{\circ}{\rho}} = \frac{1}{k_1 k_2 k_3},$$

$$I_1(\overset{(n)}{\mathbf{G}}) = \frac{1}{n - \text{III}} \sum_{\alpha=1}^{3} k_\alpha^{n-\text{III}}, \quad I_3(\overset{(n)}{\mathbf{G}}) = \frac{1}{(n - \text{III})^3} (k_1 k_2 k_3)^{n-\text{III}}.$$

Using the results of Exercise 2.3.2, show that the tensors of energetic and quasienergetic equivalence, determined by (4.38) and (4.68), for the problem on tension of a beam have the form

$${}^4\overset{(n)}{\mathbf{E}} = {}^4\overset{(n)}{\mathbf{Q}} = \sum_{\alpha,\beta=1}^{3} \overset{(n)}{E}_{\alpha\beta} \bar{\mathbf{e}}_\alpha \otimes \bar{\mathbf{e}}_\beta \otimes \bar{\mathbf{e}}^\beta \otimes \bar{\mathbf{e}}^\alpha.$$

$$\overset{\text{I}}{E}_{\alpha\beta} = \frac{1}{k_\alpha k_\beta}, \quad \overset{\text{II}}{E}_{\alpha\beta} = \frac{2}{k_\alpha + k_\beta}, \quad \overset{\text{III}}{E}_{\alpha\beta} = 1, \quad \overset{\text{IV}}{E}_{\alpha\beta} = \frac{2k_\alpha k_\beta}{k_\alpha + k_\beta}, \quad \overset{\text{V}}{E}_{\alpha\beta} = k_\alpha k_\beta.$$

**Table 4.6** For Exercise 4.2.14

| $\overset{(n)}{c}_{\alpha G}$ | $n=\mathrm{I}$ | $n=\mathrm{II}$ | $n=\mathrm{IV}$ | $n=\mathrm{V}$ |
|---|---|---|---|---|
| $\overset{(n)}{c}_{0A}$ | $-a^2/2$ | $1-U_2/\Delta$ | $U_0-1$ | $0$ |
| $\overset{(n)}{c}_{1A}$ | $a/2$ | $U_1/\Delta$ | $U_1$ | $a/2$ |
| $\overset{(n)}{c}_{2A}$ | $0$ | $1-U_0/\Delta$ | $U_2-1$ | $a^2/2$ |
| $\overset{(n)}{c}_{0B}$ | $-(1+a^2)/2$ | $-U_2/\Delta$ | $U_0$ | $1/2$ |
| $\overset{(n)}{c}_{1B}$ | $a/2$ | $U_1/\Delta$ | $U_1$ | $a/2$ |
| $\overset{(n)}{c}_{2B}$ | $-1/2$ | $-U_0/\Delta$ | $U_2$ | $(1+a^2)/2$ |
| $\overset{(n)}{c}_{0C}$ | $0$ | $1-U_0/\Delta$ | $U_2-1$ | $a^2/2$ |
| $\overset{(n)}{c}_{1C}$ | $a/2$ | $U_1/\Delta$ | $U_1$ | $a/2$ |
| $\overset{(n)}{c}_{2C}$ | $-a^2/2$ | $1-U_2/\Delta$ | $U_0-1$ | $0$ |
| $\overset{(n)}{c}_{0D}$ | $-1/2$ | $-U_0/\Delta$ | $U_2$ | $(1+a^2)/2$ |
| $\overset{(n)}{c}_{1D}$ | $a/2$ | $U_1/\Delta$ | $U_1$ | $a/2$ |
| $\overset{(n)}{c}_{2D}$ | $-(1+a^2)/2$ | $-U_2/\Delta$ | $U_0$ | $1/2$ |

**4.2.14.** Using the results of Exercises 2.2.2 and 2.3.3, show that for the problem on simple shear (see Example 2.2) the tensors $\overset{(n)}{\mathbf{C}}$, $\overset{(n)}{\mathbf{G}}$, $\overset{(n)}{\mathbf{A}}$ and $\overset{(n)}{\mathbf{g}}$ have the form

$$\overset{(n)}{\mathbf{C}}_G = \overset{(n)}{c}_{0G}\bar{\mathbf{e}}_1^2 + \overset{(n)}{c}_{1G}\mathbf{O}_3 + \overset{(n)}{c}_{2G}\bar{\mathbf{e}}_2^2, \quad G = A,\ B,\ C,\ D,$$

where the coefficients $\overset{(n)}{c}_{\alpha G}$ are determined by Table 4.6.
Here $\Delta = U_0U_2 - U_1^2$.

**4.2.15.** Using the results of Exercise 2.2.3, show that for the problem on rotation of a beam with tension (see Example 2.3) the tensors $\overset{(n)}{\mathbf{C}}$, $\overset{(n)}{\mathbf{G}}$, $\overset{(n)}{\mathbf{A}}$ and $\overset{(n)}{\mathbf{g}}$ have the forms

$$\overset{(n)}{\mathbf{C}} = \frac{1}{n-\mathrm{III}}\sum_{\alpha=1}^{3}(k_\alpha^{n-\mathrm{III}}-1)\bar{\mathbf{e}}_\alpha\otimes\bar{\mathbf{e}}_\alpha, \quad \overset{(n)}{\mathbf{G}} = \frac{1}{n-\mathrm{III}}\sum_{\alpha=1}^{3}k_\alpha^{n-\mathrm{III}}\bar{\mathbf{e}}_\alpha\otimes\bar{\mathbf{e}}_\alpha,$$

$$\overset{(n)}{\mathbf{A}} = \frac{1}{n-\mathrm{III}}\sum_{\alpha=1}^{3}(k_\alpha^{n-\mathrm{III}}-1)\mathbf{p}_\alpha\otimes\mathbf{p}_\alpha, \quad \overset{(n)}{\mathbf{g}} = \frac{1}{n-\mathrm{III}}\sum_{\alpha=1}^{3}k_\alpha^{n-\mathrm{III}}\mathbf{p}_\alpha\otimes\mathbf{p}_\alpha,$$

$$n = \mathrm{I},\ \mathrm{II},\ \mathrm{IV},\ \mathrm{V};$$

where $\mathbf{p}_\alpha = \mathbf{O}_0\cdot\bar{\mathbf{e}}_\alpha$.

**4.2.16.** Show that if the tensors $\mathbf{F}$ and $\nabla \otimes \mathbf{v}$ are diagonal (for example, as in the problem on tension of a beam):

$$\mathbf{F} = \sum_{\alpha=1}^{3} k_\alpha \bar{\mathbf{e}}_\alpha \otimes \bar{\mathbf{e}}_\alpha, \quad \nabla \otimes \mathbf{v} = \mathbf{D} = \sum_{\alpha=1}^{3} D_\alpha \bar{\mathbf{e}}_\alpha \otimes \bar{\mathbf{e}}_\alpha,$$

then expressions (4.102) for ${}^4\overset{(n)}{\mathbf{X}}$ take the forms

$${}^4\overset{\mathrm{I}}{\mathbf{X}} = (\mathbf{F}^{-1} \otimes \mathbf{F}^{-1\mathrm{T}})^{(1432)} = \sum_{\alpha,\beta=1}^{3} k_\alpha^{-1} k_\beta^{-1} \boldsymbol{\Delta}_{\alpha\beta},$$

$${}^4\overset{\mathrm{II}}{\mathbf{X}} = (\mathbf{F}^{-1} \otimes \mathbf{E})^{(1432)} = \sum_{\alpha,\beta=1}^{3} k_\alpha^{-1} \boldsymbol{\Delta}_{\alpha\beta},$$

$${}^4\overset{\mathrm{III}}{\mathbf{X}} = \boldsymbol{\Delta}_{\mathrm{III}} = \sum_{\alpha,\beta=1}^{3} \boldsymbol{\Delta}_{\alpha\beta}, \quad {}^4\overset{\mathrm{IV}}{\mathbf{X}} = (\mathbf{F}^{\mathrm{T}} \otimes \mathbf{E})^{(1432)} = \sum_{\alpha,\beta=1}^{3} k_\alpha \boldsymbol{\Delta}_{\alpha\beta},$$

$${}^4\overset{\mathrm{V}}{\mathbf{X}} = (\mathbf{F}^{\mathrm{T}} \otimes \mathbf{F})^{(1432)} = \sum_{\alpha,\beta=1}^{3} k_\alpha k_\beta \boldsymbol{\Delta}_{\alpha\beta},$$

where $\boldsymbol{\Delta}_{\alpha\beta} \equiv \bar{\mathbf{e}}_\alpha \otimes \bar{\mathbf{e}}_\beta \otimes \bar{\mathbf{e}}_\beta \otimes \bar{\mathbf{e}}_\alpha$.

**4.2.17.** Prove that all the quasienergetic stress tensors $\overset{(n)}{\mathbf{S}}$ can be expressed in the form (4.66).

**4.2.18.** Show that for $n = \mathrm{III},\ \mathrm{IV},\ \mathrm{V}$ formulae (4.38) and (4.41) hold.

**4.2.19.** Prove formula (4.68) for different values of $n$.

**4.2.20.** Introducing components of the eigenvectors $\mathbf{p}_\alpha$ and $\overset{\circ}{\mathbf{p}}_\alpha$ with respect to the basis $\widetilde{\mathbf{r}}_i$:

$$\mathbf{p}_i = \widetilde{Q}_i^k \widetilde{\mathbf{r}}_k, \quad \overset{\circ}{\mathbf{p}}_i = \overset{\circ}{\widetilde{Q}}{}_i^k \widetilde{\mathbf{r}}_k,$$

show that tensors ${}^4\overset{(n)}{\mathbf{E}}$ (4.38) and ${}^4\overset{(n)}{\mathbf{Q}}$ (4.68) in this basis have the components

$${}^4\overset{(n)}{\mathbf{E}} = \overset{(n)}{\widetilde{E}}{}^{ijkl} \widetilde{\mathbf{r}}_i \otimes \widetilde{\mathbf{r}}_j \otimes \widetilde{\mathbf{r}}_k \otimes \widetilde{\mathbf{r}}_l, \quad {}^4\overset{(n)}{\mathbf{Q}} = \overset{(n)}{\widetilde{Q}}{}^{ijkl} \widetilde{\mathbf{r}}_i \otimes \widetilde{\mathbf{r}}_j \otimes \widetilde{\mathbf{r}}_k \otimes \widetilde{\mathbf{r}}_l,$$

$$\overset{(n)}{\widetilde{E}}{}^{ijkl} = \sum_{\alpha,\beta=1}^{3} E_{\alpha\beta} \widetilde{Q}_\alpha^i \widetilde{Q}_\beta^j \overset{\circ}{\widetilde{Q}}{}_\beta^k \overset{\circ}{\widetilde{Q}}{}_\alpha^l, \quad \overset{(n)}{\widetilde{Q}}{}^{ijkl} = \sum_{\alpha,\beta=1}^{3} E_{\alpha\beta} \widetilde{Q}_\alpha^i \widetilde{Q}_\beta^j \widetilde{Q}_\beta^k \widetilde{Q}_\alpha^l.$$

**4.2.21.** Show that for $n = \text{I}$ and V the tensors ${}^4\overset{(\text{n})}{\mathbf{Q}}$ and ${}^4\overset{(\text{n})}{\mathbf{E}}$ can be represented in the explicit form in terms of **F** and **V**:

$$ {}^4\overset{\text{I}}{\mathbf{E}} = (\mathbf{F}^{-1\text{T}} \otimes \mathbf{F}^{-1})^{(1432)}, \quad {}^4\overset{\text{V}}{\mathbf{E}} = (\mathbf{F} \otimes \mathbf{F}^{\text{T}})^{(1432)}, $$

$$ {}^4\overset{\text{I}}{\mathbf{Q}} = (\mathbf{V}^{-1} \otimes \mathbf{V}^{-1})^{(1432)}, \quad {}^4\overset{\text{V}}{\mathbf{Q}} = (\mathbf{V} \otimes \mathbf{V})^{(1432)}. $$

**4.2.22.** Using Eqs. (2.283), (2.284), (2.311), and (2.321), prove formulae (4.98), (4.101) and (4.102) for $G = A$ and $B$.

**4.2.23.** Using formulae (4.42), (4.43), (4.65), (4.71), and (2.143), show that the energetic and quasienergetic deformation tensors and measures can be represented in terms of the corresponding power of the deformation gradient:

$$ \overset{(\text{n})}{\mathbf{G}} = \frac{1}{n - \text{III}}(\mathbf{F}^{\text{T}} \cdot \mathbf{F})^{(n-\text{III})/2}, \quad \overset{(\text{n})}{\mathbf{C}} = \frac{1}{n - \text{III}}((\mathbf{F}^{\text{T}} \cdot \mathbf{F})^{(n-\text{III})/2} - \mathbf{E}), $$

$$ \overset{(\text{n})}{\mathbf{g}} = \frac{1}{n - \text{III}}(\mathbf{F} \cdot \mathbf{F}^{\text{T}})^{(n-\text{III})/2}, \quad \overset{(\text{n})}{\mathbf{A}} = \frac{1}{n - \text{III}}((\mathbf{F} \cdot \mathbf{F}^{\text{T}})^{(n-\text{III})/2} - \mathbf{E}), $$

$$ n = \text{I, II, IV, V}. $$

**4.2.24.** Prove that symmetric tensors commuting with each other are coaxial and vice versa (see Sect. 4.2.21).

## 4.3 The Principal Thermodynamic Identity

### *4.3.1 Different Forms of the Principal Thermodynamic Identity*

Let us return to the question on finding constitutive equations in continuum mechanics. Write out the thermodynamic laws (3.124) and (3.166):

$$ \rho \, (de/dt) = \mathbf{T} \cdot \cdot (\nabla \otimes \mathbf{v})^{\text{T}} + \rho q_m - \nabla \cdot \mathbf{q}, \tag{4.107} $$

$$ \rho\theta \, (d\eta/dt) = \rho q_m - \nabla \cdot \mathbf{q} + w^*. \tag{4.108} $$

Eliminating $\nabla \cdot \mathbf{q}$ among these equations, we get

$$ \rho \left( \frac{de}{dt} - \theta \frac{d\eta}{dt} \right) - w_{(i)} + w^* = 0. \tag{4.109} $$

On substituting the expression (4.92) for the stress power $w_{(i)}$ into (4.109), we obtain

$$e\text{-form}: \ \rho\frac{de}{dt} - \rho\theta\frac{d\eta}{dt} - \overset{(n)}{\mathbf{T}}_G \cdot\cdot \frac{d\overset{(n)}{\mathbf{C}}_G}{dt} - \mathbf{S}_G \cdot\cdot \frac{d\mathbf{O}^{\mathrm{T}}}{dt} + w^* = 0, \quad (4.110)$$

$$n = \mathrm{I}, \ldots, \mathrm{V}; \qquad G = A,\ B,\ C,\ D.$$

This relationship is called the *principal thermodynamic identity* (PTI) written in the *generalized e-form.*

Substitution of expressions (4.87)–(4.90) into (4.109) in place of $w_{(i)}$ gives the principal thermodynamic identity in the forms $A_n^e$, $B_n^e$, $C_n^e$ and $D_n^e$:

$$A_n^e: \quad \rho\frac{de}{dt} - \rho\theta\frac{d\eta}{dt} - \overset{(n)}{\mathbf{T}} \cdot\cdot \frac{d\overset{(n)}{\mathbf{C}}}{dt} + w^* = 0, \quad (4.111)$$

$$B_n^e: \quad \rho\frac{de}{dt} - \rho\theta\frac{d\eta}{dt} - \overset{(n)}{\mathbf{T}} \cdot\cdot \frac{d\overset{(n)}{\mathbf{G}}}{dt} + w^* = 0, \quad (4.112)$$

$$C_n^e: \quad \rho\frac{de}{dt} - \rho\theta\frac{d\eta}{dt} - \overset{(n)}{\mathbf{S}} \cdot\cdot \frac{d\overset{(n)}{\mathbf{A}}}{dt} - \overset{\circ}{\mathbf{S}} \cdot\cdot \frac{d\mathbf{O}^{\mathrm{T}}}{dt} + w^* = 0, \quad (4.113)$$

$$D_n^e: \quad \rho\frac{de}{dt} - \rho\theta\frac{d\eta}{dt} - \overset{\circ}{\mathbf{S}} \cdot\cdot \frac{d\overset{(n)}{\mathbf{g}}}{dt} - \overset{(n)}{\mathbf{S}} \cdot\cdot \frac{d\mathbf{O}^{\mathrm{T}}}{dt} + w^* = 0. \quad (4.114)$$

The remarkable property of the principal thermodynamic identity consists in the fact that the identity does not include the gradients with respect to coordinates (i.e. values of the type $\nabla \cdot \mathbf{q}$), and, therefore, it may be considered as some relation connecting the changes in three main values: $e$, $\theta$ and $\overset{(n)}{\mathbf{C}}$ (or $e$, $\theta$ and $\overset{(n)}{\mathbf{G}}$; or $e$, $\theta$, $\overset{(n)}{\mathbf{A}}$ and $\mathbf{O}^{\mathrm{T}}$; or $e$, $\theta$, $\overset{(n)}{\mathbf{g}}$ and $\mathbf{O}^{\mathrm{T}}$) at a local point of a continuum.

The principal thermodynamic identity is a basis for derivation of different constitutive equations of continua.

Using the expression (4.8) for the stress power $w_{(i)}$, from (4.109) we get the relation

$$\overset{\circ}{\rho}\frac{de}{dt} - \overset{\circ}{\rho}\theta\frac{d\eta}{dt} - \mathbf{P} \cdot\cdot \frac{d\mathbf{F}}{dt} + \overset{\circ}{w}{}^* = 0, \quad (4.115)$$

which is called the *principal thermodynamic identity in the $\overset{\circ}{e}$-form in the material description* and which contains the density $\overset{\circ}{\rho}$ and the Piola–Kirchhoff tensor $\mathbf{P}$ determined in $\overset{\circ}{\mathcal{K}}$.

### 4.3.2 The Clausius–Duhem Inequality

Substituting the expression (3.167) for the dissipation function $w^*$ into the principal thermodynamic identity (4.110) and taking into account that $\rho q^*$ is always non-negative due to the Planck inequality (3.158a), we obtain the relation

$$\rho\frac{de}{dt} - \rho\theta\frac{d\eta}{dt} - \overset{(n)}{\mathbf{T}}_G \cdot\cdot \frac{d\overset{(n)}{\mathbf{C}}_G}{dt} - \mathbf{S}_G \cdot\cdot \frac{d\mathbf{O}^{\mathrm{T}}}{dt} + \frac{1}{\theta}\mathbf{q}\cdot\nabla\theta \leqslant 0, \tag{4.116}$$

which is called the *Clausius–Duhem inequality in the e-form* in the spatial description. In a similar way, from (4.115) and (3.175) we obtain the Clausius–Duhem inequality in the material description:

$$\overset{\circ}{\rho}\frac{de}{dt} - \overset{\circ}{\rho}\theta\frac{d\eta}{dt} - \mathbf{P}\cdot\cdot\frac{d\mathbf{F}}{dt} + \frac{1}{\theta}\overset{\circ}{\mathbf{q}}\cdot\overset{\circ}{\nabla}\theta \leqslant 0. \tag{4.117}$$

### 4.3.3 The Helmholtz Free Energy

Introduce a new thermodynamic function $\psi$ called the *Helmholtz free energy* as follows:

$$\psi = e - \theta\eta. \tag{4.118}$$

It is clear that

$$\frac{d}{dt}\psi = \frac{de}{dt} - \frac{d\theta}{dt}\eta - \theta\frac{d\eta}{dt}. \tag{4.119}$$

On determining the derivative $de/dt$ from (4.119) and substituting its expression into (4.109), we obtain

$$\rho\frac{d\psi}{dt} + \rho\eta\frac{d\theta}{dt} - w_{(i)} + w^* = 0. \tag{4.120}$$

Substitution of (4.119) into (4.110) yields

$$\rho\frac{d\psi}{dt} + \rho\eta\frac{d\theta}{dt} - \overset{(n)}{\mathbf{T}}_G \cdot\cdot \frac{d\overset{(n)}{\mathbf{C}}_G}{dt} - \mathbf{S}_G \cdot\cdot \frac{d\mathbf{O}^{\mathrm{T}}}{dt} + w^* = 0. \tag{4.120a}$$

This relation is called the *principal thermodynamic identity in the generalized ψ-form*.

Taking different values of index $G = A, B, C$ or $D$, from (4.120) we obtain the following four $\psi$-forms of the principal thermodynamic identity:

$$\rho\frac{d\psi}{dt} + \rho\eta\frac{d\theta}{dt} - \overset{(n)}{\mathbf{T}} \cdot\cdot \frac{d\overset{(n)}{\mathbf{C}}}{dt} + w^* = 0, \tag{4.121}$$

$$\rho\frac{d\psi}{dt} + \rho\eta\frac{d\theta}{dt} - \overset{(n)}{\mathbf{T}} \cdot\cdot \frac{d\overset{(n)}{\mathbf{G}}}{dt} + w^* = 0, \tag{4.122}$$

$$\rho\frac{d\psi}{dt} + \rho\eta\frac{d\theta}{dt} - \overset{(n)}{\mathbf{S}} \cdot\cdot \frac{d\overset{(n)}{\mathbf{A}}}{dt} - \overset{\circ}{\mathbf{S}}\cdot\cdot \frac{d\mathbf{O}^{\mathrm{T}}}{dt} + w^* = 0, \tag{4.123}$$

$$\rho\frac{d\psi}{dt} + \rho\eta\frac{d\theta}{dt} - \overset{(n)}{\mathbf{S}} \cdot\cdot \frac{d\overset{(n)}{\mathbf{g}}}{dt} - \overset{\circ}{\mathbf{S}}\cdot\cdot \frac{d\mathbf{O}^{\mathrm{T}}}{dt} + w^* = 0; \tag{4.124}$$

where (4.121) is called the *principal thermodynamic identity in the form* $A_n$, (4.122) – *in the form* $B_n$, (4.123) – *in the form* $C_n$, and (4.124) – *in the form* $D_n$.

In a similar way, from (4.115) and (4.119) we obtain the principal thermodynamic identity in the $\overset{\circ}{\psi}$-form in the material description:

$$\overset{\circ}{\rho}\frac{d\psi}{dt} + \overset{\circ}{\rho}\eta\frac{d\theta}{dt} - \mathbf{P}\cdot\cdot \frac{d\mathbf{F}}{dt} + \overset{\circ}{w}{}^* = 0. \tag{4.125}$$

### 4.3.4 The Gibbs Free Energy

Introduce one more thermodynamic function

$$\overset{(n)}{\zeta}_A = \psi - \frac{1}{\rho}\overset{(n)}{\mathbf{T}}\cdot\cdot\overset{(n)}{\mathbf{C}}, \qquad n = \mathrm{I}, \ldots, \mathrm{V}, \tag{4.126}$$

called the *Gibbs free energy in the form* $A_n$, and determine its derivative with respect to time:

$$\frac{d}{dt}\overset{(n)}{\zeta}_A = \frac{d\psi}{dt} - \frac{d}{dt}\left(\frac{\overset{(n)}{\mathbf{T}}}{\rho}\right)\cdot\cdot\overset{(n)}{\mathbf{C}} - \frac{\overset{(n)}{\mathbf{T}}}{\rho}\cdot\cdot\frac{d}{dt}\overset{(n)}{\mathbf{C}}. \tag{4.127}$$

On determining the derivative $d\psi/dt$ from this equation, and then substituting its expression into (4.121), we obtain the principal thermodynamic identity *in the form* $A_n^{\zeta}$:

$$\rho\frac{d\overset{(n)}{\zeta}_A}{dt} + \rho\eta\frac{d\theta}{dt} + \rho\overset{(n)}{\mathbf{C}}\cdot\cdot\frac{d}{dt}\left(\frac{\overset{(n)}{\mathbf{T}}}{\rho}\right) + w^* = 0, \qquad n = \mathrm{I}, \ldots, \mathrm{V}. \tag{4.128}$$

This identity connects changes in the functions $\overset{(n)}{\zeta}_A$, $\theta$ and $\overset{(n)}{\mathbf{T}}/\rho$.

For the remaining forms $B_n$, $C_n$ and $D_n$, the Gibbs free energy is introduced as follows:

$$\text{in the form } B_n: \quad \overset{(n)}{\zeta}_B = \psi - \frac{1}{\rho}\overset{(n)}{\mathbf{T}}\cdot\cdot\overset{(n)}{\mathbf{G}}, \tag{4.129}$$

$$\text{in the form } C_n: \quad \overset{(n)}{\zeta}_C = \psi - \frac{1}{\rho}\overset{(n)}{\mathbf{S}}\cdot\cdot\overset{(n)}{\mathbf{A}}, \tag{4.130}$$

$$\text{in the form } D_n: \quad \overset{(n)}{\zeta}_D = \psi - \frac{1}{\rho}\overset{(n)}{\mathbf{S}}\cdot\cdot\overset{(n)}{\mathbf{g}}. \tag{4.131}$$

Introduce also the generalized Gibbs free energy

$$\overset{(n)}{\zeta}_G = \begin{cases} \overset{(n)}{\zeta}_A, & \text{if } G = A, \\ \overset{(n)}{\zeta}_B, & \text{if } G = B, \\ \overset{(n)}{\zeta}_C, & \text{if } G = C, \\ \overset{(n)}{\zeta}_D, & \text{if } G = D. \end{cases} \tag{4.132}$$

Then, by simple manipulations similar to (4.127), from (4.122)–(4.124) we get the principal thermodynamic identity *in the generalized $\zeta$-form*:

$$\rho\frac{d\overset{(n)}{\zeta}_G}{dt} + \rho\eta\frac{d\theta}{dt} + \rho\overset{(n)}{\mathbf{C}}_G\cdot\cdot\frac{d}{dt}\Big(\frac{\overset{(n)}{\mathbf{T}}_G}{\rho}\Big) - \mathbf{S}_G\cdot\cdot\frac{d\mathbf{O}^{\mathrm{T}}}{dt} + w^* = 0, \tag{4.133}$$

$$n = \mathrm{I}, \ldots, \mathrm{V}; \qquad G = A,\ B,\ C,\ D.$$

For different values of index $G$, the principal thermodynamic identity takes the forms $A_n^{\zeta}$, $B_n^{\zeta}$, $C_n^{\zeta}$ and $D_n^{\zeta}$.

The Gibbs free energy can also be introduced formally with the help of the tensors $\mathbf{P}$ and $\mathbf{F}$:

$$\zeta = \psi - \frac{1}{\overset{\circ}{\rho}}\mathbf{P}\cdot\cdot\mathbf{F}, \tag{4.134}$$

that is called the *Gibbs free energy in the material description*; and for the function $\zeta$ we have the principal thermodynamic identity *in the $\overset{\circ}{\zeta}$-form* in the material description:

$$\overset{\circ}{\rho}\frac{d\zeta}{dt} + \overset{\circ}{\rho}\eta\frac{d\theta}{dt} + \mathbf{F}\cdot\cdot\frac{d\mathbf{P}}{dt} + \overset{\circ}{w}{}^* = 0. \tag{4.135}$$

### 4.3.5 Enthalpy

Finally, introduce one more thermodynamic function

$$\overset{(n)}{i}_G = e - \frac{1}{\rho}\overset{(n)}{\mathbf{T}}_G \cdot\cdot \overset{(n)}{\mathbf{C}}_G, \quad n = \mathrm{I}, \dots, \mathrm{V}; \quad G = A,\ B,\ C,\ D, \tag{4.136}$$

called the *enthalpy*. It follows from (4.126) and (4.129)–(4.131) that $\overset{(n)}{i}_G$ is connected to $\overset{(n)}{\zeta}_G$ by the relation

$$\overset{(n)}{i}_G = \overset{(n)}{\zeta}_G + \eta\theta. \tag{4.137}$$

Taking different values of index $G$ in (4.136), we get the following four different forms of the enthalpy:

$$\begin{aligned} \overset{(n)}{i}_A &= e - \frac{1}{\rho}\overset{(n)}{\mathbf{T}} \cdot\cdot \overset{(n)}{\mathbf{C}}, & \overset{(n)}{i}_B &= e - \frac{1}{\rho}\overset{(n)}{\mathbf{T}} \cdot\cdot \overset{(n)}{\mathbf{G}}, \\ \overset{(n)}{i}_C &= e - \frac{1}{\rho}\overset{(n)}{\mathbf{S}} \cdot\cdot \overset{(n)}{\mathbf{A}}, & \overset{(n)}{i}_D &= e - \frac{1}{\rho}\overset{(n)}{\mathbf{S}} \cdot\cdot \overset{(n)}{\mathbf{g}}. \end{aligned} \tag{4.138}$$

From (4.137) we find the rate

$$\frac{d\overset{(n)}{i}_G}{dt} = \frac{d\overset{(n)}{\zeta}_G}{dt} + \eta\frac{d\theta}{dt} + \theta\frac{d\eta}{dt}. \tag{4.139}$$

On substituting this expression into the principal thermodynamic identity in the $\zeta$-form (4.133), we get

$$\frac{d\overset{(n)}{i}_G}{dt} - \theta\frac{d\eta}{dt} + \overset{(n)}{\mathbf{C}}_G \cdot\cdot \frac{d}{dt}\left(\frac{\overset{(n)}{\mathbf{T}}_G}{\rho}\right) - \frac{\mathbf{S}_G}{\rho} \cdot\cdot \frac{d\mathbf{O}^{\mathrm{T}}}{dt} + \frac{w^*}{\rho} = 0, \tag{4.140}$$

$$n = \mathrm{I}, \dots, \mathrm{V}, \qquad G = A,\ B,\ C,\ D;$$

that is the *principal thermodynamic identity in the generalized i-form.*

Taking different values of index $G$, we obtain the principal thermodynamic identity in the forms $A_n^i$, $B_n^i$, $C_n^i$ and $D_n^i$.

Similarly to (4.136), introduce the *enthalpy in the material description*:

$$i = e - (1/\overset{\circ}{\rho})\mathbf{P} \cdot\cdot \mathbf{F} = \zeta + \eta\theta, \tag{4.141}$$

with the help of which from (4.135) and (4.141) we obtain the principal thermodynamic identity *in the i-form in the material description*:

$$\frac{di}{dt} - \theta\frac{d\eta}{dt} + \frac{\mathbf{F}}{\overset{\circ}{\rho}}\cdot\cdot\frac{d\mathbf{P}}{dt} + \frac{\overset{\circ}{w}{}^*}{\overset{\circ}{\rho}} = 0. \qquad (4.142)$$

## 4.3.6 Universal Form of the Principal Thermodynamic Identity

We can write the following universal expression of the principal thermodynamic identity for all the four forms (4.110), (4.120), (4.133) and (4.140):

$$\frac{d}{dt}\pi + \mu\frac{d\lambda}{dt} - \mathbf{N}\cdot\cdot\frac{d}{dt}\mathbf{K} - \mathbf{M}\cdot\cdot\frac{d\mathbf{O}^{\mathrm{T}}}{dt} + \frac{w^*}{\rho} = 0. \qquad (4.143)$$

Here we have introduced scalar functions: $\pi$ called the *thermodynamic potential*, $\mu$ and $\lambda$; and also symmetric tensors **N**, **K** and **M**. Table 4.7 shows their values for different forms of the principal thermodynamic identity.

The relationship (4.143) is called the *universal form of the principal thermodynamic identity*.

**Table 4.7** Different variants of values of $\pi$, $\mu$, $\lambda$, **N**, **K** and **M** in the universal form of PTI (4.143) in the spatial description

| $N$ | Generalized form of PTI | Forms of PTI | $\pi$ | $\mu$ | $\lambda$ | N | K | M |
|---|---|---|---|---|---|---|---|---|
| 1 | $\psi$-form (4.120) | $A_n$, $B_n$, $C_n$, $D_n$ | $\psi$ | $\eta$ | $\theta$ | $\overset{(n)}{\mathbf{T}}_G/\rho$ | $\overset{(n)}{\mathbf{C}}_G$ | $\mathbf{S}_G/\rho$ |
| 2 | $e$-form (4.110) | $A_n^e$, $B_n^e$, $C_n^e$, $D_n^e$ | $e$ | $-\theta$ | $\eta$ | $\overset{(n)}{\mathbf{T}}_G/\rho$ | $\overset{(n)}{\mathbf{C}}_G$ | $\mathbf{S}_G/\rho$ |
| 3 | $e$-form (4.110) | $A_n^e$, $B_n^e$, $C_n^e$, $D_n^e$ | $\eta$ | $-1/\theta$ | $e$ | $\overset{(n)}{\mathbf{T}}_G/\rho\theta$ | $\overset{(n)}{\mathbf{C}}_G$ | $\mathbf{S}_G/\rho\theta$ |
| 4 | $\psi$-form (4.120) | $A_n$, $B_n$, $C_n$, $D_n$ | $\theta$ | $1/\eta$ | $\psi$ | $\overset{(n)}{\mathbf{T}}_G/\rho$ | $\overset{(n)}{\mathbf{C}}_G$ | $\mathbf{S}_G/\rho\eta$ |
| 5 | $\zeta$-form (4.133) | $A_n^\zeta$, $B_n^\zeta$, $C_n^\zeta$, $D_n^\zeta$ | $\overset{(n)}{\zeta}_G$ | $\eta$ | $\theta$ | $-\overset{(n)}{\mathbf{C}}_G$ | $\overset{(n)}{\mathbf{T}}_G/\rho$ | $\mathbf{S}_G/\rho$ |
| 6 | $\zeta$-form (4.133) | $A_n^\zeta$, $B_n^\zeta$, $C_n^\zeta$, $D_n^\zeta$ | $\theta$ | $1/\eta$ | $\overset{(n)}{\zeta}_G$ | $-\overset{(n)}{\mathbf{C}}_G/\eta$ | $\overset{(n)}{\mathbf{T}}_G/\rho$ | $\mathbf{S}_G/\rho\eta$ |
| 7 | $i$-form (4.140) | $A_n^i$, $B_n^i$, $C_n^i$, $D_n^i$ | $\overset{(n)}{i}_G$ | $-\theta$ | $\eta$ | $-\overset{(n)}{\mathbf{C}}_G$ | $\overset{(n)}{\mathbf{T}}_G/\rho$ | $\mathbf{S}_G/\rho$ |
| 8 | $i$-form (4.140) | $A_n^i$, $B_n^i$, $C_n^i$, $D_n^i$ | $\eta$ | $-1/\theta$ | $\overset{(n)}{i}_G$ | $-\overset{(n)}{\mathbf{C}}_G/\theta$ | $\overset{(n)}{\mathbf{T}}_G/\rho$ | $\mathbf{S}_G/\rho\theta$ |

**Table 4.8** Different variants of values of $\pi$, $\mu$, $\lambda$, **N**, **K** and **M** in the universal form of PTI (4.143) in the material description

| $N$ | Form of PTI | $\pi$ | $\mu$ | $\lambda$ | **N** | **K** | **M** |
|---|---|---|---|---|---|---|---|
| 1 | $\overset{\circ}{\psi}$-form (4.125) | $\psi$ | $\eta$ | $\theta$ | $\mathbf{P}/\overset{\circ}{\rho}$ | **F** | **0** |
| 2 | $\overset{\circ}{e}$-form (4.115) | $e$ | $-\theta$ | $\eta$ | $\mathbf{P}/\overset{\circ}{\rho}$ | **F** | **0** |
| 3 | $\overset{\circ}{e}$-form (4.115) | $\eta$ | $-1/\theta$ | $e$ | $\mathbf{P}/\overset{\circ}{\rho}\theta$ | **F** | **0** |
| 4 | $\overset{\circ}{\psi}$-form (4.125) | $\theta$ | $1/\eta$ | $\psi$ | $\mathbf{P}/\overset{\circ}{\rho}\eta$ | **F** | **0** |
| 5 | $\overset{\circ}{\zeta}$-form (4.135) | $\zeta$ | $\eta$ | $\theta$ | $-\mathbf{F}$ | $\mathbf{P}/\overset{\circ}{\rho}$ | **0** |
| 6 | $\overset{\circ}{\zeta}$-form (4.135) | $\theta$ | $1/\eta$ | $\zeta$ | $-\mathbf{F}/\eta$ | $\mathbf{P}/\overset{\circ}{\rho}$ | **0** |
| 7 | $\overset{\circ}{i}$-form (4.142) | $i$ | $-\theta$ | $\eta$ | $-\mathbf{F}$ | $\mathbf{P}/\overset{\circ}{\rho}$ | **0** |
| 8 | $\overset{\circ}{i}$-form (4.142) | $\eta$ | $-1/\theta$ | $i$ | $-\mathbf{F}/\theta$ | $\mathbf{P}/\overset{\circ}{\rho}$ | **0** |

It should be noted that for each of the four forms of PTI (4.110), (4.120), (4.133), and (4.140) there are two variants of a choice of the thermodynamic potential. For example, for the $e$-form (4.110) we may put either $\pi = e$, $\lambda = \eta$ and $\mu = -\theta$, or $\pi = \eta$, $\lambda = e$ and $\mu = -1/\theta$. Both the variants are given in Table 4.7. The same holds for other forms of PTI as well. Thus, for the four forms of PTI, there are eight different variants of a choice of the thermodynamic potential.

Notice that the choice of temperature $\theta$ as a thermodynamic potential yields $\mu = 1/\eta$. But the thermodynamic laws do not exclude the situation when $\eta \to 0$; in this case $\mu = 1/\eta \to +\infty$, that should be taken into consideration when temperature $\theta$ is chosen as the potential $\pi$.

All the four forms of PTI in the material description (4.115), (4.125), (4.135), and (4.142) can also be written in the *single form* (4.143). Values of the functions $\pi$, $\mu$, $\lambda$, and of the tensors **N**, **K** and **M** for these forms are given in Table 4.8. Notice that in the material description the tensor **M** is zero-tensor, and the tensors **N** and **K** are not symmetric.

For each of the four forms of PTI in the material description, there are also two variants of a choice of the thermodynamic potential; thus, we have eight different sets of the functions $\pi$, $\lambda$, $\mu$, **N**, **K** and **M** in the universal form of PTI (4.143).

### *4.3.7 Representation of the Principal Thermodynamic Identity in Terms of Co-rotational Derivatives*

**Theorem 4.12.** *Let the tensors* $\overset{(n)}{\mathbf{T}}$ *and* $\overset{(n)}{\mathbf{C}}$ *(or the tensors* $\overset{(n)}{\mathbf{S}}$ *and* $\overset{(n)}{\mathbf{A}}$*) be coaxial in pairs for each n; then the principal thermodynamic identity (4.143) takes the form*

$$\pi^h + \mu\lambda^h - \mathbf{N}\cdot\cdot\mathbf{K}^h - \mathbf{M}\cdot\cdot\dot{\mathbf{O}}^{\mathrm{T}} + \frac{w^*}{\rho} = 0, \qquad h = \{\cdot,\ U,\ V,\ J,\ S\}. \quad (4.144)$$

▼ Since $\pi$ and $\lambda$ are scalar functions, their co-rotational derivatives coincide with the total derivative with respect to time: $\dot{\pi} = \pi^h$, $\dot{\lambda} = \lambda^h$. The second and third summands in (4.143) for the $\psi$- and $e$-forms of PTI, according to the notation introduced in Table 4.7 and formula (4.92), give the generalized form of the stress power

$$w_{(i)} = \overset{(n)}{\mathbf{T}}_G \cdot\cdot \overset{(n)}{\mathbf{G}}{}^{\bullet}_G + \mathbf{S}_G \cdot\cdot \dot{\mathbf{O}}^{\mathrm{T}} = \mathbf{N} \cdot\cdot \dot{\mathbf{K}} + \mathbf{M} \cdot\cdot \dot{\mathbf{O}}^{\mathrm{T}}. \tag{4.145}$$

Then for $w_{(i)}$ Theorem 4.11 is valid; and, due to this theorem, we can really replace the total derivative $\dot{\mathbf{K}}$ by the co-rotational derivative $\mathbf{K}^h$, if the tensors $\mathbf{N}$ and $\mathbf{K}$, i.e. the tensors $\overset{(n)}{\mathbf{T}}$ and $\overset{(n)}{\mathbf{C}}$ (or the tensors $\overset{(n)}{\mathbf{S}}$ and $\overset{(n)}{\mathbf{A}}$), are coaxial.

For the $\zeta$-form of PTI, the second and third summands in (4.143) do not coincide with $w_{(i)}$, but they can be written with the help of Table 4.7:

$$\mathbf{N} \cdot\cdot \dot{\mathbf{K}} + \mathbf{M} \cdot\cdot \dot{\mathbf{O}}^{\mathrm{T}} = \rho \overset{(n)}{\mathbf{C}}_G \cdot\cdot (\overset{(n)}{\mathbf{T}}_G/\rho)^{\bullet} + \mathbf{S}_G \cdot\cdot \dot{\mathbf{O}}^{\mathrm{T}}. \tag{4.146}$$

If the tensors $\overset{(n)}{\mathbf{T}}$ and $\overset{(n)}{\mathbf{C}}$ (or the tensors $\overset{(n)}{\mathbf{S}}$ and $\overset{(n)}{\mathbf{A}}$) are coaxial in pairs for each $n$, then tensors $\overset{(n)}{\mathbf{C}}_G$ and $\overset{(n)}{\mathbf{T}}_G/\rho$ are coaxial as well. Then, following the proof of Theorem 4.11, we get

$$\mathbf{N} \cdot\cdot \dot{\mathbf{K}} = \mathbf{N} \cdot\cdot \mathbf{K}^h, \tag{4.147}$$

as was to be shown. ▲

## Exercises for 4.3

**4.3.1.** Using formulae (4.43) and (4.70), show that the Gibbs free energies $\overset{(n)}{\zeta}_A$, $\overset{(n)}{\zeta}_B$, $\overset{(n)}{\zeta}_C$ and $\overset{(n)}{\zeta}_D$, determined by formulae (4.126), (4.129)–(4.131), are not coincident; and they are connected by the following relations:

$$\overset{(n)}{\zeta}_A = \overset{(n)}{\zeta}_B + \frac{1}{n - \mathrm{III}} I_1(\overset{(n)}{\mathbf{T}}), \qquad \overset{(n)}{\zeta}_C = \overset{(n)}{\zeta}_D + \frac{1}{n - \mathrm{III}} I_1(\overset{(n)}{\mathbf{S}}).$$

**4.3.2.** Using the results of Exercises 4.2.3 and 4.2.8, show that $\overset{(n)}{\zeta}_B$ and $\overset{(n)}{\zeta}_D$, and also $\overset{(n)}{\zeta}_A$ and $\overset{(n)}{\zeta}_C$ coincide in pairs: $\overset{(n)}{\zeta}_A = \overset{(n)}{\zeta}_C$, $\overset{(n)}{\zeta}_B = \overset{(n)}{\zeta}_D$, $n = \mathrm{I, II, IV, V}$.

**4.3.3.** Using the definition (3.56) of the tensor $\mathbf{P}$, show that the Gibbs free energy $\overset{\circ}{\zeta}$ in the material description (4.134) can be written in the form

$$\zeta = \psi - (1/\rho)\, I_1(\mathbf{T}).$$

# 4.4 Principles of Thermodynamically Consistent Determinism, Equipresence and Local Action

## 4.4.1 Active and Reactive Variables

Let us consider now the principal thermodynamic identity in the form $A_n$ (4.121). Split the functions $\psi$, $\theta$, $\eta$, $\overset{(n)}{\mathbf{T}}/\rho$, $\overset{(n)}{\mathbf{C}}$ and $w^*/\rho$ included in (4.121) into the two groups: functions included in (4.121) through their derivatives, except $\psi$, are called the *reactive variables*:

$$\mathcal{R} = \{\theta, \overset{(n)}{\mathbf{C}}\}, \tag{4.148}$$

the remaining functions are called the *active variables*:

$$\Lambda = \{\psi, \eta, \overset{(n)}{\mathbf{T}}/\rho, w^*/\rho\}. \tag{4.149}$$

This separation has the following meaning: with the help of active variables we describe some generalized internal forces caused by external actions on a material point $\mathcal{M}$ of a continuum considered; and with the help of reactive variables we describe the response of a continuum to these external actions.

The choice of reactive and active variables is not unique: considering the principal thermodynamic identity in the form $C_n$ (4.123), as reactive variables we take the functions

$$\mathcal{R} = \{\theta, \overset{(n)}{\mathbf{A}}, \mathbf{O}\}, \tag{4.150}$$

and as active variables – the functions

$$\Lambda = \{\psi, \eta, \overset{(n)}{\mathbf{S}}/\rho, \overset{\circ}{\mathbf{S}}/\rho, w^*/\rho\}. \tag{4.151}$$

Similarly, for the principal thermodynamic identity in the form $B_n$ (4.122) the variables $\mathcal{R}$ and $\Lambda$ are chosen as follows:

$$\mathcal{R} = \{\theta, \overset{(n)}{\mathbf{G}}\}, \qquad \Lambda = \{\psi, \eta, \overset{(n)}{\mathbf{T}}/\rho, w^*/\rho\}, \tag{4.152}$$

and for PTI in the form $D_n$ (4.124):

$$\mathcal{R} = \{\theta, \overset{(n)}{\mathbf{g}}, \mathbf{O}\}, \qquad \Lambda = \{\psi, \eta, \overset{(n)}{\mathbf{S}}/\rho, \overset{\circ}{\mathbf{S}}/\rho, w^*/\rho\}. \tag{4.153}$$

We can write out the active and reactive variables for the universal form of PTI (4.136) in the spatial description as follows:

$$\mathcal{R} = \{\lambda, \mathbf{K}, \mathbf{O}\}, \qquad \Lambda = \{\pi, \mu, \mathbf{N}, \mathbf{M}, w^*/\rho\}, \tag{4.154}$$

where all the generalized thermodynamic functions are determined by Tables 4.7 and 4.8.

### 4.4.2 The Principle of Thermodynamically Consistent Determinism

**Axiom 11 (The principle of thermodynamically consistent determinism).** *For every continuum, active variables $\Lambda$ are completely defined by reactive variables $\mathcal{R}$; in other words, there is a mapping of the generalized space of reactive variables $\mathcal{X}_{\mathcal{R}}$ into the space of active variables $\mathcal{X}_{\Lambda}$:*

$$\breve{f}: \quad \mathcal{X}_{\mathcal{R}} \to \mathcal{X}_{\Lambda}, \tag{4.155}$$

*which is called the operator relation (or the operator) and written in the form*

$$\Lambda = \breve{f}(\mathcal{R}), \quad \mathcal{R} \in \mathcal{X}_{\mathcal{R}}, \quad \Lambda \in \mathcal{X}_{\Lambda}; \tag{4.156}$$

*the relation is consistent with thermodynamics, i.e. satisfies identically the principal thermodynamic identity (4.143).*

A particular case of the operator (4.155) is the mapping determined by the ordinary function

$$\Lambda(t) = f(\mathcal{R}(t)), \tag{4.157}$$

where the active variables $\Lambda(t)$ depend on the reactive ones $\mathcal{R}(t)$ considered at the same times. Other examples of the operators will be given below.

If PTI is considered in the form $A_n$ (4.121), then the operator (4.156) with use of (4.148) and (4.149) takes the form

$$\begin{cases} \psi = \breve{\psi}\,(\overset{(n)}{\mathbf{C}},\ \theta), & \eta = \breve{\eta}\,(\overset{(n)}{\mathbf{C}},\ \theta), \\ \overset{(n)}{\mathbf{T}} = \breve{\mathcal{F}}\,(\overset{(n)}{\mathbf{C}},\ \theta), & w^* = \breve{w}^*\,(\overset{(n)}{\mathbf{C}},\ \theta). \end{cases} \tag{4.158}$$

Here we have taken into consideration that density $\rho$ can always be represented as a function of $\overset{(n)}{\mathbf{C}}$ (see Sect. 4.2.19).

Relations (4.158) are *constitutive equations* of a continuum, they are also called the *model $A_n$* of a continuum.

Notice that, according to Axiom 11, the operators (4.156) are not entirely arbitrary, because after their substitution into PTI (4.121) we get the equation

$$\rho\frac{d}{dt}\breve{\psi}(\overset{(n)}{\mathbf{C}},\theta) + \rho\breve{\eta}(\breve{\mathbf{C}},\ \theta)\frac{d\theta}{dt} - \breve{\mathcal{F}}(\mathbf{C},\ \theta)\cdot\cdot\frac{d\overset{(n)}{\mathbf{C}}}{dt} + w^*(\overset{(n)}{\mathbf{C}},\ \theta) = 0,$$

which should be satisfied identically for all processes of changing the arguments $\overset{(n)}{\mathbf{C}}(\tau), \theta(\tau), 0 \leqslant \tau \leqslant t$.

If PTI is considered in the form $B_n$ (4.122), then the constitutive equations (4.158) with use of (4.152) are written as follows:

$$\begin{cases} \psi = \breve{\psi}\,(\overset{(n)}{\mathbf{G}},\ \theta), \\ \eta = \breve{\eta}\,(\overset{(n)}{\mathbf{G}},\ \theta), \\ \overset{(n)}{\mathbf{T}} = \breve{\mathcal{F}}\,(\overset{(n)}{\mathbf{G}},\ \theta), \\ w^* = \breve{w}^*\,(\overset{(n)}{\mathbf{G}},\ \theta), \quad n = \mathrm{I},\ \mathrm{II},\ \mathrm{IV},\ \mathrm{V}, \end{cases} \tag{4.159}$$

because the density $\rho$ can be expressed as a function of $\overset{(n)}{\mathbf{G}}$ (see Sect. 4.2.19). The relations (4.159) are called the *model* $B_n$.

Similarly, using PTI in the form $C_n$ and the relation (4.156), we get the constitutive equations

$$\begin{cases} \psi = \breve{\psi}(\overset{(n)}{\mathbf{A}},\ \mathbf{O},\ \theta), \\ \eta = \breve{\eta}(\overset{(n)}{\mathbf{A}},\ \mathbf{O},\ \theta), \\ \overset{(n)}{\mathbf{S}} = \breve{\mathcal{F}}(\overset{(n)}{\mathbf{A}},\ \mathbf{O},\ \theta), \\ w^* = \breve{w}^*(\overset{(n)}{\mathbf{A}},\ \mathbf{O},\ \theta), \\ \overset{\circ}{\mathbf{S}} = \breve{\mathcal{F}}^0(\overset{(n)}{\mathbf{A}},\ \mathbf{O},\ \theta), \quad n = \mathrm{I}, \ldots, \mathrm{V}, \end{cases} \tag{4.160}$$

which are called the *model* $C_n$ of a continuum.

According to PTI in the form $D_n$, from (4.153) and (4.156) we obtain the constitutive equations

$$\begin{cases} \psi = \breve{\psi}(\overset{(n)}{\mathbf{g}},\ \mathbf{O},\ \theta), \\ \eta = \breve{\eta}(\overset{(n)}{\mathbf{g}},\ \mathbf{O},\ \theta), \\ \overset{(n)}{\mathbf{S}} = \breve{\mathcal{F}}(\overset{(n)}{\mathbf{g}},\ \mathbf{O},\ \theta), \\ w^* = \breve{w}^*(\overset{(n)}{\mathbf{g}},\ \mathbf{O},\ \theta), \\ \overset{\circ}{\mathbf{S}} = \breve{\mathcal{F}}^0(\overset{(n)}{\mathbf{g}},\ \mathbf{O},\ \theta), \end{cases} \tag{4.161}$$

which are called the *model* $D_n$ of a continuum.

Here we have taken into account that density $\rho$ can be represented as a function of $\overset{(n)}{\mathbf{A}}$ or of $\overset{(n)}{\mathbf{g}}$ (see Sect. 4.2.19).

Notice that the tensor operators $\breve{\mathcal{F}}$ in all the relations (4.156)–(4.159) are different.

If PTI is considered in the universal form (4.143), then Axiom 11 gives the constitutive equations

$$\begin{cases} \pi = \breve{\pi}(\lambda,\ \mathbf{K},\ \mathbf{O}), \\ \mu = \breve{\mu}(\lambda,\ \mathbf{K},\ \mathbf{O}), \\ \mathbf{N} = \breve{\mathbf{N}}(\lambda,\ \mathbf{K},\ \mathbf{O}), \\ w^* = \breve{w}^*(\lambda,\ \mathbf{K},\ \mathbf{O}), \quad \mathbf{M} = \breve{\mathbf{M}}(\lambda,\ \mathbf{K},\ \mathbf{O}). \end{cases} \tag{4.162}$$

Considering different forms of PTI, by Tables 4.7 and 4.8, we obtain a large collection of models of continuum mechanics: 20 models for each of the $\psi$-, $e$-, $\eta$- and $\zeta$-forms (and each of the forms has two representations), and also four forms: $\overset{\circ}{\psi}$, $\overset{\circ}{e}$, $\overset{\circ}{\eta}$ and $\overset{\circ}{\zeta}$ (each of the forms has two representations as well). We finally get $20 \times (4 \times 2 + 4 \times 2) = 320$ (!) models. It should be noted that the models $A_n$, $B_n$, $C_n$ and $D_n$ are most widely used in continuum mechanics.

### 4.4.3 *The Principle of Equipresence*

Analyzing the constitutive equations (4.158)–(4.162), one can notice that for each model all the operator relations include the same reactive variables. This fact is expressed by the following axiom.

**Axiom 12 (The principle of equipresence).** *Constitutive equations must be operators of the same reactive variables.*

In particular, it follows from the principle that since the models $A_n$ do not contain the rotation tensor $\mathbf{O}$, the corresponding forms of constitutive equations (4.158) cannot include the tensor $\mathbf{O}$ as an argument.

Another considerable corollary of the principle of equipresence is the following assertion: since the temperature gradient $\nabla\theta$ does not appear in PTI (4.143), the gradient $\nabla\theta$ cannot be involved in the set of active and reactive variables $\Lambda$ and $\mathcal{R}$ in constitutive equations (4.156).

This conclusion will be used in Sect. 4.11.1.

### 4.4.4 *The Principle of Local Action*

Notice that the general definition of operator (4.156) admits the situation when the process of changing the active variables $\Lambda(X^i, t)$ at a material point $X^i$ depends on the process of changing the reactive variables $\mathcal{R}(X'^i, \tau)$ at other material points $X'^i \neq X^i$, where $X^i, X'^i \in \overset{\circ}{V}$.

To prohibit such dependences, we introduce one more basic principle.

**Axiom 13 (The principle of local action).** *For any continuum, operators of constitutive equations in each of the forms (4.156) are such that active variables at every material point $X^i \in \overset{\circ}{V}$ depend only on reactive variables at the same point $X^i$ $\forall t \geqslant 0$.*

*Remark 1.* The principle of local action allows us to construct constitutive equations describing adequately the behavior of most real continuous media. Nevertheless, the principle is not an universal physical law, because there are continua, which should be described not by local constitutive equations. □

Continua satisfying the principle of local action are sometimes called *simple.*

*Remark 2.* Notice that the principle of local action admits the dependence of constitutive equations (4.156) at a material point $\mathcal{M}$ on the $\mathcal{M}$-point coordinate $X^i$. This dependence has the explicit form

$$\Lambda = \breve{f}(\mathcal{R}, X^i), \tag{4.163}$$

or, for example, for the models $A_n$:

$$\begin{cases} \psi = \breve{\psi}(\overset{(n)}{\mathbf{C}}, \theta, X^i), \\ \eta = \breve{\eta}(\overset{(n)}{\mathbf{C}}, \theta, X^i), \\ \overset{(n)}{\mathbf{T}} = \breve{\mathcal{F}}(\overset{(n)}{\mathbf{C}}, \theta, X^i), \\ w^* = \breve{w}^*(\overset{(n)}{\mathbf{C}}, \theta, X^i). \end{cases} \tag{4.164}$$

Such dependence on $X^i$ is physically justified and can be caused either by the influence of nonmechanical factors (for example, electromagnetic field etc.): in this case the dependence on $X^i$ is continuous; or by nonhomogeneity of a continuum (such continua are, for example, composite materials, multiphase continua etc.): in this case the dependence (4.163) is a discontinuous (usually piecewise continuous) function of $X^i$. □

A continuum, whose constitutive equations (4.156) are the same for all material points $X^i \in \overset{\circ}{V}$, is called *homogeneous*; otherwise, a continuum is called *inhomogeneous*.

## 4.5 Definition of Ideal Continua

### *4.5.1 Classification of Types of Continua*

Let us consider now the most widely used forms of the operators (4.156). Depending on the operator form, we distinguish the following main *types of continua*:

- ideal continua
- viscoelastic continua of the differential type
- viscoelastic continua of the integral type
- plastic continua

In Sects. 4.5–4.11 and Chap. 6 we will consider the most widely used type of continuous media, namely ideal continua. Chaps. 7 and 8 deal with viscoelastic continua of the differential and integral types, respectively; and finally, Chap. 9 investigates plastic continua. All the mentioned types of continua will be considered for the case of large deformations.

### *4.5.2 General Form of Constitutive Equations for Ideal Continua*

**Definition 4.2.** A continuum is called *ideal*, if its constitutive equations (4.156) are ordinary functions of active variables, i.e. the relations (4.157), where values of $\Lambda(t)$ and $\mathcal{R}(t)$ are determined at the same times $t$, hold.

The *model* $A_n$ *of an ideal continuum* is said to be considered, if the model $A_n$ of a continuum has been chosen and the corresponding operator constitutive equations (4.158) are functions of the indicated arguments; in particular, the free energy

$$\psi(t) = \psi(\overset{(n)}{\mathbf{C}}(t), \theta(t)) \tag{4.165}$$

is a scalar function of one tensor argument $\overset{(n)}{\mathbf{C}}$ and one scalar argument $\theta$.

Below, if the symbol $\breve{}$ of operator is absent, then a relation of the form (4.165) will be considered as a function of current values of arguments at time $t$: $\psi(t) = \psi(\overset{(n)}{\mathbf{C}}(t), \theta(t))$, and the symbol $t$ will be omitted.

All the functions (4.157) are assumed to be continuously differentiable, then we can determine, for example, the total derivative of the free energy $d\psi/dt$. Using the rule of differentiation of a scalar function with respect to tensor argument (see [12]), for models $A_n$ of an ideal continuum we get

$$\frac{d\psi}{dt} = \frac{\partial\psi}{\partial\overset{(n)}{\mathbf{C}}} \cdot\cdot \frac{d\overset{(n)}{\mathbf{C}}}{dt} + \frac{\partial\psi}{\partial\theta}\frac{d\theta}{dt}, \tag{4.166}$$

where $\partial\psi/\partial\overset{(n)}{\mathbf{C}}$ is the symmetric second-order tensor.

Since the constitutive equations (4.156) must be consistent with thermodynamics, i.e. should satisfy PTI (4.121), so, substituting (4.166) into PTI (4.121) and collecting like terms, we obtain

$$\left(\frac{\partial\psi}{\partial\overset{(n)}{\mathbf{C}}} - \frac{\overset{(n)}{\mathbf{T}}}{\rho}\right) \cdot\cdot\, d\overset{(n)}{\mathbf{C}} + \left(\frac{\partial\psi}{\partial\theta} + \eta\right) d\theta + \frac{w^*}{\rho}\, dt = 0. \tag{4.167}$$

The differentials $d\overset{(n)}{\mathbf{C}}$, $d\theta$ and $dt$ are independent, therefore the identity (4.167) holds if and only if coefficients of the differentials vanish, i.e. the following relations are valid:

$$\begin{cases} \overset{(n)}{\mathbf{T}} = \rho(\partial\psi/\partial\overset{(n)}{\mathbf{C}}) \equiv \mathcal{F}(\overset{(n)}{\mathbf{C}},\ \theta), \\ \eta = -\partial\psi/\partial\theta, \\ w^* = 0. \end{cases} \tag{4.168}$$

In other words, relations (4.168) are equivalent to the identity (4.167). The relations (4.168) are the *constitutive equations for the model $A_n$ of ideal continua*, and they are the particular case of relations (4.156). Analyzing the relations (4.168), we conclude that:

(1) Ideal continua are *nondissipative*, i.e. their dissipation function $w^*$ is identically zero.
(2) The model of an ideal continuum is specified by the only scalar function (4.165), and the remaining constitutive equations are determined by differentiation of this function, according to formulae (4.168).

*Remark.* Notice that the tensor function $\mathcal{F}(\overset{(n)}{\mathbf{C}},\theta)$ in (4.168) is not potential. Since the density $\rho$ is a scalar function of $\overset{(n)}{\mathbf{C}}$ (see Sect. 4.2.19), the function $\mathcal{F}(\overset{(n)}{\mathbf{C}},\theta)$ would be potential if there were some scalar function $\psi' = \psi'(\overset{(n)}{\mathbf{C}},\theta)$ such that $\overset{(n)}{\mathbf{T}} = \mathcal{F}(\overset{(n)}{\mathbf{C}},\theta) = \partial\psi'/\partial\overset{(n)}{\mathbf{C}}$. Tensor function $\mathcal{F}(\overset{(n)}{\mathbf{C}},\theta)$, satisfying the relations (4.168), is called *quasipotential.* □

One can say that this is the *model $B_n$ of an ideal continuum*, if some model $B_n$ of a continuum has been chosen and the corresponding constitutive equations (4.157) are functions of the indicated arguments, in particular:

$$\psi = \psi(\overset{(n)}{\mathbf{G}},\ \theta). \tag{4.169}$$

Determine the derivative of (4.169) with respect to $t$, substitute the result into PTI (4.122) and then collect like terms. Since the differentials $d\overset{(n)}{\mathbf{G}}$, $d\theta$ and $dt$ are independent, we obtain that the relation (4.122) is equivalent to the following system of equations:

$$\begin{cases} \overset{(n)}{\mathbf{T}} = \rho(\partial\psi/\partial\overset{(n)}{\mathbf{G}}) \equiv \mathcal{F}(\overset{(n)}{\mathbf{G}},\ \theta), \\ \eta = -\partial\psi/\partial\theta, \\ w^* = 0, \end{cases} \tag{4.170}$$

which are called the *constitutive equations for the model $B_n$ of an ideal continuum.* The models $B_n$ of ideal continua are also nondissipative and have the potential (4.169).

If we apply the models $C_n$ to an ideal continuum, then all the active variables $\Lambda(t)$ in (4.160) depend on values of the reactive variables $\mathcal{R}(t)$ at the same time; in particular, the free energy $\psi$ is a function in the form

$$\psi = \psi(\overset{(n)}{\mathbf{A}}, \mathbf{O}, \theta). \tag{4.171}$$

Differentiating the function with respect to $t$, we obtain

$$\frac{d\psi}{dt} = \frac{\partial\psi}{\partial\overset{(n)}{\mathbf{A}}} \cdot\cdot \frac{d\overset{(n)}{\mathbf{A}}}{dt} + \frac{\partial\psi}{\partial\mathbf{O}} \cdot\cdot \frac{d\mathbf{O}^{\mathrm{T}}}{dt} + \frac{\partial\psi}{\partial\theta}\frac{d\theta}{dt}. \tag{4.172}$$

On substituting this expression into PTI (4.123), we get the identity

$$\left(\frac{\partial\psi}{\partial\overset{(n)}{\mathbf{A}}} - \frac{\overset{(n)}{\mathbf{S}}}{\rho}\right)\cdot\cdot\, d\overset{(n)}{\mathbf{A}} + \left(\frac{\partial\psi}{\partial\mathbf{O}} - \frac{\overset{\circ}{\mathbf{S}}}{\rho}\right)\cdot\cdot\, d\mathbf{O}^{\mathrm{T}} + \left(\frac{\partial\psi}{\partial\theta} + \eta\right) d\theta + \frac{w^*}{\rho}\, dt = 0. \tag{4.173}$$

All the differentials $d\overset{(n)}{\mathbf{A}}$, $d\mathbf{O}^{\mathrm{T}}$, $d\theta$ and $dt$ are independent, therefore the identity is equivalent to the following set of relations

$$\begin{cases} \overset{(n)}{\mathbf{S}} = \rho\,(\partial\psi/\partial\overset{(n)}{\mathbf{A}}), \\ \overset{\circ}{\mathbf{S}} = \rho\,(\partial\psi/\partial\mathbf{O}), \\ \eta = -\partial\psi/\partial\theta, \\ w^* = 0, \end{cases} \tag{4.174}$$

which are the *constitutive equations for the model $C_n$ of ideal continua*. Models $C_n$ of ideal continua, as well as models $A_n$, are nondissipative and potential; however, for $C_n$ there is an additional relation, connecting the rotation tensor of stresses $\overset{\circ}{\mathbf{S}}$ to $\psi$.

If we consider the models $D_n$ of an ideal continuum, then all the relations (4.161) are functions of the indicated arguments; in particular, $\psi$ has the form

$$\psi = \psi(\overset{(n)}{\mathbf{g}}, \mathbf{O}, \theta). \tag{4.175}$$

Substituting the function (4.175) into PTI (4.124), due to independence of the differentials $d\overset{(n)}{\mathbf{g}}$, $d\theta$ and $dt$, we get the following set of relations

$$\begin{cases} \overset{(n)}{\mathbf{S}} = \rho\, (\partial\psi/\partial \overset{(n)}{\mathbf{g}}), \\ \overset{\circ}{\mathbf{S}} = \rho\, (\partial\psi/\partial \mathbf{O}), \\ \eta = -\partial\psi/\partial\theta, \\ w^* = 0, \end{cases} \tag{4.176}$$

which are called the *constitutive equations for the model $D_n$ of ideal continua.*

Finally, if for an ideal continuum we consider PTI in the universal form (4.143), then all the relations (4.162) are functions of the indicated arguments; in particular, the thermodynamic potential $\pi$ has the form

$$\pi = \pi(\lambda,\ \mathbf{K},\ \mathbf{O}). \tag{4.177}$$

Its derivative with respect to $t$ is

$$\dot{\pi} = \frac{\partial\pi}{\partial\mathbf{K}}\cdot\cdot\frac{d\mathbf{K}}{dt} + \frac{\partial\pi}{\partial\mathbf{O}}\cdot\cdot\frac{d\mathbf{O}^{\mathrm{T}}}{dt} + \frac{\partial\pi}{\partial\lambda}\frac{d\lambda}{dt}. \tag{4.178}$$

By Axiom 11, the relations (4.162) should identically satisfy PTI in the form (4.143). Substituting (4.178) into (4.143) and collecting like terms, due to independence of the differentials $d\mathbf{K}$, $d\mathbf{O}$ and $d\lambda$, we get the following system of constitutive equations:

$$\begin{cases} \mu = -\partial\pi/\partial\lambda, \\ \mathbf{N} = \partial\pi/\partial\mathbf{K}, \\ \mathbf{M} = \partial\pi/\partial\mathbf{O}, \\ w^* = 0. \end{cases} \tag{4.179}$$

## 4.6 The Principle of Material Symmetry

### *4.6.1 Different Reference Configurations*

Let us consider now the principle of material symmetry mentioned in Sect. 4.1. This principle introduces several significant concepts of continuum mechanics: anisotropy (and isotropy) of a continuum, invariants of tensors, and also indifferent tensor functions and functionals. The principle of material symmetry briefly consists in that the constitutive equations of a continuum (4.156) must remain unchanged under certain transformations of a reference configuration $\overset{\circ}{\mathcal{K}}$.

Let us give a mathematical formulation of the principle.

Constitutive equations (4.156) correspond to the motion of a continuum from $\overset{\circ}{\mathcal{K}}$ to $\mathcal{K}$. If we choose another reference configuration $\overset{*}{\mathcal{K}}$, then the motion of the continuum $\overset{*}{\mathcal{K}} \to \mathcal{K}$ is described, in general, by other constitutive equations, which can be written in the form

$$\overset{*}{\Lambda} = \breve{f}^{*}(\overset{*}{\mathcal{R}}), \tag{4.180}$$

where $\overset{*}{\Lambda}$ and $\overset{*}{\mathcal{R}}$ are active and reactive variables (4.148), (4.149) or (4.150), (4.151), or (4.154) corresponding to the motion $\overset{*}{\mathcal{K}} \to \mathcal{K}$. The operator $\breve{f}^{*}$, in general, can differ from $\breve{f}$ (4.156). The main problem is to find reference configurations $\overset{*}{\mathcal{K}}$ such that the motion $\overset{*}{\mathcal{K}} \to \mathcal{K}$ does not change the operator $\breve{f}$, i.e. in place of (4.180) we have the relation

$$\overset{*}{\Lambda} = \breve{f}(\overset{*}{\mathcal{R}}). \tag{4.181}$$

Let us find such configurations $\overset{*}{\mathcal{K}}$.

Positions of the same point $\mathcal{M}$ in $\overset{\circ}{\mathcal{K}}$ and $\overset{*}{\mathcal{K}}$ are described by the radius-vectors $\overset{\circ}{\mathbf{x}}$ and $\overset{*}{\mathbf{x}}$, respectively (Fig. 4.1):

$$\overset{\circ}{\mathbf{x}} = \overset{\circ}{\mathbf{x}}(X^i), \quad \overset{*}{\mathbf{x}} = \overset{*}{\mathbf{x}}(X^i), \tag{4.182}$$

where $\overset{\circ}{\mathbf{x}}$ and $\overset{*}{\mathbf{x}}$ are connected by the differentiable transformation

$$\overset{\circ}{\mathbf{x}} = \overset{\circ}{\mathbf{x}}(\overset{*}{x}{}^i). \tag{4.183}$$

For the same point $\mathcal{M}$, we can introduce local bases and metric matrices in $\overset{\circ}{\mathcal{K}}$ and $\overset{*}{\mathcal{K}}$:

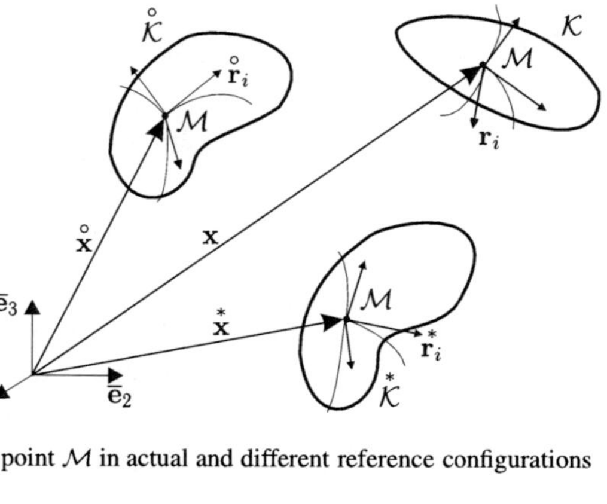


**Fig. 4.1** Positions of point $\mathcal{M}$ in actual and different reference configurations

$$\overset{\circ}{\mathbf{r}}_i = \frac{\partial \overset{\circ}{\mathbf{x}}}{\partial X^i}, \qquad \overset{*}{\mathbf{r}}_i = \frac{\partial \overset{*}{\mathbf{x}}}{\partial X^i}, \tag{4.184}$$

$$\overset{\circ}{g}_{ij} = \overset{\circ}{\mathbf{r}}_i \cdot \overset{\circ}{\mathbf{r}}_j, \qquad \overset{*}{g}_{ij} = \overset{*}{\mathbf{r}}_i \cdot \overset{*}{\mathbf{r}}_j, \tag{4.185}$$

$$\overset{\circ}{\mathbf{r}}{}^i = \overset{\circ}{g}{}^{ij}\overset{\circ}{\mathbf{r}}_j, \qquad \overset{*}{\mathbf{r}}{}^i = \overset{*}{g}{}^{ij}\overset{*}{\mathbf{r}}_j, \tag{4.186}$$

and also the deformation gradients at the transformations $\overset{\circ}{\mathcal{K}} \to \mathcal{K}$, $\overset{*}{\mathcal{K}} \to \mathcal{K}$ and $\overset{\circ}{\mathcal{K}} \to \overset{*}{\mathcal{K}}$:

$$\mathbf{F} = \mathbf{r}_i \otimes \overset{\circ}{\mathbf{r}}{}^i, \qquad \overset{*}{\mathbf{F}} = \mathbf{r}_i \otimes \overset{*}{\mathbf{r}}{}^i, \qquad \mathbf{H} = \overset{*}{\mathbf{r}}_i \otimes \overset{\circ}{\mathbf{r}}{}^i. \tag{4.187}$$

The tensor $\mathbf{H}$, describing the motion from $\overset{\circ}{\mathcal{K}}$ to $\overset{*}{\mathcal{K}}$, connects the deformation gradients $\mathbf{F}$ and $\overset{*}{\mathbf{F}}$:

$$\mathbf{F} = \overset{*}{\mathbf{F}} \cdot \mathbf{H}, \qquad \overset{*}{\mathbf{F}} = \mathbf{F} \cdot \mathbf{H}^{-1}. \tag{4.188}$$

Thus, each reference configuration $\overset{*}{\mathcal{K}}$ is assigned to its local basis $\overset{*}{\mathbf{r}}_i$ and its tensor $\mathbf{H}$. And the *transformation from one reference configuration* $\overset{\circ}{\mathcal{K}}$ *to another reference configuration* $\overset{*}{\mathcal{K}}$ is the passage from description of a continuum in the basis $\overset{\circ}{\mathbf{r}}_i$ to description in the basis $\overset{*}{\mathbf{r}}_i$.

## 4.6.2 *H*-indifferent and *H*-invariant Tensors

With the help of the tensor $\mathbf{H}$ and relationship (4.188), we can connect any tensor $\boldsymbol{\Omega}$, which has a physical meaning and is defined for the transformation $\overset{\circ}{\mathcal{K}} \to \mathcal{K}$, and a corresponding tensor $\overset{*}{\boldsymbol{\Omega}}$ defined for the transformation $\overset{*}{\mathcal{K}} \to \mathcal{K}$.

**Definition 4.3.** A tensor $\boldsymbol{\Omega}$ is called *H-indifferent*, if its components remain unchanged under the transformation from one reference configuration $\overset{*}{\mathcal{K}}$ to another $\overset{\circ}{\mathcal{K}}$:

$$\overset{\circ}{\Omega}{}^{i_1 \ldots i_n}(\overset{\circ}{\mathbf{x}}) = \overset{*}{\Omega}{}^{i_1 \ldots i_n}(\overset{*}{\mathbf{x}}). \tag{4.189}$$

Here $\overset{\circ}{\Omega}{}^{i_1 \ldots i_n}$ and $\overset{*}{\Omega}{}^{i_1 \ldots i_n}$ are components of corresponding tensors $\boldsymbol{\Omega}$ and $\overset{*}{\boldsymbol{\Omega}}$ in configurations $\overset{\circ}{\mathcal{K}}$ and $\overset{*}{\mathcal{K}}$ with radius-vectors $\overset{\circ}{\mathbf{x}}$ and $\overset{*}{\mathbf{x}}$, respectively:

$$\boldsymbol{\Omega}(\overset{\circ}{\mathbf{x}}) = \overset{\circ}{\Omega}{}^{i_1 \ldots i_n}(\overset{\circ}{\mathbf{x}})\overset{\circ}{\mathbf{r}}_{i_1} \otimes \ldots \otimes \overset{\circ}{\mathbf{r}}_{i_n}, \quad \overset{*}{\boldsymbol{\Omega}}(\overset{*}{\mathbf{x}}) = \overset{*}{\Omega}{}^{i_1 \ldots i_n}(\overset{*}{\mathbf{x}})\overset{*}{\mathbf{r}}_{i_1} \otimes \ldots \otimes \overset{*}{\mathbf{r}}_{i_n}. \tag{4.190}$$

**Definition 4.4.** A tensor $\boldsymbol{\Omega}$ is called *H-invariant*, if it remains unchanged under the transformation from $\overset{\circ}{\mathcal{K}}$ to $\overset{*}{\mathcal{K}}$, i.e.

$$\boldsymbol{\Omega}(\overset{\circ}{\mathbf{x}}) = \overset{*}{\boldsymbol{\Omega}}(\overset{*}{\mathbf{x}}). \tag{4.191}$$

Local bases vectors $\mathbf{r}^i$ and $\mathbf{r}_i$ remain unchanged under the transformation $\overset{\circ}{\mathcal{K}} \to \overset{*}{\mathcal{K}}$, therefore they are $H$-invariant.

Local basis vectors $\overset{\circ}{\mathbf{r}}_i$ change in going from $\overset{\circ}{\mathcal{K}}$ to $\overset{*}{\mathcal{K}}$, therefore they are not $H$-invariant but they can be resolved for their basis as well as the vectors $\overset{*}{\mathbf{r}}_i$:

$$\overset{\circ}{\mathbf{r}}_i = \delta_i^j \overset{\circ}{\mathbf{r}}_j, \quad \overset{*}{\mathbf{r}}_i = \delta_i^j \overset{*}{\mathbf{r}}_j, \tag{4.192}$$

i.e. components $\delta_i^j$ of vectors $\overset{\circ}{\mathbf{r}}_i$ remain unchanged under the transformation $\overset{\circ}{\mathcal{K}} \to \overset{*}{\mathcal{K}}$; hence, the vectors $\overset{\circ}{\mathbf{r}}_i$ are $H$-indifferent (in a similar way, we can show that $\overset{\circ}{\mathbf{r}}^i$ are $H$-indifferent vectors).

Besides, due to formulae (4.184) and (4.187), we get the relations

$$\overset{*}{\mathbf{r}}_i = \mathbf{H} \cdot \overset{\circ}{\mathbf{r}}_i = \overset{\circ}{\mathbf{r}}_i \cdot \mathbf{H}^{\mathrm{T}}. \tag{4.193}$$

and also

$$\overset{*}{\mathbf{r}}^i = \mathbf{H}^{-1\mathrm{T}} \cdot \overset{\circ}{\mathbf{r}}^i = \overset{\circ}{\mathbf{r}}^i \cdot \mathbf{H}^{-1}. \tag{4.194}$$

Substituting (4.193) into (4.190) and taking (4.189) into account, we obtain that any $H$-indifferent tensor satisfies the relation

$$\overset{*}{\boldsymbol{\Omega}}(\overset{*}{\mathbf{x}}) = \overset{*}{\Omega}{}^{i_1 \ldots i_n}(\overset{*}{\mathbf{x}})\overset{*}{\mathbf{r}}_{i_1} \otimes \ldots \otimes \overset{*}{\mathbf{r}}_{i_n} = \overset{\circ}{\Omega}{}^{i_1 \ldots i_n}(\overset{\circ}{\mathbf{x}})\overset{\circ}{\mathbf{r}}_{i_1} \cdot \mathbf{H}^{\mathrm{T}} \otimes \ldots \otimes \overset{\circ}{\mathbf{r}}_{i_n} \cdot \mathbf{H}^{\mathrm{T}}. \tag{4.195}$$

For a scalar $\varphi$, $H$-indifference means that

$$\varphi(\overset{*}{x}) = \varphi(x), \tag{4.196}$$

for a vector **a**:

$$\overset{*}{\mathbf{a}}(\overset{*}{\mathbf{x}}) = \mathbf{a}(\overset{\circ}{\mathbf{x}}) \cdot \mathbf{H}^{\mathrm{T}} = \mathbf{H} \cdot \mathbf{a}(\overset{\circ}{\mathbf{x}}), \tag{4.197}$$

and for a second-order tensor:

$$\overset{*}{\boldsymbol{\Omega}}(\overset{*}{\mathbf{x}}) = \mathbf{H} \cdot \boldsymbol{\Omega}(\overset{\circ}{\mathbf{x}}) \cdot \mathbf{H}^{\mathrm{T}}. \tag{4.198}$$

Figure 4.2a shows an example of $H$-invariant vector **a**, which remains unchanged under the transformation $\overset{\circ}{\mathcal{K}} \to \overset{*}{\mathcal{K}}$. Figure 4.2b shows an example of $H$-indifferent vector **b**, which moves together with the basis $\overset{\circ}{\mathbf{r}}_i$: vector $\overset{*}{\mathbf{b}}$ in basis $\overset{*}{\mathbf{r}}_i$ has the same coordinates as vector **b** in basis $\overset{\circ}{\mathbf{r}}_i$.

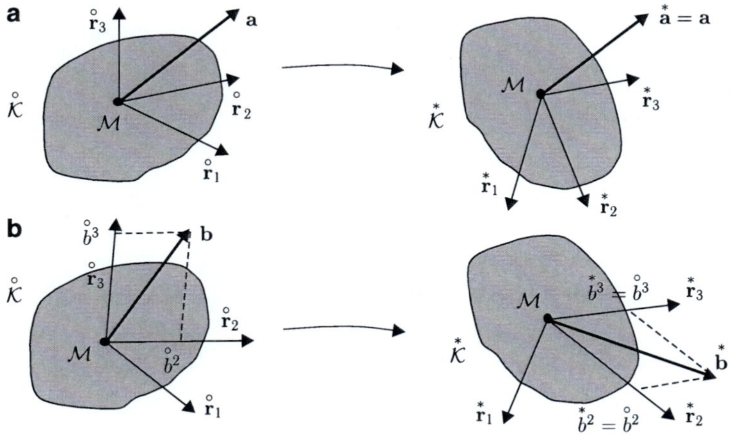


**Fig. 4.2** (**a**) – $H$-invariant vector **a**; (**b**) – $H$-indifferent vector **b**

The metric tensor **E** is the same in all the configurations: $\overset{\circ}{\mathcal{K}}$, $\overset{*}{\mathcal{K}}$ and $\mathcal{K}$, i.e. the tensor is $H$-invariant. Indeed, according to formulae (4.193) and (4.194), we have

$$\overset{*}{\mathbf{E}} = \overset{*}{\mathbf{r}}_i \otimes \overset{*}{\mathbf{r}}{}^i = \mathbf{H}\cdot\overset{\circ}{\mathbf{r}}_i \otimes \overset{\circ}{\mathbf{r}}{}^i \cdot \mathbf{H}^{-1} = \mathbf{H}\cdot\mathbf{E}\cdot\mathbf{H}^{-1} = \mathbf{H}\cdot\mathbf{H}^{-1} = \mathbf{E}. \tag{4.199}$$

The deformation gradient **F** is neither $H$-invariant nor $H$-indifferent, because during the passage from $\overset{\circ}{\mathcal{K}}$ to $\overset{*}{\mathcal{K}}$ the tensor is transformed by the law (4.188), which is different from both (4.191) and (4.198).

Similarly to the tensors $\overset{(\mathrm{n})}{\mathbf{C}}$, $\overset{(\mathrm{n})}{\mathbf{A}}$, $\overset{(\mathrm{n})}{\mathbf{G}}$ and $\overset{(\mathrm{n})}{\mathbf{g}}$ describing the deformation of a local point $\mathcal{M}$ under the transformation from $\overset{\circ}{\mathcal{K}}$ to $\mathcal{K}$, we can define the deformation tensors and measures during the passage from $\overset{*}{\mathcal{K}}$ to $\mathcal{K}$, for example:

$$\overset{\vee}{\mathbf{C}}{}^* = \overset{*}{\mathbf{C}} = \frac{1}{2}(\overset{*}{\mathbf{F}}{}^{\mathrm{T}}\cdot\overset{*}{\mathbf{F}} - \mathbf{E}), \quad \overset{\vee}{\mathbf{G}}{}^* = \frac{1}{2}\overset{*}{\mathbf{F}}{}^{\mathrm{T}}\cdot\overset{*}{\mathbf{F}}. \tag{4.200}$$

But each of the tensors $\overset{(\mathrm{n})}{\mathbf{C}}$, $\overset{(\mathrm{n})}{\mathbf{A}}$, $\overset{(\mathrm{n})}{\mathbf{G}}$ and $\overset{(\mathrm{n})}{\mathbf{g}}$ as well as **F** is neither $H$-indifferent nor $H$-invariant. For example, the tensors $\overset{\vee}{\mathbf{G}}$ and $\overset{\vee}{\mathbf{C}}$ during the passage from $\overset{\circ}{\mathcal{K}}$ to $\overset{*}{\mathcal{K}}$ are transformed as follows:

$$\overset{\vee}{\mathbf{G}}{}^* = \frac{1}{2}\overset{*}{\mathbf{G}} = \frac{1}{2}\overset{*}{\mathbf{F}}{}^{\mathrm{T}}\cdot\overset{*}{\mathbf{F}} = \frac{1}{2}\mathbf{H}^{-1\mathrm{T}}\cdot\mathbf{F}^{\mathrm{T}}\cdot\mathbf{F}\cdot\mathbf{H}^{-1} = \mathbf{H}^{-1\mathrm{T}}\cdot\overset{\vee}{\mathbf{G}}\cdot\mathbf{H}^{-1},$$

$$\overset{\vee}{\mathbf{C}}{}^* = \mathbf{H}^{-1\mathrm{T}}\cdot\overset{\vee}{\mathbf{C}}\cdot\mathbf{H}^{-1} + \frac{1}{2}(\mathbf{H}^{-1\mathrm{T}}\cdot\mathbf{H}^{-1} - \mathbf{E}).$$

Only the right Almansi deformation measure $\mathbf{G}^{-1}$ (2.75) and hence $\overset{\mathrm{I}}{\mathbf{G}}$ are $H$-indifferent, because

$$\overset{\mathrm{I}}{\mathbf{G}}{}^{*} = -\frac{1}{2}\overset{*}{\mathbf{G}}{}^{-1} = -\frac{1}{2}\overset{*}{\mathbf{F}}{}^{-1} \cdot \overset{*}{\mathbf{F}}{}^{-1\mathrm{T}} = -\frac{1}{2}\mathbf{H} \cdot \mathbf{F}^{-1} \cdot \mathbf{F}^{-1\mathrm{T}} \cdot \mathbf{H}^{\mathrm{T}}$$
$$= -\frac{1}{2}\mathbf{H} \cdot \mathbf{G}^{-1} \cdot \mathbf{H}^{\mathrm{T}} = \mathbf{H} \cdot \overset{\mathrm{I}}{\mathbf{G}} \cdot \mathbf{H}^{\mathrm{T}}; \quad (4.201)$$

but $\overset{\mathrm{I}}{\mathbf{C}}$ is neither $H$-indifferent nor $H$-invariant:

$$\overset{\mathrm{I}}{\mathbf{C}}{}^{*} = \overset{\mathrm{I}}{\mathbf{G}}{}^{*} - \frac{1}{2}\mathbf{E} = \mathbf{H} \cdot \overset{\mathrm{I}}{\mathbf{C}} \cdot \mathbf{H}^{\mathrm{T}} - \frac{1}{2}(\mathbf{H} \cdot \mathbf{H}^{\mathrm{T}} - \mathbf{E}). \quad (4.201a)$$

Since the actual configuration $\mathcal{K}$ remains unchanged under the transformation $\overset{\circ}{\mathcal{K}} \to \overset{*}{\mathcal{K}}$ and the Cauchy stress tensor $\mathbf{T}$ is defined in $\mathcal{K}$, so the tensor remains unchanged under the transformation $\overset{\circ}{\mathcal{K}} \to \overset{*}{\mathcal{K}}$, i.e. $\mathbf{T}$ is $H$-invariant:

$$\overset{*}{\mathbf{T}} = \mathbf{T}. \quad (4.202)$$

However, other stress tensors $\overset{(\mathrm{n})}{\mathbf{T}}$, $\overset{(\mathrm{n})}{\mathbf{S}}$ and $\mathbf{P}$ are not $H$-invariant, because they depend not only on $\mathbf{T}$ but also on the deformation gradient $\mathbf{F}$. Among the tensors $\overset{(\mathrm{n})}{\mathbf{S}}$, $\overset{(\mathrm{n})}{\mathbf{T}}$ and $\mathbf{P}$, only the tensor $\overset{\mathrm{V}}{\mathbf{T}}$ is $H$-indifferent, because, according to (4.188) and (4.202), we have

$$\overset{\mathrm{V}}{\mathbf{T}}{}^{*} = \overset{*}{\mathbf{F}}{}^{-1} \cdot \overset{*}{\mathbf{T}} \cdot \overset{*}{\mathbf{F}}{}^{-1\mathrm{T}} = \mathbf{H} \cdot \mathbf{F}^{-1} \cdot \mathbf{T} \cdot \mathbf{F}^{-1\mathrm{T}} \cdot \mathbf{H}^{\mathrm{T}} = \mathbf{H} \cdot \overset{\mathrm{V}}{\mathbf{T}} \cdot \mathbf{H}^{\mathrm{T}}; \quad (4.203)$$

but, for example, the tensor $\overset{\mathrm{I}}{\mathbf{T}}$ is transformed during the passage from $\overset{\circ}{\mathcal{K}}$ to $\overset{*}{\mathcal{K}}$ as follows:

$$\overset{\mathrm{I}}{\mathbf{T}}{}^{*} = \overset{*}{\mathbf{F}}{}^{\mathrm{T}} \cdot \overset{*}{\mathbf{T}} \cdot \overset{*}{\mathbf{F}} = \mathbf{H}^{-1\mathrm{T}} \cdot \mathbf{F}^{\mathrm{T}} \cdot \mathbf{T} \cdot \mathbf{F} \cdot \mathbf{H}^{-1} = \mathbf{H}^{-1\mathrm{T}} \cdot \overset{\mathrm{I}}{\mathbf{T}} \cdot \mathbf{H}^{-1}. \quad (4.203a)$$

*Remark.* The type of transformation (4.200a) and (4.203a) during the passage $\overset{\circ}{\mathcal{K}} \to \overset{*}{\mathcal{K}}$ proves to be useful for continuum mechanics, although this type is neither $H$-indifferent nor $H$-invariant. Therefore tensors $\boldsymbol{\Omega}$ transforming under $\overset{\circ}{\mathcal{K}} \to \overset{*}{\mathcal{K}}$ by the type:

$$\overset{*}{\boldsymbol{\Omega}} = \mathbf{H}^{-1\mathrm{T}} \cdot \boldsymbol{\Omega} \cdot \mathbf{H}^{-1}, \quad (4.203b)$$

will be called *H-pseudoindifferent.* □

### 4.6.3 Symmetry Groups of Continua

Suppose that there exists a set of transformations of the reference configuration $H : \overset{\circ}{\mathcal{K}} \to \overset{*}{\mathcal{K}}$ and each of the transformations does not change the constitutive equations (4.156); i.e. if the relation (4.156) is valid, then relation (4.181) holds at these transformations as well. Such transformations of the reference configuration are called *H-transformations*, and each transformation is put in correspondence with a tensor **H**.

All $H$-transformations $\overset{\circ}{\mathcal{K}} \to \overset{*}{\mathcal{K}}$ occur without changes in the density $\overset{\circ}{\rho}$:

$$\overset{*}{\rho} = \overset{\circ}{\rho}, \tag{4.204}$$

because the density $\overset{\circ}{\rho}$ is the property of a continuum, and $H$-transformations, by definition, do not change properties of a continuum.

Density $\rho$ in actual configuration $\mathcal{K}$ is independent of transformations $\overset{\circ}{\mathcal{K}} \to \overset{*}{\mathcal{K}}$, and hence the density $\rho$ is both $H$-indifferent and $H$-invariant.

**Theorem 4.13.** *The set of all H-transformations $\overset{\circ}{\mathcal{K}} \to \overset{*}{\mathcal{K}}$ of the reference configuration $\overset{\circ}{\mathcal{K}}$ of a continuum (or the set of* **H***-tensors) with constitutive equations (4.156), for which the relations (4.181) remain valid, constitutes a group called the symmetry group $\overset{\circ}{G}_s$ of the continuum.*

The group is sometimes called the *group of equivalence*.

▼ Show that the set of all $H$-transformations actually has the properties of a group (see the definition of a group in [12]); i.e. in this set the operation of multiplication, which is superposition of $H$-transformations, is defined, and the set includes the unit and inverse elements.

Indeed, superposition of two $H$-transformations is also $H$-transformation with the tensor $\mathbf{H}_2 \cdot \mathbf{H}_1$. This assertion can be proved in the following way. Let there be transformation $\mathbf{H}_1 : \overset{\circ}{\mathcal{K}} \to \overset{*}{\mathcal{K}}$ and transformation $\mathbf{H}_2 : \overset{*}{\mathcal{K}} \to \widehat{\mathcal{K}}$ with the deformation gradients:

$$\overset{*}{\mathbf{F}} = \mathbf{F} \cdot \mathbf{H}_1^{-1}, \quad \widehat{\mathbf{F}} = \overset{*}{\mathbf{F}} \cdot \mathbf{H}_2^{-1}. \tag{4.205}$$

Notice that since the energetic and quasienergetic deformation tensors are uniquely expressed in terms of the deformation gradient **F** (see Exercise 4.2.23), and the energetic and quasienergetic stress tensors – in terms of **T** and **F** by formulae (4.36) and (4.66), so the constitutive equations (4.156) can be written in the universal form

$$\Lambda = \check{f}(\mathcal{R}), \quad \mathcal{R} = \{\theta, \mathbf{F}\}, \quad \Lambda = \{\psi, \eta, \mathbf{T}/\rho, w^*/\rho\}, \tag{4.206}$$

for all the models $A_n$, $B_n$, $C_n$ and $D_n$.

By the definition of $H$-transformation, from the constitutive equations (4.206) after $\mathbf{H}_1$-transformation we get the relation (4.181):

$$\overset{*}{\Lambda} = \breve{f}(\overset{*}{\mathcal{R}}), \quad \overset{*}{\mathcal{R}} = \{\theta^*, \mathbf{F}^*\} = \{\theta, \mathbf{F}\cdot\mathbf{H}_1^{-1}\},$$
$$\overset{*}{\Lambda} = \{\psi^*, \eta^*, \mathbf{T}^*/\rho, (w^*)^*/\rho\} = \{\psi, \eta, \mathbf{T}/\rho, w^*/\rho\}, \tag{4.207}$$

and from the constitutive equations (4.207) after $\mathbf{H}_2$-transformation we obtain

$$\widehat{\Lambda} = \breve{f}(\widehat{\mathcal{R}}), \tag{4.208}$$

where

$$\widehat{\mathcal{R}} = \{\widehat{\theta}, \widehat{\mathbf{F}}\}, \quad \widehat{\Lambda} = \{\widehat{\psi}, \widehat{\eta}, \widehat{\mathbf{T}}/\rho, \widehat{w}^*/\rho\}. \tag{4.209}$$

Thus, from the relations (4.206) after the transformation $\mathbf{H}_2\cdot\mathbf{H}_1$ we get relations (4.208), because $\widehat{\mathcal{R}}$ and $\widehat{\Lambda}$ (4.209), included in the relations, can be represented in the form

$$\widehat{\mathcal{R}} = \{\theta, \mathbf{F}\cdot(\mathbf{H}_2\cdot\mathbf{H}_1)^{-1}\}, \quad \widehat{\Lambda} = \{\psi, \eta, \mathbf{T}/\rho, w^*/\rho\}. \tag{4.210}$$

Here we have taken into account that all the scalars under $H$-transformations remain unchanged: $\psi = \overset{*}{\psi} = \widehat{\psi}$, $\eta = \overset{*}{\eta} = \widehat{\eta}$ etc., and the tensor $\mathbf{T}$ is $H$-invariant: $\mathbf{T} = \overset{*}{\mathbf{T}} = \widehat{\mathbf{T}}$.

The tensor $\mathbf{H} = \mathbf{E}$ is the unit element of the set of $H$-transformations.

Using the expressions (4.206) for reactive and active variables and choosing $\mathbf{H}_1 = \mathbf{H}$ and $\mathbf{H}_2 = \mathbf{H}^{-1}$, from the relations (4.210) one can easily get that $\widehat{\mathcal{R}} = \mathcal{R}$ and $\widehat{\Lambda} = \Lambda$. Thus, after the transformation $\mathbf{H}_1\cdot\mathbf{H}_2 = \mathbf{H}\cdot\mathbf{H}^{-1}$ we get constitutive equations (4.156) with $\mathcal{R}, \Lambda$ in the form (4.206). Hence, for any $H$-transformation, the transformation with the tensor $\mathbf{H}^{-1}$ is $H$-transformation as well. ▲

### 4.6.4 The Statement of the Principle of Material Symmetry

We have shown that if there is a set of $H$-transformations satisfying (4.181), then the set is a group. However, the question, whether the set always exists, remains open. The following axiom gives a positive answer to the question.

**Axiom 14 (The principle of material symmetry).** *For every continuum with the constitutive equations (4.156) in the motion $\overset{\circ}{\mathcal{K}} \to \mathcal{K}$ with an arbitrary reference configuration $\overset{\circ}{\mathcal{K}}$, there exists a corresponding symmetry group $\overset{\circ}{G}_s$ being the group* *of H-transformations of the reference configuration* $H : \overset{\circ}{\mathcal{K}} \to \overset{*}{\mathcal{K}}$, *which do not change the constitutive equations, i.e. relation (4.181) holds in any motion* $\overset{*}{\mathcal{K}} \to \mathcal{K}$:

$$\overset{*}{\Lambda} = \breve{f}(\overset{*}{\mathcal{R}}). \tag{4.211}$$

# 4.7 Definition of Fluids and Solids

## 4.7.1 Fluids and Solids

Let us consider examples of symmetry groups $\overset{\circ}{G}_s$.

**Theorem 4.14.** *Every* **H**-*tensor corresponding to H-transformation is unimodular, i.e. it satisfies the relation*

$$\det \mathbf{H} = \pm 1. \tag{4.212}$$

▼ Indeed, the continuity equation (3.8) and relations (4.188) yield

$$\frac{\overset{\circ}{\rho}}{\rho} = |\det \mathbf{F}| = |\det (\overset{*}{\mathbf{F}} \cdot \mathbf{H})| = |\det \overset{*}{\mathbf{F}} \cdot \det \mathbf{H}| = \frac{\overset{*}{\rho}}{\rho} |\det \mathbf{H}|. \tag{4.213}$$

Due to (4.204), from (4.213) we get Eq. (4.212). ▲

Notice that $H$-transformations may be not continuous (such transformations cannot be represented as a sequence of transformations convergent to the identical one). An example of such transformation is the $H$-transformation of reflection with respect to some plane. For such $H$-transformations, in the continuity equation, unlike Eq. (3.8) written for continuous transformations, one should take the magnitude of $\det \mathbf{F}$ in order that the expression $\overset{\circ}{\rho}/\rho$ be positive. This has been made in (4.213).

The set of all unimodular tensors constitutes the full *unimodular group* $U$ [12]. Thus, we get the following theorem.

**Theorem 4.15.** *A symmetry group* $\overset{\circ}{G}_s$ *of any continuum is a subgroup of the full unimodular group:*

$$\overset{\circ}{G}_s \subset U. \tag{4.214}$$

**Definition 4.5.** A continuum, which for any reference configuration $\overset{\circ}{\mathcal{K}}$ has a symmetry group $\overset{\circ}{G}_s$ coinciding with the full unimodular group:

$$\overset{\circ}{G}_s = U, \tag{4.215}$$

is called a *fluid*.

Consider $H$-transformations of a reference configuration $\overset{\circ}{\mathcal{K}}$ with orthogonal tensors $\mathbf{H}$ satisfying the condition

$$\mathbf{H} \cdot \mathbf{H}^{\mathrm{T}} = \mathbf{E}. \tag{4.216}$$

The set of all tensors satisfying the condition (4.216) constitutes the *full orthogonal group* $I$ [12].

**Definition 4.6.** A continuum, for which there exists a reference configuration $\widehat{\mathcal{K}}$ such that its symmetry group $\widehat{G}_s$ is a subgroup of the full orthogonal group:

$$\widehat{G}_s \subset I, \tag{4.217}$$

is called a *solid* (a solid body).

The configuration $\widehat{\mathcal{K}}$ of a solid, for which the condition (4.217) is satisfied, is called *undistorted*.

### 4.7.2 Isomeric Symmetry Groups

Remind that a symmetry group $\overset{\circ}{G}_s$ of a continuum (a group of $H$-transformations) is defined for some fixed configuration $\overset{\circ}{\mathcal{K}}$. For another reference configuration $\widehat{\mathcal{K}}$, according to Axiom 14, there is a symmetry group $\widehat{G}_s$ as well, but the group is, in general, not coincident with $\overset{\circ}{G}_s$.

**Theorem 4.16 (of Noll).** *For every continuum, symmetry groups $\overset{\circ}{G}_s$ and $\widehat{G}_s$ corresponding to different reference configurations $\overset{\circ}{\mathcal{K}}$ and $\widehat{\mathcal{K}}$ are isomeric, i.e. they are connected by the relation*

$$\widehat{G}_s = \mathbf{S}\, \overset{\circ}{G}_s\, \mathbf{S}^{-1}. \tag{4.218}$$

*Here* $\mathbf{S}$ *is the nonsingular deformation gradient corresponding to the transformation* $\overset{\circ}{\mathcal{K}} \to \widehat{\mathcal{K}}$, *which connects local bases* $\overset{\circ}{\mathbf{r}}_i$ *and* $\widehat{\mathbf{r}}_i$ *in* $\overset{\circ}{\mathcal{K}}$ *and* $\widehat{\mathcal{K}}$*:*

$$\mathbf{S} = \widehat{\mathbf{r}}_i \otimes \overset{\circ}{\mathbf{r}}{}^i, \quad \widehat{\mathbf{r}}_i = \mathbf{S} \cdot \overset{\circ}{\mathbf{r}}_i, \quad \widehat{\mathbf{r}}^i = \overset{\circ}{\mathbf{r}}{}^i \cdot \mathbf{S}^{-1}. \tag{4.219}$$

▼ If the relationship (4.218) really holds, this means that we can put every transformation tensor $\mathbf{H}$ of a symmetry group $\overset{\circ}{G}_s$ in correspondence with the tensor

$$\widehat{\mathbf{H}} = \mathbf{S} \cdot \mathbf{H} \cdot \mathbf{S}^{-1} \tag{4.220}$$

of a group $\widehat{G}_s$ and vice versa.

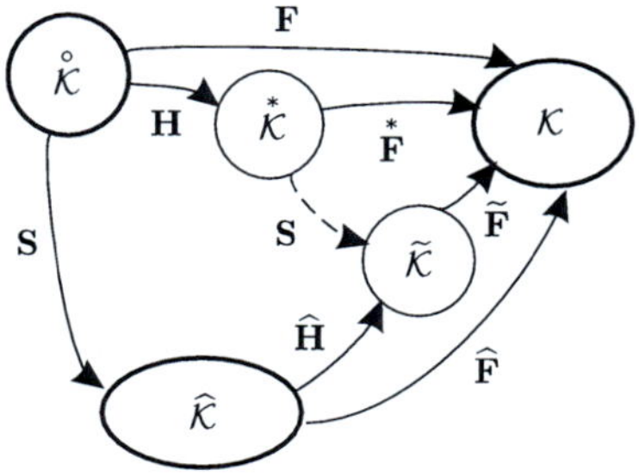


**Fig. 4.3** The connection of different configurations

Such groups $\overset{\circ}{G}_s$ and $\widehat{G}_s$ are called *isomeric*. Figure 4.3 shows reference configurations $\overset{\circ}{\mathcal{K}}$, $\widehat{\mathcal{K}}$ and their corresponding configurations $\overset{*}{\mathcal{K}}$, $\widetilde{\mathcal{K}}$ after $H$- and $\widehat{H}$-transformations. The deformation gradients $\mathbf{F}$, $\widehat{\mathbf{F}}$, $\overset{*}{\mathbf{F}}$ and $\widetilde{\mathbf{F}}$ transform configurations $\overset{\circ}{\mathcal{K}}$, $\widehat{\mathcal{K}}$, $\overset{*}{\mathcal{K}}$ and $\widetilde{\mathcal{K}}$ into $\mathcal{K}$, respectively.

According to the representation (4.206) of constitutive equations, we see that for all $H$- and $\widehat{H}$-transformations the following relations hold:

$$\Lambda = \breve{f}(\theta, \mathbf{F}) = \breve{f}(\theta, \mathbf{F}\cdot\mathbf{H}^{-1})\ \forall\mathbf{F}, \tag{4.221}$$

$$\Lambda = \widehat{\breve{f}}(\theta, \widehat{\mathbf{F}}) = \widehat{\breve{f}}(\theta, \widehat{\mathbf{F}}\cdot\widehat{\mathbf{H}}^{-1})\ \forall\widehat{\mathbf{F}}. \tag{4.222}$$

Here the hat $\widehat{\ }$ in $\widehat{\breve{f}}$ means that the dependence of $\breve{f}$ on $\mathbf{F}$ and $\widehat{\mathbf{F}}$ may be different in different configurations.

Since $\mathbf{F}$ and $\widehat{\mathbf{F}}$ are connected by the tensor $\mathbf{S}$:

$$\widehat{\mathbf{F}} = \mathbf{r}_i \otimes \widehat{\mathbf{r}}^i = \mathbf{r}_i \otimes \overset{\circ}{\mathbf{r}}{}^i \cdot \overset{\circ}{\mathbf{r}}_j \otimes \widehat{\mathbf{r}}^j = \mathbf{F}\cdot\mathbf{S}^{-1}, \tag{4.223}$$

we get the relation

$$\breve{f}(\theta,\ \mathbf{F}) = \widehat{\breve{f}}(\theta,\ \widehat{\mathbf{F}}) = \widehat{\breve{f}}(\theta,\ \mathbf{F}\cdot\mathbf{S}^{-1}). \tag{4.224}$$

Since this relation must be satisfied for any deformation gradient $\mathbf{F}$, we can apply the relation to $\mathbf{F}\cdot\mathbf{H}^{-1}$ with substitution of the product in place of $\mathbf{F}$:

$$\breve{f}(\theta,\ \mathbf{F}\cdot\mathbf{H}^{-1}) = \widehat{\breve{f}}(\theta,\ \mathbf{F}\cdot\mathbf{H}^{-1}\cdot\mathbf{S}^{-1}). \tag{4.225}$$

Since $\mathbf{H} \subset \overset{\circ}{G}_s$, so, by (4.221), two equalities (4.224) and (4.225) prove to be coincident, i.e.

$$\widehat{\breve{f}}(\mathbf{F}\cdot\mathbf{S}^{-1}) = \widehat{\breve{f}}(\mathbf{F}\cdot\mathbf{H}^{-1}\cdot\mathbf{S}^{-1}). \tag{4.226}$$

Returning from $\mathbf{F}$ to $\widehat{\mathbf{F}}$ by the relation $\mathbf{F} = \widehat{\mathbf{F}} \cdot \mathbf{S}$, from (4.226) we get

$$\Lambda = \check{f}(\widehat{\mathbf{F}}) = \check{f}(\widehat{\mathbf{F}} \cdot \mathbf{S} \cdot \mathbf{H}^{-1} \cdot \mathbf{S}^{-1}). \tag{4.227}$$

Thus, the transformation $\mathbf{S} \cdot \mathbf{H}^{-1} \cdot \mathbf{S}^{-1}$ retains the form of the tensor function $\check{f}$ without changes. This means that the tensor $\mathbf{S} \cdot \mathbf{H}^{-1} \cdot \mathbf{S}^{-1}$ belongs to the group $\widehat{G}_s$. Then there is a tensor $\widehat{\mathbf{H}} \in \widehat{G}_s$ such that $\widehat{\mathbf{H}}^{-1} = \mathbf{S} \cdot \mathbf{H}^{-1} \cdot \mathbf{S}^{-1}$. From this relation we get formula (4.220). ▲

Let us resolve now tensors $\mathbf{H}$ and $\widehat{\mathbf{H}}$ for bases $\mathring{\mathbf{r}}_i$ and $\widehat{\mathbf{r}}_i$:

$$\mathbf{H} = H^i_j \mathring{\mathbf{r}}_i \otimes \mathring{\mathbf{r}}^j, \qquad \widehat{\mathbf{H}} = \widehat{H}^i_j \widehat{\mathbf{r}}_i \otimes \widehat{\mathbf{r}}^j. \tag{4.228}$$

On substituting (4.228) into (4.220) and taking (4.219) into account, we get

$$\widehat{H}^i_j \widehat{\mathbf{r}}_i \otimes \widehat{\mathbf{r}}^j = H^i_j \mathbf{S} \cdot \mathbf{r}_i \otimes \mathbf{r}^j \cdot \mathbf{S}^{-1} = H^i_j \, \widehat{\mathbf{r}}_i \otimes \widehat{\mathbf{r}}^j. \tag{4.229}$$

Thus, due to uniqueness of a tensor resolution for a dyadic basis, we find that $\widehat{H}^i_j = H^i_j$. Hence, the tensors $\mathbf{H}$ and $\widehat{\mathbf{H}}$ have the same components but with respect to different bases. Such tensors are called *isomeric* (see [12]).

Theorem 4.16 as well as Definitions 4.5 and 4.6 have been suggested by Noll.

*Remark 1.* The conditions of Theorem 4.16 imply that as a symmetry group $\mathring{G}_s$ corresponding to a configuration $\mathring{\mathcal{K}}$, we take a subgroup of the unimodular group $U$, which is maximum permissible by the constitutive equations (4.206); i.e. if for $\mathring{\mathcal{K}}$ there is one more symmetry group $G'_s$, then $G'_s \subset \mathring{G}_s$. The same holds for $\widehat{G}_s$. This condition is assumed to be always satisfied below.

*Remark 2.* Since during the passage from one reference configuration $\mathring{\mathcal{K}}$ to another $\widehat{\mathcal{K}}$, due to Theorem 4.16, we obtain, in general, different symmetry groups $\mathring{G}_s$ or $\widehat{G}_s$, so the concept of a symmetry group of a continuum is relative and dependent on the choice of reference configuration. For example, if a group $\mathring{G}_s$ is orthogonal in $\mathring{\mathcal{K}}$, then in configuration $\widehat{\mathcal{K}}$ the group may be no longer orthogonal.

Let us apply Theorem 4.16 to solids. According to Definition 4.6, an undistorted configuration $\mathring{\mathcal{K}}$ is not uniquely defined. The following theorem establishes a relation between symmetry groups of two undistorted configurations.

**Theorem 4.17.** *Symmetry groups $\mathring{G}_s$ and $\widehat{G}_s$ of a solid, which correspond to two undistorted configurations $\mathring{\mathcal{K}}$ and $\widehat{\mathcal{K}}$, are connected by the relation*

$$\widehat{G}_s = \mathbf{O}_0 \mathring{G}_s \mathbf{O}_0^{\mathrm{T}}, \tag{4.230}$$

*where* $\mathbf{O}_0$ *is the orthogonal rotation tensor accompanying the transformation* $\mathbf{S}$ : $\overset{\circ}{\mathcal{K}} \to \widehat{\mathcal{K}}$.

▼ Rewrite formula (4.220) in the form

$$\widehat{\mathbf{H}} \cdot \mathbf{S} = \mathbf{S} \cdot \mathbf{H}, \tag{4.231}$$

where $\widehat{\mathbf{H}} \in \widehat{G}_s$ and $\mathbf{H} \in G_s$ are orthogonal tensors.

Since the tensor $\mathbf{S}$ is nonsingular, we can write its polar decomposition

$$\mathbf{S} = \mathbf{O}_0 \cdot \mathbf{U}_0 = \mathbf{V}_0 \cdot \mathbf{O}_0, \tag{4.232}$$

where the tensors $\mathbf{U}_0$ and $\mathbf{V}_0$ are connected by the relation

$$\mathbf{V}_0 = \mathbf{O}_0 \cdot \mathbf{U}_0 \cdot \mathbf{O}_0^{\mathrm{T}}. \tag{4.233}$$

On substituting (4.232) and (4.233) into (4.231), we get

$$\widehat{\mathbf{H}} \cdot \mathbf{O}_0 \cdot \mathbf{U}_0 = \mathbf{V}_0 \cdot \mathbf{O}_0 \cdot \mathbf{H}. \tag{4.234}$$

The tensor $\mathbf{B} \equiv \widehat{\mathbf{H}} \cdot \mathbf{O}_0 \cdot \mathbf{U}_0$ included in (4.234) is nonsingular, because $\det \mathbf{B} = \det \widehat{\mathbf{H}} \cdot \det \mathbf{S} = \pm \det \mathbf{S} \neq 0$. Then the tensors $\mathbf{U}_0$ and $\mathbf{V}_0$ are symmetric and positive-definite, and the tensors $\widehat{\mathbf{H}} \cdot \mathbf{O}_0$ and $\mathbf{O}_0 \cdot \mathbf{H}$ are orthogonal. Thus, formula (4.234) is a polar decomposition of the tensor $\mathbf{B}$. But, due to Theorem 2.9, this decomposition is unique, therefore the relation $\widehat{\mathbf{H}} \cdot \mathbf{O}_0 = \mathbf{O}_0 \cdot \mathbf{H}$ is valid and can be rewritten in the form

$$\widehat{\mathbf{H}} = \mathbf{O}_0 \cdot \mathbf{H} \cdot \mathbf{O}_0^{\mathrm{T}}. \tag{4.235}$$

This equation has proved formula (4.230). ▲

Consider the relationship (4.234). In the equation, replace $\mathbf{V}_0$ by $\mathbf{U}_0$ with use of formula (4.233) and then multiply both sides of (4.234) by $\mathbf{O}_0^{\mathrm{T}} \cdot \widehat{\mathbf{H}}^{\mathrm{T}}$ from the left. As a result, we obtain

$$\mathbf{U}_0 = \mathbf{O}_0^{\mathrm{T}} \cdot \widehat{\mathbf{H}}^{\mathrm{T}} \cdot \mathbf{O}_0 \cdot \mathbf{U}_0 \cdot \mathbf{O}_0^{\mathrm{T}} \cdot \mathbf{O}_0 \cdot \mathbf{H} = \mathbf{H}^{\mathrm{T}} \cdot \mathbf{U}_0 \cdot \mathbf{H}. \tag{4.236}$$

Here we have used formula (4.235).

Thus, from (4.236) we establish that the tensors $\mathbf{U}_0$ and $\mathbf{H}$ are commutative:

$$\mathbf{H} \cdot \mathbf{U}_0 = \mathbf{U}_0 \cdot \mathbf{H} \qquad \forall \mathbf{H} \in \overset{\circ}{G}_s. \tag{4.237}$$

### 4.7.3 Definition of Anisotropic Solids

Consider a solid body. According to the principle of material symmetry, the solid possesses some orthogonal symmetry group $\widehat{G}_s$ in a corresponding undistorted configuration $\widehat{\mathcal{K}}$. Such solid body is called the *anisotropic body of the class* $\widehat{G}_s$.

**Definition 4.7.** A solid is called *isotropic*, if its symmetry group $\widehat{G}_s$ in an undistorted configuration $\widehat{\mathcal{K}}$ coincides with $I$:

$$\widehat{G}_s = I. \tag{4.238}$$

It should be noted that the condition (4.238) is satisfied, in general, only in some undistorted configuration $\widehat{\mathcal{K}}$. Under the transformation to another reference configuration, for example to $\overset{\circ}{\mathcal{K}}$, the solid may be no longer isotropic.

There are 39 subgroups $\widehat{G}_s$ of the full orthogonal group $I$ (including the group $I$ itself); wherein 32 subgroups contain a finite number of elements $\mathbf{H} \in \widehat{G}_s$ (such groups are called *point groups*), and 7 subgroups include a continuous set of elements $\mathbf{H}$ (they are called *continuous*). The description of all elements of the groups one can find in [12].

Consider below only the following four subgroups $\widehat{G}_s \subset I$, which are most widely used in continuum mechanics:

- The triclinic group $\widehat{G}_s = E$ (point group, which consists of two elements).
- The orthotropy group $\widehat{G}_s = O$ (point group, which consists of eight elements).
- The transverse isotropy group $\widehat{G}_s = T_3$ (continuous group).
- The isotropy group (which is full orthogonal) $\widehat{G}_s = I$ (continuous group).

As noted above, to each configuration $\overset{\circ}{\mathcal{K}}$ (and $\widehat{\mathcal{K}}$) of a solid its local basis $\mathbf{r}_i$ (and $\widehat{\mathbf{r}}_i$), and also the symmetry group $\overset{\circ}{G}_s$ (and $\widehat{G}_s$) correspond. Orthogonal group $\widehat{G}_s$ corresponding to the undistorted configuration $\widehat{\mathcal{K}}$ is characterized by its orthonormal basis $\widehat{\mathbf{c}}_i$, which is called the *principal basis of anisotropy*, and coordinate axes oriented along the vectors $\widehat{\mathbf{c}}_i$ are called the *principal axes of anisotropy*.

Figure 4.4 shows the undistorted configuration $\widehat{\mathcal{K}}$ and the reference configuration $\overset{\circ}{\mathcal{K}}$, bases $\mathbf{r}_i$ and $\overset{\circ}{\mathbf{r}}_i$ directed tangentially to the coordinate lines $X^i$ (dashed lines),

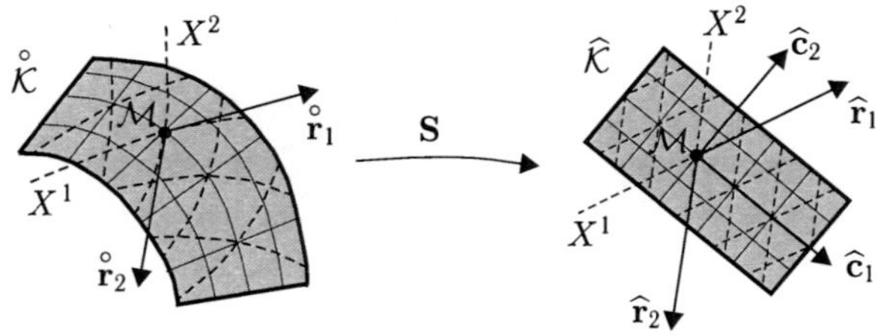


**Fig. 4.4** Distorted reference configuration and undistorted configuration of a solid

and also basis $\widehat{\mathbf{c}}_i$ directed along the curves (continuous lines) characterizing the geometric structure of a solid and its anisotropy. It should be noted that anisotropic properties of solids are caused, as a rule, by a specific geometric structure of the bodies, where not all directions have the same properties. Such materials are, for example, composites, rubber-cord materials and tires reinforced by fibres, metals and alloys, where a grained microstructure is oriented along the rolled direction, rocks and grounds having a layered structure, etc. Notice that the basis $\widehat{\mathbf{c}}_i$ can depend, in general, on coordinates $\mathbf{x}$ of a considered material point $\mathcal{M}$, in this case a solid is said to possess curvilinear anisotropy of the class $\widehat{G}_s$.

If tensors of $H$-transformations from group $\widehat{G}_s$ are represented in the principal anisotropy basis:

$$\widehat{\mathbf{H}} = \widehat{H}^i_j \, \widehat{\mathbf{c}}_i \otimes \widehat{\mathbf{c}}_j, \qquad \widehat{\mathbf{H}} \in \widehat{G}_s. \tag{4.239}$$

then matrices of components of the tensors $\widehat{H}^i_j$ are orthogonal:

$$\widehat{H}^i_j \, \widehat{H}^j_k = \delta^i_k, \tag{4.240}$$

(in orthonormal bases the matrices $\widehat{H}^i_j$ and $\widehat{H}^j_i$ coincide).

For the orthotropy group $\widehat{G}_s$, there are eight matrices of the standard form $\widehat{H}^i_j$ (see [12]):

$$\widehat{H}^i_j \in \{E,\ C,\ R_\alpha,\ D_\alpha\}, \qquad \alpha = 1, 2, 3, \tag{4.241}$$

where

$$\begin{gathered}
E = \begin{pmatrix} 1 & 0 & 0 \\ 0 & 1 & 0 \\ 0 & 0 & 1 \end{pmatrix}, \quad C = \begin{pmatrix} -1 & 0 & 0 \\ 0 & -1 & 0 \\ 0 & 0 & -1 \end{pmatrix}, \\
R_1 = \begin{pmatrix} -1 & 0 & 0 \\ 0 & 1 & 0 \\ 0 & 0 & 1 \end{pmatrix}, \quad R_2 = \begin{pmatrix} 1 & 0 & 0 \\ 0 & -1 & 0 \\ 0 & 0 & 1 \end{pmatrix}, \quad R_3 = \begin{pmatrix} 1 & 0 & 0 \\ 0 & 1 & 0 \\ 0 & 0 & -1 \end{pmatrix}, \\
D_1 = \begin{pmatrix} 1 & 0 & 0 \\ 0 & -1 & 0 \\ 0 & 0 & -1 \end{pmatrix}, \quad D_2 = \begin{pmatrix} -1 & 0 & 0 \\ 0 & 1 & 0 \\ 0 & 0 & -1 \end{pmatrix}, \quad D_3 = \begin{pmatrix} -1 & 0 & 0 \\ 0 & -1 & 0 \\ 0 & 0 & 1 \end{pmatrix}.
\end{gathered} \tag{4.242}$$

Here $E$ is the unit matrix, $C$ is the matrix of centrally symmetric transformation, $R_\alpha$ are the matrices of mirror image relative to the planes being orthogonal to vectors $\widehat{\mathbf{c}}_\alpha$, and $D_\alpha$ are the matrices of rotation by angle $\pi$ about the axes oriented along vectors $\widehat{\mathbf{c}}_\alpha$.

For the triclinic group $\widehat{G}_s = E$, there are only two matrices $\widehat{H}^i_j$:

$$\widehat{H}^i_j \in \{E,\ C\}, \tag{4.243}$$

corresponding to the identical and centrally symmetric transformations.

For the transverse isotropy group $\widehat{G}_s = T_3$, there is one axis of rotation called the *axis of transverse isotropy*; in this case it is the axis oriented along the vector $\widehat{\mathbf{c}}_3$ and matrices $\widehat{H}^i_j$ have the form

$$\widehat{H}^i_j \in \{Q^\phi_3,\ D_\gamma Q^\phi_3,\ R_\alpha Q^\phi_3\},\quad \alpha = 1,2,3;\quad \gamma = 1,2. \tag{4.244}$$

Here $Q^\phi_3$ is the matrix of rotation by angle $\phi$ about the axis oriented along the vector $\widehat{\mathbf{c}}_3$:

$$Q^\phi_3 = \begin{pmatrix} \cos\phi & \sin\phi & 0 \\ -\sin\phi & \cos\phi & 0 \\ 0 & 0 & 1 \end{pmatrix},\qquad 0 \leqslant \phi < 2\pi. \tag{4.245}$$

For the groups $\widehat{G}_s = T_1$ and $\widehat{G}_s = T_2$, the axes of transverse isotropy are oriented along the vectors $\widehat{\mathbf{c}}_1$ and $\widehat{\mathbf{c}}_2$, respectively.

The principal basis of anisotropy $\widehat{\mathbf{c}}_i$ and the basis $\widehat{\mathbf{r}}_i$ of undistorted configuration $\widehat{\mathcal{K}}$ may be not coincident (see Fig. 4.4). If we introduce a tensor $\widehat{\mathbf{Q}}$ (in general, not orthogonal), which connects these two bases: $\widehat{\mathbf{c}}_i = \widehat{\mathbf{Q}} \cdot \widehat{\mathbf{r}}_i$, $\widehat{\mathbf{Q}} = \widehat{Q}^i_j\, \widehat{\mathbf{r}}_i \otimes \widehat{\mathbf{r}}_j$, then the tensors of $H$-transformations (4.239) from group $\widehat{G}_s$ can be represented in the basis $\widehat{\mathbf{r}}_i$ as well:

$$\widehat{\mathbf{H}} = \widehat{H}^i_j \widehat{\mathbf{c}}_i \otimes \widehat{\mathbf{c}}^j = \widehat{H}^i_j\, \widehat{Q}^k_i \widehat{Q}^j_l\, \widehat{\mathbf{r}}_k \otimes \widehat{\mathbf{r}}^l. \tag{4.246}$$

### 4.7.4 *H-indifference and H-invariance of Tensors Describing the Motion of a Solid*

Let us consider a solid continuum. A reference configuration $\overset{\circ}{\mathcal{K}}$ is assumed to be undistorted, then a symmetry group $\overset{\circ}{G}_s$ of the continuum corresponding to the configuration $\overset{\circ}{\mathcal{K}}$ is orthogonal; and $\mathbf{H}$-tensors of transformations involved in the group $\overset{\circ}{G}_s$ are orthogonal. According to formula (4.230), there exists an orthogonal tensor $\mathbf{O}_0$ connecting elements of the groups $\overset{\circ}{G}_s$ and $\widehat{G}_s$:

$$\widehat{\mathbf{H}} = \mathbf{O}_0 \cdot \mathbf{H} \cdot \mathbf{O}_0^{\mathrm{T}} \quad \forall \mathbf{H} \in \overset{\circ}{G}_s,\quad \widehat{\mathbf{H}} \in \widehat{G}_s. \tag{4.247}$$

For undistorted configuration $\overset{\circ}{\mathcal{K}}$, there is its principal anisotropy basis; the basis will be denoted below by $\widehat{\mathbf{c}}_i$. Although this basis may differ from the basis introduced in Sect. 4.7.3 for undistorted configuration $\widehat{\mathcal{K}}$, below they will not be considered simultaneously, so there will be no confusion for the notation $\widehat{\mathbf{c}}_i$. Notice that all constructions made in Sects. 4.7 and 4.8 can be written in a similar way for undistorted configuration $\widehat{\mathcal{K}}$. But since for formulating the problems in continuum mechanics we need the constitutive equations in a reference configuration $\overset{\circ}{\mathcal{K}}$ and the passage from

$\widehat{\mathcal{K}}$ to $\overset{\circ}{\mathcal{K}}$ is described by the formulae of tensor transformation from Sect. 4.6.2, so the final form of the constitutive equations can prove to be rather awkward. Therefore we will consider below only the case of undistorted reference configuration $\overset{\circ}{\mathcal{K}}$.

Since for undistorted configuration $\overset{\circ}{\mathcal{K}}$ all **H**-tensors are orthogonal, we introduce the notation

$$\mathbf{H}^{\mathrm{T}} = \overset{*}{\mathbf{Q}}, \tag{4.248}$$

where $\overset{*}{\mathbf{Q}}$ is the orthogonal tensor describing the transformation of a solid from $\overset{\circ}{\mathcal{K}}$ to $\overset{*}{\mathcal{K}}$.

**Definition 4.8.** A tensor $\boldsymbol{\Omega}$ specified in configuration $\overset{\circ}{\mathcal{K}}$ is called *$H$-indifferent relative to group $\overset{\circ}{G}_s \subset I$*, if for any orthogonal tensor of transformation $\overset{*}{\mathbf{Q}} : \overset{\circ}{\mathcal{K}} \to \overset{*}{\mathcal{K}}$ from this group, the tensor $\boldsymbol{\Omega}$ satisfies the condition

$$\overset{*}{\boldsymbol{\Omega}} = \overset{*}{\mathbf{Q}}{}^{\mathrm{T}} \cdot \boldsymbol{\Omega} \cdot \overset{*}{\mathbf{Q}}, \qquad \forall\, \overset{*}{\mathbf{Q}} \in \overset{\circ}{G}_s. \tag{4.249}$$

In other words, there exist tensors, which are $H$-indifferent relative only to certain subgroups of the orthogonal group; tensors, which are $H$-indifferent relative to any $H$-transformation, are called *absolutely $H$-indifferent.*

The definition (4.188) yields

$$\overset{*}{\mathbf{F}} = \mathbf{F} \cdot \overset{*}{\mathbf{Q}} = \mathbf{F} \cdot \mathbf{H}^{\mathrm{T}}, \tag{4.250}$$

i.e. the deformation gradient is neither $H$-indifferent nor $H$-invariant even relative to subgroups $\overset{\circ}{G}_s \subset I$.

The following combinations for **F** are neither $H$-indifferent nor $H$-invariant as well, because

$$\overset{*}{\mathbf{F}}{}^{\mathrm{T}} = \overset{*}{\mathbf{Q}}{}^{\mathrm{T}} \cdot \mathbf{F}^{\mathrm{T}}, \qquad \overset{*}{\mathbf{F}}{}^{-1} = \overset{*}{\mathbf{Q}} \cdot \mathbf{F}^{-1}, \qquad \overset{*}{\mathbf{F}}{}^{-1\mathrm{T}} = \mathbf{F}^{-1\mathrm{T}} \cdot \overset{*}{\mathbf{Q}}. \tag{4.251}$$

The metric tensor is both $H$-indifferent and $H$-invariant relative to any orthogonal group $\overset{\circ}{G}_s \subset I$, because the relation (4.199) yields

$$\overset{*}{\mathbf{E}} = \overset{*}{\mathbf{Q}}{}^{\mathrm{T}} \cdot \mathbf{E} \cdot \overset{*}{\mathbf{Q}} = \overset{*}{\mathbf{Q}}{}^{\mathrm{T}} \cdot \mathbf{Q} = \mathbf{E}. \tag{4.252}$$

Notice that for arbitrary $H$-transformations (which are not orthogonal), the metric tensor is only $H$-invariant (see (4.199)).

**Theorem 4.18.** *All the quasienergetic deformation tensors* $\overset{(n)}{\mathbf{A}}$ *and deformation measures* $\overset{(n)}{\mathbf{g}}$ *are* $H$*-invariant, and all the energetic deformation tensors* $\overset{(n)}{\mathbf{C}}$ *and deformation measures* $\overset{(n)}{\mathbf{G}}$ *are* $H$*-indifferent relative to any orthogonal group* $\overset{\circ}{G}_s \subset I$.

▼ Prove that the Cauchy–Green deformation tensor $\mathbf{C}$ is $H$-indifferent, and the Almansi deformation tensor $\mathbf{A}$ is $H$-invariant. The relations (4.200) and (4.251) give

$$\overset{\mathrm{V}}{\mathbf{C}}{}^{*} = \overset{*}{\mathbf{C}} = \frac{1}{2}(\overset{*}{\mathbf{F}}{}^{\mathrm{T}} \cdot \overset{*}{\mathbf{F}} - \mathbf{E}) = \frac{1}{2}(\overset{*}{\mathbf{Q}}{}^{\mathrm{T}} \cdot \mathbf{F}^{\mathrm{T}} \cdot \mathbf{F} \cdot \overset{*}{\mathbf{Q}} - \mathbf{E}) = \overset{*}{\mathbf{Q}}{}^{\mathrm{T}} \cdot \overset{\mathrm{V}}{\mathbf{C}} \cdot \overset{*}{\mathbf{Q}}, \tag{4.253}$$

and also

$$\overset{\mathrm{I}}{\mathbf{A}}{}^{*} = \overset{*}{\mathbf{A}} = \frac{1}{2}(\mathbf{E} - \overset{*}{\mathbf{F}}{}^{-1\mathrm{T}} \cdot \overset{*}{\mathbf{F}}{}^{-1}) = \frac{1}{2}(\mathbf{E} - \mathbf{F}^{-1\mathrm{T}} \cdot \overset{*}{\mathbf{Q}} \cdot \overset{*}{\mathbf{Q}}{}^{\mathrm{T}} \cdot \mathbf{F}^{-1}) = \overset{\mathrm{I}}{\mathbf{A}}. \tag{4.254}$$

In a similar way, we can prove that the tensor $\overset{\mathrm{I}}{\mathbf{C}} = \mathbf{\Lambda}$ is $H$-indifferent and the tensor $\overset{\mathrm{V}}{\mathbf{A}} = \mathbf{J}$ is $H$-invariant.

Show that the right stretch tensor $\mathbf{U}$ is $H$-indifferent, the left stretch tensor $\mathbf{V}$ is $H$-invariant, and the rotation tensor $\mathbf{O}$ accompanying deformation is neither $H$-indifferent nor $H$-invariant.

On substituting the polar decomposition of $\mathbf{F}$ into (4.250), we get

$$\overset{*}{\mathbf{F}} = \mathbf{V} \cdot (\mathbf{O} \cdot \overset{*}{\mathbf{Q}}) = \overset{*}{\mathbf{V}} \cdot \overset{*}{\mathbf{O}}. \tag{4.255}$$

Here, on the one hand, we have obtained the new orthogonal tensor $\mathbf{O} \cdot \overset{*}{\mathbf{Q}}$, and, on the other hand, we have found the polar decomposition for $\overset{*}{\mathbf{F}}$. Since the polar decomposition is unique, we get

$$\overset{*}{\mathbf{V}} = \mathbf{V} \quad \text{and} \quad \overset{*}{\mathbf{O}} = \mathbf{O} \cdot \overset{*}{\mathbf{Q}}. \tag{4.256}$$

Formula (2.137) for the polar decomposition yields

$$\overset{*}{\mathbf{U}} = \overset{*}{\mathbf{O}}{}^{\mathrm{T}} \cdot \overset{*}{\mathbf{F}} = \overset{*}{\mathbf{Q}}{}^{\mathrm{T}} \cdot \mathbf{O} \cdot \mathbf{F} \cdot \overset{*}{\mathbf{Q}} = \overset{*}{\mathbf{Q}}{}^{\mathrm{T}} \cdot \mathbf{U} \cdot \overset{*}{\mathbf{Q}}, \tag{4.257}$$

i.e. the tensor $\mathbf{U}$ is $H$-indifferent.

In a similar way, one can prove the theorem for the remaining tensors $\overset{(n)}{\mathbf{C}}$ and $\overset{(n)}{\mathbf{A}}$. ▲

As shown in Sect. 4.6.2, only the deformation measure $\overset{\mathrm{I}}{\mathbf{G}}$ is absolutely $H$-indifferent (i.e. $H$-indifferent relative to any but not only to orthogonal $H$-transformations).

**Theorem 4.19.** *All the energetic stress tensors* $\overset{(\mathrm{n})}{\mathbf{T}}$ *are* $H$*-indifferent, and all the quasienergetic stress tensors* $\overset{(\mathrm{n})}{\mathbf{S}}$ *are* $H$*-invariant relative to any orthogonal group* $\overset{\circ}{G}_s \subset I$.

▼ The energetic stress tensors $\overset{(\mathrm{n})}{\mathbf{T}}$ are $H$-indifferent, because

$$\overset{(\mathrm{I})*}{\mathbf{T}} = \overset{*}{\mathbf{F}}{}^{\mathrm{T}} \cdot \overset{*}{\mathbf{T}} \cdot \overset{*}{\mathbf{F}} = \overset{*}{\mathbf{Q}}{}^{\mathrm{T}} \cdot \mathbf{F}^{\mathrm{T}} \cdot \mathbf{T} \cdot \mathbf{F} \cdot \overset{*}{\mathbf{Q}} = \overset{*}{\mathbf{Q}} \cdot \overset{\mathrm{I}}{\mathbf{T}} \cdot \mathbf{Q},$$

$$\begin{aligned} \overset{(\mathrm{II})*}{\mathbf{T}} &= \frac{1}{2}\left(\overset{*}{\mathbf{F}}{}^{\mathrm{T}} \cdot \overset{*}{\mathbf{T}} \cdot \overset{*}{\mathbf{O}} + \overset{*}{\mathbf{O}}{}^{\mathrm{T}} \cdot \overset{*}{\mathbf{T}} \cdot \overset{*}{\mathbf{F}}\right) \\ &= \frac{1}{2}\left(\overset{*}{\mathbf{Q}}{}^{\mathrm{T}} \cdot \mathbf{F}^{\mathrm{T}} \cdot \mathbf{O} \cdot \overset{*}{\mathbf{Q}} + \overset{*}{\mathbf{Q}}{}^{\mathrm{T}} \cdot \mathbf{O} \cdot \mathbf{T} \cdot \mathbf{F} \cdot \overset{*}{\mathbf{Q}}\right) = \overset{*}{\mathbf{Q}} \cdot \overset{\mathrm{II}}{\mathbf{T}} \cdot \mathbf{Q} \quad \text{etc.} \end{aligned} \tag{4.258}$$

Tensors $\overset{(\mathrm{n})}{\mathbf{S}}$ are $H$-invariant, because they are formed by the contraction of $H$-invariant tensors $\mathbf{T}$ and $\mathbf{V}$. ▲

As shown in Sect. 4.6.2, the only tensor $\overset{\mathrm{V}}{\mathbf{T}}$ is absolutely $H$-indifferent, and the only tensor $\overset{\mathrm{III}}{\mathbf{S}} = \mathbf{T}$ is absolutely $H$-invariant.

The rotation stress tensor $\overset{\circ}{\mathbf{S}}$ is neither $H$-invariant nor $H$-indifferent, and it is transformed similarly to the rotation tensor $\mathbf{O}$:

$$\overset{\circ}{\mathbf{S}}{}^{*} = \frac{1}{2}(\overset{*}{\mathbf{V}} \cdot \overset{*}{\mathbf{T}} \cdot \overset{*}{\mathbf{V}}{}^{-1} - \overset{*}{\mathbf{V}}{}^{-1} \cdot \overset{*}{\mathbf{T}} \cdot \overset{*}{\mathbf{V}}) \cdot \overset{*}{\mathbf{O}} = \overset{\circ}{\mathbf{S}} \cdot \overset{*}{\mathbf{O}}. \tag{4.259}$$

### 4.7.5 *H-invariance of Rate Characteristics of a Solid*

The velocity $\mathbf{v}$ is absolutely $H$-invariant (for any transformations $\overset{\circ}{\mathcal{K}} \to \overset{*}{\mathcal{K}}$), because this vector is defined in actual configuration $\mathcal{K}$.

For the same reason, the velocity gradient

$$\mathbf{L} = (\nabla \otimes \mathbf{v})^{\mathrm{T}} = (\partial \mathbf{v}/\partial X^i) \otimes \mathbf{r}^i \tag{4.260}$$

is absolutely $H$-invariant as well. Therefore the tensors $\mathbf{D}$ and $\mathbf{W}$ are also absolutely $H$-invariant:

$$\overset{*}{\mathbf{D}} = \mathbf{D}, \qquad \overset{*}{\mathbf{W}} = \mathbf{W}. \tag{4.261}$$

**Theorem 4.20.** *The total derivative with respect to time* $\dot{\mathbf{B}}$*, Oldroyd* $\mathbf{B}^{\mathrm{Ol}}$*, Cotter–Rivlin* $\mathbf{B}^{\mathrm{CR}}$*, mixed co-rotational* $\mathbf{B}^{d}$ *and* $\mathbf{B}^{D}$ *relative to the eigenbasis of the left stretch tensor* $\mathbf{B}^{V}$*, Jaumann* $\mathbf{B}^{J}$ *derivatives are absolutely* $H$*-invariant, if they are applied to an absolutely* $H$*-invariant tensor* $\mathbf{B}$.

The same co-rotational derivatives, and also derivative in the eigenbasis $\mathbf{B}^{V}$ and the spin derivative $\mathbf{B}^{S}$ are $H$-invariant relative to group $\overset{\circ}{G}_s \subset I$, if they are applied to a tensor $\mathbf{B}$, which is $H$-invariant relative to the group $\overset{\circ}{G}_s \subset I$.

▼ Suppose that a tensor $\mathbf{B}$ is $H$-invariant: $\mathbf{B} = \overset{*}{\mathbf{B}}$. Then $H$-invariance of the derivatives $\dot{\mathbf{B}}$, $\mathbf{B}^{\mathrm{Ol}}$, $\mathbf{B}^{\mathrm{CR}}$, $\mathbf{B}^{d}$, $\mathbf{B}^{D}$ and $\mathbf{B}^{J}$ follows from $H$-invariance of the tensors $\mathbf{L}$ and $\mathbf{W}$, for example,

$$(\overset{*}{\mathbf{B}}{}^{J}) = \dot{\overset{*}{\mathbf{B}}} + \overset{*}{\mathbf{B}} \cdot \overset{*}{\mathbf{W}} - \overset{*}{\mathbf{W}} \cdot \overset{*}{\mathbf{B}} = \dot{\mathbf{B}} + \mathbf{B} \cdot \mathbf{W} - \mathbf{W} \cdot \mathbf{B} = \mathbf{B}^{J}. \tag{4.262}$$

If a tensor $\mathbf{B}$ is $H$-invariant only relative to a subgroup $\overset{\circ}{G}_s$ of the full orthogonal group, then from (4.262) it follows that $\overset{*}{\mathbf{B}}{}^{J}$ is $H$-invariant relative to the subgroup $\overset{\circ}{G}_s$ as well.

We should consider particularly only $\mathbf{B}^{V}$. Since the left stretch tensor $\mathbf{V}$ is $H$-invariant (see (4.256)) for all $\mathbf{H} = \overset{*}{\mathbf{Q}}{}^{\mathrm{T}} \in \overset{\circ}{G}_s$ (involved in orthogonal groups $\overset{\circ}{G}_s$), its eigenbasis $\mathbf{p}_i$ and the total derivative of the eigenbasis $\dot{\mathbf{p}}_i$ are $H$-invariant as well:

$$\overset{*}{\mathbf{p}}_i = \mathbf{p}_i, \qquad \dot{\overset{*}{\mathbf{p}}}_i = \dot{\mathbf{p}}_i. \tag{4.263}$$

Then the spin $\boldsymbol{\Omega}_V$ and rotation tensor $\mathbf{O}_V$ are $H$-invariant:

$$\overset{*}{\boldsymbol{\Omega}}_V = \dot{\overset{*}{\mathbf{p}}}_i \otimes \overset{*}{\mathbf{p}}{}^i = \dot{\mathbf{p}}_i \otimes \mathbf{p}^i = \boldsymbol{\Omega}_V, \tag{4.264}$$

$$\overset{*}{\mathbf{Q}}_V = \overset{*}{\mathbf{p}}_i \otimes \bar{\mathbf{e}}^i = \mathbf{p}_i \otimes \bar{\mathbf{e}}^i = \mathbf{O}_V. \tag{4.265}$$

From these relations we get that the tensor $\mathbf{B}^{V}$ is $H$-invariant:

$$(\overset{*}{\mathbf{B}})^V = \dot{\overset{*}{\mathbf{B}}} - \overset{*}{\boldsymbol{\Omega}}_V \cdot \overset{*}{\mathbf{B}} + \overset{*}{\mathbf{B}} \cdot \overset{*}{\boldsymbol{\Omega}}_V = \dot{\mathbf{B}} - \boldsymbol{\Omega} \cdot \mathbf{B} + \mathbf{B} \cdot \boldsymbol{\Omega}_V = \mathbf{B}^{V}. \tag{4.266}$$

Let us consider now the spin tensor $\boldsymbol{\Omega} = \overset{*}{\dot{\mathbf{O}}} \cdot \mathbf{O}^{\mathrm{T}}$. According to the property (4.256) of transformation of the rotation tensor $\mathbf{O}$, we get

$$\overset{*}{\boldsymbol{\Omega}} = \overset{*}{\mathbf{O}}{}^{\bullet} \cdot \overset{*}{\mathbf{O}}{}^{\mathrm{T}} = \dot{\mathbf{O}} \cdot \overset{*}{\mathbf{Q}} \cdot \overset{*}{\mathbf{Q}}{}^{\mathrm{T}} \cdot \mathbf{O}^{\mathrm{T}} = \dot{\mathbf{O}} \cdot \mathbf{O}^{\mathrm{T}} = \boldsymbol{\Omega},$$

i.e. the spin $\boldsymbol{\Omega}$ is $H$-invariant relative to any orthogonal subgroup $\overset{\circ}{G}_s \subset I$.

Then the spin derivative (2.371) in configuration $\overset{*}{\mathcal{K}}$ has the form

$$\overset{*}{\mathbf{B}}{}^{S} = \overset{*}{\mathbf{B}}{}^{\bullet} - \overset{*}{\boldsymbol{\Omega}} \cdot \overset{*}{\mathbf{B}} + \overset{*}{\mathbf{B}} \cdot \overset{*}{\boldsymbol{\Omega}} = \dot{\mathbf{B}} - \boldsymbol{\Omega} \cdot \mathbf{B} + \mathbf{B} \cdot \boldsymbol{\Omega} = \mathbf{B}^{S}, \tag{4.267}$$

if a tensor $\mathbf{B}$ is $H$-invariant relative to $\overset{\circ}{G}_s \subset I$.

Notice that the derivatives $\mathbf{B}^S$ and $\mathbf{B}^V$ are not absolutely $H$-invariant even if a tensor $\mathbf{B}$ is absolutely $H$-invariant. ▲

The theorem certainly remains valid for an arbitrary $H$-invariant vector $\mathbf{a}$ (see Exercise 4.7.3).

A corollary of this theorem is $H$-invariance of the total and co-rotational derivative of the quasienergetic deformation tensors:

$$(\overset{(n)}{\mathbf{A}}{}^{*})^{\bullet} = \overset{(n)}{\mathbf{A}}{}^{\bullet} \quad \text{and} \quad (\overset{(n)}{\mathbf{A}}{}^{*})^{V} = \overset{(n)}{\mathbf{A}}{}^{V}. \tag{4.268}$$

**Theorem 4.21.** *The total derivative $\dot{\mathbf{B}}$ of a $H$-indifferent tensor $\mathbf{B}$ with respect to time is $H$-indifferent as well.*

▼ Since a transformation tensor $\mathbf{H}$ is independent of time $t$, so, differentiating the indifference condition

$$\overset{*}{\mathbf{B}} = \mathbf{H} \cdot \mathbf{B} \cdot \mathbf{H}^{\mathrm{T}}, \tag{4.269}$$

we get

$$(\overset{*}{\mathbf{B}})^{\bullet} = \mathbf{H} \cdot \dot{\mathbf{B}} \cdot \mathbf{H}^{\mathrm{T}}. \quad ▲ \tag{4.270}$$

The remaining co-rotational derivatives, mentioned in Theorem 4.20, are neither $H$-indifferent nor $H$-invariant, if they are applied to a $H$-indifferent tensor.

A corollary of this theorem is absolute $H$-indifference of the total derivative of the energetic deformation measure $\overset{\mathrm{I}}{\mathbf{G}}$:

$$(\overset{\mathrm{I}}{\mathbf{G}}{}^{*})^{\bullet} = \mathbf{H} \cdot \overset{\mathrm{I}}{\mathbf{G}}{}^{\bullet} \cdot \mathbf{H}^{\mathrm{T}}, \tag{4.271}$$

and also $H$-indifference of the derivatives $\overset{(n)}{\mathbf{G}}{}^{\bullet}$ and $\overset{(n)}{\mathbf{C}}{}^{\bullet}$ relative to any orthogonal group $\overset{\circ}{G}_s$:

$$(\overset{(n)}{\mathbf{G}}{}^{*})^{\bullet} = \overset{*}{\mathbf{Q}}{}^{\mathrm{T}} \cdot \overset{(n)}{\mathbf{G}}{}^{\bullet} \cdot \overset{*}{\mathbf{Q}}, \quad \mathbf{H} = \overset{*}{\mathbf{Q}}{}^{\mathrm{T}}. \tag{4.271a}$$

**Theorem 4.22.** *The co-rotational derivative in the eigenbasis of the right stretch tensor $\mathbf{B}^U$ is $H$-indifferent if it is applied to a tensor $\mathbf{B}$, which is $H$-indifferent relative to group $\overset{\circ}{G}_s \subset I$.*

▼ At first, let us consider how the tensors $\mathbf{O}_U$ and $\boldsymbol{\Omega}_U$ are modified in $H$-transformations. According to the relation (2.276) between $\mathbf{O}$, $\mathbf{O}_U$ and $\mathbf{O}_V$, and also formula (4.256) for the tensor $\mathbf{O}$, we get

$$\overset{*}{\mathbf{O}} = \overset{*}{\mathbf{O}}_V \cdot \overset{*}{\mathbf{O}}{}^{\mathrm{T}}_U, \tag{4.272}$$

or

$$\mathbf{O} \cdot \overset{*}{\mathbf{Q}} = \mathbf{O}_V \cdot \overset{*}{\mathbf{O}}{}^{\mathrm{T}}_U, \quad \mathbf{O} = \mathbf{O}_V \cdot \overset{*}{\mathbf{O}}{}^{\mathrm{T}}_U \cdot \overset{*}{\mathbf{Q}}{}^{\mathrm{T}} = \mathbf{O}_V \cdot \mathbf{O}^{\mathrm{T}}_U. \tag{4.273}$$

Thus, the rotation tensor $\mathbf{O}_U$ is neither $H$-indifferent nor $H$-invariant:

$$\overset{*}{\mathbf{O}}_U = \overset{*}{\mathbf{Q}}{}^{\mathrm{T}} \cdot \mathbf{O}_U. \tag{4.274}$$

Using the definition (2.269) of the tensor $\boldsymbol{\Omega}_U$, we obtain

$$\overset{*}{\boldsymbol{\Omega}}_U = \overset{*}{\mathbf{O}}{}^{\bullet}_U \cdot \overset{*}{\mathbf{O}}{}^{\mathrm{T}}_U = \overset{*}{\mathbf{O}}{}^{\mathrm{T}} \cdot \dot{\mathbf{O}}_U \cdot \mathbf{O}^{\mathrm{T}}_U \cdot \overset{*}{\mathbf{Q}} = \overset{*}{\mathbf{Q}}{}^{\mathrm{T}} \cdot \boldsymbol{\Omega}_U \cdot \overset{*}{\mathbf{Q}}, \tag{4.275}$$

i.e. the spin of the right stretch tensor $\boldsymbol{\Omega}_U$ is $H$-indifferent.

Let a tensor $\mathbf{B}$ be $H$-indifferent. According to formula (2.359), we get

$$\begin{aligned}\overset{*}{\mathbf{B}}{}^{U} &= \overset{*}{\mathbf{B}}{}^{\bullet} - \overset{*}{\boldsymbol{\Omega}}_U \cdot \overset{*}{\mathbf{B}} + \overset{*}{\mathbf{B}} \cdot \overset{*}{\boldsymbol{\Omega}}_U = \overset{*}{\mathbf{Q}}{}^{\mathrm{T}} \cdot \mathbf{B} \cdot \overset{*}{\mathbf{Q}} - \overset{*}{\mathbf{Q}}{}^{\mathrm{T}} \cdot \boldsymbol{\Omega}_U \cdot \overset{*}{\mathbf{Q}} \cdot \overset{*}{\mathbf{Q}}{}^{\mathrm{T}} \cdot \mathbf{B} \cdot \overset{*}{\mathbf{Q}} \\ &+ \overset{*}{\mathbf{O}}{}^{\mathrm{T}} \cdot \mathbf{B} \cdot \overset{*}{\mathbf{Q}} \cdot \overset{*}{\mathbf{Q}}{}^{\mathrm{T}} \cdot \boldsymbol{\Omega}_U \cdot \overset{*}{\mathbf{Q}} = \overset{*}{\mathbf{Q}}{}^{\mathrm{T}} \cdot \mathbf{B}^{U} \cdot \overset{*}{\mathbf{Q}}. \ \blacktriangle\end{aligned} \tag{4.276}$$

This theorem is valid for any $H$-indifferent vector as well (see Exercise 4.7.4).

*Remark.* If a tensor $\mathbf{B}$ is $H$-invariant, then its derivative $\mathbf{B}^U$ is neither $H$-invariant nor $H$-indifferent. □

Tables 4.9 and 4.10 give the information on $H$-indifference and $H$-invariance of basic vectors and tensors.

## Exercises for 4.7

**4.7.1.** Show that if undistorted configuration $\widehat{\mathcal{K}}$ can be obtained from configuration $\overset{\circ}{\mathcal{K}}$ with the help of volume expansion transformation, i.e. the transformation tensor $\mathbf{S}$ (4.219) is spherical:

$$\mathbf{S} = k\mathbf{E}, \quad k \neq 0,$$

then $\widehat{G}_s = \overset{\circ}{G}_s$.

**Table 4.9** $H$-indifference and $H$-invariance of basic vectors and tensors

| Tensors and vectors | Absolute $H$-indifference | $H$-pseudo-indifference | $H$-indifference relative to orthogonal subgroup $G_s \subset I$ | Absolute $H$-invariance | $H$-invariance relative to orthogonal subgroup $G_s \subset I$ |
|---|---|---|---|---|---|
| $\mathbf{r}_i, \mathbf{r}^i$ | − | | − | + | + |
| $\overset{\circ}{\mathbf{r}}_i, \overset{\circ}{\mathbf{r}}{}^i$ | + | | + | − | − |
| $\mathbf{E}$ | − | | + | + | + |
| $\mathbf{F}$ | − | − | − | − | − |
| $\overset{(n)}{\mathbf{C}}$, $n = \mathrm{I}, \ldots, \mathrm{V}$ | − | − | + | − | − |
| $\overset{\mathrm{I}}{\mathbf{G}}$ | + | − | + | − | − |
| $\overset{\mathrm{II}}{\mathbf{G}}, \overset{\mathrm{IV}}{\mathbf{G}}$ | − | − | + | − | − |
| $\overset{\mathrm{V}}{\mathbf{G}}$ | − | + | + | − | − |
| $\overset{(n)}{\mathbf{A}}, \overset{(n)}{\mathbf{g}}$, $n = \mathrm{I}, \ldots, \mathrm{V}$ | − | − | − | − | + |
| $\mathbf{T} = \overset{\mathrm{III}}{\mathbf{S}}$ | − | − | − | + | + |
| $\mathbf{P}$ | − | − | − | − | − |
| $\overset{\mathrm{I}}{\mathbf{T}}$ | − | + | + | − | − |
| $\overset{\mathrm{II}}{\mathbf{T}}, \overset{\mathrm{III}}{\mathbf{T}}, \overset{\mathrm{IV}}{\mathbf{T}}$ | − | − | + | − | − |
| $\overset{\mathrm{V}}{\mathbf{T}}$ | + | − | + | − | − |
| $\overset{(n)}{\mathbf{S}}$, $n = \mathrm{I}, \mathrm{II}, \mathrm{IV}, \mathrm{V}$ | − | − | − | − | + |
| $\overset{\circ}{\mathbf{S}}$ | − | − | − | − | − |
| $\boldsymbol{\Omega}_V, \boldsymbol{\Omega} = \dot{\mathbf{O}} \cdot \mathbf{O}^{\mathrm{T}}$ | − | − | − | − | + |
| $\boldsymbol{\Omega}_U$ | − | − | + | − | − |
| $\mathbf{v}, \mathbf{L}, \mathbf{D}, \mathbf{W}$ | − | − | − | + | + |

**4.7.2.** Show that if $H$-transformations $H : \overset{\circ}{\mathcal{K}} \to \overset{*}{\mathcal{K}}$ are not orthogonal, then the tensor $\mathbf{V}$ is not $H$-invariant, and the tensor $\mathbf{U}$ is not $H$-indifferent.

**4.7.3.** Prove Theorem 4.20 for a $H$-invariant vector $\mathbf{a}$.

**4.7.4.** Prove Theorems 4.21 and 4.22 for a $H$-indifferent vector $\mathbf{a}$.

**Table 4.10** $H$-indifference and $H$-invariance of basic vectors and tensors

| Tensors and vectors | Absolute $H$-indifference | $H$-pseudo-indifference | $H$-indifference relative to orthogonal subgroup $G_s \subset I$ | Absolute $H$-invariance | $H$-invariance relative to orthogonal subgroup $G_s \subset I$ |
|---|---|---|---|---|---|
| $\mathbf{B}^h$, $h = \{\cdot, \text{OI}, \text{CR}, d, D, J, S\}$ if $\mathbf{B}$ is absolutely $H$-invariant | − | − | − | + | + |
| $\mathbf{B}^h$, $h = \{\cdot, \text{OI}, \text{CR}, d, D, J, S, V\}$, if $\mathbf{B}$ is $H$-invariant relative to $\overset{\circ}{G}_s \subset I$ | − | − | − | − | + |
| $\dot{\mathbf{B}}$, if $\mathbf{B}$ is absolutely $H$-indifferent | + | − | + | − | − |
| $\mathbf{B}^h$, $h = \{\cdot, U\}$ if $\mathbf{B}$ is $H$-indifferent relative to $\overset{\circ}{G}_s \subset I$ | − | − | + | − | − |

## 4.8 Corollaries of the Principle of Material Symmetry and Constitutive Equations for Ideal Continua

Let us apply now the principle of material symmetry to ideal solids and consider corollaries of the principle. Ideal solids are usually called *elastic continua.*

### *4.8.1 Corollary of the Principle of Material Symmetry for Models $A_n$ of Ideal (Elastic) Solids*

Let there be an anisotropic ideal solid of class $\overset{\circ}{G}_s$ with undistorted reference configuration $\overset{\circ}{\mathcal{K}}$. Suppose that constitutive equations of the anisotropic ideal solid correspond to models $A_n$ (4.165), (4.168). Apply the principle of material symmetry in the form (4.181) to these models. Then we obtain that for any $\overset{*}{\mathbf{Q}}$-transformation from $\overset{\circ}{\mathcal{K}}$ to $\overset{*}{\mathcal{K}}$, the relations

$$\psi = \psi(\overset{(n)}{\mathbf{C}}{}^*,\ \theta), \tag{4.277}$$

$$\overset{(n)}{\mathbf{T}}{}^* = \rho(\partial\psi/\partial\overset{(n)}{\mathbf{C}}{}^*) \equiv \mathcal{F}(\overset{(n)}{\mathbf{C}}{}^*,\ \theta), \quad \eta = -\partial\psi/\partial\theta \tag{4.278}$$

hold at the motion $\overset{*}{\mathcal{K}} \to \mathcal{K}$, if for $\overset{\circ}{\mathcal{K}} \to \mathcal{K}$ Eqs. (4.165), (4.168) are satisfied. Here we have taken into account that all scalar functions are $H$-invariants, i.e. $\overset{*}{\theta} = \theta$, $\overset{*}{\psi} = \psi$, $\overset{*}{\eta} = \eta$.

Since the tensors $\overset{(n)}{\mathbf{T}}$ and $\overset{(n)}{\mathbf{C}}$ are $H$-indifferent relative to group $\overset{\circ}{G}_s$, from (4.278) we get the following assertion: the principle of material symmetry holds if the constitutive equations of ideal models $A_n$ of anisotropic continua satisfy the conditions

$$\overset{*}{\mathbf{Q}}^{\mathrm{T}} \cdot \mathcal{F}(\overset{(n)}{\mathbf{C}}, \theta) \cdot \overset{*}{\mathbf{Q}} = \mathcal{F}(\overset{*}{\mathbf{Q}}^{\mathrm{T}} \cdot \overset{(n)}{\mathbf{C}} \cdot \overset{*}{\mathbf{Q}}, \theta), \quad \forall \overset{*}{\mathbf{Q}} \in \overset{\circ}{G}_s, \tag{4.279a}$$

$$\frac{\partial \psi}{\partial \theta}(\overset{(n)}{\mathbf{C}}, \theta) = \frac{\partial \psi}{\partial \theta}(\overset{*}{\mathbf{Q}}^{\mathrm{T}} \cdot \overset{(n)}{\mathbf{C}} \cdot \overset{*}{\mathbf{Q}}, \theta), \tag{4.279b}$$

$$\psi = \Psi(\overset{(n)}{\mathbf{C}}, \theta) = \Psi(\overset{*}{\mathbf{Q}}^{\mathrm{T}} \cdot \overset{(n)}{\mathbf{C}} \cdot \overset{*}{\mathbf{Q}}, \theta). \tag{4.279c}$$

**Theorem 4.23.** *The principle of material symmetry holds for models $A_n$ of ideal solids if and only if the condition (4.279c) for the potential $\psi$ is satisfied.*

▼ The condition (4.279c) is necessary, because if the principle of material symmetry holds, then all the three conditions (4.279) are satisfied.

Prove that the condition (4.279c) is sufficient. Let only the condition (4.279c) be satisfied. On substituting the expression for the potential $\psi$ into (4.279a) and differentiating the scalar with respect to the composite tensor argument (see [12, 14]), we get

$$\rho \overset{*}{\mathbf{Q}}^{\mathrm{T}} \cdot \frac{\partial}{\partial \overset{(n)}{\mathbf{C}}} \psi(\overset{*}{\mathbf{Q}}^{\mathrm{T}} \cdot \overset{(n)}{\mathbf{C}} \cdot \overset{*}{\mathbf{Q}}, \theta) \cdot \overset{*}{\mathbf{Q}}$$
$$= \rho \overset{*}{\mathbf{Q}}^{\mathrm{T}} \cdot \overset{*}{\mathbf{Q}} \cdot \frac{\partial \psi}{\partial \overset{(n)*}{\mathbf{C}}} (\overset{*}{\mathbf{Q}}^{\mathrm{T}} \cdot \overset{(n)}{\mathbf{C}} \cdot \overset{*}{\mathbf{Q}}, \theta) \cdot \overset{*}{\mathbf{Q}}^{\mathrm{T}} \cdot \overset{*}{\mathbf{Q}} = \rho \frac{\partial \psi}{\partial \overset{(n)*}{\mathbf{C}}}. \tag{4.280}$$

In a similar way, one can prove that the condition (4.279b) is satisfied as well. ▲

## 4.8.2 Scalar Indifferent Functions of Tensor Argument

Consider in detail the condition (4.279c) imposed on the function $\psi(\overset{(n)}{\mathbf{C}}, \theta)$ for satisfying the principle of material symmetry.

**Definition 4.9.** A scalar function $\varphi(\boldsymbol{\Omega})$ of a tensor argument is called *indifferent relative to symmetry group* $\overset{\circ}{G}_s \subset I$, if for each tensor $\overset{*}{\mathbf{Q}} \in \overset{\circ}{G}_s$ the following condition is satisfied:

$$\varphi(\boldsymbol{\Omega}) = \varphi(\overset{*}{\mathbf{Q}}^{\mathrm{T}} \cdot \boldsymbol{\Omega} \cdot \overset{*}{\mathbf{Q}}). \tag{4.281}$$

With the help of the concept of an indifferent function, the results of Sect. 4.8.1 can be formulated as the following theorem.

**Theorem 4.24.** *For the model $A_n$ of an ideal solid continuum with symmetry group $\overset{\circ}{G}_s$ and undistorted configuration $\overset{\circ}{\mathcal{K}}$, the principle of material symmetry holds if and only if the free energy $\psi(\overset{(n)}{\mathbf{C}}, \theta)$ is an indifferent scalar function, i.e. it satisfies the condition (4.279c).*

Consider a solid with undistorted configuration $\overset{\circ}{\mathcal{K}}$ and symmetry group $\overset{\circ}{G}_s$. According to Sect. 4.7.4, the principal anisotropy basis in the configuration $\overset{\circ}{\mathcal{K}}$ is denoted by $\widehat{\mathbf{c}}_i$, and an arbitrary transformation tensor $\overset{*}{\mathbf{Q}} \in \overset{\circ}{G}_s$ can be resolved for this basis (see (4.239)):

$$\overset{*}{\mathbf{Q}} = \mathbf{H}^{\mathrm{T}} = \widehat{H}^i_j \widehat{\mathbf{c}}^j \otimes \widehat{\mathbf{c}}_i, \tag{4.282}$$

where $\widehat{H}^i_j$ are components of the transformation tensor $\mathbf{H} = \overset{*}{\mathbf{O}}^{\mathrm{T}}$ (for example, of the forms (4.241) and (4.244)). Let there be a scalar function $\varphi(\mathbf{\Omega})$, which is indifferent relative to the same group $\overset{\circ}{G}_s$, then the function can be written in the *component form*

$$\varphi = \varphi(\mathbf{\Omega}) = \varphi(\widehat{\Omega}^{ij}), \tag{4.283}$$

i.e. in terms of components of the tensor

$$\mathbf{\Omega} = \widehat{\Omega}^{ij} \widehat{\mathbf{c}}_i \otimes \widehat{\mathbf{c}}_j. \tag{4.284}$$

The condition (4.281) of indifference of a scalar function $\varphi(\mathbf{\Omega})$ in this case takes the form

$$\varphi(\widehat{\Omega}^{ij}) = \varphi(\Omega'^{ij}), \quad \Omega'^{ij} = \widehat{H}^i_k \widehat{H}^j_l \widehat{\Omega}^{kl} \quad \forall\, H^i_j \in \widehat{G}_s. \tag{4.285}$$

In other words, the form of a scalar indifferent function remains unchanged at replacing $\widehat{\Omega}^{ij}$ by $\Omega'^{ij}$ for each matrix $H^{ij}$ of the group $\overset{\circ}{G}_s$.

As an example, consider a scalar linear function of a symmetric tensor argument

$$\varphi(\mathbf{\Omega}) = \mathbf{M} \cdot\cdot\, \mathbf{\Omega} + \varphi_0, \tag{4.286}$$

where $\mathbf{M}$ is a symmetric second-order tensor, which is independent of $\mathbf{\Omega}$ and defining the function $\varphi(\mathbf{\Omega})$, and $\varphi_0 = \text{const}$. If the function (4.286) is indifferent, then the condition (4.281) for the function has the form

$$\mathbf{M} \cdot\cdot\, \mathbf{\Omega} = \mathbf{M} \cdot\cdot\, (\overset{*}{\mathbf{Q}}^{\mathrm{T}} \cdot \mathbf{\Omega} \cdot \overset{*}{\mathbf{Q}}). \tag{4.287}$$

Using the rules (4.4) and (4.5) of rearrangement of multipliers in the scalar product and taking the symmetry of tensors $\mathbf{\Omega}$ and $\mathbf{M}$ into account, we get

$$(\mathbf{M} - \overset{*}{\mathbf{Q}}^{\mathrm{T}} \cdot \mathbf{M} \cdot \overset{*}{\mathbf{Q}}) \cdot\cdot\, \boldsymbol{\Omega} = 0. \tag{4.288}$$

Since this relation must be satisfied for any $\boldsymbol{\Omega}$, so the relation holds if and only if the tensor $\mathbf{M}$ satisfies the condition

$$\mathbf{M} = \overset{*}{\mathbf{Q}}^{\mathrm{T}} \cdot \mathbf{M} \cdot \overset{*}{\mathbf{Q}} \quad \forall\, \overset{*}{\mathbf{Q}} \in \overset{\circ}{G}_s. \tag{4.289}$$

**Definition 4.10.** The second-order tensor $\mathbf{M}$ satisfying the condition (4.289) for each orthogonal tensor $\overset{*}{\mathbf{O}} \in \overset{\circ}{G}_s$ is called *indifferent relative to the group* $\overset{\circ}{G}_s$.

Comparing formula (4.289) with (4.249) and (4.191), we obtain that the tensor $\mathbf{M}$ is indifferent relative to group $\overset{\circ}{G}_s$ if and only if the tensor is both $H$-indifferent and $H$-invariant relative the group $\overset{\circ}{G}_s$:

$$\mathbf{M} = \overset{*}{\mathbf{M}}, \qquad \overset{*}{\mathbf{M}} = \overset{*}{\mathbf{Q}}^{\mathrm{T}} \cdot \mathbf{M} \cdot \mathbf{Q} \quad \forall \overset{*}{\mathbf{Q}} \in \overset{\circ}{G}_s. \tag{4.290}$$

On comparing this condition with (4.249), we get that the linear scalar function (4.286) is indifferent relative to group $\overset{\circ}{G}_s$ if and only if the tensor $\mathbf{M}$ defining this function is indifferent relative to the same group $\overset{\circ}{G}_s$.

This example exhibits that indifferent scalar functions (and as will be shown below, not only scalar functions) are constructed with the help of indifferent tensors.

### *4.8.3 Producing Tensors of Groups*

Let us consider now the set $S_3^{(2)}$ of all symmetric second-order tensors $\mathbf{M}$, which are indifferent relative to some group $\overset{\circ}{G}_s \subset I$.

Let $\widehat{\mathbf{c}}_i$ be the principal anisotropy basis, in which all orthogonal tensors $\overset{*}{\mathbf{Q}} \in \overset{\circ}{G}_s$ has the form (4.282), where $\widehat{H}^i_j$ are orthogonal matrices of the standard form (4.241), (4.244), etc.

The set $S_3^{(2)}$ originates a finite-dimensional space of dimension $k$ (see [12]), where we can choose a basis being a collection of symmetric tensors $\mathbf{O}_{s(\gamma)}$ ($\gamma = 1, \ldots, k$) having the following properties:

- Each of the tensors $\mathbf{O}_{s(\gamma)}$ is indifferent relative to the group $\overset{\circ}{G}_s$.
- Any other tensor $\mathbf{M} \in S_3^{(2)}$ indifferent relative to the same group is a linear combination of the tensors $\mathbf{O}_{s(\gamma)}$:

$$\mathbf{M} = \sum_{\gamma=1}^{k} a_\gamma \mathbf{O}_{s(\gamma)}, \tag{4.291}$$

where not all $a_\gamma$ are zero.

**Table 4.11** The producing tensors of groups

| Symmetry group | Producing tensors $\mathbf{O}_{s(\gamma)}$, $\gamma = 1,\dots,k$ | Number $k$ |
|---|---|---|
| Orthotropy $O$ | $\widehat{\mathbf{c}}^2_\gamma$, $\gamma = 1, 2, 3$ | 3 |
| Transverse isotropy $T_3$ | $\widehat{\mathbf{c}}^2_3$, $\mathbf{E}$ | 2 |
| Isotropy $I$ | $\mathbf{E}$ | 1 |

The tensors $\mathbf{O}_{s(\gamma)}$ are called the *producing tensors of the group* $\overset{\circ}{G}_s$.

For the groups of transverse isotropy $T_3$, orthotropy $O$ (see Sect. 4.7.3) and also for the full orthogonal group $I$, the tensors $\mathbf{O}_{s(\gamma)}$ are given in Table 4.11 (see [12]). Here we have introduced the notation

$$\widehat{\mathbf{c}}^2_\gamma = \widehat{\mathbf{c}}_\gamma \otimes \widehat{\mathbf{c}}_\gamma. \tag{4.292}$$

Then the representation (4.291) for the groups $O$, $T_3$ and $I$ takes the form

$$\begin{aligned} \mathbf{M} &= \sum_{\gamma=1}^{3} a_\gamma \widehat{\mathbf{c}}^2_\gamma \quad \text{for } \overset{\circ}{G}_s = O, \\ \mathbf{M} &= a_1 \widehat{\mathbf{c}}^2_3 + a_2 \mathbf{E} \quad \text{for } \overset{\circ}{G}_s = T_3, \\ \mathbf{M} &= a\,\mathbf{E} \quad \text{for } \overset{\circ}{G}_s = I, \end{aligned} \tag{4.293}$$

i.e. all second-order tensors indifferent relative to the full orthogonal group $I$ are proportional to the metric tensor.

### 4.8.4 Scalar Invariants of a Second-Order Tensor

Let us show how indifferent scalar functions (not only linear) are constructed with the help of indifferent tensors. At first, we give two definitions.

**Definition 4.11.** Let there be a second-order tensor $\mathbf{\Omega}$ having in the orthonormal basis $\widehat{\mathbf{c}}_i$ the following form: $\mathbf{\Omega} = \widehat{\Omega}^{ij} \widehat{\mathbf{c}}_i \otimes \widehat{\mathbf{c}}_j$, and also some symmetry group $\overset{\circ}{G}_s \subset I$, each transformation tensor of which $\overset{*}{\mathbf{Q}} \in \widehat{G}_s$ has the form (4.282), then any scalar function of the tensor $\mathbf{\Omega}$

$$I^{(s)}(\mathbf{\Omega}) = I^{(s)}(\widehat{\Omega}^{ij})$$

being indifferent relative to the group $\overset{\circ}{G}_s$ is called a *scalar invariant* of the tensor $\mathbf{\Omega}$ relative to the group $\overset{\circ}{G}_s$.

**Definition 4.12.** The system of $r$ scalar invariants $I_\gamma^{(s)}(\boldsymbol{\Omega})$ (where $\gamma = 1, \ldots, r$) of tensor $\boldsymbol{\Omega}$ relative to transformation group $\overset{\circ}{G}_s$ is called the *functional basis* of independent invariants of the tensor relative to the group $\overset{\circ}{G}_s$, if

- the system is functionally independent;
- any other scalar invariant $I^{(s)}(\boldsymbol{\Omega})$ (not involved in the system) of the tensor $\boldsymbol{\Omega}$ relative to the same group $\overset{\circ}{G}_s$ can be represented as a function of the invariants $I_\gamma^{(s)}$:
$$I^{(s)}(\boldsymbol{\Omega}) = f(I_1^{(s)}(\boldsymbol{\Omega}), \ldots, I_r^{(s)}(\boldsymbol{\Omega})), \tag{4.294}$$
where $f(I_1, \ldots, I_r)$ is a scalar function of $r$ variables, i.e. $f : \mathbb{R}^r \to \mathbb{R}^1$.

Below, unless otherwise stated, the function $f$ is assumed to be continuously differentiable in $\mathbb{R}^r$ and the invariants $I^{(s)}(\widehat{\Omega}^{ij})$ – continuously differentiable functions $I^{(s)} : \mathbb{R}^6 \to \mathbb{R}^1$ in $\mathbb{R}^6$ (the number of independent components of a symmetric second-order tensor $\boldsymbol{\Omega}$ is equal to six). The superscript $s$ of the invariants means that for different groups $\overset{\circ}{G}_s$ invariants of the same tensor are different. The definitions of functional dependence and independence of the invariant system are given in [12].

In [12] one can find a proof of the following theorem on the number of elements in the functional basis.

**Theorem 4.24a.** *For any symmetric second-order tensor $\boldsymbol{\Omega}$, its functional basis of independent invariants relative to group $\overset{\circ}{G}_s \subset I$ consists of $r$ elements, where*

- $r = 3$ *for the isotropy group* $\overset{\circ}{G}_s = I$;
- $r = 5$ *for the transverse isotropy group* $\overset{\circ}{G}_s = T_3$;
- $r = 6$ *for the orthotropy group* $\overset{\circ}{G}_s = O$ *and the triclinic group* $\overset{\circ}{G}_s = E$.

As the functional basis, we can choose different systems of invariants. In continuum mechanics, one often uses polynomial invariants $I_\gamma^{(s)}(\boldsymbol{\Omega})$ formed by contraction of the producing tensors $\mathbf{O}_{s(\gamma)}$ of the corresponding group $\overset{\circ}{G}_s$ with the tensor $\boldsymbol{\Omega}$ itself (*linear invariants*), or with its tensor powers $\boldsymbol{\Omega} \otimes \boldsymbol{\Omega}$, $\boldsymbol{\Omega} \otimes \boldsymbol{\Omega} \otimes \boldsymbol{\Omega}$ (*quadratic* and *cubic invariants*, respectively). Functional bases constructed by this method for four considered symmetry groups are shown below.

*The triclinic group* $\overset{\circ}{G}_s = E$:

$$\begin{gathered} I_\gamma^{(E)} = \boldsymbol{\Omega} \cdot\cdot\, \widehat{\mathbf{c}}_\gamma^2, \quad I_{3+\gamma}^{(E)} = \frac{1}{2}\boldsymbol{\Omega} \cdot\cdot\, (\widehat{\mathbf{c}}_\alpha \otimes \widehat{\mathbf{c}}_\beta + \widehat{\mathbf{c}}_\beta \otimes \widehat{\mathbf{c}}_\alpha), \\ \alpha, \beta, \gamma = 1, 2, 3, \quad \alpha \neq \beta \neq \gamma \neq \alpha, \end{gathered} \tag{4.295}$$

or

$$I_\gamma^{(E)} = \{\widehat{\Omega}_{11}, \widehat{\Omega}_{22}, \widehat{\Omega}_{33}, \widehat{\Omega}_{23}, \widehat{\Omega}_{13}, \widehat{\Omega}_{12}\}.$$

*The orthotropy group* $\overset{\circ}{G}_s = O$:

$$I_\alpha^{(O)} = \boldsymbol{\Omega}\cdot\cdot\widehat{\mathbf{c}}_\gamma^2,\quad \gamma=1,2,3,\quad I_4^{(O)} = (\widehat{\mathbf{c}}_2\cdot\boldsymbol{\Omega})\cdot\cdot(\widehat{\mathbf{c}}_3\cdot\boldsymbol{\Omega}),$$
$$I_5^{(O)} = (\widehat{\mathbf{c}}_1\cdot\boldsymbol{\Omega})\cdot\cdot(\widehat{\mathbf{c}}_3\cdot\boldsymbol{\Omega}),\quad I_6^{(O)} = ((\widehat{\mathbf{c}}_1\cdot\boldsymbol{\Omega})\cdot(\widehat{\mathbf{c}}_2\cdot\boldsymbol{\Omega}))\cdot\cdot(\widehat{\mathbf{c}}_3\cdot\boldsymbol{\Omega}), \tag{4.296}$$

or

$$I_\gamma^{(O)} = \{\widehat{\Omega}_{11},\ \widehat{\Omega}_{22},\ \widehat{\Omega}_{33},\ \widehat{\Omega}_{23}^2,\ \widehat{\Omega}_{13}^2,\ \widehat{\Omega}_{12}\widehat{\Omega}_{13}\widehat{\Omega}_{23}\}. \tag{4.296a}$$

*The transverse isotropy group* $\overset{\circ}{G}_s = T_3$:

$$I_1^{(3)} = (\mathbf{E}-\widehat{\mathbf{c}}_3^2)\cdot\cdot\boldsymbol{\Omega},\quad I_2^{(3)} = \boldsymbol{\Omega}\cdot\cdot\widehat{\mathbf{c}}_3^2,$$
$$I_3^{(3)} = ((\mathbf{E}-\widehat{\mathbf{c}}_3^2)\cdot\boldsymbol{\Omega})\cdot\cdot(\widehat{\mathbf{c}}_3^2\cdot\boldsymbol{\Omega}), \tag{4.297}$$
$$I_4^{(3)} = \boldsymbol{\Omega}^2\cdot\cdot\mathbf{E} - I_2^{(3)2} - 2I_3^{(3)},\quad I_5^{(3)} = \det\boldsymbol{\Omega},$$

or

$$I_\gamma^{(3)} = \{\widehat{\Omega}_{11}+\widehat{\Omega}_{22},\ \widehat{\Omega}_{33},\ \widehat{\Omega}_{13}^2+\widehat{\Omega}_{23}^2,\ \widehat{\Omega}_{11}^2+\widehat{\Omega}_{22}^2+2\widehat{\Omega}_{12}^2,\ \det\boldsymbol{\Omega}\}. \tag{4.297a}$$

*The isotropy group* $\overset{\circ}{G}_s = I$:

$$I_\gamma^{(I)} = I_\gamma(\boldsymbol{\Omega}),\qquad \gamma=1,2,3. \tag{4.298}$$

Here $I_\gamma(\boldsymbol{\Omega})$ are the *principal invariants of tensor* $\boldsymbol{\Omega}$, which are defined by formulae (4.48):

$$I_1(\boldsymbol{\Omega}) = \boldsymbol{\Omega}\cdot\cdot\mathbf{E},\quad I_2(\boldsymbol{\Omega}) = \frac{1}{2}(I_1^2(\boldsymbol{\Omega}) - I_1(\boldsymbol{\Omega}^2)),\quad I_3(\boldsymbol{\Omega}) = \det\boldsymbol{\Omega}. \tag{4.299}$$

Let us prove that the principal invariants of a tensor are indifferent, for example, for $I_1$:

$$I_1(\boldsymbol{\Omega}) = \mathbf{E}\cdot\cdot\boldsymbol{\Omega} = (\overset{*}{\mathbf{Q}}\cdot\mathbf{E}\cdot\overset{*}{\mathbf{Q}}{}^{\mathrm{T}})\cdot\cdot\boldsymbol{\Omega} = \mathbf{E}\cdot\cdot(\overset{*}{\mathbf{Q}}{}^{\mathrm{T}}\cdot\boldsymbol{\Omega}\cdot\overset{*}{\mathbf{Q}})$$
$$= I_1(\overset{*}{\mathbf{Q}}{}^{\mathrm{T}}\cdot\boldsymbol{\Omega}\cdot\overset{*}{\mathbf{Q}})\quad \forall\,\widehat{\mathbf{Q}}\in\overset{\circ}{G}_s = I. \tag{4.300}$$

In a similar way, we can prove that $I_1(\boldsymbol{\Omega}^2)$ satisfies the condition (4.281); and hence $I_2(\boldsymbol{\Omega})$ satisfies the condition as well. For $I_3(\boldsymbol{\Omega})$, we have

$$I_3(\overset{*}{\mathbf{Q}}{}^{\mathrm{T}}\cdot\boldsymbol{\Omega}\cdot\overset{*}{\mathbf{Q}}) = \det(\overset{*}{\mathbf{Q}}{}^{\mathrm{T}}\cdot\boldsymbol{\Omega}\cdot\overset{*}{\mathbf{Q}}) = \det\overset{*}{\mathbf{Q}}{}^{\mathrm{T}}\cdot\det\boldsymbol{\Omega}\cdot\det\overset{*}{\mathbf{Q}} = \det\boldsymbol{\Omega} = I_3(\boldsymbol{\Omega}), \tag{4.301}$$

because $\overset{*}{\mathbf{Q}}$ is an orthogonal tensor of the group $\overset{\circ}{G}_s = I$.

One can find a proof of indifference of the invariants $I_\alpha^{(O)}$ and $I_\alpha^{(3)}$ in [12].

In braces, formulae (4.295)–(4.297) show *component forms* of the invariants $I_\gamma^{(s)}$ in the basis $\widehat{\mathbf{c}}_i$.

Notice once more that functional bases of invariants may be different; their choice, in general, is defined by a considered problem. So for the triclinic group $\overset{\circ}{G}_s = E$, besides (4.295), one can use another basis

$$\widetilde{I}_\gamma^{(E)} = I_\gamma(\boldsymbol{\Omega}), \quad \gamma = 1,\ 2,\ 3, \quad \widetilde{I}_{3+\gamma}^{(E)} = I_{3+\gamma}^{(E)}. \tag{4.302}$$

Here we have taken into account that the principal invariants $I_\gamma(\boldsymbol{\Omega})$ are invariants relative to any subgroup $\overset{\circ}{G}_s \subset I$.

## 4.8.5 *Representation of a Scalar Indifferent Function in Terms of Invariants*

According to Theorem 4.24a, any scalar function of tensor argument $\varphi(\boldsymbol{\Omega})$, which is indifferent relative to group $\overset{\circ}{G}_s$, can be represented as a function of invariants $I_\alpha^{(s)}(\boldsymbol{\Omega})$ of the tensor, i.e.

$$\varphi(\boldsymbol{\Omega}) = \varphi(I_1^{(s)}(\boldsymbol{\Omega}), \ldots, I_r^{(s)}(\boldsymbol{\Omega})). \tag{4.303}$$

Let us return now to models $A_n$ of solids and apply this result to the free energy $\psi(\overset{(n)}{\mathbf{C}}, \theta)$. According to the principle of material symmetry, this function must be indifferent relative to some group $\overset{\circ}{G}_s \subset I$; then the free energy can be represented as a function of the functional basis of independent invariants $I_\alpha^{(s)}(\overset{(n)}{\mathbf{C}})$ of the tensor $\overset{(n)}{\mathbf{C}}$, i.e.

$$\psi(\overset{(n)}{\mathbf{C}},\ \theta) = \psi(I_1^{(s)}(\overset{(n)}{\mathbf{C}}), \ldots, I_r^{(s)}(\overset{(n)}{\mathbf{C}}),\ \theta). \tag{4.304}$$

The function (4.304) indifferent relative to the orthotropy group $O$ is called the *orthotropic scalar function*; the function depends on six invariants ($r = 6$), for example (4.296).

The function (4.297) indifferent relative to the transverse isotropy group $T_3$ is called the *transversely isotropic scalar function*; the function depends on five invariants ($r = 5$), for example (4.303).

The function (4.304) indifferent relative to the isotropy group $I$ is called the *isotropic scalar function*; the function depends on three principal invariants:

$$\psi(\overset{(n)}{\mathbf{C}},\ \theta) = \psi(I_\alpha(\overset{(n)}{\mathbf{C}}),\ \theta), \quad \alpha = 1, 2, 3. \tag{4.305}$$

### 4.8.6 *Indifferent Tensor Functions of Tensor Argument and Invariant Representation of Constitutive Equations for Elastic Continua*

Let us return now to the conditions (4.279) and consider in detail the first condition (4.279a) imposed on the tensor function of the tensor argument $\mathcal{F}(\overset{(n)}{\mathbf{C}}, \theta)$.

**Definition 4.13.** The tensor function of tensor argument $\mathcal{F}(\boldsymbol{\Omega})$ is called *indifferent relative to a symmetry group* $\overset{\circ}{G}_s \subset I$, if each tensor $\overset{*}{\mathbf{Q}} \in \overset{\circ}{G}_s$ satisfies the condition

$$\overset{*}{\mathbf{Q}}^{\mathrm{T}} \cdot \mathcal{F}(\boldsymbol{\Omega}) \cdot \overset{*}{\mathbf{Q}} = \mathcal{F}(\overset{*}{\mathbf{Q}}^{\mathrm{T}} \cdot \boldsymbol{\Omega} \cdot \overset{*}{\mathbf{Q}}). \tag{4.306}$$

Resolving the tensor $\overset{*}{\mathbf{Q}}$ for basis $\widehat{\mathbf{c}}_i$ (see (4.282)), and also the tensor $\boldsymbol{\Omega}$ and tensor function $\mathcal{F}$ for the basis $\widehat{\mathbf{c}}_i$:

$$\mathcal{F}(\boldsymbol{\Omega}) = \widehat{\mathcal{F}}^{ij}\widehat{\mathbf{c}}_i \otimes \widehat{\mathbf{c}}_j, \quad \widehat{\mathcal{F}}^{ij} = \widehat{\mathcal{F}}^{ij}(\widehat{\Omega}^{kl}), \tag{4.307}$$

$$\boldsymbol{\Omega} = \widehat{\Omega}^{kl}\widehat{\mathbf{c}}_k \otimes \widehat{\mathbf{c}}_l, \tag{4.308}$$

we can write the *component form* of the condition (4.306) of indifference of the tensor function:

$$\widehat{H}^{i_1}_{j_1} \widehat{H}^{i_2}_{j_2} \widehat{\mathcal{F}}^{j_1 j_2}(\widehat{\Omega}^{kl}) = \widehat{\mathcal{F}}^{i_1 i_2}(\widehat{H}^{k_1}_{l_1} \widehat{H}^{k_2}_{l_2} \widehat{\Omega}^{l_1 l_2}). \tag{4.309}$$

The tensor function $\mathcal{F}(\overset{(n)}{\mathbf{C}}, \theta)$ in the models $A_n$ is *quasipotential*, i.e the function satisfies the first condition of (4.168):

$$\mathcal{F}(\overset{(n)}{\mathbf{C}}, \theta) = \rho(\partial\psi/\partial\overset{(n)}{\mathbf{C}}). \tag{4.310}$$

where the scalar function $\psi(\overset{(n)}{\mathbf{C}}, \theta)$ is indifferent relative to the group $\overset{\circ}{G}_s$, and hence this function can always be represented in the form (4.304) of a function of tensor invariants $I^{(s)}_\alpha(\overset{(n)}{\mathbf{C}})$. Then for the quasipotential function $\mathcal{F}(\overset{(n)}{\mathbf{C}}, \theta)$ we have the representation

$$\overset{(n)}{\mathbf{T}} = \mathcal{F}(\overset{(n)}{\mathbf{C}}, \theta) = \sum_{\gamma=1}^{r} \varphi_\gamma I^{(s)}_{\gamma \mathbf{C}}, \tag{4.311}$$

where

$$\varphi_\gamma = \varphi_\gamma(I^{(s)}_1(\overset{(n)}{\mathbf{C}}), \ldots, I^{(s)}_r(\overset{(n)}{\mathbf{C}}),\ \theta) = \rho(\partial\psi/\partial I^{(s)}_\gamma), \quad I^{(s)}_{\gamma \mathbf{C}} = \partial I^{(s)}_\gamma/\partial\overset{(n)}{\mathbf{C}}. \tag{4.312}$$

Functions $\varphi_\gamma$ depend only on invariants $I_\gamma^{(s)}(\overset{(n)}{\mathbf{C}})$ and $\theta$, because the density $\rho$ is uniquely expressed in terms of three principal invariants of the tensor $\overset{(n)}{\mathbf{C}}$ (see Sect. 4.2.19).

Formula (4.311) is called the *representation of an indifferent quasipotential tensor function in the tensor basis* or the *invariant representation of constitutive equations for the model $A_n$ of elastic continua.*

On substituting different collections (4.295)–(4.298) in place of invariants $I_\gamma^{(s)}$ into (4.311), we obtain the representations of tensor functions indifferent relative to the groups $E$, $O$, $T_3$ and $I$.

For the triclinic group $E$, the invariant representation (4.311) has the form

$$\overset{(n)}{\mathbf{T}} = \sum_{\gamma=1}^{3}\left(\varphi_\gamma \widehat{\mathbf{c}}_\gamma^2 + \frac{1}{2}\varphi_{\gamma+3}\mathbf{O}_\gamma\right), \tag{4.313a}$$

and the component form of this function for the basis $\widehat{\mathbf{c}}_\gamma$ is written as follows:

$$\widehat{\overset{(n)}{T}}{}^{ij} = \sum_{\gamma=1}^{3}\left(\varphi_\gamma \delta_\gamma^i \delta_\gamma^j + \frac{1}{2}\varphi_{\gamma+3}(\delta_\alpha^i\delta_\beta^j + \delta_\beta^i\delta_\alpha^j)\right), \tag{4.313b}$$

where $\varphi_\gamma$ are scalar functions of invariants $I_\gamma^{(E)}$ (4.295):

$$\varphi_\gamma = \varphi_\gamma(\widehat{\overset{(n)}{C}}_{11}, \widehat{\overset{(n)}{C}}_{22}, \widehat{\overset{(n)}{C}}_{33}, \widehat{\overset{(n)}{C}}_{23}, \widehat{\overset{(n)}{C}}_{13}, \widehat{\overset{(n)}{C}}_{12}, \theta). \tag{4.313c}$$

For the orthotropy group $O$, the invariant representation has the form

$$\overset{(n)}{\mathbf{T}} = \sum_{\gamma=1}^{3}\varphi_\gamma \widehat{\mathbf{c}}_\gamma^2 + \frac{1}{2}(\varphi_4\mathbf{O}_1\otimes\mathbf{O}_1 + \varphi_5\mathbf{O}_2\otimes\mathbf{O}_2 + \varphi_7\mathbf{O}_3\otimes\mathbf{O}_3)\cdot\cdot\overset{(n)}{\mathbf{C}}$$
$$+3\varphi_6\,{}^6\mathbf{O}_m\cdot\cdot\cdot\cdot\,\overset{(n)}{\mathbf{C}}\otimes\overset{(n)}{\mathbf{C}}, \tag{4.314a}$$

and the corresponding component form of the function with respect to the basis $\widehat{\mathbf{c}}_i$ is written as follows:

$$\widehat{\overset{(n)}{T}}{}^{ij} = \sum_{\gamma=1}^{3}\varphi_\gamma\delta_\gamma^i\delta_\gamma^j + \frac{1}{2}(\varphi_4\widehat{\overset{(n)}{C}}_{23} + \widehat{\overset{(n)}{C}}_{12}\widehat{\overset{(n)}{C}}_{13}\varphi_6)(\delta_2^i\delta_3^j + \delta_3^i\delta_2^j)$$
$$+\frac{1}{2}(\varphi_5\widehat{\overset{(n)}{C}}_{13} + \varphi_6\widehat{\overset{(n)}{C}}_{12}\widehat{\overset{(n)}{C}}_{23})(\delta_1^i\delta_3^j + \delta_3^i\delta_1^j)$$
$$+\frac{1}{2}(\varphi_7\widehat{\overset{(n)}{C}}_{12} + \varphi_6\widehat{\overset{(n)}{C}}_{13}\widehat{\overset{(n)}{C}}_{23})(\delta_1^i\delta_2^j + \delta_2^i\delta_1^j). \tag{4.314b}$$

Here $\overset{(n)}{\widehat{T}}{}^{ij}$ and $\overset{(n)}{\widehat{C}}{}^{ij}$ are components of the tensors $\overset{(n)}{\mathbf{T}}$ and $\overset{(n)}{\mathbf{C}}$ with respect to the basis $\widehat{\mathbf{c}}_i$, and $\mathbf{O}_\gamma$, $\gamma = 1, 2, 3$, and ${}^6\mathbf{O}_m$ are tensors indifferent relative to the group $O$, which are constructed with the help of the producing tensors $\mathbf{O}_{s(\gamma)}$ of the group (see Table 4.11 and [12]):

$$\mathbf{O}_\gamma = \widehat{\mathbf{c}}_\alpha \otimes \widehat{\mathbf{c}}_\beta + \widehat{\mathbf{c}}_\beta \otimes \widehat{\mathbf{c}}_\alpha, \quad \alpha \neq \beta \neq \gamma \neq \alpha, \quad \alpha, \beta, \gamma = 1, 2, 3, \qquad (4.315)$$

$${}^6\mathbf{O}_m = \frac{1}{48} \sum_{\substack{\alpha,\beta,\gamma=1 \\ \alpha\neq\beta\neq\gamma\neq\alpha}}^{3} (\mathbf{O}_\alpha \otimes \mathbf{O}_\beta + \mathbf{O}_\beta \otimes \mathbf{O}_\alpha) \otimes \mathbf{O}_\gamma. \qquad (4.316)$$

When deriving the expression (4.314a) from the representation (4.311), one should use the rules of differentiation of the invariants being scalar functions of a tensor argument (see [12]). In order for formula (4.314a) to be symmetric in indices 1, 2, and 3, we must complement the system of invariants $I_\gamma^{(O)}(\overset{(n)}{\mathbf{C}})$ $(\gamma = 1, \dots, 6)$ with the seventh invariant

$$I_7^{(O)} = (\widehat{\mathbf{c}}_1^2 \cdot \overset{(n)}{\mathbf{C}}) \cdot\cdot (\widehat{\mathbf{c}}_2^2 \cdot \overset{(n)}{\mathbf{C}}) = \overset{(n)}{\widehat{C}}_{12}, \qquad (4.317)$$

which is not independent. In this case, functions $\varphi_\gamma$ in (4.314a) depend on all these seven invariants

$$\varphi_\gamma = \varphi_\gamma(\overset{(n)}{\widehat{C}}_{11}, \overset{(n)}{\widehat{C}}_{22}, \overset{(n)}{\widehat{C}}_{33}, \overset{(n)}{\widehat{C}}_{23}, \overset{(n)}{\widehat{C}}_{13}, \overset{(n)}{\widehat{C}}_{12}\overset{(n)}{\widehat{C}}_{23}\overset{(n)}{\widehat{C}}_{13}, \overset{(n)}{\widehat{C}}_{12}, \theta). \qquad (4.318)$$

The function (4.314a) is called the *orthotropic tensor function*.

For the transverse isotropy group $T_3$, the representation (4.311) takes the form

$$\overset{(n)}{\mathbf{T}} = (\varphi_1 + I_2\varphi_5)(\mathbf{E} - \widehat{\mathbf{c}}_3^2) + (\varphi_2 + I_2\varphi_5)\widehat{\mathbf{c}}_3^2 + \Bigg(\frac{1}{2}\varphi_3(\mathbf{O}_1 \otimes \mathbf{O}_1 + \mathbf{O}_2 \otimes \mathbf{O}_2)$$
$$+2\varphi_4\Big(\mathbf{\Delta} - \frac{1}{2}(\mathbf{O}_1 \otimes \mathbf{O}_1 + \mathbf{O}_2 \otimes \mathbf{O}_2) - \widehat{\mathbf{c}}_3^2 \otimes \widehat{\mathbf{c}}_3^2\Big) - \varphi_5 I_1 \mathbf{\Delta}\Bigg) \cdot\cdot \overset{(n)}{\mathbf{C}} + \varphi_5 \overset{(n)}{\mathbf{C}}{}^2, \qquad (4.319)$$

and the component form with respect to the basis $\widehat{\mathbf{c}}_i$ is written as follows:

$$\overset{(n)}{\widehat{T}}{}^{ij} = (\varphi_1 + I_2\varphi_5)\delta^{ij} + (\varphi_2 - \varphi_1 - 2\varphi_4\overset{(n)}{\widehat{C}}_{33})\delta_3^i\delta_3^j$$
$$+(\varphi_3 - 2\varphi_4)\big((\delta_2^i\delta_3^j + \delta_3^i\delta_2^j)\overset{(n)}{\widehat{C}}_{23} + (\delta_1^i\delta_3^j + \delta_3^i\delta_1^j)\overset{(n)}{\widehat{C}}_{13}\big)$$
$$+(2\varphi_4 - I_1\varphi_5)\overset{(n)}{\widehat{C}}{}^{ij} + \varphi_5\overset{(n)}{\widehat{C}}{}^{ik}\overset{(n)}{\widehat{C}}{}^{kj}, \qquad (4.320)$$

where

$$\varphi_\gamma = \varphi_\gamma\big(\overset{\frown}{\overset{(n)}{C}}_{11} + \overset{\frown}{\overset{(n)}{C}}_{22},\ \overset{\frown}{\overset{(n)}{C}}_{33},\ \overset{\frown}{\overset{(n)}{C}}_{13}{}^2 + \overset{\frown}{\overset{(n)}{C}}_{23}{}^2,\ \overset{\frown}{\overset{(n)}{C}}_{11}{}^2 + \overset{\frown}{\overset{(n)}{C}}_{22}{}^2 + 2\overset{\frown}{\overset{(n)}{C}}_{12}{}^2,\ \det \overset{\frown}{\overset{(n)}{\mathbf{C}}},\ \theta\big). \tag{4.321}$$

Here $\boldsymbol{\Delta}$ is the symmetric fourth-order unit tensor (2.22a).

For the full orthogonal (isotropic) group $I$, the representation (4.311) takes the form

$$\overset{(n)}{\mathbf{T}} = \psi_1 \mathbf{E} + \psi_2 \overset{(n)}{\mathbf{C}} + \psi_3 \overset{(n)}{\mathbf{C}}{}^2, \tag{4.322}$$

or

$$\overset{\frown}{\overset{(n)}{T}}{}^{ij} = \psi_1 \delta^{ij} + \psi_2 \overset{\frown}{\overset{(n)}{C}}{}^{ij} + \psi_3 \overset{\frown}{\overset{(n)}{C}}{}^{i}_{k} \overset{\frown}{\overset{(n)}{C}}{}^{kj},$$

where

$$\psi_1 = \varphi_1 + \varphi_2 I_1 + \varphi_3 I_2, \qquad \psi_2 = -(\varphi_2 + I_1\varphi_3), \qquad \psi_3 = \varphi_3, \tag{4.322a}$$

$$\varphi_\gamma = \varphi_\gamma\big(I_1(\overset{(n)}{\mathbf{C}}),\ I_2(\overset{(n)}{\mathbf{C}}),\ I_3(\overset{(n)}{\mathbf{C}}),\ \theta\big), \quad \psi = \psi\big(I_1(\overset{(n)}{\mathbf{C}}),\ I_2(\overset{(n)}{\mathbf{C}}),\ I_3(\overset{(n)}{\mathbf{C}}),\ \theta\big).$$

When deriving formula (4.322), we used the rules of differentiation of the principal invariants of a tensor (see [12]):

$$\frac{\partial I_1(\boldsymbol{\Omega})}{\partial \boldsymbol{\Omega}} = \mathbf{E}, \qquad \frac{\partial I_2(\boldsymbol{\Omega})}{\partial \boldsymbol{\Omega}} = \mathbf{E} I_1(\boldsymbol{\Omega}) - \boldsymbol{\Omega}^{\mathrm{T}},$$
$$\frac{\partial I_3(\boldsymbol{\Omega})}{\partial \boldsymbol{\Omega}} = I_3(\boldsymbol{\Omega})\boldsymbol{\Omega}^{-1} = \boldsymbol{\Omega}^{\mathrm{T}2} - I_1(\boldsymbol{\Omega})\boldsymbol{\Omega}^{\mathrm{T}} + \mathbf{E} I_2(\boldsymbol{\Omega}). \tag{4.323}$$

*Remark 1.* For all groups $\widehat{G}_s$, the indifferent quasipotential tensor functions (4.311) can be represented in the single tensor form

$$\overset{(n)}{\mathbf{T}} = {}^4\mathbf{M} \cdot\cdot \overset{(n)}{\mathbf{C}} + \xi \overset{(n)}{\mathbf{C}}{}^2 + {}^6\mathbf{L} \cdot\cdot\cdot\cdot (\overset{(n)}{\mathbf{C}} \otimes \overset{(n)}{\mathbf{C}}). \tag{4.324}$$

The tensor ${}^4\mathbf{M}$ is called the *quasilinear elasticity tensor*, $\xi$ is called the *parameter of quadratic elasticity*, and ${}^6\mathbf{L}$ – the *tensor of quadratic elasticity*.

For the isotropy group $I$, from (4.322) it follows that

$${}^4\mathbf{M} = \frac{\psi_1}{I_1(\overset{(n)}{\mathbf{C}})} \mathbf{E} \otimes \mathbf{E} + \psi_2 \boldsymbol{\Delta}, \qquad \xi = \psi_3, \qquad {}^6\mathbf{L} = 0. \tag{4.325a}$$

For the transverse isotropy group $T_3$, from (4.319) it follows that

$$
\begin{aligned}
{}^4\mathbf{M} = \frac{\varphi_1 + I_2\varphi_5}{I_1}\mathbf{E}\otimes\mathbf{E} + \left(\frac{\varphi_2 - \varphi_1}{I_2} - 2\varphi_4\right)\widehat{\mathbf{c}}_3^2 \otimes \widehat{\mathbf{c}}_3^2 \\
+ \left(\frac{\varphi_3}{2} - \varphi_4\right)(\mathbf{O}_1\otimes\mathbf{O}_1 + \mathbf{O}_2\otimes\mathbf{O}_2) + (2\varphi_4 + \varphi_5)I_1\boldsymbol{\Delta}, \\
\xi = \varphi_5, \qquad {}^6\mathbf{L} = 0.
\end{aligned} \tag{4.325b}
$$

For the orthotropy group $O$, from (4.314a) it follows that

$$
\begin{aligned}
{}^4\mathbf{M} = \sum_{\gamma=1}^{3} \frac{\varphi_\gamma}{I_\gamma^{(0)}}\widehat{\mathbf{c}}_\gamma^2 \otimes \widehat{\mathbf{c}}_\gamma^2 + \frac{\varphi_4}{2}\mathbf{O}_1\otimes\mathbf{O}_1 + \frac{\varphi_5}{2}\mathbf{O}_2\otimes\mathbf{O}_2 + \frac{\varphi_7}{2}\mathbf{O}_3\otimes\mathbf{O}_3, \\
{}^6\mathbf{L} = 3\varphi_6\,{}^6\mathbf{O}_m, \qquad \xi = 0.
\end{aligned} \tag{4.325c}
$$

The main result of this section is the derivation of representations (4.311), (4.313a), (4.314a), (4.319), and (4.322) for indifferent tensor functions. These representations show that tensor nonlinearity for indifferent tensor functions cannot be higher than the second power, i.e. the tensor powers $\overset{(n)}{\mathbf{C}}{}^3$, $\overset{(n)}{\mathbf{C}} \otimes \overset{(n)}{\mathbf{C}} \otimes \overset{(n)}{\mathbf{C}}$ and higher powers are not involved in these functions; but the scalar functions (4.312), (4.318), (4.321), and (4.322a) of invariants are nonlinear in a more complicated form.

The obtained representations of tensor functions are the consequences of applying the principle of material indifference to the models $A_n$ of solids; and they demonstrate how this principle seeming not sufficiently analytical allows us to derive analytical forms of constitutive equations.

Notice that the representations (4.313a), (4.314a), (4.319), and (4.322) reject attempts to construct indifferent tensor functions in the form of generalized tensor Taylor series. □

*Remark 2.* In continuum mechanics, one sometimes uses the concept of a *natural state* of a continuum, that means the configuration $\mathcal{K}^e$ where the deformation gradient $\mathbf{F}$ coincides with the metric tensor, the Cauchy stress tensor $\mathbf{T}$ take some values (for example, the tensor is spherical: $\mathbf{T}^e = -p^e\mathbf{E}$, where $p^e$ is a constant having the sense of a pressure in the natural state), and the temperature and free energy $\psi$ are equal to some natural values:

$$
\mathcal{K}^e: \quad \mathbf{T} = \mathbf{T}^e, \quad \mathbf{F} = \mathbf{E}, \quad \theta = \theta^e, \quad \psi = \psi^e. \tag{4.326}
$$

In the natural state, all the energetic and quasienergetic stress tensors $\overset{(n)}{\mathbf{T}}$ and $\overset{(n)}{\mathbf{S}}$ coincide with $\mathbf{T}^e$, all the deformation tensors $\overset{(n)}{\mathbf{A}}$ and $\overset{(n)}{\mathbf{C}}$ are zero-tensors, and the deformation measures $\overset{(n)}{\mathbf{g}}$ and $\overset{(n)}{\mathbf{G}}$ coincide with $\mathbf{E}$ to within the factor:

$$\mathcal{K}^e: \quad \overset{(n)}{\mathbf{T}} = \overset{(n)}{\mathbf{S}} = \mathbf{T}^e, \quad \overset{(n)}{\mathbf{A}} = \overset{(n)}{\mathbf{C}} = \mathbf{0}, \quad \overset{(n)}{\mathbf{g}} = \overset{(n)}{\mathbf{G}} = \frac{1}{n - \mathrm{III}}\mathbf{E}. \tag{4.327}$$

For solids, one often assumes that in the reference configuration $\overset{\circ}{\mathcal{K}}$ a continuum is at its *natural unstressed state*, i.e. $\mathbf{T}^e = \mathbf{0}$ and $\psi^e = \psi_0$; then for the solid we have the relations

$$\overset{\circ}{\mathcal{K}}: \quad \overset{(n)}{\mathbf{T}} = \overset{(n)}{\mathbf{S}} = \mathbf{T} = \mathbf{0}, \quad \overset{(n)}{\mathbf{A}} = \overset{(n)}{\mathbf{C}} = \mathbf{0}, \quad \overset{(n)}{\mathbf{g}} = \overset{(n)}{\mathbf{G}} = \frac{1}{n - \mathrm{III}}\mathbf{E}, \quad \theta = \theta_0, \quad \psi = \psi_0. \tag{4.328}$$

There is a class of problems (for example, problems on appearing technological or initial stresses in structures), for which the conditions (4.328) are replaced by (4.326), (4.327).

If in $\overset{\circ}{\mathcal{K}}$ a continuum is at an unstressed state, then in each reference configuration $\overset{*}{\mathcal{K}}$ obtained from $\overset{\circ}{\mathcal{K}}$ by $H$-transformation involved in the corresponding symmetry group $\overset{\circ}{G}_s$ the continuum is at the unstressed state as well. In this case, we impose additional conditions on the constitutive equations (4.311) or (4.324), namely the conditions (4.328) in $\overset{\circ}{\mathcal{K}}$, that lead to certain constraints on the potential $\psi$. For example, for the models $A_n$ of isotropic ideal continua, from (4.322) it follows that in $\overset{\circ}{\mathcal{K}} = \mathcal{K}^e$ we have the relations

$$\varphi_1 = \varphi_1(0, 0, 0, \theta_0) = 0, \quad \psi(0, 0, 0, \theta_0) = 0. \quad \square \tag{4.329}$$

### 4.8.7 *Quasilinear and Linear Models $A_n$ of Elastic Continua*

For the *quasilinear models* $A_n$ of elastic continua, in formulae (4.324) we should assume that

$$\xi = 0, \qquad {}^6\mathbf{L} = 0, \tag{4.330}$$

then we obtain the relation

$$\overset{(n)}{\mathbf{T}} = {}^4\mathbf{M} \cdot\cdot \overset{(n)}{\mathbf{C}}. \tag{4.330a}$$

For an isotropic continuum, the relations (4.330) mean that the scalar function $\psi_3 = \varphi_3$ in (4.322) vanishes: $\varphi_3(I_1, I_2, I_3) = 0$. This relation is the additional condition for the invariants $I_1, I_2, I_3$. On expressing one of the invariants from this relation, we can eliminate this invariant between arguments of the functions $\varphi_\gamma$ (4.322a). In continuum mechanics, one usually expresses the cubic invariant $I_3(\overset{(n)}{\mathbf{C}})$ in this way, in order that only linear and quadratic invariants remain as arguments of the functions (4.322a). Then relations (4.322) for isotropic quasilinear elastic continua take the form

$$\overset{(n)}{\mathbf{T}} = \psi_1 \mathbf{E} - \varphi_2 \overset{(n)}{\mathbf{C}},$$

$$\psi_1 = \varphi_1 + \varphi_2 I_1, \quad \varphi_\gamma = \varphi_\gamma(I_1(\overset{(n)}{\mathbf{C}}),\ I_2(\overset{(n)}{\mathbf{C}}),\ \theta), \quad \gamma = 1,\ 2.$$

For the transverse isotropy group $T_3$, relations (4.330) lead to one additional condition

$$\varphi_5(I_1^{(3)},\ \ldots,\ I_5^{(3)}) = 0,$$

which allows us to eliminate the cubic invariant $I_5^{(3)}$ among arguments of the functions $\varphi_\gamma$ (4.321); and the number of arguments of the functions $\varphi_\gamma$ reduces to four.

For the orthotropy group $O$, relations (4.330) lead to the condition

$$\varphi_6(I_1^{(O)},\ \ldots,\ I_7^{(O)}) = 0,$$

that also allows us to eliminate the cubic invariant $I_6^{(O)}$ between arguments of the scalar functions $\varphi_\gamma$ (4.312).

For *linear models* $A_n$ of solids, we assume that the free energy (4.304) is a quadratic function of invariants $I_\gamma^{(s)}(\overset{(n)}{\mathbf{C}})$, i.e. the function is a quadratic form of linear invariants $I_\gamma^{(s)}$ $(\gamma = 1, \ldots, r_1)$ and a linear form of quadratic invariants $I_\gamma^{(s)}$ $(\gamma = r_1 + 1,\ \ldots,\ r_2)$ relative to the corresponding group:

$$\overset{\circ}{\rho}\psi = \overset{\circ}{\rho}\psi_0 + \overset{\circ}{\rho}\bar{\psi}^0 + \sum_{\gamma=1}^{r_1} \bar{m}_\gamma I_\gamma^{(s)} + \frac{1}{2}\sum_{\gamma,\beta=1}^{r_1} l_{\gamma\beta} I_\gamma^{(s)} I_\beta^{(s)} + \sum_{\gamma=r_1+1}^{r_2} l_{\gamma\gamma} I_\gamma^{(s)}, \quad (4.331)$$

where $\psi_0, \bar{\psi}^0, \bar{m}_\gamma, l_{\gamma\beta}$ are the constants; the cubic invariants $I_\gamma^{(s)}, \gamma = r_2{+}1, \ldots, r$, are not involved in the expression (4.331), therefore for linear models $A_n$ both the conditions (4.330) hold.

Notice that the total number of different constants $l_{\gamma\beta}$ does not exceed 21, because the matrix $(l_{\gamma\beta})$ consisting of the constants has a maximum size $6 \times 6$ (6 is the maximum number of invariants $I_\gamma^{(s)}$) and the matrix is symmetric; and a symmetric $n \times n$ matrix has $n(n+1)/2$ different components. The functions $\varphi_\gamma$ have the forms

$$\varphi_\gamma = J \sum_{\beta=1}^{r_1} l_{\gamma\beta} I_\beta^{(s)} + J m_\gamma, \quad \gamma = 1, \ldots, r_1,$$
$$\varphi_\gamma = J l_{\gamma\gamma}, \quad \gamma = r_1 + 1, \ldots, r_2; \quad J = \rho/\overset{\circ}{\rho}, \quad I_{\gamma\mathbf{C}}^{(s)} = \partial I_\gamma^{(s)}/\partial \overset{(n)}{\mathbf{C}}, \quad (4.332)$$

and the constitutive equations (4.311) and (4.324) for linear models take the form

$$\overset{(n)}{\mathbf{T}} = J\sum_{\gamma=1}^{r_1} \bar{m}_\gamma \mathbf{O}_\gamma^{(s)} + J\sum_{\gamma,\beta=1}^{r_1} l_{\gamma\beta} I_\gamma^{(s)} \mathbf{O}_\beta^{(s)} + J\sum_{\gamma=r_1+1}^{r_2} l_{\gamma\gamma} I_{\gamma\mathbf{C}}^{(s)}. \quad (4.333)$$

If in $\overset{\circ}{\mathcal{K}}$ a continuum is at an unstressed state, then substituting (4.331) and (4.333) into the normalization equation (4.328), we find that the constants $\bar{\psi}^0$ and $\bar{m}_\gamma$ may be only zero:

$$\bar{\psi}^0 = 0, \quad \bar{m}_\gamma = 0, \quad \gamma = 1, \dots, r_1. \quad (4.334)$$

Thus, for linear models $A_n$ there are no purely linear summands.

(1) For isotropic elastic continua, there is only one linear invariant and one quadratic invariant ($r_1 = 1, r_2 = 2$); therefore for *isotropic linear models* $A_n$ the free energy $\psi$ (4.331) has the form

$$\overset{\circ}{\rho}\psi = \overset{\circ}{\rho}\psi_0 + \left(\frac{1}{2}l_1 + l_2\right) I_1^2(\overset{(n)}{\mathbf{C}}) - 2l_2 I_2(\overset{(n)}{\mathbf{C}}) = \overset{\circ}{\rho}\psi_0 + \frac{l_1}{2} I_1^2(\overset{(n)}{\mathbf{C}}) + l_2 I_1(\overset{(n)}{\mathbf{C}}{}^2), \quad (4.335)$$

where $l_1$ and $l_2$ are the constants ($l_{11} = l_1 + 2l_2, l_{22} = -2l_2$).

For the isotropic linear model, the functions $\varphi_\gamma$ (4.312) and $\psi_\gamma$ (4.322a) have the forms

$$\begin{gathered} \varphi_1 = J(l_1 + 2l_2) I_1, \quad \varphi_2 = -2l_2 J, \quad \varphi_3 = 0, \\ \psi_1 = l_1 I_1 J, \quad \psi_2 = 2l_2 J, \quad J = \rho/\overset{\circ}{\rho}, \end{gathered} \quad (4.336)$$

the tensor ${}^4\mathbf{M}$ (4.325a) has only two independent components:

$${}^4\mathbf{M} = J(l_1 \mathbf{E} \otimes \mathbf{E} + 2l_2 \mathbf{\Delta}), \quad (4.337)$$

and the constitutive equations (4.322) are written as follows:

$$\overset{(n)}{\mathbf{T}} = J(l_1 I_1(\overset{(n)}{\mathbf{C}})\mathbf{E} + 2l_2 \overset{(n)}{\mathbf{C}}). \quad (4.338)$$

Since the relations (4.338) contain the factor $J$, they are not linear; therefore the models (4.335) and (4.338) are often called *semilinear*. The semilinear model $A_{\mathrm{IV}}$ is called *John's model*, and the model $A_{\mathrm{V}}$ – *Murnaghan's model.*

(2) For transversely isotropic continua, there are two linear invariants $I_\gamma^{(s)}$ and two quadratic invariants ($r_1 = 2, r_2 = 4$); therefore, for transversely isotropic linear models $A_n$, the free energy $\psi$ (4.331) has the form

$$\overset{\circ}{\rho}\psi = \overset{\circ}{\rho}\psi_0 + \frac{1}{2}(l_{11} I_1^{(3)2} + 2l_{12} I_1^{(3)} I_2^{(3)} + l_{22} I_2^{(3)2}) + l_{33} I_3^{(3)} + l_{44} I_4^{(3)}, \quad (4.339)$$

and contains five constants $l_{\gamma\beta}$.

The functions $\varphi_\gamma$ (4.312) for this model have the forms

$$\varphi_1 = J(l_{11}I_1^{(3)} + l_{12}I_2^{(3)}), \qquad \varphi_2 = J(l_{22}I_2^{(3)} + l_{12}I_1^{(3)}),$$
$$\varphi_3 = Jl_{33}, \qquad \varphi_4 = Jl_{44}, \qquad \varphi_5 = 0, \tag{4.340}$$

the tensor ${}^4\mathbf{M}$ (4.325b) has five independent components:

$${}^4\mathbf{M} = J\Big(l_{11}\mathbf{E}\otimes\mathbf{E} + \widetilde{l}_{22}\widehat{\mathbf{c}}_3^2\otimes\widehat{\mathbf{c}}_3^2 + (l_{12}-l_{11})(\mathbf{E}\otimes\widehat{\mathbf{c}}_3^2 + \widehat{\mathbf{c}}_3^2\otimes\mathbf{E})$$
$$+\Big(\frac{l_{33}}{2} - l_{44}\Big)(\mathbf{O}_1\otimes\mathbf{O}_1 + \mathbf{O}_2\otimes\mathbf{O}_2) + 2l_{44}\boldsymbol{\Delta}\Big), \tag{4.341}$$
$$\widetilde{l}_{22} = l_{22} - 2l_{44} - 2l_{12} + l_{11},$$

and the constitutive equations (4.319) are written as follows:

$$\overset{(n)}{\mathbf{T}} = J\Big((l_{11}I_1^{(3)} + l_{12}I_2^{(3)})(\mathbf{E} - \widehat{\mathbf{c}}_3^2) + ((l_{22} - 2l_{44})I_2^{(3)} + l_{12}I_1^{(3)})\widehat{\mathbf{c}}_3^2$$
$$+\Big(\frac{l_{33}}{2} - l_{44}\Big)(\mathbf{O}_1\otimes\mathbf{O}_1 + \mathbf{O}_2\otimes\mathbf{O}_2)\cdot\cdot\overset{(n)}{\mathbf{C}} + 2l_{44}\overset{(n)}{\mathbf{C}}\Big). \tag{4.342}$$

(3) For orthotropic continua, there are three linear invariants and three quadratic invariants ($r_1 = 3,\ r_2 = r = 6$); therefore, for orthotropic linear models $A_n$, the free energy $\psi$ (4.331) has the form

$$\overset{\circ}{\rho}\psi = \overset{\circ}{\rho}\psi_0 + \frac{1}{2}\sum_{\gamma,\beta=1}^{3} l_{\gamma\beta}I_\gamma^{(O)}I_\beta^{(O)}\sum_{\gamma=1}^{3} l_{3+\gamma,3+\gamma}I_{3+\gamma}^{(O)} \tag{4.343}$$

and contains nine constants ($l_{11}, l_{22}, \ldots, l_{66}, l_{12}, l_{13}, l_{23}$), the tensor ${}^4\mathbf{M}$ (4.325c) has nine independent components:

$${}^4\mathbf{M} = J\sum_{\gamma,\beta=1}^{3} l_{\gamma\beta}\widehat{\mathbf{c}}_\gamma^2\otimes\widehat{\mathbf{c}}_\beta^2 + J\sum_{\gamma=1}^{3} l_{3+\gamma,3+\gamma}\mathbf{O}_\gamma\otimes\mathbf{O}_\gamma, \tag{4.344}$$

and the constitutive equations (4.314a) take the form

$$\overset{(n)}{\mathbf{T}} = \sum_{\gamma,\beta=1}^{3} Jl_{\gamma\beta}I_\beta^{(O)}\widehat{\mathbf{c}}_\gamma^2 + \sum_{\gamma=1}^{3} Jl_{3+\gamma,3+\gamma}\mathbf{O}_\gamma(\mathbf{O}_\gamma\cdot\cdot\overset{(n)}{\mathbf{C}}). \tag{4.345}$$

### 4.8.8 Constitutive Equations for Models $B_n$ of Elastic Continua

The theory stated in Sects. 4.8.1–4.8.6 is formally valid for models $B_n$ with the only difference that the energetic deformation tensors $\overset{(n)}{\mathbf{C}}$ in formulae of these sections must be replaced by the corresponding energetic deformation measures $\overset{(n)}{\mathbf{G}}$, because $\overset{(n)}{\mathbf{G}}$ as well as $\overset{(n)}{\mathbf{C}}$ are $H$-indifferent relative to the orthogonal group $I$. In particular, relations (4.304), (4.311), and (4.312) take the forms

$$\begin{aligned}
&\psi(\overset{(n)}{\mathbf{G}},\, \theta) = \psi(I_1^{(s)}(\overset{(n)}{\mathbf{G}}), \ldots, I_r^{(s)}(\overset{(n)}{\mathbf{G}}),\, \theta),\\
&\overset{(n)}{\mathbf{T}} = \mathcal{F}(\overset{(n)}{\mathbf{G}},\, \theta) = \sum_{\gamma=1}^{r} \varphi_\gamma \frac{\partial I_\gamma^{(s)}}{\partial \overset{(n)}{\mathbf{G}}},\\
&\varphi_\gamma = \varphi_\gamma(I_1^{(s)}(\overset{(n)}{\mathbf{G}}), \ldots, I_r^{(s)}(\overset{(n)}{\mathbf{G}}),\, \theta) = \rho \frac{\partial \psi}{\partial I_\gamma^{(s)}}.
\end{aligned} \tag{4.346}$$

In a similar way, we modify formulae (4.313), (4.319), and (4.322); for example, for the group $I$ the representation (4.346) takes the form

$$\begin{aligned}
&\psi = \psi(I_1(\overset{(n)}{\mathbf{G}}), \ldots, I_3(\overset{(n)}{\mathbf{G}}),\, \theta), \quad \overset{(n)}{\mathbf{T}} = \psi_1 \mathbf{E} + \psi_2 \overset{(n)}{\mathbf{G}} + \psi_3 \overset{(n)}{\mathbf{G}}{}^2,\\
&\psi_1 = \varphi_1 + \varphi_2 I_1 + \varphi_3 I_2, \quad \psi_2 = -(\varphi_2 + \varphi_3 I_1), \quad \psi_3 = \varphi_3,\\
&\varphi_\gamma = \varphi_\gamma(I_1(\overset{(n)}{\mathbf{G}}), \ldots, I_3(\overset{(n)}{\mathbf{G}}),\, \theta) = \rho(\partial\psi/\partial I_\gamma).
\end{aligned} \tag{4.347}$$

If in $\overset{\circ}{\mathcal{K}}$ a continuum is at an unstressed state, then relations (4.347) must satisfy the conditions (4.328), which lead to the following conditions of normalization for functions $\varphi$:

$$\varphi_1(a_1, a_2, a_3) + \varphi_2(a_1, a_2, a_3)\frac{2}{n - \mathrm{III}} + \varphi_3(a_1, a_2, a_3)\frac{1}{(n - \mathrm{III})^2} = 0. \tag{4.347a}$$

Here we have used that in $\overset{\circ}{\mathcal{K}}$

$$\begin{aligned}
&I_\gamma(\overset{(n)}{\mathbf{C}}) = I_\gamma(\mathbf{0}) = \mathbf{0}, \quad I_1(\overset{(n)}{\mathbf{G}}) = \frac{3}{n - \mathrm{III}} \equiv a_1,\\
&I_2(\overset{(n)}{\mathbf{G}}) = \frac{3}{(n - \mathrm{III})^2} \equiv a_2, \quad I_3(\overset{(n)}{\mathbf{G}}) = \frac{1}{(n - \mathrm{III})^3} \equiv a_3.
\end{aligned} \tag{4.347b}$$

For linear models $B_n$, the free energy is also chosen in the form (4.331), where $I_\gamma = I_\gamma(\overset{(n)}{\mathbf{G}})$; in particular, for an isotropic linear continuum we have

$$\overset{\circ}{\rho}\psi = \overset{\circ}{\rho}\psi_0 + \overset{\circ}{\rho}\psi^0 + \left(\frac{l_1}{2} + l_2\right) I_1^2(\overset{(n)}{\mathbf{G}}) - 2l_2 I_2(\overset{(n)}{\mathbf{G}}) + m I_1(\overset{(n)}{\mathbf{G}}), \tag{4.348}$$

and the constitutive equations take the form similar to (4.338):

$$\overset{(n)}{\mathbf{T}} = J(m + l_1 I_1(\overset{(n)}{\mathbf{G}}))\mathbf{E} + 2Jl_2\overset{(n)}{\mathbf{G}}. \tag{4.349}$$

If the continuum in $\overset{\circ}{\mathcal{K}}$ is at an unstressed state, then from (4.328) we obtain the following relations between the constants $\psi^0$, $l_1$, $l_2$ and $m$:

$$m = -\frac{3l_1 + 2l_2}{n - \mathrm{III}}, \quad \psi^0 = \frac{3(3l_1 + 2l_2)}{2\overset{\circ}{\rho}(n - \mathrm{III})^2},$$

hence relations (4.348) and (4.349) take the forms

$$\overset{\circ}{\rho}\psi = \overset{\circ}{\rho}\psi_0 + \frac{3(3l_1 + 2l_2)}{2(n - \mathrm{III})^2} - \frac{3l_1 + 2l_2}{n - \mathrm{III}} I_1(\overset{(n)}{\mathbf{G}}) + \frac{l_1}{2} I_1^2(\overset{(n)}{\mathbf{G}}) + l_2 I_1(\overset{(n)}{\mathbf{G}}{}^2),$$

$$\overset{(n)}{\mathbf{T}} = J\Big((l_1 I_1(\overset{(n)}{\mathbf{G}}) - \frac{3l_1 + 2l_2}{n - \mathrm{III}})\mathbf{E} + 2l_2\overset{(n)}{\mathbf{G}}\Big). \tag{4.350}$$

The same relations can be derived from (4.335) and (4.338) by using the substitution $\overset{(n)}{\mathbf{C}} = \overset{(n)}{\mathbf{G}} - (1/(n - \mathrm{III}))\mathbf{E}$ (see Exercise 4.8.11), i.e. the models $B_n$ (4.350) and $A_n$ (4.335), (4.338) are equivalent; they have only different forms.

In Sect. 4.9 we will show that for some types of ideal continua there exist models $A_n$ and $B_n$ which are not equivalent.

### *4.8.9 Corollaries to the Principle of Material Symmetry for Models $C_n$ and $D_n$ of Elastic Continua*

Establish corollaries to the principle of material symmetry for models $C_n$ of solids.

Let there be a solid with undistorted reference configuration $\overset{\circ}{\mathcal{K}}$ and symmetry group $\overset{\circ}{G}_s$. If constitutive equations of the solid correspond to the models $C_n$ (4.171), (4.174) for each motion $\overset{\circ}{\mathcal{K}} \to \mathcal{K}$, then, according to the principle of material symmetry (4.181), the following relations must hold:

$$
\begin{aligned}
&\overset{(n)}{\mathbf{S}}{}^{*} = \rho(\partial\psi/\partial\overset{(n)}{\mathbf{A}}{}^{*}) \equiv \mathbf{\Phi}(\overset{(n)}{\mathbf{A}}{}^{*}, \overset{*}{\mathbf{O}}, \theta),\\
&\psi = \psi(\overset{(n)}{\mathbf{A}}{}^{*}, \overset{*}{\mathbf{O}}, \theta), \quad \eta = -\partial\psi/\partial\theta,\\
&\overset{\circ}{\mathbf{S}}{}^{*} = \rho(\partial\psi/\partial\overset{*}{\mathbf{O}}) \equiv \overset{\circ}{\mathbf{\Phi}}(\overset{(n)}{\mathbf{A}}{}^{*}, \overset{*}{\mathbf{O}}, \theta)
\end{aligned}
\tag{4.351}
$$

in the motion $\overset{*}{\mathcal{K}} \to \mathcal{K}$ for each configuration $\overset{*}{\mathcal{K}}$ obtained from $\overset{\circ}{\mathcal{K}}$ by $\overset{*}{\mathbf{Q}}$-transformation involved in the group $\overset{\circ}{G}_s$.

Since $\overset{(n)}{\mathbf{S}}$ and $\overset{(n)}{\mathbf{A}}$ are $H$-invariant tensors relative to any orthogonal transformations (see Theorems 4.18 and 4.19) and the tensors $\overset{\circ}{\mathbf{S}}$ and $\mathbf{O}$ are transformed by the laws (4.256) and (4.259), so from (4.351) we obtain that the principle of material symmetry holds, if constitutive equations of the models $C_n$ of solids satisfy the conditions

$$
\mathbf{\Phi}(\overset{(n)}{\mathbf{A}}, \mathbf{O}, \theta) = \mathbf{\Phi}(\overset{(n)}{\mathbf{A}}, \mathbf{O}\cdot\overset{*}{\mathbf{Q}}, \theta), \tag{4.352a}
$$

$$
\frac{\partial\psi}{\partial\theta}(\overset{(n)}{\mathbf{A}}, \mathbf{O}, \theta) = \frac{\partial\psi}{\partial\theta}(\overset{(n)}{\mathbf{A}}, \mathbf{O}\cdot\overset{*}{\mathbf{Q}}, \theta), \tag{4.352b}
$$

$$
\overset{\circ}{\mathbf{\Phi}}(\overset{(n)}{\mathbf{A}}, \mathbf{O}, \theta)\cdot\overset{*}{\mathbf{Q}} = \overset{\circ}{\mathbf{\Phi}}(\overset{(n)}{\mathbf{A}}, \mathbf{O}\cdot\overset{*}{\mathbf{Q}}, \theta), \tag{4.352c}
$$

$$
\psi(\overset{(n)}{\mathbf{A}}, \mathbf{O}, \theta) = \psi(\overset{(n)}{\mathbf{A}}, \mathbf{O}\cdot\overset{*}{\mathbf{Q}}, \theta), \quad \forall\, \overset{*}{\mathbf{Q}} \in \overset{\circ}{G}_s. \tag{4.352d}
$$

**Theorem 4.25.** *The principle of material symmetry holds for the models $C_n$ of ideal solids if and only if the condition (4.352d) for the potential $\psi$ is satisfied.*

▼ The condition (4.352d) is necessary, because if the principle of material symmetry holds, then (4.352d) follows from (4.351).

Prove that the condition (4.352d) is sufficient. The relations (4.352a) and (4.352b) can be proved in the same way as in Theorem 4.23, by differentiating the function $\psi$ in (4.352d) with respect to its arguments $\overset{(n)}{\mathbf{A}}$ and $\theta$.

To prove the relation (4.352c), one should differentiate the function (4.352d) with respect to $\mathbf{O}$:

$$
\begin{aligned}
\overset{\circ}{\mathbf{\Phi}}(\overset{(n)}{\mathbf{A}}, \mathbf{O}, \theta) &= \frac{\partial\psi}{\partial\mathbf{O}}(\overset{(n)}{\mathbf{A}}, \mathbf{O}, \theta) = \frac{\partial\psi}{\partial\mathbf{O}}(\overset{(n)}{\mathbf{A}}, \overset{*}{\mathbf{O}}, \theta)\\
&= \frac{\partial\psi}{\partial\overset{*}{\mathbf{O}}}(\overset{(n)}{\mathbf{A}}, \overset{*}{\mathbf{O}}, \theta)\cdot\overset{*}{\mathbf{Q}}{}^{\mathrm{T}} = \overset{\circ}{\mathbf{\Phi}}(\overset{(n)}{\mathbf{A}}, \mathbf{O}\cdot\overset{*}{\mathbf{Q}}, \theta)\cdot\overset{*}{\mathbf{Q}}{}^{\mathrm{T}}.
\end{aligned}
$$

Here we have used the relation between derivatives of $\psi$ with respect to $\mathbf{O}$ and $\overset{*}{\mathbf{O}}$:

$$\frac{\partial\psi}{\partial\mathbf{O}} = \frac{\partial\psi}{\partial\overset{*}{\mathbf{O}}}\cdot\overset{*}{\mathbf{Q}}^{\mathrm{T}}, \quad \overset{*}{\mathbf{O}} = \mathbf{O}\cdot\overset{*}{\mathbf{Q}}, \tag{4.353}$$

which follows from the differential form

$$d'\psi = \frac{\partial\psi}{\partial\overset{*}{\mathbf{O}}}\cdot\cdot\, d\overset{*}{\mathbf{O}}^{\mathrm{T}} = \frac{\partial\psi}{\partial\overset{*}{\mathbf{O}}}\cdot\cdot\,(\overset{*}{\mathbf{Q}}^{\mathrm{T}}\cdot d\mathbf{O}^{\mathrm{T}}) = \frac{\partial\psi}{\partial\overset{*}{\mathbf{O}}}\cdot\overset{*}{\mathbf{Q}}^{\mathrm{T}}\cdot\cdot\, d\mathbf{O}^{\mathrm{T}} = \frac{\partial\psi}{\partial\mathbf{O}}\cdot\cdot\, d\mathbf{O}^{\mathrm{T}},$$

where $d'\psi$ is the partial differential at fixed $\overset{(n)}{\mathbf{A}}$. ▲

The condition (4.352d) imposed on the function $\psi$ differs considerably from the condition of indifference (4.279c), which has been obtained for models $A_n$.

**Definition 4.14.** A scalar function $P(\overset{(n)}{\mathbf{A}},\mathbf{O})$ of two tensor arguments $\overset{(n)}{\mathbf{A}}$ and $\mathbf{O}$ is called a *pseudoinvariant relative to group* $\overset{\circ}{G}_s$, if

(1) the function is rotary-indifferent relative to the symmetry group $\overset{\circ}{G}_s \subset I$, i.e. for each tensor $\overset{*}{\mathbf{Q}} \in \overset{\circ}{G}_s$ and for every $\overset{(n)}{\mathbf{A}}$ and $\mathbf{O}$ the following condition is satisfied:

$$P(\overset{(n)}{\mathbf{A}},\ \mathbf{O}) = P(\overset{(n)}{\mathbf{A}},\ \mathbf{O}\cdot\overset{*}{\mathbf{Q}}), \tag{4.354}$$

(2) for every $\overset{(n)}{\mathbf{A}}$ and $\mathbf{O}$ the following identity holds:

$$\overset{(n)}{\mathbf{A}}\cdot\frac{\partial P}{\partial\overset{(n)}{\mathbf{A}}} - \frac{\partial P}{\partial\overset{(n)}{\mathbf{A}}}\cdot\overset{(n)}{\mathbf{A}} = \frac{\partial P}{\partial\mathbf{O}}\cdot\mathbf{O}^{\mathrm{T}}. \tag{4.355}$$

The function $\psi(\overset{(n)}{\mathbf{A}},\mathbf{O},\theta)$ satisfying the principle of material symmetry is a pseudoinvariant, because the function satisfies the condition (4.352d) of rotary indifference; and the function satisfies the relation (4.355), because from (4.174) and (4.76) it follows that

$$0 = \overset{\circ}{\mathbf{S}} - \rho\frac{\partial\psi}{\partial\mathbf{O}} = (\overset{(n)}{\mathbf{A}}\cdot\overset{(n)}{\mathbf{S}} - \overset{(n)}{\mathbf{S}}\cdot\overset{(n)}{\mathbf{A}})\cdot\mathbf{O} - \rho\frac{\partial\psi}{\partial\mathbf{O}} = \rho\left(\overset{(n)}{\mathbf{A}}\cdot\frac{\partial\psi}{\partial\overset{(n)}{\mathbf{A}}} - \frac{\partial\psi}{\partial\overset{(n)}{\mathbf{A}}}\cdot\overset{(n)}{\mathbf{A}}\right)\cdot\mathbf{O} - \rho\frac{\partial\psi}{\partial\mathbf{O}}. \tag{4.356}$$

**Theorem 4.26.** *Let* $I_\alpha^{(s)}(\overset{(n)}{\mathbf{C}})$ *be an invariant of the tensor* $\overset{(n)}{\mathbf{C}}$ *relative to* $\overset{\circ}{G}_s$*, then the scalar function*

$$P_\alpha^{(s)}(\overset{(n)}{\mathbf{A}},\ \mathbf{O}) = I_\alpha^{(s)}(\mathbf{O}^{\mathrm{T}}\cdot\overset{(n)}{\mathbf{A}}\cdot\mathbf{O}) \tag{4.357}$$

*is a pseudoinvariant relative to the same group* $\overset{\circ}{G}_s$.

▼ (1) Indeed, if $I_\alpha^{(s)}$ is an invariant of the tensor $\overset{(n)}{\mathbf{C}}$ relative to group $\overset{\circ}{G}_s$, then, by definition, the invariant satisfies the condition of $H$-indifference relative to $\overset{\circ}{G}_s$:

$$I_\alpha^{(s)}(\overset{(n)}{\mathbf{C}}) = I_\alpha^{(s)}(\overset{*}{\mathbf{Q}}^{\mathrm{T}} \cdot \overset{(n)}{\mathbf{C}} \cdot \overset{*}{\mathbf{Q}}) \quad \forall \overset{*}{\mathbf{Q}} \in \overset{\circ}{G}_s. \tag{4.358}$$

But the tensors $\overset{(n)}{\mathbf{C}}$ can always be expressed in terms of $\overset{(n)}{\mathbf{A}}$ by the formula (see Exercise 4.2.8)

$$\overset{(n)}{\mathbf{C}} = \mathbf{O}^{\mathrm{T}} \cdot \overset{(n)}{\mathbf{A}} \cdot \mathbf{O}. \tag{4.359}$$

Substitution of (4.359) into (4.358) yields

$$I_\alpha^{(s)}(\mathbf{O}^{\mathrm{T}} \cdot \overset{(n)}{\mathbf{A}} \cdot \mathbf{O}) = I_\alpha^{(s)}(\overset{*}{\mathbf{Q}}^{\mathrm{T}} \cdot \mathbf{O}^{\mathrm{T}} \cdot \overset{(n)}{\mathbf{A}} \cdot \mathbf{O} \cdot \overset{*}{\mathbf{Q}}) \quad \forall \overset{*}{\mathbf{Q}} \in \overset{\circ}{G}_s. \tag{4.360}$$

On defining the function $P_\alpha^{(s)}$ by formula (4.357), from (4.360) we obtain that the function satisfies the condition

$$P_\alpha^{(s)}(\overset{(n)}{\mathbf{A}},\ \mathbf{O}) = P_\alpha^{(s)}(\mathbf{A},\ \mathbf{O} \cdot \overset{*}{\mathbf{Q}}) \quad \forall \overset{*}{\mathbf{Q}} \in \overset{\circ}{G}_s, \tag{4.361}$$

i.e. the function $P_\alpha^{(s)}$ (4.357) is rotary-indifferent relative to the group $\overset{\circ}{G}_s$.
(2) Show that each scalar function (4.357) satisfies the condition (4.355). At first, determine the derivative $\partial P_\alpha / \partial \overset{(n)}{\mathbf{A}}$ by using the concept of the partial differential at fixed $\mathbf{O}$:

$$\begin{aligned} d' P_\alpha^{(s)} &= \frac{\partial P_\alpha^{(s)}}{\partial \overset{(n)}{\mathbf{A}}} \cdot\cdot\, d\overset{(n)}{\mathbf{A}}^{\mathrm{T}} = \frac{dP_\alpha^{(s)}}{d\overset{(n)}{\mathbf{C}}} \cdot\cdot\, d\overset{(n)}{\mathbf{C}}^{\mathrm{T}} \\ &= \frac{dP_\alpha^{(s)}}{d\overset{(n)}{\mathbf{C}}} \cdot\cdot\, (\mathbf{O}^{\mathrm{T}} \cdot d\overset{(n)}{\mathbf{A}}^{\mathrm{T}} \cdot \mathbf{O}) = \left(\mathbf{O} \cdot \frac{dP_\alpha^{(s)}}{d\overset{(n)}{\mathbf{C}}} \cdot \mathbf{O}^{\mathrm{T}}\right) \cdot\cdot\, d\overset{(n)}{\mathbf{A}}. \end{aligned} \tag{4.362}$$

Here we have taken into account that $P_\alpha^{(s)}$, by (4.357) and (4.359), can be considered as a function of $\overset{(n)}{\mathbf{C}}$. Comparing the second and last equalities, we get

$$\frac{\partial P_\alpha^{(s)}}{\partial \overset{(n)}{\mathbf{A}}} = \mathbf{O} \cdot \frac{dP_\alpha^{(s)}}{d\overset{(n)}{\mathbf{C}}} \cdot \mathbf{O}^{\mathrm{T}}. \tag{4.363}$$

In a similar way, using the partial differential at fixed $\overset{(n)}{\mathbf{A}}$, we obtain

$$\begin{aligned} d'P_\alpha^{(s)} &= \frac{\partial P_\alpha^{(s)}}{\partial \mathbf{O}} \cdot\cdot\, d\mathbf{O}^{\mathrm{T}} = \frac{dP_\alpha^{(s)}}{d\overset{(n)}{\mathbf{C}}} \cdot\cdot\, d\overset{(n)}{\mathbf{C}}{}^{\mathrm{T}} \\ &= \frac{dP_\alpha^{(s)}}{d\overset{(n)}{\mathbf{C}}} \cdot\cdot\, (d\mathbf{O}^{\mathrm{T}} \cdot \overset{(n)}{\mathbf{A}} \cdot \mathbf{O} + \mathbf{O}^{\mathrm{T}} \cdot \overset{(n)}{\mathbf{A}} \cdot d\mathbf{O}) \\ &= \overset{(n)}{\mathbf{A}} \cdot \mathbf{O} \cdot \frac{dP_\alpha^{(s)}}{d\overset{(n)}{\mathbf{C}}} \cdot\cdot\, \mathbf{O}^{\mathrm{T}} + \frac{dP_\alpha^{(s)}}{d\overset{(n)}{\mathbf{C}}} \cdot \mathbf{O}^{\mathrm{T}} \cdot \overset{(n)}{\mathbf{A}} \cdot\cdot\, d\mathbf{O} \\ &= \left( \overset{(n)}{\mathbf{A}} \cdot \mathbf{O} \cdot \frac{dP_\alpha^{(s)}}{d\overset{(n)}{\mathbf{C}}} - \mathbf{O} \cdot \frac{dP_\alpha^{(s)}}{d\overset{(n)}{\mathbf{C}}} \cdot \mathbf{O}^{\mathrm{T}} \cdot \overset{(n)}{\mathbf{A}} \cdot \mathbf{O} \right) \cdot d\mathbf{O}^{\mathrm{T}}. \end{aligned} \tag{4.364}$$

Here we have taken into account that $d\mathbf{O} = -\mathbf{O} \cdot d\mathbf{O}^{\mathrm{T}} \cdot \mathbf{O}$. From (4.364) we get

$$\frac{\partial P_\alpha^{(s)}}{\partial \mathbf{O}} = \overset{(n)}{\mathbf{A}} \cdot \mathbf{O} \cdot \frac{dP_\alpha^{(s)}}{d\overset{(n)}{\mathbf{C}}} - \mathbf{O} \cdot \frac{dP_\alpha^{(s)}}{d\overset{(n)}{\mathbf{C}}} \cdot \mathbf{O}^{\mathrm{T}} \cdot \overset{(n)}{\mathbf{A}} \cdot \mathbf{O}. \tag{4.365}$$

On substituting (4.363) and (4.365) into (4.355), one can verify that the relation (4.355) is identically satisfied:

$$\begin{aligned} &\overset{(n)}{\mathbf{A}} \cdot \frac{\partial P_\alpha^{(s)}}{\partial \overset{(n)}{\mathbf{A}}} - \frac{\partial P_\alpha^{(s)}}{\partial \overset{(n)}{\mathbf{A}}} \cdot \overset{(n)}{\mathbf{A}} - \frac{\partial P_\alpha^{(s)}}{\partial \mathbf{O}} \cdot \mathbf{O}^{\mathrm{T}} = \overset{(n)}{\mathbf{A}} \cdot \mathbf{O} \cdot \frac{dP_\alpha^{(s)}}{d\overset{(n)}{\mathbf{C}}} \cdot \mathbf{O}^{\mathrm{T}} \\ &-\mathbf{O} \cdot \frac{dP_\alpha^{(s)}}{d\overset{(n)}{\mathbf{C}}} \cdot \mathbf{O}^{\mathrm{T}} \cdot \overset{(n)}{\mathbf{A}} - \mathbf{A} \cdot \mathbf{O} \cdot \frac{dP_\alpha^{(s)}}{d\overset{(n)}{\mathbf{C}}} \cdot \mathbf{O}^{\mathrm{T}} + \mathbf{O} \cdot \frac{dP_\alpha^{(s)}}{d\overset{(n)}{\mathbf{C}}} \cdot \mathbf{O}^{\mathrm{T}} \cdot \overset{(n)}{\mathbf{A}} \equiv 0. \end{aligned} \tag{4.366}$$

Thus, each scalar function (4.357) is a pseudoinvariant. ▲

Theorem 4.26 gives the method of construction of pseudoinvariants $P_\alpha^{(s)}$: they can be formed by the corresponding invariants $I_\alpha^{(s)}(\overset{(n)}{\mathbf{C}})$ with the help of formula (4.357). The number $z$ of elements in a functional basis of pseudoinvariants $P_\alpha^{(s)}(\overset{(n)}{\mathbf{A}}, \mathbf{O})$ is determined by the following theorem.

**Theorem 4.27.** *A functional basis of pseudoinvariants $P_\alpha^{(s)}(\overset{(n)}{\mathbf{A}}, \mathbf{O})$ relative to group $\overset{\circ}{G}_s$ consists of $z$ elements, where*

- $z = 3$ *for the isotropy group* $\overset{\circ}{G}_s = I$,
- $z = 5$ *for the transverse isotropy group* $\overset{\circ}{G}_s = T_3$,

- $z = 6$ *for the orthotropy group* $\overset{\circ}{G}_s = O$,
- $z \leqslant 6$ *for each subgroup* $\overset{\circ}{G}_s \subset I$.

▼ (1) Since the symmetric tensor $\overset{(n)}{\mathbf{A}}$ has six independent components and the orthogonal tensor $\mathbf{O}$ has three independent components, the total number of functional pseudoinvariants $P_\alpha^{(s)}$ cannot exceed nine. However, due to the additional constraint (4.355) imposed on the functions $P_\alpha^{(s)}$, the number of independent pseudoinvariants $P_\alpha^{(s)}$ diminishes. The left-hand side of (4.355) is a skew-symmetric tensor, therefore the set (4.355) contains only three independent relations. As a result, we find that the number $z$ of independent pseudoinvariants does not exceed $9 - 3 = 6$ for arbitrary orthogonal subgroup $\overset{\circ}{G}_s \subset I$.
(2) Show that for the isotropy group $I$ the number of pseudoinvariants $P_\alpha^{(s)}$ is equal to three. Perform a proof in two stages.

*Stage 1.* For isotropic continua, the condition (4.354) must be satisfied for the full orthogonal group, i.e. $\forall\, \overset{*}{\mathbf{Q}} \in I$. Since the tensor $\mathbf{O}$ is also orthogonal, the condition (4.354) in the basis $\widehat{\mathbf{c}}_i$ can be rewritten as follows:

$$P_\alpha^{(s)}(\widehat{\overset{(n)}{A}}{}^{ij},\ \widehat{O}^{kl}) = P_\alpha^{(s)}(\widehat{\overset{(n)}{A}}{}^{ij},\ \widehat{O}'^{kl}) \quad \forall \mathbf{O}', \mathbf{O} \in I,\ \forall \overset{(n)}{\mathbf{A}}, \tag{4.367}$$

where $\mathbf{O}' = \mathbf{O} \cdot \overset{*}{\mathbf{O}} \in I$. The component form $P_\alpha^{(s)}(\widehat{\overset{(n)}{A}}{}^{ij},\ \widehat{O}^{kl})$ can be considered as a function of nine variables, which is given on the set $\mathbb{R}^6 \times V_0$, where $V_0 \subset \mathbb{R}^3$ is the range of components of the orthogonal tensors $\widehat{O}^{kl}$. Since in the condition (4.367) the components $\widehat{O}^{kl}$ and $\widehat{O}'^{kl}$ take all values of $V_0$ ($I$ is the full orthogonal group), so from the condition (4.367) it follows that the function $\psi$ remains unchanged on the set $V_0$, i.e. the function is a constant; and within the domain of definition $\mathbb{R}^6 \times V_0$ the function is independent of $\widehat{O}^{kl}$, i.e. $P_\alpha^{(s)}$ are independent of the rotation tensor $\mathbf{O}$:

$$P_\alpha^{(s)} = P_\alpha^{(s)}(\overset{(n)}{\mathbf{A}}) \quad \text{or} \quad P_\alpha^{(s)} = P_\alpha^{(s)}(\widehat{\overset{(n)}{A}}{}^{ij}). \tag{4.368}$$

*Stage 2.* Notice that although the indifference condition (4.281) has not been imposed on the function (4.368), the function $P_\alpha^{(s)}$ is nevertheless indifferent, however, possibly relative to not the whole group $I$ but at least relative to the triclinic group $\overset{\circ}{G}_s = E \subset I$ (this group contains only the identical $\overset{*}{\mathbf{Q}} = \mathbf{E}$ and centrally symmetric $\overset{*}{\mathbf{Q}} = -\mathbf{E}$ transformations, and the condition (4.281) is always satisfied for each function $P_\alpha^{(s)}(\overset{(n)}{\mathbf{A}})$).

The problem consists in finding a maximum subgroup $\overset{\circ}{G}_s$, relative to which the function (4.368) is indifferent.

We know that in the group $E$ there exists the special functional basis of independent invariants (4.302). Then we can apply formula (4.304) to $P^{(s)}_\alpha(\overset{(n)}{\mathbf{A}},\theta)$ and consider $P^{(s)}_\alpha$ as a function of the functional basis $\widetilde{I}^{(E)}_\alpha(\overset{(n)}{\mathbf{A}})$ (4.302):

$$P^{(s)}_\alpha = P^{(s)}_\alpha(I_1(\overset{(n)}{\mathbf{A}}),\dots, I_3(\overset{(n)}{\mathbf{A}}),\ \widetilde{I}_4(\overset{(n)}{\mathbf{A}}),\ \dots\ \widetilde{I}_6(\overset{(n)}{\mathbf{A}}),\ \theta). \tag{4.369}$$

Substituting the potential (4.369) into (4.355), we obtain

$$\sum_{\gamma=1}^{6}\varphi_\gamma\left(\overset{(n)}{\mathbf{A}}\cdot\frac{\partial \widetilde{I}^{(E)}_\gamma}{\partial \overset{(n)}{\mathbf{A}}} - \frac{\partial I^{(E)}_\gamma}{\partial \overset{(n)}{\mathbf{A}}}\cdot\overset{(n)}{\mathbf{A}}\right) = 0, \tag{4.370}$$

where $\varphi_\gamma = \partial P^{(s)}_\alpha/\partial I^{(E)}_\gamma$.

According to formula (4.323) of differentiation of the principal invariants $I_1(\overset{(n)}{\mathbf{A}})$ and the formula of differentiation of the remaining three invariants:

$$\frac{\partial \widetilde{I}^{(E)}_{\gamma+3}}{\partial \overset{(n)}{\mathbf{A}}}(\overset{(n)}{\mathbf{A}}) = \widehat{\mathbf{O}}_\gamma,\quad \alpha,\ \beta,\ \gamma = 1,\ 2,\ 3;\quad \alpha\neq\beta\neq\gamma\neq\alpha, \tag{4.371}$$

$$\widehat{\mathbf{O}}_\gamma \equiv \frac{1}{2}(\widehat{\mathbf{c}}_\alpha\otimes\widehat{\mathbf{c}}_\beta + \widehat{\mathbf{c}}_\beta\otimes\widehat{\mathbf{c}}_\alpha), \tag{4.372}$$

Eq. (4.370) takes the form

$$\begin{aligned}&\psi_1\mathbf{E} + \psi_2\overset{(n)}{\mathbf{A}}{}^2 + \varphi_3\overset{(n)}{\mathbf{A}}{}^3 + \frac{1}{2}\sum_{\gamma=1}^{3}\varphi_{\gamma+3}\mathbf{O}_\gamma\cdot\overset{(n)}{\mathbf{A}}\\ &= \psi_1\mathbf{E} + \psi_2\overset{(n)}{\mathbf{A}}{}^2 + \varphi_3\overset{(n)}{\mathbf{A}}{}^3 + \frac{1}{2}\sum_{\gamma=1}^{3}\varphi_{\gamma+3}\overset{(n)}{\mathbf{A}}\cdot\mathbf{O}_\gamma,\end{aligned} \tag{4.373}$$

where $\psi_1$ and $\psi_2$ are determined by formulae (4.322a).

Hence the following identity must hold:

$$\sum_{\gamma=1}^{3}\varphi_{\gamma+3}(\mathbf{O}_\gamma\cdot\overset{(n)}{\mathbf{A}} - \overset{(n)}{\mathbf{A}}\cdot\mathbf{O}_\gamma) = 0. \tag{4.374}$$

Since the product of the tensors $\mathbf{O}_\gamma$ and $\overset{(n)}{\mathbf{A}}$ is not commutative (one can check this fact with multiplying $\mathbf{O}_\gamma$ (4.315) by $\overset{(n)}{\mathbf{A}} = \widehat{\overset{(n)}{A}}{}^{ij}\widehat{\mathbf{c}}_i\otimes\widehat{\mathbf{c}}_j$), we obtain that the identity (4.374) holds if and only if the coefficients $\varphi_{3+\gamma}$ $(\gamma = 1, 2, 3)$ identically vanish:

$$\varphi_\gamma = \partial P^{(s)}_\alpha/\partial \widetilde{I}^{(E)}_\gamma \equiv 0,\quad \gamma = 4, 5, 6. \tag{4.375}$$

Thus, for an isotropic continuum, each pseudoinvariant $P$ (4.369) cannot depend on the invariants $\widetilde{I}_\gamma^{(E)}(\overset{(n)}{\mathbf{A}})$, $\gamma = 4, 5, 6$; but $P$ depends only on three principal invariants $I_\gamma(\overset{(n)}{\mathbf{A}})$, i.e. for isotropic continua a basis of pseudoinvariants consists of three elements.
(3) For a transversely isotropic continuum, a proof is analogous (see Exercise 4.8.7). ▲

Consider several important corollaries of the theorem.

For isotropic continua, the potential $\psi$ (see (4.171)) is independent of the tensor $\mathbf{O}$ (because $\psi$ is a pseudoinvariant), and the rotation stress tensor $\overset{\circ}{\mathbf{S}}$, due to this fact, is identically zero (with taking formula (4.176) into account):

$$\psi = \psi(\overset{(n)}{\mathbf{A}}, \theta), \quad \overset{\circ}{\mathbf{S}} = \rho \frac{\partial \psi(\overset{(n)}{\mathbf{A}}, \theta)}{\partial \mathbf{O}} = \mathbf{0}. \tag{4.376}$$

Since for orthogonal groups $\overset{\circ}{G}_s = T_3$, $O$ and $I$ the number $z$ of elements in the functional basis of pseudoinvariants $P_\alpha^{(s)}(\overset{(n)}{\mathbf{A}}, \mathbf{O})$ is equal to the number $r$ of elements in the basis of invariants $I_\alpha^{(s)}(\overset{(n)}{\mathbf{C}})$ of the corresponding group $\overset{\circ}{G}_s$ (see Theorems 4.24 and 4.27), and for each invariant $I_\alpha^{(s)}(\overset{(n)}{\mathbf{C}})$ we can construct a pseudoinvariant by formulae (4.357) and (4.359), so the basis of pseudoinvariants $P_\alpha^{(s)}$ can be constructed only with the help of invariants $I_\alpha^{(s)}(\overset{(n)}{\mathbf{C}})$ (4.357). Hence, for the models $C_n$, the potential $\psi$ can always be written in the form

$$\psi = \psi(\overset{(n)}{\mathbf{A}}, \mathbf{O}, \theta) = \psi(P_\alpha^{(s)}(\overset{(n)}{\mathbf{A}}, \mathbf{O}), \theta) = \psi(I_\alpha^{(s)}(\mathbf{O}^{\mathrm{T}} \cdot \overset{(n)}{\mathbf{A}} \cdot \mathbf{O}), \theta). \tag{4.377}$$

On substituting (4.377) into (4.174), we obtain the general representation of constitutive equations for models $C_n$ of ideal continua in the tensor basis (this is an analog of (4.311)):

$$\overset{(n)}{\mathbf{S}} = \mathbf{\Phi}(\overset{(n)}{\mathbf{A}}, \mathbf{O}, \theta) = \sum_{\gamma=1}^{r} \varphi_\gamma \frac{\partial I_\gamma^{(s)}}{\partial \overset{(n)}{\mathbf{A}}}, \tag{4.378}$$

$$\varphi_\gamma = \rho(\partial \psi / \partial I_\gamma^{(s)}) = \varphi_\gamma(I_\alpha^{(s)}(\mathbf{O}^{\mathrm{T}} \cdot \overset{(n)}{\mathbf{A}} \cdot \mathbf{O}), \theta). \tag{4.379}$$

In particular, for an isotropic solid, the relations (4.378) and (4.379) of the model $C_n$ have standard forms, which are similar to (4.322):

$$\overset{(n)}{\mathbf{S}} = \psi_1 \mathbf{E} + \psi_2 \overset{(n)}{\mathbf{A}} + \varphi_3 \overset{(n)}{\mathbf{A}}{}^2, \tag{4.380}$$

$$\psi_1 = \varphi_1 + \varphi_2 I_1 + \varphi_3 I_2, \quad -\psi_2 = \varphi_2 + I_1 \varphi_3, \quad \varphi_\gamma = \rho \frac{\partial \psi}{\partial I_\gamma}, \quad \psi = \psi(I_\gamma(\overset{(n)}{\mathbf{A}}), \theta).$$

Let us establish a connection between the relation (4.378) of models $C_n$ and the relation (4.311) of models $A_n$ with taking into account that the quasienergetic stress tensors and quasienergetic deformation tensors are connected to the corresponding energetic tensors with the help of the rotation tensor $\mathbf{O}$ (see Exercise 4.2.3):

$$\overset{(n)}{\mathbf{S}} = \mathbf{O} \cdot \overset{(n)}{\mathbf{T}} \cdot \mathbf{O}^{\mathrm{T}}. \tag{4.381}$$

Substitution of (4.381) and (4.359) into (4.378) yields

$$\mathbf{O} \cdot \overset{(n)}{\mathbf{T}} \cdot \mathbf{O}^{\mathrm{T}} = \sum_{\gamma=1}^{r} \varphi_\gamma \mathbf{O} \cdot \frac{\partial I_\gamma^{(s)}}{\partial \overset{(n)}{\mathbf{C}}} \cdot \mathbf{O}^{\mathrm{T}}. \tag{4.382}$$

Comparing (4.382) with (4.311), we conclude that the models $A_n$ and $C_n$ of ideal continua differ from each other only by multiplication of the constitutive equations by the rotation tensor $\mathbf{O}$. Thus, we can consider these models as different forms of the same relations.

However, we have shown that the form $C_n$ (4.378) of constitutive equations is the only possible form with the rotation tensor $\mathbf{O}$, which satisfies the principle of material symmetry.

In addition, the forms $A_n$ and $C_n$ prove to be rather different, when they have been differentiated with respect to time (that will be shown in Sect. 4.8.12); therefore, we will distinguish the models $A_n$ and $C_n$ from each other. Notice that for nonideal continua a distinction between the models $A_n$ and $C_n$ is more considerable, that will be shown in Chapters 7 and 8.

Since the quasienergetic measures $\overset{(n)}{\mathbf{g}}$ as well as the quasienergetic deformation tensors $\overset{(n)}{\mathbf{A}}$ are $H$-invariant relative to the group $I$, so all the formulae of Sect. 4.8.8 still hold for the models $D_n$ of ideal solids (4.161) with substitution $\overset{(n)}{\mathbf{A}} \to \overset{(n)}{\mathbf{g}}$. In particular, the constitutive equation (4.380) for an isotropic solid in the models $D_n$ takes the form

$$\overset{(n)}{\mathbf{S}} = \psi_1 \mathbf{E} + \psi_2 \overset{(n)}{\mathbf{g}} + \varphi_3 \overset{(n)}{\mathbf{g}}{}^2, \quad \psi = \psi(I_\gamma(\overset{(n)}{\mathbf{g}}), \theta). \tag{4.383}$$

## 4.8.10 General Representation of Constitutive Equations for All Models of Elastic Continua

The constitutive equations (4.311), (4.346), and (4.378) for models $A_n$, $B_n$, $C_n$ and $D_n$ of elastic continua with the help of the generalized energetic tensors $\overset{(n)}{\mathbf{T}}_G$ and $\overset{(n)}{\mathbf{C}}_G$ (4.91) can be written in the unified form

$$\overset{(n)}{\mathbf{T}}_G = \mathcal{F}_G(\overset{(n)}{\mathbf{C}}_G, \mathbf{O}, \theta) = \sum_{\gamma=1}^{r} \varphi_\gamma \mathbf{I}_{\gamma G}^{(s)}, \tag{4.384}$$

where we have denoted the derivative tensors

$$\mathbf{I}^{(s)}_{\gamma G} = \partial I^{(s)}_{\gamma G}/\partial \overset{(\mathrm{n})}{\mathbf{C}}_G, \quad \varphi_\gamma = \rho(\partial\psi/\partial I^{(s)}_{\gamma G}), \quad \psi = \psi(I^{(s)}_{\gamma G}, \theta),$$
$$I^{(s)}_{\gamma G} = h_G I^{(s)}_\gamma(\overset{(\mathrm{n})}{\mathbf{C}}_G) + (1-h_G)I^{(s)}_\gamma(\mathbf{O}^{\mathrm{T}}\cdot \overset{(\mathrm{n})}{\mathbf{C}}_G \cdot \mathbf{O}), \qquad (4.385)$$
$$G = A,\ B,\ C,\ D; \qquad n = \mathrm{I}, \ldots, \mathrm{V}.$$

Here we have introduced the following functions, which are indicators of the model type:

$$h_G = \begin{cases} 1, & G = A,\ B, \\ 0, & G = C,\ D, \end{cases} \qquad \bar{h}_G = \begin{cases} 1, & G = A,\ C, \\ 0, & G = B,\ D. \end{cases} \qquad (4.386)$$

The constitutive equation (4.384) for all the models can always be reduced to the dependence of the Cauchy stress tensor $\mathbf{T}$ on the deformation gradient $\mathbf{F}$. For the purpose we should use relations (4.37) and (4.67) expressing $\mathbf{T}$ in terms of $\overset{(\mathrm{n})}{\mathbf{T}}$ and $\overset{(\mathrm{n})}{\mathbf{S}}$. Substituting the constitutive equations (4.311) of models $A_n$, (4.346) of models $B_n$, (4.378) of models $C_n$ and $D_n$ into relations (4.37) and (4.67), we obtain the *general representation of constitutive equations for all the models of elastic continua*:

$$\mathbf{T} = \overset{(\mathrm{n})}{\mathcal{F}}_G(\mathbf{F},\ \theta), \qquad G = A,\ B,\ C,\ D, \qquad (4.387)$$

where

$$\overset{(\mathrm{n})}{\mathcal{F}}_G(\mathbf{F},\ \theta) \equiv {}^4\overset{(\mathrm{n})}{\mathbf{E}}_G \cdot\cdot\, \mathcal{F}_G(\overset{(\mathrm{n})}{\mathbf{C}}_G,\ \theta) = \sum_{\gamma=1}^{r} \varphi_\gamma\, {}^4\overset{(\mathrm{n})}{\mathbf{E}}_G \cdot\cdot\, \mathbf{I}^{(s)}_{\gamma G}, \qquad (4.388)$$

and

$${}^4\overset{(\mathrm{n})}{\mathbf{E}}_G = h_G\, {}^4\overset{(\mathrm{n})}{\mathbf{E}} + (1-h_G)\, {}^4\overset{(\mathrm{n})}{\mathbf{Q}}$$
$$= \sum_{\alpha,\beta=1}^{3} E_{\alpha\beta}\mathbf{p}_\alpha \otimes \mathbf{p}_\beta \otimes (h_G \overset{\circ}{\mathbf{p}}_\beta \otimes \overset{\circ}{\mathbf{p}}_\alpha + (1-h_G)\mathbf{p}_\beta \otimes \mathbf{p}_\alpha) \qquad (4.389)$$

are the generalized tensors of energetic equivalence.

For the models at $n = \mathrm{I}$ and V, the tensors ${}^4\overset{(\mathrm{n})}{\mathbf{E}}_G$ can be written in the explicit form (see Exercise 4.2.21) in terms of $\mathbf{F}$ and $\mathbf{V}$, therefore for models $A_n$, $B_n$, $C_n$ and $D_n$ ($n = \mathrm{I}$ and V) the tensor function (4.387) has the explicit form

$$\overset{\mathrm{I}}{\mathcal{F}}_A = \mathbf{F}^{-1\mathrm{T}} \cdot \mathcal{F}(\overset{\mathrm{I}}{\mathbf{C}}, \theta) \cdot \mathbf{F}^{-1}, \quad \overset{\mathrm{I}}{\mathcal{F}}_B = \mathbf{F}^{-1\mathrm{T}} \cdot \mathcal{F}(\overset{\mathrm{I}}{\mathbf{G}}, \theta) \cdot \mathbf{F}^{-1},$$
$$\overset{\mathrm{V}}{\mathcal{F}}_A = \mathbf{F} \cdot \mathcal{F}(\overset{\mathrm{V}}{\mathbf{C}}, \theta) \cdot \mathbf{F}^{\mathrm{T}}, \quad \overset{\mathrm{V}}{\mathcal{F}}_B = \mathbf{F} \cdot \mathcal{F}(\overset{\mathrm{V}}{\mathbf{G}}, \theta) \cdot \mathbf{F}^{\mathrm{T}},$$
$$\overset{\mathrm{I}}{\mathcal{F}}_C = \mathbf{V}^{-1} \cdot \boldsymbol{\Phi}(\overset{\mathrm{I}}{\mathbf{A}}, \mathbf{O}, \theta) \cdot \mathbf{V}^{-1}, \quad \overset{\mathrm{I}}{\mathcal{F}}_D = \mathbf{V}^{-1} \cdot \boldsymbol{\Phi}(\overset{\mathrm{I}}{\mathbf{g}}, \mathbf{O}, \theta) \cdot \mathbf{V}^{-1}, \quad (4.390)$$
$$\overset{\mathrm{V}}{\mathcal{F}}_C = \mathbf{V} \cdot \boldsymbol{\Phi}(\overset{\mathrm{V}}{\mathbf{A}}, \mathbf{O}, \theta) \cdot \mathbf{V}, \quad \overset{\mathrm{V}}{\mathcal{F}}_D = \mathbf{V} \cdot \boldsymbol{\Phi}(\overset{\mathrm{V}}{\mathbf{g}}, \mathbf{O}, \theta) \cdot \mathbf{V}.$$

The generalized energetic tensors $\overset{(n)}{\mathbf{C}}_G$ can be expressed in terms of $\mathbf{F}$ by formulae (4.41), (4.42), (4.65), and (4.71) (see Exercise 4.2.23). These formulae can be written in the unified form

$$\overset{(n)}{\mathbf{C}}_G(\mathbf{F}) = \frac{1}{n-\mathrm{III}}((h_G \mathbf{F}^{\mathrm{T}} \cdot \mathbf{F} + (1-h_G)\mathbf{F} \cdot \mathbf{F}^{\mathrm{T}})^{(n-\mathrm{III})/2} - \bar{h}_G \mathbf{E}, \quad (4.391a)$$
$$n = \mathrm{I}, \mathrm{II}, \mathrm{IV}, \mathrm{V}.$$

For these models at $n = \mathrm{III}$, in place of relations (4.391a) Eq. (4.387) is complemented with the differential relations (2.317) and (4.43b), (4.62), (4.71b) for determining the tensors $\overset{\mathrm{III}}{\mathbf{C}}_G$ in terms of $\mathbf{F}$. These relations can also be written in the generalized form

$$\overset{\mathrm{III}}{\mathbf{C}}{}^{\bullet}_G = \frac{h_G}{2}\mathbf{U}^{-1} \cdot (\dot{\mathbf{F}}^{\mathrm{T}} \cdot \mathbf{F} + \mathbf{F}^{\mathrm{T}} \cdot \dot{\mathbf{F}}) \cdot \mathbf{U}^{-1} + \frac{1-h_G}{2}\mathbf{V}^{-1} \cdot (\dot{\mathbf{F}} \cdot \mathbf{F}^{\mathrm{T}} + \mathbf{F} \cdot \dot{\mathbf{F}}^{\mathrm{T}}) \cdot \mathbf{V}^{-1},$$
$$\overset{\mathrm{III}}{\mathbf{C}}_G(0) = \mathbf{0}, \quad G = A, C; \quad \overset{\mathrm{III}}{\mathbf{C}}_G(0) = \mathbf{E}, \quad G = B, D. \quad (4.391b)$$

Fractional and negative powers of tensors of the type $\mathbf{F}^{\mathrm{T}} \cdot \mathbf{F}$ are determined with the help of the passage to the eigenbases $\mathbf{p}_i$ and $\overset{\circ}{\mathbf{p}}_i$. According to the representation (2.140) of tensors $\mathbf{F}^{\mathrm{T}} \cdot \mathbf{F}$ and $\mathbf{F} \cdot \mathbf{F}^{\mathrm{T}}$ in the eigenbases, we can rewrite formulae (4.391a) in the equivalent form

$$\overset{(n)}{\mathbf{C}}_G = \frac{1}{n-\mathrm{III}} \sum_{\alpha=1}^{3} (\lambda_\alpha^{n-\mathrm{III}}(h_G \overset{\circ}{\mathbf{p}}_\alpha \otimes \overset{\circ}{\mathbf{p}}_\alpha + (1-h_G)\mathbf{p}_\alpha \otimes \mathbf{p}_\alpha)) - \frac{\bar{h}_G}{n-\mathrm{III}}\mathbf{E}, \quad (4.391c)$$
$$n = \mathrm{I}, \mathrm{II}, \mathrm{IV}, \mathrm{V}.$$

*Remark.* The constitutive equation (4.387) can be represented as a function of the inverse gradient $\mathbf{F}^{-1}$, that is sometimes convenient for formulation of elasticity problems. For the purpose, the tensors of energetic equivalence ${}^4\overset{(n)}{\mathbf{E}}$ and ${}^4\overset{(n)}{\mathbf{Q}}$, and $\overset{(n)}{\mathbf{C}}_G$ should be expressed in terms of $\mathbf{F}^{-1}$:

$$\overset{(n)}{\mathbf{C}}_G(\mathbf{F}^{-1}) = \frac{1}{n-\mathrm{III}}((h_G \mathbf{F}^{-1} \cdot \mathbf{F}^{-1\mathrm{T}} + (1-h_G)\mathbf{F}^{-1\mathrm{T}} \cdot \mathbf{F}^{-1})^{(\mathrm{III}-n)/2} - \bar{h}_G \mathbf{E}), \quad (4.392)$$

where eigenvalues $\lambda_\alpha$ and eigenvectors $\overset{\circ}{\mathbf{p}}_\alpha$, $\mathbf{p}_\alpha$ must be expressed in terms of $\mathbf{F}^{-1}$, i.e. in place of (2.140) we must consider the equation

$$\mathbf{F}^{-1\mathrm{T}} \cdot \mathbf{F}^{-1} = \sum_{\alpha=1}^{3} \lambda_\alpha^{-2} \mathbf{p}_\alpha \otimes \mathbf{p}_\alpha$$

to determine $\lambda_\alpha$ and $\mathbf{p}_\alpha$. Then in place of (4.387) we have the constitutive equation

$$\mathbf{T} = \overset{(\mathrm{n})}{\mathcal{F}}_G(\mathbf{F}^{-1}, \theta), \tag{4.393}$$

where $\overset{(\mathrm{n})}{\mathcal{F}}_G$ is expressed by the same formulae (4.388), (4.389) and (4.385), and $\overset{(\mathrm{n})}{\mathbf{C}}_G$ is considered as a function in the form (4.392). □

## 4.8.11 Representation of Constitutive Equations of Isotropic Elastic Continua in Eigenbases

The constitutive equations (4.322), (4.347), (4.380), and (4.383) of models $A_n$, $B_n$, $C_n$, and $D_n$ for isotropic elastic continua can be expressed in the eigenbases $\overset{\circ}{\mathbf{p}}_\alpha$ or $\mathbf{p}_\alpha$. According to the results of Exercise 4.2.4, the tensors $\overset{(\mathrm{n})}{\mathbf{C}}$ and $\overset{(\mathrm{n})}{\mathbf{G}}$ can be resolved for the basis $\overset{\circ}{\mathbf{p}}_\alpha$, and the tensors $\overset{(\mathrm{n})}{\mathbf{A}}$ and $\overset{(\mathrm{n})}{\mathbf{g}}$ – for the basis $\mathbf{p}_\alpha$. On substituting the resolutions into the constitutive equations (4.322), (4.347), (4.380) and (4.383), we obtain that the tensors $\overset{(\mathrm{n})}{\mathbf{T}}$ and $\overset{(\mathrm{n})}{\mathbf{S}}$ at $n$ = I, II, IV, V are diagonal in the bases $\overset{\circ}{\mathbf{p}}_\alpha$ and $\mathbf{p}_\alpha$:

$$\overset{(\mathrm{n})}{\mathbf{T}} = \sum_{\alpha=1}^{3} \overset{\circ(n)}{\sigma}_\alpha \overset{\circ}{\mathbf{p}}_\alpha \otimes \overset{\circ}{\mathbf{p}}_\alpha, \quad \overset{(\mathrm{n})}{\mathbf{S}} = \sum_{\alpha=1}^{3} \overset{(\mathrm{n})}{\sigma}_\alpha \mathbf{p}_\alpha \otimes \mathbf{p}_\alpha, \tag{4.394}$$

Here $\overset{\circ(n)}{\sigma}_\alpha$ and $\overset{(\mathrm{n})}{\sigma}_\alpha$ are eigenvalues of the stress tensors, which for models $B_n$ and $D_n$ have the forms

$$\begin{aligned} \overset{(\mathrm{n})}{\sigma}_\alpha &= \psi_1 + \frac{\psi_2}{n - \mathrm{III}} \lambda_\alpha^{n-\mathrm{III}} + \frac{\varphi_3}{(n-\mathrm{III})^2} \lambda_\alpha^{2(n-\mathrm{III})}, \quad \varphi_\alpha = \rho \frac{\partial \psi}{\partial I_\alpha(\overset{(\mathrm{n})}{\mathbf{G}})}, \\ \overset{\circ(n)}{\sigma}_\alpha &= \psi_1 + \frac{\psi_2}{n - \mathrm{III}} \lambda_\alpha^{n-\mathrm{III}} + \frac{\varphi_3}{(n-\mathrm{III})^2} \lambda_\alpha^{2(n-\mathrm{III})}, \quad \varphi_\alpha = \rho \frac{\partial \psi}{\partial I_\alpha(\overset{(\mathrm{n})}{\mathbf{g}})}, \end{aligned} \tag{4.395}$$

$$\psi_1 = \varphi_1 + \varphi_2 I_1 + \varphi_3 I_2, \quad -\psi_2 = \varphi_2 + I_1 \varphi_3. \tag{4.396}$$

For models $A_n$ and $C_n$, analogous relations are given in Exercise 4.8.9. Since the principal invariants of the tensors $\overset{(n)}{\mathbf{G}}$ and $\overset{(n)}{\mathbf{g}}$ are coincident (see Exercise 4.8.2), i.e. $I_\alpha(\overset{(n)}{\mathbf{G}}) = I_\alpha(\overset{(n)}{\mathbf{g}})$, the eigenvalues $\overset{(n)}{\sigma}_\alpha$ and $\overset{\circ(n)}{\sigma}_\alpha$ of the tensors $\overset{(n)}{\mathbf{S}}$ and $\overset{(n)}{\mathbf{T}}$ for isotropic continua coincide as well:

$$\overset{(n)}{\sigma}_\alpha = \overset{\circ(n)}{\sigma}_\alpha, \quad \alpha = 1, 2, 3. \tag{4.397}$$

**Theorem 4.28.** *For models $A_n$, $B_n$, $C_n$ and $D_n$ of isotropic elastic continua, eigenvalues $\overset{(n)}{\sigma}_\alpha$ of the tensors $\overset{(n)}{\mathbf{T}}$ and $\overset{(n)}{\mathbf{S}}$ can always be represented in the form*

$$\overset{(n)}{\sigma}_\alpha = \rho\lambda_\alpha^{\mathrm{IV}-n}\frac{\partial\psi}{\partial\lambda_\alpha}, \quad \alpha = 1,2,3, \quad n = \mathrm{I,\ II,\ IV,\ V}; \quad \rho = \overset{\circ}{\rho}\lambda_1\lambda_2\lambda_3, \tag{4.398}$$

*and $\psi$ can be considered as a function of $\lambda_\alpha$:*

$$\psi = \psi(\lambda_1, \lambda_2, \lambda_3, \theta). \tag{4.399}$$

▼ Since, for models $B_n$ and $D_n$ of isotropic continua, $\psi$ is a function of $I_\alpha(\overset{(n)}{\mathbf{G}}) = I_\alpha(\overset{(n)}{\mathbf{g}})$ and $\theta$ and the principal invariants $I_\alpha(\overset{(n)}{\mathbf{G}})$ are uniquely expressed in terms of $\lambda_\alpha$ (see Exercise 4.8.3), so the potential $\psi$ can always be considered as a function in the form (4.399).

To prove formula (4.398), we differentiate the composite function $\psi = \psi(I_\beta, \theta)$ with respect to $\lambda_\alpha$ and multiply the result by $\lambda_\alpha^{\mathrm{IV}-n}$:

$$\begin{aligned}
\lambda_\alpha^{\mathrm{IV}-n}\frac{\partial\psi}{\partial\lambda_\alpha} &= \lambda_\alpha^{\mathrm{IV}-n}\sum_{\beta=1}^{3}\frac{\partial\psi}{\partial I_\beta}\frac{\partial I_\beta}{\partial\lambda_\alpha} = \lambda_\alpha^{\mathrm{IV}-n}\sum_{\alpha=1}^{3}\varphi_\beta\frac{\partial I_\beta}{\partial\lambda_\alpha} \\
&= \lambda_\alpha^{\mathrm{IV}-n}\Big(\varphi_1\lambda_\alpha^{n-\mathrm{IV}} + \frac{\varphi_2}{n-\mathrm{III}}\lambda_\alpha^{n-\mathrm{IV}}(\lambda_\beta^{n-\mathrm{III}} + \lambda_\gamma^{n-\mathrm{III}}) \\
&\quad + \frac{\varphi_3}{(n-\mathrm{III})^2}\lambda_\alpha^{n-\mathrm{IV}}(\lambda_\beta\lambda_\gamma)^{n-\mathrm{III}}\Big) \\
&= \varphi_1 + \frac{\varphi_2}{n-\mathrm{III}}(\lambda_\beta^{n-\mathrm{III}} + \lambda_\gamma^{n-\mathrm{III}}) + \frac{\varphi_3}{(n-\mathrm{III})^2}(\lambda_\beta\lambda_\gamma)^{n-\mathrm{III}}.
\end{aligned} \tag{4.400}$$

Here we have used the results of Exercise 4.8.4.

On the other hand, substitution of (4.396) into (4.395) yields

$$\overset{(n)}{\sigma}_{\alpha} = \varphi_1 + \varphi_2\left(I_1 - \frac{\lambda_\alpha^{n-\text{III}}}{n-\text{III}}\right) + \varphi_3\left(I_2 - I_1\frac{\lambda_\alpha^{n-\text{III}}}{n-\text{III}} + \frac{\lambda_\alpha^{2(n-\text{III})}}{(n-\text{III})^2}\right). \quad (4.401)$$

According to the relations between $I_1$, $I_2$ and $\lambda_\alpha$ (see Exercise 4.8.3), we see that the expressions (4.400) and (4.401) exactly coincide as was to be proved. A proof of the theorem for models $A_n$ and $C_n$ should be performed as Exercise 4.8.10. ▲

Formulae (4.394), (4.398) and (4.399) are called the *representation of the constitutive equations of isotropic continua in the eigenbasis.*

Let us consider now the case when $n = \text{III}$.

**Theorem 4.28a.** *For isotropic elastic continua, the tensor* $\mathbf{B}$ *defined by (2.317) is energetically equivalent to the logarithmic deformation tensor* $\overset{\circ}{\mathbf{H}}$ *(2.167); the tensor* $\overset{\text{III}}{\mathbf{G}}$ *defined by (4.43a) is energetically equivalent to the logarithmic deformation measure* $\overset{\circ}{\mathbf{H}}_1$ *(2.167); the tensor* $\mathbf{Y}$ *(4.62) – to the tensor* $\tilde{\mathbf{H}}$*; the measure* $\overset{\text{III}}{\mathbf{g}}$ *(4.71a) – to the measure* $\mathbf{H}_1$*, in the sense that the following relations hold:*

$$\begin{aligned} w_{(i)} &= \overset{\text{III}}{\mathbf{T}}\cdot\cdot\dot{\mathbf{B}} = \overset{\text{III}}{\mathbf{T}}\cdot\cdot\overset{\circ}{\mathbf{H}}{}^{\bullet} = \mathbf{S}\cdot\cdot\dot{\mathbf{Y}} = \mathbf{S}\cdot\cdot\dot{\mathbf{H}} = \overset{\text{III}}{\mathbf{T}}\cdot\cdot\overset{\text{III}}{\mathbf{G}}{}^{\bullet} \\ &= \overset{\text{III}}{\mathbf{T}}\cdot\cdot\overset{\circ}{\mathbf{H}}{}_1^{\bullet} = \overset{\text{III}}{\mathbf{S}}\cdot\cdot\overset{\text{III}}{\mathbf{g}}{}^{\bullet} = \overset{\text{III}}{\mathbf{S}}\cdot\cdot\dot{\mathbf{H}}_1 = \sum_{\alpha=1}^{3}\frac{\overset{\text{III}}{\sigma}_\alpha}{\lambda_\alpha}\dot{\lambda}_\alpha, \end{aligned} \quad (4.402)$$

*where* $\overset{\text{III}}{\sigma}_\alpha = \sigma_\alpha$ *are eigenvalues of the tensors* $\overset{\text{III}}{\mathbf{T}}$ *and* $\mathbf{T}$*, which are connected to eigenvalues* $\lambda_\alpha$ *of the tensors* $\mathbf{U}$ *and* $\mathbf{V}$ *by the relations*

$$\sigma_\alpha = \overset{\text{III}}{\sigma}_\alpha = \rho\lambda_\alpha\frac{\partial\psi}{\partial\lambda_\alpha}, \quad \psi = \psi(\lambda_1, \lambda_2, \lambda_3, \theta). \quad (4.403)$$

▼ (a) For all the models $C_n$ $(n = \text{I}, \ldots, \text{V})$ of isotropic continua, the rotation tensor $\overset{\circ}{\mathbf{S}}$ is zero (see (4.376)), therefore, by the definition of the tensor (4.52), for these models including $C_{\text{III}}$, we obtain

$$\mathbf{V}\cdot\mathbf{T}\cdot\mathbf{V}^{-1} - \mathbf{V}^{-1}\cdot\mathbf{T}\cdot\mathbf{V} = 0, \quad (4.404)$$

or $\mathbf{V}^2\cdot\mathbf{T} - \mathbf{T}\cdot\mathbf{V}^2 = 0$, i.e. the tensor $\mathbf{T} = \overset{\text{III}}{\mathbf{S}}$ commutes with $\mathbf{V}^2$, and hence the tensors are coaxial (see Sect. 4.2.21). Thus, $\mathbf{T}$ and $\mathbf{V}^2$ have a common eigenbasis $\mathbf{p}_\alpha$ and can be written in the form

$$\mathbf{V}^2 = \sum_{\alpha=1}^{3}\lambda_\alpha^2\mathbf{p}_\alpha\otimes\mathbf{p}_\alpha, \quad \mathbf{T} = \overset{\text{III}}{\mathbf{S}} = \sum_{\alpha=1}^{3}\overset{\text{III}}{\sigma}_\alpha\mathbf{p}_\alpha\otimes\mathbf{p}_\alpha. \quad (4.405)$$

Represent the derivative $\dot{\mathbf{Y}}$ in the eigenbasis $\mathbf{p}_\alpha$ with use of its definition (4.62):

$$\dot{\mathbf{Y}} = \frac{1}{2}\left(\sum_{\alpha=1}^{3} \dot{\lambda}_\alpha \mathbf{p}_\alpha \otimes \mathbf{p}_\alpha + \lambda_\alpha(\dot{\mathbf{p}}_\alpha \otimes \mathbf{p}_\alpha + \mathbf{p}_\alpha \otimes \dot{\mathbf{p}}_\alpha)\right) \cdot \sum_{\beta=1}^{3} \frac{1}{\lambda_\beta}\mathbf{p}_\beta \otimes \mathbf{p}_\beta$$
$$+\frac{1}{2}\left(\sum_{\alpha=1}^{3} \frac{1}{\lambda_\alpha}\mathbf{p}_\alpha \otimes \mathbf{p}_\alpha\right) \cdot \sum_{\beta=1}^{3}(\dot{\lambda}_\beta \mathbf{p}_\beta \otimes \mathbf{p}_\beta + \lambda_\beta(\dot{\mathbf{p}}_\beta \otimes \mathbf{p}_\beta + \mathbf{p}_\beta \otimes \dot{\mathbf{p}}_\beta))$$
$$= \sum_{\alpha=1}^{3}\left(\left(\frac{\dot{\lambda}_\alpha}{\lambda_\alpha}\delta_{\alpha\beta} + \frac{1}{2}\sum_{\beta=1}^{3}\left(\frac{\lambda_\alpha}{\lambda_\beta} - \frac{\lambda_\beta}{\lambda_\alpha}\right)(\dot{\mathbf{p}}_\alpha \cdot \mathbf{p}_\beta)\right)\mathbf{p}_\alpha \otimes \mathbf{p}_\beta\right.$$
$$\left.+\frac{1}{2}(\dot{\mathbf{p}}_\alpha \otimes \mathbf{p}_\alpha + \mathbf{p}_\alpha \otimes \dot{\mathbf{p}}_\alpha)\right). \tag{4.406}$$

Take the properties of orthonormal eigenbasis $\mathbf{p}_\alpha$ into account:

$$\mathbf{p}_\alpha \cdot \mathbf{p}_\beta = \delta_{\alpha\beta}, \quad \dot{\mathbf{p}}_\alpha \cdot \mathbf{p}_\beta + \mathbf{p}_\alpha \cdot \dot{\mathbf{p}}_\beta = 0,$$
$$\mathbf{p}_\alpha \cdot \dot{\mathbf{p}}_\alpha = \frac{1}{2}(\mathbf{p}_\alpha \cdot \mathbf{p}_\alpha)^\bullet = \frac{1}{2}\frac{d}{dt}|\mathbf{p}_\alpha|^2 = 0, \tag{4.407}$$

and contract the tensor $\mathbf{T}$ (4.405) with $\dot{\mathbf{Y}}$ (4.406):

$$\mathbf{T} \cdot\cdot\, \dot{\mathbf{Y}} = \sum_{\alpha=1}^{3} \sigma_\alpha \frac{\dot{\lambda}_\alpha}{\lambda_\alpha}. \tag{4.408}$$

But the same result can be obtained if we consider the Hencky tensor $\tilde{\mathbf{H}}$ (2.167), determine its derivative:

$$\tilde{\mathbf{H}} = \sum_{\alpha=1}^{3} \lg \lambda_\alpha \mathbf{p}_\alpha \otimes \mathbf{p}_\alpha,$$
$$\dot{\tilde{\mathbf{H}}} = \sum_{\alpha=1}^{3}\left(\frac{\dot{\lambda}_\alpha}{\lambda_\alpha}\mathbf{p}_\alpha \otimes \mathbf{p}_\alpha + \lg \lambda_\alpha(\dot{\mathbf{p}}_\alpha \otimes \mathbf{p}_\alpha + \mathbf{p}_\alpha \otimes \dot{\mathbf{p}}_\alpha)\right), \tag{4.409}$$

and contract the derivative with $\mathbf{T}$ (4.405):

$$w_{(i)} = \mathbf{T} \cdot\cdot\, \dot{\tilde{\mathbf{H}}} = \sum_{\alpha=1}^{3} \sigma_\alpha \dot{\lambda}_\alpha / \lambda_\alpha. \tag{4.410}$$

Comparing (4.408) with (4.410), we get that for isotropic continua the tensors $\tilde{\mathbf{H}}$ and $\mathbf{Y}$ are energetically equivalent, and the contraction $\mathbf{T} \cdot\cdot\, \dot{\mathbf{Y}} = \mathbf{T} \cdot\cdot\, \dot{\tilde{\mathbf{H}}}$ gives the stress power, because for isotropic continua $\overset{\circ}{\mathbf{S}} = 0$.

Since the logarithmic deformation measure $\mathbf{H}_1$ (2.167) satisfies the relation $\dot{\mathbf{H}}_1 = \dot{\tilde{\mathbf{H}}}$, and $\overset{\mathrm{III}}{\mathbf{g}}{}^{\bullet} = \dot{\mathbf{Y}}$, Eq. (4.402) still holds for the measures:

$$\overset{\mathrm{III}}{\mathbf{S}} \cdot\cdot \overset{\mathrm{III}}{\mathbf{g}}{}^{\bullet} = \overset{\mathrm{III}}{\mathbf{S}} \cdot\cdot \dot{\mathbf{H}}_1.$$

(b) Consider the model $A_{\mathrm{III}}$ for an isotropic continuum. For this case, the constitutive equation (4.322) holds. From the equation it follows that the tensors $\overset{\mathrm{III}}{\mathbf{T}}$ and $\mathbf{B}$ are coaxial. To prove this fact it is sufficient to resolve the tensor $\mathbf{B}$ for its eigenbasis $\widetilde{\mathbf{p}}_\alpha$ and to substitute the result into (4.322); then we get that the tensor $\overset{\mathrm{III}}{\mathbf{T}}$ is also diagonal in the same basis $\widetilde{\mathbf{p}}_\alpha$:

$$\overset{\mathrm{III}}{\mathbf{T}} = \sum_{\alpha=1}^{3} \overset{\mathrm{III}}{\sigma}_\alpha \widetilde{\mathbf{p}}_\alpha \otimes \widetilde{\mathbf{p}}_\alpha, \quad \mathbf{B} = \sum_{\alpha=1}^{3} B_\alpha \widetilde{\mathbf{p}}_\alpha \otimes \widetilde{\mathbf{p}}_\alpha. \tag{4.411}$$

Unlike the model $C_{\mathrm{III}}$, in this case there is no initial information that $\mathbf{T}$ is coaxial with $\mathbf{U}$, and $\widetilde{\mathbf{p}}_\alpha$, in general, may be not coincident with $\overset{\circ}{\mathbf{p}}_\alpha$. However, for the model $A_{\mathrm{III}}$, there is an implicit additional condition that Eq. (2.317) for the tensor $\dot{\mathbf{B}}$ must be integrable or the tensor $\dot{\mathbf{B}}$ can be replaced by its energetically equivalent integrable tensor.

Take into account that the tensor $\dot{\mathbf{B}}$ in the eigenbasis $\overset{\circ}{\mathbf{p}}_\alpha$ has the same form as (4.406):

$$\dot{\mathbf{B}} = \sum_{\alpha=1}^{3} \left( \left( \frac{\dot{\lambda}_\alpha}{\lambda_\alpha} \delta_{\alpha\beta} + \frac{1}{2} \sum_{\beta=1}^{3} \left( \frac{\lambda_\alpha}{\lambda_\beta} - \frac{\lambda_\beta}{\lambda_\alpha} \right) (\overset{\circ}{\mathbf{p}}{}^{\bullet}_\alpha \cdot \overset{\circ}{\mathbf{p}}_\beta) \right) \overset{\circ}{\mathbf{p}}_\alpha \otimes \overset{\circ}{\mathbf{p}}_\beta \right.$$
$$\left. + \frac{1}{2} (\overset{\circ}{\mathbf{p}}{}^{\bullet}_\alpha \otimes \overset{\circ}{\mathbf{p}}_\alpha + \overset{\circ}{\mathbf{p}}_\alpha \otimes \overset{\circ}{\mathbf{p}}{}^{\bullet}_\alpha) \right). \tag{4.412}$$

On the other hand, from (4.411) it follows that

$$\dot{\mathbf{B}} = \sum_{\alpha=1}^{3} (\dot{B}_\alpha \widetilde{\mathbf{p}}_\alpha \otimes \widetilde{\mathbf{p}}_\alpha + B_\alpha (\dot{\widetilde{\mathbf{p}}}_\alpha \otimes \widetilde{\mathbf{p}}_\alpha + \widetilde{\mathbf{p}}_\alpha \otimes \dot{\widetilde{\mathbf{p}}}_\alpha)). \tag{4.413}$$

Contract the tensor $\dot{\mathbf{B}}$ with $\overset{\mathrm{III}}{\mathbf{T}}$ and use both the expressions (4.412) and (4.413):

$$\overset{\mathrm{III}}{\mathbf{T}} \cdot\cdot \dot{\mathbf{B}} = \sum_{\alpha=1}^{3} \overset{\mathrm{III}}{\sigma}_\alpha \left( \sum_{\gamma=1}^{3} \left( \frac{\dot{\lambda}_\alpha}{\lambda_\alpha} p_{\alpha\gamma}^2 + p_{\alpha\gamma} \widetilde{p}_{\alpha\gamma} + \sum_{\beta=1}^{3} \lambda_{\alpha\beta} p_{\alpha\gamma} p_{\gamma\beta} \right) \right) = \sum_{\alpha=1}^{3} \overset{\mathrm{III}}{\sigma}_\alpha \dot{B}_\alpha, \tag{4.414}$$

where

$$p_{\alpha\gamma} = \overset{\circ}{\mathbf{p}}_\alpha \cdot \widetilde{\mathbf{p}}_\gamma, \quad \widetilde{p}_{\alpha\gamma} = \overset{\circ}{\mathbf{p}}{}^{\bullet}_\alpha \cdot \widetilde{\mathbf{p}}_\gamma, \quad \lambda_{\alpha\beta} = \Big(\frac{\lambda_\alpha}{\lambda_\beta} - \frac{\lambda_\beta}{\lambda_\alpha}\Big)(\overset{\circ}{\mathbf{p}}{}^{\bullet}_\alpha \cdot \overset{\circ}{\mathbf{p}}_\beta).$$

According to the relation (4.414), the coefficient of $\overset{\mathrm{III}}{\sigma}_\alpha$ must form the total derivative; that occurs if and only if $p_{\alpha\gamma} = 0$ at $\alpha \neq \gamma$ and $p_{\alpha\alpha} = 1$, i.e. the bases $\widetilde{\mathbf{p}}_\alpha$ and $\overset{\circ}{\mathbf{p}}_\alpha$ coincide: $\widetilde{\mathbf{p}}_\alpha = \overset{\circ}{\mathbf{p}}_\alpha$. In this case

$$w_{(i)} = \overset{\mathrm{III}}{\mathbf{T}} \cdot\cdot \dot{\mathbf{B}} = \sum_{\alpha=1}^{3} \overset{\mathrm{III}}{\sigma}_\alpha \dot{B}_\alpha = \sum_{\alpha=1}^{3} \overset{\mathrm{III}}{\sigma}_\alpha (\lg \lambda_\alpha)^{\bullet}. \tag{4.415}$$

The contraction of the tensor $\overset{\mathrm{III}}{\mathbf{T}}$ with $\overset{\circ}{\mathbf{H}}$ gives the same expression, where

$$\overset{\circ}{\mathbf{H}} = \sum_{\alpha=1}^{3} \lg \lambda_\alpha \overset{\circ}{\mathbf{p}}_\alpha \otimes \overset{\circ}{\mathbf{p}}_\alpha,$$
$$\overset{\circ}{\mathbf{H}}{}^{\bullet} = \sum_{\alpha=1}^{3} ((\lg \lambda_\alpha)^{\bullet} \overset{\circ}{\mathbf{p}}_\alpha \otimes \overset{\circ}{\mathbf{p}}_\alpha + \lg \lambda_\alpha (\overset{\circ}{\mathbf{p}}{}^{\bullet}_\alpha \otimes \overset{\circ}{\mathbf{p}}_\alpha + \overset{\circ}{\mathbf{p}}_\alpha \otimes \overset{\circ}{\mathbf{p}}{}^{\bullet}_\alpha)). \tag{4.415a}$$

Hence, the tensor $\mathbf{B}$ can be replaced by its energetically equivalent tensor $\overset{\circ}{\mathbf{H}}$.

Since the logarithmic deformation measure $\overset{\circ}{\mathbf{H}}_1$ (2.167) satisfies the relation $\overset{\circ}{\mathbf{H}}_1{}^{\bullet} = \overset{\circ}{\mathbf{H}}{}^{\bullet}$, and $\overset{\mathrm{III}}{\mathbf{G}}{}^{\bullet} = \dot{\mathbf{B}}$, so relationship (4.402) still holds for the measures: $\overset{\mathrm{III}}{\mathbf{T}} \cdot\cdot \overset{\mathrm{III}}{\mathbf{G}}{}^{\bullet} = \overset{\mathrm{III}}{\mathbf{T}} \cdot\cdot \dot{\overset{\circ}{\mathbf{H}}}_1$.

(c) Since the left-hand sides of Eqs. (4.408) and (4.415) are coincident, the right-hand sides must coincide as well; this means that eigenvalues of the tensors $\overset{\mathrm{III}}{\mathbf{T}}$ and $\mathbf{T}$ are coincident: $\sigma_\alpha = \overset{\mathrm{III}}{\sigma}_\alpha$.

Substitution of formulae (4.402) into (4.120) yields

$$\sum_{\alpha=1}^{3} \left(\rho \frac{\partial \psi}{\partial \lambda_\alpha} - \sigma_\alpha \lambda_\alpha\right) d\lambda_\alpha + \rho \left(\frac{\partial \psi}{\partial \theta} + \eta\right) d\theta = 0, \tag{4.416}$$

because for an isotropic continuum, $\psi$ can always be considered as a function of $\lambda_\alpha$ and $\theta$: $\psi = \psi(\lambda_\alpha, \theta)$. From (4.416) we get the constitutive equations (4.403). ▲

Uniting the assertions of both theorems (4.398) and (4.403), we obtain that for all the models $A_n$, $B_n$, $C_n$ and $D_n$ ($n = \mathrm{I}, \ldots, \mathrm{V}$) constitutive equations of isotropic elastic continua take the form

$$\overset{(n)}{\sigma}_\alpha = \lambda_\alpha^{\mathrm{III}-n} \sigma_\alpha, \quad \sigma_\alpha = \rho \lambda_\alpha \frac{\partial \psi}{\partial \lambda_\alpha}, \quad n = \mathrm{I}, \ldots, \mathrm{V},$$
$$\psi = \psi(\lambda_\alpha, \theta), \quad \rho = \overset{\circ}{\rho} \lambda_1 \lambda_2 \lambda_3. \tag{4.417}$$

The stress power $w_{(i)}$ defined by (4.2), with use of relations (4.87)–(4.90), (4.394), (4.402), and expressions for $\overset{(n)}{\mathbf{C}}$, $\overset{(n)}{\mathbf{G}}$, $\overset{(n)}{\mathbf{A}}$ and $\overset{(n)}{\mathbf{g}}$ in terms of $\lambda_\alpha$ and $\overset{\circ}{\mathbf{p}}_\alpha$ (see Exercise 4.2.4) for all the models $A_n$, $B_n$, $C_n$ and $D_n$, can be represented in the form

$$w_{(i)} = \sum_{\alpha=1}^{3} \overset{(n)}{\sigma}_\alpha \lambda_\alpha^{n-\mathrm{IV}} \dot{\lambda}_\alpha, \quad n = \mathrm{I}, \ldots, \mathrm{V}. \tag{4.417a}$$

Thus, for isotropic elastic continua, the generalized constitutive equations (4.387) for all the models $A_n$, $B_n$, $C_n$ and $D_n$ reduce to the single form

$$\mathbf{T} = \mathcal{F}_B(\mathbf{F}, \theta), \tag{4.418a}$$

$$\mathcal{F}_B(\mathbf{F}, \theta) = \overset{\circ}{\rho}(\lambda_1\lambda_2\lambda_3) \sum_{\alpha=1}^{3} \lambda_\alpha \frac{\partial \psi}{\partial \lambda_\alpha} \mathbf{p}_\alpha \otimes \mathbf{p}_\alpha, \quad \psi = \psi(\lambda_\alpha, \theta), \quad \lambda_\alpha, \mathbf{p}_\alpha \parallel \mathbf{F}.$$

If $\lambda_\alpha$ and $\mathbf{p}_\alpha$ are considered as functions of $\mathbf{F}^{-1}$, then the constitutive equations (4.418a) can be written in the form (4.393):

$$\mathbf{T} = \mathcal{F}_B(\mathbf{F}^{-1}, \theta), \tag{4.418b}$$

where the tensor function $\mathcal{F}_B(\mathbf{F}^{-1}, \theta)$ has the same form as in (4.418a).

Since, for all the models $A_n$, $B_n$, $C_n$ and $D_n$ ($n = \mathrm{I}, \ldots, \mathrm{V}$) of isotropic elastic continua, all the tensors $\overset{(n)}{\mathbf{T}}$, $\overset{(n)}{\mathbf{C}}$ and $\overset{(n)}{\mathbf{G}}$, and also $\overset{(n)}{\mathbf{S}}$, $\overset{(n)}{\mathbf{A}}$ and $\overset{(n)}{\mathbf{g}}$ are coaxial, so these tensors are commutative; then, according to Theorem 4.11, the stress power can be expressed with the help of the co-rotational derivative (see Sect. 4.2.21):

$$w_{(i)} = \overset{(n)}{\mathbf{T}} \cdot\cdot \overset{(n)}{\mathbf{C}}{}^h = \overset{(n)}{\mathbf{T}} \cdot\cdot \overset{(n)}{\mathbf{G}}{}^h = \overset{(n)}{\mathbf{S}} \cdot\cdot \overset{(n)}{\mathbf{A}}{}^h = \overset{(n)}{\mathbf{S}} \cdot\cdot \overset{(n)}{\mathbf{g}}{}^h, \quad n = \mathrm{I}, \ldots, \mathrm{V}, \quad h = \{\cdot, U, V, J, S\}. \tag{4.419}$$

Here $\overset{\mathrm{III}}{\mathbf{C}} = \mathbf{H}$, $\overset{\mathrm{III}}{\mathbf{G}} = \overset{\circ}{\mathbf{H}}_1$, $\overset{\mathrm{III}}{\mathbf{A}} = \overset{\circ}{\tilde{\mathbf{H}}}$ and $\overset{\mathrm{III}}{\mathbf{g}} = \mathbf{H}_1$.

### 4.8.12 *Representation of Constitutive Equations of Isotropic Elastic Continua 'in Rates'*

For some problems in solid mechanics (usually for dynamic processes), it is convenient to represent the constitutive equations 'in rates'.

At first, consider the models $A_n$ and $B_n$ of elastic continua, which correspond to the constitutive equations (4.384) (at $G = A, B$) independent of the tensor $\mathbf{O}$:

$$\overset{(n)}{\mathbf{T}}_G = \overset{(n)}{\mathcal{F}}_G(\overset{(n)}{\mathbf{C}}_G, \theta) = \sum_{\gamma=1}^{r} \varphi_\gamma \mathbf{I}^{(s)}_{\gamma G}. \tag{4.420}$$

**Theorem 4.29.** *For the models $A_n$ and $B_n$ of ideal solids, the constitutive equations (4.420) at $G = A, B$ can be represented 'in rates':*

$$\overset{(n)}{\mathbf{T}}{}^{\bullet} = {}^4\mathbf{P}_G(\overset{(n)}{\mathbf{C}}_G, \theta) \cdot\cdot \overset{(n)}{\mathbf{C}}{}^{\bullet}_G + \overset{(n)}{\mathbf{T}}_{\theta G}\dot{\theta}, \quad G = A, B. \tag{4.421}$$

▼ Differentiating the relation (4.420) with respect to $t$ and using the definition of the derivative tensors (4.385), we get

$$\overset{(n)}{\mathbf{T}}{}^{\bullet} = \sum_{\gamma,\beta=1}^{r} \left( \frac{\partial \varphi_\gamma}{\partial I^{(s)}_\beta} \mathbf{I}^{(s)}_{\beta G} \cdot\cdot \overset{(n)}{\mathbf{C}}{}^{\bullet}_G + \frac{\partial \varphi_\gamma}{\partial \theta}\dot{\theta} \right) \mathbf{I}^{(s)}_{\gamma G} + \sum_{\gamma=1}^{r} \varphi_\gamma \frac{\partial \mathbf{I}^{(s)}_{\gamma G}}{\partial \overset{(n)}{\mathbf{C}}_G} \cdot\cdot \overset{(n)}{\mathbf{C}}{}^{\bullet}_G. \tag{4.422}$$

Introducing the fourth-order tensor ${}^4\mathbf{P}_G$ and tensor $\overset{(n)}{\mathbf{T}}_{\theta G}$ as follows:

$${}^4\mathbf{P}_G = \sum_{\gamma,\beta=1}^{r} (\varphi_{\gamma\beta} \mathbf{I}^{(s)}_{\gamma G} \otimes \mathbf{I}^{(s)}_{\beta G} + \varphi_\gamma \delta_{\gamma\beta} {}^4\mathbf{I}^{(s)}_{\gamma G}), \quad \overset{(n)}{\mathbf{T}}_{\theta G} = \sum_{\beta=1}^{r} \frac{\partial \varphi_\gamma}{\partial \theta} \mathbf{I}^{(s)}_{\gamma G}, \tag{4.423}$$

where

$$\varphi_{\gamma\beta} = \frac{\partial \varphi_\gamma}{\partial I^{(s)}_\beta} = \frac{\partial}{\partial I^{(s)}_\beta}\left( \rho \frac{\partial \psi}{\partial I^{(s)}_\gamma} \right), \quad {}^4\mathbf{I}^{(s)}_{\gamma G} = \frac{\partial \mathbf{I}^{(s)}_{\gamma G}}{\partial \overset{(n)}{\mathbf{C}}_G} = \frac{\partial^2 I^{(s)}_\gamma}{\partial \overset{(n)}{\mathbf{C}}_G \partial \overset{(n)}{\mathbf{C}}_G}, \tag{4.424}$$

we get that (4.422) can be represented in the form (4.421). ▲

Let us consider now the models $C_n$ and $D_n$ of ideal isotropic solids. As shown in Sect. 4.8.9, for isotropic continua the constitutive equations (4.384) are independent of the tensor $\mathbf{O}$ and can also be represented in the form (4.420) at $G = C, D$.

**Theorem 4.30.** *For the models $C_n$ and $D_n$ of isotropic ideal solids, the constitutive equations (4.420) can be represented 'in rates' as follows:*

$$\overset{(n)}{\mathbf{S}}{}^{h} = {}^4\mathbf{P}_G(\overset{(n)}{\mathbf{C}}_G, \theta) \cdot\cdot \overset{(n)}{\mathbf{C}}{}^{h}_G + \overset{(n)}{\mathbf{S}}_{\theta G}\dot{\theta}, \quad G = C, D, \tag{4.425}$$

*where we have introduced the notation $h$ for the co-rotational derivatives*

$$h = \{U, V, J, S\}.$$

▼ The co-rotational derivatives of a second-order tensor has the general form (formula (2.373))

$$\overset{(n)}{\mathbf{S}}{}^{h} = \overset{(n)}{\mathbf{S}}{}^{\bullet} - \mathbf{Z}_h \cdot \overset{(n)}{\mathbf{S}} - \overset{(n)}{\mathbf{S}} \cdot \mathbf{Z}_h^{\mathrm{T}},$$
$$\mathbf{Z}_h = \{\boldsymbol{\Omega}_U,\ \boldsymbol{\Omega}_V,\ \mathbf{W},\ \boldsymbol{\Omega}\},\quad h = \{U,\ V,\ J,\ S\}. \tag{4.426}$$

On substituting the right-hand side of (4.420) into (4.426), we obtain

$$\overset{(n)}{\mathbf{S}}{}^{h} = \sum_{\gamma=1}^{3} (\dot{\varphi}_\gamma \mathbf{I}_{\gamma G} + \varphi_\gamma \mathbf{I}_{\gamma G}^{h}), \tag{4.427}$$

because $\dot{\varphi}_\gamma = {\varphi_\gamma}^h$. Here we have taken into account that for isotropic continua the relation (4.420) contains only three principal invariants $I_\gamma(\overset{(n)}{\mathbf{C}}_G)$.

Let us show now that $\dot{\varphi}_\gamma$ can be represented with the help of co-rotational derivatives as follows:

$$\dot{\varphi}_\alpha = \frac{d}{dt}\varphi_\alpha(I_\gamma(\overset{(n)}{\mathbf{C}}_G),\ \theta) = \sum_{\gamma=1}^{3}\Big(\varphi_{\alpha\gamma}\mathbf{I}_{\gamma G} \cdot\cdot \overset{(n)}{\mathbf{C}}{}_G^{h} + \frac{\partial \varphi_\alpha}{\partial \theta}\dot{\theta}\Big). \tag{4.428}$$

Here we have used the following notation for derivative tensors of the principal invariants: $\mathbf{I}_{\gamma G} = \partial I_\gamma(\overset{(n)}{\mathbf{C}}_G)/\partial \overset{(n)}{\mathbf{C}}_G$..

We must show only that the total derivative in the double contractions can be replaced by the co-rotational one:

$$\mathbf{I}_{\gamma G} \cdot\cdot \overset{(n)}{\mathbf{C}}{}_G^{\bullet} = \mathbf{I}_{\gamma G} \cdot\cdot \overset{(n)}{\mathbf{C}}{}_G^{h}. \tag{4.429}$$

Indeed, for isotropic continua, the derivative tensors $\mathbf{I}_{\gamma G}$ have the form (4.323), and they are proportional to tensors $\mathbf{E}$, $\overset{(n)}{\mathbf{C}}_G$ and $\overset{(n)}{\mathbf{C}}{}_G^2$; thus, all the tensors $\mathbf{I}_{\gamma G}$ are coaxial with $\overset{(n)}{\mathbf{C}}_G$, and hence $\mathbf{I}_{\gamma G}$ and $\overset{(n)}{\mathbf{C}}_G$ are commutative. According to the representation (4.426), we obtain

$$\mathbf{I}_{\gamma G} \cdot\cdot \overset{(n)}{\mathbf{C}}_G{}^{h} = \mathbf{I}_{\gamma G} \cdot\cdot \overset{(n)}{\mathbf{C}}{}_G^{\bullet} - \mathbf{I}_{\gamma G} \cdot\cdot \mathbf{Z}_h \cdot \overset{(n)}{\mathbf{C}}_G - \mathbf{I}_{\gamma G} \cdot\cdot \overset{(n)}{\mathbf{C}}_G \cdot \mathbf{Z}_h^{\mathrm{T}}. \tag{4.430}$$

Since $\mathbf{I}_{\gamma G}$ and $\overset{(n)}{\mathbf{C}}_G$ are commutative, the second and third summands on the right-hand side of (4.430) are canceled in pair; thus, formulae (4.429) and (4.428) really hold.

Determine the co-rotational derivatives $\mathbf{I}^h_{\gamma G}$ in relation (4.427) with use of formula (4.426) (the formula still holds after substitution $\overset{(n)}{\mathbf{S}} \to \mathbf{I}_{\gamma G}$) and also the formula of differentiation of the principal invariants (4.323):

$$\mathbf{I}^h_{1G} = \mathbf{0},\quad \mathbf{I}^h_{2G} = \left(\frac{\partial I_2}{\partial \overset{(n)}{\mathbf{C}}_G}\right)^h = (\mathbf{E}I_1 - \overset{(n)}{\mathbf{C}}_G)^h = \mathbf{E}\otimes\mathbf{E}\cdot\cdot\overset{(n)}{\mathbf{C}}{}^h_G - \overset{(n)}{\mathbf{C}}{}^h_G,$$

$$\mathbf{I}^h_{3G} = (\overset{(n)}{\mathbf{C}}{}^2_G - I_1\overset{(n)}{\mathbf{C}}_G + \mathbf{E}I_2)^h = \overset{(n)}{\mathbf{C}}{}^h_G\cdot\overset{(n)}{\mathbf{C}}_G + \overset{(n)}{\mathbf{C}}_G\cdot\overset{(n)}{\mathbf{C}}{}^h_G - I_1\overset{(n)}{\mathbf{C}}{}^h_G - \dot{I}_1\overset{(n)}{\mathbf{C}}_G + \dot{I}_2\mathbf{E}$$

$$= \overset{(n)}{\mathbf{C}}{}^h_G\cdot\overset{(n)}{\mathbf{C}}_G + \overset{(n)}{\mathbf{C}}_G\cdot\overset{(n)}{\mathbf{C}}{}^h_G - I_1\overset{(n)}{\mathbf{C}}{}^h_G - (\overset{(n)}{\mathbf{C}}_G\otimes\mathbf{E})\cdot\cdot\overset{(n)}{\mathbf{C}}{}^h_G + \mathbf{E}\otimes(\mathbf{E}I_1 - \overset{(n)}{\mathbf{C}}_G)\cdot\cdot\overset{(n)}{\mathbf{C}}{}^h_G. \tag{4.431}$$

Here we have applied the rule (4.429) of substitution of the co-rotational derivative in place of the total one.

On substituting (4.428) and (4.431) into (4.427) and collecting like terms, we get the representation (4.425) of the constitutive equations, where the tensors ${}^4\mathbf{P}_G$ and $\overset{(n)}{\mathbf{S}}_{\theta G}$ have the forms

$${}^4\mathbf{P}_G = \sum_{\gamma,\beta=1}^{3}(\varphi_{\gamma\beta}\mathbf{I}_{\gamma G}\otimes\mathbf{I}_{\gamma G} + \varphi_\gamma{}^4\widetilde{\mathbf{P}}_\gamma\delta_{\gamma\beta}),\quad \overset{(n)}{\mathbf{S}}_{\theta G} = \sum_{\gamma=1}^{3}\frac{\partial\varphi_\gamma}{\partial\theta}\mathbf{I}_{\gamma G}, \tag{4.432}$$

where

$${}^4\widetilde{\mathbf{P}}_1 = \mathbf{0},\quad {}^4\widetilde{\mathbf{P}}_2 = \mathbf{E}\otimes\mathbf{E} - \mathbf{\Delta},$$

$${}^4\widetilde{\mathbf{P}}_3 = (\mathbf{E}\otimes\overset{(n)}{\mathbf{C}}_G)^{(3214)} + (\overset{(n)}{\mathbf{C}}_G\otimes\mathbf{E})^{(1423)} + I_1(\mathbf{E}\otimes\mathbf{E} - \mathbf{\Delta})$$
$$-(\mathbf{E}\otimes\overset{(n)}{\mathbf{C}}_G + \overset{(n)}{\mathbf{C}}_G\otimes\mathbf{E}).\ \blacktriangle \tag{4.433}$$

*Remark 1.* As will be shown in Sect. 4.10.7, the use of co-rotational derivatives in relations (4.425) ensures their correctness, that cannot be achieved by applying the usual derivative with respect to time to models $C_n$ and $D_n$. □

*Remark 2.* The relations (4.421) and (4.425) can be applied formally to problems of continuum mechanics without relations (4.420). In this case, there appears a question whether the relations satisfy the principle of material symmetry. This principle holds true for the models $C_n$ and $D_n$ in rates (4.425), because all the tensors $\overset{(n)}{\mathbf{S}}$ and $\overset{(n)}{\mathbf{C}}_G = \{\overset{(n)}{\mathbf{A}}, \overset{(n)}{\mathbf{g}}\}$ are $H$-invariant.

One should prove that relations (4.421) 'in rates' for the models $A_n$ and $B_n$ satisfy the principle of material symmetry as well (Exercise 4.8.8). □

Using the representations (4.380) and (4.383) of constitutive equations for the models $C_n$ and $D_n$ of isotropic continua and differentiating them, we get the relation ‘in rates’

$$\overset{(n)}{\mathbf{S}}{}^h = \dot{\psi}_1\mathbf{E} + \dot{\psi}_2\overset{(n)}{\mathbf{C}}_G + \dot{\varphi}_3\overset{(n)}{\mathbf{C}}{}^2_G + \psi_2\overset{(n)}{\mathbf{C}}{}^h_G + \varphi_3(\overset{(n)}{\mathbf{C}}{}^h_G\cdot\overset{(n)}{\mathbf{C}}_G + \overset{(n)}{\mathbf{C}}_G\cdot\overset{(n)}{\mathbf{C}}{}^h_G), \quad (4.434)$$

which is another form of relations (4.425). Here $\psi_\alpha$ are expressed by the formulae

$$\psi_1 = \varphi_1 + \varphi_2 I_1 + \varphi_3 I_2, \quad -\psi_2 = \varphi_2 + I_1\varphi_3, \quad I_\gamma = I_\gamma(\overset{(n)}{\mathbf{C}}_G). \quad (4.435)$$

*Remark 3.* The relations ‘in rates’ for the models $A_n$, $B_n$ (4.420) and for the models $C_n$, $D_n$ (4.425) can be written in the single form

$$\overset{(n)}{\mathbf{T}}{}^h_G = {}^4\mathbf{P}_G(\overset{(n)}{\mathbf{C}}_G, \theta)\cdot\cdot\,\overset{(n)}{\mathbf{C}}{}^h_G + \overset{(n)}{\mathbf{T}}_{\theta G}\dot{\theta}, \quad G = A,\ B,\ C,\ D, \quad (4.436)$$

if for $G = A, B$ the parameter $h$ takes on the only value $h = \{\cdot\}$ (the total derivative with respect to $t$), and for $G = C, D$ the parameter $h$ takes on the values $h = \{U,\ V,\ J,\ S\}$ but only isotropic continua are considered.

Then we can use the generalized relations (4.105) between $\overset{(n)}{\mathbf{C}}{}^h_G$ and $\mathbf{L}$ and rewrite Eqs. (4.436) as follows:

$$\overset{(n)}{\mathbf{T}}{}^h_G = {}^4\overset{(n)}{\mathbf{P}}_{Gh}(\mathbf{F},\ \theta)\cdot\cdot\,\mathbf{L} + \overset{(n)}{\mathbf{T}}_{\theta G}\dot{\theta}, \quad (4.437)$$

Where

$${}^4\overset{(n)}{\mathbf{P}}_{Gh} = {}^4\mathbf{P}_G\cdot\cdot\,{}^4\overset{(n)}{\mathbf{B}}_{Gh}, \quad (4.438)$$

and ${}^4\overset{(n)}{\mathbf{B}}_{Gh}$ are determined by (4.106).

According to formula (2.373) for the co-rotational derivatives $\overset{(n)}{\mathbf{T}}{}^h_G$ of a symmetric second-order tensor, we can rewrite the relation (4.437) in the form

$$\overset{(n)}{\mathbf{T}}{}^\bullet_G - \mathbf{Z}_{Gh}\cdot\overset{(n)}{\mathbf{T}}_G - \overset{(n)}{\mathbf{T}}_G\cdot\mathbf{Z}^{\mathrm{T}}_{Gh} = {}^4\overset{(n)}{\mathbf{P}}_{Gh}\cdot\cdot\,\mathbf{L} + \overset{(n)}{\mathbf{T}}_{\theta G}\dot{\theta}. \quad (4.439)$$

Here we have introduced the notation

$$\mathbf{Z}_{Gh} = \begin{cases} l\mathbf{0}, & G = A, B, \\ \mathbf{Z}_h, & G = C, D, \end{cases} \quad (4.440)$$

and taken into account that for $G = A, B$, by definition, $\overset{(n)}{\mathbf{T}}{}^h_G = \overset{(n)}{\mathbf{T}}{}^\bullet_G$. The tensors $\mathbf{Z}_h$ are defined by formulae (2.374).

According to the continuity equation, the total derivative $\overset{(n)}{\mathbf{T}}{}^{\bullet}_{G}$ takes the form (see formulae (3.20))

$$\frac{\partial \rho \overset{(n)}{\mathbf{T}}_G}{\partial t} + \nabla \cdot (\rho \mathbf{v} \otimes \overset{(n)}{\mathbf{T}}_G) = \rho \overset{(n)}{\mathbf{T}}{}^{\bullet}_{G}. \tag{4.441}$$

On substituting (4.441) into (4.439), we obtain

$$\frac{\partial \rho \overset{(n)}{\mathbf{T}}_G}{\partial t} + \nabla \cdot (\rho \mathbf{v} \otimes \overset{(n)}{\mathbf{T}}_G) = \rho^4 \overset{(n)}{\mathbf{P}}_{Gh} \cdot\cdot \mathbf{L} + \rho \mathbf{Z}_{Gh} \cdot \overset{(n)}{\mathbf{T}}_G + \rho \overset{(n)}{\mathbf{T}}_G \cdot \mathbf{Z}^{\mathrm{T}}_{Gh} + \rho \overset{(n)}{\mathbf{T}}_{\theta G} \dot{\theta},$$
$$G = A,\ B,\ C,\ D, \quad h = \{\cdot,\ U,\ V,\ J,\ S\}, \quad n = \mathrm{I}, \ldots, \mathrm{V}, \tag{4.442}$$

– the generalized divergence form of the constitutive equations 'in rates' for ideal solids. □

### 4.8.13 *Application of the Principle of Material Symmetry to Fluids*

Up to now in this section we considered only solids. Applying the principle of material symmetry to fluids, we obtain the following important theorem.

**Theorem 4.31.** *For the models $A_n$ ($n = \mathrm{I, II, IV, V}$) of ideal fluids, the constitutive equations (4.165), (4.168), satisfying the principle of material symmetry, can be written in the form*

$$\overset{(n)}{\mathbf{T}} = -\frac{p}{n - \mathrm{III}} \overset{(n)}{\mathbf{G}}{}^{-1}, \tag{4.443a}$$

$$p = p(I_3, \theta) = \overset{\circ}{\rho} I_3 (\partial \psi / \partial I_3) \left((n - \mathrm{III})^3 I_3\right)^{1/(\mathrm{III}-n)} (\mathrm{III} - n), \tag{4.443b}$$

$$\eta = \partial \psi / \partial \theta, \tag{4.443c}$$

$$\psi = \psi(I_3(\overset{(n)}{\mathbf{G}}),\ \theta), \tag{4.443d}$$

*where*

$$\mathbf{T} = -p\mathbf{E}, \tag{4.444a}$$

$$p = p(\rho,\ \theta) = \rho^2 (\partial \psi / \partial \rho), \tag{4.444b}$$

$$\psi = \Psi(\rho,\ \theta), \tag{4.444c}$$

$$\eta = -\partial \psi / \partial \theta. \tag{4.444d}$$

The function $p$ introduced by formulae (4.443b) and (4.444b) is called the *pressure*. The tensor $\mathbf{T}$ proportional to the unit tensor (see relation (4.444a)) is called *spherical*.

▼ According to Definition 4.5, in each reference configuration, including $\overset{\circ}{\mathcal{K}}$, a fluid has the unimodular symmetry group $\overset{\circ}{G}_s = U$. Therefore, the relations (4.278) must hold for each $\mathbf{H}$-transformation from $\overset{\circ}{\mathcal{K}}$ to $\overset{*}{\mathcal{K}}$, where $\mathbf{H} \in \overset{\circ}{G}_s = U$ (i.e. $\det \mathbf{H} = 1$).

Notice that the tensors $\overset{(n)}{\mathbf{T}}$ and $\overset{(n)}{\mathbf{C}}$ are not $H$-indifferent relative to the unimodular group (except $\overset{\mathrm{I}}{\mathbf{T}}$, which is absolutely $H$-indifferent), therefore relations (4.279) do not hold for a fluid, and for the function $\psi$ in place of the indifference condition (4.279c) we have a more general condition

$$\psi(\overset{(n)}{\mathbf{C}},\ \theta) = \psi(\overset{(n)}{\mathbf{C}}{}^*,\ \theta) \quad \forall \mathbf{H} \in \overset{\circ}{G}_s = U. \tag{4.445}$$

The unique solution of Eq. (4.445) is the scalar function $\psi$ depending on the determinant of the corresponding energetic deformation measure $\overset{(n)}{\mathbf{G}}$, i.e.

$$\psi = \psi\left(\det\,(\overset{(n)}{\mathbf{C}} + \frac{1}{n - \mathrm{III}}\mathbf{E}),\ \theta\right) = \psi(I_3(\overset{(n)}{\mathbf{G}}),\ \theta). \tag{4.446}$$

Indeed, one can immediately verify that the third principal invariant $I_3(\overset{(n)}{\mathbf{G}})$ always satisfies the condition (4.445):

$$I_3(\overset{(n)}{\mathbf{G}}{}^*) = I_3(\overset{(n)}{\mathbf{G}}) \qquad \forall \mathbf{H} \in U, \tag{4.447}$$

because at $n = \mathrm{I, II, IV, V}$ this invariant can be expressed in terms of the densities ratio $\rho/\overset{\circ}{\rho}$ (see formulae (4.80)):

$$I_3(\overset{(n)}{\mathbf{G}}) = \frac{1}{(n - \mathrm{III})^2}(\rho/\overset{\circ}{\rho})^{\mathrm{III}-n}, \tag{4.448}$$

and densities $\rho$ and $\overset{\circ}{\rho}$ are always absolutely $H$-indifferent (see Sect. 4.6.3). Moreover, the unimodular $H$-transformations have been introduced as transformations of a reference configuration, which hold the only scalar function – density $\overset{\circ}{\rho}$, therefore the functions $I_3(\overset{(n)}{\mathbf{G}})$ are unique solutions of Eq. (4.445). All other solutions are only functions of $I_3(\overset{(n)}{\mathbf{G}})$, in particular the function (4.446). Thus, we have shown that for a fluid the free energy $\psi$ really has the form (4.446).

On substituting (4.448) into (4.446), we obtain that $\psi$ for all the models $A_n$ can be expressed only in terms of $\rho$ and $\theta$:

$$\psi = \psi(I_3(\overset{(n)}{\mathbf{G}}),\ \theta) = \psi\left(\frac{1}{(n - \mathrm{III})^3}(\rho/\overset{\circ}{\rho})^{\mathrm{III}-n},\ \theta\right) \equiv \Psi(\rho,\ \theta). \tag{4.449}$$

Substituting (4.446) into (4.168), we determine the tensors $\overset{(n)}{\mathbf{T}}$

$$\overset{(n)}{\mathbf{T}} = \rho\frac{\partial\psi}{\partial\overset{(n)}{\mathbf{C}}} = \rho\frac{\partial\psi}{\partial I_3}\frac{\partial I_3}{\partial\overset{(n)}{\mathbf{G}}}. \tag{4.450}$$

Here we have taken into account that, by formula (4.43), from the derivative with respect to $\overset{(n)}{\mathbf{C}}$ one can pass to the derivative with respect to $\overset{(n)}{\mathbf{G}}$, because $\partial\overset{(n)}{\mathbf{G}}/\partial\overset{(n)}{\mathbf{C}} = \mathbf{\Delta}$.

According to formula (4.323) for $\partial I_3/\partial\overset{(n)}{\mathbf{G}}$:

$$\partial I_3(\overset{(n)}{\mathbf{G}})/\partial\overset{(n)}{\mathbf{G}} = I_3(\overset{(n)}{\mathbf{G}})\overset{(n)}{\mathbf{G}}^{-1}, \tag{4.451}$$

we can rewrite the constitutive equations (4.450) in the form

$$\overset{(n)}{\mathbf{T}} = \rho\frac{\partial\psi}{\partial I_3}I_3\overset{(n)}{\mathbf{G}}^{-1}. \tag{4.452}$$

Introducing the notation for pressure:

$$p = \rho\frac{\partial\psi}{\partial I_3}I_3(\overset{(n)}{\mathbf{G}})(\mathrm{III}-n) \tag{4.453}$$

and substituting the expression (4.448) for $\rho$ in terms of $I_3(\overset{(n)}{\mathbf{G}})$ into (4.453), from (4.452) we obtain the exact representations (4.443a) and (4.443b) of constitutive equations for a fluid.

Let us show that all four relations (4.443a) can be represented in the form (4.444a). To see this, one should rewrite the expression for $\overset{(n)}{\mathbf{G}}^{-1}$ with the help of formulae (4.31) and (4.47):

$$\overset{\mathrm{I}}{\mathbf{G}}^{-1} = -2\mathbf{G} = -2\mathbf{F}^{\mathrm{T}}\cdot\mathbf{F}, \quad \overset{\mathrm{II}}{\mathbf{G}}^{-1} = -\mathbf{U} = -\frac{1}{2}(\mathbf{F}^{\mathrm{T}}\cdot\mathbf{O}+\mathbf{O}^{\mathrm{T}}\cdot\mathbf{F}), \tag{4.454}$$
$$\overset{\mathrm{IV}}{\mathbf{G}}^{-1} = \mathbf{U}^{-1} = \frac{1}{2}(\mathbf{F}^{-1}\cdot\mathbf{O}+\mathbf{O}^{\mathrm{T}}\cdot\mathbf{F}^{-1\mathrm{T}}), \quad \overset{\mathrm{V}}{\mathbf{G}}^{-1} = 2\mathbf{G}^{-1} = 2\mathbf{F}^{-1}\cdot\mathbf{F}^{-1\mathrm{T}},$$

and then substitute these expressions and the expressions for energetic tensors $\overset{(n)}{\mathbf{T}}$ (see Table 4.1) into relations (4.443a). As a result, we obtain a linear equation with respect to the tensor $\mathbf{T}$ for each $n$:

$$n=\mathrm{I}: \quad \mathbf{F}^{\mathrm{T}}\cdot\mathbf{T}\cdot\mathbf{F} = -p\mathbf{F}^{\mathrm{T}}\cdot\mathbf{F},$$
$$n=\mathrm{II}: \quad \frac{1}{2}(\mathbf{F}^{\mathrm{T}}\cdot\mathbf{T}\cdot\mathbf{O}+\mathbf{O}^{\mathrm{T}}\cdot\mathbf{T}\cdot\mathbf{F}) = -\frac{1}{2}p(\mathbf{F}^{\mathrm{T}}\cdot\mathbf{O}+\mathbf{O}^{\mathrm{T}}\cdot\mathbf{F}), \tag{4.455}$$
$$n=\mathrm{IV}: \quad \frac{1}{2}(\mathbf{F}^{-1}\cdot\mathbf{T}\cdot\mathbf{O}+\mathbf{O}^{\mathrm{T}}\cdot\mathbf{T}\cdot\mathbf{F}^{-1\mathrm{T}}) = -\frac{1}{2}p(\mathbf{F}^{-1\mathrm{T}}\cdot\mathbf{O}+\mathbf{O}^{\mathrm{T}}\cdot\mathbf{F}^{-1\mathrm{T}}),$$
$$n=\mathrm{V}: \quad \mathbf{F}^{-1}\cdot\mathbf{T}\cdot\mathbf{F}^{-1\mathrm{T}} = -p\mathbf{F}^{-1}\cdot\mathbf{F}^{-1\mathrm{T}}.$$

A solution of each of the linear equations is the exact expression (4.444a), and this solution is unique (indeed, each equation in (4.455) can be rewritten in components, for example, with respect to the basis $\mathbf{r}_i$; then we get a linear nonsingular equation for components of the tensor $T^{ij}$ and the equation has a unique solution).

Formula (4.444b) for pressure follows from (4.453), if we express invariant $I_3(\overset{(n)}{\mathbf{G}})$ in terms of $\rho$ according to (4.448):

$$\begin{aligned} p &= \rho \frac{\partial \Psi(\rho,\theta)}{\partial \rho} \frac{\partial \rho}{\partial I_3} I_3(\mathrm{III}-n) \\ &= \rho \frac{\partial \Psi}{\partial \rho} \frac{\overset{\circ}{\rho}}{(\mathrm{III}-n)} I_3^{\frac{1}{\mathrm{III}-n}-1} (n-\mathrm{III})^{3/(\mathrm{III}-n)} I_3(\mathrm{III}-n) = \rho^2\,(\partial\psi/\partial\rho). \end{aligned} \quad (4.456)$$

Finally, we should show that the derived constitutive equations (4.443a) satisfy the material symmetry principle, i.e. from (4.443a) we can get the relationship

$$\overset{(n)}{\mathbf{T}}{}^{*} = -\frac{p}{n-\mathrm{III}} \overset{(n)}{\mathbf{G}}{}^{*-1} \qquad \forall \mathbf{H} \in U. \quad (4.457)$$

Notice that the pressure $p$, according to (4.443b), is a function only of $I_3(\overset{(n)}{\mathbf{G}})$ and $\theta$, and hence $p$ is always $H$-indifferent: $p^* = p$. And since the stress tensor $\mathbf{T}$ is also absolutely $H$-indifferent, so for all the models $A_n$ the relation (4.444a) holds in any configuration $\overset{*}{\mathcal{K}}$:

$$\overset{*}{\mathbf{T}} = -p^*\mathbf{E} \qquad \forall \mathbf{H} \in U,\ \mathbf{H} : \overset{\circ}{\mathcal{K}} \to \overset{*}{\mathcal{K}}. \quad (4.458)$$

On substituting (4.458) into relations (4.455) written in an arbitrary configuration $\overset{*}{\mathcal{K}}$:

$$\begin{aligned} n = \mathrm{I}: \quad & \overset{*}{\mathbf{F}}{}^{\mathrm{T}} \cdot \overset{*}{\mathbf{T}} \cdot \overset{*}{\mathbf{F}} = -p^* \overset{*}{\mathbf{F}}{}^{\mathrm{T}} \cdot \overset{*}{\mathbf{F}}, \quad (4.459) \\ n = \mathrm{II}: \ & \frac{1}{2}(\overset{*}{\mathbf{F}}{}^{\mathrm{T}} \cdot \overset{*}{\mathbf{T}} \cdot \overset{*}{\mathbf{O}} + \overset{*}{\mathbf{O}}{}^{\mathrm{T}} \cdot \overset{*}{\mathbf{T}} \cdot \overset{*}{\mathbf{F}}) = -\frac{1}{2} p (\overset{*}{\mathbf{F}}{}^{\mathrm{T}} \cdot \overset{*}{\mathbf{O}} + \overset{*}{\mathbf{O}}{}^{\mathrm{T}} \cdot \overset{*}{\mathbf{F}}) \text{ etc. } \forall \mathbf{H} \in U, \end{aligned}$$

we find that the relations are identically satisfied (independently of transformations of the tensors $\mathbf{F}$ and $\mathbf{O}$ during the passage from $\overset{\circ}{\mathcal{K}}$ to $\overset{*}{\mathcal{K}}$!).

Relations (4.459) are exactly equivalent to (4.457), hence the relations (4.457) really hold for any unimodular $H$-transformations $\mathbf{H} : \overset{\circ}{\mathcal{K}} \to \overset{*}{\mathcal{K}}$. ▲

*Remark 1.* Theorem 4.31 shows that in place of energetic deformation tensors $\overset{(n)}{\mathbf{C}}$ for fluids we use only the corresponding measures $\overset{(n)}{\mathbf{G}}$; in other words, the models $A_n$ and $B_n$ for fluids coincide. For solids, the models $A_n$ and $B_n$ prove to be considerably distinct. □

**Theorem 4.32.** *For the models $C_n$ ($n = \mathrm{I}, \mathrm{II}, \mathrm{IV}, \mathrm{V}$) of ideal fluids, the constitutive equations (4.165), (4.168) satisfying the principle of material symmetry can be written in the form*

$$\begin{cases} \overset{(n)}{\mathbf{S}} = -\dfrac{p}{n - \mathrm{III}}\overset{(n)}{\mathbf{g}}^{-1}, & (4.460a)\\ p = p(I_3, \theta) = \overset{\circ}{\rho} I_3 \dfrac{\partial \psi}{\partial I_3}\left((n - \mathrm{III})^3 I_3\right)^{1/(\mathrm{III}-n)}(\mathrm{III} - n), & (4.460b)\\ \eta = -\partial\psi/\partial\theta, & (4.460c)\\ \psi = \psi(I_3(\overset{(n)}{\mathbf{g}}),\ \theta), & (4.460d) \end{cases}$$

*where expressions of the tensor* **T** *for all the models are the same, and they have the form (4.444).*

The third invariant in (4.460) is determined by tensor $\overset{(n)}{\mathbf{g}}$: $I_3 = I_3(\overset{(n)}{\mathbf{g}})$.

▼ A proof of the theorem is similar to the proof of the preceding theorem.

According to the principle of material indifference (4.211) for the potential $\psi$, the following relation must hold for the models $C_n$ of an ideal fluid:

$$\psi(\overset{(n)}{\mathbf{A}},\ \mathbf{O},\ \theta) = \psi(\overset{(n)}{\mathbf{A}}{}^*,\ \overset{*}{\mathbf{O}},\ \theta) \quad \forall \mathbf{H} \in \overset{\circ}{G}_s = U. \tag{4.461}$$

Tensors $\overset{(n)}{\mathbf{A}}$ are not $H$-invariant under unimodular transformations, therefore relation (4.352d) does not hold.

The unique solution of Eq. (4.461) is the scalar function depending on the third principal invariant of the quasienergetic deformation measure $\overset{(n)}{\mathbf{g}}$, which is connected to $\overset{(n)}{\mathbf{A}}$ by relation (4.71):

$$\psi = \psi(I_3(\overset{(n)}{\mathbf{g}}), \theta). \tag{4.462}$$

From (4.448) it follows that the invariant $I_3(\overset{(n)}{\mathbf{g}})$ can be expressed in terms of the densities ratio $\rho/\overset{\circ}{\rho}$:

$$I_3(\overset{(n)}{\mathbf{g}}) = I_3(\overset{(n)}{\mathbf{G}}) = \frac{1}{(n - \mathrm{III})^3}(\rho/\overset{\circ}{\rho})^{\mathrm{III}-n}. \tag{4.463}$$

Since the densities $\overset{\circ}{\rho}$ and $\rho$ are always $H$-indifferent, the function $I_3(\overset{(n)}{\mathbf{g}})$ is also $H$-indifferent under any unimodular transformations, i.e.

$$I_3(\overset{(n)}{\mathbf{g}}{}^*) = I_3(\overset{(n)}{\mathbf{g}}) \quad \forall \mathbf{H} \in \overset{\circ}{G}_s = U, \tag{4.464}$$

hence, the function (4.462) is a solution of Eq. (4.461). Uniqueness of the solution (4.462) follows from formula (4.463), because unimodular $H$-transformations have been defined as transformations retaining the only scalar function being the density $\rho$.

On substituting the function (4.462) into the constitutive equations (4.174), we find the quasienergetic stress tensors $\overset{(n)}{\mathbf{S}}$ and tensor $\overset{\circ}{\mathbf{S}}$:

$$\overset{(n)}{\mathbf{S}} = \rho \frac{\partial \psi}{\partial \overset{(n)}{\mathbf{A}}} = \rho \frac{\partial \psi}{\partial I_3} \frac{\partial I_3(\overset{(n)}{\mathbf{g}})}{\partial \overset{(n)}{\mathbf{g}}}, \tag{4.465}$$

$$\overset{\circ}{\mathbf{S}} = \rho(\partial \psi / \partial \mathbf{O}) \equiv \mathbf{0}. \tag{4.466}$$

Here we have taken into account that $\partial \overset{(n)}{\mathbf{g}} / \partial \overset{(n)}{\mathbf{A}} = \mathbf{\Delta}$.

With use of formula (4.323) for the derivative of the tensor determinant, we can rewrite the constitutive equations (4.465) in the form

$$\overset{(n)}{\mathbf{S}} = \rho \frac{\partial \psi}{\partial I_3} I_3 \overset{(n)}{\mathbf{g}}{}^{-1}. \tag{4.467}$$

Introducing in formula (4.467) the notation for pressure

$$\widetilde{p} = \rho \frac{\partial \psi}{\partial I_3} I_3(\overset{(n)}{\mathbf{g}})(\mathrm{III} - n) = p, \tag{4.468}$$

we obtain the representation (4.460a) of constitutive equations for the models $C_n$ of an ideal fluid.

Since, by (4.463), the invariants $I_3(\overset{(n)}{\mathbf{g}})$ and $I_3(\overset{(n)}{\mathbf{G}})$ are coincident, the pressure $p$ introduced by (4.453) coincides with $\widetilde{p}$ introduced by (4.468). On substituting the expression (4.463) for $\rho$ in terms of $I_3(\overset{(n)}{\mathbf{g}})$ into (4.468), we get formula (4.460c).

Let us show now that all four constitutive equations (4.460a) are really equivalent to the representation (4.444a) for the Cauchy stress tensor $\mathbf{T}$. According to the definitions of quasienergetic tensors $\overset{(n)}{\mathbf{S}}$ (see Table 4.2) and quasienergetic measures $\overset{(n)}{\mathbf{g}}$ (4.71), constitutive equations (4.460a) can be rewritten in the form

$$\begin{aligned}
n = \mathrm{I}: &\quad \mathbf{V} \cdot \mathbf{T} \cdot \mathbf{V} = -p \mathbf{V} \cdot \mathbf{V}, \\
n = \mathrm{II}: &\quad \frac{1}{2}(\mathbf{V} \cdot \mathbf{T} + \mathbf{T} \cdot \mathbf{V}) = -p \mathbf{V}, \\
n = \mathrm{IV}: &\quad \frac{1}{2}(\mathbf{V}^{-1} \cdot \mathbf{T} + \mathbf{T} \cdot \mathbf{V}^{-1}) = -p \mathbf{V}^{-1}, \\
n = \mathrm{V}: &\quad \mathbf{V}^{-1} \cdot \mathbf{T} \cdot \mathbf{V}^{-1} = -p \mathbf{V}^{-1} \cdot \mathbf{V}^{-1}.
\end{aligned} \tag{4.469}$$

One can verify that an unique solution of each of the linear equations for the tensor $\mathbf{T}$ is the expression (4.444a).

Since the rotation tensor of stresses $\overset{\circ}{\mathbf{S}}$ for an ideal fluid is identically zero (see (4.466)), so, by the definition (4.52) of the tensor $\overset{\circ}{\mathbf{S}}$, the following relation must be satisfied for an ideal fluid:

$$\mathbf{V} \cdot \mathbf{T} \cdot \mathbf{V}^{-1} = \mathbf{V}^{-1} \cdot \mathbf{T} \cdot \mathbf{V}. \tag{4.470}$$

But the relation is always satisfied, because we have shown that the stress tensor for an ideal fluid is spherical (see (4.444a)).

It remains to show that the derived constitutive equations (4.460a) satisfy the principle of material symmetry, i.e., according to (4.460a), the following relations must hold:

$$\overset{(n)}{\mathbf{S}}{}^{*} = -\frac{p^*}{n - \mathrm{III}} \overset{(n)}{\mathbf{g}}{}^{*-1} \quad \forall \mathbf{H} \in U, \tag{4.471}$$

Or

$$\begin{aligned} &\text{at } n = \mathrm{I}: && \overset{*}{\mathbf{V}} \cdot \overset{*}{\mathbf{T}} \cdot \overset{*}{\mathbf{V}} = -p^* \overset{*}{\mathbf{V}}{}^2, \\ &\text{at } n = \mathrm{II}: && \frac{1}{2}(\overset{*}{\mathbf{V}} \cdot \overset{*}{\mathbf{T}} + \overset{*}{\mathbf{T}} \cdot \overset{*}{\mathbf{V}}) = -p^* \overset{*}{\mathbf{V}} \quad \text{etc.} \end{aligned} \tag{4.472}$$

But the relations (4.472) are really satisfied, because the tensor $\overset{*}{\mathbf{T}}$, by (4.458), is also spherical under any $H$-transformations $\mathbf{H} : \overset{\circ}{\mathcal{K}} \to \overset{*}{\mathcal{K}}$. ▲

*Remark 2.* As follows from Theorem 4.32, the models $C_n$ and $D_n$ for ideal fluids coincide, because the energetic deformation tensors $\overset{(n)}{\mathbf{A}}$ do not appear in the constitutive equations (they are included only in the corresponding measure $\overset{(n)}{\mathbf{g}}$). □

As a result of the section, it should be noted that

- all the models $A_n$, $B_n$, $C_n$ and $D_n$ of ideal fluids are equivalent and can be written in the form (4.444);
- nevertheless, for a fluid, the constitutive equations (4.443) and (4.460) can also be applied to different problems of continuum mechanics.

*Remark 3.* Proofs of Theorems 4.31 and 4.32 show that in order for the principal of material symmetry to hold, it is not necessary that the tensors involved in the constitutive equations be either $H$-indifferent or $H$-invariant; the principal requirement consists in that they must be transformed in the *consistent way* during the passage from $\overset{\circ}{\mathcal{K}}$ to $\overset{*}{\mathcal{K}}$ (see formulae (4.459) and (4.472)). □

Let us note another property. The principle of material symmetry has allowed us to find explicit forms of the constitutive equations for ideal solids and fluids; these are formulae (4.444), (4.311), (4.313a), (4.314a), (4.322), and (4.319), as distinct from the implicit forms (4.168) and (4.174).

### 4.8.14 Functional Energetic Couples of Tensors

Let us return now to the representation (4.13) of energetic couples of tensors and explain a method of construction of new energetic couples with the help of the theory of invariants suggested above.

Consider an arbitrary tensor function of tensor $\mathbf{U}$

$$\overset{(e)}{\mathbf{C}} = \overset{(e)}{\mathbf{C}}(\mathbf{U}), \tag{4.473}$$

such that:

(a) it forms a symmetric second-order tensor $\overset{(e)}{\mathbf{C}} = \overset{(e)}{\mathbf{C}}{}^{\mathrm{T}}$,
(b) it is indifferent relative to the full orthogonal group $I$ (i.e. it is an isotropic function),
(c) it is differentiable.

For such tensor functions, we have the theorem (see, for example [12]) that the functions can be represented in terms of two tensor powers $\mathbf{U}$ and $\mathbf{U}^2$ in the form

$$\overset{(e)}{\mathbf{C}} = \omega_1 \mathbf{E} + \omega_2 \mathbf{U} + \omega_2 \mathbf{U}^2, \tag{4.474}$$

where $\omega_\alpha$ are scalar functions of the principal invariants

$$\omega_\alpha = \omega_\alpha(I_1(\mathbf{U}),\ I_2(\mathbf{U}),\ I_3(\mathbf{U})), \tag{4.475}$$

being differentiable with respect to $I_\alpha$ within the whole domain of definition.

In a particular case, functions $\omega_\alpha$ may be the derivatives of one scalar function: $\omega_\alpha = \frac{\partial}{\partial I_\alpha}\omega(I_1,\ I_2,\ I_3)$, but this is not necessary.

Determine the derivative of (4.474) with respect to $t$:

$$\overset{(e)}{\mathbf{C}}{}^{\bullet} = {}^4\overset{(e)}{\mathbf{C}}_{\mathrm{U}} \cdot\cdot \dot{\mathbf{U}}. \tag{4.476}$$

Here ${}^4\overset{(e)}{\mathbf{C}}_{\mathrm{U}}$ is a fourth-order tensor being the derivative tensor of $\overset{(e)}{\mathbf{C}}$ with respect to the tensor argument $\mathbf{U}$:

$${}^4\overset{(e)}{\mathbf{C}}_{\mathrm{U}} = \omega_2 \boldsymbol{\Delta} + \omega_3(\mathbf{U}\otimes\mathbf{E}+\mathbf{E}\otimes\mathbf{U}) + \sum_{\alpha=1}^{3}(\omega_{1\alpha}\mathbf{E}+\omega_{2\alpha}\mathbf{U}+\omega_{3\alpha}\mathbf{U}^2)\otimes I_{\alpha\mathrm{U}}, \tag{4.477}$$

where

$$\omega_{\beta\alpha} = \frac{\omega_\beta}{\partial I_\alpha}, \quad I_{\alpha\mathrm{U}} = \frac{\partial I_{\alpha\mathrm{U}}}{\partial \mathbf{U}}, \quad \alpha = 1,\ 2,\ 3. \tag{4.478}$$

The stress tensor $\overset{(e)}{\mathbf{T}}$ being coupled to $\overset{(e)}{\mathbf{C}}$ must satisfy the equation

$$\overset{(e)}{\mathbf{T}} \cdot\cdot \overset{(e)}{\mathbf{C}}{}^{\bullet} = \overset{IV}{\mathbf{T}} \cdot\cdot \dot{\mathbf{U}}. \tag{4.479}$$

Substituting (4.476) into (4.479), we obtain the equation for $\overset{(e)}{\mathbf{T}}$:

$$\overset{IV}{\mathbf{T}} = {}^4\overset{(e)}{\mathbf{C}}_{\mathrm{U}} \cdot\cdot \overset{(e)}{\mathbf{T}}. \tag{4.480}$$

Thus, we have found the family of new energetic couples of tensors $(\overset{(e)}{\mathbf{T}}, \overset{(e)}{\mathbf{C}})$, they will be called the *functional energetic couples*.

The couples $(\overset{(n)}{\mathbf{T}}, \overset{(n)}{\mathbf{C}})$ when $n \geqslant \mathrm{IV}$ are particular cases of the functional energetic couples, because according to the Hamilton–Cayley theorem (see [12])

$$\mathbf{U}^3 = I_1(\mathbf{U})\mathbf{U}^2 - I_2(\mathbf{U})\mathbf{U} + I_3(\mathbf{U})\mathbf{E}, \tag{4.481}$$

we can express any integer power $\mathbf{U}^n$ in terms of only the first two tensor powers: $\mathbf{E}$, $\mathbf{U}$ and $\mathbf{U}^2$ in the form (4.474), where the coefficients $\omega_\alpha$ are functions of the principal invariants (4.475).

In a similar way, we can introduce the functional energetic couples $(\overset{(-e)}{\mathbf{C}}, \overset{(-e)}{\mathbf{T}})$ as follows:

$$\overset{(-e)}{\mathbf{C}} = \overset{(-e)}{\mathbf{C}}(\mathbf{U}^{-1}),$$
$$\overset{II}{\mathbf{T}} = {}^4\overset{(-e)}{\mathbf{C}}_{\mathrm{U}^{-1}} \cdot\cdot \overset{(-e)}{\mathbf{T}}, \tag{4.482}$$

where ${}^4\overset{(-e)}{\mathbf{C}}_{\mathrm{U}^{-1}}$ is the derivative tensor of $\overset{(-e)}{\mathbf{C}}$ with respect to the tensor argument $\mathbf{U}^{-1}$ in the form (4.477).

The energetic couples $(\overset{(n)}{\mathbf{T}}, \overset{(n)}{\mathbf{C}})$ when $n \leqslant \mathrm{II}$ are particular cases of the functional couples $(\overset{(-e)}{\mathbf{T}}, \overset{(-e)}{\mathbf{C}})$.

We will say that functional energetic couples are of the $A$-type, if the following initial conditions hold:

$$\overset{(e)}{\mathbf{C}}(\mathbf{E}) = \mathbf{0}, \quad \overset{(-e)}{\mathbf{C}}(\mathbf{E}) = \mathbf{0} \tag{4.483}$$

(the tensors $\overset{(e)}{\mathbf{C}}$ and $\overset{(-e)}{\mathbf{C}}$ are generalizations of energetic deformation tensors $\overset{(n)}{\mathbf{C}}$).

If the initial conditions

$$\overset{(e)}{\mathbf{C}}(\mathbf{E}) = \mathbf{E}, \quad \overset{(-e)}{\mathbf{C}}(\mathbf{E}) = \mathbf{E} \tag{4.484}$$

are satisfied, then we will say that the functional couples are of the $B$-type (such tensors $\overset{(e)}{\mathbf{C}}$ and $\overset{(-e)}{\mathbf{C}}$ are generalizations of energetic deformation measures $\overset{(n)}{\mathbf{G}}$).

We can also introduce *functional quasienergetic couples* of tensors of the $C$-type as follows:

$$\overset{(e)}{\mathbf{A}} = \overset{(e)}{\mathbf{A}}(\mathbf{V}), \qquad \overset{(-e)}{\mathbf{A}} = \overset{(-e)}{\mathbf{A}}(\mathbf{V}^{-1}),$$
$$\overset{(e)}{\mathbf{A}}(\mathbf{E}) = \mathbf{0}, \qquad \overset{(-e)}{\mathbf{A}}(\mathbf{E}) = \mathbf{0}, \tag{4.485}$$
$$\overset{\mathrm{IV}}{\mathbf{S}} = {}^4\overset{(e)}{\mathbf{A}}_{\mathbf{V}} \cdot\cdot \overset{(e)}{\mathbf{S}}, \qquad \overset{\mathrm{II}}{\mathbf{S}} = {}^4\overset{(-e)}{\mathbf{A}}_{\mathbf{V}^{-1}} \cdot\cdot \overset{(-e)}{\mathbf{S}}.$$

The derivative tensors ${}^4\overset{(e)}{\mathbf{A}}_{\mathbf{V}}$ and ${}^4\overset{(-e)}{\mathbf{A}}_{\mathbf{V}^{-1}}$ have the form (4.477) with substitutions $\mathbf{C} \to \mathbf{A}$ and $\mathbf{U} \to \mathbf{V}$.

Quasienergetic couples $(\overset{(n)}{\mathbf{S}}, \overset{(n)}{\mathbf{A}})$ when $n \geqslant \mathrm{IV}$ are particular cases of the functional quasienergetic couples.

Functional quasienergetic tensor couples of the $D$-type differ from ones of the $C$-type by the initial conditions

$$\overset{(e)}{\mathbf{A}}(\mathbf{E}) = \mathbf{E}, \qquad \overset{(-e)}{\mathbf{A}}(\mathbf{E}) = \mathbf{E}. \tag{4.486}$$

The use of the functional couples of tensors considerably extends the scope for constructions of advanced models of continuous media.

However, methods of construction of these models are the same as the ones considered above for models $A_n$, $B_n$, $C_n$, and $D_n$. Therefore we will not consider functional couples of tensors in this book, but they are kept in mind for future investigations.

## Exercises for 4.8

**4.8.1.** Using the relation (2.31) between the tensors $\mathbf{U}$ and $\mathbf{V}$, show that principal invariants of the tensors always coincide:

$$I_\alpha(\mathbf{U}) = I_\alpha(\mathbf{V}), \quad \alpha = 1, 2, 3.$$

**4.8.2.** Using the results of Exercise 4.8.1, show that principal invariants of the tensors $\overset{(n)}{\mathbf{G}}$ and $\overset{(n)}{\mathbf{g}}$ coincide:

$$I_\alpha(\overset{(n)}{\mathbf{G}}) = I_\alpha(\overset{(n)}{\mathbf{g}}), \quad \alpha = 1, 2, 3, \quad n = \mathrm{I}, \mathrm{II}, \mathrm{IV}, \mathrm{V},$$

where

$$I_3(\overset{(n)}{\mathbf{G}}) = \frac{1}{(n-\mathrm{III})^3}(I_3(\mathbf{U}))^{n-\mathrm{III}}.$$

Show that principal invariants of the tensors $\overset{(n)}{\mathbf{A}}$ and $\overset{(n)}{\mathbf{C}}$ are coincident as well:

$$I_\alpha(\overset{(n)}{\mathbf{A}}) = I_\alpha(\overset{(n)}{\mathbf{C}}), \quad \alpha = 1, 2, 3.$$

**4.8.3.** Using the results of Exercise 4.2.4, show that principal invariants of the tensors $\overset{(n)}{\mathbf{G}}$ and $\overset{(n)}{\mathbf{g}}$ can be expressed in terms of eigenvalues $\lambda_\alpha$ of the tensors $\mathbf{U}$ and $\mathbf{V}$ as follows:

$$I_1(\overset{(n)}{\mathbf{G}}) = I_1(\overset{(n)}{\mathbf{g}}) = \frac{1}{n-\mathrm{III}}\sum_{\alpha=1}^{3}\lambda_\alpha^{n-\mathrm{III}}, \quad I_3(\overset{(n)}{\mathbf{G}}) = I_3(\overset{(n)}{\mathbf{g}}) = \frac{1}{(n-\mathrm{III})^3}(\lambda_1\lambda_2\lambda_3)^{n-\mathrm{III}},$$

$$I_2(\overset{(n)}{\mathbf{G}}) = I_2(\overset{(n)}{\mathbf{g}}) = \frac{1}{(n-\mathrm{III})^2}((\lambda_1\lambda_2)^{n-\mathrm{III}} + (\lambda_2\lambda_3)^{n-\mathrm{III}} + (\lambda_1\lambda_3)^{n-\mathrm{III}}).$$

**4.8.4.** Using the results of Exercise 4.8.3, show that

$$\frac{\partial I_1(\overset{(n)}{\mathbf{G}})}{\partial\lambda_\alpha} = \lambda_\alpha^{n-\mathrm{IV}}, \quad \frac{\partial I_2(\overset{(n)}{\mathbf{G}})}{\partial\lambda_\alpha} = \frac{1}{n-\mathrm{III}}\lambda_\alpha^{n-\mathrm{IV}}(\lambda_\beta^{n-\mathrm{III}} + \lambda_\gamma^{n-\mathrm{III}}),$$

$$\frac{\partial I_3(\overset{(n)}{\mathbf{G}})}{\partial\lambda_\alpha} = \frac{1}{n-\mathrm{III}}\lambda_\alpha^{n-\mathrm{IV}}(\lambda_\beta\lambda_\gamma)^{n-\mathrm{III}}, \quad \alpha \neq \beta \neq \gamma \neq \alpha.$$

**4.8.5.** Show that for quasilinear models $A_n$ and $B_n$ (see Sect. 4.8.7) the tensors ${}^4\mathbf{P}_G$ $(G = A, B)$ coincide with the quasilinear elasticity tensor ${}^4\mathbf{M}$ introduced in Sect. 4.8.6:

$${}^4\mathbf{P}_A = {}^4\mathbf{P}_B = {}^4\mathbf{M},$$

and for linear models $A_n$ the tensor ${}^4\mathbf{M}$ has the form

$${}^4\mathbf{M} = J\,{}^4\widetilde{\mathbf{M}}, \quad J = \rho/\overset{\circ}{\rho},$$

where the constant-tensor ${}^4\mathbf{M}$ depends only on producing tensors of the corresponding symmetry group $G_s$.

**4.8.6.** Using the results (4.444) of Theorems 4.31 and 4.32, show that the Gibbs free energy $\overset{(n)}{\zeta}_G$ (4.126) and enthalpy $\overset{(n)}{i}_G$ (4.137) of an ideal fluid have the forms

$$\overset{(n)}{\zeta}_D = \overset{(n)}{\zeta}_B = \psi + \frac{p}{\rho(n - \text{III})}, \quad \overset{(n)}{\zeta}_A = \psi + \frac{p}{\rho(n - \text{III})}\left(1 - \frac{I_1(\overset{(n)}{\mathbf{G}}{}^{-1})}{n - \text{III}}\right),$$

$$\overset{(n)}{\zeta}_C = \psi + \frac{p}{\rho(n - \text{III})}\left(1 - \frac{I_1(\overset{(n)}{\mathbf{g}}{}^{-1})}{n - \text{III}}\right).$$

**4.8.7.** Prove Theorem 4.27 for a transversely isotropic continuum.

**4.8.8.** Using properties of $H$-indifference of the tensors $\overset{(n)}{\mathbf{C}}$, $\overset{(n)}{\mathbf{T}}$, $\overset{(n)}{\mathbf{G}}$, $\overset{(n)}{\mathbf{T}}{}^{\bullet}$, $\overset{(n)}{\mathbf{C}}{}^{\bullet}$ and $\overset{(n)}{\mathbf{G}}{}^{\bullet}$ (see Sect. 4.7.4), prove that the relations (4.421) satisfy the principle of material symmetry.

**4.8.9.** Show that for the models $A_n$ and $C_n$ the relations (4.395) and (4.396) between eigenvalues of the stress tensors $\overset{\circ(n)}{\sigma}_{\alpha}$, $\overset{(n)}{\sigma}_{\alpha}$ and $\lambda_\alpha$ have the forms

$$\overset{(n)}{\sigma}_{\alpha} = \overset{\circ(n)}{\sigma}_{\alpha} = \psi_1 + \frac{\psi_2}{n - \text{III}}(\lambda_\alpha^{n-\text{III}} - 1) + \frac{\varphi_3}{(n - \text{III})^2}(\lambda_\alpha^{(n-\text{III})} - 1)^2,$$

$$\psi_1 = \varphi_1 + \varphi_2 I_1 + \varphi_3 I_2, \quad -\psi_2 = \varphi_2 + I_1\varphi_3,$$

where $\varphi_\alpha = \rho(\partial\psi/\partial I_\alpha(\overset{(n)}{\mathbf{C}}))$ for the models $A_n$ and $\varphi_\alpha = \rho(\partial\psi/\partial I_\alpha(\overset{(n)}{\mathbf{A}}))$ for the models $C_n$.

**4.8.10.** Using the results of Exercise 4.8.9, prove Theorem 4.28 for the models $A_n$ and $C_n$ of isotropic elastic continua.

**4.8.11.** Show that relationships (4.350) for the models $B_n$ can be derived from relations (4.335) and (4.338) for the models $A_n$ after substitution of formula (4.43) into the relations.

## 4.9 Incompressible Continua

### *4.9.1 Definition of Incompressible Continua*

Consider a particular case of continua, that is frequently used in practice.

**Definition 4.15.** A continuum is called *incompressible*, if in each actual configuration $\mathcal{K}$ its density $\rho$ coincides with the density $\overset{\circ}{\rho}$ in a reference configuration:

$$\rho = \overset{\circ}{\rho} = \text{const} \quad \forall \mathcal{K} \quad \forall t \geqslant 0. \tag{4.487}$$

There are both incompressible fluids and incompressible solids. An incompressible fluid is often called simply a *fluid*; and if a fluid is not incompressible, then it is called a *compressible fluid* or *gas*.

As follows from the continuity equation in Lagrangian variables (3.8), for incompressible continua, a volume of each elementary domain $dV$ remains unchanged:

$$\rho = \overset{\circ}{\rho} \quad \Longrightarrow \quad dV = d\overset{\circ}{V}, \tag{4.488}$$

and, hence, a volume $|V(t)|$ of the domain occupied by an incompressible continuum, remains unchanged as well: $|V(t)| = |\overset{\circ}{V}| = \text{const}$.

It follows from the continuity equation (3.8) that for an incompressible continuum there is an additional condition imposed on the deformation gradient:

$$\det \mathbf{F} = 1. \tag{4.489}$$

According to the polar decomposition (2.137), we find that for an incompressible continuum the stretch tensors always have the unit determinant:

$$\det \mathbf{V} = \det \mathbf{F} \cdot \det \mathbf{O}^{\mathrm{T}} = \det \mathbf{F} = 1, \quad \det \mathbf{U} = 1. \tag{4.490}$$

Substituting the formulae into (4.42) and (4.71), we get that determinants of all the energetic and quasienergetic deformation measures are always equal to a constant value:

$$\det \overset{(\mathrm{n})}{\mathbf{G}} = \det \left( \frac{1}{n - \mathrm{III}} \mathbf{U}^{n-\mathrm{III}} \right) = \frac{1}{(n - \mathrm{III})^3} \det \mathbf{U}^{n-\mathrm{III}} = \frac{1}{(n - \mathrm{III})^3}, \tag{4.491}$$

$$\det \overset{(\mathrm{n})}{\mathbf{g}} = \det \left( \frac{1}{n - \mathrm{III}} \mathbf{V}^{n-\mathrm{III}} \right) = \frac{1}{(n - \mathrm{III})^3}, \quad n = \mathrm{I, II, IV, V}. \tag{4.492}$$

## 4.9.2 *The Principal Thermodynamic Identity for Incompressible Continua*

For incompressible continua, not all components of the tensors $\mathbf{F}$, $\mathbf{U}$, $\mathbf{V}$, $\overset{(\mathrm{n})}{\mathbf{C}}$ and $\overset{(\mathrm{n})}{\mathbf{A}}$ are independent, because there are additional conditions (4.489)–(4.492) imposed on the tensors. Thus, for incompressible continua, the constitutive equations (4.168) and (4.174) do not hold, because their derivation was based on independence of the differentials $d\overset{(\mathrm{n})}{\mathbf{C}}$ or $d\overset{(\mathrm{n})}{\mathbf{A}}$.

In order to find a general form of constitutive equations for incompressible continua we should transform formulae (4.121) and (4.123) of the principal thermodynamic identity. Introduce two additional scalar functions, values of which are identically zero:

$$\gamma(\overset{(n)}{\mathbf{G}}) = \det \overset{(n)}{\mathbf{G}} - \frac{1}{(n-\mathrm{III})^3} = 0, \quad \gamma(\overset{(n)}{\mathbf{g}}) = \det \overset{(n)}{\mathbf{g}} - \frac{1}{(n-\mathrm{III})^3} = 0, \quad (4.493)$$

and, according to formulae (4.491) and (4.492), derivatives of the functions with respect to their tensor arguments have the forms

$$\frac{\partial \gamma}{\partial \overset{(n)}{\mathbf{G}}} = \det \overset{(n)}{\mathbf{G}}\, \overset{(n)}{\mathbf{G}}{}^{-1} = \frac{1}{(n-\mathrm{III})^3} \overset{(n)}{\mathbf{G}}{}^{-1}, \quad \frac{\partial \gamma}{\partial \overset{(n)}{\mathbf{g}}} = \frac{1}{(n-\mathrm{III})^3} \overset{(n)}{\mathbf{g}}{}^{-1}. \quad (4.494)$$

Since $\gamma(\overset{(n)}{\mathbf{G}})$ and $\gamma(\overset{(n)}{\mathbf{g}})$ are identically zero, their total differentials are identically zero as well: $d\gamma = 0$. Rewrite the principal thermodynamic identity in forms (4.121) and (4.123) with adding the zero summand $\widetilde{p}\, d\gamma$:

$$\rho\, d\psi - \widetilde{p}\, d\gamma - \overset{(n)}{\mathbf{T}} \cdot\cdot\, d\overset{(n)}{\mathbf{C}} + \rho\eta\, d\theta + w^*\, dt = 0, \quad (4.495)$$

and

$$\rho\, d\psi - \widetilde{p}\, d\gamma - \overset{(n)}{\mathbf{S}} \cdot\cdot\, d\overset{(n)}{\mathbf{A}} - \overset{\circ}{\mathbf{S}} \cdot\cdot\, d\mathbf{O}^{\mathrm{T}} + \rho\eta\, d\theta + w^*\, dt = 0, \quad (4.496)$$

where

$$d\gamma = (\partial\gamma/\partial\overset{(n)}{\mathbf{G}}) \cdot\cdot\, d\overset{(n)}{\mathbf{G}} = (\partial\gamma/\partial\overset{(n)}{\mathbf{g}}) \cdot\cdot\, d\overset{(n)}{\mathbf{g}} = 0. \quad (4.497)$$

Here we have introduced the non-zero function $\widetilde{p}$ called the *indeterminate Lagrange multiplier*.

### *4.9.3 Constitutive Equations for Ideal Incompressible Continua*

If we consider the model $A_n$ of ideal continua, then $\psi$ is a function in the form (4.165):

$$\psi = \psi(\overset{(n)}{\mathbf{C}}, \theta). \quad (4.498)$$

Determining the differential of the function $\psi$, substituting the result into PTI (4.495) and taking formula (4.497) into account, we get

$$\left( \frac{\partial \psi}{\partial \overset{(n)}{\mathbf{C}}} - \frac{\overset{(n)}{\mathbf{T}}}{\rho} - \frac{\widetilde{p}}{\rho} \frac{\partial \gamma}{\partial \overset{(n)}{\mathbf{G}}} \right) \cdot\cdot\, d\overset{(n)}{\mathbf{C}} + \left( \frac{\partial \psi}{\partial \theta} + \eta \right) d\theta + \frac{w^*}{\rho}\, dt = 0. \quad (4.499)$$

Due to the choice of Lagrange multiplier $\widetilde{p}$, increments of the tensors $\overset{(n)}{\mathbf{C}}$ may be considered again as independent ones and the expressions at $d\overset{(n)}{\mathbf{C}}$, $d\theta$ and $dt$ in (4.499) can be equated to zero; then we get the *constitutive equations for the models* $A_n$ *of incompressible ideal continua*

$$\begin{cases} \overset{(n)}{\mathbf{T}} = -(p/(n - \mathrm{III}))\, \overset{(n)}{\mathbf{G}}{}^{-1} + \overset{\circ}{\rho}(\partial\psi/\partial\overset{(n)}{\mathbf{C}}), \\ \eta = -\partial\psi/\partial\theta, \\ w^* = 0, \qquad \psi = \psi(\overset{(n)}{\mathbf{C}}, \theta), \end{cases} \tag{4.500}$$

where

$$p = \widetilde{p}/(n - \mathrm{III})^2. \tag{4.500a}$$

Replacing the tensors $\overset{(n)}{\mathbf{C}}$ by $\overset{(n)}{\mathbf{G}}$ in (4.499), we obtain *constitutive equations for the models* $B_n$ *of incompressible ideal continua*:

$$\begin{cases} \overset{(n)}{\mathbf{T}} = -(p/(n - \mathrm{III}))\, \overset{(n)}{\mathbf{G}}{}^{-1} + \overset{\circ}{\rho}(\partial\psi/\partial\overset{(n)}{\mathbf{G}}), \\ \eta = -\partial\psi/\partial\theta, \\ w^* = 0, \qquad \psi = \psi(\overset{(n)}{\mathbf{G}}, \theta). \end{cases} \tag{4.501}$$

If we consider the models $C_n$ of ideal continua, then $\psi$ is a function in the form (4.171):

$$\psi = \psi(\overset{(n)}{\mathbf{A}}, \mathbf{O}, \theta), \tag{4.502}$$

and PTI (4.496) takes the form

$$\left(\frac{\partial\psi}{\partial\overset{(n)}{\mathbf{A}}} - \frac{\overset{(n)}{\mathbf{S}}}{\rho} - \frac{\widetilde{p}}{\rho}\frac{\partial\gamma}{\partial\overset{(n)}{\mathbf{g}}}\right)\cdot\cdot\, d\overset{(n)}{\mathbf{A}} + \left(\frac{\partial\psi}{\partial\mathbf{O}} - \frac{\overset{\circ}{\mathbf{S}}}{\rho}\right)\cdot\cdot\, d\mathbf{O}^{\mathrm{T}} + \left(\frac{\partial\psi}{\partial\theta} + \eta\right) d\theta + \frac{w^*}{\rho}\, dt = 0. \tag{4.503}$$

The introduction of Lagrange multiplier $\widetilde{p}$ allows us to consider the increments $d\overset{(n)}{\mathbf{A}}$, $d\mathbf{O}$, $d\theta$ and $dt$ as independent; then from (4.503) we get the *constitutive equations for the models* $C_n$ *of incompressible ideal continua*

$$\begin{cases} \overset{(n)}{\mathbf{S}} = -(p/(n - \mathrm{III}))\, \overset{(n)}{\mathbf{g}}{}^{-1} + \overset{\circ}{\rho}(\partial\psi/\partial\overset{(n)}{\mathbf{A}}), \\ \overset{\circ}{\mathbf{S}} = \overset{\circ}{\rho}(\partial\psi/\partial\mathbf{O}), \\ \eta = -\partial\psi/\partial\theta, \\ w^* = 0, \qquad \psi = \psi(\overset{(n)}{\mathbf{A}}, \mathbf{O}, \theta). \end{cases} \tag{4.504}$$

In a similar way, the substitution $\overset{(n)}{\mathbf{A}} \to \overset{(n)}{\mathbf{g}}$ yields the *constitutive equations for the models $D_n$ of incompressible ideal continua*:

$$\begin{cases} \overset{(n)}{\mathbf{S}} = -(p/(n-\mathrm{III}))\, \overset{(n)}{\mathbf{g}}{}^{-1} + \overset{\circ}{\rho}(\partial\psi/\partial \overset{(n)}{\mathbf{g}}), \\ \overset{\circ}{\mathbf{S}} = \overset{\circ}{\rho}(\partial\psi/\partial \mathbf{O}), \\ \eta = -\partial\psi/\partial\theta, \\ w^* = 0, \qquad \psi = \psi(\overset{(n)}{\mathbf{g}}, \mathbf{O}, \theta). \end{cases} \tag{4.504a}$$

Just as for a fluid (see formulae (4.443)), the function $p$ is called the *hydrostatic pressure*. Notice that for incompressible continua the pressure $p$ as well as the indeterminate Lagrange multiplier $\widetilde{p}$ is an additional unknown function that cannot be expressed in terms of $\rho$ and $\overset{(n)}{\mathbf{C}}$, or $\overset{(n)}{\mathbf{A}}, \overset{(n)}{\mathbf{G}}, \overset{(n)}{\mathbf{g}}$.

## 4.9.4 Corollaries to the Principle of Material Symmetry for Incompressible Fluids

Relations (4.500) and (4.504) hold for both incompressible solids and fluids. Let us consider incompressible fluids.

**Theorem 4.33.** *For the models $A_n$ ($n = \mathrm{I}, \mathrm{II}, \mathrm{IV}, \mathrm{V}$) of ideal incompressible fluids, the constitutive equations (4.500) satisfying the principle of material symmetry can be written in the form*

$$\begin{cases} \overset{(n)}{\mathbf{T}} = -(p/(n-\mathrm{III}))\, \overset{(n)}{\mathbf{G}}{}^{-1}, \\ \eta = -\partial\psi/\partial\theta, \\ \psi = \psi(\theta), \end{cases} \tag{4.505}$$

*and the equation for the Cauchy stress tensor for all the models $A_n$ has the form*

$$\mathbf{T} = -p\mathbf{E}, \tag{4.506}$$

*where $p$ is an unknown function being independent of $\rho$.*

▼ Since a considered continuum is a fluid, so, as shown in the proof of Theorem 4.31, its free energy $\psi$ can depend only on the third invariant $I_3(\overset{(n)}{\mathbf{G}})$ and $\theta$ (formula (4.446)):

$$\psi = \psi(\overset{(n)}{\mathbf{C}}, \theta) = \psi(I_3(\overset{(n)}{\mathbf{G}}), \theta). \tag{4.507}$$

This assertion holds for all fluids: compressible and incompressible. But for incompressible fluids the invariant $I_3(\overset{(n)}{\mathbf{G}})$ is always a constant (formula (4.491)); thus, in this case $\psi$ may change only with changing temperature, i.e. the function (4.507) can be written in the form

$$\psi = \psi(\theta). \tag{4.508}$$

Then $\partial\psi/\partial\overset{(n)}{\mathbf{C}} \equiv 0$, and relations (4.500) for incompressible fluids really take the form (4.505). Just as in Theorem 4.31, one can prove that formula (4.506) follows from (4.505).

Notice that for incompressible fluids, formula (4.444b) connecting the pressure $p$ to the density $\rho$ does not hold, and $p$ is an unknown function. ▲

**Theorem 4.34.** *For the models $C_n$ ($n = \mathrm{I, II, IV, V}$) of ideal incompressible fluids, the constitutive equations (4.504) satisfying the principle of material symmetry can be written in the form*

$$\begin{cases} \overset{(n)}{\mathbf{S}} = -(p/(n-\mathrm{III}))\,\overset{(n)}{\mathbf{g}}{}^{-1}, \\ \eta = -\partial\psi/\partial\theta, \\ \psi = \psi(\theta), \end{cases} \tag{4.509}$$

*where equations for the tensor* **T** *for all the models are the same and have the form (4.506), and the pressure $p$ is an unknown function.*

▼ A proof of the theorem is analogous to the proof of Theorem 4.33. ▲

### *4.9.5 Representation of Constitutive Equations for Incompressible Solids in Tensor Bases*

Let us consider now an incompressible ideal solid, which is called an *incompressible elastic continuum*. According to the principle of material symmetry, such a continuum must have some symmetry group $\widehat{G}_s$ in undistorted configuration $\widehat{\mathcal{K}}$. As before, a reference configuration $\overset{\circ}{\mathcal{K}}$ is assumed to be undistorted. If we consider the models $A_n$ of incompressible solids (4.500), then the free energy $\psi(\overset{(n)}{\mathbf{C}},\theta)$, according to formula (4.304), is a function of the invariants $I_\alpha^{(s)}(\overset{(n)}{\mathbf{C}})$ of the tensor $\overset{(n)}{\mathbf{C}}$ relative to the group $\overset{\circ}{G}_s$:

$$\psi = \psi(I_1^{(s)}(\overset{(n)}{\mathbf{C}}),\ldots,I_r^{(s)}(\overset{(n)}{\mathbf{C}}),\ \theta). \tag{4.510}$$

However, for incompressible solids, the density $\rho$ does not vary ($\rho = \overset{\circ}{\rho}$), and hence, according to formula (4.81), the principal invariants $I_\alpha(\overset{(n)}{\mathbf{C}})$ are connected by one relation

$$I_3(\overset{(n)}{\mathbf{C}}) + \frac{1}{n-\mathrm{III}} I_2(\overset{(n)}{\mathbf{C}}) + \frac{I_1(\overset{(n)}{\mathbf{C}})}{(n-\mathrm{III})^2} = 0, \tag{4.511}$$

i.e. among three invariants $I_\alpha(\overset{(n)}{\mathbf{C}})$ ($\alpha = 1, 2, 3$), only two ones are independent. Since, as shown in Sect. 4.8.4, the principal invariants $I_\alpha(\overset{(n)}{\mathbf{C}})$ can always be included in the functional basis $I_\alpha^{(s)}(\overset{(n)}{\mathbf{C}})$ of any orthogonal group $\overset{\circ}{G}_s$, so for incompressible continua the functional basis contains only $(r-1)$ invariants and the function (4.510) has the form

$$\psi = \psi(I_1^{(s)}(\overset{(n)}{\mathbf{C}}), \ldots, I_{r-1}^{(s)}(\overset{(n)}{\mathbf{C}}),\ \theta). \tag{4.512}$$

For incompressible isotropic and transversely isotropic continua, the third principal invariant $I_3(\overset{(n)}{\mathbf{C}})$ is usually eliminated from the complete functional basis $I_\alpha(\overset{(n)}{\mathbf{C}})$ (see formulae (4.297) and (4.298)), and for orthotropic incompressible continua, the invariant $I_6^{(O)}(\overset{(n)}{\mathbf{C}})$ is excluded from the set (4.296).

On substituting the free energy (4.512) into (4.500), we obtain the following *general representation of the constitutive equations for incompressible elastic continua in the tensor basis*:

$$\overset{(n)}{\mathbf{T}} = -\frac{p}{n-\mathrm{III}} \overset{(n)}{\mathbf{G}}{}^{-1} + \sum_{\gamma=1}^{r-1} \varphi_\gamma(I_1^{(s)}(\overset{(n)}{\mathbf{C}}), \ldots, I_{r-1}^{(s)}(\overset{(n)}{\mathbf{C}}),\ \theta) \frac{\partial I_\gamma^{(s)}}{\partial \overset{(n)}{\mathbf{C}}}, \tag{4.513}$$

where

$$\varphi_\gamma = \overset{\circ}{\rho}(\partial\psi/\partial I_\gamma^{(s)}), \tag{4.514}$$

which is similar to the representation (4.311), (4.312).

In particular, the constitutive equations for an isotropic incompressible elastic continuum has the form (compare with (4.322)):

$$\begin{gathered}\overset{(n)}{\mathbf{T}} = -\frac{p}{n-\mathrm{III}} \overset{(n)}{\mathbf{G}}{}^{-1} + \psi_1 \mathbf{E} - \psi_2 \overset{(n)}{\mathbf{C}}, \\ \psi_1 = \varphi_1 + \varphi_2 I_1, \qquad \psi_2 = \varphi_2, \\ \varphi_\gamma = \overset{\circ}{\rho}(\partial\psi/\partial I_\gamma),\quad \gamma = 1, 2;\quad \psi = \psi(I_1(\overset{(n)}{\mathbf{C}}),\ I_2(\overset{(n)}{\mathbf{C}}),\ \theta).\end{gathered} \tag{4.515}$$

If we assume the normalization conditions (4.326) and (4.327), where $\mathbf{T}^e = -p^e$, then the potential $\psi$ is subject to the additional conditions (compare them with (4.329)):

$$\varphi_1(0,\ 0,\ \theta_e) = p_0 - p^e, \quad \psi(0,\ 0,\ \theta_e) = 0, \tag{4.515a}$$

where $p_0$ is a value of the hydrostatic pressure $p$ in a natural state.

The models $B_n$ of incompressible elastic continua give constitutive equations, which are similar to (4.513) and differ from them only by the substitution $\overset{(\mathrm{n})}{\mathbf{C}} \to \overset{(\mathrm{n})}{\mathbf{G}}$:

$$\begin{aligned} \overset{(\mathrm{n})}{\mathbf{T}} &= -\frac{p}{n - \mathrm{III}}\overset{(\mathrm{n})}{\mathbf{G}}{}^{-1} + \sum_{\gamma=1}^{r-1} \varphi_\gamma \frac{\partial I_\gamma^{(s)}}{\partial \overset{(\mathrm{n})}{\mathbf{G}}}, \\ \varphi_\gamma(I_1^{(s)}(\overset{(\mathrm{n})}{\mathbf{G}}),\ \ldots,\ & I_{r-1}^{(s)}(\overset{(\mathrm{n})}{\mathbf{G}}),\ \theta) = \overset{\circ}{\rho}(\partial\psi/\partial I_\gamma^{(s)}). \end{aligned} \tag{4.516}$$

In particular, for the models $B_n$ of isotropic incompressible elastic continua, we have

$$\begin{aligned} \overset{(\mathrm{n})}{\mathbf{T}} &= -\frac{p}{n - \mathrm{III}}\overset{(\mathrm{n})}{\mathbf{G}}{}^{-1} + \psi_1\mathbf{E} - \varphi_2\overset{(\mathrm{n})}{\mathbf{G}}, \\ \psi_1 = \varphi_1 + \varphi_2 I_1,\ \varphi_\gamma &= \overset{\circ}{\rho}(\partial\psi/\partial I_\gamma) = \varphi_\gamma(I_1, I_2, \theta),\ \psi = \psi(I_1(\overset{(\mathrm{n})}{\mathbf{G}}),\ I_2(\overset{(\mathrm{n})}{\mathbf{G}}),\ \theta). \end{aligned} \tag{4.517}$$

If we assume the normalization conditions (4.326) and (4.327), then the potential $\psi$ is subject to the additional conditions (compare with (4.347a))

$$\varphi_1(a_1,\ a_2,\ \theta_e) + (2/(n - \mathrm{III}))\varphi_2(a_1,\ a_2,\ \theta_e) = p_0 - p^e, \quad \psi(a_1,\ a_2,\ \theta_e) = 0. \tag{4.518}$$

Here we have used the notation (4.347b).

For the models $C_n$ of incompressible ideal solids, we can use the theorems of Sect. 4.8.9 and represent the free energy in the form (4.378), then we get the following representation of the constitutive equations (4.504) in the tensor basis:

$$\begin{aligned} \overset{(\mathrm{n})}{\mathbf{S}} &= -\frac{p}{n - \mathrm{III}}\overset{(\mathrm{n})}{\mathbf{g}}{}^{-1} + \sum_{\gamma=1}^{r-1} \varphi_\gamma \frac{\partial I_\gamma^{(s)}}{\partial \overset{(\mathrm{n})}{\mathbf{A}}}, \\ \varphi_\gamma &= \varphi_\gamma(I_\alpha^{(s)}(\mathbf{O}^{\mathrm{T}} \cdot \overset{(\mathrm{n})}{\mathbf{A}} \cdot \mathbf{O}),\ \theta) = \overset{\circ}{\rho}\frac{\partial\psi}{\partial I_\gamma^{(s)}}, \quad \alpha = 1, \ldots, r-1. \end{aligned} \tag{4.519}$$

In this case the relation (4.504) for the rotation tensor $\overset{\circ}{\mathbf{S}}$ is automatically satisfied.

If we consider the models $C_n$ of incompressible isotropic continua, then, according to (4.380), the free energy $\psi$ can be represented as a function of three principal invariants $I_\alpha(\overset{(n)}{\mathbf{A}})$. However, for incompressible continua, these three invariants are also connected by the relation (4.511):

$$I_3(\overset{(n)}{\mathbf{A}}) + \frac{1}{n-\mathrm{III}} I_2(\overset{(n)}{\mathbf{A}}) + \frac{1}{(n-\mathrm{III})^2} I_1(\overset{(n)}{\mathbf{A}}) = 0,$$

i.e. only two ones of them are independent, therefore $\psi$ can be considered to depend only on two invariants; for example, on $I_1$ and $I_2$:

$$\psi = \psi(I_1(\overset{(n)}{\mathbf{A}}),\ I_2(\overset{(n)}{\mathbf{A}}),\ \theta). \tag{4.520}$$

Then the constitutive equations (4.509) take the form

$$\begin{aligned} \overset{(n)}{\mathbf{S}} &= -\frac{p}{n-\mathrm{III}} \overset{(n)}{\mathbf{g}}{}^{-1} + \psi_1 \mathbf{E} - \psi_2 \overset{(n)}{\mathbf{A}}, \\ \varphi_\gamma &= \varphi_\gamma(I_1(\overset{(n)}{\mathbf{A}}),\ I_2(\overset{(n)}{\mathbf{A}}),\ \theta). \end{aligned} \tag{4.521}$$

Just as for compressible continua, the constitutive equations for the models $A_n$ (4.515) and for the models $C_n$ (4.521) can be connected only by the rotation tensor $\mathbf{O}$, because $I_\alpha(\overset{(n)}{\mathbf{C}}) = I_\alpha(\overset{(n)}{\mathbf{A}})$ (see Exercise 4.8.2).

The constitutive equations for the models $D_n$ (4.504a) in the tensor basis are written as follows:

$$\begin{aligned} \overset{(n)}{\mathbf{S}} &= -\frac{p}{n-\mathrm{III}} \overset{(n)}{\mathbf{g}}{}^{-1} + \sum_{\gamma=1}^{r-1} \varphi_\gamma \frac{\partial I_\gamma^{(s)}}{\partial \overset{(n)}{\mathbf{g}}}, \\ \varphi_\gamma &= \varphi_\gamma(I_\alpha^{(s)}(\mathbf{O}^{\mathrm{T}} \cdot \overset{(n)}{\mathbf{G}} \cdot \mathbf{O}),\ \theta) = \overset{\circ}{\rho} \frac{\partial \psi}{\partial I_\gamma^{(s)}}, \quad \alpha = 1, \ldots, r-1. \end{aligned} \tag{4.522}$$

These equations are equivalent to relations (4.516) and can be obtained from them with the help of multiplication by the tensor $\mathbf{O}$ (see (4.382)).

### *4.9.6 General Representation of Constitutive Equations for All the Models of Incompressible Ideal Solids*

Just as in Sect. 4.8.10, constitutive equations (4.513) for the models $A_n$, (4.516) for the models $B_n$, (4.519) for the models $C_n$ and (4.522) for the models $D_n$ can be written in a general form – with the help of the generalized energetic tensors $\overset{(n)}{\mathbf{T}}_G$, $\overset{(n)}{\mathbf{C}}_G$ and $\overset{(n)}{\mathbf{G}}_G$ (4.91):

$$\overset{(n)}{\mathbf{T}}_G = -\frac{p}{n - \mathrm{III}} \overset{(n)}{\mathbf{G}}{}_G^{-1} + \sum_{\gamma=1}^{r-1} \varphi_\gamma \mathbf{I}_{\gamma G}^{(s)}. \tag{4.523}$$

Here we have used the notation (4.385) and (4.386).

These relations can be reduced to the form (4.387) by expressing the Cauchy stress tensor in the explicit form:

$$\mathbf{T} = -p\mathbf{E} + \overset{(n)}{\underset{\sim}{\mathcal{F}}}_G(\mathbf{F},\ \theta), \tag{4.524}$$

where

$$\overset{(n)}{\underset{\sim}{\mathcal{F}}}_G(\mathbf{F},\ \theta) \equiv \sum_{\gamma=1}^{r-1} \varphi_\gamma \overset{(n)}{\mathbf{H}}_{\gamma G}. \tag{4.525}$$

The remaining notation has been introduced in Sect. 4.8.10.

### 4.9.7 Linear Models of Ideal Incompressible Elastic Continua

As an example, let us consider linear models of ideal elastic isotropic continua, for which the free energy $\psi$ (4.515) or (4.516) is a quadratic function of the corresponding invariants $I_\gamma(\overset{(n)}{\mathbf{C}})$ or $I_\gamma(\overset{(n)}{\mathbf{A}})$. For isotropic linear models $A_n$ of incompressible continua, $\psi$ has the same form (4.335) as well as for corresponding compressible isotropic continua; but constitutive equations change due to the appearance of hydrostatic pressure $p$;

$$\begin{aligned} \overset{\circ}{\rho}\psi &= \overset{\circ}{\rho}\psi_0 + \overset{\circ}{\rho}\bar{\psi}^0 + \frac{l_1 + 2l_2}{2} I_1^2(\overset{(n)}{\mathbf{C}}) - 2l_2 I_2(\overset{(n)}{\mathbf{C}}) + \bar{m} I_1(\overset{(n)}{\mathbf{C}}), \\ \overset{(n)}{\mathbf{T}} &= -\frac{p}{n - \mathrm{III}} \overset{(n)}{\mathbf{G}}{}^{-1} + (\bar{m} + l_1 I_1(\overset{(n)}{\mathbf{C}}))\mathbf{E} + 2l_2 \overset{(n)}{\mathbf{C}}. \end{aligned} \tag{4.526}$$

Since we consider incompressible continua, so $J = \rho/\overset{\circ}{\rho} = 1$.

In order that the constitutive equations satisfy the conditions (4.326), (4.327) in the natural configuration $\mathcal{K}^e$ at $\mathbf{T}^e = -p^e$, in $\mathcal{K}^e$ the relations (4.515a) must hold; in this case they take the form

$$p_0 = p^e + \bar{m}, \qquad \bar{\psi}^0 = 0, \tag{4.527}$$

where $p_0$ is the value of the function $p$ in $\mathcal{K}^e$.

For isotropic linear models $B_n$ of incompressible continua, the free energy $\psi$ has the form (4.348) as well as for compressible continua, and the constitutive equations are similar to (4.526):

$$\begin{aligned} \overset{\circ}{\rho}\psi &= \overset{\circ}{\rho}\psi_0 + \overset{\circ}{\rho}\psi^0 + \frac{l_1 + 2l_2}{2} I_1^2(\overset{(n)}{\mathbf{G}}) - 2l_2 I_2(\overset{(n)}{\mathbf{G}}) + m I_1(\overset{(n)}{\mathbf{G}}), \\ \overset{(n)}{\mathbf{T}} &= -\frac{p}{n - \mathrm{III}}\overset{(n)}{\mathbf{G}}{}^{-1} + (m + l_1 I_1(\overset{(n)}{\mathbf{G}}))\mathbf{E} + 2l_2\overset{(n)}{\mathbf{G}}. \end{aligned} \tag{4.528}$$

The normalization conditions (4.326) and (4.327) lead to the restrictions (4.518) on the form of the potential $\psi$, which in this case yield relations between $\psi^0$, $m$, $l_1$, $l_2$ and $p_0$:

$$p_0 = p^e + m + \frac{3l_1 + 2l_2}{n - \mathrm{III}}, \qquad \psi^0 = -\frac{3(3l_1 + 2l_2)}{2\overset{\circ}{\rho}(n - \mathrm{III})^2} - \frac{3m}{(n - \mathrm{III})\overset{\circ}{\rho}}. \tag{4.529}$$

Relations (4.526), (4.527) for models $A_n$ and (4.528), (4.529) for models $B_n$ are completely equivalent (just as for compressible continua), and the models can be obtained one from another with the help of substitution $\overset{(n)}{\mathbf{C}} = \overset{(n)}{\mathbf{G}} - (1/(n - \mathrm{III}))\mathbf{E}$ (see Exercise 4.9.1). Each of the models includes three independent constants $m$, $l_1$ and $l_2$ that is by one constant more than for the corresponding models $A_n$ (4.335) and $B_n$ (4.350) of compressible continua.

Considering for the models (4.526) and (4.528) special cases of values of the constants, for example, assuming $m = 0$ and $\bar{m} = 0$, from the normalization conditions (4.527) and (4.529) we find new values of $p_0$, $\bar{\psi}^0$ and $\psi^0$, while the models $A_n$ and $B_n$ are not equivalent.

The models $B_n$ (see (4.528)) are most widely used in practice; in these models we assume that

$$l_1 + 2l_2 = 0, \quad l_2 = -(\mu/2)(1-\beta)(n-\mathrm{III})^2, \quad m = \mu(1+\beta)(n-\mathrm{III}), \tag{4.530}$$

where $\mu$ and $\beta$ are two new independent constants; and from the normalization conditions (4.529) we have

$$\psi^0 = -6\mu/\overset{\circ}{\rho}, \qquad p_0 = p^e + \mu(3 - \beta)(n - \mathrm{III})^2. \tag{4.531}$$

Relations (4.528) in this case can be written in the following form (see Exercise 4.9.2):

$$\begin{aligned} \overset{\circ}{\rho}\psi &= \overset{\circ}{\rho}\psi_0 + \mu\Big((1+\beta)((n - \mathrm{III})I_1(\overset{(n)}{\mathbf{G}}) - 3) + (1-\beta)((n - \mathrm{III})^2 I_2(\overset{(n)}{\mathbf{G}}) - 3)\Big), \\ \overset{(n)}{\mathbf{T}} &= -\frac{p}{n - \mathrm{III}}\overset{(n)}{\mathbf{G}}{}^{-1} + \mu(n - \mathrm{III})^2\left(\left(\frac{1+\beta}{n - \mathrm{III}} + (1-\beta)I_1(\overset{(n)}{\mathbf{G}})\right)\mathbf{E} - (1-\beta)\overset{(n)}{\mathbf{G}}\right). \end{aligned} \tag{4.532}$$

The set of relations (4.532) at $n = \text{IV}$ is called the *model of K.F. Chernykh*, which in the particular case when $\beta = 1$ reduces to the *Bartenev–Hazanovich model*. The set of relations (4.532) at $n = \text{V}$ is called *Mooney's model*, which in the special case when $\beta = 1$ reduces to *Treloar's model*.

### 4.9.8 Representation of Models $B_n$ and $D_n$ of Incompressible Isotropic Elastic Continua in Eigenbasis

For incompressible continua, three eigenvalues $\lambda_i$ are not independent, because they are connected by the relations (4.490):

$$\gamma = I_3(\mathbf{U}) - 1 = \lambda_1\lambda_2\lambda_3 - 1 = 0. \tag{4.533}$$

As shown in Sect. 4.9.2, we can introduce the Lagrange multiplier $p$ and rewrite the principal thermodynamic identity (4.495) in the form

$$\rho\, d\psi - p\, d\gamma - \sum_{\alpha=1}^{3} \overset{(\mathrm{n})}{\sigma}_{\alpha}\lambda_{\alpha}^{n-\mathrm{IV}}\, d\lambda_{\alpha} + \rho\eta\, d\theta = 0, \tag{4.534}$$

where

$$d\gamma = \sum_{\alpha=1}^{3} \frac{\partial\gamma}{\partial\lambda_{\alpha}}\, d\lambda_{\alpha} = 0. \tag{4.535}$$

Here we have used the expression (4.417a) for the stress power $w_{(i)}$. Then $d\lambda_{\alpha}$ ($\alpha = 1, 2, 3$) can be considered as independent.

Determining the derivatives of functions (4.533):

$$\frac{\partial\gamma}{\partial\lambda_{\alpha}} = \lambda_{\beta}\lambda_{\gamma} = \frac{1}{\lambda_{\alpha}}, \quad \alpha = 1, 2, 3, \quad \alpha \neq \beta \neq \gamma \neq \alpha, \tag{4.536}$$

and substituting them into (4.534), we obtain

$$\sum_{\alpha=1}^{3}\left(\rho\frac{\partial\psi}{\partial\lambda_{\alpha}} - \frac{p}{\lambda_{\alpha}} - \overset{\circ(n)}{\sigma}_{\alpha}\lambda_{\alpha}^{n-\mathrm{IV}}\right) d\lambda_{\alpha} + \rho\left(\frac{\partial\psi}{\partial\theta} + \eta\right) d\theta = 0. \tag{4.537}$$

As a result, we find the following constitutive equations for incompressible continua (the models $B_n$ and $D_n$):

$$\overset{(\mathrm{n})}{\sigma}_{\alpha} = -p\lambda_{\alpha}^{\mathrm{III}-n} + \overset{\circ}{\rho}\lambda_{\alpha}^{\mathrm{IV}-n}\frac{\partial\psi}{\partial\lambda_{\alpha}}, \quad \alpha = 1, 2, 3, \quad n = \mathrm{I}, \ldots, \mathrm{V}, \tag{4.538}$$

or with taking (4.417) into account:

$$\sigma_\alpha = -p + \overset{\circ}{\rho}\lambda_\alpha \frac{\partial \psi}{\partial \lambda_\alpha}, \quad \overset{(n)}{\sigma}_\alpha = \lambda_\alpha^{\mathrm{III}-n} \sigma_\alpha. \tag{4.539}$$

Thus, we have proved the following theorem.

**Theorem 4.35.** *The constitutive equations (4.515) and (4.521) for incompressible isotropic continua can be written in the form of three scalar relations (4.539) between $\sigma_\alpha$ and $\lambda_\alpha$.*

As an example of constitutive equations in the eigenbasis, consider the potentials (4.532) for incompressible continua and pass from $I_\alpha$ to $\lambda_\alpha$ by the formula from Exercise 4.8.3:

$$B_{\mathrm{I}},\ B_{\mathrm{V}}: \quad \psi = \psi_0 + \mu(1+\beta)(\lambda_1^{-2} + \lambda_2^{-2} + \lambda_3^{-2} - 3) \\ + \mu(1-\beta)(\lambda_1^2 + \lambda_2^2 + \lambda_3^2 - 3), \tag{4.540}$$

$$B_{\mathrm{II}},\ B_{\mathrm{IV}}: \quad \psi = \psi_0 + \mu(1+\beta)(\lambda_1^{-1} + \lambda_2^{-1} + \lambda_3^{-1} - 3) \\ + \mu(1-\beta)(\lambda_1 + \lambda_2 + \lambda_3 - 3). \tag{4.541}$$

Here we have used the incompressibility condition $\lambda_1\lambda_2\lambda_3 = 1$. The second invariant $I_2$ can be expressed as follows:

$$4I_2(\overset{\mathrm{I}}{\mathbf{G}}) = \lambda_1^{-2}\lambda_2^{-2} + \lambda_2^{-2}\lambda_3^{-2} + \lambda_1^{-2}\lambda_3^{-2} = \lambda_1^2 + \lambda_2^2 + \lambda_3^2. \tag{4.542}$$

On substituting (4.540) and (4.541) into (4.539), we get the corresponding constitutive equations in the eigenbasis:

$$B_{\mathrm{I}}, B_{\mathrm{V}}: \quad \sigma_\alpha = -p + 2\mu((1-\beta)(\lambda_1^2 + \lambda_2^2 + \lambda_3^2) - (1+\beta)(\lambda_1^{-2} + \lambda_2^{-2} + \lambda_3^{-2})), \tag{4.543}$$

$$B_{\mathrm{II}}, B_{\mathrm{IV}}: \quad \sigma_\alpha = -p + \mu((1-\beta)(\lambda_1 + \lambda_2 + \lambda_3) - (1-\beta)(\lambda_1^{-1} + \lambda_2^{-1} + \lambda_3^{-1})). \tag{4.544}$$

## Exercises for 4.9

**4.9.1.** Show that the linear models $A_n$ (4.526), (4.527) and $B_n$ (4.528), (4.529) of incompressible continua are equivalent.

**4.9.2.** Prove that relations (4.532) follow from (4.528) under the assumptions of the model (4.530).

## 4.10 The Principle of Material Indifference

### *4.10.1 Rigid Motion*

Let us consider now one more basic principle mentioned in Sect. 4.1, namely the principle of material indifference. This principle consists in the assertion that constitutive equations of a continuum cannot depend on the choice of a frame of reference, where they are considered.

The principle of material symmetry supposes a change of reference configuration: $\overset{\circ}{\mathcal{K}} \to \overset{*}{\mathcal{K}}$, but, according to the principle of material indifference, we consider special motions of actual configuration: $\mathcal{K} \to \mathcal{K}'$, namely rigid motions.

Define the *rigid motion* of a continuum. Take in configuration $\mathcal{K}$ (Fig. 4.5) an arbitrary point $\mathcal{M}_0$ with its radius-vector $\mathbf{x}_0$ (it is not to be confused with $\overset{\circ}{\mathbf{x}}$) and pass to configuration $\widehat{\mathcal{K}}$, which differs from $\mathcal{K}$ only by rotation as a rigid whole about the point $\mathcal{M}_0$. Then an arbitrary vector $\mathbf{x}-\mathbf{x}_0$ of $\mathcal{K}$ becomes the vector $\widehat{\mathbf{x}}-\mathbf{x}_0$ in the configuration $\widehat{\mathcal{K}}$; and

$$\widehat{\mathbf{x}} - \mathbf{x}_0 = (\mathbf{x} - \mathbf{x}_0) \cdot \mathbf{Q}(t). \tag{4.545}$$

Here $\mathbf{x}$ and $\widehat{\mathbf{x}}$ are coordinates of the same point $\mathcal{M}$ in $\mathcal{K}$ and $\widehat{\mathcal{K}}$, respectively, and $\mathbf{Q}$ is the orthogonal rotation tensor:

$$\mathbf{Q} = Q^{ij}\bar{\mathbf{e}}_i \otimes \bar{\mathbf{e}}_j, \qquad \mathbf{Q}^{\mathrm{T}} = \mathbf{Q}^{-1}, \quad \mathbf{Q}^{\mathrm{T}} \cdot \mathbf{Q} = \mathbf{E}, \tag{4.546}$$

where $Q^{ij}$ is the orthogonal matrix of rotation.

Pass from $\widehat{\mathcal{K}}$ to the following configuration $\mathcal{K}'$, which differs from $\widehat{\mathcal{K}}$ only by a parallel carry with vector $\mathbf{a}$, then a radius-vector of the point $\mathcal{M}$ in $\mathcal{K}'$ is determined as follows:

$$\mathbf{x}' = \widehat{\mathbf{x}} + \mathbf{a} = \mathbf{x}_0 + \mathbf{a} + (\mathbf{x} - \mathbf{x}_0) \cdot \mathbf{Q}(t), \tag{4.547}$$

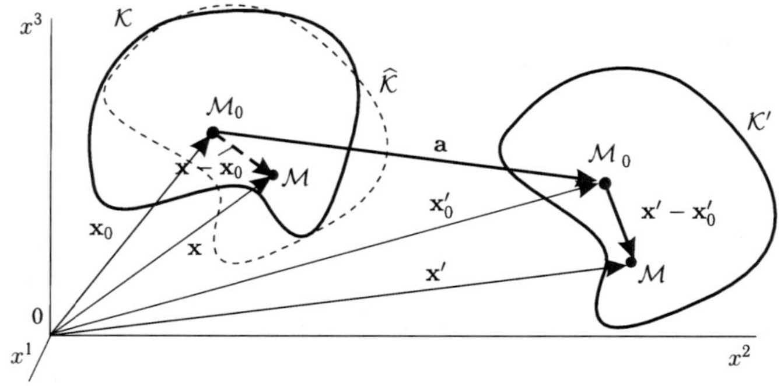


**Fig. 4.5** The rigid motion of a continuum

or

$$\mathbf{x}'(X^i,t) = \mathbf{x}'_0(X^i_0,t) + \big(\mathbf{x}(X^i,t) - \mathbf{x}_0(X^i_0,t)\big)\cdot\mathbf{Q}(t), \tag{4.548}$$

where $X^i$ and $X^j_0$ are Lagrangian coordinates of the points $\mathcal{M}$ and $\mathcal{M}_0$, respectively, and $\mathbf{x}'_0 = \mathbf{x}_0 + \mathbf{a}$ is the radius-vector of the point $\mathcal{M}_0$ in $\mathcal{K}'$, which is the instantaneous center of rotation. The rotation tensor $\mathbf{Q}(t)$ and radius-vectors $\overset{\circ}{\mathbf{x}}(t)$ and $\overset{\circ}{\mathbf{x}}{}'(t)$ depend on time $t$.

The motion, satisfying Eq. (4.548), is called the *rigid motion* of a continuum.

## *4.10.2 R-indifferent and R-invariant Tensors*

Let an arbitrary tensor $\mathbf{A}(\mathbf{x})$ be given at a point $\mathbf{x}$ in configuration $\mathcal{K}$:

$$\mathbf{A}(\mathbf{x},t) = A^{i_1\ldots i_n}\mathbf{r}_{i_1}\otimes\ldots\otimes\mathbf{r}_{i_n}, \tag{4.549}$$

where $\mathbf{r}_i$ is the local basis in $\mathcal{K}$.

A tensor $\mathbf{A}$ is called *indifferent* relative to the rigid motion (4.548) or *R-indifferent*, if under the transformation from $\mathcal{K}$ to $\mathcal{K}'$ the corresponding tensor $\mathbf{A}'$ is determined by the same rules as the tensor $\mathbf{A}$, but only in $\mathcal{K}'$:

$$\mathbf{A}'(\mathbf{x}',t) = A'^{i_1\ldots i_n}\mathbf{r}'_{i_1}\otimes\ldots\otimes\mathbf{r}'_{i_n};$$

for $\mathbf{A}'$ we have

$$A^{i_1\ldots i_n}(\mathbf{x},\ t) = A'^{i_1\ldots i_n}(\mathbf{x}',\ t), \tag{4.550}$$

where $\mathbf{r}'_i$ is the local basis in $\mathcal{K}'$.

Components of $R$-indifferent tensor $\mathbf{A}$ move together with the local basis vectors, i.e. they are 'frozen' into the continuum.

A tensor $\mathbf{A}$ is called *invariant relative to the rigid motion* (4.548) or *R-invariant*, if

$$\mathbf{A}'(\mathbf{x}',\ t) = \mathbf{A}(\mathbf{x},\ t). \tag{4.551}$$

The concept of invariance of a tensor relative to the rigid motion, does not coincide with the definition of a tensor as an invariant object, because under the transformation $\mathcal{K}\to\mathcal{K}'$ we obtain, in general, different tensors $\mathbf{A}$ and $\mathbf{A}'$, and their coincidence is a certain condition imposed on the tensor $\mathbf{A}$.

Differentiation (4.548) with respect to $X^i$ yields

$$\frac{\partial\mathbf{x}'}{\partial X^i} = \mathbf{r}'_i = \frac{\partial\mathbf{x}}{\partial X^i}\cdot\mathbf{Q} = \mathbf{r}_i\cdot\mathbf{Q}. \tag{4.552}$$

Thus, in the rigid motion, vectors of a local basis are transformed by the law (4.552), and the definition of $R$-indifferent tensor (4.550) takes the form

$$\mathbf{A}'(\mathbf{x}', t) = A'^{i_1\ldots i_n}(\mathbf{x}', t)\mathbf{r}'_{i_1} \otimes \ldots \otimes \mathbf{r}'_{i_n} = A^{i_1\ldots i_n}(\mathbf{x}, t)\mathbf{r}_{i_1} \cdot \mathbf{Q} \otimes \ldots \mathbf{r}_{i_n} \cdot \mathbf{Q}. \quad (4.553)$$

For a scalar $\varphi$, $R$-indifference means that

$$\varphi'(\mathbf{x}',\ t) = \varphi(\mathbf{x},\ t), \quad (4.554)$$

for a vector $\mathbf{b}$:

$$\mathbf{b}'(\mathbf{x}',\ t) = \mathbf{b}(\mathbf{x},\ t) \cdot \mathbf{Q} = \mathbf{Q}^{\mathrm{T}} \cdot \mathbf{b}(\mathbf{x},\ t), \quad (4.555)$$

and for a second-order tensor $\mathbf{A}$:

$$\begin{aligned}\mathbf{A}'(\mathbf{x}',\ t) = A'^{i_1 i_2}\mathbf{r}'_{i_1} \otimes \mathbf{r}'_{i_2} &= A^{i_1 i_2}\mathbf{r}_{i_1} \cdot \mathbf{Q} \otimes \mathbf{r}_{i_2} \cdot \mathbf{Q} \\ &= \mathbf{Q}^{\mathrm{T}} \cdot A^{i_1 i_2}\mathbf{r}_{i_1} \otimes \mathbf{r}_{i_2} \cdot \mathbf{Q} = \mathbf{Q}^{\mathrm{T}} \cdot \mathbf{A} \cdot \mathbf{Q}.\end{aligned} \quad (4.556)$$

According to (4.548), the vector $\widetilde{\mathbf{x}} = \mathbf{x} - \mathbf{x}_0$ is $R$-indifferent.
Consider other examples.

### 4.10.3 Density and Deformation Gradient in Rigid Motion

Density $\rho$ is a $R$-indifferent scalar. Indeed, in going from $\mathcal{K}$ to $\mathcal{K}'$ Eq. (3.6) yields

$$\rho\, dV = \rho'\, dV', \quad (4.557)$$

where $\rho'$ and $dV'$ are the density and elementary volume in the configuration $\mathcal{K}'$; but in the rigid motion $dV' = dV$ and, hence,

$$\rho'(\mathbf{x}',\ t) = \rho(\mathbf{x},\ t). \quad (4.558)$$

Local basis vectors $\mathbf{r}_i$ are $R$-indifferent due to (4.552).
The deformation gradient $\mathbf{F}'$ in $\mathcal{K}'$ is determined as well as $\mathbf{F}$ in $\mathcal{K}$:

$$\mathbf{F}' = \mathbf{r}'_i \otimes \overset{\circ}{\mathbf{r}}{}^{i\prime} = \mathbf{Q}^{\mathrm{T}} \cdot \mathbf{r}'_i \otimes \overset{\circ}{\mathbf{r}}{}^{i} = \mathbf{Q}^{\mathrm{T}} \cdot \mathbf{F}. \quad (4.559)$$

The deformation gradient is not a $R$-indifferent tensor, because in the rigid motion a reference configuration remains unchanged: $\overset{\circ}{\mathbf{r}}{}^{\prime i} = \overset{\circ}{\mathbf{r}}{}^{i}$.
The transpose deformation gradient is not $R$-indifferent as well, because

$$\mathbf{F}^{\mathrm{T}\prime} = \overset{\circ}{\mathbf{r}}{}^{i\prime} \otimes \mathbf{r}'_i = \overset{\circ}{\mathbf{r}}{}^{i} \otimes \mathbf{r}_i \cdot \cdot\, \mathbf{Q} = \mathbf{F}^{\mathrm{T}} \cdot \mathbf{Q}. \quad (4.560)$$

The inverse gradients are also not $R$-indifferent:

$$\mathbf{F}^{-1\mathrm{T}\prime} = (\mathbf{F}^{\mathrm{T}\prime})^{-1} = (\mathbf{F}^{\mathrm{T}} \cdot \mathbf{Q})^{-1} = \mathbf{Q}^{-1} \cdot \mathbf{F}^{\mathrm{T}-1} = \mathbf{Q}^{\mathrm{T}} \cdot \mathbf{F}^{\mathrm{T}-1}, \tag{4.561a}$$

$$\mathbf{F}^{-1\prime} = (\mathbf{F}')^{-1} = (\mathbf{Q}^{\mathrm{T}} \cdot \mathbf{F})^{-1} = \mathbf{F}^{-1} \cdot \mathbf{Q}^{\mathrm{T}-1} = \mathbf{F}^{-1} \cdot \mathbf{Q}. \tag{4.561b}$$

Local vectors of a reciprocal basis are $R$-indifferent, because

$$\mathbf{r}^{i\prime} = \mathbf{r}^{j\prime}\delta^i_j = \mathbf{r}^{j\prime} \otimes \mathring{\mathbf{r}}'_j \cdot \mathring{\mathbf{r}}^{i\prime} = \mathbf{F}^{-1\mathrm{T}\prime} \cdot \mathring{\mathbf{r}}^{i\prime} = \mathbf{Q}^{\mathrm{T}} \cdot \mathbf{F}^{-1\mathrm{T}} \cdot \mathring{\mathbf{r}}^{i} = \mathbf{Q}^{\mathrm{T}} \cdot \mathbf{r}^{i}. \tag{4.562}$$

The unit tensor

$$\mathbf{E} = \mathbf{Q}^{\mathrm{T}} \cdot \mathbf{E} \cdot \mathbf{Q} = \mathbf{Q}^{\mathrm{T}} \cdot \mathbf{Q} = \mathbf{E}$$

is both $R$-indifferent and $R$-invariant.

With the help of (4.562) we can determine the nabla-operator $\nabla'$ in $\mathcal{K}'$:

$$\nabla' = \mathbf{r}'^{i}(\partial/\partial X^i). \tag{4.563}$$

It follows from (4.562) and (4.563) that

$$\nabla' = \mathbf{Q}^{\mathrm{T}} \cdot \nabla, \qquad \nabla = \mathbf{r}^{i}\frac{\partial}{\partial X^i} = \bar{\mathbf{e}}^{i}\frac{\partial}{\partial x^i}. \tag{4.563a}$$

## 4.10.4 Deformation Tensors in Rigid Motion

The left Almansi deformation tensor is $R$-indifferent, because

$$\mathbf{A}' = \frac{1}{2}(\mathbf{E}' - \mathbf{F}^{-1\mathrm{T}\prime} \cdot \mathbf{F}^{-1\prime}) = \frac{1}{2}(\mathbf{E}' - \mathbf{Q}^{\mathrm{T}} \cdot \mathbf{F}^{-1\mathrm{T}} \cdot \mathbf{F}^{-1} \cdot \mathbf{Q}) = \mathbf{Q}^{\mathrm{T}} \cdot \mathbf{A} \cdot \mathbf{Q}. \tag{4.564}$$

The right Almansi deformation tensor is $R$-invariant:

$$\mathbf{\Lambda}' = \frac{1}{2}(\mathbf{E}' - \mathbf{F}^{-1\prime} \cdot \mathbf{F}^{-1\mathrm{T}\prime}) = \frac{1}{2}(\mathbf{E} - \mathbf{F}^{-1} \cdot \mathbf{Q} \cdot \mathbf{Q}^{\mathrm{T}} \cdot \mathbf{F}^{-1\mathrm{T}}) = \mathbf{\Lambda}. \tag{4.565}$$

The left Cauchy–Green deformation tensor is $R$-invariant, because

$$\mathbf{C}' = \frac{1}{2}(\mathbf{F}^{\mathrm{T}\prime} \cdot \mathbf{F}' - \mathbf{E}') = \frac{1}{2}(\mathbf{F}^{\mathrm{T}} \cdot \mathbf{Q} \cdot \mathbf{Q}^{\mathrm{T}} \cdot \mathbf{F} - \mathbf{E}) = \mathbf{C}. \tag{4.566}$$

The right Cauchy–Green deformation tensor is $R$-indifferent:

$$\mathbf{J}' = \frac{1}{2}(\mathbf{F}' \cdot \mathbf{F}^{\mathrm{T}\prime} - \mathbf{E}') = \frac{1}{2}(\mathbf{Q}^{\mathrm{T}} \cdot \mathbf{F} \cdot \mathbf{F}^{\mathrm{T}} \cdot \mathbf{O} - \mathbf{Q}^{\mathrm{T}} \cdot \mathbf{Q}) = \mathbf{Q}^{\mathrm{T}} \cdot \mathbf{J} \cdot \mathbf{Q}. \tag{4.567}$$

The right stretch tensor $\mathbf{U}$ is $R$-invariant. Indeed, by the definition of the polar decomposition, formula (4.560) of transformation of $\mathbf{F}$ has the form

$$\mathbf{F}' = \mathbf{Q}^{\mathrm{T}} \cdot \mathbf{F} = \mathbf{Q}^{\mathrm{T}} \cdot \mathbf{O} \cdot \mathbf{U} = \mathbf{O}' \cdot \mathbf{U}';$$

but since $\mathbf{Q}^{\mathrm{T}} \cdot \mathbf{O}$ is orthogonal and the polar decomposition is unique, so

$$\mathbf{U} = \mathbf{U}' \quad \text{and} \quad \mathbf{O}' = \mathbf{Q}^{\mathrm{T}} \cdot \mathbf{O}. \tag{4.568}$$

Thus, the tensor $\mathbf{U}$ is $R$-invariant.

The rotation tensor $\mathbf{O}$ accompanying deformation is neither $R$-indifferent nor $R$-invariant.

In a similar way, using the definition of $\mathbf{V}$ and the connection between $\mathbf{O}'$ and $\mathbf{Q}$, we have

$$\mathbf{F}' = \mathbf{Q}^{\mathrm{T}} \cdot \mathbf{F} = \mathbf{Q}^{\mathrm{T}} \cdot \mathbf{V} \cdot \mathbf{O} = \mathbf{V}' \cdot \mathbf{O}' = \mathbf{V}' \cdot \mathbf{Q}^{\mathrm{T}} \cdot \mathbf{O},$$

i.e. we obtain that

$$\mathbf{Q}^{\mathrm{T}} \cdot \mathbf{V} = \mathbf{V}' \cdot \mathbf{Q}^{\mathrm{T}}.$$

Therefore, the tensor $\mathbf{V}$ is $R$-indifferent:

$$\mathbf{V}' = \mathbf{Q}^{\mathrm{T}} \cdot \mathbf{V} \cdot \mathbf{Q}. \tag{4.569}$$

From (4.568) and (4.569) we derive the following theorem.

**Theorem 4.36.** *All the energetic deformation tensors* $\overset{(n)}{\mathbf{C}}$ *and energetic deformation measures* $\overset{(n)}{\mathbf{G}}$*, which are proportional to* $\mathbf{U}^{n-\mathrm{III}}$*, are R-invariant, and all the quasienergetic deformation tensors* $\overset{(n)}{\mathbf{A}}$ *and quasienergetic deformation measures* $\overset{(n)}{\mathbf{g}}$*, which are proportional to* $\mathbf{V}^{n-\mathrm{III}}$*, are R-indifferent (*$n = \mathrm{I},\ \mathrm{II},\ \mathrm{IV},\ \mathrm{V}$*).*

The tensors $\overset{\mathrm{III}}{\mathbf{C}}$ and $\overset{\mathrm{III}}{\mathbf{A}}$ will be considered in Sect. 4.10.7.

### 4.10.5 Stress Tensors in Rigid Motion

The Cauchy stress tensor is $R$-indifferent. To prove the assertion, use the Cauchy relation (3.47): $\mathbf{t}_n = \mathbf{n} \cdot \mathbf{T}$. By definition, the force vector $\mathbf{t}_n$ is 'tracking', i.e. the vector moves together with the corresponding surface element $d\Sigma$ in any motion; and, hence, $\mathbf{t}_n$ is $R$-indifferent. The vector $\mathbf{n}$ also moves with the surface element; thus, $\mathbf{n}$ is $R$-indifferent as well. Then

$$\mathbf{t}_n' = \mathbf{n}' \cdot \mathbf{T}' = \mathbf{n} \cdot \mathbf{Q} \cdot \mathbf{T}' = \mathbf{t}_n \cdot \mathbf{Q} = \mathbf{n} \cdot \mathbf{T} \cdot \mathbf{Q}, \tag{4.570}$$

and we finally get

$$\mathbf{Q} \cdot \mathbf{T}' = \mathbf{T} \cdot \mathbf{Q} \quad \text{and} \quad \mathbf{T}' = \mathbf{Q}^{\mathrm{T}} \cdot \mathbf{T} \cdot \mathbf{Q}, \tag{4.571}$$

i.e. **T** is $R$-indifferent.

The Piola–Kirchhoff tensor **P** is neither $R$-indifferent nor $R$-invariant, because

$$\mathbf{P}' = \sqrt{g/\overset{\circ}{g}}\, \mathbf{F}^{-1\prime} \cdot \mathbf{T}' = \sqrt{g/\overset{\circ}{g}}\, \mathbf{F}^{-1} \cdot \mathbf{Q} \cdot \mathbf{Q}^{\mathrm{T}} \cdot \mathbf{T}' \cdot \mathbf{Q} = \mathbf{P} \cdot \mathbf{Q}. \tag{4.572}$$

**Theorem 4.37.** *All the principal energetic stress tensors* $\overset{(n)}{\mathbf{T}}$ *are R-invariant, and all the principal quasienergetic stress tensors* $\overset{(n)}{\mathbf{S}}$ *are R-indifferent:*

$$\overset{(n)}{\mathbf{T}}{}' = \overset{(n)}{\mathbf{T}}, \quad \overset{(n)}{\mathbf{S}}{}' = \mathbf{Q}^{\mathrm{T}} \cdot \overset{(n)}{\mathbf{S}} \cdot \mathbf{Q}, \quad n = \mathrm{I}, \ldots, \mathrm{V}. \tag{4.573}$$

▼ By the definition of the tensors $\overset{(n)}{\mathbf{T}}$ (see Table 4.1), we have

$$\begin{aligned}
\overset{\mathrm{I}}{\mathbf{T}}{}' &= \mathbf{F}^{\mathrm{T}\prime} \cdot \mathbf{T}' \cdot \mathbf{F}' = \mathbf{F}^{\mathrm{T}} \cdot \mathbf{Q} \cdot \mathbf{Q}^{\mathrm{T}} \cdot \mathbf{T} \cdot \mathbf{Q} \cdot \mathbf{Q}^{\mathrm{T}} \cdot \mathbf{F} = \overset{\mathrm{I}}{\mathbf{T}},\\
\overset{\mathrm{II}}{\mathbf{T}}{}' &= \frac{1}{2}\Big(\mathbf{F}^{\mathrm{T}\prime} \cdot \mathbf{T}' \cdot \mathbf{O}' + \mathbf{O}^{\mathrm{T}\prime} \cdot \mathbf{T}' \cdot \mathbf{F}'\Big)\\
&= \frac{1}{2}\Big(\mathbf{F}^{\mathrm{T}} \cdot \mathbf{Q} \cdot \mathbf{Q}^{\mathrm{T}} \cdot \mathbf{T} \cdot \mathbf{Q} \cdot \mathbf{Q}^{\mathrm{T}} \cdot \mathbf{O} + \mathbf{O}^{\mathrm{T}} \cdot \mathbf{Q} \cdot \mathbf{Q}^{\mathrm{T}} \cdot \mathbf{T} \cdot \mathbf{Q} \cdot \mathbf{Q}^{\mathrm{T}} \cdot \mathbf{F}\Big) = \overset{\mathrm{II}}{\mathbf{T}}, \qquad (4.574)\\
\overset{\mathrm{III}}{\mathbf{T}}{}' &= \mathbf{O}^{\mathrm{T}\prime} \cdot \mathbf{T}' \cdot \mathbf{O}' = \mathbf{O}^{\mathrm{T}} \cdot \mathbf{Q} \cdot \mathbf{Q}^{\mathrm{T}} \cdot \mathbf{T} \cdot \mathbf{Q} \cdot \mathbf{Q}^{\mathrm{T}} \cdot \mathbf{O} = \overset{\mathrm{III}}{\mathbf{T}} \quad \text{etc.,}
\end{aligned}$$

and, by the definition of the tensors $\overset{(n)}{\mathbf{S}}$ (see Table 4.3), we get

$$\begin{aligned}
\overset{\mathrm{I}}{\mathbf{S}}{}' &= \mathbf{V}' \cdot \mathbf{T}' \cdot \mathbf{V}' = \mathbf{Q}^{\mathrm{T}} \cdot \mathbf{V} \cdot \mathbf{Q} \cdot \mathbf{Q}^{\mathrm{T}} \cdot \mathbf{T} \cdot \mathbf{Q} \cdot \mathbf{Q}^{\mathrm{T}} \cdot \mathbf{V} \cdot \mathbf{Q} = \mathbf{Q}^{\mathrm{T}} \cdot \overset{\mathrm{I}}{\mathbf{S}} \cdot \mathbf{Q},\\
\overset{\mathrm{II}}{\mathbf{S}}{}' &= \frac{1}{2}\Big(\mathbf{V}' \cdot \mathbf{T}' + \mathbf{T}' \cdot \mathbf{V}'\Big)\\
&= \frac{1}{2}\Big(\mathbf{Q}^{\mathrm{T}} \cdot \mathbf{V} \cdot \mathbf{Q} \cdot \mathbf{Q}^{\mathrm{T}} \cdot \mathbf{T} \cdot \mathbf{Q} + \mathbf{Q}^{\mathrm{T}} \cdot \mathbf{T} \cdot \mathbf{Q} \cdot \mathbf{Q}^{\mathrm{T}} \cdot \mathbf{V} \cdot \mathbf{Q}\Big) = \mathbf{Q}^{\mathrm{T}} \cdot \overset{\mathrm{II}}{\mathbf{S}} \cdot \mathbf{Q}, \qquad (4.575)\\
\overset{\mathrm{III}}{\mathbf{S}}{}' &= \mathbf{T}' = \mathbf{Q}^{\mathrm{T}} \cdot \overset{\mathrm{III}}{\mathbf{S}} \cdot \mathbf{Q} \quad \text{etc. } \blacktriangle
\end{aligned}$$

### 4.10.6 The Velocity in Rigid Motion

**Theorem 4.38.** *The velocity* $\mathbf{v}$ *is not* $R$*-indifferent.*

▼ According to the definition (2.190) and formula (4.548), we get

$$\mathbf{v}' = \frac{d\mathbf{x}'}{dt} = \frac{d\mathbf{x}'_0}{dt} + (\mathbf{v} - \mathbf{v}_0) \cdot \mathbf{Q} + \dot{\mathbf{Q}}^{\mathrm{T}} \cdot (\mathbf{x} - \mathbf{x}_0), \tag{4.576}$$

where $\mathbf{v} = d\mathbf{x}/dt$, and $\mathbf{v}_0 = d\mathbf{x}_0/dt$ is the velocity of the point $\mathcal{M}_0$ (which is independent of $X^i$). It follows from (4.576) that the vector $\mathbf{v}$ is not $R$-indifferent. ▲

### 4.10.7 The Deformation Rate Tensor and the Vorticity Tensor in Rigid Motion

**Theorem 4.38a.** *The deformation rate tensor* $\mathbf{D}$ *is* $R$*-indifferent, and the vorticity tensor* $\mathbf{W}$ *is neither* $R$*-indifferent nor* $R$*-invariant.*

▼ Due to (4.576), we get

$$\mathbf{L}^{\mathrm{T}\prime} = (\nabla \otimes \mathbf{v})' = \mathbf{r}^{i\prime} \otimes \frac{\partial \mathbf{v}'}{\partial X^i} = \mathbf{Q}^{\mathrm{T}} \cdot \mathbf{r}^i \otimes \left( \frac{\partial \mathbf{v}}{\partial X^i} \cdot \mathbf{Q} + \dot{\mathbf{Q}}^{\mathrm{T}} \cdot \mathbf{r}_i \right)$$

$$= \mathbf{Q}^{\mathrm{T}} \cdot \nabla \otimes \mathbf{v} \cdot \mathbf{Q} + \mathbf{Q}^{\mathrm{T}} \cdot \mathbf{r}^i \otimes \mathbf{r}_i \cdot \dot{\mathbf{Q}} = \mathbf{Q}^{\mathrm{T}} \cdot \nabla \otimes \mathbf{v} \cdot \mathbf{Q} + \mathbf{Q}^{\mathrm{T}} \cdot \dot{\mathbf{Q}}, \tag{4.577}$$

$$\mathbf{L}' = (\nabla \otimes \mathbf{v})^{\mathrm{T}\prime} = \frac{\partial \mathbf{v}'}{\partial X^i} \otimes \mathbf{r}^{i\prime} = \left( \mathbf{Q}^{\mathrm{T}} \otimes \frac{\partial \mathbf{v}}{\partial X^i} + \dot{\mathbf{Q}}^{\mathrm{T}} \cdot \mathbf{r}_i \right) \otimes \mathbf{r}^i \cdot \mathbf{Q}$$

$$= \mathbf{Q}^{\mathrm{T}} \cdot (\nabla \otimes \mathbf{v})^{\mathrm{T}} \cdot \mathbf{Q} + \dot{\mathbf{Q}}^{\mathrm{T}} \cdot \mathbf{Q}. \tag{4.578}$$

On summing (4.577) and (4.578), we find

$$\mathbf{D}' = \frac{1}{2}(\nabla \otimes \mathbf{v})' + (\nabla \otimes \mathbf{v})^{\mathrm{T}\prime} = \mathbf{Q}^{\mathrm{T}} \cdot \mathbf{D} \cdot \mathbf{Q}. \tag{4.579}$$

The vorticity tensor is transformed as follows:

$$\mathbf{W}' = \frac{1}{2}((\nabla' \otimes \mathbf{v}')^{\mathrm{T}} - \nabla' \otimes \mathbf{v}') = \mathbf{Q}^{\mathrm{T}} \cdot \mathbf{W} \cdot \mathbf{Q} + \dot{\mathbf{Q}}^{\mathrm{T}} \cdot \mathbf{Q}$$

$$= \mathbf{Q}^{\mathrm{T}} \cdot \mathbf{W} \cdot \mathbf{Q} - \mathbf{Q}^{\mathrm{T}} \cdot \dot{\mathbf{Q}}. \; ▲ \tag{4.580}$$

### 4.10.8 Co-rotational Derivatives in Rigid Motion

**Theorem 4.39.** *The total derivative of any R-invariant tensor with respect to time is R-invariant.*

▼ Let a tensor $\mathbf{\Omega}$ be $R$-invariant, then

$$\mathbf{\Omega}' = \mathbf{\Omega} \quad \text{and} \quad \dot{\mathbf{\Omega}}' = \frac{d}{dt}\mathbf{\Omega}. \ \blacktriangle \tag{4.581}$$

Hence, the total derivatives of the energetic deformation tensors $\overset{(n)}{\mathbf{C}}$, $\overset{(n)}{\mathbf{G}}$ and of the energetic stress tensors $\overset{(n)}{\mathbf{T}}$ with respect to time are $R$-invariant.

**Theorem 4.40.** *The derivative of any R-indifferent tensor* $\mathbf{\Omega}$ *with respect to time is neither R-indifferent nor R-invariant.*

▼ For the tensor $\mathbf{\Omega}$ we have

$$(\mathbf{\Omega}')^{\bullet} = (\mathbf{Q}^{\mathrm{T}} \cdot \mathbf{\Omega} \cdot \mathbf{Q})^{\bullet} = \dot{\mathbf{Q}}^{\mathrm{T}} \cdot \mathbf{\Omega} \cdot \mathbf{Q} + \mathbf{Q}^{\mathrm{T}} \cdot \dot{\mathbf{\Omega}} \cdot \mathbf{Q} + \mathbf{Q}^{\mathrm{T}} \cdot \mathbf{\Omega} \cdot \mathbf{Q} \neq \mathbf{Q}^{\mathrm{T}} \cdot \dot{\mathbf{\Omega}} \cdot \mathbf{Q}. \ \blacktriangle \tag{4.582}$$

Since the tensors $\mathbf{U}$ and $\dot{\mathbf{U}}$ are $R$-invariant, the tensor $\overset{\mathrm{III}}{\mathbf{C}} = \mathbf{B}$, defined by relations (2.317), is also $R$-invariant. But the tensor $\overset{\mathrm{III}}{\mathbf{A}} = \mathbf{Y}$, defined by (2.318), is neither $R$-invariant nor $R$-indifferent, because $\overset{\mathrm{III}}{\mathbf{A}}$ depends on the tensor $\dot{\mathbf{V}}$, which is not $R$-indifferent. In place of the tensor $\overset{\mathrm{III}}{\mathbf{A}} = \mathbf{Y}$, for isotropic continua one usually use the Hencky tensor $\mathbf{H}$ (2.167), which is energetically equivalent to $\overset{\mathrm{III}}{\mathbf{A}} = \mathbf{Y}$ (see Theorem 4.28a)) and which is $R$-indifferent.

Indeed, the eigenvalues $\lambda_\alpha$ of the tensor $\mathbf{U}$ do not change in the motion $\mathcal{K} \to \mathcal{K}'$, because the tensor $\mathbf{U}$ remains unchanged itself; but since $\lambda_\alpha$ are the eigenvalues of the tensor $\mathbf{V}$ as well, so $\mathbf{V}$ can be represented as follows:

$$\mathbf{V}' = \sum_{\alpha=1}^{3} \lambda_\alpha \mathbf{p}'_\alpha \otimes \mathbf{p}'_\alpha = \mathbf{Q}^{\mathrm{T}} \cdot \sum_{\alpha=1}^{3} \lambda_\alpha \mathbf{p}_\alpha \otimes \mathbf{p}_\alpha \cdot \mathbf{Q}.$$

From this equation it follows that

$$\mathbf{p}'_\alpha = \mathbf{Q}^{\mathrm{T}} \cdot \mathbf{p}_\alpha,$$

and the eigenvectors $\mathbf{p}_\alpha$ are $R$-indifferent.

The Hencky tensor (2.167) in $\mathcal{K}'$ has the following form:

$$\mathbf{H}' = \sum_{\alpha=1}^{3} \lg \lambda_\alpha \mathbf{p}'_\alpha \otimes \mathbf{p}'_\alpha = \mathbf{Q}^{\mathrm{T}} \cdot \sum_{\alpha=1}^{3} \lg \lambda_\alpha \mathbf{p}_\alpha \otimes \mathbf{p}_\alpha \cdot \mathbf{Q} = \mathbf{Q}^{\mathrm{T}} \cdot \mathbf{H} \cdot \mathbf{Q}.$$

Thus, the tensor $\mathbf{H}$ is really $R$-indifferent.

The Hencky tensor $\overset{\circ}{\mathbf{H}}$ (2.167) is clear to be $R$-invariant.

To obtain $R$-indifferent rate characteristics of a $R$-indifferent tensor, one should use the co-rotational (objective) derivatives with respect to time, which have been introduced in Sect. 2.5.

**Theorem 4.41.** *The Oldroyd derivatives of any $R$-indifferent vector* $\mathbf{a}$ *and of any $R$-indifferent tensor* $\mathbf{T}$ *are $R$-indifferent as well.*

▼ From (2.350) we get

$$\begin{aligned}(\mathbf{a}')^{\mathrm{Ol}} &= \dot{\mathbf{a}}' - \mathbf{a}' \cdot \mathbf{L}^{\mathrm{T}\prime} = (\mathbf{a} \cdot \mathbf{Q})^{\bullet} - \mathbf{a} \cdot \mathbf{Q} \cdot \mathbf{Q}^{\mathrm{T}} \cdot \mathbf{L} \cdot \mathbf{Q} - \mathbf{a} \cdot \mathbf{Q} \cdot \mathbf{Q}^{\mathrm{T}} \cdot \dot{\mathbf{Q}} \\ &= \mathbf{Q}^{\mathrm{T}} \cdot \dot{\mathbf{a}} + \mathbf{a} \cdot \dot{\mathbf{Q}} - \mathbf{a} \cdot \mathbf{L} \cdot \mathbf{Q} - \mathbf{a} \cdot \dot{\mathbf{Q}} = \mathbf{Q}^{\mathrm{T}} \cdot (\dot{\mathbf{a}} - \mathbf{L}^{\mathrm{T}} \cdot \mathbf{a}) = \mathbf{Q}^{\mathrm{T}} \cdot \mathbf{a}^{\mathrm{Ol}},\end{aligned} \tag{4.583}$$

and

$$\begin{aligned}(\mathbf{T}')^{\mathrm{Ol}} &= \dot{\mathbf{T}}' - \mathbf{T}' \cdot \mathbf{L}^{\mathrm{T}\prime} - \mathbf{L}' \cdot \mathbf{T}' = (\mathbf{Q}^{\mathrm{T}} \cdot \mathbf{T} \cdot \mathbf{Q})^{\bullet} - \mathbf{Q}^{\mathrm{T}} \cdot \mathbf{T} \cdot \mathbf{Q} \cdot \mathbf{Q}^{\mathrm{T}} \cdot \mathbf{L}^{\mathrm{T}} \cdot \mathbf{Q} \\ &\quad - \mathbf{Q}^{\mathrm{T}} \cdot \mathbf{T} \cdot \mathbf{Q} \cdot \mathbf{Q}^{\mathrm{T}} \cdot \dot{\mathbf{Q}} - \mathbf{Q}^{\mathrm{T}} \cdot \mathbf{L} \cdot \mathbf{Q} \cdot \mathbf{Q}^{\mathrm{T}} \cdot \mathbf{T} \cdot \mathbf{Q} - \dot{\mathbf{Q}}^{\mathrm{T}} \cdot \mathbf{Q} \cdot \mathbf{Q}^{\mathrm{T}} \cdot \mathbf{T} \cdot \mathbf{Q} \\ &= \mathbf{Q}^{\mathrm{T}} \cdot \dot{\mathbf{T}} \cdot \mathbf{Q} - \mathbf{Q}^{\mathrm{T}} \cdot \mathbf{T} \cdot \mathbf{L}^{\mathrm{T}} \cdot \mathbf{Q} - \mathbf{Q}^{\mathrm{T}} \cdot \mathbf{L} \cdot \mathbf{T} \cdot \mathbf{Q} + \dot{\mathbf{Q}}^{\mathrm{T}} \cdot \mathbf{T} \cdot \mathbf{Q} + \mathbf{Q}^{\mathrm{T}} \cdot \mathbf{T} \cdot \dot{\mathbf{Q}} \\ &\quad - \mathbf{Q}^{\mathrm{T}} \cdot \mathbf{T} \cdot \dot{\mathbf{Q}} - \dot{\mathbf{Q}} \cdot \mathbf{T} \cdot \mathbf{Q} = \mathbf{Q}^{\mathrm{T}} \cdot (\dot{\mathbf{T}} - \mathbf{T} \cdot \mathbf{L}^{\mathrm{T}} - \mathbf{L} \cdot \mathbf{T}) \cdot \mathbf{Q} = \mathbf{Q}^{\mathrm{T}} \cdot \mathbf{T}^{\mathrm{Ol}} \cdot \mathbf{Q}. \ \blacktriangle\end{aligned} \tag{4.584}$$

The Oldroyd derivatives of any $R$-invariant vector $\mathbf{a}$ and of any $R$-invariant tensor $\mathbf{\Pi}$ are neither $R$-invariant nor $R$-indifferent.

**Theorem 4.42.** *The Cotter—Rivlin derivatives of any $R$-indifferent vector* $\mathbf{a}$ *and of any $R$-indifferent tensor* $\mathbf{T}$ *are $R$-indifferent as well.*

▼ From (2.351) we obtain

$$\begin{aligned}(\mathbf{a}')^{\mathrm{CR}} &= \dot{\mathbf{a}}' + \mathbf{L}^{\mathrm{T}\prime} \cdot \mathbf{a}' = (\mathbf{a}^{\mathrm{T}} \cdot \mathbf{Q})^{\bullet} + \mathbf{Q}^{\mathrm{T}} \cdot \mathbf{L}^{\mathrm{T}} \cdot \mathbf{Q} \cdot \mathbf{Q}^{\mathrm{T}} \cdot \mathbf{a} + \mathbf{Q}^{\mathrm{T}} \cdot \dot{\mathbf{Q}} \cdot \mathbf{Q}^{\mathrm{T}} \cdot \mathbf{a} \\ &= \dot{\mathbf{Q}}^{\mathrm{T}} \cdot \mathbf{a} + \mathbf{Q}^{\mathrm{T}} \cdot \dot{\mathbf{a}} + \mathbf{Q}^{\mathrm{T}} \cdot \mathbf{L}^{\mathrm{T}} \cdot \mathbf{a} - \dot{\mathbf{Q}}^{\mathrm{T}} \cdot \mathbf{a} = \mathbf{a}^{\mathrm{T}} \cdot (\dot{\mathbf{a}} + \mathbf{L}^{\mathrm{T}} \cdot \mathbf{a}) = \mathbf{Q}^{\mathrm{T}} \cdot \mathbf{a}^{\mathrm{CR}},\end{aligned} \tag{4.585}$$

and from (2.352):

$$(\mathbf{T}')^{\mathrm{CR}} = \dot{\mathbf{T}}' + \mathbf{L}^{\mathrm{T}\prime}\cdot\mathbf{T}' + \mathbf{T}'\cdot\mathbf{L}' = \mathbf{Q}^{\mathrm{T}}\cdot\dot{\mathbf{T}}\cdot\mathbf{Q} + \dot{\mathbf{Q}}^{\mathrm{T}}\cdot\mathbf{T}\cdot\mathbf{Q}$$
$$+\mathbf{Q}^{\mathrm{T}}\cdot\mathbf{T}\cdot\dot{\mathbf{Q}} + \mathbf{Q}^{\mathrm{T}}\cdot\mathbf{L}^{\mathrm{T}}\cdot\mathbf{Q}\cdot\mathbf{Q}^{\mathrm{T}}\cdot\mathbf{T}\cdot\mathbf{Q} + \mathbf{Q}^{\mathrm{T}}\cdot\dot{\mathbf{Q}}\cdot\mathbf{Q}^{\mathrm{T}}\cdot\mathbf{T}\cdot\mathbf{Q}$$
$$+\mathbf{Q}^{\mathrm{T}}\cdot\mathbf{T}\cdot\mathbf{Q}\cdot\mathbf{Q}^{\mathrm{T}}\cdot\mathbf{L}\cdot\mathbf{Q} + \mathbf{Q}^{\mathrm{T}}\cdot\mathbf{T}\cdot\mathbf{Q}\cdot\cdot\,\mathbf{Q}^{\mathrm{T}}\cdot\mathbf{Q}$$
$$= \mathbf{Q}^{\mathrm{T}}\cdot(\dot{\mathbf{T}} + \mathbf{L}^{\mathrm{T}}\cdot\mathbf{T} + \mathbf{T}\cdot\mathbf{L})\cdot\mathbf{Q} = \mathbf{Q}^{\mathrm{T}}\cdot\mathbf{T}^{\mathrm{CR}}\cdot\mathbf{Q}.\ \blacktriangle \tag{4.586}$$

The Cotter–Rivlin derivative of any $R$-invariant vector (or of any $R$-invariant tensor) is neither $R$-invariant nor $R$-indifferent.

**Theorem 4.43.** *The mixed left and right co-rotational derivatives of any $R$-indifferent tensor* $\mathbf{T}$ *are $R$-indifferent.*

▼ From (2.355) we get

$$(\mathbf{T}')^{d} = \dot{\mathbf{T}}' - \mathbf{L}'\cdot\mathbf{T}' + \mathbf{T}'\cdot\mathbf{L}' = \mathbf{Q}^{\mathrm{T}}\cdot\dot{\mathbf{T}}\cdot\mathbf{Q} + \dot{\mathbf{Q}}^{\mathrm{T}}\cdot\mathbf{T}\cdot\mathbf{Q} + \mathbf{Q}^{\mathrm{T}}\cdot\mathbf{T}\cdot\dot{\mathbf{Q}}$$
$$-\mathbf{Q}^{\mathrm{T}}\cdot\mathbf{L}\cdot\mathbf{Q}\cdot\mathbf{Q}^{\mathrm{T}}\cdot\mathbf{T}\cdot\mathbf{Q} - \dot{\mathbf{Q}}^{\mathrm{T}}\cdot\mathbf{Q}\cdot\mathbf{Q}^{\mathrm{T}}\cdot\mathbf{T}\cdot\mathbf{Q} + \mathbf{Q}^{\mathrm{T}}\cdot\mathbf{T}\cdot\mathbf{Q}\cdot\mathbf{Q}^{\mathrm{T}}\cdot\mathbf{L}\cdot\mathbf{Q}$$
$$+\mathbf{Q}^{\mathrm{T}}\cdot\mathbf{T}\cdot\mathbf{Q}\cdot\dot{\mathbf{Q}}^{\mathrm{T}}\cdot\mathbf{Q} = \mathbf{Q}^{\mathrm{T}}\cdot(\dot{\mathbf{T}} - \mathbf{L}\cdot\mathbf{T} + \mathbf{T}\cdot\mathbf{L})\cdot\mathbf{Q} = \mathbf{Q}^{\mathrm{T}}\cdot\mathbf{T}^{d}\cdot\mathbf{Q} \tag{4.587}$$

and

$$(\mathbf{T}')^{D} = \dot{\mathbf{T}}' + \mathbf{L}'^{\mathrm{T}}\cdot\mathbf{T}' - \mathbf{T}'\cdot\mathbf{L}'^{\mathrm{T}} = \mathbf{Q}^{\mathrm{T}}\cdot\dot{\mathbf{T}}\cdot\mathbf{Q} + \mathbf{Q}^{\mathrm{T}}\cdot\mathbf{L}^{\mathrm{T}}\cdot\mathbf{Q}\cdot\mathbf{Q}^{\mathrm{T}}\cdot\mathbf{Q}$$
$$+\mathbf{Q}^{\mathrm{T}}\cdot\dot{\mathbf{Q}}\cdot\mathbf{Q}^{\mathrm{T}}\cdot\mathbf{T}\cdot\mathbf{Q} - \mathbf{Q}^{\mathrm{T}}\cdot\mathbf{T}\cdot\mathbf{Q}\cdot\mathbf{Q}^{\mathrm{T}}\cdot\mathbf{L}^{\mathrm{T}}\cdot\mathbf{Q} - \mathbf{Q}^{\mathrm{T}}\cdot\mathbf{T}\cdot\mathbf{Q}\cdot\mathbf{Q}^{\mathrm{T}}\cdot\dot{\mathbf{Q}}$$
$$+\dot{\mathbf{Q}}^{\mathrm{T}}\cdot\mathbf{T}\cdot\mathbf{Q}+\mathbf{Q}^{\mathrm{T}}\cdot\mathbf{T}\cdot\dot{\mathbf{Q}}=\mathbf{Q}^{\mathrm{T}}\cdot(\dot{\mathbf{T}}+\mathbf{L}^{\mathrm{T}}\cdot\mathbf{T}-\mathbf{T}\cdot\mathbf{L}^{\mathrm{T}})\cdot\mathbf{Q}=\mathbf{Q}^{\mathrm{T}}\cdot\mathbf{T}^{D}\cdot\mathbf{Q}.\ \blacktriangle \tag{4.588}$$

The mixed co-rotational derivatives of any $R$-invariant tensor are neither $R$-invariant nor $R$-indifferent.

**Theorem 4.44.** *The rotation tensor* $\mathbf{O}_U$ *and the spin of the right stretch tensor* $\mathbf{\Omega}_U$ *defined by formulae (2.269) are $R$-invariant.*

▼ Since the right stretch tensor $\mathbf{U}$ is $R$-invariant, i.e. it remains unchanged in the rigid motion, its eigenvectors $\overset{\circ}{\mathbf{p}}_i$ remain without changes as well:

$$\overset{\circ}{\mathbf{p}}{}'_i = \overset{\circ}{\mathbf{p}}_i. \tag{4.589}$$

Due to (4.581), the total derivative of the $R$-invariant vector $\overset{\circ}{\mathbf{p}}_\alpha$ with respect to time is $R$-invariant as well:

$$\dot{\overset{\circ}{\mathbf{p}}}{}'_i = \dot{\overset{\circ}{\mathbf{p}}}_i. \tag{4.590}$$

Thus, we get

$$\boldsymbol{\Omega}'_U = \overset{\circ\bullet}{\mathbf{p}}'_i \otimes \overset{\circ}{\mathbf{p}}'^i = \overset{\circ\bullet}{\mathbf{p}}_i \otimes \overset{\circ}{\mathbf{p}}^i = \boldsymbol{\Omega}_U, \qquad \mathbf{O}'_U = \overset{\circ}{\mathbf{p}}'_i \otimes \bar{\mathbf{e}}^i = \overset{\circ}{\mathbf{p}}_i \otimes \bar{\mathbf{e}}^i = \mathbf{O}_U. \blacktriangle \quad (4.591)$$

**Theorem 4.45.** *The derivative relative to the eigenbasis of the right stretch tensor, applied to any R-invariant vector* **a** *or tensor* $\boldsymbol{\Pi}$, *is R-invariant.*

▼ From (2.357) and (2.358) we get

$$(\mathbf{a}')^U = \dot{\mathbf{a}}' + \mathbf{a}' \cdot \boldsymbol{\Omega}'_U = \dot{\mathbf{a}} + \mathbf{a} \cdot \boldsymbol{\Omega}_U = \mathbf{a}^U,$$
$$(\boldsymbol{\Pi}')^U = \dot{\boldsymbol{\Pi}}' - \boldsymbol{\Omega}'_U \cdot \boldsymbol{\Pi}' + \boldsymbol{\Pi}' \cdot \boldsymbol{\Omega}'_U = \dot{\boldsymbol{\Pi}} - \boldsymbol{\Omega}_U \cdot \boldsymbol{\Pi} + \boldsymbol{\Pi} \cdot \boldsymbol{\Omega}_U = \boldsymbol{\Pi}^U. \blacktriangle \quad (4.592)$$

**Theorem 4.46.** *The rotation tensor* $\mathbf{O}_V$ *and the spin of the left stretch tensor* $\boldsymbol{\Omega}_V$*, defined by formulae (2.271), are neither R-indifferent nor R-invariant.*

▼ From the relation (2.276) between $\mathbf{O}$, $\mathbf{O}_U$ and $\mathbf{O}_V$ we get

$$\mathbf{O}' = \mathbf{O}'_V \cdot \mathbf{O}^{\mathrm{T}\prime}_U. \quad (4.593)$$

Since the rotation tensor $\mathbf{O}$ is transformed by formula (4.568), we find that

$$\mathbf{Q}^{\mathrm{T}} \cdot \mathbf{O} = \mathbf{O}'_V \cdot \mathbf{O}^{\mathrm{T}}_U, \quad \mathbf{O} = \mathbf{Q} \cdot \mathbf{O}'_V \cdot \mathbf{O}^{\mathrm{T}}_U = \mathbf{O}_V \cdot \mathbf{O}^{\mathrm{T}}_U.$$

Thus, the rotation tensor $\mathbf{O}_V$ is neither $R$-indifferent nor $R$-invariant:

$$\mathbf{O}'_V = \mathbf{Q}^{\mathrm{T}} \cdot \mathbf{O}_V. \quad (4.594)$$

The spin of the left stretch tensor $\boldsymbol{\Omega}_V$, defined by formula (2.271) with the help of the rotation tensor $\mathbf{O}_V$, is transformed in the rigid motion as follows:

$$\boldsymbol{\Omega}'_V = \dot{\mathbf{O}}'_V \cdot \mathbf{O}^{\mathrm{T}\prime}_V = \mathbf{Q}^{\mathrm{T}} \cdot \dot{\mathbf{O}}_V \cdot \mathbf{O}^{\mathrm{T}}_V \cdot \mathbf{Q} + \dot{\mathbf{Q}}^{\mathrm{T}} \cdot \mathbf{O}_V \cdot \mathbf{O}^{\mathrm{T}}_V \cdot \mathbf{Q}$$
$$= \mathbf{Q}^{\mathrm{T}} \cdot \boldsymbol{\Omega}_V \cdot \mathbf{Q} + \dot{\mathbf{Q}}^{\mathrm{T}} \cdot \mathbf{Q}. \quad (4.595)$$

Thus, $\boldsymbol{\Omega}'_V$ is neither $R$-indifferent nor $R$-invariant. ▲

**Theorem 4.47.** *The derivative of any R-indifferent vector* **a** *or of any R-indifferent tensor* **T** *relative to the eigenbasis of the left stretch tensor is R-indifferent.*

▼ Due to (2.361), we get

$$(\mathbf{a}')^V = (\mathbf{a} \cdot \mathbf{Q})^{\bullet} + \mathbf{a} \cdot \mathbf{Q} \cdot \boldsymbol{\Omega}'_V$$
$$= \dot{\mathbf{a}} \cdot \mathbf{Q} + \mathbf{a} \cdot \dot{\mathbf{Q}} + \mathbf{a} \cdot \mathbf{Q} \cdot \mathbf{Q}^{\mathrm{T}} \cdot \boldsymbol{\Omega}_V \cdot \mathbf{Q} - \mathbf{a} \cdot \mathbf{Q} \cdot \mathbf{Q}^{\mathrm{T}} \cdot \dot{\mathbf{Q}}$$
$$= (\dot{\mathbf{a}} + \mathbf{a} \cdot \boldsymbol{\Omega}_V) \cdot \mathbf{Q} = \mathbf{a}^V \cdot \mathbf{Q}, \quad (4.596)$$

and

$$
\begin{aligned}
(\mathbf{T}')^V &= \dot{\mathbf{T}}' - \boldsymbol{\Omega}'_V \cdot \mathbf{T}' + \mathbf{T}' \cdot \boldsymbol{\Omega}'_V = \mathbf{Q}^{\mathrm{T}} \cdot \dot{\mathbf{T}} \cdot \mathbf{Q} + \dot{\mathbf{Q}}^{\mathrm{T}} \cdot \mathbf{T} \cdot \mathbf{Q} \\
&\quad + \mathbf{Q}^{\mathrm{T}} \cdot \mathbf{Q} \cdot \dot{\mathbf{Q}} - \mathbf{Q}^{\mathrm{T}} \cdot \boldsymbol{\Omega}_V \cdot \mathbf{Q} \cdot \mathbf{Q}^{\mathrm{T}} \cdot \mathbf{T} \cdot \mathbf{Q} - \dot{\mathbf{Q}}^{\mathrm{T}} \cdot \mathbf{Q} \cdot \mathbf{Q}^{\mathrm{T}} \cdot \mathbf{T} \cdot \mathbf{Q} \\
&\quad + \mathbf{Q}^{\mathrm{T}} \cdot \mathbf{T} \cdot \mathbf{Q} \cdot \mathbf{Q}^{\mathrm{T}} \cdot \boldsymbol{\Omega}_V \cdot \mathbf{Q} + \mathbf{Q}^{\mathrm{T}} \cdot \mathbf{T} \cdot \mathbf{Q} \cdot \dot{\mathbf{Q}}^{\mathrm{T}} \cdot \mathbf{Q} \\
&= \mathbf{Q}^{\mathrm{T}} \cdot (\dot{\mathbf{T}} - \boldsymbol{\Omega}_V \cdot \mathbf{T} + \mathbf{T} \cdot \boldsymbol{\Omega}_V) \cdot \mathbf{Q} = \mathbf{Q}^{\mathrm{T}} \cdot \mathbf{T}^V \cdot \mathbf{Q}. \blacktriangle
\end{aligned} \tag{4.597}
$$

**Theorem 4.48.** *The spin $\boldsymbol{\Omega} = \dot{\mathbf{O}} \cdot \mathbf{O}^{\mathrm{T}}$ accompanying deformation is neither $R$-indifferent nor $R$-invariant.*

▼ According to (4.568), we obtain

$$
\begin{aligned}
\boldsymbol{\Omega}' = \dot{\mathbf{O}}' \cdot \mathbf{O}'^{\mathrm{T}} &= \dot{\mathbf{Q}}^{\mathrm{T}} \cdot \mathbf{O} \cdot \mathbf{O}^{\mathrm{T}} \cdot \mathbf{Q} + \mathbf{Q}^{\mathrm{T}} \cdot \dot{\mathbf{O}} \cdot \mathbf{O}^{\mathrm{T}} \cdot \mathbf{Q} \\
&= \mathbf{Q}^{\mathrm{T}} \cdot \boldsymbol{\Omega} \cdot \mathbf{Q} + \dot{\mathbf{Q}}^{\mathrm{T}} \cdot \mathbf{Q}. \ \blacktriangle
\end{aligned} \tag{4.598}
$$

**Theorem 4.49.** *The spin derivative, defined by formulae (2.368) and (2.369), of any $R$-indifferent object is $R$-indifferent.*

▼ For any $R$-indifferent vector, from (2.370) we get

$$
\begin{aligned}
\mathbf{a}'^S &= \dot{\mathbf{a}}' + \mathbf{a}' \cdot \boldsymbol{\Omega}' = \dot{\mathbf{a}} \cdot \mathbf{Q} + \mathbf{a} \cdot \dot{\mathbf{Q}} + \mathbf{a} \cdot \mathbf{Q} \cdot \mathbf{Q}^{\mathrm{T}} \cdot \boldsymbol{\Omega} \cdot \mathbf{Q} \\
&\quad + \mathbf{a} \cdot \mathbf{Q} \cdot \dot{\mathbf{Q}}^{\mathrm{T}} \cdot \mathbf{Q} = (\dot{\mathbf{a}} + \mathbf{a} \cdot \boldsymbol{\Omega}) \cdot \mathbf{Q} = \mathbf{a}^S \cdot \mathbf{Q}.
\end{aligned} \tag{4.599}
$$

In a similar way, for any $R$-indifferent tensor from (2.371) we obtain

$$
\mathbf{T}^{S\prime} = \mathbf{Q}^{\mathrm{T}} \cdot \mathbf{T}^S \cdot \mathbf{Q}. \ \blacktriangle \tag{4.600}
$$

**Theorem 4.50.** *The Jaumann derivative is $R$-indifferent, if it is applied to any $R$-indifferent vector or tensor.*

▼ For any $R$-indifferent vector, from (2.364) we get

$$
\begin{aligned}
(\mathbf{a}')^J &= (\mathbf{a} \cdot \mathbf{Q})^J = (\mathbf{a} \cdot \mathbf{Q})^{\bullet} + \mathbf{a} \cdot \mathbf{Q} \cdot \mathbf{W}' = \dot{\mathbf{a}} \cdot \mathbf{Q} + \mathbf{a} \cdot \dot{\mathbf{Q}} \\
&\quad + \mathbf{a} \cdot \mathbf{Q} \cdot (\mathbf{Q}^{\mathrm{T}} \cdot \mathbf{W} \cdot \mathbf{Q} - \mathbf{Q}^{\mathrm{T}} \cdot \mathbf{Q}) = (\dot{\mathbf{a}} + \mathbf{a} \cdot \mathbf{W}) \cdot \mathbf{Q} = \mathbf{a}^J \cdot \mathbf{Q} = (\mathbf{a}^J)'.
\end{aligned} \tag{4.601}
$$

In a similar way, we can verify that the derivative $\mathbf{T}^J$ (2.365) is $R$-indifferent:

$$
(\mathbf{T}')^J = \mathbf{Q}^{\mathrm{T}} \cdot \mathbf{T}^J \cdot \mathbf{Q}. \ \blacktriangle \tag{4.602}
$$

Table 4.12 gives summary data on $R$-indifference and $R$-invariance of basic vectors and tensors.

**Table 4.12** List of basic $R$-invariant and $R$-indifferent tensors

| Tensors and vectors | $R$-indifference | $R$-invariance |
|---|---|---|
| $\overset{\circ}{\mathbf{r}}_i, \overset{\circ}{\mathbf{r}}^i$ | − | + |
| $\mathbf{r}_i, \mathbf{r}^i$ | + | − |
| $\mathbf{E}$ | + | + |
| $\mathbf{F}$ | − | − |
| $\overset{(n)}{\mathbf{C}}, \overset{(n)}{\mathbf{G}}, n = \text{I}, \ldots, \text{V}$ | − | + |
| $\overset{(n)}{\mathbf{A}}, \overset{(n)}{\mathbf{g}}, n = \text{I}, \ldots, \text{V}$ | + | − |
| $\mathbf{T}$ | + | − |
| $\mathbf{P}$ | − | − |
| $\overset{(n)}{\mathbf{T}}, n = \text{I}, \ldots, \text{V}$ | − | + |
| $\overset{(n)}{\mathbf{S}}, n = \text{I}, \ldots, \text{V}$ | + | − |
| $\mathbf{v}, \mathbf{L}, \mathbf{W}$ | − | − |
| $\mathbf{D}$ | + | − |
| $\mathbf{B}^h, h = \{\cdot, U\}$ if $\mathbf{B}$ is $R$-invariant | − | + |
| $\mathbf{B}^h, h = \{\text{Ol}, \text{CR}, d, D, V, S, J\}$, if $\mathbf{B}$ is $R$-indifferent | + | − |
| $\mathbf{O}_U, \boldsymbol{\Omega}_U$ | − | + |
| $\mathbf{O}_V, \boldsymbol{\Omega}_V, \boldsymbol{\Omega}$ | − | − |

## 4.10.9 The Statement of the Principle of Material Indifference

**Axiom 15 (The principle of material indifference (or the principle of objectivity)).** *In the rigid motion, all the universal laws of continuum mechanics in the differential form (3.303) and also the constitutive equations (4.156), connecting the active* $\Lambda$ *and reactive* $\mathcal{R}$ *variables, remain unchanged. In other words, if the relations (3.303) and the constitutive equations*

$$\Lambda = \breve{f}(\mathcal{R}) \tag{4.603}$$

*hold in configuration* $\mathcal{K}$, *then in* $\mathcal{K}'$ *the following relations must hold:*

$$\rho' \frac{d'\bar{A}'_\alpha}{dt} = \nabla' \cdot \bar{B}'_\alpha + \rho' C'_\alpha, \quad \alpha = 1, \ldots, 6, \tag{4.604}$$

*and also*

$$\Lambda' = \breve{f}(\mathcal{R}'), \tag{4.605}$$

*where* $\breve{f}$ *is the same operator.*

### 4.10.10 Material Indifference of the Continuity Equation

Let us establish what constraints are imposed by the principle of material indifference on the continuity equation (see the system (3.303) at $\alpha = 1$).

Due to (4.558), the density $\rho$ is always $R$-invariant. Thus, by Theorem 4.39, we obtain

$$\rho' \frac{d}{dt}(1/\rho') = \rho \frac{d}{dt}(1/\rho). \tag{4.606}$$

According to the results of Exercise 4.10.1,

$$\nabla' \cdot \mathbf{v}' = \nabla \cdot \mathbf{v}. \tag{4.607}$$

Thus, we have proved the following theorem.

**Theorem 4.51.** *The principle of material indifference always remains valid for the continuity equation, i.e. from (3.303) at $\alpha = 1$ we always get that*

$$\rho' \frac{d}{dt}\left(\frac{1}{\rho'}\right) = \nabla' \cdot \mathbf{v}'. \tag{4.608}$$

### 4.10.11 Material Indifference for the Momentum Balance Equation

Consider the momentum balance equation (3.303) at $\alpha = 2$. At first, prove the following auxiliary assertion.

**Theorem 4.52.** *The acceleration vector $d\mathbf{v}/dt = d^2\mathbf{x}/dt^2 = \ddot{\mathbf{x}}$ of a material point is neither $R$-indifferent nor $R$-invariant, and in the rigid motion it is transformed as follows:*

$$\ddot{\mathbf{x}}' = \mathbf{Q}^{\mathrm{T}} \cdot (\ddot{\mathbf{x}} - \dot{\mathbf{v}}_0) + \ddot{\mathbf{x}}'_0 + (\dot{\boldsymbol{\Omega}}'_s - \boldsymbol{\Omega}'^2_s) \cdot (\mathbf{x}' - \mathbf{x}'_0) + 2\boldsymbol{\Omega}'_s \cdot (\dot{\mathbf{x}}' - \dot{\mathbf{x}}'_0), \tag{4.609}$$

*where*

$$\boldsymbol{\Omega}'_s = \dot{\mathbf{Q}}^{\mathrm{T}} \cdot \mathbf{Q}. \tag{4.610}$$

▼ From formula (4.548), for the rigid motion we get

$$\mathbf{x} - \mathbf{x}_0 = (\mathbf{x}' - \mathbf{x}'_0) \cdot \mathbf{Q}^{\mathrm{T}}. \tag{4.611}$$

Then formula (4.576) for $\mathbf{v}'$ with taking (4.611) into account yields

$$\mathbf{v}' = (\mathbf{v} - \mathbf{v}_0) \cdot \mathbf{Q} + \dot{\mathbf{x}}'_0 + \boldsymbol{\Omega}'_s \cdot (\mathbf{x}' - \mathbf{x}'_0), \tag{4.612}$$

or

$$\mathbf{v} = \mathbf{v}_0 + (\mathbf{v}' - \dot{\mathbf{x}}'_0) \cdot \mathbf{Q}^{\mathrm{T}} - \mathbf{Q} \cdot \boldsymbol{\Omega}'_s \cdot (\mathbf{x}' - \mathbf{x}'_0). \tag{4.613}$$

Differentiation of (4.612) with respect to time $t$ gives

$$\ddot{\mathbf{x}}' = \dot{\mathbf{v}}' = (\ddot{\mathbf{x}} - \dot{\mathbf{v}}_0) \cdot \mathbf{Q} + (\mathbf{v} - \mathbf{v}_0) \cdot \dot{\mathbf{Q}} + \ddot{\mathbf{x}}'_0 + \boldsymbol{\Omega}'_s \cdot (\dot{\mathbf{x}}' - \dot{\mathbf{x}}'_0) + \dot{\boldsymbol{\Omega}}'_s \cdot (\mathbf{x}' - \mathbf{x}'_0). \tag{4.614}$$

Going from $\mathbf{v}$ to $\mathbf{v}'$ by formula (4.613), we find that

$$\begin{aligned} \ddot{\mathbf{x}}' &= (\ddot{\mathbf{x}} - \dot{\mathbf{v}}_0) \cdot \mathbf{Q} + (\mathbf{v}' - \dot{\mathbf{x}}'_0) \cdot \mathbf{Q}^{\mathrm{T}} \cdot \dot{\mathbf{Q}} - \dot{\mathbf{Q}}^{\mathrm{T}} \cdot \mathbf{Q} \cdot \boldsymbol{\Omega}'_s \cdot (\mathbf{x}' - \mathbf{x}'_0) \\ &+ \ddot{\mathbf{x}}'_0 + \boldsymbol{\Omega}'_s \cdot (\dot{\mathbf{x}}' - \dot{\mathbf{x}}'_0) + \dot{\boldsymbol{\Omega}}'_s \cdot (\mathbf{x}' - \mathbf{x}'_0) = (\ddot{\mathbf{x}} - \dot{\mathbf{v}}_0) \cdot \mathbf{Q} + \ddot{\mathbf{x}}'_0 \\ &+ (\dot{\boldsymbol{\Omega}}'_s - \boldsymbol{\Omega}'^2_s) \cdot (\mathbf{x}' - \mathbf{x}'_0) + 2\boldsymbol{\Omega}'_s \cdot (\dot{\mathbf{x}}' - \dot{\mathbf{x}}'_0). \end{aligned} \tag{4.615}$$

The presence of the last three summands in the formula means that the acceleration vector $\ddot{\mathbf{x}}$ is not $R$-indifferent. It is clear that $\ddot{\mathbf{x}}$ is not $R$-invariant as well. ▲

Let us consider how the momentum balance equation (3.72) is transformed during the passage from $\mathcal{K}$ to $\mathcal{K}'$.

**Theorem 4.53.** *The momentum balance equation in Eulerian variables (3.72), written in $\mathcal{K}$ in the form*

$$\rho\mathbf{a} = \nabla \cdot \mathbf{T} + \rho\mathbf{f}, \tag{4.616}$$

*where*

*(1) the vector* $\mathbf{a}$ *is defined as follows: in $\mathcal{K}$ it coincides with the total acceleration* $\ddot{\mathbf{x}}$*, and during the passage to $\mathcal{K}'$ it is transformed as a $R$-indifferent vector:*

$$\mathcal{K}: \quad \mathbf{a} = \ddot{\mathbf{x}}; \qquad \mathcal{K}': \quad \mathbf{a}' = \ddot{\mathbf{x}} \cdot \mathbf{Q}; \tag{4.617}$$

*(2) during the passage from $\mathcal{K}$ to $\mathcal{K}'$ the vector* $\mathbf{f}$ *is assumed to be $R$-indifferent:*

$$\mathbf{f}' = \mathbf{f} \cdot \mathbf{Q}, \tag{4.618}$$

*satisfies the principle of material indifference, i.e. in $\mathcal{K}'$ the momentum balance equation has the form*

$$\rho'\mathbf{a}' = \nabla' \cdot \mathbf{T}' + \rho'\mathbf{f}'. \tag{4.619}$$

▼ Indeed, let the vectors $\mathbf{f}$ and $\mathbf{a}$ be transformed during the passage from $\mathcal{K}$ to $\mathcal{K}'$ by formulae (4.617) and (4.618), then we get

$$\rho'\mathbf{a}' - \nabla' \cdot \mathbf{T}' - \rho'\mathbf{f}' = \rho'\ddot{\mathbf{x}} \cdot \mathbf{Q} - \nabla' \cdot \mathbf{T}' - \rho'\mathbf{f} \cdot \mathbf{Q}.$$

Since the tensor $\mathbf{T}$ is $R$-indifferent, so, according to the results of Exercise 4.10.2, the divergence $\nabla \cdot \mathbf{T}$ is also a $R$-indifferent vector. Then, taking the equality $\rho = \rho'$ and formula (4.616) into account, we get

$$\rho'\mathbf{a}' - \nabla' \cdot \mathbf{T}' - \rho'\mathbf{f}' = (\rho\mathbf{a} - \nabla \cdot \mathbf{T} - \rho\mathbf{f}) \cdot \mathbf{Q} = 0.$$

Thus, formula (4.619) really follows from (4.616). ▲

*Remark 1.* If one uses the traditional form of the momentum balance equation (3.72), then, according to the principle of material indifference, during the passage to $\mathcal{K}'$ the equation should be written in the form

$$\rho'(d\mathbf{v}'/dt) = \nabla' \cdot \mathbf{T}' + \rho'\mathbf{f}';$$

but since the velocity $\mathbf{v}$ and the acceleration $d\mathbf{v}/dt$ are not $R$-indifferent vectors, this equation does not hold. This fact is one of the paradoxes of continuum mechanics: the traditional form of the momentum balance equation (3.72) does not satisfy the principle of objectivity, i.e. this depends on the choice of a coordinate system. To eliminate this paradox we should define the acceleration vector $\mathbf{a}$ in the special $R$-indifferent way (4.617). The definition (4.617) includes the axiomatic assumption on the existence of an *inertial system* $O\bar{\mathbf{e}}_i$ such that in configuration $\mathcal{K}$ the acceleration $\mathbf{a}$ is the second derivative of radius-vector $\ddot{\mathbf{x}}$. The existence of the inertial system $O\bar{\mathbf{e}}_i$ is guaranteed by Axiom 5 (see Sect. 3.2.1). □

*Remark 2.* The expression for acceleration $\mathbf{a}'$ in $\mathcal{K}'$ with taking formula (4.615) into account can be written in the form

$$\mathbf{a}' = \ddot{\mathbf{x}}' + \mathbf{v}_0 \cdot \mathbf{Q} - \ddot{\mathbf{x}}_0' - (\dot{\mathbf{\Omega}}_s' - \mathbf{\Omega}_s'^2) \cdot (\mathbf{x}' - \mathbf{x}_0') - 2\mathbf{\Omega}_s' \cdot (\dot{\mathbf{x}}' - \dot{\mathbf{x}}_0'). \square \qquad (4.617a)$$

Taking the conclusion of Theorem 4.53 and Remark 1 for the theorem into account, we should return to Axiom 5 (see Sect. 3.2.1) and reformulate this axiom in the following way.

**Axiom 5a.**

(1) *For every continuum $\mathcal{B}$ $\forall t \geqslant 0$ there exists a vector $\mathcal{F}_{(i)}(\mathcal{B}, t)$, called the body inertia force, which in the inertial system $O\bar{\mathbf{e}}_i$ for configuration $\mathcal{K}$ has the form*

$$\mathcal{K}: \quad \mathcal{F}_{(i)} = d\mathbf{I}/dt, \qquad (4.620a)$$

*where $\mathbf{I}$ is the momentum vector of the body (3.26), and the vector $\mathcal{F}_{(i)}$ is transformed during the passage to $\mathcal{K}'$ in the $R$-indifferent way:*

$$\mathcal{K}': \quad \mathcal{F}'_{(i)} = \mathcal{F}_{(i)} \cdot \mathbf{Q}. \qquad (4.620b)$$

(2) *For any pair of bodies $\mathcal{B}$ and $\mathcal{B}_1$ $\forall t \geqslant 0$ there exists a vector of bodies interaction force $\mathcal{F}(\mathcal{B}, \mathcal{B}_1, t)$, which may be zero-valued and has the following properties:*

- *the vector is additive:*

$$\mathcal{F}(\mathcal{B}' + \mathcal{B}'',\ \mathcal{B}_1,\ t) = \mathcal{F}(\mathcal{B}',\ \mathcal{B}_1,\ t) + \mathcal{F}(\mathcal{B}'',\ \mathcal{B}_1,\ t),$$
$$\mathcal{F}(\mathcal{B},\ \mathcal{B}_1' + \mathcal{B}_1'',\ t) = \mathcal{F}(\mathcal{B},\ \mathcal{B}_1',\ t) + \mathcal{F}(\mathcal{B},\ \mathcal{B}_1'',\ t),$$

*where $\mathcal{B} = \mathcal{B}' \cup \mathcal{B}''$, $\mathcal{B}_1 = \mathcal{B}_1' \cup \mathcal{B}_1''$;*

- *the vector is* $R$*-indifferent during the passage to* $\mathcal{K}'$*:*

$$\mathcal{F}' = \mathcal{F} \cdot \mathbf{Q};$$

- *the vector is inertial, i.e. the summarized vector of external forces acting on the body* $\mathcal{B}$*,* $\mathcal{F}(\mathcal{B}, t) = \mathcal{F}(\mathcal{B}, \mathcal{B}^e, t)$ *(where* $\mathcal{B}^e = \mathcal{U} \setminus \mathcal{B}$ *is the surroundings of the body* $\mathcal{B}$*) in* $\mathcal{K}$ *is equal to the body inertia force vector:*

$$\mathcal{K}: \ \mathcal{F}_{(i)} = \mathcal{F}. \tag{4.621}$$

A corollary of this axiom is the following theorem.

**Theorem 4.53a.** *If Axiom 5a is assumed to be valid, then under the transformation from configuration* $\mathcal{K}$ *to* $\mathcal{K}'$ *the momentum balance law retains its form:*

$$\mathcal{K}': \ \mathcal{F}'_{(i)} = \mathcal{F}'. \tag{4.621a}$$

This theorem is an integral analog of Theorem 4.53. Axiom 5a ensures that the conditions involved in statements of Theorems 4.53 and 4.53a be satisfied. Thus, due to the acceptance of Axiom 5a, the main purpose of this section (to ensure that the momentum balance equations satisfy the principle of material indifference) has been achieved. Axiom 5 introduced in Sect. 3.2.1 is entirely contained in the generalized Axiom 5a; therefore, all the conclusions, obtained above according to Axiom 5, remain valid.

### *4.10.12 Material Indifference of the Thermodynamic Laws*

The following axiom asserts material indifference of the thermodynamic functions.

**Axiom 7a (Objectivity of thermodynamic functions).** *The thermodynamic functions* $U$*,* $Q$*,* $H$*,* $\bar{Q}^*$ *and temperature* $\theta$ *are* $R$*-indifferent scalars:*

$$U' = U, \quad Q' = Q, \quad H' = H, \quad \bar{Q}^{*\prime} = \bar{Q}^*, \quad \theta' = \theta. \tag{4.622}$$

It follows from (4.622) that the specific internal energy and specific entropy are $R$-indifferent scalars:

$$e' = e, \qquad \eta' = \eta; \tag{4.622a}$$

and from (4.622), (3.106), and (3.159) we get that the heat fluxes from mass and surface sources, and also the specific internal entropy production are $R$-indifferent scalars:

$$q'_m = q_m, \qquad q'_\Sigma = q_\Sigma, \qquad q^{*\prime} = q^*, \tag{4.622b}$$

because the mass $m$ is independent of the rigid motion and $d\Sigma$ is always defined in the $R$-indifferent way.

The normal $\mathbf{n}$ and relations (3.111) are $R$-indifferent (see Sect. 4.10.5), then we obtain that the heat flux vector $\mathbf{q}$ is $R$-indifferent:

$$\mathbf{q}' = \mathbf{q} \cdot \mathbf{Q}, \tag{4.623}$$

because the following relation holds:

$$-q'_{\Sigma} = -\mathbf{n}' \cdot \mathbf{q}' = -\mathbf{n} \cdot \mathbf{Q} \cdot \mathbf{q}' = \mathbf{n} \cdot \mathbf{q} = q_{\Sigma}.$$

**Theorem 4.54.** *The principle of material indifference holds for the first thermodynamic law, i.e. from the energy balance equation (3.118), written for $\mathcal{K}$ in the form*

$$\rho\Big(\frac{de}{dt} + \mathbf{v} \cdot \mathbf{a}\Big) - \nabla \cdot (\mathbf{T} \cdot \mathbf{v}) + \nabla \cdot \mathbf{q} - \rho q_m - \rho \mathbf{f} \cdot \mathbf{v} = 0, \tag{4.624}$$

*we get the following energy balance equation in $\mathcal{K}'$:*

$$\rho'\Big(\frac{de'}{dt} + \mathbf{v}' \cdot \mathbf{a}'\Big) - \nabla' \cdot (\mathbf{T}' \cdot \mathbf{v}') + \nabla' \cdot \mathbf{q}' - \rho' q'_m - \rho' \mathbf{f}' \cdot \mathbf{v}' = 0, \tag{4.625}$$

*where $\mathbf{a}'$ is defined by (4.617).*

▼ Consider transformations of separate summands on the left-hand side of equation (4.625). Due to Axiom 5a, the momentum balance equation (4.619) holds in $\mathcal{K}'$; then, multiplying the left and right sides of this equation by $\mathbf{v}'$, we get

$$\rho' \mathbf{v}' \cdot \mathbf{a}' = \mathbf{v}' \cdot \nabla' \cdot \mathbf{T}' + \rho' \mathbf{f}' \cdot \mathbf{v}', \tag{4.626a}$$

or

$$\rho' \mathbf{v}' \cdot \mathbf{a}' - \nabla' \cdot (\mathbf{T}' \cdot \mathbf{v}') + \mathbf{T}' \cdot\cdot (\nabla' \otimes \mathbf{v}')^{\mathrm{T}} - \rho' \mathbf{f}' \cdot \mathbf{v}' = 0. \tag{4.626b}$$

Due to Axiom 7a and the results of Exercises 4.10.2 and 4.10.5, we have

$$\nabla' \cdot \mathbf{q}' = \nabla \cdot \mathbf{q}, \quad \rho' \frac{de'}{dt} = \rho \frac{de}{dt}. \tag{4.627}$$

Since the tensor $\mathbf{T}'$ is symmetric (see Exercise 4.10.6), the results of Exercise 4.10.4 and Theorem 4.38 yield

$$\mathbf{T}' \cdot\cdot (\nabla' \otimes \mathbf{v}')^{\mathrm{T}} = \mathbf{T}' \cdot\cdot \mathbf{D}' = \mathbf{T} \cdot\cdot \mathbf{D} = \mathbf{T} \cdot\cdot (\nabla \otimes \mathbf{v})^{\mathrm{T}}. \tag{4.628}$$

Modify the left-side expression in (4.625) with taking formulae (4.626b), (4.627) and (4.628) into account:

$$\rho'\left(\frac{de'}{dt}+\mathbf{v}'\cdot\mathbf{a}'\right)-\nabla'\cdot(\mathbf{T}'\cdot\mathbf{v}')+\nabla'\cdot\mathbf{q}'-\rho'\mathbf{f}'\cdot\mathbf{v}'-\rho' q'_m$$
$$=\rho\frac{de}{dt}-\mathbf{T}'\cdot\cdot(\nabla'\otimes\mathbf{v}')^{\mathrm{T}}+\nabla'\cdot\mathbf{q}'-\rho q'_m$$
$$=\rho\frac{de}{dt}-\mathbf{T}\cdot\cdot(\nabla\otimes\mathbf{v})^{\mathrm{T}}+\nabla\cdot\mathbf{q}-\rho q_m. \quad (4.629)$$

Since Eq. (4.624) is satisfied, the heat influx equation (3.124) being its consequence holds as well.

On comparing (4.629) with (3.124), we conclude that the expression (4.629) must vanish, and therefore the equation (4.625) really holds. ▲

**Theorem 4.55.** *The entropy balance equation (the second thermodynamic law) satisfies the principle of material indifference, i.e. from (3.303) at* $\alpha = 4$ *we get the equation*

$$\rho'\frac{d\eta'}{dt}-\nabla'\cdot\left(\frac{\mathbf{q}'}{\theta'}\right)-\frac{\rho'}{\theta'}(q'_m+q^{*\prime})=0. \quad (4.630)$$

▼ A proof of the theorem should be performed as Exercise 4.10.7. ▲

### 4.10.13 Material Indifference of the Compatibility Equations

Let us verify whether equations (3.303) at $\alpha = 5$ and 6 always satisfy the principle of material indifference.

**Theorem 4.56.** *The kinematic equation (3.303) at* $\alpha = 5$ *and the dynamic compatibility equation (3.303) at* $\alpha = 6$ *always satisfy the principle of material indifference.*

▼ Since $\mathbf{u}' = \mathbf{x}' - \overset{\circ}{\mathbf{x}}$, we have $d\mathbf{u}'/dt = d\mathbf{x}'/dt = \mathbf{v}'$; and hence the kinematic equation is always satisfied in $\mathcal{K}'$:

$$\rho'(d\mathbf{u}'/dt)=\rho'\mathbf{v}'. \quad (4.631)$$

Let the compatibility equation (3.303) at $\alpha = 6$ be satisfied in $\mathcal{K}$:

$$\rho\frac{d\mathbf{F}^{\mathrm{T}}}{dt}-\nabla\cdot(\rho\mathbf{F}\otimes\mathbf{v})=0. \quad (4.632)$$

Consider the similar expression in $\mathcal{K}'$:

$$\rho' \frac{d\mathbf{F}^{\mathrm{T}\prime}}{dt} - \nabla' \cdot (\rho' \mathbf{F}' \otimes \mathbf{v}')$$
$$= \rho' \frac{d\mathbf{F}^{\mathrm{T}}}{dt} \cdot \mathbf{Q} + \rho \mathbf{F}^{\mathrm{T}} \cdot \dot{\mathbf{Q}} - \rho' \mathbf{F}^{\mathrm{T}\prime} \cdot \nabla' \otimes \mathbf{v}' - (\nabla' \cdot \rho' \mathbf{F}') \otimes \mathbf{v}'. \quad (4.633)$$

Here we have used formula (4.559). With taking this equation and (4.578) into account, the third summand takes the form

$$\begin{aligned} \rho' \mathbf{F}^{\mathrm{T}\prime} \cdot \nabla' \otimes \mathbf{v}' &= \rho \mathbf{F}^{\mathrm{T}} \cdot \mathbf{Q} \cdot \mathbf{Q}^{\mathrm{T}} \cdot \nabla \otimes \mathbf{v} \cdot \mathbf{Q} + \rho \mathbf{F}^{\mathrm{T}} \cdot \mathbf{Q} \cdot \mathbf{Q}^{\mathrm{T}} \cdot \dot{\mathbf{Q}} \\ &= \rho \mathbf{F}^{\mathrm{T}} \cdot \nabla \otimes \mathbf{v} \cdot \mathbf{Q} + \rho \mathbf{F}^{\mathrm{T}} \cdot \dot{\mathbf{Q}}. \end{aligned} \quad (4.634)$$

The fourth summand in (4.633) is zero, because

$$\nabla' \cdot \rho' \mathbf{F}' = \mathbf{r}'^{i} \cdot \frac{\partial \rho' \mathbf{F}'}{\partial X^{i}} = \mathbf{r}^{i} \cdot \mathbf{Q} \cdot \mathbf{Q}^{\mathrm{T}} \cdot \frac{\partial \rho \mathbf{F}}{\partial X^{i}} = \nabla \cdot \rho \mathbf{F} = 0. \quad (4.635)$$

The last equality in (4.635) is valid due to Theorem 3.24.

On substituting (4.634) and (4.635) into (4.633), we finally get

$$\begin{aligned} \rho' \frac{d\mathbf{F}^{\mathrm{T}\prime}}{dt} - \nabla' \cdot (\rho' \mathbf{F}' \otimes \mathbf{v}') &= \left( \rho \frac{d\mathbf{F}^{\mathrm{T}}}{dt} - \rho \mathbf{F}^{\mathrm{T}} \cdot \nabla \otimes \mathbf{v} - (\nabla \cdot \rho \mathbf{F}) \otimes \mathbf{v} \right) \cdot \mathbf{Q} \\ &= \left( \rho \frac{d\mathbf{F}^{\mathrm{T}}}{dt} - \nabla \cdot (\rho \mathbf{F} \otimes \mathbf{v}) \right) \cdot \mathbf{Q} = 0, \end{aligned} \quad (4.636)$$

because Eq. (4.632) is satisfied by condition. ▲

## 4.10.14 *Material Indifference of Models $A_n$ and $B_n$ of Ideal Continua*

Consider the constitutive equations (4.165) and (4.167) of models $A_n$ of ideal continua (both solids and fluids) in actual configuration $\mathcal{K}$. Then in configuration $\mathcal{K}'$ these equations, according to the principle of material indifference (4.603)–(4.605), must be written in the form

$$\begin{cases} \overset{(\mathrm{n})}{\mathbf{T}}{}' = \rho(\partial \psi / \partial \overset{(\mathrm{n})}{\mathbf{C}}) \equiv \mathcal{F}(\overset{(\mathrm{n})}{\mathbf{C}}{}', \theta'), \\ \psi = \psi(\overset{(\mathrm{n})}{\mathbf{C}}{}', \theta'), \\ \eta = -\partial \psi / \partial \theta, \end{cases} \quad (4.637)$$

where $\psi$ and $\mathcal{F}(\overset{(\mathrm{n})}{\mathbf{C}}{}', \theta')$ are the same tensor functions as in (4.165) and (4.167).

But Eqs. (4.637) really hold, if the constitutive equations (4.165) and (4.167) are satisfied, because all the tensors $\overset{(n)}{\mathbf{T}}$ and $\overset{(n)}{\mathbf{C}}$ are $R$-invariant, i.e. $\overset{(n)}{\mathbf{T}}{}' = \overset{(n)}{\mathbf{T}}$, $\overset{(n)}{\mathbf{C}}{}' = \overset{(n)}{\mathbf{C}}$; and the temperature $\theta'$, due to (4.622), is also $R$-invariant: $\theta' = \theta$.

For models $B_n$, the situation is analogous.

Thus, we can formulate the following theorem.

**Theorem 4.57.** *For models $A_n$ and $B_n$ of ideal continua (both solids and fluids), the principle of material indifference is always identically satisfied for any potentials $\psi(\overset{(n)}{\mathbf{C}}, \theta)$ and $\psi(\overset{(n)}{\mathbf{G}}, \theta)$.*

### *4.10.15 Material Indifference for Models $C_n$ and $D_n$ of Ideal Continua*

The situation is different when the principle of material indifference is applied to models $C_n$ of ideal continua.

Let in configuration $\mathcal{K}$ the relations (4.171) and (4.174) be satisfied, then in configuration $\mathcal{K}'$, according to the principle of material indifference (4.603)–(4.605), the following equations must hold:

$$\begin{cases} \overset{(n)}{\mathbf{S}}{}' = \rho(\partial\psi/\partial\overset{(n)}{\mathbf{A}}{}') \equiv \mathbf{\Phi}(\overset{(n)}{\mathbf{A}}{}', \mathbf{O}', \theta), \\ \psi = \psi(\overset{(n)}{\mathbf{A}}{}', \mathbf{O}', \theta), \\ \eta = -\partial\psi/\partial\theta, \\ \overset{\circ}{\mathbf{S}}{}' = \rho\partial\psi/\partial\mathbf{O}' \equiv \overset{\circ}{\mathbf{\Phi}}(\overset{(n)}{\mathbf{A}}{}', \mathbf{O}', \theta), \end{cases} \tag{4.638}$$

where $\mathbf{\Phi}$, $\overset{\circ}{\mathbf{\Phi}}$ and $\psi$ are the same tensor functions as in (4.171) and (4.174).

Since the tensors $\overset{(n)}{\mathbf{S}}$ and $\overset{(n)}{\mathbf{A}}$ are $R$-indifferent (see Theorems 4.36 and 4.37) and the tensor $\mathbf{O}$ is transformed during the passage from $\mathcal{K}$ to $\mathcal{K}'$ by formula (4.568), so the relations (4.638) hold if and only if the functions $\mathbf{\Phi}$, $\overset{\circ}{\mathbf{\Phi}}$, and $\psi$ satisfy the following equations for any orthogonal tensor $\mathbf{Q}$:

$$\begin{cases} \mathbf{Q}^{\mathrm{T}} \cdot \mathbf{\Phi}(\overset{(n)}{\mathbf{A}}, \mathbf{O}, \theta) \cdot \mathbf{Q} = \mathbf{\Phi}(\mathbf{Q}^{\mathrm{T}} \cdot \overset{(n)}{\mathbf{A}} \cdot \mathbf{Q}, \mathbf{Q}^{\mathrm{T}} \cdot \mathbf{O}, \theta), & (4.639a) \\ (1/2)(\mathbf{Q}^{\mathrm{T}} \cdot \overset{\circ}{\mathbf{\Phi}}(\overset{(n)}{\mathbf{A}}, \mathbf{O}, \theta) + \overset{\circ}{\mathbf{\Phi}}(\overset{(n)}{\mathbf{A}}, \mathbf{O}, \theta) \cdot \mathbf{Q}) & \\ \qquad = \overset{\circ}{\mathbf{\Phi}}(\mathbf{Q}^{\mathrm{T}} \cdot \overset{(n)}{\mathbf{A}} \cdot \mathbf{Q}, \mathbf{Q}^{\mathrm{T}} \cdot \mathbf{O}, \theta), & (4.639b) \\ \psi(\overset{(n)}{\mathbf{A}}, \mathbf{O}, \theta) = \psi(\mathbf{Q}^{\mathrm{T}} \cdot \overset{(n)}{\mathbf{A}} \cdot \mathbf{Q}, \mathbf{Q}^{\mathrm{T}} \cdot \mathbf{O}, \theta) \ \ \forall \mathbf{Q}. & (4.639c) \end{cases}$$

**Theorem 4.58.** *For models $C_n$ of ideal continua (both solids and fluids), the principle of material indifference remains valid if and only if the scalar function* $\psi(\overset{(n)}{\mathbf{A}}, \mathbf{O}, \theta)$ *satisfies the condition (4.639c) for any orthogonal tensor* $\mathbf{Q}$.

▼ If the principle of material indifference is valid, then all the relations (4.639), including (4.639c), hold.

Prove the converse assertion. If the condition (4.639c) is satisfied, then, differentiating the function $\psi(\overset{(n)}{\mathbf{A}}, \mathbf{O}, \theta)$ with respect to tensor arguments $\overset{(n)}{\mathbf{A}}$ and $\mathbf{O}$ and taking the relations $\mathbf{\Phi} = \rho\partial\psi/\partial\overset{(n)}{\mathbf{A}}$ and $\overset{\circ}{\mathbf{\Phi}} = \rho\partial\psi/\partial\mathbf{O}$ into account, we get that the two remaining equations (4.639a) and (4.639b) hold as well; and hence the equations (4.638) remain valid. ▲

On comparing the condition (4.639c) with conditions (4.281) and (4.354), we obtain that for satisfaction of the principle of material indifference, the function $\psi(\overset{(n)}{\mathbf{A}}, \mathbf{O}, \theta)$ must be both indifferent under the full group of orthogonal transformations $G_s = I$ with respect to the first argument $\overset{(n)}{\mathbf{A}}$ and rotation-indifferent under the same group $I$ with respect to the second argument $\mathbf{O}$.

**Theorem 4.59.** *For models $C_n$ of ideal solids, the principle of material indifference is always satisfied.*

▼ For models $C_n$, the potential $\psi$ can always be represented in the form (4.377):

$$\psi(\overset{(n)}{\mathbf{A}}, \mathbf{O}, \theta) = \psi(I_\alpha^{(s)}(\mathbf{O}^{\mathrm{T}} \cdot \overset{(n)}{\mathbf{A}} \cdot \mathbf{O}), \theta). \tag{4.640a}$$

To verify that the condition (4.639c) is satisfied, we should substitute the tensor $\mathbf{Q}^{\mathrm{T}} \cdot \overset{(n)}{\mathbf{A}} \cdot \mathbf{Q}$ in place of $\overset{(n)}{\mathbf{A}}$ and the tensor $\mathbf{Q}^{\mathrm{T}} \cdot \mathbf{O}$ in place of $\mathbf{O}$ on the right-hand side of the relation. After these substitutions we get

$$\begin{aligned}\psi(\mathbf{Q}^{\mathrm{T}} \cdot \overset{(n)}{\mathbf{A}} \cdot \mathbf{Q}, \mathbf{Q}^{\mathrm{T}} \cdot \mathbf{O}, \theta) &= \psi(I_\alpha^{(s)}(\mathbf{O}^{\mathrm{T}} \cdot \mathbf{Q} \cdot \mathbf{Q}^{\mathrm{T}} \cdot \overset{(n)}{\mathbf{A}} \cdot \mathbf{Q} \cdot \mathbf{Q}^{\mathrm{T}} \cdot \mathbf{O}), \theta) \\ &= \psi(I_\alpha^{(s)}(\mathbf{O}^{\mathrm{T}} \cdot \overset{(n)}{\mathbf{A}} \cdot \mathbf{O}), \theta). \end{aligned} \tag{4.640b}$$

On comparing the expressions (4.640a) with (4.640b), we find that they coincide; hence the relations (4.639c) hold for any orthogonal tensor $\mathbf{Q}$. ▲

**Theorem 4.60.** *Constitutive equations in the form of models $C_n$ (4.460) for ideal fluid always satisfy the principle of material indifference.*

▼ Tensors $\overset{(n)}{\mathbf{S}}$ are $R$-indifferent, and tensors $\overset{(n)}{\mathbf{g}}{}^{-1}$ are $R$-indifferent as well (since $\overset{(n)}{\mathbf{g}}$ is $R$-indifferent, the following relation holds:

$$\overset{(n)}{\mathbf{g}}{}'^{-1} = (\mathbf{Q}^{\mathrm{T}} \cdot \overset{(n)}{\mathbf{g}} \cdot \mathbf{Q})^{-1} = \mathbf{Q}^{-1} \cdot \overset{(n)}{\mathbf{g}}{}^{-1} \cdot \mathbf{Q}^{-1\mathrm{T}} = \mathbf{Q}^{\mathrm{T}} \cdot \overset{(n)}{\mathbf{g}}{}^{-1} \cdot \mathbf{Q}.$$

Then, multiplying equation (4.460a) by $\mathbf{Q}^{\mathrm{T}}$ from the left and by $\mathbf{Q}$ from the right, we get the following relation in configuration $\mathcal{K}'$:

$$\mathbf{S}' = -\frac{p}{n - \mathrm{III}} \overset{(\mathrm{n})}{\mathbf{g}}{}'^{-1}. \tag{4.641}$$

The function $\psi(I_3(\overset{(\mathrm{n})}{\mathbf{g}}), \theta)$ (4.460d) always satisfies the condition (4.639c), because $I_3(\overset{(\mathrm{n})}{\mathbf{g}})$ is an invariant under any orthogonal transformation; therefore the pressure $p$ (4.460b) is a $R$-invariant function: $p' = p$, and hence the constitutive equations (4.460) satisfy the principle of material indifference. ▲

For models $B_n$, the situation is analogous.

Thus, the constitutive equations for models $A_n$, $B_n$, $C_n$ and $D_n$ of ideal continua, which have been derived in Sect. 4.8, satisfy the principle of material indifference.

### *4.10.16 Material Indifference for Incompressible Continua*

All the constitutive equations derived above for incompressible fluids (4.505), (4.509) and incompressible solids (4.513), (4.521) satisfy the principle of material indifference. Indeed, the constitutive equations for incompressible continua differ from the corresponding constitutive equations for compressible continua by the additional summand $-(p/(n - \mathrm{III}))\overset{(\mathrm{n})}{\mathbf{G}}{}^{-1}$ or $-(p/(n - \mathrm{III}))\overset{(\mathrm{n})}{\mathbf{g}}{}^{-1}$. But this summand, similarly to the model of ideal compressible fluid (see Theorem 4.59), is really consistent with the principle of material indifference.

### *4.10.17 Material Indifference for Models of Solids 'in Rates'*

**Theorem 4.61.** *The models $A_n$ and $B_n$ (4.421) of ideal solids 'in rates' automatically satisfy the principle of material indifference, and the models $C_n$ and $D_n$ (4.425) 'in rates' for isotropic continua satisfy the principle only if $h$ is a corotational derivative from the following set:*

$$h = \{V,\ J,\ S\}. \tag{4.642}$$

*The principle of material indifference does not hold for the derivative $h = U$ in eigenbasis of the right stretch tensor and for the total derivative with respect to time $d/dt$.*

▼ (1) Since all the tensors $\overset{(\mathrm{n})}{\mathbf{T}}$ and $\overset{(\mathrm{n})}{\mathbf{C}}_G = \{\overset{(\mathrm{n})}{\mathbf{C}},\ \overset{(\mathrm{n})}{\mathbf{G}}\}$ are $R$-invariant, the relations (4.421) remain unchanged in rigid motions $\mathcal{K} \to \mathcal{K}'$ and really satisfy the principle of material indifference.

(2) Consider constitutive equations (4.425) for models $C_n$ and $D_n$ in the form (4.434). Under the transformation from $\mathcal{K}$ to $\mathcal{K}'$, these equations take the form

$$(\overset{(\mathrm{n})}{\mathbf{S}}{}^{h})' = \varphi_1^{h\,\prime}\mathbf{E} + \varphi_2^{h\,\prime}\overset{(\mathrm{n})}{\mathbf{C}}{}'_G + \varphi_3^{h\,\prime}\overset{(\mathrm{n})}{\mathbf{C}}{}'^2_G + \varphi_2'(\overset{(\mathrm{n})}{\mathbf{C}}{}^h_G)'$$
$$+\varphi_3'((\overset{(\mathrm{n})}{\mathbf{C}}{}^h_G)'\cdot\overset{(\mathrm{n})}{\mathbf{C}}{}'_G + \overset{(\mathrm{n})}{\mathbf{C}}{}'_G\cdot(\overset{(\mathrm{n})}{\mathbf{C}}{}^h_G)'),\quad \overset{(\mathrm{n})}{\mathbf{C}}_G = \{\overset{(\mathrm{n})}{\mathbf{A}},\ \overset{(\mathrm{n})}{\mathbf{g}}\}. \quad (4.643)$$

Tensors $\overset{(\mathrm{n})}{\mathbf{S}}$ and $\overset{(\mathrm{n})}{\mathbf{C}}_G$ are $R$-indifferent (see Theorems 4.36 and 4.37), then, due to Theorems 4.43, 4.49, and 4.50, the co-rotational derivatives $\overset{(\mathrm{n})}{\mathbf{S}}{}^h$ and $\overset{(\mathrm{n})}{\mathbf{C}}{}^h_G$ are also $R$-indifferent, where $h$ is included in the set (4.642):

$$(\overset{(\mathrm{n})}{\mathbf{S}}{}^h)' = \mathbf{Q}^{\mathrm{T}}\cdot\overset{(\mathrm{n})}{\mathbf{S}}{}^h\cdot\mathbf{Q},\quad (\overset{(\mathrm{n})}{\mathbf{C}}{}^h_G)' = \mathbf{Q}^{\mathrm{T}}\cdot\overset{(\mathrm{n})}{\mathbf{C}}{}^h_G\cdot\mathbf{Q}. \quad (4.644)$$

The functions $\varphi_\alpha$ depend on invariants $I_\alpha(\overset{(\mathrm{n})}{\mathbf{C}}_G)$ (see (4.380)), and, according to the results of Exercise 4.10.5, invariants $I_\alpha(\overset{(\mathrm{n})}{\mathbf{C}}_G)$ are $R$-invariant scalar functions. Therefore the functions $\varphi_\alpha$ are also $R$-invariant, i.e. $\varphi'_\alpha = \varphi_\alpha$; and, by Theorem 4.39, we get

$$(\dot{\varphi}_\alpha)' = (\varphi^h_\alpha)' = \varphi^h_\alpha. \quad (4.645)$$

Substitution of (4.644) into (4.434) yields

$$\mathbf{Q}^{\mathrm{T}}\cdot\overset{(\mathrm{n})}{\mathbf{S}}{}^h\cdot\mathbf{Q} = \mathbf{Q}^{\mathrm{T}}\Big(\varphi_1^h\mathbf{E} + \varphi_2^h\overset{(\mathrm{n})}{\mathbf{C}}_G + \varphi_3^h\overset{(\mathrm{n})}{\mathbf{C}}{}^2_G$$
$$+\varphi_2\overset{(\mathrm{n})}{\mathbf{C}}{}^h_G + \varphi_3(\overset{(\mathrm{n})}{\mathbf{C}}{}^h_G\cdot\overset{(\mathrm{n})}{\mathbf{C}}_G + \overset{(\mathrm{n})}{\mathbf{C}}_G\cdot\overset{(\mathrm{n})}{\mathbf{C}}{}^h_G)\Big)\cdot\mathbf{Q}. \quad (4.646)$$

Thus, the relations (4.644) hold in $\mathcal{K}'$.

For the co-rotational derivative $h = U$ and the total derivative with respect to time, the relations (4.644) and hence (4.646) are not valid, and the principle of material indifference does not hold. ▲

## Exercises for 4.10

**4.10.1.** Using formulae (4.578) and (4.560), show that the divergence of the velocity is both $R$-indifferent and $R$-invariant:

$$\nabla'\cdot\mathbf{v}' = \mathbf{r}'^i\cdot\frac{\partial\mathbf{v}'}{\partial X^i} = \mathbf{E}\cdot\cdot\nabla'\otimes\mathbf{v}' = \nabla\cdot\mathbf{v}.$$

**4.10.2.** Show that the divergence of any $R$-indifferent second-order tensor $\mathbf{B}$ (and of any $R$-indifferent vector $\mathbf{b}$) is $R$-indifferent:

$$\nabla' \cdot \mathbf{B}' = \mathbf{r}'^i \cdot \frac{\partial \mathbf{B}'}{\partial X^i} = \mathbf{Q}^{\mathrm{T}} \cdot \nabla \cdot \mathbf{B}, \quad \nabla' \cdot \mathbf{b}' = \nabla \cdot \mathbf{b}.$$

**4.10.3.** Show that for any $R$-indifferent vector $\mathbf{b}$, its curl is also $R$-indifferent:

$$\nabla' \times \mathbf{b}' = \mathbf{Q}^{\mathrm{T}} \cdot \nabla \times \mathbf{b}.$$

**4.10.4.** Prove that the contraction of any two $R$-indifferent tensors $\mathbf{A}$ and $\mathbf{B}$ gives a scalar being both $R$-indifferent and $R$-invariant:

$$\mathbf{A}' \cdot\cdot \mathbf{B}' = \mathbf{A} \cdot\cdot \mathbf{B}.$$

**4.10.5.** Show that the principal invariants of any symmetric $R$-indifferent tensor $\mathbf{A}$ are both $R$-indifferent and $R$-invariant:

$$I'_\alpha(\mathbf{A}') = I_\alpha(\mathbf{A}') = I_\alpha(\mathbf{A}), \qquad \mathbf{A}' = \mathbf{Q}^{\mathrm{T}} \cdot \mathbf{A} \cdot \mathbf{Q}.$$

**4.10.6.** Show that if a second-order tensor $\mathbf{B}$ is symmetric and $R$-indifferent, then the tensor $\mathbf{B}'$ is symmetric as well.

**4.10.7.** Prove Theorem 4.55 with use of the method given in the proof of Theorem 4.54.

**4.10.8.** Show that from (4.622b) we get $R$-indifference of the dissipation rate $w^*$ (3.167):

$$w^{*\prime} = w^*.$$

**4.10.9.** Let there be a scalar function of a tensor argument:

$$\psi = \Psi(\mathbf{A}) \quad \text{and} \quad \mathbf{A}' = \mathbf{O}^{\mathrm{T}} \cdot \mathbf{A} \cdot \mathbf{O},$$

where $\mathbf{O}$ is a second-order tensor being independent of $\mathbf{A}$. Prove that

$$\frac{\partial \psi(\mathbf{A})}{\partial \mathbf{A}'} = \mathbf{O}^{\mathrm{T}} \cdot \frac{\partial \psi(\mathbf{A})}{\partial \mathbf{A}} \cdot \mathbf{O}.$$

## 4.11 Relationships in a Moving System

### *4.11.1 A Moving Reference System*

In continuum mechanics, one frequently considers not only a rigid motion of an actual configuration, investigated in Sect. 4.10, but also a rigid motion of a

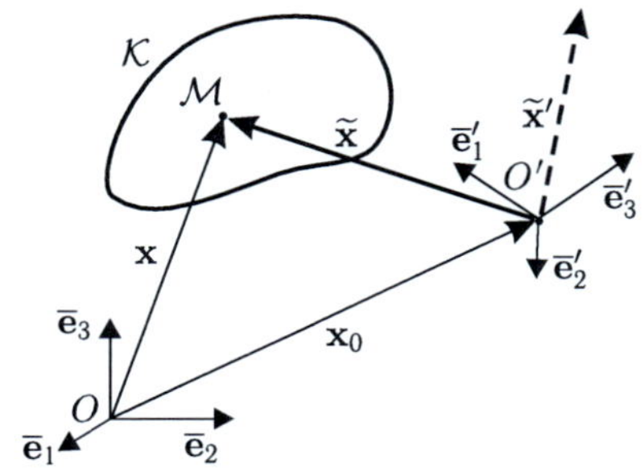


**Fig. 4.6** The moving reference system $O'\bar{\mathbf{e}}'_i$ for $\mathcal{K}$

coordinate system, methods of description for which are analogous. Let us consider such a motion.

In Sect. 2.1.7 we have introduced a *moving system* $O'\bar{\mathbf{e}}'_i$ (2.39), which is characterized by the radius-vector $\mathbf{x}_0(t)$ of the point $O'$ (the center of rotation) and the orthogonal tensor $\mathbf{Q}(t)$ of rotation of the system $O'\bar{\mathbf{e}}'_i$ about a fixed (immovable) system $O\bar{\mathbf{e}}_i$ (Fig. 4.6), where

$$\bar{\mathbf{e}}'_i = \mathbf{Q} \cdot \bar{\mathbf{e}}_i = Q_i^j \bar{\mathbf{e}}_j. \tag{4.647}$$

The position of a point $\mathcal{M}$ in the system $O\bar{\mathbf{e}}_i$ is defined by its radius-vector $\mathbf{x}$, and in the system $O'\bar{\mathbf{e}}'_i$ – by the radius-vector

$$\widetilde{\mathbf{x}} = \mathbf{x} - \mathbf{x}_0 = \widetilde{x}^i \bar{\mathbf{e}}'_i, \tag{4.648}$$

(see (2.26)).

Introduce the vector

$$\widetilde{\mathbf{x}}' = \mathbf{Q}^{\mathrm{T}} \cdot \widetilde{\mathbf{x}}, \tag{4.649}$$

whose coordinates $\widetilde{x}^i$ in the basis $\bar{\mathbf{e}}_i$ are the same as coordinates of the vector $\widetilde{\mathbf{x}}$ in the moving basis $\bar{\mathbf{e}}'_i$:

$$\widetilde{\mathbf{x}}' = \widetilde{\mathbf{x}} \cdot \mathbf{Q} = \widetilde{x}^i \bar{\mathbf{e}}'_i \cdot \mathbf{Q} = \widetilde{x}^i \bar{\mathbf{e}}_i = \widetilde{x}'^i \bar{\mathbf{e}}_i, \tag{4.650}$$

i.e.

$$\widetilde{x}'^i = \widetilde{x}^i. \tag{4.651}$$

Equations (4.647) and (4.648) yield (see (2.27))

$$\begin{gathered} \widetilde{\mathbf{x}} = \mathbf{x} - \mathbf{x}_0 = (x^i - x_0^i)\bar{\mathbf{e}}_i = \widetilde{x}^i \bar{\mathbf{e}}'_i = \widetilde{x}^i Q_i^j \bar{\mathbf{e}}_j, \\ x^i - x_0^i(t) = Q_j^i(t)\widetilde{x}^j, \quad \frac{\partial x^i}{\partial \widetilde{x}^j} = Q_j^i(t). \end{gathered} \tag{4.652}$$

### 4.11.2 The Euler Formula

Let us differentiate formula (4.648):

$$\dot{\widetilde{\mathbf{x}}} = \dot{\mathbf{x}} - \dot{\mathbf{x}}_0. \tag{4.653}$$

Since $\dot{\mathbf{x}} = \mathbf{v}$ and $\dot{\mathbf{x}}_0 = \mathbf{v}_0$ are velocities of the point $\mathcal{M}$ and the point $O'$, respectively, relative to the fixed system $O\bar{\mathbf{e}}_i$, according to formula (4.649) for $\widetilde{\mathbf{x}}$ we obtain

$$\mathbf{v} = \mathbf{v}_0 + (\widetilde{\mathbf{x}}' \cdot \mathbf{Q}^{\mathrm{T}})^{\bullet}. \tag{4.654}$$

Differentiating the product $\widetilde{\mathbf{x}}' \cdot \mathbf{Q}$ and using formula (4.649) once more, we find that

$$\mathbf{v} = \mathbf{v}_0 + \dot{\widetilde{\mathbf{x}}}' \cdot \mathbf{Q}^{\mathrm{T}} - \widetilde{\mathbf{x}} \cdot (\dot{\mathbf{Q}} \cdot \mathbf{Q}^{\mathrm{T}}). \tag{4.655}$$

Since the tensor $\mathbf{Q}$ is orthogonal, then $\dot{\mathbf{Q}} \cdot \mathbf{Q}^{\mathrm{T}}$ is skew-symmetric (see Sect. 2.4.10) and it can be represented in terms of the corresponding vector $\boldsymbol{\omega}_s$ in the form

$$\boldsymbol{\Omega}_s \equiv \dot{\mathbf{Q}} \cdot \mathbf{Q}^{\mathrm{T}} = \boldsymbol{\omega}_s \times \mathbf{E} = -\mathbf{Q} \cdot \dot{\mathbf{Q}}^{\mathrm{T}}. \tag{4.656}$$

The vector $\boldsymbol{\omega}_s$ is in general not coincident with the vorticity vector $\boldsymbol{\omega}$ introduced by formula (2.227); they coincide in the case when during the passage from $\overset{\circ}{\mathcal{K}}$ to $\mathcal{K}$ a continuum executes only a rigid motion, whose rotation tensor coincides with the tensor $\mathbf{Q}$ of a moving system.

According to (4.656), formula (4.655) takes the form

$$\mathbf{v} = \mathbf{v}_0 + \dot{\widetilde{\mathbf{x}}}' \cdot \mathbf{Q}^{\mathrm{T}} - \widetilde{\mathbf{x}} \cdot \boldsymbol{\Omega}_s; \tag{4.656a}$$

and the following relations hold:

$$-\widetilde{\mathbf{x}} \cdot \boldsymbol{\Omega}_s = -\boldsymbol{\Omega}_s^{\mathrm{T}} \cdot \widetilde{\mathbf{x}} = \boldsymbol{\Omega}_s \cdot \widetilde{\mathbf{x}} = (\boldsymbol{\omega}_s \times \mathbf{E}) \cdot \widetilde{\mathbf{x}} = \boldsymbol{\omega}_s \times \widetilde{\mathbf{x}}. \tag{4.657}$$

According to formulae (4.650), we can introduce the vector

$$\mathbf{v}_r \equiv \dot{\widetilde{\mathbf{x}}}' \cdot \mathbf{Q}^{\mathrm{T}} = \dot{\widetilde{x}}^i \bar{\mathbf{e}}_i \cdot \mathbf{Q}^{\mathrm{T}} = \dot{\widetilde{x}}^i \bar{\mathbf{e}}'_i, \tag{4.658}$$

which is called the *relative velocity* of a point $\mathcal{M}$ in the moving system $O'\bar{\mathbf{e}}'_i$. Components of the vector $\mathbf{v}_r$ in basis $\bar{\mathbf{e}}'_i$ are the derivatives $\dot{\widetilde{x}}^i$ of components $\widetilde{x}^i$ of the radius-vector $\widetilde{\mathbf{x}}$ in the same moving basis $\bar{\mathbf{e}}'_i$; in other words, $\mathbf{v}_r$ is determined by the same rules as the velocity $\mathbf{v} = \dot{\mathbf{x}} = \dot{x}^i \bar{\mathbf{e}}_i$ in a fixed system $O\bar{\mathbf{e}}_i$.

According to (4.656)–(4.658), formula (4.655) takes the form

$$\mathbf{v} = \mathbf{v}_0 + \boldsymbol{\omega}_s \times \widetilde{\mathbf{x}} + \mathbf{v}_r. \tag{4.659}$$

The velocity $\mathbf{v}$ is called the *total velocity of a point* $\mathcal{M}$, and $\mathbf{v}_e = \mathbf{v}_0 + \boldsymbol{\omega}_s \times \widetilde{\mathbf{x}}$ – the *translational velocity*. Formula (4.659) called the *Euler formula* means that the total velocity of a material point is equal to a sum of the translational and relative velocities.

### 4.11.3 The Coriolis Formula

Similarly to the acceleration vector of a material point $\mathcal{M}$ in a fixed system $O\bar{\mathbf{e}}_i$:

$$\ddot{\mathbf{x}} = \ddot{x}^i \bar{\mathbf{e}}_i, \tag{4.660}$$

let us introduce the relative acceleration vector of a point $\mathcal{M}$ in the moving system $O'\bar{\mathbf{e}}'_i$:

$$\mathbf{a}_r = \ddot{\widetilde{x}}^i \bar{\mathbf{e}}'_i. \tag{4.661}$$

By (4.647) and (4.650), the vector $\mathbf{a}_r$ can be written as follows:

$$\mathbf{a}_r = \ddot{\widetilde{x}}^i \mathbf{Q} \cdot \bar{\mathbf{e}}_i = \ddot{\widetilde{\mathbf{x}}}' \cdot \mathbf{Q}^{\mathrm{T}}. \tag{4.662}$$

**Theorem 4.62 (of Coriolis).** *The acceleration* $\ddot{\mathbf{x}}$ *of a material point* $\mathcal{M}$ *in a fixed coordinate system and the relative acceleration* $\mathbf{a}_r$ *in a moving system are connected by the Coriolis formula*

$$\ddot{\mathbf{x}} = \mathbf{a}_r + \mathbf{a}_e + \mathbf{a}_c, \tag{4.663}$$

*where*

$$\mathbf{a}_e = \dot{\mathbf{v}}_0 + \dot{\boldsymbol{\omega}}_s \times \widetilde{\mathbf{x}} + \boldsymbol{\omega}_s \times (\boldsymbol{\omega}_s \times \widetilde{\mathbf{x}}), \tag{4.664}$$

$$\mathbf{a}_c = 2\boldsymbol{\omega}_s \times \mathbf{v}_r = 2\boldsymbol{\omega}_s \times (\dot{\widetilde{\mathbf{x}}}' \cdot \mathbf{Q}^{\mathrm{T}}), \tag{4.665}$$

$\mathbf{a}_e$ *is called the translational acceleration,* $\mathbf{a}_c$ *– the Coriolis acceleration, and* $\ddot{\mathbf{x}}$ *– the total acceleration.*

▼ Consider formula (4.655) for the velocity $\mathbf{v}$ and differentiate this expression with respect to $t$:

$$\ddot{\mathbf{x}} = \dot{\mathbf{v}} = \dot{\mathbf{v}}_0 + \ddot{\widetilde{\mathbf{x}}}' \cdot \mathbf{Q}^{\mathrm{T}} + \dot{\widetilde{\mathbf{x}}}' \cdot \dot{\mathbf{Q}}^{\mathrm{T}} + \dot{\boldsymbol{\Omega}}_s \cdot \widetilde{\mathbf{x}} + \boldsymbol{\Omega}_s \cdot \dot{\widetilde{\mathbf{x}}}. \tag{4.666}$$

Here we have used formula (4.656) for the spin $\boldsymbol{\Omega}_s$. Modify the third summand on the right-hand side

$$\dot{\widetilde{\mathbf{x}}}' \cdot \dot{\mathbf{Q}}^{\mathrm{T}} = \dot{\widetilde{\mathbf{x}}}' \cdot \mathbf{Q}^{\mathrm{T}} \cdot \mathbf{Q} \cdot \dot{\mathbf{Q}}^{\mathrm{T}} = (\dot{\mathbf{Q}} \cdot \mathbf{Q}^{\mathrm{T}}) \cdot (\dot{\widetilde{\mathbf{x}}}' \cdot \mathbf{Q}^{\mathrm{T}}) = \boldsymbol{\Omega}_s \cdot \mathbf{v}_r = \boldsymbol{\omega}_s \times \mathbf{v}_r, \tag{4.667}$$

and the fourth summand according to (4.656)

$$\dot{\boldsymbol{\Omega}}_s \cdot \widetilde{\mathbf{x}} = (\boldsymbol{\omega}_s \times \mathbf{E})^{\bullet} \cdot \widetilde{\mathbf{x}} = \dot{\boldsymbol{\omega}}_s \times \mathbf{E} \cdot \widetilde{\mathbf{x}} = \dot{\boldsymbol{\omega}}_s \times \widetilde{\mathbf{x}}. \tag{4.668}$$

The last summand on the right-hand side of (4.666) can be transformed according to (4.649) and (4.658):

$$\begin{aligned}
\boldsymbol{\Omega}_s \cdot \dot{\widetilde{\mathbf{x}}} &= \boldsymbol{\Omega}_s \cdot (\widetilde{\mathbf{x}}' \cdot \mathbf{Q}^{\mathrm{T}})^{\bullet} = \boldsymbol{\Omega}_s \cdot (\dot{\widetilde{\mathbf{x}}}' \cdot \mathbf{Q}^{\mathrm{T}}) + \boldsymbol{\Omega}_s \cdot (\widetilde{\mathbf{x}}' \cdot \dot{\mathbf{Q}}^{\mathrm{T}}) \\
&= \boldsymbol{\Omega}_s \cdot \mathbf{v}_r + \boldsymbol{\Omega}_s \cdot (\widetilde{\mathbf{x}} \cdot (\mathbf{Q} \cdot \dot{\mathbf{Q}}^{\mathrm{T}})) = \boldsymbol{\omega}_s \times \mathbf{v}_r + \boldsymbol{\Omega}_s \cdot (\boldsymbol{\Omega}_s \cdot \widetilde{\mathbf{x}}) \\
&= \boldsymbol{\omega}_s \times \mathbf{v}_r + \boldsymbol{\omega}_s \times (\boldsymbol{\omega}_s \times \widetilde{\mathbf{x}}).
\end{aligned} \tag{4.669}$$

On substituting formulae (4.667)–(4.669) into (4.666), we obtain Eq. (4.663). ▲

*Remark 1.* Due to the definition (2.336) of co-rotational derivatives $\mathbf{a}^h$, expressions (4.658) and (4.662) for the relative velocity $\mathbf{v}_r$ and the relative acceleration $\mathbf{a}_r$ can be represented as co-rotational derivatives $\mathbf{a}^Q$ of the vector $\widetilde{\mathbf{x}} = \widetilde{x}^i \bar{\mathbf{e}}_i{}'$ in the orthonormal moving basis $\mathbf{h}_i = \bar{\mathbf{e}}_i{}'$:

$$\mathbf{v}_r = \widetilde{\mathbf{x}}^{Q} \equiv \dot{\widetilde{x}}^{i} \bar{\mathbf{e}}_i', \quad \mathbf{a}_r = \widetilde{\mathbf{x}}^{QQ} = \mathbf{v}_r^{Q} \equiv \ddot{\widetilde{x}}^{i} \bar{\mathbf{e}}_i'. \tag{4.670}$$

In Sect. 2.1.7 we have derived the relations (2.50) and (2.53) between the moving basis $\bar{\mathbf{e}}_i'$ and bases $\widetilde{\mathbf{r}}_i$ and $\widetilde{\mathbf{r}}^i$:

$$\begin{gathered}
\bar{\mathbf{e}}_i' = \widetilde{P}_i^j \widetilde{\mathbf{r}}_j, \quad \widetilde{\mathbf{r}}^i = \widetilde{P}_j^i \bar{\mathbf{e}}'^j, \quad \widetilde{\mathbf{r}}_i = \widetilde{Q}_i^j \, \bar{\mathbf{e}}_j', \\
\widetilde{Q}_j^i = \partial \widetilde{x}^i / \partial \widetilde{X}^j, \quad \widetilde{P}_j^i = \partial \widetilde{X}^i / \partial \widetilde{x}^j,
\end{gathered} \tag{4.671}$$

where the matrices $\widetilde{Q}_j^i$ and $\widetilde{P}_j^i$, according to (2.44), are independent of $t$; then formulae (4.670) can be written in the basis $\widetilde{\mathbf{r}}_i$ as well:

$$\begin{gathered}
\mathbf{v}_r = \widetilde{v}_r^i \widetilde{\mathbf{r}}_i = \widetilde{v}_{ri} \widetilde{\mathbf{r}}^i, \quad \mathbf{a}_r = \widetilde{a}_r^i \widetilde{\mathbf{r}}_i = \widetilde{a}_{ri} \widetilde{\mathbf{r}}^i, \\
\widetilde{v}_r^i = \frac{d}{dt} (\widetilde{x}^j \widetilde{P}_j^i), \quad \widetilde{a}_r^i = \frac{d^2}{dt^2} (\widetilde{x}^j \widetilde{P}_j^i) = \frac{d \widetilde{v}_{ri}}{dt}.
\end{gathered} \tag{4.672}$$

These formulae are exactly analogous to representations of the velocity and acceleration in the fixed basis $\bar{\mathbf{e}}_i$ and in the basis $\overset{\circ}{\mathbf{r}}_i$ of Lagrangian coordinates:

$$\mathbf{v} = \frac{d\mathbf{x}}{dt} = \dot{x}^i \bar{\mathbf{e}}_i = \bar{v}^i \bar{\mathbf{e}}_i = \overset{\circ}{v}{}^i \overset{\circ}{\mathbf{r}}_i, \quad \mathbf{a} = \frac{d^2 \mathbf{x}}{dt^2} = \ddot{x}^i \bar{\mathbf{e}}_i = \bar{a}^i \bar{\mathbf{e}}_i = \overset{\circ}{a}{}^i \overset{\circ}{\mathbf{r}}_i, \tag{4.673}$$

$$\overset{\circ}{v}{}^i = \frac{d}{dt} x^i = \frac{d}{dt} \left( x^i \left( \frac{\partial X^j}{\partial x^i} \right) \right), \quad \overset{\circ}{a}{}^i = \frac{d^2}{dt^2} x^i = \frac{d}{dt} \overset{\circ}{v}{}^i = \frac{d^2}{dt^2} \left( x^i \left( \frac{\partial X^j}{\partial x^i} \right) \right). \quad \square$$

*Remark 2.* For moving bases of the type $\bar{\mathbf{e}}'_i$, which are connected to $\bar{\mathbf{e}}_i$ with the help of the orthogonal tensor $\mathbf{Q}$ by formula (4.647), in Sect. 2.4.10 we have derived formula (2.266); this formula applied to $\bar{\mathbf{e}}'_i$ takes the form

$$\dot{\bar{\mathbf{e}}}'_i = \boldsymbol{\omega}_s \times \bar{\mathbf{e}}'_i. \tag{4.674}$$

Resolving the vector $\boldsymbol{\omega}_s$ for the basis $\bar{\mathbf{e}}'_i$: $\boldsymbol{\omega}_s = \omega_s^{\prime i}\bar{\mathbf{e}}'_i$, we get

$$\dot{\boldsymbol{\omega}}_s = (\omega_s^{\prime i}\bar{\mathbf{e}}'_i)^{\bullet} = \dot{\omega}_s^{\prime i}\bar{\mathbf{e}}'_i + \omega_s^{\prime i}\dot{\bar{\mathbf{e}}}'_i = \boldsymbol{\omega}_s \times (\omega_s^{\prime i}\bar{\mathbf{e}}'_i). \tag{4.675}$$

Since $\boldsymbol{\omega}_s \times \boldsymbol{\omega}_s = \mathbf{0}$, we obtain

$$\dot{\boldsymbol{\omega}}_s = \boldsymbol{\omega}_s^{Q}, \quad \boldsymbol{\omega}_s^{Q} \equiv \dot{\omega}_s^{\prime i}\bar{\mathbf{e}}'_i, \tag{4.676}$$

where $\boldsymbol{\omega}_s^{Q}$ is the co-rotational derivative (2.336) in the basis $\bar{\mathbf{e}}'_i$. □

### 4.11.4 The Nabla-Operator in a Moving System

In Sect. 2.1.7 we have proved Theorem 2.2 that the nabla-operators $\nabla$ and $\widetilde{\nabla}$ in a moving coordinate system are coincident:

$$\nabla = \widetilde{\nabla}, \quad \widetilde{\nabla} = \widetilde{\mathbf{r}}^i \frac{\partial}{\partial \widetilde{X}^i}, \quad \nabla = \mathbf{r}^i \frac{\partial}{\partial X^i}, \tag{4.677}$$

where $\widetilde{X}^i$ are curvilinear spatial coordinates (2.45), and $\widetilde{\mathbf{r}}_i$ and $\widetilde{\mathbf{r}}^i$ are local bases in $\mathcal{K}$ relative to the coordinates $\widetilde{X}^i$ introduced by (2.48) and (2.53).

According to relations (4.671), the nabla-operator $\nabla$ can be represented in bases $\bar{\mathbf{e}}_i$ and $\bar{\mathbf{e}}'_i$ as follows:

$$\begin{aligned}\nabla = \bar{\mathbf{e}}^i \frac{\partial}{\partial x^i} &= \delta^{ik}\delta_{ms}P_k^m \bar{\mathbf{e}}'^s \left(\frac{\partial \widetilde{x}^j}{\partial x^i}\right)\frac{\partial}{\partial \widetilde{x}^j} \\ &= \delta^{ik}\delta_{ms}P_k^m P_i^j \bar{\mathbf{e}}'^s \frac{\partial}{\partial \widetilde{x}^j} = \delta_s^j \bar{\mathbf{e}}'^s \frac{\partial}{\partial \widetilde{x}^j} = \bar{\mathbf{e}}'^j \frac{\partial}{\partial \widetilde{x}^j}.\end{aligned} \tag{4.678}$$

Here we have used the formulae of passage from one orthonormal basis to another:

$$\bar{\mathbf{e}}^i = \delta^{ik}\bar{\mathbf{e}}_k, \quad \bar{\mathbf{e}}'_m = \delta_{ms}\bar{\mathbf{e}}'^s, \tag{4.679}$$

and taken into account that both the matrices $Q_j^i$ and $P_j^i$ are orthogonal, and

$$P_k^i P_l^j \delta^{kl} = \delta^{ij}. \tag{4.680}$$

As follows from (4.678) and (4.672), (4.673), in a moving system $O'\bar{\mathbf{e}}'_i$ its orthonormal basis $\bar{\mathbf{e}}'_i$ plays the same role as the basis $\bar{\mathbf{e}}_i$ in a fixed system, and local bases $\widetilde{\mathbf{r}}^i$ and $\widetilde{\mathbf{r}}_i$ are similar to bases $\mathbf{r}^i$ and $\mathbf{r}_i$ relative to the nabla-operator, because

$$\nabla = \bar{\mathbf{e}}^i \frac{\partial}{\partial x^i} = \mathbf{r}^i \frac{\partial}{\partial X^i} = \widetilde{\nabla} = \bar{\mathbf{e}}'^i \frac{\partial}{\partial \widetilde{x}^i} = \widetilde{\mathbf{r}}^i \frac{\partial}{\partial \widetilde{X}^i}, \tag{4.681}$$

and they are analogous to bases $\mathring{\mathbf{r}}_i$ and $\mathring{\mathbf{r}}^i$ relative to the derivative with respect to time.

## 4.11.5 The Velocity Gradient in a Moving System

Applying the relations (4.681) to formula (4.656a), we obtain

$$\nabla \otimes \mathbf{v} = \widetilde{\nabla} \otimes \mathbf{v}_r - (\widetilde{\nabla} \otimes \widetilde{\mathbf{x}}) \cdot \boldsymbol{\Omega}_s. \tag{4.682}$$

Here we have taken into account that $\mathbf{v}_0$ and $\boldsymbol{\Omega}_s$ are independent of coordinates.

Since

$$\widetilde{\nabla} \otimes \widetilde{\mathbf{x}} = \bar{\mathbf{e}}'^i \frac{\partial}{\partial \widetilde{x}^i} \otimes (\widetilde{x}^j \bar{\mathbf{e}}'_j) = \bar{\mathbf{e}}'^i \otimes \bar{\mathbf{e}}'_i = \mathbf{E} \tag{4.683}$$

(similarly to $\nabla \otimes \mathbf{x} = \mathbf{E}$), formula (4.682) takes the form

$$\nabla \otimes \mathbf{v} = \widetilde{\nabla} \otimes \mathbf{v}_r - \mathbf{E} \times \boldsymbol{\omega}_s. \tag{4.684}$$

In particular, since the tensor $\boldsymbol{\Omega}_s$ is skew-symmetric, so

$$\nabla \cdot \mathbf{v} = \widetilde{\nabla} \cdot \mathbf{v}_r. \tag{4.685}$$

## 4.11.6 The Continuity Equation in a Moving System

On substituting the relation (4.685) into the continuity equation in Eulerian variables (3.15), we obtain

$$d\rho/dt = -\rho \widetilde{\nabla} \cdot \mathbf{v}_r; \tag{4.686}$$

or, since the co-rotational derivative $\rho^{\mathcal{Q}}$ of a scalar coincides with $\dot{\rho}$, so

$$\rho \left(\frac{1}{\rho}\right)^{\mathcal{Q}} = \widetilde{\nabla} \cdot \mathbf{v}_r \tag{4.687}$$

is the continuity equation in a moving system $O'\bar{\mathbf{e}}'_i$, which has the same form as the corresponding Eq. (3.15) in a fixed system $O\bar{\mathbf{e}}_i$ if all the values are considered relative the system $O'\bar{\mathbf{e}}'_i$, i.e. with substitutions $\mathbf{v} \to \mathbf{v}_r$ and $\nabla \to \widetilde{\nabla}$.

### *4.11.7 The Momentum Balance Equation in a Moving System*

For the momentum balance equation, the situation is different: substitution of the Coriolis formula (4.663) and the relation (4.681) into (3.72) yields

$$\rho(\ddot{\widetilde{\mathbf{x}}}' \cdot \mathbf{Q}^{\mathrm{T}} + \mathbf{a}_e + \mathbf{a}_c) = \widetilde{\nabla} \cdot \mathbf{T} + \rho\mathbf{f}. \tag{4.688}$$

Here we have taken into account that the stress tensor $\mathbf{T}$ is defined in actual configuration $\mathcal{K}$ and is independent of the choice of a coordinate system (but components of the tensor $\widetilde{T}^{ij}$ with respect to the basis $\bar{\mathbf{e}}'_i$: $\mathbf{T} = \widetilde{T}^{ij}\bar{\mathbf{e}}'_i \otimes \bar{\mathbf{e}}'_j$ depend on this choice).

According to representation (4.670) for the acceleration $\mathbf{a}_r$, we can rewrite equation (4.688) in the form

$$\rho\widetilde{\mathbf{x}}^{QQ} = \widetilde{\nabla} \cdot \mathbf{T} + \rho\widetilde{\mathbf{f}}, \tag{4.689}$$

where

$$\widetilde{\mathbf{f}} = \mathbf{f} - \mathbf{a}_e - \mathbf{a}_c = \mathbf{f} - \dot{\mathbf{v}}_0 - \dot{\boldsymbol{\omega}}_s \times \widetilde{\mathbf{x}} - \boldsymbol{\omega}_s \times (\boldsymbol{\omega}_s \times \widetilde{\mathbf{x}}) - 2\boldsymbol{\omega}_s \times \mathbf{v}_r \tag{4.690}$$

is the specific mass force in a moving system, which includes the translational and Coriolis accelerations. If $\mathbf{v}_0$ and $\boldsymbol{\omega}_s$ are assumed to be known, then the translational acceleration $\mathbf{a}_e$ is a known value, but the Coriolis acceleration, due to the presence of the velocity $\mathbf{v}_r$, is an unknown vector.

According to (4.670), the momentum balance equation (4.689) takes the form

$$\rho\mathbf{v}_r^Q = \widetilde{\nabla} \cdot \mathbf{T} + \rho\widetilde{\mathbf{f}}. \tag{4.691}$$

This equation is similar to the momentum balance equation (3.72) in a fixed system $O\bar{\mathbf{e}}_i$; the distinction consists only in that the nabla-operator $\widetilde{\nabla}$, the co-rotational derivative $\mathbf{v}_r^Q$ and the velocity $\mathbf{v}_r$ are considered in a moving system $O'\bar{\mathbf{e}}'_i$, and the specific mass force $\widetilde{\mathbf{f}}$ involves additional terms, namely the translational $\mathbf{a}_e$ and Coriolis $\mathbf{a}_c$ accelerations.

### *4.11.8 The Thermodynamic Laws in a Moving System*

The scalar thermodynamic functions $e$, $\eta$, $\theta$, and $q_m$ are independent of the choice a coordinate system, because they were introduced in an actual configuration. The heat flux vector $\mathbf{q}$, as well as $\mathbf{T}$, is also defined in $\mathcal{K}$; therefore, it remains unchanged during the passage to the system $O'\bar{\mathbf{e}}'_i$, and only its components with respect to basis $\bar{\mathbf{e}}'_i$ become different. The specific kinetic energy determined by the velocity magnitude squared also changes; indeed, according to the Euler formula (4.659) we have

$$\frac{1}{2}|\mathbf{v}|^2 = \frac{1}{2}\mathbf{v}\cdot\mathbf{v} = \frac{1}{2}(|\mathbf{v}_r|^2 + |\mathbf{v}_0|^2 + |\boldsymbol{\omega}_s \times \widetilde{\mathbf{x}}|) \\ +\mathbf{v}_0\cdot(\boldsymbol{\omega}_s \times \widetilde{\mathbf{x}}) + \mathbf{v}_0\cdot\mathbf{v}_r + \mathbf{v}_r\cdot(\boldsymbol{\omega}_s \times \widetilde{\mathbf{x}}).$$

However, if the heat influx equation (3.124) is considered as a statement of the first thermodynamic law, then it does not include $(1/2)|\mathbf{v}|^2$; and, thus, we obtain the *heat influx equation in a moving system* $O'\bar{\mathbf{e}}'_i$:

$$\rho\frac{de}{dt} = \mathbf{T}\cdot\cdot(\widetilde{\nabla}\otimes\mathbf{v}_r)^{\mathrm{T}} + \rho q_m - \widetilde{\nabla}\cdot\mathbf{q} \tag{4.692}$$

Here we have used formula (4.684) and also the fact that $\mathbf{T}\cdot\cdot\boldsymbol{\Omega}_s = 0$, because the tensor $\mathbf{T}$ is symmetric, and $\boldsymbol{\Omega}_s = \mathbf{E}\times\boldsymbol{\omega}_s$ is skew-symmetric.

In order to find an analog of the energy balance law in the form (3.118), one should consider the contraction

$$\mathbf{v}_r^{\mathcal{Q}}\cdot\mathbf{v}_r = (\ddot{\widetilde{x}}^i\bar{\mathbf{e}}'_i)\cdot(\dot{\widetilde{x}}^j\bar{\mathbf{e}}'_j) = \ddot{\widetilde{x}}^i\dot{\widetilde{x}}^j\delta_{ij} = \frac{1}{2}(\dot{\widetilde{x}}^i\dot{\widetilde{x}}^j\delta_{ij})^{\bullet} = \frac{1}{2}(\mathbf{v}_r\cdot\mathbf{v}_r)^{\bullet} = |\mathbf{v}_r|^{\bullet}/2. \tag{4.693}$$

Multiplying Eq. (4.691) by $\mathbf{v}_r$, from (4.693) we obtain the *kinetic energy balance equation in a moving system*

$$\rho\frac{d}{dt}\frac{|\mathbf{v}_r|}{2} = \mathbf{v}_r\cdot\widetilde{\nabla}\cdot\mathbf{T} + \rho\widetilde{\mathbf{v}}_r\cdot\widetilde{\mathbf{f}}. \tag{4.694}$$

On summing (4.694) and (4.692), we finally get the *energy balance law in a moving coordinate system*

$$\rho\frac{d}{dt}\left(e + \frac{|\mathbf{v}_r|^2}{2}\right) = \widetilde{\nabla}\cdot(\mathbf{T}\cdot\mathbf{v}_r - \mathbf{q}) + \rho\widetilde{\mathbf{f}}\cdot\mathbf{v}_r + \rho q_m. \tag{4.695}$$

The second thermodynamic law (3.166) during the passage to $O'\bar{\mathbf{e}}'_i$ retains its form

$$\rho\theta\frac{d\eta}{dt} = -\widetilde{\nabla}\cdot\mathbf{q} + \rho q_m + w^*. \tag{4.696}$$

### *4.11.9 The Equation of Deformation Compatibility in a Moving System*

Let us consider now the dynamic compatibility equation in the form (3.277):

$$\frac{d\mathbf{F}^{\mathrm{T}}}{dt} = \mathbf{F}^{\mathrm{T}}\cdot\nabla\otimes\mathbf{v}. \tag{4.697}$$

By analogy with the Cartesian coordinates $\overset{\circ}{x}{}^i$ for $\overset{\circ}{\mathcal{K}}$ in the system $O\bar{\mathbf{e}}_i$, we introduce Cartesian coordinates $\overset{\circ}{\widetilde{x}}{}^i$ in the moving system $O'\bar{\mathbf{e}}'_i$ also for the configuration $\overset{\circ}{\mathcal{K}}$ (according to (4.652)):

$$\overset{\circ}{\widetilde{x}}{}^i = \overset{\circ}{P}{}^i_j(\overset{\circ}{x}{}^j - x^j_{00}), \tag{4.698}$$

where $\overset{\circ}{P}{}^i_j$, $\overset{\circ}{Q}{}^i_j$, $x^j_{00}$ are values of $P^i_j$, $Q^i_j$ and $x^j_0$ at $t = 0$. Since $x^i$ and $\overset{\circ}{x}{}^i$ are connected by $x^i = x^i(\overset{\circ}{x}{}^j, t)$, so from (4.652) and (4.698) we find the relation between $\widetilde{x}^i$ and $\overset{\circ}{\widetilde{x}}{}^j$:

$$\begin{aligned}\widetilde{x}^i &= P^i_j(x^j - x^j_0) = P^i_j(x^j(\overset{\circ}{x}{}^k, t) - x^j_0)\\ &= P^i_j(x^j(x^k_{00} + \overset{\circ}{Q}{}^k_m \overset{\circ}{\widetilde{x}}{}^m,\ t) - x^j_0) \equiv \widetilde{x}^i(\overset{\circ}{\widetilde{x}}{}^m,\ t).\end{aligned} \tag{4.699}$$

Introduce the basis $\overset{\circ}{\bar{\mathbf{e}}}{}'_i = \bar{\mathbf{e}}'_i(0)$ that is in general not coincident with $\bar{\mathbf{e}}_i$; for this basis, according to (4.647), we have

$$\begin{aligned}\overset{\circ}{\bar{\mathbf{e}}}{}'_j &= \overset{\circ}{Q}{}^i_j\bar{\mathbf{e}}_i = \left(\frac{\partial x^i}{\partial \widetilde{x}^j}\right)_{t=0}\bar{\mathbf{e}}_i = (\partial\overset{\circ}{x}{}^i/\partial\overset{\circ}{\widetilde{x}}{}^j)\bar{\mathbf{e}}_i = \overset{\circ}{\mathbf{Q}}\cdot\bar{\mathbf{e}}_i = \overset{\circ}{\mathbf{Q}}\cdot\mathbf{Q}^{\mathrm{T}}\cdot\bar{\mathbf{e}}'_i,\\ \overset{\circ}{\bar{\mathbf{e}}}{}'^j &= \overset{\circ}{\mathbf{Q}}\cdot\mathbf{Q}^{\mathrm{T}}\cdot\bar{\mathbf{e}}'^i, \quad \bar{\mathbf{e}}^i = \overset{\circ}{P}{}^i_j\overset{\circ}{\bar{\mathbf{e}}}{}'^j = \left(\frac{\partial\overset{\circ}{\widetilde{x}}{}^i}{\partial\overset{\circ}{x}{}^j}\right)\overset{\circ}{\bar{\mathbf{e}}}{}'^j,\end{aligned} \tag{4.700}$$

where $\overset{\circ}{\mathbf{Q}} = \overset{\circ}{Q}{}^i_j\,\bar{\mathbf{e}}_i\otimes\bar{\mathbf{e}}^j$ is the value of the orthogonal tensor $\mathbf{Q}$ at $t = 0$ (if $\overset{\circ}{\bar{\mathbf{e}}}{}'_i = \bar{\mathbf{e}}_i$, then $\overset{\circ}{\mathbf{Q}} = \mathbf{E}$).

Remind the representation of the deformation gradient $\mathbf{F}$ in the Cartesian basis $\bar{\mathbf{e}}_i$ (see Exercise 2.1.5) and find its representation in the basis $\bar{\mathbf{e}}'_i$:

$$\mathbf{F} = \frac{\partial x^i}{\partial\overset{\circ}{x}{}^j}\bar{\mathbf{e}}_i\otimes\bar{\mathbf{e}}^j = \frac{\partial x^i}{\partial\widetilde{x}^k}\frac{\partial\widetilde{x}^k}{\partial\overset{\circ}{\widetilde{x}}{}^m}\frac{\partial\overset{\circ}{\widetilde{x}}{}^m}{\partial\overset{\circ}{x}{}^j}\bar{\mathbf{e}}_i\otimes\bar{\mathbf{e}}^j = \frac{\partial\widetilde{x}^k}{\partial\overset{\circ}{\widetilde{x}}{}^m}\bar{\mathbf{e}}'_k\otimes\overset{\circ}{\bar{\mathbf{e}}}{}'^m \equiv \widetilde{F}^k_m\bar{\mathbf{e}}'_k\otimes\overset{\circ}{\bar{\mathbf{e}}}{}'^m, \tag{4.701}$$

$$\widetilde{F}^k_m = (\partial\widetilde{x}^k/\partial\overset{\circ}{\widetilde{x}}{}^m).$$

Introduce two more tensors, which have the same components $\widetilde{F}^k_m$ but with respect to other dyadic bases:

$$\widetilde{\mathbf{F}} = \widetilde{F}^k_m\bar{\mathbf{e}}'_k\otimes\bar{\mathbf{e}}'^m, \quad \mathbf{F}' = \widetilde{F}^k_m\bar{\mathbf{e}}_k\otimes\bar{\mathbf{e}}^m. \tag{4.702}$$

According to (4.700) and (4.647), we get

$$\mathbf{F} = \widetilde{\mathbf{F}}\cdot\mathbf{Q}\cdot\overset{\circ}{\mathbf{Q}}{}^{\mathrm{T}}, \quad \mathbf{F}' = \mathbf{Q}^{\mathrm{T}}\cdot\widetilde{\mathbf{F}}\cdot\mathbf{Q}. \tag{4.703}$$

Using the representation (4.701), we can determine the derivative with respect to time

$$\frac{d\mathbf{F}^{\mathrm{T}}}{dt} = \frac{d\widetilde{F}^{k}_{m}}{dt}\overset{\circ}{\bar{\mathbf{e}}}{}'^{m} \otimes \bar{\mathbf{e}}'_{k} + \widetilde{F}^{k}_{m}\overset{\circ}{\bar{\mathbf{e}}}{}'^{m} \otimes \dot{\bar{\mathbf{e}}}'_{k}$$
$$= \overset{\circ}{\mathbf{Q}} \cdot \mathbf{Q}^{\mathrm{T}} \cdot \frac{d\widetilde{F}^{k}_{m}}{dt}\bar{\mathbf{e}}'^{m} \otimes \bar{\mathbf{e}}'_{k} - \overset{\circ}{\mathbf{Q}} \cdot \mathbf{Q}^{\mathrm{T}} \cdot \widetilde{F}^{k}_{m}\bar{\mathbf{e}}'^{m} \otimes \bar{\mathbf{e}}'_{k} \times \boldsymbol{\omega}_{s}. \quad (4.704)$$

Here we have taken into account that $\overset{\circ}{\bar{\mathbf{e}}}{}'^{m}$ is independent of $t$ and used formulae (4.700) and (4.674) for $\dot{\bar{\mathbf{e}}}'_{k}$.

If in the first summand we replace the basis $\bar{\mathbf{e}}'_{i}$ by $\mathbf{e}_{i}$ according to (4.647) and take into account that $d\mathbf{F}'/dt = (d\widetilde{F}^{k}_{m}/dt)\bar{\mathbf{e}}_{k} \otimes \bar{\mathbf{e}}^{m}$, then we obtain

$$\frac{d\mathbf{F}^{\mathrm{T}}}{dt} = \overset{\circ}{\mathbf{Q}} \cdot \mathbf{Q}^{\mathrm{T}} \cdot \left(\mathbf{Q} \cdot \frac{d\mathbf{F}'^{\mathrm{T}}}{dt} \cdot \mathbf{Q}^{\mathrm{T}} - \widetilde{\mathbf{F}} \times \boldsymbol{\omega}_{s}\right). \quad (4.705)$$

Rearrange the right-hand side of the compatibility equation (4.697) with the help of (4.703) and (4.684):

$$\mathbf{F}^{\mathrm{T}} \cdot \nabla \otimes \mathbf{v} = \overset{\circ}{\mathbf{Q}} \cdot \mathbf{Q}^{\mathrm{T}} \cdot (\widetilde{\mathbf{F}}^{\mathrm{T}} \cdot \widetilde{\nabla} \otimes \mathbf{v}_{r} - \widetilde{\mathbf{F}}^{\mathrm{T}} \times \boldsymbol{\omega}_{s}). \quad (4.706)$$

On substituting (4.705) and (4.706) into (4.697), we finally get the *dynamic compatibility equation in a moving coordinate system* $O'\bar{\mathbf{e}}'_{i}$:

$$\mathbf{Q} \cdot \frac{d\mathbf{F}'^{\mathrm{T}}}{dt} \cdot \mathbf{Q}^{\mathrm{T}} = \widetilde{\mathbf{F}}^{\mathrm{T}} \cdot \widetilde{\nabla} \otimes \mathbf{v}_{r}. \quad (4.707)$$

Since

$$\mathbf{Q} \cdot \frac{d\mathbf{F}'^{\mathrm{T}}}{dt} \cdot \mathbf{Q}^{\mathrm{T}} = \frac{d\widetilde{F}^{k}_{m}}{dt}\bar{\mathbf{e}}'^{m} \otimes \bar{\mathbf{e}}'_{k} = \widetilde{\mathbf{F}}^{\mathrm{T}\mathcal{Q}} \quad (4.708)$$

(here we have used the definition (2.338) of the co-rotational derivative in the moving orthonormal basis $\bar{\mathbf{e}}'_{i}$), so Eq. (4.707) can be written in the equivalent form

$$\widetilde{\mathbf{F}}^{\mathrm{T}\mathcal{Q}} = \widetilde{\mathbf{F}}^{\mathrm{T}} \cdot \widetilde{\nabla} \otimes \mathbf{v}_{r}, \quad (4.709)$$

which is a formal analog of the compatibility equation (4.695) in a fixed coordinate system.

Equation (4.707) can also be written in the form similar to (3.303), $\alpha = 5$. To do this, one should determine the divergence of the tensor $\rho\widetilde{\mathbf{F}}$ according to formulae (4.701) and (3.272):

$$\widetilde{\nabla} \cdot (\rho\widetilde{\mathbf{F}}) = \nabla \cdot (\rho\mathbf{F} \cdot \overset{\circ}{\mathbf{Q}} \cdot \mathbf{Q}^{\mathrm{T}}) = (\nabla \cdot (\rho\mathbf{F}) \cdot \overset{\circ}{\mathbf{Q}} \cdot \mathbf{Q}^{\mathrm{T}} + \rho\mathbf{F}^{\mathrm{T}} \cdot\cdot \nabla \otimes (\overset{\circ}{\mathbf{Q}} \cdot \mathbf{Q}^{\mathrm{T}})$$
$$= (\nabla \cdot \rho\mathbf{F}) \cdot \overset{\circ}{\mathbf{Q}} \cdot \mathbf{Q}^{\mathrm{T}} = 0. \quad (4.710)$$

Here we have taken into account that the rotation tensors $\overset{\circ}{\mathbf{Q}}$ and $\mathbf{Q}^{\mathrm{T}}$ are independent of coordinates $x^{i}$.

It follows from (4.710) that $(\widetilde{\nabla} \cdot \rho\widetilde{\mathbf{F}}) \otimes \mathbf{v}_r = \mathbf{0}$. Multiplying (4.709) by $\rho$ and adding the result to the last equation, we obtain

$$\rho \widetilde{\mathbf{F}}^{\mathrm{T}Q} = \widetilde{\nabla} \cdot (\rho \widetilde{\mathbf{F}} \otimes \widetilde{\mathbf{v}}_r). \tag{4.711}$$

### 4.11.10 The Kinematic Equation in a Moving System

An analog of the kinematic equation (3.303) ($\alpha = 6$) in a fixed system $O\bar{\mathbf{e}}_i$:

$$\dot{\mathbf{u}} = \mathbf{v}, \tag{4.712}$$

during the passage to a moving system $O'\bar{\mathbf{e}}_i'$ is the first relation of (4.670) for the relative velocity $\mathbf{v}_r$, which is expressed in terms of the co-rotational derivative.

### 4.11.11 The Complete System of Continuum Mechanics Laws in a Moving Coordinate System

Equations (4.686), (4.691), (4.695), (4.696), (4.711), and (4.670) can be written in the single form similar to (3.303):

$$\rho \bar{A}_\alpha^Q = \widetilde{\nabla} \cdot \bar{B}_\alpha + \rho C_\alpha, \quad \alpha = 1, \ldots, 6, \tag{4.713}$$

where we have introduced the generalized vectors

$$\bar{A}_\alpha = \begin{pmatrix} 1/\rho \\ \mathbf{v}_r \\ e + (|\mathbf{v}_r|^2/2) \\ \eta \\ \widetilde{\mathbf{x}} \\ \widetilde{\mathbf{F}}^{\mathrm{T}} \end{pmatrix}, \quad \bar{B}_\alpha = \begin{pmatrix} \mathbf{v}_r \\ \mathbf{T} \\ \mathbf{T} \cdot \mathbf{v}_r - \mathbf{q} \\ -\mathbf{q}/\theta \\ 0 \\ \rho \widetilde{\mathbf{F}} \otimes \mathbf{v}_r \end{pmatrix}, \quad C_\alpha = \begin{pmatrix} 0 \\ \widetilde{\mathbf{f}} \\ \widetilde{\mathbf{f}} \cdot \mathbf{v}_r + q_m \\ (q_m + q^*)/\theta \\ \mathbf{v}_r \\ 0 \end{pmatrix}, \tag{4.714}$$

which are similar to the corresponding generalized vectors of (3.304).

### 4.11.12 Constitutive Equations in a Moving System

Consider the case when $\overset{\circ}{\bar{\mathbf{e}}}{}_i' = \bar{\mathbf{e}}_i$, i.e. the moving basis $\bar{\mathbf{e}}_i'$ at $t = 0$ coincides with $\bar{\mathbf{e}}_i$, then $\overset{\circ}{\mathbf{Q}} = \mathbf{E}$, and from (4.703) we get

$$\widetilde{\mathbf{F}} = \mathbf{F} \cdot \mathbf{Q}^{\mathrm{T}}. \tag{4.715}$$

Comparing (4.715) with (4.188), we can establish an analogy between the transformation of coordinate system $O\bar{\mathbf{e}}_i \to O'\bar{\mathbf{e}}'_i$ described by the rotation tensor $\mathbf{Q} = \bar{\mathbf{e}}'_i \otimes \bar{\mathbf{e}}^i$ and the orthogonal transformation of reference configuration $\overset{\circ}{\mathcal{K}} \to \overset{*}{\mathcal{K}}$, which, according to Sect. 4.6, is determined by the tensor $\mathbf{H} = \overset{*}{\mathbf{r}}_i \otimes \overset{\circ}{\mathbf{r}}{}^i = \overset{*}{\mathbf{Q}}{}^{\mathrm{T}}$. The tensor $\mathbf{Q}$ is an analog of the tensor $\overset{*}{\mathbf{Q}}{}^{\mathrm{T}} = \mathbf{H}$.

Drawing on the analogy, we can notice that all the results of Sect. 4.6 for the tensors $\overset{(n)}{\mathbf{C}}$, $\overset{(n)}{\mathbf{A}}$ and $\overset{(n)}{\mathbf{T}}$, $\overset{(n)}{\mathbf{S}}$ under orthogonal $H$-transformations still hold for orthogonal transformations of a coordinate system. Thus, according to Theorems 4.18 and 4.19, all the energetic tensors $\overset{(n)}{\mathbf{C}}$ and $\overset{(n)}{\mathbf{T}}$ are $H$-invariant, i.e.

$$\widetilde{\overset{(n)}{\mathbf{T}}} = \mathbf{Q} \cdot \overset{(n)}{\mathbf{T}} \cdot \mathbf{Q}^{\mathrm{T}}, \qquad \widetilde{\overset{(n)}{\mathbf{C}}} = \mathbf{Q} \cdot \overset{(n)}{\mathbf{C}} \cdot \mathbf{Q}^{\mathrm{T}}, \tag{4.716}$$

where $\widetilde{\overset{(n)}{\mathbf{T}}}$ and $\widetilde{\overset{(n)}{\mathbf{C}}}$ are the tensors determined in $O'\bar{\mathbf{e}}'_i$.

Tensors $\overset{(n)}{\mathbf{A}}$ and $\overset{(n)}{\mathbf{S}}$ are $H$-invariant, and the tensor $\mathbf{O}$ is transformed by the formula similar to (4.256)

$$\widetilde{\overset{(n)}{\mathbf{A}}} = \overset{(n)}{\mathbf{A}}, \qquad \widetilde{\overset{(n)}{\mathbf{S}}} = \overset{(n)}{\mathbf{S}}, \qquad \widetilde{\mathbf{O}} = \mathbf{O} \cdot \mathbf{Q}^{\mathrm{T}}. \tag{4.717}$$

Therefore, the constitutive equations for models $A_n$ of ideal solids (4.311) during the passage to $O'\bar{\mathbf{e}}'_i$ take the form

$$\mathbf{Q}^{\mathrm{T}} \cdot \widetilde{\overset{(n)}{\mathbf{T}}} \cdot \mathbf{Q} = \sum_{\gamma=1}^{r} \varphi_\gamma I^{(s)}_{\gamma \mathbf{C}}, \tag{4.718}$$

$$\varphi_\gamma = \rho \frac{\partial \psi}{\partial I^{(s)}_\gamma}, \quad I^{(s)}_{\gamma \mathbf{C}} = \frac{\partial I^{(s)}_\gamma}{\partial \overset{(n)}{\mathbf{C}}} = \mathbf{Q}^{\mathrm{T}} \cdot \frac{\partial I^{(s)}_\gamma}{\partial \widetilde{\overset{(n)}{\mathbf{C}}}} \cdot \mathbf{Q}, \quad I^{(s)}_\gamma(\overset{(n)}{\mathbf{C}}) = I^{(s)}_\gamma(\mathbf{Q}^{\mathrm{T}} \cdot \widetilde{\overset{(n)}{\mathbf{C}}} \cdot \mathbf{Q}). \tag{4.719}$$

As follows from (4.716), the tensors $\widetilde{\overset{(n)}{\mathbf{T}}}$ and $\widetilde{\overset{(n)}{\mathbf{C}}}$ in basis $\widetilde{\widehat{\mathbf{c}}}_i = \mathbf{Q} \cdot \widehat{\mathbf{c}}_i$ have the same components $\widehat{\overset{(n)}{T}}{}^{ij}$, $\widehat{\overset{(n)}{C}}{}^{ij}$ as the tensors $\overset{(n)}{\mathbf{T}}$ and $\overset{(n)}{\mathbf{C}}$ in basis $\widehat{\mathbf{c}}_i$ of principal anisotropy axes of a body. The basis $\widetilde{\widehat{\mathbf{c}}}_i$ is orthonormal, and it is called the *basis of moving anisotropy axes*. As follows from (4.718) and (4.719), in the basis $\widetilde{\widehat{\mathbf{c}}}_i$ the invariants $I^{\widetilde{(s)}}_\gamma(\widetilde{\overset{(n)}{\mathbf{C}}})$ of the tensor $\widetilde{\overset{(n)}{\mathbf{C}}}$ relative to the group $\widetilde{G}_s = \mathbf{Q}^{\mathrm{T}} \cdot G_s \cdot \mathbf{Q}$ have the same form as the invariants $I^{(s)}_\gamma(\overset{(n)}{\mathbf{C}})$ relative to the group $G_s$ while they are written in basis $\widehat{\mathbf{c}}_i$:

$$I^{\widetilde{(s)}}_\gamma(\widetilde{\overset{(n)}{\mathbf{C}}}) = I^{\widetilde{(s)}}_\gamma(\widehat{\overset{(n)}{C}}{}^{ij}) = I^{(s)}_\gamma(\widehat{\overset{(n)}{C}}{}^{ij}) = I^{(s)}_\gamma(\overset{(n)}{\mathbf{C}}) = I^{(s)}_\gamma(\mathbf{Q}^{\mathrm{T}} \cdot \widetilde{\overset{(n)}{\mathbf{C}}} \cdot \mathbf{Q}), \tag{4.720}$$

(the index $\widetilde{(s)}$ means that the invariant is considered relative to the group $\widetilde{G}_s$ with the basis of moving anisotropy axes $\widetilde{\widetilde{\mathbf{c}}}_i$). According to (4.719) and (4.720), the constitutive equations (4.718) take the form

$$\overset{(n)}{\widetilde{\mathbf{T}}} = \sum_{\gamma=1}^{r} \varphi_\gamma I^{\widetilde{(s)}}_{\gamma\widetilde{\mathbf{C}}}, \quad \varphi_\gamma = \rho(\partial\psi/\partial I_\gamma^{\widetilde{(s)}}), \quad I^{\widetilde{(s)}}_{\gamma\widetilde{\mathbf{C}}} = \partial I_\gamma^{\widetilde{(s)}}/\partial \overset{(n)}{\widetilde{\mathbf{C}}}. \tag{4.721}$$

For isotropic continua ($G_s = I$), the principal invariants $I_\gamma(\overset{(n)}{\mathbf{C}})$ coincide with $I_\gamma(\overset{(n)}{\widetilde{\mathbf{C}}})$; therefore the constitutive equations (4.721) exactly coincide with the corresponding relations in a fixed system $O\bar{\mathbf{e}}_i$ and have the form (4.322)

$$\overset{(n)}{\widetilde{\mathbf{T}}} = \psi_1 \mathbf{E} + \psi_2 \overset{(n)}{\widetilde{\mathbf{C}}} + \psi_3 \overset{(n)}{\widetilde{\mathbf{C}}}{}^2, \quad \psi_\alpha = \psi_\alpha(I_\gamma(\overset{(n)}{\widetilde{\mathbf{C}}}), \theta). \tag{4.722}$$

For transversely isotropic continua with the group $G_s = T_3$, whose invariants $I_\gamma^{(3)}(\overset{(n)}{\mathbf{C}}) = I_\gamma^{(3)}(\overset{(n)}{\widehat{C}}{}^{ij})$ are chosen in the form (4.297), the corresponding invariants $I_\gamma^{\widetilde{(3)}}(\overset{(n)}{\widetilde{\mathbf{C}}})$ in basis $\widehat{\widetilde{\mathbf{c}}}_i$ have the same form, and they are determined by the rule (4.720):

$$\begin{aligned} I_1^{\widetilde{(3)}} &= (\mathbf{E} - \widetilde{\widetilde{\mathbf{c}}}{}_3^2) \cdot\cdot \overset{(n)}{\widetilde{\mathbf{C}}}, \quad I_2^{\widetilde{(3)}} = \widetilde{\widetilde{\mathbf{c}}}{}_3^2 \cdot\cdot \overset{(n)}{\widetilde{\mathbf{C}}}, \quad I_5^{\widetilde{(3)}} = \det \overset{(n)}{\widetilde{\mathbf{C}}}, \\ I_3^{\widetilde{(3)}} &= ((\mathbf{E} - \widetilde{\widetilde{\mathbf{c}}}{}_3^2) \cdot \overset{(n)}{\widetilde{\mathbf{C}}}) \cdot\cdot (\widetilde{\widetilde{\mathbf{c}}}{}_3^2 \cdot \overset{(n)}{\widetilde{\mathbf{C}}}), \quad I_4^{\widetilde{(3)}} = \overset{(n)}{\widetilde{\mathbf{C}}}{}^2 \cdot\cdot \mathbf{E} - I_2^{\widetilde{(3)}2} - 2I_3^{\widetilde{(3)}}. \end{aligned} \tag{4.723}$$

In a similar way, we can determine invariants for other groups $G_s \subset I$.

For models $B_n$ of ideal solids, the constitutive equations (4.346) are transformed during the passage to $O'\bar{\mathbf{e}}'_i$ in a similar way and take the form

$$\begin{gathered} \overset{(n)}{\widetilde{\mathbf{T}}} = \sum_{\gamma=1}^{r} \varphi_\gamma I^{\widetilde{(s)}}_{\gamma\widetilde{\mathbf{G}}}, \\ \varphi_\gamma = \rho \frac{\partial\psi}{\partial I_\gamma^{\widetilde{(s)}}}, \quad I^{\widetilde{(s)}}_{\gamma\widetilde{\mathbf{G}}} = \partial I_\gamma^{\widetilde{(s)}}/\partial \overset{(n)}{\widetilde{\mathbf{G}}}, \quad I_\gamma^{\widetilde{(s)}}(\overset{(n)}{\widetilde{\mathbf{G}}}) = I_\gamma^{(s)}(\mathbf{Q}^{\mathrm{T}} \cdot \overset{(n)}{\widetilde{\mathbf{G}}} \cdot \mathbf{Q}). \end{gathered} \tag{4.724}$$

For models $C_n$ of ideal solids, the constitutive equations (4.378) and (4.379) in a moving system $O'\bar{\mathbf{e}}'_i$ with the help of (4.717) are written as follows:

$$\begin{gathered} \overset{(n)}{\widetilde{\mathbf{S}}} = \sum_{\gamma=1}^{r} \varphi_\gamma I^{\widetilde{(s)}}_{\gamma\widetilde{\mathbf{A}}}, \quad \varphi_\gamma = \rho \frac{\partial\psi}{\partial I_\gamma^{\widetilde{(s)}}}, \quad I^{\widetilde{(s)}}_{\gamma\widetilde{\mathbf{A}}} = \frac{\partial I_\gamma^{\widetilde{(s)}}}{\partial \overset{(n)}{\widetilde{\mathbf{A}}}}, \\ I_\gamma^{\widetilde{(s)}}(\widetilde{\mathbf{O}}^{\mathrm{T}} \cdot \overset{(n)}{\widetilde{\mathbf{A}}} \cdot \widetilde{\mathbf{O}}) = I_\gamma^{(s)}(\mathbf{Q}^{\mathrm{T}} \cdot \mathbf{O}^{\mathrm{T}} \cdot \overset{(n)}{\widetilde{\mathbf{A}}} \cdot \mathbf{O} \cdot \mathbf{Q}). \end{gathered} \tag{4.725}$$

For ideal fluids, the constitutive equations (4.444) retain their form during the passage to $O'\bar{\mathbf{e}}'_i$, because the tensor $\mathbf{T}$ is $H$-invariant, and $\rho$ and $\theta$ are independent of orthogonal transformations of a reference configuration.

### *4.11.13 General Remarks*

The above reasoning allows us to make the following conclusion: during the passage to a moving coordinate system the balance equations (4.713) and (4.714) and the constitutive equations (4.721) are written in the same form as for a fixed system except that: (1) the additional terms $\mathbf{a}_e$ and $\mathbf{a}_c$ should be included in the expression for the mass force $\mathbf{f}$, and (2) the derivatives with respect to time should be replaced by the co-rotational derivatives.

The presence of the translational $\mathbf{a}_e$ and Coriolis $\mathbf{a}_c$ accelerations in the set (4.713) is usually quite considerable. However, for the special case of the accelerationless rectilinear motion of a moving system $O'\bar{\mathbf{e}}'_i$, for example, along the axis $O\bar{\mathbf{e}}_1$ when

$$\mathbf{Q} = \mathbf{E}, \quad \mathbf{v}_0 = \bar{v}_0^1\bar{\mathbf{e}}_1, \quad \bar{v}_0^1 = \text{const}, \tag{4.726}$$

we have

$$\boldsymbol{\omega}_s = \mathbf{0}, \quad \mathbf{a}_e = \mathbf{0}, \quad \mathbf{a}_c = \mathbf{0}, \tag{4.727}$$

i.e. there are no translational and Coriolis accelerations; and, thus, Eqs. (4.713) exactly coincide with Eqs. (3.303) in a fixed coordinate system. In this case all processes and phenomena described by the equation sets will be coincident as well. Therefore, an observer moving with such a moving coordinate system cannot recognize whether he moves rectilinearly and without acceleration or he is at rest.

This conclusion in a less general formulation was made by Galilei and Newton when they were studying only kinematics of nondeformable bodies; therefore, the motion (4.726) is called *Galilei's motion.*

## Exercises for 4.11

**4.11.1.** Show that if a moving system $O'\bar{\mathbf{e}}'_i$ rotates with a constant angular speed $\omega$ about the fixed axis $O\bar{\mathbf{e}}_3$ and $O' = O$, then for the Coriolis formula (4.663) we have

$$\mathbf{v}_0 \equiv 0, \quad \dot{\boldsymbol{\omega}}_s \equiv 0, \quad \boldsymbol{\omega}_s = \omega\bar{\mathbf{e}}_3, \quad \widetilde{\mathbf{x}}_3 \equiv x^1\bar{\mathbf{e}}_1 + x^2\bar{\mathbf{e}}_2 = r\mathbf{e}_r,$$
$$-\mathbf{a}_e = \mathbf{f}_{\text{cp}} \equiv \omega^2\widetilde{\mathbf{x}}_3 = \omega^2 r\mathbf{e}_r,$$

and expression (4.690) for $\widetilde{\mathbf{f}}$ takes the form

$$\widetilde{\mathbf{f}} = \mathbf{f} + \mathbf{f}_{\text{cp}} - \mathbf{a}_c,$$

where the vector $\mathbf{f}_{\text{cp}}$ is called the specific centripetal force, the direction of which coincides with the direction of vector $\mathbf{e}_r$ of the cylindrical coordinate system $O\mathbf{e}_r\mathbf{e}_\varphi\bar{\mathbf{e}}_3$ with the axis $O\bar{\mathbf{e}}_3$.

## 4.12 The Onsager Principle

### *4.12.1 The Onsager Principle and the Fourier Law*

Let us pay attention to the fact that the principal thermodynamic identity forms (4.121)–(4.124), which were the foundation of derivation of constitutive equations for the models $A_n$, $B_n$, $C_n$ and $D_n$, do not include the heat flux vector $\mathbf{q}$. However, the vector $\mathbf{q}$ appears in the set of balance laws (3.303), and without some relations for $\mathbf{q}$ this system is not closed even with taking account of constitutive equations (4.158), (4.159), (4.160) or (4.161).

To derive the relations for $\mathbf{q}$, one should use the Onsager principle [45, 46]. Formula (3.169) for specific internal entropy production yields

$$\rho q^* = w^* - \frac{\mathbf{q}}{\theta} \cdot \nabla\theta \geqslant 0. \tag{4.728}$$

The value $\rho q^*$ is nonnegative due to the Planck inequality (3.158a).

**Axiom 16 (The Onsager Principle).** *In order for the Planck inequality to hold, the specific internal entropy production $q^*$ must be representable in the quadratic form*

$$\rho q^* = \sum_{\beta=1} Q_\beta X_\beta = \sum_{\beta,\gamma=1} L_{\beta\gamma} X_\beta X_\gamma \geqslant 0. \tag{4.729}$$

Here

$$Q_\beta = \sum_{\beta=1} L_{\beta\gamma} X_\gamma \tag{4.730}$$

are functions called *thermodynamic fluxes*, and $X_\beta$ are functions called *thermodynamic forces*.

As follows from (4.729), the matrix $L_{\beta\gamma}$ is symmetric:

$$L_{\beta\gamma} = L_{\gamma\beta}. \tag{4.730a}$$

*Remark.* As shown in Sect. 4.4.3, the principles of thermodynamically consistent determinism and equipresence result in the fact that all the active variables $\Lambda$ included in constitutive equations (4.158)–(4.161) are independent of the temperature gradient $\nabla\theta$. Hence, the dissipation function $w^*$ being among the active variables is independent of $\nabla\theta$ as well. Thus, if we assume $\nabla\theta \equiv 0 \ \forall \mathcal{M} \in V$ and $\forall t \geqslant 0$, then the value of $w^*$ remains unchanged and the summand $\mathbf{q} \cdot \nabla\theta$ vanishes; and from (4.728) we obtain

$$w^* \geqslant 0. \tag{4.731}$$

This inequality is known (see (3.171)) as the *dissipation inequality*.

The reasoning has proved the fact that the dissipation inequality is a consequence of the Planck inequality (4.728) and principles of thermodynamically consistent determinism and equipresence.

For ideal continua, $w^* \equiv 0$, then the Planck inequality (4.728) yields

$$\mathbf{q} \cdot \nabla\theta \leqslant 0, \tag{4.732}$$

this relation is called the *Fourier inequality*.

For nonideal continua, the Fourier inequality (4.732) does not follow from the Planck inequality (4.728) and it should be assumed to be an independent axiom (see Axiom 10 in Sect. 3.5.2). □

The Onsager principle is applied as follows: expression (4.728) for $\rho q^*$ is represented as the sum $\sum_\beta Q_\beta X_\beta$; for example, for ideal continua we choose the temperature gradient $X_1 = \nabla\theta$ as a thermodynamic force $X_1$ and the value $Q_1 = -(1/\theta)\mathbf{q}$ as a corresponding thermodynamic flux.

Then, according to the Onsager principle, the thermodynamic flux must be a linear function of $X_1$, i.e.

$$-\frac{1}{\theta}\mathbf{q} = Q_1 = L_{11}X_1 = L_{11}\nabla\theta. \tag{4.733}$$

The coefficient $L_{11}$ in this case is a second-order tensor, and the product $\theta L_{11} = \boldsymbol{\lambda}$ is called the *heat conductivity tensor*; then relation (4.733) takes the form

$$\mathbf{q} = -\boldsymbol{\lambda} \cdot \nabla\theta, \tag{4.734}$$

which is called the *Fourier law* in the spatial description.

Due to the Fourier inequality, the heat conductivity tensor $\boldsymbol{\lambda}$ is positive-definite:

$$\nabla\theta \cdot \boldsymbol{\lambda} \cdot \nabla\theta \geqslant 0. \tag{4.735}$$

The Fourier law can be formulated in the material description as well. To do this, one should replace the vector $\mathbf{q}$ in (4.734) by the vector $\overset{\circ}{\mathbf{q}}$ according to formula (3.128) and the gradient $\nabla\theta$ by $\overset{\circ}{\nabla}\theta$ according to the formula similar to formulae (2.23): $\nabla\theta = \mathbf{F}^{-1\mathrm{T}} \cdot \overset{\circ}{\nabla}\theta$. Then we obtain the relation

$$\overset{\circ}{\mathbf{q}} = -\overset{\circ}{\boldsymbol{\lambda}} \cdot \overset{\circ}{\nabla}\theta, \tag{4.736}$$

where $\overset{\circ}{\boldsymbol{\lambda}}$ is the heat conductivity tensor in a reference configuration:

$$\overset{\circ}{\boldsymbol{\lambda}} = \sqrt{g/\overset{\circ}{g}}\,\mathbf{F}^{-1} \cdot \boldsymbol{\lambda} \cdot \mathbf{F}^{-1\mathrm{T}}. \tag{4.737}$$

The Fourier law (4.734) is included in the general system of constitutive equations (4.158), (4.159), (4.160) or (4.161) of a continuum. This law can also be written in the form of operator relation (4.156), where the vector $\mathbf{q}$ is an active variable and the vector $\nabla\theta$ is a reactive one.

Then, depending on the considered model $A_n$, $B_n$, $C_n$ or $D_n$, the heat conductivity tensors $\boldsymbol{\lambda}$ and $\overset{\circ}{\boldsymbol{\lambda}}$ should be assumed to be operators of the corresponding set of reactive variables $\mathcal{R}$:

$$\begin{aligned}
&\text{models } A_n: \quad \boldsymbol{\lambda} = \breve{\boldsymbol{\lambda}}(\overset{(n)}{\mathbf{C}},\ \theta), && \overset{\circ}{\boldsymbol{\lambda}} = \overset{\circ}{\breve{\boldsymbol{\lambda}}}(\overset{(n)}{\mathbf{C}},\ \theta),\\
&\text{models } B_n: \quad \boldsymbol{\lambda} = \breve{\boldsymbol{\lambda}}(\overset{(n)}{\mathbf{G}},\ \theta), && \overset{\circ}{\boldsymbol{\lambda}} = \overset{\circ}{\breve{\boldsymbol{\lambda}}}(\overset{(n)}{\mathbf{G}},\ \theta),\\
&\text{models } C_n: \quad \boldsymbol{\lambda} = \breve{\boldsymbol{\lambda}}(\overset{(n)}{\mathbf{A}},\ \mathbf{O},\ \theta), && \overset{\circ}{\boldsymbol{\lambda}} = \overset{\circ}{\breve{\boldsymbol{\lambda}}}(\overset{(n)}{\mathbf{A}},\ \mathbf{O},\ \theta),\\
&\text{models } D_n: \quad \boldsymbol{\lambda} = \breve{\boldsymbol{\lambda}}(\overset{(n)}{\mathbf{g}},\ \mathbf{O},\ \theta), && \overset{\circ}{\boldsymbol{\lambda}} = \overset{\circ}{\breve{\boldsymbol{\lambda}}}(\overset{(n)}{\mathbf{g}},\ \mathbf{O},\ \theta).
\end{aligned} \tag{4.737a}$$

### *4.12.2 Corollaries of the Principle of Material Symmetry for the Fourier Law*

Consider the Fourier law (4.736) in a reference configuration $\overset{\circ}{\mathcal{K}}$ with use of one of the models $A_n, \ldots, D_n$; for definiteness, we choose the model $B_n$. Since the Fourier law as well as the constitutive equations (4.158)–(4.161) describes some physical properties of a continuum, so this law must satisfy the principle of material symmetry (4.180)–(4.181) if the vector $\overset{\circ}{\mathbf{q}}$ is considered as $\Lambda$ and the vector $\overset{\circ}{\nabla}\theta$ – as $\mathcal{R}$. In other words, if Eq. (4.736) holds in a configuration $\overset{\circ}{\mathcal{K}}$, then there exists a symmetry group $\overset{\circ}{G}_s$ such that for each $H$-transformation from this group ($\mathbf{H} : \overset{\circ}{\mathcal{K}} \to \overset{*}{\mathcal{K}}$) the form of function (4.736) remains unchanged:

$$\overset{\circ}{\mathbf{q}} = -\overset{\circ}{\boldsymbol{\lambda}}(\overset{(n)}{\mathbf{G}},\ \theta) \cdot \overset{\circ}{\nabla}\theta. \tag{4.738a}$$

and

$$\overset{*}{\mathbf{q}} = -\overset{\circ}{\boldsymbol{\lambda}}(\overset{(n)}{\mathbf{G}}{}^{*},\ \theta) \cdot \overset{*}{\nabla}\theta. \tag{4.738b}$$

**Theorem 4.63.** *The principle of material symmetry (4.738a), (4.738b) holds for the Fourier law in the material description if and only if in* $\overset{\circ}{\mathcal{K}}$ *there exists a symmetry group* $\overset{\circ}{G}_s$ *such that the heat conductivity tensor is transformed as follows:*

$$\overset{\circ}{\boldsymbol{\lambda}}(\overset{(n)}{\mathbf{G}},\ \theta) = \sqrt{\overset{*}{g}/\overset{\circ}{g}}\, \mathbf{H}^{-1} \cdot \overset{\circ}{\boldsymbol{\lambda}}(\overset{(n)}{\mathbf{G}}{}^{*},\ \theta) \cdot \mathbf{H}^{-1\mathrm{T}} \quad \forall \mathbf{H} \in \overset{\circ}{G}_s. \tag{4.739}$$

▼ Since the vector $\mathbf{q}$ and temperature $\theta$ are defined in $\mathcal{K}$, they are $H$-invariant. Then $\overset{\circ}{\mathbf{q}}$ and $\overset{\circ}{\nabla}\theta$ under $H$-transformations change as follows:

$$\overset{*}{\mathbf{q}} = \sqrt{g/\overset{*}{g}}\, \overset{*}{\mathbf{F}}{}^{-1} \cdot \mathbf{q} = \sqrt{g/\overset{*}{g}}\, \mathbf{H} \cdot \mathbf{F}^{-1} \cdot \mathbf{q} = \sqrt{g/\overset{*}{g}}\, \mathbf{H} \cdot \overset{\circ}{\mathbf{q}},$$

$$\overset{*}{\nabla}\theta = \overset{*}{\mathbf{r}}{}^{i} \frac{\partial \theta}{\partial X^i} = \overset{*}{\mathbf{r}}{}^{i} \otimes \overset{\circ}{\mathbf{r}}_i \cdot \overset{\circ}{\mathbf{r}}{}^{k} \frac{\partial \theta}{\partial X^k} = \mathbf{H}^{-1\mathrm{T}} \cdot \overset{\circ}{\nabla}\theta. \tag{4.740}$$

Let the principle of material symmetry be valid, then relations (4.738a) and (4.738b) hold. On substituting (4.740) into (4.738b), we obtain

$$\sqrt{\overset{\circ}{g}/\overset{*}{g}}\, \mathbf{H} \cdot \overset{\circ}{\mathbf{q}} = -\overset{\circ}{\boldsymbol{\lambda}} \cdot \mathbf{H}^{-1\mathrm{T}} \cdot \overset{\circ}{\nabla}\theta, \tag{4.741}$$

or

$$\overset{\circ}{\mathbf{q}} = -\sqrt{\overset{*}{g}/\overset{\circ}{g}}\, (\mathbf{H}^{-1} \cdot \overset{\circ}{\boldsymbol{\lambda}} \cdot \mathbf{H}^{-1\mathrm{T}}) \cdot \overset{\circ}{\nabla}\theta. \tag{4.742}$$

Comparing (4.738a) with (4.742), we see that in this case the relation (4.739) must hold for the heat conductivity tensor $\overset{\circ}{\boldsymbol{\lambda}}$.

Prove the converse assertion. Let the condition (4.739) be satisfied. Substituting the condition into (4.736), we obtain formula (4.742). Performing the transformations in the reverse order, we get formulae (4.738a) and (4.738b). ▲

Theorem 4.63 still holds for the remaining models $A_n$, $C_n$ and $D_n$. For them the relation (4.739) has the forms

$$\overset{\circ}{\boldsymbol{\lambda}}(\overset{(n)}{\mathbf{C}}{}^{*},\ \theta) = \sqrt{\overset{\circ}{g}/\overset{*}{g}}\, \mathbf{H} \cdot \overset{\circ}{\boldsymbol{\lambda}}(\overset{(n)}{\mathbf{C}},\ \theta) \cdot \mathbf{H}^{\mathrm{T}},$$

$$\overset{\circ}{\boldsymbol{\lambda}}(\overset{(n)}{\mathbf{A}}{}^{*},\ \mathbf{O}^{*},\ \theta) = \sqrt{\overset{\circ}{g}/\overset{*}{g}}\, \mathbf{H} \cdot \overset{\circ}{\boldsymbol{\lambda}}(\overset{(n)}{\mathbf{A}},\ \mathbf{O},\ \theta) \cdot \mathbf{H}^{\mathrm{T}},$$

$$\overset{\circ}{\boldsymbol{\lambda}}(\overset{(n)}{\mathbf{g}}{}^{*},\ \mathbf{O}^{*},\ \theta) = \sqrt{\overset{\circ}{g}/\overset{*}{g}}\, \mathbf{H} \cdot \overset{\circ}{\boldsymbol{\lambda}}(\overset{(n)}{\mathbf{g}},\ \mathbf{O},\ \theta) \cdot \mathbf{H}^{\mathrm{T}}. \tag{4.743}$$

**Theorem 4.64.** *The Fourier law (4.734) in the spatial description identically satisfies the principle of material symmetry for each symmetry group* $G_s$.

▼ Indeed, for any $H$-transformation of a reference configuration $\overset{\circ}{\mathcal{K}} \to \overset{*}{\mathcal{K}}$, since the vectors $\mathbf{q}$ and $\nabla\theta$ are always $H$-invariant, relation (4.734) retains its form with the same heat conductivity tensor $\boldsymbol{\lambda}$ for each tensor $\mathbf{H}$. ▲

Certainly, there is no discrepancy between relation (4.739) for the tensor $\overset{\circ}{\boldsymbol{\lambda}}$ during the passage from $\overset{\circ}{\mathcal{K}}$ to $\overset{*}{\mathcal{K}}$ and invariability of the tensor $\boldsymbol{\lambda}$, because formula (4.739) together with (4.737) and formula (4.188) of transformation of the deformation gradient $\mathbf{F}$ ensure that the tensor $\boldsymbol{\lambda}$ be independent of the choice of reference configuration $\overset{*}{\mathcal{K}}$:

$$\boldsymbol{\lambda} = \sqrt{\overset{\circ}{g}/g}\,\mathbf{F}\cdot\overset{\circ}{\boldsymbol{\lambda}}\cdot\mathbf{F}^{\mathrm{T}} = \sqrt{\overset{\circ}{g}/g}\,\overset{*}{\mathbf{F}}\cdot\mathbf{H}\cdot\overset{\circ}{\boldsymbol{\lambda}}\cdot\mathbf{H}^{\mathrm{T}}\cdot\overset{*}{\mathbf{F}}{}^{\mathrm{T}} = \sqrt{\overset{*}{g}/g}\,\overset{*}{\mathbf{F}}\cdot\overset{*}{\boldsymbol{\lambda}}\cdot\overset{*}{\mathbf{F}}{}^{\mathrm{T}} = (\boldsymbol{\lambda})^{*}.$$

Here $(\boldsymbol{\lambda})^{*}$ is the tensor $\boldsymbol{\lambda}$ defined in configuration $\overset{*}{\mathcal{K}}$, which, as it was expected, coincides with the tensor $\boldsymbol{\lambda}$ itself (it is not to be confused with $\overset{*}{\boldsymbol{\lambda}}$, which is the tensor $\overset{\circ}{\boldsymbol{\lambda}}$ defined in configuration $\overset{*}{\mathcal{K}}$).

### 4.12.3 *Corollary of the Principle of Material Indifference for the Fourier Law*

The heat flux vector $\overset{\circ}{\mathbf{q}}$ and temperature gradient $\overset{\circ}{\nabla}\theta$ are defined in a reference configuration $\overset{\circ}{\mathcal{K}}$, therefore they remain unchanged at rigid motions from $\mathcal{K}$ to $\mathcal{K}'$ (i.e. $\overset{\circ}{\mathbf{q}}{}' = \overset{\circ}{\mathbf{q}}$, $(\overset{\circ}{\nabla}\theta)' = \overset{\circ}{\nabla}\theta$). Hence, the Fourier law (4.736) in the material description also remains unchanged under the transformation from $\mathcal{K}$ to $\mathcal{K}'$:

$$\overset{\circ}{\mathbf{q}}{}' = -\overset{\circ}{\boldsymbol{\lambda}}\cdot(\overset{\circ}{\nabla}\theta)' \tag{4.744}$$

for any orthogonal rotation tensor $\mathbf{Q}$. This means that relation (4.736) of the Fourier law satisfies the principle of material indifference (4.603), (4.605) with the same heat conductivity tensor $\overset{\circ}{\boldsymbol{\lambda}}$, i.e.

$$\overset{\circ}{\boldsymbol{\lambda}}{}' = \overset{\circ}{\boldsymbol{\lambda}}. \tag{4.745}$$

For the Fourier law (4.734) in the spatial description, the situation is different.

**Theorem 4.65.** *The Fourier law (4.734) in the spatial description satisfies the principle of material indifference (4.603), (4.605) (i.e. the relation*

$$\mathbf{q}' = -\boldsymbol{\lambda}\cdot(\nabla\theta)' \tag{4.746}$$

*follows from (4.734) for any rigid motion $\mathcal{K} \longrightarrow \mathcal{K}'$) if and only if the heat conductivity tensor is both R-invariant and R-indifferent:*

$$\boldsymbol{\lambda}' = \boldsymbol{\lambda}, \qquad \boldsymbol{\lambda}' = \mathbf{Q}^{\mathrm{T}}\cdot\boldsymbol{\lambda}\cdot\mathbf{Q}. \tag{4.747}$$

Remind that tensors satisfying the condition (4.747) for any orthogonal tensor $\mathbf{Q}$, according to (4.290), are called *indifferent relative to the full orthogonal group*.
▼ The heat flux vector $\mathbf{q}$ is $R$-indifferent (just as the tensor $\mathbf{T}$). To prove the fact, one should use relation (3.111) and take into account that the scalar function of heat influx due to surface sources $q_\Sigma$ is $R$-invariant: $q'_\Sigma = q_\Sigma$, and the normal vector $\mathbf{n}$ is $R$-indifferent, then

$$-q'_\Sigma = \mathbf{q}' \cdot \mathbf{n}' = \mathbf{q}' \cdot \mathbf{Q}^{\mathrm{T}} \cdot \mathbf{n} = -q_\Sigma = \mathbf{q} \cdot \mathbf{n}. \tag{4.748}$$

Thus, we obtain

$$\mathbf{q}' = \mathbf{Q}^{\mathrm{T}} \cdot \mathbf{q}. \tag{4.749}$$

The gradient of a scalar is also $R$-indifferent:

$$(\nabla\theta)' = \mathbf{r}^{i\prime} \frac{\partial\theta}{\partial X^i} = \mathbf{Q}^{\mathrm{T}} \cdot \mathbf{r}^i \frac{\partial\theta}{\partial X^i} = \mathbf{Q}^{\mathrm{T}} \cdot \nabla\theta. \tag{4.750}$$

Multiplying the relation (4.734) by $\mathbf{Q}^{\mathrm{T}}$, we obtain

$$\mathbf{Q}^{\mathrm{T}} \cdot \mathbf{q} = -\mathbf{Q}^{\mathrm{T}} \cdot \boldsymbol{\lambda} \cdot \nabla\theta = -\mathbf{Q}^{\mathrm{T}} \cdot \boldsymbol{\lambda} \cdot \mathbf{Q} \cdot \mathbf{Q}^{\mathrm{T}} \cdot \nabla\theta, \tag{4.751}$$

or

$$\mathbf{q}' = -\boldsymbol{\lambda}' \cdot \nabla'\theta, \tag{4.752}$$

$$\boldsymbol{\lambda}' = \mathbf{Q}^{\mathrm{T}} \cdot \boldsymbol{\lambda} \cdot \mathbf{Q}. \tag{4.753}$$

As follows from (4.752), the principle of material indifference holds if and only if the tensor $\boldsymbol{\lambda}$ is indifferent, i.e. $\boldsymbol{\lambda}' = \boldsymbol{\lambda}$. ▲

Notice that relations (4.747) are consistent with (4.745): substituting formulae (4.747), (4.559), and (4.560) into (4.737), we obtain

$$\begin{aligned} \overset{\circ}{\boldsymbol{\lambda}}{}' &= \sqrt{g/\overset{\circ}{g}}\, \mathbf{F}^{-1\prime} \cdot \boldsymbol{\lambda}' \cdot \mathbf{F}^{-1\mathrm{T}\prime} = \sqrt{g/\overset{\circ}{g}}\, \mathbf{F}^{-1} \cdot \mathbf{Q} \cdot \boldsymbol{\lambda}' \cdot \mathbf{Q}^{\mathrm{T}} \cdot \mathbf{F}^{-1\mathrm{T}} \\ &= \sqrt{g/\overset{\circ}{g}}\, \mathbf{F}^{-1} \cdot \boldsymbol{\lambda} \cdot \mathbf{F}^{-1\mathrm{T}} = \overset{\circ}{\boldsymbol{\lambda}}, \end{aligned} \tag{4.754}$$

that is exactly the relation (4.745).

### 4.12.4 The Fourier Law for Fluids

For fluids, one usually use the Fourier law in the spatial description (4.734).

According to Theorem 4.64, the relation (4.734) for any tensor $\boldsymbol{\lambda}$ satisfies identically the principle of material symmetry for each symmetry group $\overset{\circ}{G}_s$, including the unimodular group $\overset{\circ}{G}_s = U$ corresponding to a fluid.

To satisfy the principle of material indifference, the tensor $\boldsymbol{\lambda}$ in (4.734) must be indifferent relative to the full orthogonal group $G_s = I$, i.e. the relations (4.747) must hold.

As shown in Sect. 4.8.3 (see (4.293)), Eq. (4.747) for the group $I$ has an unique solution being the spherical tensor $\boldsymbol{\lambda}$:

$$\boldsymbol{\lambda} = \lambda \mathbf{E}, \qquad \lambda > 0, \tag{4.755}$$

where $\lambda$ is the constant called the *heat conductivity coefficient.*

Substitution of (4.755) into (4.734) gives the Fourier law for fluids

$$\mathbf{q} = -\lambda \nabla \theta. \tag{4.756}$$

With the help of Eq. (4.737) we can write the heat conductivity tensor $\overset{\circ}{\boldsymbol{\lambda}}$ for fluids as follows:

$$\overset{\circ}{\boldsymbol{\lambda}} = \lambda \sqrt{g/\overset{\circ}{g}}\, \mathbf{F}^{-1} \cdot \mathbf{F}^{-1\mathrm{T}} = \lambda \sqrt{g/\overset{\circ}{g}}\, \mathbf{G}^{-1}. \tag{4.757}$$

This tensor is not spherical, but it satisfies Eq. (4.739).

The Fourier law (4.736) in the material description for fluids has the form

$$\overset{\circ}{\mathbf{q}} = -\lambda \sqrt{g/\overset{\circ}{g}}\, \mathbf{G}^{-1} \cdot \overset{\circ}{\nabla} \theta. \tag{4.758}$$

### *4.12.5 The Fourier Law for Solids*

For solids, one usually apply the Fourier law in the material description (4.736).

As noted in Sect. 4.12.3, for any tensor $\overset{\circ}{\boldsymbol{\lambda}}$ this relation satisfies identically the principle of material indifference. To satisfy the principle of material symmetry, by Theorem 4.63, the tensor $\overset{\circ}{\boldsymbol{\lambda}}$ must be indifferent relative to a group $\overset{\circ}{G}_s$ corresponding to the symmetry group of a considered solid, i.e. the relation (4.739) must hold.

According to the results of Sect. 4.8.3 (see formulae (4.293)), we find that:

- if $\overset{\circ}{G}_s = I$ (an isotropic solid), then the tensor $\overset{\circ}{\boldsymbol{\lambda}}$ has the only independent component $\overset{\circ}{\lambda}$:

$$\overset{\circ}{\boldsymbol{\lambda}} = \overset{\circ}{\lambda} \mathbf{E}, \tag{4.759}$$

- if $\overset{\circ}{G}_s = \widehat{T}_3$ (a transversely isotropic solid with the transverse isotropy vector $\widehat{\mathbf{c}}_3$), then the tensor $\overset{\circ}{\boldsymbol{\lambda}}$ has two independent components – $\overset{\circ}{\lambda}_1$ and $\overset{\circ}{\lambda}_2$:

$$\overset{\circ}{\boldsymbol{\lambda}} = \overset{\circ}{\lambda}_1 \widehat{\mathbf{c}}_3^2 + \overset{\circ}{\lambda}_2 \mathbf{E}, \tag{4.760}$$

- if $\overset{\circ}{G}_s = \widehat{O}$ (an orthotropic solid with the principal orthotropy basis $\widehat{\mathbf{c}}_\gamma$), then the tensor $\overset{\circ}{\boldsymbol{\lambda}}$ has three independent components – $\overset{\circ}{\lambda}_\gamma$:

$$\overset{\circ}{\boldsymbol{\lambda}} = \sum_{\gamma=1}^{3} \overset{\circ}{\lambda}_\gamma \widehat{\mathbf{c}}_\gamma^2. \qquad (4.761)$$

To find the tensor $\boldsymbol{\lambda}$ for solids one should substitute formulae (4.759)–(4.761) into (4.737). For example, for an isotropic continuum the tensor $\boldsymbol{\lambda}$ has the form

$$\boldsymbol{\lambda} = \sqrt{\overset{\circ}{g}/g}\, \mathbf{F} \cdot \overset{\circ}{\boldsymbol{\lambda}} \cdot \mathbf{F}^{\mathrm{T}} = \overset{\circ}{\lambda} \sqrt{\overset{\circ}{g}/g}\, \mathbf{F} \cdot \mathbf{F}^{\mathrm{T}}. \qquad (4.762)$$

On substituting expressions (4.759)–(4.761) into (4.736), we get the Fourier law for isotropic solids

$$\overset{\circ}{\mathbf{q}} = -\overset{\circ}{\lambda} \mathbf{E}, \qquad (4.763)$$

the Fourier law for transversely isotropic solids

$$\overset{\circ}{\mathbf{q}} = -(\overset{\circ}{\lambda}_1 \widehat{\mathbf{c}}_3^2 + \overset{\circ}{\lambda}_2 \mathbf{E}), \qquad (4.764)$$

and the Fourier law for orthotropic solids

$$\overset{\circ}{\mathbf{q}} = -\sum_{\gamma=1}^{3} \overset{\circ}{\lambda}_\gamma \widehat{\mathbf{c}}_\gamma^2. \qquad (4.765)$$

# Chapter 5
# Relations at Singular Surfaces

## 5.1 Relations at a Singular Surface in the Material Description

### 5.1.1 Singular Surfaces

Up to now we considered the case when all functions appearing in the balance laws: $\rho$, $\mathbf{u}$, $\mathbf{v}$, $\mathbf{T}$, $\mathbf{F}$, $\mathbf{f}$ etc. are continuously differentiable functions of coordinates $X^i$ (or of $x^i$) and time $t$. However, in practice one often meets with problems, where this condition is violated. For example, for the phenomena of impact, explosion, combustion etc., a part of the indicated functions in a domain $V$ considered can suffer a jump discontinuity across some surface $S$ (Fig. 5.1).

As a *singular surface* we can consider an interface $S$ between two contacting media (Fig. 5.2).

Besides differential equations in a domain $V$, mathematical formulation of problems in continuum mechanics should involve conditions on the surface bounding the domain $V$ (*boundary conditions*). This boundary can also be considered as a singular surface for the functions.

The integral balance laws still hold for a volume $V$ containing a singular surface $S$; but differential equations of the corresponding laws are not valid across such surface $S$, because in deriving the equations we have essentially used the condition of continuous differentiability of the functions inside $V$. Therefore, we need to establish appropriate consequences of the balance laws for the case when there is a singular surface for the functions. This is a purpose of the chapter.

### 5.1.2 The First Classification of Singular Surfaces

Consider a continuum occupying a volume $V$ in $\mathcal{K}$. We assume that there is a surface $S$ dividing the volume $V$ into two subdomains $V_+$ and $V_-$ such that inside $V_+$ and $V_-$ all functions ($\rho$, $\mathbf{v}$, $\mathbf{F}$ etc.) considered are continuously differentiable. However, during the passage from $V_+$ to $V_-$ across $S$ some of the functions can suffer a jump discontinuity. Consider the set $M_t$ of material points belonging to the singular

Y.I. Dimitrienko, *Nonlinear Continuum Mechanics and Large Inelastic Deformations*, Solid Mechanics and Its Applications 174, DOI 10.1007/978-94-007-0034-5_5,

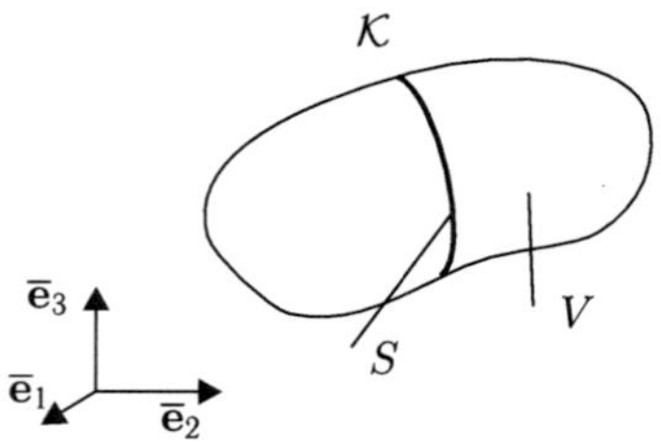


**Fig. 5.1** The singular surface $S$ in domain $V$

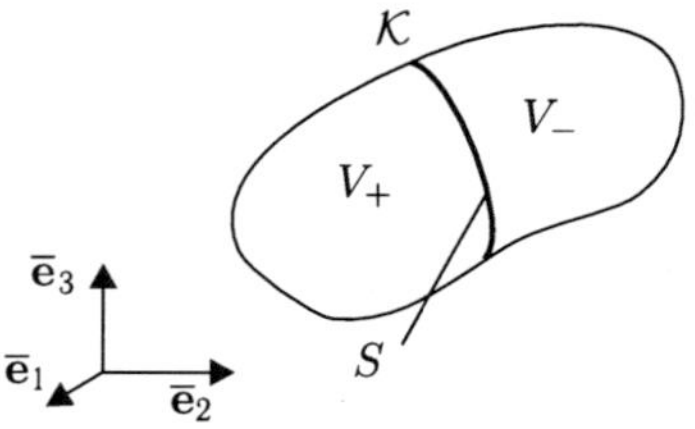


**Fig. 5.2** The interface $S$ between two media $V_+$ and $V_-$

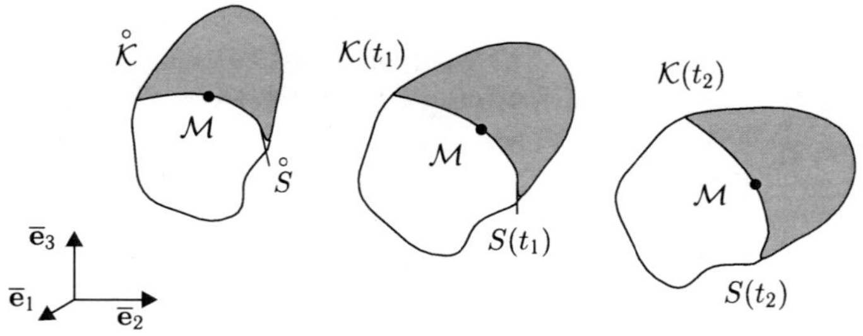


**Fig. 5.3** There is no transition of material points across a singular surface

surface $S(t)$ at a time $t > 0$. This set $M_t$ in a reference configuration $\mathring{\mathcal{K}}$ belongs to a surface $\mathring{S}(t)$, which, in general, depends on $t$ and divides $\mathring{V}$ into two subdomains $V_+$ and $V_-$.

At different times $t$ ($t_1 \leqslant t \leqslant t_2$), positions of the singular surface $S(t)$ in the actual configuration may be distinct, and there is an alternative:

- Sets $M_t$, corresponding to surfaces $S(t)$ $\forall t \in [t_1, t_2]$, coincide
- Not all sets $M_t$, corresponding to surfaces $S(t)$ at considered times $t$ ($t \in [t_1, t_2]$), are coincident

In the first case, one can say that material points *do not cross* the singular surface $S(t)$ $\forall t \in [t_1, t_2]$; in the second case, there occurs a *transition of material points across the singular surface*.

If there is no transition of material points across the interface $S(t)$, $\forall t \in [t_1, t_2]$, then the position of the surface $\mathring{S}(t)$ in $\mathring{\mathcal{K}}$ $\forall t \in [t_1, t_2]$ remains unchanged (Fig. 5.3). While material points cross a singular surface at $t \in [t_1, t_2]$, the location of the surface $\mathring{S}(t)$ in the reference configuration varies (Fig. 5.4).

Let us give now a classification of singular surfaces (Fig. 5.5).

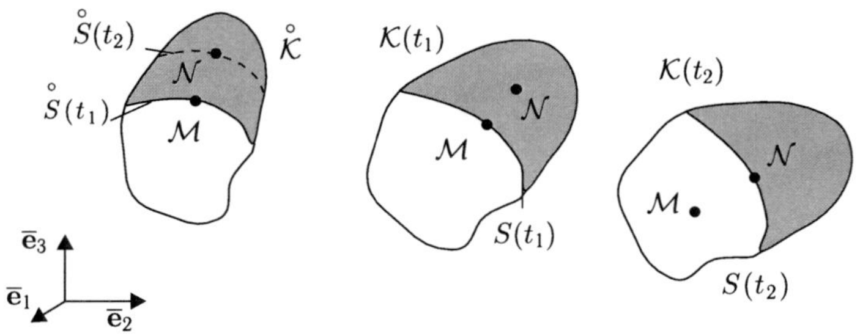


**Fig. 5.4** Material points cross a singular surface

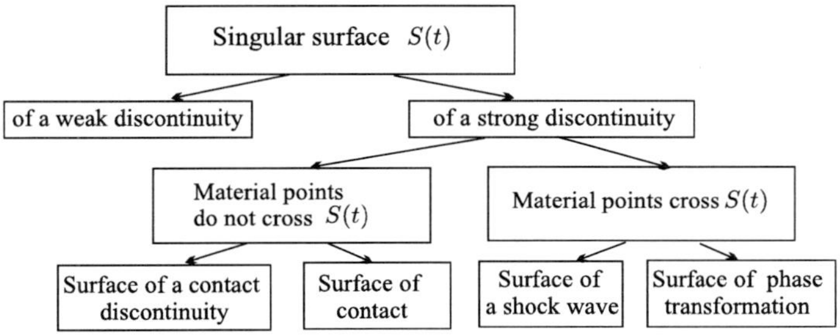


**Fig. 5.5** Classification of singular surfaces

**Definition 5.1.** A singular surface $S(t)$, across which unknown functions $\rho$, $\mathbf{u}$, $\mathbf{v}$, $\mathbf{T}$, $\mathbf{F}$ etc. suffer jump discontinuities, is called a *surface of a strong discontinuity*. If these functions remain continuous across $S(t)$ and only their first derivatives with respect to time $t$ and coordinates ($\boldsymbol{\nabla}\rho$, $\boldsymbol{\nabla}\otimes\mathbf{u}$, $\boldsymbol{\nabla}\otimes\mathbf{v}$ etc.) suffer jump discontinuities, then $S(t)$ is called a *surface of a weak discontinuity*.

The transition of material points can occur only across surfaces of a strong discontinuity.

**Definition 5.2.** A surface of a strong discontinuity $S(t)$, across which there is no transition of material points $\forall t \in [t_1, t_2]$, is called a *surface of a contact discontinuity* if constitutive equations in domains $V_+(t)$ and $V_-(t)$ are the same at all times $t$ considered; otherwise, it is called a *surface of contact*.

**Definition 5.3.** A surface of a strong discontinuity $S(t)$, across which there is a transition of material points at $t \in [t_1, t_2]$, is called a *surface of a shock wave* if constitutive equations are the same at all times $t$ considered; otherwise, $S(t)$ is called a *surface of phase transformation*.

Besides the classification introduced above, singular surfaces can be split into *coherent* and *incoherent* ones.

**Definition 5.4.** A singular surface $S(t)$, across which the radius-vectors $\mathbf{x}$ of material points $\mathcal{M} \in S(t)$ suffer jump discontinuities, is called *incoherent*; otherwise, the surface is called *coherent*.

For an incoherent surface $S(t)$, a local neighborhood $dV$ of every point $\mathcal{M}$, belonging to the surface $S(t)$ at some time $t$, at the time $t + \Delta t$ is divided into two half-neighborhoods $dV_+$ and $dV_-$, moving apart one from another, because if in $\overset{\circ}{\mathcal{K}}$ the radius-vector $\overset{\circ}{\mathbf{x}}(\mathcal{M})$ of the point $\mathcal{M}$ is continuous, then the radius-vector $\mathbf{x}(\mathcal{M}, t + \Delta t)$ of this point is no longer continuous:

$$[\overset{\circ}{\mathbf{x}}] = 0, \qquad [\mathbf{x}] \neq 0. \tag{5.1}$$

For a coherent surface $S(t)$, local half-neighborhoods $dV_+$ and $dV_-$ of its every point $\mathcal{M} \in S(t)$ at any time $t$ do not move apart one from another, and the jump of the radius-vector $\mathbf{x}$ is zero if $[\overset{\circ}{\mathbf{x}}] = 0$:

$$[\overset{\circ}{\mathbf{x}}] = 0, \qquad [\mathbf{x}] = 0. \tag{5.2}$$

The example of an incoherent surface is a surface of contact (an interface) between an ideal fluid and a solid: an ideal fluid slips along a solid without grip.

If at time $t$ we choose two contacting points $\mathcal{M}_+$ and $\mathcal{M}_-$ of an incoherent surface $S(t)$, then their radius-vectors $\mathbf{x}_+$ and $\mathbf{x}_-$ coincide but the radius-vectors $\overset{\circ}{\mathbf{x}}_+$ and $\overset{\circ}{\mathbf{x}}_-$ in $\overset{\circ}{\mathcal{K}}$ may be not coincident, i.e. instead of (5.1) the following relations may hold

$$[\overset{\circ}{\mathbf{x}}] \neq 0, \qquad [\mathbf{x}] = 0.$$

It is evident that an incoherent surface is a surface of a strong discontinuity.

### 5.1.3 *Axiom on the Class of Functions across a Singular Surface*

Let us consider now a surface of a strong discontinuity $\overset{\circ}{S}(t)$ dividing a volume $\overset{\circ}{V}$ in $\overset{\circ}{\mathcal{K}}$ into two subdomains $\overset{\circ}{V}_+$ and $\overset{\circ}{V}_-$.

**Definition 5.5.** Let there be a function $A(\overset{\circ}{\mathbf{x}}, t)$ defined in a domain $\overset{\circ}{V}$ which contains a singular surface $\overset{\circ}{S}$. The value

$$[A] = A_+\big|_{\overset{\circ}{S}} - A_-\big|_{\overset{\circ}{S}}, \tag{5.3}$$

where

$$A_{\pm}\big|_{\overset{\circ}{S}} = \lim_{\substack{\overset{\circ}{\mathbf{x}} \to \overset{\circ}{\mathbf{x}}_{\Sigma} \\ \overset{\circ}{\mathbf{x}} \in \overset{\circ}{V}_{\pm},\ \overset{\circ}{\mathbf{x}}_{\Sigma} \in \overset{\circ}{S}}} A_{\pm}(\overset{\circ}{\mathbf{x}}, t),$$

is called the *jump of the function across the surface* $\overset{\circ}{S}$.

If a surface $S(t)$ is coherent, then from (5.2) it follows that the functions $x^i(X^i, t)$ do not suffer jump discontinuities across the surface $\overset{\circ}{S}(t)$:

$$x^i(X^j\big|_{\overset{\circ}{S}_+}, t) = x^i(X^j\big|_{\overset{\circ}{S}_-}, t).$$

Then in (5.3) we can always pass from coordinates $\overset{\circ}{x}{}^i$ to $x^i$ and consider the jump of a function across a corresponding singular surface $S(t)$ in $\mathcal{K}$:

$$[A] = A_+\big|_S - A_-\big|_S, \tag{5.4}$$

where

$$A_{\pm}\big|_S = \lim_{\substack{x^i \to x^i_{\Sigma} \\ x^i \in V_{\pm},\ x^i_{\Sigma} \in S(t)}} A(x^i, t).$$

**Axiom 17.** *For a continuum containing in* $\overset{\circ}{\mathcal{K}}$ *a singular surface* $\overset{\circ}{S}(t)$, *which divides the volume* $\overset{\circ}{V}$ *into two parts* $\overset{\circ}{V}_+$ *and* $\overset{\circ}{V}_-$, *the functions* $\overset{\circ}{\rho}$, $\overset{\circ}{A}_{\alpha}$ *and* $\overset{\circ}{B}_{\alpha}$ *appearing in the balance equation system (3.310) are assumed to be smooth in* $\overset{\circ}{V}_+$ *and* $\overset{\circ}{V}_-$:

$$\overset{\circ}{A}_{\alpha} = \begin{cases} \overset{\circ}{A}_{\alpha+}, & \overset{\circ}{\mathbf{x}} \in \overset{\circ}{V}_+, \\ \overset{\circ}{A}_{\alpha-}, & \overset{\circ}{\mathbf{x}} \in \overset{\circ}{V}_-, \end{cases} \qquad \overset{\circ}{B}_{\alpha} = \begin{cases} \overset{\circ}{B}_{\alpha+}, & \overset{\circ}{\mathbf{x}} \in \overset{\circ}{V}_+, \\ \overset{\circ}{B}_{\alpha-}, & \overset{\circ}{\mathbf{x}} \in \overset{\circ}{V}_-, \end{cases} \tag{5.5}$$

*and across the singular surface* $\overset{\circ}{S}(t)$ *they can suffer a finite jump discontinuity, i.e.*

$$[\overset{\circ}{A}_{\alpha}] < +\infty, \qquad [\overset{\circ}{B}_{\alpha}] < +\infty. \tag{5.6}$$

*The functions* $C_{\alpha}$ *included in the system (3.310) are also assumed to be smooth in* $\overset{\circ}{V}_+$ *and* $\overset{\circ}{V}_-$, *and across the singular surface they can suffer an unbounded jump discontinuity of the type as the* $\delta$*-function has:*

$$\overset{\circ}{\rho} C_{\alpha} = \overset{\circ}{\rho}\widetilde{C}_{\alpha} + \overset{\circ}{C}_{\alpha\Sigma}\delta(\overset{\circ}{\mathbf{x}}_{\Sigma} - \overset{\circ}{\mathbf{x}}), \quad \widetilde{C}_{\alpha} = \begin{cases} C_{\alpha+}, & \overset{\circ}{\mathbf{x}} \in \overset{\circ}{V}_+, \\ C_{\alpha-}, & \overset{\circ}{\mathbf{x}} \in \overset{\circ}{V}_-. \end{cases} \tag{5.7}$$

*Here $\overset{\circ}{C}_{\alpha\Sigma}(\overset{\circ}{\mathbf{x}}_\Sigma, t)$ is a finite, smooth across $\overset{\circ}{S}(t)$ function satisfying the condition*

$$\int\limits_{\overset{\circ}{V}} \overset{\circ}{C}_{\alpha\Sigma}\, \delta(\overset{\circ}{\mathbf{x}}_\Sigma - \overset{\circ}{\mathbf{x}})\, d\overset{\circ}{V} = \int\limits_{\overset{\circ}{S}} \overset{\circ}{C}_{\alpha\Sigma}\, d\overset{\circ}{\Sigma}. \tag{5.8}$$

*The function $\overset{\circ}{C}_{4\Sigma}$ consists of the two summands: the jump of entropy production by external sources $\overset{\circ}{\bar{C}}_{4\Sigma}$ and the jump of entropy production by internal sources $\overset{\circ}{C}{}^*_{4\Sigma}$, which is assumed to be always non-negative:*

$$\overset{\circ}{C}_{4\Sigma} = \overset{\circ}{\bar{C}}_{4\Sigma} + \overset{\circ}{C}{}^*_{4\Sigma}, \quad \overset{\circ}{C}{}^*_{4\Sigma} \geqslant 0. \tag{5.8a}$$

To apply Axiom 17 to a continuum containing not one but several singular surfaces $\overset{\circ}{S}_\alpha$, one should divide the whole domain $\overset{\circ}{V}$ into subdomains $\overset{\circ}{V}_\alpha$ ($\alpha = 1, 2, \ldots$), each of which contains only one singular surface $\overset{\circ}{S}_\alpha$.

### *5.1.4 The Rule of Differentiation of a Volume Integral in the Presence of a Singular Surface*

Let us establish a rule of differentiation of a volume integral for the case when the integration domain $\overset{\circ}{V}$ contains a singular surface $\overset{\circ}{S}(t)$ for the functions $\overset{\circ}{\rho}\overset{\circ}{A}_\alpha(\overset{\circ}{x}{}^i, t)$ defined in $\overset{\circ}{V}$.

According to Axiom 17, determine this derivative for each of the volumes $\overset{\circ}{V}_+(t)$ and $\overset{\circ}{V}_-(t)$:

$$\frac{\partial}{\partial t}\int\limits_{\overset{\circ}{V}} \overset{\circ}{\rho}\overset{\circ}{A}_\alpha\, d\overset{\circ}{V} = \frac{\partial}{\partial t}\int\limits_{\overset{\circ}{V}_-(t)} \overset{\circ}{\rho}_-\overset{\circ}{A}_{\alpha-}\, d\overset{\circ}{V} + \frac{\partial}{\partial t}\int\limits_{\overset{\circ}{V}_+(t)} \overset{\circ}{\rho}_+\overset{\circ}{A}_{\alpha+}\, d\overset{\circ}{V}. \tag{5.9}$$

Let us use the definition of the derivative of a scalar function with respect to a scalar argument (time):

$$\frac{\partial}{\partial t}\int\limits_{\overset{\circ}{V}_-(t)} \overset{\circ}{\rho}_-\overset{\circ}{A}_{\alpha-}(\overset{\circ}{\mathbf{x}}, t)\, d\overset{\circ}{V}$$

$$= \lim_{\Delta t\to 0}\frac{1}{\Delta t}\left(\int\limits_{\overset{\circ}{V}_-+\Delta\overset{\circ}{V}} \overset{\circ}{\rho}_-\overset{\circ}{A}_{\alpha-}(\overset{\circ}{\mathbf{x}}, t+\Delta t)\, d\overset{\circ}{V} - \int\limits_{\overset{\circ}{V}_-} \overset{\circ}{\rho}_-\overset{\circ}{A}_{\alpha-}(\overset{\circ}{\mathbf{x}}, t)\, d\overset{\circ}{V}\right)$$

$$= \lim_{\Delta t \to 0} \left( \int\limits_{\overset{\circ}{V}_-} \overset{\circ}{\rho}_- \frac{\overset{\circ}{A}_{\alpha-}(\overset{\circ}{\mathbf{x}}, t + \Delta t) - \overset{\circ}{A}_{\alpha-}(\overset{\circ}{\mathbf{x}}, t)}{\Delta t} d\overset{\circ}{V} + \frac{1}{\Delta t} \int\limits_{\Delta \overset{\circ}{V}} \overset{\circ}{\rho}_- \overset{\circ}{A}_{\alpha-}(\overset{\circ}{\mathbf{x}}, t + \Delta t) d\overset{\circ}{V} \right)$$

$$= \int\limits_{\overset{\circ}{V}_-(t)} \overset{\circ}{\rho}_- \frac{\partial \overset{\circ}{A}_{\alpha-}}{\partial t} d\overset{\circ}{V} + \lim_{\Delta t \to 0} \int\limits_{\Delta \overset{\circ}{V}} \overset{\circ}{\rho}_- \overset{\circ}{A}_{\alpha-}(\overset{\circ}{\mathbf{x}}, t + \Delta t) \frac{1}{\Delta t} d\overset{\circ}{V}. \tag{5.10}$$

Here $\Delta \overset{\circ}{V}$ is the change in the volume $\overset{\circ}{V}_-$ due to transition of material points across the singular surface in $\overset{\circ}{\mathcal{K}}$: $\Delta \overset{\circ}{V} = \Delta \overset{\circ}{\mathbf{x}}_\Sigma \cdot \overset{\circ}{\mathbf{n}} \overset{\circ}{S}$, and $\Delta \overset{\circ}{\mathbf{x}}_\Sigma$ is the change in the radius-vector $\overset{\circ}{\mathbf{x}}$ of points belonging to the surface $\overset{\circ}{S}$.

The corresponding elementary volume $d\overset{\circ}{V}$ is determined as follows (Fig. 5.6):

$$d\overset{\circ}{V} = d\overset{\circ}{\mathbf{x}}_\Sigma \cdot \overset{\circ}{\mathbf{n}}\, d\overset{\circ}{\Sigma}, \tag{5.11}$$

where $\overset{\circ}{\mathbf{n}}$ is the normal directed from $\overset{\circ}{V}_-$ to $\overset{\circ}{V}_+$, and $d\overset{\circ}{\mathbf{x}}_\Sigma = d\overset{\circ}{\mathbf{x}}(X^i|_{\overset{\circ}{\Sigma}}, t)$ is the infinitesimal increment of the radius-vector $\overset{\circ}{\mathbf{x}}$ of points belonging to the singular surface. Then the vector

$$\overset{\circ}{\mathbf{c}} = d\overset{\circ}{\mathbf{x}}_\Sigma / dt \tag{5.12}$$

is the *velocity of the singular surface* $\overset{\circ}{S}$.

Due to the construction, the vector $\overset{\circ}{\mathbf{c}}$ is different from zero if and only if material points cross the singular surface $\overset{\circ}{S}(t)$ (and hence $S(t)$).

According to (5.11) and (5.12), the second summand in the last line of formula (5.10) can be written as follows:

$$\lim_{\Delta t \to 0} \int\limits_{\Delta \overset{\circ}{V}} \overset{\circ}{\rho}_- \overset{\circ}{A}_{\alpha-}(\overset{\circ}{\mathbf{x}}, t + \Delta t) \frac{d\overset{\circ}{V}}{\Delta t}$$

$$= \lim_{\Delta t \to 0} \int\limits_{\Delta \overset{\circ}{V}} \overset{\circ}{\rho}_- \overset{\circ}{A}_{\alpha-}(\overset{\circ}{\mathbf{x}}, t + \Delta t) \frac{d\overset{\circ}{\mathbf{x}}_\Sigma}{\Delta t} \cdot \overset{\circ}{\mathbf{n}}\, d\overset{\circ}{\Sigma} = \int\limits_{\overset{\circ}{\Sigma}} \overset{\circ}{\rho}_- \overset{\circ}{A}_{\alpha-} \overset{\circ}{\mathbf{v}}_\Sigma \cdot \overset{\circ}{\mathbf{n}}\, d\overset{\circ}{\Sigma}. \tag{5.13}$$

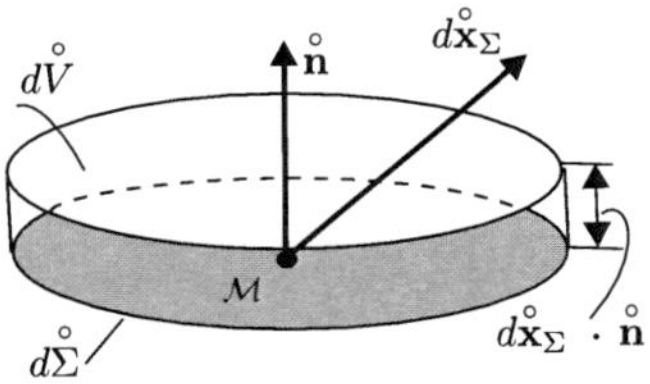


**Fig. 5.6** The infinitesimal change in a volume due to transition of material points across a singular surface

Then the expression for the derivative of the first integral in (5.9) takes the form

$$\frac{\partial}{\partial t}\int\limits_{\overset{\circ}{V}_-} \overset{\circ}{\rho}_- \overset{\circ}{A}_{\alpha-}(\overset{\circ}{\mathbf{x}}, t)\, d\overset{\circ}{V} = \int\limits_{\overset{\circ}{V}_-} \overset{\circ}{\rho}_- \frac{\partial \overset{\circ}{A}_{\alpha-}}{\partial t}\, d\overset{\circ}{V} + \int\limits_{\overset{\circ}{S}} \overset{\circ}{\rho}_- \overset{\circ}{A}_{\alpha-} \overset{\circ}{\mathbf{c}} \cdot \overset{\circ}{\mathbf{n}}\, d\overset{\circ}{\Sigma}. \qquad (5.14)$$

The derivative of the second integral in (5.9) is determined in a similar way:

$$\frac{\partial}{\partial t}\int\limits_{\overset{\circ}{V}_+(t)} \overset{\circ}{\rho}_+ \overset{\circ}{A}_{\alpha+}(\overset{\circ}{\mathbf{x}}, t)\, d\overset{\circ}{V}$$

$$= \lim_{\Delta t \to 0} \frac{1}{\Delta t}\left( \int\limits_{\overset{\circ}{V}_+ - \Delta \overset{\circ}{V}} \overset{\circ}{\rho}_+ \overset{\circ}{A}_{\alpha+}(\overset{\circ}{\mathbf{x}}, t+\Delta t)\, d\overset{\circ}{V} - \int\limits_{\overset{\circ}{V}_+} \overset{\circ}{\rho}_+ \overset{\circ}{A}_{\alpha+}(\overset{\circ}{\mathbf{x}}, t)\, d\overset{\circ}{V} \right)$$

$$= \int\limits_{\overset{\circ}{V}_+} \overset{\circ}{\rho}_+ \frac{\partial \overset{\circ}{A}_{\alpha+}}{\partial t}\, d\overset{\circ}{V} - \int\limits_{\overset{\circ}{S}} \overset{\circ}{\rho}_+ \overset{\circ}{A}_{\alpha+} \overset{\circ}{\mathbf{c}} \cdot \overset{\circ}{\mathbf{n}}\, d\overset{\circ}{\Sigma}. \qquad (5.15)$$

Here we have taken into account that normals $\overset{\circ}{\mathbf{n}}$ to the surface $\overset{\circ}{S}$ for the volumes $\overset{\circ}{V}_+$ and $\overset{\circ}{V}_-$ are opposite in direction.

On substituting (5.14) and (5.15) into (5.9), we obtain the following theorem.

**Theorem 5.1.** *For any functions $\overset{\circ}{A}_\alpha$ defined in a domain $\overset{\circ}{V}$, which contains a singular surface $\overset{\circ}{S}$, and satisfying Axiom 17, the rule of differentiation of a volume integral holds:*

$$\frac{\partial}{\partial t}\int\limits_{\overset{\circ}{V}} \overset{\circ}{\rho} \overset{\circ}{A}_\alpha\, d\overset{\circ}{V} = \int\limits_{\overset{\circ}{V}} \overset{\circ}{\rho} \frac{\partial \overset{\circ}{A}_\alpha}{\partial t}\, d\overset{\circ}{V} - \int\limits_{\overset{\circ}{S}} (\overset{\circ}{\rho}_+ \overset{\circ}{A}_{\alpha+} - \overset{\circ}{\rho}_- \overset{\circ}{A}_{\alpha-}) \overset{\circ}{D}\, d\overset{\circ}{\Sigma}, \qquad (5.16)$$

*where*

$$\overset{\circ}{D} = \overset{\circ}{\mathbf{c}} \cdot \overset{\circ}{\mathbf{n}} \qquad (5.16a)$$

*is the normal speed of propagation of the singular surface.*

According to Definition 5.4 and the fact that in $\overset{\circ}{\mathcal{K}}$ the total and partial derivatives with respect to $t$ coincide, Eq. (5.16) can be rewritten in the form

$$\frac{d}{dt}\int_{\overset{\circ}{V}} \overset{\circ}{\rho}\overset{\circ}{A}_\alpha \, d\overset{\circ}{V} = \int_{\overset{\circ}{V}} \overset{\circ}{\rho}\frac{d\overset{\circ}{A}_\alpha}{dt} \, d\overset{\circ}{V} - \int_{\overset{\circ}{S}} [\overset{\circ}{\rho}\overset{\circ}{A}_\alpha]\overset{\circ}{D} \, d\overset{\circ}{\Sigma}. \qquad (5.17)$$

### 5.1.5 Relations at a Coherent Singular Surface in $\overset{\circ}{\mathcal{K}}$

Let a singular surface $S(t)$ be coherent. Consider its image in $\overset{\circ}{\mathcal{K}}$, being the singular surface $\overset{\circ}{S}$ and a point $\mathcal{M} \in \overset{\circ}{S}$. Let us construct a special domain $\overset{\circ}{V}_h$ called the *neighborhood of the singular surface* and containing the point $\mathcal{M}$ (Fig. 5.7).

To construct the domain $\overset{\circ}{V}_h$ we choose the part $\overset{\circ}{\Sigma}_h$ of the surface $\overset{\circ}{S}$ so that $\mathcal{M} \in \overset{\circ}{\Sigma}_h$ and introduce curvilinear coordinates $X^i$ so that the curves $X^1, X^2$ belong to the surface $\overset{\circ}{S}$ and the curves $X^3$ be oriented normally to $\overset{\circ}{S}$.

The domain $\overset{\circ}{V}_h$ is bounded by the surface $\partial\overset{\circ}{V}_h$ consisting of the lateral surfaces $\overset{\circ}{\Sigma}_{h\pm}$ (their equations are $X^3 = \pm h/2$), the end surfaces $\overset{\circ}{\widetilde{\Sigma}}_{h\pm}$ (their equations are $0 < X^3 < h/2$ and $-h/2 < X^3 < 0$, and $X^I$, $I = 1, 2$, satisfy the equation of the contour $\overset{\circ}{\mathcal{L}}_\Sigma$ bounding the surface $\overset{\circ}{\Sigma}_h$: $X^I = X^I_\Sigma(s)$, $0 \leqslant s \leqslant s_0$).

Let us use the integral form of the balance laws (3.313), which hold for an arbitrary finite volume, including $\overset{\circ}{V}_h$:

$$\frac{\partial}{\partial t}\int_{\overset{\circ}{V}_h} \overset{\circ}{\rho}\overset{\circ}{A}_\alpha \, d\overset{\circ}{V} = \int_{\overset{\circ}{V}_h} \overset{\circ}{\rho}C_\alpha \, d\overset{\circ}{V} + \int_{\partial\overset{\circ}{V}_h} \overset{\circ}{\mathbf{n}} \cdot \overset{\circ}{B}_\alpha \, d\overset{\circ}{\Sigma}. \qquad (5.18)$$

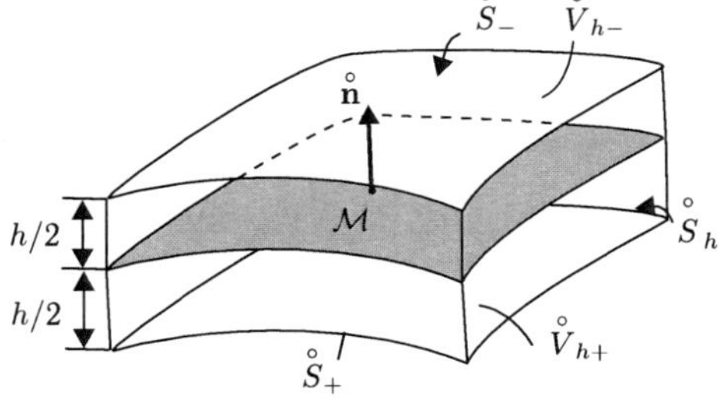


**Fig. 5.7** The domain for derivation of conditions at a singular surface

According to theorem (5.16) and conditions (5.7) and (5.8) for the functions $C_\alpha$, the balance equations (5.18) can be written in the form

$$\int\limits_{\overset{\circ}{V}_h} \overset{\circ}{\rho}\frac{\partial \overset{\circ}{A}_\alpha}{\partial t}\, d\overset{\circ}{V} - \int\limits_{\overset{\circ}{\Sigma}_h} [\overset{\circ}{\rho}\overset{\circ}{A}_\alpha]\overset{\circ}{D}\, d\overset{\circ}{\Sigma} - \int\limits_{\partial\overset{\circ}{V}_h} \overset{\circ}{\mathbf{n}}\cdot\overset{\circ}{B}_\alpha\, d\overset{\circ}{\Sigma} - \int\limits_{\overset{\circ}{V}_h} \overset{\circ}{\rho}\widetilde{C}_\alpha\, d\overset{\circ}{V} - \int\limits_{\overset{\circ}{\Sigma}_h} \overset{\circ}{C}_{\alpha\Sigma}\, d\overset{\circ}{\Sigma} = 0. \tag{5.19}$$

Here $[\overset{\circ}{\rho}\overset{\circ}{A}_\alpha]$ is the jump of the functions at the singular surface $\overset{\circ}{\Sigma}_h$:

$$[\overset{\circ}{\rho}\overset{\circ}{A}_\alpha] = (\overset{\circ}{\rho}_+\overset{\circ}{A}_{\alpha+} - \overset{\circ}{\rho}_-\overset{\circ}{A}_{\alpha-})_{\overset{\circ}{\Sigma}_h}. \tag{5.20}$$

Notice that the domain $\overset{\circ}{V}_h$ has been constructed in the special way, so we can consider the one-parametric family of such domains, numbering them by the parameter $h \in (0, h_0)$. Then, since, according to Axiom 17, the functions $\overset{\circ}{\rho}(\partial\overset{\circ}{A}_\alpha/\partial t)$, $\overset{\circ}{\mathbf{n}}\cdot\overset{\circ}{B}_\alpha$ and $C_\alpha$ are continuous in domains $\overset{\circ}{V}_{h\pm}$ and can suffer only a finite discontinuity across $\overset{\circ}{\Sigma}_h$ (a field of the normal $\overset{\circ}{\mathbf{n}}(X^i)$ at every point $X^i$ of the domain $\overset{\circ}{V}_{h\pm}$ is defined as follows: $\overset{\circ}{\mathbf{n}}(X^i) = \pm\overset{\circ}{\mathbf{n}}(X^I)$), and the functions $[\overset{\circ}{\rho}\overset{\circ}{A}_\alpha]D$ and $\overset{\circ}{C}_{\alpha\Sigma}$ are continuous across $\overset{\circ}{\Sigma}_h$, we can pass to the limit as $h \to 0$ (i.e. the domain $\overset{\circ}{V}_h$ shrinks to a point). So we have the equations

$$\lim_{h\to 0}\frac{1}{|\overset{\circ}{\Sigma}_h|}\int\limits_{\overset{\circ}{V}_h} \overset{\circ}{\rho}\frac{\partial A_\alpha}{\partial t}\, d\overset{\circ}{V} = 0, \tag{5.21}$$

$$\lim_{h\to 0}\frac{1}{|\overset{\circ}{\Sigma}_h|}\int\limits_{\partial\overset{\circ}{V}_h} \overset{\circ}{\mathbf{n}}\cdot\overset{\circ}{B}_\alpha\, d\overset{\circ}{\Sigma} = \overset{\circ}{\mathbf{n}}_+\cdot\overset{\circ}{B}_{\alpha+} + \overset{\circ}{\mathbf{n}}_-\cdot\overset{\circ}{B}_{\alpha-} = \overset{\circ}{\mathbf{n}}\cdot(\overset{\circ}{B}_{\alpha+} - \overset{\circ}{B}_{\alpha-}) = \overset{\circ}{\mathbf{n}}\cdot[\overset{\circ}{B}_\alpha], \tag{5.22}$$

$$\lim_{h\to 0}\frac{1}{|\overset{\circ}{\Sigma}_h|}\int\limits_{\overset{\circ}{\Sigma}_h} [\overset{\circ}{\rho}\overset{\circ}{A}_\alpha]\overset{\circ}{D}\, d\overset{\circ}{\Sigma} = [\overset{\circ}{\rho}\overset{\circ}{A}_\alpha]\overset{\circ}{D}, \tag{5.23}$$

$$\lim_{h\to 0}\frac{1}{|\overset{\circ}{\Sigma}_h|}\int\limits_{\overset{\circ}{V}_h} \overset{\circ}{\rho}\widetilde{C}_\alpha\, d\overset{\circ}{V} = 0, \quad \lim_{h\to 0}\frac{1}{|\overset{\circ}{\Sigma}_h|}\int\limits_{\overset{\circ}{\Sigma}_h} \overset{\circ}{C}_{\alpha\Sigma}\, d\overset{\circ}{\Sigma} = \overset{\circ}{C}_{\alpha\Sigma}. \tag{5.24}$$

Dividing each summand in (5.19) by $|\overset{\circ}{\Sigma}_h|$ and passing to the limit as $h \to 0$, with the help of formulae (5.21)–(5.24) we obtain the following theorem.

**Theorem 5.2.** *For the functions $\overset{\circ}{\rho}$, $\overset{\circ}{A}_\alpha$, $\overset{\circ}{B}_\alpha$ and $C_\alpha$:*

- *Defined in a domain $\overset{\circ}{V}$ which contains a coherent singular surface $\overset{\circ}{S}$,*
- *Satisfying Axiom 17,*
- *Satisfying the integral balance laws (3.313),*

*the following relations at the singular surface $\overset{\circ}{S}(t)$ hold:*

$$-[\overset{\circ}{\rho}\overset{\circ}{A}_\alpha]\overset{\circ}{D} = \overset{\circ}{\mathbf{n}}\cdot[\overset{\circ}{B}_\alpha] + \overset{\circ}{C}_{\alpha\Sigma}, \qquad \alpha = 1,\ldots,6. \tag{5.25}$$

### 5.1.6 Relation Between Velocities of a Singular Surface in $\overset{\circ}{\mathcal{K}}$ and $\mathcal{K}$

Define the velocity of a singular surface $S(t)$ in $\mathcal{K}$ as follows:

$$\mathbf{c} = \frac{d\mathbf{x}_\Sigma}{dt}\left(X^i|_{\overset{\circ}{S}(t)}, t\right), \qquad \mathbf{x}_\Sigma \in S(t). \tag{5.26}$$

Even if there is no transition of material points across a singular surface (i.e. $\overset{\circ}{S}$ is independent of $t$), the vector $\mathbf{c}$ may be different from zero due to the motion of material points $\mathcal{M}$ of the singular surface under the transformation from $\overset{\circ}{\mathcal{K}}$ to $\mathcal{K}$.

Let us establish now a relation between the velocities $\overset{\circ}{\mathbf{c}}$ and $\mathbf{c}$ of a singular surface in $\overset{\circ}{\mathcal{K}}$ and $\mathcal{K}$, respectively, defined by Eqs. (5.12) and (5.26). Consider the relations (5.2) at a coherent surface with use of the motion law (2.4) for material points of the surface $\overset{\circ}{S}(t)$. During the passage from coordinates $X^i$ to $\overset{\circ}{x}{}^i$ this law takes the form

$$\mathbf{x}_\Sigma\left(X^i|_{\overset{\circ}{S}}, t\right) = \mathbf{x}\left(\overset{\circ}{x}{}^i_\Sigma(X^j|_{\overset{\circ}{S}_+}, t), t\right) = \mathbf{x}\left(\overset{\circ}{x}{}^i_\Sigma(X^j|_{\overset{\circ}{S}_-}, t), t\right), \tag{5.27}$$

where $\overset{\circ}{\mathbf{x}}_\Sigma = \overset{\circ}{x}{}^i_\Sigma\bar{\mathbf{e}}_i$ are the radius-vectors of points of the singular surface $\overset{\circ}{S}$ in $\overset{\circ}{\mathcal{K}}$, depending not only on $X^j$ but also on $t$. Here we have used the condition of coherence of the surface $S(t)$: $X^i|_{\overset{\circ}{S}_+} = X^i|_{\overset{\circ}{S}_-}$.

On substituting (5.27) into the definition of $\mathbf{c}$ (5.26), we find that

$$\mathbf{c} = \frac{\partial\mathbf{x}_\Sigma}{\partial t}\left(\overset{\circ}{x}{}^i_\Sigma(X^j|_{\overset{\circ}{S}_+}, t), t\right) + \frac{\partial\mathbf{x}_+}{\partial\overset{\circ}{x}{}^i}\frac{\partial\overset{\circ}{x}{}^i_\Sigma}{\partial t} = \mathbf{v}_{\Sigma+} + \mathbf{F}_+\cdot\overset{\circ}{\mathbf{c}} = \mathbf{v}_{\Sigma-} + \mathbf{F}_-\cdot\overset{\circ}{\mathbf{c}}. \tag{5.28}$$

Here we have used the definition (2.181) of the velocity of a material point $\mathcal{M}$ at the singular surface:

$$\mathbf{v}_{\Sigma\pm} = \frac{\partial \mathbf{x}_\Sigma}{\partial t}\left(X^i\big|_{\overset{\circ}{S}_\pm}, t\right),$$

and also the definition (5.12) of the velocity of a singular surface in $\overset{\circ}{\mathcal{K}}$:

$$\overset{\circ}{\mathbf{c}} = \frac{\partial \overset{\circ}{\mathbf{x}}_\Sigma}{\partial t} = \frac{\partial \overset{\circ}{x}{}^i_\Sigma}{\partial t}\bar{\mathbf{e}}_i,$$

and introduced the notation for the deformation gradient at the singular surface $\overset{\circ}{S}(t)$:

$$\mathbf{F}_\pm = \left(\frac{\partial x^m}{\partial \overset{\circ}{x}{}^i}\right)\Big|_{\overset{\circ}{S}_\pm} \bar{\mathbf{e}}_m \otimes \bar{\mathbf{e}}^i. \tag{5.29}$$

Thus, we have proved the following theorem.

**Theorem 5.3.** *Velocities* $\overset{\circ}{\mathbf{c}}$ *and* $\mathbf{c}$ *of a coherent singular surface* $S(t)$ *in reference and actual configurations, which are determined by (5.12) and (5.26), respectively, are connected by the relation*

$$\mathbf{c} = \mathbf{v}_{\Sigma\pm} + \mathbf{F}_\pm \cdot \overset{\circ}{\mathbf{c}}, \tag{5.30}$$

*or*

$$\overset{\circ}{\mathbf{c}} = \mathbf{F}_\pm^{-1} \cdot (\mathbf{c} - \mathbf{v}_{\Sigma\pm}). \tag{5.31}$$

## 5.2 Relations at a Singular Surface in the Spatial Description

### *5.2.1 Relations at a Coherent Singular Surface in the Spatial Description*

Let us establish now relations between jumps of the functions $\rho_\alpha$, $A_\alpha$, $B_\alpha$ and $C_\alpha$ at a coherent singular surface $S(t)$ corresponding to the singular surface $\overset{\circ}{S}(t)$ in $\overset{\circ}{\mathcal{K}}$. To do this, multiply Eq. (5.25) by the surface element $d\overset{\circ}{\Sigma}$:

$$-[\overset{\circ}{\rho}\overset{\circ}{A}_\alpha]\overset{\circ}{\mathbf{c}} \cdot \overset{\circ}{\mathbf{n}}\, d\overset{\circ}{\Sigma} = \overset{\circ}{\mathbf{n}} \cdot [\overset{\circ}{B}_\alpha]\, d\overset{\circ}{\Sigma} + \overset{\circ}{C}_{\alpha\Sigma}\, d\overset{\circ}{\Sigma}. \tag{5.32}$$

**Theorem 5.4.** *The functions* $\overset{\circ}{B}_\alpha$ *and* $B_\alpha$ *defined by formulae (3.308) and (3.311) satisfy the relations*

$$\overset{\circ}{\mathbf{n}} \cdot [\overset{\circ}{B}_\alpha]\, d\overset{\circ}{\Sigma} = \mathbf{n} \cdot [B_\alpha]\, d\Sigma, \quad \alpha = 1, \ldots, 6. \tag{5.33}$$

▼ When $\alpha = 2$ we have $B_2 = \mathbf{T}$ and $\overset{\circ}{B}_2 = \mathbf{P}$; then, according to (3.55), we get

$$\overset{\circ}{\mathbf{n}} \cdot [\mathbf{P}]\, d\overset{\circ}{\Sigma} = \mathbf{n} \cdot [\mathbf{T}]\, d\Sigma. \tag{5.34}$$

When $\alpha = 3$ we have

$$B_3 = \mathbf{T} \cdot \mathbf{v} - \mathbf{q}, \quad \overset{\circ}{B}_3 = \mathbf{P} \cdot \mathbf{v} - \overset{\circ}{\mathbf{q}},$$

and hence

$$\overset{\circ}{\mathbf{n}} \cdot [\mathbf{P} \cdot \mathbf{v} - \overset{\circ}{\mathbf{q}}]\, d\overset{\circ}{\Sigma} = \mathbf{n} \cdot [\mathbf{T} \cdot \mathbf{v} - \mathbf{q}]\, d\Sigma. \tag{5.35}$$

When $\alpha = 4$ we have $B_4 = \mathbf{q}/\theta$, and $\overset{\circ}{B}_4 = \overset{\circ}{\mathbf{q}}/\theta$. Therefore, according to (3.129), the following relation holds:

$$\overset{\circ}{\mathbf{n}} \cdot [\overset{\circ}{\mathbf{q}}/\theta]\, d\overset{\circ}{\Sigma} = \mathbf{n} \cdot [\mathbf{q}/\theta]\, d\Sigma. \tag{5.36}$$

When $\alpha = 6$ we have

$$B_6 = \rho \mathbf{F} \otimes \mathbf{v}, \quad \overset{\circ}{B}_6 = \overset{\circ}{\rho} \mathbf{E} \otimes \mathbf{v}.$$

According to formula (2.116), we get

$$\mathbf{n} \cdot [\rho \mathbf{F} \otimes \mathbf{v}]\, d\Sigma = \overset{\circ}{\mathbf{n}} \cdot \left[\rho \sqrt{g/\overset{\circ}{g}}\mathbf{E} \otimes \mathbf{v}\right] d\overset{\circ}{\Sigma} = \overset{\circ}{\mathbf{n}} \cdot [\overset{\circ}{\rho} \mathbf{E} \otimes \mathbf{v}]\, d\overset{\circ}{\Sigma}, \tag{5.37}$$

as was to be shown. ▲

To modify the left-hand side of (5.32), we use Eq. (5.31); then we find the relations

$$\begin{aligned} \overset{\circ}{\rho}_+ \overset{\circ}{D}\, d\overset{\circ}{\Sigma} &= \overset{\circ}{\rho}_+ \overset{\circ}{\mathbf{c}} \cdot \overset{\circ}{\mathbf{n}}\, d\overset{\circ}{\Sigma} = \overset{\circ}{\rho}_+ \overset{\circ}{\mathbf{n}} \cdot \mathbf{F}_+^{-1} \cdot (\mathbf{c} - \mathbf{v}_{\Sigma+})\, d\overset{\circ}{\Sigma} \\ &= \overset{\circ}{\rho}_- \overset{\circ}{\mathbf{n}} \cdot \mathbf{F}_-^{-1} \cdot (\mathbf{c} - \mathbf{v}_{\Sigma-})\, d\overset{\circ}{\Sigma}. \end{aligned} \tag{5.38}$$

Using the property of transformation of oriented surface elements (see (2.116)) applied to the ones on the singular surface:

$$\overset{\circ}{\rho}_\pm \overset{\circ}{\mathbf{n}} \cdot \mathbf{F}_\pm^{-1}\, d\overset{\circ}{\Sigma} = \rho_\pm \mathbf{n}\, d\Sigma, \tag{5.39}$$

we get

$$\overset{\circ}{\rho}_\pm \overset{\circ}{D}\, d\overset{\circ}{\Sigma} = \rho_\pm \mathbf{n} \cdot \mathbf{c}\, d\Sigma - \rho_\pm \mathbf{n} \cdot \mathbf{v}_{\Sigma\pm}\, d\Sigma = \rho_\pm (D - \mathbf{n} \cdot \mathbf{v}_{\Sigma\pm})\, d\Sigma. \tag{5.40}$$

Here we have denoted the normal speed of the singular surface in $\mathcal{K}$ by

$$D = \mathbf{c} \cdot \mathbf{n}. \tag{5.41}$$

On substituting (5.40) and (5.33) into (5.32), we obtain the following theorem.

**Theorem 5.5.** *Equations (5.25) at a coherent singular surface* $\overset{\circ}{S}$ *in* $\overset{\circ}{\mathcal{K}}$ *are equivalent to the relations*

$$-[\rho A_\alpha]D + \mathbf{n} \cdot [\rho \mathbf{v} \otimes A_\alpha] - \mathbf{n} \cdot [B_\alpha] = C_{\alpha\Sigma}, \tag{5.42}$$

*which hold at the corresponding singular surface* $S(t)$ *in* $\mathcal{K}$.

Here the jumps of functions across the singular surface $S$, according to (5.2), are determined by the equations

$$[\rho A_\alpha] = (\rho_+ A_{\alpha+} - \rho_- A_{\alpha-})_S. \tag{5.43}$$

In relations (5.42) we have introduced singular sources in $\mathcal{K}$ connected to $\overset{\circ}{C}_{\alpha\Sigma}$ by the relations

$$C_{\alpha\Sigma}\, d\Sigma = \overset{\circ}{C}_{\alpha\Sigma}\, d\overset{\circ}{\Sigma}. \tag{5.44}$$

### 5.2.2 The Rule of Differentiation of an Integral over a Moving Volume Containing a Singular Surface

Let us derive now a rule of differentiation of an integral over a moving volume $V(t)$. This rule is a generalization of formula (3.12) for the case when the volume $V(t)$ contains a singular surface $S(t)$.

Consider the function $A_\alpha(\overset{\circ}{x}{}^i, t)$ (where $\overset{\circ}{x}{}^i \in \overset{\circ}{V}$) satisfying Axiom 17. Going from $\overset{\circ}{\mathcal{K}}$ to $\mathcal{K}$, according to the motion law $x^i = x^i(\overset{\circ}{x}{}^j, t)$, we find that the function $A_\alpha(x^j, t) = A_\alpha(x^i(\overset{\circ}{x}{}^j, t), t) = A_\alpha(\overset{\circ}{x}{}^j, t)$ is continuously differentiable within the volumes $V_+$ and $V_-$ and can suffer a finite jump discontinuity across the interface $S(t)$ between the volumes, i.e.

$$A_\alpha = \begin{cases} A_{\alpha+}, & x^i \in V_+, \\ A_{\alpha-}, & x^i \in V_-, \end{cases} \qquad [A_\alpha] < +\infty, \tag{5.45}$$

where

$$[A_\alpha] = A_+|_S - A_-|_S. \tag{5.46}$$

According to Theorem 5.1, the differentiation rule (5.17) holds for the function $\overset{\circ}{A}_\alpha$ in $\mathcal{K}$. Transform the integral on the left-hand side of (5.17) as follows:

$$\frac{d}{dt}\int_{\overset{\circ}{V}} \overset{\circ}{\rho} A_\alpha \, d\overset{\circ}{V} = \frac{d}{dt}\int_{\overset{\circ}{V}_+} A_{\alpha+}\overset{\circ}{\rho}_+ \, d\overset{\circ}{V} + \frac{d}{dt}\int_{\overset{\circ}{V}_-} A_{\alpha-}\overset{\circ}{\rho}_- \, d\overset{\circ}{V}$$
$$= \frac{d}{dt}\int_{V_+} A_{\alpha+}\rho_+ \, dV + \frac{d}{dt}\int_{V_-} A_{\alpha-}\rho_- \, dV = \frac{d}{dt}\int_V \rho A_\alpha \, dV. \tag{5.47}$$

Here we have used the fact that inside the volumes $\overset{\circ}{V}_+$ and $\overset{\circ}{V}_-$, and also inside $V_+$ and $V_-$, the functions $\overset{\circ}{\rho}_\pm$ and $A_{\alpha\pm}$ are smooth (differentiable) and satisfy the continuity equation (3.6) allowing us to go from integrals over $\overset{\circ}{V}_\pm$ to integrals over $V_\pm$.

In a similar way, modify the first integral on the right-hand side of formula (5.17):

$$\int_{\overset{\circ}{V}} \overset{\circ}{\rho}\frac{dA_\alpha}{dt} \, d\overset{\circ}{V} = \int_V \rho\frac{dA_\alpha}{dt} \, dV. \tag{5.48}$$

The second integral on the right-hand side of formula (5.17) can be transformed by using the relation (5.40) between surface elements $d\overset{\circ}{\Sigma}$ and $d\Sigma$:

$$\int_{\overset{\circ}{S}} [\overset{\circ}{\rho} A_\alpha]\overset{\circ}{D} \, d\Sigma = \int_{\overset{\circ}{S}} (\overset{\circ}{\rho}_+ A_+ \overset{\circ}{D} \, d\overset{\circ}{\Sigma} - \overset{\circ}{\rho}_- A_- \overset{\circ}{D} \, d\overset{\circ}{\Sigma})$$
$$= \int_S (\rho_+ A_+ (D - \mathbf{n}\cdot\mathbf{v}_+) \, d\Sigma - \rho_- A_- (D - \mathbf{n}\cdot\mathbf{v}_-)) \, d\Sigma$$
$$= \int_S [\rho A (D - \mathbf{n}\cdot\mathbf{v})] \, d\Sigma. \tag{5.49}$$

On substituting formulae (5.47)–(5.49) into relation (5.17), we obtain the following theorem.

**Theorem 5.6.** *For any function $A_\alpha$, defined in a volume $V(t)$, which contains a coherent singular surface $S(t)$, and satisfying the corollary (5.45) of Axiom 17, the following rule of differentiation of an integral over a moving volume, which contains a singular surface, holds:*

$$\frac{d}{dt}\int_V \rho A_\alpha \, dV = \int_V \rho\frac{dA_\alpha}{dt} \, dV - \int_S [\rho A_\alpha (D - v_n)] \, d\Sigma. \tag{5.50}$$

Formula (5.50) may be written in another form. Splitting the first integral on the right-hand side of (5.50) into integrals over volumes $V_+$ and $V_-$ and using formula (3.20), we get

$$\int_{V_\pm} \left(\rho \frac{dA_\alpha}{dt}\right)_\pm dV = \int_{V_\pm} \left(\frac{\partial}{\partial t}(\rho A_\alpha) + \nabla \cdot (\rho \mathbf{v} \otimes A_\alpha)\right)_\pm dV. \tag{5.51}$$

On substituting (5.51) into (5.50), we obtain two more representations of the rule of differentiation of an integral over a moving volume, which contains a singular surface $S(t)$:

$$\frac{d}{dt}\int_V \rho A_\alpha \, dV = \int_V \left(\frac{\partial \rho A_\alpha}{\partial t} + \nabla \cdot (\rho \mathbf{v} \otimes A_\alpha)\right) dV - \int_S [\rho A_\alpha (D - v_n)] \, d\Sigma \tag{5.52}$$

and

$$\frac{d}{dt}\int_V \rho A_\alpha \, dV = \int_V \frac{\partial \rho A_\alpha}{\partial t} dV + \int_\Sigma \mathbf{n} \cdot (\rho \mathbf{v} \otimes A_\alpha) \, d\Sigma - \int_S [\rho A_\alpha (D - v_n)] \, d\Sigma. \tag{5.53}$$

## 5.3 Explicit Form of Relations at a Singular Surface

### *5.3.1 Explicit Form of Relations at a Surface of a Strong Discontinuity in a Reference Configuration*

Let us write now relations (5.25) in the explicit form. Using the notations $\overset{\circ}{A}_\alpha$, $\overset{\circ}{B}_\alpha$ and $C_\alpha$ (see (3.304), (3.308), (3.311)) and substituting them into (5.25), we obtain the explicit form of the relations between jumps of the functions at a surface of a strong discontinuity in $\overset{\circ}{\mathcal{K}}$:

$$\begin{aligned}
&[\overset{\circ}{\rho}]\overset{\circ}{D} = 0,\\
&[\overset{\circ}{\rho}\mathbf{v}]\overset{\circ}{D} + \overset{\circ}{\mathbf{n}} \cdot [\mathbf{P}] + \overset{\circ}{C}_{2\Sigma} = 0,\\
&[\overset{\circ}{\rho}(e + v^2/2)]\overset{\circ}{D} + \overset{\circ}{\mathbf{n}} \cdot [\mathbf{P} \cdot \mathbf{v} - \overset{\circ}{\mathbf{q}}] + \overset{\circ}{C}_{3\Sigma} = 0,\\
&[\overset{\circ}{\rho}\eta]\overset{\circ}{D} - \overset{\circ}{\mathbf{n}} \cdot [\overset{\circ}{\mathbf{q}}/\theta] + \overset{\circ}{C}_{4\Sigma} = 0,\\
&[\overset{\circ}{\rho}\mathbf{u}]\overset{\circ}{D} + \overset{\circ}{C}_{5\Sigma} = 0,\\
&[\overset{\circ}{\rho}\mathbf{F}^{\mathrm{T}}]\overset{\circ}{D} + \overset{\circ}{\mathbf{n}} \otimes [\overset{\circ}{\rho}\mathbf{v}] + \overset{\circ}{C}_{6\Sigma} = 0.
\end{aligned} \tag{5.54}$$

### 5.3.2 Explicit Form of Relations at a Surface of a Strong Discontinuity in an Actual Configuration

Substitution of the notations $A_\alpha$, $B_\alpha$ and $C_\alpha$ (see (3.304) and (3.308)) into relations (5.42) gives their explicit forms

$$\begin{aligned}
&[\rho(D - v_n)] = 0, \\
&[\rho\mathbf{v}(D - v_n)] + \mathbf{n}\cdot[\mathbf{T}] + C_{2\Sigma} = 0, \\
&[\rho(e + v^2/2)(D - v_n)] + \mathbf{n}\cdot[\mathbf{T}\cdot\mathbf{v} - \mathbf{q}] + C_{3\Sigma} = 0, \\
&[\rho\eta(D - v_n)] - \mathbf{n}\cdot[\mathbf{q}/\theta] + C_{4\Sigma} = 0, \\
&[\rho\mathbf{u}(D - v_n)] + C_{5\Sigma} = 0, \\
&[\rho\mathbf{F}^{\mathrm{T}}(D - v_n)] + \mathbf{n}\cdot[\rho\mathbf{F}\otimes\mathbf{v}] + C_{6\Sigma} = 0,
\end{aligned} \tag{5.55}$$

where

$$v_{n\pm} = \mathbf{v}_{\Sigma\pm}\cdot\mathbf{n}. \tag{5.56}$$

The functions $\overset{\circ}{C}_{\alpha\Sigma}$ and $C_{\alpha\Sigma}$ in these relations, in general, are independent. Expressions for them can be given with the help of additional assumptions on singular surfaces $\overset{\circ}{S}$ and $S$.

### 5.3.3 Mass Rate of Propagation of a Singular Surface

Denote a *mass rate of propagation of a singular surface* in $\overset{\circ}{\mathcal{K}}$ by

$$\overset{\circ}{M} = \overset{\circ}{D}\overset{\circ}{\rho}_{+}, \tag{5.57}$$

then the first relation of (5.54) yields

$$\overset{\circ}{\rho}_{+}\overset{\circ}{D} = \overset{\circ}{\rho}_{-}\overset{\circ}{D} = \overset{\circ}{M}. \tag{5.58}$$

The first equation of (5.55) gives

$$\rho_{+}(D - \mathbf{v}_{\Sigma+}\cdot\mathbf{n}) = \rho_{-}(D - \mathbf{v}_{\Sigma-}\cdot\mathbf{n}) \equiv M, \tag{5.59}$$

where $M$ is the *mass rate* of propagation of the singular surface in actual configuration $\mathcal{K}$.

**Theorem 5.7.** *Mass rates $\overset{\circ}{M}$ and $M$ are connected by the relation*

$$M = \frac{k_+\rho_+}{\overset{\circ}{\rho}_+}\overset{\circ}{M}, \quad k_\pm = (\mathbf{n}\cdot\mathbf{g}_\pm^{-1}\cdot\mathbf{n})^{1/2} = (\mathbf{n}\cdot\mathbf{F}_\pm\cdot\mathbf{F}_\pm^{\mathrm{T}}\cdot\mathbf{n})^{1/2}, \tag{5.60}$$

*where*

$$k_+\rho_+/\overset{\circ}{\rho}_+ = k_-\rho_-/\overset{\circ}{\rho}_-, \tag{5.61}$$

*i.e.* $[k\rho/\overset{\circ}{\rho}] = 0$ *or* $[\rho/(\overset{\circ}{k}\,\overset{\circ}{\rho})] = 0.$

▼ Substitution of expression (5.41) and formula (5.30) into the definition (5.59) yields

$$M = \rho_+(D - \mathbf{v}_{\Sigma+}\cdot\mathbf{n}) = \rho_+(\mathbf{c} - \mathbf{v}_{\Sigma\pm})\cdot\mathbf{n} = \rho_+\mathbf{n}\cdot\mathbf{F}_+\cdot\overset{\circ}{\mathbf{c}}. \tag{5.62}$$

Apply formula (2.124), connecting the normals $\mathbf{n}$ and $\overset{\circ}{\mathbf{n}}$, to material points at the singular surface:

$$\mathbf{n}\cdot\mathbf{F}_+ = k_+\overset{\circ}{\mathbf{n}}, \qquad \text{or} \qquad \mathbf{n}\cdot\mathbf{F}_- = k_-\overset{\circ}{\mathbf{n}}. \tag{5.63}$$

On substituting (5.63) into (5.62), we finally obtain

$$M = k_+\rho_+\overset{\circ}{\mathbf{c}}\cdot\overset{\circ}{\mathbf{n}} = \frac{k_+\rho_+}{\overset{\circ}{\rho}_+}\overset{\circ}{M}. \tag{5.64}$$

In a similar way, we can show that

$$M = \frac{k_-\rho_-}{\overset{\circ}{\rho}_-}\overset{\circ}{M}. \quad \blacktriangle \tag{5.65}$$

One frequently introduces also the *speed of propagation of a singular surface*

$$D_0 = D - \mathbf{v}_{\Sigma+}\cdot\mathbf{n}, \tag{5.66}$$

which, according to (5.30) and (5.41), is connected to $\overset{\circ}{\mathbf{c}}$ by the relation

$$D_0 = \mathbf{n}\cdot\mathbf{F}_+\cdot\overset{\circ}{\mathbf{c}}. \tag{5.67}$$

If there is no transition of material points across a singular surface, then $\overset{\circ}{\mathbf{c}} = 0$ and hence

$$\overset{\circ}{D} = D_0 = 0, \qquad \overset{\circ}{M} = M = 0. \tag{5.68}$$

Relation (5.59) with use of (5.66) can be written in the form

$$[\rho]D_0 = \rho_-[v_n]. \tag{5.69}$$

The introduction of the mass rate $\overset{\circ}{M}$ allows us to rewrite the relations (5.25) for jumps as follows:

$$\overset{\circ}{M}[A_\alpha] + \overset{\circ}{\mathbf{n}} \cdot [\overset{\circ}{B}_\alpha] + \overset{\circ}{C}_{\alpha\Sigma} = 0, \quad \alpha = 2, \ldots, 6, \tag{5.70}$$

or in the explicit form

$$[\overset{\circ}{\rho}]\overset{\circ}{D} = 0, \tag{5.70a}$$

$$\overset{\circ}{M}[\mathbf{v}] + \overset{\circ}{\mathbf{n}} \cdot [\mathbf{P}] + \overset{\circ}{\mathbf{C}}_{2\Sigma} = 0, \tag{5.70b}$$

$$\overset{\circ}{M}\left[e + \frac{v^2}{2}\right] + \overset{\circ}{\mathbf{n}} \cdot [\mathbf{P} \cdot \mathbf{v} - \overset{\circ}{\mathbf{q}}] + \overset{\circ}{C}_{3\Sigma} = 0, \tag{5.70c}$$

$$\overset{\circ}{M}[\eta] - \overset{\circ}{\mathbf{n}} \cdot [\overset{\circ}{\mathbf{q}}/\theta] + \overset{\circ}{C}_{4\Sigma} = 0, \tag{5.70d}$$

$$\overset{\circ}{M}[\mathbf{u}] + \overset{\circ}{\mathbf{C}}_{5\Sigma} = 0, \tag{5.70e}$$

$$\overset{\circ}{M}[\mathbf{F}^{\mathrm{T}}] + \overset{\circ}{\mathbf{n}} \otimes [\overset{\circ}{\rho}\mathbf{v}] + \overset{\circ}{\mathbf{C}}_{6\Sigma} = 0. \tag{5.70f}$$

In a similar way, relations (5.42) with use of (5.59) can be rewritten as follows:

$$M[A_\alpha] + \mathbf{n} \cdot [B_\alpha] + C_{\alpha\Sigma} = 0, \quad \alpha = 2, \ldots, 6, \tag{5.71}$$

or in the explicit form

$$[\rho(D - \mathbf{v} \cdot \mathbf{n})] = 0, \tag{5.72}$$

$$M[\mathbf{v}] + \mathbf{n} \cdot [\mathbf{T}] + \mathbf{C}_{2\Sigma} = 0, \tag{5.73}$$

$$M\left[e + \frac{v^2}{2}\right] + \mathbf{n} \cdot [\mathbf{T} \cdot \mathbf{v} - \mathbf{q}] + C_{3\Sigma} = 0, \tag{5.74}$$

$$M[\eta] - \mathbf{n} \cdot [\mathbf{q}/\theta] + C_{4\Sigma} = 0, \tag{5.75}$$

$$M[\mathbf{u}] + \mathbf{C}_{5\Sigma} = 0, \tag{5.76}$$

$$M[\mathbf{F}^{\mathrm{T}}] + \mathbf{n} \cdot [\rho\mathbf{F} \otimes \mathbf{v}] + \mathbf{C}_{6\Sigma} = 0. \tag{5.77}$$

### *5.3.4 Relations at a Singular Surface Without Transition of Material Points*

If there is no transition of material points across a singular surface, i.e. $\overset{\circ}{M} = M = 0$, then from (5.71) it follows that at $\overset{\circ}{S}$ and $S$ the following relations hold:

$$\overset{\circ}{\mathbf{n}} \cdot [\overset{\circ}{B}_\alpha] + \overset{\circ}{C}_{\alpha\Sigma} = 0, \quad \overset{\circ}{\mathbf{x}} \in \overset{\circ}{S}, \tag{5.78}$$

$$\mathbf{n} \cdot [B_\alpha] + C_{\alpha\Sigma} = 0, \quad \mathbf{x} \in S, \quad \alpha = 2, \ldots, 6; \tag{5.79}$$

or

$$\begin{cases} \overset{\circ}{\mathbf{n}} \cdot [\mathbf{P}] + \overset{\circ}{\mathbf{C}}_{2\Sigma} = 0, \\ \overset{\circ}{\mathbf{n}} \cdot [\mathbf{P} \cdot \mathbf{v} - \overset{\circ}{\mathbf{q}}] + \overset{\circ}{C}_{3\Sigma} = 0, \\ \overset{\circ}{\mathbf{n}} \cdot [\mathbf{q}/\theta] + \overset{\circ}{C}_{4\Sigma} = 0, \\ \overset{\circ}{C}_{5\Sigma} = 0, \\ \overset{\circ}{\mathbf{n}} \otimes [\overset{\circ}{\rho}\mathbf{v}] + \overset{\circ}{\mathbf{C}}_{6\Sigma} = 0, \qquad \overset{\circ}{\mathbf{x}} \in \overset{\circ}{S}; \end{cases} \tag{5.80}$$

and also

$$\begin{cases} \mathbf{n} \cdot [\mathbf{T}] + \mathbf{C}_{2\Sigma} = 0, \\ \mathbf{n} \cdot [\mathbf{T} \cdot \mathbf{v} - \mathbf{q}] + C_{3\Sigma} = 0, \\ \mathbf{n} \cdot [\mathbf{q}/\theta] + C_{4\Sigma} = 0, \\ C_{5\Sigma} = 0, \\ \mathbf{n} \cdot [\rho\mathbf{F} \otimes \mathbf{v}] + \mathbf{C}_{6\Sigma} = 0, \qquad \mathbf{x} \in S. \end{cases} \tag{5.81}$$

## 5.4 The Main Types of Singular Surfaces

Let us analyze now a possible form of the surface functions $C_{\alpha\Sigma}$ and $\overset{\circ}{C}_{\alpha\Sigma}$ for a singular surface $S(t)$, which, in general, may be incoherent.

### 5.4.1 *Jump of Density*

At any singular surface $S(t)$ there is no jump of mass (as $C_1 = \overset{\circ}{C}_1 \equiv 0$), hence

$$\overset{\circ}{C}_{1\Sigma} = C_{1\Sigma} = 0. \tag{5.82}$$

As follows from (5.70a), the density jump $[\overset{\circ}{\rho}]$ may be different from zero only if there is no transition of material points across the singular surface (the velocity $\overset{\circ}{\mathbf{c}}$ and hence the speed $\overset{\circ}{D}$ are zero):

$$\overset{\circ}{D} = 0, \qquad [\overset{\circ}{\rho}] \neq 0. \tag{5.83}$$

If material points cross a singular surface, then the density jump $\overset{\circ}{\rho}$ is always zero:

$$\overset{\circ}{D} \neq 0, \quad [\overset{\circ}{\rho}] = 0. \tag{5.84}$$

Notice that the density jump $[\rho]$ in this case is non-zero (see formula (5.72)).

## 5.4.2 Jumps of Radius-Vector and Displacement Vector

The made assumptions on differentiability of all functions $\rho_{\pm}$, $A_{\alpha\pm}$ and $B_{\alpha\pm}$ in domains $V_{\pm}$ up to a boundary $S$ mean the absence of discontinuities within a medium in an actual configuration, i.e. the impossibility of the motion of two media so as it is shown in Fig. 5.8. This fact can be described mathematically as follows: for all points of a surface $S$ in $\mathcal{K}$ their radius-vectors $\mathbf{x}$ on the both sides of the surface are coincident:

$$\mathbf{x}_{\Sigma+} - \mathbf{x}_{\Sigma-} = 0 \quad \text{or} \quad [\mathbf{x}] = 0. \tag{5.85}$$

However, the displacement vector $\mathbf{u}$ may suffer a jump discontinuity across $S$.

A surface $\Sigma$ is called a *surface without singular displacements*, if the following conditions are satisfied:

$$\overset{\circ}{C}_{5\Sigma} = C_{5\Sigma} = 0. \tag{5.86}$$

At a surface without singular displacements, from (5.72)–(5.77) we have:

- If there is no transition of material points across $S$ ($D_0 = \overset{\circ}{D} = 0$ and $M = \overset{\circ}{M} = 0$), then a jump of the displacement vector may be nonzero:

$$M = 0, \quad [\mathbf{u}] \neq 0. \tag{5.87}$$

- If material points cross the surface $S$ ($D_0 \neq 0$ and $M \neq 0$), then there is no jump of the displacement vector:

$$M \neq 0, \quad [\mathbf{u}] = 0. \tag{5.88}$$

For coherent surfaces $S$ (see Definition 5.4), besides $\mathbf{x}$, the radius-vector $\overset{\circ}{\mathbf{x}}$ in a reference configuration is also continuous:

$$[\overset{\circ}{\mathbf{x}}(\mathbf{x}, t)] = 0. \tag{5.89}$$

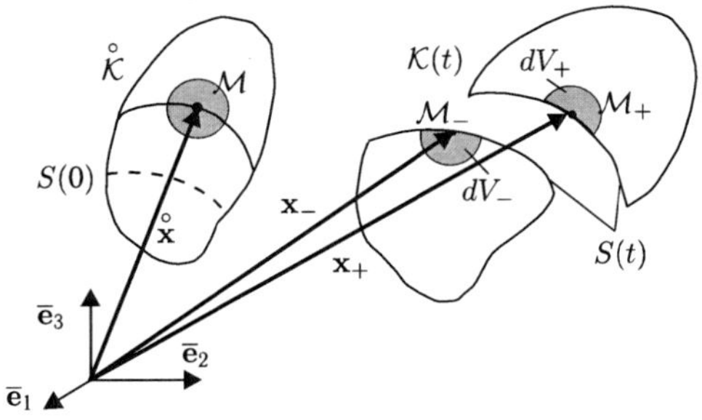


**Fig. 5.8** Such motion of a continuum is impossible under the made assumptions on differentiability of all functions $\rho_{\pm}$, $A_{\alpha\pm}$ and $B_{\alpha\pm}$

Examples of *incoherent surfaces* are a surface of interaction of a solid and an ideal gas continua, on which there is slip of gas, and also a surface $S$ with singular displacements, on which the following condition holds:

$$C_{5\Sigma} \neq 0. \tag{5.90}$$

At an incoherent surface, material points $\mathcal{M}_+$ and $\mathcal{M}_-$ of a continuum, being neighboring on different sides of the surface in a reference configuration, can move apart within an arbitrary distance from one another without leaving the surface, so there will be no discontinuities at the surface $S$ (Fig. 5.9). The displacement vector in this case suffers a jump discontinuity

$$M \neq 0, \qquad [\mathbf{u}] \neq 0. \tag{5.91}$$

Figure 5.10 shows a logical scheme of coherent and incoherent surfaces.

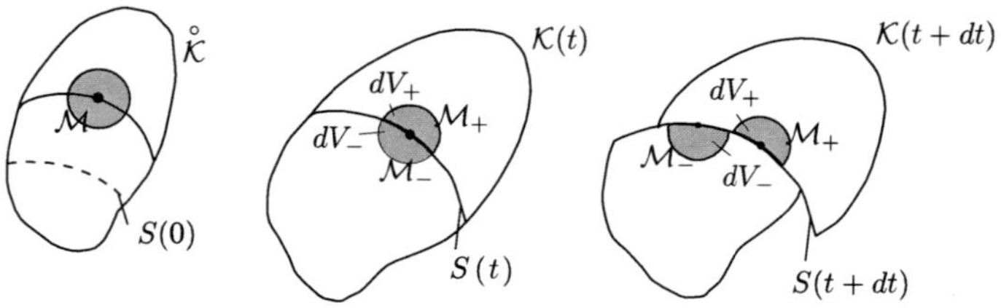


**Fig. 5.9** Displacements of neighboring points at an incoherent surface

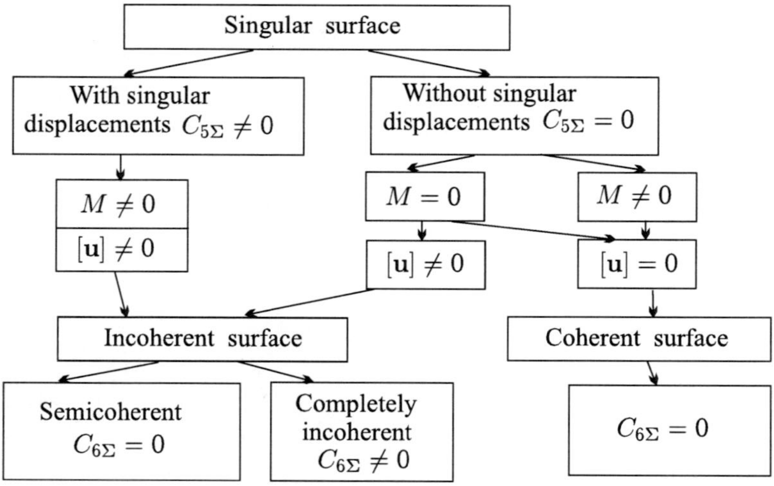


**Fig. 5.10** Logical scheme of singular surfaces

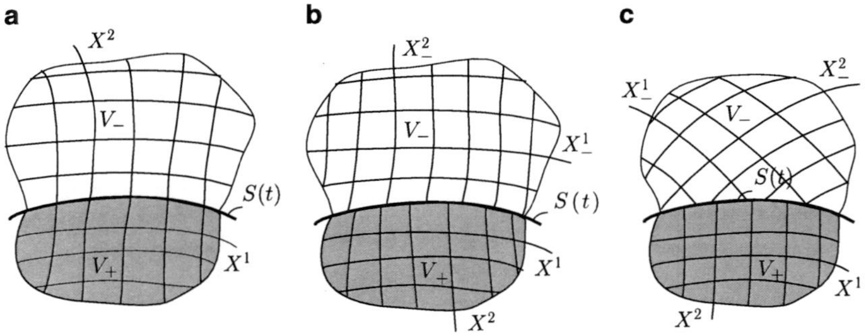


**Fig. 5.11** Scheme of a singular surface $S(t)$ in $\mathcal{K}$: (**a**) coherent, (**b**) semicoherent, and (**c**) incoherent. Here $X^i$ and $X^i_-$ are Lagrangian coordinates in subdomains $V_+$ and $V_-$

## 5.4.3 Semicoherent and Completely Incoherent Singular Surfaces

An incoherent singular surface $S(t)$ is called *semicoherent*, if the following conditions are satisfied at this surface:

$$\overset{\circ}{C}_{6\Sigma} = 0, \qquad C_{6\Sigma} = 0. \tag{5.92}$$

If at an incoherent surface $S(t)$ the conditions

$$\overset{\circ}{C}_{6\Sigma} \neq 0, \qquad C_{6\Sigma} \neq 0 \tag{5.93}$$

are satisfied, then the surface is called *completely incoherent*.

There are no singular jumps of the deformation gradient **F** across semicoherent surfaces (Fig. 5.11).

## 5.4.4 Nondissipative and Homothermal Singular Surfaces

A singular surface is *nondissipative*, if there are no singular sources of entropy production at the surface:

$$\overset{\circ}{C}_{4\Sigma} = 0, \qquad C_{4\Sigma} = 0. \tag{5.94}$$

Surfaces of melting, solidification, sublimation and precipitation of gas on solids, and also many surfaces of phase transformations are usually nondissipative. Surfaces of shock waves, surfaces of chemical reactions etc. are examples of dissipative surfaces.

Notice that the function $C_{4\Sigma}$, according to (5.8a), is the jump of entropy production by external heat sources and by internal sources. If there are no external heat sources (such processes of motion are called *adiabatic*), then $\mathbf{q} = 0$, $q_m = 0$ and $\widetilde{C}_4 = 0$, and $C_{4\Sigma}$ is non-negative:

$$C_{4\Sigma} = C^*_{4\Sigma} \geqslant 0, \tag{5.94a}$$

The vector of surface forces $\mathbf{C}_{2\Sigma}$ and the energy of surface forces $C_{3\Sigma}$ are usually connected with the phenomenon of surface tension:

$$\mathbf{C}_{2\Sigma} = -P_\Sigma \mathbf{n}, \qquad P_\Sigma = \sigma_\Sigma \left(\frac{1}{R_1} + \frac{1}{R_2}\right). \tag{5.95}$$

Here $R_1$ and $R_2$ are the principal curvature radii of the surface $S(t)$ (see [12]), and $\sigma_\Sigma$ is the *coefficient of surface tension* being a characteristic of the medium surface, which is determined in experiments.

A corresponding value of the surface forces energy is determined as follows:

$$C_{3\Sigma} = (1/2)\mathbf{C}_{2\Sigma} \cdot (\mathbf{v}_{\Sigma+} + \mathbf{v}_{\Sigma-}). \tag{5.96}$$

When there are electromagnetic effects, the functions $\mathbf{C}_{2\Sigma}$ and $C_{3\Sigma}$ are connected also with surface electromagnetic forces and energy.

When there are no electromagnetic effects and surface tension, we have

$$\overset{\circ}{\mathbf{C}}_{2\Sigma} = \mathbf{C}_{2\Sigma} = 0 \quad \text{and} \quad \overset{\circ}{C}_{3\Sigma} = C_{3\Sigma} = 0. \tag{5.97}$$

A surface $\Sigma$ is called *homothermal*, if there is no jump of temperature across this surface:

$$[\theta] = 0 \quad \text{at } \mathbf{x} \in \Sigma. \tag{5.98}$$

This condition is used in many problems on phase transformations, chemical reactions, shock waves etc.

## 5.4.5 Surfaces with Ideal Contact

If a surface $S(t)$, across which there is no transition of material points, is homothermal, coherent and nondissipative, then it is called a *surface with ideal contact*. For this surface,

$$\overset{\circ}{C}_{\alpha\Sigma} = 0, \qquad C_{\alpha\Sigma} = 0, \qquad \alpha = 1, \ldots, 6, \tag{5.99}$$

and

$$[\overset{\circ}{\mathbf{x}}] = 0, \qquad M = 0, \qquad [\theta] = 0, \qquad [\mathbf{u}] = 0, \qquad [\mathbf{v}] = 0. \tag{5.100}$$

Since material points do not cross a surface with ideal contact, the velocity of the surface $S(t)$ coincides with the velocity $\mathbf{v}$ of material points $\mathcal{M}$ of $S(t)$; and the jump of the velocity is zero that follows from (5.59) or (5.70f).

Relations (5.99) and (5.100) should be complemented with relations (5.81), which in this case have the forms

$$\mathbf{n}\cdot[\mathbf{T}] = 0, \tag{5.101}$$

$$\mathbf{n}\cdot[\mathbf{T}\cdot\mathbf{v}] = \mathbf{n}\cdot[\mathbf{q}], \tag{5.102}$$

$$\mathbf{n}\cdot[\mathbf{q}/\theta] = 0, \tag{5.103}$$

$$\mathbf{n}\cdot[\rho\mathbf{F}\otimes\mathbf{v}] = 0. \tag{5.104}$$

As follows from (5.100) and (5.103), the jump of the normal component of the heat flux vector is zero:

$$\mathbf{n}\cdot[\mathbf{q}] = 0, \tag{5.105}$$

and relation (5.102) is not independent. This relation is identically satisfied if equations (5.100) and (5.103) hold. Since $[\mathbf{v}] = 0$, Eq. (5.104) is equivalent to the relation

$$\mathbf{n}\cdot[\rho\mathbf{F}] = 0.$$

According to (2.116), this relation is a geometric condition for the normal vector in $\overset{\circ}{\mathcal{K}}$:

$$0 = [\overset{\circ}{\mathbf{n}}]d\overset{\circ}{\Sigma} = [\rho\mathbf{n}\cdot\mathbf{F}]d\Sigma/\overset{\circ}{\rho}.$$

Thus, in the system (5.100)–(5.105) the following relations are independent:

$$\begin{cases} [\mathbf{u}] = 0, \\ \mathbf{n}\cdot[\mathbf{T}] = 0, \\ [\theta] = 0, \\ \mathbf{n}\cdot[\mathbf{q}] = 0. \end{cases} \tag{5.106}$$

These equations are called the *relations at a surface with ideal contact.*

In a similar way, in $\overset{\circ}{\mathcal{K}}$ from (5.100) and (5.80) we obtain the following independent relations at a surface with ideal contact:

$$\begin{cases} [\mathbf{u}] = 0, \\ \overset{\circ}{\mathbf{n}}\cdot[\mathbf{P}] = 0, \\ [\theta] = 0, \\ \overset{\circ}{\mathbf{n}}\cdot[\overset{\circ}{\mathbf{q}}] = 0. \end{cases} \tag{5.107}$$

### 5.4.6 On Boundary Conditions

Before we have considered the case when in both subdomains $V_+$ and $V_-$ the functions $\rho_\pm$, $A_{\alpha\pm}$ and $B_{\alpha\pm}$ are unknown and should be found. However, Eq. (5.71) can also be used for another case: when in one of the subdomains, for example in $V_+$, the functions $\rho_+$, $A_{\alpha+}$ and $B_{\alpha+}$ are given, and in the other subdomain (i.e. in $V_-$) the functions $\rho_-$, $A_{\alpha-}$ and $B_{\alpha-}$ are unknown. In this case, relations (5.71) at the interface $S$ between $V_+$ and $V_-$ serve to obtain *boundary conditions* at $S$ for formulation of initial boundary-value problems. Notice that, as a rule, not all the functions $\rho_+$, $A_{\alpha+}$ and $B_{\alpha+}$ in the domain $V_+$ can be given in an arbitrary way; in formulating a problem of continuum mechanics, some of the functions should usually remain unknown and be determined in its solving.

### 5.4.7 Equation of a Singular Surface in $\overset{\circ}{\mathcal{K}}$

An equation of any smooth surface, including a singular surface $\overset{\circ}{S}(t)$, can be given *implicitly*, namely with the help of one scalar equation in coordinates $\overset{\circ}{x}{}^i$:

$$\overset{\circ}{S}: \quad \overset{\circ}{f}(\overset{\circ}{x}{}^i, t) = 0. \tag{5.108}$$

The argument $t$ shows that the surface function $\overset{\circ}{f}$ may vary with time.

Introducing curvilinear coordinates $\tilde{X}^I$ $(I = 1, 2)$ on the surface $\overset{\circ}{S}(t)$ [12], we can give the surface $\overset{\circ}{S}$ in the *parametric form* in coordinates $\overset{\circ}{x}{}^i$:

$$\overset{\circ}{x}{}^i_\Sigma = \overset{\circ}{x}{}^i(\tilde{X}^I, t), \tag{5.109}$$

where $\overset{\circ}{\mathbf{x}}_\Sigma = \overset{\circ}{x}{}^i_\Sigma \bar{\mathbf{e}}_i$ is the radius-vector of a point of the surface $\overset{\circ}{S}$.

Introduce vectors tangent to the surface $\overset{\circ}{S}$ and the unit normal vector being orthogonal to them:

$$\overset{\circ}{\boldsymbol{\rho}}_I = \frac{\partial \overset{\circ}{\mathbf{x}}_\Sigma}{\partial \tilde{X}^I}, \quad \overset{\circ}{\mathbf{n}} = \frac{\overset{\circ}{\boldsymbol{\rho}}_1 \times \overset{\circ}{\boldsymbol{\rho}}_2}{|\overset{\circ}{\boldsymbol{\rho}}_1 \times \overset{\circ}{\boldsymbol{\rho}}_2|}, \quad \overset{\circ}{\mathbf{n}} \cdot \overset{\circ}{\boldsymbol{\rho}}_I = 0, \quad I = 1, 2. \tag{5.110}$$

By tangent vectors $\overset{\circ}{\boldsymbol{\rho}}_I$, we can always construct orthonormal tangent vectors $\overset{\circ}{\boldsymbol{\tau}}_I$; therefore, at surface $\overset{\circ}{S}$ there is an orthonormal basis $\overset{\circ}{\boldsymbol{\tau}}_I, \overset{\circ}{\mathbf{n}}$ $(I = 1, 2)$ such that

$$\overset{\circ}{\mathbf{n}} \cdot \overset{\circ}{\boldsymbol{\tau}}_I = 0, \quad \overset{\circ}{\boldsymbol{\tau}}_1 \cdot \overset{\circ}{\boldsymbol{\tau}}_2 = 0, \quad I = 1, 2. \tag{5.110a}$$

By substituting (5.109) into (5.108), the equation of a singular surface can be represented as a composite function of curvilinear coordinates on the surface: $\overset{\circ}{f}(\overset{\circ}{x}^i(\tilde{X}^I, t), t) = 0$. Differentiating this equation with respect to $\tilde{X}^I$, we get

$$\frac{\partial \overset{\circ}{f}}{\partial \overset{\circ}{x}^i} \frac{\partial \overset{\circ}{x}^i_\Sigma}{\partial \tilde{X}^I} = 0. \tag{5.111}$$

Introducing the vector $\overset{\circ}{\nabla} \overset{\circ}{f} = \frac{\partial \overset{\circ}{f}}{\partial \overset{\circ}{x}^i} \bar{\mathbf{e}}^i$ called the *surface gradient* and using formulae (5.110), we can represent Eq. (5.111) as the scalar product $\overset{\circ}{\nabla} \overset{\circ}{f} \cdot \overset{\circ}{\boldsymbol{\rho}}_I = 0$. In other words, the surface gradient $\overset{\circ}{\nabla} \overset{\circ}{f}$ is orthogonal to the tangent vectors; hence, the vector $\overset{\circ}{\nabla} \overset{\circ}{f}$ is collinear to the vector $\overset{\circ}{\mathbf{n}}$. Then, normalizing the vector $\overset{\circ}{\nabla} \overset{\circ}{f}$, we find the following important formula:

$$\overset{\circ}{\mathbf{n}} = \overset{\circ}{\nabla} \overset{\circ}{f} / |\overset{\circ}{\nabla} \overset{\circ}{f}|, \tag{5.112}$$

which allows us to calculate the normal $\overset{\circ}{\mathbf{n}}$ to the surface $\overset{\circ}{S}$ given implicitly by equation (5.108).

Determine the total differential of Eq. (5.108) with respect to $t$:

$$d\overset{\circ}{f} = \frac{\partial \overset{\circ}{f}}{\partial t} dt + \frac{\partial \overset{\circ}{f}}{\partial \overset{\circ}{x}^i} d\overset{\circ}{x}^i_\Sigma = 0.$$

This equation can be rewritten as follows:

$$\frac{\partial \overset{\circ}{f}}{\partial t} + \left( \frac{\partial \overset{\circ}{f}}{\partial \overset{\circ}{x}^i} \bar{\mathbf{e}}^i \right) \cdot \left( \frac{\partial \overset{\circ}{x}^j_\Sigma}{\partial t} \bar{\mathbf{e}}_j \right) = 0. \tag{5.113}$$

According to the definitions (5.12) of the velocity $\overset{\circ}{\mathbf{c}}$ of a singular surface and the surface gradient, we get

$$\frac{\partial \overset{\circ}{f}}{\partial t} + \overset{\circ}{\mathbf{c}} \cdot \overset{\circ}{\nabla} \overset{\circ}{f} = 0. \tag{5.114}$$

With the help of definition (5.16a) for the normal speed $\overset{\circ}{D}$ of a singular surface and formula (5.112) for the normal vector, we can rewrite Eq. (5.114) as follows:

$$\frac{\partial \overset{\circ}{f}}{\partial t} + \overset{\circ}{D} |\overset{\circ}{\nabla} \overset{\circ}{f}| = 0. \tag{5.115}$$

This equation is called the *differential equation of the singular surface* $\overset{\circ}{S}(t)$; it allows us to find Eq. (5.108) of the surface $\overset{\circ}{S}(t)$ if only the normal speed $\overset{\circ}{D}$ and initial position of the surface $\overset{\circ}{S}(0)$ are given:

$$\overset{\circ}{S}(0): \quad \overset{\circ}{f}(\overset{\circ}{x}{}^i,\, 0) = 0. \tag{5.116}$$

## 5.4.8 Equation of a Singular Surface in $\mathcal{K}$

Let us derive now an equation of a singular surface $S(t)$ in an actual configuration $\mathcal{K}$. This equation has the *implicit form* similar to (5.108)

$$S: \quad f(x^i,\, t) = 0; \tag{5.117}$$

and the parametric equation of a singular surface is analogous to (5.109):

$$x^i_\Sigma = x^i(\tilde{X}^I,\, t), \quad I = 1,\, 2. \tag{5.118}$$

Here $\mathbf{x}_\Sigma = x^i_\Sigma \bar{\mathbf{e}}_i$ is the radius-vector of a point of the surface $S$. Differentiating this radius-vector with respect to curvilinear coordinates on the surface, we find the expression for tangent vectors to $S(t)$ and for the normal vector:

$$\boldsymbol{\rho}_I = \frac{\partial \mathbf{x}_\Sigma}{\partial \tilde{X}^I} \quad \mathbf{n} = \frac{\boldsymbol{\rho}_1 \times \boldsymbol{\rho}_2}{|\boldsymbol{\rho}_1 \times \boldsymbol{\rho}_2|}, \quad \mathbf{n} \cdot \boldsymbol{\rho}_I = 0, \quad I = 1, 2. \tag{5.119}$$

By tangent vectors $\boldsymbol{\rho}_I$, we can always construct orthonormal tangent vectors $\boldsymbol{\tau}_I$; therefore, at the surface $S$ there is an orthonormal basis $\boldsymbol{\tau}_I, \mathbf{n}$ $(I = 1, 2)$ such that

$$\mathbf{n} \cdot \boldsymbol{\tau}_I = 0, \quad \boldsymbol{\tau}_1 \cdot \boldsymbol{\tau}_2 = 0, \quad I = 1,\, 2. \tag{5.119a}$$

Substituting the function (5.118) into (5.117), we represent the equation of a singular surface as a composite function of curvilinear coordinates on the surface: $f(x^i(\tilde{X}^I, t),\ t) = 0$. Differentiation of this equation with respect to $X^I$ yields

$$\frac{\partial f}{\partial x^i} \frac{\partial x^i_\Sigma}{\partial \tilde{X}^I} = 0. \tag{5.120}$$

Introducing the surface gradient $\nabla f = (\partial f / \partial x^i) \bar{\mathbf{e}}^i$ and using formulae (5.119), we can rewrite Eq. (5.120) in the form $\nabla f \cdot \boldsymbol{\rho}_I = 0$. On comparing this equation with formula (5.119) of mutual orthogonality of tangent vectors and a normal vector, we obtain the following connection between the unit normal vector $\mathbf{n}$ and the surface gradient $\nabla f$:

$$\mathbf{n} = \frac{\nabla f}{|\nabla f|}, \tag{5.121}$$

being an analog of formula (5.112).

Determine the total differential of Eq. (5.117) with respect to $t$:

$$df = \frac{\partial f}{\partial t} dt + \frac{\partial f}{\partial x^i} dx^i_\Sigma = 0.$$

According to (5.26), this equation takes the form

$$\frac{\partial f}{\partial t} + \mathbf{c} \cdot \nabla f = 0. \tag{5.122}$$

With use of the definition (5.41) of the normal speed $D$ of a singular surface $S(t)$ in $\mathcal{K}$, Eq. (5.122) becomes

$$\frac{\partial f}{\partial t} + D|\nabla f| = 0. \tag{5.123}$$

This equation is called the *differential equation of the singular surface* $S(t)$ in $\mathcal{K}$. Passing from $D$ to the speed $D_0$ determined by (5.66), Eq. (5.123) can be written in the form

$$\frac{\partial f}{\partial t} + (D_0 + \mathbf{v}_{\Sigma +} \cdot \mathbf{n})|\nabla f| = 0. \tag{5.124}$$

This equation allows us to find an equation of the surface $S(t)$ in the form (5.117) if only the normal speed $D$ and initial position of the surface $S(0) = \overset{\circ}{S}(0)$ are given:

$$S(0): \qquad f(x^i, 0) = \overset{\circ}{f}(x^i, 0) = 0. \tag{5.125}$$

## Exercises for 5.4

**5.4.1.** Show that Eq. (5.124) of propagation of a singular surface can be rewritten in the form

$$\frac{\partial f}{\partial t} + D_0|\nabla f| + \mathbf{v}_{\Sigma +} \cdot \nabla f = 0,$$

and when there is no transition of material points across the singular surface, the equation reduces to

$$\frac{\partial f}{\partial t} + \mathbf{v}_{\Sigma +} \cdot \nabla f = 0.$$

# Chapter 6
# Elastic Continua at Large Deformations

## 6.1 Closed Systems in the Spatial Description

### 6.1.1 $\theta RUVF$-system of Thermoelasticity

Let us formulate the problem for the general case of elastic (i.e. ideal) solids at large deformations; their constitutive equations have been derived in Sects. 4.5–4.9.

The general system of balance laws (3.307) in the spatial (Eulerian) description consists of the continuity equation ($\alpha = 1$), the momentum balance equation ($\alpha = 2$), the energy balance equation ($\alpha = 3$), the dynamic compatibility equation ($\alpha = 5$) and the kinematic equation ($\alpha = 6$):

$$\frac{\partial \rho}{\partial t} + \nabla \cdot \rho \mathbf{v} = 0, \tag{6.1a}$$

$$\frac{\partial \rho \mathbf{v}}{\partial t} + \nabla \cdot \rho \mathbf{v} \otimes \mathbf{v} = \nabla \cdot \mathbf{T} + \rho \mathbf{f}, \tag{6.1b}$$

$$\frac{\partial \rho \varepsilon}{\partial t} + \nabla \cdot (\rho \mathbf{v} \varepsilon + \mathbf{q}) = \nabla \cdot (\mathbf{T} \cdot \mathbf{v}) + \rho \mathbf{f} \cdot \mathbf{v} + \rho q_m, \tag{6.1c}$$

$$\frac{\partial \rho \mathbf{F}^{\mathrm{T}}}{\partial t} + \nabla \cdot \rho (\mathbf{v} \otimes \mathbf{F}^{\mathrm{T}} - \mathbf{F} \otimes \mathbf{v}) = 0, \tag{6.1d}$$

$$\frac{\partial \rho \mathbf{u}}{\partial t} + \nabla \cdot (\rho \mathbf{v} \otimes \mathbf{u}) = \rho \mathbf{v}. \tag{6.1e}$$

As noted above, the balance laws system (3.307) consists of six groups of equations ($\alpha = 1, \ldots 6$); the total number of scalar equations is $1+3+1+1+9+3 = 18$. However, after complementing the system with the constitutive equations, we should exclude one thermodynamic equation from the set. Indeed, remind (see Sect. 4.4) that the part of the constitutive equations is equivalent to the principal thermodynamic identity, which has been obtained by summation of two thermodynamic equations (see Sect. 4.3); therefore, the principal thermodynamic identity can replace one of them. One usually eliminates the entropy balance equation (3.307)

Y.I. Dimitrienko, *Nonlinear Continuum Mechanics and Large Inelastic Deformations*, Solid Mechanics and Its Applications 174, DOI 10.1007/978-94-007-0034-5_6,

at $\alpha = 4$. In this case the system (6.1a)–(6.1e) consists of 17 scalar equations and contains 27 scalar unknowns:

$$\rho,\ \mathbf{v},\ \mathbf{u},\ \mathbf{T},\ e,\ \theta,\ \mathbf{q},\ \mathbf{F}\ \|\ \mathbf{x},\ t. \tag{6.1f}$$

Each of the vector fields **v**, **u** and **q** is equivalent to three scalar functions, the tensor field **T** – to six scalar functions, and the field **F** – to nine scalar functions.

To close the system we should complement it with constitutive equations. A part of the equations has the same form for all ideal solids, namely the Fourier law (4.734) and the expressions for specific total energy $\varepsilon$ and internal energy $e$ (see (4.118) and (4.168)):

$$\mathbf{q} = -\boldsymbol{\lambda}\cdot\nabla\theta, \tag{6.2a}$$

$$\varepsilon = e + \frac{|\mathbf{v}|^2}{2} = \psi - \theta\frac{\partial\psi}{\partial\theta} + \frac{v^2}{2}. \tag{6.2b}$$

The remaining constitutive equations for ideal solids depend on the choice of a continuum model. Forms of the constitutive equations have been considered in Sect. 4.8. Remind the general representation of constitutive equations for elastic continua in the form (4.387) (see Sect. 4.8.10):

$$\mathbf{T} = \overset{(\mathrm{n})}{\mathcal{F}}_G(\mathbf{F},\ \theta), \tag{6.3a}$$

$$\psi = \psi(I^{(s)}_{\gamma G}(\overset{(\mathrm{n})}{\mathbf{C}}_G(\mathbf{F})),\ \theta) \equiv \psi(\mathbf{F},\ \theta), \tag{6.3b}$$

where the tensor functions $\overset{(\mathrm{n})}{\mathcal{F}}_G$ and the potential $\psi$ depend on the choice of model $A_n$, $B_n$, $C_n$ or $D_n$ and are defined by relations (4.385)–(4.391):

$$\begin{gathered}
\overset{(\mathrm{n})}{\mathcal{F}}_G(\mathbf{F},\theta) = {}^4\overset{(\mathrm{n})}{\mathbf{E}}_G\cdot\cdot\overset{(\mathrm{n})}{\mathbf{T}}_G = {}^4\overset{(\mathrm{n})}{\mathbf{E}}_G\cdot\cdot\mathcal{F}_G(\mathbf{F},\ \theta) = \sum_{\gamma=1}^{r}\varphi_\gamma\,{}^4\overset{(\mathrm{n})}{\mathbf{E}}_G\cdot\cdot\mathbf{I}^{(s)}_{\gamma G},\\
\mathbf{I}^{(s)}_{\gamma G} = \partial I^{(s)}_{\gamma G}/\partial\overset{(\mathrm{n})}{\mathbf{C}}_G,\quad \varphi_\gamma = \rho(\partial\varphi/\partial I^{(s)}_{\gamma G}),\\
I^{(s)}_{\gamma G} = h_G I^{(s)}_\gamma(\overset{(\mathrm{n})}{\mathbf{C}}_G) + (1-h_G)I^{(s)}_\gamma(\mathbf{O}^{\mathrm{T}}\cdot\overset{(\mathrm{n})}{\mathbf{C}}_G\cdot\mathbf{O}),\\
\overset{(\mathrm{n})}{\mathbf{C}}_G = \frac{1}{n-\mathrm{III}}\sum_{\alpha=1}^{3}(\lambda_\alpha^{n-\mathrm{III}}(h_G\mathring{\mathbf{p}}_\alpha\otimes\mathring{\mathbf{p}}_\alpha + (1-h_G)\mathbf{p}_\alpha\otimes\mathbf{p}_\alpha)) - \frac{\bar{h}_G}{n-\mathrm{III}}\mathbf{E},\\
{}^4\overset{(\mathrm{n})}{\mathbf{E}}_G = \sum_{\alpha,\beta=1}^{3}E_{\alpha\beta}\mathbf{p}_\alpha\otimes\mathbf{p}_\beta\otimes(h_G\mathring{\mathbf{p}}_\beta\otimes\mathring{\mathbf{p}}_\alpha + (1-h_G)\mathbf{p}_\beta\otimes\mathbf{p}_\alpha).
\end{gathered} \tag{6.4}$$

Here $I_\gamma^{(s)}$ are invariants of the tensors $\overset{(\mathrm{n})}{\mathbf{C}}_G$ relative to the symmetry group $\overset{\circ}{G}_s$ of a considered elastic continuum; and

$$\lambda_\alpha,\ \mathbf{p}_\alpha,\ \overset{\circ}{\mathbf{p}}_\alpha \ \| \ \mathbf{F} \tag{6.5}$$

are eigenvalues and eigenvectors, which are functions of the tensor $\mathbf{F}$; the coefficients $E_{\alpha\beta}$ are expressed in terms only of $\lambda_\alpha$ (see (4.39)); $\mathbf{O} = \mathbf{p}_1 \otimes \overset{\circ}{\mathbf{p}}{}^1$ is the rotation tensor accompanying the deformation, which also depends only on $\mathbf{F}$; and $h_G$ and $\bar{h}_G$ are function-indicators of the model type:

$$h_G = \begin{cases} 1, & G = A,\ B, \\ 0, & G = C,\ D, \end{cases} \qquad \bar{h}_G = \begin{cases} 1, & G = A,\ C, \\ 0, & G = B,\ D. \end{cases} \tag{6.4a}$$

The internal energy $e$ can be considered as a function of $\mathbf{F}$ and $\theta$, i.e.

$$e = \psi - \theta \frac{\partial \psi}{\partial \theta} = e(\mathbf{F},\ \theta). \tag{6.6}$$

Then we obtain the system of 10 scalar constitutive equations (the vector relation (6.2a) is equivalent to three scalar equations, relation (6.2b) where (6.3b) has been substituted is one scalar equation, and (6.3a) is equivalent to six scalar equations), which together with (6.1) constitute the closed system of 27 scalar equations for 27 scalar unknown functions (6.1f).

The number of equations and unknowns in the system can be decreased by substituting the constitutive equations (6.2) and (6.3) into (6.1b) and (6.1c); as a result, we obtain the system of 17 scalar equations for 17 unknowns:

$$\theta,\ \rho,\ \nu,\ \mathbf{v},\ \mathbf{F}\ \|\ \mathbf{x},\ t. \tag{6.7}$$

The set (6.1) is called the *$\theta RUVF$-system of thermoelasticity.*

*Remark 1.* Choosing the function $\overset{(\mathrm{n})}{\mathbf{T}}_G = \mathcal{F}_G(\overset{(\mathrm{n})}{\mathbf{C}}_G, \theta) = \sum_{\gamma=1}^{r} \varphi_\gamma I_{\gamma G}^{(s)}$ for the models $A_n$, $B_n$, $C_n$ or $D_n$ in a certain form (one of the forms considered in Sect. 4.8), we get special representations of the tensor functions (6.4). So if the model $A_n$ is considered, then the general representation of the function has the form (see (4.324)):

$$\mathcal{F}_A(\overset{(\mathrm{n})}{\mathbf{C}}, \theta) = {}^4\mathbf{M} \cdot\cdot \overset{(\mathrm{n})}{\mathbf{C}} + \xi \overset{(\mathrm{n})}{\mathbf{C}}{}^2 + {}^6\mathbf{L} \cdot\cdot\cdot\cdot (\overset{(\mathrm{n})}{\mathbf{C}} \otimes \overset{(\mathrm{n})}{\mathbf{C}}), \tag{6.8}$$

where ${}^4\mathbf{M}$ is the quasilinear elasticity tensor, $\xi$ is the parameter of quadratic elasticity, ${}^6\mathbf{L}$ is the tensor of quadratic elasticity; they depend, in general, on the invariants $I_\gamma^{(s)}(\overset{(n)}{\mathbf{C}})$. For quasilinear models $A_n$, we assume (see (4.330)) that $\xi = 0$, ${}^6\mathbf{L} = 0$:

$$\mathcal{F}_A(\overset{(n)}{\mathbf{C}}, \theta) = {}^4\mathbf{M} \cdot\cdot \overset{(n)}{\mathbf{C}}; \tag{6.8a}$$

and for linear models, in addition, ${}^4\mathbf{M} = J\,{}^4\overset{\circ}{\mathbf{M}}$, where ${}^4\overset{\circ}{\mathbf{M}}$ is the tensor of elastic moduli, which is independent of $I_\gamma^{(s)}(\overset{(n)}{\mathbf{C}})$.

On substituting the given expression for $\mathcal{F}_A$ into (6.4), we obtain the corresponding representation of the tensor function $\overset{(n)}{\mathcal{F}}_A$:

$$\mathbf{T} = \overset{(n)}{\mathcal{F}}_A(\mathbf{F}, \theta) = {}^4\overset{(n)}{\mathbf{E}} \cdot\cdot \mathcal{F}_A = {}^4\overset{(n)}{\mathbf{E}} \cdot\cdot \left({}^4\mathbf{M} \cdot\cdot \overset{(n)}{\mathbf{C}} + \xi \overset{(n)}{\mathbf{C}}{}^2 + {}^6\mathbf{L} \cdot\cdot\cdot\cdot (\overset{(n)}{\mathbf{C}} \otimes \overset{(n)}{\mathbf{C}})\right),$$

$$\overset{(n)}{\mathbf{C}} = \frac{1}{n - \mathrm{III}} \sum_{\alpha=1}^{3} \lambda_\alpha^{n-\mathrm{III}} \overset{\circ}{\mathbf{p}}_\alpha \otimes \overset{\circ}{\mathbf{p}}_\alpha, \quad {}^4\overset{(n)}{\mathbf{E}} = \sum_{\alpha,\beta=1}^{3} E_{\alpha\beta} \mathbf{p}_\alpha \otimes \mathbf{p}_\beta \otimes \overset{\circ}{\mathbf{p}}_\beta \otimes \overset{\circ}{\mathbf{p}}_\alpha.$$

□

*Remark 2.* In place of the energy balance equation (6.1c), we can use its equivalent – the heat influx equation (3.125):

$$\frac{\partial}{\partial t}\rho e + \nabla \cdot (\rho \mathbf{v} e + \mathbf{q}) = \mathbf{T} \cdot\cdot \nabla \otimes \mathbf{v}^{\mathrm{T}} + \rho q_m, \tag{6.9a}$$

or the entropy balance equation (3.307) at $\alpha = 4$, which can be written in the form (3.166):

$$\theta\rho\frac{d\eta}{dt} = -\nabla \cdot \mathbf{q} + \rho q_m. \tag{6.9b}$$

The constitutive equations (6.2b) in this case are replaced by

$$\eta = -\partial\psi/\partial\theta = \eta(\mathbf{F}, \theta). \tag{6.9c}$$

Equations (6.9a) or (6.9b) sometimes prove to be more convenient than the energy balance equation (6.1c), because they contain a smaller number of summands. However, in numerical solving Eq. (6.1c), having a completely divergence form (Eq. (6.9a) has a divergence form up to the summand $\mathbf{T} \cdot\cdot \nabla \otimes \mathbf{v}^{\mathrm{T}}$), is frequently more preferable.

The entropy balance equation (6.9b) can be rewritten by using the principal thermodynamic identity in the form (4.120):

$$\frac{d\psi}{dt} + \eta\frac{d\theta}{dt} - \frac{1}{\rho}\mathbf{T} \cdot\cdot \nabla \otimes \mathbf{v}^{\mathrm{T}} = 0. \tag{6.10}$$

Here we have used the definition (4.2) for the stress power $w_{(i)}$ and taken into account that for elastic media $w^* = 0$.

Differentiating (6.10) with respect to $\theta$ and taking into account that $\rho$ and $\nabla \otimes \mathbf{v}^{\mathrm{T}} = \dot{\mathbf{F}} \cdot \mathbf{F}^{-1}$ are expressed only in terms of $\mathbf{F}$ and are independent of $\theta$, and $\mathbf{F}$ and $\theta$ are independent variables, we obtain

$$\frac{d}{dt}\frac{\partial \psi}{\partial \theta} = -\frac{\partial \eta}{\partial \theta}\frac{d\theta}{dt} + \frac{1}{\rho}\frac{\partial \mathbf{T}}{\partial \theta} \cdot\cdot \nabla \otimes \mathbf{v}^{\mathrm{T}}.$$

On substituting the relation (6.9c) into the last equation, we find the following expression for the rate of changing specific entropy:

$$\frac{d\eta}{dt} = -\frac{\partial^2 \psi}{\partial \theta^2}\frac{d\theta}{dt} - \frac{1}{\rho}\frac{\partial \mathbf{T}}{\partial \theta} \cdot\cdot \nabla \otimes \mathbf{v}^{\mathrm{T}}. \tag{6.10a}$$

Substitution of (6.2a), (6.3a), and (6.10a) into (6.9b) gives the equation

$$\rho c_{\varepsilon}\left(\frac{\partial \theta}{\partial t} + \mathbf{v} \cdot \nabla \theta\right) = \nabla \cdot (\boldsymbol{\lambda} \cdot \nabla \theta) + \frac{\partial}{\partial \theta}\overset{(\mathrm{n})}{\mathcal{F}}(\mathbf{F}, \theta) \cdot\cdot \nabla \mathbf{v}^{\mathrm{T}} + \rho q_m, \tag{6.11}$$

which is called the *heat conduction equation for an elastic continuum in the spatial description*. Here we have introduced the function

$$c_{\varepsilon}(\mathbf{F}, \theta) = -\theta \frac{\partial^2 \psi}{\partial \theta^2}, \tag{6.11a}$$

that is called the *heat capacity of an elastic continuum* at fixed deformations. □

### 6.1.2 *θRVF-, θRUV-, and θUV-Systems of Dynamic Equations of Thermoelasticity*

Notice that the displacement vector $\mathbf{u}$ formally appears only in the kinematic equation (6.5), and this equation can be excluded from the general system. In this case the system is considered to be the *θRVF-system of thermoelasticity*. However, due to boundary conditions, Eq. (6.5) cannot always be excluded from the general system (see Sect. 6.3).

The number of equations and unknowns in the system (6.1)–(6.3) can be reduced by eliminating the deformation gradient $\mathbf{F}$ between the unknowns (6.1f). To do this, one should recall that the inverse gradient has a vector potential (see (2.77)):

$$\mathbf{F}^{-1} = \mathbf{E} - \nabla \otimes \mathbf{u}^{\mathrm{T}}. \tag{6.12}$$

This relation is, in essence, a compatibility equation: the dynamic compatibility equation (6.1d) has been derived from this equation (see Sect. 3.7.1). Therefore, relation (6.12) can be considered in place of Eq. (6.1d). Then, substituting (6.12) into the constitutive equations (6.3), we represent them as functions of the inverse gradient (see Sect. 4.8.10, formulae (4.392) and (4.393)):

$$\begin{aligned}
\mathbf{T} &= \overset{(n)}{\mathcal{F}}_G(\mathbf{F}^{-1},\ \theta),\\
\psi &= \psi(I^{(s)}_{\gamma G}(\overset{(n)}{\mathbf{C}}_G(\mathbf{F}^{-1})),\ \theta) \equiv \theta(\mathbf{F}^{-1},\ \theta),\\
e &= \psi - \theta\frac{\partial\psi}{\partial\theta} = e(\mathbf{F}^{-1},\ \theta).
\end{aligned} \tag{6.13a}$$

Here the tensor function $\overset{(n)}{\mathcal{F}}_G$ has the same form (6.4) but $\lambda_\alpha$, $\mathbf{p}_\alpha$ and $\overset{\circ}{\mathbf{p}}_\alpha$ are considered to depend on $\mathbf{F}^{-1}$:

$$\lambda_\alpha,\ \overset{\circ}{\mathbf{p}}_\alpha,\ \mathbf{p}_\alpha\ \parallel\ \mathbf{F}^{-1}. \tag{6.13b}$$

Substituting the constitutive equations (6.13a) and (6.2a) into Eqs. (6.1b) and (6.1c), we obtain the *$\theta RUV$-system of thermoelasticity*

$$\frac{\partial\rho}{\partial t} + \nabla\cdot\rho\mathbf{v} = 0, \tag{6.14a}$$

$$\frac{\partial\rho\mathbf{v}}{\partial t} + \nabla\cdot\rho\mathbf{v}\otimes\mathbf{v} = \nabla\cdot\overset{(n)}{\mathcal{F}}_G(\mathbf{E}-\nabla\otimes\mathbf{u}^{\mathrm{T}},\theta) + \rho\mathbf{f}, \tag{6.14b}$$

$$\frac{\partial\rho\varepsilon}{\partial t} + \nabla\cdot\rho\mathbf{v}\varepsilon = \nabla\cdot(\boldsymbol{\lambda}\cdot\nabla\theta + \overset{(n)}{\mathcal{F}}_G(\mathbf{E}-\nabla\otimes\mathbf{u}^{\mathrm{T}},\theta)\cdot\mathbf{v}) + \rho\mathbf{f}\cdot\mathbf{v} + \rho q_m, \tag{6.14c}$$

$$\frac{\partial\rho\mathbf{u}}{\partial t} + \nabla\cdot(\rho\mathbf{v}\otimes\mathbf{u}) = \rho\mathbf{v}, \tag{6.14d}$$

$$\varepsilon = e(\mathbf{E}-\nabla\otimes\mathbf{u}^{\mathrm{T}},\ \theta) + \frac{v^2}{2}, \tag{6.14e}$$

which consists of eight scalar equations for eight scalar unknowns:

$$\theta,\ \rho,\ \mathbf{u},\ \mathbf{v}\ \parallel\ \mathbf{x},\ t. \tag{6.15}$$

*Remark 3.* The dependence of tensors ${}^4\overset{(n)}{\mathbf{E}}_G$ and $\overset{(n)}{\mathbf{C}}_G$ on $\mathbf{F}^{-1}$ (see Eqs. (6.4) and (6.13b)) is sufficiently complex; and, in general, it even hasn't got an explicit analytic expression (eigenvalues $\lambda_\alpha$ and eigenvectors $\mathbf{p}_\alpha$ and $\overset{\circ}{\mathbf{p}}_\alpha$, in general, cannot be expressed in terms of $\mathbf{F}^{-1}$ analytically, but only in the form of a computing algorithm (see Sect. 2.3.2)). Due to this, the dependence of the tensor $\mathbf{T}$ on the displacement gradient $\nabla\otimes\mathbf{u}$ in the theory of finite elastic deformations has a rather complex form. Even for linear models (see Sect. 4.8.7 and (6.8)), where the tensors $\overset{(n)}{\mathbf{T}}$ and $\overset{(n)}{\mathbf{C}}_G$ are linearly connected, the dependences $\overset{(n)}{\mathbf{C}}_G(\mathbf{F}^{-1})$ and ${}^4\overset{(n)}{\mathbf{E}}_G(\mathbf{F}^{-1})$ lead to a complicated nonlinear connection between $\mathbf{T}$ and $\nabla\otimes\mathbf{u}$. However, the models $A_{\mathrm{I}}$ and $B_{\mathrm{I}}$ are the exceptions; for them the tensors $\overset{\mathrm{I}}{\mathbf{G}}_A = \overset{\mathrm{I}}{\mathbf{C}} = \mathbf{\Lambda}$, $\overset{\mathrm{I}}{\mathbf{C}}_B = \overset{\mathrm{I}}{\mathbf{G}}$ and ${}^4\overset{\mathrm{I}}{\mathbf{E}}_A = {}^4\overset{\mathrm{I}}{\mathbf{E}}_B = {}^4\overset{\mathrm{I}}{\mathbf{E}}$ can be expressed in terms of $\mathbf{F}^{-1}$ in a sufficiently simple form (see Exercise 4.2.21 and (2.73a)), and the tensor dependences $\overset{\mathrm{I}}{\mathbf{C}}_G(\mathbf{F}^{-1})$ and ${}^4\overset{\mathrm{I}}{\mathbf{E}}_G(\mathbf{F}^{-1})$ are quadratic. The $\theta RUV$-system of dynamic equations of elasticity theory (6.14) for these models is essentially simplified and has the form (for the model $A_{\mathrm{I}}$)

$$\begin{cases} \partial\rho/\partial t + \nabla\cdot\rho\mathbf{v} = 0, \\ \partial\rho\mathbf{v}/\partial t + \nabla\cdot\rho\mathbf{v}\otimes\mathbf{v} = \nabla\cdot\mathbf{T} + \rho\mathbf{f}, \\ \partial\rho\varepsilon/\partial t + \nabla\cdot\rho\mathbf{v}\varepsilon = \nabla\cdot(\boldsymbol{\lambda}\cdot\nabla\theta + \mathbf{T}\cdot\mathbf{v}) + \rho\mathbf{f}\cdot\mathbf{v} + \rho q_m, \\ \partial\rho\mathbf{u}/\partial t + \nabla\cdot(\rho\mathbf{v}\otimes\mathbf{u}) = \rho\mathbf{v}, \end{cases} \tag{6.16a}$$

$$\begin{cases} \mathbf{T} = \sum_{\gamma=1}^{r}\varphi_\gamma \mathbf{F}^{-1\mathrm{T}}\cdot\mathbf{I}_\gamma^{(s)}\cdot\mathbf{F}^{-1}, \\ \mathbf{I}_\gamma^{(s)} = \partial I_\gamma^{(s)}/\partial\mathbf{\Lambda}, \quad \varphi_\gamma = \rho(\partial\psi/\partial I_\gamma^{(s)}), \quad \psi = \psi(I_\gamma^{(s)}(\mathbf{\Lambda}),\theta), \\ \mathbf{\Lambda} = (1/2)(\nabla\otimes\mathbf{u} + \nabla\otimes\mathbf{u}^{\mathrm{T}} - \nabla\otimes\mathbf{u}^{\mathrm{T}}\cdot\nabla\otimes\mathbf{u}), \\ \mathbf{F}^{-1} = \mathbf{E} - \nabla\otimes\mathbf{u}^{\mathrm{T}}, \\ \varepsilon = e + v^2/2, \quad e = \psi - \theta(\partial\psi/\partial\theta). \end{cases} \tag{6.16b}$$

The substitution $\mathbf{\Lambda} \to \overset{\mathrm{I}}{\mathbf{G}}$ gives a similar system for the model $B_{\mathrm{I}}$. There is no need to find eigenbases and eigenvalues $\lambda_\alpha$ for the models $A_{\mathrm{I}}$ and $B_{\mathrm{I}}$, that discerns these models among the remaining ones and explains their wide application to the spatial description.

The models $A_{\mathrm{V}}$ and $B_{\mathrm{V}}$ are of a higher level of complexity. For the model $A_{\mathrm{V}}$, constitutive equations (6.16b) are replaced by

$$\begin{cases} \mathbf{T} = \sum_{\gamma=1}^{r}\varphi_\gamma \mathbf{F}\cdot\mathbf{I}_\gamma^{(s)}\cdot\mathbf{F}^{\mathrm{T}}, \\ \mathbf{I}_\gamma^{(s)} = \partial I_\gamma^{(s)}/\partial\mathbf{C}, \quad \varphi_\gamma = \rho(\partial\psi/\partial I_\gamma^{(s)}), \quad \psi = \psi(I_\gamma^{(s)}(\mathbf{C}),\theta), \\ \mathbf{C} = (1/2)(\mathbf{F}^{\mathrm{T}}\cdot\mathbf{F} - \mathbf{E}), \quad \mathbf{F} = (\mathbf{E} - \nabla\otimes\mathbf{u}^{\mathrm{T}})^{-1}, \\ \varepsilon = e + (v^2/2), \quad e = \psi - \theta(\partial\psi/\partial\theta). \end{cases} \tag{6.16c}$$

For this model, there is no need to determine eigenvectors and eigenvalues as well; but, unlike the models $A_{\mathrm{I}}$ and $B_{\mathrm{I}}$, one should invert the inverse deformation gradient in order to restore the tensor $\mathbf{F}$.

The models $A_{\mathrm{II}}$, $B_{\mathrm{II}}$ and $A_{\mathrm{IV}}$, $B_{\mathrm{IV}}$ are the most complicated for numerical solving the problems with arbitrary geometry of the domain $V$; for them we need to calculate $\lambda_\alpha$, $\mathbf{p}_\alpha$ and $\overset{\circ}{\mathbf{p}}_\alpha$. □

*Remark 4.* Notice that the density $\rho$ in the systems (6.14) and (6.16) is considered to be an independent unknown value, therefore one should retain the multiplier $J = \rho/\overset{\circ}{\rho}$ in the constitutive equations (see Sect. 4.8.7) in its form not expressing the factor in terms of the tensors $\overset{(n)}{\mathbf{C}}$, $\overset{(n)}{\mathbf{A}}$, $\overset{(n)}{\mathbf{G}}$ or $\overset{(n)}{\mathbf{g}}$ by formulae (4.81) and (4.82). But if such substitution has been performed, then we obtain that $\rho$ can be expressed in terms of $\mathbf{F}$, and the relation $\rho = \rho(\mathbf{F})$ is the well-known continuity equation in Lagrangian variables (3.8):

$$\rho = \overset{\circ}{\rho} \det \mathbf{F}^{-1} = \overset{\circ}{\rho} \det (\mathbf{E} - \nabla \otimes \mathbf{u}^{\mathrm{T}}). \tag{6.17}$$

Since this equation is exactly equivalent to the continuity equation (6.14a) (see Sect. 3.1), so in this case the continuity equation should be excluded from the general system (6.14) or (6.16), and Eq. (6.17) should be substituted into the remaining equations of the systems including the constitutive equations. Thus, we obtain the $\theta UV$-system (6.14b)–(6.14e), (6.17) consisting of seven equations for the seven unknowns

$$\theta,\ \mathbf{u},\ \mathbf{v}\ \|\ \mathbf{x},\ t.\ \square \tag{6.18}$$

*Remark 5.* Reduction of the number of equations and unknown functions usually makes the system more complicated. For example, the $\theta RUVF$-system (6.1) involves the first-order derivatives with respect to $\mathbf{x}$ and $t$, but the $\theta UV$-system is of the mixed order: for $\mathbf{v}$ – the first order, and for $\mathbf{u}$ – the second order with respect to $\mathbf{x}$ and the first order with respect to $t$. Therefore, often it is more convenient to consider the $\theta RUVF$-system; although it contains more equations, all the equations are of the same type. □

### 6.1.3 *TθRUVF-system of Dynamic Equations of Thermoelasticity*

Sometimes in solving special problems for solids it is more convenient to increase the number of equations even in comparison with the $\theta RUVF$-system (6.1) by introducing the tensor $\mathbf{T}$ or the generalized energetic stress tensor $\overset{(n)}{\mathbf{T}}_G$ (see (4.91)) as an additional unknown tensor.

To do this, consider the constitutive equations 'in rates' (4.442). Expressing the tensor $\mathbf{T}$ in terms of the energetic stress tensors $\overset{(n)}{\mathbf{T}}_G$ by relations (4.37), (4.67), and (4.387):

$$\mathbf{T} = {}^4\overset{(n)}{\mathbf{E}} \cdot\cdot \overset{(n)}{\mathbf{T}} = {}^4\overset{(n)}{\mathbf{Q}} \cdot\cdot \mathbf{S} = {}^4\overset{(n)}{\mathbf{E}}_G \cdot\cdot \overset{(n)}{\mathbf{T}}_G, \tag{6.19}$$

we obtain the *$T\theta RUVF$-system of thermoelasticity*:

$$\frac{\partial \rho}{\partial t} + \nabla \cdot \rho \mathbf{v} = 0, \tag{6.20}$$

$$\frac{\partial \rho \mathbf{v}}{\partial t} + \nabla \cdot \rho \mathbf{v} \otimes \mathbf{v} = \nabla \cdot ({}^4\overset{(n)}{\mathbf{E}}_G \cdot\cdot \overset{(n)}{\mathbf{T}}_G) + \rho \mathbf{f}, \tag{6.21}$$

$$\frac{\partial \rho \varepsilon}{\partial t} + \nabla \cdot \rho \mathbf{v} \varepsilon = \nabla \cdot (\boldsymbol{\lambda} \cdot \nabla \theta + ({}^4\overset{(n)}{\mathbf{E}}_G \cdot \cdot \overset{(n)}{\mathbf{T}}_G) \cdot \mathbf{v}) + \rho \mathbf{f} \cdot \mathbf{v} + \rho q_m, \tag{6.22}$$

$$\frac{\partial \rho \mathbf{u}}{\partial t} + \nabla \cdot (\rho \mathbf{v} \otimes \mathbf{u}) = \rho \mathbf{v}, \tag{6.23}$$

$$\frac{\partial \rho \overset{(n)}{\mathbf{T}}_G}{\partial t} + \nabla \cdot (\rho \mathbf{v} \otimes \overset{(n)}{\mathbf{T}}_G) = \rho^4 \overset{(n)}{\mathbf{P}}_{Gh}(\mathbf{F}, \theta) \cdot \cdot \nabla \otimes \mathbf{v}^{\mathrm{T}}$$
$$+ \rho \mathbf{Z}_{Gh} \cdot \overset{(n)}{\mathbf{T}}_G + \rho \overset{(n)}{\mathbf{T}}_G \cdot \mathbf{Z}_{Gh} + \rho \overset{(n)}{\mathbf{T}}_{\theta G} \dot{\theta}, \tag{6.24}$$

$$\frac{\partial \rho \mathbf{F}^{\mathrm{T}}}{\partial t} + \nabla \cdot \rho (\mathbf{v} \otimes \mathbf{F}^{\mathrm{T}} - \mathbf{F} \otimes \mathbf{v}) = 0, \tag{6.25}$$

which consists of 23 equations for $6 + 1 + 1 + 3 + 3 + 9 = 23$ scalar unknowns:

$$\overset{(n)}{\mathbf{T}}_G, \ \theta, \ \rho, \ \mathbf{u}, \ \mathbf{v}, \ \mathbf{F} \ \| \ \mathbf{x}, \ t. \tag{6.26}$$

An advantage of the system is that all its equations have the same divergence form.

The tensors appearing in Eq. (6.24) have been determined in Sect. 4.8.12: the tensor ${}^4\overset{(n)}{\mathbf{P}}_{Gh}$ – by formulae (4.423), (4.438), (4.106), (4.99), and (2.375), the tensor $\mathbf{Z}_{Gh}$ – by (4.440) and (2.374), and the tensor $\mathbf{T}_{\theta G}$ – by (4.423). All the tensors are functions of $\mathbf{F}$ and $\nabla \otimes \mathbf{v}$.

*Remark 6.* Equation (6.24) has been derived in Sect. 4.8.12 for elastic continua models not containing the rotation tensor $\mathbf{O}$, i.e. for models $A_n$, $B_n$, and also for isotropic materials described by models $C_n$ and $D_n$.

Therefore, for these models the $T\theta RUVF$-system (6.20)–(6.25) holds. □

### *6.1.4 Component Form of the Dynamic Equation System of Thermoelasticity in the Spatial Description*

To rewrite the equation systems given above in components, one usually uses fixed bases $\widetilde{\mathbf{r}}_i$ and $\widetilde{\mathbf{r}}^i$ (see Sect. 2.1.7). Local bases $\mathbf{r}_i$ and $\mathbf{r}^i$ are rarely applied for the reason mentioned in Sect. 3.1.2.

The $\theta RUVF$-system (6.1) in basis $\widetilde{\mathbf{r}}_i$ has the component form

$$\frac{\partial \rho}{\partial t} + \widetilde{\nabla}_j(\rho \widetilde{v}^j) = 0,$$

$$\frac{\partial \rho \widetilde{v}^i}{\partial t} + \widetilde{\nabla}_j(\rho \widetilde{v}^i \widetilde{v}^j - \widetilde{T}^{ji}) = \rho \widetilde{f}^{\,i},$$

$$\frac{\partial \rho \varepsilon}{\partial t} + \widetilde{\nabla}_j(\widetilde{v}^i(\rho\varepsilon\delta_i^j - \widetilde{T}^{jk}\widetilde{g}_{ki}) + \widetilde{q}^j) = \rho\widetilde{f}^i\widetilde{v}^j\widetilde{g}_{ij} + \rho q_m,$$

$$\frac{\partial \rho \widetilde{F}^i_j}{\partial t} + \widetilde{\nabla}_k(\rho\widetilde{v}^k\widetilde{F}^i_j - \rho\widetilde{F}^k_j\widetilde{v}^i) = 0, \tag{6.27}$$

$$\frac{\partial \rho \widetilde{u}^i}{\partial t} + \widetilde{\nabla}_j \rho\widetilde{v}^j\widetilde{u}^i = \rho\widetilde{v}^i.$$

The constitutive equations (6.2a) and (6.2b) in components are written as follows:

$$\widetilde{q}^i = -\widetilde{\lambda}^{ij}\widetilde{\nabla}_j\theta, \quad \varepsilon = \psi - \theta\frac{\partial\psi}{\partial\theta} + \frac{1}{2}\widetilde{v}^i\widetilde{v}^j\widetilde{g}_{ij}. \tag{6.28}$$

The constitutive equations (6.3) and (6.4) have the component forms

$$\widetilde{T}^{ij} = \overset{(\mathrm{n})}{\widetilde{\mathcal{F}}}{}^{ij}_G(\widetilde{F}^k_l, \theta),$$

$$\psi = \psi(I^{(s)}_{\gamma G}(\overset{(\mathrm{n})}{\widetilde{\mathcal{F}}}{}^{ij}_G(\widetilde{F}^k_l)), \theta) \equiv \psi(\widetilde{F}^k_l, \theta),$$

$$\overset{(\mathrm{n})}{\widetilde{\mathcal{F}}}{}^{ij}_G = \sum_{\gamma=1}^{3}\varphi_\gamma \overset{(\mathrm{n})}{\widetilde{E}}{}^{ijkl}_G \widetilde{I}^{(s)}_{\gamma G,kl},$$

$$\widetilde{I}^{(s)}_{\gamma G,kl} = \partial I^{(s)}_{\gamma G}/\partial \overset{(\mathrm{n})}{\widetilde{C}}{}^{kl}_G, \quad \varphi_\gamma = \rho(\partial\psi/\partial I^{(s)}_{\gamma G}), \tag{6.29}$$

$$I^{(s)}_{\gamma G} = h_G I^{(s)}_\gamma(\overset{(\mathrm{n})}{\widetilde{C}}{}^{kl}_G) + (1-h_G)I^{(s)}_\gamma(\overset{(\mathrm{n})}{\widetilde{C}}{}^{ij}_G\widetilde{O}_{ik}\widetilde{O}_{jl}),$$

$$\overset{(\mathrm{n})}{\widetilde{C}}{}^{ij}_G = \frac{1}{n-\mathrm{III}}\sum_{\gamma=1}^{3}\lambda_\alpha^{n-\mathrm{III}}(h_G\overset{\circ}{\widetilde{Q}}{}^i_\alpha\overset{\circ}{\widetilde{Q}}{}^j_\alpha + (1-h_G)\widetilde{Q}^i_\alpha\widetilde{Q}^j_\alpha) - \frac{\bar{h}_G}{n-\mathrm{III}}\widetilde{g}^{ij},$$

$$\widetilde{E}^{ijkl}_G = \sum_{\alpha,\beta=1}^{3} E_{\alpha\beta}\widetilde{Q}^i_\alpha\widetilde{Q}^j_\alpha(h_G\overset{\circ}{\widetilde{Q}}{}^k_\beta\overset{\circ}{\widetilde{Q}}{}^l_\alpha + (1-h_G)\widetilde{Q}^k_\beta\widetilde{Q}^l{}_\alpha).$$

Here eigenvalues $\lambda_\alpha$ and Jacobian matrices $\overset{\circ}{\widetilde{Q}}{}^i_\alpha$, $\widetilde{Q}^i_\alpha$ of eigenvectors $\overset{\circ}{\mathbf{p}}_\alpha$, $\mathbf{p}_\alpha$ (see Exercise 4.2.20), according to (6.5), are functions of components $\widetilde{F}^i_j$ of the deformation gradient $\mathbf{F}$ in the basis $\widetilde{\mathbf{r}}_i$:

$$\lambda_\alpha, \ \overset{\circ}{\widetilde{Q}}{}^i_\alpha, \ \widetilde{Q}^i_\alpha \ \| \ \widetilde{F}^i_j. \tag{6.30}$$

To rewrite the $\theta RUV$-system (6.14) in components, one should only exclude the dynamic compatibility equation from (6.27) and replace it by the component form of relations (6.13a):

$$(\widetilde{F}^{-1})^k_l = (\widetilde{F}^k_l)^{-1} = \delta^k_l - \widetilde{\nabla}_l\widetilde{u}^k. \tag{6.31}$$

As a result, we obtain the system

$$\begin{aligned}
&\frac{\partial \rho}{\partial t} + \widetilde{\nabla}_j(\rho \widetilde{v}^j) = 0,\\
&\frac{\partial \rho \widetilde{v}^i}{\partial t} + \widetilde{\nabla}_j(\rho \widetilde{v}^i \widetilde{v}^j - \widetilde{T}^{ji}) = \rho \widetilde{f}^{\,i},\\
&\frac{\partial \rho \varepsilon}{\partial t} + \widetilde{\nabla}_j(\widetilde{v}^i(\rho\varepsilon\delta_i^j - \widetilde{g}_{mi}\widetilde{T}^{jm})) = \widetilde{\nabla}_j(\widetilde{\lambda}^{ij}\widetilde{\nabla}_i\theta) + \rho \widetilde{f}^{\,i}\widetilde{v}^j\widetilde{g}_{ij} + \rho q_m,\\
&\frac{\partial \rho \widetilde{u}^i}{\partial t} + \widetilde{\nabla}_j \rho \widetilde{v}^j \widetilde{u}^i = \rho \widetilde{v}^i,\\
&\varepsilon = \psi - \theta\frac{\partial \psi}{\partial \theta} + \frac{1}{2}\widetilde{v}^i\widetilde{v}^j\widetilde{g}_{ij},\\
&\overset{(n)}{\widetilde{T}}{}^{ij} = \overset{(n)}{\widetilde{\mathcal{F}}}{}^{ij}_{G}(\delta_l^k - \widetilde{\nabla}_l \widetilde{u}^k,\, \theta), \quad \psi = \psi(\delta_k^l - \widetilde{\nabla}_l\widetilde{u}^k,\, \theta).
\end{aligned} \tag{6.32}$$

Here, just as in (6.13b), we have turned to the inverse gradient $\mathbf{F}^{-1}$ in the constitutive equations (6.29).

The component form of the system (6.16) coincides with (6.32), and the tensor function $\overset{(n)}{\widetilde{\mathcal{F}}}_{ij}(F_l^k)$ corresponding to this system (the model $A_{\mathrm{I}}$) is given in Exercise 6.1.3.

The $T\theta RUVF$-system has the component form

$$\begin{aligned}
&\frac{\partial \rho}{\partial t} + \widetilde{\nabla}_j(\rho \widetilde{v}^j) = 0,\\
&\frac{\partial \rho \widetilde{v}^i}{\partial t} + \widetilde{\nabla}_j(\rho \widetilde{v}^i \widetilde{v}^j - \overset{(n)}{\widetilde{E}}{}^{ijkl}_{G}\overset{(n)}{\widetilde{T}}_{Gkl}) = \rho \widetilde{f}^{\,i},\\
&\frac{\partial \rho \varepsilon}{\partial t} + \widetilde{\nabla}_j(\widetilde{v}^i(\rho\varepsilon\delta_i^j - \overset{(n)}{\widetilde{E}}{}^{kjml}_{G}\overset{(n)}{\widetilde{T}}_{Gml}\widetilde{g}_{ki})) = \widetilde{\nabla}_j(\widetilde{\lambda}^{ij}\widetilde{\nabla}_i\theta) + \rho \widetilde{f}^{\,i}\widetilde{v}^j\widetilde{g}_{ij} + \rho q_m,\\
&\frac{\partial \rho \widetilde{F}_j^i}{\partial t} + \widetilde{\nabla}_k(\rho \widetilde{v}^k \widetilde{F}_j^i - \rho \widetilde{F}_j^k \widetilde{v}^i) = 0,\\
&\frac{\partial \rho \widetilde{u}^i}{\partial t} + \widetilde{\nabla}_j(\rho \widetilde{v}^j \widetilde{u}^i) = \rho \widetilde{v}^i,\\
&\frac{\partial \rho \overset{(n)}{\widetilde{T}}_{Gij}}{\partial t} + \widetilde{\nabla}_k(\rho \widetilde{v}^k \overset{(n)}{\widetilde{T}}_{Gij})\\
&\quad = \rho \overset{(n)}{\widetilde{P}}{}_{Ghij}{}^{k}{}_{l}\widetilde{\nabla}_k\widetilde{v}^l + \rho \widetilde{Z}{}_{Ghi}^{k}\overset{(n)}{\widetilde{T}}_{Gkj} + \rho \widetilde{Z}{}_{Gh}{}^{k}{}_{j}\overset{(n)}{\widetilde{T}}_{Gik} + \rho \overset{(n)}{\widetilde{T}}_{\theta Gij}\frac{\partial \theta}{\partial t},
\end{aligned} \tag{6.33}$$

where $\overset{(n)}{\widetilde{P}}{}_{Ghij}{}^{k}{}_{l}$ and $\widetilde{Z}{}_{Ghi}^{k}$ are components of the tensors $\overset{(n)}{\mathbf{P}}_{Gh}$ and $\mathbf{Z}_{Gh}$ with respect to the basis $\widetilde{\mathbf{r}}_i$.

### 6.1.5 The Model of Quasistatic Processes in Elastic Solids at Large Deformations

Continuum mechanics widely uses the concept of models of processes (i.e. motions of media under some assumptions). One of them is the model of quasistatic processes (i.e., in a certain sense, very slow motions from $\overset{\circ}{\mathcal{K}}$ to $\mathcal{K}$).

**Definition 6.1.** One can say that this is the *model of quasistatic processes* in a solid, if the terms containing the velocity **v** can be neglected in Eqs. (6.1b) and (6.1c) of the $\theta RUVF$-system (6.1) in comparison with the other terms; i.e. in these equations one assumes that

$$\mathbf{v} \equiv 0. \tag{6.34}$$

Figure 6.1 shows the schematic graph of varying all known and unknown functions $\Omega$ with time for quasistatic and dynamic processes (for the latter motions we cannot assume the model of quasistatic processes).

Due to (6.34), the momentum balance equation (6.1b) and the heat conduction equation (6.11) used in place of the energy balance equation (6.1c) take the forms

$$\begin{aligned} &\nabla \cdot \mathbf{T} + \rho \mathbf{f} = 0, \\ &\rho c_\varepsilon \frac{\partial \theta}{\partial t} = \nabla \cdot (\boldsymbol{\lambda} \cdot \nabla \theta) + \rho q_m. \end{aligned} \tag{6.35}$$

Complementing this system with constitutive equations (6.12) and (6.13):

$$\mathbf{T} = \overset{(\mathrm{n})}{\mathcal{F}}_G(\mathbf{E} - \nabla \otimes \mathbf{u}^{\mathrm{T}},\ \theta), \tag{6.36}$$

we obtain the closed system of four scalar equations for four scalar functions:

$$\theta,\ \mathbf{u}\ \|\ \mathbf{x},\ t. \tag{6.37}$$

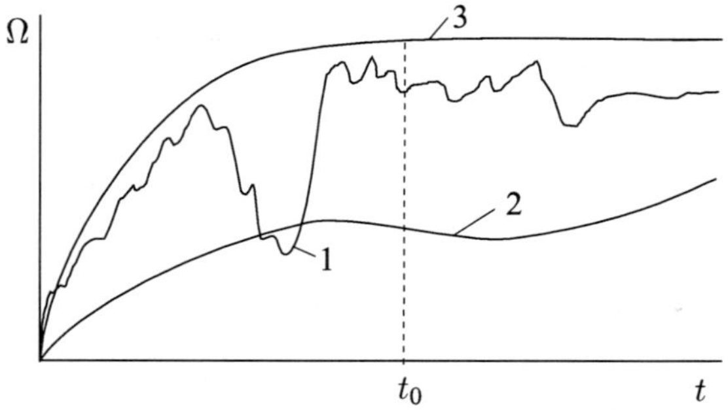


**Fig. 6.1** The typical graph of functions of time for dynamic (*1*), quasistatic (*2*), and static (*3*) processes

This system is called the *quasistatic equation system of thermoelasticity in the spatial description*. The density $\rho$ in the system is assumed to be expressed in terms of displacements by the continuity equation (6.17).

The remaining equations of the system (6.1) for the model of quasistatic processes hold in general only approximately, and, therefore, they are not considered. However, if one considers the equation system (6.35) for *static processes*, for which starting from a certain time $t_0$ the functions $\theta$ and $\mathbf{u}$ are independent of $t$ (see Fig. 6.1), then Eqs. (6.1a), (6.1d), and (6.1e) take the forms

$$\partial\rho/\partial t = 0, \quad \partial\mathbf{F}/\partial t = 0, \quad \partial\mathbf{u}/\partial t = 0, \tag{6.38}$$

and hold exactly at $t > 0$; and their solution is written as follows:

$$\rho = \rho(\mathbf{x}), \quad \mathbf{F} = \mathbf{F}(\mathbf{x}), \quad \mathbf{u} = \mathbf{u}(\mathbf{x}). \tag{6.39}$$

Here the right-hand sides of the relations can be found after solving the system (6.35), (6.36).

## Exercises for 6.1

**6.1.1.** Show that for the quasilinear models $A_{\mathrm{I}}$ (see Remark 1 or Sect. 4.8.7) the constitutive equations (6.8a) have the explicit analytic form similar to (6.16b):

$$\mathbf{T} = \mathbf{F}^{-1\mathrm{T}} \cdot ({}^4\mathbf{M} \cdot\cdot\, \mathbf{\Lambda}) \cdot \mathbf{F}^{-1},$$
$$\mathbf{\Lambda} = \frac{1}{2}(\nabla \otimes \mathbf{u} + \nabla \otimes \mathbf{u}^{\mathrm{T}} - \nabla \otimes \mathbf{u}^{\mathrm{T}} \cdot \nabla \otimes \mathbf{u}), \quad \mathbf{F}^{-1} = \mathbf{E} - \nabla \otimes \mathbf{u}^{\mathrm{T}},$$

or

$$\mathbf{T} = (\mathbf{E} - \nabla \otimes \mathbf{u}) \cdot \left({}^4\mathbf{M} \cdot\cdot \left(\mathbf{E} - \frac{1}{2}\nabla \otimes \mathbf{u}^{\mathrm{T}}\right) \cdot \nabla \otimes \mathbf{u}\right) \cdot (\mathbf{E} - \nabla \otimes \mathbf{u}^{\mathrm{T}}).$$

**6.1.2.** Using formula (4.322), show that for isotropic continua the relations (6.9a) can be written as follows:

$$\mathbf{T} = \mathbf{F}^{-1\mathrm{T}} \cdot (\psi_1 \mathbf{E} + \psi_2 \mathbf{\Lambda} + \psi_3 \mathbf{\Lambda}^2) \cdot \mathbf{F}^{-1},$$
$$\psi_1 = \varphi_1 + \varphi_2 I_1 + \varphi_3 I_2, \quad -\psi_2 = \varphi_2 + I_1 \varphi_3,$$
$$\psi_3 = \varphi_3, \quad I_\alpha = I_\alpha(\mathbf{\Lambda}), \ \alpha = 1, 2, 3.$$

Using formula (4.337), show that for the linear model $A_{\mathrm{I}}$ of isotropic continua this relation takes the form

$$\mathbf{T} = J(l_1 I_1(\mathbf{\Lambda})\mathbf{F}^{-1\mathrm{T}} \cdot \mathbf{F}^{-1} + 2l_2 \mathbf{F}^{-1\mathrm{T}} \cdot \mathbf{\Lambda} \cdot \mathbf{F}^{-1}),$$

or

$$\mathbf{T} = J\Big(l_1\Big(\nabla\cdot\mathbf{u} - \frac{1}{2}\nabla\otimes\mathbf{u}^{\mathrm{T}}\cdot\cdot\,\nabla\otimes\mathbf{u}\Big)(\mathbf{E}-\nabla\otimes\mathbf{u})\cdot(\mathbf{E}-\nabla\otimes\mathbf{u}^{\mathrm{T}})$$
$$+2l_2(\mathbf{E}-\nabla\otimes\mathbf{u})\cdot(\mathbf{E}-\frac{1}{2}\nabla\otimes\mathbf{u}^{\mathrm{T}})\cdot\nabla\otimes\mathbf{u}\cdot(\mathbf{E}-\nabla\otimes\mathbf{u})^{\mathrm{T}}\Big),$$

where $l_1$ and $l_2$ are the constants.

**6.1.3.** Show that constitutive equations (6.16b) for the model $A_{\mathrm{I}}$ in basis $\widetilde{\mathbf{r}}_i$ have the component form

$$\widetilde{T}^{ij} = \sum_{\gamma=1}^{r}\varphi_\gamma \widetilde{g}^{ik}\widetilde{g}^{jl}(\widetilde{F}^{-1})_k^m(\widetilde{F}^{-1})_l^s\widetilde{I}^{(s)}_{\gamma ms},$$
$$\widetilde{I}^{(s)}_{\gamma ms} = \partial I^{(s)}_\gamma/\partial\widetilde{\Lambda}^{ms},\quad \varphi_\gamma = \rho(\partial\psi/\partial I^{(s)}_\gamma),\quad \psi = \psi(I^{(s)}_\gamma(\widetilde{\Lambda}^{ms}),\theta),$$
$$\widetilde{\Lambda}^{ms} = \frac{1}{2}(\widetilde{\nabla}^m\widetilde{u}^s + \widetilde{\nabla}^s\widetilde{u}^m - \widetilde{\nabla}^k\widetilde{u}^m\widetilde{\nabla}_k\widetilde{u}_s),\quad (\widetilde{F}^{-1})_k^m = \delta_k^m - \widetilde{\nabla}_k\widetilde{u}^m.$$

Using the result of Exercise 6.1.1, show that for the quasilinear model $A_{\mathrm{I}}$ this component representation takes the form

$$\widetilde{T}^{ij} = \widetilde{g}^{ik}\widetilde{g}^{jl}(\widetilde{F}^{-1})_k^p(\widetilde{F}^{-1})_l^q\widetilde{M}_{pqms}\widetilde{\Lambda}^{ms},$$

and for the model $A_{\mathrm{I}}$ of isotropic continua – the form

$$\widetilde{T}^{ij} = \widetilde{g}^{ik}\widetilde{g}^{jl}(\widetilde{F}^{-1})_k^p(\widetilde{F}^{-1})_l^q\Big(\psi_1\widetilde{g}_{pq} + \psi_2\widetilde{g}_{pm}\widetilde{g}_{qs}\widetilde{\Lambda}^{ms} + \psi_3\widetilde{g}_{pm}\widetilde{g}_{qs}\widetilde{g}_{uv}\widetilde{\Lambda}^{mu}\widetilde{\Lambda}^{sv}\Big).$$

## 6.2 Closed Systems in the Material Description

### 6.2.1 *θUVF-system of Dynamic Equations of Thermoelasticity in the Material Description*

For the material description, we should consider the balance law system (3.310) having the explicit form

$$\rho = \overset{\circ}{\rho}\det\mathbf{F}^{-1}, \tag{6.40a}$$

$$\overset{\circ}{\rho}(\partial\mathbf{v}/\partial t) = \overset{\circ}{\nabla}\cdot\mathbf{P} + \overset{\circ}{\rho}\mathbf{f}, \tag{6.40b}$$

$$\overset{\circ}{\rho}\theta(\partial\eta/\partial t) = -\overset{\circ}{\nabla}\cdot\overset{\circ}{\mathbf{q}} + \overset{\circ}{\rho}q_m, \tag{6.40c}$$

$$\partial\mathbf{F}^{\mathrm{T}}/\partial t = \overset{\circ}{\nabla}\otimes\mathbf{v}, \tag{6.40d}$$

$$\partial\mathbf{u}/\partial t = \mathbf{v}. \tag{6.40e}$$

Here we have used the entropy balance equation (3.174) in place of the energy balance equation.

The system (6.40) as well as (6.1) becomes closed after complementing it with the constitutive equations consisting of the two equations

$$\overset{\circ}{\mathbf{q}} = -\overset{\circ}{\boldsymbol{\lambda}} \cdot \overset{\circ}{\nabla}\theta, \qquad \eta = -\partial\psi/\partial\theta, \tag{6.41}$$

and the universal constitutive equations (6.3) and (6.4) written for the Piola–Kirchhoff tensor:

$$\mathbf{P} = \overset{(n)}{\mathcal{F}}{}^{\circ}_{G}(\mathbf{F},\theta), \qquad \psi = \psi(I^{(s)}_{\gamma G}(\overset{(n)}{\mathbf{C}}_G(\mathbf{F})),\theta) \equiv \psi(\mathbf{F},\theta). \tag{6.42}$$

Here we have introduced the tensor function

$$\overset{(n)}{\mathcal{F}}{}^{\circ}_{G}(\mathbf{F},\theta) \equiv (\overset{\circ}{\rho}/\rho)\mathbf{F}^{-1} \cdot \overset{(n)}{\mathcal{F}}_{G}(\mathbf{F},\theta). \tag{6.43}$$

The system (6.40)–(6.42) contains 16 scalar unknowns (the density $\rho$ is assumed to be expressed in terms of $\mathbf{F}$):

$$\theta,\ \mathbf{u},\ \mathbf{v},\ \mathbf{F}\ \|\ X^i,\ t, \tag{6.44}$$

which are functions of Lagrangian coordinates $X^i$ and time $t$, and consists of 16 scalar equations (after substitution of (6.41) and (6.42) into (6.40)). This system is called the *$\theta UVF$-system of thermoelasticity in the material description.*

For solids, this system proves to be more preferable, because a domain of definition of the functions (6.44) is known: this is $\overset{\circ}{V} \times [0, t_{\max}]$. The exceptions are problems with phase transformations, where $\overset{\circ}{V}$ varies with time and is determined in the process of solving. However, in this case the system (6.40)–(6.42) also proves to be more preferable than the corresponding system (6.1)–(6.6) in the spatial description.

*Remark 1.* Although the definition (6.43) of the tensor function $\overset{(n)}{\mathcal{F}}{}^{\circ}_{G}$ includes $\mathbf{F}^{-1}$, this function can be represented as a function of $\mathbf{F}$. To do this, we should use the representation of $\mathbf{F}^{-1}$ in the eigenbasis (see (2.171)) and the expression (6.4) of the function $\overset{(n)}{\mathcal{F}}_{G}$; then we obtain

$$\mathbf{P} = \overset{(n)}{\mathcal{F}}{}^{\circ}_{G}(\mathbf{F},\theta) = \sum_{\gamma=1}^{r} \overset{\circ}{\varphi}_{\gamma}\, {}^4\overset{(n)}{\mathbf{E}}{}^{\circ}_{G} \cdot\cdot\, \mathbf{I}^{(s)}_{\gamma G}, \tag{6.45}$$

$${}^4\overset{(n)}{\mathbf{E}}{}^{\circ}_{G} = \mathbf{F}^{-1} \cdot {}^4\overset{(n)}{\mathbf{E}}_{G} = \sum_{\alpha,\beta=1}^{3} \overset{\circ}{E}_{\alpha\beta}\overset{\circ}{\mathbf{p}}_{\alpha} \otimes \mathbf{p}_{\beta} \otimes (h_G\overset{\circ}{\mathbf{p}}_{\beta} \otimes \overset{\circ}{\mathbf{p}}_{\alpha} + (1-h_G)\mathbf{p}_{\beta} \otimes \mathbf{p}_{\alpha}),$$

$$\overset{\circ}{E}_{\alpha\beta} = E_{\alpha\beta}/\lambda_{\alpha}, \quad \overset{\circ}{\varphi}_{\gamma} = (\overset{\circ}{\rho}/\rho)\varphi_{\gamma}, \quad G = A, B, C, D.$$

The form of the tensors $\mathbf{I}^{(s)}_{\gamma G}$ coincides with (6.9c). The tensors ${}^4\overset{(n)}{\mathbf{E}}{}^{\circ}_G$ are called *Lagrangian tensors of energetic equivalence*; they connect the Piola–Kirchhoff stress tensor to $\overset{(n)}{\mathbf{T}}_G$:

$$\mathbf{P} = {}^4\overset{(n)}{\mathbf{E}}{}^{\circ}_G \cdot\cdot \overset{(n)}{\mathbf{T}}_G. \tag{6.46}$$

For example, choosing the tensor functions $\overset{(n)}{\mathbf{T}}_G = \mathcal{F}_G(\overset{(n)}{\mathbf{C}}_G, \theta)$ for the models $A_n$ in the form (6.8), we obtain tensor function $\overset{(n)}{\mathcal{F}}{}^{\circ}_G$ in the form

$$\mathbf{P} = {}^4\overset{(n)}{\mathbf{E}}{}^{\circ} \cdot\cdot ({}^4\mathbf{M} \cdot\cdot \overset{(n)}{\mathbf{C}} + \xi \overset{(n)}{\mathbf{C}}{}^2 + {}^6\mathbf{L} \cdot\cdot\cdot\cdot (\overset{(n)}{\mathbf{C}} \otimes \overset{(n)}{\mathbf{C}})), \tag{6.47}$$

where ${}^4\overset{(n)}{\mathbf{E}}{}^{\circ} = \mathbf{F}^{-1} \cdot {}^4\overset{(n)}{\mathbf{E}}_G$.

Notice that although the equation system (6.40) contains the Piola–Kirchhoff stress tensor $\mathbf{P}$, the final (after solving the system (6.40) with boundary and initial conditions) analysis of a stress field in a solid usually needs the Cauchy stress tensor $\mathbf{T}$ to be known; therefore, besides constitutive equations (6.45), the material description involves the relations (6.4). □

*Remark 2.* Although the tensor function (6.45) has a rather complex form, which in general cannot be expressed analytically, this form may be simplified for two exceptional models $A_{\mathrm{V}}$ and $B_{\mathrm{V}}$, because

$$\mathbf{P} = \frac{\overset{\circ}{\rho}}{\rho}\mathbf{F}^{-1} \cdot \mathbf{T} = \frac{\overset{\circ}{\rho}}{\rho}\mathbf{F}^{-1} \cdot (\mathbf{F} \cdot \overset{\mathrm{V}}{\mathbf{T}} \cdot \mathbf{F}^{\mathrm{T}}) = \frac{\overset{\circ}{\rho}}{\rho}\overset{\mathrm{V}}{\mathbf{T}} \cdot \mathbf{F}^{\mathrm{T}}. \tag{6.48}$$

On substituting the expression (6.4) for the tensor function $\overset{\mathrm{V}}{\mathbf{T}} = \mathcal{F}_A(\mathbf{F}, \theta)$ into (6.48), we obtain the following tensor function having the explicit analytical representation:

$$\mathbf{P} = \overset{\mathrm{V}}{\mathcal{F}}{}^{\circ}_A(\mathbf{F}, \theta) = \sum_{\gamma=1}^{r} \mathbf{I}^{(s)}_{\gamma} \cdot \mathbf{F}^{\mathrm{T}}, \tag{6.49}$$

$$\overset{\circ}{\varphi}_\gamma = \overset{\circ}{\rho}(\partial\psi/\partial I^{(s)}_\gamma), \quad \psi = \psi(I^{(s)}_\gamma(\mathbf{C}), \theta), \quad \mathbf{I}^{(s)}_\gamma = \partial I^{(s)}_\gamma/\partial\mathbf{C},$$
$$\mathbf{C} = (1/2)(\mathbf{F}^{\mathrm{T}} \cdot \mathbf{F} - \mathbf{E}).$$

The substitution $\mathbf{C} \to \overset{\mathrm{V}}{\mathbf{G}}$ in (6.49) yields the tensor function $\overset{\mathrm{V}}{\mathcal{F}}{}^{\circ}_B(\mathbf{F}, \theta)$ for the model $B_{\mathrm{V}}$.

For the models $A_{\mathrm{V}}$ and $B_{\mathrm{V}}$, there is no need to determine eigenbases and eigenvalues in constructing the functions (6.49). The mentioned advantages discern the models $A_{\mathrm{V}}$ and $B_{\mathrm{V}}$ among others while using the material description as well as the models $A_{\mathrm{I}}$ and $B_{\mathrm{I}}$ while using the spatial description (see Remark 3 in Sect. 6.1). These models are the most frequently applied to computations of problems of elasticity theory at large deformations.

The models $A_{\mathrm{I}}$ and $B_{\mathrm{I}}$ are of a higher level of complexity in the material description. For the model $A_{\mathrm{I}}$, constitutive equations (6.49) are replaced by the relations

$$\mathbf{P} = \overset{\mathrm{I}}{\mathcal{F}}{}^{\circ}_{A}(\mathbf{F},\theta) = \sum_{\gamma=1}^{r} \overset{\circ}{\varphi}_{\gamma} \mathbf{F}^{-1} \cdot \mathbf{F}^{-1\mathrm{T}} \cdot \mathbf{I}^{(s)}_{\gamma} \cdot \mathbf{F}^{-1}, \tag{6.50}$$

$$\overset{\circ}{\varphi}_{\gamma} = \overset{\circ}{\rho}(\partial\psi/\partial I^{(s)}_{\gamma}), \quad \psi = \psi(I^{(s)}_{\gamma}(\mathbf{\Lambda}),\theta), \quad \mathbf{I}^{(s)}_{\gamma} = \partial I^{(s)}_{\gamma}/\partial\mathbf{\Lambda},$$

$$\mathbf{\Lambda} = \frac{1}{2}(\mathbf{E} - \mathbf{F}^{-1}\cdot\mathbf{F}^{-1\mathrm{T}}).$$

For this model, although there is no need to determine eigenvalues and eigenvectors, however, in comparison with the models $A_{\mathrm{V}}$ and $B_{\mathrm{V}}$, one should invert the tensor $\mathbf{F}^{-1}$.

Just as in the spatial description, the models $A_{\mathrm{II}}$, $B_{\mathrm{II}}$ and $A_{\mathrm{IV}}$, $B_{\mathrm{IV}}$ are the most complicated for computations. □

The entropy balance equation in the system (6.40) can be modified just as in Sect. 6.1.1 by using the principal thermodynamic identity in the material description (4.125):

$$\overset{\circ}{\rho}\frac{d\psi}{dt} + \overset{\circ}{\rho}\eta\frac{d\theta}{dt} - \mathbf{P}\cdot\cdot\overset{\circ}{\nabla}\otimes\mathbf{v}^{\mathrm{T}} = 0. \tag{6.51}$$

Differentiating the identity with respect to $\theta$ and using the second relation of (6.41), we obtain the following expression for the rate of changing the specific entropy in the material description:

$$\overset{\circ}{\rho}\frac{\partial\eta}{\partial t} = -\overset{\circ}{\rho}\frac{\partial^2\psi}{\partial\theta^2}\frac{\partial\theta}{\partial t} - \frac{\partial\mathbf{P}}{\partial\theta}\cdot\cdot\overset{\circ}{\nabla}\otimes\mathbf{v}^{\mathrm{T}}. \tag{6.52}$$

On substituting (6.52) and (6.41), (6.42) into the entropy balance equation of the system (6.40), we get the desired heat conduction equation for an elastic continuum in the material description:

$$\overset{\circ}{\rho}c_{\varepsilon}\frac{\partial\theta}{\partial t} = \overset{\circ}{\nabla}\cdot(\boldsymbol{\lambda}\cdot\overset{\circ}{\nabla}\theta) + \frac{\overset{\circ}{\rho}}{\rho}({}^{4}\overset{(\mathrm{n})}{\mathbf{E}}{}^{\circ}_{G}\cdot\cdot\overset{(\mathrm{n})}{\mathbf{T}}_{\theta G})\cdot\cdot\overset{\circ}{\nabla}\otimes\mathbf{v}^{\mathrm{T}} + \overset{\circ}{\rho}q_m, \tag{6.53}$$

where the heat capacity $c_{\varepsilon}$ is determined by (6.11a). Here we have taken into account that equations (6.45) give the following expression for the derivative with respect to $\theta$:

$$\frac{\partial\mathbf{P}}{\partial\theta} = \frac{\partial}{\partial\theta}\overset{(\mathrm{n})}{\mathcal{F}}{}^{\circ}_{G}(\mathbf{F},\theta) = {}^{4}\overset{(\mathrm{n})}{\mathbf{E}}{}^{\circ}_{G}\cdot\cdot\sum_{\gamma=1}^{r}\frac{\partial\overset{\circ}{\varphi}_{\gamma}}{\partial\theta}\mathbf{I}^{(s)}_{\gamma G} = \frac{\overset{\circ}{\rho}}{\rho}\,{}^{4}\overset{(\mathrm{n})}{\mathbf{E}}{}^{\circ}_{G}\cdot\cdot\overset{(\mathrm{n})}{\mathbf{T}}_{\theta G}, \tag{6.53a}$$

where the tensor $\overset{(n)}{\mathbf{T}}_{\theta G}$ has been introduced by formula (4.423):

$$\overset{(n)}{\mathbf{T}}_{\theta G} = \sum_{\gamma=1}^{r} \frac{\partial \varphi_\gamma}{\partial \theta} \mathbf{I}^{(s)}_{\gamma G}. \tag{6.53b}$$

## 6.2.2 $\theta UV$- and $\theta U$-systems of Thermoelasticity in the Material Description

Return to the general system (6.40)–(6.42) and notice that the deformation gradient $\mathbf{F}$ can be excluded from the system with the help of Eq. (2.77):

$$\mathbf{F} = \mathbf{E} + \overset{\circ}{\nabla} \otimes \mathbf{u}^{\mathrm{T}}. \tag{6.54}$$

Then we obtain the *$\theta UV$-system of thermoelasticity in the material description*:

$$\begin{aligned}
\overset{\circ}{\rho}\frac{\partial \mathbf{v}}{\partial t} &= \overset{\circ}{\nabla} \cdot \overset{(n)}{\mathcal{F}}{}^{\circ}_{G}((\mathbf{E} + \overset{\circ}{\nabla} \otimes \mathbf{u}^{\mathrm{T}}), \theta) + \overset{\circ}{\rho}\mathbf{f}, \\
\overset{\circ}{\rho} c_\varepsilon \frac{\partial \theta}{\partial t} &= \overset{\circ}{\nabla} \cdot (\overset{\circ}{\boldsymbol{\lambda}} \cdot \overset{\circ}{\nabla}\theta) + \frac{\partial}{\partial \theta}\overset{(n)}{\mathcal{F}}{}^{\circ}_{G}(\mathbf{E} + \overset{\circ}{\nabla} \otimes \mathbf{u}^{\mathrm{T}}, \theta) \cdot\cdot \overset{\circ}{\nabla} \otimes \mathbf{v}^{\mathrm{T}} + \overset{\circ}{\rho} q_m, \\
\partial \mathbf{u}/\partial t &= \mathbf{v},
\end{aligned} \tag{6.55}$$

consisting of seven equations for seven scalar unknowns:

$$\theta,\ \mathbf{u},\ \mathbf{v}\ \parallel\ X^i,\ t. \tag{6.56}$$

The density $\rho$ does not appear in this system explicitly, and it can always be evaluated with the help of the continuity equation (6.17):

$$\rho = \overset{\circ}{\rho} \det (\mathbf{E} + \overset{\circ}{\nabla} \otimes \mathbf{u}^{\mathrm{T}}). \tag{6.57}$$

Since in the material description the velocity $\mathbf{v}$ is connected to the displacement vector $\mathbf{u}$ by an explicit kinematic relation, we can eliminate the velocity $\mathbf{v}$ between the unknowns. As a result, we obtain the *$\theta U$-system of thermoelasticity in the material description*:

$$\begin{aligned}
\overset{\circ}{\rho}\frac{\partial^2 \mathbf{u}}{\partial t^2} &= \overset{\circ}{\nabla} \cdot \overset{(n)}{\mathcal{F}}{}^{\circ}_{G}((\mathbf{E} + \overset{\circ}{\nabla} \otimes \mathbf{u}^{\mathrm{T}}), \theta) + \overset{\circ}{\rho}\mathbf{f}, \\
\overset{\circ}{\rho} c_\varepsilon \frac{\partial \theta}{\partial t} &= \overset{\circ}{\nabla} \cdot (\overset{\circ}{\boldsymbol{\lambda}} \cdot \overset{\circ}{\nabla}\theta) + \frac{\partial}{\partial \theta}\overset{(n)}{\mathcal{F}}{}^{\circ}_{G}((\mathbf{E} + \overset{\circ}{\nabla} \otimes \mathbf{u}^{\mathrm{T}}), \theta) \cdot\cdot \overset{\circ}{\nabla} \otimes \frac{\partial \mathbf{u}}{\partial t} + \overset{\circ}{\rho} q_m,
\end{aligned} \tag{6.58}$$

for four scalar unknowns:

$$\theta,\ \mathbf{u}\ \parallel\ X^i,\ t. \tag{6.59}$$

In particular, for the exceptional model $A_{\mathrm{V}}$, the $\theta U$-system (6.58) with account of (6.49) and (2.78) is written in the form

$$\begin{aligned}
&\overset{\circ}{\rho}\frac{\partial^2\mathbf{u}}{\partial t^2} = \overset{\circ}{\nabla}\cdot\mathbf{P} + \overset{\circ}{\rho}\mathbf{f},\\
&\overset{\circ}{\rho}c_\varepsilon\frac{\partial\theta}{\partial t} = \overset{\circ}{\nabla}\cdot(\overset{\circ}{\boldsymbol{\lambda}}\cdot\overset{\circ}{\nabla}\theta) + \left(\sum_{\gamma=1}^{r}\varphi_{\gamma\theta}\mathbf{I}^{(s)}_{\gamma\mathbf{C}}\right)\cdot\cdot\,(\mathbf{F}^{\mathrm{T}}\cdot\overset{\circ}{\nabla}\otimes\mathbf{v}^{\mathrm{T}}) + \overset{\circ}{\rho}q_m,\\
&\varphi_{\gamma\theta} = \overset{\circ}{\rho}\frac{\partial^2\psi}{\partial\theta\,\partial I^{(s)}_\gamma},\quad \overset{\circ}{\varphi}_\gamma = \overset{\circ}{\rho}\frac{\partial\psi}{\partial I^{(s)}_\gamma},\\
&\mathbf{P} = \sum_{\gamma=1}^{r}\overset{\circ}{\varphi}_\gamma\mathbf{I}^{(s)}_{\gamma\mathbf{C}}\cdot\mathbf{F}^{\mathrm{T}},\quad \psi = \psi(I^{(s)}_\gamma(\mathbf{C}),\ \theta),\quad \mathbf{I}^{(s)}_{\gamma\mathbf{C}} = \partial I^{(s)}_\gamma/\partial\mathbf{C},\\
&\mathbf{C} = \frac{1}{2}(\overset{\circ}{\nabla}\otimes\mathbf{u} + \overset{\circ}{\nabla}\otimes\mathbf{u}^{\mathrm{T}} + \overset{\circ}{\nabla}\otimes\mathbf{u}\cdot\overset{\circ}{\nabla}\otimes\mathbf{u}^{\mathrm{T}}),\quad \mathbf{F} = \mathbf{E} + \overset{\circ}{\nabla}\otimes\mathbf{u}^{\mathrm{T}}.
\end{aligned} \tag{6.60}$$

### *6.2.3 $T\theta UVF$-system of Thermoelasticity in the Material Description*

Using the relations (6.46) between the tensors $\mathbf{P}$ and $\overset{(\mathrm{n})}{\mathbf{T}}_G$ and complementing the system (6.40) with constitutive equations (4.442) 'in rates', we obtain the $T\theta UVF$-system of thermoelasticity in the material description:

$$\begin{aligned}
&\overset{\circ}{\rho}\frac{\partial\mathbf{v}}{\partial t} = \overset{\circ}{\nabla}\cdot({}^4\overset{(\mathrm{n})}{\mathbf{E}}{}^{\circ}_G\cdot\cdot\,\overset{(\mathrm{n})}{\mathbf{T}}_G) + \overset{\circ}{\rho}\mathbf{f},\\
&\overset{\circ}{\rho}c_\varepsilon\frac{\partial\theta}{\partial t} = \overset{\circ}{\nabla}\cdot(\overset{\circ}{\boldsymbol{\lambda}}\cdot\overset{\circ}{\nabla}\theta) + (\overset{\circ}{\rho}/\rho)({}^4\overset{(\mathrm{n})}{\mathbf{E}}{}^{\circ}_G\cdot\cdot\,\overset{(\mathrm{n})}{\mathbf{T}}_{\theta G})\cdot\cdot\,\overset{\circ}{\nabla}\otimes\mathbf{v}^{\mathrm{T}} + \overset{\circ}{\rho}q_m,\\
&\partial\mathbf{u}/\partial t = \mathbf{v},\quad \partial\mathbf{F}/\partial t = \overset{\circ}{\nabla}\otimes\mathbf{v}^{\mathrm{T}},\\
&\frac{\partial\overset{(\mathrm{n})}{\mathbf{T}}_G}{\partial t} = {}^4\overset{(\mathrm{n})}{\mathbf{P}}{}^{\circ}_{Gh}(\mathbf{F},\theta)\cdot\cdot\,\overset{\circ}{\nabla}\otimes\mathbf{v}^{\mathrm{T}} + \mathbf{Z}_{Gh}\cdot\overset{(\mathrm{n})}{\mathbf{T}}_G + \overset{(\mathrm{n})}{\mathbf{T}}_G\cdot\mathbf{Z}_{Gh} + \overset{(\mathrm{n})}{\mathbf{T}}_{\theta G}\frac{\partial\theta}{\partial t},
\end{aligned} \tag{6.61}$$

for 22 scalar unknowns:

$$\overset{(\mathrm{n})}{\mathbf{T}}_G,\ \theta,\ \mathbf{u},\ \mathbf{v},\ \mathbf{F}\ \parallel\ X^i,\ t. \tag{6.62}$$

Here we have introduced the notation for the fourth-order tensor

$$ {}^4\overset{(\mathrm{n})}{\mathbf{P}}{}^{\circ}_{Gh} \equiv \big({}^4\overset{(\mathrm{n})}{\mathbf{P}}{}^{(1243)}_{Gh} \cdot \mathbf{F}^{-1\mathrm{T}}\big)^{(1243)}. \tag{6.63}$$

The remaining notations are the same as in Sect. 6.1.3. The heat conduction equation in (6.61) has been written with the help of (6.53).

Component forms of equation systems of elasticity theory in the material description are usually used in the basis $\overset{\circ}{\mathbf{r}}_i$ of a reference configuration $\overset{\circ}{\mathcal{K}}$ (see Exercise 6.2.2 and 6.2.3).

### 6.2.4 *The Equation System of Thermoelasticity for Quasistatic Processes in the Material Description*

The model of quasistatic processes introduced in Sect. 6.1.5 can be considered for the material description as well. In this case the equation system (6.40)–(6.42) with account of the assumption (6.34) takes the form

$$\overset{\circ}{\nabla} \cdot \mathbf{P} + \overset{\circ}{\rho}\mathbf{f} = 0, \tag{6.64}$$

$$\rho c_{\varepsilon} \frac{\partial \theta}{\partial t} = \overset{\circ}{\nabla} \cdot (\overset{\circ}{\boldsymbol{\lambda}} \cdot \overset{\circ}{\nabla}\theta) + \overset{\circ}{\rho} q_m, \tag{6.65}$$

$$\mathbf{P} = \overset{(\mathrm{n})}{\mathcal{F}}{}^{\circ}_{G}(\mathbf{E} + \overset{\circ}{\nabla} \otimes \mathbf{u}, \theta). \tag{6.66}$$

The system consists of four scalar equations for four scalar unknown functions:

$$\theta,\ \mathbf{u}\ \|\ X^i,\ t, \tag{6.67}$$

and it is called the *quasistatic equation system of thermoelasticity in the material description*.

In this case the dynamic compatibility equation and the kinematic equation in system (6.40) are satisfied only approximately, and they are not considered. These equations are satisfied exactly for static processes starting from some time $t_0$.

## Exercises for 6.2

**6.2.1.** Show that constitutive equations (6.49) for the quasilinear model $A_{\mathrm{V}}$ can be written in the form

$$\mathbf{P} = (\overset{\circ}{\rho}/\rho)({}^4\mathbf{M} \cdot\cdot\, \mathbf{C}) \cdot \mathbf{F}^{\mathrm{T}}, \quad \mathbf{T} = \mathbf{F} \cdot ({}^4\mathbf{M} \cdot\cdot\, \mathbf{C}) \cdot \mathbf{F}^{\mathrm{T}},$$

or in terms of the displacement gradient:

$$\mathbf{P} = \frac{\overset{\circ}{\rho}}{\rho}\left({}^4\mathbf{M}\cdot\cdot\left(\mathbf{E} + \frac{1}{2}\overset{\circ}{\nabla}\otimes\mathbf{u}\right)\cdot\overset{\circ}{\nabla}\otimes\mathbf{u}^{\mathrm{T}}\right)\cdot(\mathbf{E} + \overset{\circ}{\nabla}\otimes\mathbf{u}).$$

**6.2.2.** Using the results of Exercise 6.1.1, show that for isotropic media the relations (6.49) have the forms

$$\mathbf{P} = (\overset{\circ}{\psi}_1\mathbf{E} + \overset{\circ}{\psi}_2\mathbf{C} + \overset{\circ}{\psi}_3\mathbf{C}^2)\cdot\mathbf{F}^{\mathrm{T}}, \quad \mathbf{T} = J\mathbf{F}\cdot(\overset{\circ}{\psi}_1 + \overset{\circ}{\psi}_2\mathbf{C} + \overset{\circ}{\psi}_3\mathbf{C}^2)\cdot\mathbf{F}^{\mathrm{T}},$$

$$\overset{\circ}{\psi}_1 = \overset{\circ}{\varphi}_1 + \overset{\circ}{\varphi}_2 I_1 + \overset{\circ}{\varphi}_3 I_2, \quad -\overset{\circ}{\psi}_2 = \overset{\circ}{\varphi}_2 + I_1\overset{\circ}{\varphi}_3, \quad \overset{\circ}{\psi}_3 = \overset{\circ}{\varphi}_3,$$

$$\overset{\circ}{\varphi}_\alpha = \overset{\circ}{\rho}(\partial\psi/\partial I_\alpha), \quad I_\alpha = I_\alpha(\mathbf{C}), \quad \alpha = 1,2,3.$$

Show that for the linear model $A_{\mathrm{V}}$ of isotropic continua these relations take the forms

$$\mathbf{P} = l_1 I_1(\mathbf{C})\mathbf{F}^{\mathrm{T}} + 2l_2\mathbf{C}\cdot\mathbf{F}^{\mathrm{T}}, \quad \mathbf{T} = J(l_1 I_1(\mathbf{C})\mathbf{F}\cdot\mathbf{F}^{\mathrm{T}} + 2l_2\mathbf{F}\cdot\mathbf{C}\cdot\mathbf{F}^{\mathrm{T}}),$$

or in terms of the displacement gradient:

$$\mathbf{P} = l_1\left(\overset{\circ}{\nabla}\cdot\mathbf{u} + \frac{1}{2}\overset{\circ}{\nabla}\otimes\mathbf{u}\cdot\cdot\overset{\circ}{\nabla}\otimes\mathbf{u}^{\mathrm{T}}\right)(\mathbf{E} + \overset{\circ}{\nabla}\otimes\mathbf{u})$$
$$+l_2(\overset{\circ}{\nabla}\otimes\mathbf{u} + \overset{\circ}{\nabla}\otimes\mathbf{u}^{\mathrm{T}} + \overset{\circ}{\nabla}\otimes\mathbf{u}\cdot\overset{\circ}{\nabla}\otimes\mathbf{u}^{\mathrm{T}})\cdot(\mathbf{E} + \overset{\circ}{\nabla}\otimes\mathbf{u}).$$

**6.2.3.** Using the result of Exercise 6.2.1, show that the constitutive equations for the quasilinear model $A_{\mathrm{V}}$ in basis $\overset{\circ}{\mathbf{r}}_i$ have the component form

$$\overset{\circ}{P}{}_i^j = (\overset{\circ}{\rho}/\rho)\overset{\circ}{M}{}^{jkms}\varepsilon_{ms}\overset{\circ}{F}{}^l_k\overset{\circ}{g}_{li};$$

and for the model $A_{\mathrm{V}}$ of isotropic media:

$$\overset{\circ}{P}{}_i^j = (\overset{\circ}{\psi}_1\overset{\circ}{g}{}^{jk} + \overset{\circ}{\psi}_2\overset{\circ}{g}{}^{jm}\overset{\circ}{g}{}^{ks}\varepsilon_{ms} + \overset{\circ}{\psi}_3\overset{\circ}{g}{}^{jm}\overset{\circ}{g}{}^{ks}\overset{\circ}{g}{}^{uv}\varepsilon_{mu}\varepsilon_{sv})\overset{\circ}{F}{}^l_k\overset{\circ}{g}_{li},$$

$$T^{ij} = \overset{\circ}{\psi}_1\overset{\circ}{g}{}^{ij} + \overset{\circ}{\psi}_2\overset{\circ}{g}{}^{im}\overset{\circ}{g}{}^{js}\varepsilon_{ms} + \overset{\circ}{\psi}_3\overset{\circ}{g}{}^{im}\overset{\circ}{g}{}^{js}\overset{\circ}{g}{}^{uv}\varepsilon_{mu}\varepsilon_{sv}.$$

**6.2.4.** Show that the $\theta UVF$-system (6.40)–(6.42) and (6.45) in basis $\overset{\circ}{\mathbf{r}}_i$ in the material description has the component form

$$\begin{cases} \overset{\circ}{\rho}(\partial \overset{\circ}{v}^i/\partial t) = \overset{\circ}{\nabla}_j \overset{\circ}{P}^{ji} + \overset{\circ}{\rho} \overset{\circ}{f}^i, \\ \overset{\circ}{\rho}(\partial e/\partial t) = \overset{\circ}{\nabla}_j(\overset{\circ}{\lambda}^{ji} \overset{\circ}{\nabla}_i \theta) + \overset{\circ}{P}^{ij} \overset{\circ}{\nabla}_i \overset{\circ}{v}^k \overset{\circ}{g}_{kj} + \overset{\circ}{\rho} q_m, \\ \partial \overset{\circ}{F}{}^i_j/\partial t = \overset{\circ}{\nabla}_j \overset{\circ}{u}^i, \\ \partial \overset{\circ}{u}^i/\partial t = \overset{\circ}{v}^i, \end{cases}$$

$$\begin{cases} \overset{\circ}{P}^{ij} = \overset{(n)}{\mathcal{F}}{}^{\circ ij}_{G}(\overset{\circ}{F}{}^k_l, \theta), \quad e = \psi - \theta(\partial\psi/\partial\theta), \\ \psi = \psi(I^{(s)}_{\gamma G}(\overset{(n)}{C}{}^{\circ ij}_{G}(\overset{\circ}{F}{}^k_l)), \theta) \equiv \psi(\overset{\circ}{F}{}^k_l, \theta), \end{cases}$$

$$\begin{cases} \overset{(n)}{\mathcal{F}}{}^{\circ ij}_{G} = \sum_{\gamma=1}^{r} \overset{\circ}{\varphi}_\gamma \overset{(n)}{E}{}^{\circ ijkl}_{G} \overset{\circ}{I}{}^{(s)}_{\gamma Gkl}, \\ \overset{\circ}{I}{}^{(s)}_{\gamma Gkl} = \partial I^{(s)}_{\gamma G}/\partial \overset{(n)}{C}{}^{\circ kl}_{G}, \quad \overset{\circ}{\varphi}_\gamma = \overset{\circ}{\rho}(\partial\psi/\partial I^{(s)}_{\gamma G}), \\ I^{(s)}_{\gamma G} = h_G I^{(s)}_\gamma(\overset{(n)}{C}{}^{\circ kl}_{G}) + (1-h_G) I^{(s)}_\gamma(\overset{(n)}{C}{}^{\circ ij}_{G} \overset{\circ}{O}_{ik} \overset{\circ}{O}_{jl}), \\ \overset{(n)}{C}{}^{\circ kl}_{G} = \frac{1}{n-\mathrm{III}} \sum_{\alpha=1}^{3} \lambda_\alpha^{n-\mathrm{III}}(h_G \overset{\circ}{Q}{}^i{}_\alpha \overset{\circ}{Q}{}^j_\alpha + (1-h_G) Q^i_\alpha Q^j_\alpha) - \frac{h_G}{n-\mathrm{III}} \overset{\circ}{g}{}^{ij}, \\ \overset{(n)}{E}{}^{\circ ijkl}_{G} = \sum_{\alpha,\beta=1}^{3} \overset{\circ}{E}_{\alpha\beta} Q^i_\alpha \overset{\circ}{Q}{}^j_\beta (h_G \overset{\circ}{Q}{}^k{}_\beta \overset{\circ}{Q}{}^l_\alpha + (1-h_G) Q^k_\beta Q^l{}_\alpha). \end{cases}$$

Here $\overset{\circ}{\mathbf{p}}_\alpha = \overset{\circ}{Q}{}^i{}_\alpha \overset{\circ}{\mathbf{r}}_i$, $\mathbf{p}_\alpha = Q^i_\alpha \mathbf{p}_i$.

**6.2.5.** Show that the $\theta U$-system (6.60) for the model $A_{\mathrm{V}}$ in basis $\overset{\circ}{\mathbf{r}}_i$ has the component form

$$\begin{cases} \overset{\circ}{\rho}(\partial^2 \overset{\circ}{u}_i/\partial t^2) = \overset{\circ}{\nabla}_j \overset{\circ}{P}{}^j_i + \overset{\circ}{\rho} \overset{\circ}{f}_i, \\ \overset{\circ}{\rho}(\partial e/\partial t) = \overset{\circ}{\nabla}_j(\overset{\circ}{\lambda}^{ji} \overset{\circ}{\nabla}_i \theta) + \overset{\circ}{P}{}^j_i \overset{\circ}{\nabla}_j(\partial \overset{\circ}{u}_k/\partial t) \overset{\circ}{g}{}^{ik} + \overset{\circ}{\rho} q_m, \\ \overset{\circ}{P}{}^j_i = \sum_{\gamma=1}^{r} \overset{\circ}{\varphi}_\gamma (\partial I^{(s)}_{\gamma G}/\partial \varepsilon_{jk}) \overset{\circ}{F}{}^l_k \overset{\circ}{g}_{li}, \\ \overset{\circ}{\varphi}_\gamma = \overset{\circ}{\rho}(\partial\psi/\partial I^{(s)}_\gamma), \quad \psi = \psi(I^{(s)}_\gamma(\varepsilon_{jk}), \theta), e = \psi - \theta(\partial\psi/\partial\theta), \\ \varepsilon_{jk} = (1/2)(\overset{\circ}{\nabla}_j \overset{\circ}{u}_k + \overset{\circ}{\nabla}_k \overset{\circ}{u}_j + \overset{\circ}{\nabla}_j \overset{\circ}{u}_m \overset{\circ}{\nabla}_k \overset{\circ}{u}_i \overset{\circ}{g}{}^{mi}), \\ \overset{\circ}{F}{}^i_k = \delta^i_k + \overset{\circ}{\nabla}_k \overset{\circ}{u}_m \overset{\circ}{g}{}^{mi}. \end{cases}$$

Here $\varepsilon_{jk} = \overset{\circ}{C}_{jk}$ are components of the deformation tensor $\mathbf{C} = \varepsilon_{ij} \overset{\circ}{\mathbf{r}}{}^i \otimes \overset{\circ}{\mathbf{r}}{}^j$ (see (2.68)).

**6.2.6.** Using the fact that components of the tensor $\overset{\mathrm{V}}{\mathbf{T}}$ with respect to basis $\overset{\circ}{\mathbf{r}}_i$ coincide with components $T^{ij}$ of the Cauchy tensor $\mathbf{T}$ with respect to basis $\mathbf{r}_i$ (see Exercise 4.2.1), show that for the model $A_{\mathrm{V}}$ components $T^{ij}$ can be evaluated with the help of constitutive equations (6.3a):

$$T^{ij} = \mathcal{F}^{ij}_A = \sum_{\gamma=1}^{r} \varphi_\gamma \frac{\partial I^{(s)}_{\gamma G}}{\partial \varepsilon_{kl}} \overset{\circ}{g}{}^{ik} \overset{\circ}{g}{}^{jl}.$$

## 6.3 Statements of Problems for Elastic Continua at Large Deformations

### 6.3.1 Boundary Conditions in the Spatial Description

Remind that the closed systems of differential equations given in Sects. 6.1.1–6.1.4 hold only in a domain $V$ not containing singular surfaces $S(t)$. In the presence of such surfaces, in place of these systems one should consider the relations for the jumps (5.72)–(5.77).

(1) For a homothermal coherent surface $S_1(t)$, the relations have the form

$$\begin{cases} [\rho(D - \mathbf{v}\cdot\mathbf{n})] = 0, \\ M[\mathbf{v}] + \mathbf{n}\cdot[\mathbf{T}] + \mathbf{C}_{2\Sigma} = 0, \\ M[e + v^2/2] + \mathbf{n}\cdot[\mathbf{T}\cdot\mathbf{v} - \mathbf{q}] + C_{3\Sigma} = 0, \\ [\theta] = 0, \\ [\mathbf{u}] = 0, \\ M[\mathbf{F}^{\mathrm{T}}] + \mathbf{n}\cdot[\rho\mathbf{F}\otimes\mathbf{v}] = 0. \end{cases} \tag{6.68}$$

Here $[\mathbf{v}] = \mathbf{v}' - \mathbf{v}''$ are the jumps of functions across the singular surface $S_1(t)$ (see Fig. 6.2), and

$$M = \rho'(D - \mathbf{v}'\cdot\mathbf{n}) = \rho''(D - \mathbf{v}''\cdot\mathbf{n})$$

is the mass rate of propagation of the singular surface (see (5.59)).

For quasistatic processes in an elastic continuum (see Sect. 6.1.5), according to the definition (6.34), we should assume that $\mathbf{v} = 0$ in relations (6.68), then in place of (6.68) we have

$$\begin{cases} [\rho] = 0, \\ \mathbf{n}\cdot[\mathbf{T}] + \mathbf{C}_{2\Sigma} = 0, \\ [\mathbf{u}] = 0, \\ M[e] - \mathbf{n}\cdot[\mathbf{q}] + C_{3\Sigma} = 0, \\ [\theta] = 0. \end{cases} \tag{6.68a}$$

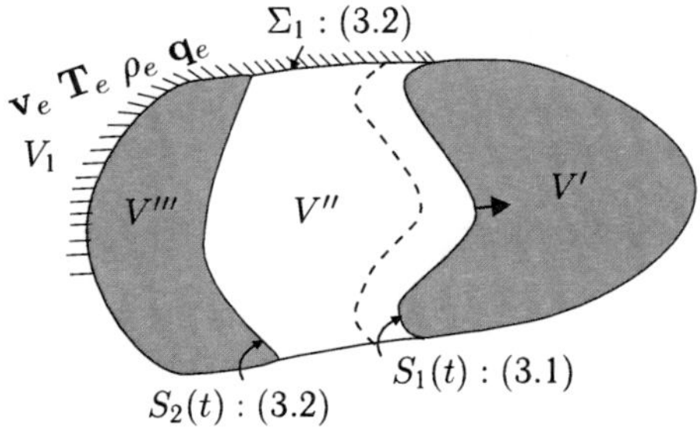


**Fig. 6.2** Boundary conditions for solids in configuration $\mathcal{K}$

One can neglect the last equation in (6.68) as well as the deformation compatibility equation in the quasistatic equation system.

(2) In the special case when a surface $S_2(t)$ is a surface with ideal contact, the relations (5.106) hold at this surface (Fig. 6.2):

$$\begin{cases} [\mathbf{u}] = 0, \\ \mathbf{n} \cdot [\mathbf{T}] = 0, \\ [\theta] = 0, \\ \mathbf{n} \cdot [\mathbf{q}] = 0. \end{cases} \tag{6.69}$$

The boundary $\Sigma$ of a domain $V$ is also a singular surface; therefore, we should formulate boundary conditions at $\Sigma$ by using the relations for jumps (6.68). Let us specify several main special cases of boundary conditions.

(3) When material points cross the surface part $\Sigma_1$, boundary conditions at $\Sigma_1$ for a solid have the form

$$\begin{aligned} &M[\mathbf{v}] + \mathbf{n} \cdot [\mathbf{T}] + \mathbf{C}_{2\Sigma} = 0, \\ &M\left[e + \frac{v^2}{2}\right] + \mathbf{n} \cdot [\mathbf{T} \cdot \mathbf{v} - \mathbf{q}] + C_{3\Sigma} = 0. \end{aligned} \tag{6.70}$$

Here $[\mathbf{v}] = \mathbf{v}_e - \mathbf{v}$ etc., where functions without subscripts correspond to a domain $V$ considered and are unknown, and functions with subscript $e$ correspond to the surroundings of the domain $V$ and are assumed to be known. The remaining relations for jumps in the system (6.68) take no part in formulating boundary conditions, they only impose constraints on parameters of the surroundings.

If the model of quasistatic processes is considered for an elastic continuum, then, according to the definition (6.34), we assume that $\mathbf{v} \equiv 0$ in the boundary conditions (6.70). In this case the conditions take the form

$$\begin{cases} \mathbf{n} \cdot [\mathbf{T}] + \mathbf{C}_{2\Sigma} = 0, \\ M[e] - \mathbf{n} \cdot [\mathbf{q}] + C_{3\Sigma} = 0. \end{cases} \tag{6.70a}$$

(4) If the surroundings on the part $\Sigma_2$ is an ideal fluid, then for the fluid the Cauchy stress tensor is spherical: $\mathbf{T}_e = -p_e$ (see (4.444a)), and relations (6.70) take the form (see Exercise 6.3.5)

$$\begin{cases} M[v_n] - T_n = p_e - C_{n\Sigma}, \\ M[v_{\tau_I}] - T_{\tau_I} = 0, \quad I = 1, 2, \\ M[e + v_n^2/2] - T_n v_n = p_e v_{ne} - \widetilde{C}'_{3\Sigma}, \end{cases} \tag{6.71}$$

$$\widetilde{C}'_{3\Sigma} = C_{3\Sigma} - \frac{1}{2}\sum_{I=1}^{2}(v_{\tau_I} + v_{\tau_I e})^2 - [q_n].$$

Here $e$, $T_n = \mathbf{n}\cdot\mathbf{T}\cdot\mathbf{n}$, $v_n = \mathbf{n}\cdot\mathbf{v}$ and $T_{\tau_I} = \boldsymbol{\tau}_I\cdot\mathbf{T}\cdot\mathbf{n}$ are unknown functions referred to the solid; $e_0$, $p_e$, $v_{\tau_I e}$ and $v_{ne}$, $q_{ne}$ are given functions referred to the fluid.

For quasistatic processes, the relations (6.71) take the form

$$\begin{cases} T_n = -p_e + C_{n\Sigma}, \\ T_{\tau_I} = 0, \quad I = 1,2, \\ M[e] - [q_n] = C_{3\Sigma}. \end{cases} \tag{6.71a}$$

(5) Across the surface part $\Sigma_3$ of a domain $V$, the transition of material points may be absent ($M = 0$). If this surface is in addition coherent, then the following boundary conditions hold at the surface:

$$\mathbf{n}\cdot\mathbf{T} = \widetilde{\mathbf{t}}_{ne}, \quad \mathbf{n}\cdot\mathbf{q} = \widetilde{q}_{ne}, \tag{6.72}$$

where $\widetilde{\mathbf{t}}_{ne} = \mathbf{n}\cdot\mathbf{T}_e + \mathbf{C}_{2\Sigma}$, $\widetilde{q}_{ne} = \mathbf{n}\cdot\mathbf{q}_e - C_{3\Sigma}$ are given values.

(6) If $M = 0$ and the surrounding medium is an ideal fluid, then although such a surface $\Sigma_4$ is not coherent (see Sects. 5.1.2 and 5.4.2), the boundary conditions (6.72) hold at the surface; and the stress vector $\widetilde{\mathbf{t}}_{ne}$ is collinear to the normal vector:

$$\widetilde{\mathbf{t}}_{ne} = -p_e\mathbf{n}. \tag{6.73}$$

The boundary condition (6.73) is frequently called the *tracking loading*, because it follows the change of the normal to the surface $\Sigma_4$ in the actual configuration. On the contrary, the boundary condition (6.72) admits the existence of the stresses vector $\widetilde{\mathbf{t}}_{ne}$ with its fixed value and direction in any actual configuration; such a boundary condition is called the *fixed loading* (Fig. 6.3).

(7) When $M = 0$, at the surface part $\Sigma_5$ instead of the stress vector $\mathbf{t}_n = \mathbf{n}\cdot\mathbf{T}$ we can give the displacement vector (or the radius-vector $\mathbf{x}$) of a material point $\mathcal{M}$ and temperature $\theta$:

$$\mathbf{u} = \mathbf{u}_e, \quad \theta = \theta_e. \tag{6.74}$$

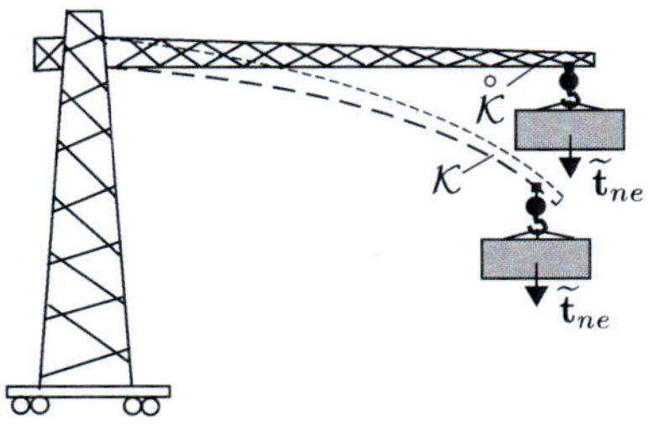


**Fig. 6.3** Fixed loading of the crane boom under the action of the gravity force of a suspended freight

In this case the relations for jumps (5.81) take no part in formulation of boundary conditions, they only impose constraints on parameters of the surroundings.

(8), (9) In addition, two more combinations of boundary conditions at the surface parts $\Sigma_6$ and $\Sigma_7$ of a domain $V$ are possible:

$$\begin{cases} \mathbf{u} = \mathbf{u}_e, \\ \mathbf{n}\cdot\mathbf{q} = q_{ne}, \end{cases} \qquad \begin{cases} \mathbf{n}\cdot\mathbf{T} = \mathbf{t}_{ne}, \\ \theta = \theta_e. \end{cases} \tag{6.75}$$

(10) For solids, boundary conditions can be given at artificial boundaries. The *symmetry conditions* are the most widely used conditions of this type; they remain valid when:

(a) there is a fixed symmetry plane $\Pi_\Sigma$ in the domain $V$, which divides the domain into two parts $V'$ and $V''$
(b) at outer boundaries $\Sigma'$ and $\Sigma''$ of the domains $V'$ and $V''$ boundary conditions are symmetric with respect to $\Pi_\Sigma$, and mass forces and heat sources are symmetric too
(c) constitutive equations of a solid and values of the unknown functions at the initial time are symmetric with respect to $\Pi_\Sigma$

In this case, fields of all the unknown functions must remain symmetric with respect to $\Pi_\Sigma$ at any time $t \geqslant 0$. This means that the derivatives of all the scalar unknowns with respect to the normal to $\Pi_\Sigma$ must vanish, the motion of points along the normal to $\Pi_\Sigma$ must be absent (otherwise, the symmetry is violated), and the stress vector $\mathbf{t}_n$ must be collinear to the normal vector $\mathbf{n}$ to $\Pi_\Sigma$, i.e. the following relations at $\Pi_\Sigma$ must hold:

$$\begin{gathered} \partial\rho/\partial\mathbf{n} = 0, \quad \mathbf{u}\cdot\mathbf{n} = 0, \quad \mathbf{v}\cdot\mathbf{n} = 0, \quad \mathbf{q}\cdot\mathbf{n} = 0, \\ \mathbf{n}\cdot\mathbf{T}\cdot\boldsymbol{\tau}_I = T_{\tau_I} = 0, \quad I = 1,2. \end{gathered} \tag{6.76}$$

### 6.3.2 *Boundary Conditions in the Material Description*

All the considered types of boundary conditions correspond to an actual configuration $\mathcal{K}$; however, they can be rewritten in terms of a reference configuration $\overset{\circ}{\mathcal{K}}$ and then used to formulate problems in $\overset{\circ}{\mathcal{K}}$.

(1) For a singular surface $\overset{\circ}{S}_1(t)$ inside a domain $\overset{\circ}{V}$, when the surface $\overset{\circ}{S}_1(t)$ is homothermal and coherent, the conditions (5.70) should be satisfied:

$$\begin{cases} [\overset{\circ}{\rho}]\overset{\circ}{D} = 0, \\ \overset{\circ}{M}[\mathbf{v}] + \overset{\circ}{\mathbf{n}} \cdot \mathbf{P} + \overset{\circ}{\mathbf{C}}_{2\Sigma} = 0, \\ \theta \overset{\circ}{M}[\eta] - \overset{\circ}{\mathbf{n}} \cdot [\overset{\circ}{\mathbf{q}}] + \theta \overset{\circ}{C}_{4\Sigma} = 0, \\ [\mathbf{u}] = 0, \\ \overset{\circ}{M}[\mathbf{F}^{\mathrm{T}}] + \overset{\circ}{\mathbf{n}} \otimes [\overset{\circ}{\rho}\mathbf{v}] = 0, \\ [\theta] = 0. \end{cases} \tag{6.77}$$

Here $\overset{\circ}{M} = \overset{\circ}{\rho}'\overset{\circ}{D} = \overset{\circ}{\rho}''\overset{\circ}{D}$ is the mass rate of propagation of the singular surface in $\overset{\circ}{\mathcal{K}}$ (see (5.58)).

Notice that the set (6.77) includes the condition for the entropy jump (the third relation), because the balance equation system (6.40) contains just the entropy balance equation. If instead of this equation the system (6.40) involved the energy balance equation, then the set (6.77) should contain a relation for the energy jump (the relation (5.70c)).

For quasistatic processes, according to (6.34), the system (6.77) takes the form

$$\begin{gathered} [\overset{\circ}{\rho}] = 0, \\ \overset{\circ}{\mathbf{n}} \cdot \mathbf{P} + \overset{\circ}{\mathbf{C}}_{2\Sigma} = 0, \\ [\mathbf{u}] = 0, \\ \theta \overset{\circ}{M}[\eta] - \overset{\circ}{\mathbf{n}} \cdot [\overset{\circ}{\mathbf{q}}] + \theta \overset{\circ}{C}_{4\Sigma} = 0, \\ [\theta] = 0. \end{gathered} \tag{6.77a}$$

(For the quasistatic statement, the deformation compatibility equation is not considered.)

(2) If the surface $\overset{\circ}{S}_2(t)$ inside the domain $\overset{\circ}{V}$ is a surface with ideal contact, then the relations (5.107) hold at the surface:

$$[\mathbf{u}] = 0, \quad \overset{\circ}{\mathbf{n}} \cdot [\mathbf{P}] = 0, \quad [\theta] = 0, \quad \overset{\circ}{\mathbf{n}} \cdot [\overset{\circ}{\mathbf{q}}] = 0. \tag{6.78}$$

(3) At the outer surface part $\overset{\circ}{\Sigma}_1$ of the domain $\overset{\circ}{V}$ (corresponding to the surface $\Sigma_1$), where $M \neq 0$ and $\overset{\circ}{M} \neq 0$, the following conditions should be satisfied:

$$\begin{cases} \overset{\circ}{M}[\mathbf{v}] + \overset{\circ}{\mathbf{n}} \cdot [\mathbf{P}] + \overset{\circ}{\mathbf{C}}_{2\Sigma} = 0, \\ \overset{\circ}{M}[\eta] - \overset{\circ}{\mathbf{n}} \cdot [\overset{\circ}{\mathbf{q}}/\theta] + \overset{\circ}{C}_{4\Sigma} = 0. \end{cases} \tag{6.79}$$

Here $[\mathbf{v}] = \mathbf{v}_e - \mathbf{v}$ etc., where functions without subscripts correspond to the considered domain $V$, and functions with subscript $e$ – to the surroundings (given values).

For quasistatic processes, these relations take the forms

$$\overset{\circ}{\mathbf{n}} \cdot [\mathbf{P}] + \overset{\circ}{\mathbf{C}}_{2\Sigma} = 0,$$
$$\overset{\circ}{M}[\eta] - \overset{\circ}{\mathbf{n}} \cdot [\overset{\circ}{\mathbf{q}}/\theta] + \overset{\circ}{C}_{4\Sigma} = 0. \tag{6.79a}$$

(4) At the part $\overset{\circ}{\Sigma}_2$, corresponding to the surface part $\Sigma_2$, where $\overset{\circ}{M} \neq 0$ and $M \neq 0$, and where the surrounding medium is an ideal fluid, boundary conditions have the same form (6.79), but the Piola–Kirchhoff stress vector in the surroundings $\overset{\circ}{\mathbf{n}} \cdot \mathbf{P}_e = \overset{\circ}{\mathbf{t}}_{ne}$ is determined by the formula

$$\overset{\circ}{\mathbf{t}}_{ne} = -p_e\,(\overset{\circ}{\rho}/\rho)\,\overset{\circ}{\mathbf{n}} \cdot \mathbf{F}^{-1}. \tag{6.80}$$

Here $p_e$ is the pressure in the fluid (referred to actual configuration $\mathcal{K}$), and $\overset{\circ}{\rho}$, $\rho$ and $\mathbf{F}^{-1}$ are parameters of the solid.

(5) At the part $\overset{\circ}{\Sigma}_3$, corresponding to the homothermal coherent surface $\Sigma_3$, where $M = 0$ and $\overset{\circ}{M} = 0$, the boundary conditions (6.79) become

$$\overset{\circ}{\mathbf{n}} \cdot \mathbf{P} = \overset{\circ}{\mathbf{t}}_{ne}, \qquad \overset{\circ}{\mathbf{n}} \cdot \overset{\circ}{\mathbf{q}} = \overset{\circ}{q}_{ne}, \tag{6.81}$$

where $\overset{\circ}{\mathbf{t}}_{ne}$ and $\overset{\circ}{q}_{ne}$ are given values.

(6) At the part $\overset{\circ}{\Sigma}_4$, corresponding to the surface part $\Sigma_4$, where parameters of the ideal fluid are given and $M = 0$, $\overset{\circ}{M} = 0$, we have the conditions (6.81). In these conditions the stress vector $\overset{\circ}{\mathbf{t}}_{ne}$ has the form (6.80), and $\overset{\circ}{q}_{ne}$ is determined by $q_{ne}$:

$$\overset{\circ}{q}_{ne} = \mathbf{q}_e \cdot \mathbf{n}(d\Sigma/d\overset{\circ}{\Sigma}) = q_{ne}\overset{\circ}{k}(\overset{\circ}{\rho}/\rho),$$

where $\overset{\circ}{k}$ is evaluated by formula (2.121). Therefore, at $\overset{\circ}{\Sigma}_4$ the following boundary conditions must be satisfied:

$$\overset{\circ}{\mathbf{n}} \cdot \mathbf{P} = -p_e(\overset{\circ}{\rho}/\rho)\overset{\circ}{\mathbf{n}} \cdot \mathbf{F}^{-1},$$
$$\overset{\circ}{\mathbf{n}} \cdot \overset{\circ}{\mathbf{q}} = q_{ne}\overset{\circ}{k}(\overset{\circ}{\rho}/\rho), \tag{6.82}$$
$$\overset{\circ}{k} = (\overset{\circ}{\mathbf{n}} \cdot \mathbf{G}^{-1} \cdot \overset{\circ}{\mathbf{n}})^{-1/2}.$$

(7) At the part $\overset{\circ}{\Sigma}_5$, corresponding to the surface $\Sigma_5$, the boundary conditions (6.74) formally remain unchanged.

(8), (9) At the part $\overset{\circ}{\Sigma}_6$, corresponding to the surface $\Sigma_6$, the boundary conditions (6.75) become

$$\begin{cases} \mathbf{u} = \mathbf{u}_e, \\ \overset{\circ}{\mathbf{n}} \cdot \overset{\circ}{\mathbf{q}} = \overset{\circ}{q}_{ne}, \end{cases} \qquad \begin{cases} \overset{\circ}{\mathbf{n}} \cdot \mathbf{P} = \overset{\circ}{\mathbf{t}}_{ne}, \\ \theta = \theta_e. \end{cases} \tag{6.83}$$

(10) The symmetry conditions (6.76), defined in $\mathcal{K}$ at the plane $\Pi_\Sigma$ which is fixed at any time $t \geqslant 0$, during the passage to $\overset{\circ}{\mathcal{K}}$ take the form

$$\mathbf{u} \cdot \overset{\circ}{\mathbf{n}} = 0, \quad \mathbf{v} \cdot \overset{\circ}{\mathbf{n}} = 0, \quad \overset{\circ}{\mathbf{q}} \cdot \overset{\circ}{\mathbf{n}} = 0, \quad \overset{\circ}{\mathbf{n}} \cdot \mathbf{P} \cdot \overset{\circ}{\boldsymbol{\tau}}_I = 0, \tag{6.84}$$

because at $\Pi_\Sigma$: $\mathbf{n} = \overset{\circ}{\mathbf{n}}$ and $\boldsymbol{\tau}_I = \overset{\circ}{\boldsymbol{\tau}}_I$.

### 6.3.3 *Statements of Main Problems of Thermoelasticity at Large Deformations in the Spatial Description*

Let us complement now each of the closed systems formulated in Sect. 6.1 with the boundary conditions given in Sect. 6.3.1 and also with initial conditions (the number of which is equal to the number of derivatives with respect to $t$ in the closed system) and Eq. (5.124) of the moving surface $\Sigma$ of a body $V$ considered. Then we obtain statements of the corresponding initial boundary-value problems for elastic continua at large deformations.

Thus, the statement of the *dynamic $\theta RUVF$-problem of thermoelasticity in the spatial description* consists of:

- Equation system (6.1) defined in $V \times (0, t_{\max})$.
- Constitutive equations (6.2)–(6.4) given in $\overline{V} \times (0, t_{\max})$.
- Equation (5.124) of the moving surface $\Sigma$ of a body $V$:

$$\partial f/\partial t + D_0|\nabla f| + \mathbf{v} \cdot \nabla f = 0. \tag{6.85}$$

- Boundary conditions (6.68)–(6.75) given at $\Sigma \times (0, t_{\max})$.
- The initial conditions given in $V$

$$t = 0: \quad \rho = \overset{\circ}{\rho}, \quad \mathbf{v} = \mathbf{v}_0, \quad \theta = \theta_0, \quad \mathbf{F} = \mathbf{E}, \quad \mathbf{u} = \mathbf{0}, \quad f = f^0. \tag{6.86}$$

The statement of the *dynamic $\theta RUV$-problem of thermoelasticity* consists of:

- Equation system (6.14) defined in $V \times (0, t_{\max})$.
- Equation (6.85) of the moving surface $\Sigma$ of a body $V$.
- Boundary conditions (6.68)–(6.75) at $\Sigma \times (0, t_{\max})$, where the constitutive equations (6.2)–(6.4) have been substituted.

- The initial conditions in $V$

$$t = 0: \quad \rho = \overset{\circ}{\rho}, \quad \mathbf{v} = \mathbf{v}_0, \quad \theta = \theta_0, \quad \mathbf{u} = 0, \quad f = f^0. \qquad (6.87)$$

In a similar way, we can give statements of the *dynamic $\theta UV$-problem of thermoelasticity* and the *dynamic $\theta U$-problem.*

The statement of the *dynamic $T\theta RUVF$-problem of thermoelasticity* consists of:

- Equation system (6.20)–(6.25) defined in $V \times (0, t_{\max})$.
- Equation (6.85) of the moving surface $\Sigma$ of a body $V$.
- Boundary conditions (6.68)–(6.75) at $\Sigma \times (0, t_{\max})$, where relations (6.2)–(6.4) and (6.19) have been substituted.
- The initial conditions in $V$

$$t = 0: \quad \mathbf{T} = \mathbf{T}_0, \quad \theta = \theta_0, \quad \rho = \overset{\circ}{\rho}, \quad \mathbf{u} = 0, \quad \mathbf{v} = \mathbf{v}_0, \quad \mathbf{F} = \mathbf{E}, \quad f = f^0. \qquad (6.88)$$

By a given value of the Cauchy stress tensor $\mathbf{T}_0$ and $\mathbf{F} = \mathbf{E}$ at $t = 0$, one can evaluate $\overset{(n)}{\mathbf{T}}_G$ at $t = 0$.

The statement of the *quasistatic problem of thermoelasticity in the spatial description* consists of:

- Equation system (6.35) defined in $V \times (0, t_{\max})$.
- Constitutive equations (6.36) in $\overline{V} \times (0, t_{\max})$.
- Equation (6.85) of the moving surface of a body.
- Boundary conditions (6.68a), (6.69), (6.70a), (6.71a), (6.72)–(6.75) (the velocity $\mathbf{v}$ does not appear in the boundary conditions (6.75) for this statement).
- The initial conditions

$$t = 0: \quad \theta = \theta_0, \quad f = f^0.$$

*Remark 1.* For all the statements considered, the constitutive equations hold both inside the domain $V$ and at its boundary $\Sigma$. □

*Remark 2.* Equation (6.85) of the moving surface $\Sigma$ of a body $V$ holds for all its parts $\Sigma_\alpha$ mentioned in Sect. 6.3.1, where at the parts $\Sigma_1$ and $\Sigma_2$: $M \neq 0$, and at the remaining parts $M = 0$. Moreover, Eq. (6.85) can be written for singular surfaces $S_1(t)$ and $S_2(t)$ moving inside the volume $V$. For surfaces $\Sigma_\alpha$, across which there is no transition of material points ($M = 0$), Eq. (6.85) may be replaced by formula (2.76):

$$\mathbf{x} = \overset{\circ}{\mathbf{x}}(x^i|_{\Sigma_\alpha}, t) + \mathbf{u}(x^i|_{\Sigma_\alpha}, t), \qquad (6.89)$$

where $x^i|_{\Sigma_\alpha}$ are coordinates of points at the surface $\Sigma_\alpha$. □

*Remark 3.* In all the problem statements considered above, the rate $D_0$ of phase transformation is assumed to be a given function in the form

$$D_0 = D_0(\mathbf{F},\ \theta,\ \Omega_e), \tag{6.90}$$

where $\Omega_e = \{\rho_e,\ \mathbf{F}_e,\ \theta_e,\ \mathbf{v}_e\}$ are parameters of the surroundings. □

*Remark 4.* All the problem statements considered above are, in general, *coupled*; i.e. the heat conduction problem consisting of the energy balance equation (6.1c) with the corresponding boundary conditions (6.68)–(6.75) for $q_n$ or $\theta$ and the initial condition (6.86) for $\theta$ cannot be solved separately from the elasticity problem at large deformations. Indeed, if we consider, for example, the dynamic $\theta RUVF$-problem of thermoelasticity and if in place of the energy equation (6.1c) we use the heat conduction equation (6.11), then there are six causes of the connection:

(1) Equation (6.11) involves the convective term $\mathbf{v} \cdot \nabla\theta$ containing the velocity;
(2) this equation also includes the term $\frac{\partial}{\partial\theta}\overset{(n)}{\mathcal{F}} \cdot\cdot \nabla \otimes \mathbf{v}^{\mathrm{T}}$ depending on $\mathbf{F}$ and $\nabla \otimes \mathbf{v}$;
(3) the heat conductivity tensor for a solid is usually defined in a reference configuration $\overset{\circ}{\mathcal{K}}$ (this is the tensor $\overset{\circ}{\boldsymbol{\lambda}}$), and in $\mathcal{K}$ the tensor is determined by formula (4.737) $\boldsymbol{\lambda} = (\rho/\overset{\circ}{\rho})\mathbf{F} \cdot \overset{\circ}{\boldsymbol{\lambda}} \cdot \mathbf{F}^{\mathrm{T}}$, and hence $\boldsymbol{\lambda}$ depends upon $\mathbf{F}$;
(4) the domain $V$ of integration of the system (6.1) is unknown and determined by solving Eq. (6.85) or by formula (6.87), which contain the mechanical unknowns: the velocity $\mathbf{v}$ and the displacement vector $\mathbf{u}$;
(5) the rate $D_0$ of phase transformation (6.90), in general, may depend on the mechanical unknowns $\mathbf{F}$ and $\mathbf{T}$;
(6) in turn, constitutive equations of elasticity (6.3a) depend upon the temperature $\theta$.

A contribution of the term $\frac{\partial}{\partial\theta}\overset{(n)}{\mathcal{F}} \cdot\cdot \nabla \otimes \mathbf{v}^{\mathrm{T}}$ to the heat conduction equation can often be neglected, the rate $D_0$ is usually assumed to depend only on the temperature: $D_0 = D_0(\theta, \Omega_e)$. However, the remaining causes 1, 3, and 4, ensuring the system (6.1a), (6.1b), (6.11), (6.12), and (6.19) to be coupled, in general, cannot be neglected. So the statement of the coupled *dynamic $\theta RUVF$-problem of thermoelasticity at large deformations* differs fundamentally from the thermoelasticity problem under infinitesimal deformations, where the connection, as a rule, can be neglected. Due to this, problems of thermoelasticity at infinitesimal deformations are called *weakly coupled*, and the $\theta RUVF$-problem is called *strongly coupled*.

The dynamic $\theta RUV$-, $\theta UV$- and $T\theta RUVF$-problems of thermoelasticity in the spatial description are also strongly coupled. □

### 6.3.4 *Statements of Thermoelasticity Problems in the Material Description*

Statements of all the problems in the material description, when there are no phase transformations ($\overset{\circ}{M} = 0$), are formulated in the domain $\overset{\circ}{V}$ of a reference configuration, which is known; this is their main advantage over the problem statements in the spatial description.

If a part of the surface $\overset{\circ}{\Sigma}$ of domain $\overset{\circ}{V}$ is moving due to phase transformations ($\overset{\circ}{M} \neq 0$), then the domain $\overset{\circ}{V}$ becomes unknown too, and the system (6.40) of balance laws should be complemented with Eq. (5.115) for determining a shape of the surface $\overset{\circ}{\Sigma}(t)$ of the domain $\overset{\circ}{V}$.

The statement of the *dynamic $\theta UVF$-problem of thermoelasticity in the material description* consists of:

- Equation system (6.40) defined in $\overset{\circ}{V} \times (0, t_{\max})$.
- Constitutive equations (6.41)–(6.43) in $\overline{\overset{\circ}{V}} \times (0, t_{\max})$.
- Equation (5.115) of the motion of a surface $\overset{\circ}{\Sigma}$ due to phase transformations:

$$\partial \overset{\circ}{f}/\partial t + \overset{\circ}{D}|\overset{\circ}{\nabla}\overset{\circ}{f}| = 0. \tag{6.91}$$

- Boundary conditions (6.77)–(6.84) (a part of which may be absent) at $\overset{\circ}{\Sigma} \times (0, t_{\max})$, where the constitutive equations (6.41) and (6.42) have been substituted.
- The initial conditions in $V$

$$t = 0: \quad \theta = \theta_0, \quad \mathbf{u} = 0, \quad \mathbf{v} = \mathbf{v}_0, \quad \mathbf{F} = \mathbf{E}, \quad \overset{\circ}{f} = 0. \tag{6.92}$$

The statement of the *dynamic $\theta UV$-problem of thermoelasticity in the material description* consists of:

- Equation system (6.55) defined in $\overset{\circ}{V} \times (0, t_{\max})$.
- Equation (6.91) of the moving surface $\overset{\circ}{\Sigma}$ of a body.
- Boundary conditions (6.77)–(6.84) at $\overset{\circ}{\Sigma} \times (0, t_{\max})$, where the constitutive equations (6.41), (6.42), and (6.54) have been substituted.
- The initial conditions in $\overset{\circ}{V}$

$$t = 0: \quad \theta = \theta_0, \quad \mathbf{u} = 0, \quad \mathbf{v} = \mathbf{v}_0, \quad \overset{\circ}{f} = 0. \tag{6.93}$$

In a similar way, we can give a statement of the dynamic $\theta U$-problem.

The statement of the *dynamic $T\theta UVF$-problem of thermoelasticity in the material description* consists of:

- Equation system (6.61) defined in $\overset{\circ}{V} \times (0, t_{\max})$.
- Equation (6.91) of the moving surface $\overset{\circ}{\Sigma}$ of a body.
- Boundary conditions (6.77)–(6.84) at $\overset{\circ}{\Sigma} \times (0, t_{\max})$, where the constitutive equations (6.41) and (6.46) have been substituted.
- The initial conditions in $\overset{\circ}{V}$

$$t = 0: \quad \overset{(n)}{\mathbf{T}}_G = \overset{(n)}{\mathbf{T}}_{G0}, \quad \theta = \theta_0, \quad \mathbf{u} = 0, \quad \mathbf{v} = \mathbf{v}_0, \quad \mathbf{F} = \mathbf{E}, \quad \overset{\circ}{f} = 0. \quad (6.94)$$

The statement of the *quasistatic problem of thermoelasticity in the material description* consists of:

- Equation system (6.64), (6.65) defined in $\overset{\circ}{V} \times (0, t_{\max})$.
- Constitutive equations (6.66) in $\overline{\overset{\circ}{V}} \times (0, t_{\max})$.
- Boundary conditions (6.77a), (6.78), (6.79a), and (6.80)–(6.84) at $\overset{\circ}{\Sigma}$ (the equation of (6.84) for $\mathbf{v}$ does not appear in the statement).
- Equation (6.91) of the moving surface $\overset{\circ}{\Sigma}$ of a body.
- The initial conditions

$$t = 0: \ \theta = \theta_0, \qquad \overset{\circ}{f} = 0. \qquad (6.95)$$

The speed $\overset{\circ}{D}$ in all the statements mentioned above is assumed to be a given function in the form

$$\overset{\circ}{D} = \overset{\circ}{D}(\mathbf{F}, \ \theta, \ \Omega_e), \qquad (6.96)$$

where $\Omega_e = \{\rho_e, \ \mathbf{F}_e, \ \theta_e, \ \mathbf{v}_e\}$ are parameters of the surroundings.

*Remark 5.* The statements of thermoelasticity problems in the material description, which have been formulated above, are *weakly coupled* when there are no phase transformations or when the rate of phase transformations depends only on temperature: $\overset{\circ}{D} = \overset{\circ}{D}(\theta, \Omega_e)$, because in this case the heat conduction problem (for example, in the $\theta UVF$-statement this is Eq. (6.53) with the corresponding heat boundary and initial conditions from (6.77)–(6.84), (6.86), and also (6.85)) and the mechanics problem (the remaining equations in the $\theta UVF$-problem) are connected only by the entropy term $(\overset{\circ}{\rho}/\rho)({}^4\overset{(n)}{\mathbf{E}}{}^{\circ}_{G} \cdot\cdot \overset{(n)}{\mathbf{T}}_{\theta G}) \cdot\cdot \overset{\circ}{\nabla} \otimes \mathbf{v}^{\mathrm{T}}$ in (6.53) and by the dependence of the entropy jump $[\eta]$ upon $\mathbf{F}$. These effects in most problems of thermoelasticity can be neglected; as a result, the problems of heat conduction and mechanics turn out to be uncoupled. Usually, one first solves the problem of heat conduction and finds the temperature field $\theta(\overset{\circ}{\mathbf{x}}, t)$ in $\overset{\circ}{V}$, and with the known temperature field one then solves the problem of mechanics.

If the phase transformation rate $\overset{\circ}{D}$ essentially depends on the temperature $\theta$ and on the mechanical unknowns $\mathbf{F}$ and $\mathbf{T}$ (in some problems on solids with phase transformations there is such a situation), then problems of thermoelasticity in the material description turn out to be *strongly coupled* (i.e. the connection cannot be neglected). □

## 6.3.5 Statements of Quasistatic Problems of Elasticity Theory at Large Deformations

An important particular case of the motion of elastic continua, occurring in practice, is the case when the temperature is a constant:

$$\theta = \theta_0 = \text{const} \qquad \forall t \geqslant 0.$$

In this case, one can say that this is the model of *isothermal processes* in solids.

For this model, the energy balance equation (or the entropy balance equation) is eliminated among the balance equations of the general system and the temperature $\theta$ is eliminated between the unknowns for all the statements considered in Sects. 6.3.3 and 6.3.4. The corresponding problems are called *problems of elasticity at large deformations*. For example, the statement of the *quasistatic problem of elasticity in the spatial description* has the form

$$\begin{cases} \nabla \cdot \mathbf{T} + \rho \mathbf{f} = 0 \quad \text{in } V, \\ \mathbf{T} = \overset{(n)}{\mathcal{F}}_G(\mathbf{E} - \nabla \otimes \mathbf{u}^{\mathrm{T}}) \quad \text{in } \overline{V}, \\ \mathbf{n} \cdot \mathbf{T} = \widetilde{\mathbf{t}}_{ne}, \quad \text{at } \Sigma_1,\ \Sigma_2,\ \Sigma_3,\ \Sigma_4,\ \Sigma_7, \\ \mathbf{u} = \mathbf{u}_e, \quad \text{at } \Sigma_5,\ \Sigma_6, \\ \mathbf{u} \cdot \mathbf{n} = 0, \quad \mathbf{n} \cdot \mathbf{T} \cdot \boldsymbol{\tau}_I = 0 \quad \text{at } \Sigma_8, \quad G = A,\ B,\ C,\ D, \end{cases} \tag{6.97}$$

and the statement of the quasistatic problem of elasticity in the material description is written as follows:

$$\begin{cases} \overset{\circ}{\nabla} \cdot \mathbf{P} + \overset{\circ}{\rho} \mathbf{f} = 0 \quad \text{in } \overset{\circ}{V}, \\ \mathbf{P} = \overset{(n)}{\mathcal{F}}{}^{\circ}_{G}(\mathbf{E} + \overset{\circ}{\nabla} \otimes \mathbf{u}^{\mathrm{T}}) \quad \text{in } \overline{\overset{\circ}{V}}, \\ \overset{\circ}{\mathbf{n}} \cdot \mathbf{P} = \overset{\circ}{\mathbf{t}}_{ne}, \quad \text{at } \overset{\circ}{\Sigma}_1,\ \ldots,\ \overset{\circ}{\Sigma}_4,\ \overset{\circ}{\Sigma}_7, \\ \mathbf{u} = \mathbf{u}_e, \quad \text{at } \overset{\circ}{\Sigma}_5,\ \overset{\circ}{\Sigma}_6, \\ \mathbf{u} \cdot \overset{\circ}{\mathbf{n}} = 0, \quad \overset{\circ}{\mathbf{n}} \cdot \mathbf{P} \cdot \overset{\circ}{\boldsymbol{\tau}}_I = 0 \quad \text{at } \overset{\circ}{\Sigma}_8. \end{cases} \tag{6.98}$$

Notice that the boundary conditions (6.70) and (6.71) at the surface parts $\Sigma_1$ and $\Sigma_2$, across which there is a transition of material points ($M \neq 0$), for quasistatic processes coincide with the conditions (6.72) and (6.73) at surfaces without such transition, because under these conditions the term $M[\mathbf{v}]$ can be neglected in comparison with $\mathbf{n} \cdot [\mathbf{T}]$.

For incompressible continua, a statement of the *quasistatic problem of elasticity theory in the spatial description* includes the constitutive equations from Sect. 4.9.6:

$$\begin{cases} \nabla \cdot \mathbf{T} + \rho \mathbf{f} = 0 \quad \text{in } V, \\ \det \mathbf{F}^{-1} = 1 \quad \text{in } V, \\ \mathbf{T} = -p\mathbf{E} + \overset{(n)}{\widetilde{\mathcal{F}}}_G(\mathbf{F}^{-1}) \quad \text{in } \overline{V}, \\ \mathbf{F}^{-1} = \mathbf{E} - \nabla \otimes \mathbf{u}^{\mathrm{T}} \quad \text{in } \overline{V}, \\ \mathbf{n} \cdot \mathbf{T} = \widetilde{\mathbf{t}}_{ne}, \quad \text{at } \Sigma_1, \ \ldots, \ \Sigma_4, \ \Sigma_7, \\ \mathbf{u} = \mathbf{u}_e, \quad \text{at } \Sigma_5, \ \Sigma_6, \\ \mathbf{u} \cdot \mathbf{n} = 0, \quad \mathbf{n} \cdot \mathbf{T} \cdot \boldsymbol{\tau}_I = 0 \quad \text{at } \Sigma_8. \end{cases} \tag{6.99}$$

The second equation in this system is the equation of incompressibility.

The tensor function $\overset{(n)}{\widetilde{\mathcal{F}}}_G(\mathbf{F}^{-1}, \theta)$ has the same form as the corresponding function $\overset{(n)}{\mathcal{F}}_G(\mathbf{F}^{-1}, \theta)$ in (6.13a) and (6.4), and differs from the last one only by the smaller number $(r-1)$ of invariants $I_\gamma^{(s)}(\overset{(n)}{\mathbf{C}}_G)$ in the potential (6.3b) $\psi = \psi(I_{\gamma G}^{(s)}(\overset{(n)}{\mathbf{C}}_G), \theta)$ due to the incompressibility condition.

The problem (6.99) is solved for the unknown functions

$$\mathbf{u}, p \ \parallel \ x^i.$$

The corresponding statement of the *quasistatic problem of elasticity for an incompressible continuum in the material description* has the form

$$\begin{cases} \overset{\circ}{\nabla} \cdot \mathbf{P} + \overset{\circ}{\rho} \mathbf{f} = 0 \quad \text{in } \overset{\circ}{V}, \\ \det \mathbf{F} = 1 \quad \text{in } \overset{\circ}{V}, \\ \mathbf{P} = -p\mathbf{F}^{-1} + \overset{(n)}{\widetilde{\mathcal{F}}}{}^{\circ}_G(\mathbf{F}) \quad \text{in } \overline{\overset{\circ}{V}}, \\ \mathbf{F} = \mathbf{E} + \overset{\circ}{\nabla} \otimes \mathbf{u} \quad \text{in } \overline{\overset{\circ}{V}}, \\ \overset{\circ}{\mathbf{n}} \cdot \mathbf{P} = \overset{\circ}{\mathbf{t}}_{ne}, \quad \text{at } \overset{\circ}{\Sigma}_1, \ \ldots, \ \overset{\circ}{\Sigma}_4, \ \overset{\circ}{\Sigma}_7, \\ \mathbf{u} = \mathbf{u}_e, \quad \text{at } \overset{\circ}{\Sigma}_5, \ \overset{\circ}{\Sigma}_6, \\ \mathbf{u} \cdot \overset{\circ}{\mathbf{n}} = 0, \quad \overset{\circ}{\mathbf{n}} \cdot \mathbf{P} \cdot \boldsymbol{\tau}_\alpha = 0 \quad \text{at } \overset{\circ}{\Sigma}_8, \end{cases} \tag{6.100}$$

and is solved for the same unknown functions but depending on the coordinates $X^i$:

$$\mathbf{u},\ p\ \|\ X^i.$$

The tensor function $\overset{(n)}{\widetilde{\mathcal{F}}}{}^{\circ}_G(\mathbf{F},\ \theta)$ differs from $\overset{(n)}{\mathcal{F}}{}^{\circ}_G$ (6.45) only by the number of invariants being smaller by 1 (this number is equal to $(r-1)$).

The inverse gradient in (6.100) is determined by the formula (see [12])

$$\mathbf{F}^{-1} = \mathbf{F}^2 - I_1(\mathbf{F})\mathbf{F} + I_2(\mathbf{F})\mathbf{E}.$$

*Remark 1.* Since in actual configuration $\mathcal{K}$ the solution domain $V$ is unknown, the systems (6.97) and (6.99) must be complemented with relations (6.89), which allow us to find the unknown geometry of the domain $V$. For the problems (6.98) and (6.100) in the material description, the solution domain $\overset{\circ}{V}$ is known. □

*Remark 2.* A form of the tensor functions $\overset{(n)}{\mathcal{F}}_G$ and $\overset{(n)}{\mathcal{F}}{}^{\circ}_G$ in the statements (6.30)–(6.33) is determined by Eqs. (6.4), (6.13a) and (6.45), and in general, as noted in Remark 1 of Sect. 6.1.1, it is rather complicated and has no explicit analytical representation. The exceptions are the models $A_{\mathrm{I}}$ and $B_{\mathrm{I}}$ for the problems (6.30) and (6.32) in the spatial description and the models $A_{\mathrm{V}}$ and $B_{\mathrm{V}}$ for the problems (6.31) and (6.33) in the material description; they admit an explicit analytical representation of the tensor functions mentioned above (see (6.16b), (6.16c) and Exercises 6.1.1 and 6.1.2).

If we consider a special model of an elastic continuum, for example, the quasi-linear model $A_n$, then as the tensor functions $\overset{(n)}{\mathcal{F}}_G$ and $\overset{(n)}{\mathcal{F}}{}^{\circ}_G$ in (6.30) and (6.31) we can use the representations (6.8a) and (6.47) under the assumption that $\xi = 0$ and ${}^6\mathbf{L} \equiv 0$. For linear models $A_n$, the tensor ${}^4\mathbf{M}$ has the form ${}^4\mathbf{M} = J\ {}^4\widetilde{\mathbf{M}}$, where ${}^4\widetilde{\mathbf{M}}$ is the elastic moduli tensor, which for different symmetry groups $\overset{\circ}{G}_s$ of an elastic continuum is chosen in accordance with the representations from Sect. 4.8.7. □

### *6.3.6 Conditions on External Forces in Quasistatic Problems*

Remind (see Sect. 3.3.1) that besides the motion equations (or the equilibrium equations) having a local character, in continuum mechanics there are integral laws: the momentum balance law (3.32) and the angular momentum balance law (3.82). For quasistatic processes, inertial forces in these laws can be neglected; thus, we obtain the equations

$$\int_V \rho\mathbf{f}\,dV + \int_\Sigma \widetilde{\mathbf{t}}_n\,d\Sigma = 0, \quad \int_V \mathbf{x}\times\rho\mathbf{f}\,dV + \int_\Sigma \mathbf{x}\times\widetilde{\mathbf{t}}_n\,d\Sigma = 0. \qquad (6.101)$$

These equations are additional conditions on external forces $\mathbf{f}$ and $\widetilde{\mathbf{t}}_{ne}$ and displacements $\mathbf{u}_e$ in the quasistatic problem (6.97). Using the boundary conditions of (6.97), we can rewrite the conditions (6.101) in the form

$$\begin{gathered}\int_V \rho\mathbf{f}\,dV + \sum_{\alpha=1,\dots,4,7}\int_{\Sigma_\alpha}\widetilde{\mathbf{t}}_{ne}\,d\Sigma + \sum_{\alpha=5,6}\int_{\Sigma_\alpha}\mathbf{t}_n\,d\Sigma = 0,\\ \int_V \mathbf{x}\times\rho\mathbf{f}\,dV + \sum_{\alpha=1,\dots,4,7}\int_{\Sigma_\alpha}\mathbf{x}\times\widetilde{\mathbf{t}}_{ne}\,d\Sigma + \sum_{\alpha=5,6}\int_{\Sigma_\alpha}(\mathring{\mathbf{x}}+\mathbf{u}_e)\times\mathbf{t}_n\,d\Sigma = 0.\end{gathered} \tag{6.102}$$

These conditions can be expressed in terms of a reference configuration as follows:

$$\begin{gathered}\int_{\mathring{V}} \mathring{\rho}\mathbf{f}\,d\mathring{V} + \sum_{\alpha=1,\dots,4,7}\int_{\mathring{\Sigma}_\alpha}\widetilde{\mathring{\mathbf{t}}}_{ne}\,d\mathring{\Sigma} + \sum_{\alpha=5,6}\int_{\mathring{\Sigma}_\alpha}\mathring{\mathbf{t}}_n\,d\mathring{\Sigma} = 0,\\ \int_{\mathring{V}} \mathbf{x}\times\mathring{\rho}\mathbf{f}\,d\mathring{V} + \sum_{\alpha=1,\dots,4,7}\int_{\mathring{\Sigma}_\alpha}\mathbf{x}\times\widetilde{\mathring{\mathbf{t}}}_{ne}\,d\mathring{\Sigma} + \sum_{\alpha=5,6}\int_{\mathring{\Sigma}_\alpha}(\mathring{\mathbf{x}}+\mathbf{u}_e)\times\mathring{\mathbf{t}}_n\,d\mathring{\Sigma} = 0.\end{gathered} \tag{6.103}$$

As a result, we find conditions on the vectors $\widetilde{\mathring{\mathbf{t}}}_{ne}$, $\mathbf{f}$ and $\mathbf{u}_e$ for the quasistatic problem in the material description.

### 6.3.7 *Variational Statement of the Quasistatic Problem in the Spatial Description*

In solving problems of continuum mechanics at large deformations, variational statements of the problems are widely used.

To state these problems we introduce the concept of a *kinematically admissible* vector field $\mathbf{u}(\mathbf{x})$, which is defined in a domain $V \cup \Sigma$, two times continuously differentiable in this domain and satisfies the boundary conditions of the problem (6.97) at the surface parts $\Sigma_u = \Sigma_5 \cup \Sigma_6$ and $\Sigma_8$, where displacements are given:

$$\mathbf{u}\big|_{\Sigma_u} = \mathbf{u}_e, \quad \mathbf{u}\cdot\mathbf{n}\big|_{\Sigma_8} = 0. \tag{6.104}$$

If the vector field $\mathbf{u}(\mathbf{x})$ in addition satisfies all the remaining equations of the system (6.97), then it is called *real*. So a real vector field $\mathbf{u}$ is the desired solution of the problem (6.97).

Introduce the concept of the *variation* $\delta\mathbf{u}$ of a vector field as the difference of two kinematically admissible fields. The variation $\delta\mathbf{u}$ satisfies the zero boundary conditions

$$\delta\mathbf{u}\big|_{\Sigma_u} = 0, \quad \delta\mathbf{u}\cdot\mathbf{n}\big|_{\Sigma_8} = 0. \tag{6.105}$$

Introduce integral characteristics of the motion of a continuum: $A^e$ is the work done by external forces acting over the displacements $\mathbf{u}$, which consists of the work $A^e_\Sigma$ done by external surface forces and the work $A^e_m$ done by external mass forces:

$$A^e = A^e_\Sigma + A^e_m, \quad A^e_\Sigma = \int_{\Sigma_\sigma} \widetilde{\mathbf{t}}_{ne} \cdot \mathbf{u}\, d\Sigma, \quad A^e_m = \int_V \rho \mathbf{f} \cdot \mathbf{u}\, dV. \tag{6.106}$$

Here $\Sigma_\sigma = \Sigma_1 \cup \Sigma_2 \cup \Sigma_3 \cup \Sigma_4 \cup \Sigma_7$ is the part of the body surface, where the stresses vector $\widetilde{\mathbf{t}}_{ne}$ included in system (6.97) is given.

In addition, introduce the *potential energy* $\Pi$ of a body as follows:

$$\Pi = \int_V \rho \psi \, dV, \tag{6.107}$$

where $\psi(\overset{(\mathrm{n})}{\mathbf{C}}_G, \mathbf{O}, \theta)$ is the free energy (potential) (6.3b) in isothermal processes, which depends on the generalized energetic deformation tensor $\overset{(\mathrm{n})}{\mathbf{C}}_G$ and the rotation tensor $\mathbf{O}$ in accordance with the models $A_n$, $B_n$, $C_n$ and $D_n$.

Write the functional

$$L(\mathbf{u}) = \Pi - A^e, \tag{6.107a}$$

called *Lagrangian*, and introduce the concept of *the variation of functionals $\delta L$, $\delta\Pi$ and $dA^e$* over displacements $\mathbf{u}$, which can be calculated by the same rules as the differential $df$ of function $f(t)$. Then for $\delta\Pi$ we obtain the expression

$$\delta\Pi = \delta \int_V \rho\psi \, dV = \delta \int_{\overset{\circ}{V}} \overset{\circ}{\rho}\psi \, d\overset{\circ}{V} = \int_{\overset{\circ}{V}} \overset{\circ}{\rho}\delta\psi \, d\overset{\circ}{V} = \int_V \rho\delta\psi \, dV. \tag{6.108}$$

Here we have used the rule of differentiation of an integral over a moving volume (see Exercise 3.1.2).

Let us apply now formula (4.178) of differentiation of the scalar function $\psi(\overset{(\mathrm{n})}{\mathbf{C}}_G(t), \mathbf{O}(t), \theta)$ with respect to the argument $t$ to the case of isothermal processes (when $\theta = \mathrm{const}$) and ideal continua (when $w^* = 0$). According to this formula, we can determine the variation $\delta\psi$:

$$\rho\delta\psi(\overset{(\mathrm{n})}{\mathbf{C}}_G, \mathbf{O}, \theta) = \rho\frac{\delta\psi}{\partial \overset{(\mathrm{n})}{\mathbf{C}}_G} \cdot\cdot\, \delta\overset{(\mathrm{n})}{\mathbf{C}}_G + \rho\frac{\delta\psi}{\partial\mathbf{O}} \cdot\cdot\, \delta\mathbf{O}^{\mathrm{T}} = \overset{(\mathrm{n})}{\mathbf{T}}_G \cdot\cdot\, \delta\overset{(\mathrm{n})}{\mathbf{C}}_G + \mathbf{S}_G \cdot\cdot\, \delta\mathbf{O}^{\mathrm{T}}. \tag{6.109}$$

Since formula (6.109) holds for all $n = \mathrm{I}, \ldots, \mathrm{V}$ and $G = A, B, C, D$, then, choosing $n = \mathrm{V}$ and $G = A$, we obtain

$$\rho\delta\psi = \overset{\mathrm{V}}{\mathbf{T}} \cdot\cdot\, \delta\mathbf{C} = \frac{1}{2}\overset{\mathrm{V}}{\mathbf{T}} \cdot\cdot\, \delta(\mathbf{F}^{\mathrm{T}} \cdot \mathbf{F} + \mathbf{F}^{\mathrm{T}} \cdot \delta\mathbf{F})$$

$$= \frac{1}{2}\mathbf{F}\cdot\overset{\mathrm{V}}{\mathbf{T}}\cdot\mathbf{F}^{\mathrm{T}}\cdot\cdot(\mathbf{F}^{-1\mathrm{T}}\cdot\delta\mathbf{F}^{\mathrm{T}}+\delta\mathbf{F}\cdot\mathbf{F}^{-1})$$

$$= \mathbf{T}\cdot\cdot\frac{1}{2}(\mathbf{F}^{-1\mathrm{T}}\cdot\overset{\circ}{\nabla}\otimes\delta\mathbf{u}+(\overset{\circ}{\nabla}\otimes\mathbf{u}^{\mathrm{T}})\cdot\mathbf{F}^{-1}) = \mathbf{T}\cdot\cdot\boldsymbol{\varepsilon}(\delta\mathbf{u}). \tag{6.110}$$

Here we have used the relation (2.73) between $\mathbf{C}$ and $\mathbf{F}$, the relation (4.25) between $\overset{\mathrm{V}}{\mathbf{T}}$ and $\mathbf{T}$, and also formulae (2.79) and (2.37) which yield the relations

$$\delta\mathbf{F}^{\mathrm{T}} = \overset{\circ}{\nabla}\otimes\delta\mathbf{u}, \qquad \mathbf{F}^{-1\mathrm{T}}\cdot\delta\mathbf{F}^{\mathrm{T}} = \nabla\otimes\delta\mathbf{u}. \tag{6.111}$$

In Eqs. (6.110), we have denoted the linear deformation tensor $\boldsymbol{\varepsilon}$ and its variation:

$$\boldsymbol{\varepsilon}(\mathbf{u}) = \frac{1}{2}(\nabla\otimes\mathbf{u}+\nabla\otimes\mathbf{u}^{\mathrm{T}}), \quad \delta\boldsymbol{\varepsilon}(\mathbf{u}) = \boldsymbol{\varepsilon}(\delta\mathbf{u}) = \frac{1}{2}(\nabla\otimes\delta\mathbf{u}+\nabla\otimes\delta\mathbf{u}^{\mathrm{T}}). \tag{6.112}$$

Equating the right-hand sides of (6.109) and (6.110), we find that

$$\overset{(\mathrm{n})}{\mathbf{T}}_G\cdot\cdot\,\delta\overset{(\mathrm{n})}{\mathbf{C}}_G + \mathbf{S}_G\cdot\cdot\,\delta\mathbf{O}^{\mathrm{T}} = \mathbf{T}\cdot\cdot\,\boldsymbol{\varepsilon}(\delta\mathbf{u}). \tag{6.113}$$

This equation is an analog of formulae (4.92) and (4.9) for the stress power when the tensor $\mathbf{T}$ is symmetric. The tensor $\boldsymbol{\varepsilon}(\delta\mathbf{u})$ is an analog of the tensor $\mathbf{D}dt = \boldsymbol{\varepsilon}(\mathbf{v}dt)$ in (6.113).

Substituting (6.109) and (6.113) into (6.108) and evaluating the variation $\delta A^e$, we find the expression for the variation of Lagrangian

$$\delta L = \int_V \mathbf{T}\cdot\cdot\,\boldsymbol{\varepsilon}(\delta\mathbf{u})\,dV - \int_V \rho\mathbf{f}\cdot\delta\mathbf{u}\,dV - \int_{\Sigma_\sigma}\widetilde{\mathbf{t}}_{ne}\cdot\delta\mathbf{u}\,d\Sigma. \tag{6.114}$$

Let us formulate now the main theorem.

**Theorem 6.1 (Lagrange's variational principle).** *Among all kinematically admissible fields* $\mathbf{u}(\mathbf{x})$*, a real field is distinguished by the fact that its and only its Lagrangian* $L$ *has a stationary value:*

$$\delta L = 0. \tag{6.115}$$

▼ Let the field $\mathbf{u}$ satisfy Eq. (6.114). Then, substituting (6.114) into (6.115), we obtain

$$\int_V \mathbf{T}\cdot\cdot\,\boldsymbol{\varepsilon}(\delta\mathbf{u})\,dV - \int_V \rho\mathbf{f}\cdot\delta\mathbf{u}\,dV - \int_{\Sigma_\sigma}\widetilde{\mathbf{t}}_{ne}\cdot\delta\mathbf{u}\,d\Sigma = 0. \tag{6.116}$$

According to (6.112) and the properties of the product of the divergence of a tensor and a vector (see [12]), we can represent the first integral of (6.116) in the form

$$\int_V \mathbf{T}\cdot\cdot\,\boldsymbol{\varepsilon}(\delta\mathbf{u})\,dV = \int_V \mathbf{T}\cdot\cdot\,\nabla\otimes\delta\mathbf{u}^{\mathrm{T}}\,dV = \int_V \nabla\cdot(\mathbf{T}\cdot\mathbf{u})\,dV - \int_V \delta\mathbf{u}\cdot\nabla\cdot\mathbf{T}\,dV$$

$$= \int_{\Sigma_\sigma} \mathbf{n}\cdot\mathbf{T}\cdot\delta\mathbf{u}\,d\Sigma + \sum_{\alpha=1}^{2}\int_{\Sigma_8} \mathbf{n}\cdot\mathbf{T}\cdot\boldsymbol{\tau}_\alpha\delta u_{\tau_\alpha}\,d\Sigma - \int_V \delta\mathbf{u}\cdot\nabla\cdot\mathbf{T}\,dV. \quad (6.117)$$

Here we have taken into account that the tensor $\mathbf{T}$ is symmetric and applied the Gauss–Ostrogradskii formula, and also used that $\delta\mathbf{u} = 0$ at $\Sigma_u$, and at the part $\Sigma_8$ the following relationship holds:

$$\mathbf{n}\cdot\mathbf{T}\cdot\delta\mathbf{u} = \mathbf{t}_n\cdot\delta\mathbf{u} = \left(t_{nn}\mathbf{n} + \sum_{\alpha=1}^{2} t_{n\tau_\alpha}\boldsymbol{\tau}_\alpha\right)\cdot\left(\delta u_n\mathbf{n} + \sum_{\beta=1}^{2}\delta u_{\tau_\alpha}\boldsymbol{\tau}_\alpha\right)$$

$$= t_{nn}\delta u_n + \sum_{\alpha=1}^{2} t_{n\tau_\alpha}\delta u_{\tau_\alpha} = \sum_{\alpha=1}^{2} t_{n\tau_\alpha}\delta u_{\tau_\alpha},$$

as $\delta u_n = \mathbf{n}\cdot\delta\mathbf{u} = 0$ at $\Sigma_8$. Here $\mathbf{n}$, $\boldsymbol{\tau}_\alpha$ is the orthonormal basis; and $t_{nn}$, $t_{n\tau_\alpha}$ and $u_n$, $u_{\tau_\alpha}$ are projections of the vectors $\mathbf{t}_n$ and $\mathbf{u}$ onto the basis vectors.

On substituting (6.117) into (6.116), we obtain

$$\sum_{\alpha=1}^{2}\int_{\Sigma_8}(\mathbf{n}\cdot\mathbf{T}\cdot\boldsymbol{\tau}_\alpha)\delta u_{\tau_\alpha}\,d\Sigma + \int_{\Sigma_\sigma}(\mathbf{n}\cdot\mathbf{T} - \widetilde{\mathbf{t}}_{ne})\cdot\delta\mathbf{u}\,d\Sigma - \int_V \delta\mathbf{u}\cdot(\nabla\cdot\mathbf{T} - \rho\mathbf{f})\,dV = 0. \quad (6.118)$$

Since functions $\delta\mathbf{u}$ are arbitrary, from Eq. (6.118) it follows that all expressions in parentheses must vanish. Thus, the equilibrium equations in the system (6.97) actually hold, and the boundary conditions at $\Sigma_\sigma$ and $\Sigma_8$ are satisfied. At the remaining surface part $\Sigma_u$, the boundary conditions are satisfied due to the choice of functions $\mathbf{u}$. Thus, $\mathbf{u}$ is a solution of the problem (6.97).

Conversely, let $\mathbf{u}(\mathbf{x})$ be a real field of displacements, which satisfies the system (6.97). Then, multiplying the equilibrium equations and the boundary conditions at $\Sigma_\sigma$ and $\Sigma_8$ by $\delta\mathbf{u}$, we get Eq. (6.118). Following in the reverse order all manipulations from (6.117) to (6.114), we verify that Eq. (6.114) actually holds. ▲

Equation (6.116) is called the *variational equation*, and the variational statement of the quasistatic problem (6.97) serves for finding a kinematically admissible displacement field $\mathbf{u}(\mathbf{x})$ satisfying the variational equation (6.116).

### 6.3.8 Variational Statement of Quasistatic Problem in the Material Description

In a similar way, we can formulate a variational statement of the quasistatic problem (6.98). With the help of formulae (6.110), (2.37), and (4.5) we can transform the variation $\delta\psi$ as follows:

$$\begin{aligned}\rho\delta\psi &= \frac{1}{2}(\mathbf{T}^{\mathrm{T}}\cdot\cdot\nabla\otimes\delta\mathbf{u}+\mathbf{T}\cdot\cdot\nabla\otimes\delta\mathbf{u}^{\mathrm{T}})\\ &= \frac{1}{2}(\mathbf{T}^{\mathrm{T}}\cdot\mathbf{F}^{-1\mathrm{T}}\cdot\cdot\overset{\circ}{\nabla}\otimes\delta\mathbf{u}+\mathbf{F}^{-1}\cdot\mathbf{T}\cdot\cdot\overset{\circ}{\nabla}\otimes\delta\mathbf{u}^{\mathrm{T}})\\ &= \frac{\rho}{2\overset{\circ}{\rho}}(\mathbf{P}^{\mathrm{T}}\cdot\cdot\overset{\circ}{\nabla}\otimes\delta\mathbf{u}+\mathbf{P}\cdot\cdot\overset{\circ}{\nabla}\otimes\delta\mathbf{u}^{\mathrm{T}}) = \frac{\rho}{\overset{\circ}{\rho}}\mathbf{P}\cdot\cdot\overset{\circ}{\nabla}\otimes\delta\mathbf{u}^{\mathrm{T}}.\end{aligned} \tag{6.119}$$

Then using formulae (3.55) of passage from $\mathcal{K}$ to $\overset{\circ}{\mathcal{K}}$ and substituting (6.119) into (6.116), we obtain the variational equation in $\overset{\circ}{\mathcal{K}}$

$$\int_{\overset{\circ}{V}}\mathbf{P}\cdot\cdot\overset{\circ}{\nabla}\otimes\delta\mathbf{u}^{\mathrm{T}}\,d\overset{\circ}{V}-\int_{\overset{\circ}{V}}\overset{\circ}{\rho}\mathbf{f}\cdot\delta\mathbf{u}\,d\overset{\circ}{V}-\int_{\overset{\circ}{\Sigma}_{\sigma}}\overset{\circ}{\mathbf{t}}_{ne}\cdot\delta\mathbf{u}\,d\overset{\circ}{\Sigma}=0. \tag{6.120}$$

### 6.3.9 Variational Statement for Incompressible Continua in the Material Description

Let us formulate now a variational statement of the quasistatic problem (6.99) of elasticity for incompressible materials.

Introduce a field of *possible pressures* $p(\mathbf{x})$ being an arbitrary scalar field defined and continuously differentiable in $V\cup\Sigma$, a *real field of pressures* $p(\mathbf{x})$ satisfying (together with a real field of displacements $\mathbf{u}(\mathbf{x})$) the system (6.99), and also the variation of pressures $\delta p$ being the difference of two possible pressures.

The potential energy for an incompressible continuum is introduced as follows:

$$\Pi=\int_V(\rho\psi-\widetilde{p}\gamma)\,dV. \tag{6.121}$$

Here $\widetilde{p}$ is the indeterminate Lagrange multiplier: $\widetilde{p}=p(n-\mathrm{III})^2$ (see (4.500a)), and $\gamma$ is a scalar function of the tensor $\overset{(n)}{\mathbf{G}}_G$, whose value is zero (see (4.493)):

$$\gamma(\overset{(n)}{\mathbf{G}}_G)=\det\overset{(n)}{\mathbf{G}}_G-\frac{1}{(1-\mathrm{III})^3}=0. \tag{6.122}$$

The variation of the function $\gamma$ is calculated by the same rules as the differential $d\gamma$ (see (4.494)):

$$\delta\gamma = \frac{\partial\gamma}{\partial\overset{(n)}{\mathbf{G}}_G}\cdot\cdot\,\delta\overset{(n)}{\mathbf{G}}_G = \frac{1}{(n-\mathrm{III})^3}\overset{(n)}{\mathbf{G}}{}_G^{-1}\cdot\cdot\,\delta\overset{(n)}{\mathbf{G}}_G. \tag{6.123}$$

Since constitutive equations for an incompressible elastic continuum can be written in the form (4.523)

$$\overset{(n)}{\mathbf{T}}_G = -\frac{p}{n-\mathrm{III}}\overset{(n)}{\mathbf{G}}{}_G^{-1} + \widetilde{\mathcal{F}}_G, \quad \widetilde{\mathcal{F}}_G = \rho\frac{\partial\psi}{\partial\overset{(n)}{\mathbf{G}}_G} = \sum_{\gamma=1}^{r-1}\varphi_\gamma\mathbf{I}^{(s)}_{\gamma G}, \tag{6.124}$$

the variation $\delta\psi$, by analogy with formula (6.109), can be represented as follows:

$$\rho\delta\psi(\overset{(n)}{\mathbf{C}}_G,\ \mathbf{O},\ \theta) = \widetilde{\mathcal{F}}_G\cdot\cdot\,\delta\overset{(n)}{\mathbf{C}}_G + \mathbf{S}_G\cdot\cdot\,\delta\mathbf{O}^{\mathrm{T}}. \tag{6.125}$$

According to (6.123) and (6.125), we find

$$\begin{aligned}\delta\Pi &= \int_V(\rho\delta\psi - \widetilde{p}\delta\gamma - \gamma\delta\widetilde{p})\,dV = \int_V\Big(\widetilde{\mathcal{F}}_G - \frac{p}{n-\mathrm{III}}\overset{(n)}{\mathbf{G}}{}_G^{-1}\Big)\cdot\cdot\,\delta\overset{(n)}{\mathbf{G}}_G\,dV\\ &+\int_V\mathbf{S}_G\cdot\cdot\,\delta\mathbf{O}^{\mathrm{T}}\,dV - \int_V\Big(\det\overset{(n)}{\mathbf{G}}_G - \frac{1}{(n-\mathrm{III})^3}\Big)\delta\widetilde{p}\,dV\\ &= \int_V(\overset{(n)}{\mathbf{T}}_G\cdot\cdot\,\delta\overset{(n)}{\mathbf{C}}_G + \mathbf{S}_G\cdot\cdot\,\delta\mathbf{O}^{\mathrm{T}})\,dV - \int_V\Big(\det\overset{(n)}{\mathbf{G}}_G - \frac{1}{(n-\mathrm{III})^3}\Big)\delta\widetilde{p}\,dV\\ &= \int_V\mathbf{T}\cdot\cdot\,\boldsymbol{\varepsilon}(\delta\mathbf{u})\,dV - \frac{1}{n-\mathrm{III}}\int_V((\det\mathbf{F})^{n-\mathrm{III}}-1)\delta p\,dV.\end{aligned} \tag{6.126}$$

Here we have used formulae (6.124) and (6.113), which hold for incompressible materials too, and taken into account that

$$\begin{aligned}\det\overset{(n)}{\mathbf{G}} &= \frac{1}{(n-\mathrm{III})^3}\det\mathbf{U}^{n-\mathrm{III}} = \frac{1}{(n-\mathrm{III})^3}(\det\mathbf{U})^{n-\mathrm{III}}\\ &= \frac{1}{(n-\mathrm{III})^3}(\det\mathbf{F})^{n-\mathrm{III}} = \det\overset{(n)}{\mathbf{g}}\end{aligned}$$

and $\det\mathbf{F} = \det\mathbf{U} = \det\mathbf{V}$.

According to constitutive equations for incompressible elastic continua in the form (6.99), we finally obtain

$$\delta\Pi = \int_V(\overset{(n)}{\widetilde{\mathcal{F}}}_G - p\mathbf{E})\cdot\cdot\,\boldsymbol{\varepsilon}(\delta\mathbf{u})\,dV - \frac{1}{n-\mathrm{III}}\int_V((\det\mathbf{F})^{n-\mathrm{III}}-1)\delta p\,dV. \tag{6.127}$$

Writing Lagrangian $L(\mathbf{u}, p)$ according to formulae (6.107a), (6.121) and (6.106), we can formulate Lagrange's variational principle for incompressible continua.

**Theorem 6.2.** *Among all kinematically admissible fields* $\mathbf{u}(\mathbf{x})$ *and possible pressures* $p$, *real fields of displacements and pressures for incompressible continua are distinguished by the fact that their and only their Lagrangian* $L(\mathbf{u}, p)$ *has a stationary value:*

$$\delta L(\mathbf{u},\ p) = 0. \tag{6.128}$$

Substituting the expressions (6.107a), (6.121) and (6.106) into (6.128) and taking into account that the variations $\delta\mathbf{u}$ and $\delta p$ are mutually independent, we obtain the following *system of variational equations for incompressible elastic continua*:

$$\begin{cases} \int_V (\overset{(n)}{\widetilde{\boldsymbol{\mathcal{F}}}}_G - p\mathbf{E}) \cdot\cdot\, \boldsymbol{\varepsilon}(\delta\mathbf{u})\, dV - \int_V \rho\mathbf{f} \cdot \delta\mathbf{u}\, dV - \int_{\Sigma_\sigma} \widetilde{\mathbf{t}}_{ne} \cdot \delta\mathbf{u}\, d\Sigma = 0, \\ \int_V (\det(\mathbf{E} - \nabla \otimes \mathbf{u}) - 1)\delta p\, dV = 0. \end{cases} \tag{6.129}$$

A proof of the theorem is similar to the proof of Theorem 6.1 (see Exercise 6.3.6).

A variational statement of the quasistatic problem (6.100) of elasticity for an incompressible continuum can be obtained from (6.129) with the help of transformations (6.119):

$$\begin{cases} \int_{\overset{\circ}{V}} (\overset{(n)}{\boldsymbol{\mathcal{F}}}{}^0_G - p\mathbf{F}^{-1}) \cdot\cdot\, \overset{\circ}{\nabla} \otimes \delta\mathbf{u}\, d\overset{\circ}{V} - \int_{\overset{\circ}{V}} \overset{\circ}{\rho}\mathbf{f} \cdot \delta\mathbf{u}\, d\overset{\circ}{V} - \int_{\overset{\circ}{\Sigma}_\sigma} \overset{\circ}{\widetilde{\mathbf{t}}}_{ne} \cdot \delta\mathbf{u}\, d\overset{\circ}{\Sigma} = 0, \\ \int_{\overset{\circ}{V}} (\det(\mathbf{E} + \overset{\circ}{\nabla} \otimes \mathbf{u}) - 1)\delta p\, d\overset{\circ}{V} = 0. \end{cases} \tag{6.130}$$

## Exercises for 6.3

**6.3.1.** Show that the statement (6.99) of the quasistatic problem of elasticity theory in the spatial description for the model $A_{\mathrm{I}}$ of incompressible media has the form (by using the relation (6.16b)):

$$\nabla \cdot \mathbf{T} + \rho\mathbf{f} = 0, \quad \det \mathbf{F}^{-1} = 1 \ \text{ in } V,$$

$$\begin{cases} \mathbf{T} = -p\mathbf{E} + \sum_{\gamma=1}^{r-1} \varphi_\gamma \mathbf{F}^{-1\mathrm{T}} \cdot \mathbf{I}_\gamma^{(s)} \cdot \mathbf{F}^{-1}, \\ \mathbf{I}_\gamma^{(s)} = \partial I_\gamma^{(s)}/\partial\boldsymbol{\Lambda}, \quad \varphi_\gamma = \rho(\partial\psi/\partial I_\gamma^{(s)}), \quad \psi = \psi(I_\gamma^{(s)}(\boldsymbol{\Lambda}),\ \theta), \\ \boldsymbol{\Lambda} = (1/2)(\nabla \otimes \mathbf{u} + \nabla \otimes \mathbf{u}^{\mathrm{T}} - \nabla \otimes \mathbf{u}^{\mathrm{T}} \cdot \nabla \otimes \mathbf{u}), \\ \mathbf{F}^{-1} = \mathbf{E} - \nabla \otimes \mathbf{u}^{\mathrm{T}} \ \text{ in } \overline{V}, \end{cases}$$

$$\mathbf{n} \cdot \mathbf{T}\big|_{\Sigma_\sigma} = \widetilde{\mathbf{t}}_{ne}, \qquad \mathbf{u}\big|_{\Sigma_u} = \mathbf{u}_e,$$

$$\mathbf{u}\big|_{\Sigma_8} \cdot \mathbf{n} = 0, \qquad \mathbf{n} \cdot \mathbf{T}\big|_{\Sigma_8} \cdot \boldsymbol{\tau}_I = 0.$$

**6.3.2.** Show that the quasistatic problem statement (6.97) of elasticity theory for the models $B_n$ and $D_n$ of isotropic media can be written in the form (by using the constitutive equations (4.418a)):

$$\nabla \cdot \mathbf{T} + \rho \mathbf{f} = 0, \quad \mathbf{T} = \overset{\circ}{\rho}\lambda_1\lambda_2\lambda_3 \sum_{\alpha=1}^{3} \lambda_\alpha \frac{\partial \psi(\lambda_\beta)}{\partial \lambda_\alpha} \mathbf{p}_\alpha \otimes \mathbf{p}_\alpha,$$

$$\lambda_\alpha,\ \mathbf{p}_\alpha \ \| \ \mathbf{F}^{-1}, \quad \mathbf{F}^{-1} = \mathbf{E} - \nabla \otimes \mathbf{u}^{\mathrm{T}},$$

$$\mathbf{n} \cdot \mathbf{T}\big|_{\Sigma_\sigma} = \widetilde{\mathbf{t}}_{ne}, \quad \mathbf{u}\big|_{\Sigma_u} = \mathbf{u}_e, \quad \mathbf{u}\big|_{\Sigma_8} \cdot \mathbf{n} = 0, \quad \mathbf{n} \cdot \mathbf{T}\big|_{\Sigma_8} \cdot \boldsymbol{\tau}_I = 0.$$

Here for incompressible continua, constitutive equations take the form

$$\begin{cases} \mathbf{T} = -p\mathbf{E} + \overset{\circ}{\rho} \sum_{\alpha=1}^{3} \lambda_\alpha \dfrac{\partial \psi(\lambda_\beta)}{\partial \lambda_\alpha} \mathbf{p}_\alpha \otimes \mathbf{p}_\alpha, \\ \det \mathbf{F}^{-1} = 1. \end{cases}$$

**6.3.3.** Show that the quasistatic problem statement (6.98) of elasticity theory in the material description for the models $B_n$ and $D_n$ of isotropic media can be written in the form (by using the results of Exercise 6.3.2):

$$\overset{\circ}{\nabla} \cdot \mathbf{P} + \overset{\circ}{\rho} \mathbf{f} = 0, \quad \mathbf{P} = \overset{\circ}{\rho} \sum_{\alpha=1}^{3} \frac{\partial \psi(\lambda_\beta)}{\partial \lambda_\alpha} \overset{\circ}{\mathbf{p}}_\alpha \otimes \mathbf{p}_\alpha,$$

$$\lambda_\alpha, \mathbf{p}_\alpha,\ \overset{\circ}{\mathbf{p}}_\alpha \ \| \ \mathbf{F}^{-1}, \quad \mathbf{F}^{-1} = \mathbf{E} - \overset{\circ}{\nabla} \otimes \mathbf{u}^{\mathrm{T}},$$

$$\overset{\circ}{\mathbf{n}} \cdot \mathbf{P}\big|_{\overset{\circ}{\Sigma}_\sigma} = \overset{\circ}{\mathbf{t}}_{ne}, \quad \mathbf{u}\big|_{\overset{\circ}{\Sigma}_u} = \mathbf{u}_e, \quad \mathbf{u}\big|_{\overset{\circ}{\Sigma}_8} \cdot \overset{\circ}{\mathbf{n}} = 0, \quad \overset{\circ}{\mathbf{n}} \cdot \mathbf{P}\big|_{\overset{\circ}{\Sigma}_8} \cdot \overset{\circ}{\boldsymbol{\tau}}_I = 0,$$

where for incompressible continua, constitutive equations take the form

$$\begin{cases} \mathbf{P} = -p\mathbf{F}^{-1} + \overset{\circ}{\rho} \sum_{\alpha=1}^{3} \dfrac{\partial \psi(\lambda_\beta)}{\partial \lambda_\alpha} \overset{\circ}{\mathbf{p}}_\alpha \otimes \overset{\circ}{\mathbf{p}}_\alpha, \\ \det \mathbf{F}^{-1} = 1. \end{cases}$$

**6.3.4.** Show that the quasistatic problem statement (6.100) of elasticity theory in the material description for the model $A_{\mathrm{V}}$ of incompressible continua has the form (by using the relation (6.49)):

$$\overset{\circ}{\nabla} \cdot \mathbf{P} + \overset{\circ}{\rho} \mathbf{f} = 0, \ \det \mathbf{F} = 1 \ \text{ in } V,$$

$$\begin{cases} \mathbf{P} = -p\mathbf{F}^{-1} + \sum_{\gamma=1}^{r-1} \overset{\circ}{\varphi}_\gamma \mathbf{I}_\gamma^{(s)} \cdot \mathbf{F}^{\mathrm{T}}, \\ \mathbf{I}_\gamma^{(s)} = \partial I_\gamma^{(s)}/\partial \mathbf{C}, \quad \overset{\circ}{\varphi}_\gamma = \overset{\circ}{\rho}(\partial \psi/\partial I_\gamma^{(s)}), \quad \psi = \psi(I_\gamma^{(s)}(\mathbf{C}), \theta), \\ \mathbf{C} = (1/2)(\overset{\circ}{\nabla} \otimes \mathbf{u} + \overset{\circ}{\nabla} \otimes \mathbf{u}^{\mathrm{T}} + \overset{\circ}{\nabla} \otimes \mathbf{u} \cdot \overset{\circ}{\nabla} \otimes \mathbf{u}^{\mathrm{T}}), \\ \mathbf{F} = \mathbf{E} + \overset{\circ}{\nabla} \otimes \mathbf{u}^{\mathrm{T}} \ \text{ in } \overline{V}, \end{cases}$$

$$\overset{\circ}{\mathbf{n}} \cdot \mathbf{P}\big|_{\overset{\circ}{\Sigma}_\sigma} = \overset{\circ}{\mathbf{t}}_{ne}, \quad \mathbf{u}\big|_{\overset{\circ}{\Sigma}_u} = \mathbf{u}_e, \quad \mathbf{u}\big|_{\overset{\circ}{\Sigma}_8} \cdot \overset{\circ}{\mathbf{n}} = 0, \quad \overset{\circ}{\mathbf{n}} \cdot \mathbf{P}\big|_{\overset{\circ}{\Sigma}_8} \cdot \overset{\circ}{\boldsymbol{\tau}}_I = 0.$$

Show that the problem for basis $\overset{\circ}{\mathbf{r}}_i$ can be written in components:

$$\overset{\circ}{\nabla}_j \overset{\circ}{P}{}^j_i + \overset{\circ}{\rho}\overset{\circ}{f}_i = 0, \quad \det\,(\overset{\circ}{F}{}^k_l) = 1,$$

$$\begin{cases} \overset{\circ}{P}{}^j_i = -p(\overset{\circ}{F}{}^{-1})^j_i + \sum_{\gamma=1}^{r-1} \overset{\circ}{\varphi}_\gamma (\partial I_{\gamma G}/\partial \varepsilon_{jk}) \overset{\circ}{F}{}^l_k \overset{\circ}{g}_{il}, \\ \overset{\circ}{\varphi}_\gamma = \overset{\circ}{\rho}(\partial\psi/\partial I_\gamma^{(s)}), \quad \psi = \psi(I_\gamma^{(s)}(\varepsilon_{jk}), \theta), \\ \varepsilon_{jk} = (1/2)(\overset{\circ}{\nabla}_j \overset{\circ}{u}_k + \overset{\circ}{\nabla}_k \overset{\circ}{u}_j + \overset{\circ}{\nabla}_j \overset{\circ}{u}_m \overset{\circ}{\nabla}_k \overset{\circ}{u}_i \overset{\circ}{g}{}^{mi}), \\ \overset{\circ}{F}{}^i_k = \delta^i_k + \overset{\circ}{\nabla}_k \overset{\circ}{u}_m \overset{\circ}{g}{}^{mi}, \\ (\overset{\circ}{F}{}^{-1})^j_i = (1/2)\epsilon^{jmk}\epsilon_{iqs} \overset{\circ}{F}{}^q_m \overset{\circ}{F}{}^s_k, \end{cases}$$

$$\overset{\circ}{n}_j \overset{\circ}{P}{}^j_i\big|_{\overset{\circ}{\Sigma}_\sigma} = \overset{\circ}{t}_{nei}, \quad \overset{\circ}{u}_i\big|_{\overset{\circ}{\Sigma}_u} = \overset{\circ}{u}_{ei}, \quad \overset{\circ}{u}_i\big|_{\overset{\circ}{\Sigma}_g} \overset{\circ}{n}{}^i = 0, \quad \overset{\circ}{n}_j \overset{\circ}{P}{}^j{}_k\big|_{\overset{\circ}{\Sigma}_g} \overset{\circ}{\tau}{}^k_I = 0.$$

(One should derive an expression for $(\overset{\circ}{F}{}^{-1})^j_i$ by using the relation (2.125) between components of a matrix and its inverse.)

**6.3.5.** Prove relations (6.71).

**6.3.6.** Prove Theorem 6.2.

## 6.4 The Problem on an Elastic Beam in Tension

### *6.4.1 Semi-Inverse Method*

Consider several classical examples of solving the elasticity problems (6.97) and (6.98). Practically all main analytical solutions of quasistatic problems of elasticity theory at large deformations can be obtained by using the *semi-inverse method*, where proceeding from the geometric shape of a body, in a reference configuration one selects (guesses) a law of its motion $\mathbf{x}(X^i, t)$, according to which the deformation gradient field $\mathbf{F}(X^i, t)$ is calculated. Then one can evaluate the stress tensor fields $\mathbf{T}$ and $\mathbf{P}$ by constitutive equations. Further, one should verify that the tensor fields $\mathbf{P}(X^i, t)$ and $\mathbf{T}(X^i, t)$ obtained satisfy the equilibrium equations and boundary conditions of the systems (6.97) and (6.98). If the motion law is guessed right, then these equations and conditions are satisfied identically.

### *6.4.2 Deformation of a Beam in Tension*

Consider the problem on a beam in tension; then the motion law has the form (2.6)

$$x^\alpha = k_\alpha(t)\, X^\alpha. \tag{6.131}$$

**Table 6.1** Sequential calculation of the deformation tensors

| | Tension | Shear | Rotation with tension |
|---|---|---|---|
| The motion law | (2.6) | (2.7) | (2.8) |
| Local basis vectors and metric matrices | Exercise 2.1.1 | Exercise 2.1.2 | Exercise 2.1.3 |
| Deformation gradient **F**, **C**, **G**, **A**, **J** | Exercise 2.2.1 | Exercise 2.2.2 | Exercise 2.2.3 |
| Tensors **U**, **V**, **O** | Exercise 2.3.2 | Exercise 2.3.3 | Exercise 2.3.4 |
| Tensors $\overset{(n)}{\mathbf{C}}$, $\overset{(n)}{\mathbf{A}}$, $\overset{(n)}{\mathbf{G}}$, $\overset{(n)}{\mathbf{g}}$ | Exercise 4.2.13 | Exercise 4.2.14 | Exercise 4.2.15 |

The deformation gradient **F** for this problem becomes (see Exercise 2.2.1)

$$\mathbf{F} = \sum_{\alpha=1}^{3} k_\alpha \bar{\mathbf{e}}_\alpha \otimes \bar{\mathbf{e}}_\alpha. \tag{6.132}$$

The deformation tensors for this problem were calculated above (this procedure is shown in Table 6.1).

In particular, the tensors $\overset{(n)}{\mathbf{C}}$ and $\overset{(n)}{\mathbf{A}}$ have the form

$$\overset{(n)}{\mathbf{C}} = \overset{(n)}{\mathbf{A}} = \sum_{\alpha=1}^{3} \overset{(n)}{C}_\alpha \bar{\mathbf{e}}_\alpha \otimes \bar{\mathbf{e}}_\alpha, \tag{6.133}$$

$$\overset{(n)}{C}_\alpha = \frac{1}{n-\mathrm{III}}(k_\alpha^{n-\mathrm{III}} - 1), \quad n = \mathrm{I},\ \mathrm{II},\ \mathrm{IV},\ \mathrm{V}; \qquad \overset{(n)}{C}_\alpha = \ln k_\alpha, \quad n = \mathrm{III}. \tag{6.134}$$

### 6.4.3 Stresses in a Beam

Assume that a beam is made of elastic (ideal, solid) isotropic material described by linear models $A_n$ (4.338):

$$\overset{(n)}{\mathbf{T}} = J(l_1 I_1(\overset{(n)}{\mathbf{C}})\mathbf{E} + 2l_2 \overset{(n)}{\mathbf{C}}), \quad J = \rho/\overset{\circ}{\rho}. \tag{6.135}$$

Let us calculate stresses in the beam. On substituting the expressions (6.133) and (6.134) for the energetic deformation tensors $\overset{(n)}{\mathbf{C}}$ into (6.135), we find components $\overset{(n)}{T}_{\alpha\alpha}$ of the energetic stress tensors

$$\overset{(n)}{\mathbf{T}} = \sum_{\alpha=1}^{3} \overset{(n)}{T}_{\alpha\alpha} \bar{\mathbf{e}}_\alpha \otimes \bar{\mathbf{e}}_\alpha,$$

$$\overset{(n)}{T}_{\alpha\alpha} = \frac{1}{(n-\mathrm{III})k_1k_2k_3}\left(l_1\left(\sum_{\beta=1}^{3} k_\beta^{n-\mathrm{III}} - 3\right) + 2l_2(k_\alpha^{n-\mathrm{III}} - 1)\right), \tag{6.136}$$

$$n = \mathrm{I}, \mathrm{II}, \mathrm{IV}, \mathrm{V},$$

$$\overset{\mathrm{III}}{T}_{\alpha\alpha} = \frac{1}{k_1k_2k_3}\left(l_1 \sum_{\alpha=1}^{3} \lg k_\alpha + 2l_2 \lg k_\alpha\right).$$

Since in this problem the rotation tensor $\mathbf{O}$ is the unit tensor (see Exercise 2.3.2), the quasienergetic stress tensors $\overset{(n)}{\mathbf{S}}$ coincide with $\overset{(n)}{\mathbf{T}}$ (see Exercise 4.2.3): $\overset{(n)}{\mathbf{S}} = \overset{(n)}{\mathbf{T}}$.

To determine the Cauchy stress tensor $\mathbf{T}$ we use the tensors of energetic equivalence ${}^4\overset{(n)}{\mathbf{E}}$ and formulae (4.37) having the following forms in this problem:

$$\begin{aligned}
n=\mathrm{I}: \quad & \mathbf{T} = \mathbf{F}^{-1\mathrm{T}} \cdot \overset{\mathrm{I}}{\mathbf{T}} \cdot \mathbf{F}^{-1},\\
n=\mathrm{II}: \quad & \mathbf{T} = (1/2)(\mathbf{F}^{-1\mathrm{T}} \cdot \overset{\mathrm{II}}{\mathbf{T}} + \overset{\mathrm{II}}{\mathbf{T}} \cdot \mathbf{F}^{-1}),\\
n=\mathrm{III}: \quad & \mathbf{T} = \overset{\mathrm{III}}{\mathbf{T}},\\
n=\mathrm{IV}: \quad & \mathbf{T} = (1/2)(\mathbf{F} \cdot \overset{\mathrm{IV}}{\mathbf{T}} + \overset{\mathrm{IV}}{\mathbf{T}} \cdot \mathbf{F}^{\mathrm{T}}),\\
n=\mathrm{V}: \quad & \mathbf{T} = \mathbf{F} \cdot \overset{\mathrm{V}}{\mathbf{T}} \cdot \mathbf{F}^{\mathrm{T}}.
\end{aligned} \tag{6.137}$$

On substituting the expression (6.132) for $\mathbf{F}$ and formulae (6.133) into (6.137), we find components $\sigma_{\alpha\alpha}$ of the Cauchy stress tensor $\mathbf{T}$ with respect to the Cartesian basis for different models $A_n$:

$$\mathbf{T} = \sum_{\alpha=1}^{3} \sigma_{\alpha\alpha} \bar{\mathbf{e}}_\alpha \otimes \bar{\mathbf{e}}_\alpha, \quad \sigma_{\alpha\alpha} = k_\alpha^{n-\mathrm{III}} \overset{(n)}{T}_{\alpha\alpha}, \tag{6.138}$$

$$\sigma_{\alpha\alpha} = \frac{k_\alpha^{n-\mathrm{III}}}{(n-\mathrm{III})k_1k_2k_3}\left(l_1\left(\sum_{\beta=1}^{3} k_\beta^{n-\mathrm{III}} - 3\right) + 2l_2(k_\alpha^{n-\mathrm{III}} - 1)\right),$$

$$n = \mathrm{I},\ \mathrm{II},\ \mathrm{IV},\ \mathrm{V}, \tag{6.139a}$$

$$\sigma_{\alpha\alpha} = \overset{\mathrm{III}}{T}_\alpha, \qquad n = \mathrm{III}. \tag{6.139b}$$

Since the expression for $\mathbf{T}$ obtained is independent of the coordinates $X^i$, the tensor $\mathbf{T}$ (6.138) automatically satisfies the equilibrium equations of the system (6.97): $\nabla \cdot \mathbf{T} = 0$ when there are no mass forces ($\mathbf{f} = 0$).

### 6.4.4 The Boundary Conditions

Boundary conditions of the system (6.97) applying to the problem on a beam in tension become

$$\begin{aligned} x^1 &= 0: \quad \mathbf{n}\cdot\mathbf{u} = 0, \quad \mathbf{n}\cdot\mathbf{T}\cdot\boldsymbol{\tau}_\alpha = 0, \quad \alpha = 2,\ 3; \\ x^1 &= h_1: \quad \mathbf{n}\cdot\mathbf{u} = u_e^1, \quad \mathbf{n}\cdot\mathbf{T}\cdot\boldsymbol{\tau}_\alpha = 0, \quad \alpha = 2,3; \\ x^\alpha &= \pm h_\alpha/2: \quad \mathbf{n}\cdot\mathbf{T} = 0, \quad \alpha = 2,\ 3. \end{aligned} \tag{6.140}$$

In other words, according to the problem statement from Example 2.1, at one end of the beam ($x^1 = 0$) the symmetry condition is given, at the other end ($x^1 = h_1$) – the displacement $u_e^1$, and the lateral surfaces ($x^I = \pm h_I/2$) are free of loads. Since at the surface $x^1 = 0$ we have $\mathbf{n} = -\bar{\mathbf{e}}_1$, $\boldsymbol{\tau}_\alpha = \bar{\mathbf{e}}_\alpha$ ($\alpha = 2,\ 3$) and at the lateral surface $x^\alpha = \pm h_\alpha/2$: $\mathbf{n} = \bar{\mathbf{e}}_\alpha$, the boundary conditions (6.140) take the form

$$x^1 = 0: \quad u^1 = 0, \quad \bar{\mathbf{e}}_1\cdot\mathbf{T}\cdot\bar{\mathbf{e}}_\alpha = 0, \quad \alpha = 2,3, \tag{6.141a}$$

$$x^1 = h_1: \quad u^1 = u_e^1, \quad \bar{\mathbf{e}}_1\cdot\mathbf{T}\cdot\bar{\mathbf{e}}_\alpha = 0, \quad \alpha = 2,3; \tag{6.141b}$$

$$x^\alpha = \pm h_\alpha/2: \quad \sigma_{\alpha\alpha} = 0, \quad \alpha = 2,3. \tag{6.141c}$$

As follows from formula (6.138), there are no tangential stresses in this problem; therefore, the conditions $\bar{\mathbf{e}}_1\cdot\mathbf{T}\cdot\bar{\mathbf{e}}_\alpha = 0$ are automatically satisfied in the whole beam. The boundary condition $u^1 = 0$ at $x^1 = 0$ is satisfied due to (6.131), and from the boundary condition at $x^1 = h_1$ we express $k_1$ in terms of $u_e^1$:

$$u_e^1 = h_1 - h_1^0 = (k_1 - 1)h_1^0, \qquad k_1 = 1 + u_e^1/h_1^0. \tag{6.142}$$

### 6.4.5 Resolving Relation $\sigma_1 \sim k_1$

On substituting (6.139a) into the boundary condition (6.141c), we find that the components $\sigma_{22}$ and $\sigma_{33}$ of the stress tensor $\mathbf{T}$ in this problem must be zero, and only the component $\sigma_{11}$ is nonzero. As a result, uniting (6.139b) and (6.141c), we obtain three equations

$$\sigma_{11} = \frac{k_1^{n-\mathrm{III}}}{(n-\mathrm{III})k_1k_2k_3}\left(l_1\left(\sum_{\beta=1}^{3} k_\beta^{n-\mathrm{III}} - 3\right) + 2l_2(k_1^{n-\mathrm{III}} - 1)\right),$$
$$n = \mathrm{I}, \mathrm{II}, \mathrm{IV}, \mathrm{V}, \tag{6.143}$$

$$0 = l_1\left(\sum_{\beta=1}^{3} k_\beta^{n-\mathrm{III}} - 3\right) + 2l_2(k_\alpha^{n-\mathrm{III}} - 1), \quad \alpha = 2,\ 3, \tag{6.144}$$

for the three unknowns: $k_1$, $k_2$ and $\sigma_{11}$ ($k_1$ is determined by (6.142)).

Notice that the system (6.143)–(6.144) admits the solution $k_2 = k_3$; then with the help of (6.144) we can express $k_2$ and $k_3$ in terms of $k_1$:

$$k_\alpha^{n-\mathrm{III}} - 1 = -\nu(k_1^{n-\mathrm{III}} - 1), \quad \alpha = 2,\ 3, \tag{6.145}$$

where

$$\nu = \frac{l_1}{2(l_1 + l_2)} \tag{6.146}$$

is the *Poisson ratio*. Calculating the ratio of components $\overset{(n)}{C}_2$ and $\overset{(n)}{C}_1$ of the energetic deformation tensors, for which Eqs. (6.134) hold, and using formula (6.145), we find that Poisson's ratio is the ratio of these components with the opposite sign:

$$\overset{(n)}{C}_2/\overset{(n)}{C}_1 = \overset{(n)}{C}_3/\overset{(n)}{C}_1 = -\nu. \tag{6.147}$$

This relationship holds for all models $A_n$. For most real isotropic elastic materials, values of the Poisson ratio are within the interval $0 < \nu < 0.5$. Therefore, for a beam in longitudinal tension along the axis $Ox^1$ when $\overset{(n)}{C}_1 > 0$, the transverse deformations are negative: $\overset{(n)}{C}_2 = \overset{(n)}{C}_3 < 0$, i.e. there occurs a transverse compression of the beam. This phenomenon is called the *Poisson effect*.

On substituting the expression (6.145) into (6.143), we find the resolving relation for the problem: the dependence of the stress $\sigma_{11}$ on the elongation ratio $k_1$:

$$\sigma_{11} = 2l_2(1+\nu)\frac{k_1^{n-\mathrm{III}-1}}{n-\mathrm{III}}(k_1^{n-\mathrm{III}} - 1)(1 - \nu(k_1^{n-\mathrm{III}} - 1))^{-2/(n-\mathrm{III})}, \tag{6.148}$$
$$n = \mathrm{I},\ \mathrm{II},\ \mathrm{IV},\ \mathrm{V}.$$

By the same method, we obtain the resolving relation for the model $A_{\mathrm{III}}$ (see Exercise 6.4.1):

$$k_2 = k_3 = 1/k_1^\nu, \quad \sigma_{11} = 2l_2(1+\nu)\frac{\lg k_1}{k_1^{1-2\nu}}. \tag{6.149}$$

### 6.4.6 Comparative Analysis of Different Models $A_n$

Compare the dependences $k_2(k_1)$ and $\sigma_{11}(k_1)$ obtained for different models $A_n$.

Notice that for all models $A_n$, the dependences $k_2(k_1)$ (6.145) and (6.149) are monotonically decreasing, i.e. the extension of the beam along the axis $Ox^1$ is accompanied by its compression along the axes $Ox^2$ and $Ox^3$ (remind that the values $k_\alpha > 1$ correspond to extension along the axis $Ox^\alpha$, and the values $0 < k_\alpha < 1$ – to compression along $Ox^\alpha$). This property is called the *Poisson effect*.

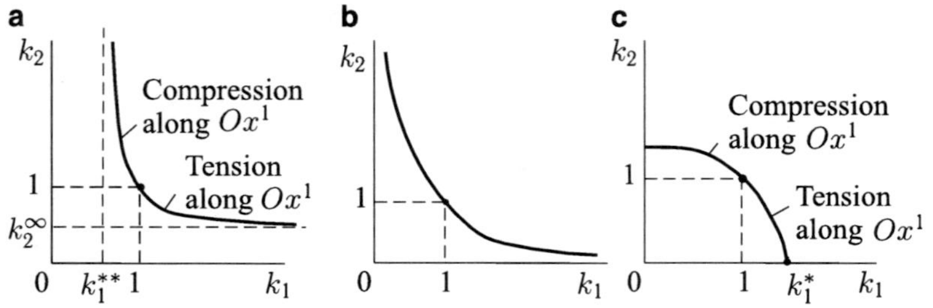


**Fig. 6.4** Dependence $k_2(k_1)$ for different models $A_n$: **(a)** – $n$ = I and $n$ = II, **(b)** – $n$ = III, **(c)** – $n$ = IV and $n$ = V

However, for the models $A_{\mathrm{I}}$ and $A_{\mathrm{II}}$, the function $k_2(k_1)$ has horizontal and vertical asymptotes (Fig. 6.4):

$$k_1 = k_1^{**} \equiv \left(\frac{\nu}{1+\nu}\right)^{1/(\mathrm{III}-n)} \quad \text{and} \quad k_2 = k_2^\infty \equiv \left(\frac{1}{1+\nu}\right)^{1/(\mathrm{III}-n)}, \quad n = \mathrm{I},\ \mathrm{II},$$

and for the models $A_{\mathrm{IV}}$ and $A_{\mathrm{V}}$ there are no asymptotes, but there is ultimate value $k_1^* = (1+\nu/\nu)^{1/(n-\mathrm{III})}$, at which the solution degenerates with loss of the physical meaning.

For the model $A_{\mathrm{III}}$, according to (6.149), the function $k_2(k_1)$ also has horizontal and vertical asymptotes which are coincident with the abscissa and ordinate axes, respectively (Fig. 6.4).

Comparing graphs of the functions $\sigma_{11}(k_1)$ (6.148) and (6.149), we can see that for all models $A_n$ ($n = \mathrm{I}, \dots, \mathrm{V}$), values of the stresses $\sigma_{11}$ are positive for the beam in tension along the axis $Ox^1$ ($k_1 > 1$) and negative for the beam in compression along the axis $Ox^1$ ($0 < k_1 < 1$), that agrees with actual stresses observed in practice. Notice that values of $k_1$ can be found in experiments by displacements $u_e^1$ according to formula (6.142), and stresses $\sigma_1$ – according to formula (3.64), which for the beam takes the form

$$\sigma_{11} = \widehat{T}^{11} = \mathcal{F}_1^1/\Sigma_1. \tag{6.150}$$

Here $\mathcal{F}_1^1$ is the component referred to the axis $Ox^1$ of the external force vector $\boldsymbol{\mathcal{F}}^1 = \mathcal{F}_1^1\bar{\mathbf{e}}_1$ applied to the beam side $\Sigma_1$ orthogonal to $Ox^1$.

However, for the models $A_{\mathrm{I}}$ and $A_{\mathrm{II}}$, the function $\sigma_{11}(k_1)$ in tension (when $k_1 > 1$) first grows and then diminishes, and in compression (when $k_1 < 1$) the magnitude $|\sigma_{11}(k_1)|$ also first increases and then decreases, while $\sigma_{11}(k_1^{**}) = 0$ (see Fig. 6.5, where we have denoted the dimensionless stress $\bar{\sigma}_{11} = \sigma_{11}/(2l_2(1+\nu))$). The monotone increase of the function $\sigma_{11}(k_1)$, that is usually observed in experiments, occurs only within the interval $k_1^{**} < k_1^{\min} \leqslant k_1 \leqslant k_1^{\max}$. The extremum values $k_1^{\min}$ and $k_1^{\max}$, at which monotonicity of the function $\sigma_{11}(k_1)$ is violated (i.e. $\frac{\partial}{\partial k_1}\sigma_{11}(k_1)\big|_{k_1=k_1^{\min},k_1^{\max}} = 0$), are sometimes interpreted as ultimate values of loss of stability of the material in compression or in tension, respectively. However, this

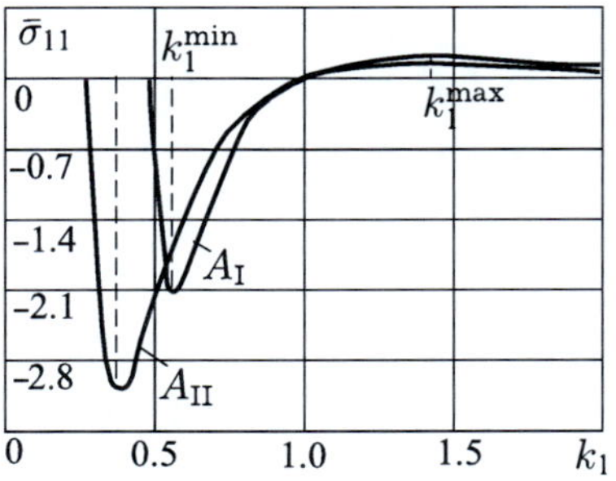


**Fig. 6.5** The function $\sigma_{11}(k_1)$ for the models $A_I$ and $A_{II}$, $\nu = 0.3$

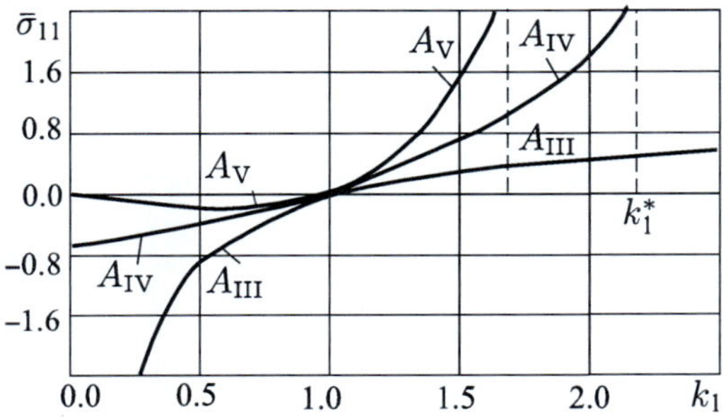


**Fig. 6.6** The function $\sigma_{11}(k_1)$ for the models $A_{III}$, $A_{IV}$ and $A_V$, $\nu = 0.3$

effect cannot be identified with the phenomenon of loss of stability for structures of solids in compression, that is connected, as a rule, not with non-linearity of the constitutive equations but with the nonlinear dependence of deformations upon displacements.

For the models $A_{III}$ and $A_{IV}$, the function $\sigma_{11}(k_1)$ is monotonically increasing at all values of $k_1$. For the model $A_{IV}$, the function is defined only within the interval $0 < k_1 < k_1^*$, and $\sigma_{11} \to +\infty$ as $k_1 \to k_1^*$. For the model $A_V$, the function $\sigma_{11}(k_1)$ in tension ($k_1 > 1$) behaves as well as it does in the model $A_{IV}$; and in compression the function has an extremum point ($k_1 = k_{min}$, see Fig. 6.6).

The model $A_{III}$ is the only one of models $A_n$, for which the function $\sigma_{11}(k_1)$ exists and is monotone along the whole semiaxis $k > 0$, and $\sigma_{11}(k) \to -\infty$ as $k \to 0$.

Thus, the problem on a beam in tension shows that models $A_n$ at different $n$ yield distinct (even in kind) dependences of the stress on the elongation ratio $\sigma_{11}(k)$. The question what model $A_n$ should be applied to a specific elastic continuum is solved by comparing experimental data with computed results determined by different models; in particular, by comparing the functions $\sigma_{11} = \sigma_{11}(k_1)$ (6.148) and (6.149) with corresponding experimental dependences.

In experimental investigations, instead of $k_1$ one usually uses the relative elongation

$$\delta_1 = k_1 - 1 = \frac{u_e^1}{h_1^0} = \frac{h_1 - h_1^0}{h_1^0}, \qquad (6.151)$$

and the dependence $\sigma_{11} = \sigma_{11}(\delta_1)$ is called the *diagram of deforming*.

Notice also that in experiment one frequently determines the Piola–Kirchhoff stress $P_{11}$, that is the ratio of component $\mathcal{F}_1^1$ of the external force vector (measured in experiments) to the undistorted surface element $\overset{\circ}{\Sigma}_1$:

$$P_{11} = \mathcal{F}_1^1/\overset{\circ}{\Sigma}_1. \tag{6.152}$$

In the problem on tension, the areas $\overset{\circ}{\Sigma}_1$ and $\Sigma_1$ are connected by the relation

$$\Sigma_1/\overset{\circ}{\Sigma}_1 = k_2k_3 = k_2^2, \tag{6.153}$$

therefore, $P_{11}$ and $\sigma_{11}$ are connected by

$$\sigma_{11} = P_{11}/k_2^2. \tag{6.154}$$

According to (6.148), (6.149) and (6.154), we find the following expressions for the function $P_{11}(k_1)$:

$$P_{11} = 2l_2(1+\nu)\frac{k_1^{n-\mathrm{III}-1}}{n-\mathrm{III}}(k_1^{n-\mathrm{III}}-1), \quad n = \mathrm{I, II, IV, V},$$
$$P_{11} = 2l_2(1+\nu)\frac{\lg k_1}{k_1}, \quad n = \mathrm{III}. \tag{6.155}$$

Notice that the functions $P_{11}(k_1)$ are independent of the Poisson ratio. Thus, if we have the experimental diagram $P_{11}(k_1)$, then we cannot calculate values of $\nu$ by (6.155); they may be determined by changing the thickness of a specimen in tension or in compression according to formula (6.146).

Figure 6.7 shows the experimental diagram of deforming $P_{11}^{(\mathrm{ex})}(\delta_1)$ obtained by the method mentioned above for filled rubber, where $\delta_1 = k_1 - 1$, and also

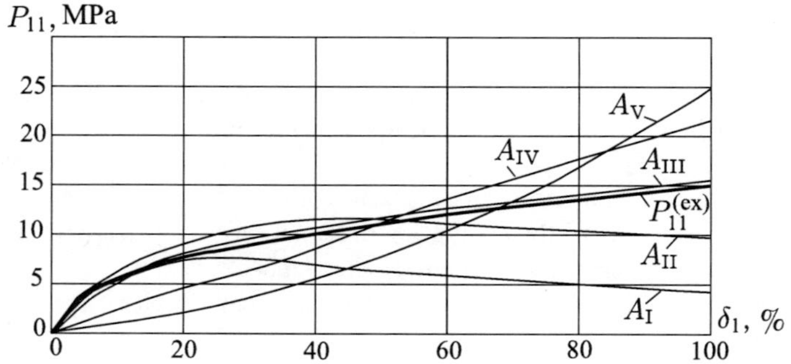


**Fig. 6.7** Experimental diagram of deforming $P_{11}^{(\mathrm{ex})}(\delta_1)$ for filled rubber in tension and its approximation by linear models $A_n$ at $\nu = 0.4$

approximation of this diagram with the help of models $A_n$ by formulae (6.155). A value of the constant $l_2$ for these models was determined by minimizing the mean-square distance $\delta_{1(i)}$ between computed and experimental curves at $N$ points:

$$\Delta = \left( \frac{1}{N} \sum_{i=1}^{N} \left| 1 - \frac{P_{11}(\delta_{1(i)})}{P_{11}^{(\text{ex})}(\delta_{1(i)})} \right|^2 \right)^{1/2} \to \min. \tag{6.156}$$

The Poisson ratio for filled rubber proved to be close to $\nu = 0.4$.

Values of the constant $l_2$, obtained by the method mentioned above, are shown in Table 6.2; this table gives also the minimum mean-square deviations corresponding to these optimal values of $l_2$.

The values found allow us to conclude that the model $A_{\text{III}}$ gives the best approximation to the experimental diagram of deforming (see Fig. 6.7). The models $A_{\text{I}}$ and $A_{\text{II}}$ satisfactorily approximate the diagram $P_{11}^{(\text{ex})}(\delta_1)$, when $\delta_1$ ranges from 0 to 22 and 45%, respectively. At higher values of $\delta_1$ the condition $k_1 > k_{\max}$ holds, and functions $P_{11}(\delta_1)$ for these models give the non-physical effect of lowering the stresses $P_{11}$.

Figure 6.8 shows diagrams of deforming $\sigma_{11}^{(\text{ex})}(\delta_1)$ for the same rubber, which were recalculated from the experimental diagram $P_{11}^{(\text{ex})}(\delta_1)$ by formulae (6.154), where $k_2$ were determined by (6.145) and (6.149). Since the functions $k_2(k_1)$ are distinct for different models $A_n$, the diagrams $\sigma_{11}^{(\text{ex})}(\delta_1)$ are also distinct for these models.

**Table 6.2** Values of the constant $l_2$ for rubber in different models $A_n$

| $n$ | $l_2$, MPa | $\Delta$, % |
|---|---|---|
| I | 29.6 | 37 |
| II | 27.1 | 17 |
| III | 16.4 | 9 |
| IV | 7.5 | 42 |
| V | 3.0 | 58 |

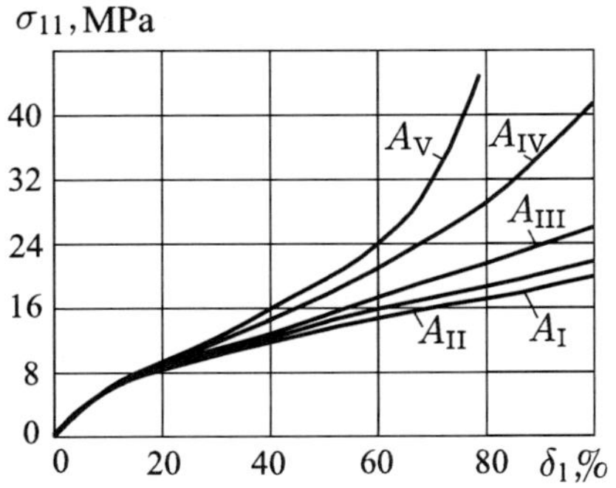


**Fig. 6.8** Experimental diagram of deforming in tension for filled rubber, recalculated in terms of the Cauchy stresses by models $A_n$ when $\nu = 0.4$

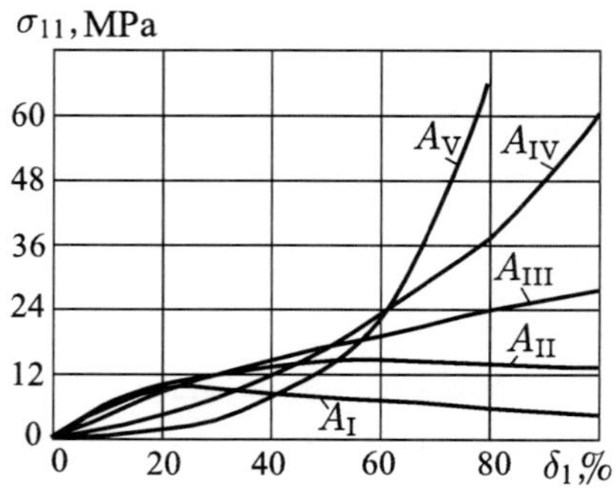


**Fig. 6.9** Diagrams of deforming for filled rubber in tension, computed in terms of the Cauchy stresses by models $A_n$ when $\nu = 0.4$

Figure 6.9 exhibits diagrams $\sigma_{11}(\delta_1)$ for models $A_n$, computed by formulae (6.148) and (6.149), where the constant $l_2$ was found with the help of approximation to the experimental diagram $P_{11}^{(\mathrm{ex})}(\delta_1)$ by functions (6.155) for rubber (see Table 6.2).

The situation is also possible when in experiment one determines not $P_{11}^{(\mathrm{ex})}(\delta_1)$ but the diagram of deforming $\sigma_{11}^{(\mathrm{ex})}(\delta_1)$. Then the constant $l_2$ can be found from the condition of the best approximation to the function $\sigma_{11}^{(\mathrm{ex})}(\delta_1)$ by the dependence (6.148) or (6.149), while the coefficient $\nu$ is usually determined by Eq. (6.147). For the considered class of filled rubbers, the qualitative relative position of the functions $\sigma_{11}(\delta_1)$ and $\sigma_{11}^{(\mathrm{ex})}(\delta_1)$ is the same as for the Piola–Kirchhoff stresses; in this case the model $A_{\mathrm{III}}$ gives the best results too.

## Exercises for 6.4

**6.4.1.** Prove the relations (6.148) for the linear model $A_{\mathrm{III}}$.

## 6.5 Tension of an Incompressible Beam

### *6.5.1 Deformation of an Incompressible Elastic Beam*

Consider the problem on tension of a beam (see Example 2.1 and Sect. 6.4.2), but the beam is assumed to be an incompressible elastic (ideal) solid.

Let boundary conditions for the beam be the same conditions (6.140), and a law of the beam motion will be sought in the form (6.131) too. Then the deformation gradient $\mathbf{F}$ is expressed by formula (6.132), and the deformation measures and tensors have the same forms as the ones for a compressible continuum and are determined by Table 6.1. In particular, the energetic deformation measures $\overset{(n)}{\mathbf{G}}$ have the form (see Exercise 4.2.13):

$$\overset{(n)}{\mathbf{G}} = \overset{(n)}{\mathbf{g}} = \frac{1}{n - \text{III}} \sum_{\alpha=1}^{3} k_\alpha^{n-\text{III}} \bar{\mathbf{e}}_\alpha \otimes \bar{\mathbf{e}}_\alpha, \quad n = \text{I, II, IV, V}. \tag{6.157}$$

Since the solid was assumed to be incompressible, the condition of incompressibility (4.489) must be satisfied. Then substituting the expression (6.132) for $\mathbf{F}$ into (4.489), we obtain the relation

$$1 = \det \mathbf{F} = k_1 k_2 k_3, \tag{6.158}$$

from which we can express $k_2$ and $k_3$ in terms of $k_1$ (in this problem $k_2 = k_3$):

$$k_3 = k_2 = 1/\sqrt{k_1}. \tag{6.159}$$

This simple relation is an analog of relations (6.145) and (6.149) for a compressible material; from (6.159) it follows that for incompressible continua the Poisson effect also occurs (see Sect. 6.4.5) but the dependence of $k_2$ and $k_3$ upon $k_1$ has an universal character: $k_2$ and $k_3$ are independent of the constants $\mu$ and $\beta$ of an incompressible continuum (for a compressible medium, the situation is different).

Since in the problem on tension of a beam $k_1$ is immediately evaluated from the boundary condition (6.142), values of $k_2$ and $k_3$ can be found by formula (6.159) without use of the constitutive equations too.

### 6.5.2 Stresses in an Incompressible Beam for Models $B_n$

Let constitutive equations of an elastic incompressible beam correspond to linear models $B_n$ (4.532):

$$\overset{(n)}{\mathbf{T}} = -\frac{p}{n - \text{III}} \overset{(n)}{\mathbf{G}}{}^{-1} + \mu(n - \text{III})^2 \left( \left( \frac{1 + \beta}{n - \text{III}} + (1 - \beta) I_1(\overset{(n)}{\mathbf{G}}) \right) \mathbf{E} - (1 - \beta) \overset{(n)}{\mathbf{G}} \right), \tag{6.160}$$

which contain two constants: $\mu$ and $\beta$.

On substituting relation (6.157) into this expression, we find that the tensors $\overset{(n)}{\mathbf{T}}$ in this problem, just as for compressible media, do not contain tangential stresses:

$$\overset{(n)}{\mathbf{T}} = \sum_{\alpha=1}^{3} \overset{(n)}{T}_{\alpha\alpha} \bar{\mathbf{e}}_\alpha \otimes \bar{\mathbf{e}}_\alpha, \tag{6.161}$$

$$\overset{(n)}{T}_{\alpha\alpha} = -pk_\alpha^{\mathrm{III}-n} + \mu(n - \mathrm{III})\Big(1 + \beta + (1-\beta)\Big(\sum_{\gamma=1}^{3} k_\gamma^{n-\mathrm{III}} - k_\alpha^{n-\mathrm{III}}\Big)\Big),$$

$$\alpha = 1,\ 2,\ 3, \quad n = \mathrm{I},\ \mathrm{II},\ \mathrm{IV},\ \mathrm{V}. \tag{6.162}$$

Substitution of these expressions into (6.137) yields a representation of the Cauchy stress tensor $\mathbf{T}$ for models $B_n$:

$$\mathbf{T} = \sum_{\alpha=1}^{3} \sigma_{\alpha\alpha} \bar{\mathbf{e}}_\alpha \otimes \bar{\mathbf{e}}_\alpha, \tag{6.163}$$

$$\sigma_{\alpha\alpha} = -p + \mu(n - \mathrm{III})k_\alpha^{n-\mathrm{III}}\Big(1 + \beta + (1-\beta)\Big(\sum_{\gamma=1}^{3} k_\gamma^{n-\mathrm{III}} - k_\alpha^{n-\mathrm{III}}\Big)\Big),$$

$$\alpha = 1,\ 2,\ 3. \tag{6.164}$$

### 6.5.3 Resolving Relation $\sigma_1(k_1)$

Since $k_2$ and $k_3$ are coincident, in (6.164) there are only two independent relations: when $\alpha = 1$ and $\alpha = 2$, and from the boundary condition (6.141c) it follows that $\sigma_{22} = 0$. As a result, we have the two equations

$$\begin{aligned} \sigma_{11} &= -p + \mu(n - \mathrm{III})k_1^{n-\mathrm{III}}(1 + \beta + 2(1-\beta)k_2^{n-\mathrm{III}}), \\ 0 &= -p + \mu(n - \mathrm{III})k_2^{n-\mathrm{III}}(1 + \beta + (1-\beta)(k_1^{n-\mathrm{III}} + k_2^{n-\mathrm{III}})) \end{aligned} \tag{6.165}$$

for the two unknowns: $\sigma_1$ and $p$. Eliminating $p$ and using Eq. (6.159), we find the resolving relation between the stress $\sigma_{11}$ and the longitudinal elongation ratio $k_1$:

$$\begin{aligned} \sigma_{11} = \mu(n - \mathrm{III})\Big((1+\beta)\Big(k_1^{n-\mathrm{III}} - \frac{1}{k_1^{(n-\mathrm{III})/2}}\Big) \\ +(1-\beta)\Big(k_1^{(n-\mathrm{III})/2} - \frac{1}{k_1^{n-\mathrm{III}}}\Big)\Big), \quad n = \mathrm{I},\ \mathrm{II},\ \mathrm{IV},\ \mathrm{V}. \end{aligned} \tag{6.166}$$

### 6.5.4 Comparative Analysis of Models $B_n$

Figure 6.10 exhibits graphs of the functions $\frac{\sigma_{11}}{\mu}(k_1)$ determined by formula (6.166) for different values of the parameter $\beta$ and different $n$. For the models $B_{\mathrm{II}}$ and $B_{\mathrm{IV}}$, all the functions are convex upwards at all values of the constant $\beta$ within the interval $-1 \leqslant \beta \leqslant 1$. For the models $B_{\mathrm{I}}$ and $B_{\mathrm{V}}$, the functions $\frac{\sigma_{11}}{\mu}(k_1)$ have a point

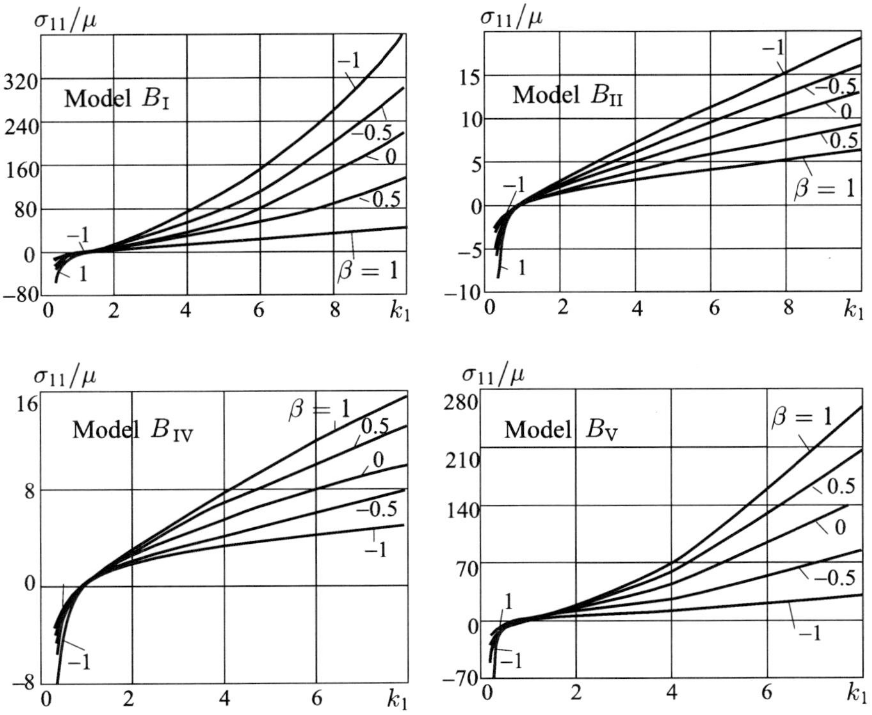


**Fig. 6.10** The functions $\sigma_{11}/\mu$ determined by formula (6.166) for different values of $\beta$ and $n$

of inflection at $\beta < 1$ and $\beta > -1$, respectively: in the case of compression (when $0 < k < 1$) these functions are convex upwards, and in the case of tension (when $k \geqslant 1$) they are convex downwards. For the model $B_{\mathrm{I}}$ with $\beta = 1$ and the model $B_{\mathrm{V}}$ with $\beta = 1$ (Treloar's model), the graphs of the functions $\frac{\sigma_{11}}{\mu}(k)$ have no point of inflection, they are convex upwards.

When $\beta > 1$ and $\beta < -1$, graphs of the function $\frac{\sigma_{11}}{\mu}(k_1)$ have the physical meaning only within some interval $k^*_{\min} < k_1 < k^*_{\max}$, because for other values of $k_1$ monotonicity of the function is violated.

Figures 6.11 and 6.13 show experimental diagrams of deforming $\sigma_{11}^{(\mathrm{ex})}(\delta_1)$ for rubber and polyurethane elastomer and also the dependences $\sigma_{11}(\delta_1)$ (6.166) approximating these experimental curves. For each model $B_n$, the constants $\mu$ and $\beta$ have been chosen in the optimal way by minimizing the mean-square distance between the computed and experimental diagrams of deforming in tension, calculated at $N$ points:

$$\Delta = \left( \frac{1}{N} \sum_{i=1}^{N} \left( 1 - \frac{\sigma_{11}(\delta_{1(i)})}{\sigma_{11}^{(\mathrm{ex})}(\delta_{1(i)})} \right)^2 \right)^{1/2} \to \min. \qquad (6.167)$$

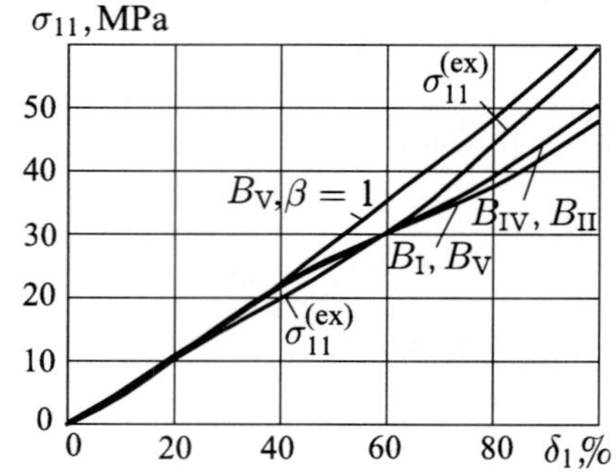


**Fig. 6.11** Experimental diagram of deforming for rubber in tension and its approximation by models $B_n$ and Treloar's model

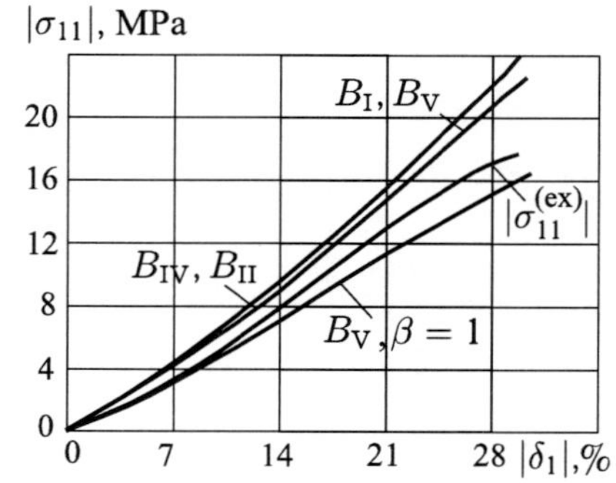


**Fig. 6.12** Experimental diagram of deforming for rubber in compression and its approximation by models $B_n$ and Treloar's model

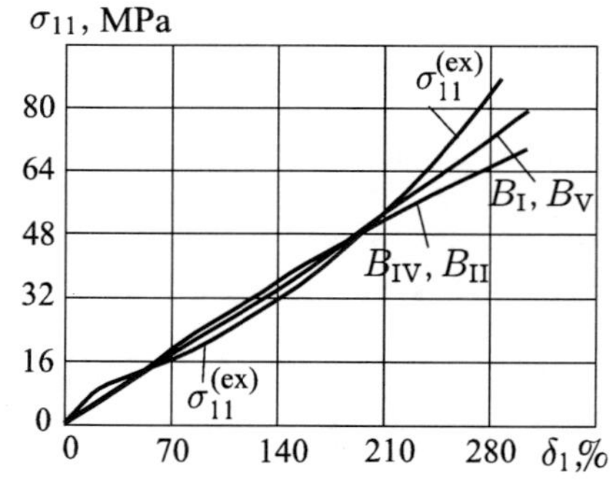


**Fig. 6.13** Experimental diagram of deforming for polyurethane in tension and its approximation by models $B_n$

Values of the stresses $\sigma_{11}^{(\text{ex})}(\delta_1)$ have been determined by experimental values of the Piola–Kirchhoff stresses $P_{11}^{(\text{ex})}(\delta_1)$ with the help of formula (6.154) for incompressible continua:

$$\sigma_{11}^{(\text{ex})} = P_{11}^{(\text{ex})} k_1. \tag{6.168}$$

The models $B_{\text{IV}}$ and $B_{\text{V}}$ with $\beta = 1$ were also considered (the Bartenev–Hazanovich and Treloar's models, see Sect. 4.9.7).

**Table 6.3** Values of the constants $\mu$ and $\beta$ for rubber and polyurethane

| Rubber | | | | Polyurethane | | | |
|---|---|---|---|---|---|---|---|
| $n$ | $\mu$, MPa | $\beta$ | $\Delta$, % | $n$ | $\mu$, MPa | $\beta$ | $\Delta$, % |
| I | 5.145 | 0.13 | 7.8 | I | 3.15 | 0.616 | 11.3 |
| II | 19.11 | 1 | 8.7 | II | 11.5 | −0.45 | 14.5 |
| IV | 19.11 | −1 | 8.7 | IV | 11.56 | 0.45 | 14.5 |
| V | 5.145 | −0.13 | 8.7 | V | 3.15 | −0.616 | 11.3 |
| V* | 4.41 | 1 | 11 | | | | |

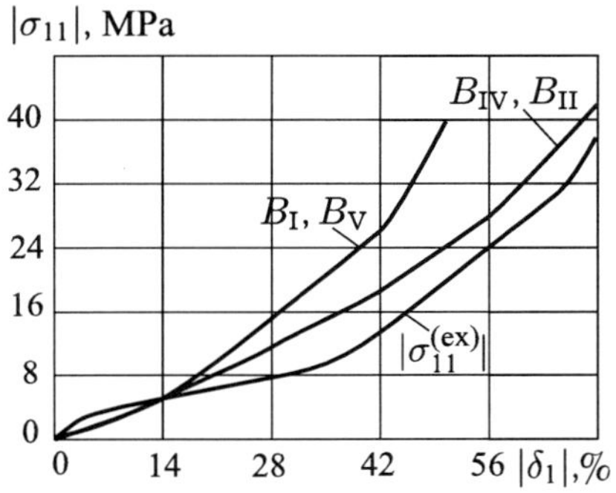


**Fig. 6.14** Experimental and computed diagrams of deforming for polyurethane in compression

Table 6.3 shows values of the constants $\mu$ and $\beta$, calculated by the method mentioned above. This table gives also values of the mean-square distance $\Delta$ between experimental and computed curves.

All the models $B_n$ exhibit practically the same accuracy of approximation to the experimental curves; the models $B_{\mathrm{I}}$ and $B_{\mathrm{V}}$ are some more accurate.

Figures 6.12 and 6.14 show computed and experimental diagrams of deforming for rubber and polyurethane in compression, where the computed diagrams were obtained by formulae (6.166), in which the constants $\mu$ and $\beta$ were predetermined by diagrams of deforming in tension. For the materials considered, the models $B_{\mathrm{II}}$ and $B_{\mathrm{IV}}$ forecast more accurately a behavior of polyurethane in compression than the models $B_{\mathrm{I}}$ and $B_{\mathrm{V}}$. It should be noted that experimental results in compression depend considerably on the conditions of testing, in particular, on the shape of a specimen and the method of its fastening. For example, in compression of specimens with a cylindrical shape the diagrams $\sigma_{11}(\delta_1)$ depend essentially on the ratio of their initial diameter and thickness. Moreover, at a certain value of $\delta_1$ there usually occurs a loss of stability of a shape of the specimen; for example, the cylindrical shape often becomes the barrel-type one. The circumstances indicated appear to be main causes of the deviation of the computed diagrams $\sigma_{11}(\delta_1)$ from the experimental one $\sigma_{11}^{(\mathrm{ex})}(\delta_1)$ in compression.

### 6.5.5 Stresses in an Incompressible Beam for Models $A_n$

To describe the process of deforming of an incompressible elastic beam, we can choose other models stated in Sect. 4.9. For comparison, let us consider the linear models $A_n$ (4.526)

$$\overset{(n)}{\mathbf{T}} = -\frac{p}{n-\mathrm{III}}\overset{(n)}{\mathbf{G}}{}^{-1} + (\bar{m} + l_1 I_1(\overset{(n)}{\mathbf{C}}))\mathbf{E} + 2l_2\overset{(n)}{\mathbf{C}}, \quad n = \mathrm{I},\ \mathrm{II},\ \mathrm{IV},\ \mathrm{V}, \tag{6.169}$$

which contain three constants: $\bar{m}$, $l_1$ and $l_2$.

The algorithm of determining the diagram of deforming $\sigma_{11}(\delta_1)$ for these models is the same as the one for models $B_n$. The stresses $\overset{(n)}{T}_{\alpha\alpha}$ in the problem on tension have the form

$$\overset{(n)}{T}_{\alpha\alpha} = -pk_\alpha^{\mathrm{III}-n} + \bar{m} + \frac{l_1}{n-\mathrm{III}}(k_1^{n-\mathrm{III}} + 2k_1^{\frac{\mathrm{III}-n}{2}} - 3) + \frac{2l_2}{n-\mathrm{III}}(k_\alpha^{n-\mathrm{III}} - 1),$$
$$\alpha = 1,\ 2,\ 3. \tag{6.170}$$

Cartesian components $\sigma_{\alpha\alpha}$ of the Cauchy stress tensor are determined by formulae (6.138): $\sigma_{\alpha\alpha} = k_\alpha^{n-\mathrm{III}}\overset{(n)}{T}_{\alpha\alpha}$; as a result, we obtain the equations

$$\sigma_{11} = -p + \left(\bar{m} + \frac{l_1}{n-\mathrm{III}}\left(k_1^{n-\mathrm{III}} + 2k_1^{\frac{\mathrm{III}-n}{2}} - 3\right) + \frac{2l_2}{n-\mathrm{III}}(k_1^{n-\mathrm{III}} - 1)\right)k_1^{n-\mathrm{III}}, \tag{6.171}$$

$$\sigma_{22} = 0 = -p + \left(\bar{m} + \frac{l_1}{n-\mathrm{III}}(k_1^{n-\mathrm{III}} + 2k_1^{\frac{\mathrm{III}-n}{2}} - 3) + \frac{2l_2}{n-\mathrm{III}}(k_1^{\frac{\mathrm{III}-n}{2}} - 1)\right)k_1^{\frac{\mathrm{III}-n}{2}}.$$

Eliminating $p$ among these equations, we find the desired function $\sigma_{11}(k_1)$, which can be written in the form

$$\sigma_{11} = \bar{m}\overset{(n)}{Q}(k_1) + l_1\overset{(n)}{M}(k_1) + l_2\overset{(n)}{N}(k_1), \tag{6.172}$$

where

$$\overset{(n)}{Q} = k_1^{n-\mathrm{III}} - k_1^{\frac{\mathrm{III}-n}{2}},$$
$$\overset{(n)}{M} = \frac{1}{n-\mathrm{III}}(k_1^{n-\mathrm{III}} + 2k_1^{\frac{\mathrm{III}-n}{2}} - 3)(k_1^{n-\mathrm{III}} - k_1^{\frac{\mathrm{III}-n}{2}}), \tag{6.173}$$
$$\overset{(n)}{N} = \frac{2}{n-\mathrm{III}}((k_1^{n-\mathrm{III}} - 1)k_1^{n-\mathrm{III}} - (k_1^{\frac{\mathrm{III}-n}{2}} - 1)k_1^{\frac{\mathrm{III}-n}{2}}).$$

Figures 6.15 and 6.16 exhibit results of approximation to the experimental diagrams of deforming $\sigma_{11}^{(\mathrm{ex})}(\delta_1)$ for rubber and polyurethane with the help of the

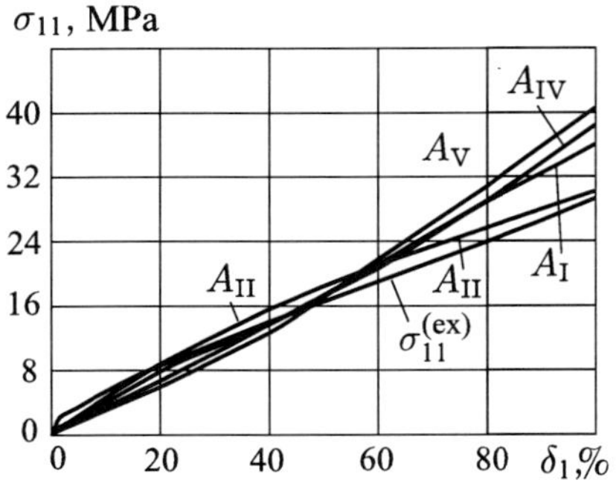


**Fig. 6.15** Experimental diagram of deforming for rubber and its approximation by models $A_n$ of incompressible continua

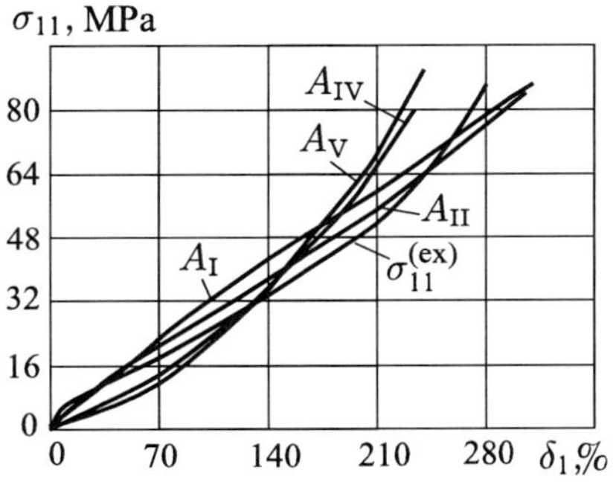


**Fig. 6.16** Experimental diagram of deforming for polyurethane and its approximation by models $A_n$ of incompressible continua

**Table 6.4** Values of the constants $\bar{m}$, $l_1$, and $l_2$ for rubber and polyurethane

| Rubber | | | | | Polyurethane | | | | |
|---|---|---|---|---|---|---|---|---|---|
| $n$ | $\bar{m}$, MPa | $l_1$, MPa | $l_2$, MPa | $\Delta$, % | $n$ | $\bar{m}$, MPa | $l_1$, MPa | $l_2$, MPa | $\Delta$, % |
| I | 0.2 | 0.2 | 17.6 | 21.9 | I | 0.2 | 0.2 | 10.6 | 21 |
| II | 2.8 | 0.2 | 19.8 | 16.1 | II | 25 | 0.2 | 29.4 | 9 |
| IV | 19.8 | 0.2 | 2.4 | 28 | IV | 0.2 | 0.2 | 9 | 34 |
| V | 10.6 | 0.2 | 0.2 | 31 | V | 5.4 | 0.2 | 0.2 | 38 |

functions (6.172) for different $n$ =I, II, IV and V. The constants $\bar{m}$, $l_1$ and $l_2$ were calculated by minimizing the functional of the mean-square distance between the functions $\sigma_{11}(\delta_1)$ and $\sigma_{11}^{(\mathrm{ex})}(\delta_1)$ at $N$ points:

$$\Delta = \left( \frac{1}{N} \sum_{i=1}^{N} \left( 1 - \frac{\sigma_{11}(\delta_i)}{\sigma_{11}^{(\mathrm{ex})}(\delta_i)} \right)^2 \right)^{1/2} \to \min. \tag{6.174}$$

Values of the constants $\bar{m}$, $l_1$, and $l_2$ calculated by this method are shown in Table 6.4.

The model $A_{\mathrm{II}}$ demonstrates the best approximation to the experimental diagram of deforming for the two considered types of materials (see Figs. 6.15 and 6.16 and Table 6.4); values of $\Delta$ for this model do not exceed 16.1% for rubber and 9% for polyurethane. Notice that for the model $A_{\mathrm{II}}$, unlike other models $A_n$ and $B_n$ considered above, the function $\sigma_{11}(\delta_1)$ has a point of inflection as well as the experimental diagram $\sigma_{11}^{(\mathrm{ex})}(\delta_1)$.

For other solids, the situation may be different.

Consideration of the whole complex of energetic (or quasienergetic) models is appropriate, because, on performing calculations by all the models, one can choose the model giving the best results for a specific elastic continuum.

## Exercises for 6.5

**6.5.1.** Consider the problem on an incompressible elastic beam in tension with the help of the linear models $A_n$ (4.526) when $\bar{m} = p_0 = 0$:

$$\overset{(\mathrm{n})}{\mathbf{T}} = -\frac{p}{n-\mathrm{III}}\mathbf{G}^{-1} + l_1 I_1(\overset{(\mathrm{n})}{\mathbf{C}})\mathbf{E} + 2l_2\overset{(\mathrm{n})}{\mathbf{C}}.$$

Show that in this case the final expression for the component $\overset{(\mathrm{n})}{\sigma}_1$ of the Cauchy stress tensor has the form

$$\overset{(\mathrm{n})}{\sigma}_1 = \frac{l_1}{n-\mathrm{III}}(k_1^{n-\mathrm{III}} + 2k_1^{(\mathrm{III}-n)/2} - 3)(k_1^{n-\mathrm{III}} - k_1^{(\mathrm{III}-n)/2})$$
$$+\frac{2l_2}{n-\mathrm{III}}(k_1^{n-\mathrm{III}}(k_1^{n-\mathrm{III}}-1)-k_1^{(\mathrm{III}-n)/2}(k_1^{(\mathrm{III}-n)/2}-1)), \quad n = \mathrm{I},\ \mathrm{II},\ \mathrm{IV},\ \mathrm{V}.$$

## 6.6 Simple Shear

### 6.6.1 Deformations in Simple Shear

Consider the problem on an elastic solid body, whose shape is a parallelepiped, in simple shear. The motion of the body is determined by Eq. (2.7) (see Example 2.2 in Sect. 2.1.1):

$$x^i = X^i + a\delta_1^i X^2, \quad i = 1,\ 2,\ 3. \qquad (6.175)$$

The deformation tensors for this problem have been calculated above (see Table 6.1).

### 6.6.2 Stresses in the Problem on Shear

Assume that the body considered is isotropic and complies with the linear model $A_n$ (4.338). Calculate the stress tensors $\overset{(n)}{\mathbf{T}}$ by substituting the tensors $\overset{(n)}{\mathbf{C}}$ from Exercise 4.2.14 into Eq. (4.338) or (6.135):

$$\overset{(n)}{\mathbf{T}} = \sum_{\alpha=1}^{3} \overset{(n)}{T}_{\alpha\alpha}\bar{\mathbf{e}}_{\alpha}^{2} + \overset{(n)}{T}_{12}\mathbf{O}_3, \tag{6.176}$$

$$\overset{(n)}{T}_{11} = (l_1 + 2l_2)\overset{(n)}{c}_{0A} + l_1\overset{(n)}{c}_{2A}, \quad \overset{(n)}{T}_{12} = 2l_2\overset{(n)}{c}_{1A},$$

$$\overset{(n)}{T}_{22} = (l_1 + 2l_2)\overset{(n)}{c}_{2A} + l_1\overset{(n)}{c}_{0A}, \quad \overset{(n)}{T}_{33} = l_1(\overset{(n)}{c}_{0A} + \overset{(n)}{c}_{2A}),$$

$$n = \mathrm{I},\ \mathrm{II},\ \mathrm{IV},\ \mathrm{V}.$$

Here $\overset{(n)}{c}_{0A}$, $\overset{(n)}{c}_{1A}$ and $\overset{(n)}{c}_{2A}$ are determined by Table 4.6 from Exercise 4.2.14.

Since the shear stresses $\overset{(n)}{T}_{12}$ are nonzero, the tensors $\overset{(n)}{\mathbf{T}}$ in this problem are not diagonal: each of them has four nonzero components. To evaluate the Cauchy stress tensor $\mathbf{T}$, one should use formulae (4.37) and (4.38). Since in the problem on shear the rotation tensor $\mathbf{O}$ is different from $\mathbf{E}$, calculating the tensors of energetic equivalence ${}^4\overset{(n)}{\mathbf{E}}$ is more complicated than in the problem on tension, because we need relations (4.37) to be represented in the eigenbases:

$$\mathbf{T} = {}^4\overset{(n)}{\mathbf{E}} \cdot\cdot \overset{(n)}{\mathbf{T}} = \sum_{\alpha,\beta=1}^{3} \overset{(n)}{E}_{\alpha\beta}\mathbf{p}_{\alpha} \otimes \mathbf{p}_{\beta} \otimes \overset{\circ}{\mathbf{p}}_{\beta} \otimes \overset{\circ}{\mathbf{p}}_{\alpha} \cdot\cdot \left( \sum_{\gamma=1}^{3} \overset{(n)}{T}_{\gamma\gamma}\bar{\mathbf{e}}_{\gamma}^{2} + \overset{(n)}{T}_{12}\mathbf{O}_3 \right)$$

$$= \sum_{\alpha,\beta=1}^{3} T_{\alpha\beta}^{(p)}\mathbf{p}_{\alpha} \otimes \mathbf{p}_{\beta} \tag{6.177}$$

Here we have denoted components of the Cauchy stress tensor with respect to the eigenbasis $\mathbf{p}_{\alpha}$ by

$$T_{\alpha\beta}^{(p)} = \overset{(n)}{E}_{\alpha\beta} \left( \sum_{\gamma=1}^{3} \overset{(n)}{T}_{\gamma\gamma}\overset{\circ}{\bar{p}}_{\beta\gamma}\overset{\circ}{\bar{p}}_{\alpha\gamma} + \overset{(n)}{T}_{12}(\overset{\circ}{\bar{p}}_{\alpha 1}\overset{\circ}{\bar{p}}_{\beta 2} + \overset{\circ}{\bar{p}}_{\alpha 2}\overset{\circ}{\bar{p}}_{\beta 1}) \right), \tag{6.178}$$

and also introduced components $\overset{\circ}{\bar{p}}_{\alpha\gamma}$ of resolution of the eigenvectors $\overset{\circ}{\mathbf{p}}_\alpha$ for the basis $\bar{\mathbf{e}}_\gamma$:

$$\overset{\circ}{\mathbf{p}}_\alpha = \overset{\circ}{\bar{p}}_{\alpha i}\bar{\mathbf{e}}^i, \quad \overset{\circ}{\bar{p}}_{\alpha\gamma} = \overset{\circ}{\mathbf{p}}_\alpha \cdot \bar{\mathbf{e}}_\gamma = \begin{pmatrix} s_1 & s_1 b_1 & 0 \\ 0 & s_2 b_2 & 0 \\ 0 & 0 & 1 \end{pmatrix}, \tag{6.179}$$

$$s_\alpha = (1+b_\alpha^2)^{-1/2}, \qquad b_\alpha = \frac{a}{2} - (-1)^\alpha \sqrt{1+a^2/4}.$$

Here we have used the expression for $\overset{\circ}{\mathbf{p}}_\alpha$ from Exercise 2.3.3:

$$\overset{\circ}{\mathbf{p}}_\alpha = s_\alpha(\bar{\mathbf{e}}_1 + b_\alpha \bar{\mathbf{e}}_2), \quad \alpha = 1,\ 2; \quad \overset{\circ}{\bar{\mathbf{p}}}_3 = 1.$$

With the help of (6.179) we can rewrite formulae (6.178) in the explicit form

$$\begin{aligned} T_{11}^{(p)} &= \overset{(n)}{E}_{11}(\overset{(n)}{T}_{11}s_1^2 + \overset{(n)}{T}_{22}s_1^2 b_1^2 + 2\overset{(n)}{T}_{12}s_1^2 b_1), \\ T_{22}^{(p)} &= \overset{(n)}{E}_{22}(\overset{(n)}{T}_{11}s_2^2 + \overset{(n)}{T}_{22}s_2^2 b_2^2 + 2\overset{(n)}{T}_{12}s_2^2 b_2), \\ T_{12}^{(p)} &= \overset{(n)}{E}_{12}(\overset{(n)}{T}_{11} + \overset{(n)}{T}_{22}b_1 b_2 + \overset{(n)}{T}_{12}(b_1+b_2))s_1 s_2, \\ T_{13}^{(p)} &= T_{23}^{(p)} = 0, \quad T_{33}^{(p)} = \overset{(n)}{T}_{33}. \end{aligned} \tag{6.180}$$

The matrices $\overset{(n)}{E}_{\alpha\beta}$ are determined by formulae (4.39); and eigenvalues $\lambda_\alpha$ appearing in these formulae, according to the result of Exercise 2.3.3, have the forms

$$\lambda_1 = \sqrt{1+b_1|a|}, \quad \lambda_2 = \lambda_1^{-1}, \quad \lambda_3 = 1. \tag{6.181}$$

Using the expression for the eigenvectors (see Exercise 2.3.3)

$$\mathbf{p}_\alpha = (-1)^{\alpha+1}s_\alpha(b_\alpha \bar{\mathbf{e}}_1 + \bar{\mathbf{e}}_2),$$

from (6.177) we obtain the following representation for the Cauchy stress tensor:

$$\mathbf{T} = \sum_{\alpha=1}^{3} \sigma_{\alpha\alpha}\bar{\mathbf{e}}_\alpha^2 + \sigma_{12}\mathbf{O}_3. \tag{6.182}$$

Here $\sigma_{\alpha\beta}$ are components of the Cauchy stress tensor with respect to the Cartesian basis $\bar{\mathbf{e}}_i$, which in the problem on shear have the forms

$$\begin{aligned} \sigma_{11} &= T_{11}^{(p)}s_1^2 b_1^2 - 2T_{12}^{(p)}s_1 s_2 b_1 b_2 + T_{22}^{(p)}s_2^2 b_2^2, \\ \sigma_{22} &= T_{11}^{(p)}s_1^2 - 2T_{12}^{(p)}s_1 s_2 + T_{22}^{(p)}s_2^2, \\ \sigma_{12} &= T_{11}^{(p)}s_1^2 b_1 - T_{12}^{(p)}s_1 s_2(b_1+b_2) + T_{22}^{(p)}s_2^2 b_2, \quad \sigma_{33} = T_{33}^{(p)}. \end{aligned} \tag{6.183}$$

Thus, the Cauchy stress tensor (6.182) in the problem on simple shear also proves to be independent of coordinates; therefore, the tensor automatically satisfies the equilibrium equations in a Cartesian basis:

$$\nabla \cdot \mathbf{T} = \sum_{\alpha,\beta=1}^{3} (\partial \sigma_{\alpha\beta} / \partial x^{\beta}) \bar{\mathbf{e}}_{\alpha} = 0. \tag{6.184}$$

### 6.6.3 Boundary Conditions in the Problem on Shear

In the problem on simple shear, boundary conditions corresponding to the motion law (6.175) have the form

$$x^2 = h^2: \quad u^1 = u^1_e, \quad u^2 = 0, \quad u^3 = 0, \tag{6.185a}$$

$$x^2 = 0: \quad u^\alpha = 0, \quad \alpha = 1, 2, 3, \tag{6.185b}$$

$$x^3 = 0, h^3: \quad u^3 = 0, \quad \bar{\mathbf{e}}_1 \cdot \mathbf{T} \cdot \bar{\mathbf{e}}_3 = 0, \quad \bar{\mathbf{e}}_2 \cdot \mathbf{T} \cdot \bar{\mathbf{e}}_3 = 0, \tag{6.185c}$$

$$x^1 = 0, h^1: \quad \bar{\mathbf{e}}_1 \cdot \mathbf{T} \cdot \bar{\mathbf{e}}_3 = 0, \quad u^2 = 0, \quad u^2 = u^1_e X^2 / h_2. \tag{6.185d}$$

Thus, the side $x^2 = 0$ of the parallelepiped remains fixed, and at the opposite side $x^2 = h^2$ we give the displacement $u^1_e$ shifting the parallelepiped towards the axis $Ox^1$, at the sides $x^3 = 0$ and $x^3 = h^3$ there is no displacement along the axis $Ox^3$ and both tangential stresses. At the inclined sides $x^1 = 0, h^1$ we give the zero tangential stress $\sigma_{13} = 0$, the zero displacement $u = 0$ and the longitudinal displacement $u^1$ varying by a linear law along the coordinate $X^2$.

We can verify that the motion law (6.175) and the stress tensor $\mathbf{T}$ (6.182) automatically satisfy the boundary conditions (6.185a)–(6.185c), because Eqs. (6.175) and (6.182) yield

$$u^i = x^i - X^i = a\delta^i_1 X^2, \quad \alpha = 1, 2, 3, \tag{6.186a}$$

$$\sigma_{\alpha 3} = \bar{\mathbf{e}}_\alpha \cdot \mathbf{T} \cdot \bar{\mathbf{e}}_3 \equiv 0, \quad \alpha = 1, 2. \tag{6.186b}$$

From (6.185d) and (6.186a) we can find values of the function $a$ through $u^1_e$:

$$a = u^1_e / h^2. \tag{6.187}$$

According to (6.187), the first boundary condition in (6.185d) is satisfied. The second and third boundary conditions in (6.185d) are satisfied identically due to formulae (6.186) and (6.187).

The problem on a simple shear with the boundary conditions (6.185) is realized approximately in the experiment on longitudinal shear of a thin strip of considered material (for example, of rubber), placed between two rigid (for example, steel) sheets displaced one parallel to another (Fig. 6.17).

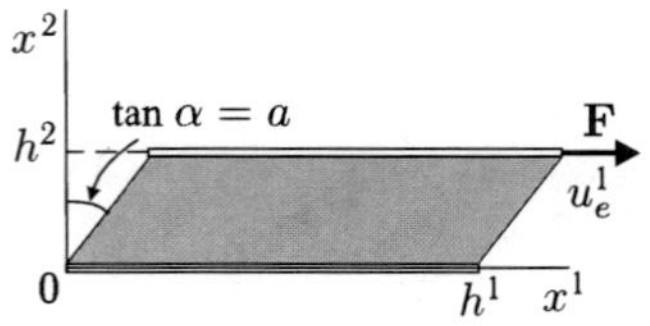


**Fig. 6.17** The experimental scheme, where the solution of the problem on a simple shear is approximately realized

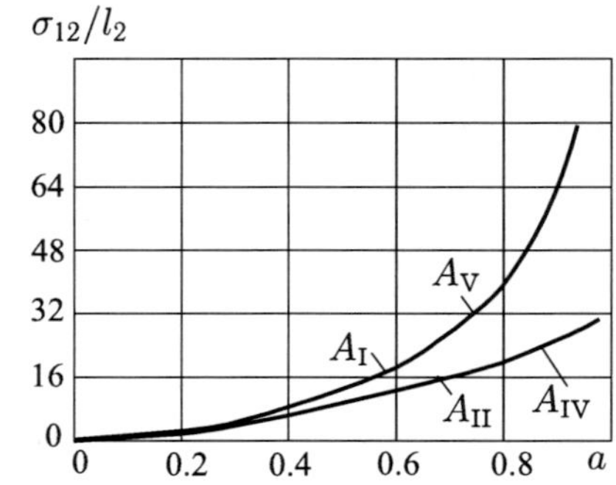


**Fig. 6.18** Diagrams of deforming in shear for linear models $A_n$ and $\nu = 0.4$

At some distance from the ends $x^1 = 0, h^1$ and $x^3 = 0, h^3$ in such a plate the motion law (6.175) holds and there is a uniform stress state.

### *6.6.4 Comparative Analysis of Different Models $A_n$ for the Problem on Shear*

Notice that in the problem on simple shear, besides the shear stresses $\sigma_{12}$, all normal stresses $\sigma_{11}$, $\sigma_{22}$ and $\sigma_{33}$ are nonzero; this phenomenon, which is typical for large deformations, is called the *Poynting effect*.

The dependence $\sigma_{12}(a)$ expressed by formulae (6.183) is called the *diagram of deforming in simple shear* (remind that $a = \tan\alpha = u_e^1/h^2$ is the tangent of shear angle for the parallelepiped).

Figures 6.18 and 6.19 exhibit diagrams of deforming in shear for different linear models $A_n$ and different values of the Poisson ratio (the stress $\sigma_{12}$ is referred to $l_2$, and the constant $l_1$ is expressed in terms of $\nu$ by formula (6.146)). The increase in value of $\nu$ from 0 to 0.5 leads to displacing the diagram $\sigma_{12}(a)$ into the domain of higher values. A peculiarity of the problem on a simple shear is the fact that the diagrams $\sigma_{12}(a)$ coincide for the models $A_{\mathrm{I}}$ and $A_{\mathrm{V}}$, and also for the models $A_{\mathrm{II}}$ and $A_{\mathrm{IV}}$ (Fig. 6.18). However, corresponding diagrams of the normal stresses $\sigma_{11}(a)$ and $\sigma_{22}(a)$ determined by (6.183) are essentially distinct: for the models $A_{\mathrm{I}}$ and $A_{\mathrm{II}}$ the stresses $\sigma_{11}$ and $\sigma_{22}$ are negative, and for the models $A_{\mathrm{IV}}$ and $A_{\mathrm{V}}$ they are positive (Figs. 6.20 and 6.21).

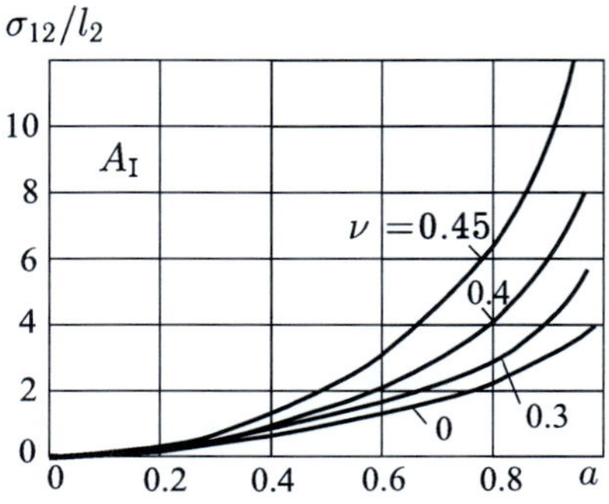


**Fig. 6.19** Diagrams of deforming in shear for the linear model $A_I$ with different values of the Poisson ratio

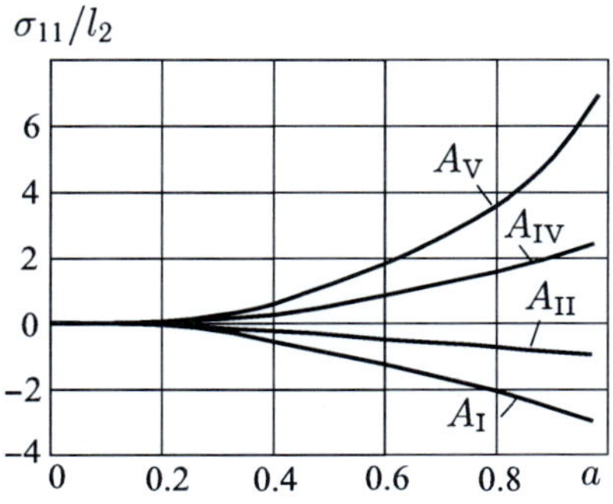


**Fig. 6.20** Dependence of the stress $\sigma_{11}$ on the shear angle $a$ in the problem on a simple shear for linear models $A_n$ and $\nu = 0.4$

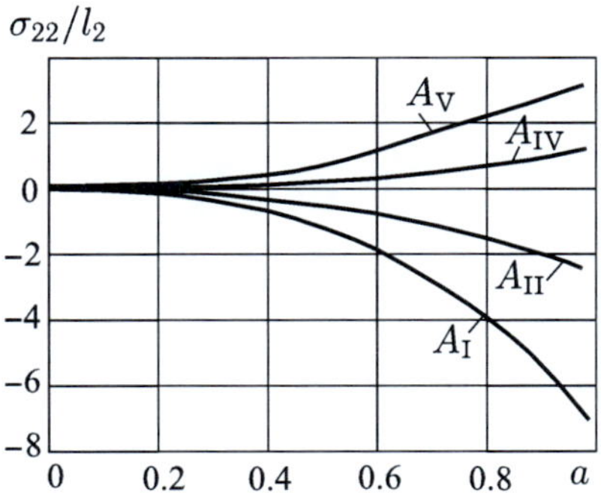


**Fig. 6.21** Dependence of the stress $\sigma_{22}$ on the shear angle $a$ in the problem on a simple shear for linear models $A_n$ and $\nu = 0.4$

### *6.6.5 Shear of an Incompressible Elastic Continuum*

Consider a simple shear for linear models $B_n$ of incompressible continua (4.532). Notice that the motion law (6.175) even for compressible continua conserves their volume, because $\det \mathbf{F} = 1$ (see Exercise 4.2.14). Therefore, for incompressible continua a simple shear is also described by the law (6.175) and the boundary conditions (6.185a), (6.185b) and (6.185d), and in place of conditions (6.185c) one should consider the case of free ends:

$$x^3 = 0, h^3: \quad \sigma_{33} = 0, \quad \sigma_{13} = 0, \quad \sigma_{23} = 0. \tag{6.188}$$

On substituting the expression for the tensors $\overset{(n)}{\mathbf{G}}$ in the problem on shear (see Exercise 4.2.14) into (4.532), we find

$$\overset{(n)}{\mathbf{T}} = -\frac{p}{n-\mathrm{III}}\overset{(n)}{\mathbf{G}}{}^{-1} + \sum_{\alpha=1}^{3} \overset{(n)}{T}_{\alpha\alpha}\bar{\mathbf{e}}_{\alpha}^{2} + \overset{(n)}{T}_{12}\mathbf{O}_3, \tag{6.189}$$

$$\overset{(n)}{T}_{11} = \mu(n-\mathrm{III})^2\left(\frac{1+\beta}{n-\mathrm{III}} + (1-\beta)\left(\overset{(n)}{c}_{2B} + \frac{1}{n-\mathrm{III}}\right)\right),$$

$$\overset{(n)}{T}_{22} = \mu(n-\mathrm{III})^2\left(\frac{1+\beta}{n-\mathrm{III}} + (1-\beta)\left(\overset{(n)}{c}_{0B} + \frac{1}{n-\mathrm{III}}\right)\right),$$

$$\overset{(n)}{T}_{33} = \mu(n-\mathrm{III})^2\left(\frac{1+\beta}{n-\mathrm{III}} + (1-\beta)(\overset{(n)}{c}_{0B} + \overset{(n)}{c}_{2B})\right),$$

$$\overset{(n)}{T}_{12} = -\mu(n-\mathrm{III})^2\overset{(n)}{c}_{1B},$$

where $\overset{(n)}{c}_{0B}$, $\overset{(n)}{c}_{1B}$ and $\overset{(n)}{c}_{2B}$ are determined by Table 4.6 from Exercise 4.2.14.

With the help of the tensors of energetic equivalence ${}^4\overset{(n)}{\mathbf{E}}$ by analogy with (6.177) we obtain the expression for $\mathbf{T}$:

$$\mathbf{T} = -p\mathbf{E} + \sum_{\alpha,\beta=1}^{3} T_{\alpha\beta}^{(p)}\mathbf{p}_\alpha \otimes \mathbf{p}_\beta = \sum_{\alpha=1}^{3} \sigma_{\alpha\alpha}\bar{\mathbf{e}}_{\alpha}^{2} + \sigma_{12}\mathbf{O}_3, \tag{6.190}$$

where $T_{\alpha\beta}^{(p)}$ are determined by the same formulae (6.180). Here we have used that ${}^4\overset{(n)}{\mathbf{E}} \cdot\cdot \overset{(n)}{\mathbf{G}}{}^{-1} = (n-\mathrm{III})\mathbf{E}$. Cartesian components of the Cauchy stress tensor $\sigma_{\alpha\beta}$ can be determined with the help of expressions for the eigenvectors $\mathbf{p}_\alpha$ in the Cartesian basis $\bar{\mathbf{e}}_i$; as a result, we obtain relations similar to (6.183):

$$\sigma_{11} = -p + T_{11}^{(p)}s_1^2b_1^2 - 2T_{12}^{(p)}s_1s_2b_1b_2 + T_{22}^{(p)}s_2^2b_2^2,$$

$$\sigma_{22} = -p + T_{11}^{(p)}s_1^2 - 2T_{12}^{(p)}s_1s_2 + T_{22}^{(p)}s_2^2,$$

$$\sigma_{12} = T_{11}^{(p)}s_1^2b_1 - T_{12}^{(p)}s_1s_2(b_1+b_2) + T_{22}^{(p)}s_2^2b_2, \quad \sigma_{33} = -p + T_{33}^{(p)}. \tag{6.191}$$

Just as for a compressible material, in the problem on a simple shear $\sigma_{13} = \sigma_{23} = 0$. Then, substituting the formula for $\sigma_{33}$ into the boundary condition (6.188), we find that

$$p = T_{33}^{(p)} = \overset{(n)}{T}_{33} = \mu(n-\mathrm{III})^2\left(\frac{1+\beta}{n-\mathrm{III}} + (1-\beta)(\overset{(n)}{c}_{0B} + \overset{(n)}{c}_{2B})\right). \tag{6.192}$$

Here we have used formulae (6.189). Substitution of (6.192) into (6.191) yields a final expression for the normal stresses $\sigma_{11}$ and $\sigma_{22}$, and the stress $\sigma_{33}$ in this case proves to be zero in the whole body: $\sigma_{33} = 0$.

Figures 6.22–6.25 exhibit diagrams of deforming $\sigma_{12}(a)$ in shear, and also the functions $\sigma_{11}(a)$ and $\sigma_{22}(a)$ for linear models $B_n$ of incompressible materials.

For incompressible continua, not only the diagrams of deforming $\sigma_{12}(a)$ for the models $B_{\mathrm{I}}$ and $B_{\mathrm{V}}$, and also for the models $B_{\mathrm{II}}$ and $B_{\mathrm{IV}}$ are coincident, but also the normal stresses $\sigma_{11}(a)$ and $\sigma_{22}(a)$ (Figs. 6.22–6.24); here for all the models $B_n$ $\sigma_{11} > 0$, while $\sigma_{22} < 0$. With growing the parameter $\beta$ from $-1$ to 1, all the diagrams $|\sigma_{11}(a)|$ and $|\sigma_{22}(a)|$ are displaced into the domain of smaller values (Fig. 6.25). The diagrams of deforming $\sigma_{12}(a)$ are independent of the parameter $\beta$.

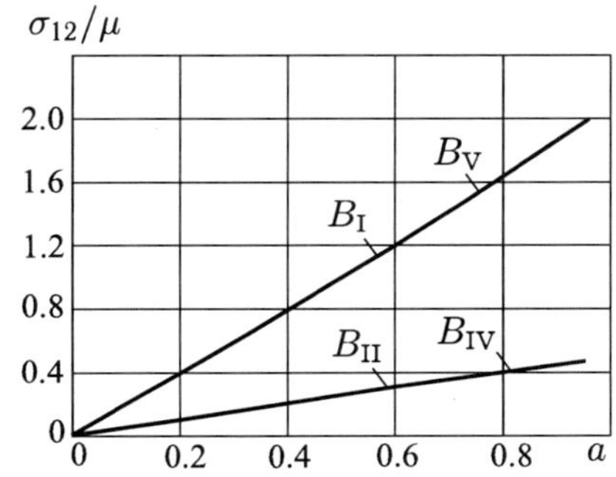


**Fig. 6.22** Diagrams of deforming in shear for models $B_n$ of incompressible continua

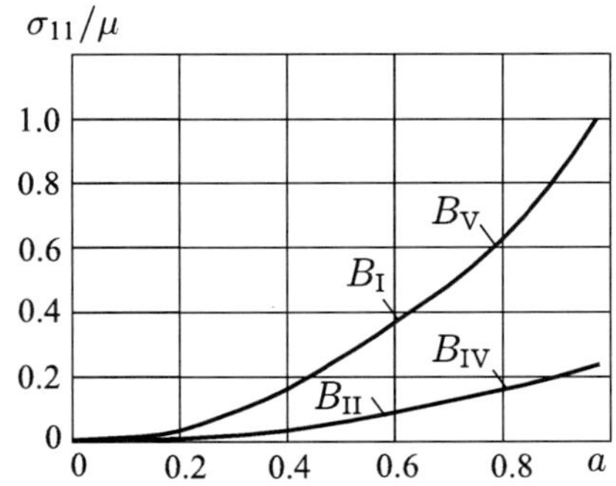


**Fig. 6.23** Dependence of the stress $\sigma_{11}$ on the shear angle $a$ in the problem on a simple shear for models $B_n$ of incompressible continua

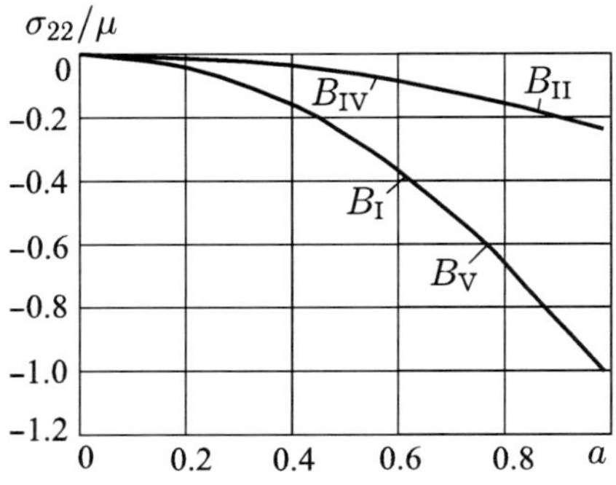


**Fig. 6.24** Dependence of the stress $\sigma_{22}$ on the shear angle $a$ in the problem on a simple shear for models $B_n$ of incompressible continua

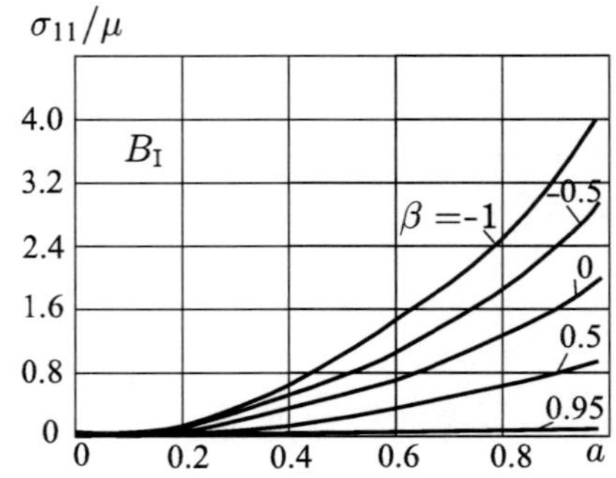


**Fig. 6.25** Dependence of the stress $\sigma_{11}$ on the shear angle $a$ in the problem on a simple shear for the model $B_{\mathrm{I}}$ of incompressible continua with different values of the coefficient $\beta$

## Exercises for 6.6

**6.6.1.** Consider the problem on shear for the models $A_{\mathrm{I}}$ and $A_{\mathrm{V}}$ with replacing the general equations (4.37) by the explicit relations between the Cauchy stress tensor and energetic stress tensors:

$$\mathbf{T} = \mathbf{F}^{-1\mathrm{T}} \cdot \overset{\mathrm{I}}{\mathbf{T}} \cdot \mathbf{F}^{-1} = \mathbf{F} \cdot \overset{\mathrm{V}}{\mathbf{T}} \cdot \mathbf{F}^{\mathrm{T}}.$$

Show that in this case formulae (6.183) for stresses $\sigma_{\alpha\beta}$ take the form

$$\sigma_{\alpha\alpha} = \overset{\mathrm{I}}{T}_{\alpha\alpha}, \quad \alpha = 1,\ 3, \quad \sigma_{22} = \overset{\mathrm{I}}{T}_{22} - 2a\overset{\mathrm{I}}{T}_{12} + a^2\overset{\mathrm{I}}{T}_{11}, \quad \sigma_{12} = \overset{\mathrm{I}}{T}_{12} - a\overset{\mathrm{I}}{T}_{11}$$

for the model $A_{\mathrm{I}}$, and

$$\sigma_{11} = \overset{\mathrm{V}}{T}_{11} + 2a\overset{\mathrm{V}}{T}_{12} + a^2\overset{\mathrm{V}}{T}_{22}, \quad \sigma_{22} = \overset{\mathrm{V}}{T}_{22}, \quad \sigma_{33} = \overset{\mathrm{V}}{T}_{33}, \quad \sigma_{12} = \overset{\mathrm{V}}{T}_{12} + a\overset{\mathrm{V}}{T}_{22}$$

for the model $A_{\mathrm{V}}$.

## 6.7 The Lamé Problem

### *6.7.1 The Motion Law for a Pipe in the Lamé Problem*

In the problems on tension and simple shear considered above, a stress tensor field $\mathbf{T}(X^i)$ was uniform, that ensured the equilibrium equations to be satisfied. With the help of the semi-inverse method we can find an analytical solution also for the problem on a pipe under external and internal pressures (*the Lamé problem*), where the field $\mathbf{T}(X^i)$ is not uniform.

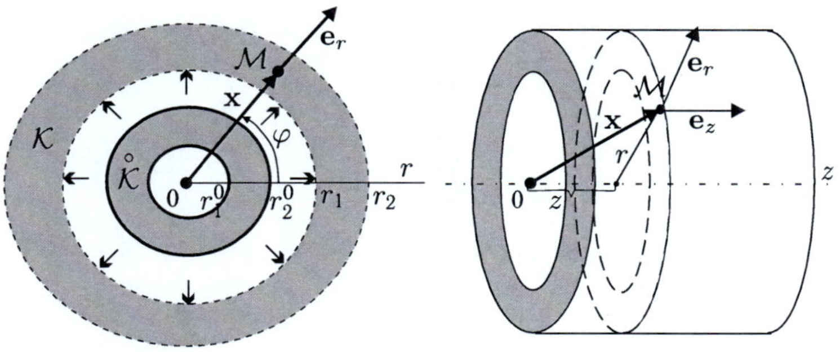


**Fig. 6.26** For the Lamé problem

In this problem, a body $\mathcal{B}$ in $\overset{\circ}{\mathcal{K}}$ and $\mathcal{K}$ is a pipe (a thick-walled cylinder) of finite length $h^3$ (Fig. 6.26), which under the action of pressures $p_{e2}$ and $p_{e1}$ on the outer $r = r_2$ and inner $r = r_1$ surfaces, respectively, has the same symmetry axis $Ox^3 = Oz$, i.e. remains self-similar. At the end $z = h^3$ of the cylinder the displacement $u_3^e$ is given, and at the second end $z = 0$ the symmetry condition is assumed to be satisfied.

Let us introduce cylindrical coordinates $r, \varphi, z$ (Fig. 6.26); and as Lagrangian coordinates $X^i$ we choose $X^1 = \overset{\circ}{r}$, $X^2 = \overset{\circ}{\varphi}$ and $X^3 = \overset{\circ}{z}$, which are the values of cylindrical coordinates of material points at the initial time $t = 0$ in $\overset{\circ}{\mathcal{K}}$. Then the radius-vector $\overset{\circ}{\mathbf{x}}$ of a material point $\mathcal{M}$ in $\overset{\circ}{\mathcal{K}}$ can be resolved for basis vectors of the cylindrical coordinate system $\mathbf{e}_r, \mathbf{e}_\varphi, \mathbf{e}_z$ (i.e. for the physical basis) as follows:

$$\overset{\circ}{\mathbf{x}} = \overset{\circ}{r}\mathbf{e}_r + \overset{\circ}{z}\mathbf{e}_z. \tag{6.193}$$

Since the pipe retains its axial symmetry in $\mathcal{K}$ under the action of pressures and axial displacement, a law of the motion is sought in the form

$$\mathbf{x} = r\mathbf{e}_r + z\mathbf{e}_z. \tag{6.194}$$

Here $r$ and $z$ are functions in the forms

$$r = f(\overset{\circ}{r}, t), \qquad z = k(t)\overset{\circ}{z}, \tag{6.195}$$

where

$$f(\overset{\circ}{r}, 0) = \overset{\circ}{r}, \qquad k(0) = 1. \tag{6.196}$$

### 6.7.2 The Deformation Gradient and Deformation Tensors in the Lamé Problem

Determine the local basis vectors $\mathbf{r}_i$ and $\overset{\circ}{\mathbf{r}}_i$ by using formulae (2.10), (6.194), and (6.195):

$$\begin{aligned}
\mathbf{r}_1 &= \frac{\partial \mathbf{x}}{\partial X^1} = \frac{\partial}{\partial \overset{\circ}{r}}(f(\overset{\circ}{r},t)\mathbf{e}_r(\overset{\circ}{\varphi}) + \overset{\circ}{z}k\mathbf{e}_z) = f'\mathbf{e}_r, \quad f' \equiv \frac{\partial f}{\partial \overset{\circ}{r}}, \\
\mathbf{r}_2 &= \frac{\partial \mathbf{x}}{\partial X^2} = \frac{\partial}{\partial \overset{\circ}{\varphi}}(f\,\mathbf{e}_r(\overset{\circ}{\varphi}) + \overset{\circ}{z}k\mathbf{e}_z) = f\frac{\partial \mathbf{e}_r}{\partial \overset{\circ}{\varphi}} = f\,\mathbf{e}_\varphi, \\
\mathbf{r}_3 &= \frac{\partial \mathbf{x}}{\partial X^3} = \frac{\partial}{\partial \overset{\circ}{z}}(f\,\mathbf{e}_r + \overset{\circ}{z}k\mathbf{e}_z) = k\mathbf{e}_z, \\
\overset{\circ}{\mathbf{r}}_1 &= \mathbf{e}_r, \quad \overset{\circ}{\mathbf{r}}_2 = \overset{\circ}{r}\mathbf{e}_\varphi, \quad \overset{\circ}{\mathbf{r}}_3 = \mathbf{e}_z.
\end{aligned} \tag{6.197}$$

Here we have taken into account the formulae of differentiation of physical basis vectors of a cylindrical coordinate system (see [12]).

Let us form the metric matrices $g_{ij}$, $g^{ij}$ and $\overset{\circ}{g}_{ij}$, $\overset{\circ}{g}^{ij}$:

$$\begin{aligned}
g_{ij} &= \mathbf{r}_i \cdot \mathbf{r}_j = \begin{pmatrix} f'^2 & 0 & 0 \\ 0 & f^2 & 0 \\ 0 & 0 & k^2 \end{pmatrix}, \quad g^{ij} = \begin{pmatrix} f'^{-2} & 0 & 0 \\ 0 & f^{-2} & 0 \\ 0 & 0 & k^{-2} \end{pmatrix}, \\
\overset{\circ}{g}_{ij} &= \overset{\circ}{\mathbf{r}}_i \cdot \overset{\circ}{\mathbf{r}}_j = \begin{pmatrix} 1 & 0 & 0 \\ 0 & \overset{\circ}{r}{}^2 & 0 \\ 0 & 0 & 1 \end{pmatrix}, \quad \overset{\circ}{g}{}^{ij} = \begin{pmatrix} 1 & 0 & 0 \\ 0 & \overset{\circ}{r}{}^{-2} & 0 \\ 0 & 0 & 1 \end{pmatrix}, \\
g &= (f'fk)^2, \quad \overset{\circ}{g} = \overset{\circ}{r}{}^2
\end{aligned} \tag{6.198}$$

and find local vectors of the reciprocal bases by formulae (2.13):

$$\begin{aligned}
\mathbf{r}^1 &= \frac{\mathbf{e}_r}{f'}, \quad \mathbf{r}^2 = \frac{\mathbf{e}_\varphi}{f}, \quad \mathbf{r}^3 = \frac{1}{k}\mathbf{e}_z, \\
\overset{\circ}{\mathbf{r}}{}^1 &= \mathbf{e}_r, \quad \overset{\circ}{\mathbf{r}}{}^2 = \frac{\mathbf{e}_\varphi}{\overset{\circ}{r}}, \quad \overset{\circ}{\mathbf{r}}{}^3 = \mathbf{e}_z.
\end{aligned} \tag{6.199}$$

According to (6.198) and (6.199), we calculate the deformation gradient

$$\begin{aligned}
\mathbf{F} &= \mathbf{r}_i \otimes \overset{\circ}{\mathbf{r}}{}^i = f'\mathbf{e}_r \otimes \mathbf{e}_r + \frac{f}{\overset{\circ}{r}}\mathbf{e}_\varphi \otimes \mathbf{e}_\varphi + k\mathbf{e}_z \otimes \mathbf{e}_z = \mathbf{F}^{\mathrm{T}}, \\
\mathbf{F}^{-1} &= \overset{\circ}{\mathbf{r}}_i \otimes \mathbf{r}^i = \frac{1}{f'}\mathbf{e}_r \otimes \mathbf{e}_r + \frac{\overset{\circ}{r}}{f}\mathbf{e}_\varphi \otimes \mathbf{e}_\varphi + \frac{1}{k}\mathbf{e}_z \otimes \mathbf{e}_z, \\
J &= \rho/\overset{\circ}{\rho} = \det \mathbf{F}^{-1} = \frac{\overset{\circ}{r}}{f'fk},
\end{aligned} \tag{6.200}$$

which, just as for the problem on a beam in tension, is a diagonal tensor but its components depend on Lagrangian coordinates. Since $f = r$ is the radius and $k$ is the elongation ratio of a cylinder along its axis, values of $f$ and $k$ are always positive; and the last relation of (6.200) yields that $f'$ is positive too. Thus, there are constraints on signs of $f'$, $f$ and $k$:

$$f' > 0, \quad f > 0, \quad k > 0. \tag{6.201}$$

Since $\mathbf{F}$ is diagonal, one can readily find the stretch tensors $\mathbf{U}$ and $\mathbf{V}$ and the rotation tensor $\mathbf{O}$:

$$\mathbf{U} = \mathbf{V} = \mathbf{F}, \qquad \mathbf{O} = \mathbf{E}. \tag{6.202}$$

With the help of formulae (4.25) and (4.42) we find the energetic deformation tensors $\overset{(n)}{\mathbf{C}}$ and deformation measures $\overset{(n)}{\mathbf{G}}$:

$$\overset{(n)}{\mathbf{C}} = \frac{1}{n - \text{III}}\Big(((f')^{n-\text{III}} - 1)\mathbf{e}_r \otimes \mathbf{e}_r + ((f/\overset{\circ}{r})^{n-\text{III}} - 1)\mathbf{e}_\varphi \otimes \mathbf{e}_\varphi + (k^{n-\text{III}} - 1)\mathbf{e}_z \otimes \mathbf{e}_z\Big), \tag{6.203}$$

$$\overset{(n)}{\mathbf{G}} = \frac{1}{n - \text{III}}\big(f'^{n-\text{III}}\mathbf{e}_r \otimes \mathbf{e}_r + (f/\overset{\circ}{r})^{n-\text{III}}\mathbf{e}_\varphi \otimes \mathbf{e}_\varphi + k^{n-\text{III}}\mathbf{e}_z \otimes \mathbf{e}_z\big).$$

### 6.7.3 *Stresses in the Lamé Problem for Models $A_n$*

Assume that constitutive equations of the cylinder correspond to the linear model $A_n$ (6.135) of an isotropic elastic continuum. Then substituting the expression (6.203) into (6.135), we find the energetic stress tensors

$$\overset{(n)}{\mathbf{T}} = \overset{(n)}{T}_r \mathbf{e}_r \otimes \mathbf{e}_r + \overset{(n)}{T}_\varphi \mathbf{e}_\varphi \otimes \mathbf{e}_\varphi + \overset{(n)}{T}_z \mathbf{e}_z \otimes \mathbf{e}_z, \tag{6.204}$$

$$\overset{(n)}{T}_r = \frac{\overset{\circ}{r}}{(n - \text{III})f'fk}(l_1 I_1 + 2l_2(f'^{n-\text{III}} - 1)),$$

$$\overset{(n)}{T}_\varphi = \frac{\overset{\circ}{r}}{(n - \text{III})f'fk}(l_1 I_1 + 2l_2((f/\overset{\circ}{r})^{n-\text{III}} - 1)),$$

$$\overset{(n)}{T}_z = \frac{\overset{\circ}{r}}{(n - \text{III})f'fk}(l_1 I_1 + 2l_2(k^{n-\text{III}} - 1)),$$

$$I_1 = (f')^{n-\text{III}} + (f/\overset{\circ}{r})^{n-\text{III}} + k^{n-\text{III}} - 3.$$

Since the tensors $\mathbf{F}$ and $\overset{(n)}{\mathbf{T}}$ are diagonal, from Eqs. (6.137), (6.200), and (6.204) we obtain the expression for the Cauchy stress tensor

$$\mathbf{T} = \sigma_r \mathbf{e}_r \otimes \mathbf{e}_r + \sigma_\varphi \mathbf{e}_\varphi \otimes \mathbf{e}_\varphi + \sigma_z \mathbf{e}_z \otimes \mathbf{e}_z = \mathbf{F}^{n-\mathrm{III}} \cdot \overset{(n)}{\mathbf{T}}, \tag{6.205}$$

$$\sigma_r = \frac{\overset{\circ}{r}(f')^{n-\mathrm{III}-1}}{(n-\mathrm{III}) f k}(l_1 I_1 + 2l_2(f'^{n-\mathrm{III}} - 1)),$$

$$\sigma_\varphi = \frac{1}{(n-\mathrm{III}) f' k}(f/\overset{\circ}{r})^{n-\mathrm{III}-1}(l_1 I_1 + 2l_2((f/\overset{\circ}{r})^{n-\mathrm{III}} - 1)),$$

$$\sigma_z = \frac{\overset{\circ}{r} k^{n-\mathrm{III}-1}}{(n-\mathrm{III}) f' f}(l_1 I_1 + 2l_2(k^{n-\mathrm{III}} - 1)).$$

Components of the stress tensor in this problem prove to depend on the coordinate $\overset{\circ}{r}$ and time $t$, therefore, the equilibrium equations are not automatically satisfied. In this case, as a rule, it is convenient to use the equilibrium equations (6.100) in the material description. To do this, we should calculate the Piola–Kirchhoff stress tensor by formulae (3.56):

$$\mathbf{P} = \frac{1}{J_0}\mathbf{F}^{-1} \cdot \mathbf{T} = \frac{1}{J}\mathbf{F}^{n-\mathrm{III}-1} \cdot \overset{(n)}{\mathbf{T}} = \overset{\circ}{\sigma}_r \mathbf{e}_r \otimes \mathbf{e}_r + \overset{\circ}{\sigma}_\varphi \mathbf{e}_\varphi \otimes \mathbf{e}_\varphi + \overset{\circ}{\sigma}_z \mathbf{e}_z \otimes \mathbf{e}_z, \tag{6.206}$$

where

$$\overset{\circ}{\sigma}_r = \frac{(f')^{n-\mathrm{III}-1}}{n-\mathrm{III}}((l_1 + 2l_2) f'^{n-\mathrm{III}} + l_1 (f/\overset{\circ}{r})^{n-\mathrm{III}} - (3l_1 + 2l_2) + l_1 k^{n-\mathrm{III}}),$$

$$\begin{aligned}\overset{\circ}{\sigma}_\varphi = \frac{1}{n-\mathrm{III}}(f/\overset{\circ}{r})^{n-\mathrm{III}-1}&((l_1 + 2l_2)(f/\overset{\circ}{r})^{n-\mathrm{III}} \\ &+ l_1 f'^{n-\mathrm{III}} - (3l_1 + 2l_2) + l_1 k^{n-\mathrm{III}}),\end{aligned} \tag{6.207}$$

$$\overset{\circ}{\sigma}_z = \frac{k^{n-\mathrm{III}-1}}{n-\mathrm{III}}((l_1 + 2l_2) k^{n-\mathrm{III}} + l_1 (f'^{n-\mathrm{III}} + (1/\overset{\circ}{r})^{n-\mathrm{III}}) - (3l_1 + 2l_2)).$$

### 6.7.4 *Equation for the Function f*

Writing components of the divergence $\overset{\circ}{\nabla} \cdot \mathbf{P}$ with respect to the physical basis $\mathbf{e}_r, \mathbf{e}_\varphi, \mathbf{e}_z$ (see [12]), we can represent the equilibrium equation projected onto the axis $O\mathbf{e}_r$ as follows:

$$\frac{\partial \overset{\circ}{\sigma}_r}{\partial \overset{\circ}{r}} + \frac{\overset{\circ}{\sigma}_r - \overset{\circ}{\sigma}_\varphi}{\overset{\circ}{r}} = 0. \tag{6.208}$$

Here we have taken into account that $\overset{\circ}{\sigma}_r$, $\overset{\circ}{\sigma}_\varphi$ and $\overset{\circ}{\sigma}_z$ are independent of $\overset{\circ}{\varphi}$ and $\overset{\circ}{z}$. The other two projections of the equilibrium equation onto the axes $O\mathbf{e}_\varphi$ and $O\mathbf{e}_z$ are satisfied identically.

On substituting the expressions (6.207) into (6.208), we obtain an ordinary differential equation of the second order for the function $f(\overset{\circ}{r}, t)$. The function $f(\overset{\circ}{r}, t)$ is determined up to the two constants of integration $C_1(t)$ and $C_2(t)$ being functions only of time; to evaluate them one should use boundary conditions.

### 6.7.5 *Boundary Conditions of the Weak Type*

Boundary conditions at inner and outer surfaces of the cylinder have the form (6.80) (gas or fluid pressures $p_{e1}$ and $p_{e2}$ are given). According to formula (6.206) and the fact that at the surfaces $\overset{\circ}{r} = \overset{\circ}{r}_\alpha$ ($\alpha = 1, 2$) the normal vectors have the form $\overset{\circ}{\mathbf{n}} = \mp\mathbf{e}_r$, the boundary condition (6.80) for this problem becomes

$$\overset{\circ}{r} = \overset{\circ}{r}_\alpha : \quad \overset{\circ}{\sigma}_r = -p_{e\alpha}\frac{f(\overset{\circ}{r}_\alpha, t)k}{\overset{\circ}{r}_\alpha}, \quad \alpha = 1, 2. \tag{6.209}$$

Boundary conditions at the end surfaces $\overset{\circ}{z} = 0, h_3$ serve for determining the function $k$. For example, if at the surface $z = 0$ the symmetry conditions are given, and at $\overset{\circ}{z} = h_3^0$ – the pressure $p_{e3}$, then we have

$$\overset{\circ}{z} = 0 : \quad u_z = (z - \overset{\circ}{z})\big|_{\overset{\circ}{z}=0} = 0, \tag{6.210}$$

$$\overset{\circ}{z} = \overset{\circ}{h}_3 : \quad \sigma_z = -p_{e3}. \tag{6.211}$$

On substituting the expression (6.194) and (6.205) into (6.210) and (6.211), we verify that the first boundary condition is satisfied identically, and the second one is an additional differential equation for the function $f$ and the constant $k$.

The solution found above cannot satisfy this boundary condition completely, therefore we should consider other variants of boundary conditions.

One of such variants weakens the boundary condition (6.211) by replacing it with the integral one

$$\overset{\circ}{z} = \overset{\circ}{h}_3 : \quad \int_{r_1}^{r_2} \sigma_z r\,dr = -\frac{1}{2}p_{e3}(r_2^2 - r_1^2). \tag{6.212}$$

One can say that such a condition is of the weak type. The method of replacing exact boundary conditions by the integral ones is called the *Saint-Venant method*.

According to (6.195), we have $rdr = fdf = ff'd\overset{\circ}{r}$. Then, substituting the expression (6.205) for $\sigma_z$ into (6.212) and passing to the reference configuration, we rewrite the condition (6.212) as follows:

$$\frac{k^{n-\text{III}-1}}{n-\text{III}}\left(\frac{2l_1}{\overset{\circ}{r}{}_2^2-\overset{\circ}{r}{}_1^2}\int_{\overset{\circ}{r}_1}^{\overset{\circ}{r}_2}\left((f')^{n-\text{III}}+\left(\frac{f}{\overset{\circ}{r}}\right)^{n-\text{III}}\right)\overset{\circ}{r}\,d\overset{\circ}{r}\right.$$
$$\left.+(l_1+2l_2)k^{n-\text{III}}-(3l_1+2l_2)\right)=-p_{e3}\left(\frac{f^2(\overset{\circ}{r}_2)-f^2(\overset{\circ}{r}_1)}{\overset{\circ}{r}{}_2^2-\overset{\circ}{r}{}_1^2}\right). \tag{6.213}$$

Expressing the function $k$ by Eq. (6.213) and substituting the result into (6.207) and then formulae (6.207) into (6.208) and (6.209), we obtain an integro-differential equation for determining the function $f(\overset{\circ}{r},t)$.

For the case when $n = \text{IV}$ (John's model $A_{\text{IV}}$), Eqs. (6.208) and (6.209) have a simple analytical solution:

$$f(\overset{\circ}{r},t) = C_1(t)\overset{\circ}{r} + \frac{C_2(t)}{\overset{\circ}{r}}. \tag{6.214}$$

Indeed, in this case Eqs. (6.207) become

$$\begin{aligned}\overset{\circ}{\sigma}_r &= 2(l_1+l_2)C_1 - 2l_2\frac{C_2}{\overset{\circ}{r}{}^2} + l_1k - (3l_1+2l_2),\\ \overset{\circ}{\sigma}_\varphi &= 2(l_1+l_2)C_1 + 2l_2\frac{C_2}{\overset{\circ}{r}{}^2} + l_1k - (3l_1+2l_2),\end{aligned} \tag{6.215}$$

and relations (6.205) take the forms

$$\begin{aligned}\sigma_r &= \frac{\overset{\circ}{r}{}^2}{(C_1\overset{\circ}{r}{}^2+C_2)k}\left(l_1(2C_1+k)+2l_2\left(C_1-\frac{C_2}{\overset{\circ}{r}{}^2}\right)-(3l_1+2l_2)\right),\\ \sigma_\varphi &= \frac{\overset{\circ}{r}{}^2}{(C_1\overset{\circ}{r}{}^2-C_2)k}\left(l_1(2C_1+k)+2l_2\left(C_1+\frac{C_2}{\overset{\circ}{r}{}^2}\right)-(3l_1+2l_2)\right),\\ \sigma_z &= \frac{\overset{\circ}{r}{}^4}{(C_1^2\overset{\circ}{r}{}^4-C_2^2)}\left(l_1(2C_1+k)+2l_2k-(3l_1+2l_2)\right).\end{aligned} \tag{6.216}$$

Substituting the expressions (6.215) into (6.208), one can readily verify that the equilibrium equation is identically satisfied.

Substitution of (6.215) into the boundary conditions (6.209) and (6.211) yields

$$2(l_1+l_2)C_1 - \frac{2l_2}{\overset{\circ}{r}{}_1^2}C_2 + l_1 k - (3l_1+2l_2) = -p_{e1}\left(C_1 + \frac{C_2}{\overset{\circ}{r}{}_1^2}\right),$$
$$2(l_1+l_2)C_1 - \frac{2l_2}{\overset{\circ}{r}{}_2^2}C_2 + l_1 k - (3l_1+2l_2) = -p_{e2}\left(C_1 + \frac{C_2}{\overset{\circ}{r}{}_2^2}\right). \quad (6.217)$$

After substitution of (6.214), the boundary condition (6.213) for the model $A_{\mathrm{IV}}$ takes the form

$$2C_1 l_1 + (l_1+2l_2)k - (3l_1+2l_2) = -p_{e3}\left(C_1^2 - \frac{C_2^2}{\overset{\circ}{r}{}_1^2\overset{\circ}{r}{}_2^2}\right). \quad (6.218)$$

Solving the system of three algebraic equations (6.217), (6.218), we find $C_1$, $C_2$, and $k$.

## 6.7.6 *Boundary Conditions of the Rigid Type*

In place of (6.212) we can consider another boundary condition, namely the condition of the rigid type when displacements $u_{ez}$ along the axis $Oz$ are given:

$$\overset{\circ}{z} = \overset{\circ}{h}_3: \qquad u_z = (z - \overset{\circ}{z})\big|_{\overset{\circ}{z}=\overset{\circ}{h}_3} = u_{ez}. \quad (6.219)$$

Then we obtain the following simple expression for $k$:

$$k = 1 + (u_{ez}/\overset{\circ}{h}_3), \quad (6.220)$$

which is similar to the corresponding expression of (6.142) in the problem on a beam in tension. For the model $A_{\mathrm{IV}}$, this relation takes the place of the condition (6.218) and the system (6.217), (6.218) for $C_1$, $C_2$, and $k$ becomes linear; its solution has the form

$$C_1 = \gamma C_2, \qquad C_2 = \overset{\circ}{r}{}_1^2 \frac{1-2\nu(k-1)}{(1+(1-2\nu)\widetilde{p}_1)\gamma - (1-2\nu)(1-\widetilde{p}_1)},$$
$$\gamma = \frac{\widetilde{p}_1 - \widetilde{p}_2\beta_0^2 - 1 + \beta_0^2}{\widetilde{p}_2 - \widetilde{p}_1}, \qquad \beta_0 = \overset{\circ}{r}_1/\overset{\circ}{r}_2. \quad (6.221)$$

Here $\widetilde{p}_\alpha = p_{e\alpha}/2l_2$, and the Poisson ratio (6.146) has been introduced.

Expressions for outer $r_2$ and inner $r_1$ radii of the cylinder in $\mathcal{K}$ can be found with the help of formulae (6.194) and (6.214):

$$r_\alpha/\overset{\circ}{r}_\alpha = f(\overset{\circ}{r}_\alpha, t)/\overset{\circ}{r}_\alpha = C_1 + (C_2/\overset{\circ}{r}{}^2_\alpha), \quad \alpha = 1, 2. \tag{6.222}$$

For other models $A_n$, we can also consider the boundary condition (6.219), which yields the expression (6.220) for $k$. Then, substituting the expressions (6.207) into (6.208) and (6.209), we obtain one nonlinear differential second-order equation for the function $f$ with two boundary conditions.

## 6.8 The Lamé Problem for an Incompressible Continuum

### *6.8.1 Equation for the Function $f$*

Consider the Lamé problem on a cylindrical pipe under internal and external pressures (see Sect. 6.7), but let the pipe be made of isotropic incompressible material described by linear models $B_n$ (see (4.532) or (6.160)). In this case the motion law for the pipe is also sought by the semi-inverse method in the form (6.194) and (6.195); therefore, all the relations (6.197)–(6.203) hold. From the condition of incompressibility of a material considered $\det \mathbf{F} = 1$ and from (6.201) it follows that the function $f(\overset{\circ}{r}, t)$ must satisfy the equation

$$f' f k = \overset{\circ}{r}. \tag{6.223}$$

Rewriting this equation in the form $f\ df = \frac{1}{k}\overset{\circ}{r}\ d\overset{\circ}{r}$, one can easily find its solution (choosing a positive root)

$$f^2 = \frac{\overset{\circ}{r}{}^2}{k} + C, \tag{6.224}$$

where $C$ is the constant of integration.

### *6.8.2 Stresses in the Lamé Problem for an Incompressible Continuum*

On substituting Eq. (6.223) into (6.203), we find

$$\overset{(n)}{\mathbf{G}} = \frac{1}{n - \mathrm{III}} \left( \left( \frac{\overset{\circ}{r}}{fk} \right)^{n-\mathrm{III}} \mathbf{e}_r \otimes \mathbf{e}_r + \left( \frac{f}{\overset{\circ}{r}} \right)^{n-\mathrm{III}} \mathbf{e}_\varphi \otimes \mathbf{e}_\varphi + k^{n-\mathrm{III}} \mathbf{e}_z \otimes \mathbf{e}_z \right). \tag{6.225}$$

Having substituted this expression int the constitutive equation (6.160), we obtain that the energetic stress tensors $\overset{(n)}{\mathbf{T}}$ have the diagonal form (6.204) in the case of an incompressible continuum too, but their components are different from the ones for compressible materials:

$$\overset{(n)}{T}_r = -p\left(\frac{fk}{\overset{\circ}{r}}\right)^{n-\mathrm{III}} + \mu(n-\mathrm{III})\left(1+\beta+(1-\beta)\left(\left(\frac{f}{\overset{\circ}{r}}\right)^{n-\mathrm{III}} + k^{n-\mathrm{III}}\right)\right),$$

$$\overset{(n)}{T}_\varphi = -p\left(\frac{\overset{\circ}{r}}{f}\right)^{n-\mathrm{III}} + \mu(n-\mathrm{III})\left(1+\beta+(1-\beta)\left(\left(\frac{\overset{\circ}{r}}{fk}\right)^{n-\mathrm{III}} + k^{n-\mathrm{III}}\right)\right),$$

$$\overset{(n)}{T}_z = -pk^{n-\mathrm{III}} + \mu(n-\mathrm{III})\left(1+\beta+(1-\beta)\left(\left(\frac{\overset{\circ}{r}}{fk}\right)^{n-\mathrm{III}} + \left(\frac{f}{\overset{\circ}{r}}\right)^{n-\mathrm{III}}\right)\right),$$

$$n = \mathrm{I},\ \mathrm{II},\ \mathrm{IV},\ \mathrm{V}. \tag{6.226}$$

For the Cauchy stress tensor, relations (6.205) hold as well, and their components are written as follows:

$$\sigma_r = -p + \widetilde{\sigma}_r, \quad \sigma_\varphi = -p + \widetilde{\sigma}_\varphi, \quad \sigma_z = -p + \widetilde{\sigma}_z, \tag{6.227}$$

where

$$\widetilde{\sigma}_r = \mu(n-\mathrm{III})\left(\frac{\overset{\circ}{r}}{fk}\right)^{n-\mathrm{III}}\left(1+\beta+(1-\beta)\left(\left(\frac{f}{\overset{\circ}{r}}\right)^{n-\mathrm{III}} + k^{n-\mathrm{III}}\right)\right),$$

$$\widetilde{\sigma}_\varphi = \mu(n-\mathrm{III})\left(\frac{f}{\overset{\circ}{r}}\right)^{n-\mathrm{III}}\left(1+\beta+(1-\beta)\left(\left(\frac{\overset{\circ}{r}}{fk}\right)^{n-\mathrm{III}} + k^{n-\mathrm{III}}\right)\right),$$

$$\widetilde{\sigma}_z = \mu(n-\mathrm{III})k^{n-\mathrm{III}}\left(1+\beta+(1-\beta)\left(\left(\frac{\overset{\circ}{r}}{fk}\right)^{n-\mathrm{III}} + \left(\frac{f}{\overset{\circ}{r}}\right)^{n-\mathrm{III}}\right)\right). \tag{6.228}$$

With the help of the relation $\mathbf{P} = \mathbf{F}^{-1}\cdot\mathbf{T}$, we find components of the Piola–Kirchhoff tensor:

$$\overset{\circ}{\sigma}_r = \sigma_r\frac{fk}{\overset{\circ}{r}}, \quad \overset{\circ}{\sigma}_\varphi = \frac{\overset{\circ}{r}}{f}\sigma_\varphi, \quad \overset{\circ}{\sigma}_z = \frac{\sigma_z}{k}. \tag{6.229}$$

### 6.8.3 Equation for Hydrostatic Pressure p

Substitution of the Piola–Kirchhoff tensor components (6.229) into the equilibrium equation (6.208) yields an ordinary differential equation for unknown function $p(\overset{\circ}{r},t)$ being the hydrostatic pressure in an incompressible continuum:

$$\left(\frac{pfk}{\overset{\circ}{r}}\right)' + \frac{p(f^2k-\overset{\circ}{r}{}^2)}{\overset{\circ}{r}{}^2 f} = \frac{\widetilde{h}}{\overset{\circ}{r}}, \quad \widetilde{h} = \frac{\partial}{\partial\overset{\circ}{r}}(fk\widetilde{\sigma}_r) - \frac{\overset{\circ}{r}}{f}\widetilde{\sigma}_\varphi. \tag{6.230}$$

Rearranging the left-hand side of Eq. (6.230) with the help of (6.223) and (6.224), we reduce the equation to the form

$$\frac{dp}{d\overset{\circ}{r}}\frac{fk}{\overset{\circ}{r}} = \frac{\widetilde{h}}{\overset{\circ}{r}}. \tag{6.231}$$

Integration of this equation yields

$$p = p_0 + \frac{1}{k}\int_{\overset{\circ}{r}_1}^{\overset{\circ}{r}} \frac{\widetilde{h}}{f}\, d\overset{\circ}{r}. \tag{6.232}$$

Here $p_0$ is the constant of integration, which together with $C$ can be found from the boundary conditions (6.209) after substitution of expressions (6.229) and (6.232) into them:

$$p_0 + \frac{1}{k}\int_{\overset{\circ}{r}_1}^{\overset{\circ}{r}_2} \frac{\widetilde{h}}{f}\, d\overset{\circ}{r} = p_{e2} + \widetilde{\sigma}_r(\overset{\circ}{r}_2), \quad p_0 = p_{e1} + \widetilde{\sigma}_r(\overset{\circ}{r}_1). \tag{6.233}$$

For the constant $C$, from (6.233) we obtain the nonlinear algebraic equation:

$$F(C) \equiv \widetilde{\sigma}_r(\overset{\circ}{r}_2) - \widetilde{\sigma}_r(\overset{\circ}{r}_1) - \frac{1}{k}\int_{\overset{\circ}{r}_1}^{\overset{\circ}{r}_2} \frac{\widetilde{h}\, d\overset{\circ}{r}}{f} - \Delta p = 0, \quad \Delta p = p_{e1} - p_{e2}, \tag{6.234}$$

into which the expressions (6.224), (6.228), and (6.230) for $\widetilde{\sigma}_r$, $f$ and $\widetilde{h}$ should be substituted.

### 6.8.4 Analysis of the Problem Solution

Figure 6.27 shows the graph of the function $F(C)$.

In the general case, for all $n$ and $\beta$ the equation $F(C) = 0$ may have two roots $C_1$ and $C_2$, among which one should choose the least root $C_1$, because just the one satisfies the normalization condition $F(C_1) = 0$ at $k = 1$ and $\Delta p = 0$ when there is no loading and deformations of the cylinder. For this case

$$\begin{gathered} C = 0, \quad f = \overset{\circ}{r}, \\ \widetilde{\sigma}_r = \widetilde{\sigma}_\varphi = \widetilde{\sigma}_z \equiv \sigma_0 = \mu(n - \mathrm{III})(3 - \beta) = \text{const}, \\ p = p_0 = \text{const}, \qquad p_0 = p_{e_1} + \sigma_0, \\ \overset{\circ}{\sigma}_r = \overset{\circ}{\sigma}_\varphi = \overset{\circ}{\sigma}_z = -p_0 + \sigma_0 = p_{e1} = \text{const}. \end{gathered} \tag{6.235}$$

The stresses $\sigma_r$, $\sigma_\varphi$ and $\sigma_z$ in the cylinder are zero when $p_{e1} = p_{e2} = 0$; and they are equal to each other but are nonzero when $p_{e1} = p_{e2} \neq 0$.

When $k = 1$ and $\Delta p > 0$ (excess internal pressure), the least root $C_1$ of the function $F(C)$ is positive; and when $\Delta p < 0$ (excess external pressure) the root $C_1$ is negative (Fig. 6.27). With growing the value $k > 1$ (longitudinal extension), the root $C_1$ is displaced into the domain of negative values (there occurs transverse compression of the cylinder); and with decreasing the value $k < 1$ (longitudinal compression), on the contrary, the root $C_1$ is displaced into the domain of positive values (there occurs lateral dilatation of the cylinder).

For the function $F(C)$, a peculiarity of interest is the existence of the ultimate value $\Delta p_*$: while $\Delta p > \Delta p_*$, there are no roots of the function $F(C)$ (see Fig. 6.27). This means that at such values of $\Delta p$ the nonlinear Lamé problem has no solution (unlike the Lamé problem in linear elasticity theory, which has a solution at all values of $\Delta p$). If we consider the process of monotone increasing the pressure difference $\Delta p$ from 0 up to $\Delta p_*$, then the cylinder radius also monotonically grows; and when $\Delta p > \Delta p_*$ a solution does not exist. This effect is called the loss of stability of a material in tension.

Figure 6.28 shows the dependence of the dilatation coefficient $y_R = (r_1 - \overset{\circ}{r}_1)/\overset{\circ}{r}_1$ of the cylinder upon the dimensionless pressure difference $\Delta p/\mu$ (when $\Delta p > 0$)

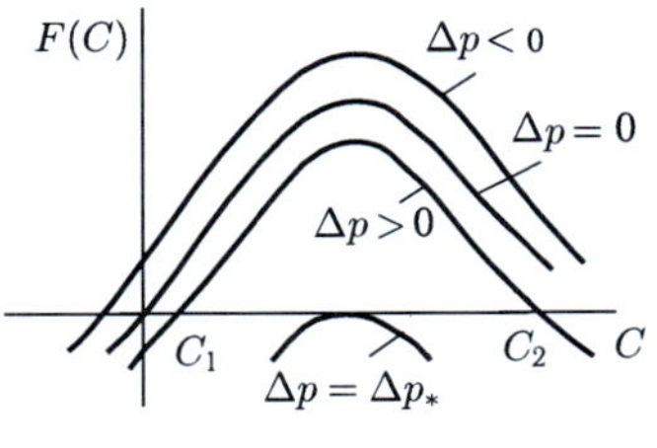


**Fig. 6.27** The function $F(C)$

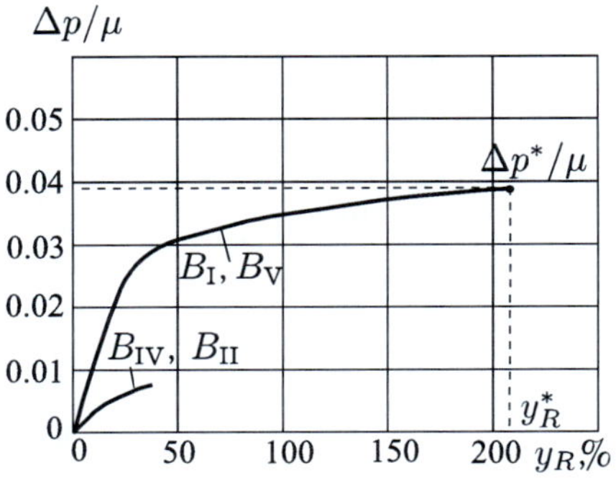


**Fig. 6.28** Dependence of the coefficient of relative dilatation of a thin-walled cylinder on internal excess pressure for different models of incompressible materials ($k = 1$, $r_2/r_1 = 1.01$)

for the cylinder with a very thin wall: $\overset{\circ}{h}/\overset{\circ}{r}_1 = 0.01$, where $\overset{\circ}{h} = \overset{\circ}{r}_2 - \overset{\circ}{r}_1$ is the initial thickness of the wall. The function $\Delta p(y_R)$ is nonlinear and exists only if $\Delta p \leqslant \Delta p^*$. A fact of interest is a very weak dependence of the function $\widetilde{h}(\overset{\circ}{r})$ on the coefficient $\beta$ when $k = 1$; due to this, the functions $F(C)$ and $\Delta p(y_R)$ are also practically independent of the values of $\beta$ within the interval $[-1, 1]$. Moreover, values of the function $\Delta p(y_R)$ at $n = $ I and V, and also at $n = $ II and IV are practically not distinguishable in pairs. Therefore, when $k = 1$ there are only two essentially distinct functions $\Delta p(y_R)$ at $n = $ I and II (Fig. 6.28). Their corresponding ultimate values $\Delta p^*/\mu$ are 0.04 and 0.0072 while $\overset{\circ}{h}/\overset{\circ}{r}_1 = 0.01$. Ultimate magnitudes of the dilatation coefficient $y_R^*$ are 218 and 41%. If we consider the cylinder with a thicker wall $\overset{\circ}{h}$, then the ultimate values $y_R^*$ prove to be smaller.

In the case of compression when $\Delta p < 0$, the root $C_1$ is negative, and from Eq. (6.224) it follows that there exists a limiting value

$$C_1^* = -\overset{\circ}{r}{}_1^2/k, \tag{6.236}$$

such that the root $C_1$ cannot take on values smaller than $C_1^*$. Hence there also exists a limiting negative value of the pressure difference $\Delta p^{**}$ such that when $\Delta p < \Delta p^{**}$ there is no solution of the Lamé problem. Graphs of the function $\Delta p(y_R)$ in compression are shown in Fig. 6.29. Just as in tension, for the case when $k = 1$, the functions $\Delta p(y_R)$ at different values of the parameter $\beta$ are practically coincident, and they are distinct only for the models $n = $ I, V and $n = $ II, IV. In the case when $\overset{\circ}{h}/\overset{\circ}{r}_1 = 0.01$, we found the limiting value $\Delta p^{**}/\mu = -3.2$ for $n = $ I, V, and for $n = $ II, IV: $\Delta p^{**}/\mu = -0.33$. It should be noted that for real thin-walled structures at essentially smaller values of pressure in compression such that $|\Delta p|/\mu \ll |\Delta p^{**}|/\mu$, there occurs a loss of stability of the structure itself; thus, the values $\Delta p^{**}/\mu$ are usually not realized.

Figures 6.30 and 6.31 exhibit distributions of the stresses $\sigma_r/\mu$ and $\sigma_\varphi/\mu$ *versus* the cylinder thickness at different values of the parameter $\beta$ and different $n$ when $p_{e2} = 0$. The radial stress $\sigma_r(\bar{r})$, where $\bar{r} = \overset{\circ}{r}/\overset{\circ}{r}_1$, depends weakly on $\beta$ and

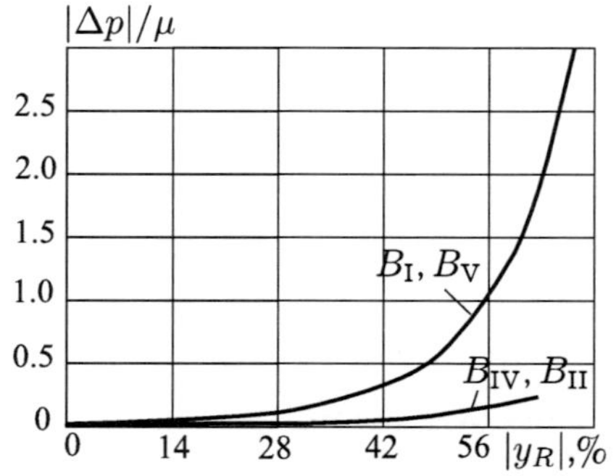


**Fig. 6.29** Dependence of the coefficient of relative compression of a thin-walled cylinder on external excess pressure for different models of incompressible materials ($k = 1, r_2/r_1 = 1.01$)

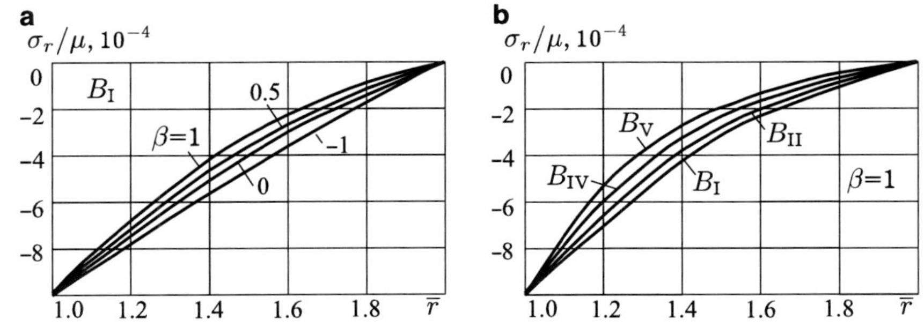


**Fig. 6.30** Distribution of the radial stresses *versus* the thickness of a thick-walled cylinder for the model $B_I$ at different values of $\beta$ (**a**) and for different models of incompressible material ($\beta = 1$) (**b**)

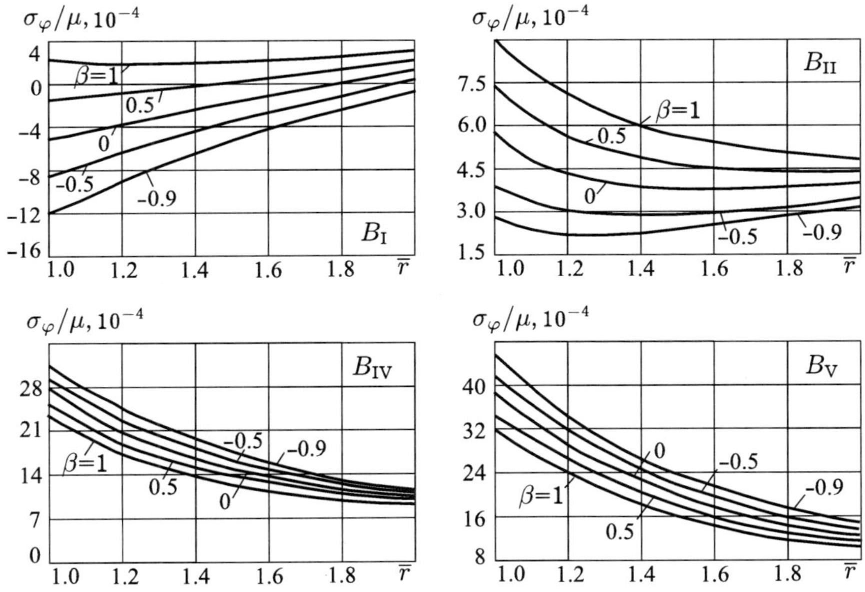


**Fig. 6.31** Distribution of the tangential stresses *versus* the thickness of a thick-walled cylinder for different models of incompressible material and different values of the parameter $\beta$

$n$; it monotonically decreases from the value $-p_{e1}$ to 0. The tangential stress $\sigma_\varphi(r)$ depends considerably more on $\beta$ and $n$ especially for thin-walled cylinders. So when $r_2/r_1 = 2$, for the model $B_I$ and $\beta = 1$ the stress $\sigma_\varphi$ is positive everywhere, and it reaches its minimum value at the interior point $r/r_1 \approx 1.2$ of the cylinder. When $\beta \leqslant 0.6$, the stress $\sigma_\varphi$ on the inner surface of the cylinder becomes negative, and when $\beta = -1$, the stress $\sigma_\varphi$ is negative in the whole cylinder. For the models $B_{II}$, $B_{IV}$ and $B_V$, the stress $\sigma_\varphi$ is always positive. For the model $B_{II}$, when $\beta = -1$,

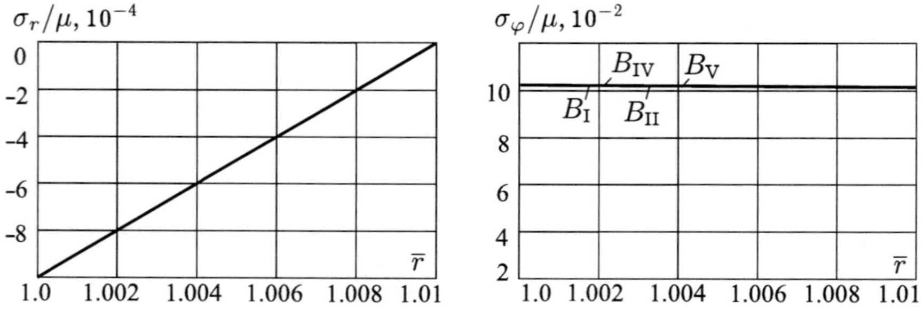


**Fig. 6.32** Distributions of radial and tangential stresses *versus* the thickness of a thin-walled cylinder for different models of incompressible material ($\beta = 1$)

the stress $\sigma_\varphi$ reaches its maximum on the outer surface of the cylinder; and for the models $B_{IV}$ and $B_V$, at all $\beta$ the function $\sigma_\varphi(\bar{r})$ is always monotonically decreasing and reaches its maximum on the outer surface.

For thin-walled cylinders, the stress $\sigma_\varphi$ is positive and practically constant *versus* the cylinder thickness; it is almost independent of $\beta$ and $n$, and its value is close to the value determined by the theory of thin linear-elastic shells at small deformations: $\sigma_\varphi \approx \Delta p \mathring{r}_1/\mathring{h}$ (Fig. 6.32).

# Chapter 7
# Continua of the Differential Type

## 7.1 Models $A_n$ and $B_n$ of Continua of the Differential Type

### *7.1.1 Constitutive Equations for Models $A_n$ of Continua of the Differential Type*

Let us consider now nonideal continua. Practically all real bodies are nonideal media, and they can be considered as ideal ones in a certain approximation. According to the general theory of constitutive equations stated in Sect. 4.4.2, a continuum is nonideal if its operator constitutive equations (4.156) include the dissipation function $w^*$ being nonzero.

Models of nonideal materials, which are widely used in practice, are models of *continua of the differential type*.

**Definition 7.1.** A continuum is called a *continuum of the differential type*, if corresponding operator relations (4.156) are usual functions of active variables $\mathcal{R}(t)$ and their derivatives $\dot{\mathcal{R}}(t)$, i.e.

$$\Lambda(t) = f\left(\mathcal{R}(t),\ \dot{\mathcal{R}}(t)\right). \tag{7.1}$$

Functions (7.1) are assumed to be continuously differentiable. One can say that this is the *model $A_n$ of a continuum of the differential type*, if some model $A_n$ of a continuum has been chosen and its corresponding operator constitutive equations (4.156) are simply functions of the arguments indicated above and their rates. In particular, the Helmholtz free energy has the form

$$\psi(t) = \psi(\overset{(n)}{\mathbf{C}}(t),\ \overset{(n)}{\mathbf{C}}{}^{\bullet}(t),\ \theta(t)). \tag{7.2}$$

(In continuum mechanics, the set of arguments of the function (7.2) does not involve the derivative $\dot{\theta}$.)

Y.I. Dimitrienko, *Nonlinear Continuum Mechanics and Large Inelastic Deformations*, Solid Mechanics and Its Applications 174, DOI 10.1007/978-94-007-0034-5_7, 

The total derivative of $\psi$ with respect to time has the form

$$\frac{d\psi}{dt} = \frac{\partial \psi}{\partial \overset{(n)}{\mathbf{C}}} \cdot\cdot \frac{d\overset{(n)}{\mathbf{C}}}{dt} + \frac{\partial \psi}{\partial \overset{(n)}{\mathbf{C}}{}^{\bullet}} \cdot\cdot \frac{d^2\overset{(n)}{\mathbf{C}}}{dt^2} + \frac{\partial \psi}{\partial \theta}\frac{d\theta}{dt}. \tag{7.3}$$

The partial derivatives $\partial\psi/\partial\overset{(n)}{\mathbf{C}}$ and $\partial\psi/\partial\overset{(n)}{\mathbf{C}}{}^{\bullet}$ are symmetric second-order tensors.

The remaining constitutive equations (4.158), connecting the active variables to the reactive ones, for the model $A_n$ of a continuum of the differential type have the forms

$$\overset{(n)}{\mathbf{T}} = \overset{(n)}{\mathbf{T}}(\overset{(n)}{\mathbf{C}}, \overset{(n)}{\mathbf{C}}{}^{\bullet}, \theta), \quad \eta = \eta(\overset{(n)}{\mathbf{C}}, \overset{(n)}{\mathbf{C}}{}^{\bullet}, \theta), \quad w^* = w^*(\overset{(n)}{\mathbf{C}}, \overset{(n)}{\mathbf{C}}{}^{\bullet}, \theta). \tag{7.4}$$

By analogy with the tensor function connecting the energetic tensors of stresses and deformations, introduce the two tensor functions

$$\overset{(n)}{\mathbf{T}}_e(\overset{(n)}{\mathbf{C}}, \theta) = \overset{(n)}{\mathbf{T}}(\overset{(n)}{\mathbf{C}}, \mathbf{0}, \theta), \tag{7.5}$$

and

$$\overset{(n)}{\mathbf{T}}_v = \overset{(n)}{\mathbf{T}} - \overset{(n)}{\mathbf{T}}_e, \tag{7.6}$$

called the *function of equilibrium stresses* and the *function of viscous stresses*, respectively.

Substituting the expressions (7.3) and (7.6) into PTI (4.121) and collecting like terms, we get

$$\left(\frac{\partial \psi}{\partial \overset{(n)}{\mathbf{C}}} - \frac{\overset{(n)}{\mathbf{T}}_e}{\rho}\right) \cdot\cdot\, d\overset{(n)}{\mathbf{C}} + \frac{\partial \psi}{\partial \overset{(n)}{\mathbf{C}}{}^{\bullet}} \cdot\cdot\, d\overset{(n)}{\mathbf{C}}{}^{\bullet} + \left(\frac{\partial \psi}{\partial \theta} + \eta\right) d\theta + \frac{1}{\rho}(w^* - \overset{(n)}{\mathbf{T}}_v \cdot\cdot \overset{(n)}{\mathbf{C}}{}^{\bullet})\, dt = 0. \tag{7.7}$$

Since the differentials $d\overset{(n)}{\mathbf{C}}$, $d\overset{(n)}{\mathbf{C}}{}^{\bullet}$, $d\theta$ and $dt$ are mutually independent, the identity (7.7) is equivalent to the equation system

$$\begin{cases} \overset{(n)}{\mathbf{T}}_e = \rho(\partial\psi/\partial\overset{(n)}{\mathbf{C}}) \equiv \mathcal{F}(\overset{(n)}{\mathbf{C}}, \theta), & (7.8a) \\ \partial\psi/\partial\overset{(n)}{\mathbf{C}}{}^{\bullet} = 0, & (7.8b) \\ \eta = -\partial\psi/\partial\theta, & (7.8c) \\ w^* = \overset{(n)}{\mathbf{T}}_v \cdot\cdot \overset{(n)}{\mathbf{C}}{}^{\bullet}. & (7.8d) \end{cases}$$

Just the fact that terms within the first parentheses of (7.7) are independent of $\overset{(n)}{\mathbf{C}}{}^{\bullet}$ ensures that the differentials $d\overset{(n)}{\mathbf{C}}$ and $d\overset{(n)}{\mathbf{C}}{}^{\bullet}$ be mutually independent. In turn, the presence of $\overset{(n)}{\mathbf{T}}_e$ instead of $\overset{(n)}{\mathbf{T}}$ within the first parentheses of (7.7) and the validity of Eq. (7.8b) ensure that terms within these parentheses be independent of $\overset{(n)}{\mathbf{C}}{}^{\bullet}$.

Thus, continua of the differential type have the following properties:

(1) They are dissipative (i.e. nonideal), because for them the dissipation function $w^*$ is not identically zero.

(2) The tensor function of equilibrium stresses $\overset{(n)}{\mathbf{T}}_e$ (but not $\overset{(n)}{\mathbf{T}}$) is quasipotential (see Sect. 4.5.2).

(3) The quasipotential $\psi$ depends only on $\overset{(n)}{\mathbf{C}}$ and $\theta$:

$$\psi = \psi(\overset{(n)}{\mathbf{C}}, \theta), \tag{7.9}$$

the remaining functions $\eta$, $\overset{(n)}{\mathbf{T}}_v$, $w^*$ depend on $\overset{(n)}{\mathbf{C}}$, $\overset{(n)}{\mathbf{C}}{}^{\bullet}$ and $\theta$.

According to the property (2), for models of the differential type it is not sufficient to specify only one function $\psi$ (7.9); in addition we need the viscous stresses function (7.6) to be given:

$$\overset{(n)}{\mathbf{T}}_v = \mathcal{F}_v(\overset{(n)}{\mathbf{C}}, \overset{(n)}{\mathbf{C}}{}^{\bullet}, \theta). \tag{7.10}$$

According to formulae (7.4)–(7.6), this function depends also on the deformation tensor rates $\overset{(n)}{\mathbf{C}}{}^{\bullet}$.

Relationships (7.8)–(7.10) are *constitutive equations for models* $A_n$ of *continua of the differential type*.

## 7.1.2 *Corollary of the Onsager Principle for Models $A_n$ of Continua of the Differential Type*

The tensor function of viscous stresses (7.10) is not quite arbitrary: it satisfies the conditions (7.5) and (7.6), i.e.

$$\mathcal{F}_v(\overset{(n)}{\mathbf{C}}, \mathbf{0}, \theta) = \mathbf{0}. \tag{7.11}$$

In addition, the function must satisfy the Onsager principle (see Sect. 4.12.1, Axiom 16). This principle is applied to continua of the differential type as follows.

First, we should form the scalar function (4.728) being the specific internal entropy production:

$$\rho q^* = w^* - \frac{\mathbf{q}}{\theta}\cdot\nabla\theta = \mathcal{F}_v(\overset{(n)}{\mathbf{C}}, \overset{(n)}{\mathbf{C}}{}^\bullet, \theta)\cdot\cdot\overset{(n)}{\mathbf{C}}{}^\bullet - \frac{\mathbf{q}}{\theta}\cdot\nabla\theta \geqslant 0. \tag{7.12}$$

Here we have used expression (7.8d) for the dissipation function $w^*$ of continua of the differential type. Then we represent the expression (7.12) in the form (4.729) with thermodynamic forces

$$\mathbf{X}_1 = \nabla\theta, \quad \mathbf{X}_2 = \overset{(n)}{\mathbf{C}}{}^\bullet \tag{7.13}$$

(where $\mathbf{X}_1$ is a vector, and $\mathbf{X}_2$ is a second-order tensor) and thermodynamic fluxes

$$\mathbf{Q}_1 = -\frac{1}{\theta}\mathbf{q}, \quad \mathbf{Q}_2 = \overset{(n)}{\mathbf{T}}_v = \mathcal{F}_v(\overset{(n)}{\mathbf{C}}, \overset{(n)}{\mathbf{C}}{}^\bullet, \theta). \tag{7.14}$$

According to the Onsager principle, thermodynamic fluxes $Q_\beta$ must be linear (tensor-linear) functions of $X_\beta$ in the form (4.730), i.e.

$$\begin{cases} -(1/\theta)\mathbf{q} = \mathbf{L}_{11}\cdot\nabla\theta + \mathbf{L}_{12}\overset{(n)}{\mathbf{C}}{}^\bullet, & (7.15)\\ \overset{(n)}{\mathbf{T}}_v = \mathcal{F}_v(\overset{(n)}{\mathbf{C}}, \overset{(n)}{\mathbf{C}}{}^\bullet, \theta) = \mathbf{L}_{12}\nabla\theta + \mathbf{L}_{22}\cdot\cdot\overset{(n)}{\mathbf{C}}{}^\bullet. & (7.16)\end{cases}$$

Here $\mathbf{L}_{11}$ is a second-order tensor, $\mathbf{L}_{12}$ is a third-order tensor, and $\mathbf{L}_{22}$ is a fourth-order tensor. Notice that since the viscous stresses function $\mathcal{F}_v$ must depend only on $\overset{(n)}{\mathbf{C}}{}^\bullet$ and $\overset{(n)}{\mathbf{C}}$ and may not depend upon the temperature gradient $\nabla\theta$ (because the opposite conflicts with the principle of equipresence), so in (7.16) we have $\mathbf{L}_{12} \equiv \mathbf{0}$; and hence we obtain the relationships

$$\mathbf{q} = -\boldsymbol{\lambda}\cdot\nabla\theta, \quad \boldsymbol{\lambda} \equiv \theta\mathbf{L}_{11}, \tag{7.17}$$

$$\mathcal{F}_v(\overset{(n)}{\mathbf{C}}, \overset{(n)}{\mathbf{C}}{}^\bullet, \theta) = {}^4\mathbf{L}_v\cdot\cdot\overset{(n)}{\mathbf{C}}{}^\bullet, \quad {}^4\mathbf{L}_v \equiv \mathbf{L}_{22}. \tag{7.18}$$

Equation (7.17) is the known Fourier law, which exactly coincides with the Fourier law (4.734) for ideal continua; and Eq. (7.18) is the quasilinear *Stokes law* for viscous stresses. The fourth-order tensor ${}^4\mathbf{L}_v$ is called the *viscosity tensor.*

Since $\mathbf{q}$ is independent of $\overset{(n)}{\mathbf{C}}{}^\bullet$ and $\mathcal{F}_v$ is independent of $\nabla\theta$, the Planck inequality (7.12) is equivalent to the two inequalities

$$-\mathbf{q}\cdot\nabla\theta = \nabla\cdot\boldsymbol{\lambda}\cdot\nabla\theta \geqslant 0, \tag{7.19}$$

$$w^* = \mathcal{F}_v\cdot\cdot\overset{(n)}{\mathbf{C}}{}^\bullet = \overset{(n)}{\mathbf{C}}{}^\bullet\cdot\cdot{}^4\mathbf{L}_v\cdot\cdot\overset{(n)}{\mathbf{C}}{}^\bullet \geqslant 0, \tag{7.20}$$

which are the Fourier inequality and the dissipation inequality for continua of the differential type, respectively. From these inequalities it follows that the tensors $\boldsymbol{\lambda}$ and ${}^4\mathbf{L}_v$ are nonnegative-definite. In addition, from (7.20) it follows that the dissipation function $w^*$ for continua of the differential type is a quadratic scalar function of $\overset{(n)}{\mathbf{C}}{}^\bullet$, and the viscosity tensor ${}^4\mathbf{L}_v$ has the following symmetry in components:

$$\begin{gathered}{}^4\mathbf{L}_v = \widehat{L}^{ijkl}\widehat{\mathbf{c}}_i \otimes \widehat{\mathbf{c}}_j \otimes \widehat{\mathbf{c}}_k \otimes \widehat{\mathbf{c}}_l, \\ \widehat{L}^{ijkl} = \widehat{L}^{ijlk}, \quad \widehat{L}^{ijkl} = \widehat{L}^{jikl}, \quad \widehat{L}^{ijkl} = \widehat{L}^{klij},\end{gathered} \tag{7.21}$$

where $\widehat{\mathbf{c}}_i$ is an orthonormal basis.

The total number of components of the tensor ${}^4\mathbf{L}_v$ is 81, and the number of independent components, according to Eqs. (7.21), does not exceed 21 (see [12]).

Notice that the viscosity tensor ${}^4\mathbf{L}_v$ and its components, in general, depend on the tensor $\overset{(n)}{\mathbf{C}}$ and also on $\overset{(n)}{\mathbf{C}}{}^\bullet$ (the Onsager principle (7.18) does not prohibit this dependence, it only requires that $\mathcal{F}_v$ be a quasilinear tensor function (see Sect. 4.8.7 and [12])).

If the tensor ${}^4\mathbf{L}_v$ is independent of $\overset{(n)}{\mathbf{C}}{}^\bullet$ and depends only on $\overset{(n)}{\mathbf{C}}$, then the function of viscous stresses $\mathcal{F}_v$ is *pseudopotential*, i.e. satisfies the relation

$$\overset{(n)}{\mathbf{T}}_v = \mathcal{F}_v(\overset{(n)}{\mathbf{C}}, \overset{(n)}{\mathbf{C}}{}^\bullet, \theta) = \frac{1}{2}(\partial w^* / \partial \overset{(n)}{\mathbf{C}}{}^\bullet). \tag{7.22}$$

Such function $\mathcal{F}_v$ is also called *potential with respect to the second tensor argument*.

Equation (7.22) can readily be verified by differentiating the scalar function $w^*$ (7.20) with respect to the argument $\overset{(n)}{\mathbf{C}}{}^\bullet$.

### *7.1.3 The Principle of Material Symmetry for Models $A_n$ of Continua of the Differential Type*

Notice that the principle of material symmetry (Axiom 14), and also Definitions 4.5 and 4.6 of fluids and solids, respectively, have been formulated for arbitrary operator constitutive equations (4.156); i.e. they can be applied not only to ideal materials but also to continua of the differential type and other types of nonideal materials.

According to this principle, models $A_n$ of continua of the differential type with constitutive equations (7.8)–(7.10) in some undistorted reference configuration $\widehat{\mathcal{K}}_s$ (as this configuration we will choose $\overset{\circ}{\mathcal{K}}$) have a symmetry group $\overset{\circ}{G}_s$ such that for each transformation tensor $\mathbf{H} \in \overset{\circ}{G}_s$ ($\mathbf{H} : \overset{\circ}{\mathcal{K}} \to \overset{*}{\mathcal{K}}$) the constitutive equations (7.8)–(7.10) in the reference configuration $\overset{*}{\mathcal{K}}$ have the form

$$\begin{cases} \overset{(n)}{\mathbf{T}}{}^*_e = \mathcal{F}(\overset{(n)}{\mathbf{C}}{}^*,\ \theta) \equiv \rho(\partial\psi/\partial\overset{(n)}{\mathbf{C}}{}^*), & (7.23a)\\ \psi = \psi(\overset{(n)}{\mathbf{C}}{}^*,\ \theta), & (7.23b)\\ \eta = -\partial\psi/\partial\theta, & (7.23c)\\ (w^*)^* = \overset{(n)}{\mathbf{T}}{}^*_v \cdot\cdot\ \overset{(n)}{\mathbf{C}}{}^{*\bullet}, & (7.23d)\\ \overset{(n)}{\mathbf{T}}{}^*_v = \mathcal{F}_v(\overset{(n)}{\mathbf{C}}{}^*,\ \overset{(n)}{\mathbf{C}}{}^{*\bullet},\ \theta). & (7.23e) \end{cases}$$

Let us consider first the *models $A_n$ of solids of the differential type*, whose symmetry group $\overset{\circ}{G}_s$ is a subgroup of the full orthogonal group $I$. For these materials, the further construction is the same as in Sect. 4.8 for ideal solids: since the tensors $\overset{(n)}{\mathbf{T}}$ and $\overset{(n)}{\mathbf{C}}$ are $H$-indifferent relative to the group $\overset{\circ}{G}_s$, tensors $\overset{(n)}{\mathbf{T}}_e$ (7.5), $\overset{(n)}{\mathbf{T}}_v$ (7.6) and $\overset{(n)}{\mathbf{C}}{}^\bullet$ are also $H$-indifferent, according to Theorem 4.21. Then Eqs. (7.23) are equivalent to the relations

$$\overset{*}{\mathbf{Q}}{}^{\mathrm{T}} \cdot \mathcal{F}(\overset{(n)}{\mathbf{C}},\ \theta) \cdot \overset{*}{\mathbf{Q}} = \mathcal{F}(\overset{*}{\mathbf{Q}}{}^{\mathrm{T}} \cdot \overset{(n)}{\mathbf{C}} \cdot \overset{*}{\mathbf{Q}},\ \theta), \tag{7.24a}$$

$$\psi(\overset{(n)}{\mathbf{C}}, \theta) = \psi(\overset{*}{\mathbf{Q}}{}^{\mathrm{T}} \cdot \overset{(n)}{\mathbf{C}} \cdot \overset{*}{\mathbf{Q}},\ \theta), \tag{7.24b}$$

$$\overset{*}{\mathbf{Q}}{}^{\mathrm{T}} \cdot \mathcal{F}_v(\overset{(n)}{\mathbf{C}},\ \overset{(n)}{\mathbf{C}}{}^\bullet,\ \theta) \cdot \overset{*}{\mathbf{Q}} = \mathcal{F}_v(\overset{*}{\mathbf{Q}}{}^{\mathrm{T}} \cdot \overset{(n)}{\mathbf{C}} \cdot \overset{*}{\mathbf{Q}},\ \overset{*}{\mathbf{Q}}{}^{\mathrm{T}} \cdot \overset{(n)}{\mathbf{C}}{}^\bullet \cdot \overset{*}{\mathbf{Q}},\ \theta), \tag{7.24c}$$

for any $\overset{*}{\mathbf{Q}} \in \overset{\circ}{G}_s$.

Since

$$(w^*)^* = \overset{(n)}{\mathbf{T}}{}^*_v \cdot\cdot\ \overset{(n)}{\mathbf{C}}{}^{*\bullet} = \overset{*}{\mathbf{Q}}{}^{\mathrm{T}} \cdot \overset{(n)}{\mathbf{T}}_v \cdot \overset{*}{\mathbf{Q}} \cdot\cdot\ \overset{*}{\mathbf{Q}}{}^{\mathrm{T}} \cdot \overset{(n)}{\mathbf{C}}{}^\bullet \cdot \overset{*}{\mathbf{Q}} = \overset{(n)}{\mathbf{T}}_v \cdot\cdot \overset{(n)}{\mathbf{C}}{}^\bullet = w^*, \tag{7.25}$$

Eq. (7.23d) is always satisfied.

**Theorem 7.1.** *For models $A_n$ of solids of the differential type, the principle of material symmetry (7.24) holds if and only if the two conditions (7.24b) and (7.24c) for the function $\psi$ and the viscous stresses function $\mathcal{F}_v$ are satisfied.*

▼ The conditions (7.24b) and (7.24c) are necessary, because if the principle of material symmetry holds, then all conditions (7.24) are satisfied.

Prove that the conditions (7.24b) and (7.24c) are sufficient. If the condition (7.24b) is satisfied, then (7.24a) follows from (7.24b); a proof of this fact is similar to the proof of formula (4.280). ▲

### 7.1.4 Representation of Constitutive Equations for Models $A_n$ of Solids of the Differential Type in Tensor Bases

In comparison with ideal continua, for solids of the differential type not only the condition (7.24b), being the condition of indifference of the scalar function $\psi(\overset{(n)}{\mathbf{C}}, \theta)$ relative to some group $\overset{\circ}{G}_s$ (see Definition 4.9), must be satisfied but also the condition (7.24c) for the tensor function $\mathcal{F}_v(\overset{(n)}{\mathbf{C}}, \overset{(n)}{\mathbf{C}}{}^{\bullet}, \theta)$ of two tensor arguments.

For the scalar function $\psi(\overset{(n)}{\mathbf{C}}, \theta)$, the same representations as were derived in Sect. 4.8 remain valid; in particular, these are representations (4.304) as functions of invariants $I_\gamma^{(s)}(\overset{(n)}{\mathbf{C}})$ relative to group $\overset{\circ}{G}_s$:

$$\psi = \psi(I_1^{(s)}(\overset{(n)}{\mathbf{C}}), \ldots, I_r^{(s)}(\overset{(n)}{\mathbf{C}}), \theta), \tag{7.26}$$

and also representations (4.311) and (4.324) for the tensor function $\mathcal{F}(\overset{(n)}{\mathbf{C}}, \theta)$ (7.8a).

For example, for an isotropic solid of the differential type, from (4.322) we obtain the expression for equilibrium stresses

$$\overset{(n)}{\mathbf{T}}_e = \psi_1 \mathbf{E} + \psi_2 \overset{(n)}{\mathbf{C}} + \psi_3 \overset{(n)}{\mathbf{C}}{}^2, \tag{7.27}$$

where $\psi_\gamma$ are determined by formulae (4.322a).

By analogy with Definition 4.13 (see Sect. 4.8.6) for a function of one tensor argument, the function $\mathcal{F}_v(\overset{(n)}{\mathbf{C}}, \overset{(n)}{\mathbf{C}}{}^{\bullet}, \theta)$ of two tensor arguments, which satisfies the condition (7.24c), is called *indifferent relative to the group* $\overset{\circ}{G}_s$ (see [12]). The following theorem gives representations of this function in tensor bases.

**Theorem 7.2.** *Any tensor function of viscous stresses (7.10), which is*

- *indifferent relative to some orthogonal group $\overset{\circ}{G}_s \subset I$, i.e. satisfies the condition (7.24c),*
- *quasilinear in the second argument $\overset{(n)}{\mathbf{C}}{}^{\bullet}$, i.e. satisfies the conditions (7.18) and (7.20),*

*can be represented in the tensor basis of the corresponding group $\overset{\circ}{G}_s$:*

$$\overset{(n)}{\mathbf{T}}_v = \mathcal{F}_v(\overset{(n)}{\mathbf{C}}, \overset{(n)}{\mathbf{C}}{}^{\bullet}, \theta) = (\mu_1 \mathbf{E} \otimes \mathbf{E} + 2\mu_2 \mathbf{\Delta}) \cdot\cdot \overset{(n)}{\mathbf{C}}{}^{\bullet}, \tag{7.28}$$

*or*

$$\overset{(n)}{\mathbf{T}}_v = \mu_1 I_1(\overset{(n)}{\mathbf{C}}{}^{\bullet})\mathbf{E} + 2\mu \overset{(n)}{\mathbf{C}}{}^{\bullet}, \tag{7.29}$$

—*for an isotropic continuum of the differential type* ($\overset{\circ}{G}_s = I$),

$$\overset{(n)}{\mathbf{T}}_v = (\mu_1 \mathbf{E} \otimes \mathbf{E} + \mu_2 \widehat{\mathbf{c}}_3^2 \otimes \widehat{\mathbf{c}}_3 + \mu_3(\mathbf{E} \otimes \widehat{\mathbf{c}}_3^2 + \widehat{\mathbf{c}}_3^2 \otimes \mathbf{E})$$
$$+ \mu_4(\mathbf{O}_1 \otimes \mathbf{O}_1 + \mathbf{O}_2 \otimes \mathbf{O}_2) + 2\mu_5 \boldsymbol{\Delta}) \cdot\cdot \overset{(n)}{\mathbf{C}}{}^{\bullet} \quad (7.30)$$

—*for a transversely isotropic continuum of the differential type* ($\overset{\circ}{G}_s = T_3$),

$$\overset{(n)}{\mathbf{T}}_v = \sum_{\alpha=1}^{3} (\mu_\alpha \widehat{\mathbf{c}}_\alpha^2 \otimes \widehat{\mathbf{c}}_\alpha + \mu_{3+\alpha}(\widehat{\mathbf{c}}_\beta^2 \otimes \widehat{\mathbf{c}}_\gamma^2 + \widehat{\mathbf{c}}_\gamma^2 \otimes \widehat{\mathbf{c}}_\beta^2)$$
$$+ \mu_{6+\alpha} \mathbf{O}_\alpha \otimes \mathbf{O}_\alpha) \cdot\cdot \overset{(n)}{\mathbf{C}}{}^{\bullet}, \quad \alpha \neq \beta \neq \gamma \neq \alpha, \; \alpha, \; \beta, \; \gamma = 1, \; 2, \; 3, \quad (7.31)$$

—*for an orthotropic continuum of the differential type* ($\overset{\circ}{G}_s = O$).

The tensors $\widehat{\mathbf{c}}_\alpha^2$ and $\mathbf{O}_\alpha$ are determined by formulae (4.292) and (4.315).
▼ Substituting relation (7.8) of quasilinearity of the viscous stresses function $\mathcal{F}_v$ into (7.24c), we obtain that the following condition must be satisfied:

$$\overset{*}{\mathbf{Q}}{}^{\mathrm{T}} \cdot ({}^4\mathbf{L}_v \cdot\cdot \overset{(n)}{\mathbf{C}}{}^{\bullet}) \cdot \overset{*}{\mathbf{Q}} = {}^4\mathbf{L}_v \cdot\cdot (\overset{*}{\mathbf{Q}}{}^{\mathrm{T}} \cdot \overset{(n)}{\mathbf{C}}{}^{\bullet} \overset{*}{\mathbf{Q}}) \quad \forall \overset{(n)}{\mathbf{C}}{}^{\bullet}. \quad (7.32)$$

Hence the viscosity tensor ${}^4\mathbf{L}_v$ must satisfy the equation (see Exercise 7.1.1)

$${}^4\mathbf{L}_v = {}^4\mathbf{L}_v \cdot\cdot\cdot\cdot (\overset{*}{\mathbf{Q}} \otimes \overset{*}{\mathbf{Q}} \otimes \overset{*}{\mathbf{Q}} \otimes \overset{*}{\mathbf{Q}})^{(57312468)} \quad \forall \overset{*}{\mathbf{Q}} \in \overset{\circ}{G}_s. \quad (7.32a)$$

This condition is an analog of the condition (4.289) for fourth-order tensors. Therefore, a fourth-order tensor ${}^4\mathbf{L}_v$ satisfying Eq. (7.32a) is called *indifferent relative to the group* $\overset{\circ}{G}_s$. All indifferent fourth-order tensors are formed with the help of the operations of tensor and scalar products by producing tensors of a corresponding group $\overset{\circ}{G}_s$ (see Sect. 4.8.3) and can be resolved for the tensor basis ${}^4\mathbf{O}_{s(\gamma)}$ by analogy with the resolution (4.291) of second-order tensors: ${}^4\mathbf{L}_v = \sum_{\gamma=1}^{k} a_\gamma {}^4\mathbf{O}_{s(\gamma)}$. The number $k$ of elements of this basis coincides with the number of independent components of the tensor ${}^4\mathbf{L}_v$; it is equal to 2, 5, and 9 for the groups $\overset{\circ}{G}_s = I, \; T_3, \; O$, respectively (see [12]). The elastic moduli tensors ${}^4\mathbf{M}$ defined by formulae (4.337), (4.341), and (4.344) are also indifferent relative to the groups $\overset{\circ}{G}_s = I, \; T_3, \; O$, respectively; and the representations (4.337), (4.341), and (4.344) are the desired resolutions for the tensor bases, being the same for each group [12]. Hence, the viscosity tensor can always be represented in the similar form for the corresponding group $\overset{\circ}{G}_s = I, \; T_3, \; O$ too. Thus, the representations (7.28)–(7.31) actually hold. ▲

*Remark.* The coefficients $\mu_\gamma$ in representations (7.28)–(7.31) are, in general, scalar functions of corresponding invariants of the tensors $\overset{(n)}{\mathbf{C}}$ and $\overset{(n)}{\mathbf{C}}{}^{\bullet}$:

$$\mu_\gamma = \mu_\gamma(\overset{(n)}{\mathbf{C}},\ \overset{(n)}{\mathbf{C}}{}^{\bullet},\ \theta). \tag{7.33}$$

But since we have assumed everywhere that $\mu_\gamma$ remain unchanged under $H$-transformations in a corresponding group, so the coefficients $\mu_\gamma$ must satisfy the relations

$$\mu_\gamma(\overset{(n)}{\mathbf{C}},\ \overset{(n)}{\mathbf{C}}{}^{\bullet},\ \theta) = \mu_\gamma(\overset{*}{\mathbf{Q}}{}^{\mathrm{T}} \cdot \overset{(n)}{\mathbf{C}} \cdot \overset{*}{\mathbf{Q}},\ \overset{*}{\mathbf{Q}}{}^{\mathrm{T}} \cdot \overset{(n)}{\mathbf{C}}{}^{\bullet} \cdot \overset{*}{\mathbf{Q}},\ \theta) \quad \forall \overset{*}{\mathbf{Q}} \in \overset{\circ}{G}_s. \tag{7.34}$$

Such scalar functions are called *simultaneous invariants of two tensor arguments relative to a group* $\overset{\circ}{G}_s$ considered. According to the theorem proved in [12], for each group $\overset{\circ}{G}_s$ there is a functional basis of independent simultaneous invariants $J_\gamma^{(s)}(\overset{(n)}{\mathbf{C}}, \overset{(n)}{\mathbf{C}}{}^{\bullet})$ $(\gamma = 1, \ldots, z)$, where

- $z = 9$ for the full orthogonal group $I$,
- $z = 11$ for the transverse isotropy group $T_3$,
- $z = 12$ for the orthotropy group $O$.

As a functional basis of simultaneous invariants $J_\gamma^{(s)}(\overset{(n)}{\mathbf{C}},\ \overset{(n)}{\mathbf{C}}{}^{\bullet})$ of two tensors we can choose the following sets [12]:
for the isotropy group $\overset{\circ}{G}_s = I$:

$$J_\alpha^{(I)} = I_\alpha(\overset{(n)}{\mathbf{C}}), \quad J_{3+\alpha}^{(I)} = I_\alpha(\overset{(n)}{\mathbf{C}}{}^{\bullet}), \quad \alpha = 1,\ 2,\ 3,$$
$$J_7^{(I)} = \overset{(n)}{\mathbf{C}} \cdot\cdot \overset{(n)}{\mathbf{C}}{}^{\bullet}, \quad J_8^{(I)} = \overset{(n)}{\mathbf{C}}{}^2 \cdot\cdot \overset{(n)}{\mathbf{C}}{}^{\bullet}, \quad J_9^{(I)} = \overset{(n)}{\mathbf{C}} \cdot\cdot (\overset{(n)}{\mathbf{C}}{}^{\bullet})^2; \tag{7.35}$$

for the group $\overset{\circ}{G}_s = T_3$:

$$J_\alpha^{(3)} = I_\alpha^{(3)}(\overset{(n)}{\mathbf{C}}), \quad \alpha = 1, \ldots, 5, \quad J_{5+\beta}^{(3)} = I_\beta^{(3)}(\overset{(n)}{\mathbf{C}}{}^{\bullet}), \quad \beta = 1, \ldots, 4,$$
$$J_{10}^{(3)} = ((\mathbf{E} - \widetilde{\mathbf{c}}_3^2) \cdot \overset{(n)}{\mathbf{C}}) \cdot\cdot (\widetilde{\mathbf{c}}_3^2 \cdot \overset{(n)}{\mathbf{C}}{}^{\bullet}), \quad J_{11}^{(3)} = \overset{(n)}{\mathbf{C}} \cdot\cdot \overset{(n)}{\mathbf{C}}{}^{\bullet} - 2J_{10}^{(3)} - J_2^{(3)} J_8^{(3)}; \tag{7.36}$$

for the group $\overset{\circ}{G}_s = O$:

$$J_\alpha^{(O)} = I_\alpha^{(O)}(\overset{(n)}{\mathbf{C}}), \quad \alpha = 1, \ldots, 6; \quad J_{6+\beta}^{(O)} = I_\beta^{(O)}(\overset{(n)}{\mathbf{C}}{}^{\bullet}), \quad \beta = 1,\ 2,\ 3,\ 6;$$
$$J_{10}^{(O)} = (\widetilde{\mathbf{c}}_2^2 \cdot \overset{(n)}{\mathbf{C}}) \cdot\cdot (\widetilde{\mathbf{c}}_3^2 \cdot \overset{(n)}{\mathbf{C}}{}^{\bullet}), \quad J_{11}^{(O)} = (\widetilde{\mathbf{c}}_1^2 \cdot \overset{(n)}{\mathbf{C}}) \cdot\cdot (\widetilde{\mathbf{c}}_3^2 \cdot \overset{(n)}{\mathbf{C}}{}^{\bullet}). \tag{7.37}$$

Invariants $I_\alpha^{(s)}$ of corresponding groups are determined by formulae (4.295)–(4.297). Then the *viscous coefficients* can always be represented as functions of simultaneous invariants:

$$\mu_\gamma = \mu_\gamma(J_1^{(s)}(\overset{(n)}{\mathbf{C}}, \overset{(n)}{\mathbf{C}}{}^\bullet), \ldots, J_z^{(s)}(\overset{(n)}{\mathbf{C}}, \overset{(n)}{\mathbf{C}}{}^\bullet), \theta). \tag{7.38}$$

The dissipation function (7.20) for an isotropic medium of the differential type, according to (7.29), has the form

$$w^* = \mu_1 I_1^2(\overset{(n)}{\mathbf{C}}{}^\bullet) + \mu_2 I_1(\overset{(n)}{\mathbf{C}}{}^{\bullet 2}). \ \square \tag{7.39}$$

### 7.1.5 Models $B_n$ of Solids of the Differential Type

For solids of the differential type, models $B_n$ can be obtained formally from corresponding models $A_n$ by replacing the tensors $\overset{(n)}{\mathbf{C}}$ with the measures $\overset{(n)}{\mathbf{G}}$. In particular, constitutive equations (7.8)–(7.10) become

$$\begin{cases} \overset{(n)}{\mathbf{T}}_e = \mathcal{F}(\overset{(n)}{\mathbf{G}}, \theta) = \rho(\partial\psi/\partial\overset{(n)}{\mathbf{G}}), \\ \psi = \psi(\overset{(n)}{\mathbf{G}}, \theta), \\ w^* = \overset{(n)}{\mathbf{T}}_v \cdot\cdot \overset{(n)}{\mathbf{G}}{}^\bullet, \\ \overset{(n)}{\mathbf{T}}_v = \mathcal{F}_v(\overset{(n)}{\mathbf{G}}, \overset{(n)}{\mathbf{G}}{}^\bullet, \theta). \end{cases} \tag{7.40}$$

For models $B_n$ of isotropic continua of the differential type, the tensors of equilibrium and viscous stresses have the forms

$$\overset{(n)}{\mathbf{T}}_e = \psi_1 \mathbf{E} + \psi_2 \overset{(n)}{\mathbf{G}} + \psi_3 \overset{(n)}{\mathbf{G}}{}^2, \tag{7.41}$$

$$\overset{(n)}{\mathbf{T}}_v = \mu_1 I_1(\overset{(n)}{\mathbf{G}}{}^\bullet)\mathbf{E} + 2\mu_2 \overset{(n)}{\mathbf{G}}{}^\bullet, \tag{7.42}$$

where the coefficients $\psi_\gamma$ are expressed by formulae (4.322), and $\mu_\gamma$ are functions of invariants:

$$\mu_\gamma = \mu_\gamma(J_1(\overset{(n)}{\mathbf{G}}, \overset{(n)}{\mathbf{G}}{}^\bullet, \theta), \ldots, J_{10}(\overset{(n)}{\mathbf{G}}, \overset{(n)}{\mathbf{G}}{}^\bullet, \theta)). \tag{7.43}$$

In a similar way, from (7.28) and (7.29) we get equations for models $B_n$ of transversely isotropic and orthotropic materials of the differential type.

### 7.1.6 Models $B_n$ of Incompressible Continua of the Differential Type

Similarly to models $B_n$ for elastic incompressible continua (see Sect. 4.9), we can introduce models $B_n$ for incompressible continua of the differential type. As shown in Sect. 4.9, for these models the potential $\psi$ in (7.40) depends only on $r-1$ linear and quadratic invariants $I_\alpha^{(s)}(\overset{(n)}{\mathbf{G}})$. In particular, for *an isotropic incompressible continuum of the differential type*, $\psi$ depends only on $I_1(\overset{(n)}{\mathbf{G}})$ and $I_2(\overset{(n)}{\mathbf{G}})$, and $\overset{(n)}{\mathbf{T}}_e$ is a quasilinear function of $\overset{(n)}{\mathbf{G}}$:

$$\overset{(n)}{\mathbf{T}} = -\frac{p}{n-\mathrm{III}}\overset{(n)}{\mathbf{G}}{}^{-1} + \psi_1\mathbf{E} + \psi_2\overset{(n)}{\mathbf{G}} + \mu_1 I_1(\overset{(n)}{\mathbf{G}}{}^{\bullet})\mathbf{E} + 2\mu_2\overset{(n)}{\mathbf{G}}{}^{\bullet},$$

$$\psi = \psi(I_1(\overset{(n)}{\mathbf{G}}),\ I_2(\overset{(n)}{\mathbf{G}}),\ \theta),\quad \psi_1 = \varphi_1 + \varphi_2 I_1,\quad \psi_2 = \varphi_2,\quad \varphi_\alpha = \overset{\circ}{\rho}(\partial\psi/\partial I_\alpha). \tag{7.44}$$

For the *simplest model* $B_n$ of isotropic incompressible continua of the differential type, we assume that the viscous coefficients are connected by the relation

$$\mu_1 = -\frac{2}{3}\mu_2. \tag{7.45}$$

Then the constitutive equation (7.44) takes the form

$$\overset{(n)}{\mathbf{T}} = -\frac{p}{n-\mathrm{III}}\overset{(n)}{\mathbf{G}}{}^{-1} + \psi_1\mathbf{E} + \psi_2\overset{(n)}{\mathbf{G}} + 2\mu_2\,\mathrm{dev}\,\overset{(n)}{\mathbf{G}}{}^{\bullet}, \tag{7.46}$$

where

$$\mathrm{dev}\,\overset{(n)}{\mathbf{G}} = \overset{(n)}{\mathbf{G}} - \frac{1}{3}I_1(\overset{(n)}{\mathbf{G}})\mathbf{E} \tag{7.47}$$

is the deviator of the tensor (the more detailed information on deviators can be found in Sect. 8.2.13 and [12]).

For the simplest models $B_n$, the first principal invariants of the stress tensor $\overset{(n)}{\mathbf{T}}$ and the deformation measures $\overset{(n)}{\mathbf{G}}$ are connected by the elastic relations (see Exercise 7.1.4)

$$I_1(\overset{(n)}{\mathbf{T}}) = -\frac{p}{n-\mathrm{III}}I_1(\overset{(n)}{\mathbf{G}}{}^{-1}) + 3\psi_1 + \psi_2 I_1(\overset{(n)}{\mathbf{G}}), \tag{7.48}$$

being independent of the deformation rates.

When solving problems in practice, one usually applies *models $B_n$ of incompressible continua of the differential type with steady creep*, where the potential $\psi$ is independent of the invariants $I_\alpha(\overset{(n)}{\mathbf{G}})$:

$$\overset{(n)}{\mathbf{T}} = -\frac{p}{n-\mathrm{III}}\overset{(n)}{\mathbf{G}}{}^{-1} + 2\mu_2 \operatorname{dev} \overset{(n)}{\mathbf{G}}{}^{\bullet}, \quad \psi = \psi(\theta), \tag{7.49}$$

This model describes the phenomenon of *creep* being a change in deformation of a body with time at constant stresses (see Sect. 7.4).

*Remark.* The consistency conditions (4.328), which must be satisfied by constitutive equations at a natural unstressed state, for materials of the differential type should be complemented by the requirement that the rates of the deformation tensors and measures vanish in $\overset{*}{\mathcal{K}}$:

$$\overset{\circ}{\mathcal{K}}: \quad \overset{(n)}{\mathbf{T}} = \overset{(n)}{\mathbf{S}} = \mathbf{T} = \mathbf{0}, \quad \overset{(n)}{\mathbf{A}} = \overset{(n)}{\mathbf{C}} = \mathbf{0}, \quad \overset{(n)}{\mathbf{g}} = \overset{(n)}{\mathbf{G}} = \frac{1}{n-\mathrm{III}}\mathbf{E},$$

$$\overset{(n)}{\mathbf{A}}{}^{\bullet} = \overset{(n)}{\mathbf{C}}{}^{\bullet} = \mathbf{0}, \quad \overset{(n)}{\mathbf{g}}{}^{\bullet} = \overset{(n)}{\mathbf{G}}{}^{\bullet} = \mathbf{0}, \quad \theta = \theta_0, \quad \psi = \psi_0. \tag{7.50}$$

All the constitutive equations (7.29), (7.30), (7.31), (7.41), (7.42), (7.44), (7.46), and (7.49) derived above satisfy these conditions. □

### 7.1.7 *The Principle of Material Indifference for Models $A_n$ and $B_n$ of Continua of the Differential Type*

Since all the energetic tensors $\overset{(n)}{\mathbf{T}}$, $\overset{(n)}{\mathbf{C}}$ and $\overset{(n)}{\mathbf{G}}$ are $R$-invariant, the constitutive equations (7.8)–(7.10) and also (7.28)–(7.31) for solids of the differential type are the same in actual configurations $\mathcal{K}$ and $\mathcal{K}'$ obtained one from another by a rigid motion. Therefore, the principle of material indifference for models $A_n$ and $B_n$ of solids of the differential type (as well as for elastic continua) is satisfied identically.

## Exercises for 7.1

**7.1.1.** Show equivalence of relations (7.32) and (7.32a).

**7.1.2.** Show that the component representations of functional bases of simultaneous invariants (7.35)–(7.37) in the basis $\widehat{\mathbf{c}}_\alpha$ have the forms for the transverse isotropy group $\overset{\circ}{G}_s = T_3$:

$$J^{(3)}_\alpha = \{\overset{\widehat{(n)}}{C}_{11} + \overset{\widehat{(n)}}{C}_{22},\ \overset{\widehat{(n)}}{C}_{33},\ \overset{\widehat{(n)}}{C}{}^2_{13} + \overset{\widehat{(n)}}{C}{}^2_{23},\ \overset{\widehat{(n)}}{C}{}^2_{11} + \overset{\widehat{(n)}}{C}{}^2_{22} + 2\overset{\widehat{(n)}}{C}{}^2_{12},\ \det \mathbf{C},$$

$$\overset{\widehat{(n)}}{C}{}^\bullet_{11} + \overset{\widehat{(n)}}{C}{}^\bullet_{22},\ \overset{\widehat{(n)}}{C}{}^\bullet_{33},\ (\overset{\widehat{(n)}}{C}{}^\bullet_{13})^2 + (\overset{\widehat{(n)}}{C}{}^\bullet_{23})^2,\ (\overset{\widehat{(n)}}{C}{}^\bullet_{11})^2 + (\overset{\widehat{(n)}}{C}{}^\bullet_{22})^2 + 2(\overset{\widehat{(n)}}{C}{}^\bullet_{12})^2,$$

$$\overset{\widehat{(n)}}{C}_{13}\overset{\widehat{(n)}}{C}{}^\bullet_{13} + \overset{\widehat{(n)}}{C}_{23}\overset{\widehat{(n)}}{C}{}^\bullet_{23},\ \overset{\widehat{(n)}}{C}_{11}\overset{\widehat{(n)}}{C}{}^\bullet_{11} + \overset{\widehat{(n)}}{C}_{22}\overset{\widehat{(n)}}{C}{}^\bullet_{22} + 2\overset{\widehat{(n)}}{C}_{12}\overset{\widehat{(n)}}{C}{}^\bullet_{12}\};$$

for the orthotropy group $\overset{\circ}{G}_s = O$:

$$J^{(O)}_\alpha = \{\overset{\widehat{(n)}}{C}_{11},\ \overset{\widehat{(n)}}{C}_{22},\ \overset{\widehat{(n)}}{C}_{33},\ \overset{\widehat{(n)}}{C}{}^2_{23},\ \overset{\widehat{(n)}}{C}{}^2_{13},\ \overset{\widehat{(n)}}{C}_{12}\overset{\widehat{(n)}}{C}_{13}\overset{\widehat{(n)}}{C}_{23},\ \overset{\widehat{(n)}}{C}{}^\bullet_{11},\ \overset{\widehat{(n)}}{C}{}^\bullet_{22},\ \overset{\widehat{(n)}}{C}{}^\bullet_{33},$$

$$\overset{\widehat{(n)}}{C}_{23}\overset{\widehat{(n)}}{C}{}^\bullet_{23},\ \overset{\widehat{(n)}}{C}_{13}\overset{\widehat{(n)}}{C}{}^\bullet_{13},\ \overset{\widehat{(n)}}{C}{}^\bullet_{12}\overset{\widehat{(n)}}{C}{}^\bullet_{13}\overset{\widehat{(n)}}{C}{}^\bullet_{23}\}.$$

**7.1.3.** Show that for the simplest model $B_n$ of isotropic incompressible continua of the differential type (7.46), the relation (7.48) between the first principal invariants holds.

## 7.2 Models $A_n$ and $B_n$ of Fluids of the Differential Type

### 7.2.1 *Tensor of Equilibrium Stresses for Fluids of the Differential Type*

Let us use the principle of material symmetry (7.23) to derive constitutive equations for fluids of the differential type.

For fluids, the symmetry group $\overset{\circ}{G}_s$, relative to which Eqs. (7.23) hold, is the unimodular group $\overset{\circ}{G}_s = U$. Tensors $\overset{(n)}{\mathbf{T}}$, $\overset{(n)}{\mathbf{C}}$ and $\overset{(n)}{\mathbf{C}}{}^\bullet$ are no longer $H$-indifferent relative to this group, except $\underset{V}{\mathbf{T}}$ and the tensor $\underset{I}{\mathbf{G}} = \underset{I}{\mathbf{C}} - (1/2)\mathbf{E}$ (see Table 4.9); therefore, relations (7.24) for $\overset{\circ}{G}_s = U$ do not hold and one should use their general form (7.23).

Due to Theorem 4.31, from these relations it follows that the function $\mathcal{F}(\overset{(n)}{\mathbf{C}}, \theta)$ has the form (4.443a)

$$\overset{(n)}{\mathbf{T}}_e = \mathcal{F}(\overset{(n)}{\mathbf{C}}, \theta) = -\frac{p}{n - \mathrm{III}} \overset{(n)}{\mathbf{G}}{}^{-1}, \tag{7.51}$$

$$p = p(I_3(\overset{(n)}{\mathbf{G}}),\ \theta), \tag{7.52}$$

where $p$ is the pressure being a scalar function only of the third invariant of the tensor $\overset{(n)}{\mathbf{G}}$. Thus, for fluids, the tensor $\overset{(n)}{\mathbf{C}}$ appears in models $A_n$ only in the combination $\overset{(n)}{\mathbf{G}} = \overset{(n)}{\mathbf{C}} + \frac{1}{n-\mathrm{III}}\,\mathbf{E}$, i.e. in fact, models $A_n$ and $B_n$ are coincident.

For the tensor of viscous stresses, the situation is analogous.

### 7.2.2 *The Tensor of Viscous Stresses in the Model $A_{\mathrm{I}}$ of a Fluid of the Differential Type*

Let us consider further only the models $A_{\mathrm{I}}$ and $A_{\mathrm{V}}$. The tensors $\overset{\mathrm{I}}{\mathbf{T}}$ and $\overset{\mathrm{I}}{\mathbf{G}}$ are transformed during the passage to configuration $\overset{*}{\mathcal{K}}$ according to formulae (4.201) and (4.203a). On substituting these formulae into (7.23e), we obtain that for the group $\overset{\circ}{G}_s = U$ the following equation must hold:

$$\mathbf{H}^{-1\mathrm{T}} \cdot \mathcal{F}_v(\overset{\mathrm{I}}{\mathbf{G}},\ \overset{\mathrm{I}}{\mathbf{G}}{}^{\bullet},\ \theta) \cdot \mathbf{H}^{-1} = \mathcal{F}_v(\mathbf{H} \cdot \overset{\mathrm{I}}{\mathbf{G}} \cdot \mathbf{H}^{\mathrm{T}},\ \mathbf{H} \cdot \overset{\mathrm{I}}{\mathbf{G}}{}^{\bullet} \cdot \mathbf{H}^{\mathrm{T}},\ \theta) \quad \forall \mathbf{H} \in U. \tag{7.53}$$

Here we have gone from the argument $\overset{\mathrm{I}}{\mathbf{C}}$ to $\overset{\mathrm{I}}{\mathbf{G}}$, because just the measure $\overset{\mathrm{I}}{\mathbf{G}}$ satisfies the relation (7.53), and the tensor $\overset{\mathrm{I}}{\mathbf{C}}$ is transformed during the passage to $\overset{*}{\mathcal{K}}$ in another way, namely according to (4.201a), and its expression involves the additional term: $-(1/2)(\mathbf{H} \cdot \mathbf{H}^{\mathrm{T}} - \mathbf{E})$.

The tensor function $\mathcal{F}_v(\overset{\mathrm{I}}{\mathbf{G}},\ \overset{\mathrm{I}}{\mathbf{G}}{}^{\bullet},\ \theta)$ satisfying the condition (7.53) is called $A_{\mathrm{I}}$-*unimodular* or $A_{\mathrm{I}}$-*indifferent relative to the group* $U$.

Notice that the group $U$ contains a subgroup being the full orthogonal group $I$; therefore, if $\mathbf{H} = \mathbf{O}^{\mathrm{T}}$ and $\mathbf{Q} \in I$, then formula (7.53) becomes (7.24c). In other words, if a tensor function is $A_{\mathrm{I}}$-unimodular, then the function is isotropic (i.e. indifferent relative to the group $I$). Then, since the Onsager principle holds for all groups and the function $\mathcal{F}_v(\overset{\mathrm{I}}{\mathbf{G}},\ \overset{\mathrm{I}}{\mathbf{G}}{}^{\bullet},\ \theta)$ is quasilinear:

$$\overset{\mathrm{I}}{\mathbf{T}}_v = \mathcal{F}_v(\overset{\mathrm{I}}{\mathbf{G}},\ \overset{\mathrm{I}}{\mathbf{G}}{}^{\bullet},\ \theta) = {}^4\mathbf{L}_v \cdot\cdot\ \overset{\mathrm{I}}{\mathbf{G}}{}^{\bullet}, \tag{7.54}$$

so for this function one of the representations of an isotropic function in the tensor basis (7.29) with two independent constants must hold. But such a representation may be not $A_{\mathrm{I}}$-unimodular; in particular, the representation (7.29) does not satisfy the condition (7.53). The following isotropic tensor function proves to be appropriate:

$$\overset{\mathrm{I}}{\mathbf{T}}_v = \mathcal{F}_v = \frac{1}{4}(\mu_1 I_1 \overset{\mathrm{I}}{\mathbf{G}}^{-1} + 2\mu_2 \overset{\mathrm{I}}{\mathbf{G}}^{-1} \cdot \overset{\mathrm{I}}{\mathbf{G}}^{\bullet} \cdot \overset{\mathrm{I}}{\mathbf{G}}^{-1}), \tag{7.55}$$

where

$$I_1(\overset{\mathrm{I}}{\mathbf{G}}^{-1} \cdot \overset{\mathrm{I}}{\mathbf{G}}^{\bullet}) = \overset{\mathrm{I}}{\mathbf{G}}^{-1} \cdot\cdot\, \overset{\mathrm{I}}{\mathbf{G}}^{\bullet}. \tag{7.56}$$

We can verify that the function (7.55) is $A_{\mathrm{I}}$-unimodular, i.e. satisfies the Eq. (7.53):

$$\begin{aligned}
4\mathcal{F}(\overset{\mathrm{I}}{\mathbf{G}}^{*}, \overset{\mathrm{I}}{\mathbf{G}}^{*\bullet}, \theta) &= \mu_1(\overset{\mathrm{I}}{\mathbf{G}}^{*-1} \cdot\cdot\, \overset{\mathrm{I}}{\mathbf{G}}^{*\bullet})\overset{\mathrm{I}}{\mathbf{G}}^{*-1} + 2\mu_2 \overset{\mathrm{I}}{\mathbf{G}}^{*-1} \cdot \overset{\mathrm{I}}{\mathbf{G}}^{*\bullet} \cdot \overset{\mathrm{I}}{\mathbf{G}}^{*-1} \\
&= \mu_1(\mathbf{H}^{-1\mathrm{T}}\overset{\mathrm{I}}{\mathbf{G}}^{-1} \cdot \mathbf{H}^{-1} \cdot\cdot\, \mathbf{H} \cdot \overset{\mathrm{I}}{\mathbf{G}}^{\bullet} \cdot \mathbf{H}^{\mathrm{T}})\mathbf{H}^{-1\mathrm{T}} \cdot \overset{\mathrm{I}}{\mathbf{G}}^{-1} \cdot \mathbf{H}^{-1} \\
&\quad + 2\mu_2 \mathbf{H}^{-1\mathrm{T}} \cdot \overset{\mathrm{I}}{\mathbf{G}}^{-1} \cdot \mathbf{H}^{-1} \cdot \mathbf{H} \cdot \overset{\mathrm{I}}{\mathbf{G}}^{\bullet} \cdot \mathbf{H}^{\mathrm{T}} \cdot \mathbf{H}^{-1\mathrm{T}} \cdot \overset{\mathrm{I}}{\mathbf{G}}^{-1} \cdot \mathbf{H}^{-1} \\
&= \mathbf{H}^{-1\mathrm{T}} \cdot (\mu_1 J_1 \overset{\mathrm{I}}{\mathbf{G}}^{-1} + 2\mu_2 \overset{\mathrm{I}}{\mathbf{G}}^{-1} \cdot \overset{\mathrm{I}}{\mathbf{G}}^{\bullet} \cdot \mathbf{G}^{-1}) \cdot \mathbf{H}^{-1} \\
&= 4\mathbf{H}^{-1\mathrm{T}} \cdot \mathcal{F}(\overset{\mathrm{I}}{\mathbf{G}}, \overset{\mathrm{I}}{\mathbf{G}}^{\bullet}, \theta) \cdot \mathbf{H}^{-1}.
\end{aligned} \tag{7.57}$$

As follows from (7.57), the function (7.55) satisfies also the condition (7.24c) for $G_s = I$, i.e. it is isotropic. Thus, the isotropic tensor function (7.55) constructed is $A_{\mathrm{I}}$-unimodular.

Thus, we obtain the following theorem.

**Theorem 7.3.** *Any quasilinear $A_{\mathrm{I}}$-unimodular tensor function of viscous stresses (7.54) can be represented in the form (7.55) or, with the help of the Cauchy stress tensor, in the form*

$$\mathbf{T}_v = \mu_1 I_1(\mathbf{D})\mathbf{E} + 2\mu_2\mathbf{D}, \tag{7.58}$$

*where* $\mathbf{D}$ *is the tensor of deformation rates (2.225).*

▼ The first assertion of the theorem (formula (7.55)) has been proved above. Let us prove that the representations (7.55) and (7.58) are equivalent.

Go from the energetic tensor of viscous stresses $\overset{\mathrm{I}}{\mathbf{T}}_v$ to the tensor of viscous stresses $\mathbf{T}$:

$$\mathbf{T}_v = \mathbf{F}^{-1\mathrm{T}} \cdot \overset{\mathrm{I}}{\mathbf{T}}_v \cdot \mathbf{F}^{-1}. \tag{7.59}$$

Substitution of (7.55) into (7.59) yields

$$\begin{aligned}
\mathbf{T}_v &= \frac{1}{4}\mathbf{F}^{-1\mathrm{T}} \cdot (\mu_1 \overset{\mathrm{I}}{\mathbf{G}}^{-1} J_1 + 2\mu_2 \overset{\mathrm{I}}{\mathbf{G}}^{-1} \cdot \overset{\mathrm{I}}{\mathbf{G}}^{\bullet} \cdot \overset{\mathrm{I}}{\mathbf{G}}^{-1}) \cdot \mathbf{F}^{-1} \\
&= \frac{1}{4}(-2\mu_1 J_1 \mathbf{E} - 4\mu_2 \mathbf{F} \cdot \dot{\mathbf{G}}^{-1} \cdot \mathbf{F}^{\mathrm{T}}).
\end{aligned} \tag{7.60}$$

Here we have taken into account that $\overset{\mathrm{I}}{\mathbf{G}} = -\frac{1}{2}\mathbf{G}^{-1} = -\frac{1}{2}\mathbf{F}^{-1}\cdot\mathbf{F}^{-1\mathrm{T}}$ and $\overset{\mathrm{I}}{\mathbf{G}}{}^{-1} = -2\mathbf{G} = -2\mathbf{F}^{\mathrm{T}}\cdot\mathbf{F}$. According to formula (2.284), we can express the tensor $\overset{\mathrm{I}}{\mathbf{G}}{}^{\bullet}$ in terms of the deformation rate tensor $\mathbf{D}$:

$$-2\overset{\mathrm{I}}{\mathbf{G}}{}^{\bullet} = \dot{\mathbf{G}}^{-1} = -2\mathbf{F}^{-1}\cdot\mathbf{D}\cdot\mathbf{F}^{-1\mathrm{T}}.$$

Taking into account that

$$\begin{aligned} I_1(\overset{\mathrm{I}}{\mathbf{G}}{}^{-1}\cdot\overset{\mathrm{I}}{\mathbf{G}}{}^{\bullet}) &= \overset{\mathrm{I}}{\mathbf{G}}{}^{-1}\cdot\cdot\,\overset{\mathrm{I}}{\mathbf{G}}{}^{\bullet} = \mathbf{G}\cdot\cdot\,\dot{\mathbf{G}}^{-1} \\ &= -2\mathbf{F}^{\mathrm{T}}\cdot\mathbf{F}\cdot\cdot\,\mathbf{F}^{-1}\cdot\mathbf{D}\cdot\mathbf{F}^{-1\mathrm{T}} = -2\mathbf{D}\cdot\cdot\,\mathbf{E} = -2I_1(\mathbf{D}), \end{aligned} \tag{7.61}$$

from (7.60) we really get formula (7.58). ▲

## 7.2.3 *Simultaneous Invariants for Fluids of the Differential Type*

Notice that the coefficients $\mu_1$ and $\mu_2$ in formula (7.55) may depend on $\overset{\mathrm{I}}{\mathbf{G}}$ and $\overset{\mathrm{I}}{\mathbf{G}}{}^{\bullet}$:

$$\mu_\gamma = \mu_\gamma(\overset{\mathrm{I}}{\mathbf{G}},\ \overset{\mathrm{I}}{\mathbf{G}}{}^{\bullet},\ \theta). \tag{7.62}$$

However, they cannot vary under unimodular transformations:

$$\mu_\gamma(\overset{\mathrm{I}}{\mathbf{G}},\ \overset{\mathrm{I}}{\mathbf{G}}{}^{\bullet},\ \theta) = \mu_\gamma(\mathbf{H}\cdot\overset{\mathrm{I}}{\mathbf{G}}\cdot\mathbf{H}^{\mathrm{T}},\ \mathbf{H}\cdot\overset{\mathrm{I}}{\mathbf{G}}{}^{\bullet}\cdot\mathbf{H}^{\mathrm{T}},\ \theta), \quad \forall\mathbf{H}\in U, \tag{7.63}$$

i.e. $\mu_\gamma$ must be scalar $A_{\mathrm{I}}$-unimodular functions of two tensor arguments. Such functions are called *simultaneous $A_{\mathrm{I}}$-invariants* relative to the group $U$.

**Theorem 7.4.** *A functional basis of independent simultaneous $A_{\mathrm{I}}$-invariants $J_\gamma^{(U)}(\overset{\mathrm{I}}{\mathbf{G}},\ \overset{\mathrm{I}}{\mathbf{G}}{}^{\bullet})$ of the tensors $\overset{\mathrm{I}}{\mathbf{G}}$ and $\overset{\mathrm{I}}{\mathbf{G}}{}^{\bullet}$ relative to the unimodular group $U$ consists of not more than five elements, which can be chosen as follows:*

$$J_\gamma^{(U)} = I_\gamma(\overset{\mathrm{I}}{\mathbf{G}}{}^{-1}\cdot\overset{\mathrm{I}}{\mathbf{G}}{}^{\bullet}),\ \gamma = 1,\ 2,\ 3;\ \ J_4^{(U)} = I_3(\overset{\mathrm{I}}{\mathbf{G}}),\ \ J_5^{(U)} = I_3(\overset{\mathrm{I}}{\mathbf{G}}{}^{\bullet}). \tag{7.64}$$

▼ Each simultaneous invariant $J_\gamma^{(U)}(\overset{\mathrm{I}}{\mathbf{G}},\ \overset{\mathrm{I}}{\mathbf{G}}{}^{\bullet})$ is an $A_{\mathrm{I}}$-unimodular function of two tensor arguments, i.e. it satisfies Eq. (7.63). Then the invariant is an isotropic tensor function of two arguments if $\mathbf{H} = \mathbf{Q}^{\mathrm{T}}$ and $\mathbf{Q}\in I\subset U$; i.e. it is a simultaneous invariant relative to the group $I$.

In the group $I$, a functional basis of simultaneous invariants can be formed by contractions of powers of the tensors: $\overset{\mathrm{I}}{\mathbf{G}}{}^{-1} \cdot \overset{\mathrm{I}}{\mathbf{G}}{}^{\bullet}$, $(\overset{\mathrm{I}}{\mathbf{G}}{}^{-1})^2 \cdot\cdot\, \overset{\mathrm{I}}{\mathbf{G}}{}^{\bullet}$ etc. However, among all the contractions, only contractions of the tensor

$$\boldsymbol{\Omega}_G = \overset{\mathrm{I}}{\mathbf{G}}{}^{-1} \cdot \overset{\mathrm{I}}{\mathbf{G}}{}^{\bullet} \tag{7.65}$$

are invariants relative to the group $U$ (see Exercise 7.2.1). The number $r$ of independent invariants of this tensor cannot exceed three (invariants of the tensor relative to the group $U$ are its invariants relative to the group $I$ too, and for the group $I$: $r = 3$). As these invariants we can choose $I_\gamma(\boldsymbol{\Omega}_G)$, $\gamma = 1, 2, 3$. Moreover, each of the tensors $\overset{\mathrm{I}}{\mathbf{G}}$ and $\overset{\mathrm{I}}{\mathbf{G}}{}^{\bullet}$ has one unimodular invariant (see Theorem 4.31), these are $\det \overset{\mathrm{I}}{\mathbf{G}}$ and $\det \overset{\mathrm{I}}{\mathbf{G}}{}^{\bullet}$, respectively. There are no other independent simultaneous invariants of the tensors $\overset{\mathrm{I}}{\mathbf{G}}$ and $\overset{\mathrm{I}}{\mathbf{G}}{}^{\bullet}$ relative to this group. ▲

*Remark 1.* Notice that in the theorem the number $r$ does not exceed 5; but the invariant $I_3(\boldsymbol{\Omega}_G)$ is not independent: it may be expressed in terms of $I_3(\overset{\mathrm{I}}{\mathbf{G}})$ and $I_3(\overset{\mathrm{I}}{\mathbf{G}}{}^{\bullet})$. Let us show this fact.

Consider the representation (7.58), then simultaneous $A_{\mathrm{I}}$-invariants $J_\gamma^{(U)}$ (7.64) can be written as functions of the principal invariants of the tensor $\mathbf{D}$ and of the density $\rho$:

$$\begin{aligned}
J_1^{(U)}(\overset{\mathrm{I}}{\mathbf{G}},\ \overset{\mathrm{I}}{\mathbf{G}}{}^{\bullet}) &= I_1(\boldsymbol{\Omega}_G) = -2I_1(\mathbf{D}),\\
J_2^{(U)}(\overset{\mathrm{I}}{\mathbf{G}},\ \overset{\mathrm{I}}{\mathbf{G}}{}^{\bullet}) = I_2(\boldsymbol{\Omega}_G) &= \frac{1}{2}(I_1^2(\boldsymbol{\Omega}_G) - I_1(\boldsymbol{\Omega}_G^2)) = 4I_2(\mathbf{D}),\\
J_3^{(U)}(\overset{\mathrm{I}}{\mathbf{G}},\ \overset{\mathrm{I}}{\mathbf{G}}{}^{\bullet}) = \det \boldsymbol{\Omega}_G &= \det \mathbf{G} \cdot \det \dot{\mathbf{G}}^{-1} = -8I_3(\mathbf{D}),\\
J_4^{(U)} = -\frac{1}{8}(\overset{\circ}{\rho}/\rho)^2, &\qquad J_5^{(U)} = (\overset{\circ}{\rho}/\rho)^2 I_3(\mathbf{D}).
\end{aligned} \tag{7.66}$$

Hence $J_5^{(U)}$ is not independent, because it is expressed in terms of $J_3^{(U)}$ and $J_4^{(U)}$:

$$J_5^{(U)} = J_3^{(U)}\ J_3^{(U)}. \tag{7.67}$$

Due to Theorem 7.4, the viscous coefficients $\mu_\gamma$ (7.62) can be represented as functions of the simultaneous invariants (7.64):

$$\mu_\gamma = \mu_\gamma(J_1^{(U)}, \ldots, J_4^{(U)},\ \theta), \tag{7.68}$$

or in the form

$$\mu_\gamma = \mu_\gamma(I_\gamma(\mathbf{D}),\ \rho,\ \theta), \qquad \gamma = 1,\ 2,\ 3. \ \square \tag{7.69}$$

### *7.2.4 Tensor of Viscous Stresses in Model $A_{\mathrm{V}}$ of Fluids of the Differential Type*

Let us consider now the model $A_{\mathrm{V}}$.

The tensors $\overset{\mathrm{V}}{\mathbf{T}}$ and $\overset{\mathrm{V}}{\mathbf{G}}$ are transformed during the passage to $\overset{*}{\mathcal{K}}$ by formulae (4.200a) and (4.203). On substituting these formulae into (7.23e), we obtain that the function of viscous stresses must satisfy the relation

$$\mathbf{H}\cdot\mathcal{F}_{\nu}(\overset{\mathrm{V}}{\mathbf{G}},\ \overset{\mathrm{V}}{\mathbf{G}}{}^{\bullet},\ \theta)\cdot\mathbf{H}^{\mathrm{T}} = \mathcal{F}_{\nu}(\mathbf{H}^{-1\mathrm{T}}\cdot\overset{\mathrm{V}}{\mathbf{G}}\cdot\mathbf{H}^{-1},\ \mathbf{H}^{-1\mathrm{T}}\cdot\overset{\mathrm{V}}{\mathbf{G}}{}^{\bullet}\cdot\mathbf{H}^{-1},\ \theta) \tag{7.70}$$

$\forall\mathbf{H}\in U$. Such a tensor function $\mathcal{F}_{\nu}(\overset{\mathrm{V}}{\mathbf{G}},\ \overset{\mathrm{V}}{\mathbf{G}}{}^{\bullet},\ \theta)$ is called *$A_{\mathrm{V}}$-unimodular* or *$A_{\mathrm{V}}$-indifferent relative to the group $U$.*

According to the Onsager principle, this function must be quasilinear:

$$\overset{\mathrm{V}}{\mathbf{T}}_{\nu} = \mathcal{F}_{\nu}(\overset{\mathrm{V}}{\mathbf{G}},\ \overset{\mathrm{V}}{\mathbf{G}}{}^{\bullet},\ \theta) = {}^{4}\mathbf{L}_{\nu}\cdot\cdot\overset{\mathrm{V}}{\mathbf{G}}{}^{\bullet}. \tag{7.71}$$

The condition for $A_{\mathrm{V}}$-unimodularity (7.70) goes into the condition for the isotropy of a tensor function when $\mathbf{H}=\mathbf{Q}^{\mathrm{T}}$ and $\mathbf{Q}\in I\subset U$; therefore, for the function (7.71) one of the representations of an isotropic function in the tensor basis (7.29) with two independent constants must hold. The following isotropic tensor function is appropriate:

$$\overset{\mathrm{V}}{\mathbf{T}}_{\nu} = \mathcal{F}_{\nu} = \frac{1}{4}(\mu_1 I_1\overset{\mathrm{V}}{\mathbf{G}}{}^{-1} + \mu_2\overset{\mathrm{V}}{\mathbf{G}}{}^{-1}\cdot\overset{\mathrm{V}}{\mathbf{G}}{}^{\bullet}\cdot\overset{\mathrm{V}}{\mathbf{G}}{}^{-1}), \tag{7.72}$$

where $I_1 = I_1(\overset{\mathrm{V}}{\mathbf{G}}{}^{-1}\cdot\overset{\mathrm{V}}{\mathbf{G}}{}^{\bullet})$. We can immediately verify that this function is $A_{\mathrm{V}}$-unimodular, i.e. satisfies the relation (7.70) (Exercise 7.2.3). Thus, the following theorem holds.

**Theorem 7.5.** *Any quasilinear $A_{\mathrm{V}}$-unimodular tensor function of viscous stresses (7.71) can be represented in the form (7.72) or, with the help of the Cauchy stress tensor, in the form (7.58).*

▼ The first part of the theorem has been proved above, therefore we will show only equivalence of representations (7.72) and (7.58).

Going from $\overset{\mathrm{V}}{\mathbf{T}}_{\nu}$ to $\mathbf{T}_{\nu}$ by the formula

$$\mathbf{T}_{\nu} = \mathbf{F}\cdot\overset{\mathrm{V}}{\mathbf{T}}_{\nu}\cdot\mathbf{F}^{\mathrm{T}}, \tag{7.73}$$

from (7.72) we obtain

$$\begin{aligned}\mathbf{T}_v &= \frac{1}{4}\mathbf{F}\cdot(\mu_1\overset{\mathrm{V}}{\mathbf{G}}{}^{-1}I_1 + 2\mu_2\overset{\mathrm{V}}{\mathbf{G}}{}^{-1}\cdot\overset{\mathrm{V}}{\mathbf{G}}{}^{\bullet}\cdot\overset{\mathrm{V}}{\mathbf{G}}{}^{-1})\cdot\mathbf{F}^{\mathrm{T}} \\ &= \frac{1}{4}(2\mu_1 I_1\mathbf{E} + 4\mu_2\mathbf{F}^{-1\mathrm{T}}\cdot\dot{\mathbf{G}}\cdot\mathbf{F}^{-1}).\end{aligned} \tag{7.74}$$

Here we have taken into account that $\overset{\mathrm{V}}{\mathbf{G}} = \frac{1}{2}\mathbf{G} = \frac{1}{2}\mathbf{F}^{\mathrm{T}}\cdot\mathbf{F}$ and $\overset{\mathrm{V}}{\mathbf{G}}{}^{-1} = 2\mathbf{G}^{-1} = 2\mathbf{F}^{-1}\cdot\mathbf{F}^{-1\mathrm{T}}$. With the help of formula (2.284) we can express $\dot{\mathbf{G}}$ in terms of $\mathbf{D}$:

$$\dot{\mathbf{G}} = 2\overset{\mathrm{V}}{\mathbf{G}}{}^{\bullet} = 2\mathbf{F}^{\mathrm{T}}\cdot\mathbf{D}\cdot\mathbf{F}. \tag{7.75}$$

Using the relations

$$I_1(\overset{\mathrm{V}}{\mathbf{G}}{}^{-1}\cdot\overset{\mathrm{V}}{\mathbf{G}}{}^{\bullet}) = \mathbf{G}^{-1}\cdot\cdot\,\dot{\mathbf{G}} = 2\mathbf{F}^{-1}\cdot\mathbf{F}^{-1\mathrm{T}}\cdot\cdot\,\mathbf{F}^{\mathrm{T}}\cdot\mathbf{D}\cdot\mathbf{F} = 2I_1(\mathbf{D}), \tag{7.76}$$

from (7.74) and (7.75) we actually obtain representation (7.58). ▲

### 7.2.5 *Viscous Coefficients in Model $A_{\mathrm{V}}$ of a Fluid of the Differential Type*

Coefficients $\mu_\gamma$ in (7.72) are functions of $\overset{\mathrm{V}}{\mathbf{G}}$ and $\overset{\mathrm{V}}{\mathbf{G}}{}^{\bullet}$:

$$\mu_\gamma = \mu_\gamma(\overset{\mathrm{V}}{\mathbf{G}},\ \overset{\mathrm{V}}{\mathbf{G}}{}^{\bullet},\ \theta) \tag{7.77}$$

and remain unchanged under unimodular transformations:

$$\mu_\gamma(\overset{\mathrm{V}}{\mathbf{G}},\ \overset{\mathrm{V}}{\mathbf{G}}{}^{\bullet},\ \theta) = \mu_\gamma(\mathbf{H}^{-1\mathrm{T}}\cdot\overset{\mathrm{V}}{\mathbf{G}}\cdot\mathbf{H}^{-1},\ \mathbf{H}^{-1\mathrm{T}}\cdot\overset{\mathrm{V}}{\mathbf{G}}{}^{\bullet}\cdot\mathbf{H}^{-1},\ \theta), \quad \forall\mathbf{H}\in U; \tag{7.78}$$

i.e. they are scalar $A_{\mathrm{V}}$-unimodular functions. In other words, they are *simultaneous $A_{\mathrm{V}}$-invariants* of the tensors $\overset{\mathrm{V}}{\mathbf{G}}$ and $\overset{\mathrm{V}}{\mathbf{G}}{}^{\bullet}$ relative to the group $U$.

For these invariants, Theorem 7.4 and Remark 1 still remain valid. Therefore, $\mu_\gamma$ (7.77) can be represented as functions

$$\mu_\gamma = \mu_\gamma(J_1^{(U)},\ldots,J_4^{(U)},\ \theta) \tag{7.79}$$

of simultaneous $A_{\mathrm{V}}$-invariants formed by formulae (7.64):

$$J_\gamma^{(U)} = I_\gamma(\overset{\mathrm{V}}{\mathbf{G}}{}^{-1}\cdot\overset{\mathrm{V}}{\mathbf{G}}{}^{\bullet}), \qquad \gamma = 1,\ 2,\ 3, \quad J_4^{(U)} = \det\overset{\mathrm{V}}{\mathbf{G}}. \tag{7.80}$$

These invariants can be expressed in terms of the principal invariants of the tensor $\mathbf{D}$ and $\rho$:

$$J_1^{(U)}(\overset{\text{V}}{\mathbf{G}}{}^{-1} \cdot \overset{\text{V}}{\mathbf{G}}{}^{\bullet}) = 2I_1(\mathbf{D}), \quad J_2^{(U)} = 4I_2(\mathbf{D}),$$
$$J_3^{(U)} = 8I_3(\mathbf{D}), \quad J_4^{(U)} = \frac{1}{8}(\rho/\overset{\circ}{\rho})^2, \tag{7.81}$$

hence the viscous coefficients (7.79) can be written in the form (7.69) as well.

### 7.2.6 General Representation of Constitutive Equations for Fluids of the Differential Type

Since for both the models $A_{\text{I}}$ and $A_{\text{V}}$ constitutive equations can be written in the single generalized form (7.58) and (7.68), it is appropriate to give a common name for the models of fluids of the differential type.

**Definition 7.2.** The models $A_{\text{I}}$ and $A_{\text{V}}$ (and also $B_{\text{I}}$ and $B_{\text{V}}$ coincident with them) of fluids of the differential type, whose constitutive equations (7.51), (7.55), and (7.72) can be represented in the single form (7.58), (7.68), i.e.

$$\begin{cases} \mathbf{T} = \mathbf{T}_e + \mathbf{T}_v, \\ \mathbf{T}_e = -p\mathbf{E}, \quad \mathbf{T}_v = \mu_1 I_1(\mathbf{D})\mathbf{E} + 2\mu_2\mathbf{D}, \\ \mu_\gamma = \mu_\gamma(I_\alpha(\mathbf{D}),\ \rho,\ \theta), \\ \psi = \psi(\rho,\ \theta), \quad p = p(\rho,\ \theta) = \rho^2(\partial\psi/\partial\rho), \quad \eta = \partial\psi/\partial\theta, \end{cases} \tag{7.82}$$

are called the *model of a viscous fluid*.

According to (7.20), the dissipation function $w^*$ for a viscous fluid takes the form

$$w^* = \overset{(\text{n})}{\mathbf{T}}_v \cdot\cdot \overset{(\text{n})}{\mathbf{C}}{}^{\bullet} = \mathbf{T}_v \cdot\cdot \mathbf{D} = \mu_1 I_1^2(\mathbf{D}) + 2\mu_2 I_1(\mathbf{D}^2). \tag{7.83}$$

The Fourier law (7.17) for a viscous fluid has the same form as the one for an ideal fluid:

$$\mathbf{q} = -\boldsymbol{\lambda}\ \nabla\theta, \quad \boldsymbol{\lambda} = \lambda\mathbf{E}. \tag{7.84}$$

If the viscous coefficients $\mu_\gamma$ are independent of the invariants $I_\gamma(\mathbf{D})$ and the density $\rho$, then the model (7.82) is called the *model of a linear-viscous fluid* or the model of a *Newtonian fluid*.

### 7.2.7 Constitutive Equations for Incompressible Viscous Fluids

Similarly to Sect. 4.9.4 where the model of incompressible ideal fluids has been stated, we can introduce a model of an incompressible viscous fluid.

**Theorem 7.6.** *The models $A_{\mathrm{I}}$ and $A_{\mathrm{V}}$ (and also $B_{\mathrm{I}}$ and $B_{\mathrm{V}}$) of incompressible fluids of the differential type (7.51), (7.55), (7.68) and (7.72), (7.79) can be represented in the single form*

$$\begin{cases} \mathbf{T} = \mathbf{T}_e + \mathbf{T}_v, \\ \mathbf{T}_e = -p\mathbf{E}, \quad \mathbf{T}_v = \mu_1 I_1(\mathbf{D})\mathbf{E} + 2\mu_2\mathbf{D}, \\ \mu_\gamma = \mu_\gamma(I_\alpha(\mathbf{D}),\ \theta), \\ \psi = \psi(\theta), \qquad \eta = \partial\psi/\partial\theta, \end{cases} \tag{7.85}$$

*where $p$ is a function independent of $\rho$.*

▼ A proof of the theorem is similar to the proof of Theorem 4.33. ▲

The models $A_{\mathrm{I}}$ and $A_{\mathrm{V}}$ (and also $B_{\mathrm{I}}$ and $B_{\mathrm{V}}$) of an incompressible fluid of the differential type, represented in the form (7.85), are called the *model of an incompressible viscous fluid*; and if $\mu_\gamma$ are independent of $I_\gamma(\mathbf{D})$, then – the model of a *linear-viscous incompressible fluid.*

### 7.2.8 The Principle of Material Indifference for Models $A_n$ and $B_n$ of Fluids of the Differential Type

Since all the energetic tensors $\overset{(n)}{\mathbf{T}}$, $\overset{(n)}{\mathbf{C}}$ and $\overset{(n)}{\mathbf{G}}$ are $R$-invariant, the constitutive equations (7.51), (7.55), and (7.72) for fluids of the differential type are the same in actual configurations $\mathcal{K}$ and $\mathcal{K}'$ differing from one another by a rigid motion. Therefore, the principle of material indifference for models $A_n$ and $B_n$ of fluids of the differential type (as well as for ideal fluids) is identically satisfied.

We should consider separately only representation (7.85) for fluids, which in configuration $\mathcal{K}'$ takes the form

$$\mathbf{T}_v' = \mu_1 I_1(\mathbf{D}')\mathbf{E} + 2\mu_2\mathbf{D}', \quad \mu_\gamma = \mu_\gamma(I_1(\mathbf{D}'), \ldots, I_3(\mathbf{D}'),\ \rho,\ \theta). \tag{7.86}$$

This relation actually holds, because both the tensors $\mathbf{D}$ and $\mathbf{T}_v$ are $R$-indifferent (see Table 4.12), and the invariants $I_\gamma(\mathbf{D})$ and $I_\gamma(\mathbf{D}')$ are coincident.

## Exercises for 7.2

**7.2.1.** Prove that contractions of the tensor (7.65) give scalar invariants relative to the group $U$.

**7.2.2.** Show that tensor ${}^4\mathbf{L}_v$ for the $A_{\mathrm{I}}$-unimodular function (7.54) has the form

$$ {}^4\mathbf{L}_v = \frac{1}{4}(\mu_1 \overset{\mathrm{I}}{\mathbf{G}}{}^{-1} \otimes \overset{\mathrm{I}}{\mathbf{G}}{}^{-1} + 2\mu_2 (\overset{\mathrm{I}}{\mathbf{G}}{}^{-1} \otimes \overset{\mathrm{I}}{\mathbf{G}}{}^{-1})^{(1432)}), $$

and for the $A_{\mathrm{V}}$-unimodular function (7.71) – the form

$$ {}^4\mathbf{L}_v = \frac{1}{4}(\mu_1 \overset{\mathrm{V}}{\mathbf{G}}{}^{-1} \otimes \overset{\mathrm{V}}{\mathbf{G}}{}^{-1} + 2\mu_2 (\overset{\mathrm{V}}{\mathbf{G}}{}^{-1} \otimes \overset{\mathrm{V}}{\mathbf{G}}{}^{-1})^{(1432)}). $$

**7.2.3.** Show that the tensor function (7.72) satisfies Eq. (7.70).

**7.2.4.** Show that the dissipation function (7.40) for models $B_{\mathrm{I}}$ and $B_{\mathrm{V}}$ of a fluid of the differential type has the form

$$ w^* = \mu_1 I_1^2(\mathbf{D}) + 2\mu_2 I_1(\mathbf{D}^2). \tag{7.87} $$

## 7.3 Models $C_n$ and $D_n$ of Continua of the Differential Type

### 7.3.1 Models $C_n$ of Continua of the Differential Type

According to *models* $C_n$ of *continua of the differential type*, the free energy and the quasienergetic stress tensors are functions in the form

$$ \psi = \psi(\overset{(n)}{\mathbf{A}}, \overset{(n)}{\mathbf{A}}{}^{\bullet}, \mathbf{O}, \dot{\mathbf{O}}, \theta), \tag{7.88} $$

$$ \overset{(n)}{\mathbf{S}} = \overset{(n)}{\mathbf{S}}(\overset{(n)}{\mathbf{A}}, \overset{(n)}{\mathbf{A}}{}^{\bullet}, \mathbf{O}, \dot{\mathbf{O}}, \theta), \quad \overset{\circ}{\mathbf{S}} = \overset{\circ}{\mathbf{S}}(\overset{(n)}{\mathbf{A}}, \overset{(n)}{\mathbf{A}}{}^{\bullet}, \mathbf{O}, \dot{\mathbf{O}}, \theta). \tag{7.89} $$

$$ w^* = w^*(\overset{(n)}{\mathbf{A}}, \overset{(n)}{\mathbf{A}}{}^{\bullet}, \mathbf{O}, \dot{\mathbf{O}}, \theta). \tag{7.90} $$

Introducing the tensors of equilibrium stresses $\overset{(n)}{\mathbf{S}}_e$, $\overset{\circ}{\mathbf{S}}_e$ and viscous stresses $\overset{(n)}{\mathbf{S}}_v$, $\overset{\circ}{\mathbf{S}}_v$:

$$ \overset{(n)}{\mathbf{S}}_v = \overset{(n)}{\mathbf{S}} - \overset{(n)}{\mathbf{S}}_e, \quad \overset{\circ}{\mathbf{S}}_v = \overset{\circ}{\mathbf{S}} - \overset{\circ}{\mathbf{S}}_e, \tag{7.91} $$

$$ \overset{(n)}{\mathbf{S}}_e = \overset{(n)}{\mathbf{S}}(\overset{(n)}{\mathbf{A}}, \mathbf{0}, \mathbf{O}, \mathbf{0}, \theta), \quad \overset{\circ}{\mathbf{S}}_e = \overset{\circ}{\mathbf{S}}(\overset{(n)}{\mathbf{A}}, \mathbf{0}, \mathbf{O}, \mathbf{0}, \theta), \tag{7.92} $$

and substituting them into PTI (4.123), we obtain the identity

$$\left(\frac{\partial\psi}{\partial \overset{(n)}{\mathbf{A}}} - \frac{\overset{(n)}{\mathbf{S}}_e}{\rho}\right)\cdot\cdot\, d\overset{(n)}{\mathbf{A}} + \frac{\partial\psi}{\partial \overset{(n)}{\mathbf{A}}{}^{\bullet}}\cdot\cdot\, d\overset{(n)}{\mathbf{A}}{}^{\bullet} + \left(\frac{\partial\psi}{\partial \mathbf{O}} - \frac{\overset{\circ}{\mathbf{S}}_e}{\rho}\right)\cdot\cdot\, d\mathbf{O}^{\mathrm{T}} + \frac{\partial\psi}{\partial \dot{\mathbf{O}}}\cdot\cdot\, d\dot{\mathbf{O}}^{\mathrm{T}}$$

$$+\left(\frac{\partial\psi}{\partial\theta} + \eta\right) d\theta + \frac{1}{\rho}(w^* - \overset{(n)}{\mathbf{S}}_v\cdot\cdot\,\overset{(n)}{\mathbf{A}}{}^{\bullet} - \overset{\circ}{\mathbf{S}}_v\cdot\cdot\,\dot{\mathbf{O}})\, dt = 0. \tag{7.93}$$

Since all differentials $d\overset{(n)}{\mathbf{A}}$, $d\overset{(n)}{\mathbf{A}}{}^{\bullet}$, $d\mathbf{O}^{\mathrm{T}}$, $d\dot{\mathbf{O}}^{\mathrm{T}}$, $d\theta$ and $dt$ are independent, the identity (7.93) is equivalent to the equation system

$$\begin{cases} \overset{(n)}{\mathbf{S}}_e = \rho(\partial\psi/\partial\overset{(n)}{\mathbf{A}}) \equiv \mathbf{\Phi}(\overset{(n)}{\mathbf{A}},\ \mathbf{O},\ \theta), \\ \overset{\circ}{\mathbf{S}}_e = \rho(\partial\psi/\partial\mathbf{O}) = \overset{\circ}{\mathbf{\Phi}}(\overset{(n)}{\mathbf{A}},\ \mathbf{O},\ \theta), \\ \eta = -\partial\psi/\partial\theta, \quad w^* = \overset{(n)}{\mathbf{S}}_v\cdot\cdot\,\overset{(n)}{\mathbf{A}}{}^{\bullet} + \overset{\circ}{\mathbf{S}}_v\cdot\cdot\,\dot{\mathbf{O}}^{\mathrm{T}}, \\ \psi = \psi(\overset{(n)}{\mathbf{A}},\ \mathbf{O},\ \theta), \quad \partial\psi/\partial\dot{\mathbf{O}} = 0, \quad \partial\psi/\partial\overset{(n)}{\mathbf{A}}{}^{\bullet} = 0, \end{cases} \tag{7.94}$$

i.e. $\psi$, just as for models $A_n$, cannot depend on the rates $\overset{(n)}{\mathbf{A}}{}^{\bullet}$ and $\dot{\mathbf{O}}$. Just as in the case of models $A_n$, Eqs. (7.94) should be complemented by relations connecting the tensors of viscous quasienergetic stresses $\overset{(n)}{\mathbf{S}}_v$, $\overset{\circ}{\mathbf{S}}_v$ to the reactive variables

$$\overset{(n)}{\mathbf{S}}_v = \mathbf{\Phi}_v(\overset{(n)}{\mathbf{A}},\ \overset{(n)}{\mathbf{A}}{}^{\bullet},\ \mathbf{O},\ \dot{\mathbf{O}},\ \theta), \qquad \overset{\circ}{\mathbf{S}}_v = \overset{\circ}{\mathbf{\Phi}}_v(\overset{(n)}{\mathbf{A}},\ \overset{(n)}{\mathbf{A}}{}^{\bullet},\ \mathbf{O},\ \dot{\mathbf{O}},\ \theta). \tag{7.95}$$

The relations (7.94) and (7.95) are *constitutive equations for models $C_n$ of continua of the differential type*.

However, these constitutive equations prove to be not correct: they do not satisfy the principle of material indifference. Indeed, since the tensors $\overset{(n)}{\mathbf{S}}$, $\overset{(n)}{\mathbf{S}}_e$, $\overset{(n)}{\mathbf{S}}_v$ and $\overset{(n)}{\mathbf{A}}$ are $R$-indifferent and the tensors $\overset{(n)}{\mathbf{A}}{}^{\bullet}$ are not $R$-indifferent, relations (7.95) are transformed during the passage from $\mathcal{K}$ to $\mathcal{K}'$ as follows (even if there is no tensor $\mathbf{O}$ being not $R$-indifferent in these relations):

$$\begin{aligned} \overset{(n)}{\mathbf{S}}{}'_v &= \mathbf{Q}^{\mathrm{T}}\cdot\mathbf{\Phi}_v(\overset{(n)}{\mathbf{A}},\ \overset{(n)}{\mathbf{A}}{}^{\bullet},\ \theta)\cdot\mathbf{Q} \\ &= \mathbf{\Phi}_v(\mathbf{Q}^{\mathrm{T}}\cdot\overset{(n)}{\mathbf{A}}\cdot\mathbf{Q},\ \mathbf{Q}^{\mathrm{T}}\cdot\overset{(n)}{\mathbf{A}}{}^{\bullet}\cdot\mathbf{Q} + \dot{\mathbf{Q}}^{\mathrm{T}}\cdot\overset{(n)}{\mathbf{A}}\cdot\mathbf{Q} + \mathbf{Q}^{\mathrm{T}}\cdot\overset{(n)}{\mathbf{A}}\cdot\dot{\mathbf{Q}},\ \theta). \end{aligned} \tag{7.96}$$

One can show that for isotropic continua of the differential type satisfying the Onsager principle, the tensor of viscous stresses $\mathbf{\Phi}_v$ has the form

$$\mathbf{\Phi}_v = \mu_1 I_1(\overset{(n)}{\mathbf{A}}{}^{\bullet})\mathbf{E} + 2\mu_2 \overset{(n)}{\mathbf{A}}{}^{\bullet}. \tag{7.97}$$

Derivation of this equation is performed by the same method as the one used for models $A_n$ (see Sect. 7.1.2).

On substituting (7.97) into (7.96), we obtain that this equation does not hold. Therefore, models $C_n$ of continua of the differential type cannot describe adequately a behavior of materials.

### 7.3.2 Models $C_n^h$ of Solids with Co-rotational Derivatives

In order to obtain a correct model $C_n$, we should replace the total derivative by some co-rotational one in Eqs. (7.88)–(7.89).

One can say that this is the *model* $C_n^h$ of *continua with co-rotational derivatives,* if active variables $\Lambda = \{\psi, \overset{(n)}{\mathbf{S}}, \overset{\circ}{\mathbf{S}}\}$ depend on the following reactive ones:

$$\psi = \psi(\overset{(n)}{\mathbf{A}}, \overset{(n)}{\mathbf{A}}{}^h, \mathbf{O}, \dot{\mathbf{O}}, \theta), \tag{7.98}$$

$$\overset{(n)}{\mathbf{S}} = \overset{(n)}{\mathbf{S}}(\overset{(n)}{\mathbf{A}}, \overset{(n)}{\mathbf{A}}{}^h, \mathbf{O}, \dot{\mathbf{O}}, \theta), \tag{7.99}$$

$$\overset{\circ}{\mathbf{S}} = \overset{\circ}{\mathbf{S}}(\overset{(n)}{\mathbf{A}}, \overset{(n)}{\mathbf{A}}{}^h, \mathbf{O}, \dot{\mathbf{O}}, \theta), \tag{7.100}$$

where

$$h = \{\, V,\ S,\ J \,\}. \tag{7.101}$$

Substitution of (7.98)–(7.100) into PTI (4.123) yields

$$\left(\frac{\partial \psi}{\partial \overset{(n)}{\mathbf{A}}} - \frac{\overset{(n)}{\mathbf{S}}_e}{\rho}\right) \cdot\cdot\, \overset{(n)}{\mathbf{A}}{}^{\bullet} + \frac{\partial \psi}{\partial \overset{(n)}{\mathbf{A}}{}^h} \cdot\cdot\, \overset{(n)}{\mathbf{A}}{}^{h\bullet} + \left(\frac{\partial \psi}{\partial \mathbf{O}} - \frac{\overset{\circ}{\mathbf{S}}_e}{\rho}\right) \cdot\cdot\, \dot{\mathbf{O}}^{\mathrm{T}}$$

$$+ \frac{\partial \psi}{\partial \dot{\mathbf{O}}} \cdot\cdot\, \ddot{\mathbf{O}}^{\mathrm{T}} + \left(\frac{\partial \psi}{\partial \theta} + \eta\right)\dot{\theta} + \frac{1}{\rho}(w^* - \overset{(n)}{\mathbf{S}}_v \cdot\cdot\, \overset{(n)}{\mathbf{A}}{}^{\bullet} - \overset{\circ}{\mathbf{S}}_v \cdot\cdot\, \dot{\mathbf{O}}^{\mathrm{T}}) = 0. \tag{7.102}$$

Since all derivatives $\overset{(n)}{\mathbf{A}}{}^{\bullet}$, $\overset{(n)}{\mathbf{A}}{}^{h\bullet}$, $\dot{\mathbf{O}}^{\mathrm{T}}$, $\ddot{\mathbf{O}}^{\mathrm{T}}$ and $\dot{\theta}$ are independent, the identity (7.102) is equivalent to the equation system

$$
\begin{cases}
\psi = \psi(\overset{(n)}{\mathbf{A}}, \mathbf{O}, \theta), & (7.103\text{a}) \\
\partial\psi/\partial\dot{\mathbf{O}} = 0, \quad \partial\psi/\partial\dot{\overset{(n)}{\mathbf{A}}}{}^{h} = 0, & (7.103\text{b}) \\
\overset{(n)}{\mathbf{S}}_e = \rho(\partial\psi/\partial\overset{(n)}{\mathbf{A}}) \equiv \mathbf{\Phi}(\overset{(n)}{\mathbf{A}}, \mathbf{O}, \theta), & (7.103\text{c}) \\
\overset{\circ}{\mathbf{S}}_e = \rho(\partial\psi/\partial\mathbf{O}) = \overset{\circ}{\mathbf{\Phi}}(\overset{(n)}{\mathbf{A}}, \mathbf{O}, \theta), & (7.103\text{d}) \\
w^* = \overset{(n)}{\mathbf{S}}_v \cdot\cdot \overset{(n)}{\mathbf{A}}{}^{\bullet} + \overset{\circ}{\mathbf{S}}_v \cdot\cdot \dot{\mathbf{O}}^{\mathrm{T}}. & (7.103\text{e})
\end{cases}
$$

For viscous stresses, according to the principle of equipresence, we should give the functions

$$\overset{(n)}{\mathbf{S}}_v = \mathbf{\Phi}_v(\overset{(n)}{\mathbf{A}}, \overset{(n)}{\mathbf{A}}{}^{h}, \mathbf{O}, \dot{\mathbf{O}}, \theta), \tag{7.104}$$

$$\overset{\circ}{\mathbf{S}}_v = \overset{\circ}{\mathbf{\Phi}}_v(\overset{(n)}{\mathbf{A}}, \overset{(n)}{\mathbf{A}}{}^{h}, \mathbf{O}, \dot{\mathbf{O}}, \theta). \tag{7.105}$$

Notice that the expression (7.103e) for the dissipation function, which has been derived formally, conflicts with the principle of equipresence, because $w^*$ depends on the following arguments:

$$w^* = w^*(\overset{(n)}{\mathbf{A}}, \overset{(n)}{\mathbf{A}}{}^{\bullet}, \overset{(n)}{\mathbf{A}}{}^{h}, \mathbf{O}, \dot{\mathbf{O}}, \theta), \tag{7.106}$$

that is inconsistent with the dependences (7.98)–(7.100). However, we will show below that this inconsistence may be avoided.

### 7.3.3 Corollaries of the Principle of Material Symmetry for Models $C_n^h$ of Solids

Consider the principle of material symmetry for models $C_n^h$ of solids. According to the principle, the following relations must hold during the passage to reference configuration $\overset{*}{\mathcal{K}}$:

$$\overset{(n)}{\mathbf{S}}{}_e^* = \mathbf{\Phi}(\overset{(n)}{\mathbf{A}}{}^*, \overset{*}{\mathbf{O}}, \theta), \tag{7.107a}$$

$$\overset{\circ}{\mathbf{S}}{}_e^* = \overset{\circ}{\mathbf{\Phi}}(\overset{(n)}{\mathbf{A}}{}^*, \overset{*}{\mathbf{O}}, \theta), \tag{7.107b}$$

$$\psi = \psi(\overset{(n)}{\mathbf{A}}{}^*, \overset{*}{\mathbf{O}}, \theta), \tag{7.107c}$$

$$\overset{(n)}{\mathbf{S}}{}^*_v = \mathbf{\Phi}_v(\overset{(n)}{\mathbf{A}}{}^*,\ \overset{(n)}{\mathbf{A}}{}^{*h},\ \overset{*}{\mathbf{O}},\ \dot{\mathbf{O}}^*,\ \theta), \tag{7.107d}$$

$$\overset{\circ}{\mathbf{S}}{}^*_v = \overset{\circ}{\mathbf{\Phi}}_v(\overset{(n)}{\mathbf{A}}{}^*,\ \overset{(n)}{\mathbf{A}}{}^{*h},\ \overset{*}{\mathbf{O}},\ \dot{\mathbf{O}}^*,\ \theta), \tag{7.107e}$$

$$w^* = \overset{(n)}{\mathbf{S}}{}^*_v \cdot\cdot \overset{(n)}{\mathbf{A}}{}^{*\bullet} + \overset{\circ}{\mathbf{S}}{}^*_v \cdot\cdot \dot{\mathbf{O}}^*, \tag{7.107f}$$

for each configuration $\overset{*}{\mathcal{K}}$ obtained from $\overset{\circ}{\mathcal{K}}$ with the help of $H$-transformation $(\overset{\circ}{\mathcal{K}} \to \overset{*}{\mathcal{K}})$ from group $\overset{\circ}{G}_s$.

From Eqs. (7.103e) and (7.107f) it follows that the dissipation function (7.106) must satisfy the relation

$$w^*(\overset{(n)}{\mathbf{A}}{}^*,\ \overset{(n)}{\mathbf{A}}{}^{*\bullet},\ \overset{(n)}{\mathbf{A}}{}^{*h},\ \overset{*}{\mathbf{O}},\ \dot{\mathbf{O}}^*,\ \theta) = w^*(\overset{(n)}{\mathbf{A}},\ \overset{(n)}{\mathbf{A}}{}^{\bullet},\ \overset{(n)}{\mathbf{A}}{}^{h},\ \mathbf{O},\ \dot{\mathbf{O}},\ \theta) \quad \forall \mathbf{H} \in \overset{\circ}{G}_s. \tag{7.108}$$

Below, we will consider solids of the differential type, i.e. the case when $\overset{\circ}{G}_s \subset I$.

Since for all orthogonal groups $\overset{\circ}{G}_s$ the tensors $\overset{(n)}{\mathbf{A}}$ and $\overset{(n)}{\mathbf{A}}{}^h$ are $H$-invariant (see Tables 4.9 and 4.10) and the tensor $\mathbf{O}$ is transformed by formula (4.256) during the passage from $\overset{\circ}{\mathcal{K}}$ to $\overset{*}{\mathcal{K}}$, relation (7.108) can be written in the form

$$w^*(\overset{(n)}{\mathbf{A}},\ \overset{(n)}{\mathbf{A}}{}^{\bullet},\ \overset{(n)}{\mathbf{A}}{}^{h},\ \mathbf{O}\cdot\overset{*}{\mathbf{Q}},\ \dot{\mathbf{O}}\cdot\overset{*}{\mathbf{Q}},\ \theta) = w^*(\overset{(n)}{\mathbf{A}},\ \overset{(n)}{\mathbf{A}}{}^{\bullet},\ \overset{(n)}{\mathbf{A}}{}^{h},\ \mathbf{O},\ \dot{\mathbf{O}},\ \theta) \ \forall \overset{*}{\mathbf{Q}} \in \overset{\circ}{G}_s \subset I. \tag{7.109}$$

The similar relation follows from (7.103a) and (7.107c) for the Helmholtz free energy $\psi$:

$$\psi(\overset{(n)}{\mathbf{A}},\ \mathbf{O}\cdot\overset{*}{\mathbf{Q}},\ \theta) = \psi(\overset{(n)}{\mathbf{A}},\ \mathbf{O},\ \theta) \quad \forall \overset{*}{\mathbf{Q}} \in \overset{\circ}{G}_s \subset I. \tag{7.110}$$

In other words, according to Definition 4.14, the dissipation function (7.106) and the free energy (7.103a) must be rotation-indifferent relative to $\overset{\circ}{G}_s \subset I$.

**Theorem 7.7.** *For models $C_n^h$ of isotropic continua, the dissipation function (7.106) and the free energy (7.103a), which satisfy the principle of material symmetry (i.e. Eqs. (7.109) and (7.110)), are independent of the tensors* $\mathbf{O}$ *and* $\dot{\mathbf{O}}$*; i.e.*

$$\psi = \psi(\overset{(n)}{\mathbf{A}},\ \theta), \tag{7.111}$$

$$w^* = w^*(\overset{(n)}{\mathbf{A}},\ \overset{(n)}{\mathbf{A}}{}^{\bullet},\ \overset{(n)}{\mathbf{A}}{}^{h},\ \theta). \tag{7.112}$$

▼ A proof of the theorem is similar to the proof of the first stage of Theorem 4.27 (see Exercise 7.3.3). ▲

Since the tensors $\overset{(n)}{\mathbf{A}}{}^{\bullet}$ and $\dot{\mathbf{O}}$ are independent, from (7.112) and (7.103e) we obtain that for isotropic solids the following relations must hold:

$$\overset{\circ}{\mathbf{S}}_v \equiv \overset{\circ}{\mathbf{\Phi}}_v \equiv 0, \tag{7.113a}$$

$$\overset{(n)}{\mathbf{S}}_v = \mathbf{\Phi}_v(\overset{(n)}{\mathbf{A}}, \overset{(n)}{\mathbf{A}}{}^h, \theta); \tag{7.113b}$$

i.e. the rotation tensor of viscous stresses is identically zero, and the tensor of viscous stresses $\overset{(n)}{\mathbf{S}}_v$ is independent of $\mathbf{O}$ and $\dot{\mathbf{O}}$.

Moreover, since for isotropic continua $\psi$ is independent of $\mathbf{O}$, from (7.103d) it follows that the rotation tensor of equilibrium stresses is identically zero too:

$$\overset{\circ}{\mathbf{S}}_e = \overset{\circ}{\mathbf{\Phi}} \equiv 0. \tag{7.114}$$

Since both the rotation tensors $\overset{\circ}{\mathbf{S}}_v$ and $\overset{\circ}{\mathbf{S}}_e$ are identically zero, so, according to Theorem 4.9 of Sect. 4.2.18, the corresponding tensors $\overset{(n)}{\mathbf{S}}_v$ and $\overset{(n)}{\mathbf{S}}_e$ for isotropic continua are commutative with $\overset{(n)}{\mathbf{A}}$:

$$\overset{(n)}{\mathbf{S}}_v \cdot \overset{(n)}{\mathbf{A}} = \overset{(n)}{\mathbf{A}} \cdot \overset{(n)}{\mathbf{S}}_v, \tag{7.115}$$

$$\overset{(n)}{\mathbf{S}}_e \cdot \overset{(n)}{\mathbf{A}} = \overset{(n)}{\mathbf{A}} \cdot \overset{(n)}{\mathbf{S}}_e. \tag{7.116}$$

Then, according to Theorem 4.11, the stress power $w_{(i)}$ and hence the viscous stress power can be represented in terms of the co-rotational derivatives:

$$w_v \equiv \overset{(n)}{\mathbf{S}}_v \cdot\cdot \overset{(n)}{\mathbf{A}}{}^{\bullet} = \overset{(n)}{\mathbf{S}}_v \cdot\cdot \overset{(n)}{\mathbf{A}}{}^h = w^*, \quad h = \{U, V, S, J\}. \tag{7.117}$$

Since the tensor of viscous stresses depends only on $\overset{(n)}{\mathbf{A}}$, $\overset{(n)}{\mathbf{A}}{}^h$ and $\theta$, from (7.117) and (7.103e) it follows that the dissipation function (7.112) for isotropic continua is independent of $\overset{(n)}{\mathbf{A}}{}^{\bullet}$, i.e.

$$w^* = w^*(\overset{(n)}{\mathbf{A}}, \overset{(n)}{\mathbf{A}}{}^h, \theta), \tag{7.118}$$

that ensures that the principle of equipresence be satisfied.

### 7.3.4 Viscosity Tensor in Models $C_n^h$

Let us apply now the Onsager principle to models $C_n^h$ of isotropic continua. Due to this principle, the function of viscous stresses (7.113b) is quasilinear with respect to $\overset{(n)}{\mathbf{A}}{}^h$:

$$\overset{(n)}{\mathbf{S}}_\nu = \mathbf{\Phi}_\nu(\overset{(n)}{\mathbf{A}},\ \overset{(n)}{\mathbf{A}}{}^h,\ \theta) = {}^4\mathbf{L}_\nu \cdot\cdot \overset{(n)}{\mathbf{A}}{}^h. \tag{7.119}$$

Here ${}^4\mathbf{L}_\nu$ is the viscosity tensor being positive-definite:

$$w^* = \overset{(n)}{\mathbf{S}}_\nu \cdot\cdot \overset{(n)}{\mathbf{A}}{}^h = \overset{(n)}{\mathbf{A}}{}^h \cdot\cdot {}^4\mathbf{L}_\nu \cdot\cdot \overset{(n)}{\mathbf{A}}{}^h \geqslant 0, \tag{7.120}$$

and having the symmetry (7.21).

**Theorem 7.8.** *For models $C_n$ of isotropic solids, the viscosity tensor ${}^4\mathbf{L}_\nu$ has the form*

$${}^4\mathbf{L}_\nu = \mu_1 \mathbf{E} \otimes \mathbf{E} + 2\mu_2 \mathbf{\Delta}, \tag{7.121}$$

*and the function of viscous stresses becomes*

$$\overset{(n)}{\mathbf{S}}_\nu = \mathbf{\Phi}_\nu = \mu_1 I_1(\overset{(n)}{\mathbf{A}}{}^h)\mathbf{E} + 2\mu_2 \overset{(n)}{\mathbf{A}}{}^h, \tag{7.122}$$

*where $\mu_\gamma$ are the viscous coefficients.*

▼ Notice that the quasilinear function (7.119) of the argument $\overset{(n)}{\mathbf{A}}{}^h$ is indifferent relative to the triclinic group $E$, i.e. satisfies the condition

$$\overset{*}{\mathbf{Q}}{}^{\mathrm{T}} \cdot ({}^4\mathbf{L}_\nu \cdot\cdot \overset{(n)}{\mathbf{A}}{}^h) \cdot \overset{*}{\mathbf{Q}} = {}^4\mathbf{L}_\nu \cdot\cdot (\overset{*}{\mathbf{Q}}{}^{\mathrm{T}} \cdot \overset{(n)}{\mathbf{A}}{}^h \cdot \mathbf{Q}) \tag{7.123}$$

when $\overset{*}{\mathbf{Q}} \in \{\mathbf{E}, -\mathbf{E}\}$.

Just as in Theorem 4.27, we want to find a maximum orthogonal subgroup $\overset{\circ}{G}_s \subset I$, relative to which the tensor function (7.119) is indifferent.

But in the triclinic group there is a tensor basis consisting of six tensors (see [12]), for which we can resolve any tensor function (7.119); this is the set $\{\mathbf{E},\ \overset{(n)}{\mathbf{A}}{}^h,\ \overset{(n)}{\mathbf{A}}{}^{h2},\ \mathbf{O}_1, \mathbf{O}_2, \mathbf{O}_3\}$, i.e.

$$\overset{(n)}{\mathbf{S}}_\nu = {}^4\mathbf{L}_\nu \cdot\cdot \overset{(n)}{\mathbf{A}}{}^h = \varphi_1 \mathbf{E} + \varphi_2 \overset{(n)}{\mathbf{A}}{}^h + \varphi_3 \overset{(n)}{\mathbf{A}}{}^{h\,2} + \sum_{\gamma=1}^{3} \varphi_{3+\gamma} \mathbf{O}_\gamma, \tag{7.124}$$

where $\varphi_\gamma$ are the resolution coefficients.

As shown above (see (7.115)), for isotropic solids of the differential type the rotation viscosity tensor $\overset{\circ}{\mathbf{S}}_\nu$ commutes with the tensor $\overset{(n)}{\mathbf{A}}$. Substitution of (7.124) into (7.115) yields

$$\sum_{\gamma=1}^{3} \varphi_{3+\gamma}(\mathbf{O}_\gamma \cdot \overset{(n)}{\mathbf{A}} - \overset{(n)}{\mathbf{A}} \cdot \mathbf{O}_\gamma) = 0. \tag{7.125}$$

Hence the coefficients $\varphi_{3+\gamma}$ $(\gamma = 1, 2, 3)$ in the representation (7.124) must always vanish.

Moreover, since the function (7.119) is quasilinear in $\overset{(n)}{\mathbf{A}}{}^h$, so $\varphi_3 = 0$; and then from (7.124) we obtain that the function (7.119) for an isotropic continuum can always be written in the form

$$\overset{(n)}{\mathbf{S}}_\nu = \varphi_1 \mathbf{E} + \varphi_2 \overset{(n)}{\mathbf{A}}{}^h = {}^4\mathbf{L}_\nu \cdot\cdot \overset{(n)}{\mathbf{A}}{}^h. \tag{7.126}$$

Introducing the notation

$$\varphi_1 = \mu_1 I_1(\overset{(n)}{\mathbf{A}}{}^h), \qquad \varphi_2 = 2\mu_2, \tag{7.127}$$

from (7.126) and (7.127) we really get representations (7.121) and (7.122). ▲

### 7.3.5 *Final Representation of Constitutive Equations for Model $C_n^h$ of Isotropic Solids*

Let us establish now a final form of constitutive equations for the models $C_n^h$ of isotropic solids.

**Theorem 7.9.** *For models $C_n^h$ of isotropic solids, constitutive equations (7.111) and (7.103c) for equilibrium stresses can be represented as functions of three principal invariants $I_\alpha(\overset{(n)}{\mathbf{A}})$:*

$$\psi(\overset{(n)}{\mathbf{A}}, \theta) = \psi(I_\alpha(\overset{(n)}{\mathbf{A}}), \theta), \qquad \alpha = 1, 2, 3, \tag{7.128}$$

$$\overset{(n)}{\mathbf{S}}_e = \mathbf{\Phi} \equiv \rho(\partial\psi/\partial\overset{(n)}{\mathbf{A}}) = \psi_1 \mathbf{E} + \psi_2 \overset{(n)}{\mathbf{A}} + \psi_3 \overset{(n)}{\mathbf{A}}{}^2, \tag{7.129}$$

*where $\psi_\alpha$ are expressed by formulae (4.336).*

▼ A proof of the theorem is similar to the proof of Theorem 4.27. ▲

Formulae (7.122) and (7.129) give the possibility to derive a final representation of constitutive equations for the model $C_n^h$ of an isotropic solid:

$$\overset{(n)}{\mathbf{S}} = \psi_1 \mathbf{E} + \psi_2 \overset{(n)}{\mathbf{A}} + \psi_3 \overset{(n)}{\mathbf{A}}{}^2 + \mu_1 I_1(\overset{(n)}{\mathbf{A}}{}^h)\mathbf{E} + 2\mu_2 \overset{(n)}{\mathbf{A}}{}^h. \tag{7.130}$$

The relation obtained is correct, because it satisfies the principle of material indifference.

Indeed, in the rigid motion from actual configuration $\mathcal{K}$ to $\mathcal{K}'$ the relationship (7.130) is transformed as follows:

$$\psi_1 \mathbf{E} + \psi_2 \overset{(n)}{\mathbf{A}}{}' + \psi_3 \overset{(n)}{\mathbf{A}}{}'^2 + \mu_1 I_1(\overset{(n)}{\mathbf{A}}{}^{h\,\prime})\mathbf{E} + 2\mu_2 \overset{(n)}{\mathbf{A}}{}^{h\,\prime}$$
$$= \mathbf{Q}^{\mathrm{T}} \cdot (\psi_1 \mathbf{E} + \psi_2 \overset{(n)}{\mathbf{A}} + \psi_3 \overset{(n)}{\mathbf{A}}{}^2 + \mu_1 I_1(\overset{(n)}{\mathbf{A}}{}^h)\mathbf{E} + 2\mu_2 \overset{(n)}{\mathbf{A}}{}^h) \cdot \mathbf{Q}. \tag{7.131}$$

*Remark 1.* Notice that although the representation (7.117) for the viscous stress power still holds for the co-rotational derivatives included in (7.117), however, for the final result (7.130) we need to return to the list (7.101) with the excluded right derivative in the eigenbasis $h = U$, because this derivative, according to Table 4.12, does not maintain $R$-indifference of the derivative $\overset{(n)}{\mathbf{A}}{}^h$ for the $R$-indifferent tensor $\overset{(n)}{\mathbf{A}}$. But all co-rotational derivatives occurring in the list (7.101) satisfy this requirement, that has been used in deriving the formula (7.131). □

*Remark 2.* The coefficients $\psi_\gamma$ occurring in (7.130) are functions of the invariants $I_\gamma(\overset{(n)}{\mathbf{A}})$ (see formulae (4.336)), and the viscous coefficients depend on $\overset{(n)}{\mathbf{A}}$, $\overset{(n)}{\mathbf{A}}{}^h$ and $\theta$:

$$\mu_\gamma = \mu_\gamma(\overset{(n)}{\mathbf{A}},\ \overset{(n)}{\mathbf{A}}{}^h,\ \theta). \tag{7.132}$$

Since the functions (7.132) cannot vary in rigid motions of actual configuration $\mathcal{K} \longrightarrow \mathcal{K}'$, they must satisfy the relation

$$\mu_\gamma(\overset{(n)}{\mathbf{A}},\ \overset{(n)}{\mathbf{A}}{}^h,\ \theta) = \mu_\gamma(\mathbf{Q}^{\mathrm{T}} \cdot \overset{(n)}{\mathbf{A}} \cdot \mathbf{Q},\ \mathbf{Q}^{\mathrm{T}} \cdot \overset{(n)}{\mathbf{A}}{}^h \cdot \mathbf{Q},\ \theta) \quad \forall \mathbf{Q} \in I \tag{7.133}$$

(due to $R$-indifference of $\overset{(n)}{\mathbf{A}}$ and $\overset{(n)}{\mathbf{A}}{}^h$). But this means that $\mu_\gamma$ must be $H$-indifferent relative to the full orthogonal group.

Then, according to the results of Sect. 7.1.4, the functions (7.132) must depend on simultaneous invariants $J_\alpha$ of the tensors $\overset{(n)}{\mathbf{A}}$ and $\overset{(n)}{\mathbf{A}}{}^h$ relative to the group $I$:

$$\mu_\gamma = \mu_\gamma(J_1, \ldots, J_{10},\ \theta), \quad \gamma = 1,\ 2; \qquad J_\alpha = J_\alpha(\overset{(n)}{\mathbf{A}},\ \overset{(n)}{\mathbf{A}}{}^h), \qquad \alpha = 1, \ldots, 10. \tag{7.134}$$

Simultaneous invariants $J_\alpha$ satisfy the condition (7.133) and can be formed as shown in (7.35).

Due to $H$-invariance of the tensors $\overset{(n)}{\mathbf{A}}$ and $\overset{(n)}{\mathbf{A}}{}^h$, under orthogonal transformations of reference configuration $\overset{\circ}{\mathcal{K}} \to \overset{*}{\mathcal{K}}$ the viscous coefficients (7.132) always remain unchanged, that has been used in proving the formula (7.131). □

### 7.3.6 Models $D_n^h$ of Isotropic Solids

Models $D_n^h$ of solids of the differential type can be obtained from Eqs. (7.98)–(7.100) for models $C_n^h$ by replacing the deformation tensors $\overset{(n)}{\mathbf{A}}$ with the measures $\overset{(n)}{\mathbf{g}}$.

Due to $H$-invariance of the tensors $\overset{(n)}{\mathbf{g}}$, all manipulations performed for models $C_n^h$ remain valid for models $D_n^h$ too. In particular, the final relations (7.130) for isotropic solids in the model $D_n^h$ take the form

$$\overset{(n)}{\mathbf{S}} = \psi_1\mathbf{E} + \psi_2\overset{(n)}{\mathbf{g}} + \psi_3\overset{(n)}{\mathbf{g}}{}^2 + \mu_1 I_1(\overset{(n)}{\mathbf{g}}{}^h)\mathbf{E} + 2\mu_2\overset{(n)}{\mathbf{g}}{}^h, \tag{7.135}$$

where the coefficients $\psi_\gamma$ are functions of the invariants $I_\gamma(\overset{(n)}{\mathbf{g}})$ (see formulae (4.380)), and the viscous coefficients $\mu_\gamma$ depend on simultaneous invariants of the tensors $\overset{(n)}{\mathbf{g}}$ and $\overset{(n)}{\mathbf{g}}{}^h$:

$$\mu_\gamma = \mu_\gamma(J_1,\dots,J_{10},\theta),\quad \gamma = 1, 2;\qquad J_\alpha = J_\alpha(\overset{(n)}{\mathbf{g}}, \overset{(n)}{\mathbf{g}}{}^h),\quad \alpha = 1,\dots,10. \tag{7.136}$$

Relations (7.135) satisfy the principles of material symmetry (since all the tensors $\overset{(n)}{\mathbf{S}}$, $\overset{(n)}{\mathbf{g}}$ and $\overset{(n)}{\mathbf{g}}{}^h$ are $H$-invariant under orthogonal transformations) and material indifference (since all the tensors $\overset{(n)}{\mathbf{S}}$, $\overset{(n)}{\mathbf{g}}$, and $\overset{(n)}{\mathbf{g}}{}^h$ are $R$-indifferent).

## Exercises for 7.3

**7.3.1.** Prove Theorem 7.9.

**7.3.2.** Prove that the Onsager principle yields representation (7.97) for the functions $\mathbf{\Phi}_\gamma$.

**7.3.3.** Prove Theorem 7.7.

**7.3.4.** Using relations (7.120) and (7.122), show that for models of isotropic continua of the differential type the dissipation function has the form

$$w^* = \mu_1 I_1^2(\overset{(n)}{\mathbf{A}}{}^h) + 2\mu_2 I_1((\overset{(n)}{\mathbf{A}}{}^h)^2).$$

## 7.4 The Problem on a Beam in Tension

### 7.4.1 Rate Characteristics of a Beam

Consider the problem on a beam in tension, which has been investigated in Sect. 6.4.1, but the beam is made of material described by the model $B_n$ of incompressible continua of the differential type with steady creep (see (7.49)). The motion law and the tensors $\mathbf{F}$, $\overset{(n)}{\mathbf{C}}$ in this case retain their forms (6.131)–(6.134), and the velocity $\mathbf{v}$ and the velocity gradient $\mathbf{L}$, according to (2.223), become

$$\mathbf{v} = \sum_{\alpha=1}^{3} \dot{k}_\alpha X^\alpha \bar{\mathbf{e}}_\alpha^2, \quad \mathbf{L} = \frac{\partial \mathbf{v}}{\partial X^i} \otimes \mathbf{r}^i = \sum_{\alpha=1}^{3} \frac{\dot{k}_\alpha}{k_\alpha} \bar{\mathbf{e}}_\alpha^2. \tag{7.137}$$

Here we have used that $\mathbf{r}^\alpha = \bar{\mathbf{e}}_\alpha / k_\alpha$ (see Exercise 2.1.1).

The rates $\overset{(n)}{\mathbf{C}}{}^\bullet$ of the deformation tensors and their deviators can be found by differentiating the expression (6.133):

$$\overset{(n)}{\mathbf{C}}{}^\bullet = \sum_{\alpha=1}^{3} k_\alpha^{n-\mathrm{III}-1} \dot{k}_\alpha \bar{\mathbf{e}}_\alpha^2, \quad \operatorname{dev} \overset{(n)}{\mathbf{G}}{}^\bullet = \sum_{\alpha=1}^{3} \overset{(n)}{f}_\alpha{}^\bullet(k_1) \bar{\mathbf{e}}_\alpha^2, \tag{7.138}$$

$$\overset{(n)}{f}_1(k_1) = \frac{2}{3(n-\mathrm{III})}\left(k_1^{n-\mathrm{III}} - k_1^{\frac{n-\mathrm{III}}{2}}\right), \quad \overset{(n)}{f}_2 = \overset{(n)}{f}_3 = -\frac{1}{2}\overset{(n)}{f}_1. \tag{7.139}$$

Here we have taken into account that for incompressible materials $k_2 = k_3 = 1/\sqrt{k_1}$ (see Sect. 6.5.1).

### 7.4.2 Stresses in the Beam

On substituting the expression (7.138) for $\operatorname{dev} \overset{(n)}{\mathbf{C}}{}^\bullet$ into constitutive equations (7.49), we find components $\overset{(n)}{T}_{\alpha\alpha}$ of the energetic stress tensors

$$\overset{(n)}{\mathbf{T}} = \sum_{\alpha=1}^{3} \overset{(n)}{T}_{\alpha\alpha} \bar{\mathbf{e}}_\alpha^2, \quad \overset{(n)}{T}_{\alpha\alpha} = -p k_\alpha^{\mathrm{III}-n} + 2\mu_2 \overset{(n)}{f}_\alpha{}^\bullet. \tag{7.140}$$

The Cauchy stress tensor **T** is calculated by the same relations (6.138) just as for elastic materials:

$$\mathbf{T} = \sum_{\alpha=1}^{3} \sigma_{\alpha\alpha}\bar{\mathbf{e}}_{\alpha}^{2}, \quad \sigma_{\alpha\alpha} = k_{\alpha}^{n-\mathrm{III}} \overset{(\mathrm{n})}{T}_{\alpha\alpha}, \tag{7.141}$$

$$\sigma_{\alpha\alpha} = -p + 2\mu_2 \overset{(\mathrm{n})}{f}_{\alpha}{}^{\bullet} k_{\alpha}^{n-\mathrm{III}}. \tag{7.142}$$

Just as in the case of an elastic beam, the Cauchy stress tensor (7.141) is independent of coordinates; therefore, the equilibrium equations for the beam, when there are no mass forces, are identically satisfied.

### 7.4.3 Resolving Relation $\sigma(k_1, \dot{k}_1)$

On substituting (7.142) into the boundary condition (6.141c), we find that, just as for an elastic beam, the components $\sigma_{22}$ and $\sigma_{33}$ must be zero, and the only component $\sigma_{11}$ is nonzero. As a result, we obtain two relations

$$\sigma_{11} = -p + 2\mu_2 \overset{(\mathrm{n})}{f}_{1}{}^{\bullet} k_1^{n-\mathrm{III}} \quad \text{and} \quad 0 = -p + \mu_2 \overset{(\mathrm{n})}{f}_{1}{}^{\bullet} k_1^{\frac{\mathrm{III}-n}{2}}, \tag{7.143}$$

for the three unknowns: $k_1$, $p$ and $\sigma_{11}$. Excluding $p$ from this system, we get the following resolving relation for the problem:

$$\sigma_{11} = \frac{2\mu_2}{3k_1} \overset{(\mathrm{n})}{L}{}^{2}(k_1)\dot{k}_1, \tag{7.144}$$

where

$$\overset{(\mathrm{n})}{L}(k_1) = k_1^{n-\mathrm{III}} + \frac{1}{2} k_1^{\frac{\mathrm{III}-n}{2}}. \tag{7.145}$$

### 7.4.4 Comparative Analysis of Creep Curves for Different Models $B_n$

Consider a *stepwise process of loading* (Fig. 7.1) when the stress $\sigma_{11}(t)$ is given in the form

$$\sigma_{11}(t) = \sigma^0 h(t), \tag{7.146}$$

where $\sigma^0$ is the constant, and $h(t)$ is the *Heaviside function*

$$h(t) = \begin{cases} 0, & t \leqslant 0, \\ 1, & t > 0. \end{cases} \tag{7.147}$$

The elongation function $\delta_1(t) = k_1(t) - 1$ (measured in experiments) for a beam of material of the differential type under such loading is usually called the *creep*

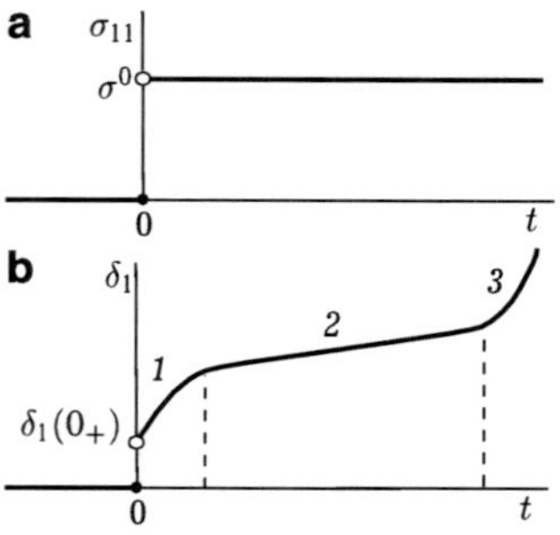


**Fig. 7.1** The stepwise process of loading (**a**) and the typical curve of creep (**b**)

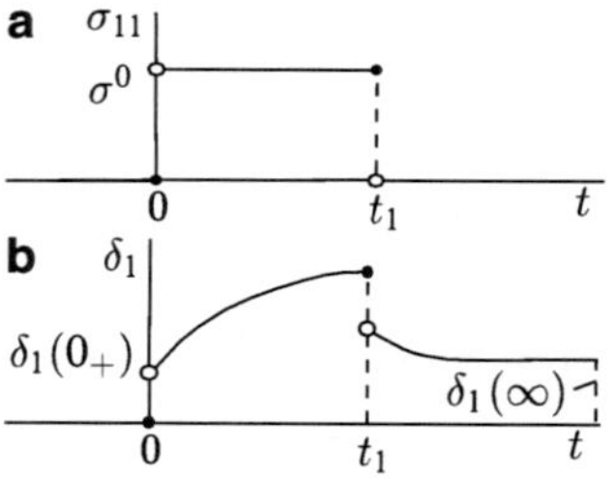


**Fig. 7.2** The process of loading–unloading (**a**) and creep curve in this process (**b**)

*curve*. A typical creep curve for materials, whose properties depend on the deformation rate, is shown in Fig. 7.1. There are three typical sections along this curve: (1) the initial section, (2) the section of steady creep, and (3) the section of unsteady creep. The value of elongation $\delta_1(0_+)$ at $t \to 0_+$ is called the instantly elastic elongation. The creep effect is observed for metals and alloys under high temperatures.

If at the time $t_1$ unloading occurs (Fig. 7.2):

$$\sigma_{11}(t) = \sigma^0(h(t) - h(t - t_1)), \tag{7.148}$$

then the elongation $\delta_1(t)$ decreases, and for materials of the differential type the elongation does not return to its initial zero value as $t \to \infty$ (Fig. 7.2), i.e. there appears a *residual deformation of creep* $\delta_1(\infty)$.

If the process of loading $\sigma_{11}(t)$ is given, then Eq. (7.144) is a nonlinear ordinary differential equation for the functions $k_1$. Solving the equation, we obtain

$$k_1(t) = H^{-1}\left(\int_0^t \frac{\sigma_{11}(\tau)}{\mu_2} d\tau\right), \quad H(k_1) \equiv \frac{2}{3}\int_1^{k_1} \frac{\overset{(n)}{L}{}^2(k)}{k} dk. \tag{7.149}$$

On substituting (7.146) into (7.149), we find that under a constant load the elongation of the beam grows with time:

$$\delta_1(t) = H^{-1}(\sigma^0 t/\mu_2) - 1, \tag{7.150}$$

and under unloading the elongation remains constant and does not decrease: $\delta_1(t) = \delta_1(t_1)$ at $t > t_1$.

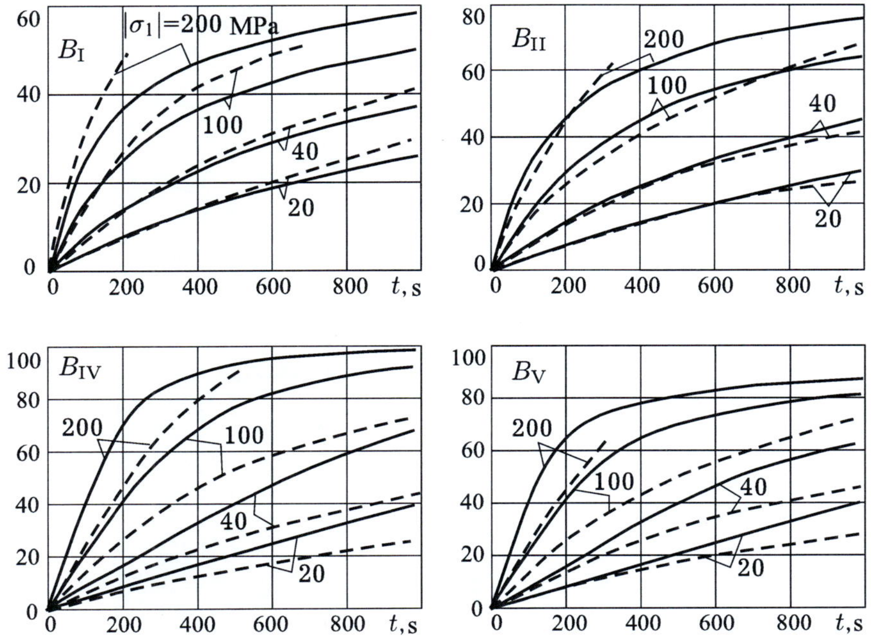


**Fig. 7.3** Creep curves for nickel alloy (solid curves are computations, dashed lines are experimental data)

Figure 7.3 shows computed and experimental creep curves $\delta_1(t)$ for nickel alloy at temperature 1100°C in compression for different values of the stress $\sigma^0 \leqslant 0$. Theoretical curves have been computed according to formula (7.149) for different models $B_n$, and the creep curves $\delta_{1\rm p}(t)$ are the difference in elongation between the experimental values $\delta_1(t)$ and the initial value: $\delta_{1\rm p}(t) = \delta_1(t) - \delta_1(0_+)$. One of the experimental creep curves $\delta_{1\rm p}(t)$ at the smallest value $\sigma^0$ was used for determining the constant $\mu_2$, that was calculated by minimizing the mean-square distance between computed and experimental results at $N$ points being times $t_i$:

$$\Delta = \left(\frac{1}{N}\sum_{i=1}^{N}\left(1 - \frac{\delta_1(t_i)}{\delta_{1\rm p}(t_i)}\right)^2\right)^{1/2} \to \min. \qquad (7.151)$$

Table 7.1 shows values of the constant $\mu_2$ for nickel alloy, computed by the method mentioned above for different models $B_n$. The model $B_{\rm II}$ exhibits the best approximation to the experimental data for the case considered (Fig. 7.3). It should be noted that for many metals, values of the high-temperature creep elongation $\delta_{1\rm p}(t)$ at sufficiently great $t$ considerably exceed instantly elastic values $\delta_1(0_+)$, therefore the last ones are often neglected in computations of creep problems. The considered models $B_n$ of continua of the differential type with steady creep fall into this class of models.

**Table 7.1** Values of the constant $\mu_2$ and the relative error $\delta$ of approximation to the creep curves for models $B_n$ of continua of the differential type for nickel alloy at temperature 1100°C

| $n$ | I | II | IV | V |
|---|---|---|---|---|
| $\mu_2$, GPa·s | 30 | 30 | 32 | 35 |
| $\delta$, % | 18 | 17 | 19 | 20 |

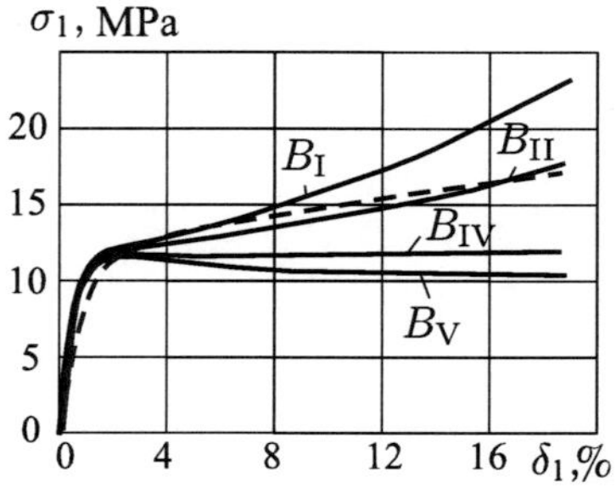


**Fig. 7.4** Diagrams of deforming for nickel in compression

## 7.4.5 *Analysis of Deforming Diagrams for Different Models $B_n$ of Continua of the Differential Type*

Consider one more regime of deforming, when the deforming process of a beam is given by the law

$$k_1(t) = 1 + \begin{cases} bt^2/2t_1, & t < t_1, \\ b(t - t_1/2), & t \geqslant t_1. \end{cases} \quad (7.152)$$

where $b$ = const is the rate of deforming, and $t_1$ is the beginning of deforming with a constant rate (the initial interval $0 \leqslant t \leqslant t_1$ is necessary for the consistency conditions (7.50) to be satisfied). The rate $\dot{k}_1$ is determined by the expression

$$\dot{k}_1 = \sqrt{(2k_1 b/t_1)} \quad \text{when } t < t_1$$

and

$$\dot{k}_1 = b \quad \text{when } t \geqslant t_1.$$

Substituting (7.152) into (7.144), we obtain the relation $\sigma_{11}(k_1)$ called the diagram of deforming for the beam considered.

Figure 7.4 exhibits the experimental diagram of deforming for nickel in compression at 1100 °C and $b = -0.00025\,\text{s}^{-1}$ and computed diagrams, for which the viscosity coefficient $\mu_2$ has been evaluated with the help of the creep curves (see Sect. 7.4.4 and Table 7.1). The model $B_{\text{II}}$ gives the best approximation to the experimental data.

# Chapter 8
# Viscoelastic Continua at Large Deformations

## 8.1 Viscoelastic Continua of the Integral Type

### *8.1.1 Definition of Viscoelastic Continua*

Besides models of the differential type considered in Chap. 7, in continuum mechanics there are other types of nonideal media. One widely uses models of viscoelastic materials, which are also called continua of the integral type, or hereditarily elastic continua. Models of viscoelastic continua most adequately describe the mechanical properties of polymer materials, composites based on polymers, different elastomers, rubbers and biomaterials, in particular, human muscular tissues.

Further, viscoelastic continua of the differential type will be called *continua of the differential type*, and viscoelastic continua of the integral type – simply *viscoelastic materials*.

**Definition 8.1.** A medium is called a *viscoelastic continuum of the integral type* (or simply a viscoelastic continuum), if any of the models $A_n$, $B_n$, $C_n$ or $D_n$ is assumed for the medium, and corresponding operator constitutive equations (4.156) or (4.158)–(4.161) are *functionals* of time $t$:

$$\Lambda(t) = \underset{\tau=0}{\overset{\tau=t}{f}} (\mathcal{R}(t),\ \mathcal{R}^t(\tau)), \tag{8.1}$$

i.e. values of active variables $\Lambda(t)$ depend not only on values of reactive variables $\mathcal{R}(t)$ at the same instant of time but also on their prehistory $\mathcal{R}^t(\tau) \equiv \mathcal{R}(t-\tau)$, i.e. on their values at all preceding instants of time $0 < \tau \leqslant t$, starting from initial one $\tau = 0$.

Due to such specific dependence, viscoelastic continua are also called *continua with memory*.

For viscoelastic continua:

1. "The present can depend only on the past but not on the future"; therefore, all the functionals (8.1) depend only on their previous history (*prehistory*) $\mathcal{R}(t-\tau)$, $0 < \tau \leqslant t$.

Y.I. Dimitrienko, *Nonlinear Continuum Mechanics and Large Inelastic Deformations*, Solid Mechanics and Its Applications 174, DOI 10.1007/978-94-007-0034-5_8,

2. "The past is not infinite", i.e. the times $\tau > t$ do not affect the functionals (8.1). This means that

$$\mathcal{R}(\tau) \equiv 0 \text{ at } \tau < 0; \quad \mathcal{R}^t(\tau) = \mathcal{R}(t-\tau) \equiv 0 \text{ at } t > \tau. \tag{8.2}$$

## 8.1.2 Tensor Functional Space

To perform operations with functionals (8.1), we need some additional knowledge of functional analysis.

Consider the set of previous histories of $k$-order tensors ${}^k\mathbf{T}^t(\tau) = {}^k\mathbf{T}(t-\tau)$ $(0 < \tau \leqslant t)$, and define for two arbitrary prehistories ${}^k\mathbf{T}_1^t$ and ${}^k\mathbf{T}_2^t$ their scalar product as follows:

$$({}^k\mathbf{T}_1^t, {}^k\mathbf{T}_2^t)_t = \int_0^t {}^k\mathbf{T}_1^t(\tau) \underbrace{\cdots\cdots}_{k} ({}^k\mathbf{T}_2^t(\tau))^{(k\ldots1)} \gamma^2(\tau)\, d\tau. \tag{8.3}$$

The function $\gamma(\tau)$ is called the *function of memory*. This function is positive, continuous, monotonically decreasing, defined within the interval $[0, +\infty)$ and squared-integrable, i.e.

$$\int_0^\infty \gamma^2(\tau)\, d\tau = \gamma_0^2 < +\infty. \tag{8.4}$$

Since the memory function is monotonically decreasing, quantities of the prehistory ${}^k\mathbf{T}^t(\tau) = {}^k\mathbf{T}(t-\tau)$ at small values of $\tau$ give a greater contribution to the scalar product (8.3) than quantities of ${}^k\mathbf{T}^t(\tau)$ at large values of $\tau$. In other words, a continuum better remembers events having occurred at times closer to the current instant $t$ than the ones at more remote times. The functionals (8.1) are assumed to have this property too; therefore, viscoelastic continua of the integral type are also called continua with *fading memory*.

Let us consider now the set ${}^k\mathcal{H}_t$ of processes of changing the tensor ${}^k\mathbf{T}(\tau)$ $(0 < \tau \leqslant t)$ and assign to each process the pair $({}^k\mathbf{T}(t), {}^k\mathbf{T}(\tau))$ consisting of values of the tensor ${}^k\mathbf{T}(t)$ at time $t$ and its prehistory ${}^k\mathbf{T}(\tau) = {}^k\mathbf{T}(t-\tau)$ $(0 < \tau \leqslant t)$. Then we can introduce the scalar product of processes ${}^k\mathbf{T}_1(\tau)$ and ${}^k\mathbf{T}_2(\tau)$ included in ${}^k\mathcal{H}_t$:

$$\left({}^k\mathbf{T}_1, {}^k\mathbf{T}_2\right)_t = \left({}^k\mathbf{T}(t), {}^k\mathbf{T}(t)\right) + \left({}^k\mathbf{T}_1^t, {}^k\mathbf{T}_2^t\right)_t, \tag{8.5}$$

where

$$\left({}^k\mathbf{T}_1(t), {}^k\mathbf{T}_2(t)\right) = {}^k\mathbf{T}_1(t) \underbrace{\cdots\cdots}_{k} \left({}^k\mathbf{T}_2(t)\right)^{(k\,\ldots\,1)} \tag{8.6}$$

is the scalar product of $k$-order tensors.

The set ${}^k\mathcal{H}_t$ of all processes of changing the tensor ${}^k\mathbf{T}(\tau)$ $(0 \leqslant \tau \leqslant t)$, for which the scalar product (8.5) exists and which at each fixed $\tau$ are elements of the tensor space $\mathcal{T}_3^k(\mathcal{E}_3)$ with the operations of addition and multiplication by a number, is called the *tensor functional space* ${}^k\mathcal{H}_t$.

The space ${}^k\mathcal{H}_t$ is a Hilbert space, because we can always go into a Cartesian basis, where the components $\bar{T}^{i_1\dots i_k}(\tau)$ of tensors included in ${}^k\mathcal{H}_t$ are squared-integrable functions, i.e. they belong to the function space $L_2^{(m)}[0,t]$ (where $m = 3^k$), which is known as a Hilbert space.

Due to the property (8.2), the scalar product of processes included in ${}^k\mathcal{H}_t$ (8.5) can be written in the form

$$({}^k\mathbf{T}_1^t,\ {}^k\mathbf{T}_2^t)_t = \int\limits_0^\infty {}^k\mathbf{T}_1^t(\tau) \underbrace{\cdot \ldots \cdot}_{k} (\mathbf{T}_2^t(\tau))^{(k\dots 1)}\gamma^2(\tau)\, d\tau < +\infty, \tag{8.7}$$

which often proves to be convenient for analysis of the models of a viscoelastic continuum.

In the space ${}^k\mathcal{H}_t$, there is a natural norm of the process ${}^k\mathbf{T}(\tau)$:

$$\| {}^k\mathbf{T} \| = ({}^k\mathbf{T},\ {}^k\mathbf{T})_t^{1/2}, \tag{8.8}$$

where $(\,\cdot\,)_t$ is the scalar product (8.5).

### *8.1.3 Continuous and Differentiable Functionals*

Using the norm (8.8), we can introduce the concept of a *continuous functional* in the form (8.1)

$${}^m\mathbf{S} = {}^m\underset{\tau=0}{\overset{t}{\mathcal{F}}}({}^k\mathbf{T}(t),\ {}^k\mathbf{T}^t(\tau)), \tag{8.9}$$

which is considered as the mapping of the domain $U$ contained in the space ${}^k\mathcal{H}_t$ into the domain $V$ of the space ${}^m\mathcal{H}_t$:

$${}^m\mathcal{F}: \quad U \subset {}^k\mathcal{H}_t \to V \subset {}^m\mathcal{H}_t. \tag{8.10}$$

**Definition 8.2.** The functional (8.10) is called *continuous* in a domain $U \subset {}^k\mathcal{H}_t$, if for each process ${}^k\mathbf{T} \in U$ the following condition is satisfied: $\forall \varepsilon > 0\ \exists \delta > 0$ such that for every process ${}^k\widetilde{\mathbf{T}}$, where $({}^k\mathbf{T} + {}^k\widetilde{\mathbf{T}}) \in U$ and

$$\| {}^k\widetilde{\mathbf{T}} \|_t < \delta, \tag{8.11}$$

values of the operators in the norm (8.8) of the space ${}^k\mathcal{H}_t$ are sufficiently close:

$$\| {}^m\mathop{\mathcal{F}}_{\tau=0}^{t}\left({}^k\mathbf{T}(t)+{}^k\widetilde{\mathbf{T}}(t),\ {}^k\mathbf{T}^t(\tau)+{}^k\widetilde{\mathbf{T}}^t(\tau)\right)-{}^m\mathop{\mathcal{F}}_{\tau=0}^{t}({}^k\mathbf{T}(t),\ {}^k\mathbf{T}^t(\tau))\ \|_t<\varepsilon. \tag{8.12}$$

The functional (8.10) is called *linear* if it satisfies the two conditions

$$\mathop{\mathcal{F}}_{\tau=0}^{t}\left({}^k\mathbf{T}_1(t)+{}^k\mathbf{T}_2(t),\ {}^k\mathbf{T}_1(\tau)+{}^k\mathbf{T}_2(\tau)\right)$$
$$=\mathop{\mathcal{F}}_{\tau=0}^{t}\left({}^k\mathbf{T}_1(t),\ {}^k\mathbf{T}_1(\tau)\right)+\mathop{\mathcal{F}}_{\tau=0}^{t}\left({}^k\mathbf{T}_2(t),\ {}^k\mathbf{T}_2(\tau)\right), \tag{8.13}$$

$$\mathop{\mathcal{F}}_{\tau=0}^{t}\left(s{}^k\mathbf{T}(t),\ s{}^k\mathbf{T}(\tau)\right)=s\mathop{\mathcal{F}}_{\tau=0}^{t}\left({}^k\mathbf{T}(t),\ {}^k\mathbf{T}(\tau)\right), \tag{8.14}$$

for all processes ${}^k\mathbf{T}_1(\tau)$ and ${}^k\mathbf{T}_2(\tau)$ included in ${}^k\mathcal{H}_t$ and for every real number $s$.

In space ${}^k\mathcal{H}_t$ we can use *Riesz's theorem* that any linear functional (8.9) can be represented as the scalar product of a fixed element from ${}^k\mathcal{H}_t$ and an arbitrary process ${}^k\mathbf{T}(\tau)$ from ${}^k\mathcal{H}_t$; so a scalar linear functional in ${}^2\mathcal{H}_t$ has the form

$$f(\mathbf{T}(t),\ \mathbf{T}^t(\tau))=\widetilde{\mathbf{\Gamma}}(t,t)\cdot\cdot\,\mathbf{T}^{\mathrm{T}}(t)+\int_0^t\widetilde{\mathbf{\Gamma}}(t,\ t-\tau)\cdot\cdot\,\mathbf{T}^{\mathrm{T}}(t-\tau)\gamma^2(\tau)\,d\tau, \tag{8.15a}$$

where $\widetilde{\mathbf{\Gamma}}(t,t-\tau)$ is the prehistory and $\widetilde{\mathbf{\Gamma}}(t,t)$ is the instantaneous value at $\tau=t$ of the fixed process $\widetilde{\mathbf{\Gamma}}(t,\ \tau)$ $(0\leqslant\tau\leqslant t)$ for the given functional $f$ (the appearance of one more argument $t$ for the process $\widetilde{\mathbf{\Gamma}}(t,\ \tau)$ means that this process can vary with changing the time interval considered).

Replacing the variable $t-\tau=y$ and introducing the notation $\widetilde{\mathbf{\Gamma}}^{\mathrm{T}}(t,y)\gamma^2(\tau)=\mathbf{\Gamma}(t,y)$, $\mathbf{\Gamma}_0=\mathbf{\Gamma}^{\mathrm{T}}(t,t)$, after the reverse substitution $y\to\tau$ we obtain another representation of the linear scalar functional

$$f=\mathbf{\Gamma}_0\cdot\cdot\,\mathbf{T}(t)+\int_0^t\mathbf{\Gamma}(t,\ \tau)\cdot\cdot\,\mathbf{T}(\tau)\,d\tau, \tag{8.15b}$$

called *Volterra's representation.*

For viscoelastic continua, it is convenient to use the Dirac *δ-function* $\delta(t)$ having the following main property:

$$\int_0^t\mathbf{B}(t,\ \tau)\delta(t_0-\tau)\,d\tau=\begin{cases}\mathbf{B}(t,t_0), & t_0\in[0,t],\\ 0, & t_0\notin[0,t],\end{cases} \tag{8.16}$$

for any continuous tensor process $\mathbf{B}(t,\tau)$.

According to (8.16), the linear functional (8.15b) can be written in the form

$$f = \int_0^t \mathbf{A}(t,\ \tau) \cdot\cdot \mathbf{T}(\tau)\ d\tau, \tag{8.15c}$$

$$\mathbf{A}(t,\ \tau) = \boldsymbol{\Gamma}_0 \delta(t-\tau) + \boldsymbol{\Gamma}(t,\ \tau). \tag{8.15d}$$

**Definition 8.3.** The functional (8.9) is called *Fréchet–differentiable* at point ${}^m\mathbf{T} \in U$ of a domain $U \subset {}^m\mathcal{H}_t$, if there exist two functionals $\partial\mathcal{F}$ and $\delta\mathcal{F}$ having the following properties:

- they are defined over the Cartesian product of the space $\mathcal{H}_t$

$$\partial^m\mathcal{F}:\quad {}^k\mathcal{H}_t \times {}^k\mathcal{H}_t \to {}^m\mathcal{H}_t; \qquad \delta^m\mathcal{F}:\quad {}^k\mathcal{H}_t \times {}^k\mathcal{H}_t \to {}^m\mathcal{H}_t, \tag{8.17}$$

- they can be written in the form similar to (8.9)

$$\begin{aligned} {}^m\mathbf{P}_1 &= \partial^m \mathop{\mathcal{F}}_{\tau=0}^{t} ({}^k\mathbf{T}(t),\ {}^k\mathbf{T}^t(\tau) | {}^k\widetilde{\mathbf{T}}(t)) \\ &= \frac{\partial}{\partial^k\mathbf{T}(t)} \mathop{\mathcal{F}}_{\tau=0}^{t} ({}^k\mathbf{T}(t),\ {}^k\mathbf{T}^t(\tau)) \cdot \ldots \cdot {}^k\mathbf{T}^{(k\ldots1)}(t), \end{aligned} \tag{8.18}$$

$${}^m\mathbf{P}_2 = \delta^m \mathop{\mathcal{F}}_{\tau=0}^{t} ({}^k\mathbf{T}(t),\ {}^k\mathbf{T}^t(\tau) | {}^k\widetilde{\mathbf{T}}^t(\tau)), \tag{8.19}$$

(the vertical line separates two different arguments of the process),
- they are linear and continuous in the second argument,
- they satisfy the condition: $\forall\varepsilon > 0\ \exists\delta$ such that for any process $({}^k\widetilde{\mathbf{T}}(t),\ {}^k\widetilde{\mathbf{T}}^t(\tau))$, for which $({}^k\mathbf{T}_1(t),\ {}^k\mathbf{T}_1^t(\tau)) \subset U$ and

$$\| {}^k\mathbf{T} \|_t < \delta, \tag{8.20}$$

the following inequality holds:

$$\| \Delta^m\mathcal{F} \|_t \leqslant \varepsilon \| {}^k\widetilde{\mathbf{T}} \|_t, \tag{8.21}$$

where

$$\begin{aligned} \Delta^m\mathcal{F} = {}&{}^m \mathop{\mathcal{F}}_{\tau=0}^{t} ({}^k\mathbf{T}_1(t),\ {}^k\mathbf{T}_1^t(\tau)) - {}^m \mathop{\mathcal{F}}_{\tau=0}^{t} ({}^k\mathbf{T}(t),\ {}^k\mathbf{T}^t(\tau)) \\ &- \partial^m \mathop{\mathcal{F}}_{\tau=0}^{t} ({}^k\mathbf{T}(t),\ {}^k\mathbf{T}^t(\tau) | {}^k\widetilde{\mathbf{T}}(t)) - \delta \mathop{\mathcal{F}}_{\tau=0}^{t} ({}^k\mathbf{T}(t),\ {}^k\mathbf{T}^t(\tau) | {}^k\widetilde{\mathbf{T}}^t(\tau)), \end{aligned} \tag{8.22}$$

$$({}^k\mathbf{T}_1(t),\ {}^k\mathbf{T}_1^t(\tau)) \equiv ({}^k\mathbf{T}(t) + {}^k\widetilde{\mathbf{T}}(t),\ {}^k\mathbf{T}^t(\tau) + {}^k\widetilde{\mathbf{T}}^t(\tau)). \tag{8.23}$$

The operator (8.19) is called the *Fréchet–derivative*; and the right-hand side of expression (8.18) is the partial derivative of $\mathcal{F}$ (considered as a tensor function of $\mathbf{T}(t)$) with respect to the tensor argument $\mathbf{T}(t)$.

If the functional (8.9) is Fréchet–differentiable, then it is continuous (see Exercise 8.1.2).

The set of processes ${}^k\mathbf{T}(t) \in {}^k\mathcal{H}_{t'}$ having the first and second continuous derivatives with respect to time $t$: ${}^k\dot{\mathbf{T}}(t)$ and ${}^k\ddot{\mathbf{T}}(t)$, which belong to ${}^k\mathcal{H}_{t'}$, will be denoted by $U_{t'}$.

**Theorem 8.1.** *Let the functional (8.9) be Fréchet–differentiable in ${}^k\mathcal{H}_{t'}$, then there exists such $t$ ($t \in (0, t')$) that for all processes ${}^k\mathbf{T}(\tau) \in U_t$ the process ${}^m\mathbf{S}(t)$ is differentiable with respect to $t$ and the following rule of differentiation of the functional with respect to time holds:*

$$\frac{d}{dt}\,{}^m\mathbf{S}(t) = \frac{\partial}{\partial\, {}^k\mathbf{T}(t)} \mathop{\mathcal{F}}_{\tau=0}^{t}({}^k\mathbf{T}(t),\ {}^k\mathbf{T}^t(\tau)) \cdot \ldots \cdot \frac{d}{dt}\,{}^k\mathbf{T}^{(k\ldots1)}(t)$$
$$+\,\delta \mathop{\mathcal{F}}_{\tau=0}^{t}({}^k\mathbf{T}(t),\ {}^k\mathbf{T}^t(\tau)\big|\,{}^k\dot{\mathbf{T}}^t(\tau)). \tag{8.24}$$

Here

$${}^k\dot{\mathbf{T}}^t = \frac{d}{d(t-\tau)}\,{}^k\mathbf{T}(t-\tau) = -\frac{d}{dt}\,{}^k\mathbf{T}^t(\tau).$$

▼ A proof of the theorem can be found in [31]. ▲

*Remark.* The theorem gives the possibility to calculate the Fréchet–derivatives of the operators (8.9) by evaluating the ordinary derivative of the functions $\mathbf{S}(t)$ with respect to $t$ according to formula (8.24). □

*Example 8.1.* Determine the Fréchet–derivatives of the linear operator (8.15b) for the case when $\mathbf{\Gamma}(t,\tau) = \mathbf{\Gamma}(t-\tau)$ and $\mathbf{\Gamma}(t,t) = \mathbf{\Gamma}(0)$. According to formula (8.24), we calculate the ordinary derivative with respect to time $t$ by the rule of differentiation of an integral with a varying upper limit:

$$\frac{d}{dt} f = \mathbf{\Gamma}_0 \cdot\cdot \frac{d\mathbf{T}}{dt}(t) + \mathbf{\Gamma}(0) \cdot\cdot \mathbf{T}(t) + \int_0^t \mathbf{\Gamma}'(t-\tau) \cdot\cdot \mathbf{T}(\tau)\, d\tau, \tag{8.25}$$

where $\mathbf{\Gamma}'(y) = \partial \mathbf{\Gamma}(y)/\partial y$. Comparing (8.25) with (8.24), we find the partial derivative and the Fréchet–derivative:

$$\partial\mathcal{F} = \mathbf{\Gamma}_0 \cdot\cdot \frac{d}{dt}\mathbf{T}(t), \qquad \frac{\partial \mathcal{F}}{\partial \mathbf{T}} = \mathbf{\Gamma}_0,$$
$$\delta\mathcal{F} = \mathbf{\Gamma}(0) \cdot\cdot \mathbf{T}(t) + \int_0^t \mathbf{\Gamma}'(t-\tau) \cdot\cdot \mathbf{T}(\tau)\, d\tau. \ \square \tag{8.26}$$

### 8.1.4 Axiom of Fading Memory

For viscoelastic continua, in addition we assume the following axiom.

**Axiom 18 (of fading memory).** *The functionals (8.1) occurring in constitutive equations of viscoelastic continua are Fréchet–differentiable and hence satisfy the rule of differentiation with respect to time (8.24):*

$$\frac{d}{dt}\Lambda(t) = \frac{\partial}{\partial \mathcal{R}(t)} \overset{t}{\underset{\tau=0}{f}} (\mathcal{R}(t),\ \mathcal{R}^t(\tau))\frac{d}{dt}\mathcal{R}(t) + \delta \overset{t}{\underset{\tau=0}{f}} (\mathcal{R}(t),\ \mathcal{R}^t(\tau)\big|\dot{\mathcal{R}}^t(\tau)). \quad (8.27)$$

The interconnection between Fréchet–differentiability and fading memory of functionals is clarified by the theorem on relaxation.

Let there be a process $\mathcal{R}(\tau)$ $(0 \leqslant \tau \leqslant t)$, which is arbitrary up to some time $t_0 < t$, and when $t_0 \geqslant \tau \geqslant t$ it remains constant:

$$\mathcal{R}(\tau) = \mathcal{R}(t_0), \qquad t_0 \geqslant \tau \geqslant t. \quad (8.28)$$

Such process $\mathcal{R}(\tau)$ is called a *process with the constant extension* (Fig. 8.1).

In addition, consider a *static process*

$$\overset{*}{\mathcal{R}}(\tau) = \mathcal{R}(t_0) = \text{const}, \qquad 0 \leqslant \tau \leqslant t. \quad (8.29)$$

Then the following theorem can be formulated.

**Theorem 8.2.** *Let the functional $f$ (8.1) be Fréchet–differentiable, then its partial derivative $\partial f$ and Fréchet–derivative $\delta f$, and also the function $\Lambda(t)$ for any process $\mathcal{R}(\tau)$ with the constant extension at fixed time $t_0$ have the limits as $t \to +\infty$, which*

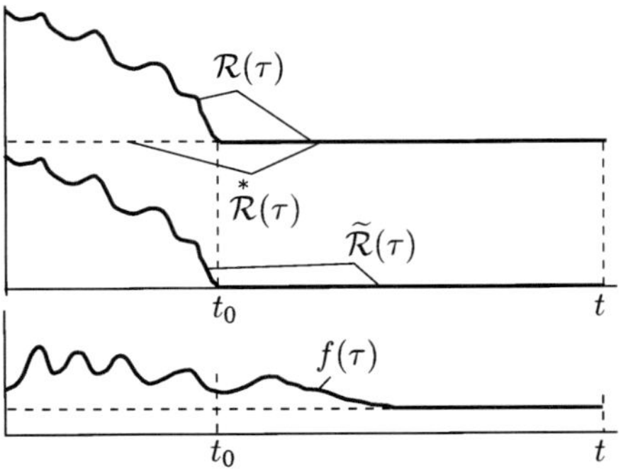


**Fig. 8.1** For Theorem 8.2

*coincide with values of the derivatives* $\partial \overset{*}{f}$, $\delta \overset{*}{f}$ *and* $\Lambda^*$, *respectively, in corresponding static processes* $\overset{*}{\mathcal{R}}(\tau)$:

$$\lim_{t\to+\infty} \overset{t}{\underset{\tau=0}{f}} (\mathcal{R}(t),\ \mathcal{R}^t(\tau)) = \Lambda^* \equiv \overset{t_0}{\underset{\tau=0}{f}} (\overset{*}{\mathcal{R}}(t_0),\ \overset{*}{\mathcal{R}}{}^{t_0}(\tau)),$$

$$\lim_{t\to+\infty} \partial \overset{t}{\underset{\tau=0}{f}} (\mathcal{R}(t),\ \mathcal{R}^t(\tau)\big|\dot{\mathcal{R}}(t)) = \partial \overset{*}{f} \equiv \partial \overset{t_0}{\underset{\tau=0}{f}} (\overset{*}{\mathcal{R}}(t_0),\ \overset{*}{\mathcal{R}}{}^{t_0}(\tau)\big|0),$$

$$\lim_{t\to+\infty} \delta \overset{t}{\underset{\tau=0}{f}} (\mathcal{R}(t),\ \mathcal{R}^t(\tau)\big|\dot{\mathcal{R}}^t(\tau)) = \delta \overset{*}{f} \equiv \delta \overset{t_0}{\underset{\tau=0}{f}} (\overset{*}{\mathcal{R}}(t_0),\ \overset{*}{\mathcal{R}}{}^{t_0}(\tau)\big|0). \qquad (8.30)$$

In simple words, for a process $\mathcal{R}(\tau)$ such that starting from some time $t_0$ the process reaches a constant level, in a certain time interval a viscoelastic continuum forgets the process $\mathcal{R}(\tau)$ up to the time $t_0$, because the continuum response, expressed by the functionals $f$, $\partial f$ and $\delta f$, at sufficiently great values of $t$ differs little from its response to a static process.

▼ Consider the process

$$\widetilde{\mathcal{R}}(\tau) = \mathcal{R}(\tau) - \overset{*}{\mathcal{R}}(\tau), \quad 0 \leqslant \tau \leqslant t. \qquad (8.31)$$

Since

$$\widetilde{\mathcal{R}}(\tau) \equiv 0 \quad \text{at } t_0 \geqslant \tau \geqslant t, \qquad (8.32)$$

so we have

$$\| \widetilde{\mathcal{R}} \|_t^2 = \widetilde{\mathcal{R}}^2(t) + \int_0^t \widetilde{\mathcal{R}}^2(t-\tau)\gamma^2(\tau)\,d\tau = \int_0^t \widetilde{\mathcal{R}}^2(\tau)\gamma^2(t-\tau)\,d\tau$$

$$= \int_0^{t_0} \widetilde{\mathcal{R}}^2(\tau)\gamma^2(t-\tau)\,d\tau \leqslant c \int_{t-t_0}^{t} \gamma^2(\tau)\,d\tau. \qquad (8.33)$$

Due to the property (8.4) and monotone decreasing the function $\gamma(\tau)$, we find that $\| \widetilde{\mathcal{R}} \|_t \to 0$ at $t \to +\infty$.

Since a Fréchet–differentiable functional $f$ is continuous as well as $\partial f$ and $\delta f$, the condition (8.12) yields the inequality

$$\| \overset{t}{\underset{\tau=0}{f}} (\overset{*}{\mathcal{R}}(t) + \widetilde{\mathcal{R}}(t),\ \overset{*}{\mathcal{R}}{}^t(\tau) + \widetilde{\mathcal{R}}^t(\tau)) - \overset{t}{\underset{\tau=0}{f}} (\overset{*}{\mathcal{R}}(t),\ \overset{*}{\mathcal{R}}{}^t(\tau)) \|_t < \varepsilon; \qquad (8.34)$$

this means that the first limit of (8.30) exists. In a similar way, we can prove that the second and third limits of (8.30) exist too. ▲

### 8.1.5 Models $A_n$ of Viscoelastic Continua

For models $A_n$ of viscoelastic continua, the free energy $\psi$ is a functional in the form (8.1), and as reactive variables one should choose the set (4.148):

$$\psi = \mathop{\psi}_{\tau=0}^{t}\left(\overset{(n)}{\mathbf{C}}(t),\ \theta(t),\ \overset{(n)}{\mathbf{C}}{}^t(\tau),\ \theta^t(\tau)\right). \tag{8.35}$$

According to the rule (8.24) of differentiation of a functional, we obtain the expression for the total derivative of $\psi$ with respect to $t$:

$$\frac{d\psi}{dt} = \frac{\partial\psi}{\partial\overset{(n)}{\mathbf{C}}}\cdot\cdot\frac{d\overset{(n)}{\mathbf{C}}}{dt} + \frac{\partial\psi}{\partial\theta}\frac{d\theta}{dt} + \delta\psi. \tag{8.36}$$

Substituting this expression into PTI (4.121) and collecting like terms, we get

$$\left(\frac{\partial\psi}{\partial\overset{(n)}{\mathbf{C}}} - \frac{\overset{(n)}{\mathbf{T}}}{\rho}\right)\cdot\cdot\, d\overset{(n)}{\mathbf{C}} + \left(\frac{\partial\psi}{\partial\theta} + \eta\right)d\theta + \left(\frac{w^*}{\rho} + \delta\psi\right)dt = 0. \tag{8.37}$$

When the prehistories $\overset{(n)}{\mathbf{C}}{}^t$, $\theta^t$ and the current values $\overset{(n)}{\mathbf{C}}(t)$ and $\theta(t)$ are fixed, the increments $d\overset{(n)}{\mathbf{C}}$, $d\theta$ and $dt$ can vary arbitrarily, therefore the identity (8.37) holds when and only when the coefficients of these increments vanish. As a result, we obtain the equation system

$$\begin{cases} \overset{(n)}{\mathbf{T}} = \rho(\partial\psi/\partial\overset{(n)}{\mathbf{C}}) = \mathop{\mathcal{F}}_{\tau=0}^{t}(\overset{(n)}{\mathbf{C}}(t),\ \theta(t),\ \overset{(n)}{\mathbf{C}}{}^t(\tau),\ \theta^t(\tau)), \\ \eta = -\partial\psi/\partial\theta, \\ w^* = -\rho\delta\psi, \end{cases} \tag{8.38}$$

that together with (8.35) is a system of *constitutive equations for models $A_n$ of viscoelastic continua.*

Just as for ideal continua, for viscoelastic materials it is sufficient to give only the free energy functional (8.35), then the remaining relations are determined by its differentiation according to formulae (8.38).

Notice that although relations (8.38) are formally similar to the corresponding relations (4.168) for models $A_n$ of ideal continua, they essentially differ by the fact that in (8.38) there is a functional dependence on $\overset{(n)}{\mathbf{C}}$ and $\theta$. Moreover, viscoelastic continua are dissipative: for them the dissipation function $w^*$ is not identically zero.

### 8.1.6 Corollaries of the Principle of Material Symmetry for Models $A_n$ of Viscoelastic Continua

According to the principle of material symmetry (Axiom 14), for each viscoelastic continuum there exists an undistorted reference configuration $\widehat{\mathcal{K}}$. Just as in Sect. 4.7.4, for simplicity, let the reference configuration $\overset{\circ}{\mathcal{K}}$ be undistorted. Then for $\overset{\circ}{\mathcal{K}}$ there is a subgroup $\overset{\circ}{G}_s \subset U$ of the unimodular group $U$ such that for each transformation tensor $\mathbf{H} \in \overset{\circ}{G}_s$ ($\mathbf{H} : \overset{\circ}{\mathcal{K}} \to \overset{*}{\mathcal{K}}$) the constitutive equations (8.38) written for $\overset{\circ}{\mathcal{K}}$ are transformed during the passage to $\overset{*}{\mathcal{K}}$ as follows:

$$\begin{cases} \overset{(n)}{\mathbf{T}}{}^* = \overset{t}{\underset{\tau=0}{\mathcal{F}}}(\overset{(n)}{\mathbf{C}}{}^*,\ \theta,\ \overset{(n)}{\mathbf{C}}{}^{t*},\ \theta^t) = \rho(\partial\psi/\partial\overset{(n)}{\mathbf{C}}{}^*), \\ \psi = \overset{t}{\underset{\tau=0}{\psi}}(\overset{(n)}{\mathbf{C}}{}^*,\ \theta,\ \overset{(n)}{\mathbf{C}}{}^{t*},\ \theta^t) \equiv \psi^*, \\ \eta = -\partial\psi^*/\partial\theta, \\ (w^*)^* = -\rho(\delta\psi)^*, \quad \forall \mathbf{H} \in \overset{\circ}{G}_s. \end{cases} \tag{8.39}$$

For viscoelastic solids, due to $H$-indifference of all the tensors $\overset{(n)}{\mathbf{T}}$, $\overset{(n)}{\mathbf{C}}(t)$ and $\overset{(n)}{\mathbf{C}}{}^t(\tau) = \overset{(n)}{\mathbf{C}}(t-\tau)$ (see Sect. 4.7.4), relations (8.39) take the forms

$$\overset{*}{\mathbf{Q}}{}^{\mathrm{T}} \cdot \overset{t}{\underset{\tau=0}{\mathcal{F}}}(\overset{(n)}{\mathbf{C}},\ \theta,\ \overset{(n)}{\mathbf{C}}{}^t,\ \theta^t) \cdot \overset{*}{\mathbf{Q}} = \overset{t}{\underset{\tau=0}{\mathcal{F}}}(\overset{*}{\mathbf{Q}}{}^{\mathrm{T}} \cdot \overset{(n)}{\mathbf{C}} \cdot \overset{*}{\mathbf{Q}},\ \theta,\ \overset{*}{\mathbf{Q}}{}^{\mathrm{T}} \cdot \overset{(n)}{\mathbf{C}}{}^t \cdot \overset{*}{\mathbf{Q}},\ \theta^t), \tag{8.40a}$$

$$\overset{t}{\underset{\tau=0}{\psi}}(\overset{(n)}{\mathbf{C}},\ \theta) = \overset{t}{\underset{\tau=0}{\psi}}(\overset{*}{\mathbf{Q}}{}^{\mathrm{T}} \cdot \overset{(n)}{\mathbf{C}} \cdot \overset{*}{\mathbf{Q}},\ \theta,\ \overset{*}{\mathbf{Q}}{}^{\mathrm{T}} \cdot \overset{(n)}{\mathbf{C}}{}^t \cdot \overset{*}{\mathbf{Q}},\ \theta^t) \ \ \forall \overset{*}{\mathbf{Q}} \in \overset{\circ}{G}_s, \tag{8.40b}$$

$$\delta\psi = \delta\psi^*. \tag{8.40c}$$

**Theorem 8.3.** *The principle of material symmetry in the form (8.40) holds for models $A_n$ of viscoelastic solids if and only if the condition (8.40b) for $\psi$ is satisfied.*

▼ It is evident that the condition (8.40b) is necessary.

Prove that this condition is sufficient. Let the condition (8.40b) be satisfied. Then, since the functional $\mathcal{F}$ is a tensor function being the derivative of $\psi$ with respect to $\overset{(n)}{\mathbf{C}}(t)$, so with the help of the method used in the proof of Theorem 4.23 we can prove that the condition (8.40b) yields (8.40a). To prove Eq. (8.40c) we should use formula (8.36) for $\delta\psi$ and rewrite it in $\overset{*}{\mathcal{K}}$:

$$\delta\psi^* = \frac{d\psi^*}{dt} - \frac{1}{\rho}\overset{(n)}{\mathbf{T}}{}^* \cdot\cdot \frac{d\overset{(n)}{\mathbf{C}}{}^*}{dt} - \frac{\partial\psi^*}{\partial\theta}\frac{d\theta}{dt}. \tag{8.41}$$

Since $\psi = \psi^*$ and the passage from $\overset{\circ}{\mathcal{K}}$ to $\overset{*}{\mathcal{K}}$ is independent of $t$, so $d\psi/dt = d\psi^*/dt$. Due to $H$-indifference of the tensors $\overset{(n)}{\mathbf{T}}$ and $\overset{(n)}{\mathbf{C}}$, we have the relation $\overset{(n)}{\mathbf{T}}{}^* \cdot\cdot (d\overset{(n)}{\mathbf{C}}{}^*/dt) = \overset{(n)}{\mathbf{T}} \cdot\cdot (d\overset{(n)}{\mathbf{C}}/dt)$, and also $\partial\psi^*/\partial\theta = \partial\psi/\partial\theta$. Thus, the right-hand side of Eq. (8.41) coincides with $\delta\psi$, and hence $(\delta\psi)^* = \delta\psi$; i.e. Eq. (8.40c) actually holds. ▲

### 8.1.7 General Representation of Functional of Free Energy in Models $A_n$

Scalar functional $\psi$ (8.35) satisfying the condition (8.40b) is called *functionally indifferent relative to the group* $\overset{\circ}{G}_s$. Let us find a general representation of such a functional in terms of invariants of tensor $\overset{(n)}{\mathbf{C}}$ of the corresponding group $\overset{\circ}{G}_s$.

In Sect. 8.1.3 we have derived a general representation of a linear scalar functional in the form (8.15d). Similarly to (8.15d), define a *quadratic scalar functional* as a double integral in the form

$$\psi_2 = \int_0^t \int_0^t {}^4\mathbf{A}(t, \tau_1, \tau_2) \cdot\cdot\cdot\cdot (\overset{(n)}{\mathbf{C}}(\tau_1) \otimes \overset{(n)}{\mathbf{C}}(\tau_2))^{(4321)} d\tau_1 d\tau_2, \tag{8.42}$$

where ${}^4\mathbf{A}(t, \tau_1, \tau_2)$ is a fixed fourth-order tensor called the *core of the functional.* We also define a *n-fold scalar functional*:

$$\psi_m = \int_0^t \ldots \int_0^t \widetilde{\psi}_m(t,\ \tau_1,\ \ldots,\ \tau_m)\, d\tau_1 \ldots d\tau_m, \tag{8.43}$$

where

$$\widetilde{\psi}_m(t, \tau_1, \ldots, \tau_m) = {}^{2n}\mathbf{A}(t, \tau_1, \ldots, \tau_m) \underbrace{\cdot \ldots \cdot}_{2m} (\overset{(n)}{\mathbf{C}}(\tau_1) \otimes \ldots \otimes \overset{(n)}{\mathbf{C}}(\tau_m))^{(m,m-1,\ldots,2,1)}. \tag{8.44}$$

Here ${}^{2n}\mathbf{A}(t,\ \tau_1, \ldots, \tau_m)$ is the core of this functional (it is a fixed tensor of order $(2n)$ depending on $m+1$ arguments).

**Theorem 8.4 (Stone–Weierstrass).** *Any continuous scalar functional (8.35) in space $\mathcal{H}_t$ can be uniformly approximated by n-fold scalar functionals (8.43):*

$$\psi = \underset{\tau=0}{\overset{t}{\psi}} (\overset{(n)}{\mathbf{C}}(t),\ \theta(t),\ \overset{(n)}{\mathbf{C}}{}^t(\tau),\ \theta^t(\tau)) = \sum_{m=1}^{\infty} \psi_m, \tag{8.45}$$

*where the equality means that the partial sum uniformly converges in the norm (8.8).*

▼ A proof of the theorem for the space $\mathcal{H}_t$ can be found in [9]. ▲

Let us consider the integrand $\psi_n(t, \tau_1, \ldots, \tau_n)$ of the $n$-fold functional (8.43). At any fixed set of values $t, \tau_1, \ldots, \tau_m$, this expression is a scalar function (but not a functional!) of $n$ tensor arguments $\overset{(n)}{\mathbf{C}}(\tau_i) \equiv \overset{(n)}{\mathbf{C}}_i$ $(i = 1, \ldots, m)$:

$$\widetilde{\psi}_m(t, \tau_1, \ldots, \tau_m) = \widetilde{\psi}_m(t, \tau_1, \ldots, \tau_m, \mathbf{C}_1, \ldots, \mathbf{C}_m). \tag{8.46}$$

Here while values of $t, \tau_1, \ldots, \tau_m$ vary, the number and the form of tensor arguments of the functions remain unchanged.

On substituting the representation (8.45) into the condition (8.40b) of functional indifference of $\psi$, we find that the functions $\widetilde{\psi}_m$ (8.44) at each fixed value $t, \tau_1, \ldots, \tau_m$ must satisfy the condition

$$\begin{aligned} &\widetilde{\psi}_m(t, \tau_1, \ldots, \tau_m, \mathbf{C}_1, \ldots, \mathbf{C}_m) \\ &\quad = \widetilde{\psi}_m(t, \tau_1, \ldots, \tau_m, \overset{*}{\mathbf{Q}}{}^{\mathrm{T}} \cdot \mathbf{C}_1 \cdot \overset{*}{\mathbf{Q}}, \ldots, \overset{*}{\mathbf{Q}}{}^{\mathrm{T}} \cdot \mathbf{C}_m \cdot \overset{*}{\mathbf{Q}}), \end{aligned} \tag{8.47}$$

i.e. they must be $H$-indifferent scalar functions relative to the group $\overset{\circ}{G}_s$.

**Theorem 8.5.** *Every scalar function (8.46) of m tensor arguments* $\mathbf{C}_1, \ldots, \mathbf{C}_m$, *which is indifferent relative to a group* $\overset{\circ}{G}_s$, *can be represented as a function of finite number z* ($z \leqslant 6m$) *of simultaneous invariants*

$$J_\gamma^{(s)} = J_\gamma^{(s)}(\mathbf{C}_1, \ldots, \mathbf{C}_m), \quad \gamma = 1, \ldots, z, \tag{8.48}$$

*relative to the group* $\overset{\circ}{G}_s$ *in the form*

$$\widetilde{\psi}_m = \widetilde{\psi}_m(t, \tau_1, \ldots, \tau_m, J_\gamma^{(s)}(\mathbf{C}_1, \ldots, \mathbf{C}_m)). \tag{8.49}$$

▼ By analogy with simultaneous invariants of two tensors, which have been considered in Sect. 7.1.4, we now introduce *simultaneous invariants of m tensors* relative to a group $\overset{\circ}{G}_s$. The simultaneous invariants are scalar functions $J_\gamma$ (8.48), which are $H$-indifferent relative to the group $\overset{\circ}{G}_s$; i.e. they satisfy the relations

$$J_\gamma^{(s)}(\mathbf{C}_1, \ldots, \mathbf{C}_m) = J_\gamma^{(s)}(\overset{*}{\mathbf{Q}}{}^{\mathrm{T}} \cdot \mathbf{C}_1 \cdot \overset{*}{\mathbf{Q}}, \ldots, \overset{*}{\mathbf{Q}}{}^{\mathrm{T}} \cdot \mathbf{C}_m \cdot \overset{*}{\mathbf{Q}}) \quad \forall \overset{*}{\mathbf{Q}} \in G_s. \tag{8.50}$$

Their functional basis consists of $z$ simultaneous invariants, where $z$ is a finite number, which cannot exceed the total number of components of all the tensors, i.e. $z \leqslant 6m$. Moreover, since $J_\gamma^{(s)}$ form a basis, any other $H$-indifferent scalar function relative to the same group $\overset{\circ}{G}_s$ can be expressed in this basis. But the function $\widetilde{\psi}$ (8.40) is just such a function due to (8.47), therefore the relation (8.49) actually holds. ▲

Substitution of the expression (8.49) into (8.43) and then into (8.45) yields the following general representation of the continuous functional (8.35), which is functionally indifferent relative to a group $\overset{\circ}{G}_s$:

$$\psi = \sum_{m=1}^{\infty} \int_0^t \cdots \int_0^t \widetilde{\psi}_m(t,\ \tau_1, \ldots, \tau_m,\ J_\gamma(\overset{(n)}{\mathbf{C}}(\tau_1), \ldots, \overset{(n)}{\mathbf{C}}(\tau_m)))\, d\tau_1 \ldots d\tau_m. \tag{8.51}$$

In deriving this formula we have used representation (8.15c) of linear functionals with the help of $\delta$-function. Let us perform now the inverse operation: segregate the $\delta$-type constituent from the cores $\psi_m$, that allows us to separate the instantaneous value of the tensor $\overset{(n)}{\mathbf{C}}(t)$ from its prehistory. Formula (8.15d) can be generalized for functions of $m+1$ arguments $t, \tau_1, \ldots, \tau_m$ as follows:

$$\widetilde{\psi}_m(t,\ \tau_1, \ldots, \tau_m,\ J_\gamma^{(s)}) = \sum_{k=0}^{m} \delta(t-\tau_1)\ldots\delta(t-\tau_k)\psi_{mk}(t,\ \tau_{k+1}, \ldots, \tau_m,\ J_\gamma^{(s)}). \tag{8.51a}$$

We assume that at $k=m$ the argument $\tau_{k+1}=\tau_{m+1}$ does not appear among arguments of the function $\psi_{mk}$: $\psi_{mm}=\psi_{mk}(t,\ J_\gamma^{(s)})$.

On substituting (8.51a) into (8.51), we get

$$\psi = \sum_{m=1}^{\infty}\sum_{k=0}^{m} \underbrace{\int_0^t \cdots \int_0^t}_{m-k} \psi_{mk}\Bigg(t,\ \tau_{k+1}, \ldots, \tau_m,\ J_\gamma^{(s)}\Bigg(\underbrace{\overset{(n)}{\mathbf{C}}(t), \ldots, \overset{(n)}{\mathbf{C}}(t)}_{k},$$
$$\overset{(n)}{\mathbf{C}}(\tau_{k+1}), \ldots, \overset{(n)}{\mathbf{C}}(\tau_m)\Bigg)\Bigg)\, d\tau_{k+1} \ldots d\tau_m. \tag{8.52}$$

Rearrange summands in the expression (8.52) and take into account that simultaneous invariants $J_\gamma^{(s)}$ of $m$ tensors, among which there are $k$ tensors $\overset{(n)}{\mathbf{C}}(t)$, can always be expressed in terms of simultaneous invariants of $(m-k+1)$ tensors $\overset{(n)}{\mathbf{C}}(\tau_{k+1}), \ldots \overset{(n)}{\mathbf{C}}(\tau_m)$ and $\overset{(n)}{\mathbf{C}}(t)$. Then we finally obtain the general form of the functional (8.35)

$$\psi = \varphi_0(t,\ J_\gamma^{(s)}(\overset{(n)}{\mathbf{C}}(t)))$$
$$+ \sum_{m=1}^{\infty} \underbrace{\int_0^t \cdots \int_0^t}_{m} \varphi_m\Big(t,\ \tau_1, \ldots, \tau_m,\ J_\gamma^{(s)}(\overset{(n)}{\mathbf{C}}(t),$$
$$\overset{(n)}{\mathbf{C}}(\tau_1), \ldots, \overset{(n)}{\mathbf{C}}(\tau_m))\Big)\, d\tau_1 \ldots d\tau_m, \tag{8.53}$$

which is functionally indifferent relative to the group $\overset{\circ}{G}_s$. Here we have introduced the notation: $\varphi_0$ – the instantly elastic part and $\varphi_m$ – cores of the functional:

$$\varphi_0 = \sum_{m=1}^{\infty} \psi_{mm}(t,\ J_\gamma^{(s)}(\mathbf{C}(t))),$$

$$\varphi_m = \sum_{k=0}^{\infty} \psi_{m+k,k}(t,\ \tau_1, \ldots, \tau_m,\ J_\gamma^{(s)}(\underbrace{\overset{(n)}{\mathbf{C}}(t) \ldots \overset{(n)}{\mathbf{C}}(t)}_{k},\ \mathbf{C}(\tau_1), \ldots, \mathbf{C}(\tau_m))). \tag{8.54}$$

Simultaneous invariants $J_\gamma^{(s)}(\overset{(n)}{\mathbf{C}}(t))$ of one tensor are simply invariants $I_\gamma^{(s)}(\overset{(n)}{\mathbf{C}}(t))$ of the tensor relative to the same group $\overset{\circ}{G}_s$.

The expression (8.53) is the desired general representation of the functional (8.35) for models $A_n$.

### 8.1.8 Model $A_n$ of Stable Viscoelastic Continua

**Definition 8.4.** One can say that this is the model $A_n$ of a *stable viscoelastic continuum*, if the functional $\psi$ of the model is invariant relative to a shift of the process of deforming and heating in time; i.e. if there are two processes $(\overset{(n)}{\mathbf{C}}(\tau), \theta(\tau))$ and $\overset{(n)}{\widetilde{\mathbf{C}}}(\tau), \widetilde{\theta}(\tau))$, $0 \leqslant \tau \leqslant t_1$, such that they are different from each other only by a shift in time:

$$(\overset{(n)}{\widetilde{\mathbf{C}}}(\tau),\ \widetilde{\theta}(\tau)) = \begin{cases} (\overset{(n)}{\mathbf{C}}(\tau - t_0),\ \theta(\tau - t_0)), & t_0 < \tau \leqslant t_1, \\ (\mathbf{0},\ \theta_0), & 0 \leqslant \tau \leqslant t_0; \end{cases} \tag{8.55}$$

then the corresponding values of the functionals $\psi$ and $\widetilde{\psi}$ are different only by a shift in time as well (Fig. 8.2):

$$\widetilde{\psi} = \begin{cases} \psi(t - t_0), & t_0 < t \leqslant t_1, \\ \psi(0), & 0 \leqslant t \leqslant t_0. \end{cases} \tag{8.56}$$

*Remark.* Since the functional $\psi$ for stable continua is invariant relative to a shift in time, its partial derivatives with respect to $\overset{(n)}{\mathbf{C}}(t)$ and $\theta(t)$ and also the Fréchet–derivative $\delta\psi$ have this property; hence the constitutive equations (8.38) are invariant relative to a shift in time too. Due to the property of invariance, stable continua are also called non-aging, and their constitutive equations do not change with time themselves when there are no deformations and variations of temperature. □

Let us consider two important models of stable continua.

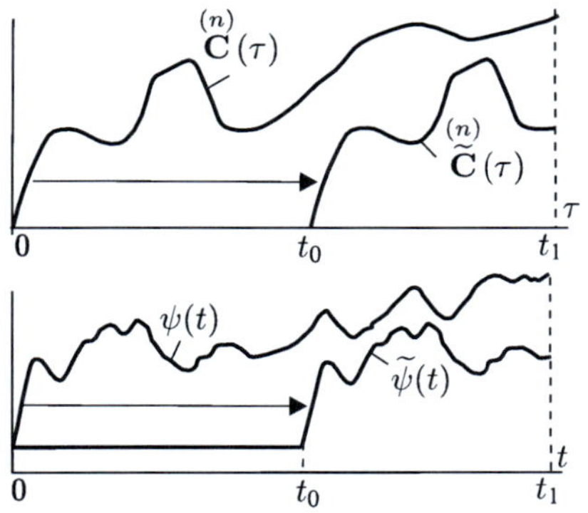


**Fig. 8.2** For the definition of a stable viscoelastic continuum

## 8.1.9 *Model $A_n$ of a Viscoelastic Continuum with Difference Cores*

One can say that this is the *model $A_n$ of a viscoelastic continuum with difference cores*, if in the general representation of the functional $\psi$ (8.53) there is no explicit dependence of cores $\varphi_m$ on the times $t$ and $\tau_i$, but there is a dependence only on their difference $t - \tau_i$ or on the temperatures $\theta(t)$ and $\theta(\tau_i)$:

$$\varphi_m = \varphi_m(t-\tau_1, \ldots, t-\tau_m, \theta(t), \theta(\tau_1), \ldots, \theta(\tau_m), J_\gamma^{(s)}(\overset{(n)}{\mathbf{C}}(t), \overset{(n)}{\mathbf{C}}(\tau_1), \ldots, \overset{(n)}{\mathbf{C}}(\tau_m))), \tag{8.57}$$

and $\varphi_0$ does not depend explicitly on temperature:

$$\varphi_0 = \varphi_0(I_\gamma^{(s)}(\overset{(n)}{\mathbf{C}}(t)), \ \theta(t)). \tag{8.58}$$

Functions $\varphi_m$ are assumed to satisfy the following conditions of normalization and symmetry with respect to any permutations of the first $m$ arguments:

$$\varphi_m(t - \tau_1, \ldots, t - \tau_m, \ \theta_0, \ldots, \theta_0, \ 0, \ldots, 0) = 0, \tag{8.59}$$

$$\varphi_m(y_1, \ldots, y_n, \ldots, y_l, \ldots, y_m, \ \theta, \ \theta_1, \ldots, \theta_m, \ J_\gamma^{(s)})$$

$$= \varphi_m(y_1, \ldots, y_l, \ldots, y_n, \ldots, y_m, \ \theta, \ \theta_1, \ldots, \theta_m, \ J_\gamma^{(s)}),$$

where $\theta_0 = \theta(0), \ \theta_n = \theta(\tau_n), \ y_n = t - \tau_n$, and $1 \leqslant n, \ l \leqslant m$.

Simultaneous invariants $J_\gamma^{(s)}$ can always be chosen to satisfy the normalization conditions

$$J_\gamma^{(s)}(0, \ldots, 0) = 0, \qquad \gamma = 1, \ldots, z. \tag{8.60}$$

For the model with difference cores, the functional (8.53) has the form

$$\psi(t) = \varphi_0(I_\gamma^{(s)}(\overset{(n)}{\mathbf{C}}(t)),\ \theta(t)) + \sum_{m=1}^{\infty} \int_0^t \dots \int_0^t \varphi_m \, d\tau_1 \dots d\tau_m, \tag{8.61}$$

where $\varphi_m$ are determined by formula (8.57).

**Theorem 8.6.** *The model $A_n$ of a viscoelastic continuum with difference cores is stable.*

▼ Let the first process be of the form $\overset{(n)}{\mathbf{C}}(\tau)$, $0 \leqslant \tau \leqslant t$, then the corresponding functional $\psi(t)$ has the form (8.61). Since the second process $\overset{(n)}{\widetilde{\mathbf{C}}}(\tau)$ when $\tau \leqslant t_0$ is identically zero, so, due to the normalization conditions (8.59) and (8.60), we have $\varphi_m \equiv 0$ when $0 \leqslant \tau_i \leqslant t_0$ ($i = 1, \dots, m$); therefore, the lower limits of the integrals in the expression (8.61) for functional $\widetilde{\psi}(t)$ can be determined as $t_0$:

$$\widetilde{\psi}(t) = \varphi_0\left(I_\gamma^{(s)}(\overset{(n)}{\widetilde{\mathbf{C}}}(t)),\ \widetilde{\theta}(t)\right)$$
$$+ \sum_{m=1}^{\infty} \int_{t_0}^t \dots \int_{t_0}^t \varphi_m\Big(t - \tau_1, \dots, t - \tau_m,\ \widetilde{\theta}(t),\ \widetilde{\theta}(\tau_1), \dots, \widetilde{\theta}(\tau_m),$$
$$J_\gamma^{(s)}(\overset{(n)}{\widetilde{\mathbf{C}}}(t),\ \overset{(n)}{\widetilde{\mathbf{C}}}(\tau_1), \dots, \overset{(n)}{\widetilde{\mathbf{C}}}(\tau_m))\Big)\, d\tau_1 \dots d\tau_m. \tag{8.62}$$

Replacing $\overset{(n)}{\widetilde{\mathbf{C}}}(\tau_i)$ by $\overset{(n)}{\mathbf{C}}(\tau_i - t_0)$ when $\tau_i > t_0$ and then substituting $\tau_i - t_0 = \widetilde{\tau}_i$, we obtain (when $t > t_0$)

$$\widetilde{\psi}(t) = \varphi_0(I_\gamma^{(s)}(\overset{(n)}{\mathbf{C}}(t - t_0)), \theta(t - t_0))$$
$$+ \sum_{m=1}^{\infty} \int_0^{t-t_0} \dots \int_0^{t-t_0} \varphi_m\Big(t - t_0 - \widetilde{\tau}_1, \dots, t - t_0 - \widetilde{\tau}_m, \theta(t - t_0),$$
$$\theta(\widetilde{\tau}_1), \dots, \theta(\widetilde{\tau}_m), J_\gamma^{(s)}(\overset{(n)}{\mathbf{C}}(t - t_0), \overset{(n)}{\mathbf{C}}(\widetilde{\tau}_1), \dots, \overset{(n)}{\mathbf{C}}(\widetilde{\tau}_m))\Big)\, d\widetilde{\tau}_1 \dots d\widetilde{\tau}_m. \tag{8.63}$$

Comparing (8.61) and (8.63), we verify that $\widetilde{\psi}(t)$ and $\psi(t)$ are connected only by the time-shift $t \to (t - t_0)$, because at $t = t_0$

$$\widetilde{\psi}(t_0) = \varphi_0(0, \theta(0)) = \psi(0),$$

i.e. the relation (8.56) holds. ▲

On substituting the functional (8.61) into (8.38), we get the general form of constitutive equations for stable continua:

$$\overset{(n)}{\mathbf{T}} = \rho \sum_{\gamma=1}^{z} \left( \frac{\partial \varphi_0}{\partial I_\gamma^{(s)}} \frac{\partial I_\gamma^{(s)}}{\partial \overset{(n)}{\mathbf{C}}(t)} + \sum_{m=1}^{\infty} \int_0^t \dots \int_0^t \frac{\partial \varphi_m}{\partial J_\gamma^{(s)}} \frac{\partial J_\gamma^{(s)}}{\partial \overset{(n)}{\mathbf{C}}(t)} \, d\tau_1 \dots d\tau_m \right),$$

$$-\eta = \frac{\partial \varphi_0}{\partial \theta} + \sum_{m=1}^{\infty} \int_0^t \dots \int_0^t \frac{\partial \varphi_m}{\partial \theta(t)} \, d\tau_1 \dots d\tau_m,$$

$$-w^* = \rho \varphi_1^0 + \rho \sum_{m=1}^{\infty} \int_0^t \dots \int_0^t \left( \frac{\partial \varphi_m}{\partial t} + \varphi_{m+1}^0 \right) d\tau_1 \dots d\tau_m. \tag{8.64}$$

Here the cores $\varphi_m$ are functions in the form (8.57), and $\partial \varphi_m / \partial t$ is the partial derivative of the function when its first arguments $(t - \tau_1), \dots, \theta(\tau_m)$ vary and the arguments $J_\gamma^{(s)}$ are fixed (i.e. there is no differentiation with respect to $J_\gamma^{(s)}$). We have introduced the following notation for a value of the function $\varphi_{m+1}$ (8.57) at $\tau_{m+1} = t$:

$$\varphi_{m+1}^0 = \varphi_{m+1} \Big( t - \tau_1, \dots, t - \tau_m, \ 0, \ \theta(t), \ \theta(\tau_1), \dots, \theta(\tau_m), \ \theta(t),$$
$$J_\gamma^{(s)} (\overset{(n)}{\mathbf{C}}(t), \ \overset{(n)}{\mathbf{C}}(\tau_1), \dots, \overset{(n)}{\mathbf{C}}(\tau_m)), \ \overset{(n)}{\mathbf{C}}(t) \Big).$$

In deriving the expression for the dissipation function we have used the conditions (8.59).

### 8.1.10 Model $A_n$ of a Thermoviscoelastic Continuum

Let us consider the most widely used method to take the dependence of cores $\varphi_m$ (8.57) on temperature $\theta$ into account.

For the *model* $A_n$ *of a thermoviscoelastic continuum with difference cores*, temperature $\theta$ appears in simultaneous invariants $J_\gamma^{(s)}$, i.e.

$$\varphi_m = \varphi_m \left( t - \tau_1, \dots, t - \tau_m, \ J_\gamma^{(s)} \left( \overset{(n)}{\mathbf{C}}_\theta(t), \ \overset{(n)}{\mathbf{C}}_\theta(\tau_1), \dots, \overset{(n)}{\mathbf{C}}_\theta(\tau_m) \right) \right), \tag{8.65}$$

where

$$\overset{(n)}{\mathbf{C}}_\theta(\tau) = \overset{(n)}{\mathbf{C}}(\tau) - \mathring{\boldsymbol{\varepsilon}}(\tau), \quad \mathring{\boldsymbol{\varepsilon}}(\tau) = \int_{\theta_0}^{\theta(\tau)} \boldsymbol{\alpha}(\widetilde{\theta}) d\widetilde{\theta}. \tag{8.66}$$

The tensor $\overset{\circ}{\boldsymbol{\varepsilon}}$ is called the tensor of *heat deformation*, and $\boldsymbol{\alpha}$ – the tensor of *heat expansion*. Both the tensors are symmetric and $H$-indifferent relative to a considered group $\overset{\circ}{G}_s$:

$$\overset{*}{\mathbf{Q}}^{\mathrm{T}} \cdot \boldsymbol{\alpha} \cdot \overset{*}{\mathbf{Q}} = \boldsymbol{\alpha} \quad \forall \overset{*}{\mathbf{Q}} \in \overset{\circ}{G}_s, \tag{8.67}$$

therefore, the functions $J_\gamma^{(s)}$ of $\overset{(\mathrm{n})}{\mathbf{C}}_\theta$ are also $H$-indifferent relative to the group $\overset{\circ}{G}_s$.

Taking the dependence of constitutive equations upon temperature as the difference between the deformation tensor and the heat deformation tensor (8.66) is called the *Duhamel–Neumann model*.

In a similar way, the Duhamel–Neumann model describes the dependence of the function $\varphi_0$ on temperature:

$$\varphi_0 = \varphi_0(I_\gamma^{(s)}\left(\overset{(\mathrm{n})}{\mathbf{C}}_\theta(t)\right),\ \theta(t)). \tag{8.68}$$

Since $\partial \overset{(\mathrm{n})}{\mathbf{C}}_\theta / \partial\theta = -\boldsymbol{\alpha}$, the derivatives with respect to $\theta(t)$ in (8.64) for this model have the form

$$\frac{\partial \varphi_0}{\partial \theta} = -\sum_{\gamma=1}^{z} \frac{\partial \varphi_0}{\partial I_\gamma} \frac{\partial I_\gamma^{(s)}}{\partial \overset{(\mathrm{n})}{\mathbf{C}}_\theta} \cdot\cdot \boldsymbol{\alpha} + \frac{\partial' \varphi_0}{\partial \theta(t)}, \quad \frac{\partial \varphi_m}{\partial \theta(t)} = -\sum_{\gamma=1}^{z} \frac{\partial \varphi_m}{\partial J_\gamma} \frac{\partial J_\gamma^{(s)}}{\partial \overset{(\mathrm{n})}{\mathbf{C}}_\theta(t)} \cdot\cdot \boldsymbol{\alpha}(\theta(t)), \tag{8.68a}$$

where $\partial'/\partial\theta$ means the derivative with respect to the second argument in formula (8.68).

Taking into account that $\partial I_\gamma^{(s)} / \partial \overset{(\mathrm{n})}{\mathbf{C}}_\theta = \partial I_\gamma^{(s)} / \partial \overset{(\mathrm{n})}{\mathbf{C}}$ and $\partial J_\gamma^{(s)} / \partial \overset{(\mathrm{n})}{\mathbf{C}}_\theta = \partial J_\gamma^{(s)} / \partial \overset{(\mathrm{n})}{\mathbf{C}}$ and substituting (8.68a) into (8.64), we obtain the expression for the specific entropy

$$\eta = -\frac{\partial' \varphi_0}{\partial \theta(t)} + \frac{1}{\rho} \boldsymbol{\alpha} \cdot\cdot \overset{(\mathrm{n})}{\mathbf{T}}. \tag{8.69}$$

### *8.1.11 Model $A_n$ of a Thermorheologically Simple Viscoelastic Continuum*

One can say that this is the *model $A_n$ of a thermorheologically simple viscoelastic continuum*, if the cores $\varphi_m$ (8.57) in constitutive equations (8.53) and (8.64) depend on temperature $\theta$ in the functional way through the so-called *reduced time*:

$$\varphi_m = \varphi_m\Big(t' - \tau_1', \ldots, t' - \tau_m',\ J_\gamma^{(s)}(\overset{(\mathrm{n})}{\mathbf{C}}(t),\ \overset{(\mathrm{n})}{\mathbf{C}}(\tau_1), \ldots,$$

$$\overset{(\mathrm{n})}{\mathbf{C}}(\tau_m))\Big) a_\theta(\theta(\tau_1)) \ldots a_\theta(\theta(\tau_m)), \tag{8.70}$$

where

$$t' = \int_0^t a_\theta(\theta(\widetilde{\tau}))\, d\widetilde{\tau}, \qquad \tau_i' = \int_0^{\tau_i} a_\theta(\theta(\widetilde{\tau}))\, d\widetilde{\tau} \tag{8.71}$$

is the reduced time being a functional of the function $a_\theta(\theta)$ called the *function of the temperature-time shift.*

The functions $\varphi_m$ (8.70) and $a_\theta$ satisfy the normalization conditions

$$\varphi_m\left(0,\ldots,0,\ J_\gamma^{(s)}\right) = 0,\ \ \varphi_m\left(t'-\tau_1',\ldots,t'-\tau_m',0\right) = 0,\ \ a_\theta(\theta_0) = 1. \tag{8.72}$$

If the process $\overset{(n)}{\mathbf{C}}(\tau)$ is considered with respect to the reduced time

$$\overset{(n)}{\underset{\sim}{\mathbf{C}}}(\tau') = \overset{(n)}{\underset{\sim}{\mathbf{C}}}\left(\int_0^\tau a_\theta d\widetilde{\tau}\right) = \overset{(n)}{\mathbf{C}}(\tau),$$

then, since $d\tau_i' = a_\theta(\theta(\tau_i))\, d\tau_i$, the functional (8.53) with the core (8.70) can be written with respect to the reduced time

$$\psi(t') = \varphi_0\left(I_\gamma^{(s)}(\overset{(n)}{\underset{\sim}{\mathbf{C}}}(t')),\ \theta(t')\right) + \sum_{m=1}^{\infty}\int_0^{t'}\cdots\int_0^{t'}\varphi_m\left(t'-\tau_1',\ldots,t'-\tau_m',\right.$$
$$\left. J_\gamma^{(s)}(\overset{(n)}{\underset{\sim}{\mathbf{C}}}(t'),\ \overset{(n)}{\underset{\sim}{\mathbf{C}}}(\tau_1'),\ldots,\overset{(n)}{\underset{\sim}{\mathbf{C}}}(\tau_m'))\right) d\tau_1'\ldots d\tau_m'. \tag{8.73}$$

Substituting the functional (8.73) into (8.38) and using the differentiation rule (8.27), we obtain constitutive equations for a thermorheologically simple viscoelastic continuum

$$\overset{(n)}{\mathbf{T}} = \rho\sum_{\gamma=1}^{z}\left(\frac{\partial\varphi_0}{\partial I_\gamma^{(s)}}\frac{\partial I_\gamma^{(s)}}{\partial\overset{(n)}{\mathbf{C}}} + \sum_{m=1}^{\infty}\int_0^{t'}\cdots\int_0^{t'}\frac{\partial\varphi_m}{\partial J_\gamma^{(s)}}\frac{\partial J_\gamma^{(s)}}{\partial\overset{(n)}{\mathbf{C}}(t')}\, d\tau_1'\ldots d\tau_m'\right),$$
$$\eta = -(\partial'\varphi_0/\partial\theta) + (1/\rho)\boldsymbol{\alpha}\cdot\cdot\overset{(n)}{\mathbf{T}}, \tag{8.74}$$
$$-w^* = \rho a_\theta\varphi_1^0 + \rho a_\theta\sum_{m=1}^{\infty}\int_0^{t'}\cdots\int_0^{t'}\left(\frac{\partial\varphi_m}{\partial t'} + \varphi_{m+1}^0\right)d\tau_1'\ldots d\tau_m'.$$

Here we have used that $\partial/\partial t = a_\theta(\partial/\partial t')$.

Notice that with the help of (8.38) and (8.74) the dissipation function $w^*$ can be represented in another equivalent form

$$w^* = \overset{(\mathrm{n})}{\mathbf{T}} \cdot\cdot \frac{d}{dt}\overset{(\mathrm{n})}{\mathbf{C}} - \rho\frac{d\psi}{dt} + \left(\rho\frac{\partial'\varphi_0}{\partial\theta} - \boldsymbol{\alpha}\cdot\cdot\overset{(\mathrm{n})}{\mathbf{T}}\right)\frac{d\theta}{dt}, \tag{8.75}$$

which proves to be useful for cyclic loading.

**Theorem 8.7.** *A thermorheologically simple continuum is stable.*

▼ The reduced time (8.71) for the shifted process of heating $\widetilde{\theta}(\tau) = \theta(\tau - t_0)$ with use of the normalization condition (8.72) can be represented in the form

$$t' = \int_0^{t_0} a_\theta\, d\tau + \int_{t_0}^{t} a_\theta\left(\widetilde{\theta}(\tau)\right)\, d\tau = t_0 + \int_{t_0}^{t} a_\theta(\theta(\tau - t_0))\, d\tau$$

$$= t_0 + \int_0^{t-t_0} a_\theta(\theta(\widetilde{\tau}))\, d\widetilde{\tau},$$

$$\tau' = t_0 + \int_0^{\tau-t_0} a_\theta(\theta(\widetilde{\tau}))\, d\widetilde{\tau}, \tag{8.76}$$

when $t_0 < \tau < t$. The further proof is the same as the one in Theorem 8.6 (see Exercise 8.1.1). ▲

## Exercises for 8.1

**8.1.1.** Complete the proof of Theorem 8.7.

**8.1.2.** Using Definitions 8.2 and 8.3, prove that if a functional is Fréchet– differentiable, then it is continuous.

## 8.2 Principal, Quadratic and Linear Models of Viscoelastic Continua

### *8.2.1 Principal Models $A_n$ of Viscoelastic Continua*

Constitutive equations containing multiple integrals of the type (8.53), (8.61) or (8.73) are very awkward, and their application in practice is considerably difficult. Therefore, special models of viscoelastic materials, in which one may retain a finite number of integrals, are widely used.

For the *principal model $A_n$ of a thermoviscoelastic continuum with difference cores*, the sum (8.61) contains only one integral ($m = 1$), i.e. $\psi$ in this model has the form

$$\psi = \varphi_0(I_\gamma^{(s)}(\overset{(n)}{\mathbf{C}}_\theta),\ \theta) - \int_0^t \varphi_1(t-\tau,\ J_\gamma^{(s)}(\overset{(n)}{\mathbf{C}}_\theta(t),\ \overset{(n)}{\mathbf{C}}_\theta(\tau))\,d\tau. \tag{8.77}$$

Here $\varphi_0$ and $\varphi_1$ are functions of the arguments indicated, and the function $\varphi_1$ is chosen with the negative sign that can always be done by simple renaming the functions.

Constitutive equations for the continuum considered have the form (8.64), where $m$ should be assumed to be equal to 1:

$$\overset{(n)}{\mathbf{T}} = \sum_{\gamma=1}^{z}\Big(\varphi_{0\gamma} J_{\gamma\mathbf{C}}^{(s)} - \int_0^t \varphi_{1\gamma} J_{\gamma\mathbf{C}}^{(s)}\,d\tau\Big). \tag{8.78}$$

Here we have denoted the partial derivatives of $\varphi_0$ and $\varphi_1$

$$\begin{aligned}\varphi_{0\gamma}(J_\alpha^{(s)}(\overset{(n)}{\mathbf{C}}_\theta(t))) &= \rho(\partial\varphi_0/\partial I_\gamma^{(s)}), \quad \gamma = 1,\dots,r,\\ \varphi_{1\gamma}(t-\tau,\ J_\alpha^{(s)}(\overset{(n)}{\mathbf{C}}_\theta(t),\ \overset{(n)}{\mathbf{C}}_\theta(\tau))) &= \rho(\partial\varphi_1/\partial J_\gamma^{(s)}), \quad \gamma = 1,\dots,z,\end{aligned} \tag{8.79}$$

and also the partial derivative tensors of $J_\gamma^{(s)}$ with respect to $\overset{(n)}{\mathbf{C}}_\theta(t)$:

$$J_{\gamma\mathbf{C}}^{(s)} = \partial J_\gamma^{(s)}/\partial\overset{(n)}{\mathbf{C}}_\theta(t) = \partial J_\gamma^{(s)}/\partial\overset{(n)}{\mathbf{C}}(t), \quad \gamma = 1,\dots,z. \tag{8.80}$$

The simultaneous invariants $J_\gamma^{(s)}(\overset{(n)}{\mathbf{C}}_\theta(t),\ \overset{(n)}{\mathbf{C}}_\theta(\tau))$ can be chosen so that the first $r$ ones form a functional basis of invariants $I_\gamma^{(s)}(\overset{(n)}{\mathbf{C}}_\theta(t))$ in the same group $\overset{\circ}{G}_s$. We will assume below that $J_\gamma^{(s)}$ are ordered in such a way; then the following equations hold:

$$J_{\gamma\mathbf{C}}^{(s)} = \partial I_\gamma^{(s)}/\partial\overset{(n)}{\mathbf{C}}_\theta(t), \quad \gamma = 1,\dots,r; \qquad \varphi_{0\gamma} \equiv 0, \quad \gamma = r+1,\dots,z. \tag{8.81}$$

These equations have been used for deriving the relation (8.78).

According to (8.64) and (8.69), the dissipation function $w^*$ and the specific entropy $\eta$ for the principal models $A_n$ have the forms

$$\begin{aligned} w^* &= \rho\varphi_1(0,\ J_\gamma^{(s)}(\overset{(n)}{\mathbf{C}}_\theta(t),\ \overset{(n)}{\mathbf{C}}_\theta(t)) + \rho\int_0^t \frac{\partial}{\partial t}\varphi_1(t-\tau,\ J_\gamma^{(s)}(\overset{(n)}{\mathbf{C}}_\theta(t),\ \overset{(n)}{\mathbf{C}}_\theta(\tau))\,d\tau \geqslant 0,\\ \eta &= -\frac{\partial'\varphi_0}{\partial\theta} + \frac{1}{\rho}\boldsymbol{\alpha}\cdot\cdot\overset{(n)}{\mathbf{T}}.\end{aligned} \tag{8.82}$$

Here $\partial/\partial t$ is the partial derivative of $\varphi_1$ with respect to the first argument, and $\partial'/\partial\theta$ is the partial derivative of $\varphi_0$ with respect to the second argument.

Simultaneous invariants $J_\gamma^{(s)}$ of two tensors can be written by analogy with the ones for continua of the differential type (see Sect. 7.1.4).

## 8.2.2 Principal Model $A_n$ of an Isotropic Thermoviscoelastic Continuum

For the principal model $A_n$ of an isotropic thermoviscoelastic continuum with difference cores, the functional basis of simultaneous invariants $J_\gamma^{(I)}$ consists of 9 invariants, which can be chosen as follows (see (7.35)):

$$J_\alpha^{(I)} = I_\alpha(\overset{(n)}{\mathbf{C}}_\theta(t)),\quad J_{\alpha+3}^{(I)} = I_\alpha(\overset{(n)}{\mathbf{C}}_\theta(\tau)),\quad \alpha = 1,2,3, \tag{8.83}$$
$$J_7^{(I)} = \overset{(n)}{\mathbf{C}}_\theta(\tau)\cdot\cdot\,\overset{(n)}{\mathbf{C}}_\theta(t),\quad J_8^{(I)} = \overset{(n)}{\mathbf{C}}{}_\theta^2(\tau)\cdot\cdot\,\overset{(n)}{\mathbf{C}}_\theta(t),\quad J_9^{(I)} = \overset{(n)}{\mathbf{C}}_\theta(\tau)\cdot\cdot\,\overset{(n)}{\mathbf{C}}{}_\theta^2(t),$$
$$r = 3 \quad \text{and} \quad z = 9.$$

The derivative tensors of these invariants are calculated by the formulae (see [12])

$$J_{1\mathbf{C}}^{(I)} = \mathbf{E},\quad J_{2\mathbf{C}}^{(I)} = \mathbf{E}I_1(\overset{(n)}{\mathbf{C}}_\theta(t)) - \overset{(n)}{\mathbf{C}}_\theta(t),\quad J_{3\mathbf{C}}^{(I)} = \overset{(n)}{\mathbf{C}}{}_\theta^2(t) - I_1\overset{(n)}{\mathbf{C}}_\theta(t) + \mathbf{E}I_2,$$
$$J_{\gamma+3,\mathbf{C}}^{(I)} = \mathbf{0},\ \gamma = 1,2,3;\quad J_{7\mathbf{C}}^{(I)} = \overset{(n)}{\mathbf{C}}_\theta(\tau),\quad J_{8\mathbf{C}}^{(I)} = \overset{(n)}{\mathbf{C}}{}_\theta^2(\tau),$$
$$J_{9\mathbf{C}}^{(I)} = \overset{(n)}{\mathbf{C}}_\theta(\tau)\cdot\overset{(n)}{\mathbf{C}}_\theta(t) + \overset{(n)}{\mathbf{C}}_\theta(t)\cdot\overset{(n)}{\mathbf{C}}_\theta(\tau). \tag{8.84}$$

Substituting these expressions into (8.77) and collecting terms with the same tensor powers, we obtain constitutive equations for the principal model $A_n$ of an isotropic thermoviscoelastic continuum:

$$\overset{(n)}{\mathbf{T}} = \breve{\varphi}_1\mathbf{E} + \breve{\varphi}_2\overset{(n)}{\mathbf{C}}_\theta + \breve{\varphi}_3\overset{(n)}{\mathbf{C}}{}_\theta^2. \tag{8.85}$$

Here we have denoted the functionals

$$\breve{\varphi}_1 \equiv \varphi_{01} + \varphi_{02}I_1(t) + \varphi_{03}I_2(t) - \int_0^t (\varphi_{11} + \varphi_{12}I_1(t) + \varphi_{13}I_2(t))\,d\tau,$$
$$-\breve{\varphi}_2\overset{(n)}{\mathbf{C}}_\theta \equiv (\varphi_{02} + \varphi_{03}I_1(t))\overset{(n)}{\mathbf{C}}_\theta(t) - \int_0^t ((\varphi_{12} + \varphi_{13}I_1(t))\overset{(n)}{\mathbf{C}}_\theta(t) - \varphi_{17}\overset{(n)}{\mathbf{C}}_\theta(\tau))\,d\tau, \tag{8.86}$$

$$\breve{\varphi}_3 \overset{(n)}{\mathbf{C}}{}^2_\theta \equiv \varphi_{03} \overset{(n)}{\mathbf{C}}{}^2_\theta(t)$$
$$- \int\limits_0^t \Big( \varphi_{13} \overset{(n)}{\mathbf{C}}{}^2_\theta(t) + \varphi_{18} \overset{(n)}{\mathbf{C}}{}^2_\theta(\tau) + \varphi_{19} (\overset{(n)}{\mathbf{C}}_\theta(t) \cdot \overset{(n)}{\mathbf{C}}_\theta(\tau) + \overset{(n)}{\mathbf{C}}_\theta(\tau) \cdot \overset{(n)}{\mathbf{C}}_\theta(t)) \Big)\, d\tau.$$

Relation (8.85) is formally similar to the corresponding relation (4.322) for an isotropic elastic continuum, but in (8.85) $\breve{\varphi}_1$, $\breve{\varphi}_2$ and $\breve{\varphi}_3$ are no longer functions of invariants of the tensor $\overset{(n)}{\mathbf{C}}$; they are functionals in the form (8.86).

### 8.2.3 *Principal Model $A_n$ of a Transversely Isotropic Thermoviscoelastic Continuum*

For the principal model $A_n$ of a transversely isotropic (relative to the group $T_3$) thermoviscoelastic continuum with difference cores, the functional basis of simultaneous invariants $J_\gamma^{(3)}$ consists of 11 invariants, which can be chosen as follows (see (7.36)):

$$J_\gamma^{(3)} = I_\gamma^{(3)}\Big(\overset{(n)}{\mathbf{C}}_\theta(t)\Big),\ \gamma = 1, \ldots, 5;\quad J_{5+\gamma}^{(3)} = I_\gamma^{(3)}(\overset{(n)}{\mathbf{C}}_\theta(\tau)),\ \gamma = 1, \ldots, 4;$$
$$J_{10}^{(3)} = ((\mathbf{E} - \widehat{\mathbf{c}}_3^2) \cdot \overset{(n)}{\mathbf{C}}_\theta(t)) \cdot\cdot \Big(\widehat{\mathbf{c}}_3^2 \cdot \overset{(n)}{\mathbf{C}}_\theta(\tau)\Big),$$
$$J_{11}^{(3)} = \overset{(n)}{\mathbf{C}}_\theta(t) \cdot\cdot \overset{(n)}{\mathbf{C}}_\theta(\tau) - 2J_{10}^{(3)} - J_2^{(3)} J_7^{(3)},$$
$$r = 5, \qquad z = 11. \tag{8.87}$$

Here the invariants $I_\gamma^{(3)}$ are determined by formulae (4.297). The partial derivatives $J_{\gamma\mathbf{C}}^{(3)}$ of these invariants have the forms (see [12])

$$J_{1\mathbf{C}}^{(3)} = \mathbf{E} - \widehat{\mathbf{c}}_3^2,\quad J_{2\mathbf{C}}^{(3)} = \widehat{\mathbf{c}}_3^2,\quad J_{3\mathbf{C}}^{(3)} = \frac{1}{2}(\mathbf{O}_1 \otimes \mathbf{O}_1 + \mathbf{O}_2 \otimes \mathbf{O}_2) \cdot\cdot \overset{(n)}{\mathbf{C}}_\theta(t),$$
$$J_{4\mathbf{C}}^{(3)} = 2\,{}^4\mathbf{O}_3 \cdot\cdot \overset{(n)}{\mathbf{C}}_\theta(t),\quad {}^4\mathbf{O}_3 \equiv \mathbf{\Delta} - \frac{1}{2}(\mathbf{O}_1 \otimes \mathbf{O}_1 + \mathbf{O}_2 \otimes \mathbf{O}_2) - \widehat{\mathbf{c}}_3^2 \otimes \widehat{\mathbf{c}}_3^2,$$
$$J_{5\mathbf{C}}^{(3)} = \overset{(n)}{\mathbf{C}}{}^2_\theta(t) - I_1 \overset{(n)}{\mathbf{C}}_\theta(t) + \mathbf{E} I_2,\quad J_{6\mathbf{C}}^{(3)} = J_{7\mathbf{C}}^{(3)} = J_{8\mathbf{C}}^{(3)} = J_{9\mathbf{C}}^{(3)} = 0,$$
$$J_{10\mathbf{C}}^{(3)} = \frac{1}{4}(\mathbf{O}_1 \otimes \mathbf{O}_1 + \mathbf{O}_2 \otimes \mathbf{O}_2) \cdot\cdot \overset{(n)}{\mathbf{C}}_\theta(\tau),\quad J_{11\mathbf{C}}^{(3)} = {}^4\mathbf{O}_3 \cdot\cdot \overset{(n)}{\mathbf{C}}_\theta(\tau). \tag{8.88}$$

Substituting these expressions into (8.77) and rearranging the summands, we obtain constitutive equations for the *principal model $A_n$ of a transversely isotropic thermoviscoelastic continuum*:

$$\overset{(n)}{\mathbf{T}} = \breve{\varphi}_1 \mathbf{E} + \breve{\varphi}_2 \widehat{\mathbf{c}}_3^2 + (\mathbf{O}_1 \otimes \mathbf{O}_1 + \mathbf{O}_2 \otimes \mathbf{O}_2) \cdot\cdot \breve{\varphi}_3 \overset{(n)}{\mathbf{C}}_\theta + \breve{\varphi}_4 \overset{(n)}{\mathbf{C}}_\theta + \breve{\varphi}_5 \overset{(n)}{\mathbf{C}}{}^2_\theta. \tag{8.89}$$

Here we have denoted the functionals

$$\breve{\varphi}_1 \equiv \varphi_{01} + \varphi_{05} I_2 - \int_0^t (\varphi_{11} + \varphi_{15} I_2(t))\, d\tau,$$

$$\breve{\varphi}_2 \equiv \varphi_{02} - \varphi_{01} - 2\varphi_{04} I_2^{(3)} - \int_0^t (\varphi_{12} - \varphi_{11} - 2\varphi_{14} I_2^{(3)}(t) - 2\varphi_{1,11} I_2^{(3)}(\tau))\, d\tau,$$

$$\breve{\varphi}_3 \overset{(n)}{\mathbf{C}}_\theta \equiv \frac{1}{2}(\varphi_{03} - 2\varphi_{04}) \overset{(n)}{\mathbf{C}}_\theta - \frac{1}{2}\int_0^t \left( \left(\frac{\varphi_{13}}{2} - \varphi_{14}\right) \overset{(n)}{\mathbf{C}}_\theta(t) + \left(\frac{\varphi_{1,10}}{2} - \varphi_{1,11}\right) \overset{(n)}{\mathbf{C}}_\theta(\tau)\right) d\tau,$$

$$\breve{\varphi}_4 \overset{(n)}{\mathbf{C}}_\theta \equiv (2\varphi_{04} - \varphi_{05} I_1) \overset{(n)}{\mathbf{C}}_\theta - \int_0^t \left( (2\varphi_{14} - \varphi_{15} I_1(t)) \overset{(n)}{\mathbf{C}}_\theta(t) + \varphi_{1,11} \overset{(n)}{\mathbf{C}}_\theta(\tau)\right) d\tau,$$

$$\breve{\varphi}_5 \overset{(n)}{\mathbf{C}}{}^2_\theta \equiv \left( \varphi_{05} - \int_0^t \varphi_{15}\, d\tau \right) \overset{(n)}{\mathbf{C}}{}^2_\theta(t). \tag{8.90}$$

## 8.2.4 Principal Model $A_n$ of an Orthotropic Thermoviscoelastic Continuum

For the principal model $A_n$ of an orthotropic thermoviscoelastic continuum with difference cores, the functional basis of simultaneous invariants consists of 12 invariants, which can be chosen as follows (see (7.37)):

$$J_\gamma^{(O)} = I_\gamma^{(O)}(\overset{(n)}{\mathbf{C}}_\theta(t)),\ \gamma = 1, \ldots, 6; \quad J_{\gamma+6}^{(O)} = I_\gamma^{(O)}(\overset{(n)}{\mathbf{C}}_\theta(\tau)),\ \gamma = 1, 2, 3, 6;$$
$$J_{10}^{(O)} = \left(\widehat{\mathbf{c}}_2^2 \cdot \overset{(n)}{\mathbf{C}}_\theta(t)\right) \cdot\cdot \left(\widehat{\mathbf{c}}_3^2 \cdot \overset{(n)}{\mathbf{C}}_\theta(\tau)\right),\ J_{11}^{(O)} = \left(\widehat{\mathbf{c}}_1^2 \cdot \overset{(n)}{\mathbf{C}}_\theta(t)\right) \cdot\cdot \left(\widehat{\mathbf{c}}_3^2 \cdot \overset{(n)}{\mathbf{C}}_\theta(\tau)\right),$$
$$r = 6, \quad z = 12. \tag{8.91}$$

This set should be complemented by two more invariants (being dependent) in order to obtain relations symmetric with respect to the vectors $\widehat{\mathbf{c}}_\alpha^2$:

$$J_{13}^{(O)} = (\widehat{\mathbf{c}}_1^2 \cdot \overset{(n)}{\mathbf{C}}_\theta(t)) \cdot\cdot (\widehat{\mathbf{c}}_2^2 \cdot \overset{(n)}{\mathbf{C}}_\theta(\tau)),$$
$$J_{14}^{(O)} = I_7^{(O)}(\overset{(n)}{\mathbf{C}}_\theta(t)) = (\widehat{\mathbf{c}}_1^2 \cdot \overset{(n)}{\mathbf{C}}_\theta(t)) \cdot\cdot (\widehat{\mathbf{c}}_2^2 \cdot \overset{(n)}{\mathbf{C}}_\theta(t)). \tag{8.92}$$

The partial derivative tensors of these invariants have the forms (see [12])

$$
\begin{aligned}
&J^{(O)}_{\gamma\mathbf{C}} = \widehat{\mathbf{c}}^2_\gamma,\ \gamma = 1,2,3; \quad J^{(O)}_{\gamma+3,\mathbf{C}} = \frac{1}{2}(\mathbf{O}_\gamma \otimes \mathbf{O}_\gamma)\cdot\cdot \overset{(n)}{\mathbf{C}}_\theta(t),\ \ \gamma = 1,2;\\
&J^{(O)}_{6\mathbf{C}} = 3\,{}^6\mathbf{O}_m \cdots\cdot \overset{(n)}{\mathbf{C}}_\theta(t) \otimes \overset{(n)}{\mathbf{C}}_\theta(t), \quad J^{(O)}_{\gamma+6,\mathbf{C}} = J^{(O)}_{12\mathbf{C}} = 0,\ \ \gamma = 1,2,3;\\
&J^{(O)}_{10\mathbf{C}} = \frac{1}{4}(\mathbf{O}_1 \otimes \mathbf{O}_1)\cdot\cdot \overset{(n)}{\mathbf{C}}_\theta(\tau), \quad J^{(O)}_{11\mathbf{C}} = \frac{1}{4}(\mathbf{O}_2 \otimes \mathbf{O}_2)\cdot\cdot \overset{(n)}{\mathbf{C}}_\theta(\tau),\\
&J^{(O)}_{13\mathbf{C}} = \frac{1}{4}(\mathbf{O}_3 \otimes \mathbf{O}_3)\cdot\cdot \overset{(n)}{\mathbf{C}}_\theta(\tau), \quad J^{(O)}_{14\mathbf{C}} = \frac{1}{2}(\mathbf{O}_3 \otimes \mathbf{O}_3)\cdot\cdot \overset{(n)}{\mathbf{C}}_\theta(t),
\end{aligned}
\tag{8.93}
$$

where the tensor ${}^6\mathbf{O}_m$ is determined by formula (4.316).

Substituting these expressions into (8.77) and grouping like terms, we obtain constitutive equations for the *principal model $A_n$ of an orthotropic thermoviscoelastic continuum*:

$$
\overset{(n)}{\mathbf{T}} = \sum_{\gamma=1}^{3}(\breve{\varphi}_\gamma \widehat{\mathbf{c}}^2_\gamma + \mathbf{O}_\gamma \otimes \mathbf{O}_\gamma \cdot\cdot\, \breve{\varphi}_{3+\gamma} \overset{(n)}{\mathbf{C}}_\theta) + \breve{\varphi}_7\, {}^6\mathbf{O}_m \cdots\cdot \overset{(n)}{\mathbf{C}}_\theta \otimes \overset{(n)}{\mathbf{C}}_\theta. \tag{8.94}
$$

Here we have denoted the functionals

$$
\begin{aligned}
&\breve{\varphi}_\gamma \equiv \varphi_{0\gamma} - \int_0^t \varphi_{1\gamma}\big(t-\tau, J^{(O)}_\alpha\big)\, d\tau, \quad \gamma = 1,\ 2,\ 3,\\
&\breve{\varphi}_{3+\gamma} \overset{(n)}{\mathbf{C}}_\theta \equiv \frac{1}{2}\varphi_{0,3+\gamma} \overset{(n)}{\mathbf{C}}_\theta - \frac{1}{2}\int_0^t \bigg(\frac{1}{2}\varphi_{1,9+\gamma}\big(t-\tau,\ J^{(O)}_\alpha\big) \overset{(n)}{\mathbf{C}}_\theta(\tau)\\
&\qquad + \varphi_{1,3+\gamma}\big(t-\tau,\ J^{(O)}_\alpha\big) \overset{(n)}{\mathbf{C}}_\theta(t)\bigg)\, d\tau, \quad \gamma = 1,\ 2,
\end{aligned}
\tag{8.95}
$$

$$
\begin{aligned}
&\breve{\varphi}_7 \equiv 3\varphi_{06} - 3\int_0^t \varphi_{16}\big(t-\tau, J^{(O)}_\alpha\big)\, d\tau,\\
&\breve{\varphi}_6 \overset{(n)}{\mathbf{C}}_\theta \equiv \frac{1}{2}\varphi_{0,14} \overset{(n)}{\mathbf{C}}_\theta - \frac{1}{2}\int_0^t \bigg(\frac{1}{2}\varphi_{1,13}\big(t-\tau, J^{(O)}_\alpha\big) \overset{(n)}{\mathbf{C}}_\theta(\tau)\\
&\qquad + \varphi_{1,14}\big(t-\tau, J^{(O)}_\alpha\big) \overset{(n)}{\mathbf{C}}_\theta(t)\bigg)\, d\tau.
\end{aligned}
$$

### 8.2.5 Quadratic Models $A_n$ of Thermoviscoelastic Continua

For the *quadratic model $A_n$ of a thermoviscoelastic continuum with difference cores*, we retain two integrals in the sum (8.61), i.e. $m = 1, 2$. A form of constitutive equations for specific symmetry groups $\overset{\circ}{G}_s$ becomes considerably more complicated, because there appear double integrals and we need to consider simultaneous invariants $J_\gamma^{(s)}$ of three tensors. Therefore, one usually considers the particular case of the quadratic model when $m = 1$ and 2, but simultaneous invariants $J_\gamma^{(s)}$ of only two tensors appear there just as in the principal model:

$$\psi = \varphi_0 \left( I_\gamma^{(s)}(\overset{(n)}{\mathbf{C}}_\theta(t)), \theta \right) - \int_0^t \varphi_1(t-\tau, J_\gamma^{(s)} \left( \overset{(n)}{\mathbf{C}}_\theta(t), \overset{(n)}{\mathbf{C}}_\theta(\tau_1) \right) d\tau$$
$$+ \int_0^t \int_0^t \varphi_2(t-\tau_1, t-\tau_2, J_\gamma^{(s)} \left( \overset{(n)}{\mathbf{C}}_\theta(\tau_1), \overset{(n)}{\mathbf{C}}_\theta(\tau_2) \right) d\tau_1\, d\tau_2. \qquad (8.96)$$

Here $\varphi_0$, $\varphi_1$ and $\varphi_2$ are functions of the arguments indicated. Since the core $\varphi_2$ in this model is independent of $\overset{(n)}{\mathbf{C}}_\theta(t)$, so $\partial\varphi_2/\partial\overset{(n)}{\mathbf{C}}_\theta(t) \equiv 0$, and constitutive equations prove to be coincident with (8.78); and hence they coincide with (8.85), (8.89) and (8.94) too. The distinction between the principal and quadratic models consists only in the forms of the functional $\psi$ and the dissipation function $w^*$. Such a situation is typical for viscoelastic continua when distinct functionals of the free energy $\psi$ correspond to the same relations between the tensors of stresses and deformations. Comparing the principal model (8.77) with the quadratic one (8.96), we can also notice that the principal model has only one core $\varphi_1$ appearing also in relation (8.78), and the quadratic model has two cores $\varphi_1$ and $\varphi_2$, one of which is not included in relation (8.78) between stresses and deformations. Thus, for the principal model we can restore the functional of the free energy $\psi$ by Eqs. (8.78) up to the entropy term $\partial\varphi_0/\partial\theta$ and the constant $\varphi_0(0, \theta_0) = \psi_0$.

Models of viscoelastic continua having such a property are called *mechanically determinate*.

The quadratic model is not mechanically determinate: due to the presence of the core $\varphi_2$ we cannot restore the form of $\psi$ by relations (8.78) between stresses and deformations. Nevertheless, this model is also used in practice due to its quadratic structure being typical for thermodynamic potentials.

### 8.2.6 Linear Models $A_n$ of Viscoelastic Continua

The quadratic model $A_n$ of a thermoviscoelastic continuum with difference cores (8.96), where the functions $\varphi_0$, $\varphi_1$ and $\varphi_2$ depend linearly upon the quadratic invariants $J_\gamma^{(s)}$ and quadratically upon the linear invariants $J_\gamma^{(s)}$, is called *linear* (cubic invariants do not occur in this model):

$$\varphi_0 = \psi_0 + \frac{1}{2\overset{\circ}{\rho}} \sum_{\gamma,\beta=1}^{r_1} l_{\gamma\beta} I_\gamma^{(s)}(t) I_\beta^{(s)}(t) + \frac{1}{\overset{\circ}{\rho}} \sum_{\gamma=r_1+1}^{r_2} l_{\gamma\gamma} I_\gamma^{(s)}(t),$$

$$\varphi_1 = \frac{1}{\overset{\circ}{\rho}} \sum_{\gamma,\beta=1}^{r_1} q_{\gamma\beta}(t-\tau) I_\gamma^{(s)}(t) I_\beta^{(s)}(\tau) + \frac{2}{\overset{\circ}{\rho}} \sum_{\gamma=r_1+1}^{r_2} q_{\gamma\gamma}(t-\tau) J_\gamma^{(s)}(t,\tau), \quad (8.97)$$

$$\varphi_2 = \frac{1}{2\overset{\circ}{\rho}} \sum_{\gamma,\beta=1}^{r_1} p_{\gamma\beta}(t-\tau_1, t-\tau_2) I_\gamma^{(s)}(\tau_1) I_\beta^{(s)}(\tau_2)$$
$$+ \frac{1}{\overset{\circ}{\rho}} \sum_{\gamma=r_1+1}^{r_2} p_{\gamma\gamma}(t-\tau_1,\ t-\tau_2) J_\gamma^{(s)}(\tau_1,\ \tau_2).$$

Here $l_{\gamma\beta}$, $l_{\gamma\gamma}$ are constants; $q_{\gamma\beta}(t-\tau)$, $q_{\gamma\gamma}(t-\tau)$ are one-instant cores (they are functions of one argument, being symmetric in $\gamma$ and $\beta$); $p_{\gamma\beta}(t-\tau_1, t-\tau_2)$ and $p_{\gamma\gamma}(t-\tau_1, t-\tau_2)$ are two-instant cores (they are functions of two variables, being symmetric in $\gamma$ and $\beta$, and also in $t-\tau_1$ and $t-\tau_2$). We have introduced the notation

$$I_\gamma^{(s)}(\tau) = I_\gamma^{(s)}\left(\overset{(n)}{\mathbf{C}}_\theta(\tau)\right), \quad J_\gamma^{(s)}(\tau_1,\ \tau_2) = J_\gamma^{(s)}\left(\overset{(n)}{\mathbf{C}}_\theta(\tau_1), \overset{(n)}{\mathbf{C}}_\theta(\tau_2)\right), \quad (8.98)$$

where $r_1$ is the number of linear invariants $I_\gamma^{(s)}(\overset{(n)}{\mathbf{C}}_\theta(t))$ in group $\overset{\circ}{G}_s$ $(r_1 \leqslant r)$, and $(r_2 - r_1)$ is the number of the quadratic simultaneous invariants $J_\gamma^{(s)}(\tau_1,\ \tau_2)$ in this group, where $r_2 \leqslant z$.

Since not all simultaneous invariants contained in the full basis $J_\gamma(\overset{(n)}{\mathbf{C}}_\theta(\tau_1),\ \overset{(n)}{\mathbf{C}}_\theta(\tau_2))$ appear in the expression for the functional $\psi$ in linear models, it is convenient to renumber these invariants in comparison with the bases (8.83), (8.87), and (8.91) by enumerating first the linear invariants $I_\gamma(\overset{(n)}{\mathbf{C}}_\theta(t))$ and then the quadratic simultaneous invariants $J_\gamma(\overset{(n)}{\mathbf{C}}_\theta(\tau_1),\ \overset{(n)}{\mathbf{C}}_\theta(\tau_2))$.

The *principal linear model* $A_n$ of viscoelastic continua can be obtained by applying similar relationships to the functions $\varphi_0$ and $\varphi_1$ of the principal model (8.77) when $\varphi_2 = 0$.

The functions $\varphi_{0\gamma}$ and $\varphi_{1\gamma}$ (8.79) for the principal linear model have the forms

$$\varphi_{0\gamma} = J \sum_{\beta=1}^{r_1} l_{\gamma\beta} I_\beta^{(s)}(t), \quad \varphi_{1\gamma} = J \sum_{\beta=1}^{r_1} q_{\gamma\beta}(t-\tau) I_\beta^{(s)}(\tau), \quad \gamma = 1, \ldots, r_1,$$
$$\varphi_{0\gamma} = J l_{\gamma\gamma}, \quad \varphi_{1\gamma} = J q_{\gamma\gamma}(t-\tau), \quad \gamma = r_1+1, \ldots, r_2,$$
$$J = \rho/\overset{\circ}{\rho}. \quad (8.99)$$

As noted above, constitutive equations (8.77) for both the models coincide; and, for the linear models, they have the form

$$\overset{(\mathrm{n})}{\mathbf{T}} = J \sum_{\gamma,\beta=1}^{r_1} \breve{l}_{\gamma\beta} I_\gamma^{(s)} \mathbf{O}_\beta^{(s)} + J \sum_{\gamma=r_1+1}^{r_2} \breve{l}_{\gamma\gamma} I_{\gamma\mathbf{C}}^{(s)}, \tag{8.100}$$

where we have denoted the linear functionals

$$\breve{l}_{\gamma\beta} I_\gamma^{(s)} \equiv l_{\gamma\beta} I_\gamma^{(s)}(t) - \int_0^t q_{\gamma\beta}(t-\tau) I_\gamma^{(s)}(\tau)\, d\tau, \quad \gamma,\ \beta = 1, \ldots, r_1,$$

$$\breve{l}_{\gamma\gamma} I_{\gamma\mathbf{C}}^{(s)} \equiv l_{\gamma\gamma} I_{\gamma\mathbf{C}}^{(s)}(t) - \int_0^t q_{\gamma\gamma}(t-\tau) I_{\gamma\mathbf{C}}^{(s)}(\tau)\, d\tau, \quad \gamma = r_1+1, \ldots, r_2. \tag{8.101}$$

Here we have taken into account that all the linear invariants have the form $I_\gamma^{(s)} = \overset{(\mathrm{n})}{\mathbf{C}}_\theta \cdot\cdot\, \mathbf{O}_\gamma^{(s)}$, where $\mathbf{O}_\gamma^{(s)}$ are producing tensors of the group, and for the quadratic invariants,

$$J_{\gamma\mathbf{C}}^{(s)}(\tau) = \frac{\partial J_\gamma(t,\ \tau)}{\partial \overset{(\mathrm{n})}{\mathbf{C}}_\theta(t)} = \frac{1}{2} \frac{\partial I_\gamma^{(s)}(\tau,\ \tau)}{\partial \overset{(\mathrm{n})}{\mathbf{C}}_\theta(\tau)} = \frac{1}{2} I_{\gamma\mathbf{C}}^{(s)}(\tau), \quad \gamma = r_1+1, \ldots, r_2. \tag{8.102}$$

(For isotropic continua, in order to satisfy this condition, as invariants $I_\gamma^{(I)}$ one should choose the invariants $I_1(\mathbf{C}_\theta)$ and $I_1(\mathbf{C}_\theta^2)$.) Notice that when the cores $q_{\gamma\beta}(t)$ and $q_{\gamma\gamma}(t)$ in (8.101) are absent, then these relations exactly coincide with relations (4.333) of linear models $A_n$ for ideal continua if in the last ones we assume that $\bar{m}_\gamma = 0$.

For principal linear models $A_n$ of viscoelastic continua, the dissipation function $w^*$ (8.82) has the form

$$w^* = J \sum_{\gamma,\beta=1}^{r_1} \left( q_{\gamma\beta}(0) I_\gamma^{(s)}(t) I_\beta^{(s)}(t) + \int_0^t \frac{\partial}{\partial t} q_{\gamma\beta}(t-\tau) I_\gamma^{(s)}(t) I_\beta^{(s)}(\tau) d\tau \right)$$
$$+ 2J \sum_{\gamma=r_1+1}^{r_2} \left( q_{\gamma\gamma}(0) J_\gamma^{(s)}(t,t) + \int_0^t \frac{\partial}{\partial t} q_{\gamma\gamma}(t-\tau) J_\gamma^{(s)}(t,\tau)\, d\tau \right), \tag{8.103a}$$

and for linear models $A_n$ of viscoelastic continua, terms with two-instant cores should be added to the expression (8.103a):

$$w^* = J \sum_{\gamma,\beta=1}^{r_1} \left( q_{\gamma\beta}(0) I_\gamma^{(s)}(t) I_\beta^{(s)}(t) + \int_0^t \frac{\partial}{\partial t} q_{\gamma\beta}(t-\tau) I_\gamma^{(s)}(t) I_\beta^{(s)}(\tau)\, d\tau \right.$$
$$\left. - \frac{1}{2} \int_0^t \int_0^t \frac{\partial}{\partial t} p_{\gamma\beta}(t-\tau_1, t-\tau_2) I_\gamma^{(s)}(\tau_1) I_\beta^{(s)}(\tau_2)\, d\tau_1\, d\tau_2 \right)$$
$$+ J \sum_{\gamma=r_1+1}^{r_2} \left( 2 q_{\gamma\gamma}(0) J_\gamma^{(s)}(t,t) + 2 \int_0^t \frac{\partial}{\partial t} q_{\gamma\gamma}(t-\tau) J_\gamma^{(s)}(t,\tau)\, d\tau \right.$$
$$\left. - \int_0^t \int_0^t \frac{\partial}{\partial t} p_{\gamma\gamma}(t-\tau_1, t-\tau_2) J_\gamma(\tau_1,\tau_2)\, d\tau_1\, d\tau_2 \right). \quad (8.103b)$$

For linear models $A_n$, the specific entropy $\eta$, according to (8.69) and (8.82), has the form

$$\eta = -\frac{\partial \psi_0}{\partial \theta} + \frac{1}{\rho} \boldsymbol{\alpha} \cdot\cdot \overset{(n)}{\mathbf{T}}, \quad (8.104)$$

where $\psi_0(\theta)$ is a function only of the temperature. The dissipation function with the help of formula (8.75) can be represented in the form

$$w^* = \overset{(n)}{\mathbf{T}} \cdot\cdot \frac{d}{dt} \overset{(n)}{\mathbf{C}} - \rho \frac{d\psi}{dt} + \left( \rho \frac{\partial \psi_0}{\partial \theta} - \boldsymbol{\alpha} \cdot\cdot \overset{(n)}{\mathbf{T}} \right) \frac{d\theta}{dt}. \quad (8.104a)$$

### 8.2.7 Representation of Linear Models $A_n$ in the Boltzmann Form

For linear models $A_n$ of viscoelastic continua, representations of the free energy $\psi$ in the form (8.96), (8.97), the constitutive equations in the form (8.100), (8.101) and the dissipation function $w^*$ in the form (8.103b) are called the model $A_n$ in the *Volterra form* (the name is connected to the fact that integral expressions of the form (8.101) occurring in this representation were considered for the first time by Volterra in 1909).

Let us give another equivalent representation for these models. Introduce new two-instant cores $\bar{\psi}_{\gamma\beta}(y,z)$ and $\bar{\psi}_{\gamma\gamma}(y,z)$ satisfying the differential equations

$$\frac{\partial^2 \bar{\psi}_{\gamma\beta}(y,z)}{\partial y \partial z} = p_{\gamma\beta}(y,z), \quad \frac{\partial \bar{\psi}_{\gamma\beta}(y,0)}{\partial y} = -q_{\gamma\beta}(y), \quad \bar{\psi}_{\gamma\beta}(0,0) = l_{\gamma\beta},$$
$$\frac{\partial^2 \bar{\psi}_{\gamma\gamma}(y,z)}{\partial y \partial z} = p_{\gamma\gamma}(y,z), \quad \frac{\partial \bar{\psi}_{\gamma\gamma}(y,0)}{\partial y} = -q_{\gamma\gamma}(y), \quad \bar{\psi}_{\gamma\gamma}(0,0) = l_{\gamma\gamma},$$
$$y = t - \tau_1, \quad z = t - \tau_2. \quad (8.105)$$

Then the following theorem holds.

**Theorem 8.8.** *Let the two-instant cores* $\bar{\psi}_{\gamma\beta}(y,z)$ *and* $\bar{\psi}_{\gamma\gamma}(y,z)$

1. *Be symmetric functions of their arguments:*

$$\bar{\psi}_{\gamma\beta}(y,z) = \bar{\psi}_{\gamma\beta}(z,y), \quad \bar{\psi}_{\gamma\gamma}(y,z) = \bar{\psi}_{\gamma\gamma}(z,y), \tag{8.105a}$$

2. *Be two times continuously differentiable functions of their arguments within the interval* $(0,t)$*,*
3. *Satisfy the conditions (8.105),*

*then we can pass from the representation of linear model* $A_n$ *in the Volterra form (Eqs. (8.96), (8.97), (8.100), (8.101), and (8.103b)) to an equivalent representation of the model* $A_n$ *in the Boltzmann form*

$$\psi = \psi_0 + \frac{1}{2\overset{\circ}{\rho}} \sum_{\gamma,\beta=1}^{r_1} \int_0^t \int_0^t \bar{\psi}_{\gamma\beta}(t-\tau_1, t-\tau_2)\, dI_\gamma^{(s)}(\tau_1)\, dI_\beta^{(s)}(\tau_2)$$
$$+ \frac{1}{\overset{\circ}{\rho}} \sum_{\gamma=r_1+1}^{r_2} \int_0^t \int_0^t \bar{\psi}_{\gamma\gamma}(t-\tau_1, t-\tau_2)\, dJ_\gamma^{(s)}(\tau_1,\ \tau_2), \tag{8.106}$$

$$\overset{(n)}{\mathbf{T}} = J \sum_{\gamma,\beta=1}^{r_1} \mathbf{O}_\beta^{(s)} \int_0^t r_{\gamma\beta}(t-\tau_1)\, dI_\gamma^{(s)}(\tau) + J \sum_{\gamma=r_1+1}^{r_2} \int_0^t r_{\gamma\gamma}(t-\tau)\, dI_{\gamma\mathbf{C}}^{(s)}(\tau), \tag{8.107}$$

$$w^* = -\frac{1}{2} J \sum_{\gamma,\beta=1}^{r_1} \int_0^t \int_0^t \frac{\partial}{\partial t} \bar{\psi}_{\gamma\beta}(t-\tau_1,\ t-\tau_2)\, dI_\gamma^{(s)}(\tau_1)\, dI_\beta^{(s)}(\tau_2)$$
$$- J \sum_{\gamma=r_1+1}^{r_2} \int_0^t \int_0^t \frac{\partial}{\partial t} \bar{\psi}_{\gamma\gamma}(t-\tau_1, t-\tau_2)\, dJ_\gamma^{(s)}(\tau_1,\ \tau_2). \tag{8.108}$$

Here we have introduced the notation

$$r_{\gamma\beta}(y) = \bar{\psi}_{\gamma\beta}(y,0), \quad r_{\gamma\gamma}(y) = \bar{\psi}_{\gamma\gamma}(y,0), \tag{8.109a}$$

$$dI_\gamma^{(s)}(\tau_1) = \dot{I}_1^{(s)}(\tau_1)\, d\tau_1, \quad dJ_\gamma^{(s)}(\tau_1,\tau_2) = J_\gamma^{(s)}\left(\overset{(n)}{\mathbf{C}}{}_\theta^\bullet(\tau_1), \overset{(n)}{\mathbf{C}}{}_\theta^\bullet(\tau_2)\right) d\tau_1\, d\tau_2. \tag{8.109b}$$

▼ To prove the theorem, it is sufficient to transform the integrals in (8.106)–(8.108) as follows:

$$\int_0^t \left( \int_0^t \bar{\psi}_{\gamma\beta}(t-\tau_1, t-\tau_2)\, dI_\gamma^{(s)}(\tau_1) \right) dI_\beta^{(s)}(\tau_2) = \int_0^t \Big( \bar{\psi}_{\gamma\beta}(0, t-\tau_2) I_\gamma^{(s)}(t)$$

$$- \int_0^t \frac{\partial \bar{\psi}_{\gamma\beta}}{\partial \tau_1}(t-\tau_1, t-\tau_2) I_\gamma^{(s)}(\tau_1)\, d\tau_1 \Big)\, dI_\beta^{(s)}(\tau_2) = I_\gamma^{(s)}(t) \Big( \bar{\psi}_{\gamma\beta}(0,0) I_\beta^{(s)}(t)$$

$$- \int_0^t \frac{\partial}{\partial \tau_2} \bar{\psi}_{\gamma\beta}(0, t-\tau_2) I_\beta^{(s)}(\tau_2)\, d\tau_2 \Big) - \int_0^t \frac{\partial \bar{\psi}_{\gamma\beta}}{\partial \tau_1}(t-\tau_1, 0) I_\gamma^{(s)}(\tau_1) d\tau_1 I_\beta^{(s)}(t)$$

$$+ \int_0^t \int_0^t \frac{\partial^2 \bar{\psi}_{\gamma\beta}}{\partial \tau_1 \partial \tau_2}(t-\tau_1, t-\tau_2) I_\gamma^{(s)}(\tau_1) I_\beta^{(s)}(\tau_2)\, d\tau_1\, d\tau_2$$

$$= l_{\gamma\beta} I_\gamma^{(s)}(t) I_\beta^{(s)}(t) - \int_0^t q_{\gamma\beta}(t-\tau) (I_\gamma^{(s)}(t) I_\beta^{(s)}(\tau) + I_\gamma^{(s)}(\tau) I_\beta^{(s)}(t))\, d\tau$$

$$+ \int_0^t \int_0^t p_{\gamma\beta}(t-\tau_1, t-\tau_2) I_\gamma^{(s)}(\tau_1) I_\beta^{(s)}(\tau_2)\, d\tau_1\, d\tau_2. \tag{8.110}$$

Here we have taken into account that $I_\gamma^{(s)}(0) = 0$ and changed the variables

$$\frac{\partial}{\partial \tau_2} \bar{\psi}_{\gamma\beta}(0, t-\tau_2) = -\frac{\partial}{\partial (t-\tau_2)} \bar{\psi}_{\gamma\beta}(0, t-\tau_2) = -\frac{\partial}{\partial y} \bar{\psi}_{\gamma\beta}(0, y) = q_{\gamma\beta}(y).$$

In a similar way, we can transform the integrals of $\bar{\psi}_{\gamma\gamma}$ with taking into account that $J_\gamma^{(s)}(\overset{(n)}{\mathbf{C}}_\theta(\tau_1), \overset{(n)}{\mathbf{C}}_\theta(\tau_2))$ is a linear function with respect to each tensor argument; therefore, the following relations hold:

$$dJ_\gamma^{(s)}(\tau_1, \tau_2) = \frac{\partial}{\partial \tau_1} J_\gamma^{(s)}(\overset{(n)}{\mathbf{C}}_\theta(\tau_1), \overset{(n)}{\mathbf{C}}{}^{\bullet}_\theta(\tau_2))\, d\tau_1\, d\tau_2$$

$$= \frac{\partial}{\partial \tau_2} J_\gamma^{(s)}(\overset{(n)}{\mathbf{C}}{}^{\bullet}_\theta(\tau_1), \overset{(n)}{\mathbf{C}}_\theta(\tau_2))\, d\tau_1\, d\tau_2$$

$$= \frac{\partial^2}{\partial \tau_1 \partial \tau_2} J_\gamma^{(s)}(\overset{(n)}{\mathbf{C}}_\theta(\tau_1), \overset{(n)}{\mathbf{C}}_\theta(\tau_2))\, d\tau_1\, d\tau_2,$$

hence

$$\int_0^t\int_0^t \bar{\psi}_{\gamma\gamma}(t-\tau_1,t-\tau_2)\,dJ_\gamma^{(s)}(\tau_1,\tau_2)$$
$$= l_{\gamma\gamma}J_\gamma^{(s)}(\overset{(n)}{\mathbf{C}}_\theta(t),\overset{(n)}{\mathbf{C}}_\theta(t)) - 2\int_0^t q_{\gamma\gamma}(t-\tau)J_\gamma^{(s)}(\overset{(n)}{\mathbf{C}}_\theta(t),\overset{(n)}{\mathbf{C}}_\theta(\tau))\,d\tau$$
$$+\int_0^t\int_0^t p_{\gamma\gamma}(t-\tau)J_\gamma^{(s)}(\overset{(n)}{\mathbf{C}}_\theta(\tau_1),\overset{(n)}{\mathbf{C}}_\theta(\tau_2))\,d\tau_1\,d\tau_2. \qquad (8.111)$$

On substituting (8.110) and (8.111) into (8.106), we actually obtain the expressions (8.96) and (8.97) for $\psi$.

The representations (8.107) and (8.108) can be proved in a similar way (see Exercise 8.2.1). ▲

*Remark.* If we consider constitutive equations in the Volterra form (8.96), (8.97), and (8.100) and pass to the limit at $t \to 0$, then all the integral summands containing the cores $q_{\gamma\beta}$ and $q_{\gamma\gamma}$ vanish. As a result, we get *instantly elastic relations* which exactly coincide with the corresponding equations (4.331), (4.333), and (4.334) of models $A_n$ of elastic continua:

$$\begin{aligned}\psi(0) &= \psi_0 + \frac{1}{2\overset{\circ}{\rho}}\sum_{\gamma,\beta=1}^{r_1} l_{\gamma\beta}I_\gamma^{(s)}(0)I_\beta^{(s)}(0) + \frac{1}{\overset{\circ}{\rho}}\sum_{\gamma=r_1+1}^{r_2} l_{\gamma\gamma}I_\gamma^{(s)}(0),\\ \overset{(n)}{\mathbf{T}}(0) &= J\sum_{\gamma,\beta=1}^{r_1} l_{\gamma\beta}I_\gamma^{(s)}(0)\mathbf{O}_\beta^{(s)} + J\sum_{\gamma=r_1+1}^{r_2} l_{\gamma\gamma}I_{\gamma\mathbf{C}}^{(s)}(0).\end{aligned} \qquad (8.112)$$

In order to obtain these relations from the Boltzmann form (8.106), (8.107), one should represent the deformation tensors in the form $\overset{(n)}{\mathbf{C}}_\theta(\tau) = \overset{(n)}{\mathbf{C}}_\theta(0)h(\tau)$, where $h(\tau)$ is the Heaviside function. Then we find that $I_\gamma^{(s)}(\tau) = I_\gamma^{(s)}(0)h(\tau)$ and $J_\gamma^{(s)}(\tau_1,\tau_2) = I_\gamma^{(s)}(0)h(\tau_1)h(\tau_2)$. Substituting these expressions into (8.106) and (8.107) and using (8.105), we obtain the desired relations (8.112) as $t \to 0_+$. □

## 8.2.8 *Mechanically Determinate Linear Models $A_n$ of Viscoelastic Continua*

As noted in Sect. 8.2.5, quadratic models $A_n$, including the linear models (8.97), are not mechanically determinate due to the presence of the two-instant cores $p_{\gamma\beta}(t-\tau_1,t-\tau_2)$ and $p_{\gamma\gamma}(t-\tau_1,t-\tau_2)$. However, the models may become mechanically determinate after introduction of the additional assumption on a form of the two-instant cores; we assume that they depend on the sum of their arguments:

$$\bar{\psi}_{\gamma\beta}(y,z) = \bar{\psi}_{\gamma\beta}(y+z), \quad \bar{\psi}_{\gamma\gamma}(y,z) = \bar{\psi}_{\gamma\gamma}(y+z). \tag{8.113}$$

In this case, the cores $\bar{\psi}_{\gamma\beta}$ and $\bar{\psi}_{\gamma\gamma}$ become one-instant, and with the help of formula (8.109a) they can be uniquely expressed in terms of the cores $r_{\gamma\beta}(y)$ and $r_{\gamma\gamma}(y)$ included in constitutive equations (8.107):

$$\bar{\psi}_{\gamma\beta}(y) = r_{\gamma\beta}(y), \quad \bar{\psi}_{\gamma\gamma}(y) = r_{\gamma\gamma}(y). \tag{8.114}$$

The functional (8.106) of the free energy takes the form

$$\psi = \psi_0 + \frac{1}{2\overset{\circ}{\rho}} \sum_{\gamma,\beta=1}^{r_1} \int_0^t \int_0^t r_{\gamma\beta}(2t-\tau_1-\tau_2)\, dI_\gamma^{(s)}(\tau_1)\, dI_\beta^{(s)}(\tau_2)$$
$$+ \frac{1}{\overset{\circ}{\rho}} \sum_{\gamma=r_1+1}^{r_2} \int_0^t \int_0^t r_{\gamma\gamma}(2t-\tau_1-\tau_2)\, dJ_\gamma^{(s)}(\tau_1,\tau_2). \tag{8.115}$$

The dissipation function $w^*$ (8.108) in this model is also determined completely by the functional (8.107) of the constitutive equations:

$$w^* = -\frac{J}{2} \sum_{\gamma,\beta=1}^{r_1} \int_0^t \int_0^t \frac{\partial}{\partial t} r_{\gamma\beta}(2t-\tau_1-\tau_2)\, dI_\gamma^{(s)}(\tau_1)\, dI_\beta^{(s)}(\tau_2)$$
$$-J \sum_{\gamma=r_1+1}^{r_2} \int_0^t \int_0^t \frac{\partial}{\partial t} r_{\gamma\gamma}(2t-\tau_1-\tau_2)\, dJ_\gamma^{(s)}(\tau_1,\tau_2). \tag{8.116}$$

The cores $r_{\gamma\beta}(y)$ and $r_{\gamma\gamma}(y)$, according to (8.114) and (8.105), are connected to the cores $q_{\gamma\beta}(y)$ and $q_{\gamma\gamma}(y)$ by the relations

$$\frac{\partial r_{\gamma\beta}(y)}{\partial y} = -q_{\gamma\beta}(y), \quad \frac{\partial r_{\gamma\gamma}(y)}{\partial y} = -q_{\gamma\gamma}(y), \quad r_{\gamma\beta}(0) = l_{\gamma\beta}, \quad r_{\gamma\gamma}(0) = l_{\gamma\gamma}. \tag{8.117}$$

The cores $q_{\gamma\beta}(y)$ and $q_{\gamma\gamma}(y)$ are called the *relaxation cores*, and the cores $r_{\gamma\beta}(y)$ and $r_{\gamma\gamma}(y)$ are called the *relaxation functions*.

### 8.2.9 Linear Models $A_n$ for Isotropic Viscoelastic Continua

Let us derive now constitutive equations for linear models $A_n$ of viscoelastic continua in the Volterra (8.100) and Boltzmann (8.107) forms for different symmetry groups $\overset{\circ}{G}_s$.

For linear models $A_n$ of viscoelastic isotropic continua, the invariants (8.98) and the derivative tensors $I^{(I)}_{\gamma\mathbf{C}}$ (8.102) have the forms

$$r = 3, \quad r_1 = 1, \quad r_2 = 2,$$
$$I_1^{(I)}(\tau) = I_1(\overset{(n)}{\mathbf{C}}_\theta(\tau)), \;\; I_2^{(I)}(\tau) = J_2^{(I)}(\tau_1, \tau_2) = \overset{(n)}{\mathbf{C}}_\theta(\tau_1) \cdot\cdot \overset{(n)}{\mathbf{C}}_\theta(\tau_2), \tag{8.118}$$
$$\mathbf{O}_1^{(I)} = I^{(I)}_{1\mathbf{C}}(t) = \mathbf{E}, \quad I^{(I)}_{2\mathbf{C}}(t) = 2\overset{(n)}{\mathbf{C}}_\theta(t).$$

Then constitutive equations (8.100), (8.106), and (8.107) become

$$\overset{\circ}{\rho}\psi = \overset{\circ}{\rho}\psi_0 + \frac{1}{2}\int_0^t\int_0^t r_1(2t - \tau_1 - \tau_2) dI_1(\overset{(n)}{\mathbf{C}}_\theta(\tau_1))\, dI_1(\overset{(n)}{\mathbf{C}}_\theta(\tau_2))$$
$$+ \int_0^t\int_0^t r_2(2t - \tau_1 - \tau_2)\, d\overset{(n)}{\mathbf{C}}_\theta(\tau_1) \cdot\cdot\, d\overset{(n)}{\mathbf{C}}_\theta(\tau_2), \tag{8.119}$$

$$\overset{(n)}{\mathbf{T}} = J(\breve{l}_1 I_1\mathbf{E} + 2\breve{l}_2\overset{(n)}{\mathbf{C}}_\theta). \tag{8.120}$$

Here we have denoted the linear functionals

$$\breve{l}_1 I_1 \equiv l_1 I_1(\overset{(n)}{\mathbf{C}}_\theta(t)) - \int_0^t q_1(t-\tau) I_1(\overset{(n)}{\mathbf{C}}_\theta(\tau))\, d\tau = \int_0^t r_1(t-\tau)\, dI_1(\overset{(n)}{\mathbf{C}}_\theta(\tau)),$$
$$\breve{l}_2\overset{(n)}{\mathbf{C}}_\theta = l_2\overset{(n)}{\mathbf{C}}_\theta(t) - \int_0^t q_2(t-\tau)\overset{(n)}{\mathbf{C}}_\theta(\tau)\, d\tau = \int_0^t r_2(t-\tau)\, d\overset{(n)}{\mathbf{C}}_\theta(\tau). \tag{8.121}$$

Thus, for an isotropic continuum, there are two independent constants $l_1, l_2$ and two cores $q_\gamma(t-\tau)$ connected to the cores $r_\gamma(y)$ by the relations (8.117)

$$\frac{\partial r_\gamma(y)}{\partial y} = -q_\gamma(y), \quad r_\gamma(0) = l_\gamma, \quad \gamma = 1,\ 2. \tag{8.122}$$

Introducing the fourth-order tensor functional similar to the tensor ${}^4\overset{\circ}{\mathbf{M}}$ (4.337) for elastic continua:

$${}^4\breve{\mathbf{R}} = \mathbf{E}\otimes\mathbf{E}\breve{l}_1 + 2\Delta\breve{l}_2, \tag{8.123}$$

we can represent constitutive equations (8.120) in the operator form

$$\overset{(n)}{\mathbf{T}} = J\,{}^4\breve{\mathbf{R}}\cdot\cdot\overset{(n)}{\mathbf{C}}_\theta, \tag{8.124}$$

which is analogous to relations (4.338) for semilinear isotropic elastic continua.

### 8.2.10 Linear Models $A_n$ of Transversely Isotropic Viscoelastic Continua

For linear models $A_n$ of viscoelastic transversely isotropic continua, from (8.87) and (8.88) we obtain

$$\begin{gathered}
r = 5, \quad r_1 = 2, \quad r_2 - r_1 = 2,\\
I_1^{(3)}(\tau) = (\mathbf{E} - \widehat{\mathbf{c}}_3^2)\cdot\cdot\overset{(n)}{\mathbf{C}}_\theta(\tau), \quad I_2^{(3)}(\tau) = \widehat{\mathbf{c}}_3^2\cdot\cdot\overset{(n)}{\mathbf{C}}_\theta(\tau),\\
J_3^{(3)}(\tau_1,\tau_2) = ((\mathbf{E} - \widehat{\mathbf{c}}_3^2)\cdot\overset{(n)}{\mathbf{C}}_\theta(\tau_1))\cdot\cdot(\widehat{\mathbf{c}}_3^2\cdot\overset{(n)}{\mathbf{C}}_\theta(\tau_2)),\\
I_\alpha^{(3)}(\tau) = J_\alpha^{(3)}(\tau,\ \tau), \ \alpha = 3,\ 4,\\
J_4^{(3)}(\tau_1,\ \tau_2) = \overset{(n)}{\mathbf{C}}_\theta(\tau_1)\cdot\cdot\overset{(n)}{\mathbf{C}}_\theta(\tau_2) - 2J_3^{(3)}(\tau_1,\ \tau_2) - I_2^{(3)}(\tau_1)I_2^{(3)}(\tau_2),\\
I_{3C}^{(3)}(\tau) = \frac{1}{2}(\mathbf{O}_1\otimes\mathbf{O}_1 + \mathbf{O}_2\otimes\mathbf{O}_2)\cdot\cdot\overset{(n)}{\mathbf{C}}_\theta(\tau), \quad I_{4C}^{(3)}(\tau) = 2\,{}^4\mathbf{O}_3\cdot\cdot\overset{(n)}{\mathbf{C}}_\theta(\tau),
\end{gathered} \tag{8.125}$$

where ${}^4\mathbf{O}_3$ is determined by formula (8.88). Relations (8.100) take the form

$$\begin{aligned}
\overset{(n)}{\mathbf{T}} = J\Bigg(&\left(\breve{l}_{11}I_1^{(3)} + \breve{l}_{12}I_2^{(3)}\right)(\mathbf{E} - \widehat{\mathbf{c}}_3^2) + \left(\left(\breve{l}_{22} - 2\breve{l}_{44}\right)I_2^{(3)} + \breve{l}_{12}I_1^{(3)}\right)\widehat{\mathbf{c}}_3^2\\
&+(\mathbf{O}_1\otimes\mathbf{O}_1 + \mathbf{O}_2\otimes\mathbf{O}_2)\cdot\cdot\left(\frac{\breve{l}_{33}}{2} - \breve{l}_{44}\right)\overset{(n)}{\mathbf{C}}_\theta + 2\breve{l}_{44}\overset{(n)}{\mathbf{C}}_\theta\Bigg).
\end{aligned} \tag{8.126}$$

Here the linear operators $\breve{l}_{\gamma\beta}I_\beta^{(3)}$ and $\breve{l}_{\gamma\gamma}\overset{(n)}{\mathbf{C}}$ are determined by expressions (8.101), which can be represented in the Boltzmann form (8.107)

$$\breve{l}_{\gamma\beta}I_\beta^{(3)} = \int_0^t r_{\gamma\beta}(t-\tau)\,dI_\beta^{(3)}(\tau), \quad \breve{l}_{\gamma\gamma}J_{\gamma C}^{(3)} = \int_0^t r_{\gamma\gamma}(t-\tau)\,dJ_{\gamma C}^{(3)}(\tau). \tag{8.127}$$

For a transversely isotropic continuum, there are five independent constants $l_{11}$, $l_{22}$, $l_{12}$, $l_{33}$, $l_{44}$ and five cores $q_{\gamma\beta}(t-\tau)$ or $r_{\gamma\beta}(t-\tau)$.

Introducing the tensor functional being analogous to the tensor of elastic moduli (4.341):

$$
\begin{aligned}
{}^4\breve{\mathbf{R}} &= \mathbf{E}\otimes\mathbf{E}\breve{l}_{11} + \widehat{\mathbf{c}}_3^2\otimes\widehat{\mathbf{c}}_3^2\breve{\breve{l}}_{22} + (\breve{l}_{12}-\breve{l}_{11})(\mathbf{E}\otimes\widehat{\mathbf{c}}_3^2 + \widehat{\mathbf{c}}_3^2\otimes\mathbf{E}) \\
&\quad + (\mathbf{O}_1\otimes\mathbf{O}_1 + \mathbf{O}_2\otimes\mathbf{O}_2)\cdot\cdot\left(\frac{\breve{l}_{33}}{2} - \breve{l}_{44}\right) + 2\Delta\breve{l}_{44}, \\
\breve{\breve{l}}_{22} &= \breve{l}_{22} - 2\breve{l}_{44} - 2\breve{l}_{12} + \breve{l}_{11},
\end{aligned} \tag{8.128}
$$

we can also represent constitutive equations (8.126) in the form (8.124).

### 8.2.11 Linear Models $A_n$ of Orthotropic Viscoelastic Continua

For linear models $A_n$ of viscoelastic orthotropic continua, due to (8.91)–(8.93), the invariants (8.98) and the derivative tensors (8.102) take the forms

$$
\begin{aligned}
& r = 6, \quad r_1 = 3, \quad r_2 = 6, \\
& I_\alpha^{(O)}(\tau) = \widehat{\mathbf{c}}_\alpha^2\cdot\cdot\overset{(n)}{\mathbf{C}}_\theta(\tau), \quad \alpha = 1,2,3, \\
& J_4^{(O)}(\tau_1,\tau_2) = \left(\widehat{\mathbf{c}}_2^2\cdot\overset{(n)}{\mathbf{C}}_\theta(\tau_1)\right)\cdot\cdot\left(\widehat{\mathbf{c}}_3^2\cdot\overset{(n)}{\mathbf{C}}_\theta(\tau_2)\right), \\
& J_5^{(O)}(\tau_1,\tau_2) = \left(\widehat{\mathbf{c}}_1^2\cdot\overset{(n)}{\mathbf{C}}_\theta(\tau_1)\right)\cdot\cdot\left(\widehat{\mathbf{c}}_3^2\cdot\overset{(n)}{\mathbf{C}}_\theta(\tau_2)\right), \\
& J_6^{(O)}(\tau_1,\tau_2) = \left(\widehat{\mathbf{c}}_1^2\cdot\overset{(n)}{\mathbf{C}}_\theta(\tau_1)\right)\cdot\cdot\left(\widehat{\mathbf{c}}_2^2\cdot\overset{(n)}{\mathbf{C}}_\theta(\tau_2)\right), \\
& I_\alpha^{(O)}(\tau) = J_\alpha^{(O)}(\tau,\tau), \quad \alpha = 4,5,6; \quad I_{4\mathrm{C}}^{(O)}(\tau) = \frac{1}{2}(\mathbf{O}_1\otimes\mathbf{O}_1)\cdot\cdot\overset{(n)}{\mathbf{C}}_\theta(\tau), \\
& I_{5\mathrm{C}}^{(O)}(\tau) = 2(\mathbf{O}_2\otimes\mathbf{O}_2)\cdot\cdot\overset{(n)}{\mathbf{C}}_\theta(\tau), \quad I_{6\mathrm{C}}^{(O)}(\tau) = \frac{1}{2}(\mathbf{O}_3\otimes\mathbf{O}_3)\cdot\cdot\overset{(n)}{\mathbf{C}}_\theta(\tau),
\end{aligned} \tag{8.129}
$$

and the relations (8.100) can be written as follows:

$$
\overset{(n)}{\mathbf{T}} = J\sum_{\gamma,\beta=1}^{3}\breve{l}_{\gamma\beta}I_\beta^{(O)}\widehat{\mathbf{c}}_\gamma^2 + J\sum_{\gamma=1}^{3}\mathbf{O}_\gamma(\mathbf{O}_\gamma\cdot\cdot\breve{l}_{3+\gamma,3+\gamma}\overset{(n)}{\mathbf{C}}_\theta). \tag{8.130}
$$

Thus, there are nine independent constants $l_{11}, l_{22}, l_{33}, l_{12}, l_{13}, l_{23}, l_{44}, l_{55}, l_{66}$ and nine cores $q_{\gamma\beta}(t-\tau)$ or $r_{\gamma\beta}(t-\tau)$.

Introducing the tensor functional being analogous to (4.344):

$$
{}^4\breve{\mathbf{R}} = \sum_{\gamma,\beta=1}^{3}\widehat{\mathbf{c}}_\gamma^2\otimes\widehat{\mathbf{c}}_\beta^2\breve{l}_{\gamma\beta} + \sum_{\gamma=1}^{3}\mathbf{O}_\gamma\otimes\mathbf{O}_\gamma\breve{l}_{3+\gamma,3+\gamma}, \tag{8.131}
$$

we can represent constitutive equations (8.130) in the operator form (8.124).

### 8.2.12 The Tensor of Relaxation Functions

According to the operator form (8.124) of constitutive equations for linear models $A_n$ of viscoelastic continua, we can introduce the fourth-order tensor ${}^4\mathbf{R}(t)$, called the *tensor of relaxation functions*, by the same formulae as for the elastic moduli tensor ${}^4\overset{\circ}{\mathbf{M}}$ in linear models $A_n$ of elastic continua (see Sect. 4.8.7) if in the corresponding formulae the elastic constants $l_{\alpha\beta}$ are replaced by the relaxation functions $r_{\alpha\beta}(t)$.

For an isotropic continuum, this tensor has the form

$$ {}^4\mathbf{R}(t) = r_1(t)\mathbf{E} \otimes \mathbf{E} + 2r_2(t)\Delta; \tag{8.132} $$

for a transversely isotropic continuum

$$ \begin{aligned} {}^4\mathbf{R}(t) &= r_{11}(t)\mathbf{E} \otimes \mathbf{E} + \widetilde{r}_{22}(t)\widehat{\mathbf{c}}_3^2 \otimes \widehat{\mathbf{c}}_3^2 + (r_{12}(t) - r_{11}(t))(\mathbf{E} \otimes \widehat{\mathbf{c}}_3^2 + \widehat{\mathbf{c}}_3^2 \otimes \mathbf{E}) \\ &\quad + \left(\frac{1}{2}r_{33}(t) - r_{44}(t)\right)(\mathbf{O}_1 \otimes \mathbf{O}_1 + \mathbf{O}_2 \otimes \mathbf{O}_2) + 2r_{44}(t)\Delta, \\ \widetilde{r}_{22}(t) &= r_{11}(t) + r_{22}(t) - 2r_{12}(t) - 2r_{44}(t), \end{aligned} \tag{8.133} $$

and for an orthotropic continuum

$$ {}^4\mathbf{R}(t) = \sum_{\alpha,\beta=1}^{3} r_{\alpha\beta}(t)\widehat{\mathbf{c}}_\gamma^2 \otimes \widehat{\mathbf{c}}_\beta^2 + \sum_{\alpha=1}^{3} r_{3+\alpha,3+\alpha}(t)\mathbf{O}_\gamma \otimes \mathbf{O}_\gamma. \tag{8.134} $$

Then the operator relations (8.124) can be represented as follows:

$$ \overset{(n)}{\mathbf{T}} = J\, {}^4\breve{\mathbf{R}} \cdot\cdot \overset{(n)}{\mathbf{C}}_\theta, \tag{8.135} $$

where

$$ {}^4\breve{\mathbf{R}} \cdot\cdot \overset{(n)}{\mathbf{C}}_\theta = \int_0^t {}^4\mathbf{R}(t-\tau) \cdot\cdot d\overset{(n)}{\mathbf{C}}_\theta(\tau) \tag{8.136} $$

is a tensor linear functional.

For instantaneous loading as $t \to 0_+$, these relations coincide with (8.112) and with the corresponding relations (4.330a) of models $A_n$ for a linear-elastic continuum, because

$$ {}^4\mathbf{R}(0) = {}^4\overset{\circ}{\mathbf{M}}. \tag{8.137} $$

The tensor

$$ {}^4\mathbf{K}(t) = -\frac{d\,{}^4\mathbf{R}}{dt}(t) \tag{8.138} $$

is called the *tensor of relaxation cores*. This tensor for different groups $\overset{\circ}{G}_s$ has the same form as the tensor ${}^4\mathbf{R}(t)$ in (8.132)–(8.134) if in these formulae the substitution $r_{\alpha\beta}(t) \to q_{\alpha\beta}(t)$ has been made.

According to (8.137) and (8.138), the constitutive equations (8.135) can be written in the Volterra form

$$\overset{(n)}{\mathbf{T}} = J({}^4\overset{\circ}{\mathbf{M}}\cdot\cdot\overset{(n)}{\mathbf{C}}_\theta - \int_0^t \mathbf{K}(t-\tau)\cdot\cdot\overset{(n)}{\mathbf{C}}_\theta(\tau)\,d\tau), \tag{8.139}$$

that is equivalent to the form (8.100). The operator (8.115) of the free energy $\psi$ for the mechanically determinate model $A_n$ with the help of the tensor of relaxation functions can be represented in the form (see Exercise 8.2.3)

$$\overset{\circ}{\rho}\psi = \overset{\circ}{\rho}\psi_0 + \frac{1}{2}\int_0^t\int_0^t d\overset{(n)}{\mathbf{C}}_\theta(\tau_1)\cdot\cdot{}^4\mathbf{R}(2t-\tau_1-\tau_2)\cdot\cdot d\overset{(n)}{\mathbf{C}}_\theta(\tau_2), \tag{8.140}$$

and the dissipation function (8.116) – in the form

$$w^* = -\frac{J}{2}\int_0^t\int_0^t d\overset{(n)}{\mathbf{C}}_\theta(\tau_1)\cdot\cdot\frac{d}{dt}{}^4\mathbf{R}(2t-\tau_1-\tau_2)\cdot\cdot d\overset{(n)}{\mathbf{C}}_\theta(\tau_2). \tag{8.141}$$

Formula (8.141) gives the following theorem.

**Theorem 8.9.** *For mechanically determinate linear models $A_n$ of viscoelastic continua, the tensors of relaxation cores ${}^4\mathbf{K}(t)$ are*

*1. Nonnegative-definite:*

$$\mathbf{h}\cdot\cdot{}^4\mathbf{K}(t)\cdot\cdot\mathbf{h} \geqslant 0, \quad \forall\mathbf{h}\neq 0,\ \forall t\geqslant 0, \tag{8.142}$$

*2. Symmetric in the following combinations of indices:*

$${}^4\mathbf{K}(t) = {}^4\mathbf{K}^{(1243)}(t) = {}^4\mathbf{K}^{(2134)}(t) = {}^4\mathbf{K}^{(3412)}(t) \quad \forall t\geqslant 0, \tag{8.143}$$

*(i.e. these tensors have the same symmetry as the elastic moduli tensor ${}^4\overset{\circ}{\mathbf{M}}$ for linear models $A_n$ of elastic continua).*

▼ The dissipation function is always nonnegative ($w^* \geqslant 0$) and vanishes for viscoelastic continua only if $\overset{(n)}{\mathbf{C}}_\theta(\tau) \equiv 0$. Then, choosing the process of deforming in the form of a step-function:

$$\overset{(n)}{\mathbf{C}}_\theta(\tau) = \mathbf{h}\,h(\tau), \quad \tau\geqslant 0, \tag{8.144}$$

where $h(\tau)$ is the Heaviside function, and $\mathbf{h}$ is a symmetric non-zero constant tensor, we obtain

$$d\overset{(n)}{\mathbf{C}}_\theta(\tau) = \mathbf{h}\,\delta(\tau)\,d\tau. \tag{8.144a}$$

Substituting (8.144a) into (8.141) and using the property (8.16) of the $\delta$-function and formula (8.138), we find that

$$\begin{aligned} w^* &= \frac{J}{2}\int_0^t\int_0^t \mathbf{h}\cdot\cdot\,{}^4\mathbf{K}(2t-\tau_1-\tau_2)\cdot\cdot\,\mathbf{h}\delta(\tau_1)\delta(\tau_2)\,d\tau_1\,d\tau_2 \\ &= \frac{J}{2}\,\mathbf{h}\cdot\cdot\,{}^4\mathbf{K}(2t)\cdot\cdot\,\mathbf{h} \geqslant 0, \quad \forall t>0, \end{aligned} \tag{8.145}$$

i.e. the tensor ${}^4\mathbf{K}(t)$ is nonnegative-definite.

The existence of the quadratic form (8.145) and the symmetry of the tensor $\mathbf{h}$ lead to symmetry of ${}^4\mathbf{K}(t)$ in the first–second and third–fourth indices and also in pairs of the indices, i.e. the relations (8.143) actually hold. ▲

As follows from (8.142) and (8.138), the tensor of relaxation functions ${}^4\mathbf{R}(t)$ generates the monotonically non-increasing form

$$\mathbf{h}\cdot\cdot\,\frac{d\,{}^4\mathbf{R}}{dt}(t)\cdot\cdot\,\mathbf{h} \leqslant 0, \quad \forall \mathbf{h}\neq 0, \ \ \forall t \geqslant 0. \tag{8.146}$$

And if the elastic moduli tensor $\overset{\circ}{\mathbf{M}} = {}^4\mathbf{R}(0)$ has the symmetry (8.143), then from (8.143) it follows that the tensor ${}^4\mathbf{R}(t)$ has the same symmetry $\forall t \geqslant 0$:

$${}^4\mathbf{R}(t) = {}^4\mathbf{R}^{(1243)}(t) = {}^4\mathbf{R}^{(2134)}(t) = {}^4\mathbf{R}^{(3412)}(t) \quad \forall t \geqslant 0. \tag{8.147}$$

### 8.2.13 *Spectral Representation of Linear Models $A_n$ of Viscoelastic Continua*

Let us apply now the theory of spectral decompositions of symmetric second-order tensors (see [12]). According to this theory, for any symmetric tensors, in particular for $\overset{(n)}{\mathbf{C}}_\theta(\tau)$ and $\overset{(n)}{\mathbf{T}}(\tau)$, we can introduce their *spectral decompositions* relative to a symmetry group $\overset{\circ}{G}_s$ chosen:

$$\overset{(n)}{\mathbf{T}} = \sum_{\alpha=1}^{\bar{n}} \mathbf{P}_\alpha^{(\mathrm{T})}, \quad \overset{(n)}{\mathbf{C}}_\theta = \sum_{\alpha=1}^{\bar{n}} \mathbf{P}_\alpha^{(\mathrm{C})}, \quad 1 < \bar{n} \leqslant 6. \tag{8.148}$$

Here $\mathbf{P}_\alpha^{(\mathrm{T})}$ and $\mathbf{P}_\alpha^{(\mathrm{C})}$ ($\alpha = 1, \ldots, \bar{n}$) are the *orthoprojectors* of the tensors $\overset{(\mathrm{n})}{\mathbf{T}}$ and $\overset{(\mathrm{n})}{\mathbf{C}}_\theta$ (their number is denoted by $\bar{n}$), which are symmetric second-order tensors having the following properties: a) they are mutually orthogonal, b) they are linear, c) they are indifferent relative to the group $\overset{\circ}{G}_s$:

$$\mathbf{P}_\alpha^{(\mathrm{T})} \cdot\cdot\, \mathbf{P}_\beta^{(\mathrm{T})} = 0, \quad \text{if } \alpha \neq \beta; \tag{8.149}$$

$$\mathbf{P}_\alpha^{(\mathrm{T})} = \mathbf{P}_\alpha(\overset{(\mathrm{n})}{\mathbf{T}}) = {}^4\boldsymbol{\Gamma}_\alpha \cdot\cdot\, \overset{(\mathrm{n})}{\mathbf{T}}, \quad \alpha = 1, \ldots, \bar{n}; \tag{8.150}$$

$$\mathbf{Q}^{\mathrm{T}} \cdot \mathbf{P}_\alpha(\overset{(\mathrm{n})}{\mathbf{T}}) \cdot \mathbf{Q} = \mathbf{P}_\alpha(\mathbf{Q}^{\mathrm{T}} \cdot \overset{(\mathrm{n})}{\mathbf{T}} \cdot \mathbf{Q}), \quad \forall \mathbf{Q} \in \overset{\circ}{G}_s. \tag{8.151}$$

Here the fourth-order tensors ${}^4\boldsymbol{\Gamma}_\alpha$ are indifferent relative to the group $\overset{\circ}{G}_s$, independent of $\overset{(\mathrm{n})}{\mathbf{T}}$ and formed only by producing tensors of the group (see Sect. 4.8.3). Among the tensors ${}^4\boldsymbol{\Gamma}_\alpha$ ($\alpha = 1, \ldots, \bar{n}$), there are $m$ *reducible* tensors, i.e. obtained with the help of the tensor product of the second-order tensor $\mathbf{a}_\alpha$ being symmetric and indifferent relative to the same group $\overset{\circ}{G}_s$ :

$${}^4\boldsymbol{\Gamma}_\alpha = \frac{1}{a_\alpha^2} \mathbf{a}_\alpha \otimes \mathbf{a}_\alpha, \quad a_\alpha^2 = \mathbf{a}_\alpha \cdot\cdot\, \mathbf{a}_\alpha, \quad \alpha = 1, \ldots, m < \bar{n}. \tag{8.152}$$

Expressions for ${}^4\boldsymbol{\Gamma}_\alpha$ and $\mathbf{a}_\alpha$ have the following forms (see [12]):
for the isotropy group $\overset{\circ}{G}_s = I$

$$\mathbf{a}_1 = \mathbf{E}, \quad {}^4\boldsymbol{\Gamma}_{(2)} = \boldsymbol{\Delta} - \frac{1}{3} \mathbf{E} \otimes \mathbf{E}, \quad m = 1, \quad \bar{n} = 2; \tag{8.153}$$

for the transverse isotropy group $\overset{\circ}{G}_s = T_3$

$$\mathbf{a}_1 = \widehat{\mathbf{c}}_3^2, \quad \mathbf{a}_2 = \mathbf{E} - \widehat{\mathbf{c}}_3^2, \quad m = 2, \quad \bar{n} = 4;$$

$${}^4\boldsymbol{\Gamma}_3 = \boldsymbol{\Delta} - \frac{1}{2}\left(\mathbf{E} - \widehat{\mathbf{c}}_3^2\right) \otimes \left(\mathbf{E} - \widehat{\mathbf{c}}_3^2\right) - \widehat{\mathbf{c}}_3^2 \otimes \widehat{\mathbf{c}}_3^2 - \frac{1}{2}(\mathbf{O}_1 \otimes \mathbf{O}_1 + \mathbf{O}_2 \otimes \mathbf{O}_2),$$

$${}^4\boldsymbol{\Gamma}_4 = \frac{1}{2}(\mathbf{O}_1 \otimes \mathbf{O}_1 + \mathbf{O}_2 \otimes \mathbf{O}_2); \tag{8.154}$$

for the orthotropy group $\overset{\circ}{G}_s = O$

$$\mathbf{a}_\alpha = \widehat{\mathbf{c}}_\alpha^2, \quad \alpha = 1, 2, 3; \qquad {}^4\boldsymbol{\Gamma}_{\alpha+3} = \frac{1}{2} \mathbf{O}_\alpha \otimes \mathbf{O}_\alpha, \quad m = 3, \quad \bar{n} = 6. \tag{8.155}$$

With the help of the orthoprojectors $\mathbf{P}_\alpha^{(\mathrm{T})}$, introduce *spectral invariants* of the tensor $\overset{(n)}{\mathbf{T}}$ and denote them by $Y_\alpha(\overset{(n)}{\mathbf{T}})$. For those $\mathbf{P}_\alpha^{(\mathrm{T})}$, for which ${}^4\boldsymbol{\Gamma}_\alpha$ is a reducible tensor, the invariant $Y_\alpha$ is introduced as follows:

$$Y_\alpha(\overset{(n)}{\mathbf{T}}) = \frac{1}{a_\alpha}\mathbf{a}_{(\alpha)} \cdot\cdot \overset{(n)}{\mathbf{T}}, \quad \alpha = 1, \ldots, m, \tag{8.156a}$$

and called the *spectral linear invariant*. For the remaining $\mathbf{P}_\alpha^{(\mathrm{T})}$ ($\alpha = m+1, \ldots, \bar{n}$), these invariants are introduced by the formula

$$Y_\alpha(\overset{(n)}{\mathbf{T}}) = \left(\mathbf{P}_\alpha^{(\mathrm{T})} \cdot\cdot \mathbf{P}_\alpha^{(\mathrm{T})}\right)^{1/2} \tag{8.156b}$$

and called the *spectral quadratic invariants*. From (8.150) and (8.152) it follows that

$$\mathbf{P}_\alpha^{(\mathrm{T})} = \frac{1}{a_\alpha} Y_\alpha \mathbf{a}_{(\alpha)}, \quad \alpha = 1, \ldots, m. \tag{8.157}$$

Notice that for the linear invariants (8.156a), formula (8.156b) also holds.

Due to (8.157), the spectral decomposition of the symmetric second-order tensor $\overset{(n)}{\mathbf{T}}$ (8.148) can be represented in the form

$$\overset{(n)}{\mathbf{T}} = \sum_{\alpha=1}^{m} \frac{\mathbf{a}_\alpha}{a_\alpha} Y_\alpha(\overset{(n)}{\mathbf{T}}) + \sum_{\alpha=m+1}^{n} \mathbf{P}_\alpha^{(\mathrm{T})}. \tag{8.158}$$

For any fourth-order tensor indifferent relative to a group $\overset{\circ}{G}_s$, including the tensor of relaxation functions ${}^4\mathbf{R}(t)$, we can also introduce the *spectral representation*

$${}^4\mathbf{R}(t) = \sum_{\alpha,\beta=1}^{m} R_{\alpha\beta}(t) \frac{\mathbf{a}_\alpha \otimes \mathbf{a}_\beta}{a_\alpha a_\beta} + \sum_{\alpha=m+1}^{\bar{n}} R_{\alpha\alpha}(t)\, {}^4\boldsymbol{\Gamma}_\alpha, \tag{8.159}$$

where $R_{\alpha\beta}(t)$ and $R_{\alpha\alpha}(t)$ are the spectral relaxation functions expressed uniquely in terms of $r_{\alpha\beta}(t)$ and $r_{\alpha\alpha}(t)$ (see Exercise 8.2.6).

With the help of the spectral decompositions (8.148) and (8.159) the constitutive equations (8.135) can be represented as relations between the spectral linear invariants and the orthoprojectors (see Exercise 8.2.9)

$$\begin{aligned} Y_\alpha(\overset{(n)}{\mathbf{T}}) &= J \sum_{\beta=1}^{m} \breve{R}_{\alpha\beta} Y_\beta(\overset{(n)}{\mathbf{C}}), \quad \alpha = 1, \ldots, m; \\ \mathbf{P}_\alpha^{(\mathrm{T})} &= J \breve{R}_{\alpha\alpha} \mathbf{P}_\alpha^{(\mathrm{C})}, \quad \alpha = m+1, \ldots, \bar{n}, \end{aligned} \tag{8.160}$$

where

$$\breve{R}_{\alpha\beta}\mathbf{P}_{\beta}^{(C)} = \int_{0}^{t} R_{\alpha\beta}(t-\tau)\, d\mathbf{P}_{\beta}^{(C)}(\tau). \tag{8.161}$$

Relations (8.160) are called the *spectral representation for linear models $A_n$ of viscoelastic continua*. If we introduce the spectral decomposition (8.148) also for the tensor $\mathbf{h}$:

$$\mathbf{h} = \sum_{\alpha=1}^{\bar{n}} \mathbf{P}_{\alpha}^{(\mathbf{h})},$$

then the inequality (8.146) with use of (8.159) takes the form

$$\sum_{\alpha,\beta=1}^{m} \frac{d}{dt} R_{\alpha\beta}(t) Y_{\alpha}(\mathbf{h}) Y_{\beta}(\mathbf{h}) + \sum_{\alpha=m+1}^{\bar{n}} \frac{d}{dt} R_{\alpha\alpha}(t) Y_{\alpha}^{2}(\mathbf{h}) \leqslant 0. \tag{8.162}$$

Values of the spectral linear invariants $Y_{\alpha}(\mathbf{h})$ can be assumed to be zero; then, since the spectral invariants are independent, from (8.162) we obtain the condition of monotone non-increasing the spectral relaxation functions:

$$\frac{dR_{\alpha\alpha}}{dt}(t) \leqslant 0, \qquad \alpha = 1, \ldots, \bar{n}. \tag{8.163}$$

With the help of the spectral relaxation functions one can formulate special cases of linear models $A_n$ for viscoelastic continua. So for the *simplest linear* model $A_n$ of an isotropic viscoelastic continuum, one of the two spectral relaxation cores is assumed to be constant:

$$R_{11}(t) = R_{11}(0) = l_1 + \frac{2}{3} l_2 = \text{const}, \qquad \frac{\partial R_{11}}{\partial t} = 0. \tag{8.164}$$

According to the results of Exercise 8.2.6 and formula (8.122), this condition can be rewritten as the relation between the cores $q_1(t)$ and $q_2(t)$

$$q_1(t) = -\frac{2}{3} q_2(t), \qquad \frac{\partial r_{\alpha}}{\partial t} = -q_{\alpha}(t). \tag{8.165}$$

Constitutive equations (8.160) in this case take the form

$$\begin{cases} I_1(\overset{(\mathrm{n})}{\mathbf{T}}) = J R_{11}(0) I_1(\overset{(\mathrm{n})}{\mathbf{C}}_{\theta}), \\ \operatorname{dev} \overset{(\mathrm{n})}{\mathbf{T}} = J \int_{0}^{t} R_{22}(t-\tau) \operatorname{dev} \frac{\partial}{\partial \tau} \overset{(\mathrm{n})}{\mathbf{C}}_{\theta}(\tau)\, d\tau. \end{cases} \tag{8.166}$$

Here we have denoted the orthoprojectors of the tensors relative to the full orthogonal group $I$, called the *deviators*

$$\mathrm{dev}\,\overset{(n)}{\mathbf{C}}_\theta = \overset{(n)}{\mathbf{C}}_\theta - \frac{1}{3}I_1(\overset{(n)}{\mathbf{C}}_\theta)\mathbf{E}, \qquad \mathrm{dev}\,\overset{(n)}{\mathbf{T}} = \overset{(n)}{\mathbf{T}} - \frac{1}{3}I_1(\overset{(n)}{\mathbf{T}})\mathbf{E}. \tag{8.167}$$

Equations (8.166) and (8.167) can be rewritten in the form

$$\overset{(n)}{\mathbf{T}} = \frac{J}{3}R_{11}(0)I_1(\overset{(n)}{\mathbf{C}}_\theta)\mathbf{E} + J\int_0^t R_{22}(t-\tau)\mathrm{dev}\,\frac{\partial \overset{(n)}{\mathbf{C}}_\theta(\tau)}{\partial \tau}\,d\tau. \tag{8.168}$$

### 8.2.14 Exponential Relaxation Functions and Differential Form of Constitutive Equations

To solve problems analytically or numerically it is convenient to have an analytical form of the spectral relaxation functions $R_{\alpha\beta}(t)$. As established in the preceding section, the property of monotone non-increasing the spectral functions $R_{\alpha\alpha}(t)$ is a consequence of the dissipation inequality $w^* \geqslant 0$. Functions having such a property can be approximated by the sum of exponents:

$$R_{\alpha\beta}(t) = R^{\infty}_{\alpha\beta} + \sum_{\gamma=1}^{N} B^{(\gamma)}_{\alpha\beta}\exp\left(-\frac{t}{\tau^{(\gamma)}_{\alpha\beta}}\right), \tag{8.169}$$

where $B^{(\gamma)}_{\alpha\beta}$ and $\tau^{(\gamma)}_{\alpha\beta}$ are the constants called the *spectrum of relaxation values* and the *spectrum of relaxation times*, respectively, and $R^{\infty}_{\alpha\beta}$ is the limiting value of the relaxation functions:

$$\lim_{t\to\infty} R_{\alpha\beta}(t) = R^{\infty}_{\alpha\beta}, \tag{8.170}$$

which may be zero: $R^{\infty}_{\alpha\beta} = 0$.

The constants $R^{\infty}_{\alpha\beta}$ and $B^{(\gamma)}_{\alpha\beta}$ satisfy the normalization condition at $t = 0$:

$$R^{\infty}_{\alpha\beta} + \sum_{\gamma=1}^{N} B^{(\gamma)}_{\alpha\beta} = \overset{\circ}{C}_{\alpha\beta}, \tag{8.171}$$

where $\overset{\circ}{C}_{\alpha\beta} = R_{\alpha\beta}(0)$ are the spectral (two-index) elastic moduli under instantaneous loading.

There are other methods of analytical approximation to the relaxation functions, however exponential functions have certain merits: (1) choosing a sufficiently large number $N$ of exponents in (8.169), we can approximate practically any function $R_{\alpha\beta}(t)$, (2) constitutive equations (8.160) and (8.161) with exponential cores admit their inversion (see Sect. 8.2.15), where cores of the inverse functionals prove to be exponential as well, and (3) the cores (8.169) allow us to represent constitutive equations (8.160), (8.161) or (8.135), (8.136) in the differential form.

Indeed, performing the subsequent substitutions (8.169) →(8.159)→(8.138), we find the expression for the tensor of relaxation cores:

$$ {}^4\mathbf{K}(t) = \sum_{\alpha,\beta=1}^{m} K_{\alpha\beta}(t)\frac{\mathbf{a}_\alpha \otimes \mathbf{a}_\beta}{a_\alpha a_\beta} + \sum_{\alpha=m+1}^{\bar{n}} K_{\alpha\alpha}(t)\,{}^4\boldsymbol{\Gamma}_\alpha, \tag{8.172} $$

$$ K_{\alpha\beta}(t) = -\frac{\partial R_{\alpha\beta}(t)}{\partial t} = \sum_{\gamma=1}^{N} \frac{B^{(\gamma)}_{\alpha\beta}}{\tau^{(\gamma)}_{\alpha\beta}} \exp\left(-\frac{t}{\tau^{(\gamma)}_{\alpha\beta}}\right). \tag{8.173} $$

Introduce the second-order tensors

$$ \mathbf{W}^{(\gamma)}_{\alpha\beta} = \int_0^t \exp\left(-\frac{t-\tau}{\tau^{(\gamma)}_{\alpha\beta}}\right) \frac{\overset{(\mathrm{n})}{\mathbf{C}}_\theta(\tau)d\tau}{\tau^{(\gamma)}_{\alpha\beta}}, \quad \gamma = 1,\ldots,N. \tag{8.174} $$

Differentiating $\mathbf{W}^{(\gamma)}_{\alpha\beta}$ with respect to $t$ and eliminating the integral, we obtain that the tensors $\mathbf{W}^{(\gamma)}_{\alpha\beta}$ satisfy the first-order differential equations

$$ \frac{d\mathbf{W}^{(\gamma)}_{\alpha\beta}}{dt} + \frac{\mathbf{W}^{(\gamma)}_{\alpha\beta}}{\tau^{(\gamma)}_{\alpha\beta}} = \frac{\overset{(\mathrm{n})}{\mathbf{C}}_\theta(t)}{\tau^{(\gamma)}_{\alpha\beta}}, \quad \gamma = 1,\ldots,N. \tag{8.175} $$

Substituting (8.172) into (8.139) and using the expressions (8.173) and (8.174), we obtain the following representation of constitutive equations:

$$ \overset{(\mathrm{n})}{\mathbf{T}} = J\left({}^4\overset{\circ}{\mathbf{M}} \cdot\cdot \overset{(\mathrm{n})}{\mathbf{C}}_\theta - \sum_{\gamma=1}^{N} \mathbf{W}^{(\gamma)}\right). \tag{8.176} $$

Here the *spectrum of viscous stresses* is denoted by $\mathbf{W}^{(\gamma)}$ being second-order tensors of the form

$$ \mathbf{W}^{(\gamma)} = \sum_{\alpha,\beta=1}^{m} B^{(\gamma)}_{\alpha\beta}\mathbf{W}^{(\gamma)}_{\alpha\beta} \cdot\cdot \frac{\mathbf{a}_\alpha \otimes \mathbf{a}_\beta}{a_\alpha a_\beta} + \sum_{\alpha=m+1}^{\bar{n}} B^{(\gamma)}_{\alpha\alpha}\mathbf{W}^{(\gamma)}_{\alpha\alpha} \cdot\cdot {}^4\boldsymbol{\Gamma}_\alpha. \tag{8.177} $$

Thus, with the help of the exponential cores (8.169) the constitutive equations for the mechanically determinate model $A_n$ of viscoelastic continua (8.139) can be represented in the *differential form* (8.175)–(8.177). A result of the passage from integral relations to differential ones is the appearance of additional unknowns, namely the tensors $\mathbf{W}^{(\gamma)}_{\alpha\beta}$, for which Eqs. (8.175) have been stated.

In computations the differential form (8.175)–(8.177), as a rule, proves to be more convenient than the integral form (8.139).

Substitution of the expressions (8.172) and (8.138) into (8.141) yields

$$w^* = \frac{J}{2}\sum_{\alpha,\beta=1}^{m}\int_0^t\int_0^t K_{\alpha\beta}(2t-\tau_1-\tau_2)\,dY_\alpha^{(C)}(\tau_1)\,dY_\beta^{(C)}(\tau_2)+$$
$$+\frac{J}{2}\sum_{\alpha=m+1}^{\bar n}\int_0^t\int_0^t K_{\alpha\alpha}(2t-\tau_1-\tau_2)\,d\overset{(n)}{\mathbf{C}}_\theta(\tau_1)\cdot\cdot\,{}^4\mathbf{\Gamma}_\alpha\cdot\cdot\,d\overset{(n)}{\mathbf{C}}_\theta(\tau_2). \tag{8.178}$$

Here we have denoted the linear spectral invariants of the tensor $\overset{(n)}{\mathbf{C}}_\theta(\tau)$ (see (8.156a)):

$$Y_\alpha^{(C)}(\tau)=Y_\alpha(\overset{(n)}{\mathbf{C}}_\theta(\tau))=\frac{1}{a_\alpha}\mathbf{a}_\alpha\cdot\cdot\,\overset{(n)}{\mathbf{C}}_\theta(\tau),\quad \alpha=1,\dots,m. \tag{8.179}$$

Substituting the exponential cores (8.173) into (8.178) and modifying the double integrals

$$\int_0^t\int_0^t \exp\left(-\frac{2t-\tau_1-\tau_2}{\tau_{\alpha\beta}}\right)dY_\alpha^{(C)}(\tau_1)\,dY_\beta^{(C)}(\tau_2)$$
$$=\int_0^t \exp\left(-\frac{t-\tau_1}{\tau_{\alpha\beta}}\right)dY_\alpha^{(C)}(\tau_1)\int_0^t \exp\left(-\frac{t-\tau_2}{\tau_{\alpha\beta}}\right)dY_\beta^{(C)}(\tau_2)$$
$$=\left(Y_\alpha^{(C)}(t)-\frac{1}{\tau_{\alpha\beta}}\int_0^t \exp\left(-\frac{t-\tau_1}{\tau_{\alpha\beta}}\right)dY_\alpha^{(C)}(\tau_1)\,d\tau_1\right)$$
$$\times\left(Y_\beta^{(C)}(t)-\frac{1}{\tau_{\alpha\beta}}\int_0^t \exp\left(-\frac{t-\tau_2}{\tau_{\alpha\beta}}\right)dY_\beta^{(C)}(\tau_2)\,d\tau_2\right), \tag{8.180}$$

with use of the notation (8.174) we can represent (8.178) in the form

$$w^*=\frac{J}{2}\sum_{\gamma=1}^{N}\left(\sum_{\alpha,\beta=1}^{m}\frac{B_{\alpha\beta}^{(\gamma)}}{a_\alpha a_\beta}\left(\mathbf{a}_\alpha\cdot\cdot\,\frac{d\mathbf{W}_{\alpha\beta}^{(\gamma)}}{dt}\right)\left(\mathbf{a}_\beta\cdot\cdot\,\frac{d\mathbf{W}_{\alpha\beta}^{(\gamma)}}{dt}\right)\right.$$
$$\left.+\sum_{\alpha=m+1}^{\bar n}B_{\alpha\alpha}^{(\gamma)}\frac{d\mathbf{W}_{\alpha\alpha}^{(\gamma)}}{dt}\cdot\cdot\,{}^4\mathbf{\Gamma}\cdot\cdot\,\frac{d\mathbf{W}_{\alpha\alpha}^{(\gamma)}}{dt}\right). \tag{8.181}$$

### 8.2.15 Inversion of Constitutive Equations for Linear Models of Viscoelastic Continua

In viscoelasticity theory one often uses constitutive equations inverse to (8.135) or (8.139). To derive the equations we should consider relationship (8.139) as a linear integral Volterra's equation of the second kind relative to the process $\overset{(n)}{\mathbf{C}}_\theta(\tau)$, $0 \leqslant \tau \leqslant t$. The core $\mathbf{K}(t)$ of this equation is assumed to be continuously differentiable and to satisfy the conditions (8.142) and (8.143). As known from the theory of integral equations, Eq. (8.139) with such core always has a solution, and this solution is written in the same form as the initial equation:

$$\overset{(n)}{\mathbf{C}}_\theta = {}^4\mathbf{\Pi} \cdot\cdot \frac{\overset{(n)}{\mathbf{T}}}{J} + \int_0^t {}^4\mathbf{N}(t-\tau) \cdot\cdot \frac{\overset{(n)}{\mathbf{T}}}{J}(\tau)\, d\tau. \tag{8.182}$$

Here ${}^4\mathbf{\Pi}$ is the tensor of elastic pliabilities, which is inverse of the tensor of elastic moduli ${}^4\overset{\circ}{\mathbf{M}}$:

$${}^4\mathbf{\Pi} \cdot\cdot {}^4\overset{\circ}{\mathbf{M}} = \mathbf{\Delta}, \tag{8.183}$$

and ${}^4\mathbf{N}(t)$ is the *tensor of creep cores* having the same form as the tensor ${}^4\mathbf{K}(t)$ (8.172):

$${}^4\mathbf{N}(t) = \sum_{\alpha,\beta=1}^{m} N_{\alpha\beta}(t) \frac{\mathbf{a}_\alpha \otimes \mathbf{a}_\beta}{a_\alpha a_\beta} + \sum_{\alpha=m+1}^{\bar{n}} N_{\alpha\alpha}\, {}^4\mathbf{\Gamma}_\alpha. \tag{8.184}$$

The functions $N_{\alpha\beta}(t)$ and $N_{\alpha\alpha}(t)$ are called the *spectral creep cores*.

To find a relation between the cores ${}^4\mathbf{N}(t)$ and ${}^4\mathbf{K}(t)$, we should substitute (8.139) into (8.182); as a result, we obtain the identity

$$\begin{aligned}\overset{(n)}{\mathbf{C}}_\theta(t) &= \overset{(n)}{\mathbf{C}}_\theta(t) + {}^4\mathbf{\Pi} \cdot\cdot \int_0^t {}^4\mathbf{K}(t-\tau) \cdot\cdot \overset{(n)}{\mathbf{C}}_\theta(\tau)\, d\tau \\ &\quad - \int_0^t {}^4\mathbf{N}(t-\tau) \cdot\cdot {}^4\overset{\circ}{\mathbf{M}} \cdot\cdot \overset{(n)}{\mathbf{C}}_\theta(\tau)\, d\tau \\ &\quad - \int_0^t {}^4\mathbf{N}(t-y) \cdot\cdot \int_0^y {}^4\mathbf{K}(y-\tau) \cdot\cdot \overset{(n)}{\mathbf{C}}_\theta(\tau)\, d\tau\, dy. \end{aligned} \tag{8.185}$$

Changing the integration order in the double integral:

$$(0 \leqslant \tau \leqslant y) \times (0 \leqslant y \leqslant t) \to (\tau \leqslant y \leqslant t) \times (0 \leqslant \tau \leqslant t)$$

**Fig. 8.3** The integration domain in the double integral

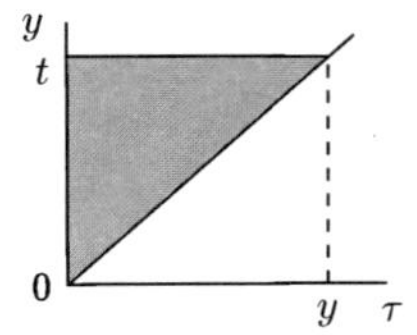


(see Fig. 8.3, where the integration domain is a shaded triangle), from (8.185) we obtain

$$\int_0^t \left\{ {}^4\boldsymbol{\Pi} \cdot\cdot {}^4\mathbf{K}(t-\tau) - {}^4\mathbf{N}(t-\tau) \cdot\cdot {}^4\overset{\circ}{\mathbf{M}} - \int_\tau^t {}^4\mathbf{N}(t-y) \cdot\cdot {}^4\mathbf{K}(y-\tau)dy \right\} \cdot\cdot \overset{(n)}{\mathbf{C}}_\theta(\tau)d\tau = 0. \tag{8.186}$$

This equation holds for any $\overset{(n)}{\mathbf{C}}_\theta(\tau)$ if and only if the expression in braces vanishes.
The substitution of variables $x = t - \tau$ in braces gives

$${}^4\boldsymbol{\Pi} \cdot\cdot {}^4\mathbf{K}(x) = {}^4\mathbf{N}(x) \cdot\cdot {}^4\overset{\circ}{\mathbf{M}} + \int_{t-x}^t {}^4\boldsymbol{\Gamma}(t-y) \cdot\cdot {}^4\mathbf{K}(y+x-t)\, dy, \quad 0 \leqslant x \leqslant t.$$

Then the substitution of variables $u = y + x - t$ under the integral sign, where $(t - x \leqslant y \leqslant t)$ and $(0 \leqslant u \leqslant x)$, yields

$${}^4\boldsymbol{\Pi} \cdot\cdot {}^4\mathbf{K}(x) = {}^4\mathbf{N}(x) \cdot\cdot {}^4\overset{\circ}{\mathbf{M}} + \int_0^x {}^4\mathbf{N}(x-u) \cdot\cdot {}^4\mathbf{K}(u)\, du, \quad 0 \leqslant u \leqslant x.$$

Reverting to the initial notation of arguments $x \to t$ and $u \to \tau$, we obtain the integral relation between the tensor of relaxation cores ${}^4\mathbf{K}(t)$ and the tensor of creep cores ${}^4\mathbf{N}(t)$:

$${}^4\boldsymbol{\Pi} \cdot\cdot {}^4\mathbf{K}(t) = {}^4\mathbf{N}(t) \cdot\cdot {}^4\overset{\circ}{\mathbf{M}} + \int_0^t {}^4\mathbf{N}(t-\tau) \cdot\cdot {}^4\mathbf{K}(\tau)\, d\tau. \tag{8.187}$$

If the core ${}^4\mathbf{K}(t)$ is known, then the relation (8.187) is a linear integral Volterra's equation of the second kind for evaluation of the core ${}^4\mathbf{N}(t)$, and vice versa.

On substituting the spectral decompositions (8.172) and (8.184) of the tensors ${}^4\mathbf{K}(t)$, ${}^4\mathbf{N}(t)$ and analogous decompositions of the tensors ${}^4\boldsymbol{\Pi}$ and ${}^4\overset{\circ}{\mathbf{M}}$ into (8.187), due to mutual orthogonality of the tensors ${}^4\boldsymbol{\Gamma}_\alpha$, $\mathbf{a}_\alpha \otimes \mathbf{a}_\beta$ (see [12]), we obtain

$$\sum_{\beta=1}^{m}\left(\Pi_{\alpha\beta}K_{\beta\varepsilon}(t)-N_{\alpha\beta}(t)C_{\beta\varepsilon}-\int_0^t N_{\alpha\beta}(t-\tau)K_{\beta\varepsilon}(\tau)\,d\tau\right)=0,\ \ \alpha,\varepsilon=1,\ldots,m, \tag{8.188a}$$

$$\Pi_{\alpha\alpha}K_{\alpha\alpha}(t)-N_{\alpha\alpha}(t)C_{\alpha\alpha}-\int_0^t N_{\alpha\alpha}(t-\tau)K_{\alpha\alpha}(t)\,d\tau=0,\ \ \alpha=m+1,\ldots,\bar{n}, \tag{8.188b}$$

– the system of scalar integral equations for determining the cores $N_{\alpha\beta}(t)$ in terms of the cores $K_{\alpha\beta}(t)$ or vice versa.

By analogy with the tensor of relaxation functions ${}^4\mathbf{R}(t)$, introduce the tensor of creep functions ${}^4\mathbf{\Pi}(t)$ satisfying the equation

$$\frac{d}{dt}{}^4\mathbf{\Pi}(t)={}^4\mathbf{N}(t),\quad {}^4\mathbf{\Pi}(0)={}^4\mathbf{\Pi}. \tag{8.189}$$

Then the inverse constitutive equation (8.182) can be written in the Boltzmann form

$$\overset{(n)}{\mathbf{C}}_\theta={}^4\breve{\mathbf{\Pi}}\cdot\cdot\frac{\overset{(n)}{\mathbf{T}}}{J}\equiv\int_0^t {}^4\mathbf{\Pi}(t-\tau)\cdot\cdot\frac{d\overset{(n)}{\mathbf{T}}}{J}(\tau). \tag{8.190}$$

The tensors of creep cores and functions have the same properties of symmetry (8.143) and (8.147) as the tensors ${}^4\mathbf{K}$ and ${}^4\mathbf{R}$ (see Exercise 8.2.11):

$${}^4\mathbf{\Pi}(t)={}^4\mathbf{\Pi}^{(1243)}(t)={}^4\mathbf{\Pi}^{(2134)}(t)={}^4\mathbf{\Pi}^{(3412)}(t),\quad \forall t\geqslant 0. \tag{8.191}$$

For the tensor of creep functions ${}^4\mathbf{\Pi}(t)$ as well as for ${}^4\mathbf{R}(t)$, we can introduce a spectral representation by formula (8.159):

$${}^4\mathbf{\Pi}(t)=\sum_{\alpha,\beta=1}^{m}\Pi_{\alpha\beta}(t)\frac{\mathbf{a}_\alpha\otimes\mathbf{a}_\beta}{a_\alpha a_\beta}+\sum_{\alpha=m+1}^{\bar{n}}\Pi_{\alpha\alpha}(t){}^4\mathbf{\Gamma}_\alpha, \tag{8.192}$$

where $\Pi_{\alpha\alpha}(t)$ and $\Pi_{\alpha\beta}(t)$ are the spectral creep functions.

**Theorem 8.10.** *If the spectral relaxation cores $K_{\alpha\beta}(t)$ are exponential, i.e. have the form (8.173), then the spectral creep cores $N_{\alpha\beta}(t)$ are exponential too:*

$$N_{\alpha\beta}(t)=\sum_{\gamma=1}^{N}\frac{A_{\alpha\beta}^{(\gamma)}}{t_{\alpha\beta}^{(\gamma)}}\exp\left(-\frac{t}{t_{\alpha\beta}^{(\gamma)}}\right), \tag{8.193}$$

*and vice versa.*

The constants $A^{(\gamma)}_{\alpha\beta}$ and $t^{(\gamma)}_{\alpha\beta}$ are called the *spectra of creep values* and *creep times*, respectively. They, in general, are not coincident with $B^{(\gamma)}_{\alpha\beta}$ and $\tau^{(\gamma)}_{\alpha\beta}$, respectively; however the number $N$ in (8.192) and (8.173) is the same.

▼ Show that if the cores $K_{\alpha\beta}(t)$ have the form (8.173), then the cores (8.193) are a solution of the integral equation (8.188). Substitution of (8.173) and (8.193) into (8.188) yields

$$\sum_{\beta=1}^{m}\left(\sum_{\gamma=1}^{N}\left(\Pi_{\alpha\beta}\frac{B^{(\gamma)}_{\beta\varepsilon}}{\tau^{(\gamma)}_{\beta\varepsilon}}\exp\left(-\frac{t}{\tau^{(\gamma)}_{\beta\varepsilon}}\right)-C_{\beta\varepsilon}\frac{A^{(\gamma)}_{\alpha\beta}}{t^{(\gamma)}_{\alpha\beta}}\exp\left(-\frac{t}{t^{(\gamma)}_{\alpha\beta}}\right)\right)\right.$$
$$\left.-\sum_{\gamma=1}^{N}\sum_{\gamma'=1}^{N}\left(\frac{A^{(\gamma)}_{\alpha\beta}B^{(\gamma')}_{\beta\varepsilon}}{t^{(\gamma)}_{\alpha\beta}\tau^{(\gamma')}_{\beta\varepsilon}}\int_0^t\exp\left(-\frac{t-\tau}{t^{(\gamma)}_{\alpha\beta}}\right)\exp\left(-\frac{\tau}{\tau^{(\gamma')}_{\beta\varepsilon}}\right)d\tau\right)\right)=0. \tag{8.194}$$

Calculating the integral in (8.194)

$$\int_0^t\exp\left(-\frac{t-\tau}{t^{(\gamma)}_{\alpha\beta}}-\frac{\tau}{\tau^{(\gamma')}_{\beta\varepsilon}}\right)d\tau=\frac{t^{(\gamma)}_{\alpha\beta}\tau^{(\gamma')}_{\beta\varepsilon}}{t^{(\gamma)}_{\alpha\beta}-\tau^{(\gamma')}_{\beta\varepsilon}}\left(\exp\left(-\frac{t}{\tau^{(\gamma')}_{\beta\varepsilon}}\right)-\exp\left(-\frac{t}{t^{(\gamma)}_{\alpha\beta}}\right)\right) \tag{8.195}$$

and equating coefficients in (8.194) at the same exponents, we get the system

$$\sum_{\beta=1}^{m}\left(\frac{\Pi_{\alpha\beta}}{\tau^{(\gamma)}_{\beta\varepsilon}}-\sum_{\gamma'=1}^{N}\frac{A^{(\gamma')}_{\alpha\beta}}{\tau^{(\gamma)}_{\beta\varepsilon}-t^{(\gamma')}_{\alpha\beta}}\right)B^{(\gamma)}_{\beta\varepsilon}=0,\ \sum_{\beta=1}^{m}\left(\frac{C_{\beta\varepsilon}}{t^{(\gamma)}_{\alpha\beta}}-\sum_{\gamma'=1}^{N}\frac{B^{(\gamma')}_{\beta\varepsilon}}{\tau^{(\gamma')}_{\beta\varepsilon}-t^{(\gamma)}_{\alpha\beta}}\right)A^{(\gamma)}_{\alpha\beta}=0 \tag{8.196}$$

for determining the constants $A^{(\gamma)}_{\alpha\beta}$ and $t^{(\gamma)}_{\alpha\beta}$ in terms of the constants $B^{(\gamma)}_{\beta\varepsilon}$ and $\tau^{(\gamma)}_{\beta\varepsilon}$ (the constants $\Pi_{\alpha\beta}$ can always be determined in terms of $C_{\alpha\beta}$ and are assumed to be known).

At those values of $B^{(\gamma)}_{\beta\varepsilon}$, $\tau^{(\gamma)}_{\beta\varepsilon}$, $\Pi_{\alpha\beta}$, at which there exists a solution of the system (8.196), the exponential representation of the creep cores (8.193) exists too. ▲

Formulae (8.196) give the method of calculation of the constants $A^{(\gamma)}_{\alpha\beta}$ and $t^{(\gamma)}_{\alpha\beta}$ in terms of $B^{(\gamma)}_{\beta\varepsilon}$, $\tau^{(\gamma)}_{\beta\varepsilon}$ and vice versa. From (8.188b) we can obtain simpler formulae for determining the constants $A^{(\gamma)}_{\alpha\alpha}$ and $t^{(\gamma)}_{\alpha\alpha}$ ($\alpha=m+1,\dots,n$):

$$\frac{\Pi_{\alpha\alpha}}{\tau^{(\gamma)}_{\alpha\alpha}}=\sum_{\gamma'=1}^{N}\frac{A^{(\gamma')}_{\alpha\alpha}}{\tau^{(\gamma)}_{\alpha\alpha}-t^{(\gamma')}_{\alpha\alpha}},\quad \frac{C_{\alpha\alpha}}{t^{(\gamma)}_{\alpha\alpha}}=\sum_{\gamma'=1}^{N}\frac{B^{(\gamma')}_{\alpha\alpha}}{\tau^{(\gamma')}_{\alpha\alpha}-t^{(\gamma)}_{\alpha\alpha}}. \tag{8.197}$$

On substituting (8.184) and (8.193) into (8.189), we find an expression for the spectral creep functions in the case of exponential cores:

$$\Pi_{\alpha\beta}(t) = \Pi_{\alpha\beta} + \sum_{\gamma=1}^{N} A_{\alpha\beta}^{(\gamma)}\left(1 - \exp\left(-\frac{t}{t_{\alpha\beta}^{(\gamma)}}\right)\right), \tag{8.198}$$

where

$$\lim_{t\to+\infty} \Pi_{\alpha\beta}(t) = \Pi_{\alpha\beta} + \sum_{\gamma=1}^{N} A_{\alpha\beta}^{(\gamma)} \equiv \Pi_{\alpha\beta}^{\infty}. \tag{8.199}$$

## Exercises for 8.2

**8.2.1.** Using the rule of differentiation of an integral with a varying upper limit (see formulae (8.15b) and (8.25)) and calculating the derivative of the functional (8.106) with respect to $t$, show that PTI (4.121) actually yields formulae (8.107) and (8.108) for $\overset{(n)}{\mathbf{T}}$ and $w^*$.

**8.2.2.** Using the definition (8.70), show that for linear models $A_n$ of thermorheologically simple viscoelastic media, relations (8.106)–(8.108) have the forms

$$\psi = \psi_0(\theta) + \frac{1}{2\overset{\circ}{\rho}} \sum_{\alpha,\beta=1}^{r_1} \int_0^t\int_0^t \bar{\psi}_{\alpha\beta}(t' - \tau_1', t' - \tau_2') dI_\alpha^{(s)}(\tau_1)\, dI_\beta^{(s)}(\tau_2)$$

$$+ \frac{1}{\overset{\circ}{\rho}} \sum_{\alpha=r_1+1}^{r_2} \int_0^t\int_0^t \bar{\psi}_{\alpha\alpha}(t' - \tau_1', t' - \tau_2')\, dJ_\alpha^{(s)}(\tau_1, \tau_2),$$

$$\overset{(n)}{\mathbf{T}} = J \sum_{\alpha,\beta=1}^{r_1} \mathbf{O}_\beta^{(s)} \int_0^t r_{\alpha\beta}(t' - \tau_1')\, dI_\alpha^{(s)}(\tau_1) + J \sum_{\alpha=r_1+1}^{r_2} \int_0^t r_{\alpha\alpha}(t' - \tau_1')\, dI_{\alpha\mathbf{C}}^{(s)}(\tau_1),$$

$$w^* = -\frac{J}{2} \sum_{\alpha,\beta=1}^{r_1} \int_0^t\int_0^t \frac{\partial}{\partial t}\bar{\psi}_{\alpha\beta}(t' - \tau_1', t' - \tau_2')\, dI_\alpha^{(s)}(\tau_1)\, dI_\beta^{(s)}(\tau_2)$$

$$-J \sum_{\alpha=r_1+1}^{r_2} \int_0^t\int_0^t \frac{\partial}{\partial t}\bar{\psi}_{\alpha\alpha}(t' - \tau_1', t' - \tau_2')\, dJ_\alpha^{(s)}(\tau_1, \tau_2),$$

where $t'$, $\tau_1'$ and $\tau_2'$ are determined by (8.71).

Taking

$$dI_\alpha(\tau) = \frac{d}{dt} I_\alpha(\tau)\, d\tau = \frac{d}{d\tau'} I_\alpha(\tau')\, d\tau' = dI_\alpha(\tau')$$

into account, show that these relations can be represented as functions of the reduced time; in particular

$$\overset{(n)}{\mathbf{T}} = J \sum_{\alpha,\beta=1}^{r_1} \mathbf{O}_\beta^{(s)} \int_0^{t'} r_{\alpha\beta}(t'-\tau_1')\, dI_\alpha^{(s)}(\tau_1') + J \sum_{\alpha=r_1+1}^{r_2} \int_0^{t'} r_{\alpha\alpha}(t'-\tau_1')\, dI_{\alpha C}^{(s)}(\tau_1').$$

**8.2.3.** Using representations (8.132)–(8.134) for the tensor of relaxation functions ${}^4\mathbf{R}(t)$, show that representations (8.140) and (8.115) for $\psi$ are equivalent.

**8.2.4.** Substituting formulae (8.132)–(8.134) for the tensor of relaxation functions ${}^4\mathbf{R}(t)$ into the expression (8.141) for $w^*$, show that formulae (8.141) exactly coincide with (8.116).

**8.2.5.** Show that the condition (8.146) causes monotone non-increasing the relaxation functions

$$\frac{\partial}{\partial t}\widehat{R}^{\alpha\alpha\alpha\alpha}(t) \leqslant 0, \quad \frac{\partial}{\partial t}\widehat{R}^{\alpha\beta\alpha\beta}(t) \leqslant 0, \quad \forall t \geqslant 0, \quad \alpha,\ \beta = 1,\ 2,\ 3, \quad \alpha \neq \beta,$$

where $\widehat{R}^{ijkl}(t)$ are components of the tensor ${}^4\mathbf{R}(t)$ with respect to the basis $\widehat{\mathbf{c}}_i$.

**8.2.6.** Using representations (8.132)–(8.134) for the tensor of relaxation functions ${}^4\mathbf{R}(t)$ and its spectral representation (8.159), and also formulae (8.152)–(8.155) for the tensors $\mathbf{a}_\alpha$ and ${}^4\mathbf{\Gamma}_\alpha$, show that the functions $r_{\alpha\beta}(t)$ and $R_{\alpha\beta}(t)$ are connected by the following relations for *isotropic continua*

$$R_{11}(t) = r_1(t) + (2/3)r_2(t), \qquad R_{22}(t) = 2r_2(t),$$

and for *transversely isotropic continua*

$$R_{11}(t) = \widetilde{R}^{3333}(t) = \widetilde{r}_{22}(t), \quad R_{22}(t) + R_{33}(t) = 2\widehat{R}^{1111}(t) = r_{11}(t) + 2r_{44}(t).$$

**8.2.7.** Show that for mechanically determinate linear models $A_n$ of thermorheologically simple continua, the constitutive equations obtained in Exercise 8.2.2 can be written in the forms (8.136), (8.140), and (8.141)

$$\overset{\circ}{\rho}\psi = \overset{\circ}{\rho}\psi_0(\theta) + \frac{1}{2}\int_0^t\int_0^t d\overset{(n)}{\mathbf{C}}_\theta(\tau_1)\cdot\cdot\,{}^4\mathbf{R}(2t'-\tau_1'-\tau_2')\cdot\cdot\, d\overset{(n)}{\mathbf{C}}_\theta(\tau_2),$$

$$\overset{(n)}{\mathbf{T}} = J\int_0^t {}^4\mathbf{R}(t'-\tau')\cdot\cdot\, d\overset{(n)}{\mathbf{C}}_\theta(\tau),$$

$$w^* = -\frac{Ja_\theta}{2}\int_0^t\int_0^t d\overset{(n)}{\mathbf{C}}_\theta(\tau_1)\cdot\cdot\,\frac{\partial}{\partial t'}{}^4\mathbf{R}(2t'-\tau_1'-\tau_2')\cdot\cdot\, d\overset{(n)}{\mathbf{C}}_\theta(\tau_2).$$

Show that the constitutive equation (8.139) for this model becomes

$$\overset{(n)}{\mathbf{T}} = J\left({}^4\overset{\circ}{\mathbf{M}} \cdot\cdot \overset{(n)}{\mathbf{C}}_\theta(t) - \int_0^t {}^4\mathbf{K}(t'-\tau') \cdot\cdot \overset{(n)}{\mathbf{C}}_\theta(\tau)a_\theta(\tau)d\tau\right), {}^4\mathbf{K}(t') = -d^4\mathbf{R}(t)/dt'.$$

**8.2.8.** Show that for the linear models $A_n$ of thermorheologically simple media with the exponential cores, the constitutive equations from Exercise 8.2.7 can be written in the form (8.175)–(8.177), (8.181)

$$\overset{(n)}{\mathbf{T}} = J\left({}^4\overset{\circ}{\mathbf{M}} \cdot\cdot \overset{(n)}{\mathbf{C}}_\theta - \sum_{\gamma=1}^{N} \mathbf{W}\right), \quad \frac{d\mathbf{W}^{(\gamma)}_{\alpha\beta}}{dt} + \frac{a_\theta(t)}{\tau^{(\gamma)}_{\alpha\beta}}(\mathbf{W}^{(\gamma)}_{\alpha\beta}(t) - \overset{(n)}{\mathbf{C}}_\theta(t)) = 0,$$

$$w^* = \frac{Ja_\theta}{2}\sum_{\gamma=1}^{N}\left(\sum_{\alpha,\beta=1}^{m} \frac{B^{(\gamma)}_{\alpha\beta}}{\tau^{(\gamma)}_{\alpha\beta}} \frac{\mathbf{a}_\alpha}{a_\alpha} \cdot\cdot \left(\overset{(n)}{\mathbf{C}}_\theta - \mathbf{W}^{(\gamma)}_{\alpha\beta}\right) \otimes \left(\overset{(n)}{\mathbf{C}}_\theta - \mathbf{W}^{(\gamma)}_{\alpha\beta}\right) \cdot\cdot \frac{\mathbf{a}_\beta}{a_\beta}\right.$$

$$\left. + \sum_{\alpha=m+1}^{\bar{n}} \frac{B^{(\gamma)}_{\alpha\alpha}}{\tau^{(\gamma)}_{\alpha\alpha}} \left(\overset{(n)}{\mathbf{C}}_\theta - \mathbf{W}^{(\gamma)}_{\alpha\alpha}\right) \cdot\cdot {}^4\mathbf{\Gamma}_\alpha \cdot\cdot \left(\overset{(n)}{\mathbf{C}}_\theta - \mathbf{W}^{(\gamma)}_{\alpha\alpha}\right)\right).$$

**8.2.9.** Prove that the spectral representations (8.148) and (8.159) lead to the spectral form (8.150) of constitutive equations for the linear models $A_n$ (8.135).

**8.2.10.** Show that in hydrostatic compression when the Cauchy stress tensor and the deformation gradient are spherical:

$$\mathbf{T} = -p\mathbf{E}, \quad \mathbf{F} = k\mathbf{E}, \quad \overset{(n)}{\mathbf{C}} = \frac{1}{n-\mathrm{III}}(k^{n-\mathrm{III}} - 1)\mathbf{E}, \quad \operatorname{dev} \overset{(n)}{\mathbf{C}} = 0,$$

a viscoelastic continuum, according to the simplest model (8.166), behaves as a purely elastic one:

$$\overset{(n)}{\mathbf{T}} = \frac{J}{3}R_{11}(0)I_1(\overset{(n)}{\mathbf{C}})\mathbf{E},$$

i.e. there are no viscoelastic deformations in hydrostatic compression for this model. Many solids actually have such properties up to high pressures $p$; therefore, the simplest model (8.166) is widely used in continuum mechanics.

**8.2.11.** Show that the tensors of creep cores and creep functions ${}^4\mathbf{N}(t)$ and ${}^4\mathbf{\Pi}(t)$ have the same properties of symmetry (8.143), (8.147) as the tensors of relaxation cores and relaxation functions ${}^4\mathbf{K}(t)$ and ${}^4\mathbf{R}(t)$, and vice versa.

## 8.3 Models of Incompressible Viscoelastic Solids and Viscoelastic Fluids

### 8.3.1 Models $A_n$ of Incompressible Viscoelastic Continua

For incompressible viscoelastic continua, as usual, there is an additional condition of incompressibility, which can be written in one of the forms (4.487)–(4.492), and the principal thermodynamic identity takes the form (4.495) for models $A_n$. For models $B_n$, its form is analogous. Substituting the functional (8.35) into (4.495) and using the rule (8.27), we get constitutive equations for incompressible viscoelastic continua

$$\begin{cases} \overset{(n)}{\mathbf{T}} = -\dfrac{p}{n-\mathrm{III}}\overset{(n)}{\mathbf{G}}{}^{-1} + \overset{\circ}{\rho}(\partial\psi/\partial\overset{(n)}{\mathbf{C}}(t)), \\ \eta = -\partial\psi/\partial\theta(t), \\ w^* = -\rho\delta\psi. \end{cases} \tag{8.200}$$

Since for incompressible continua the number $r$ of independent invariants $I_\gamma^{(s)}(\overset{(n)}{\mathbf{C}}(t))$ is smaller by 1 than that for compressible materials, in each of the representations (8.53), (8.61), (8.73), (8.77), and (8.96) of the free energy functional the subscript $\gamma$ of the function $\varphi_0(I_\gamma^{(s)}(\overset{(n)}{\mathbf{C}}), \theta)$ takes on $r-1$ values.

In the consistent way, the number $z$ of simultaneous invariants $J_\gamma^{(s)}$ occurring in arguments of the cores $\varphi_m$ decreases too.

### 8.3.2 Principal Models $A_n$ of Incompressible Isotropic Viscoelastic Continua

For the *principal models $A_n$ of incompressible isotropic viscoelastic continua*, the functional $\psi$ (8.77) depends only on five simultaneous invariants:

$$J_\alpha^{(I)} = I_\alpha\left(\overset{(n)}{\mathbf{C}}_\theta(t)\right), \quad J_{3+\alpha}^{(I)} = I_\alpha\left(\overset{(n)}{\mathbf{C}}_\theta(\tau)\right), \quad \alpha = 1,\ 2; \quad J_7^{(I)} = \overset{(n)}{\mathbf{C}}_\theta(\tau)\cdot\cdot\,\overset{(n)}{\mathbf{C}}_\theta(t), \tag{8.201}$$

and the constitutive equations (8.200) become

$$\overset{(n)}{\mathbf{T}} = -\frac{p}{n-\mathrm{III}}\overset{(n)}{\mathbf{G}}{}^{-1} + \breve{\varphi}_1\mathbf{E} + \breve{\varphi}_2\overset{(n)}{\mathbf{C}}_\theta, \tag{8.202}$$

where

$$\breve{\varphi}_1\mathbf{E} \equiv \varphi_{01} + \varphi_{02}I_1(t) - \int_0^t (\varphi_{11} + \varphi_{12}I_1(t))\,d\tau,$$
$$-\breve{\varphi}_2\overset{(\mathrm{n})}{\mathbf{C}}_\theta \equiv \varphi_{02}\overset{(\mathrm{n})}{\mathbf{C}}_\theta(t) - \int_0^t (\varphi_{12}\overset{(\mathrm{n})}{\mathbf{C}}_\theta(t) + \varphi_{17}\overset{(\mathrm{n})}{\mathbf{C}}_\theta(\tau))\,d\tau, \tag{8.203}$$

and $\varphi_{0\gamma}$ and $\varphi_{1\gamma}$ are determined by (8.79).

For the *principal linear models $A_n$ of incompressible isotropic viscoelastic continua*, the functional $\psi$ has the form (compare this with the potential for elastic incompressible materials (4.526))

$$\overset{\circ}{\rho}\psi = \overset{\circ}{\rho}\psi_0 + \overset{\circ}{\rho}\bar{\psi}^0 + \Big(\bar{m} + \frac{l_1 + 2l_2}{2}I_1(\overset{(\mathrm{n})}{\mathbf{C}}_\theta) - \int_0^t q_1(t-\tau)I_1(\overset{(\mathrm{n})}{\mathbf{C}}_\theta(\tau))d\tau\Big)I_1(\overset{(\mathrm{n})}{\mathbf{C}}_\theta)$$
$$-2\Big(l_2I_2(\overset{(\mathrm{n})}{\mathbf{C}}_\theta) + \int_0^t q_2(t-\tau)\overset{(\mathrm{n})}{\mathbf{C}}_\theta(t)\cdot\cdot\,\overset{(\mathrm{n})}{\mathbf{C}}_\theta(\tau)\,d\tau\Big), \tag{8.204}$$

where $l_1$, $l_2$ and $\bar{m}$ are the constants, and $q_1(t-\tau)$ and $q_2(t-\tau)$ are the cores.

The corresponding constitutive equations (8.202) have the form (compare with (4.526) for elastic continua)

$$\overset{(\mathrm{n})}{\mathbf{T}} = -\frac{p}{n-\mathrm{III}}\overset{(\mathrm{n})}{\mathbf{G}}{}^{-1} + (\bar{m} + \breve{l}_1I_1(\overset{(\mathrm{n})}{\mathbf{C}}_\theta))\mathbf{E} + 2\breve{l}_2\overset{(\mathrm{n})}{\mathbf{C}}_\theta. \tag{8.205}$$

Here we have denoted two linear functionals

$$\breve{l}_1I_1 = l_1I_1(\overset{(\mathrm{n})}{\mathbf{C}}_\theta) - \int_0^t q_1(t-\tau)I_1(\overset{(\mathrm{n})}{\mathbf{C}}_\theta(\tau))\,d\tau = \int_0^t r_1(t-\tau)\,dI_1(\overset{(\mathrm{n})}{\mathbf{C}}_\theta(\tau)),$$
$$\breve{l}_2\overset{(\mathrm{n})}{\mathbf{C}}_\theta = l_2\overset{(\mathrm{n})}{\mathbf{C}}_\theta - \int_0^t q_2(t-\tau)\overset{(\mathrm{n})}{\mathbf{C}}_\theta(\tau)\,d\tau = \int_0^t r_2(t-\tau)\,d\overset{(\mathrm{n})}{\mathbf{C}}_\theta(\tau). \tag{8.206}$$

The constants $\bar{\psi}^0$ and $p_0 = p(0)$ are chosen from the conditions (4.326) and (4.327) according to formulae (4.527), just as for elastic materials:

$$p_0 = p^e + \bar{m}, \quad \bar{\psi}^0 = 0, \tag{8.207}$$

where $p^e$ is the constant appearing in the initial values of the stress tensors in the natural configuration $\overset{e}{\mathcal{K}}$: $\overset{(\mathrm{n})}{\mathbf{T}} = -p^e\mathbf{E}$.

### 8.3.3 Linear Models $A_n$ of Incompressible Isotropic Viscoelastic Continua

For linear models $A_n$ of incompressible isotropic viscoelastic continua, obtained from quadratic mechanically determinate models $A_n$ (see Sects. 8.2.6–8.2.9), constitutive equations have the form similar to formulae (8.120) for compressible materials, but the functional $\psi$ should involve the summands linear in the invariant $I_1$:

$$\overset{\circ}{\rho}\psi = \overset{\circ}{\rho}\psi_0 + \overset{\circ}{\rho}\bar{\psi}^0 + \bar{m}I_1(\overset{(n)}{\mathbf{C}}_\theta) + \frac{1}{2}\int_0^t\int_0^t r_1(2t-\tau_1-\tau_2)\,dI_1(\overset{(n)}{\mathbf{C}}_\theta(\tau_1))\,dI_1(\overset{(n)}{\mathbf{C}}_\theta(\tau_2)) + \int_0^t\int_0^t r_2(2t-\tau_1-\tau_2)\,d\overset{(n)}{\mathbf{C}}_\theta(\tau_1)\cdot\cdot\,d\overset{(n)}{\mathbf{C}}_\theta(\tau_2), \tag{8.208}$$

where $\bar{\psi}^0$ and $\bar{m}$ are the constants, and $r_1(y)$ and $r_2(y)$ are the relaxation functions.

According to formulae (8.110) and (8.111) and also Theorem 8.8, the functional $\psi$ can be written in the Volterra form

$$\overset{\circ}{\rho}\psi = \overset{\circ}{\rho}\psi_0 + \overset{\circ}{\rho}\bar{\psi}^0 + \bar{m}I_1(\overset{(n)}{\mathbf{C}}_\theta) + \frac{l_1}{2}I_1^2(\overset{(n)}{\mathbf{C}}_\theta) + l_2I_1(\overset{(n)}{\mathbf{C}}{}^2_\theta) - \int_0^t q_1(t-\tau)I_1(\overset{(n)}{\mathbf{C}}_\theta(\tau))\,d\tau I_1(\overset{(n)}{\mathbf{C}}_\theta(t)) - 2\int_0^t q_2(t-\tau)\overset{(n)}{\mathbf{C}}_\theta(\tau_1)\cdot\cdot\,\overset{(n)}{\mathbf{C}}_\theta(t)\,d\tau + \frac{1}{2}\int_0^t\int_0^t p_1(2t-\tau_1-\tau_2)I_1(\overset{(n)}{\mathbf{C}}_\theta(\tau_1))I_1(\overset{(n)}{\mathbf{C}}_\theta(\tau_2))\,d\tau_1\,d\tau_2 + \frac{1}{2}\int_0^t\int_0^t p_2(2t-\tau_1-\tau_2)\overset{(n)}{\mathbf{C}}_\theta(\tau_1)\cdot\cdot\,\overset{(n)}{\mathbf{C}}_\theta(\tau_2)\,d\tau_1\,d\tau_2, \tag{8.209}$$

where

$$\frac{\partial^2 r_\gamma(y)}{\partial y^2} = p_\gamma(y), \quad \frac{\partial r_\gamma(y)}{\partial y} = -q_\gamma(y), \quad r_\gamma(0) = l_\gamma. \tag{8.210}$$

By analogy with compressible media (see Sect. 8.2.13), we can consider the *simplest model $A_n$ of isotropic incompressible continua*, in which the creep functions $q_1(y)$ and $q_2(y)$ are connected by the relation (8.165):

$$q_1(y) = -\frac{1}{3}q(y), \qquad q(y) = 2q_2(y), \tag{8.211}$$

i.e. this model involves only one core $q(y)$.

Integrating Eq. (8.211) with respect to $y$ and taking the initial condition (8.210) into account, we find the connection between $r_1(y)$ and $r_2(y)$

$$r_1(y) = -\frac{1}{3}r_2(y) + l_1 + \frac{2}{3}l_2, \quad r(y) \equiv 2r_2(y). \tag{8.211a}$$

Substituting (8.211) into (8.205) and (8.206) and grouping like terms, we obtain the following constitutive equation (compare with (8.168)):

$$\overset{(n)}{\mathbf{T}} = -\frac{p}{n - \mathrm{III}}\overset{(n)}{\mathbf{G}}{}^{-1} + (\bar{m} + l_1 I_1(\overset{(n)}{\mathbf{C}}_\theta))\mathbf{E} + 2l_2\overset{(n)}{\mathbf{C}}_\theta - \int_0^t q(t-\tau)\,\mathrm{dev}\,\overset{(n)}{\mathbf{C}}_\theta(\tau)\,d\tau, \tag{8.212}$$

where

$$\mathrm{dev}\,\overset{(n)}{\mathbf{C}}_\theta = \overset{(n)}{\mathbf{C}}_\theta - \frac{1}{3}I_1(\overset{(n)}{\mathbf{C}}_\theta)\mathbf{E} \tag{8.213}$$

is the deviator of the tensor $\overset{(n)}{\mathbf{C}}_\theta$ (see (8.167)).

If $q(t) \equiv 0$, then Eqs. (8.212) coincide with relations (4.526) for isotropic incompressible elastic continua.

## 8.3.4 *Models $B_n$ of Viscoelastic Continua*

In *models $B_n$ of viscoelastic continua*, the free energy $\psi$ is a functional in the form

$$\psi = \overset{t}{\underset{\tau=0}{\psi}}(\overset{(n)}{\mathbf{G}}(t),\ \theta(t),\ \overset{(n)}{\mathbf{G}}{}^t(\tau),\ \theta^t(\tau)), \tag{8.214}$$

and corresponding constitutive equations can be obtained with the help of the rule (8.27) of differentiation of a functional with respect to time; they have the form

$$\begin{cases} \overset{(n)}{\mathbf{T}} = \rho(\partial\psi/\partial\overset{(n)}{\mathbf{G}}(t)) \equiv \overset{t}{\underset{\tau=0}{\mathcal{F}}}(\overset{(n)}{\mathbf{G}}(t),\ \theta(t),\ \overset{(n)}{\mathbf{G}}{}^t(\tau),\ \theta^t(\tau)), \\ \eta = -\partial\psi/\partial\theta, \\ w^* = -\rho\delta\psi. \end{cases} \tag{8.215}$$

All further constructions with functionals $\overset{t}{\underset{\tau=0}{\psi}}$ and $\overset{t}{\underset{\tau=0}{\mathcal{F}}}$ can be performed for models $B_n$ as well.

Special models $B_n$ of viscoelastic continua can be obtained immediately from models $A_n$, in which one should make the substitution $\overset{(n)}{\mathbf{C}} = \overset{(n)}{\mathbf{G}} - (1/(n - \mathrm{III}))\mathbf{E}$.

In particular, for *linear mechanically determinate models $B_n$ of isotropic incompressible continua*, according to the relation $\overset{(n)}{\mathbf{C}}{}^{\bullet} = \overset{(n)}{\mathbf{G}}{}^{\bullet}$, (8.205) and (8.208), we obtain the constitutive equations

$$\overset{\circ}{\rho}\psi = \overset{\circ}{\rho}\psi_0 + \overset{\circ}{\rho}\psi^0 + m I_1(\overset{(n)}{\mathbf{G}}) + \frac{1}{2}\int_0^t\int_0^t r_1(2t-\tau_1-\tau_2)\, dI_1(\overset{(n)}{\mathbf{G}}(\tau_1))\, dI_1(\overset{(n)}{\mathbf{G}}(\tau_2))$$
$$+\int_0^t\int_0^t r_2(2t-\tau_1-\tau_2)\, d\overset{(n)}{\mathbf{G}}(\tau_1)\cdot\cdot\, d\overset{(n)}{\mathbf{G}}(\tau_2), \qquad (8.216)$$

$$w^* = \int_0^t\int_0^t q_1(2t-\tau_1-\tau_2)\, dI_1(\overset{(n)}{\mathbf{G}}(\tau_1))\, dI_2(\overset{(n)}{\mathbf{G}}(\tau_2))$$
$$+2\int_0^t\int_0^t q_2(2t-\tau_1-\tau_2)\, d\overset{(n)}{\mathbf{G}}(\tau_1)\cdot\cdot\, d\overset{(n)}{\mathbf{G}}(\tau_2), \qquad (8.217)$$

$$\overset{(n)}{\mathbf{T}} = -\frac{p}{n-\mathrm{III}}\overset{(n)}{\mathbf{G}}{}^{-1} + \left(m + \int_0^t r_1(t-\tau)\,dI_1(\overset{(n)}{\mathbf{G}}(\tau))\right)\mathbf{E} + 2\int_0^t r_2(t-\tau)\, d\overset{(n)}{\mathbf{G}}(\tau). \qquad (8.218)$$

For the simplest models $B_n$, the assumption (8.211a) on the functions $r_\gamma(t)$ yields

$$\overset{(n)}{\mathbf{T}} = -\frac{p}{n-\mathrm{III}}\overset{(n)}{\mathbf{G}}{}^{-1} + (m + l_1 I_1(\overset{(n)}{\mathbf{G}}))\mathbf{E} + 2l_2\overset{(n)}{\mathbf{G}} - \int_0^t q(t-\tau)\,\mathrm{dev}\,\overset{(n)}{\mathbf{G}}(\tau)\, d\tau, \qquad (8.219)$$

$$w^* = \int_0^t\int_0^t q(2t-\tau_1-\tau_2)\,\mathrm{dev}\,\frac{\partial\overset{(n)}{\mathbf{G}}(\tau_1)}{\partial\tau_1}\cdot\cdot\,\mathrm{dev}\,\frac{\partial\overset{(n)}{\mathbf{G}}(\tau_2)}{\partial\tau_2}\, d\tau_1\, d\tau_2 \geqslant 0. \qquad (8.220)$$

Since the function $w^*$ is nonnegative, we find that

$$q(y) = -\partial r(y)/\partial y \geqslant 0, \qquad (8.221)$$

i.e. the relaxation core $q(y)$ is always nonnegative.

Passing to the limit as $t \to 0$, in the natural configuration $\overset{e}{\mathcal{K}}$, where $\overset{(n)}{\mathbf{T}}(0) = -p^e\mathbf{E}$, $\overset{(n)}{\mathbf{G}}(0) = \mathbf{E}/(n - \mathrm{III})$ and $p = p_0$ (see Sect. 4.8.6), from (8.219) we obtain the following relations between the constants $\psi^0, m, l_1, l_2$ and $p_0$ (see (4.529)):

$$p_0 = p^e + m + \frac{3l_1 + 2l_2}{n - \mathrm{III}}, \quad \psi^0 = -\frac{3(3l_1 + 2l_2)}{2\overset{\circ}{\rho}(n - \mathrm{III})^2} - \frac{3m}{(n - \mathrm{III})\overset{\circ}{\rho}}. \tag{8.222}$$

Notice that relations (8.216) and (8.218) are entirely equivalent to Eqs. (8.208), (8.205); and (8.219) – to Eqs. (8.212) (the constants $l_1$ and $l_2$ in these equations are distinct). We can obtain new models of the class $B_n$ by taking additional assumptions on the constants $m$, $l_1$ and $l_2$. For example, if we assume just as in the corresponding elastic models $B_n$ (see (4.530)) that

$$l_1 + 2l_2 = 0, \quad l_2 = -\frac{\mu}{2}(1 - \beta)(n - \mathrm{III})^2, \quad m = \mu(1 + \beta)(n - \mathrm{III}), \tag{8.223}$$

where $\mu$ and $\beta$ are two new independent constants, then from (8.219) we obtain the constitutive equations

$$\overset{(n)}{\mathbf{T}} = -\frac{p}{n - \mathrm{III}}\overset{(n)}{\mathbf{G}}{}^{-1} + \mu(n - \mathrm{III})^2\left(\left(\frac{1 + \beta}{n - \mathrm{III}} + (1 - \beta)I_1(\overset{(n)}{\mathbf{G}})\right)\mathbf{E} - (1 - \beta)\overset{(n)}{\mathbf{G}}\right)$$
$$- \int_0^t q(t - \tau)\,\mathrm{dev}\,\overset{(n)}{\mathbf{G}}(\tau)\,d\tau, \tag{8.224}$$

which are not equivalent to the corresponding relations (8.212) of models $A_n$.

Due to (8.223), relations (8.222) become

$$p_0 = p^e + \mu(3 - \beta)(n - \mathrm{III})^2, \quad \psi^0 = -6\mu/\overset{\circ}{\rho}. \tag{8.222a}$$

### 8.3.5 Models $A_n$ and $B_n$ of Viscoelastic Fluids

Equations (8.214) and (8.215) as well as (8.35) and (8.38) hold true for both solid and fluid viscoelastic continua. However, for fluids, according to the principle of material symmetry, relations (8.39) (and the analogous relations for models $B_n$) must be satisfied:

$$\overset{(n)}{\mathbf{T}}{}^* = \mathop{\mathcal{F}}_{\tau=0}^{t}(\overset{(n)}{\mathbf{G}}{}^*(t), \theta(t), \overset{(n)}{\mathbf{G}}{}^{t*}(\tau), \theta^t(\tau)) = \rho(\partial\psi/\partial\overset{(n)}{\mathbf{G}}{}^*), \tag{8.225}$$

$$\psi = \mathop{\psi}_{\tau=0}^{t}(\overset{(n)}{\mathbf{G}}{}^*(t), \theta(t), \overset{(n)}{\mathbf{G}}{}^{t*}(\tau), \theta^t(\tau)) \equiv \psi^* \quad \forall H \in U, \tag{8.226}$$

for any $H$-transformations included in the unimodular group $U$.

**Theorem 8.11.** *For models $A_n$ and $B_n$ ($n = \mathrm{I, II, IV, V}$) of viscoelastic fluids, the constitutive equations (8.214), (8.215) and (8.35), (8.38), satisfying the principle of material symmetry and being continuous functionals in space $H_t$, can be written as follows:*

$$\overset{(n)}{\mathbf{T}} = -\frac{p}{n-\mathrm{III}}\overset{(n)}{\mathbf{G}}{}^{-1}, \tag{8.227a}$$

$$p=\rho^2\frac{\partial\psi}{\partial\rho(t)} = \overset{t}{\underset{\tau=0}{p}}\,(\rho(\tau),\theta(\tau)) = \rho^2\left(\frac{\partial\varphi_0}{\partial\rho} + \sum_{m=1}^{\infty}\int_0^t\cdots\int_0^t\frac{\partial\varphi_m}{\partial\rho(t)}\,d\tau_1\ldots d\tau_m\right), \tag{8.227b}$$

$$\psi = \overset{t}{\underset{\tau=0}{\psi}}\,(\rho(t),\ \theta(t),\ \rho^t(\tau),\ \theta^t(\tau)) = \varphi_0(\rho(t),\ \theta(t))$$
$$+\sum_{m=1}^{\infty}\int_0^t\cdots\int_0^t \varphi_m(t,\ \tau_1,\ldots,\tau_m,\ \rho(t),\ \rho(\tau_1),\ldots,\rho(\tau_m))\,d\tau_1\ldots\tau_m, \tag{8.227c}$$

$$\mathbf{T} = -p\mathbf{E}. \tag{8.227d}$$

▼ Since $\psi$ is assumed to be a continuous functional, we can apply Theorem 8.4 and expand $\psi$ in terms of $n$-fold scalar functionals:

$$\psi = \sum_{m=1}^{\infty}\int_0^t\cdots\int_0^t\widetilde{\psi}_m(t,\ \tau_1,\ldots,\tau_m)\,d\tau_1\ldots d\tau_m, \tag{8.228}$$

whose cores are scalar functions of $m$ tensor arguments:

$$\widetilde{\psi}_m(t,\ \tau_1,\ldots,\tau_m) = \widetilde{\psi}_m(t,\ \tau_1,\ldots,\tau_m,\ \overset{(n)}{\mathbf{G}}(\tau_1),\ldots,\overset{(n)}{\mathbf{G}}(\tau_m)). \tag{8.229}$$

On substituting the representation (8.228) into (8.226), we get that functions $\widetilde{\psi}_m$ must satisfy the relation

$$\widetilde{\psi}_m(t,\ \tau_1,\ldots,\tau_m,\ \overset{(n)}{\mathbf{G}}_1,\ldots,\overset{(n)}{\mathbf{G}}_m) = \widetilde{\psi}_m\left(t,\ \tau_1,\ldots,\tau_m,\ \overset{(n)}{\mathbf{G}}{}^*_1,\ldots,\overset{(n)}{\mathbf{G}}{}^*_m\right)\quad\forall H\in U, \tag{8.230}$$

where $\overset{(n)}{\mathbf{G}}_i \equiv \overset{(n)}{\mathbf{G}}(\tau_i)$, i.e. they must be $H$-indifferent relative to the unimodular group $U$.

Applying the same reasoning as we used in proving Theorem 4.31, we can show that functions of the third invariant of the tensors $\overset{(n)}{\mathbf{G}}_i$ (or, that is the same, of values of the density $\rho_i = \rho(\tau_i)$ at different times $\tau_i$) are the only functions ensuring that the condition (8.230) is satisfied:

$$\widetilde{\psi}_m=\widetilde{\psi}_m(t, \tau_1,\dots,\tau_m, I_3(\overset{(n)}{\mathbf{G}}_1),\dots,I_3(\overset{(n)}{\mathbf{G}}_m))=\widetilde{\psi}_m(t, \tau_1, \dots, \tau_m, \rho_1, \dots, \rho_m).$$

Separating $\delta$-type components from the cores in this expression by analogy with (8.51a), from (8.228) we get the representation (8.227c), where the cores $\varphi_m$ are connected to $\widetilde{\psi}_m$ by relations (8.54).

Substituting the functional (8.227c) into (8.215) and differentiating $\psi$ with respect to $\overset{(n)}{\mathbf{G}}(t)$, from (4.448)–(4.453) we actually obtain formulae (8.227a) and (8.227b).

Finally, using the transformations (4.455) and (4.456), from (8.227a) we obtain that all the relations (8.227a) for models $A_n$ and $B_n$ are equivalent to the single relation (8.227d). ▲

Notice that although Eq. (8.227d) for the Cauchy stress tensor is formally the same as the one for an ideal fluid, a viscoelastic fluid is not ideal (it is dissipative), because in this case the pressure $p$ is a functional of the density $\rho$, and the dissipation function $w^*$ is not zero due to (8.64):

$$w^* = -\rho\delta\psi = -\rho\varphi_1^0 - \rho\sum_{m=1}^{\infty}\int_0^t\dots\int_0^t\left(\frac{\partial\varphi_m}{\partial t}+\varphi_{m+1}^0\right)d\tau_1\dots d\tau_m \geqslant 0. \tag{8.231}$$

## 8.3.6 *The Principle of Material Indifference for Models $A_n$ and $B_n$ of Viscoelastic Continua*

Since all the energetic tensors $\overset{(n)}{\mathbf{T}}$, $\overset{(n)}{\mathbf{C}}$ and $\overset{(n)}{\mathbf{G}}$ are $R$-invariant in rigid motions, all the constitutive equations given in this section for solids and fluids, and also for incompressible continua, are the same in actual configurations $\mathcal{K}$ and $\mathcal{K}'$; therefore, the principle of material indifference for models $A_n$ and $B_n$ of viscoelastic media is satisfied identically.

## Exercises for 8.3

**8.3.1.** Using the stepwise loading (8.144) and passing to the limit as $t \to 0_+$, show that the instantly elastic relations obtained from (8.224) coincide with the constitutive equations (4.532) of the model $B_n$ of an elastic isotropic incompressible continuum.

**8.3.2.** Using the method applied in Sect. 8.2.14, show that for the simplest model $A_n$ of isotropic incompressible viscoelastic continua with the exponential core

$$q(t) = \sum_{\gamma=1}^{N} \frac{B^{(\gamma)}}{\tau^{(\gamma)}} \exp\left(-\frac{t}{\tau^{(\gamma)}}\right),$$

the constitutive equation (8.212) takes the form

$$\overset{(n)}{\mathbf{T}} = -\frac{p}{n-\mathrm{III}}\overset{(n)}{\mathbf{G}}{}^{-1} + (\bar{m} + l_1 I_1(\overset{(n)}{\mathbf{C}}_\theta))\mathbf{E} + 2l_2\overset{(n)}{\mathbf{C}}_\theta - \sum_{\gamma=1}^{N} \mathbf{W}^{(\gamma)} B^{(\gamma)},$$

$$\frac{d\mathbf{W}^{(\gamma)}}{dt} = \frac{1}{\tau^{(\gamma)}}(\operatorname{dev}\overset{(n)}{\mathbf{C}}_\theta - \mathbf{W}^{(\gamma)});$$

and for the simplest model $B_n$ of isotropic incompressible viscoelastic continua with the same exponential core, the constitutive equation (8.224) takes the form

$$\overset{(n)}{\mathbf{T}} = -\frac{p}{n-\mathrm{III}}\overset{(n)}{\mathbf{G}}{}^{-1}$$
$$+\mu(n-\mathrm{III})^2\left(\left(\frac{1+\beta}{n-\mathrm{III}} + (1-\beta)I_1(\overset{(n)}{\mathbf{G}})\right)\mathbf{E} - (1-\beta)\overset{(n)}{\mathbf{G}}\right) - \sum_{\gamma=1}^{N} \mathbf{W}^{(\gamma)} B^{(\gamma)},$$

$$\frac{d\mathbf{W}^{(\gamma)}}{dt} = \frac{1}{\tau^{(\gamma)}}\left(\operatorname{dev}\overset{(n)}{\mathbf{G}} - \mathbf{W}^{(\gamma)}\right).$$

**8.3.3.** Using the results of Exercise 8.2.7 and Eqs. (8.220) and (8.224), show that for the simplest linear models $B_n$ of isotropic incompressible thermorheologically simple media the following constitutive equations hold:

$$\overset{(n)}{\mathbf{T}} = -\frac{p}{n-\mathrm{III}}\overset{(n)}{\mathbf{G}}{}^{-1} + \mu(n-\mathrm{III})^2\left(\left(\frac{1+\beta}{n-\mathrm{III}} + (1-\beta)I_1(\overset{(n)}{\mathbf{G}})\right)\mathbf{E} - (1-\beta)\overset{(n)}{\mathbf{G}}\right)$$
$$-\int_0^t q(t'-\tau')\operatorname{dev}\overset{(n)}{\mathbf{G}}(\tau)a_\theta(\tau)\,d\tau,$$

$$w^* = a_\theta \int_0^t\int_0^t q(2t'-\tau_1'-\tau_2')\operatorname{dev}\frac{\partial\overset{(n)}{\mathbf{G}}(\tau_1)}{\partial\tau_1}\cdot\cdot\operatorname{dev}\frac{\partial\overset{(n)}{\mathbf{G}}(\tau_2)}{\partial\tau_2}\,d\tau_1\,d\tau_2,$$

where $t'$, $\tau_1'$ and $\tau_2'$ are determined by (8.71).

## 8.4 Statements of Problems in Viscoelasticity Theory at Large Deformations

### 8.4.1 Statements of Dynamic Problems in the Spatial Description

Statements of problems of viscoelasticity at large deformations can be obtained formally from the corresponding statements of problems of elasticity theory at large deformations (see Sect. 6.1), if in the last ones we replace the generalized constitutive equations of elasticity $\overset{(n)}{\mathbf{T}}_G = \mathcal{F}_G(\overset{(n)}{\mathbf{C}}_\theta, \theta)$ by relations of viscoelasticity (for a viscoelastic model chosen from the models considered in Sects. 8.1–8.3).

Choosing the most general representations (8.38) and (8.215) for models $A_n$ and $B_n$ of viscoelasticity and using the method mentioned above and the statement of the dynamic $\theta RUVF$-problem of elasticity theory (see Sects. 6.1.1 and 6.3.3), we obtain a statement of the *dynamic $\theta RUVF$-problem of thermoviscoelasticity in the spatial description*. This statement consists of the equation system in the domain $V \times (0, t_{\max})$:

$$\frac{\partial \rho}{\partial t} + \nabla \cdot \rho \mathbf{v} = 0, \tag{8.232a}$$

$$\frac{\partial \rho \mathbf{v}}{\partial t} + \nabla \cdot \rho \mathbf{v} \otimes \mathbf{v} = \nabla \cdot \mathbf{T} + \rho \mathbf{f}, \tag{8.232b}$$

$$\frac{\partial \rho \eta}{\partial t} + \nabla \cdot \rho \mathbf{v} \eta = -\frac{1}{\theta} \nabla \cdot \mathbf{q} + \frac{\rho q_m + w^*}{\theta}. \tag{8.232c}$$

$$\frac{\partial \rho \mathbf{F}^{\mathrm{T}}}{\partial t} + \nabla \cdot \rho (\mathbf{v} \otimes \mathbf{F}^{\mathrm{T}} - \mathbf{F} \otimes \mathbf{v}) = 0, \tag{8.232d}$$

$$\frac{\partial \rho \mathbf{u}}{\partial t} + \nabla \cdot (\rho \mathbf{v} \otimes \mathbf{u}) = \rho \mathbf{v}, \tag{8.232e}$$

the constitutive equations in the domain $\bar{V} \times (0, t_{\max})$:

$$\mathbf{q} = -\boldsymbol{\lambda} \cdot \nabla \theta, \tag{8.233a}$$

$$\mathbf{T} = {}^4\overset{(n)}{\mathbf{E}}_G \cdot\cdot \overset{(n)}{\mathbf{T}}_G, \tag{8.233b}$$

$$\overset{(n)}{\mathbf{T}}_G = \overset{t}{\underset{\tau=0}{\mathcal{F}_G}}(\overset{(n)}{\mathbf{C}}_G(t),\ \theta(t),\ \overset{(n)}{\mathbf{C}}{}^t_G(\tau),\ \theta^t(\tau)) \equiv \rho(\partial \psi / \partial \overset{(n)}{\mathbf{C}}_G(t)), \tag{8.233c}$$

$$\eta = -\partial \psi / \partial \theta(t), \quad w^* = -\rho \delta \psi, \tag{8.233d}$$

$$\psi = \overset{t}{\underset{\tau=0}{\psi}}(\overset{(n)}{\mathbf{C}}_G(t),\ \theta(t),\ \overset{(n)}{\mathbf{C}}{}^t_G(\tau),\ \theta^t(\tau)),\quad G = A,\ B, \tag{8.233e}$$

which should be complenented by expressions (4.389) and (4.391c) for tensors ${}^4\overset{(n)}{\mathbf{E}}_G$ and $\overset{(n)}{\mathbf{C}}_G$:

$$\overset{(n)}{\mathbf{C}}_G = \frac{1}{n-\mathrm{III}}\left(\sum_{\alpha=1}^{3}\lambda_\alpha^{n-\mathrm{III}}\overset{\circ}{\mathbf{p}}_\alpha \otimes \overset{\circ}{\mathbf{p}}_\alpha - \bar{h}_G\mathbf{E}\right),$$

$${}^4\overset{(n)}{\mathbf{E}} = \sum_{\alpha,\beta=1}^{3} E_{\alpha\beta}\mathbf{p}_\alpha \otimes \mathbf{p}_\beta \otimes \overset{\circ}{\mathbf{p}}_\beta \otimes \overset{\circ}{\mathbf{p}}_\alpha,$$

$$\lambda_\alpha,\ \overset{\circ}{\mathbf{p}}_\alpha,\ \mathbf{p}_\alpha \ \| \ \mathbf{F}, \tag{8.234}$$

the boundary conditions (6.68)–(6.76), which for the case when there are no phase transformations take the form

$$\begin{gathered}
\mathbf{n}\cdot\mathbf{T} = \widetilde{\mathbf{t}}_{ne}, \quad \mathbf{n}\cdot\mathbf{q} = \widetilde{q}_{ne} \quad \text{at } \Sigma_1,\ldots,\Sigma_4,\ \Sigma_7,\\
\mathbf{u} = \mathbf{u}_e, \quad \theta = \theta_e \quad \text{at } \Sigma_5,\ \Sigma_6,\\
\partial\rho/\partial\mathbf{n} = 0, \quad \mathbf{v}\cdot\mathbf{n} = 0, \quad \mathbf{n}\cdot\mathbf{q} = 0, \quad \mathbf{n}\cdot\mathbf{T}\cdot\boldsymbol{\tau}_\alpha = 0 \quad \text{at } \Sigma_8,
\end{gathered} \tag{8.235}$$

and also the initial conditions

$$t = 0: \quad \rho = \overset{\circ}{\rho}, \ \mathbf{v} = \mathbf{v}_0, \ \theta = \theta_0, \ \mathbf{F} = \mathbf{E}, \ \mathbf{u} = \mathbf{u}_0. \tag{8.236}$$

On substituting the constitutive equations (8.233) and (8.234) into (8.232), we get the system of 17 scalar equations for 17 unknown scalar functions:

$$\theta,\ \rho,\ \mathbf{u},\ \mathbf{v},\ \mathbf{F} \ \| \ \mathbf{x},\ t. \tag{8.237}$$

Since the domain $V(t)$ in the spatial description is unknown, the system obtained should be complemented with either relation (6.89) or Eq. (6.85) for the function $f(\mathbf{x},t)$ specifying a shape of the surface $\Sigma(t)$ bounding the domain $V(t)$:

$$\begin{gathered}
\frac{\partial f}{\partial t} + \mathbf{v}\cdot\nabla f = 0,\\
t = 0: \qquad f = f^0(\mathbf{x}).
\end{gathered} \tag{8.238}$$

In the second case the function $f(\mathbf{x},t)$ appears among the unknowns (8.237).

*Remark 1.* In viscoelasticity theory, in place of the energy balance equation one often uses the entropy balance equation (3.166a) (the system (8.232) has been written in this way), which contains the dissipation function $w^*$ explicitly. The entropy balance equation can be represented in the nondivergence form (3.166)

$$\rho\theta\frac{d\eta}{dt} = -\nabla\cdot\mathbf{q} + \rho q_m + w^*. \tag{8.239}$$

When the model $A_n$ of a thermoviscoelastic continuum with difference cores is considered, then $\eta$ is determined by formula (8.69). If in this formula the derivative $\partial'\varphi/\partial\theta$ is assumed to depend only on $\theta$, then after substitution of (8.69) and (8.233a) into (8.239) we obtain the following equation of heat conduction for a viscoelastic continuum in the spatial description:

$$\rho c_\varepsilon\left(\frac{\partial\theta}{\partial t}+\mathbf{v}\cdot\nabla\theta\right)=\nabla\cdot(\boldsymbol{\lambda}\cdot\nabla\theta)-\rho\theta\left(\frac{\partial}{\partial t}\left(\boldsymbol{\alpha}\cdot\cdot\frac{\overset{(n)}{\mathbf{T}}}{\rho}\right)+\mathbf{v}\cdot\nabla\left(\boldsymbol{\alpha}\cdot\cdot\frac{\overset{(n)}{\mathbf{T}}}{\rho}\right)\right)+\rho q_m+w^*. \tag{8.240}$$

Here

$$c_\varepsilon=-\theta(\partial'^2\varphi_0/\partial\theta^2) \tag{8.241}$$

is the heat capacity at fixed deformations. □

*Remark 2.* If, for example, we consider the statement of the dynamic $\theta RUVF$-problem for mechanically determinate linear models $A_n$ of thermorheologically simple viscoelastic continua with exponential cores, then the constitutive equations (8.233) has the form derived in Exercise 8.2.8:

$$\overset{(n)}{\mathbf{T}}=J\left({}^4\overset{\circ}{\mathbf{M}}\cdot\cdot\overset{(n)}{\mathbf{C}}_\theta-\sum_{\gamma=1}^{N}\mathbf{W}^{(\gamma)}\right),$$

$$\mathbf{W}^{(\gamma)}=\sum_{\alpha,\beta=1}^{m}B^{(\gamma)}_{\alpha\beta}\mathbf{W}^{(\gamma)}_{\alpha\beta}\cdot\cdot\frac{\mathbf{a}_{(\alpha)}\otimes\mathbf{a}_{(\beta)}}{a_\alpha a_\beta}+\sum_{\alpha=m+1}^{\bar{n}}B^{(\gamma)}_{\alpha\alpha}\mathbf{W}^{(\gamma)}_{\alpha\alpha}\cdot\cdot{}^4\boldsymbol{\Gamma}_\alpha,$$

$$\frac{\partial\mathbf{W}^{(\gamma)}_{\alpha\beta}}{\partial t}+\mathbf{v}\cdot\nabla\otimes\mathbf{W}^{(\gamma)}_{\alpha\beta}=a_\theta\frac{\overset{(n)}{\mathbf{C}}_\theta-\mathbf{W}^{(\gamma)}_{\alpha\beta}}{\tau^{(\gamma)}_{\alpha\beta}},$$

$$w^*=\frac{Ja_\theta}{2}\sum_{\gamma=1}^{N}\left(\sum_{\alpha,\beta=1}^{m}\frac{B^{(\gamma)}_{\alpha\beta}}{\tau^{(\gamma)}_{\alpha\beta}}\frac{\mathbf{a}_\alpha}{a_\alpha}\cdot\cdot(\overset{(n)}{\mathbf{C}}_\theta-\mathbf{W}^{(\gamma)}_{\alpha\beta})\otimes(\overset{(n)}{\mathbf{C}}_\theta-\mathbf{W}^{(\gamma)}_{\alpha\beta})\cdot\cdot\frac{\mathbf{a}_\beta}{a_\beta}\right.$$
$$\left.+\sum_{\alpha=m+1}^{\bar{n}}\frac{B^{(\gamma)}_{\alpha\alpha}}{\tau^{(\gamma)}_{\alpha\alpha}}(\overset{(n)}{\mathbf{C}}_\theta-\mathbf{W}^{(\gamma)}_{\alpha\alpha})\cdot\cdot{}^4\boldsymbol{\Gamma}_\alpha\cdot\cdot(\overset{(n)}{\mathbf{C}}_\theta-\mathbf{W}^{(\gamma)}_{\alpha\alpha})\right). \tag{8.242}$$

In this case the initial conditions (8.236) are complemented by the additional conditions

$$t=0:\quad \mathbf{W}^{(\gamma)}_{\alpha\beta}=0, \tag{8.243}$$

and the number of unknown functions (8.237) becomes greater due to adding the functions

$$\mathbf{W}^{(\gamma)}_{\alpha\beta} \parallel \mathbf{x},\ t, \quad \alpha,\ \beta = 1,\ldots,\bar{n};\ \gamma = 1,\ldots,N; \tag{8.244}$$

(here we always have $\mathbf{W}^{(\gamma)}_{\alpha\beta} \equiv 0$ when $\alpha \neq \beta$ and $\alpha,\ \beta > m$). □

By analogy with the dynamic $\theta RUV$-, $\theta RV$- and $\theta U$-problems of thermoelasticity (see Sect. 6.3.3), we can state the *dynamic $\theta RUV$-problem of viscoelasticity*, in which the deformation gradient $\mathbf{F}$ is eliminated between the unknowns, the *$\theta UV$-problem of viscoelasticity*, where in addition the density $\rho$ is eliminated, and the *dynamic $\theta U$-problem of viscoelasticity*, where only $\mathbf{u}$ and $\theta$ are unknown.

*Remark 3.* Just as the statements of thermoelasticity problems in the spatial description (see Sect. 6.3.3), the statements of thermoviscoelasticity problems mentioned above are *strongly coupled*, because they cannot be split into heat conduction problems and viscoelasticity problems even if we neglect the entropy term of the connection (i.e. the term $(\boldsymbol{\alpha}/\rho)\cdot\cdot \overset{(n)}{\mathbf{T}}$ in the equation of heat conduction (8.240)).

To the six causes of the connection in thermoviscoelasticity problems, mentioned in Remark 4 of Sect. 6.3.3, we should add one more factor: the presence of the dissipation function $w^*$ in the entropy balance equation (8.232c) or in the heat conduction equation (8.240), that is a consequence of non-ideality of viscoelastic continua. In many problems, a contribution of the dissipation function $w^*$ to the heat conduction equation (8.240) proves to be rather considerable and cannot be neglected in non-isothermal processes.

The effect of growing the temperature in viscoelastic materials without heat supplied to a body from the outside but only due to internal heat release in deforming (caused by the presence of the dissipation function $w^*$), is called *dissipative heating* of the body (see Sect. 8.6). □

Let us pay attention to the sixth cause of the connection mentioned in Remark 4 of Sect. 6.3.3: for viscoelastic continua the dependence of the constitutive equations (8.233c) on temperature can be split into the three constituents:

(1) Dependence of the heat deformation $\overset{\circ}{\boldsymbol{\varepsilon}}$ (8.66) when the Duhamel–Neumann model is used.
(2) Dependence of the elastic properties on the temperature $\theta(t)$.
(3) Dependence of the viscous properties, i.e. the integral part of Eqs. (8.233c), upon the temperature prehistory $\theta^t(\tau)$.

As established in experiments, for most viscoelastic continua, the viscous properties more considerably depend on temperature than the elastic ones. Since the dissipation function $w^*$ depends on just the viscous properties, it also depends explicitly upon the temperature (in the model $A_n$ with the exponential cores (8.242) this dependence has the form of function $a_\theta(\theta(t))$). The temperature dependence $w^*(\theta)$ leads to the intensification of dissipative heating in viscoelastic materials and, under certain conditions, can cause the effect of *heat explosion* (see Sect. 8.6.9).

### 8.4.2 Statements of Dynamic Problems in the Material Description

Using the statement of the dynamic $\theta UVF$-problem of thermoelasticity in the material description (see Sects. 6.2.1 and 6.3.4) and replacing the constitutive equations (6.42) by viscoelasticity relations in the forms (8.38) and (8.215), for models $A_n$ and $B_n$ we obtain a statement of the *dynamic $\theta UVF$-problem of thermoviscoelasticity in the material description*. This statement consists of the equation system

$$\begin{aligned}
\rho &= \overset{\circ}{\rho} \det \mathbf{F}^{-1}, \\
\overset{\circ}{\rho}(\partial \mathbf{v}/\partial t) &= \overset{\circ}{\nabla} \cdot \mathbf{P} + \overset{\circ}{\rho}\mathbf{f}, \\
\overset{\circ}{\rho}\theta(\partial \eta/\partial t) &= \overset{\circ}{\nabla} \cdot (\overset{\circ}{\boldsymbol{\lambda}} \cdot \overset{\circ}{\nabla}\theta) + \overset{\circ}{\rho} q_m + \overset{\circ}{w}{}^*, \\
\partial \mathbf{F}^{\mathrm{T}}/\partial t &= \overset{\circ}{\nabla} \otimes \mathbf{v}, \\
\partial \mathbf{u}/\partial t &= \mathbf{v}
\end{aligned} \tag{8.245}$$

defined in the domain $\overset{\circ}{V} \times (0, t_{\max})$, and the constitutive equations in the same domain $\overset{\circ}{V} \times (0, t_{\max})$:

$$\mathbf{P} = {}^4\overset{(n)}{\mathbf{E}}{}^0_G \cdot\cdot \overset{(n)}{\mathbf{T}}_G, \tag{8.246a}$$

$$\overset{(n)}{\mathbf{T}}_G = \rho(\partial\psi/\partial \overset{(n)}{\mathbf{C}}_G(t)) \equiv \underset{\tau=0}{\overset{t}{\mathcal{F}_G}}\left(\overset{(n)}{\mathbf{C}}_G(t),\ \theta(t),\ \overset{(n)}{\mathbf{C}}{}^t_G(\tau),\ \theta^t(\tau)\right), \tag{8.246b}$$

$$\eta = -\partial\psi/\partial\theta, \quad \overset{\circ}{w}{}^* = -\overset{\circ}{\rho}\delta\psi, \quad G = A,\ B, \tag{8.246c}$$

$$\psi = \underset{\tau=0}{\overset{t}{\psi}}\left(\overset{(n)}{\mathbf{C}}_G(t),\ \theta(t),\ \overset{(n)}{\mathbf{C}}{}^t_G(\tau),\ \theta^t(\tau)\right), \tag{8.246d}$$

which are complemented with expressions for the tensors ${}^4\overset{(n)}{\mathbf{E}}{}^0$ and $\overset{(n)}{\mathbf{C}}_G$:

$$\begin{aligned}
{}^4\overset{(n)}{\mathbf{E}}{}^0 &= \sum_{\alpha,\beta=1}^{3} \overset{\circ}{E}_{\alpha\beta}\overset{\circ}{\mathbf{p}}_\alpha \otimes \mathbf{p}_\beta \otimes \overset{\circ}{\mathbf{p}}_\beta \otimes \overset{\circ}{\mathbf{p}}_\alpha, \\
\overset{(n)}{\mathbf{C}}_G &= \frac{1}{n - \mathrm{III}} \sum_{\alpha=1}^{3}(\lambda_\alpha^{n-\mathrm{III}}\overset{\circ}{\mathbf{p}}_\alpha \otimes \overset{\circ}{\mathbf{p}}_\alpha - \bar{h}_G\mathbf{E}), \\
&\lambda_\alpha,\ \overset{\circ}{\mathbf{p}}_\alpha,\ \mathbf{p}_\alpha \ \| \ \mathbf{F},
\end{aligned} \tag{8.247}$$

the boundary conditions (6.77)–(6.84) taking the forms (when there are no phase transformations)

$$\overset{\circ}{\mathbf{n}} \cdot \mathbf{P} = \overset{\circ}{\mathbf{t}}_{ne}, \quad \overset{\circ}{\mathbf{n}} \cdot \overset{\circ}{\mathbf{q}} = \overset{\circ}{q}_{ne} \quad \text{at } \overset{\circ}{\Sigma}_1, \ldots, \overset{\circ}{\Sigma}_4, \ \overset{\circ}{\Sigma}_7,$$

$$\mathbf{u} = \mathbf{u}_e, \quad \theta = \theta_e \quad \text{at } \overset{\circ}{\Sigma}_5, \ \overset{\circ}{\Sigma}_6,$$

$$\mathbf{u} \cdot \overset{\circ}{\mathbf{n}} = 0, \quad \mathbf{v} \cdot \overset{\circ}{\mathbf{n}} = 0, \quad \overset{\circ}{\mathbf{q}} \cdot \overset{\circ}{\mathbf{n}} = 0, \quad \overset{\circ}{\mathbf{n}} \cdot \mathbf{P} \cdot \boldsymbol{\tau}_\alpha = 0 \quad \text{at } \overset{\circ}{\Sigma}_8, \tag{8.248}$$

and the initial conditions

$$t = 0: \quad \mathbf{v} = \mathbf{v}_0, \quad \mathbf{u} = \mathbf{u}_0, \quad \mathbf{F} = \mathbf{E}, \quad \theta = \theta_0. \tag{8.249}$$

On substituting the constitutive equations (8.246) and (8.247) into (8.245), we obtain the system for 16 unknown scalar functions being components of the following vectors and tensors:

$$\theta, \ \mathbf{u}, \ \mathbf{v}, \ \mathbf{F} \ \ \| \ \ X^i, \ t.$$

Due to the continuity equation, the density $\rho$ can be eliminated between the unknown functions.

Just as for problems of thermoelasticity in the material description, the problem (8.245)–(8.250) is formulated for a known domain $\overset{\circ}{V}$, that considerably simplifies its solving.

For particular models of viscoelastic continua, formulae (8.246b)–(8.246d) are replaced by appropriate relations derived in Sects. 8.1–8.3.

When models $A_n$ of viscoelastic continua with the difference cores and the Duhamel–Neumann model (8.66) are considered, the specific entropy $\eta$ is determined by formula (8.69). Assuming that $\partial'\varphi_0/\partial\theta$ depends only on temperature, we can rewrite the entropy balance equation of system (8.245) in the form of the heat conduction equation for a viscoelastic medium in the material description:

$$\overset{\circ}{\rho} c_\varepsilon \frac{\partial \theta}{\partial t} = \overset{\circ}{\nabla} \cdot \left( \overset{\circ}{\boldsymbol{\lambda}} \cdot \overset{\circ}{\nabla} \theta \right) - \overset{\circ}{\rho} \theta \frac{\partial}{\partial t} \left( \boldsymbol{\alpha} \cdot\cdot \frac{\overset{(n)}{\mathbf{T}}}{\rho} \right) + \overset{\circ}{\rho} q_m + \overset{\circ}{w}{}^*. \tag{8.250}$$

Here the second entropy term on the right-hand side of the equation, as a rule, can be neglected in comparison with $\overset{\circ}{w}{}^*$.

Unlike the statement of the thermoelasticity problem in the material description given in Sect. 6.3.4, the thermoviscoelasticity problem (8.245)–(8.249) is strongly coupled even if there are no phase transformations, that is caused by the presence of the dissipation function $\overset{\circ}{w}{}^*$. As noted in Sect. 8.4.1, in the general case of non-isothermal processes a contribution of the function $\overset{\circ}{w}{}^*$ to the heat conduction equation may be rather essential and cannot be neglected.

Using the statements of the dynamic $\theta UV$-, $T\theta UVF$- and $\theta U$-problems of thermoelasticity in the material description (see Sect. 6.3.4), we can formulate the corresponding dynamic problems of thermoviscoelasticity. So the statement of the *dynamic $\theta U$-problem of thermoviscoelasticity in the material description* consists of the equation system (6.58):

$$\begin{aligned} \overset{\circ}{\rho}(\partial^2\mathbf{u}/\partial t^2) &= \overset{\circ}{\nabla}\cdot\mathbf{P} + \overset{\circ}{\rho}\mathbf{f}, \\ \overset{\circ}{\rho}c_\varepsilon(\partial\theta/\partial t) &= \overset{\circ}{\nabla}\cdot(\overset{\circ}{\boldsymbol{\lambda}}\cdot\overset{\circ}{\nabla}\theta) + \overset{\circ}{\rho}q_m + \overset{\circ}{w}{}^* \end{aligned} \tag{8.251}$$

in the domain $\overset{\circ}{V}\times(0,t_{\max})$; constitutive equations (8.246); the expressions for tensors ${}^4\overset{(n)}{\mathbf{E}}{}^0$ and $\overset{(n)}{\mathbf{C}}_G$ (8.247); the kinematic equation

$$\mathbf{F} = \mathbf{E} + \overset{\circ}{\nabla}\otimes\mathbf{u}^{\mathrm{T}}; \tag{8.251a}$$

boundary conditions (8.248) and initial conditions (8.249). The problem is solved for the four scalar functions: components of the displacement vector $\mathbf{u}$ and the temperature $\theta$.

### 8.4.3 Statements of Quasistatic Problems of Viscoelasticity Theory in the Spatial Description

Statements of quasistatic problems of viscoelasticity theory can be obtained formally from the corresponding statements of quasistatic problems of elasticity theory at large deformations (see Sect. 6.3.5) by replacing the constitutive equations of elasticity with appropriate relations of viscoelasticity derived in Sects. 8.1–8.3. So the statement of the coupled *quasistatic problem of thermoviscoelasticity in the spatial description* for linear mechanically determinate models $A_n$ of thermorheologically simple continua has the form

$$\begin{cases} \nabla\cdot\mathbf{T} + \rho\mathbf{f} = 0, \\ \rho c_\varepsilon(\partial\theta/\partial t) = \nabla\cdot(\boldsymbol{\lambda}\cdot\nabla\theta) + \rho q_m + w^* \quad \text{in } V, \end{cases} \tag{8.252}$$

$$\begin{cases} \overset{(n)}{\mathbf{T}} = J\int\limits_0^t {}^4\mathbf{R}(t'-\tau')\cdot\cdot\, d\overset{(n)}{\mathbf{C}}_\theta(\tau) \quad \text{in } V\cup\Sigma, \\ w^* = -J(a_\theta/2)\int\limits_0^t\int\limits_0^t d\overset{(n)}{\mathbf{C}}_\theta(\tau_1)\cdot\cdot\,\dfrac{\partial}{\partial t'}{}^4\mathbf{R}(2t'-\tau_1'-\tau_2')\cdot\cdot\, d\overset{(n)}{\mathbf{C}}_\theta(\tau_2), \\ \overset{(n)}{\mathbf{C}}_\theta = \overset{(n)}{\mathbf{C}} - \overset{\circ}{\boldsymbol{\varepsilon}}, \quad \overset{\circ}{\boldsymbol{\varepsilon}} = \int\limits_{\theta_0}^{\theta}\boldsymbol{\alpha}(\widetilde{\theta})d\widetilde{\theta}, \quad (t',\tau') = \int\limits_0^{(t,\tau)} a_\theta(\theta(\widetilde{\tau}))\,d\widetilde{\tau}, \end{cases} \tag{8.253}$$

$$\begin{cases} \mathbf{T} = {}^4\overset{(n)}{\mathbf{E}} \cdot\cdot \overset{(n)}{\mathbf{T}}, \\ \overset{(n)}{\mathbf{C}} = \frac{1}{n-\mathrm{III}} \sum_{\alpha=1}^{3} \lambda_\alpha^{n-\mathrm{III}} \overset{\circ}{\mathbf{p}}_\alpha \otimes \overset{\circ}{\mathbf{p}}_\alpha, \\ {}^4\overset{(n)}{\mathbf{E}} = \sum_{\alpha,\beta=1}^{3} E_{\alpha\beta} \mathbf{p}_\alpha \otimes \mathbf{p}_\beta \otimes \overset{\circ}{\mathbf{p}}_\beta \otimes \overset{\circ}{\mathbf{p}}_\alpha, \\ \lambda_\alpha, \overset{\circ}{\mathbf{p}}_\alpha, \mathbf{p}_\alpha \parallel \mathbf{F}, \quad \mathbf{F} = \mathbf{E} - \nabla \otimes \mathbf{u}^{\mathrm{T}} \ \text{in } V \cup \Sigma, \end{cases} \tag{8.254}$$

$$\begin{cases} \mathbf{n} \cdot \mathbf{T}\big|_{\Sigma_\sigma} = \mathbf{t}_{ne}, \quad \mathbf{u}\big|_{\Sigma_u} = \mathbf{u}_e, \quad \mathbf{n} \cdot \mathbf{q}\big|_{\Sigma_q} = q_e, \quad \theta\big|_{\Sigma_\theta} = \theta_e, \\ \mathbf{u} \cdot \mathbf{n} = 0, \quad \mathbf{n} \cdot \mathbf{T} \cdot \boldsymbol{\tau}_\alpha = 0 \ \text{at } \Sigma_8, \\ t = 0: \quad \theta = \theta_0. \end{cases} \tag{8.255}$$

Here (8.252) is the system of the equilibrium and heat conduction equations, (8.253) are the constitutive equations, (8.254) is the set of kinematic and energetic equivalence relations, (8.255) are the boundary and initial conditions, where $\Sigma_\sigma = \Sigma_1 \cup \Sigma_2 \cup \Sigma_3 \cup \Sigma_4 \cup \Sigma_7$, $\Sigma_u = \Sigma_5 \cup \Sigma_6$, $\Sigma_q = \Sigma_1 \cup \Sigma_2 \cup \Sigma_3 \cup \Sigma_4 \cup \Sigma_6$, $\Sigma_\theta = \Sigma_5 \cup \Sigma_7$.

On substituting Eqs. (8.253) and (8.254) into (8.252), we obtain the system of four scalar equations for the four scalar unknowns: components of the displacement vector and the temperature

$$\mathbf{u}, \ \theta \ \parallel \ \mathbf{x}, \ t. \tag{8.256}$$

If the model $A_n$ with exponential cores is considered, then constitutive equations (8.253) should be replaced by relations (8.242).

In particular, one can consider *isothermal processes* when a temperature field in a body $V$ remains unchanged: $\theta(\mathbf{x}, t) = \theta_0 = \text{const}$; then the heat conduction equation can be excluded from the system (8.252). As a result, we obtain the following *statement of the quasistatic problem of viscoelasticity in the spatial description for linear models $A_n$*:

$$\nabla \cdot \mathbf{T} + \rho \mathbf{f} = 0 \ \text{in } V; \qquad \overset{(n)}{\mathbf{T}} = J \int_0^t {}^4\mathbf{R}(t-\tau) \cdot\cdot d\overset{(n)}{\mathbf{C}}(\tau),$$

$$\begin{cases} \mathbf{T} = {}^4\overset{(n)}{\mathbf{E}} \cdot\cdot \overset{(n)}{\mathbf{T}}, \\ \overset{(n)}{\mathbf{C}} = (1/(n-\mathrm{III})) \sum_{\alpha=1}^{3} \lambda_\alpha^{n-\mathrm{III}} \overset{\circ}{\mathbf{p}}_\alpha \otimes \overset{\circ}{\mathbf{p}}_\alpha, \\ {}^4\overset{(n)}{\mathbf{E}} = \sum_{\alpha,\beta=1}^{3} E_{\alpha\beta} \mathbf{p}_\alpha \otimes \mathbf{p}_\beta \otimes \overset{\circ}{\mathbf{p}}_\beta \otimes \overset{\circ}{\mathbf{p}}_\alpha, \\ \lambda_\alpha, \overset{\circ}{\mathbf{p}}_\alpha, \mathbf{p}_\alpha \ \parallel \ \mathbf{F}^{-1}, \quad \mathbf{F}^{-1} = \mathbf{E} - \nabla \otimes \mathbf{u}^{\mathrm{T}}, \end{cases} \tag{8.257}$$

$$\begin{cases} \mathbf{n} \cdot \mathbf{T}\big|_{\Sigma_\sigma} = \mathbf{t}_{ne}, \quad \mathbf{u}\big|_{\Sigma_u} = \mathbf{u}_e, \\ \mathbf{u} \cdot \mathbf{n} = 0, \quad \mathbf{n} \cdot \mathbf{T} \cdot \boldsymbol{\tau}_\alpha = 0 \ \text{at } \Sigma_8, \end{cases}$$

to solve it for three components of the displacement vector $\mathbf{u}(\mathbf{x}, t)$.

If the linear models $A_n$ (8.205) and $B_n$ (8.218) of incompressible isotropic continua are considered, then the quasistatic statement of the viscoelasticity problem in the spatial description takes the form

$$\nabla \cdot \mathbf{T} + \overset{\circ}{\rho}\mathbf{f} = 0, \quad \det \mathbf{F}^{-1} = 1,$$

$$\mathbf{T} = -p\mathbf{E} + \left( m + \int_0^t r_1(t-\tau)\, dI_1(\overset{(\mathrm{n})}{\mathbf{C}}_G(\tau)) \right) \overset{(\mathrm{n})}{\mathbf{E}} + {}^4\overset{(\mathrm{n})}{\mathbf{E}} \cdot\cdot 2 \int_0^t r_2(t-\tau)\, d\overset{(\mathrm{n})}{\mathbf{C}}_G(\tau),$$

$$G = A,\ B,$$

$$\begin{cases} \overset{(\mathrm{n})}{\mathbf{C}}_G = (1/(n-\mathrm{III})) \sum_{\alpha=1}^{3} (\lambda_\alpha^{n-\mathrm{III}} - \bar{h}_G) \overset{\circ}{\mathbf{p}}_\alpha \otimes \overset{\circ}{\mathbf{p}}_\alpha, \\ {}^4\overset{(\mathrm{n})}{\mathbf{E}} = \sum_{\alpha,\beta=1}^{3} E_{\alpha\beta} \mathbf{p}_\alpha \otimes \mathbf{p}_\beta \otimes \overset{\circ}{\mathbf{p}}_\beta \otimes \overset{\circ}{\mathbf{p}}_\alpha, \\ \lambda_\alpha, \mathbf{p}_\alpha, \overset{\circ}{\mathbf{p}}_\alpha \ \parallel\ \mathbf{F}^{-1}, \quad \mathbf{F}^{-1} = \mathbf{E} - \nabla \otimes \mathbf{u}^{\mathrm{T}}, \end{cases}$$

$$\begin{cases} \mathbf{n} \cdot \mathbf{T}\big|_{\Sigma_\sigma} = \mathbf{t}_{ne}, \quad \mathbf{u}\big|_{\Sigma_u} = \mathbf{u}_e, \\ \mathbf{u}\big|_{\Sigma_8} \cdot \mathbf{n} = 0, \quad \mathbf{n} \cdot \mathbf{T}\big|_{\Sigma_8} \cdot \boldsymbol{\tau}_\alpha = 0. \end{cases} \tag{8.258}$$

Remind that, according to (6.4a): $\bar{h}_G = 1$ if $G = A$, and $\bar{h}_G = 0$ if $G = B$. Here we have denoted the tensor $\overset{(\mathrm{n})}{\mathbf{E}} = {}^4\overset{(\mathrm{n})}{\mathbf{E}} \cdot\cdot \mathbf{E}$.

Since the domain $V$ in the spatial description is unknown, so to determine it one should complement the system (8.252)–(8.255) (and also (8.257) and (8.258)) with Eq. (6.89).

### 8.4.4 Statements of Quasistatic Problems for Models of Viscoelastic Continua in the Material Description

In a similar way, we can obtain statements of quasistatic problems in the material description. So the statement of *the coupled quasistatic problem of thermoviscoelasticity in the material description* for linear mechanically determinate models $A_n$ of thermorheologically simple continua has the form

$$\overset{\circ}{\nabla} \cdot \mathbf{P} + \overset{\circ}{\rho}\mathbf{f} = 0, \quad \overset{\circ}{\rho} c_\varepsilon \frac{\partial \theta}{\partial t} = \overset{\circ}{\nabla} \cdot (\overset{\circ}{\boldsymbol{\lambda}} \cdot \overset{\circ}{\nabla}\theta) + \overset{\circ}{\rho} q_m + \overset{\circ}{w}{}^* \ \text{ in } V, \tag{8.259}$$

$$\begin{cases} \mathbf{P} = {}^4\overset{(n)}{\mathbf{E}}{}^0 \cdot\cdot \int\limits_0^t {}^4\mathbf{R}(t' - \tau') \cdot\cdot d\overset{(n)}{\mathbf{C}}_\theta(\tau) \ \text{ in } \overset{\circ}{V} \cup \overset{\circ}{\Sigma}, \\ \overset{\circ}{w}{}^* = -(a_\theta/2) \int\limits_0^t \int\limits_0^t d\overset{(n)}{\mathbf{C}}_\theta(\tau_1) \cdot\cdot \dfrac{\partial}{\partial t'} {}^4\mathbf{R}(2t' - \tau_1' - \tau_2') \cdot\cdot d\overset{(n)}{\mathbf{C}}_\theta(\tau_2), \\ \overset{(n)}{\mathbf{C}}_\theta = \overset{(n)}{\mathbf{C}} - \overset{\circ}{\boldsymbol{\varepsilon}}, \quad \overset{\circ}{\boldsymbol{\varepsilon}} = \int\limits_{\theta_0}^{\theta} \boldsymbol{\alpha}(\widetilde{\theta})\, d\widetilde{\theta}, \quad (t', \tau') = \int\limits_0^{(t,\tau)} a_\theta(\theta(\tau))\, d\tau, \end{cases} \tag{8.260}$$

$$\begin{cases} \overset{(n)}{\mathbf{C}} = \dfrac{1}{n - \text{III}} \sum_{\alpha=1}^3 \lambda_\alpha^{n-\text{III}} \overset{\circ}{\mathbf{p}}_\alpha \otimes \overset{\circ}{\mathbf{p}}_\alpha, \\ {}^4\overset{(n)}{\mathbf{E}}{}^0 = \sum_{\alpha,\beta=1}^3 \dfrac{E_{\alpha\beta}}{\lambda_\alpha} \overset{\circ}{\mathbf{p}}_\alpha \otimes \mathbf{p}_\beta \otimes \overset{\circ}{\mathbf{p}}_\beta \otimes \overset{\circ}{\mathbf{p}}_\alpha, \\ \lambda_\alpha, \mathbf{p}_\alpha, \overset{\circ}{\mathbf{p}}_\alpha \parallel \mathbf{F}, \quad \mathbf{F} = \mathbf{E} + \overset{\circ}{\nabla} \otimes \mathbf{u}^{\mathrm{T}} \ \text{ in } \overset{\circ}{V} \cup \overset{\circ}{\Sigma}, \end{cases} \tag{8.261}$$

$$\begin{cases} \overset{\circ}{\mathbf{n}} \cdot \mathbf{P}\big|_{\overset{\circ}{\Sigma}_\sigma} = \overset{\circ}{\mathbf{t}}_{ne}, \quad \mathbf{u}\big|_{\overset{\circ}{\Sigma}_u} = \mathbf{u}_e, \quad -\overset{\circ}{\mathbf{n}} \cdot \overset{\circ}{\boldsymbol{\lambda}} \cdot \overset{\circ}{\nabla}\theta\big|_{\overset{\circ}{\Sigma}_q} = \overset{\circ}{q}_e, \quad \theta\big|_{\overset{\circ}{\Sigma}_\theta} = \theta_e, \\ \mathbf{u}\big|_{\overset{\circ}{\Sigma}_8} \cdot \overset{\circ}{\mathbf{n}} = 0, \quad \overset{\circ}{\mathbf{n}} \cdot \mathbf{P}\big|_{\overset{\circ}{\Sigma}_8} \cdot \boldsymbol{\tau}_\alpha = 0, \\ t = 0: \quad \theta = \theta_0. \end{cases} \tag{8.262}$$

Substitution of constitutive equations (8.260) and kinematic relations (8.261) into (8.259) gives a system of four scalar equations of equilibrium and heat conduction with the boundary and initial conditions (8.262) for four scalar unknowns, namely three components of the displacement vector and temperature:

$$\mathbf{u}, \theta \ \parallel \ X^i, t. \tag{8.263}$$

Considering isothermal processes when $\theta(X^i, t) = \text{const}$, from (8.259)–(8.262) we obtain the *statement of the quasistatic problem of viscoelasticity in the material description for linear models $A_n$*:

$$\overset{\circ}{\nabla} \cdot \mathbf{P} + \overset{\circ}{\rho}\mathbf{f} = 0, \quad \mathbf{P} = {}^4\overset{(n)}{\mathbf{E}}{}^0 \cdot\cdot \int\limits_0^t {}^4\mathbf{R}(t - \tau) \cdot\cdot d\overset{(n)}{\mathbf{C}}(\tau),$$

$$\begin{cases} \overset{(n)}{\mathbf{C}} = \frac{1}{n-\text{III}} \sum_{\alpha=1}^3 \lambda_\alpha^{n-\text{III}} \overset{\circ}{\mathbf{p}}_\alpha \otimes \overset{\circ}{\mathbf{p}}_\alpha, \\ {}^4\overset{(n)}{\mathbf{E}}{}^0 = \sum_{\alpha,\beta=1}^3 (E_{\alpha\beta}/\lambda_\alpha) \overset{\circ}{\mathbf{p}}_\alpha \otimes \mathbf{p}_\beta \otimes \overset{\circ}{\mathbf{p}}_\beta \otimes \overset{\circ}{\mathbf{p}}_\alpha, \\ \lambda_\alpha, \overset{\circ}{\mathbf{p}}_\alpha, \mathbf{p}_\alpha \ \parallel \ \mathbf{F}, \quad \mathbf{F} = \mathbf{E} + \overset{\circ}{\nabla} \otimes \mathbf{u}^{\mathrm{T}}, \end{cases} \tag{8.264}$$

$$\begin{cases} \overset{\circ}{\mathbf{n}} \cdot \mathbf{P}\big|_{\overset{\circ}{\Sigma}_\sigma} = \overset{\circ}{\mathbf{t}}_{ne}, \quad \mathbf{u}\big|_{\overset{\circ}{\Sigma}_u} = \mathbf{u}_e, \\ \mathbf{u}\big|_{\overset{\circ}{\Sigma}_8} \cdot \overset{\circ}{\mathbf{n}} = 0, \quad \overset{\circ}{\mathbf{n}} \cdot \mathbf{P}\big|_{\overset{\circ}{\Sigma}_8} \cdot \boldsymbol{\tau}_\alpha = 0, \end{cases}$$

to solve it for three components of the displacement vector $\mathbf{u}(X^i,\ t)$.

If the models $B_n$ of incompressible isotropic continua (8.218) are considered, then the constitutive equations in (8.264) should be replaced by the relations

$$\mathbf{P} = -p\mathbf{F}^{-1} + \left( m + \int_0^t r_1(t-\tau)\, dI_1(\overset{(n)}{\mathbf{G}}(\tau)) \right) \overset{(n)}{\mathbf{E}}{}^0 + 2\,{}^4\overset{(n)}{\mathbf{E}}{}^0 \cdot\cdot \int_0^t r_2(t-\tau)\, d\overset{(n)}{\mathbf{G}}(\tau),$$

$$\det \mathbf{F} = 1, \tag{8.265}$$

and the problem (8.264), (8.265) is solved for the four unknown scalar functions: $\mathbf{u},\ p\ \|\ X^i,\ t$.

## 8.5 The Problem on Uniaxial Deforming of a Viscoelastic Beam

### *8.5.1 Deformation of a Viscoelastic Beam in Uniaxial Tension*

As an example, consider the classical problem on a beam in uniaxial tension; this problem was studied in detail before (see Sect. 6.5). A beam is assumed to be viscoelastic, isotropic and incompressible, and its constitutive equations correspond to the simplest linear models $A_n$ or $B_n$ with exponential cores (see (8.212) and (8.224) and Exercise 8.3.2). A general statement of the quasistatic problem in the spatial description has the form (8.258).

The motion law for the beam in tension is independent of peculiarities of its mechanical properties; the law is the same for both elastic and viscoelastic continua and defined by formula (6.131). Hence all the kinematic characteristics, namely the tensors $\mathbf{F}$, $\overset{(n)}{\mathbf{C}}$ and $\overset{(n)}{\mathbf{G}}$, are the same as ones for elastic continua; they are determined according to the results of Exercises 2.2.1, 2.3.2, and 4.2.13:

$$\mathbf{F} = \sum_{\alpha=1}^{4} k_\alpha(t)\bar{\mathbf{e}}_\alpha^2, \quad \overset{\circ}{\mathbf{p}}_\alpha = \mathbf{p}_\alpha = \bar{\mathbf{e}}_\alpha, \quad \lambda_\alpha = k_\alpha, \tag{8.266}$$

$$\overset{(n)}{\mathbf{C}} = \frac{1}{n - \mathrm{III}} \sum_{\alpha=1}^{3} (k_\alpha^{n-\mathrm{III}} - 1)\bar{\mathbf{e}}_\alpha^2, \quad \operatorname{dev} \overset{(n)}{\mathbf{C}} = \operatorname{dev} \overset{(n)}{\mathbf{G}} = \sum_{\alpha=1}^{3} \overset{(n)}{f}_\alpha(k_1)\bar{\mathbf{e}}_\alpha^2,$$

$$\overset{(n)}{\mathbf{G}} = \frac{1}{n - \mathrm{III}} \sum_{\alpha=1}^{3} k_\alpha^{n-\mathrm{III}}\bar{\mathbf{e}}_\alpha^2, \quad \overset{(n)}{\mathbf{G}}{}^{-1} = (n - \mathrm{III}) \sum_{\alpha=1}^{3} k_\alpha^{\mathrm{III}-n}\bar{\mathbf{e}}_\alpha^2, \tag{8.267}$$

where we have denoted the functions of $k_1$

$$\overset{(n)}{f}_1(k_1) = \frac{2}{3(n - \mathrm{III})}\left(k_1^{n-\mathrm{III}} - k_1^{(\mathrm{III}-n)/2}\right), \quad \overset{(n)}{f}_2 = \overset{(n)}{f}_3 = -\frac{1}{2}\overset{(n)}{f}_1, \tag{8.268}$$

and $k_\alpha(t)$ – the elongation ratios of the beam along corresponding coordinate directions. Here, due to the incompressibility condition, $k_2 = k_3 = 1/\sqrt{k_1}$.

### 8.5.2 Viscous Stresses in Uniaxial Tension

On substituting the expression (8.267) into the differential equation of the constitutive relations for $\mathbf{W}^{(\gamma)}$ (see Exercise 8.3.2), we find that the tensors $\mathbf{W}^{(\gamma)}$ have a diagonal form too:

$$\mathbf{W}^{(\gamma)} = \sum_{\alpha=1}^{3} W_\alpha^{(\gamma)} \bar{\mathbf{e}}_\alpha^2. \tag{8.269}$$

Functions $W_\alpha^{(\gamma)}$ are the same for models $A_n$ and $B_n$; they satisfy the differential equations

$$\frac{d\, W_\alpha^{(\gamma)}}{dt} = \frac{1}{\tau^{(\gamma)}}\left(\overset{(n)}{f}_\alpha - W_\alpha^{(\gamma)}\right), \tag{8.269a}$$

which have the solution

$$W_\alpha^{(\gamma)} = \int_0^t \exp\left(-\frac{t-\tau}{\tau^{(\gamma)}}\right)\frac{\overset{(n)}{f}_\alpha(k_1(\tau))d\tau}{\tau^{(\gamma)}}, \quad \gamma = 1, \ldots, N. \tag{8.269b}$$

### 8.5.3 Stresses in a Viscoelastic Beam in Tension

Substituting the expressions (8.267) and (8.269) into the constitutive equations from Exercise 8.3.2, we find that the energetic stress tensors $\overset{(n)}{\mathbf{T}}$ have the diagonal form

$$\overset{(n)}{\mathbf{T}} = \sum_{\alpha=1}^{3} \overset{(n)}{T}_{\alpha\alpha} \bar{\mathbf{e}}_\alpha^2, \tag{8.270}$$

$$\overset{(n)}{T}_{\alpha\alpha} = -pk_\alpha^{\mathrm{III}-n} + \bar{m} + \frac{l_1}{n-\mathrm{III}}(k_1^{n-\mathrm{III}} + 2k_1^{(\mathrm{III}-n)/2} - 3)$$
$$+ \frac{2l_2}{n-\mathrm{III}}(k_\alpha^{n-\mathrm{III}} - 1) - \sum_{\gamma=1}^{N} W_\alpha^{(\gamma)} B^{(\gamma)} \tag{8.271}$$

– for models $A_n$ and

$$\overset{(n)}{T}_{\alpha\alpha} = -pk_\alpha^{\mathrm{III}-n} + \mu(n-\mathrm{III})(1+\beta+(1-\beta)(k_1^{n-\mathrm{III}} + 2k_1^{(\mathrm{III}-n)/2} - k_\alpha^{n-\mathrm{III}})$$
$$- \sum_{\gamma=1}^{N} W_\alpha^{(\gamma)} B^{(\gamma)} \tag{8.272}$$

– for models $B_n$. As shown in Sect. 6.4.3, the tensors $\mathbf{T}$ and $\overset{(n)}{\mathbf{T}}$ in this problem are connected by the relations

$$\overset{(n)}{\mathbf{T}} = \sum_{\alpha=1}^{3} k_\alpha^{\mathrm{III}-n} \sigma_\alpha \bar{\mathbf{e}}_\alpha^2, \quad \mathbf{T} = \sum_{\alpha=1}^{3} \sigma_\alpha \bar{\mathbf{e}}_\alpha^2, \tag{8.273}$$

$$\sigma_\alpha = k_\alpha^{n-\mathrm{III}} \overset{(n)}{T}_{\alpha\alpha}. \tag{8.273a}$$

As follows from Sect. 6.4.3, in uniaxial tension of a beam $\sigma_1 \neq 0$, and $\sigma_2 = \sigma_3 = 0$. Then, substituting (8.270)–(8.272) into (8.273a), we obtain the system of two equations

$$\begin{cases} \sigma_1 = -p + \big(\bar{m} + \frac{l_1}{n-\mathrm{III}}(k_1^{n-\mathrm{III}} + 2k_1^{(\mathrm{III}-n)/2} - 3) \\ \quad + \frac{2l_2}{n-\mathrm{III}}(k_1^{n-\mathrm{III}} - 1) - \sum_{\gamma=1}^{N} W_1^{(\gamma)} B^{(\gamma)}\big) k_1^{n-\mathrm{III}}, \\ 0 = -p + \big(\bar{m} + \frac{l_1}{n-\mathrm{III}}(k_1^{n-\mathrm{III}} + 2k_1^{(\mathrm{III}-n)/2} - 3) \\ \quad + \frac{2l_2}{n-\mathrm{III}}(k_1^{(\mathrm{III}-n)/2} - 1) - \sum_{\gamma=1}^{N} W_2^{(\gamma)} B^{(\gamma)}\big) k_1^{(\mathrm{III}-n)/2} \end{cases} \tag{8.274a}$$

– for models $A_n$, and

$$\begin{cases} \sigma_1 = -p + \big(\mu(n-\mathrm{III})(1+\beta+2(1-\beta)k_1^{n-\mathrm{III}}) \\ \qquad - \sum_{\gamma=1}^{N} W_1^{(\gamma)} B^{(\gamma)}\big) k_1^{n-\mathrm{III}}, \\ 0 = -p + \big(\mu(n-\mathrm{III})(1+\beta+(1-\beta)(k_1^{n-\mathrm{III}} + k_1^{(\mathrm{III}-n)/2})) \\ \qquad - \sum_{\gamma=1}^{N} W_2^{(\gamma)} B^{(\gamma)}\big) k_1^{(\mathrm{III}-n)/2} \end{cases} \tag{8.274b}$$

– for models $B_n$.

### 8.5.4 Resolving Relation $\sigma_1(k_1)$ for a Viscoelastic Beam

Eliminating $p$ among these systems, we obtain the relations between $\sigma_1$ and $k_1$:

$$\sigma_1 = \bar{m}\overset{(n)}{Q}(k_1) + l_1 \overset{(n)}{M}(k_1) + l_2 \overset{(n)}{N}(k_1) - \overset{(n)}{L}(k_1) \sum_{\gamma=1}^{N} W_1^{(\gamma)} B^{(\gamma)} \tag{8.275a}$$

– for models $A_n$,

$$\sigma_1 = \mu(1+\beta)\overset{(n)}{Z}(k_1) + \mu(1-\beta)\overset{(n)}{H}(k_1) - \overset{(n)}{L}(k_1)\sum_{\gamma=1}^{N} W_1^{(\gamma)} B^{(\gamma)} \qquad (8.275b)$$

– for models $B_n$. Here we have denoted the following functions of $k_1$:

$$\overset{(n)}{Q} = k_1^{n-\text{III}} - k_1^{(\text{III}-n)/2},$$

$$\overset{(n)}{M} = \frac{1}{n-\text{III}}(k_1^{n-\text{III}} + 2k_1^{(\text{III}-n)/2} - 3)(k_1^{n-\text{III}} - k_1^{(\text{III}-n)/2}),$$

$$\overset{(n)}{N} = \frac{2}{n-\text{III}}((k_1^{n-\text{III}} - 1)k_1^{n-\text{III}} - (k_1^{(\text{III}-n)/2} - 1)k_1^{(\text{III}-n)/2}),$$

$$\overset{(n)}{L} = k_1^{n-\text{III}} + (1/2)k_1^{(\text{III}-n)/2}, \quad \overset{(n)}{Z} = (n-\text{III})(k_1^{n-\text{III}} - k_1^{(\text{III}-n)/2}),$$

$$\overset{(n)}{H} = (n-\text{III})(k_1^{(n-\text{III})/2} - k_1^{\text{III}-n}). \qquad (8.276)$$

Here we have taken into account that $W_2^{(\gamma)} = -(1/2)\, W_1^{(\gamma)}$ according to (8.268) and (8.272).

When $W_1^{(\gamma)} \equiv 0$, the elasticity relations (6.166) follow from (8.275b).

### 8.5.5 *Method of Calculating the Constants $B^{(\gamma)}$ and $\tau^{(\gamma)}$*

Consider a beam under stepwise deforming, when the elongation ratio $k_1(t)$ is given in the form

$$k_1(t) = k_1^0\, h(t). \qquad (8.277)$$

Substitution of (8.277) into (8.270b) yields

$$W_1^{(\gamma)}(t) = \overset{(n)}{f}_1(k_1^0)(1 - \exp(-t/\tau^{(\gamma)})). \qquad (8.278)$$

Taking (8.278) into account, from (8.275) we find the expression for the *relaxation stresses*

$$\Delta\sigma(t) = \sigma_1(0) - \sigma_1(t) = \overset{(n)}{L}(k_1^0)\overset{(n)}{f}_1(k_1^0)(r(0) - r(t)) \qquad (8.279)$$

for all the models $A_n$ and $B_n$, where $r(t)$ is the relaxation function, which, according to (8.169), has the exponential form

$$r(t) = r^{\infty} + \sum_{\gamma=1}^{N} B^{(\gamma)} \exp(-t/\tau^{(\gamma)}), \qquad r(0) = r^{\infty} + \sum_{\gamma=1}^{N} B^{(\gamma)} = 2l_2. \quad (8.280)$$

Equation (8.279) rewritten in the form

$$\frac{\Delta\sigma(t)}{\overset{(n)}{L}(k_1^0)f_1(k_1^0)} = \sum_{\gamma=1}^{N} B^{(\gamma)}(1-\exp(-t/\tau^{(\gamma)})), \tag{8.281}$$

can be used for determining the constants $B^{(\gamma)}$ and $\tau^{(\gamma)}$. If the experimental relaxation curve $\Delta\sigma^{(ex)}(t)$ obtained at some fixed value of $k_1^0$ is known, then with the help of relation (8.281) this curve can be approximated by choosing the parameters $B^{(\gamma)}$ and $\tau^{(\gamma)}$ satisfied the condition of the minimum distance between the functions $\Delta\sigma(t)$ and $\Delta\sigma^{(ex)}(t)$ at $K$ points $t_i$ $(i = 1, \ldots, K)$:

$$\Delta^2 = \sum_{i=1}^{K}\left(1 - \frac{\Delta\sigma(t_i)}{\Delta\sigma^{(ex)}(t_i)}\right)^2 \to \min. \tag{8.282}$$

The functions $\overset{(n)}{L}(k_1^0)$ and $\overset{(n)}{f}_1(k_1^0)$, according to (8.268) and (8.276), include no material constants; therefore, at given $k_1^0$ their values are known. Sometimes to improve the convergence of the iterative procedure for minimizing the functional (8.282), the parameters $\tau^{(\gamma)}$ should be given a priori, for example, in the form $\tau^{(\gamma)} = t_{(\gamma)}$, where $t_{(\gamma)}$ are some instants of time. Substituting experimental values of the relaxation stresses $\Delta\sigma^{(ex)}(t_\gamma)$ at the times $t_\gamma$ into Eq. (8.281), we obtain a system of linear algebraic equations for determining the constants $B^{(\gamma)}$. This system can be solved numerically, for example, by Holetskii's method.

We seek further values of the parameters $t_\gamma$, for which the functional $\Delta$ (8.282) reaches its minimum value. Values $B^{(\gamma)}$ at such $t_\gamma$ are unknown.

Table 8.1 shows values of the parameters $\tau^{(\gamma)}$ and $B^{(\gamma)}$ obtained by the method mentioned above for rubber and polyurethane elastomer. The values of $\tau^{(\gamma)}$ and

**Table 8.1** Values of the constants $B^{(\gamma)}$ and $\tau^{(\gamma)}$ for rubber and polyurethane

| Rubber | | | | | Polyurethane | | | | |
|---|---|---|---|---|---|---|---|---|---|
| $B_n$ | I | II | IV | V | $B_n$ | I | II | IV | V |
| | | | | | | 5.083 | 4.819 | 2.047 | 0.993 |
| $B^{(\gamma)}$, | 2.765 | 2.598 | 2.253 | 2.078 | $B^{(\gamma)}$, | 3.568 | 3.384 | 1.827 | 0.886 |
| MPa | 16.475 | 15.486 | 15.424 | 12.385 | MPa | 2.437 | 2.311 | 0.816 | 0.396 |
| | 4.920 | 4.626 | 4.009 | 3.698 | | 0.312 | 0.296 | 0.140 | 0.068 |
| | | | | | | 3.068 | 2.909 | 0.939 | 0.456 |
| | | | | | | 0.2 | 0.2 | 1.2 | 1.2 |
| | 0.2 | 0.2 | 0.2 | 0.2 | | 5.4 | 5.4 | 24 | 24 |
| $\tau^{(\gamma)}$, s | 5.0 | 5.0 | 5.0 | 5.0 | $\tau^{(\gamma)}$, s | 66 | 66 | 300 | 300 |
| | 30 | 30 | 30 | 30 | | 300 | 300 | 7500 | 7500 |
| | | | | | | 7500 | 7500 | 14700 | 14700 |
| $\Delta$,% | 1 | 1 | 1 | 1 | $\Delta$,% | 0.3 | 0.3 | 0.7 | 0.7 |

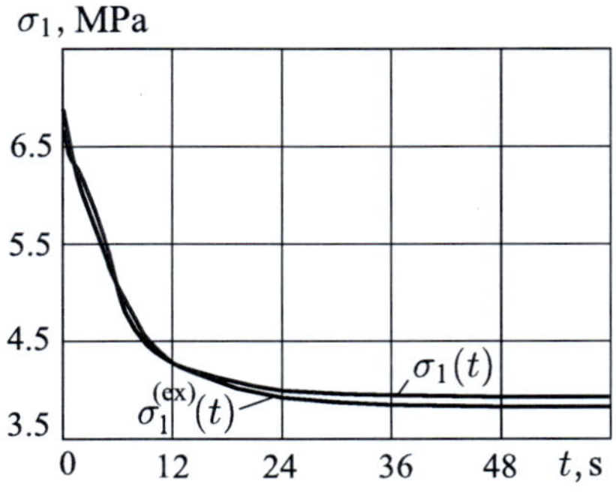


**Fig. 8.4** Graph of stress-relaxation for rubber at deformation 10% and its approximation with the help of the exponential cores (8.281)

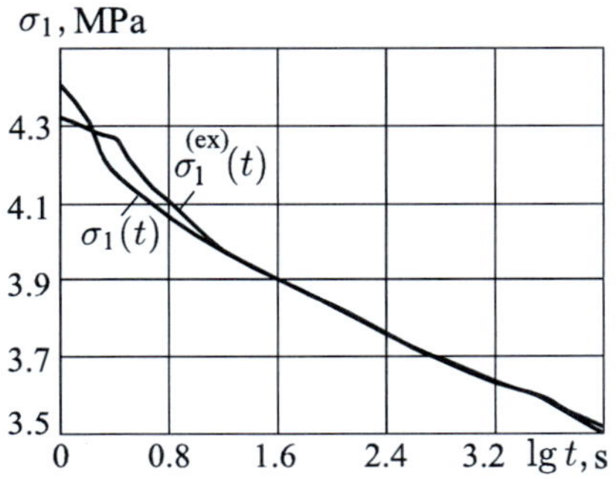


**Fig. 8.5** Graph of stress-relaxation for polyurethane at deformation 8.3% and its approximation with the help of the exponential cores (8.281)

$B^{(\gamma)}$ are the same for corresponding models $A_n$ and $B_n$ but are different for the models with different subscripts $n$.

For polyurethane, in approximating by the models $B_{\rm IV}$ and $B_{\rm V}$ we used the relaxation curve at higher values of the extension deformation $\delta_1 = 80\%$ than for the models $B_{\rm I}$ and $B_{\rm II}$ ($\delta_1 = 8.3\%$), that improved the quality of further simulation of the viscoelastic properties by these models.

Figures 8.4 and 8.5 show graphs of relaxation curves $\sigma_1(t) = \sigma_1(0) - \Delta\sigma(t)$, where the value of $\Delta\sigma(t)$ has been determined by (8.281) with the optimum constants $B^{(\gamma)}$ and $\tau^{(\gamma)}$ for rubber and polyurethane, and also their corresponding experimental relaxation curves $\sigma_1^{(\rm ex)}(t) = \sigma_1(0) - \Delta\sigma^{(\rm ex)}(t)$. The accuracy of approximation to the experimental curves with the help of the exponential cores (8.281) is sufficiently high: the mean-square deviation does not exceed 1%. Notice that the longer is a time interval of relaxation considered, the greater number $N$ of exponents in the relaxation spectrum we need to reach a high accuracy of the approximation.

For rubber, the relaxation curve was approximated within the interval from 0 to 1 min. To do this, we used three exponents; and for approximation of the relaxation curve for polyurethane within the interval up to one hour we used five exponents.

## 8.5.6 Method for Evaluating the Constants $\bar{m}$, $l_1$, $l_2$ and $\beta$, $m$

After calculation of the material constants $B^{(\gamma)}$, $\tau^{(\gamma)}$, we can determine values of the constants $\bar{m}$, $l_1$, $l_2$ or $\beta$, $m$ by Eq. (8.275) and by experimental diagrams of deforming $\sigma_1 = \sigma_1^{(\mathrm{ex})}(k_1)$ obtained, for example, in the case of deforming the beam with a constant rate $b$:

$$k_1(t) = 1 + bt, \qquad b = \text{const.} \tag{8.283}$$

The function $W_1(t)$ in these computations is known and determined by (8.269a):

$$W_1 = \sum_{\gamma=1}^{N} W_1^{(\gamma)} B^{(\gamma)}, \qquad \frac{\partial W_1^{(\gamma)}}{\partial t} + \frac{W_1^{(\gamma)}}{\tau^{(\gamma)}} = \frac{\overset{(n)}{f}_1(k_1(t))}{\tau^{(\gamma)}}. \tag{8.284}$$

The constants $l_1$, $l_2$, $\bar{m}$ or $m$, $\beta$ have been found from the condition of the best approximation to the function $\sigma_1^{(\mathrm{ex})}(k_1)$ by $\sigma_1(k_1)$, determined according to (8.275), by minimizing the functional of a mean-square error:

$$\Delta^2 = \sum_{i=1}^{k} \left(1 - \frac{\sigma_1(k_{1(i)})}{\sigma_1^{(\mathrm{ex})}(k_{1(i)})}\right)^2 \to \min.$$

For models $B_n$, the space of the optimization parameters $\mu$ and $\beta$ is two-dimensional; and for models $A_n$, the space of the parameters $l_1$, $l_2$ and $\bar{m}$ is three-dimensional. To accelerate the process of solving the optimization problem, we have used the method of gradient descent with fitting an initial point of the minimizing procedure. Values of the constants $\mu$ and $\beta$ obtained for rubber and polyurethane are given in Table 8.2, and $l_1$, $l_2$ and $\bar{m}$ – in Table 8.3.

Figures 8.6 and 8.7 show experimental diagrams of deforming $\sigma_1^{(\mathrm{ex})}(\delta_1)$ for rubber and polyurethane in tension, and also approximations to the diagrams with the help of models $A_n$ and $B_n$ of incompressible viscoelastic continua. For the materials considered, the models $A_{\mathrm{I}}$ and $B_{\mathrm{I}}$ exhibit the best approximation to the experimental data. For rubber, the model $A_{\mathrm{I}}$ approximates accurately enough the diagram $\sigma_1^{(\mathrm{ex})}(\delta_1)$ within the whole interval of deforming, including the domain of maximum deformations (higher than 100%), while the other models give a considerable error in this domain.

**Table 8.2** Values of the constants $\mu$ and $\beta$ for rubber and polyurethane in models $B_n$ of viscoelastic continua

| Rubber | | | | | Polyurethane | | | | |
|---|---|---|---|---|---|---|---|---|---|
| $n$ | I | II | IV | V | $n$ | I | II | IV | V |
| $\mu$, MPa | 5.145 | 21.31 | 23.52 | 5.88 | $\mu$, MPa | 3.68 | 13.662 | 12.61 | 2.627 |
| $\beta$ | −0.556 | −1 | 1 | 1 | $\beta$ | 0.515 | −0.46 | 1 | 0.778 |
| $\Delta$,% | 9 | 10 | 19.4 | 23.8 | $\Delta$,% | 11.8 | 14.3 | 16.8 | 23.2 |

**Table 8.3** Values of the constants $l_1$, $l_2$ and $\bar{m}$ in models $A_n$ of viscoelastic continua for rubber and polyurethane

| Rubber | | | | | Polyurethane | | | | |
|---|---|---|---|---|---|---|---|---|---|
| $n$ | I | II | IV | V | $n$ | I | II | IV | V |
| $l_1$, MPa | 1.2 | 39.8 | 0.2 | 2.8 | $l_1$, MPa | −12.2 | 0.2 | 0.2 | 0.2 |
| $l_2$, MPa | 40 | 37.2 | 11.2 | 1.8 | $l_2$, MPa | 17.8 | 29.2 | 1.6 | 0.2 |
| $\bar{m}$, MPa | 11 | 20 | 19.8 | 19 | $\bar{m}$, MPa | −2.6 | 19.6 | 20 | 8.2 |
| $\Delta$,% | 6.1 | 10 | 14.1 | 14.9 | $\Delta$,% | 13.6 | 13.5 | 25 | 35.8 |

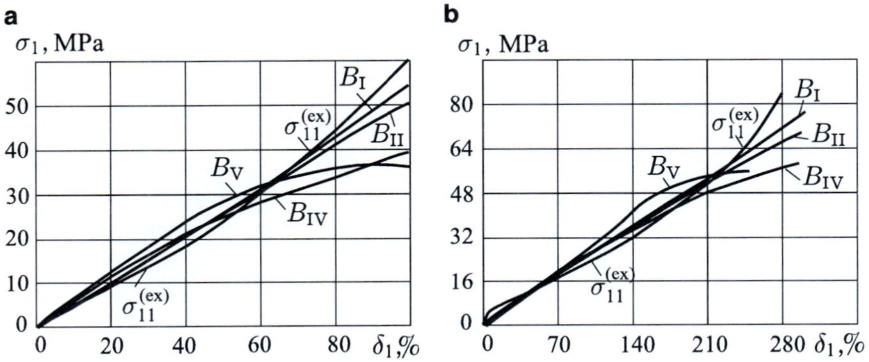


**Fig. 8.6** Approximation of the diagrams of deforming for rubber (**a**) and polyurethane (**b**) in tension by linear models $B_n$ of viscoelastic continua

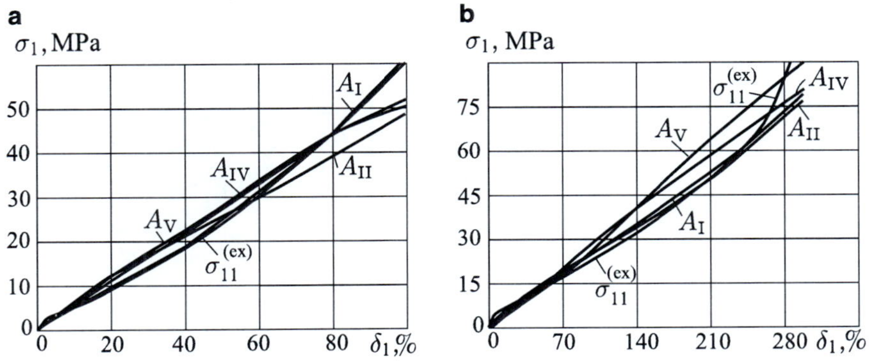


**Fig. 8.7** Approximation of the diagrams of deforming for rubber (**a**) and polyurethane (**b**) in tension by linear models $A_n$ of incompressible viscoelastic continua

## *8.5.7 Computations of Relaxation Curves*

After evaluation of all the constants $B^{(\gamma)}$, $\tau^{(\gamma)}$ and $l_1$, $l_2$, $\bar{m}$ or $m$, $\beta$, we can verify the models $A_n$ and $B_n$, for example, by performing computations of relaxation curves at different values of the parameter $k_1^0$ with use of Eq. (8.275) for the stepwise process (8.277):

$$\sigma_1(t) = \bar{m}\overset{(n)}{Q}(k_1^0) + l_1\overset{(n)}{M}(k_1^0) + l_2\overset{(n)}{N}(k_1^0)$$
$$-\overset{(n)}{L}(k_1^0)\overset{(n)}{f}_1(k_1^0)\sum_{\gamma=1}^{N} B^{(\gamma)}(1 - \exp(-t/\tau^{(\gamma)})) \qquad (8.285a)$$

– for models $A_n$, or

$$\sigma_1(t) = m(1+\beta)\overset{(n)}{Z}(k_1) + m(1-\beta)\overset{(n)}{H}(k_1)$$
$$-\overset{(n)}{L}(k_1^0)\overset{(n)}{f}_1(k_1^0)\sum_{\gamma=1}^{N} B^{(\gamma)}(1 - \exp(-t/\tau^{(\gamma)})) \qquad (8.285b)$$

– for models $B_n$.

Figures 8.8 and 8.9 show the relaxation curves $\sigma_1(t)$ obtained in experiments and in computations by the method mentioned above for polyurethane elastomer.

To verify the models $A_n$ and $B_n$ we use the relaxation curves when $k_1^0 = 1.23$ ($\delta_1 = 23\%$), $k_1^0 = 1.49$ ($\delta_1 = 49\%$) and $k_1^0 = 1.8$ ($\delta_1 = 80\%$). The computations performed show that for the considered material the models $A_{\text{I}}$ and $A_{\text{II}}$ are of the best accuracy, while the model $A_{\text{V}}$ gives the worst result of description of the relaxation curves.

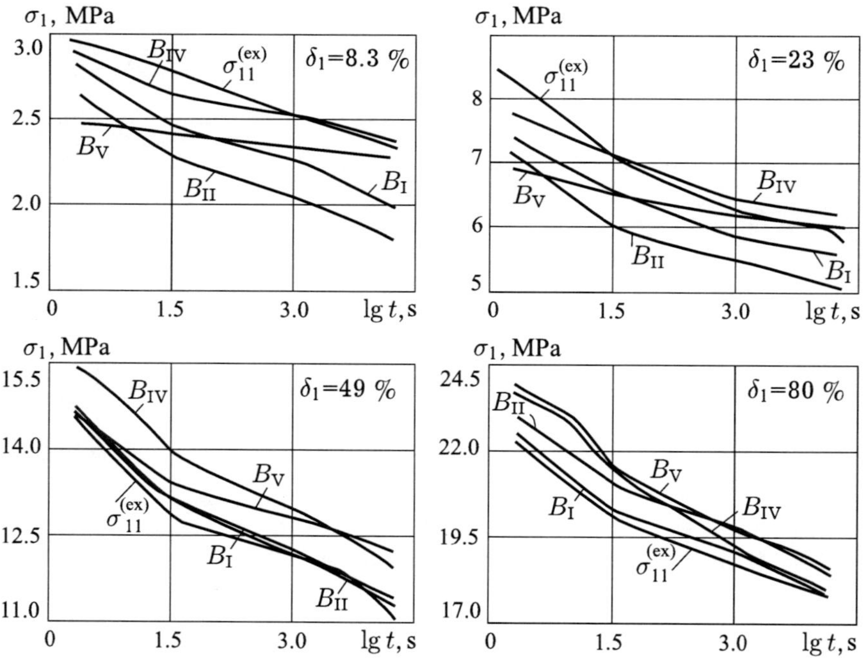


**Fig. 8.8** Computed and experimental relaxation curves for polyurethane at different values of the extension deformation; computations for different models $B_n$ of viscoelastic continua

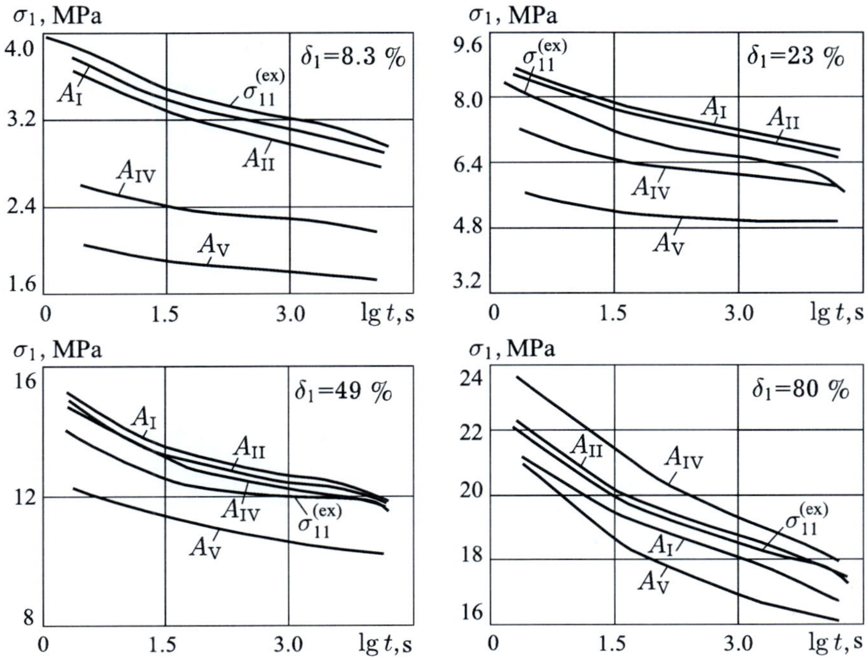


**Fig. 8.9** Computed and experimental relaxation curves for polyurethane at different values of the extension deformation; computations for different models $A_n$ of viscoelastic continua

### 8.5.8 Cyclic Deforming of a Beam

Assume that all material constants appearing in models $A_n$ or $B_n$ of viscoelastic continua are known, for example, they are determined by the method mentioned above. Consider a cyclic regime of deforming, which is widely used in practice. In this case, the function $k(t)$ first grows with a constant rate $b$ and then diminishes with the same rate, then grows again etc. (the saw-tooth regime, Fig. 8.10). Such a function $k(t)$ can be written in an analytical form with the help of the Heaviside functions:

$$k_1(t) = 1 + b \sum_{m=0}^{M} a_m (t - m t_0) h(t - m t_0). \tag{8.286}$$

Here $b$ is the rate of deforming; $h(t)$ is the Heaviside function; $a_0 = 1$, $a_m = 2(-1)^m$ and $m \geqslant 1$ are the constants; and $t_0$ is the time interval of monotonicity of the function (the semiperiod of the cycle).

Substituting (8.286) into Eq. (8.284) and then solving this equation numerically, for example, with the help of the implicit difference approximation

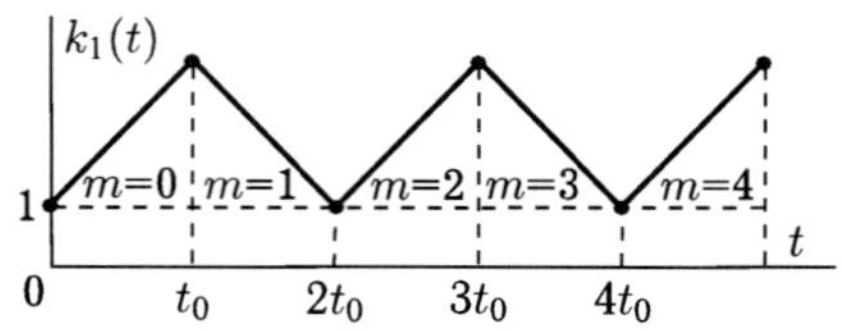


**Fig. 8.10** The cyclic process of deforming with a constant rate

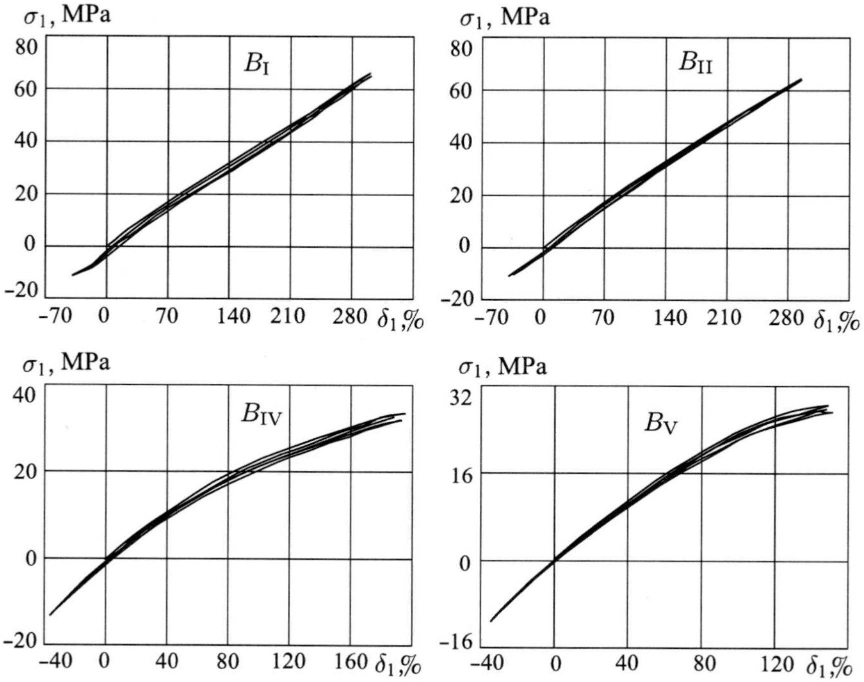


**Fig. 8.11** Cyclic diagrams of deforming of polyurethane for different models $B_n$ of viscoelastic continua

$$W^{(\gamma)}_{1,j+1} - W^{(\gamma)}_{1,j} + \frac{\Delta t}{\tau^{(\gamma)}} W^{(\gamma)}_{1,j+1} = \overset{(n)}{f}_1(k_1(t_j)), \quad j = 0, \ldots, N, \tag{8.287}$$

we can find values of the functions $W^{(\gamma)}_1(t)$ for the cyclic process of deforming. Here we have introduced the notation: $t_j$ – instants of time (nodes), $W^{(\gamma)}_{1,j} = W^{(\gamma)}_1(t_j)$ – values of functions in the nodes, $\Delta t$ – the step in time.

The stresses $\sigma_1$ in cyclic deforming are determined by the general formula (8.275a) for models $A_n$, or by (8.275b) – for models $B_n$.

If all material constants of the models are known, then, according to these formulae, we can calculate the dependences of cyclically varying stresses upon time $\sigma_1(t)$, and also plot the cyclic diagrams of deforming $(\sigma_1(t), \ \delta_1(t))$, where $\delta_1(t) = k_1(t) - 1$ is the relative elongation.

Figure 8.11 shows cyclic diagrams of deforming for polyurethane, which have been obtained by the method mentioned above for models $B_n$ of viscoelastic continua. Unlike ideal elastic media, for viscoelastic continua the cyclic diagrams of deforming do not coincide at the stages of loading ($m = 0, 2, 4, \ldots$) and unloading ($m = 1, 3, 5, \ldots$) but exhibit characteristic loops.

## 8.6 Dissipative Heating of a Viscoelastic Continuum Under Cyclic Deforming

### *8.6.1 The Problem on Dissipative Heating of a Beam Under Cyclic Deforming*

Let us remind now that viscoelastic continua are not ideal, and for them the dissipation function $w^*$ is different from identically zero. Therefore, in general, at any regime of deforming of viscoelastic continua, even under constant loads, there occurs an internal heat release in these materials, i.e. dissipative heating due to the presence of the dissipation function $w^*$ in the heat conduction equation. However, at such small-intensive regimes, dissipative heating is rather small (the temperature changes by some fractions of the Celsius degree due to $w^*$), and it can usually be neglected (although for some special problems these values can prove to be considerable). For cyclic regimes of loading when the number of cycles $M$ is rather great, the situation is different. The temperature of dissipative heating at such regimes can reach 100°C and higher, and even prove to be the cause of heat destruction of the material.

Consider a computational method for dissipative heating of a beam in longitudinal quasistatic cyclic tension by an arbitrary periodic law

$$k_1(t) = k_1(t + t_0). \tag{8.288}$$

Here $k_1(t)$ is the elongation ratio (see Sects. 6.5 and 8.5.1), $t_0$, as before, is the cycle period (Fig. 8.12), and the number of cycles is assumed to be large: $M \gg 1$. In this case one can say that the multicycle regime of deforming is considered.

Then the functions $W_1^{(\gamma)}(t)$ can be evaluated with the help of formula (8.284) and difference approximation (8.287), and the stresses $\sigma_1(t)$ – with the help of formula (8.275).

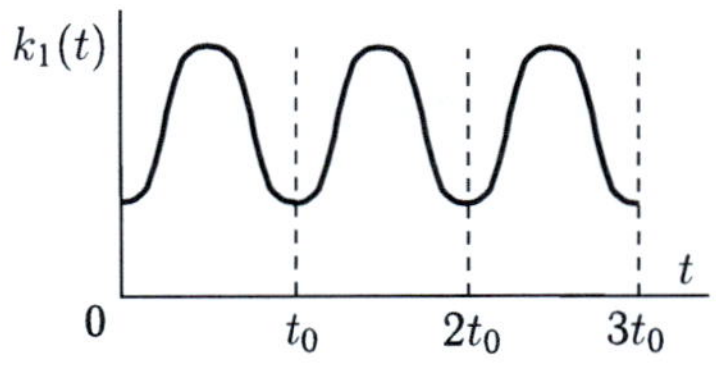


**Fig. 8.12** The cyclic regime of deforming

### 8.6.2 Fast and Slow Times in Multicycle Deforming

For the multicycle regime of deforming, we introduce the small parameter $\varkappa = 1/M \ll 1$, the new dimensionless variable $\xi = t/t_0$ having the sense of a counter of the number of oscillation cycles and called the *fast time*, and also the *slow time* $\bar{t} = t/t_1$ (where $t_1 = M t_0$), which takes on values $0 \leqslant \bar{t} \leqslant 1$ within the whole interval of cyclic deforming.

The periodic functions (8.288) can be considered as functions of the fast time with the period equal to 1:

$$k_1(t) = k_1(\xi) = k_1(\xi + 1). \tag{8.289}$$

For example, for harmonic oscillations we have

$$k_1(t) = \bar{k}_1 + k_1^0 \sin \omega t = \bar{k}_1 + k_1^0 \sin 2\pi\xi \equiv k_1(\xi), \tag{8.290}$$

because the oscillation period in this case is equal to $t_0 = 2\pi/\omega$. Here $\bar{k}_1$ is the mean value of the function $k_1$, and $k_1^0$ is the oscillation amplitude (both the values are constants).

The relaxation core $q(t)$ is considered, on the contrary, as a function of the slow time, because we assume that during one period of oscillations its changes are negligible:

$$q(t) = q(\bar{t}), \tag{8.291}$$

for example, for the exponential core of relaxation (8.280)

$$q(t) = -\frac{dr(t)}{dt} = \frac{1}{t_1} \sum_{\gamma=1}^{N} \frac{B^{(\gamma)}}{\bar{\tau}^{(\gamma)}} \exp\left(-\frac{\bar{t}}{\bar{\tau}^{(\gamma)}}\right), \quad \bar{\tau}^{(\gamma)} = \tau^{(\gamma)}/t_1. \tag{8.292}$$

The temperature $\theta$ of a viscoelastic continuum can be considered as a *quasiperiodic function* of both the fast and slow times:

$$\theta(t) = \theta(\xi, \bar{t}) = \theta(\xi + 1, \bar{t}), \tag{8.293}$$

and is a singly-periodic function of $\xi$. The arguments $\xi$ and $\bar{t}$ for such a function are assumed to be independent.

### 8.6.3 Differentiation and Integration of Quasiperiodic Functions

Differentiation of quasiperiodic functions is performed by the rule of differentiation of composite functions:

$$\frac{\partial}{\partial t}\theta(t) = \frac{\partial \theta}{\partial \bar{t}}\frac{\partial \bar{t}}{\partial t} + \frac{\partial \theta}{\partial \xi}\frac{\partial \xi}{\partial t} = \frac{1}{t_1}\left(\frac{\partial \theta}{\partial \bar{t}} + \frac{1}{\varkappa}\frac{\partial \theta}{\partial \xi}\right). \tag{8.294}$$

Integration of quasiperiodic functions $b(\tau) = b(\bar{\tau},\ \xi)$, where $\bar{\tau} = \tau/t_1$, is realized as follows:

$$\int_0^t b(\tau)\, d\tau = \int_0^t b(\bar{\tau},\xi)\, d\tau = t_1 \int_0^{\bar{t}} \left( \int_0^1 b(\bar{\tau},\xi)\, d\xi \right) d\bar{\tau} + \varkappa O(1)$$
$$= t_1 \int_0^{\bar{t}} \langle b\rangle(\bar{\tau})\, d\bar{\tau} + \varkappa O(1). \tag{8.295}$$

Here we have denoted the average value of a quasiperiodic function over the oscillation cycle, which is a function only of the slow time $\bar{\tau}$, by

$$\langle b\rangle(\bar{t}) = \int_0^1 b(\bar{t},\ \xi)\, d\xi, \tag{8.296}$$

and the value $O(1)$ contains the terms comparable in magnitude with the first term $t_1 \int_0^{\bar{t}} \langle b\rangle d\bar{\tau}$ of formula (8.295).

### *8.6.4 Heat Conduction Equation for a Thin Viscoelastic Beam*

Consider the simplest linear models $B_n$ of isotropic incompressible thermorheologically simple continua, whose constitutive equations have been derived in Exercise 8.3.3. A statement of the coupled quasistatic problem of thermoviscoelasticity in the material description has the form (8.259), (8.261), (8.262), and the heat conduction equation in this system becomes

$$\overset{\circ}{\rho} c_\varepsilon \frac{\partial \theta}{\partial t} = \overset{\circ}{\nabla} \cdot (\overset{\circ}{\boldsymbol{\lambda}} \cdot \overset{\circ}{\nabla} \theta) + \overset{\circ}{w}{}^*. \tag{8.297}$$

Here and below for simplicity we assume that $q_m = 0$.

Integrating this equation over $\overset{\circ}{V}$ and then using the Gauss–Ostrogradskii formula and the boundary conditions (8.262), we obtain

$$\overset{\circ}{\rho} c_\varepsilon \frac{\partial}{\partial t} \int_{\overset{\circ}{V}} \theta\, d\overset{\circ}{V} = -\int_{\overset{\circ}{\Sigma}} \overset{\circ}{q}_e\, d\overset{\circ}{\Sigma} + \int_{\overset{\circ}{V}} \overset{\circ}{w}{}^*\, d\overset{\circ}{V}. \tag{8.298}$$

Suppose that a beam considered is thin, i.e. its width $h_2^0$ and height $h_3^0$ are much smaller than its length $h_1^0$, so changes in the temperature $\theta$ along the coordinates $\overset{\circ}{x}{}^2$ and $\overset{\circ}{x}{}^3$ can be neglected. The beam ends $\overset{\circ}{x}{}^1 = 0$ and $\overset{\circ}{x}{}^1 = h_1^0$ are assumed to be heat-insulated (for them $\overset{\circ}{q}_e = 0$), and on the lateral surface of the beam the conditions of convective heat transfer are given:

$$\overset{\circ}{q}_e = \alpha_T(\theta - \theta_e), \tag{8.299}$$

where $\alpha_T$ is the specific *heat transfer coefficient*, and $\theta_e$ is the temperature of the surroundings (being a constant). Since in cyclic tension of the beam $\overset{\circ}{w}{}^*$ is independent of coordinates, Eq. (8.298) takes the form

$$\overset{\circ}{\rho} c_\varepsilon \frac{\partial\theta}{\partial t} = -\bar{\alpha}_T(\theta - \theta_e) + \overset{\circ}{w}{}^*, \tag{8.300}$$

where

$$\bar{\alpha}_T = \alpha_T \frac{|\overset{\circ}{\Sigma}|}{|\overset{\circ}{V}|} = \frac{2\alpha_T}{h_2^0 h_3^0}(h_2^0 + h_3^0) \tag{8.301}$$

is the integral coefficient of heat transfer.

### 8.6.5 Dissipation Function for a Viscoelastic Beam

The dissipation function $\overset{\circ}{w}{}^*$ for the beam has the form (see Exercise 8.3.3)

$$\overset{\circ}{w}{}^* = a_\theta \sum_{\alpha=1}^{3} \int_0^t \int_0^t q(2t - \tau_1' - \tau_2') \frac{d}{d\tau_1} \overset{(n)}{f}_\alpha(k_1(\tau_1)) \frac{d}{d\tau_2} \overset{(n)}{f}_\alpha(k_1(\tau_2))\, d\tau_1\, d\tau_2. \tag{8.302}$$

Integrating by parts yields

$$\begin{aligned} \overset{\circ}{w}{}^* = {} & a_\theta q(0) \sum_{\alpha=1}^{3} \overset{(n)}{f}{}_\alpha^2(k_1(t)) - 2a_\theta \sum_{\alpha=1}^{3} \overset{(n)}{f}_\alpha(k_1(t)) \int_0^t \frac{\partial}{\partial\tau_1} q(t' - \tau_1') \overset{(n)}{f}_\alpha(k_1(\tau_1))\, d\tau_1 \\ & + a_\theta \sum_{\alpha=1}^{3} \int_0^t \int_0^t \frac{\partial^2}{\partial\tau_1 \partial\tau_2} q(t' - \tau_1' - \tau_2') \overset{(n)}{f}_\alpha(k_1(\tau_1)) \overset{(n)}{f}_\alpha(k_1(\tau_2))\, d\tau_1\, d\tau_2. \end{aligned} \tag{8.303}$$

### 8.6.6 Asymptotic Expansion in Terms of a Small Parameter

Since the function $k_1(t)$ is periodic, any algebraic function of $k_1$, in particular, the function $\overset{(n)}{f}_\alpha(k_1)$ determined by (8.268), will be also periodic: $\overset{(n)}{f}_\alpha = \overset{(n)}{f}_\alpha(k_1(\xi))$. Then the temperature $\theta$, being a solution of Eq. (8.300), will be, in general, a quasiperiodic function. Let us seek a solution of Eq. (8.300) as an asymptotic expansion in terms of a small parameter $\varkappa$:

$$\theta = \bar{\theta}(\bar{t}) + \varkappa\theta^{(1)}(\bar{t},\ \xi) + \varkappa^2 O(1). \tag{8.304}$$

The first term $\bar{\theta}$ of this expansion depends only on the slow time $\bar{t}$. According to the rule (8.294), the derivative $\partial\theta/\partial t$ of the expansion (8.304) is calculated as follows:

$$\frac{\partial\theta}{\partial t} = \frac{1}{t_1}\left(\frac{\partial\bar{\theta}}{\partial\bar{t}} + \frac{\partial\theta^{(1)}}{\partial\xi}\right) + \varkappa O(1). \tag{8.305}$$

The function $a_\theta(\theta)$ and the reduced time $t'$ and $\tau'$ after substitution of the expansion (8.304) into them can also be represented in the form of asymptotic expansion:

$$a_\theta(\theta) = a_\theta(\bar{\theta}) + \varkappa a_\theta^{(1)} + \varkappa^2 O(1), \quad t' = \bar{t}' + \varkappa t'^{(1)} + \varkappa^2 O(1). \tag{8.306}$$

Here, as before, $O(1)$ means the terms comparable in magnitude with the first terms of the expansion. The functions $\bar{a}_\theta = a_\theta(\bar{\theta})$ and $\bar{t}'$ are expressed by the same formulae as the input functions $a_\theta$ and $t'$, if in them the substitution $\theta \to \bar{\theta}$ has been made:

$$\bar{a}_\theta(\bar{\theta}) = a_\theta(\bar{\theta}), \qquad \bar{t}' = \int_0^{\bar{t}} a_\theta(\bar{\theta})\, d\bar{t}. \tag{8.307}$$

Substituting the expansions (8.304) and (8.306) into formula (8.303) and using the rule (8.295) of integration of quasiperiodic functions, we find an asymptotic expansion for the dissipation function

$$\overset{\circ}{w}{}^* = \overset{\circ}{w}{}^{*(0)} + \varkappa\overset{\circ}{w}{}^{*(1)} + \varkappa^2 O(1), \tag{8.308}$$

where

$$\overset{\circ}{w}{}^{*(0)} = \bar{a}_\theta \sum_{\alpha=1}^{3}\Big(q(0)\overset{(n)}{f}{}^2_\alpha(k_1(\xi)) - 2\overset{(n)}{f}_\alpha(k_1(\xi))\langle\overset{(n)}{f}_\alpha(k_1)\rangle(q(0) - q(\bar{t}'))$$
$$+\langle\overset{(n)}{f}_\alpha(k_1)\rangle^2(q(0) - 2q(\bar{t}') + q(2\bar{t}'))\Big). \tag{8.309}$$

In deriving this expression, we have taken into account that the integrands are products of functions of the fast time $\overset{(n)}{f}_{\alpha}(k_1(\xi))$ and the slow time $\frac{\partial}{\partial\tau_1}q(\bar{t}' - \tau_1')$, and $\frac{\partial^2}{\partial\tau_1\partial\tau_2}q(2\bar{t}' - \tau_1' - \tau_2')$; hence, the integrals with respect to $\xi$ and $\bar{t}$ in formula (8.303) can be calculated independently.

On substituting (8.304) and (8.308) into (8.300), we obtain the following asymptotic expansion of the heat conduction equation:

$$\frac{\overset{\circ}{\rho} c_{\varepsilon}}{t_1}\left(\frac{\partial\bar{\theta}}{\partial\bar{t}} + \frac{\partial\theta^{(1)}}{\partial\xi}\right) = -\bar{\alpha}_T(\bar{\theta} - \theta_e) + \overset{\circ}{w}{}^{*(0)} + \varkappa O(1). \tag{8.310}$$

## 8.6.7 Averaged Heat Conduction Equation

Averaging the Eq. (8.310) over the oscillation period by (8.296) and taking periodicity of the function $\theta^{(1)}$ into account (i.e. $\langle\partial\theta^{(1)}/\partial\xi\rangle = \theta^{(1)}(1) - \theta^{(1)}(0) = 0$), we find the final form of the heat conduction equation for the temperature $\bar{\theta}$:

$$\overset{\circ}{\rho} c_{\varepsilon}\frac{\partial\bar{\theta}}{\partial t} = -\bar{\alpha}_T(\bar{\theta} - \theta_e) + \bar{w}^*, \quad t = 0: \ \bar{\theta} = \theta_0, \tag{8.311}$$

where $\langle\overset{\circ}{w}{}^{*(0)}\rangle \equiv \bar{w}^*$ is the averaged dissipation function. In deriving Eq. (8.311) we have replaced the dimensionless time by the dimensional one: $t = \bar{t}t_1$. Collecting terms with higher powers of $\varkappa$ (with $\varkappa$, $\varkappa^2$ etc.) in Eq. (8.310), we obtain an equation for determining the functions $\theta^{(1)}$, $\theta^{(2)}$ etc.; however, a contribution of these terms to the value of the temperature $\theta$ is small due to (8.304). Therefore, for the problems investigated, we can consider only the zero approximation for determining the temperature $\bar{\theta}$.

Averaging (8.309), we find the expression for the dissipation function $\bar{w}^*$ averaged over the cycle of oscillations:

$$\bar{w}^* \equiv \langle\overset{\circ}{w}{}^{*(0)}\rangle = \bar{a}_{\theta}\sum_{\alpha=1}^{3}\left(q(0)\langle\overset{(n)}{f}{}^2_{\alpha}(k_1)\rangle - \langle\overset{(n)}{f}_{\alpha}(k_1)\rangle^2(q(0) - q(2\bar{t}'))\right). \tag{8.312}$$

If the beam is elastic, then $q(\bar{t}) \equiv 0$ and $\bar{w}^* \equiv 0$, and the heat conduction problem at $\theta_0 = \theta_e$ has a trivial solution: $\theta(t) = \theta_0 = \theta_e = \text{const}$; i.e. an elastic beam in cyclic deforming does not change its temperature. For a viscoelastic beam $\bar{w}^* \geqslant 0$, and in Eq. (8.311) there appears a heat source; therefore, $\partial\bar{\theta}/\partial\bar{t} \geqslant 0$, i.e. a heat-insulated beam will always be heated with time (while $\theta_0 = \theta_e$ and $\bar{\alpha}_T = 0$). This heating is caused only by energy dissipation; therefore, it is called *dissipative heating*.

### 8.6.8 Temperature of Dissipative Heating in a Symmetric Cycle

The process of deforming is called a *symmetric cycle*, if the average value of function $\overset{(n)}{f}_1(k_1)$ over the oscillation cycle is zero: $\langle \overset{(n)}{f}_1(k_1)\rangle = 0$.

As follows from (8.268), for a symmetric cycle the conditions $\langle \overset{(n)}{f}_\alpha(k_1)\rangle = 0$ ($\alpha = 1, 2, 3$) are satisfied simultaneously. And from (8.312) it follows that the dissipation function depends only on the temperature $\bar{\theta}$:

$$\bar{w}^*(\bar{\theta}) = a_\theta(\bar{\theta})q(0)\bar{f}^2, \quad \bar{f}^2 \equiv \sum_{\alpha=1}^{3}\langle \overset{(n)}{f}{}^2_\alpha(k_1)\rangle. \tag{8.313}$$

Then the heat conduction problem (8.311) takes the form

$$\overset{\circ}{\rho} c_\varepsilon \frac{\partial\bar{\theta}}{\partial t} = -\bar{\alpha}_T(\bar{\theta} - \theta_0) + a_\theta(\bar{\theta})q(0)\bar{f}^2, \quad t = 0: \ \bar{\theta} = \theta_0. \tag{8.314}$$

Its solution has the form

$$t = H(\bar{\theta}), \quad H(\bar{\theta}) \equiv \int_{\theta_0}^{\bar{\theta}} \frac{\overset{\circ}{\rho} c_\varepsilon d\widetilde{\theta}}{a_\theta(\bar{\theta})q(0)\bar{f}^2 - \bar{\alpha}_T(\bar{\theta} - \theta_e)}. \tag{8.315}$$

### 8.6.9 Regimes of Dissipative Heating Without Heat Removal

If there is no heat removal from the beam ($\bar{\alpha}_T = 0$), then, depending on a form of the function $a_\theta(\bar{\theta})$, two basically distinct regimes of dissipative heating are possible.

(1) If the function $a_\theta(\bar{\theta})$ is such that the integral

$$H(\bar{\theta}) = \int_{\theta_0}^{\theta} \frac{\overset{\circ}{\rho} c_\varepsilon \, d\theta}{(a_\theta(\bar{\theta})q(0)\bar{f}^2} \tag{8.316}$$

becomes infinite at infinity (i.e. $H(\bar{\theta}) \to +\infty$ as $\bar{\theta} \to \infty$), then the temperature of dissipative heating in the beam gradually grows with no limit (Fig. 8.13, the curve *1*).

For many real viscoelastic materials, as the function $a_\theta(\theta)$ one frequently uses the Williams–Landel–Ferry dependence

$$a_\theta(\theta) = \exp\frac{a_1(\theta - \theta_0)}{a_2 + \theta - \theta_0}, \quad a_1, a_2 - \text{const}, \tag{8.317}$$

for which the condition $H(\bar{\theta}) \to +\infty$ as $\bar{\theta} \to +\infty$ is actually satisfied.

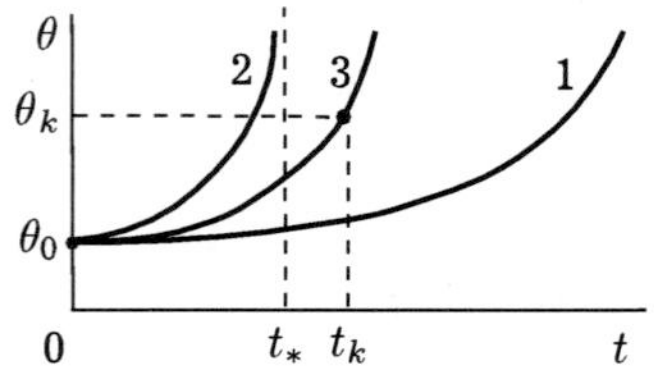


**Fig. 8.13** Different regimes of dissipative heating for a heat-insulated viscoelastic beam: 1 – unbounded growth of the temperature, 2 – heat explosion, 3 – heat pseudoexplosion

(2) If the dependence $a_\theta(\theta)$ is such that the function $H(\bar{\theta})$ is bounded at infinity (i.e. $H(+\infty) < +\infty$), then from (8.316) it follows that the temperature $\bar{\theta}(t)$ of dissipative heating reaches infinite values in the finite time $t_* = H(+\infty)$; in other words, the function $\bar{\theta}(t)$ has a vertical asymptote as $t \to t_*$ (Fig. 8.13, the curve 2). The phenomenon of sharp growth of the temperature at a certain time $t_*$ is called the *heat explosion*.

If, for example, the function $a_\theta(\theta)$ is exponential (that is typical for some elastomers):

$$a_\theta = \mathrm{e}^{a_1(\theta-\theta_0)}, \quad a_1 > 0, \tag{8.318}$$

then, calculating the integral in (8.316), we obtain the following expression for the temperature of dissipative heating:

$$\bar{\theta}(t) = \theta_0 - \frac{1}{a_1}\lg\left(1 - \frac{q(0)\bar{f}^2 a_1 t}{\overset{\circ}{\rho} c_\varepsilon}\right), \quad t_* = \frac{\overset{\circ}{\rho} c_\varepsilon}{q(0)\bar{f}^2 a_1}. \tag{8.319}$$

From this equation it really follows that $\bar{\theta} \to +\infty$ as $t \to t_*$.

(3) In practice sometimes there is an intermediate situation when the condition $H(\bar{\theta}) \to +\infty$ holds as $\bar{\theta} \to +\infty$, but the growth of temperature $\bar{\theta}(t)$ with time proves to be so sharp that it becomes similar to the effect of heat explosion: the temperature reaches its ultimate value $\theta_k$, at which there occurs a heat destruction of the material, in comparably small time $t_k$ (Fig. 8.13, the curve *3*). Such regimes are called the *heat pseudoexplosion*.

### *8.6.10 Regimes of Dissipative Heating in the Presence of Heat Removal*

When there is heat removal ($\alpha_T > 0$), the character of dissipative heating of a body changes. Since both the functions $\alpha_\theta(\bar{\theta})$ and $\alpha_T(\bar{\theta} - \theta_0)$ are non-negative, the denominator of the integrand in (8.315) at a certain finite value $\bar{\theta} = \theta_\infty < +\infty$ can vanish; in this case the function $H(\bar{\theta}) = t$ tends to infinity: $\bar{\theta} \to \theta_\infty$ as $t \to +\infty$. Thus, when there is heat removal, the regimes *1*, *2*, and *3* are impossible, because

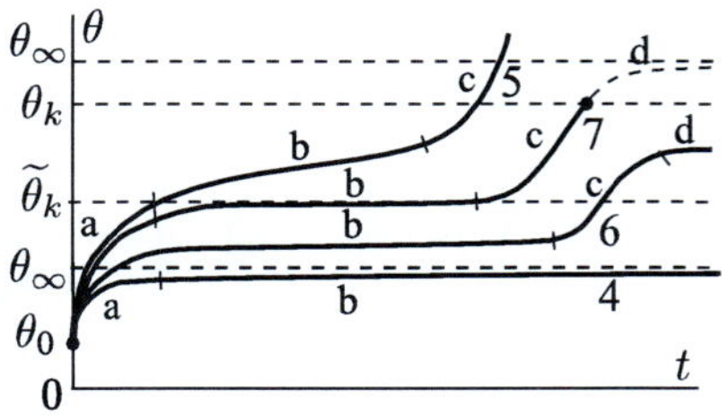


**Fig. 8.14** Typical regimes of dissipative heating for a viscoelastic beam in oscillations with heat removal

the temperature always remains bounded, in this case there are other four typical regimes *4*, *5*, *6*, and *7* (Fig. 8.14).

(4) The regime *4* is realized when the dissipation function $\bar{w}^*$ is independent of temperature ($a_\theta = 1$). In this case the heat conduction equation (8.314) yields $d^2\bar{\theta}/dt^2 \leqslant 0$ (Exercise 8.6.1); i.e. the dissipative heating curve is convex upwards. This is the most wide-spread type of a curve of dissipative heating for real practical problems.

At this curve there are two typical sections: the section 'a', where the rate of heating decreases rapidly from its maximum value $\dot{\bar{\theta}}(0)$ to practically zero $\dot{\bar{\theta}} \approx 0$, and the steady section 'b', where $\dot{\bar{\theta}} \approx 0$.

If $a_\theta(\bar{\theta}) = 1$, then the heat conduction problem (8.314) admits a solution in the explicit form (Exercise 8.6.1)

$$\bar{\theta} = \theta_0 + \frac{q(0)\bar{f}^2 \overset{\circ}{\rho} c_\varepsilon}{\bar{\alpha}_T}\left(1 - \exp\left(-\frac{\bar{\alpha}_T t}{\overset{\circ}{\rho} c_\varepsilon}\right)\right). \tag{8.320}$$

(5) If $a_\theta(\bar{\theta})$ is unbounded as $\bar{\theta} \to +\infty$ (for example, the exponential function (8.318)), then the temperature regime *5* is realized when the curve of dissipative heating can be split into the three typical sections: 'a' – initial, 'b' – steady, where $\dot{\bar{\theta}} \approx$ const, and 'c' – unsteady, where the function $\bar{\theta}(\bar{t})$ is convex downwards (Fig. 8.14) and unbounded as $t \to +\infty$.

(6) If $a_\theta(\bar{\theta})$ depends on the temperature but is bounded as $\bar{\theta} \to +\infty$ and has a point of inflection at $\bar{\theta} = \tilde{\theta}_k$ (as the function (8.317)), then the temperature regime *6* of dissipative heating occurs (Fig. 8.14), when there are four typical sections: initial – 'a', steady – 'b', unsteady – 'c', followed by steady one – 'd'. The temperature $\bar{\theta}(t)$ remains bounded as $t \to +\infty$.

(7) The regime *7* is realized under the same conditions as the regime *6*, but at the unsteady section 'c' the temperature of dissipative heating $\bar{\theta}(t)$ reaches an ultimate value $\theta_k$ such that there occurs a heat destruction of the material (by analogy with the regime *3*), and the section 'd' is not realizable.

### *8.6.11 Experimental and Computed Data on Dissipative Heating of Viscoelastic Bodies*

Figure 8.15 shows computed and experimental curves of dissipative heating for a polyurethane beam in cyclic deforming by the harmonic law (8.290). The change in temperature $\Delta\theta = \theta - \theta_0$ was found by solving the problem (8.311), (8.312) with the help of the implicit difference approximation

$$\Delta\theta^{i+1} = \frac{\Delta\theta^i + \bar{w}^* \Delta t / \overset{\circ}{\rho} c_\varepsilon}{1 + \bar{\alpha}^{\mathrm{T}} / \overset{\circ}{\rho} c_\varepsilon}, \tag{8.321}$$

where $\theta^i = \theta(t_i)$ is the value of temperature in the $i$th node at time $t_i$, and $\Delta t$ is the step in time. Values of the relaxation core $q(2t')$ and $q(0)$ appearing in expression (8.312) for the dissipation function were determined by (8.292), and $B^{(\gamma)}$ and $\tau^{(\gamma)}$ in this formula – by the relaxation curves with the help of the method given in Sect. 8.5.5. Values of the constants $B^{(\gamma)}$ and $\tau^{(\gamma)}$ for polyurethane are shown in Table 8.1. The function $a_\theta(\theta)$ was approximated by formula (8.317), and values of the constants in this formula were assumed to be $a_1 = 21$, $a_2 = 208\,\mathrm{K}$. The remaining constants in (8.311) take on the following values: $\overset{\circ}{\rho} = 10^3\ \mathrm{kg/m^3}$, $c_\varepsilon = 0.8\ \mathrm{kJ/(kg \cdot K)}$, $\bar{\alpha}^{\mathrm{T}} = 10\ \mathrm{kWt/(m^3 \cdot K)}$. The mean value $\bar{k}_1$ and the oscillation amplitude $k_1^0$ were expressed in terms of the minimum and maximum deformation values in the cycle ($\delta_{\min}$ and $\delta_{\max}$):

$$\bar{k}_1 = 1 + \frac{1}{2}(\delta_{\max} + \delta_{\min}), \quad \bar{k}_1 = \frac{1}{2}(\delta_{\max} - \delta_{\min}). \tag{8.322}$$

Figure 8.15 exhibits temperatures of dissipative heating, computed by the method mentioned above for different models $B_n$ at $\delta_{\max} = 50\%$ and $\delta_{\min} = 34\%$. Under the conditions considered, the model $B_{\mathrm{I}}$ gives the best approximation to experimental data. The distinction between different models $B_n$ is considerable: the models $B_{\mathrm{IV}}$

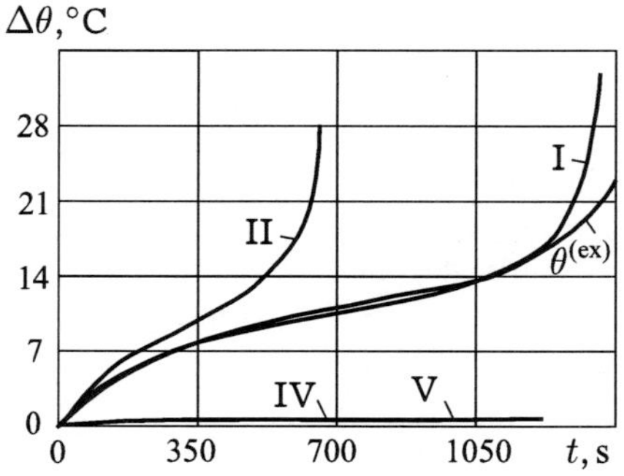


**Fig. 8.15** Curves of dissipative heating for polyurethane, computed by models $B_n$, and experimental curve of dissipative heating $\theta^{(\mathrm{ex})}$

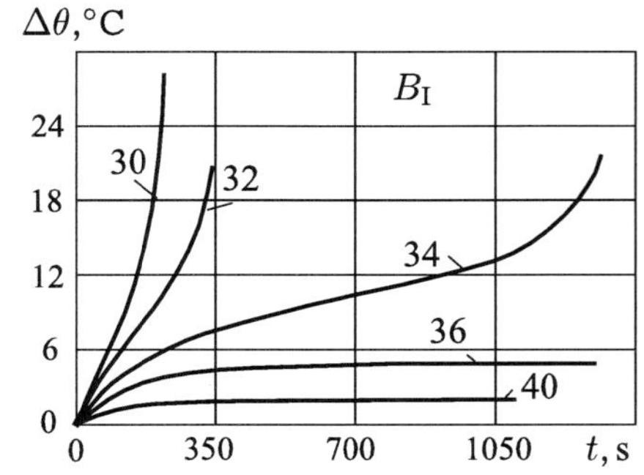


**Fig. 8.16** Curves of dissipative heating for polyurethane, computed by the model $B_I$ at different amplitudes of oscillations

and $B_V$ lead to a stationary regime of dissipative heating according to the type (4), and the models $B_I$ and $B_{II}$ forecast the regime (5) with the presence of unsteady section.

Figure 8.16 shows curves of dissipative heating for polyurethane, computed by the model $B_I$ at different values of $\delta_{min}$ (values of this parameter are given in Fig. 8.16 by numbers at curves); the value $\delta_{max} = 50\%$ has been fixed. With growing the oscillation amplitude (i.e. in this case with decreasing the value $\delta_{min}$), the intensity of dissipative heating sharply increases.

Notice that the phenomenon of dissipative heating of viscoelastic materials can essentially reduce the durability of structures under cyclic deforming.

## Exercises for 8.6

**8.6.1.** Show that if $a_\theta(\bar{\theta}) = \text{const}$, then a solution of the problem (8.314) is a function being convex upwards, i.e. $d^2\bar{\theta}/dt^2 \leqslant 0$. Prove formula (8.320).

# Chapter 9
# Plastic Continua at Large Deformations

## 9.1 Models $A_n$ of Plastic Continua at Large Deformations

### 9.1.1 Main Assumptions of the Models

While models of viscoelastic continua most adequately describe a behavior of 'soft' materials (rubbers, polymers, biomaterials), for simulation of mechanical inelastic properties of 'stiff' materials (metals and alloys) one widely uses models of plastic media.

There are different models of plastic continua, but we consider only the models falling into the class of *models of plastic yield* that is frequently used in practice. These models are convenient to be considered in the forms $A_n^\zeta$, $B_n^\zeta$, $C_n^\zeta$ and $D_n^\zeta$, where in place of $\psi$ one should use the Gibbs free energy $\overset{(n)}{\zeta}_A \equiv \zeta$ (4.126), and the principal thermodynamic identity – in the form (4.133).

**Definition 9.1.** One can say that this is the model $A_n^\zeta$ of a *plastic continuum*, if for the medium

1. The operator constitutive equations (4.156) are functionals with respect to time $t$ of reactive variables $\mathcal{R}$ and their derivatives $\dot{\mathcal{R}}$:

$$\Lambda(t) = \overset{t}{\underset{\tau=0}{\mathbf{f}}}(\mathcal{R}(t),\ \dot{\mathcal{R}}(t),\ \mathcal{R}^t(\tau),\ \dot{\mathcal{R}}^t(\tau)); \tag{9.1}$$

2. The set of reactive variables $\mathcal{R}$ in addition includes a symmetric second-order tensor $\overset{(n)}{\mathbf{C}}_p$ called the plastic deformation tensor, and the set of active variables $\Lambda$ contains a symmetric second-order tensor $\overset{(n)}{\mathbf{C}}_e$ called the elastic deformation tensor:

$$\Lambda = \{\zeta,\ \eta,\ \overset{(n)}{\mathbf{C}},\ \overset{(n)}{\mathbf{C}}_e,\ w^*\},\quad \mathcal{R} = \{\overset{(n)}{\mathbf{T}}/\rho,\ \overset{(n)}{\mathbf{C}}_p,\ \theta\}, \tag{9.2}$$

3. The tensors $\overset{(n)}{\mathbf{C}}_e$ and $\overset{(n)}{\mathbf{C}}_p$ are connected to $\overset{(n)}{\mathbf{C}}$ by the additive relation

$$\overset{(n)}{\mathbf{C}} = \overset{(n)}{\mathbf{C}}_e + \overset{(n)}{\mathbf{C}}_p. \tag{9.3}$$

Y.I. Dimitrienko, *Nonlinear Continuum Mechanics and Large Inelastic Deformations*, Solid Mechanics and Its Applications 174, DOI 10.1007/978-94-007-0034-5_9,

Unless the tensors $\overset{(n)}{\mathbf{C}}_e$ and $\overset{(n)}{\mathbf{C}}_p$ were involved in the set of variables $\mathcal{R}$ and $\Lambda$, the relation (9.1) could be considered as the single model of viscoelastic continua of the differential and integral types. However, just the appearance of the tensors $\overset{(n)}{\mathbf{C}}_e$ and $\overset{(n)}{\mathbf{C}}_p$ leads to new effects which are not usual for the continuum types considered.

Just as for continua of the differential type, the dependences (9.1) upon $\mathcal{R}(t)$ and $\dot{\mathcal{R}}(t)$ are assumed to be differentiable functions, and the functionals of prehistory $\mathcal{R}^t(\tau)$, $\dot{\mathcal{R}}^t(\tau)$ are assumed to be continuous and Fréchet–differentiable.

The tensors $\overset{(n)}{\mathbf{C}}_e$ and $\overset{(n)}{\mathbf{C}}_p$ can be introduced axiomatically; however, to justify physically the introduction of these tensors one usually considers the following model.

Introduce a reference configuration $\overset{\circ}{\mathcal{K}}$, which is assumed to be unstressed (i.e. $\mathbf{T}(0) = 0$ in $\overset{\circ}{V}$), and an actual configuration $\mathcal{K}$, where a stress tensor field $\mathbf{T}(t)$, in general, is different from identically zero. In addition, introduce one more configuration $\overset{p}{\mathcal{K}}$ (it may be not physically realizable), to which the continuum would correspond, if at the time $t_p > t$ its stress field were zero again: $\mathbf{T}(t_p) = 0$. Moreover, we impose on $\overset{p}{\mathcal{K}}$ the requirement that if the transformations $\overset{\circ}{\mathcal{K}} \to \mathcal{K} \to \overset{p}{\mathcal{K}}$ occur without plastic deformations (i.e. according to the model of an ideal continuum), then configuration $\overset{p}{\mathcal{K}}$ must coincide with $\overset{\circ}{\mathcal{K}}$. Such a configuration $\overset{p}{\mathcal{K}}$ is called *unloaded*.

Introduce in each of the configurations $\overset{\circ}{\mathcal{K}}$, $\mathcal{K}$ and $\overset{p}{\mathcal{K}}$ local basis vectors $\overset{\circ}{\mathbf{r}}_i$, $\mathbf{r}_i$ and $\overset{p}{\mathbf{r}}_i$, respectively:

$$\overset{\circ}{\mathbf{r}}_i = \frac{\partial \overset{\circ}{\mathbf{x}}}{\partial X^i}, \quad \mathbf{r}_i = \frac{\partial \mathbf{x}}{\partial X^i}, \quad \overset{p}{\mathbf{r}}_i = \frac{\partial \overset{p}{\mathbf{x}}}{\partial X^i}, \tag{9.4}$$

where $\overset{\circ}{\mathbf{x}}$, $\mathbf{x}$ and $\overset{p}{\mathbf{x}}$ are radius-vectors of a material point $\mathcal{M}$ in $\overset{\circ}{\mathcal{K}}$, $\mathcal{K}$ and $\overset{p}{\mathcal{K}}$, respectively (Fig. 9.1).

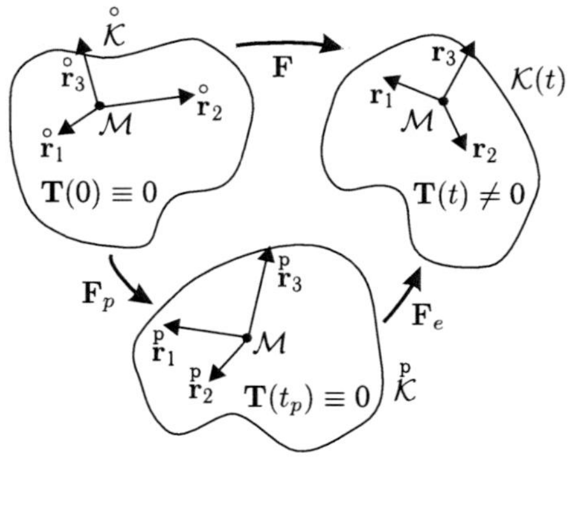


**Fig. 9.1** A scheme of transformation of configurations

As before, construct the metric matrices in $\overset{\circ}{\mathcal{K}}$, $\mathcal{K}$ and $\overset{p}{\mathcal{K}}$:

$$g_{ij} = \mathbf{r}_i \cdot \mathbf{r}_j, \quad \overset{\circ}{g}_{ij} = \overset{\circ}{\mathbf{r}}_i \cdot \overset{\circ}{\mathbf{r}}_j, \quad \overset{p}{g}_{ij} = \overset{p}{\mathbf{r}}_i \cdot \overset{p}{\mathbf{r}}_j, \tag{9.5a}$$

and their inverse matrices $g^{ij}$, $\overset{\circ}{g}{}^{ij}$ and $\overset{p}{g}{}^{ij}$, with the help of which introduce the vectors of reciprocal bases

$$\mathbf{r}^i = g^{ij}\mathbf{r}_j, \quad \overset{\circ}{\mathbf{r}}{}^i = \overset{\circ}{g}{}^{ij}\overset{\circ}{\mathbf{r}}_j, \quad \overset{p}{\mathbf{r}}_i = \overset{p}{g}{}^{ij}\overset{p}{\mathbf{r}}_j. \tag{9.5b}$$

Introduce the tensors of transformations of a local neighborhood of the point $\mathcal{M}$ from $\overset{\circ}{\mathcal{K}}$ to $\overset{p}{\mathcal{K}}$, from $\overset{p}{\mathcal{K}}$ to $\mathcal{K}$ and from $\overset{\circ}{\mathcal{K}}$ to $\mathcal{K}$:

$$\mathbf{F}_p = \overset{p}{\mathbf{r}}_i \otimes \overset{\circ}{\mathbf{r}}{}^i, \quad \mathbf{F}_e = \mathbf{r}_i \otimes \overset{p}{\mathbf{r}}{}^i, \quad \mathbf{F} = \mathbf{r}_i \otimes \overset{\circ}{\mathbf{r}}{}^i. \tag{9.6}$$

The tensors $\mathbf{F}_p$ and $\mathbf{F}_e$ are called the *gradients of plastic* and *elastic deformations*, respectively.

These tensors are connected by the relation

$$\mathbf{F} = \mathbf{F}_e \cdot \mathbf{F}_p. \tag{9.7}$$

Introduce also three right Cauchy–Green deformation tensors:

$$\mathbf{C} = \varepsilon_{ij}\overset{\circ}{\mathbf{r}}{}^i \otimes \overset{\circ}{\mathbf{r}}{}^j, \quad \mathbf{C}_p = \varepsilon^p_{ij}\overset{\circ}{\mathbf{r}}{}^i \otimes \overset{\circ}{\mathbf{r}}{}^j, \quad \mathbf{C}_e = \varepsilon^e_{ij}\overset{\circ}{\mathbf{r}}{}^i \otimes \overset{\circ}{\mathbf{r}}{}^j, \tag{9.8}$$

where we have denoted components of the deformation tensor by $\varepsilon_{ij}$, of the plastic deformation tensor – by $\varepsilon^p_{ij}$ and of the elastic deformation tensor – by $\varepsilon^e_{ij}$:

$$\varepsilon_{ij} = \frac{1}{2}\left(g_{ij} - \overset{\circ}{g}_{ij}\right), \quad \varepsilon^p_{ij} = \frac{1}{2}\left(\overset{p}{g}_{ij} - \overset{\circ}{g}_{ij}\right), \quad \varepsilon^e_{ij} = \frac{1}{2}\left(g_{ij} - \overset{p}{g}_{ij}\right). \tag{9.9}$$

In a similar way, introduce three right Almansi deformation tensors:

$$\mathbf{\Lambda} = \varepsilon^{ij}\overset{\circ}{\mathbf{r}}_i \otimes \overset{\circ}{\mathbf{r}}_j, \quad \mathbf{\Lambda}_e = \varepsilon^{ij}_e\overset{\circ}{\mathbf{r}}_i \otimes \overset{\circ}{\mathbf{r}}_j, \quad \mathbf{\Lambda}_p = \varepsilon^{ij}_p\overset{\circ}{\mathbf{r}}_i \otimes \overset{\circ}{\mathbf{r}}_j, \tag{9.10}$$

where we have denoted contravariant components of the deformation tensor by $\varepsilon^{ij}$, of the plastic deformation tensor – by $\varepsilon^{ij}_p$ and of the elastic deformation tensor – by $\varepsilon^{ij}_e$:

$$\varepsilon^{ij} = \frac{1}{2}\left(\overset{\circ}{g}{}^{ij} - g^{ij}\right), \quad \varepsilon^{ij}_p = \frac{1}{2}\left(\overset{\circ}{g}{}^{ij} - \overset{p}{g}{}^{ij}\right), \quad \varepsilon^{ij}_e = \frac{1}{2}\left(\overset{\circ}{g}{}^{ij} - g^{ij}\right). \tag{9.11}$$

From (9.8)–(9.11) it follows that three right Cauchy–Green tensors and three right Almansi tensors are connected by the additive relations

$$\mathbf{C} = \mathbf{C}_e + \mathbf{C}_p, \quad \mathbf{\Lambda} = \mathbf{\Lambda}_e + \mathbf{\Lambda}_p. \tag{9.12}$$

From these equations it follows that the relation of additivity (9.3) holds for the energetic deformation tensors $\overset{\mathrm{I}}{\mathbf{C}}$ and $\overset{\mathrm{V}}{\mathbf{C}}$ if we assume as usual that

$$\overset{\mathrm{I}}{\mathbf{C}}_e = \mathbf{\Lambda}_e, \quad \overset{\mathrm{V}}{\mathbf{C}}_e = \mathbf{C}, \quad \overset{\mathrm{I}}{\mathbf{C}}_p = \mathbf{\Lambda}_p, \quad \overset{\mathrm{V}}{\mathbf{C}}_p = \mathbf{C}_p. \tag{9.13}$$

In order to justify the additive relation (9.3) for $n = \mathrm{II}, \mathrm{IV}$, we introduce the polar decompositions for the deformation gradients (9.6):

$$\mathbf{F} = \mathbf{O} \cdot \mathbf{U}, \quad \mathbf{F}_p = \mathbf{O}_p \cdot \mathbf{U}_p, \quad \mathbf{F}_e = \mathbf{O}_e \cdot \mathbf{U}_e, \tag{9.14}$$

and represent the symmetric stretch tensors $\mathbf{U}$ and $\mathbf{U}_e$ in their eigenbases:

$$\mathbf{U} = \sum_{\alpha=1}^{3} \lambda_\alpha \overset{\circ}{\mathbf{p}}_\alpha \otimes \overset{\circ}{\mathbf{p}}_\alpha, \quad \mathbf{U}_e = \sum_{\alpha=1}^{3} \overset{e}{\lambda}_\alpha \overset{e}{\mathbf{p}}_\alpha \otimes \overset{e}{\mathbf{p}}_\alpha. \tag{9.15}$$

Here $\lambda_\alpha$ and $\overset{e}{\lambda}_\alpha$ are eigenvalues of the tensors $\mathbf{U}$ and $\mathbf{U}_e$.

Then we can introduce the tensors

$$\overset{\mathrm{IV}}{\mathbf{C}} = \mathbf{U} - \mathbf{E} = \textstyle\sum_{\alpha=1}^{3} (\lambda_\alpha - 1)\, \overset{\circ}{\mathbf{p}}_\alpha \otimes \overset{\circ}{\mathbf{p}}_\alpha,$$

$$\overset{\mathrm{IV}}{\mathbf{C}}_p = \textstyle\sum_{\alpha=1}^{3} \left(\lambda_\alpha - \overset{e}{\lambda}_\alpha\right) \overset{\circ}{\mathbf{p}}_\alpha \otimes \overset{\circ}{\mathbf{p}}_\alpha, \quad \overset{\mathrm{IV}}{\mathbf{C}}_e = \textstyle\sum_{\alpha=1}^{3} \left(\overset{e}{\lambda}_\alpha - 1\right) \overset{\circ}{\mathbf{p}}_\alpha \otimes \overset{\circ}{\mathbf{p}}_\alpha,$$

$$\overset{\mathrm{II}}{\mathbf{C}} = \mathbf{E} - \mathbf{U}^{-1} = \textstyle\sum_{\alpha=1}^{3} (1 - \lambda_\alpha^{-1})\, \overset{\circ}{\mathbf{p}}_\alpha \otimes \overset{\circ}{\mathbf{p}}_\alpha,$$

$$\overset{\mathrm{II}}{\mathbf{C}}_p = \textstyle\sum_{\alpha=1}^{3} \left(\overset{e}{\lambda}{}_\alpha^{-1} - \lambda_\alpha^{-1}\right) \overset{\circ}{\mathbf{p}}_\alpha \otimes \overset{\circ}{\mathbf{p}}_\alpha, \quad \overset{\mathrm{II}}{\mathbf{C}}_e = \textstyle\sum_{\alpha=1}^{3} \left(1 - \overset{e}{\lambda}{}_\alpha^{-1}\right) \overset{\circ}{\mathbf{p}}_\alpha \otimes \overset{\circ}{\mathbf{p}}_\alpha, \tag{9.16}$$

for which the additive relations (9.3) still hold.

Thus, we have shown how the tensors of elastic and plastic deformations satisfying the relation (9.3) can be introduced.

### 9.1.2 General Representation of Constitutive Equations for Models $A_n^\xi$ of Plastic Continua

Consider the principal thermodynamic identity in the form $A_n$ (4.121):

$$\rho\, d\psi + \rho\eta\, d\theta - \overset{(\mathrm{n})}{\mathbf{T}} \cdot\cdot\, d\overset{(\mathrm{n})}{\mathbf{C}} + w^*\, dt = 0. \tag{9.17}$$

With the help of the additive relation (9.3) this identity can be rewritten as follows:

$$\rho\, d\psi + \rho\eta\, d\theta - \overset{(\mathrm{n})}{\mathbf{T}} \cdot\cdot\, d\overset{(\mathrm{n})}{\mathbf{C}}_e - \overset{(\mathrm{n})}{\mathbf{T}} \cdot\cdot\, d\overset{(\mathrm{n})}{\mathbf{C}}_p + w^*\, dt = 0. \tag{9.18}$$

Introduce the Gibbs free energy $\zeta$ just as in (4.126):

$$\zeta = \psi - \frac{\overset{(\mathrm{n})}{\mathbf{T}}}{\rho} \cdot\cdot \overset{(\mathrm{n})}{\mathbf{C}}_e. \tag{9.19}$$

Then for $\zeta$ we obtain the identity (the principal thermodynamic identity in the form $A_n^{\zeta}$):

$$\rho\, d\zeta + \rho\eta\, d\theta + \rho\overset{(\mathrm{n})}{\mathbf{C}}_e \cdot\cdot\, d\left(\overset{(\mathrm{n})}{\mathbf{T}}/\rho\right) - \overset{(\mathrm{n})}{\mathbf{T}} \cdot\cdot\, d\overset{(\mathrm{n})}{\mathbf{C}}_p + w^*\, dt = 0. \tag{9.20}$$

According to the model $A_n^{\zeta}$, the Gibbs free energy $\zeta$ is a functional in the form (9.1):

$$\zeta = \mathop{\zeta}_{\tau=0}^{t} \left(\mathcal{R}(t),\ \dot{\mathcal{R}}(t),\ \mathcal{R}^t(\tau),\ \dot{\mathcal{R}}^t(\tau)\right), \quad \mathcal{R} = \{\overset{(\mathrm{n})}{\mathbf{T}}/\rho,\ \overset{(\mathrm{n})}{\mathbf{C}}_p,\ \theta\}. \tag{9.21}$$

Determine the total differential of this functional by using the rule of differentiation of functionals (8.24) with respect to time:

$$d\zeta = \dot{\zeta}\, dt = \frac{\partial\zeta}{\partial\overset{(\mathrm{n})}{\mathbf{T}}/\rho} \cdot\cdot\, d\left(\frac{\overset{(\mathrm{n})}{\mathbf{T}}}{\rho}\right) + \frac{\partial\zeta}{\partial\theta}\, d\theta + \frac{\partial\zeta}{\partial\overset{(\mathrm{n})}{\mathbf{C}}_p} \cdot\cdot\, d\overset{(\mathrm{n})}{\mathbf{C}}_p$$
$$+ \frac{\partial\zeta}{\partial(\overset{(\mathrm{n})}{\mathbf{T}}/\rho)^{\bullet}} \cdot\cdot\, d\left(\frac{\overset{(\mathrm{n})}{\mathbf{T}}}{\rho}\right)^{\bullet} + \frac{\partial\zeta}{\partial\dot{\theta}}\, d\dot{\theta} + \frac{\partial\zeta}{\partial\overset{(\mathrm{n})}{\mathbf{C}}{}_p^{\bullet}} \cdot\cdot\, d\overset{(\mathrm{n})}{\mathbf{C}}{}_p^{\bullet} + \delta\zeta\, dt, \tag{9.22}$$

where $\delta\zeta$ is the Fréchet–derivative.

The elastic deformation tensor $\overset{(\mathrm{n})}{\mathbf{C}}_e$ is a functional of the same type (9.1):

$$\overset{(\mathrm{n})}{\mathbf{C}}_e = \mathop{\mathbf{C}_e}_{\tau=0}^{t} \left(\mathcal{R}(t),\ \dot{\mathcal{R}}(t),\ \mathcal{R}^t(\tau),\ \dot{\mathcal{R}}^t(\tau)\right), \quad \mathcal{R} = \{\overset{(\mathrm{n})}{\mathbf{T}}/\rho,\ \overset{(\mathrm{n})}{\mathbf{C}}_p,\ \theta\}. \tag{9.23}$$

Just as for continua of the differential type (see Sect. 7.1.1), we introduce two new tensor functionals:

$$\overset{(\mathrm{n})}{\mathbf{C}}{}_e^0 = \mathop{\mathbf{C}_e}_{\tau=0}^{t} \left(\mathcal{R}(t),\ \mathbf{0},\ \mathcal{R}^t(\tau),\ \dot{\mathcal{R}}^t(\tau)\right), \tag{9.24}$$

$$\overset{(\mathrm{n})}{\mathbf{C}}{}_e^1 = \overset{(\mathrm{n})}{\mathbf{C}}_e - \overset{(\mathrm{n})}{\mathbf{C}}{}_e^0. \tag{9.25}$$

In (9.24) the arguments corresponding to the rates of changing functions $\dot{\mathcal{R}}$ are assumed to be zero. The tensors $\overset{(n)}{\mathbf{C}}{}_e^0$ and $\overset{(n)}{\mathbf{C}}{}_e^1$ are called the *equilibrium elastic deformation tensor* and *nonequilibrium elastic deformation tensor*, respectively.

Substituting (9.22), (9.24) and (9.25) into (9.20) and grouping like terms, we obtain the identity

$$\rho\Big(\frac{\partial\zeta}{\partial\overset{(n)}{\mathbf{T}}/\rho}+\overset{(n)}{\mathbf{C}}{}_e^0\Big)\cdot\cdot\, d\frac{\overset{(n)}{\mathbf{T}}}{\rho}+\rho\Big(\frac{\partial\zeta}{\partial\theta}+\eta\Big)\,d\theta+\rho\frac{\partial\zeta}{\partial(\overset{(n)}{\mathbf{T}}/\rho)^{\bullet}}\cdot\cdot\, d\Big(\frac{\overset{(n)}{\mathbf{T}}}{\rho}\Big)^{\bullet}+\rho\frac{\partial\zeta}{\partial\theta^{\bullet}}\,d\theta^{\bullet}$$

$$+\rho\frac{\partial\zeta}{\partial\overset{(n)}{\mathbf{C}}{}_p^{\bullet}}\cdot\cdot\, d\overset{(n)}{\mathbf{C}}{}_p^{\bullet}+\Big(w^{*}-(\overset{(n)}{\mathbf{T}}-\overset{(n)}{\mathbf{H}}_p)\cdot\cdot\,\overset{(n)}{\mathbf{C}}{}_p^{\bullet}+\rho\overset{(n)}{\mathbf{C}}{}_e^1\cdot\cdot\Big(\frac{\overset{(n)}{\mathbf{T}}}{\rho}\Big)^{\bullet}+\rho\delta\zeta\Big)\,dt=0. \tag{9.26}$$

Here we have introduced the *tensors of strengthening* $\overset{(n)}{\mathbf{H}}_p$ and the *reduced stress tensors* $\overset{(n)}{\mathbf{T}}_H$:

$$\overset{(n)}{\mathbf{H}}_p=\rho(\partial\zeta/\partial\overset{(n)}{\mathbf{C}}_p),\qquad \overset{(n)}{\mathbf{T}}_H=\overset{(n)}{\mathbf{T}}-\overset{(n)}{\mathbf{H}}_p. \tag{9.27}$$

Since the differentials $d(\overset{(n)}{\mathbf{T}}/\rho)$, $d\theta$, $(\overset{(n)}{\mathbf{T}}/\rho)^{\bullet}$, $d\theta^{\bullet}$, $\overset{(n)}{\mathbf{C}}{}_p^{\bullet}$ and $dt$ are independent, the identity (9.26) holds if and only if coefficients of these differentials vanish; i.e. we have the relations

$$\begin{cases}\overset{(n)}{\mathbf{C}}{}_e^0=-\partial\zeta/\partial(\overset{(n)}{\mathbf{T}}/\rho), & (9.28a)\\ \eta=-\partial\zeta/\partial\theta, & (9.28b)\\ \partial\zeta/\partial(\overset{(n)}{\mathbf{T}}/\rho)^{\bullet}=0,\quad \partial\zeta/\partial\dot{\theta}=0,\quad \partial\zeta/\partial\overset{(n)}{\mathbf{C}}{}_p^{\bullet}=0, & (9.28c)\\ w^{*}=(\overset{(n)}{\mathbf{T}}-\overset{(n)}{\mathbf{H}}_p)\cdot\cdot\,\overset{(n)}{\mathbf{C}}{}_p^{\bullet}-\rho\overset{(n)}{\mathbf{C}}{}_e^1\cdot\cdot\Big(\overset{(n)}{\mathbf{T}}/\rho\Big)^{\bullet}-\rho\delta\zeta, & (9.28d)\end{cases}$$

which are constitutive equations for models $A_n^{\zeta}$ of plastic continua. From these relations it follows that:

1. Plastic media are dissipative (for them $w^{*}\neq 0$).
2. The equilibrium deformation tensor $\overset{(n)}{\mathbf{C}}{}_e^0$ (but not $\overset{(n)}{\mathbf{C}}_e$) has the potential $\zeta$.
3. The potential $\zeta$ and hence $\overset{(n)}{\mathbf{C}}_e$, and $\eta$ are independent of the rates $(\overset{(n)}{\mathbf{T}}/\rho)^{\bullet}$, $\dot{\theta}$ and $\overset{(n)}{\mathbf{C}}{}_p^{\bullet}$:

$$\zeta=\overset{t}{\underset{\tau=0}{\zeta}}\,(\mathcal{R}(t),\,\mathcal{R}^t(\tau),\,\dot{\mathcal{R}}^t(\tau)),\quad \mathcal{R}=\{\overset{(n)}{\mathbf{T}}/\rho,\,\overset{(n)}{\mathbf{C}}_p,\,\theta\}; \tag{9.29}$$

however, the dissipation function $w^{*}$ and tensor $\overset{(n)}{\mathbf{C}}{}_e^1$ depend on $\dot{\mathcal{R}}(t)$.

Thus, the model $A_n^{\xi}$ of plastic continua is specified by three functionals: the Gibbs free energy (9.29), the nonequilibrium elastic deformation tensor

$$\overset{(n)}{\mathbf{C}}{}_e^{1} = \overset{t}{\underset{\tau=0}{\mathbf{C}_e^{1}}} \left(\mathcal{R}(t),\ \dot{\mathcal{R}}(t),\ \mathcal{R}^t(\tau),\ \dot{\mathcal{R}}^t(\tau)\right), \quad (9.30)$$

and the plastic deformation tensor $\overset{(n)}{\mathbf{C}}_p$.

### 9.1.3 Corollary of the Onsager Principle for Models $A_n^{\xi}$ of Plastic Continua

To construct the functional (9.30) and tensor $\overset{(n)}{\mathbf{C}}_p$ we use the Onsager principle (see Sect. 4.12.1, Axiom 16) and form the specific internal entropy production (4.728) with the help of (9.28d):

$$\rho q^* = w^* - \frac{\mathbf{q}}{\theta}\cdot\nabla\theta = \overset{(n)}{\mathbf{T}}_H \cdot\cdot \overset{(n)}{\mathbf{C}}{}_p^{\bullet} - \rho \overset{(n)}{\mathbf{C}}{}_e^{1} \cdot\cdot \left(\frac{\overset{(n)}{\mathbf{T}}}{\rho}\right)^{\bullet} - \rho\delta\zeta - \frac{\mathbf{q}}{\theta}\cdot\nabla\theta \geqslant 0. \quad (9.31)$$

In order for this function to be non-negative, according to the Onsager principle, it can be represented in the generalized quadratic form. For this, introduce thermodynamic forces $X_\beta$:

$$X_1 = \nabla\theta, \qquad X_2 = \overset{(n)}{\mathbf{T}}_H, \qquad X_3 = (\overset{(n)}{\mathbf{T}}/\rho)^{\bullet} \quad (9.32)$$

and thermodynamic fluxes

$$Q_1 = -\mathbf{q}/\theta, \qquad Q_2 = \overset{(n)}{\mathbf{C}}{}_p^{\bullet}, \qquad Q_3 = \rho \overset{(n)}{\mathbf{C}}{}_e^{1}. \quad (9.33)$$

Then, according to the Onsager principle, the thermodynamic fluxes $Q_\beta$ must be linear (or tensor-linear) functions of $X_\beta$:

$$\begin{cases} -\mathbf{q}/\theta = L_{11}\cdot\nabla\theta + L_{12}\cdot\cdot\overset{(n)}{\mathbf{T}}_H + L_{13}\cdot\cdot(\overset{(n)}{\mathbf{T}}/\rho)^{\bullet}, \\ \overset{(n)}{\mathbf{C}}{}_p^{\bullet} = L_{12}\cdot\nabla\theta + L_{22}\cdot\cdot\overset{(n)}{\mathbf{T}}_H + L_{23}\cdot\cdot(\overset{(n)}{\mathbf{T}}/\rho)^{\bullet}, \\ \rho\overset{(n)}{\mathbf{C}}{}_e^{1} = L_{13}\cdot\nabla\theta + L_{23}\cdot\cdot\overset{(n)}{\mathbf{T}}_H + L_{33}\cdot\cdot(\overset{(n)}{\mathbf{T}}/\rho)^{\bullet}. \end{cases} \quad (9.34)$$

Here $L_{11}$ is a second-order tensor, $L_{12}$ and $L_{13}$ are third-order tensors, and $L_{22}$, $L_{23}$ and $L_{33}$ are fourth-order tensors, which are, according to the principle of equipresence, tensor functionals of the same form as the ones appearing in the general constitutive equations (9.1) of the model $A_n^\zeta$:

$$L_{\alpha\beta} = \mathop{L_{\alpha\beta}}_{\tau=0}^{t}(\mathcal{R}(t),\ \dot{\mathcal{R}}(t),\ \mathcal{R}^t(\tau),\ \dot{\mathcal{R}}^t(\tau)). \tag{9.35}$$

Relations (9.34) are complementary constitutive equations to the system (8.28) for the model $A_n^\zeta$ of plastic continua. The first of the relations is the generalized *Fourier law*, the second one – the *law of changing plastic deformations*, and the third one – the law of changing nonequilibrium elastic deformations.

### 9.1.4 Models $A_n^\zeta$ of Plastic Yield

Many special models of plastic continua follow from the law of changing plastic deformations (9.34) after introduction of some assumptions on a form of the functionals (9.29) and (9.35).

In practice one often uses models $A_n^\zeta$ of *plastic yield*, where the functionals (9.29) and (9.35) are only functions of indicated arguments $\mathcal{R}$ and $\dot{\mathcal{R}}$ and of one scalar functional $w_p$:

$$\zeta = \zeta(\overset{(n)}{\mathbf{T}}/\rho,\ \overset{(n)}{\mathbf{C}}_p,\ \theta,\ w_p), \tag{9.36a}$$

$$L_{22} \equiv {}^4\mathbf{L}_p(\overset{(n)}{\mathbf{T}},\ \overset{(n)}{\mathbf{C}}_p,\ \theta,\ \overset{(n)}{\mathbf{T}}{}^{\bullet},\ \overset{(n)}{\mathbf{C}}{}^{\bullet}_p,\ w_p), \tag{9.36b}$$

$$L_{11}\theta = \lambda(\theta),\ L_{12} = 0,\ L_{13} = 0,\ L_{23} = 0,\ L_{33} = 0. \tag{9.36c}$$

The dependence on $\dot{\theta}$ and $\rho$ in these models is neglected; therefore, instead of the argument $\overset{(n)}{\mathbf{T}}/\rho$ the function $L_{22}$ can always have the argument $\overset{(n)}{\mathbf{T}}$ (because $\rho = \rho(\overset{(n)}{\mathbf{C}})$ and $\dot{\rho}$ can always be expressed in terms of $\overset{(n)}{\mathbf{T}}{}^{\bullet}$ and $\overset{(n)}{\mathbf{C}}{}^{\bullet}_p$). These models do not consider the cross–effects in relations (9.34), and the tensor $\mathbf{C}_e^1$ is identically zero:

$$\mathbf{C}_e^1 \equiv 0. \tag{9.37}$$

The scalar functional $w_p$ is usually chosen in the form

$$w_p = \int_0^t \overset{(n)}{\mathbf{T}}(\tau) \cdot\cdot\ \overset{(n)}{\mathbf{C}}{}^{\bullet}_p(\tau)\, d\tau, \tag{9.38}$$

called the *Taylor parameter* (the specific work done by plastic deformations), or in the form

$$w_p = \int_0^t \left( \overset{(n)}{\mathbf{C}}{}^{\bullet}_p(\tau) \cdot\cdot \overset{(n)}{\mathbf{C}}{}^{\bullet}_p(\tau) \right)^{1/2} d\tau \tag{9.39}$$

called the *Odkwist parameter*.

In this case relations (9.34) and (9.28d) have the following general form:

$$-\mathbf{q} = \boldsymbol{\lambda} \cdot \nabla\theta, \tag{9.40a}$$

$$\overset{(n)}{\mathbf{C}}{}^{\bullet}_p = {}^4\mathbf{L}_p \cdot\cdot \overset{(n)}{\mathbf{T}}_H. \tag{9.40b}$$

$$w^* = (\overset{(n)}{\mathbf{T}}_H - \rho \frac{\partial \zeta}{\partial w_p} \overset{(n)}{\mathbf{T}}) \cdot\cdot \overset{(n)}{\mathbf{C}}{}^{\bullet}_p. \tag{9.40c}$$

From the inequality (9.31) it follows that the heat conductivity tensor $\boldsymbol{\lambda}$ is symmetric and positive-definite, and the tensor ${}^4\mathbf{L}_p$ is symmetric in pairs of indices $(1, 2) \leftrightarrow (3, 4)$ (the symmetry in indices $1 \leftrightarrow 2$ and $3 \leftrightarrow 4$ follows from symmetry of the tensors $\overset{(n)}{\mathbf{C}}{}^{\bullet}_p$ and $\overset{(n)}{\mathbf{T}}$, $\overset{(n)}{\mathbf{H}}_p$), i.e. the tensor is symmetric in the form (7.21), has not more than 21 independent components and is also positive-definite.

### 9.1.5 Associated Model of Plasticity $A_n^{\zeta}$

In applications one frequently uses the *associated model of plasticity*, where the law of plastic yield (9.40b) is connected (or associated) with the concept of a yield surface. Consider this model.

Let in the six-dimensional space of components $\overset{(n)}{T}{}^{ij}$ of the stress tensor $\overset{(n)}{\mathbf{T}}$ with respect to some basis, for example, with respect to the basis $\overset{\circ}{\mathbf{r}}_i$, there be a surface whose position is given by the scalar equation system

$$f_\beta = 0, \qquad \beta = 1, \ldots, k. \tag{9.41}$$

Here $f_\beta$ are functions in the form (9.36a)

$$f_\beta = f_\beta(\overset{(n)}{\mathbf{T}}, \overset{(n)}{\mathbf{C}}_p, \theta, w_p), \tag{9.42}$$

depending parametrically on $\overset{(n)}{\mathbf{C}}_p$, $\theta$ and $w_p$, they are called the *plastic potentials*.

Denote the partial derivative of functions (9.42) with respect to time by

$$\frac{d' f_\beta}{dt} = \frac{\partial f_\beta}{\partial \overset{(n)}{\mathbf{T}}} \cdot\cdot \overset{(n)}{\mathbf{T}}{}^{\bullet}, \tag{9.43}$$

that coincides with the total derivative $f^{\bullet}_{\beta} = df_{\beta}/dt$ only if $f_{\beta}$ is independent of $\overset{(n)}{\mathbf{C}}_p$, $\theta$ and $w_p$:

$$f_{\beta} = f_{\beta}(\overset{(n)}{\mathbf{T}}). \tag{9.44}$$

In this case one can say that the *model of an ideally plastic continuum* is considered. The associated model (9.42) taking such dependence into account is called the model of a *strengthening plastic continuum.*

We assume axiomatically that:

1. Inside a domain bounded by the yield surface (9.41) plastic deformations remain unchanged, i.e.

$$\text{if all} \quad f_{\beta} < 0, \quad \text{then} \quad \overset{(n)}{\mathbf{C}}{}^{\bullet}_p = 0; \tag{9.45}$$

   if at least for one value of $\beta$ the condition $df_{\beta}/dt > 0$ is satisfied, then one can say that there occurs *active loading*, and if all $d' f_{\beta}/dt \leqslant 0$, then there occurs *passive loading* or *unloading*.
2. On the yield surface when $d' f_{\beta}/dt = 0$, plastic deformations remain also unchanged (such loading is called *neutral*), and if $d' f_{\beta}/dt > 0$, then $\overset{(n)}{\mathbf{C}}_p$ vary (one can say that there occurs *plastic loading*), i.e.

$$\text{if} \quad f_{\beta} = 0, \quad d' f_{\beta}/dt = 0, \text{ then } \overset{(n)}{\mathbf{C}}{}^{\bullet}_p = 0, \tag{9.46}$$

$$\text{if} \quad f_{\beta} = 0, \quad d' f_{\beta}/dt > 0, \text{ then } \overset{(n)}{\mathbf{C}}{}^{\bullet}_p \neq 0. \tag{9.47}$$

Notice that the state of a continuum when its tensor $\overset{(n)}{\mathbf{T}}$ is outside of the yield surface is impossible, because we assume axiomatically that the yield surface moves with changing the tensor $\overset{(n)}{\mathbf{T}}$ if $\overset{(n)}{\mathbf{C}}{}^{\bullet}_p \neq 0$; i.e. the condition $f > 0$ cannot be satisfied.

The specific expression for the rate of plastic deformation $\overset{(n)}{\mathbf{C}}{}^{\bullet}_p$ in the case (9.47) of plastic loading is given by the *Drucker model* (or the *gradient law*). According to this model, the tensor $\overset{(n)}{\mathbf{C}}{}^{\bullet}_p$ is chosen to be proportional to the gradient of the yield surface:

$$\overset{(n)}{\mathbf{C}}{}^{\bullet}_p = \sum_{\alpha=1}^{k} \dot{\varkappa}_{\alpha}(\partial f_{\alpha}/\partial \overset{(n)}{\mathbf{T}}). \tag{9.48}$$

Here $\dot{\varkappa}_{\alpha}$ are the ratio coefficients, which can be written in the convenient form of derivatives with respect to time and which are scalar functions in the form

$$\dot{\varkappa}_{\alpha} = \dot{\varkappa}_{\alpha}(\overset{(n)}{\mathbf{T}}, \overset{(n)}{\mathbf{C}}_p, \overset{(n)}{\mathbf{C}}{}^{\bullet}_p, \theta, w_p), \quad \alpha = 1, \ldots, k. \tag{9.49}$$

If $k \leqslant 6$, then one can find these functions from the gradient equation with complementing the tensor equation (9.48) by $k$ equations (9.41) of the yield surface. Thus, we have $6+k$ scalar equations (9.48) and (9.41) to determine six components of the tensor $\overset{(n)}{\mathbf{C}}{}_p^{\bullet}$ and $k$ functions $\varkappa_\alpha$ $(\alpha = 1, \ldots, k)$.

Notice that the gradient law (9.48) is written only for plastic loading. In order to derive an expression for $\overset{(n)}{\mathbf{C}}{}_p^{\bullet}$ under arbitrary loading, we should combine the relations (9.45)–(9.47). This can be done with the help of the Heaviside functions $h_+(x)$ and $h_-(x)$:

$$h_+(x) = \begin{cases} 1, & x \geqslant 0, \\ 0, & x < 0, \end{cases} \qquad h_-(x) = \begin{cases} 1, & x > 0, \\ 0, & x \leqslant 0, \end{cases} \tag{9.50}$$

and their combinations

$$h = 1 - \prod_{\beta=1}^{k} \left(1 - h_+(f_\beta) h_-\left(\frac{d' f_\beta}{dt}\right)\right). \tag{9.51}$$

Here $\prod_{\beta=1}^{k}$ is the product. One can easily verify that if the conditions (9.45) or (9.46) are satisfied, then $h = 0$, and if the condition (9.47) is satisfied, then $h = 1$. Thus, for $\overset{(n)}{\mathbf{C}}{}_p^{\bullet}$ under arbitrary loading we obtain

$$\overset{(n)}{\mathbf{C}}{}_p^{\bullet} = h \sum_{\alpha=1}^{k} \dot{\varkappa}_\alpha (\partial f_\alpha / \partial \overset{(n)}{\mathbf{T}}). \tag{9.52}$$

This relation must satisfy the corollary of the Onsager principle (9.40b) for models of plastic yield; i.e. the following equation must hold

$$h \sum_{\alpha=1}^{k} \dot{\varkappa}_\alpha (\partial f_\beta / \partial \overset{(n)}{\mathbf{T}}) = {}^4\mathbf{L}_p \cdot\cdot \overset{(n)}{\mathbf{T}}{}_H. \tag{9.53}$$

Here ${}^4\mathbf{L}_p$ is some indeterminate symmetric fourth-order tensor. Equation (9.53) means that the functions $f_\beta$ called *plastic potentials* must be quasilinear functions of $\overset{(n)}{\mathbf{T}} - \overset{(n)}{\mathbf{H}}{}_p$.

*Remark.* For the associated model of plasticity, a part of constitutive equations for $\overset{(n)}{\mathbf{C}}{}_p^{\bullet}$ is given by Eqs. (9.41) being the implicit forms of components of the tensor $\overset{(n)}{\mathbf{C}}{}_p^{\bullet}$. These equations can also be represented by the expression (9.40b) written in the implicit form

$$\Phi_\beta \cdot\cdot \overset{(n)}{\mathbf{C}}{}_p^{\bullet} - \varphi_\beta = 0, \tag{9.54}$$

where $\Phi_\beta$ are symmetric second-order tensors, and

$$\varphi_\beta = \Phi_\beta \cdot\cdot {}^4\mathbf{L}_p \cdot\cdot (\overset{(n)}{\mathbf{T}} - \overset{(n)}{\mathbf{H}}_p). \tag{9.55}$$

Indeed, differentiating (9.41) with respect to $t$, we obtain that

$$\Phi_\beta = \partial f / \partial \overset{(n)}{\mathbf{C}}_p, \quad \varphi_\beta = \frac{\partial f_\beta}{\partial \overset{(n)}{\mathbf{T}}} \cdot\cdot \overset{(n)}{\mathbf{T}}{}^{\bullet} + \frac{\partial f_\beta}{\partial \theta}\dot{\theta}. \tag{9.56}$$

Since relations (9.44) are scalar, the corollary (9.54) of the Onsager principle imposes no constraints on the form of the plastic potentials $f_\beta$, and the relation

$$\frac{\partial f_\beta}{\partial \overset{(n)}{\mathbf{T}}} \cdot\cdot \overset{(n)}{\mathbf{T}}{}^{\bullet} + \frac{\partial f_\beta}{\partial \theta}\dot{\theta} = \frac{\partial f_\beta}{\partial \overset{(n)}{\mathbf{C}}_p} \cdot\cdot {}^4\mathbf{L}_p \cdot\cdot \overset{(n)}{\mathbf{T}}_p, \tag{9.57}$$

being a consequence of (9.55) and (9.56) and an analog of formula (9.53), can always be satisfied by the proper choice of indeterminate tensor ${}^4\mathbf{L}_p$. □

For the elastic deformations $\overset{(n)}{\mathbf{C}}_e$ in the associated model, from (9.28a) and (9.37) we get the relation

$$\overset{(n)}{\mathbf{C}}_e = \mathbf{K}(\overset{(n)}{\mathbf{T}}/\rho, \overset{(n)}{\mathbf{C}}_p, \theta, w_p) = -\frac{\partial \zeta}{\partial \overset{(n)}{\mathbf{T}}/\rho}. \tag{9.58}$$

The model of a plastic continuum, where the potential $\zeta$ does not depend explicitly on the plastic deformation tensor $\overset{(n)}{\mathbf{C}}_p$, is called the *model $A_n^\zeta$ of an elastoplastic continuum*; for this model,

$$\zeta = \zeta(\overset{(n)}{\mathbf{T}}/\rho, \theta, w_p), \quad \overset{(n)}{\mathbf{H}}_p \equiv 0, \quad \overset{(n)}{\mathbf{C}}_e = \mathbf{K}(\overset{(n)}{\mathbf{T}}/\rho, \theta, w_p) = -\rho\frac{\partial \zeta}{\partial \overset{(n)}{\mathbf{T}}/\rho}. \tag{9.59}$$

This model is widely used in practice. A model with explicit dependence of $\zeta$ upon $\overset{(n)}{\mathbf{C}}_p$ is usually applied when one needs the effect of deformation anisotropy to be taken into account. This effect consists in the change of a symmetry group for a continuum considered in the process of varying plastic deformations.

The model, where the dependence of $\zeta$ upon both $\overset{(n)}{\mathbf{C}}_p$ and $w_p$ may be neglected, is called the *model of an ideally elastoplastic continuum* (it is not to be confused with an ideally plastic continuum, where, according to the definitions (9.44), the plastic potentials $f_\beta$ are independent of $\overset{(n)}{\mathbf{C}}_p$ and $w_p$).

Finally, if $\zeta$ is independent of $\overset{(n)}{\mathbf{C}}_p$ and $w_p$, and depends on $\overset{(n)}{\mathbf{T}}/\rho$ only quadratically, then one can say that this is the *model* $A_n^\zeta$ *of a plastic continuum with linear elasticity*.

### 9.1.6 Corollary of the Principle of Material Symmetry for the Associated Model $A_n^\zeta$ of Plasticity

Apply the principle of material symmetry to constitutive equations (9.53) and (9.58). Since all the tensors $\overset{(n)}{\mathbf{C}}_e$, $\overset{(n)}{\mathbf{C}}_p$ and $\overset{(n)}{\mathbf{T}}$ are $H$-indifferent relative to orthogonal $H$-transformations, from the principle of material symmetry it follows that the elastic potential $\zeta$ (9.36a) must be a function of simultaneous invariants $J_\gamma^{(s)}$ of the tensors $\overset{(n)}{\mathbf{T}}$ and $\overset{(n)}{\mathbf{C}}_p$ relative to some group $\overset{\circ}{G}_s$ in an undistorted reference configuration $\overset{\circ}{\mathcal{K}}$:

$$\zeta = \zeta\left(J_\gamma^{(s)},\ \theta,\ w_\beta^p\right). \tag{9.60}$$

Here

$$J_\gamma^{(s)} = J_\gamma^{(s)}\left(\overset{(n)}{\mathbf{T}},\ \overset{(n)}{\mathbf{C}}_p\right), \qquad \gamma = 1,\ldots,z. \tag{9.61}$$

A proof of this assertion is the same as the one for continua of the differential type (see Sect. 7.1.4).

For simplicity, a reference configuration $\overset{\circ}{\mathcal{K}}$ is assumed to be undistorted; and, as before (see Sects. 4.7.3 and 8.1.6), we will consider groups $\overset{\circ}{G}_s$ only in this configuration.

In place of one scalar functional $w_p$ (9.38), the set of arguments of the function $\zeta$ (9.60) in general may include functionals $w_\beta^p$ being integrals of such simultaneous invariants $J_\gamma^{(s)}$ that contain both the tensors $\overset{(n)}{\mathbf{T}}$ and $\overset{(n)}{\mathbf{C}}{}^{\bullet}_p$:

$$w_\beta^p = \int_0^t J_\beta^{(s)}(\overset{(n)}{\mathbf{T}}(\tau),\ \overset{(n)}{\mathbf{C}}{}^{\bullet}_p(\tau))\, d\tau, \quad \beta = 1,\ldots,z. \tag{9.62}$$

Then the constitutive equations (9.58) can be represented in the tensor basis by analogy with ideal continua:

$$\overset{(n)}{\mathbf{C}}_e = \sum_{\gamma=1}^{z} \varphi_\gamma J_{\gamma\mathbf{T}}^{(s)}. \tag{9.63}$$

Here $\varphi_\gamma$ are scalar functions in the form (9.60), and $J^{(s)}_{\gamma\mathbf{T}}$ are the derivative tensors:

$$\varphi_\gamma = \varphi_\gamma\left(J^{(s)}_\alpha,\ \theta,\ w^p_\beta\right) = -\partial\zeta/\partial J^{(s)}_\gamma,\quad J^{(s)}_{\gamma\mathbf{T}} = \frac{\partial J^{(s)}_\gamma}{\partial \overset{(n)}{\mathbf{T}}/\rho},\qquad \gamma = 1,\ldots,z. \quad (9.64)$$

As follows from the principle of material symmetry, the plasticity functions $f_\beta$ (9.42) are also functions of simultaneous invariants $J^{(s)p}_\alpha$:

$$f_\beta = f_\beta\left(J^{(s)p}_\alpha,\ \theta,\ w^p_\beta\right),\qquad \beta = 1,\ldots,k. \quad (9.65)$$

However, according to the corollary (9.53) of the Onsager principle, the derivatives $\partial f_\beta/\partial \overset{(n)}{\mathbf{T}}$ must be quasilinear functions of the tensor $\overset{(n)}{\mathbf{T}}_p$; therefore, in (9.65) $J^{(s)p}_\alpha$ must be simultaneous invariants of $\overset{(n)}{\mathbf{T}}_p$ and $\overset{(n)}{\mathbf{C}}_p$, being only linear and quadratic functions of these tensors:

$$J^{(s)p}_\alpha = J^{(s)}_\alpha(\overset{(n)}{\mathbf{T}}_H, \overset{(n)}{\mathbf{C}}_p),\qquad \alpha = 1,\ldots,z_1 \leqslant z. \quad (9.66)$$

Then constitutive equations (9.52) in the tensor basis take the form

$$\overset{(n)}{\mathbf{C}}{}^{\bullet}_p = \sum_{\alpha=1}^{z_1} \psi_\alpha J^{(s)p}_{\alpha\mathbf{T}}, \quad (9.67)$$

where

$$\psi_\alpha = h\sum_{\beta=1}^{k} \dot{\varkappa}_\beta \frac{\partial f_\beta}{\partial J^{(s)p}_\alpha},\qquad J^{(s)p}_{\alpha\mathbf{T}} = \partial J^{(s)p}_\alpha/\partial \overset{(n)}{\mathbf{T}}. \quad (9.68)$$

Relations (9.60)–(9.68) are called the *representation of the associated model of plasticity in the tensor basis.*

The derivative tensors $J^{(s)p}_{\alpha\mathbf{T}}$ and $J^{(s)}_{\alpha\mathbf{T}}$ are connected by the relation

$$J^{(s)p}_{\alpha\mathbf{T}} = \frac{\partial J^{(s)p}_\alpha}{\partial \overset{(n)}{\mathbf{T}}_H}\cdot\cdot\frac{\partial(\overset{(n)}{\mathbf{T}} - \overset{(n)}{\mathbf{H}}_p)}{\partial \overset{(n)}{\mathbf{T}}} = J^{(s)p}_{\alpha\mathbf{T}_H}\cdot\cdot(\mathbf{\Delta} - {}^4\overset{(n)}{\mathbf{H}}_{p\mathbf{T}}). \quad (9.69)$$

With the help of (9.27) and (9.59) we find an expression for the fourth-order tensor

$$\begin{aligned}{}^4\overset{(n)}{\mathbf{H}}_{p\mathbf{T}} \equiv \frac{\partial \overset{(n)}{\mathbf{H}}_p}{\partial \overset{(n)}{\mathbf{T}}} &= \sum_{\gamma=1}^{z} \frac{\partial}{\partial \overset{(n)}{\mathbf{T}}}\left(\frac{\partial\zeta}{\partial J^{(s)}_\gamma}\frac{\partial J^{(s)}_\gamma}{\partial \overset{(n)}{\mathbf{C}}_p}\right)\\ &= \sum_{\gamma,\beta=1}^{z}\left(\frac{\varphi_{\gamma\beta}}{2}\left(J^{(s)}_{\beta\mathbf{T}}\otimes J^{(s)}_{\gamma\mathbf{C}_p} + J^{(s)}_{\gamma\mathbf{C}_p}\otimes J^{(s)}_{\gamma\mathbf{T}}\right) + \delta_{\gamma\beta}\varphi_\gamma J^{(s)}_{\gamma\mathbf{T}\mathbf{C}_p}\right).\end{aligned} \quad (9.70)$$

Here we have introduced the notation

$$\varphi_{\gamma\beta} = \frac{\partial \varphi_\gamma}{\partial J_\beta^{(s)}} = \frac{\partial^2 \zeta}{\partial J_\beta^{(s)} \partial J_\gamma^{(s)}}, \qquad J_{\gamma \mathbf{T} \mathbf{C}_p}^{(s)} = \frac{\partial^2 J_\gamma^{(s)}}{\partial \overset{(n)}{\mathbf{T}} \partial \overset{(n)}{\mathbf{C}}_p}. \tag{9.71}$$

On substituting (9.68) into (9.67), we finally obtain

$$\overset{(n)}{\mathbf{C}}{}^{\bullet}_p = \sum_{\alpha=1}^{z_1} \psi_\alpha J_{\alpha \mathbf{T}_H}^{(s)p} \cdot\cdot (\boldsymbol{\Delta} - {}^4\overset{(n)}{\mathbf{H}}_{p\mathbf{T}}). \tag{9.72}$$

For the associated model of an elastoplastic continuum, $\zeta$ is independent of $\overset{(n)}{\mathbf{C}}_p$, hence

$$\overset{(n)}{\mathbf{H}}_p \equiv 0, \qquad 4\overset{(n)}{\mathbf{H}}_{p\mathbf{T}} \equiv 0, \tag{9.73}$$
$$J_\gamma^{(s)p} = J_\gamma^{(s)}, \quad J_{\gamma\mathbf{T}}^{(s)p} = J_{\gamma\mathbf{T}_H}^{(s)p} = J_{\gamma\mathbf{T}}^{(s)}, \quad J_{\gamma\mathbf{C}_p}^{(s)} = 0, \quad J_{\gamma\mathbf{T}\mathbf{C}_p}^{(s)} = 0.$$

Thus, the invariants $J_\gamma^{(s)p}$ and $J_\gamma^{(s)}$ coincide, and constitutive equations (9.60), (9.63), (9.65), and (9.72) take the forms

$$\zeta = \zeta\left(I_\gamma^{(s)}(\overset{(n)}{\mathbf{T}}/\rho),\ \theta,\ w_\beta^p\right), \quad \gamma = 1, \ldots, r, \tag{9.74a}$$

$$\overset{(n)}{\mathbf{C}}_e = \sum_{\gamma=1}^{r} \varphi_\gamma I_{\gamma\mathbf{T}}^{(s)}(\overset{(n)}{\mathbf{T}}/\rho), \tag{9.74b}$$

$$\overset{(n)}{\mathbf{C}}{}^{\bullet}_p = \sum_{\alpha=1}^{z_1} \psi_\alpha J_{\alpha\mathbf{T}}^{(s)}(\overset{(n)}{\mathbf{T}},\ \overset{(n)}{\mathbf{C}}_p), \tag{9.74c}$$

$$f_\beta = f_\beta(J_\alpha^{(s)}(\overset{(n)}{\mathbf{T}},\ \overset{(n)}{\mathbf{C}}_p),\ \theta,\ w_\beta^p), \tag{9.74d}$$

$$\varphi_\gamma = -\frac{\partial \zeta}{\partial I_\gamma^{(s)}}, \quad I_{\gamma\mathbf{T}}^{(s)} = \frac{\partial I_\gamma^{(s)}(\overset{(n)}{\mathbf{T}}/\rho)}{\partial \overset{(n)}{\mathbf{T}}/\rho}, \quad \psi_\alpha = h \sum_{\beta=1}^{k} \dot{\varkappa}_\beta \frac{\partial f_\beta}{\partial J_\alpha^{(s)}}. \tag{9.74e}$$

### 9.1.7 Associated Models of Plasticity $A_n^\zeta$ for Isotropic Continua

Let us write representations (9.74) for the three main symmetry groups $G_s$: $O$, $T_3$ and $I$, with choosing simultaneous invariants $J_\gamma^{(s)}$ in the same way as it was done for viscoelastic continua.

For an isotropic continuum, the functional basis of simultaneous invariants $J_\gamma^{(I)}(\overset{(n)}{\mathbf{T}}/\rho, \overset{(n)}{\mathbf{C}}_p)$ consists of nine invariants, which can be chosen as follows (see (7.35) and (8.83)):

$$J_\alpha^{(I)} = I_\alpha(\overset{(n)}{\mathbf{T}}/\rho), \quad J_{\alpha+3}^{(I)} = I_\alpha(\overset{(n)}{\mathbf{C}}_p), \quad \alpha = 1,\ 2,\ 3,$$
$$J_7^{(I)} = \frac{1}{\rho}\overset{(n)}{\mathbf{T}} \cdot\cdot \overset{(n)}{\mathbf{C}}_p, \quad J_8^{(I)} = \frac{1}{\rho}\overset{(n)}{\mathbf{T}} \cdot\cdot \overset{(n)}{\mathbf{C}}{}_p^2, \quad J_9^{(I)} = \frac{1}{\rho^2}\overset{(n)}{\mathbf{T}}{}^2 \cdot\cdot \overset{(n)}{\mathbf{C}}_p, \tag{9.75}$$

Then the derivative tensors $J_{\gamma\mathbf{T}}^{(s)}$ (9.64) have the forms (see (8.84))

$$J_{1\mathbf{T}}^{(I)} = \mathbf{E}, \quad J_{2\mathbf{T}}^{(I)} = \tfrac{1}{\rho}(\mathbf{E}I_1(\overset{(n)}{\mathbf{T}}) - \overset{(n)}{\mathbf{T}}), \quad J_{3\mathbf{T}}^{(I)} = \tfrac{1}{\rho^2}(\overset{(n)}{\mathbf{T}}{}^2 - I_1\overset{(n)}{\mathbf{T}} + \mathbf{E}I_2),$$
$$J_{\alpha+3,\mathbf{T}}^{(I)} = \mathbf{0}, \ \alpha = 1,\ 2,\ 3; \ J_{7\mathbf{T}}^{(I)} = \overset{(n)}{\mathbf{C}}_p, \quad J_{8\mathbf{T}}^{(I)} = \overset{(n)}{\mathbf{C}}{}_p^2,$$
$$J_{9\mathbf{T}}^{(I)} = \tfrac{1}{\rho}(\overset{(n)}{\mathbf{T}} \cdot \overset{(n)}{\mathbf{C}}_p + \overset{(n)}{\mathbf{C}}_p \cdot \overset{(n)}{\mathbf{T}}). \tag{9.76}$$

Substituting these expressions into (9.63) and collecting terms with the same tensor powers, we obtain the following representation of constitutive equations (9.63) in the tensor basis:

$$\overset{(n)}{\mathbf{C}}_e = \widetilde{\varphi}_1\mathbf{E} + \frac{\widetilde{\varphi}_2}{\rho}\overset{(n)}{\mathbf{T}} + \frac{\varphi_3}{\rho^2}\overset{(n)}{\mathbf{T}}{}^2 + \widetilde{\varphi}_4\overset{(n)}{\mathbf{C}}_p + \widetilde{\varphi}_5\overset{(n)}{\mathbf{C}}{}^2 + \frac{\widetilde{\varphi}_6}{\rho}(\overset{(n)}{\mathbf{T}} \cdot \overset{(n)}{\mathbf{C}}_p + \overset{(n)}{\mathbf{C}}_p \cdot \overset{(n)}{\mathbf{T}}). \tag{9.77}$$

Here we have denoted the scalar functions (compare with (4.322a))

$$\widetilde{\varphi}_1 = \varphi_1 + \varphi_2 I_1 + \varphi_3 I_2, \quad -\widetilde{\varphi}_2 = \varphi_2 + \varphi_3 I_1, \ \widetilde{\varphi}_{3+\gamma} = \varphi_{6+\gamma}, \ \gamma = 1,\ 2,\ 3. \tag{9.78}$$

For the model of an elastoplastic continuum (9.59), the functions $\widetilde{\varphi}_4$, $\widetilde{\varphi}_5$ and $\widetilde{\varphi}_6$, due to (9.64), are zero; and we obtain the relation

$$\overset{(n)}{\mathbf{C}}_e = \widetilde{\varphi}_1\mathbf{E} + (\widetilde{\varphi}_2/\rho)\overset{(n)}{\mathbf{T}} + (\varphi_3/\rho^2)\overset{(n)}{\mathbf{T}}{}^2, \tag{9.79}$$

which is analogous to the constitutive equation of an ideally elastic isotropic continuum (4.322) but written in the inverse form. Here

$$\varphi_\gamma = -\frac{\partial\zeta}{\partial I_\gamma}, \quad \zeta = \zeta(I_1(\overset{(n)}{\mathbf{T}}/\rho),\ I_2(\overset{(n)}{\mathbf{T}}/\rho),\ I_3(\overset{(n)}{\mathbf{T}}/\rho),\ \theta, w_p), \tag{9.80}$$

and there is only one Taylor parameter $w_p$.

For an elastoplastic continuum, according to the Onsager principle, the set of arguments of the plasticity functions $f_\beta$ includes only linear and quadratic invariants, i.e. only $J_\gamma^{(I)}(\overset{(n)}{\mathbf{T}}, \overset{(n)}{\mathbf{C}}_p)$, $\gamma = 1, 2, 4, 5, 7$, $(z_1 = 5)$, among which there is only one simultaneous invariant, namely $J_7^{(I)}$. On substituting the derivative tensors (9.76) into (9.74c), we obtain

$$\overset{(n)}{\mathbf{C}}{}^{\bullet}_p = \widetilde{\psi}_1 \mathbf{E} - \psi_2 \overset{(n)}{\mathbf{T}} + \psi_7 \overset{(n)}{\mathbf{C}}_p, \tag{9.81}$$

where $\psi_1$, $\psi_2$, and $\psi_7$ are determined by formulae (9.68):

$$\psi_\alpha = h \sum_{\beta=1}^{k} \varkappa_\beta \frac{\partial f_\beta}{\partial J_\alpha(\overset{(n)}{\mathbf{T}})}, \quad \psi_7 = h \sum_{\beta=1}^{k} \varkappa_\beta \frac{\partial f_\beta}{\partial J_7^{(I)}}, \quad \widetilde{\psi}_1 = \psi_1 + \psi_2 I_1(\overset{(n)}{\mathbf{T}}), \tag{9.82}$$

$$f_\beta = f_\gamma(I_1(\overset{(n)}{\mathbf{T}}),\ I_2(\overset{(n)}{\mathbf{T}}),\ I_1(\overset{(n)}{\mathbf{C}}_p),\ I_2(\overset{(n)}{\mathbf{C}}_p),\ J_7^{(I)}(\overset{(n)}{\mathbf{T}}, \overset{(n)}{\mathbf{C}}_p),\ \theta,\ w_p). \tag{9.83}$$

Formulae (9.78)–(9.82) give the general form of constitutive equations of the associated model of an isotropic elastoplastic continuum.

### *9.1.8 The Huber–Mises Model for Isotropic Plastic Continua*

For special models of isotropic elastoplastic continua, one usually accepts additional assumptions on a form of the elastic and plastic potentials $\zeta$ and $f_\beta$. As shown in experiments, for many elastoplastic continua, their behavior can be described adequately enough by the *Huber–Mises model*, where there is only one plastic potential $f$ depending explicitly on the only simultaneous invariant $Y_H$:

$$f = f(Y_H, \theta, w_p). \tag{9.84}$$

Here the invariant $Y_H$ has been introduced as contraction of the tensor $\mathbf{P}_H$ being the deviator of the tensor $\overset{(n)}{\mathbf{T}} - H \overset{(n)}{\mathbf{C}}_e$ (see the definition of the deviator (8.167)):

$$Y_H^2 = \frac{3}{2} \mathbf{P}_H \cdot\cdot\, \mathbf{P}_H, \quad \mathbf{P}_H = (\overset{(n)}{\mathbf{T}} - H \overset{(n)}{\mathbf{C}}_p) - \frac{1}{3} I_1(\overset{(n)}{\mathbf{T}} - H \overset{(n)}{\mathbf{C}}_p)\mathbf{E}, \tag{9.85}$$

and $H$ is the *strengthening parameter* being a scalar function in the form

$$H = H_0\, Y_p^{2n_0}, \tag{9.86}$$

where $H_0$ and $n_0$ are the constants, and $Y_p$ is the invariant of the deviator of the tensor $\overset{(n)}{\mathbf{C}}_p$ determined similarly to (9.85):

$$Y_p^2 = \frac{3}{2}\mathbf{P}_p \cdot\cdot \mathbf{P}_p, \quad \mathbf{P}_p = \overset{(n)}{\mathbf{C}}_p - \frac{1}{3}I_1(\overset{(n)}{\mathbf{C}}_p)\mathbf{E}. \tag{9.87}$$

The invariants $Y_H$ and $Y_p$ are called the *intensities* of the tensors $\overset{(n)}{\mathbf{T}} - H\overset{(n)}{\mathbf{C}}_p$ and $\overset{(n)}{\mathbf{C}}_p$, respectively.

The deviator can be constructed for any tensor; for example, for the tensor $\overset{(n)}{\mathbf{T}}$:

$$\mathbf{P}_T = \overset{(n)}{\mathbf{T}} - \frac{1}{3}I_1(\overset{(n)}{\mathbf{T}})\mathbf{E}, \quad Y_T^2 = \frac{3}{2}\mathbf{P}_T \cdot\cdot \mathbf{P}_T, \tag{9.88}$$

where $Y_T$ is the intensity of the tensor $\overset{(n)}{\mathbf{T}}$.

Each deviator of a tensor is orthogonal to the unit (metric) tensor (more detailed information on properties of the deviators can be found in [12]):

$$\mathbf{P}_H \cdot\cdot \mathbf{E} = 0, \quad \mathbf{P}_p \cdot\cdot \mathbf{E} = 0, \quad \mathbf{P}_T \cdot\cdot \mathbf{E} = 0. \tag{9.89}$$

The invariants $Y_T$ and $Y_p$ can be expressed in terms of the principal invariants of corresponding tensors (see Exercise 9.1.2):

$$Y_T^2 = I_1^2(\overset{(n)}{\mathbf{T}}) - 3I_2(\overset{(n)}{\mathbf{T}}), \quad Y_p^2 = I_1^2(\overset{(n)}{\mathbf{C}}_p) - 3I_2(\overset{(n)}{\mathbf{C}}_p). \tag{9.90}$$

We can immediately verify that the invariant $Y_H^2$ can be expressed in terms of the invariants $I_\alpha(\overset{(n)}{\mathbf{T}})$, $I_\alpha(\overset{(n)}{\mathbf{C}}_p)$ and simultaneous invariant $J_7^{(I)}$:

$$Y_H^2 = Y_T^2 + H^2Y_p^2 - 3HJ_7^{(I)} + HI_1(\overset{(n)}{\mathbf{T}})I_1(\overset{(n)}{\mathbf{C}}_p). \tag{9.91}$$

Determining the derivatives $\partial f/\partial I_\alpha(\overset{(n)}{\mathbf{T}})$ of the function (9.84):

$$\frac{\partial f}{\partial I_\alpha} = \frac{\partial f}{\partial Y_\alpha}\frac{\partial Y_H}{\partial I_\alpha} = f_Y\frac{\partial Y_H^2}{\partial I_\alpha}, \quad f_Y \equiv \frac{1}{2Y_H}\frac{\partial f}{\partial Y_H},$$

$$\frac{\partial f}{\partial I_1} = f_Y(2I_1(\overset{(n)}{\mathbf{T}}) + HI_1(\overset{(n)}{\mathbf{C}}_p)), \; \frac{\partial f}{\partial I_2} = -\,3f_Y, \; \frac{\partial f}{\partial I_7^{(I)}} = -\,3Hf_Y, \tag{9.91a}$$

and substituting them into (9.82), we find that

$$\psi_1 = \dot{\varkappa}hf_Y(2I_1(\overset{(n)}{\mathbf{T}})+HI_1(\overset{(n)}{\mathbf{C}}_p)), \quad \psi_2 = -3\dot{\varkappa}hf_Y, \quad \psi_7 = -3\dot{\varkappa}hf_YH. \tag{9.91b}$$

Then constitutive equations (9.81) take the form

$$\overset{(n)}{\mathbf{C}}{}^{\bullet}_{p} = 3\dot{\varkappa} h f_Y \mathbf{P}_H. \tag{9.92}$$

Due to the property (9.89) of the deviators, from (9.92) it follows that a continuum described by the Huber–Mises model is *plastically incompressible*, i.e.

$$\overset{(n)}{\mathbf{C}}{}^{\bullet}_{p} \cdot\cdot \mathbf{E} = 0 \quad \text{or} \quad I_1(\overset{(n)}{\mathbf{C}}_p) = 0. \tag{9.93}$$

Therefore, with the help of (9.87) Eq. (9.92) can be written as the following quasi-linear relation between the deviators:

$$\dot{\mathbf{P}}_p = 3\dot{\varkappa} h f_Y \mathbf{P}_H. \tag{9.94}$$

The scalar product of Eq. (9.94) by itself yields the following expression for $\dot{\varkappa}$ under plastic loading:

$$\dot{\varkappa} = \pm \frac{\sqrt{\dot{\mathbf{P}}_p \cdot\cdot \dot{\mathbf{P}}_p}}{\sqrt{6} f_Y Y_H}. \tag{9.95}$$

After substitution of expression (9.95) into (9.94) the number of independent equations in (9.94) reduces to four, and relationships (9.84), (9.93), and (9.94) form the complete system of constitutive equations (it consists of six independent relationships).

If the plastic potential is chosen in the Mises form

$$f = \frac{1}{3}(Y_H/\sigma_s)^2 - 1, \tag{9.96}$$

where $\sigma_s = \sigma_s(\theta, w_p)$ is the given function of $\theta$ and $w_p$ called the *yield point* or the yield strength, then $f_Y = 1/(3\sigma_s^2)$, and the final relations of plasticity (9.84), (9.92), and (9.93) become

$$\overset{(n)}{\mathbf{C}}{}^{\bullet}_{p} = \frac{\dot{\varkappa} h}{\sigma_s^2}(\mathbf{P}_T - H \overset{(n)}{\mathbf{C}}_p), \quad I_1(\overset{(n)}{\mathbf{C}}_p) = 0, \tag{9.97}$$

$$f = \frac{1}{3}(Y_H/\sigma_s)^2 - 1 = 0, \quad \mathbf{P}_T = \overset{(n)}{\mathbf{T}} - \frac{1}{3} I_1(\overset{(n)}{\mathbf{T}})\mathbf{E}, \quad H = H_0 Y_p^{2n_0}.$$

The Taylor parameter (9.38), due to plastic incompressibility (9.93), can be written in terms of the deviators:

$$w_p = \int_0^t \mathbf{P}_T \cdot\cdot \dot{\mathbf{P}}_p \, d\tau. \tag{9.98}$$

One can say that this is the *model of an isotropic plastic continuum with linear strengthening*, if the strengthening parameter $H$ (9.86) is a constant, i.e. $H = H_0$ and $n_0 = 0$.

### 9.1.9 Associated Models of Plasticity $A_n^\zeta$ for Transversely Isotropic Continua

For a *transversely isotropic* continuum, the functional basis of simultaneous invariants $J_\gamma^{(3)}(\overset{(n)}{\mathbf{T}}, \overset{(n)}{\mathbf{C}}_p)$ consists of 11 invariants, which can be chosen as follows (see (7.36) and (8.87)):

$$J_\gamma^{(3)} = I_\gamma^{(3)}(\overset{(n)}{\mathbf{T}}),\ \gamma = 1,\ldots,5;\ \ J_{5+\gamma}^{(3)} = I_\gamma^{(3)}(\overset{(n)}{\mathbf{C}}_p),\ \gamma = 1,\ldots,4;$$

$$J_{10}^{(3)} = \left((\mathbf{E} - \widehat{\mathbf{c}}_3^2)\cdot \overset{(n)}{\mathbf{T}} \cdot\cdot \left(\widehat{\mathbf{c}}_3^2 \cdot \overset{(n)}{\mathbf{C}}_p\right)\right),\ \ J_{11}^{(3)} = \overset{(n)}{\mathbf{T}} \cdot\cdot \overset{(n)}{\mathbf{C}}_p - 2J_{10}^{(3)} - J_2^{(3)}J_8^{(3)}. \tag{9.99}$$

The derivative tensors $J_{\gamma\mathbf{T}}^{(s)}$ (9.64) in this case have the forms (see (8.88))

$$J_{1\mathbf{T}}^{(3)} = \mathbf{E} - \widehat{\mathbf{c}}_3^2,\quad J_{2\mathbf{T}}^{(3)} = \widehat{\mathbf{c}}_3^2,\quad J_{3\mathbf{T}}^{(3)} = \frac{1}{2}(\mathbf{O}_1 \otimes \mathbf{O}_1 + \mathbf{O}_2 \otimes \mathbf{O}_2)\cdot\cdot \overset{(n)}{\mathbf{T}},$$

$$J_{4\mathbf{T}}^{(3)} = 2\,{}^4\mathbf{O}_3 \cdot\cdot \overset{(n)}{\mathbf{T}},\quad J_{5\mathbf{T}}^{(3)} = \overset{(n)}{\mathbf{T}}{}^2 - I_1\overset{(n)}{\mathbf{T}} + \mathbf{E}I_2,$$

$$J_{6\mathbf{T}}^{(3)} = J_{7\mathbf{T}}^{(3)} = J_{8\mathbf{T}}^{(3)} = J_{9\mathbf{T}}^{(3)} = 0, \tag{9.100}$$

$$J_{10\mathbf{T}}^{(3)} = \frac{1}{4}(\mathbf{O}_1 \otimes \mathbf{O}_1 + \mathbf{O}_2 \otimes \mathbf{O}_2)\cdot\cdot \overset{(n)}{\mathbf{C}}_p,\quad J_{11\mathbf{T}}^{(3)} = {}^4\mathbf{O}_3 \cdot\cdot \overset{(n)}{\mathbf{C}}_p.$$

Substituting these expressions into (9.63) and collecting terms with the same tensor powers, we obtain the following representation of constitutive equations (9.63) for a transversely isotropic plastic continuum in the tensor basis:

$$\overset{(n)}{\mathbf{C}}_e = \widetilde{\varphi}_1\mathbf{E} + \widetilde{\varphi}_2\widehat{\mathbf{c}}_3^2 + (\mathbf{O}_1 \otimes \mathbf{O}_1 + \mathbf{O}_2 \otimes \mathbf{O}_2)\cdot\cdot(\widetilde{\varphi}_3\overset{(n)}{\mathbf{T}} + \widetilde{\varphi}_{10}\overset{(n)}{\mathbf{C}}_p) + \widetilde{\varphi}_4\overset{(n)}{\mathbf{T}} + \varphi_5\overset{(n)}{\mathbf{T}}{}^2 + \varphi_{11}\overset{(n)}{\mathbf{C}}_p. \tag{9.101}$$

Here we have denoted the scalar functions

$$\widetilde{\varphi}_1 = \varphi_1 + \varphi_5 I_2^{(3)}(\overset{(n)}{\mathbf{T}}),\ \ \widetilde{\varphi}_2 = \varphi_2 - \varphi_1 - 2\varphi_4 I_2^{(3)}(\overset{(n)}{\mathbf{T}}) - \varphi_{11} I_2^{(3)}(\overset{(n)}{\mathbf{C}}_p),$$

$$\widetilde{\varphi}_3 = \tfrac{\varphi_3}{2} - \varphi_4,\ \ \widetilde{\varphi}_{10} = \tfrac{\varphi_{10}}{4} - \tfrac{\varphi_{11}}{2},\ \ \widetilde{\varphi}_4 = 2\varphi_4 - \varphi_5 I_1(\overset{(n)}{\mathbf{T}}). \tag{9.102}$$

For the model of an elastoplastic continuum (9.58), we have $\varphi_{10} = \varphi_{11} = 0$, and Eq. (9.101) takes the form

$$\overset{(n)}{\mathbf{C}}_e = \widetilde{\varphi}_1 \mathbf{E} + \widetilde{\varphi}_2 \widehat{\mathbf{c}}_3^2 + \widetilde{\varphi}_3 (\mathbf{O}_1 \otimes \mathbf{O}_1 + \mathbf{O}_2 \otimes \mathbf{O}_2) \cdot\cdot \overset{(n)}{\mathbf{T}} + \widetilde{\varphi}_4 \overset{(n)}{\mathbf{T}} + \varphi_5 \overset{(n)}{\mathbf{T}}{}^2. \quad (9.103)$$

Here

$$\zeta = \zeta(I_1^{(3)}(\overset{(n)}{\mathbf{T}}), \ldots, I_5^{(3)}(\overset{(n)}{\mathbf{T}}), \theta, w_1^p, \ldots, w_3^p), \quad \varphi_\gamma = -\rho(\partial \zeta / \partial I_\gamma^{(3)}), \quad (9.104)$$

and $w_\beta^p$ are the Taylor parameters (9.62) for a transversely isotropic continuum, and the number of the parameters in this case is equal to three:

$$\begin{aligned}
w_1^p &= \int_0^t (\widehat{\mathbf{c}}_3^2 \cdot\cdot \overset{(n)}{\mathbf{T}})(\widehat{\mathbf{c}}_3^2 \cdot\cdot \overset{(n)}{\mathbf{C}}{}_p^{\bullet})\, d\tau = \int_0^t \widehat{\overset{(n)}{T}}_{33} \widehat{\overset{(n)}{C}}{}_{33}^{\bullet}\, d\tau, \\
w_3^p &= \int_0^t \overset{(n)}{\mathbf{T}} \cdot\cdot \overset{(n)}{\mathbf{C}}{}_p^{\bullet} - w_2^p - w_1^p, \\
w_2^p &= \int_0^t ((\mathbf{E} - \widehat{\mathbf{c}}_3^2) \cdot \overset{(n)}{\mathbf{T}}) \cdot\cdot (\widehat{\mathbf{c}}_3^2 \cdot\cdot \overset{(n)}{\mathbf{C}}{}_p^{\bullet})\, d\tau = \int_0^t (\widehat{\overset{(n)}{T}}_{13} \widehat{\overset{(n)}{C}}{}_{13}^{\bullet} + \widehat{\overset{(n)}{T}}_{23} \widehat{\overset{(n)}{C}}{}_{23}^{\bullet})\, d\tau.
\end{aligned} \quad (9.105)$$

For an elastoplastic continuum, according to the Onsager principle, only the cubic invariant $J_3^{(3)} = I_3(\overset{(n)}{\mathbf{T}})$ does not appear in the set of arguments of the plasticity functions $f_\beta$:

$$f_\beta = f_\beta(J_1^{(3)}, J_2^{(3)}, J_4^{(3)}, \ldots, J_{11}^{(3)}, \theta, w_1^p, \ldots, w_3^p). \quad (9.106)$$

There are two simultaneous invariants in the set (9.99), namely $J_{10}^{(3)}$ and $J_{11}^{(3)}$. Then, substituting the derivative tensors (9.100) into (9.74c), we obtain

$$\begin{aligned}
\overset{(n)}{\mathbf{C}}{}_p^{\bullet} &= \psi_1 \mathbf{E} + (\psi_2 - \psi_1) \widehat{\mathbf{c}}_3^2 + \frac{1}{2} (\mathbf{O}_1 \otimes \mathbf{O}_1 + \mathbf{O}_2 \otimes \mathbf{O}_2) \cdot\cdot (\psi_3 \overset{(n)}{\mathbf{T}} + \frac{\psi_{10}}{2} \overset{(n)}{\mathbf{C}}_p) \\
&\quad + {}^4\mathbf{O}_3 \cdot\cdot (2\psi_4 \overset{(n)}{\mathbf{T}} + \psi_{11} \overset{(n)}{\mathbf{C}}_p),
\end{aligned} \quad (9.107)$$

$$\psi_\alpha = h \sum_{\beta=1}^{k} \dot{\varkappa}_\beta (\partial f_\beta / \partial J_\alpha^{(3)}). \quad (9.108)$$

### 9.1.10 Two-Potential Model of Plasticity for a Transversely Isotropic Continuum

For special models of transversely isotropic elastoplastic continua, we accept an additional assumption on a form of the potentials $\zeta$ and $f_\beta$. In the *two-potential* model, we suppose that there are two plastic potentials, one of which, namely $f_2$, depends only on those invariants $J_\gamma^{(3)}$, which in basis $\widehat{\mathbf{c}}_\gamma$ contain components with subscript 3, i.e. $\overset{\widehat{(n)}}{T}_{\alpha 3}$ and $\overset{(n)}{C}_{p\alpha 3}$, and $f_1$ depends on the remaining invariants:

$$f_1 = f_1(I_1^{(3)}(\overset{(n)}{\mathbf{T}}),\ I_4^{(3)}(\overset{(n)}{\mathbf{T}}),\ I_1^{(3)}(\overset{(n)}{\mathbf{C}}_p),\ I_4^{(3)}(\overset{(n)}{\mathbf{C}}_p),\ J_{10}^{(3)}(\overset{(n)}{\mathbf{T}},\ \overset{(n)}{\mathbf{C}}_p),\ \theta,\ w_1^p,\ w_2^p),$$

$$f_2 = f_2(I_2^{(3)}(\overset{(n)}{\mathbf{T}}),\ I_3^{(3)}(\overset{(n)}{\mathbf{T}}),\ I_2^{(3)}(\overset{(n)}{\mathbf{C}}_p),\ I_3^{(3)}(\overset{(n)}{\mathbf{C}}_p),\ J_{11}^{(3)}(\overset{(n)}{\mathbf{T}},\ \overset{(n)}{\mathbf{C}}_p),\ \theta,\ w_3^p). \quad (9.109)$$

In the *transversely isotropic two-potential Huber–Mises model*, we assume that each of the potentials $f_1$ and $f_2$ is a function of the simultaneous Huber–Mises invariants for a transversely isotropic continuum $Y_\alpha^{(3)}$:

$$f_2 = f_2(Y_2^{(3)},\ Y_3^{(3)},\ \theta,\ w_1^p,\ w_2^p), \quad f_1 = f_1(Y_1^{(3)},\ Y_4^{(3)},\ \theta,\ w_3^p). \quad (9.110)$$

Here

$$Y_\alpha^{(3)} = I_\alpha^{(3)}(\overset{(n)}{\mathbf{T}} - H_\alpha \overset{(n)}{\mathbf{C}}_p), \quad \alpha = 1, \ldots, 4; \quad H_\alpha = H_\alpha^0 (I_\alpha^{(3)}(\overset{(n)}{\mathbf{C}}_p))^{n_\alpha^0}, \quad (9.111)$$

are simultaneous invariants which are uniquely expressed in terms of the invariants (9.99) (see Exercise 9.1.5):

$$\begin{aligned} Y_\alpha^{(3)} &= I_\alpha^{(3)}(\overset{(n)}{\mathbf{T}}) - H_\alpha I_\alpha^{(3)}(\overset{(n)}{\mathbf{C}}_p), \quad \alpha = 1,\ 2, \\ Y_3^{(3)} &= I_3^{(3)}(\overset{(n)}{\mathbf{T}}) + H_2 I_3^{(3)}(\overset{(n)}{\mathbf{C}}_p) - 2H_3 J_{10}^{(3)}, \\ Y_4^{(3)} &= I_4^{(3)}(\overset{(n)}{\mathbf{T}}) + H_2 I_4^{(3)}(\overset{(n)}{\mathbf{C}}_p) - 2H_4 J_{11}^{(3)}, \end{aligned} \quad (9.112)$$

and $H_\alpha^0$ and $n_\alpha^0$ are the constants.

Calculating the derivatives $\partial f/\partial J_\alpha^{(3)}$ of the functions (9.110) (only occurring in the expression (9.67)):

$$\partial f_1/\partial I_1^{(3)} = f_{11}, \quad \partial f_1/\partial I_4^{(3)} = f_{14}, \quad \partial f_1/\partial J_{11}^{(3)} = -2 f_{14} H_4, \quad (9.113)$$

$$\partial f_2/\partial I_2^{(3)} = f_{22}, \quad \partial f_2/\partial I_3^{(3)} = f_{23}, \quad \partial f_2/\partial J_{10}^{(3)} = -2H_3 f_{23}, \quad f_{\beta\gamma} \equiv \partial f_\beta/\partial Y_\gamma^{(3)},$$

and substituting them into (9.108), we obtain

$$\begin{gathered}\psi_1 = \dot{\varkappa}_1 f_{11} h, \quad \psi_2 = \dot{\varkappa}_2 f_{22} h, \quad \psi_3 = \dot{\varkappa}_2 f_{23} h, \quad \psi_4 = \dot{\varkappa}_1 f_{14} h, \\ \psi_{10} = -2\dot{\varkappa}_2 f_{23} H_3 h, \quad \psi_{11} = -2\dot{\varkappa}_1 f_{14} H_4 h.\end{gathered} \tag{9.114}$$

Then the constitutive equations (9.107) for plastic deformations become

$$\begin{aligned}\overset{(n)}{\mathbf{C}}{}^{\bullet}_{p} &= \dot{\varkappa}_1 h (f_{11}(\mathbf{E} - \widehat{\mathbf{c}}_3^2) + 2 f_{14}\, {}^4\mathbf{O}_3 \cdot\cdot (\overset{(n)}{\mathbf{T}} - H_3 \overset{(n)}{\mathbf{C}}_p)) \\ &+ \dot{\varkappa}_2 h (f_{22} \widehat{\mathbf{c}}_3^2 + \frac{f_{23}}{2} (\mathbf{O}_1 \otimes \mathbf{O}_1 + \mathbf{O}_2 \otimes \mathbf{O}_2) \cdot\cdot (\overset{(n)}{\mathbf{T}} - H_4 \overset{(n)}{\mathbf{C}}_p)).\end{aligned} \tag{9.115}$$

We can immediately verify (see Exercise 9.1.4) that the tensors

$$\begin{aligned}\mathbf{P}_{1H} &\equiv f_{11}(\mathbf{E} - \widehat{\mathbf{c}}_3^2) + 2 f_{14}\, {}^4\mathbf{O}_3 \cdot\cdot (\overset{(n)}{\mathbf{T}} - H_3 \overset{(n)}{\mathbf{C}}_p), \\ \mathbf{P}_{1H} &\equiv f_{22} \widehat{\mathbf{c}}_3^2 + \tfrac{f_{23}}{2} (\mathbf{O}_1 \otimes \mathbf{O}_1 + \mathbf{O}_2 \otimes \mathbf{O}_2) \cdot\cdot (\overset{(n)}{\mathbf{T}} - H_4 \overset{(n)}{\mathbf{C}}_p)\end{aligned} \tag{9.116}$$

are mutually orthogonal:

$$\mathbf{P}_{1H} \cdot\cdot \mathbf{P}_{2H} = 0. \tag{9.117}$$

Therefore, rewriting Eq. (9.115) with use of notation (9.116) in the form

$$\overset{(n)}{\mathbf{C}}{}^{\bullet}_{p} = \dot{\varkappa}_1 h \mathbf{P}_{1H} + \dot{\varkappa}_2 h \mathbf{P}_{2H} \tag{9.118}$$

(this is an analog of the relationship (9.94)) and multiplying the result by $\mathbf{P}_{1H}$ and $\mathbf{P}_{2H}$, we get

$$\dot{\varkappa}_1 = \pm \sqrt{\frac{\overset{(n)}{\mathbf{C}}{}^{\bullet}_{p} \cdot\cdot \mathbf{P}_{1H}}{\mathbf{P}_{1H} \cdot\cdot \mathbf{P}_{1H}}}, \quad \dot{\varkappa}_2 = \pm \sqrt{\frac{\overset{(n)}{\mathbf{C}}{}^{\bullet}_{p} \cdot\cdot \mathbf{P}_{2H}}{\mathbf{P}_{2H} \cdot\cdot \mathbf{P}_{2H}}} \tag{9.119}$$

– the expressions for $\varkappa_1$ and $\varkappa_2$, being similar to the expression (9.95).

After substitution of (9.119) into (9.118), for determining all components of the plastic deformation tensor Eq. (9.118) should be complemented by two more scalar ones

$$f_1 = 0, \quad f_2 = 0,$$

where $f_\beta$ are expressed by formulae (9.110).

These functions are usually chosen in a quadratic form being similar to the Mises model (9.96):

$$2f_1 = \left(\frac{Y_{4H}}{\sigma_{4s}}\right)^2 + \left(\frac{|Y_{1H}| + Y_{1H}}{2\sigma_{1s}^+}\right)^2 + \left(\frac{|Y_{1H}| - Y_{1H}}{2\sigma_{1s}^-}\right)^2 - 1,$$

$$2f_2 = \left(\frac{|Y_{2H}| + Y_{2H}}{2\sigma_{2s}^+}\right)^2 + \left(\frac{|Y_{2H}| - Y_{2H}}{2\sigma_{2s}^-}\right)^2 + \left(\frac{Y_{3H}}{\sigma_{3s}}\right)^2 - 1. \quad (9.120)$$

The functions $\sigma_{1s}^{\pm}(\theta, w_3^p)$ are called the *yield points* (or the *yield strengths*) in longitudinal tension and compression, respectively, and $\sigma_{4s}(\theta,\ w_3^p)$ – the yield point in shear along the plane of transverse isotropy. The functions $\sigma_{2s}^{\pm}(\theta,\ w_1^p,\ w_2^p)$ are called the yield points in transverse tension and compression, and $\sigma_{3s}(\theta,\ w_1^p,\ w_2^p)$ – the yield point in interlayer shear. These functions are usually determined in experiments. For anisotropic continua, the distinction between the yield points in tension and compression is rather considerable; therefore, the functions $\sigma_{\alpha s}^+$ and $\sigma_{\alpha s}^-$ may be essentially distinct. Notice that although the functions $f_\beta$ depend on the sign of invariants $Y_{1H}$ and $Y_{2H}$, they are differentiable everywhere, including the case when $Y_{1H} = 0$ and $Y_{2H} = 0$, and their derivatives (9.113) take on the values

$$f_{\alpha\alpha} = \frac{|Y_{\alpha H}| + Y_{\alpha H}}{2\sigma_{\alpha s}^+} + \frac{|Y_{\alpha H}| - Y_{\alpha H}}{2\sigma_{\alpha s}^-}, \quad \alpha = 1,\ 2; \quad f_{14} = Y_{4H}/\sigma_{4s},$$

$$f_{23} = Y_{3H}/\sigma_{3s}. \quad (9.121)$$

### 9.1.11 Associated Models of Plasticity $A_n^\xi$ for Orthotropic Continua

For an *orthotropic continuum*, the functional basis of simultaneous invariants $J_\gamma^{(O)}(\overset{(n)}{\mathbf{T}}, \overset{(n)}{\mathbf{C}}_p)$ consists of twelve invariants, that should be complemented by two more, in general, dependent invariants in order to obtain the set of invariant, being symmetric relative to all the basis vectors $\widehat{\mathbf{c}}_\alpha$ (see (7.37) and (8.91), (8.92)):

$$J_\gamma^{(O)} = I_\gamma^{(O)}(\overset{(n)}{\mathbf{T}}),\ \gamma = 1, \cdots, 6; \quad J_{\gamma+6}^{(O)} = I_\gamma^{(O)}(\overset{(n)}{\mathbf{C}}_p),\ \ \gamma = 1,\ 2,\ 3,\ 6;$$

$$J_{10}^{(O)} = (\widehat{\mathbf{c}}_2^2 \cdot \overset{(n)}{\mathbf{T}}) \cdot\cdot (\widehat{\mathbf{c}}_3^2 \cdot \overset{(n)}{\mathbf{C}}_p), \quad J_{11}^{(O)} = (\widehat{\mathbf{c}}_1^2 \cdot \overset{(n)}{\mathbf{T}}) \cdot\cdot (\widehat{\mathbf{c}}_3^2 \cdot \overset{(n)}{\mathbf{C}}_p),$$

$$J_{13}^{(O)} = (\widehat{\mathbf{c}}_1^2 \cdot \overset{(n)}{\mathbf{T}}) \cdot\cdot (\widehat{\mathbf{c}}_2^2 \cdot \overset{(n)}{\mathbf{C}}_p), \quad J_{14}^{(O)} = (\widehat{\mathbf{c}}_1^2 \cdot \overset{(n)}{\mathbf{T}}) \cdot (\widehat{\mathbf{c}}_2^2 \cdot \overset{(n)}{\mathbf{T}}) = I_7^{(O)}(\overset{(n)}{\mathbf{T}}). \quad (9.122)$$

The derivative tensors in this case become (see (8.93))

$$J^{(O)}_{\gamma\mathbf{T}} = \widehat{\mathbf{c}}^2_\gamma, \quad \gamma = 1,\ 2,\ 3; \quad J^{(O)}_{\gamma+3,\mathbf{T}} = \frac{1}{2}(\mathbf{O}_\gamma \otimes \mathbf{O}_\gamma) \cdot\cdot \overset{(n)}{\mathbf{T}}, \quad \gamma = 1,\ 2;$$

$$J^{(O)}_{6\mathbf{T}} = 3\,{}^6\mathbf{O}_m \cdot\cdot\cdot\cdot \overset{(n)}{\mathbf{T}} \otimes \overset{(n)}{\mathbf{T}}, \quad J^{(O)}_{\gamma+6,\mathbf{T}} = 0, \quad \gamma = 1,\ 2,\ 3,\ 6;$$

$$J^{(O)}_{10\mathbf{T}} = \frac{1}{4}(\mathbf{O}_1 \otimes \mathbf{O}_1) \cdot\cdot \overset{(n)}{\mathbf{C}}_p, \quad J^{(O)}_{11\mathbf{T}} = \frac{1}{4}(\mathbf{O}_2 \otimes \mathbf{O}_2) \cdot\cdot \overset{(n)}{\mathbf{C}}_p, \tag{9.123}$$

$$J^{(O)}_{13\mathbf{T}} = \frac{1}{4}(\mathbf{O}_3 \otimes \mathbf{O}_3) \cdot\cdot \overset{(n)}{\mathbf{C}}_p, \quad J^{(O)}_{14\mathbf{T}} = \frac{1}{4}(\mathbf{O}_3 \otimes \mathbf{O}_3) \cdot\cdot \overset{(n)}{\mathbf{T}},$$

where the tensor ${}^6\mathbf{O}_m$ is determined by (4.316).

Substitution of these expressions into (9.63) yields the following representation of constitutive equations (9.63) for an orthotropic plastic continuum in the tensor basis:

$$\overset{(n)}{\mathbf{C}}_e = \sum_{\gamma=1}^{3} \left( \varphi_\gamma \widehat{\mathbf{c}}^2_\gamma + \frac{1}{2}\mathbf{O}_\gamma \otimes \mathbf{O}_\gamma \cdot\cdot (\widetilde{\varphi}_{3+\gamma} \overset{(n)}{\mathbf{T}} + \frac{1}{2}\widetilde{\varphi}_{6+\gamma} \overset{(n)}{\mathbf{C}}_p) \right) + 3\varphi_6\,{}^6\mathbf{O}_m \cdot\cdot\cdot\cdot \overset{(n)}{\mathbf{T}} \otimes \overset{(n)}{\mathbf{T}}. \tag{9.124}$$

Here we have denoted the scalar functions

$$\widetilde{\varphi}_4 = \varphi_4, \quad \widetilde{\varphi}_5 = \varphi_5, \quad \widetilde{\varphi}_6 = \varphi_{14}, \quad \widetilde{\varphi}_7 = \varphi_{10}, \quad \widetilde{\varphi}_8 = \varphi_{11}, \quad \widetilde{\varphi}_9 = \varphi_{13}. \tag{9.125}$$

For the model of an elastoplastic continuum (9.59): $\widetilde{\varphi}_7 = \widetilde{\varphi}_8 = \widetilde{\varphi}_9$, and equations (9.124) take the form

$$\overset{(n)}{\mathbf{C}}_e = \sum_{\gamma=1}^{3} (\varphi_\gamma \widehat{\mathbf{c}}^2_\gamma + \frac{\widetilde{\varphi}_{3+\gamma}}{2}(\mathbf{O}_\gamma \otimes \mathbf{O}_\gamma) \cdot\cdot \overset{(n)}{\mathbf{T}}) + 3\varphi_6\,{}^6\mathbf{O}_m \cdot\cdot\cdot\cdot \overset{(n)}{\mathbf{T}} \otimes \overset{(n)}{\mathbf{T}}, \tag{9.126}$$

where

$$\varphi_\gamma = -\rho(\partial\zeta/\partial I^{(O)}_\gamma), \qquad \zeta = \zeta(I^{(O)}_1(\overset{(n)}{\mathbf{T}}), \ldots, I^{(O)}_7(\overset{(n)}{\mathbf{T}}),\ \theta,\ w^p_1, \ldots, w^p_6). \tag{9.127}$$

The number of the Taylor parameters (9.62) for an orthotropic continuum is equal to six:

$$w^p_\gamma = \int_0^t (\widehat{\mathbf{c}}^2_\gamma \cdot\cdot \overset{(n)}{\mathbf{T}})(\widehat{\mathbf{c}}^2_\gamma \cdot\cdot \overset{(n)}{\mathbf{C}}{}^\bullet_p)\, d\tau, \quad w^p_{3+\gamma} = \int_0^t (\widehat{\mathbf{c}}^2_\alpha \cdot \overset{(n)}{\mathbf{T}}) \cdot\cdot (\widehat{\mathbf{c}}^2_\beta \cdot\cdot \overset{(n)}{\mathbf{C}}{}^\bullet_p)\, d\tau, \tag{9.128}$$

$$\gamma = 1,\ 2,\ 3; \quad \alpha \neq \beta \neq \gamma \neq \alpha.$$

For an orthotropic elastoplastic continuum, the plastic potentials $f_\beta$ (9.65) depend on all simultaneous invariants (9.122) except the cubic invariant $J_6^{(O)} = I_6^{(O)}(\overset{(n)}{\mathbf{T}})$:

$$f_\beta = f_\beta(J_\gamma^{(O)},\ \theta,\ w_1^p,\ldots,w_6^p), \quad \gamma = 1,\ldots,14 \text{ and } \gamma \neq 6. \tag{9.129}$$

There are three simultaneous invariants in the set (9.122), namely $J_{10}^{(O)}$, $J_{12}^{(O)}$ and $J_{13}^{(O)}$. Then, having substituted the derivative tensors (9.123) into (9.74c), we obtain the constitutive equation

$$\overset{(n)}{\mathbf{C}}{}^{\bullet}_p = \sum_{\gamma=1}^{3}\left(\psi_\gamma \widehat{\mathbf{c}}_\gamma^2 + \frac{1}{2}\mathbf{O}_\gamma \otimes \mathbf{O}_\gamma \cdot\cdot (\widetilde{\psi}_{3+\gamma}\overset{(n)}{\mathbf{T}} + \frac{1}{2}\widetilde{\psi}_{6+\gamma}\overset{(n)}{\mathbf{C}}_p)\right), \tag{9.130}$$

where

$$\widetilde{\psi}_4 = \psi_4, \quad \widetilde{\psi}_5 = \psi_5, \quad \widetilde{\psi}_6 = \psi_{14}, \quad \widetilde{\psi}_7 = \psi_{10}, \quad \widetilde{\psi}_8 = \psi_{11}, \quad \widetilde{\psi}_9 = \psi_{13},$$

$$\psi_\gamma = h\sum_{\beta=1}^{k}\dot{\varkappa}_\beta \frac{\partial f_\beta}{\partial J_\gamma^{(O)}}. \tag{9.131}$$

### *9.1.12 The Orthotropic Unipotential Huber–Mises Model for Plastic Continua*

For special models of orthotropic elastoplastic continua one should accept an additional assumption on a form of the potentials $f_\beta$. The adequacy of a model accepted is verified in experiments. In the *orthotropic unipotential Huber–Mises model* one assumes that there is only one potential $f$ depending on six simultaneous orthotropic Huber–Mises invariants $Y_\alpha^{(O)}$ ($\alpha = 1,\ldots,6$):

$$f = f(Y_1^{(O)},\ldots,Y_6^{(O)},\ \theta,\ w_1^p,\ldots,w_6^p). \tag{9.132}$$

Here

$$Y_\alpha^{(O)} = I_\alpha^{(O)}(\overset{(n)}{\mathbf{T}} - H_\alpha \overset{(n)}{\mathbf{C}}_p), \ \alpha = 1,\ldots,5, \quad Y_6^{(O)} = I_7^{(O)}(\overset{(n)}{\mathbf{T}} - H_6 \overset{(n)}{\mathbf{C}}_p),$$

$$H_\alpha = H_\alpha^0(I_\alpha^{(O)}(\overset{(n)}{\mathbf{C}}_p))^{n_\alpha^0}, \quad H_6 = H_6^0(I_7^{(O)}(\overset{(n)}{\mathbf{C}}_p))^{n_6^0}, \tag{9.133}$$

are the simultaneous invariants, which are uniquely expressed in terms of the invariants (9.122) (see Exercise 9.1.6):

$$Y_{\alpha}^{(O)} = I_{\alpha}^{(O)}(\overset{(n)}{\mathbf{T}}) - H_{\alpha} I_{\alpha}^{(O)}(\overset{(n)}{\mathbf{C}}_p), \quad \alpha = 1,\ 2,\ 3;$$

$$Y_{3+\alpha}^{(O)} = I_{3+\alpha}^{(O)}(\overset{(n)}{\mathbf{T}}) + H_{3+\alpha}^2 I_{3+\alpha}^{(O)}(\overset{(n)}{\mathbf{C}}_p) - 2H_{3+\alpha} J_{9+\alpha}^{(O)}, \quad \alpha = 1,\ 2; \tag{9.134}$$

$$Y_6^{(O)} = I_7^{(O)}(\overset{(n)}{\mathbf{T}}) + H_6^2 I_7^{(O)}(\overset{(n)}{\mathbf{C}}_p) - 2H_6 J_{13}^{(O)}.$$

Calculating the derivatives $\partial f/\partial J_{\alpha}^{(O)}$ of the function (9.132)

$$\partial f/\partial I_{\alpha}^{(O)} = f_{\alpha}, \ \alpha = 1, \ldots, 5; \quad \partial f/\partial J_{14}^{(O)} = f_6, \quad \partial f/\partial J_{10}^{(O)} = -2H_4 f_4,$$
$$\partial f/\partial J_{11}^{(O)} = -2H_5 f_5, \quad \partial f/\partial J_{13}^{(O)} = -2H_6 f_6, \quad f_{\alpha} \equiv \partial f/\partial Y_{\alpha}^{(O)} \tag{9.135}$$

and substituting them into (9.131), we find the non-zero functions

$$\psi_{\gamma} = \dot{\varkappa} h f_{\gamma}, \quad \gamma = 1, \ldots, 5, \quad \psi_{10} = -2\dot{\varkappa} h H_4 f_4,$$
$$\psi_{11} = -2\dot{\varkappa} h H_5 f_5, \quad \psi_{13} = -2\dot{\varkappa} h H_6 f_6, \quad \psi_{14} = \dot{\varkappa} h f_6. \tag{9.136}$$

Then the constitutive equation (9.130) takes the form

$$\overset{(n)}{\mathbf{C}}{}^{\bullet}_p = \dot{\varkappa} h \sum_{\gamma=1}^{3} \left( f_{\gamma} \widehat{\mathbf{c}}_{\gamma}^2 + \frac{f_{3+\gamma}}{2} \mathbf{O}_{\gamma} \otimes \mathbf{O}_{\gamma} \cdot\cdot (\overset{(n)}{\mathbf{T}} - H_{3+\gamma} \overset{(n)}{\mathbf{C}}_p) \right). \tag{9.137}$$

Multiplying this equation by itself, we obtain

$$\dot{\varkappa} = \sqrt{\frac{\overset{(n)}{\mathbf{C}}{}^{\bullet}_p \cdot\cdot \overset{(n)}{\mathbf{C}}{}^{\bullet}_p}{\mathbf{P}_H^{(O)} \cdot\cdot \mathbf{P}_H^{(O)}}}, \tag{9.138}$$

where

$$\mathbf{P}_H^{(O)} \equiv \sum_{\gamma 1}^{3} \left( f_{\gamma} \widehat{\mathbf{c}}_{\gamma}^2 + \frac{f_{3+\gamma}}{2} \mathbf{O}_{\gamma} \otimes \mathbf{O}_{\gamma} \cdot\cdot (\overset{(n)}{\mathbf{T}} - H_{3+\gamma} \overset{(n)}{\mathbf{C}}_p) \right). \tag{9.139}$$

The complete equation system for plastic deformations consists of relations (9.137), (9.138) and the following scalar equation of the yield surface:

$$f = 0, \tag{9.140}$$

where $f$ is expressed by formula (9.132).

In the quadratic model, the potential (9.132) is chosen in the form

$$2f = \sum_{\gamma=1}^{3}\left(\frac{|Y_\gamma^{(O)}| + Y_\gamma^{(O)}}{2\sigma_{\gamma s}^{+}}\right)^2 + \left(\frac{|Y_\gamma^{(O)}| - Y_\gamma^{(O)}}{2\sigma_{\gamma s}^{-}}\right)^2 + \left(\frac{Y_{3+\gamma}^{(O)}}{2\sigma_{3+\gamma,s}^{+}}\right)^2$$
$$-\frac{Y_1^{(O)}Y_2^{(O)}}{\sigma_{12s}} - \frac{Y_1^{(O)}Y_3^{(O)}}{\sigma_{13s}} - \frac{Y_2^{(O)}Y_3^{(O)}}{\sigma_{23s}} - 1. \qquad (9.141)$$

The functions $\sigma_{\gamma s}^{\pm}(\theta,\ w_\gamma^p)$ are called the *yield strengths in tension* (or in compression) *in the direction* $\gamma$; $\sigma_{3+\gamma,\,s}(\theta,\ w_{3+\gamma}^p)$ – the yield strengths in shear along the plane $(\alpha,\ \beta)$; and $\sigma_{\alpha\beta,s}(\theta,\ w_1^p,\ldots,w_3^p)$ – the mixed yield strengths. All the functions are determined in experiments.

The derivatives of the function (9.141) have the forms

$$f_\gamma = \frac{|Y_\gamma^{(O)}| + Y_\gamma^{(O)}}{2\sigma_{\gamma s}^{+}} + \frac{|Y_\gamma^{(O)}| - Y_\gamma^{(O)}}{2\sigma_{\gamma s}^{-}}, \quad f_{\gamma+3} = Y_{3+\gamma}^{(O)}/\sigma_{3+\gamma,s}, \quad \gamma = 1,\ 2,\ 3. \qquad (9.142)$$

### 9.1.13 The Principle of Material Indifference for Models $A_n^\xi$ of Plastic Continua

All the energetic tensors $\overset{(n)}{\mathbf{T}}$, $\overset{(n)}{\mathbf{C}}$, and also $\overset{(n)}{\mathbf{C}}_e$ and $\overset{(n)}{\mathbf{C}}_p$ defined by $\overset{(n)}{\mathbf{T}}$ and $\overset{(n)}{\mathbf{C}}$, are $R$-invariant. Then all the constitutive equations of models $A_n^\xi$ of plastic continua, stated in Sects. 9.1.2–9.1.6, satisfy the principle of material indifference, because they remain unchanged at the change of actual configuration $\mathcal{K} \to \mathcal{K}'$ in rigid motion.

## Exercises for 9.1

**9.1.1.** Show that the scalar invariant of the tensor $\mathbf{T}$, being its intensity $Y(\mathbf{T})$ determined by (9.88), in any basis has the following component representation:

$$Y^2(\mathbf{T}) = \frac{3}{2}\mathbf{P}\cdot\cdot\,\mathbf{P}, \quad \mathbf{P} = \mathbf{T} - \frac{1}{3}I_1(\mathbf{T})\mathbf{E},$$
$$Y^2(\mathbf{T}) = \frac{1}{2}\left((T_{11}-T_{22})^2+(T_{22}-T_{33})^2+(T_{33}-T_{11})^2+6(T_{12}^2+T_{23}^2+T_{13}^2)\right).$$

**9.1.2.** Prove that the invariant $Y$ defined in Exercise 9.1.1 can always be expressed in terms of the principal invariants as follows:

$$Y^2(\mathbf{T}) = I_1^2(\mathbf{T}) - 3I_2(\mathbf{T}).$$

**9.1.3.** Show that for the model of an isotropic elastoplastic continuum without strengthening, when $H \equiv 0$ (i.e. $H_0 = 0$), Eq. (9.97) takes the form

$$\dot{\mathbf{P}}_p = (\dot{\varkappa} h/\sigma_s^2)\mathbf{P}_T.$$

This relation is called the *Prandtl–Reuss equation*.

**9.1.4.** Show that the relation of orthogonality (9.117) holds for the tensors (9.116).

**9.1.5.** Prove relations (9.112) between the simultaneous invariants $Y_\alpha^{(3)}$ (9.111) and $J_\gamma^{(3)}$ (9.99).

**9.1.6.** Prove relations (9.134).

**9.1.7.** Consider the *plastically compressible Huber–Mises model* $A_n^\xi$ of an isotropic continuum, where the constitutive equations (9.81)–(9.83) hold and there is only one plastic potential $f$ depending, unlike (9.84), in addition on the first invariant $Y_{1H}$:

$$f = f(Y_H, Y_{1H}, \theta, w_p).$$

Here

$$Y_{1H} = (\overset{(n)}{\mathbf{T}} - H_1 \overset{(n)}{\mathbf{C}}_p) \cdot\cdot\, \mathbf{E}, \quad H_1 = H_1^0 Y_{1p}^{2n_1}, \quad Y_{1p} = I_1(\overset{(n)}{\mathbf{C}}_p), \quad f_{1Y} \equiv \partial f/\partial Y_{1H},$$

and $H_1^0$, $n_1$ are the constants.

This model describes the plastic properties of porous media, some grounds and also materials sensitive to the type of loading being volumetric tension or compression (see Sect. 9.6).

Show that for this model the constitutive equations (9.92) take the form

$$\overset{(n)}{\mathbf{C}}{}_p^\bullet = 3\dot{\varkappa} h \left( f_Y \mathbf{P}_H + \frac{1}{3} f_{1Y} \mathbf{E} \right).$$

Show that these equations written for the deviators and spherical parts of the tensors of plastic deformations and stresses have the forms (compare them with (9.94) and (9.95))

$$\begin{cases} \dot{\mathbf{P}} = 3\dot{\varkappa} h f_Y \mathbf{P}_H, \\ \dot{I}_1(\mathbf{C}_p) = \dot{\varkappa} h f_{1Y}, \\ \dot{\varkappa} = \pm\sqrt{\overset{(n)}{\mathbf{C}}{}_p^\bullet \cdot\cdot\, \overset{(n)}{\mathbf{C}}{}_p^\bullet} \Big/ \sqrt{f_Y^2 Y_H^2 + 3 f_{1Y}^2}, \quad f = 0. \end{cases}$$

Show that if the plastic potential $f$ has the form

$$f = \frac{1}{3}\left(\frac{Y_H}{\sigma_S}\right)^2 + \frac{Y_+^2}{\sigma_T'^2} + \frac{Y_-^2}{\sigma_C'^2} - 1, \quad \frac{1}{\sigma_T'^2} = \frac{1}{\sigma_T^2} - \frac{1}{3\sigma_S^2}, \quad \frac{1}{\sigma_C'^2} = \frac{1}{\sigma_C^2} - \frac{1}{3\sigma_S^2},$$

where $\sigma_S$, $\sigma_T$, and $\sigma_C$ are the yield strengths in shear, tension and compression, respectively (depend on $\theta$ and $w_p$), and $Y_+$ and $Y_-$ are invariants of constant signs

$$Y_\pm = \frac{1}{2}\left(|Y_{1H}| \pm Y_{1H}\right),$$

then

$$f_Y = \frac{1}{3\sigma_S^2} \quad \text{and} \quad f_{1Y} = \frac{2Y_+}{\sigma_T'^2} + \frac{2Y_-}{\sigma_C'^2},$$

and the constitutive equations take the form

$$\begin{cases} \overset{(n)}{\mathbf{P}}{}^{\bullet}_p = (\dot{\varkappa} h/\sigma_S^2)\mathbf{P}_H, \\ \dot{I}_1(\overset{(n)}{\mathbf{C}}_p) = 2\dot{\varkappa} h((Y_+/\sigma_T'^2) + (Y_-/\sigma_C'^2)). \end{cases}$$

**9.1.8.** Consider the *associated model of plasticity* $A_n^\zeta$ (9.74) with *quasilinear elasticity*, for which the elastic potential $\zeta$ (9.74a) depends only on linear and quadratic invariants $I_\gamma^{(s)}$ of the tensor $\overset{(n)}{\mathbf{T}}/\rho$ (by analogy with the quasilinear models $A_n$ of elastic continua (see Sect. 4.8.7)). Introducing for the functions $\varphi_\gamma(I_\gamma^{(s)}(\overset{(n)}{\mathbf{T}}/\rho), \theta, w_\beta^p)$ the representation

$$\varphi_\gamma = \overset{\circ}{\rho}\sum_{\beta=1}^{r_1} l'_{\gamma\beta} I_\gamma^{(s)}, \quad \gamma = 1,\ldots,r_1; \qquad \varphi_\gamma = \overset{\circ}{\rho} l'_{\gamma\gamma}, \quad \gamma = r_1+1,\ldots,r_2,$$

where $l'_{\gamma\beta}$ and $l'_{\gamma\gamma}$ are functions in the form

$$l'_{\gamma\beta} = l'_{\gamma\beta}(I_\alpha^{(s)}(\overset{(n)}{\mathbf{T}}/\rho),\ \theta,\ w_\alpha^p), \quad l'_{\gamma\beta} = l'_{\beta\gamma},$$

show that in this case Eq. (9.74b) may be written as follows:

$$\overset{(n)}{\mathbf{C}}_e = {}^4\mathbf{N}\cdot\cdot\overset{(n)}{\mathbf{T}}, \quad J = \rho/\overset{\circ}{\rho}.$$

Here the tensor ${}^4\mathbf{N}$ has the form

$${}^4\mathbf{N} = \frac{1}{J}\left(l'_1\mathbf{E}\otimes\mathbf{E} + 2l'_2\mathbf{\Delta}\right), \qquad l'_1 = l'_{11} + 2l'_{22}, \quad l'_{22} = -2l'_2,$$

– for isotropic media ($r_1 = 1,\ r_2 = 2$);

$$^4\mathbf{N} = \frac{1}{J}\Big(l'_{11}\mathbf{E}\otimes\mathbf{E} + \widetilde{l}'_{22}\widehat{\mathbf{c}}_3^2\otimes\widehat{\mathbf{c}}_3^2 + (l'_{12}-l'_{11})(\mathbf{E}\otimes\widehat{\mathbf{c}}_3^2 + \widehat{\mathbf{c}}_3^2\otimes\mathbf{E}) + \left(\frac{l'_{11}}{2} - l'_{44}\right)(\mathbf{O}_1\otimes\mathbf{O}_1 + \mathbf{O}_2\otimes\mathbf{O}_2) + 2l'_{44}\boldsymbol{\Delta}\Big),$$

$$\widetilde{l}'_{22} = l'_{22} - 2(l'_{44} + l'_{12}) - l'_{11},$$

– for transversely isotropic media ($r_1 = 2, r_2 = 4$);

$$^4\mathbf{N} = \frac{1}{J}\left(\sum_{\gamma,\beta=1}^{3} l'_{\gamma\beta}\widehat{\mathbf{c}}_\gamma^2\otimes\widehat{\mathbf{c}}_\beta^2 + \sum_{\gamma=1}^{2} l'_{3+\gamma,3+\gamma}\mathbf{O}_\gamma\otimes\mathbf{O}_\gamma\right)$$

– for orthotropic media ($r_1 = 3, r_2 = 6$).

Show that there exist inverse relations

$$\overset{(n)}{\mathbf{T}} = {}^4\mathbf{M}\cdot\cdot\overset{(n)}{\mathbf{C}}_e,$$

where the tensors $^4\mathbf{M}$ are inverse of $^4\mathbf{N}$:

$$^4\mathbf{M}\cdot\cdot{}^4\mathbf{N} = \boldsymbol{\Delta},$$

and have formally the same structure as the tensor $^4\mathbf{N}$, but coefficients $l_{\gamma\beta}$ of the tensor $^4\mathbf{M}$ are functions in another form

$$l_{\gamma\beta} = l_{\gamma\beta}(I_\alpha^{(s)}(\overset{(n)}{\mathbf{C}}_e),\ \theta, w_\alpha^p).$$

**9.1.9.** For the *models* $A_n^\zeta$ of an *elastoplastic continuum* (9.79) and (9.80) *with linear elasticity*, whose potential $\zeta$ (9.80) has the quadratic form

$$\zeta = \zeta_0 - \frac{\overset{\circ}{\rho}}{2}\sum_{\gamma,\beta=1}^{r_1} l'_{\gamma\beta}I_\gamma^{(s)}I_\beta^{(s)} - \overset{\circ}{\rho}\sum_{\gamma=r_1+1}^{r_2} l_{\gamma\gamma}I_\gamma^{(s)},$$

where $l'_{\gamma\beta}$ are the constants and $I_\gamma^{(s)} = I_\gamma^{(s)}(\overset{(n)}{\mathbf{T}}/\rho)$, show with use of the results of Exercise 9.1.8 that the tensors $^4\mathbf{N}$ are tensor-constants up to the factor $J$, and the constitutive equations (9.74b) between $\overset{(n)}{\mathbf{C}}_e$ and $\overset{(n)}{\mathbf{T}}$ take the form

$$\overset{(n)}{\mathbf{T}} = J(l_1 I_1(\overset{(n)}{\mathbf{C}}_e)\mathbf{E} + 2l_2\overset{(n)}{\mathbf{C}}_e),\quad l_1 = -\frac{l'_1}{2l'_2(3l'_1 + 2l'_2)},\quad l_2 = \frac{1}{4l'_2},$$

– for isotropic materials;

$$\overset{(n)}{\mathbf{T}} = J\Big((l_{11}I_1^{(3)} + l_{12}I_2^{(3)})(\mathbf{E} - \widehat{\mathbf{c}}_3^2) + ((l_{22} - 2l_{44})I_2^{(3)} + l_{12}I_1^{(3)})\,\widehat{\mathbf{c}}_3^2$$
$$+\left(\frac{l_{33}}{2} - l_{44}\right)(\mathbf{O}_1 \otimes \mathbf{O}_1 + \mathbf{O}_2 \otimes \mathbf{O}_2)\cdot\cdot\overset{(n)}{\mathbf{C}}_e + 2l_{44}\overset{(n)}{\mathbf{C}}_e\Big),$$

– for transversely isotropic materials (here $I_\gamma^{(3)} = I_\gamma^{(3)}(\overset{(n)}{\mathbf{C}}_e)$);

$$\overset{(n)}{\mathbf{T}} = J\left(\sum_{\gamma,\beta=1}^{3} l_{\gamma\beta}I_\gamma^{(O)}\widehat{\mathbf{c}}_\gamma^2 + \sum_{\gamma=1}^{3} l_{3+\gamma,3+\gamma}\mathbf{O}_\gamma(\mathbf{O}_\gamma\cdot\cdot\overset{(n)}{\mathbf{C}}_e)\right)$$

– for orthotropic materials ($I_\gamma^{(O)} = I_\gamma^{(O)}(\overset{(n)}{\mathbf{C}}_e)$).

Show that for the models $A_n^\zeta$ of elastoplastic isotropic continua with linear elasticity, whose potential $\zeta$ has the form

$$\zeta = \zeta_0 - \frac{\overset{\circ}{\rho}}{2}l_1' I_1^2(\overset{(n)}{\mathbf{T}}/\rho) - \overset{\circ}{\rho}l_2' I_1(\overset{(n)}{\mathbf{T}}{}^2/\rho^2),$$

the constitutive equations (9.74b) become

$$\overset{(n)}{\mathbf{C}}_e = \frac{1}{J}(l_1' I_1(\overset{(n)}{\mathbf{T}})\mathbf{E} + 2l_2'\overset{(n)}{\mathbf{T}}),$$
$$l_1' = -\frac{l_1}{2l_2(3l_1 + 2l_2)}, \quad l_1 = -\frac{l_1'}{2l_2'(3l_1' + 2l_2')}, \quad 2l_2 = \frac{1}{2l_2'}.$$

**9.1.10.** Using formulae (9.19) and (8.28b), show that the specific internal energy $e$ for the models $A_n^\zeta$ of plastic continua is expressed by the formula

$$e = \zeta + \frac{\overset{(n)}{\mathbf{T}}}{\rho}\cdot\cdot\overset{(n)}{\mathbf{C}}_e - \theta\frac{\partial\zeta}{\partial\theta}.$$

Show that for models $A_n^\zeta$ of an isotropic elastoplastic continuum with linear elasticity the results of Exercise 9.1.9 give

$$e = e_0 + \frac{\overset{\circ}{\rho}}{2}l_1' I_1^2\left(\frac{\overset{(n)}{\mathbf{T}}}{\rho}\right) + \overset{\circ}{\rho}l_2' I_1\left(\frac{\overset{(n)}{\mathbf{T}}{}^2}{\rho^2}\right) = e_0 + \frac{l_1}{2\overset{\circ}{\rho}}I_1^2\left(\overset{(n)}{\mathbf{C}}_e\right) + \frac{l_2}{\overset{\circ}{\rho}}I_1(\overset{(n)}{\mathbf{C}}{}_e^2),$$
$$e_0 = \zeta_0 - \theta(\partial\zeta_0/\partial\theta).$$

**9.1.11.** Show that relations (9.52) can be represented in the form independent explicitly of time:

$$\frac{d}{d\varkappa}\overset{(n)}{\mathbf{C}}_p = h\sum_{\alpha=1}^{k}\frac{d\varkappa_\alpha}{d\varkappa}\frac{\partial f_\alpha}{\partial \overset{(n)}{\mathbf{T}}},$$

where $\varkappa = \varkappa_1$.

## 9.2 Models $B_n$ of Plastic Continua

### 9.2.1 Representation of Stress Power for Models $B_n$ of Plastic Continua

Let us consider now models $B_n$ for plastic continua. Constructing these models essentially differs from constructing the models $A_n$, because the main additive relation (9.3) for the energetic measures $\overset{(n)}{\mathbf{G}}$ does not hold. The relation (9.3) is replaced by Eq. (9.7), that is axiomatically assumed in models $B_n$ and justified in Sect. 9.1.1. Since the relation (9.7) is the product of the gradients $\mathbf{F}_p$ and $\mathbf{F}_e$, models $B_n$ of plastic continua are called *multiplicative*, unlike models $A_n$ which in this case are called *additive*.

Consider the principal thermodynamic identity in the form $B_n$ (4.122)

$$\rho\, d\psi + \rho\eta\, d\theta - \overset{(n)}{\mathbf{T}} \cdot\cdot\, d\overset{(n)}{\mathbf{G}} + w^*\, dt = 0. \tag{9.143}$$

Let us show that for models $B_n$ the additive relation for the stress power derived in the following theorem is an analog of the additive relation (9.3).

**Theorem 9.1.** *The stress power $w_{(i)}$ (4.2) can always be represented in the additive form*

$$w_{(i)} = \overset{(n)}{\mathbf{T}} \cdot\cdot \overset{(n)}{\mathbf{G}}{}^{\bullet} = \overset{(n)}{\mathbf{T}}_e \cdot\cdot \overset{(n)}{\mathbf{G}}{}^{\bullet}_e + \overset{(n)}{\mathbf{T}}_p \cdot \overset{(n)}{\mathbf{D}}_p, \tag{9.144}$$

*where $\overset{(n)}{\mathbf{T}}_e$ and $\overset{(n)}{\mathbf{T}}_p$ are the symmetric tensors of elastic stresses and yield stresses, respectively, $\overset{(n)}{\mathbf{G}}_e$ are the symmetric measures of elastic deformations, and $\overset{(n)}{\mathbf{D}}_p$ are the symmetric energetic measures of plastic deformation rates; their expressions for $n = \mathrm{I}, \dots, \mathrm{V}$ are given in Table 9.1.*

Notice that the tensors $\overset{(n)}{\mathbf{T}}_e$ and $\overset{(n)}{\mathbf{G}}_e$ are entirely analogous to the tensors $\overset{(n)}{\mathbf{T}}$ and $\overset{(n)}{\mathbf{G}}$ and differ from them only by replacing $\mathbf{F} \to \mathbf{F}_e$ and $\mathbf{U} \to \mathbf{U}_e$. The tensor $\mathbf{B}_e$, as

**Table 9.1** Expressions for $\overset{(n)}{\mathbf{T}}_e$, $\overset{(n)}{\mathbf{T}}_p$, $\overset{(n)}{\mathbf{G}}_e$ and $\overset{(n)}{\mathbf{D}}_p$ at $n = \mathrm{I}, \dots, \mathrm{V}$

| $n$ | $\overset{(n)}{\mathbf{T}}_e$ | $\overset{(n)}{\mathbf{G}}_e$ | $\overset{(n)}{\mathbf{T}}_p$ | $\overset{(n)}{\mathbf{D}}_p$ |
|---|---|---|---|---|
| I | $\mathbf{F}_e^{\mathrm{T}} \cdot \mathbf{T} \cdot \mathbf{F}_e$ | $-\frac{1}{2}\mathbf{U}_e^{-2}$ | $\overset{\mathrm{I}}{\mathbf{T}}_e$ | $\frac{1}{2}(\mathbf{D}_p \cdot \mathbf{U}_e^{-2} + \mathbf{U}_e^{-2} \cdot \mathbf{D}_p^{\mathrm{T}})$ |
| II | $\frac{1}{2}(\mathbf{F}_e^{\mathrm{T}} \cdot \mathbf{T} \cdot \mathbf{O}_e + \mathbf{O}_e \cdot \mathbf{T} \cdot \mathbf{F}_e)$ | $-\mathbf{U}_e^{-1}$ | $\overset{\mathrm{I}}{\mathbf{T}}_e$ | $\overset{\mathrm{I}}{\mathbf{D}}_p$ |
| III | $\mathbf{O}_e^{\mathrm{T}} \cdot \mathbf{T} \cdot \mathbf{O}_e$ | $\mathbf{B}_e$ | $\overset{\mathrm{III}}{\mathbf{T}}_e$ | $\frac{1}{2}(\mathbf{U}_e^{\mathrm{T}} \cdot \mathbf{D}_p \cdot \mathbf{U}_e^{-1} + \mathbf{U}_e^{-1} \cdot \mathbf{D}_p^{\mathrm{T}} \cdot \mathbf{U}_e)$ |
| IV | $\frac{1}{2}(\mathbf{F}_e^{-1} \cdot \mathbf{T} \cdot \mathbf{O}_e + \mathbf{O}_e \cdot \mathbf{T} \cdot \mathbf{F}_e^{-1\mathrm{T}})$ | $\mathbf{U}_e$ | $\overset{\mathrm{V}}{\mathbf{T}}_e$ | $\overset{\mathrm{V}}{\mathbf{D}}_p$ |
| V | $\mathbf{F}_e^{-1} \cdot \mathbf{T} \cdot \mathbf{F}_e^{-1\mathrm{T}}$ | $\frac{1}{2}\mathbf{U}_e^2$ | $\overset{\mathrm{V}}{\mathbf{T}}_e$ | $\frac{1}{2}(\mathbf{U}_e^2 \cdot \mathbf{D}_p + \mathbf{D}_p^{\mathrm{T}} \cdot \mathbf{U}_e^2)$ |

well as the tensor $\mathbf{B}$, is determined by its derivative and initial value $\mathbf{B}_e^0$ (when there are no initial plastic deformations $\mathbf{B}_e^0 = \mathbf{E}$):

$$\dot{\mathbf{B}}_e = \frac{1}{2}(\dot{\mathbf{U}}_e \cdot \mathbf{U}_e^{-1} + \mathbf{U}_e^{-1} \cdot \dot{\mathbf{U}}_e), \quad \mathbf{B}_e(0) = \mathbf{B}_e^0. \tag{9.145}$$

The *plastic deformation rate measure* $\mathbf{D}_p$ is determined by the relation

$$\mathbf{D}_p = \dot{\mathbf{F}}_p \cdot \mathbf{F}_p^{-1} \tag{9.146}$$

and is a nonsymmetric tensor. It follows from Table 9.1 that

$$\overset{\mathrm{I}}{\mathbf{T}}_p = \overset{\mathrm{II}}{\mathbf{T}}_p = \overset{\mathrm{I}}{\mathbf{T}}_e, \quad \overset{\mathrm{III}}{\mathbf{T}}_p = \overset{\mathrm{III}}{\mathbf{T}}_e, \quad \overset{\mathrm{IV}}{\mathbf{T}}_p = \overset{\mathrm{V}}{\mathbf{T}}_p = \overset{\mathrm{V}}{\mathbf{T}}_e,$$

$$\overset{\mathrm{I}}{\mathbf{D}}_p = \overset{\mathrm{II}}{\mathbf{D}}_p, \quad \overset{\mathrm{IV}}{\mathbf{D}}_p = \overset{\mathrm{V}}{\mathbf{D}}_p.$$

▼ 1. Using the definition of the measure $\overset{\mathrm{I}}{\mathbf{G}}$ and introducing the measure $\overset{\mathrm{I}}{\mathbf{G}}_e$ by the similar formula $\overset{\mathrm{I}}{\mathbf{G}}_e = \mathbf{F}_e^{-1} \cdot \mathbf{F}_e^{-1\mathrm{T}}$, with the help of the multiplicative decomposition (9.7) we obtain

$$\overset{\mathrm{I}}{\mathbf{G}} = -\frac{1}{2}\mathbf{F}^{-1} \cdot \mathbf{F}^{-1\mathrm{T}} = -\frac{1}{2}\mathbf{F}_p^{-1} \cdot \mathbf{F}_e^{-1} \cdot \mathbf{F}_e^{-1\mathrm{T}} \cdot \mathbf{F}_p^{-1\mathrm{T}} = \mathbf{F}_p^{-1} \cdot \overset{\mathrm{I}}{\mathbf{G}}_e \cdot \mathbf{F}_p^{-1\mathrm{T}}, \tag{9.147}$$

then

$$\overset{\mathrm{I}}{\mathbf{T}} \cdot\cdot \overset{\mathrm{I}}{\mathbf{G}}^{\bullet} = \overset{\mathrm{I}}{\mathbf{T}} \cdot\cdot (\mathbf{F}_p^{-1} \cdot \overset{\mathrm{I}}{\mathbf{G}}{}_e^{\bullet} \cdot \mathbf{F}_p^{-1\mathrm{T}} + \dot{\mathbf{F}}_p^{-1} \cdot \overset{\mathrm{I}}{\mathbf{G}}_e \cdot \mathbf{F}_p^{-1\mathrm{T}} + \mathbf{F}_p^{-1} \cdot \overset{\mathrm{I}}{\mathbf{G}}_e \cdot \dot{\mathbf{F}}_p^{-1\mathrm{T}}). \tag{9.148}$$

According to the rule of permutations of tensors in the triple scalar product (4.4), we get

$$\overset{\mathrm{I}}{\mathbf{T}} \cdot\cdot \overset{\mathrm{I}}{\mathbf{G}}{}^{\bullet} = (\mathbf{F}_p^{-1\mathrm{T}} \cdot \overset{\mathrm{I}}{\mathbf{T}} \cdot \mathbf{F}_p^{-1}) \cdot\cdot \overset{\mathrm{I}}{\mathbf{G}}{}^{\bullet}_e + (\mathbf{F}_p^{-1\mathrm{T}} \cdot \overset{\mathrm{I}}{\mathbf{T}}) \cdot\cdot (\mathbf{F}_p^{-1} \cdot \mathbf{F}_p \cdot \dot{\mathbf{F}}_p^{-1} \cdot \overset{\mathrm{I}}{\mathbf{G}}_e)$$
$$+ (\mathbf{F}_p^{-1\mathrm{T}} \overset{\mathrm{I}}{\mathbf{T}}) \cdot\cdot \mathbf{F}_p^{-1} \cdot \overset{\mathrm{I}}{\mathbf{G}}_e \cdot \dot{\mathbf{F}}_p^{-1\mathrm{T}}) \cdot \mathbf{F}_p^{\mathrm{T}}. \qquad (9.149)$$

Since $\overset{\mathrm{I}}{\mathbf{G}}_e = -(1/2)\mathbf{U}_e^{-2}$, $\mathbf{F}_p \cdot \dot{\mathbf{F}}_p^{-1} = -\dot{\mathbf{F}}_p \cdot \mathbf{F}_p^{-1} = -\mathbf{D}_p$ and

$$\mathbf{F}_p^{-1\mathrm{T}} \cdot \overset{\mathrm{I}}{\mathbf{T}} \cdot \mathbf{F}_p^{-1} = \mathbf{F}_p^{-1\mathrm{T}} \cdot \mathbf{F}^{\mathrm{T}} \cdot \mathbf{T} \cdot \mathbf{F} \cdot \mathbf{F}_p^{-1} = \mathbf{F}_e^{\mathrm{T}} \cdot \mathbf{T} \cdot \mathbf{F}_e, \qquad (9.150)$$

so the representation (9.144) really holds at $n = \mathrm{I}$:

$$\overset{\mathrm{I}}{\mathbf{T}} \cdot\cdot \overset{\mathrm{I}}{\mathbf{G}}{}^{\bullet} = (\mathbf{F}_p^{-1\mathrm{T}} \cdot \overset{\mathrm{I}}{\mathbf{T}} \cdot \mathbf{F}_p^{-1}) \cdot\cdot (\overset{\mathrm{I}}{\mathbf{G}}{}^{\bullet}_e + \frac{1}{2}\mathbf{D}_p \cdot \mathbf{U}_e^{-2} + \frac{1}{2}\mathbf{U}_e^{-2} \cdot \mathbf{D}_p^{\mathrm{T}})$$
$$= (\mathbf{F}_e^{\mathrm{T}} \cdot \mathbf{T} \cdot \mathbf{F}_e) \cdot\cdot (\overset{\mathrm{I}}{\mathbf{G}}{}^{\bullet}_e + \overset{\mathrm{I}}{\mathbf{D}}_p). \qquad (9.151)$$

2. Since the tensors $\overset{\mathrm{I}}{\mathbf{T}}_e$ and $\overset{\mathrm{I}}{\mathbf{G}}_e$ differ from tensors $\overset{\mathrm{I}}{\mathbf{T}}$ and $\overset{\mathrm{I}}{\mathbf{G}}$ only by the substitutions $\mathbf{F} \to \mathbf{F}_e$ and $\mathbf{U} \to \mathbf{U}_e$, respectively, the first couple $(\overset{\mathrm{I}}{\mathbf{T}}_e, \overset{\mathrm{I}}{\mathbf{G}}_e)$ gives all the remaining couples $\overset{(n)}{\mathbf{T}}_e, \overset{(n)}{\mathbf{G}}_e$ in the same way as the energetic couple $(\overset{\mathrm{I}}{\mathbf{T}}, \overset{\mathrm{I}}{\mathbf{G}})$ does (the tensors $\mathbf{F}_p$ do not appear in these relations).

3. We should show only that the first couple gives the third and fifth ones:

$$\overset{\mathrm{I}}{\mathbf{T}}_e \cdot\cdot \overset{\mathrm{I}}{\mathbf{D}}_p = \overset{\mathrm{III}}{\mathbf{T}}_e \cdot\cdot \overset{\mathrm{III}}{\mathbf{D}}_p = \overset{\mathrm{V}}{\mathbf{T}}_e \cdot\cdot \overset{\mathrm{V}}{\mathbf{D}}_p \qquad (9.152)$$

(the second and fourth couples coincide with the first and fifth ones).

Indeed, according to the definitions of $\overset{\mathrm{I}}{\mathbf{T}}_e$ and $\overset{\mathrm{I}}{\mathbf{D}}_p$, we have

$$\overset{\mathrm{I}}{\mathbf{T}}_e \cdot\cdot \overset{\mathrm{I}}{\mathbf{D}}_p = (\mathbf{F}_e^{\mathrm{T}} \cdot \mathbf{T} \cdot \mathbf{F}_e) \cdot\cdot \overset{\mathrm{I}}{\mathbf{D}}_p$$
$$= (\mathbf{U}_e \cdot \mathbf{O}_e^{\mathrm{T}} \cdot \mathbf{T} \cdot \mathbf{O}_e \cdot \mathbf{U}_e) \cdot \frac{1}{2}(\mathbf{D}_p \cdot \mathbf{U}_e^{-2} + \mathbf{U}_e^{-2} \cdot \mathbf{D}_e^{\mathrm{T}})$$
$$= \frac{1}{2}\mathbf{O}_e^{\mathrm{T}} \cdot \mathbf{T} \cdot \mathbf{O}_e \cdot\cdot (\mathbf{U}_e \cdot \mathbf{D}_p \cdot \mathbf{U}_e^{-1} + \mathbf{U}_e^{-1} \cdot \mathbf{D}_p^{\mathrm{T}} \cdot \mathbf{U}_e) = \overset{\mathrm{III}}{\mathbf{T}}_e \cdot\cdot \overset{\mathrm{III}}{\mathbf{D}}_p$$
$$(9.153)$$

and

$$\overset{\text{III}}{\mathbf{T}}_e \cdot\cdot \overset{\text{III}}{\mathbf{D}}_p = \frac{1}{2}(\mathbf{O}_e^{\mathrm{T}} \cdot \mathbf{T} \cdot \mathbf{O}_e \cdot \mathbf{U}_e^{-1}) \cdot\cdot \mathbf{U}_e^2 \cdot \mathbf{D}_p \cdot \mathbf{U}_e^{-1}$$
$$+\frac{1}{2}(\mathbf{U}_e^{-1} \cdot \mathbf{O}_e^{\mathrm{T}} \cdot \mathbf{T} \cdot \mathbf{O}_e \cdot \mathbf{U}_e^{-1}) \cdot\cdot \mathbf{D}_p^{\mathrm{T}} \cdot \mathbf{U}_e^2$$
$$= \frac{1}{2}(\mathbf{F}_e^{-1} \cdot \mathbf{T} \cdot \mathbf{F}_e^{-1\mathrm{T}}) \cdot\cdot (\mathbf{U}_e^2 \cdot\cdot \mathbf{D}_p + \mathbf{D}_p^{\mathrm{T}} \cdot \mathbf{U}_e^2) = \overset{\text{V}}{\mathbf{T}}_e \cdot\cdot \overset{\text{V}}{\mathbf{D}}_p. \blacktriangle \tag{9.154}$$

**Theorem 9.2.** *The yield stress power $w_{(p)}$ can always be represented as a sum of the powers of plastic stretches and plastic rotations:*

$$w_{(p)} = \overset{(\mathrm{n})}{\mathbf{T}}_p \cdot\cdot \overset{(\mathrm{n})}{\mathbf{D}}_p = \mathbf{T}_U \cdot\cdot \dot{\mathbf{U}}_p + \mathbf{T}_o \cdot\cdot \boldsymbol{\Omega}_p. \tag{9.155}$$

*Here $\mathbf{T}_U$ is the symmetric tensor of plastic stretch stresses, $\mathbf{T}_o$ is the skew-symmetric tensor of plastic rotation stresses, which are defined as follows:*

$$\mathbf{T}_U = \frac{1}{2}(\mathbf{F}_p^{-1} \cdot \mathbf{T}_e \cdot \mathbf{O}_p + \mathbf{O}_p^{\mathrm{T}} \cdot \mathbf{T}_e^{\mathrm{T}} \cdot \mathbf{F}_p^{-1\mathrm{T}}), \tag{9.156a}$$

$$\mathbf{T}_o = \frac{1}{2}(\mathbf{T}_e - \mathbf{T}_e^{\mathrm{T}}), \tag{9.156b}$$

$$\mathbf{T}_e = \mathbf{F}_e^{-1} \cdot \mathbf{T} \cdot \mathbf{F}_e, \tag{9.157}$$

*and $\boldsymbol{\Omega}_p$ is the skew-symmetric spin tensor of plastic rotation:*

$$\boldsymbol{\Omega}_p = \dot{\mathbf{O}}_p \cdot \mathbf{O}_p^{\mathrm{T}}. \tag{9.158}$$

▼ Modify the plastic deformation rate measure $\mathbf{D}_p$ (9.146) as follows:

$$\mathbf{D}_p = \dot{\mathbf{F}}_p \cdot \mathbf{F}_p^{-1} = (\dot{\mathbf{O}}_p \cdot \mathbf{U}_p + \mathbf{O}_p \cdot \dot{\mathbf{U}}_p) \cdot \mathbf{U}_p^{-1} \cdot \mathbf{O}_p^{\mathrm{T}}$$
$$= \dot{\mathbf{O}}_p \cdot \mathbf{O}_p^{\mathrm{T}} + \mathbf{O}_p \cdot \dot{\mathbf{U}}_p \cdot \mathbf{U}_p^{-1} \cdot \mathbf{O}_p^{\mathrm{T}} = \boldsymbol{\Omega}_p + \mathbf{D}_v. \tag{9.159}$$

Here we have denoted the plastic stretch rate measure

$$\mathbf{D}_v = \mathbf{O}_p \cdot \dot{\mathbf{U}}_p \cdot \mathbf{U}_p^{-1} \cdot \mathbf{O}_p^{\mathrm{T}} = \mathbf{O}_p \cdot \dot{\mathbf{U}}_p \cdot \mathbf{F}_p^{-1}. \tag{9.160}$$

Then the first couple in (9.152) can be written in the form

$$w_{(p)} = \overset{\text{I}}{\mathbf{T}}_e \cdot\cdot \overset{\text{I}}{\mathbf{D}}_p = \frac{1}{2}\overset{\text{I}}{\mathbf{T}}_e \cdot\cdot (\boldsymbol{\Omega}_p \cdot \mathbf{U}_e^{-2} + \mathbf{U}_e^{-2} \cdot \boldsymbol{\Omega}_p^{\mathrm{T}} + \mathbf{D}_v \cdot \mathbf{U}_e^{-2} + \mathbf{U}_e^{-2} \cdot \mathbf{D}_v^{\mathrm{T}}). \tag{9.161}$$

According to the permutation rule for tensors in the triple scalar product, we obtain

$$w_{(p)} = \frac{1}{2}(\mathbf{U}_e^{-2} \cdot \overset{\mathrm{I}}{\mathbf{T}}_e - \overset{\mathrm{I}}{\mathbf{T}}_e \cdot \mathbf{U}_e^{-2}) \cdot\cdot \boldsymbol{\Omega}_p + \frac{1}{2}((\mathbf{F}_p^{-1} \cdot \mathbf{U}_e^{-2} \cdot \overset{\mathrm{I}}{\mathbf{T}} \cdot \mathbf{O}_p) \cdot\cdot \dot{\mathbf{U}}_p$$
$$+(\mathbf{O}_p^{\mathrm{T}} \cdot \overset{\mathrm{I}}{\mathbf{T}}_e \cdot \mathbf{U}_e^{-2} \cdot \mathbf{F}_p^{-1\mathrm{T}}) \cdot\cdot \dot{\mathbf{U}}_p) = \mathbf{T}_o \cdot\cdot \boldsymbol{\Omega}_p + \mathbf{T}_U \cdot\cdot \dot{\mathbf{U}}_p. \tag{9.162}$$

Here we have taken into account that the spin $\boldsymbol{\Omega}_p$ is skew-symmetric and that

$$\mathbf{U}_e^{-2} \cdot \overset{\mathrm{I}}{\mathbf{T}}_e - \overset{\mathrm{I}}{\mathbf{T}}_e \cdot \mathbf{U}_e^{-2} = \mathbf{U}_e^{-2} \cdot \mathbf{F}_e^{\mathrm{T}} \cdot \mathbf{T} \cdot \mathbf{F}_e - \mathbf{F}_e^{\mathrm{T}} \cdot \mathbf{T} \cdot \mathbf{F}_e \cdot \mathbf{U}_e^{-2}$$
$$= \mathbf{F}_e^{-1} \cdot \mathbf{T} \cdot \mathbf{F}_e - \mathbf{F}_e^{\mathrm{T}} \cdot \mathbf{T} \cdot \mathbf{F}_e^{-1\mathrm{T}} = \mathbf{T}_e - \mathbf{T}_e^{\mathrm{T}} = 2\mathbf{T}_0,$$

$$\mathbf{F}_p^{-1} \cdot \mathbf{U}_e^{-2} \cdot \overset{\mathrm{I}}{\mathbf{T}} \cdot \mathbf{O}_p + \mathbf{O}_p^{\mathrm{T}} \cdot \overset{\mathrm{I}}{\mathbf{T}} \cdot \mathbf{U}_e^{-2} \cdot \mathbf{F}_p^{-1}$$
$$= \mathbf{F}_p^{-1} \cdot \mathbf{U}_e^{-2} \cdot \mathbf{F}_e^{\mathrm{T}} \cdot \mathbf{T} \cdot \mathbf{F}_e \cdot \mathbf{O}_p + \mathbf{O}_p^{\mathrm{T}} \cdot \mathbf{F}_e^{\mathrm{T}} \cdot \mathbf{T} \cdot \mathbf{F}_e \cdot \mathbf{U}_e^{-2} \cdot \mathbf{F}_p^{-1}$$
$$= \mathbf{F}_p^{-1} \cdot \mathbf{T}_e \cdot \mathbf{O}_p + \mathbf{O}_p^{\mathrm{T}} \cdot \mathbf{T}_e^{\mathrm{T}} \cdot \mathbf{F}_p^{-1\mathrm{T}} = 2\mathbf{T}_U,$$

because $\mathbf{U}_e^{-2} \cdot \mathbf{F}_e^{\mathrm{T}} = \mathbf{F}_e^{-1}$.

The fact that other couples $\overset{(\mathrm{n})}{\mathbf{T}}_p, \overset{(\mathrm{n})}{\mathbf{D}}_p$ give the same result (9.155) follows from equivalence of contractions of the couples (9.152). ▲

On substituting the expression (9.155) into (9.144), we get representations for the stress power

$$w_{(i)} = \overset{(\mathrm{n})}{\mathbf{T}} \cdot\cdot \overset{(\mathrm{n})}{\mathbf{G}}{}^{\bullet} = \overset{(\mathrm{n})}{\mathbf{T}}_e \cdot\cdot \overset{(\mathrm{n})}{\mathbf{G}}{}_e^{\bullet} + \mathbf{T}_U \cdot\cdot \dot{\mathbf{U}}_p + \mathbf{T}_0 \cdot\cdot \boldsymbol{\Omega}_p. \tag{9.163}$$

## 9.2.2 General Representation of Constitutive Equations for Models $B_n$ of Plastic Continua

Substitution of this expression into the principal thermodynamic identity (9.143) yields

$$\rho\dot{\psi} + \rho\eta\dot{\theta} - \overset{(\mathrm{n})}{\mathbf{T}}_e \cdot\cdot \overset{(\mathrm{n})}{\mathbf{G}}{}_e^{\bullet} - \mathbf{T}_U \cdot\cdot \mathbf{U}_p^{\bullet} - \mathbf{T}_o \cdot\cdot \boldsymbol{\Omega}_p + w^* = 0. \tag{9.164}$$

Introduce the Gibbs free energy

$$\zeta = \psi - \frac{1}{\rho} \overset{(\mathrm{n})}{\mathbf{T}}_e \cdot\cdot \overset{(\mathrm{n})}{\mathbf{G}}_e, \tag{9.165}$$

then for $\zeta$ we obtain the principal thermodynamic identity in the form $B_n^\zeta$:

$$\rho\dot{\zeta} + \rho\eta\dot{\theta} + \rho\overset{(n)}{\mathbf{G}}_e\cdot\cdot(\overset{(n)}{\mathbf{T}}_e/\rho)^\bullet - \overset{(n)}{\mathbf{T}}_v\cdot\cdot\mathbf{U}_p^\bullet - \overset{(n)}{\mathbf{T}}_o\cdot\cdot\boldsymbol{\Omega}_p + w^* = 0; \qquad (9.166)$$

i.e. a change in the free energy $\zeta$ is determined only by changing the functions $\theta$, $\overset{(n)}{\mathbf{T}}_e/\rho$, $\mathbf{U}_p$ and $\mathbf{O}_p$. Therefore, in the model $B_n^\zeta$ of a plastic continuum, the Gibbs free energy is considered to be a functional in the form (similar to (9.21))

$$\zeta = \overset{t}{\underset{\tau=0}{\zeta}}(\mathcal{R}(t), \dot{\mathcal{R}}(t), \mathcal{R}^t(\tau), \dot{\mathcal{R}}^t(\tau)), \quad \mathcal{R} = (\overset{(n)}{\mathbf{T}}_e/\rho, \mathbf{U}_p, \mathbf{O}_p^{\mathrm{T}}, \theta). \qquad (9.167)$$

According to the principle of equipresence, the measure $\overset{(n)}{\mathbf{G}}_e$ is a functional (only a tensor one) in the same form

$$\overset{(n)}{\mathbf{G}}_e = \overset{t}{\underset{\tau=0}{\mathbf{G}_e}}(\mathcal{R}(t), \dot{\mathcal{R}}(t), \mathcal{R}^t(\tau), \dot{\mathcal{R}}^t(\tau)). \qquad (9.168)$$

By analogy with (9.24) and (9.25), introduce the *equilibrium elastic deformation measure* $\overset{(n)}{\mathbf{G}}{}_e^0$ and *nonequilibrium elastic deformation measure* $\overset{(n)}{\mathbf{G}}{}_e^1$ as follows:

$$\overset{(n)}{\mathbf{G}}{}_e^0 = \overset{t}{\underset{\tau=0}{\mathbf{G}_e}}(\mathcal{R}(t), \mathcal{R}^t(\tau), \dot{\mathcal{R}}^t(\tau)), \quad \overset{(n)}{\mathbf{G}}{}_e^1 = \overset{(n)}{\mathbf{G}}_e - \overset{(n)}{\mathbf{G}}{}_e^0. \qquad (9.169)$$

Substituting the functionals (9.167)–(9.169) into the principal thermodynamic identity (9.166) and grouping like terms, we obtain the identity

$$\rho\left(\frac{\partial\zeta}{\partial(\overset{(n)}{\mathbf{T}}_e/\rho)} + \overset{(n)}{\mathbf{G}}{}_e^0\right)\cdot\cdot d\frac{\overset{(n)}{\mathbf{T}}_e}{\rho} + \rho\left(\frac{\partial\zeta}{\partial\theta}\right)d\theta + \rho\frac{\partial\zeta}{\partial(\overset{(n)}{\mathbf{T}}_e/\rho)^\bullet}\cdot\cdot\left(\frac{\overset{(n)}{\mathbf{T}}_e}{\rho}\right)^\bullet$$
$$+\rho\frac{\partial\zeta}{\partial\dot{\mathbf{O}}_p^{\mathrm{T}}}\cdot\cdot d\dot{\mathbf{O}}_p + \rho\frac{\partial\zeta}{\partial\theta^\bullet}d\theta^\bullet + \rho\frac{\partial\zeta}{\partial\mathbf{U}_p^\bullet}\cdot\cdot d\mathbf{U}_p^\bullet$$
$$+(w^* - \widetilde{\mathbf{T}}_U\cdot\cdot\dot{\mathbf{U}}_p + \rho\overset{(n)}{\mathbf{G}}{}_e^1\cdot\cdot(\overset{(n)}{\mathbf{T}}_e/\rho)^\bullet + \rho\delta\zeta - \widetilde{\mathbf{T}}_o\cdot\cdot\boldsymbol{\Omega}_p)\,dt = 0, \qquad (9.170)$$

where we have introduced the notation

$$\widetilde{\mathbf{T}}_U = \mathbf{T}_U - \mathbf{N}_p, \quad \mathbf{N}_p = \rho(\partial\zeta/\partial\mathbf{U}_p), \quad \widetilde{\mathbf{T}}_o = \mathbf{T}_o - \mathbf{N}_o, \quad \mathbf{N}_o = \rho\mathbf{O}_p\cdot(\partial\zeta/\partial\mathbf{O}_p^{\mathrm{T}}).$$

Since the differentials $d(\overset{(n)}{\mathbf{T}}_e/\rho)$, $d\theta$, $d(\overset{(n)}{\mathbf{T}}_e/\rho)^\bullet$, $d\theta^\bullet$, $d\mathbf{U}_p^\bullet$, $d\mathbf{O}_p^\bullet$ and $dt$ are independent, the identity (9.170) is equivalent to the following set of relations:

$$\overset{(n)}{\mathbf{G}}{}_e^0 = -\partial\zeta/\partial(\overset{(n)}{\mathbf{T}}_e/\rho), \tag{9.171}$$

$$\eta = -\partial\zeta/\partial\theta, \tag{9.172}$$

$$\partial\zeta/\partial(\overset{(n)}{\mathbf{T}}_e/\rho)^\bullet = 0, \quad \partial\zeta/\partial\dot{\theta} = 0, \quad \partial\zeta/\partial\dot{\mathbf{U}}_p = 0, \quad \partial\zeta/\partial\dot{\mathbf{O}}_p^{\mathrm{T}} = 0, \tag{9.173}$$

$$w^* = \widetilde{\mathbf{T}}_U + \widetilde{\mathbf{T}}_o \cdot\cdot \boldsymbol{\Omega}_p - \rho\overset{(n)}{\mathbf{G}}{}_e^1 \cdot\cdot (\overset{(n)}{\mathbf{T}}_e/\rho)^\bullet - \rho\delta\zeta, \tag{9.174}$$

which are constitutive equations for models $B_n^\zeta$ of plastic continua.

As follows from relations (9.173), the free energy $\zeta$ is independent of the rates of reactive variables $\dot{\mathcal{R}}$:

$$\zeta = \underset{\tau=0}{\overset{t}{\zeta}} (\mathcal{R}(t), \mathcal{R}^t(\tau), \dot{\mathcal{R}}^t(\tau)), \qquad \mathcal{R} = (\overset{(n)}{\mathbf{T}}_e/\rho, \mathbf{U}_p, \mathbf{O}_p^{\mathrm{T}}, \theta). \tag{9.175}$$

However, this dependence is observed for the dissipation function $w^*$ and for the functional $\overset{(n)}{\mathbf{G}}{}_e^1$:

$$\overset{(n)}{\mathbf{G}}{}_e{}^1 = \overset{(n)}{\mathbf{G}}{}_e{}^1(\mathcal{R}(t), \dot{\mathcal{R}}(t), \mathcal{R}^t(\tau), \dot{\mathcal{R}}^t(\tau)). \tag{9.176}$$

Thus, the model $B_n^\zeta$ of a plastic continuum is specified by four functionals: the scalar functional (9.175) for $\zeta$, the tensor functional (9.176) and two more tensor functionals for determining the tensors $\mathbf{U}_p$ and $\mathbf{O}_p^{\mathrm{T}}$.

### *9.2.3 Corollaries of the Onsager Principle for Models $B_n^\zeta$ of Plastic Continua*

Just as for models $A_n^\zeta$, we use the Onsager principle in order to construct functionals for the tensors $\overset{(n)}{\mathbf{G}}{}_e^1$, $\mathbf{U}_p$ and $\mathbf{O}_p$. Form the specific internal entropy production (4.728) with the help of expression (9.174):

$$\rho q^* = w^* - \frac{\mathbf{q}}{\theta}\cdot\nabla\theta = \widetilde{\mathbf{T}}_U \cdot\cdot \dot{\mathbf{U}}_p + \widetilde{\mathbf{T}}_o \cdot\cdot \boldsymbol{\Omega}_p - \rho\overset{(n)}{\mathbf{G}}{}_e^1 \cdot\cdot \left(\frac{\overset{(n)}{\mathbf{T}}_e}{\rho}\right)^\bullet - \rho\delta\zeta - \frac{\mathbf{q}}{\theta}\cdot\nabla\theta \geqslant 0 \tag{9.177}$$

and introduce thermodynamic forces $X_\beta$ and fluxes $Q_\beta$ as follows:

$$X_1 = \nabla\theta, \quad X_2 = \widetilde{\mathbf{T}}_U, \quad X_3 = \widetilde{\mathbf{T}}_o, \quad X_4 = (\overset{(n)}{\mathbf{T}}_e/\rho)^{\bullet},$$
$$Q_1 = -\mathbf{q}/\theta, \quad Q_2 = \dot{\mathbf{U}}_p, \quad Q_3 = \boldsymbol{\Omega}_p, \quad Q_4 = \rho\overset{(n)}{\mathbf{G}}{}^1_e. \tag{9.178}$$

Then, according to the Onsager principle, the following tensor-linear relations between $Q_\beta$ and $X_\beta$ hold:

$$\begin{aligned}
-\mathbf{q}/\theta &= L_{11}\cdot\nabla\theta + L_{12}\cdot\cdot\widetilde{\mathbf{T}}_U + L_{13}\cdot\cdot\widetilde{\mathbf{T}}_o + L_{14}\cdot\cdot(\overset{(n)}{\mathbf{T}}_e/\rho)^{\bullet},\\
\dot{\mathbf{U}}_p &= L_{12}\cdot\nabla\theta + L_{22}\cdot\cdot\widetilde{\mathbf{T}}_U + L_{23}\cdot\cdot\widetilde{\mathbf{T}}_o + L_{24}\cdot\cdot(\overset{(n)}{\mathbf{T}}_e/\rho)^{\bullet},\\
\boldsymbol{\Omega}_p &= L_{13}\cdot\nabla\theta + L_{23}\cdot\cdot\widetilde{\mathbf{T}}_U + L_{33}\cdot\cdot\widetilde{\mathbf{T}}_o + L_{34}\cdot\cdot(\overset{(n)}{\mathbf{T}}_e/\rho)^{\bullet},\\
\rho\overset{(n)}{\mathbf{G}}_e &= L_{14}\cdot\nabla\theta + L_{24}\cdot\cdot\widetilde{\mathbf{T}}_U + L_{34}\cdot\cdot\widetilde{\mathbf{T}}_o + L_{44}\cdot\cdot(\overset{(n)}{\mathbf{T}}_e/\rho)^{\bullet}.
\end{aligned} \tag{9.179}$$

Here tensors $L_{\alpha\beta}$ are functionals with the general form (9.176) of reactive variables $\mathcal{R} = \{\mathbf{T}_e/\rho,\ \mathbf{U}_p,\ \mathbf{O}_p^{\mathrm{T}},\ \theta\}$; their special forms are given by a model of a plastic continuum considered.

For *models* $B_n^\zeta$ *of plastic yield*, all $L_{\alpha\beta}$ and $\zeta$ are only functions of $\mathcal{R}$ and $\dot{\mathcal{R}}$ and of the Taylor parameters $w_\beta^p$:

$$\zeta = \zeta(\overset{(n)}{\mathbf{T}}_e,\ \mathbf{U}_p,\ \mathbf{O}_p^{\mathrm{T}},\ \theta,\ w_\beta^p), \tag{9.180a}$$

$$L_{22} = {}^4\mathbf{L}_U(\overset{(n)}{\mathbf{T}}_e,\ \mathbf{U}_p,\ \mathbf{O}_p^{\mathrm{T}},\ \theta,\ \overset{(n)}{\mathbf{T}}{}^{\bullet}_e,\ \dot{\mathbf{U}}_p,\ \dot{\mathbf{O}}_p,\ w_\beta^p), \tag{9.180b}$$

$$L_{33} = {}^4\mathbf{L}_o(\overset{(n)}{\mathbf{T}}_e,\ \mathbf{U}_p,\ \mathbf{O}_p^{\mathrm{T}},\ \theta,\ \overset{(n)}{\mathbf{T}}{}^{\bullet}_e,\ \dot{\mathbf{U}}_p,\ \dot{\mathbf{O}}_p,\ w_\beta^p), \tag{9.180c}$$

$$L_{11}\theta = \boldsymbol{\lambda}(\theta), \quad \text{the remaining} \quad L_{\beta\gamma} = 0.$$

In this model, the cross–effects are not considered, and constitutive equations (9.179) take the form

$$-\mathbf{q} = \boldsymbol{\lambda}\cdot\nabla\theta, \tag{9.181a}$$

$$\dot{\mathbf{U}}_p = {}^4\mathbf{L}_U\cdot\cdot\widetilde{\mathbf{T}}_U, \tag{9.181b}$$

$$\boldsymbol{\Omega}_p = {}^4\mathbf{L}_o\cdot\cdot\widetilde{\mathbf{T}}_o, \tag{9.181c}$$

$$\overset{(n)}{\mathbf{G}}{}^1_e \equiv 0. \tag{9.181d}$$

The tensor ${}^4\mathbf{L}_U$ is a fourth-order symmetric tensor, and the tensor ${}^4\mathbf{L}_o$ is skew-symmetric in indices 1, 2 and 3, 4, and symmetric in pairs of indices $(1, 2) \leftrightarrow (3, 4)$.

### 9.2.4 Associated Models $B_n^\zeta$ of Plastic Continua

In the *associated model of plasticity* $B_n^\zeta$, the equations of plastic yield (9.181b) and (9.181c) are connected to the yield surface by the gradient law

$$\dot{\mathbf{U}}_p = h\sum_{\beta=1}^{k} \dot{\varkappa}_\beta \frac{\partial f_\beta}{\partial \mathbf{T}_U}, \qquad (9.182a)$$

$$\boldsymbol{\Omega}_p = h\sum_{\beta=1}^{k} \dot{\varkappa}_\beta \frac{\partial f_\beta}{\partial \mathbf{T}_o}. \qquad (9.182b)$$

Here $f_\beta$ are plastic potentials assumed to be functions in the form (9.180a), which can be considered as functions of $T_U$, $T_o$ and $\mathbf{U}_p$, $\mathbf{O}_p^{\mathrm{T}}$:

$$f_\beta = f_\beta(T_U,\ T_o,\ \mathbf{U}_p,\ \mathbf{O}_p^{\mathrm{T}},\ \theta,\ w_\alpha^p). \qquad (9.183)$$

An equation of the yield surface, just as in the model $A_n^\zeta$, is given by the set of scalar equations (9.41)

$$f_\beta = 0, \quad \beta = 1, \ldots, k. \qquad (9.184)$$

The function $h$, taking on a value 0 or 1 and determining a domain of plastic loading, is evaluated by formula (9.51), where as the partial derivative $d'f_\beta/dt$ we use an analog of formula (9.43):

$$\frac{d'f_\beta}{dt} = \frac{\partial f_\beta}{\partial \mathbf{T}_U}\cdot\cdot\,\dot{\mathbf{T}}_U + \frac{\partial f_\beta}{\partial \mathbf{T}_o}\cdot\cdot\,\dot{\mathbf{T}}_o. \qquad (9.185)$$

The system of $(9 + k)$ scalar equations (9.182) and (9.183) (Eq. (9.182a) is equivalent to six scalar equations, (9.182b) – to three scalar equations due to skew-symmetry of the tensors $\boldsymbol{\Omega}_p$ and $\mathbf{T}_o$) allows us to find the $9 + k$ scalar unknowns: six components of the tensor $\mathbf{U}_p$, three components of the tensor $\mathbf{O}_p$ and $k$ scalar functions $\dot{\varkappa}_\beta$ $(\beta = 1, \ldots k)$ which are determined by Eqs. (9.182). The parameters $\dot{\varkappa}_\beta$ are functions in the form (9.180b)

$$\dot{\varkappa}_\beta = \dot{\varkappa}_\beta(T_U,\ T_o,\ \mathbf{U}_p,\ \mathbf{O}_p^{\mathrm{T}},\ \dot{\mathbf{U}}_p,\ \boldsymbol{\Omega},\ \theta,\ w_\beta^p). \qquad (9.186)$$

In the model $B_n^\zeta$ of an *elastoplastic continuum*, the elastic potential $\zeta$ does not depend explicitly on plastic deformations; therefore, for this model the constitutive equations (9.180a) and (9.171) become

$$\zeta = \zeta(\overset{(n)}{\mathbf{T}}_e,\ \theta,\ w_\beta^p), \quad \mathbf{N}_1 \equiv 0, \quad \mathbf{N}_0 \equiv 0, \quad \overset{(n)}{\mathbf{G}}_e = \mathbf{K}(\overset{(n)}{\mathbf{T}}_e,\ \theta,\ w_\beta^p) = -\rho(\partial\zeta/\partial\overset{(n)}{\mathbf{T}}_e). \qquad (9.187)$$

### 9.2.5 Corollary of the Principle of Material Symmetry for Associated Model $B_n^\zeta$ of Plasticity

Apply the principle of material symmetry to the constitutive equations (9.182) and (9.184). To do this we should clarify how the introduced tensors with subscripts $p$ and $e$ are changed under an orthogonal transformation of the reference configuration $\overset{\circ}{\mathcal{K}} \to \overset{*}{\mathcal{K}}$.

**Theorem 9.3.** *The tensors $\mathbf{O}_e$, $\mathbf{U}_e$, and $\mathbf{V}_e$ are $H$-invariant under orthogonal $H$-transformations: $\overset{\circ}{\mathcal{K}} \to \overset{*}{\mathcal{K}}$, and all the tensors $\mathbf{F}_p$, $\mathbf{O}_p$, $\mathbf{U}_p$, $\dot{\mathbf{U}}_p$, $\mathbf{V}_p$, $\overset{(n)}{\mathbf{T}}_e$, $\overset{(n)}{\mathbf{T}}_p$, $\mathbf{T}_U$, $\mathbf{T}_o$, and $\boldsymbol{\Omega}$ are $H$-indifferent.*

▼ The unloaded configuration $\overset{p}{\mathcal{K}}$ introduced in Sect. 9.1.1 is in accord with the reference configuration $\overset{\circ}{\mathcal{K}}$. They must coincide (by definition), if loading is not plastic; for the associated model, this means that loading does not occur outside the yield surface. Therefore, the configuration $\overset{p}{\mathcal{K}}$ must be transformed in the same way as $\overset{\circ}{\mathcal{K}}$ (the local basis vectors $\overset{p}{\mathbf{r}}_i$ must be $H$-indifferent), and the elastic deformation gradient $\mathbf{F}_e$ (9.6) must be transformed during the passage $\overset{\circ}{\mathcal{K}} \to \overset{*}{\mathcal{K}}$ in the same way as $\mathbf{F}$ (see (4.188)):

$$\mathbf{F}_e = \overset{*}{\mathbf{F}}_e \cdot \mathbf{H}, \tag{9.188}$$

where

$$\overset{*}{\mathbf{F}}_e = \mathbf{r}_i \otimes \overset{p*}{\mathbf{r}}{}^i, \quad \overset{p*}{\mathbf{r}}{}^i = \mathbf{H}^{-1\mathrm{T}} \cdot \overset{p}{\mathbf{r}}{}^i, \quad \overset{p*}{\mathbf{r}}_i = \mathbf{H} \cdot \overset{p}{\mathbf{r}}_i. \tag{9.189}$$

Thus, the tensors $\mathbf{O}_e$, $\mathbf{U}_e$ and $\mathbf{V}_e$ are transformed in the same way as $\mathbf{O}$, $\mathbf{U}$ and $\mathbf{V}$ (see (4.256), (4.257)) under orthogonal transformations with the tensor $\mathbf{H} = \overset{*}{\mathbf{Q}}{}^{\mathrm{T}}$:

$$\overset{*}{\mathbf{O}}_e = \mathbf{O}_e \cdot \overset{*}{\mathbf{Q}}, \quad \overset{*}{\mathbf{V}}_e = \mathbf{V}_e, \quad \overset{*}{\mathbf{U}}_e = \overset{*}{\mathbf{Q}}{}^{\mathrm{T}} \cdot \mathbf{U}_e \cdot \overset{*}{\mathbf{Q}}. \tag{9.190}$$

Figure 9.2 shows a scheme of transformations of different configurations at the change of reference configuration $\overset{\circ}{\mathcal{K}} \to \overset{*}{\mathcal{K}}$.

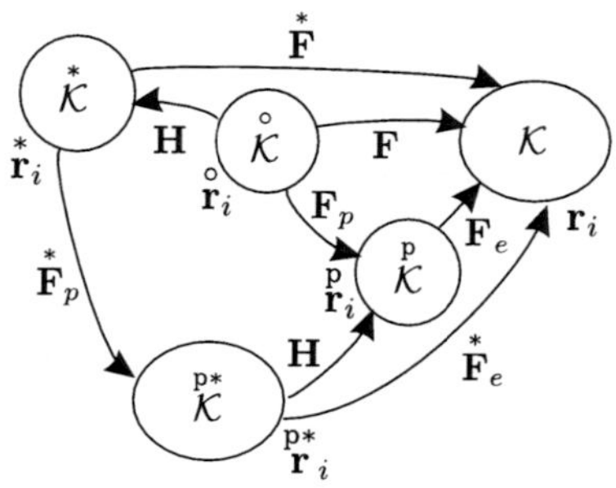

Fig. 9.2 $H$-transformations of reference and unloaded configurations

According to (9.189) and (4.194), the plastic deformation gradient $\mathbf{F}_p$ (9.6) under the transformation $\overset{\circ}{\mathcal{K}} \to \overset{*}{\mathcal{K}}$ takes the form

$$\overset{*}{\mathbf{F}}_p = \overset{p*}{\mathbf{r}}_i \otimes \overset{*}{\mathbf{r}}{}^i = \mathbf{H} \cdot \overset{p}{\mathbf{r}}_i \otimes \overset{\circ}{\mathbf{r}}{}^i \cdot \mathbf{H}^{-1} = \overset{*}{\mathbf{Q}}{}^{\mathrm{T}} \cdot \mathbf{F}_p \cdot \overset{*}{\mathbf{Q}}; \tag{9.191}$$

i.e. it is $H$-indifferent under orthogonal transformations. Using the polar decomposition for $\overset{*}{\mathbf{F}}_p$ and $\mathbf{F}_p$, we obtain

$$\overset{*}{\mathbf{F}}_p = \overset{*}{\mathbf{Q}}_p \cdot \overset{*}{\mathbf{U}}_p = \overset{*}{\mathbf{Q}}{}^{\mathrm{T}} \cdot \mathbf{O}_p \cdot \mathbf{U}_p \cdot \overset{*}{\mathbf{Q}} = (\overset{*}{\mathbf{Q}}{}^{\mathrm{T}} \cdot \mathbf{O}_p \cdot \overset{*}{\mathbf{Q}}) \cdot (\overset{*}{\mathbf{Q}}{}^{\mathrm{T}} \cdot \mathbf{U}_p \cdot \overset{*}{\mathbf{Q}}).$$

Since the tensor $(\overset{*}{\mathbf{Q}}{}^{\mathrm{T}} \cdot \mathbf{O}_p \cdot \overset{*}{\mathbf{Q}})$ is orthogonal, the tensor $(\overset{*}{\mathbf{Q}}{}^{\mathrm{T}} \cdot \mathbf{U}_p \cdot \overset{*}{\mathbf{Q}})$ is symmetric and the polar decomposition is unique, we get the following transformation formulae for $\mathbf{O}_p$ and $\mathbf{U}_p$:

$$\overset{*}{\mathbf{Q}}_p = \overset{*}{\mathbf{Q}}{}^{\mathrm{T}} \cdot \mathbf{O}_p \cdot \overset{*}{\mathbf{Q}}, \quad \overset{*}{\mathbf{U}}_p = \overset{*}{\mathbf{Q}}{}^{\mathrm{T}} \cdot \mathbf{U}_p \cdot \overset{*}{\mathbf{Q}}. \tag{9.192}$$

The tensor $\overset{*}{\mathbf{V}}_p$ is transformed as follows:

$$\overset{*}{\mathbf{V}}_p = \overset{*}{\mathbf{F}}_p \cdot \overset{*}{\mathbf{O}}{}_p^{\mathrm{T}} = \overset{*}{\mathbf{Q}}{}^{\mathrm{T}} \cdot \mathbf{F}_p \cdot \overset{*}{\mathbf{Q}} \cdot \overset{*}{\mathbf{Q}}{}^{\mathrm{T}} \cdot \mathbf{O}_p^{\mathrm{T}} \cdot \overset{*}{\mathbf{Q}} = \overset{*}{\mathbf{Q}}{}^{\mathrm{T}} \cdot \mathbf{V}_p \cdot \overset{*}{\mathbf{Q}}, \tag{9.193}$$

then the tensors $\mathbf{T}_p$, $\mathbf{O}_p$, $\mathbf{U}_p$, $\dot{\mathbf{U}}_p$ and $\mathbf{V}_p$ are $H$-indifferent under orthogonal transformations.

Due to $H$-invariance of the tensor $\mathbf{T}$, the tensors of elastic and plastic deformations ($\overset{(n)}{\mathbf{T}}_e$ and $\overset{(n)}{\mathbf{T}}_p$, respectively) defined by Table 9.1 are $H$-indifferent under orthogonal transformations as well as the tensors $\overset{(n)}{\mathbf{T}}$. Similarly, the tensors $\overset{(n)}{\mathbf{G}}_e$ are $H$-indifferent too.

The spin tensor $\boldsymbol{\Omega}_p$ (9.158) is $H$-indifferent as well as the tensors $\mathbf{T}_U$ (9.156) and $\mathbf{T}_o$ (9.157), because

$$\overset{*}{\mathbf{T}}_e = \overset{*}{\mathbf{F}}{}_e^{-1} \cdot \overset{*}{\mathbf{T}} \cdot \overset{*}{\mathbf{F}}_e = \overset{*}{\mathbf{Q}}{}^{\mathrm{T}} \cdot \mathbf{F}_e^{-1} \cdot \mathbf{T} \cdot \mathbf{F}_e \cdot \overset{*}{\mathbf{Q}} = \overset{*}{\mathbf{Q}}{}^{\mathrm{T}} \cdot \mathbf{T}_e \cdot \overset{*}{\mathbf{Q}}, \tag{9.194}$$

$$\overset{*}{\mathbf{T}}_U = \frac{1}{2}(\overset{*}{\mathbf{Q}}{}^{\mathrm{T}} \cdot \mathbf{F}_p^{-1} \cdot \overset{*}{\mathbf{Q}} \cdot \overset{*}{\mathbf{Q}}{}^{\mathrm{T}} \cdot \mathbf{T}_e \cdot \overset{*}{\mathbf{Q}} \cdot \overset{*}{\mathbf{Q}}{}^{\mathrm{T}} \cdot \mathbf{O}_p \cdot \overset{*}{\mathbf{Q}}$$
$$+ \overset{*}{\mathbf{Q}}{}^{\mathrm{T}} \cdot \mathbf{Q}_p \cdot \overset{*}{\mathbf{Q}} \cdot \overset{*}{\mathbf{Q}}{}^{\mathrm{T}} \cdot \mathbf{T}_e \cdot \overset{*}{\mathbf{Q}} \cdot \overset{*}{\mathbf{Q}}{}^{\mathrm{T}} \cdot \mathbf{F}_p^{-1\mathrm{T}} \cdot \overset{*}{\mathbf{Q}}) = \overset{*}{\mathbf{Q}}{}^{\mathrm{T}} \cdot \mathbf{T}_U \cdot \overset{*}{\mathbf{Q}}. \ \blacktriangle$$

Applying the principle of material symmetry to Eq. (9.182) and taking $H$-indifference of the tensors $\dot{\mathbf{U}}_p$, $\boldsymbol{\Omega}_p$, $\mathbf{T}_U$, $\mathbf{T}_o$, $\mathbf{U}_p$ and $\mathbf{O}_p^{\mathrm{T}}$ into account, we obtain that the equations of plastic yield (9.182) must be $H$-indifferent tensor functions relative to one or another orthogonal subgroup $\overset{\circ}{G}_s$, and the plastic potentials $f_\beta$ (9.183) must be scalar $H$-indifferent functions relative to $\overset{\circ}{G}_s$ and hence depend on simultaneous invariants $J_\gamma^{(s)p}$ in this group $\overset{\circ}{G}_s$:

$$f_\beta = f_\beta(J_\gamma^{(s)p}, \theta, w_\alpha^p), \quad J_\gamma^{(s)p} = J_\gamma^{(s)}(\mathbf{T}_U, \mathbf{T}_o, \mathbf{U}_p, \mathbf{O}_p^{\mathrm{T}}), \quad \gamma = 1, \ldots, z. \tag{9.195}$$

The scalar functionals $w_\alpha^p$ are integrals of the quadratic simultaneous invariants of the tensors $\mathbf{T}_U$, $\dot{\mathbf{U}}_p$ and $\mathbf{T}_o$, $\boldsymbol{\Omega}_p$:

$$\begin{aligned} w_\alpha^p &= \int_0^t J_\alpha^{(s)}(\mathbf{T}_U(\tau), \dot{\mathbf{U}}_p(\tau))\, d\tau, \quad \alpha = 1, \ldots, r_1, \\ w_\alpha^p &= \int_0^t J_\alpha^{(s)}(\mathbf{T}_o(\tau), \boldsymbol{\Omega}(\tau))\, d\tau, \quad \alpha = r_1 + 1, \ldots, r. \end{aligned} \tag{9.196}$$

Similarly, the elastic potential $\zeta$ (9.184a) is a $H$-indifferent scalar function too; therefore, it can be written in the form

$$\zeta = \zeta(J_\gamma^{(s)}(\overset{(n)}{\mathbf{T}}_e, \mathbf{U}_p, \mathbf{O}_p^{\mathrm{T}}), \theta, w_\alpha^p). \tag{9.197}$$

Then constitutive equations (9.171) and (9.182) can be represented in the tensor basis:

$$\overset{(n)}{\mathbf{G}}_e = \sum_{\gamma=1}^{z} \varphi_\gamma J_{\gamma\mathbf{T}}^{(s)}, \tag{9.198a}$$

$$\varphi_\gamma = -\rho(\partial\zeta/\partial J_\gamma^{(s)}), \quad J_{\gamma\mathbf{T}}^{(s)} = \partial J_\gamma^{(s)}/\partial \overset{(n)}{\mathbf{T}}_e, \tag{9.198b}$$

$$\dot{\mathbf{U}}_p = \sum_{\alpha=1}^{z_1} \psi_\alpha J_{\alpha\mathbf{T}_U}^{(s)}, \tag{9.198c}$$

$$\boldsymbol{\Omega}_p = \sum_{\alpha=1}^{z_1} \psi_\alpha J_{\alpha\mathbf{T}_o}^{(s)}, \tag{9.198d}$$

$$\psi_\alpha = h \sum_{\beta=1}^{k} \dot{\varkappa}_\beta(\partial f_\beta/\partial J_\alpha^{(s)}), \quad J_{\alpha\mathbf{T}_U}^{(s)} = \partial J_\alpha^{(s)}/\partial \mathbf{T}_U, \quad J_{\alpha\mathbf{T}_o}^{(s)} = \partial J_\alpha^{(s)}/\partial \mathbf{T}_o. \tag{9.198e}$$

### 9.2.6 Associated Models of Plasticity $B_n^\xi$ with Proper Strengthening

We will consider below only the case of *associated models* $B_n^\xi$ *of plasticity with proper strengthening*, where the simultaneous invariants $J_\gamma^{(s)p}$ (9.195) depend only on the tensors $\mathbf{T}_U$ and $\mathbf{T}_o$ and do not depend explicitly on $\mathbf{U}_p$ and $\mathbf{O}_p^{\mathrm{T}}$:

$$J_\gamma^{(s)p} = J_\gamma^{(s)}(\mathbf{T}_U, \mathbf{T}_o). \tag{9.199}$$

Since the tensors $\mathbf{T}_U$ and $\mathbf{T}_o$ defined by formulae (9.156) and (9.158) are combinations of the tensors $\overset{(n)}{\mathbf{T}}$, $\mathbf{U}_e$, $\mathbf{O}_e$, $\mathbf{U}_p$ and $\mathbf{O}_p$, the model (9.199) allows us to take account of plastic strengthening of a continuum (an increase of the yield strength after the appearance of plastic deformations) but in the special way, namely in terms of the tensors $\mathbf{T}_U$ and $\mathbf{T}_o$.

Notice that since the tensor $\mathbf{T}_o$ is skew-symmetric, it has only three independent components, and this tensor can be connected uniquely with the vorticity vector (see (2.227))

$$\boldsymbol{\omega}_0 = \frac{1}{2}\boldsymbol{\varepsilon} \cdot\cdot\, \mathbf{T}_o. \tag{9.200}$$

Then the simultaneous invariants (9.199) are scalar $H$-indifferent functions of the symmetric tensor $\mathbf{T}_U$ and the vector $\boldsymbol{\omega}_0$:

$$J_\gamma^{(s)p} = J_\gamma^{(s)}(\mathbf{T}_U,\ \boldsymbol{\omega}_0). \tag{9.201}$$

### 9.2.7 Associated Models of Plasticity $B_n^\xi$ for Isotropic Continua

Let us write the constitutive equations (9.198) for the three main symmetry groups $\overset{\circ}{G}_s$: $O$, $T_3$ and $I$ while considering only the case of elastoplastic continua, i.e. while the simultaneous invariants $J_\gamma^{(s)}$ in (9.197) coincide with the invariants $I_\gamma^{(s)}(\overset{(n)}{\mathbf{T}}_e)$.

For an *isotropic continuum*, the functional basis of invariants (9.201) consists of six elements, which can be chosen as follows:

$$J_\gamma^{(I)} = J_\gamma(\mathbf{T}_U),\ \ \gamma = 1,\ 2,\ 3, \quad J_4^{(I)} = |\boldsymbol{\omega}_0|^2,$$
$$J_5^{(I)} = \boldsymbol{\omega}_0 \cdot \mathbf{T}_U \cdot \boldsymbol{\omega}_0, \quad J_6^{(I)} = \boldsymbol{\omega}_0 \cdot \mathbf{T}_U^2 \cdot \boldsymbol{\omega}_0. \tag{9.202}$$

The derivative tensors $J^{(s)}_{\gamma\mathbf{T}_U}$ and $J^{(s)}_{\gamma\mathbf{T}_o}$ in this case become

$$
\begin{gathered}
J^{(I)}_{1\mathbf{T}_U}=\mathbf{E},\ \ J^{(I)}_{2\mathbf{T}_U}=\mathbf{E}I_1(\mathbf{T}_U)-\mathbf{T}_U,\ \ J^{(I)}_{3\mathbf{T}_U}=\mathbf{T}_U^2-I_1\mathbf{T}_U+I_2\mathbf{E},\ \ J^{(I)}_{4\mathbf{T}_U}=0,\\
J^{(I)}_{5\mathbf{T}_U}=\boldsymbol{\omega}_0\otimes\boldsymbol{\omega}_0,\ \ J^{(I)}_{6\mathbf{T}_U}=\boldsymbol{\omega}_0\otimes\mathbf{T}_U\cdot\boldsymbol{\omega}_0+\boldsymbol{\omega}_0\cdot\mathbf{T}_U\otimes\boldsymbol{\omega}_0,\\
J^{(I)}_{4\mathbf{T}_o}=\boldsymbol{\omega}_0\cdot\boldsymbol{\varepsilon},\ \ J^{(I)}_{5\mathbf{T}_o}=(\boldsymbol{\omega}_0\cdot\mathbf{T}_U+\mathbf{T}_U\cdot\boldsymbol{\omega}_0)\cdot\boldsymbol{\varepsilon},\\
J^{(I)}_{6\mathbf{T}_o}=(\boldsymbol{\omega}_0\cdot\mathbf{T}_U^2+\mathbf{T}_U^2\cdot\boldsymbol{\omega}_0)\cdot\boldsymbol{\varepsilon},\ \ J^{(I)}_{\gamma\mathbf{T}_o}=0,\ \ \gamma=1,\ 2,\ 3.
\end{gathered}
\tag{9.203}
$$

Here we have used that

$$
d\boldsymbol{\omega}_0/d\mathbf{T}_o=\frac{1}{2}\,\boldsymbol{\varepsilon}. \tag{9.204}
$$

The tensors $J^{(I)}_{\gamma\mathbf{T}_o}$ are seen to be skew-symmetric.

According to the Onsager principle, Eqs. (9.198c) and (9.198d) must be quasi-linear with respect to $\mathbf{T}_U$ and $\mathbf{T}_o$, i.e. they must have the forms (9.181b) and (9.181c); therefore, invariants $J^{(I)}_3$, $J^{(I)}_5$ and $J^{(I)}_6$ must be eliminated between arguments of the potential $f_\beta$ (9.195). Thus, $f_\beta$ depends only on three invariants $J^{(I)}_\gamma$, $\gamma=1,2,4$.

On substituting the expressions (9.202) and (9.203) into (9.198), we obtain the following constitutive equations of the *associated model* $B_n^\zeta$ *of an isotropic elasto-plastic continuum with proper strengthening*:

$$
\begin{cases}
\overset{(n)}{\mathbf{G}}_e=\widetilde{\varphi}_1\mathbf{E}+\widetilde{\varphi}_2\overset{(n)}{\mathbf{T}}_e+\varphi_3\overset{(n)}{\mathbf{T}}{}^2_e,\\
\dot{\mathbf{U}}_p=\widetilde{\psi}_1\mathbf{E}-\psi_2\mathbf{T}_U,\\
\boldsymbol{\Omega}_p=\psi_4\boldsymbol{\omega}_0\cdot\boldsymbol{\varepsilon},
\end{cases}
\tag{9.205}
$$

where

$$
\begin{gathered}
\widetilde{\varphi}_1=\varphi_1+\varphi_2I_1(\overset{(n)}{\mathbf{T}}_e)+\varphi_3I_2(\overset{(n)}{\mathbf{T}}_e),\ \ \widetilde{\varphi}_2=-\varphi_2-\varphi_3I_1(\overset{(n)}{\mathbf{T}}_e),\ \ \widetilde{\psi}_1=\psi_1+\psi_2I_1(\mathbf{T}_U),\\
\varphi_\alpha=-\rho\frac{\partial\zeta}{\partial I_\alpha(\overset{(n)}{\mathbf{T}}_e)},\quad \psi_\gamma=h\sum_{\beta=1}^{k}\dot{\varkappa}_\beta\frac{\partial f_\beta}{\partial J^{(I)}_\gamma},\\
\zeta=\zeta(I_\alpha(\overset{(n)}{\mathbf{T}}_e),\ \theta,\ w^p_\delta),\quad \alpha=1,\ 2,\ 3,\\
f_\beta=f_\beta(J^{(I)}_\gamma(\mathbf{T}_U,\boldsymbol{\omega}_0),\ \theta,\ w^p_\delta),\quad \gamma=1,\ 2,\ 4.
\end{gathered}
\tag{9.206}
$$

The Taylor parameters $w^p_\delta$ for an isotropic continuum (9.196) have the forms

$$
w^p_1=\int_0^t\mathbf{T}_U(\tau)\cdot\cdot\,\dot{\mathbf{U}}_p(\tau)\,d\tau,\quad w^p_2=\int_0^t\mathbf{T}_o(\tau)\cdot\cdot\,\boldsymbol{\Omega}(\tau)\,d\tau. \tag{9.207}
$$

The plastic potentials can be chosen in the quadratic form (9.84).

### 9.2.8 Associated Models of Plasticity $B_n^\zeta$ for Transversely Isotropic Continua

For a *transversely isotropic continuum*, the functional basis $J_\gamma^{(s)}(\mathbf{T}_U, \boldsymbol{\omega}_0)$ consists of eight simultaneous invariants. However, according to the Onsager principle, the plastic potentials $f_\beta$ must depend only on invariants being linear and quadratic in $\mathbf{T}_U$ and $\boldsymbol{\omega}_0$, which can be chosen as follows:

$$J_1^{(3)} = (\mathbf{E} - \widehat{\mathbf{c}}_3^2) \cdot\cdot\, \mathbf{T}_U, \quad J_2^{(3)} = \mathbf{T}_U \cdot\cdot\, \widehat{\mathbf{c}}_3^2, \quad J_3^{(3)} = ((\mathbf{E} - \widehat{\mathbf{c}}_3^2) \cdot \mathbf{T}_U) \cdot\cdot\, (\widehat{\mathbf{c}}_3^2 \cdot \mathbf{T}_U),$$
$$J_4^{(3)} = \mathbf{T}_U^2 \cdot\cdot\, \mathbf{E} - J_2^{(3)2} - 2J_3^{(3)}, \quad J_5^{(3)} = |\boldsymbol{\omega}_0|^2, \quad J_6^{(3)} = (\boldsymbol{\omega}_0 \cdot \widehat{\mathbf{c}}_3)^2. \quad (9.208)$$

The non-zero derivative tensors for them have the forms

$$J_{1\mathbf{T}_U}^{(3)} = \mathbf{E} - \widehat{\mathbf{c}}_3^2, \quad J_{2\mathbf{T}_U}^{(3)} = \widehat{\mathbf{c}}_3^2, \quad J_{3\mathbf{T}_U}^{(3)} = \frac{1}{2}(\mathbf{O}_1 \otimes \mathbf{O}_1 + \mathbf{O}_2 \otimes \mathbf{O}_2) \cdot\cdot\, \mathbf{T}_U,$$
$$J_{4\mathbf{T}_U}^{(3)} = 2\,{}^4\mathbf{O}_3 \cdot\cdot \mathbf{T}_U, \quad J_{5\mathbf{T}_o}^{(3)} = \boldsymbol{\omega}_0 \cdot \boldsymbol{\varepsilon}, \quad J_{6\mathbf{T}_o}^{(3)} = \boldsymbol{\omega}_0 \cdot \widehat{\mathbf{c}}_3^2 \cdot \boldsymbol{\varepsilon}. \quad (9.209)$$

Then the constitutive equations (9.198) become

$$\begin{cases} \overset{(\mathrm{n})}{\mathbf{G}}_e = \widetilde{\varphi}_1 \mathbf{E} + \widetilde{\varphi}_2 \widehat{\mathbf{c}}_3^2 + \widetilde{\varphi}_3 (\mathbf{O}_1 \otimes \mathbf{O}_1 + \mathbf{O}_2 \otimes \mathbf{O}_2) \cdot\cdot \overset{(\mathrm{n})}{\mathbf{T}}_e + \widetilde{\varphi}_4 \overset{(\mathrm{n})}{\mathbf{T}}_e + \varphi_5 \overset{(\mathrm{n})}{\mathbf{T}}{}^2, \\ \dot{\mathbf{U}}_p = \psi_1 \mathbf{E} + \widetilde{\psi}_2 \widehat{\mathbf{c}}_3^2 + \widetilde{\psi}_3 (\mathbf{O}_1 \otimes \mathbf{O}_1 + \mathbf{O}_2 \otimes \mathbf{O}_2) \cdot\cdot \mathbf{T}_U + 2\psi_4 \mathbf{T}_U, \\ \boldsymbol{\Omega}_p = (\psi_5 \boldsymbol{\omega}_0 + \psi_6 \boldsymbol{\omega}_0 \cdot \widehat{\mathbf{c}}_3^2) \cdot \boldsymbol{\varepsilon}, \end{cases} \quad (9.210)$$

where

$$\widetilde{\varphi}_1 = \varphi_1 + \varphi_5 I_2(\overset{(\mathrm{n})}{\mathbf{T}}_e), \quad \widetilde{\varphi}_2 = \varphi_2 - \varphi_1 - 2\varphi_4 I_2^{(3)}(\overset{(\mathrm{n})}{\mathbf{T}}_e), \quad \widetilde{\varphi}_3 = \frac{\varphi_3}{2} - \varphi_4,$$
$$\widetilde{\varphi}_4 = 2\varphi_4 - \varphi_5 I_1(\overset{(\mathrm{n})}{\mathbf{T}}_e), \quad \widetilde{\psi}_2 = \psi_2 - \psi_1 - 2\psi_4 I_2^{(3)}(\mathbf{T}_U), \quad \widetilde{\psi}_3 = \frac{\psi_3}{2} - \psi_4,$$
$$\varphi_\alpha = -\rho \frac{\partial \zeta}{\partial I_\alpha^{(3)}(\overset{(\mathrm{n})}{\mathbf{T}}_e)}, \quad \psi_\alpha = h \sum_{\beta=1}^{k} \dot{\varkappa}_\beta \frac{\partial f_\beta}{\partial J_\alpha^{(3)}}, \quad (9.211)$$
$$\zeta = \zeta(I_1^{(3)}(\overset{(\mathrm{n})}{\mathbf{T}}_e), \ldots, I_5^{(3)}(\overset{(\mathrm{n})}{\mathbf{T}}_e), \, \theta, \, w_1^p, \ldots, w_5^p),$$
$$f_\beta = f_\beta(J_1^{(3)}, \ldots, J_6^{(3)}, \, \theta, \, w_1^p, \ldots, w_5^p).$$

In this model, the Taylor parameters $w_\beta^p$ (9.196) have the forms

$$w_1^p = \int_0^t (\widehat{\mathbf{c}}_3^2 \cdot\cdot \mathbf{T}_U)(\widehat{\mathbf{c}}_3^2 \cdot\cdot \dot{\mathbf{U}}_p)\, d\tau, \quad w_2^p = \int_0^t ((\mathbf{E} - \widehat{\mathbf{c}}_3^2) \cdot \mathbf{T}_U) \cdot (\widehat{\mathbf{c}}_3^2 \cdot \dot{\mathbf{U}}_p)\, d\tau,$$

$$w_3^p = \int_0^t \mathbf{T}_U \cdot\cdot \dot{\mathbf{U}}_p\, d\tau - 2w_2^p - w_1^p, \quad w_4^p = \int_0^t \boldsymbol{\Omega} \cdot\cdot \mathbf{T}_o\, d\tau, \tag{9.212}$$

$$w_5^p = \int_0^t (\widehat{\mathbf{c}}_3 \cdot \boldsymbol{\omega}_0)(\widehat{\mathbf{c}}_3 \cdot \boldsymbol{\omega}_\Omega)\, d\tau,$$

where $\boldsymbol{\omega}_\Omega$ is the vorticity vector connected to the tensor $\boldsymbol{\Omega}_p$:

$$\boldsymbol{\omega}_\Omega = \frac{1}{2}\boldsymbol{\varepsilon} \cdot\cdot \boldsymbol{\Omega}_p. \tag{9.213}$$

### 9.2.9 Associated Models of Plasticity $B_n^\xi$ for Orthotropic Continua

For an *orthotropic continuum*, the functional basis $J_\gamma^{(O)}(\mathbf{T}_U, \boldsymbol{\omega}_0)$ consists of eight simultaneous invariants. According to the Onsager principle, the set of these invariants must be chosen so that it contains only invariants being linear and quadratic in $\mathbf{T}_U$ and $\boldsymbol{\omega}_0$. This requirement is satisfied by the set

$$J_\gamma^{(O)} = \mathbf{T}_U \cdot\cdot \widehat{\mathbf{c}}_\gamma^2, \quad J_4^{(O)} = (\widehat{\mathbf{c}}_2^2 \cdot \mathbf{T}_U) \cdot\cdot (\widehat{\mathbf{c}}_3^2 \cdot \mathbf{T}_U), \quad J_5^{(O)} = (\widehat{\mathbf{c}}_1^2 \cdot \mathbf{T}_U) \cdot\cdot (\widehat{\mathbf{c}}_3^2 \cdot \mathbf{T}_U),$$

$$J_6^{(O)} = (\widehat{\mathbf{c}}_1^2 \cdot \mathbf{T}_U) \cdot\cdot (\widehat{\mathbf{c}}_2^2 \cdot \mathbf{T}_U), \quad J_{6+\gamma}^{(O)} = (\boldsymbol{\omega}_0 \cdot \widehat{\mathbf{c}}_\gamma)^2, \quad \gamma = 1,\ 2,\ 3. \tag{9.214}$$

The non-zero derivative tensors have the forms

$$J_{\gamma \mathbf{T}_U}^{(O)} = \widehat{\mathbf{c}}_\gamma^2, \quad J_{3+\gamma,\mathbf{T}_U}^{(O)} = \frac{1}{2}(\mathbf{O}_\gamma \otimes \mathbf{O}_\gamma) \cdot\cdot \mathbf{T}_U, \quad J_{6+\gamma,\mathbf{T}_o}^{(O)} = 2\boldsymbol{\omega}_0 \cdot \widehat{\mathbf{c}}_\gamma^2 \cdot \boldsymbol{\varepsilon}, \quad \gamma = 1, 2, 3. \tag{9.215}$$

The constitutive equations (9.198) in this case become

$$\begin{cases} \overset{(\mathrm{n})}{\mathbf{G}}_e = \sum_{\gamma=1}^3 (\varphi_\gamma \widehat{\mathbf{c}}_\gamma^2 + (\widetilde{\varphi}_{3+\gamma}/2)(\mathbf{O}_\gamma \otimes \mathbf{O}_\gamma) \cdot\cdot \overset{(\mathrm{n})}{\mathbf{T}}_e) \\ \qquad\qquad + 3\varphi_6\, {}^6\mathbf{O}_m \cdot\cdot\cdot\cdot \overset{(\mathrm{n})}{\mathbf{T}}_e \otimes \overset{(\mathrm{n})}{\mathbf{T}}_e, \\ \dot{\mathbf{U}}_p = \sum_{\gamma=1}^3 (\psi_\gamma \widehat{\mathbf{c}}_\gamma^2 + (\psi_{3+\gamma}/2)(\mathbf{O}_\gamma \otimes \mathbf{O}_\gamma) \cdot\cdot \mathbf{T}_U), \\ \boldsymbol{\Omega}_p = 2\sum_{\gamma=1}^3 \psi_{6+\gamma} \boldsymbol{\omega}_0 \cdot \widehat{\mathbf{c}}_\gamma^2 \cdot \boldsymbol{\varepsilon}, \end{cases} \tag{9.216}$$

where

$$\varphi_\gamma = -\rho \frac{\partial \zeta}{\partial I_\gamma^{(O)}}, \quad \psi_\gamma = h \sum_{\beta=1}^{k} \dot{\varkappa}_\beta \frac{\partial f_\beta}{\partial J_\gamma^{(O)}},$$

$$\zeta = \zeta(I_1^{(O)}(\overset{(n)}{\mathbf{T}}_e), \ldots, I_7^{(O)}(\overset{(n)}{\mathbf{T}}_e),\ \theta,\ w_1^p, \ldots, w_9^p), \tag{9.217}$$
$$f_\beta = f_\beta(J_1^{(O)}, \ldots, J_9^{(O)},\ \theta,\ w_1^p, \ldots, w_9^p).$$

The Taylor parameters have the forms

$$w_\gamma^p = \int_0^t (\widehat{\mathbf{c}}_\gamma^2 \cdot\cdot\, \mathbf{T}_U)(\widehat{\mathbf{c}}_\gamma^2 \cdot\cdot\, \dot{\mathbf{U}}_p)\, d\tau, \quad w_{3+\gamma}^p = \int_0^t (\widehat{\mathbf{c}}_\alpha^2 \cdot \mathbf{T}_U) \cdot\cdot (\widehat{\mathbf{c}}_\beta^2 \cdot \dot{\mathbf{U}}_p)\, d\tau,$$
$$w_{6+\gamma}^p = \int_0^t (\widehat{\mathbf{c}}_\gamma \cdot \boldsymbol{\omega}_\Omega)(\widehat{\mathbf{c}}_\gamma \cdot \boldsymbol{\omega}_0)\, d\tau, \quad \gamma = 1, 2, 3;\ \alpha \neq \beta \neq \gamma \neq \alpha. \tag{9.218}$$

### 9.2.10 The Principle of Material Indifference for Models $B_n^\zeta$ of Plastic Continua

**Theorem 9.4.** *The tensors* $\mathbf{F}_p$, $\mathbf{O}_p$, $\mathbf{U}_p$, $\dot{\mathbf{U}}_p$, $\mathbf{V}_p$, $\overset{(n)}{\mathbf{T}}_e$, $\overset{(n)}{\mathbf{T}}_p$, $\mathbf{T}_U$, $\mathbf{T}_o$, $\mathbf{T}_e$, $\boldsymbol{\Omega}_p$, *and* $\mathbf{U}_e$ *are R-invariant, and the tensor* $\mathbf{V}_e$ *is R-indifferent in the rigid motion of actual configuration* $\mathcal{K} \to \mathcal{K}'$.

▼ Introduce with the help of the rigid motion (4.547) another actual configuration $\mathcal{K}'$ and define the deformation gradients from $\overset{\circ}{\mathcal{K}}$ to $\mathcal{K}'$ and from $\overset{p}{\mathcal{K}}$ to $\mathcal{K}'$ by the usual rule of construction of the deformation gradient:

$$\mathbf{F}' = \mathbf{r}'_i \otimes \overset{\circ}{\mathbf{r}}{}^i, \quad \mathbf{F}'_e = \mathbf{r}'_i \otimes \overset{p}{\mathbf{r}}{}^i, \quad \mathbf{Q}^{\mathrm{T}} = \mathbf{r}'_i \otimes \mathbf{r}^i. \tag{9.219}$$

Hence the following relations hold (see (4.559)):

$$\mathbf{F}' = \mathbf{Q}^{\mathrm{T}} \cdot \mathbf{F}, \quad \mathbf{F}'_e = \mathbf{Q}^{\mathrm{T}} \cdot \mathbf{F}_e, \tag{9.220}$$

i.e. the gradient $\mathbf{F}_e$ is transformed similarly to $\mathbf{F}$ when $\mathcal{K} \to \mathcal{K}'$; therefore $\mathbf{O}_e$, $\mathbf{U}_e$ and $\mathbf{V}_e$ are transformed just as $\mathbf{O}$, $\mathbf{U}$, $\mathbf{V}$ (see Sect. 4.10.4):

$$\mathbf{O}'_e = \mathbf{Q}^{\mathrm{T}} \cdot \mathbf{O}_e, \quad \mathbf{U}'_e = \mathbf{U}_e, \quad \mathbf{V}'_e = \mathbf{Q}^{\mathrm{T}} \cdot \mathbf{V}_e \cdot \mathbf{Q}. \tag{9.221}$$

The plastic deformation gradient $\mathbf{F}_p$, by definition, connects two reference configurations $\overset{\circ}{\mathcal{K}}$ and $\overset{p}{\mathcal{K}}$ (Fig. 9.3), and hence in rigid motions $\mathcal{K} \to \mathcal{K}'$ it remains

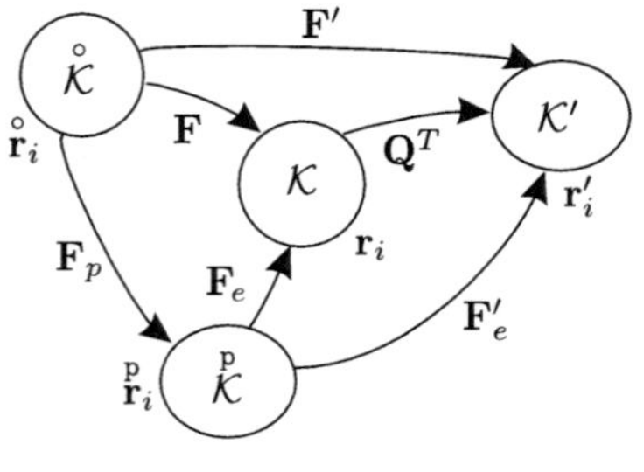


**Fig. 9.3** $R$-transformation of actual configuration $\mathcal{K}$ for a plastic continuum

unchanged, i.e. it is $R$-invariant. Then its polar decomposition also remains unchanged under transformations $\mathcal{K} \to \mathcal{K}'$, i.e. the tensors $\mathbf{O}_p$, $\mathbf{U}_p$ and $\mathbf{V}_p$ are $R$-invariant.

Due to $R$-indifference of the tensor $\mathbf{T}$, from (9.220), (9.221) and the definition of tensors $\overset{(n)}{\mathbf{T}}_e$ (see Table 9.1) it follows that the tensors $\overset{(n)}{\mathbf{T}}_e$ are transformed just as $\overset{(n)}{\mathbf{T}}$ when $\mathcal{K} \to \mathcal{K}'$, i.e. all of them are $R$-invariant. Tensors $\overset{(n)}{\mathbf{T}}_p$ coincide with $\overset{(n)}{\mathbf{T}}_e$, hence they are also $R$-invariant.

Due to the definition (9.156)–(9.158) of tensors $\mathbf{T}_U$, $\mathbf{T}_o$ and $\boldsymbol{\Omega}_p$ and $R$-invariance of tensors $\mathbf{F}_p$, $\overset{(n)}{\mathbf{T}}_e$, $\mathbf{U}_e$ and $\mathbf{O}_p$, we conclude that the tensors $\mathbf{T}_U$, $\mathbf{T}_o$ and $\boldsymbol{\Omega}_p$ are also $R$-invariant. ▲

From the theorem it follows that all the constitutive equations of models $B_n^\xi$ of plastic continua, which have been stated in Sects. 9.2.1–9.2.4, remain unchanged in rigid motions, i.e. they satisfy the principle of material indifference.

## Exercises for 9.2

**9.2.1.** Consider the *model* $B_n^\xi$ *of rotation-free plasticity*, for which the plastic potentials $f_\beta$ (9.183) are independent of $\mathbf{T}_0$. Show that in this case it follows from (9.182b) that $\mathbf{O}_p = \mathbf{E}$, and the constitutive equations of plasticity (9.182)–(9.187) are independent of $\mathbf{O}_p$.

**9.2.2.** Prove that for isotropic continua the tensor $\overset{(n)}{\mathbf{H}}_p$ commutes with $\overset{(n)}{\mathbf{A}}_p$.

## 9.3 Models $C_n$ and $D_n$ of Plastic Continua

### *9.3.1 General Representation of Constitutive Equations for Models $C_n$ of Plastic Continua*

Let us consider now models $C_n^\xi$ of plastic continua. The method of construction of these models is somewhat different from the similar one for models $A_n^\xi$.

With the help of polar decompositions of the deformation gradients $\mathbf{F}$, $\mathbf{F}_p$, and $\mathbf{F}_e$:

$$\mathbf{F} = \mathbf{V} \cdot \mathbf{O}, \qquad \mathbf{F}_e = \mathbf{V}_e \cdot \mathbf{O}_e, \qquad \mathbf{F}_p = \mathbf{V}_p \cdot \mathbf{O}_p, \tag{9.222}$$

and representations of the stretch tensors $\mathbf{V}$ and $\mathbf{V}_p$ in their eigenbases:

$$\mathbf{V} = \sum_{\alpha=1}^{3} \varkappa_\alpha \mathbf{p}_\alpha \otimes \mathbf{p}_\alpha, \qquad \mathbf{V}_p = \sum_{\alpha=1}^{3} \overset{p}{\lambda}_\alpha \overset{p}{\mathbf{p}}_\alpha \otimes \overset{p}{\mathbf{p}}_\alpha, \tag{9.223}$$

we can introduce tensors of elastic and plastic deformations $\overset{(n)}{\mathbf{A}}_e$ and $\overset{(n)}{\mathbf{A}}_p$:

$$\overset{(n)}{\mathbf{A}} = \frac{1}{n-\mathrm{III}}(\mathbf{V}^{n-\mathrm{III}} - \mathbf{E}) = \frac{1}{n-\mathrm{III}} \sum_{\alpha=1}^{3} (\lambda_\alpha^{n-\mathrm{III}} - 1)\mathbf{p}_\alpha \otimes \mathbf{p}_\alpha,$$
$$\overset{(n)}{\mathbf{A}}_p = \frac{1}{n-\mathrm{III}} \sum_{\alpha=1}^{3} (\lambda_\alpha^{n-\mathrm{III}} - \overset{p}{\lambda}_\alpha{}^{n-\mathrm{III}})\mathbf{p}_\alpha \otimes \mathbf{p}_\alpha, \tag{9.223a}$$
$$\overset{(n)}{\mathbf{A}}_e = \frac{1}{n-\mathrm{III}} \sum_{\alpha=1}^{3} (\overset{p}{\lambda}_\alpha^{n-\mathrm{III}} - 1)\mathbf{p}_\alpha \otimes \mathbf{p}_\alpha, \quad n = \mathrm{I, II, IV, V},$$

which satisfy the relation of additivity

$$\overset{(n)}{\mathbf{A}} = \overset{(n)}{\mathbf{A}}_e + \overset{(n)}{\mathbf{A}}_p. \tag{9.224}$$

The further construction of models $C_n^\zeta$ of plastic materials with the help of formal substitutions $\overset{(n)}{\mathbf{C}} \to \overset{(n)}{\mathbf{A}}$ and $\overset{(n)}{\mathbf{T}} \to \overset{(n)}{\mathbf{S}}$ in models $A_n^\zeta$, as known from the construction of models for continua of the differential type (see Sect. 7.3.1), leads to incorrect constitutive equations, which do not satisfy the principle of material indifference, as the tensors $\overset{(n)}{\mathbf{A}}{}^{\bullet}_p$ are not $R$-indifferent (but the tensors $\overset{(n)}{\mathbf{A}}_p$ are $R$-indifferent just as $\overset{(n)}{\mathbf{A}}$). The method for overcoming the difficulties is the same as for continua of the differential type, namely by using the co-rotational derivatives.

Consider the principal thermodynamic identity in the form $C_n$ (4.123) and rewrite it with taking (9.224) into account:

$$\rho\, d\psi + \rho\eta\, d\theta - \overset{(n)}{\mathbf{S}} \cdot\cdot\, d\overset{(n)}{\mathbf{A}}_e - \overset{(n)}{\mathbf{S}} \cdot\cdot\, d\overset{(n)}{\mathbf{A}}_p - \overset{\circ}{\mathbf{S}} \cdot\cdot\, d\mathbf{O}^{\mathrm{T}} + w^*\, dt = 0. \tag{9.225}$$

Introducing the Gibbs free energy

$$\zeta = \psi - (\overset{(n)}{\mathbf{S}}/\rho) \cdot\cdot\, \overset{(n)}{\mathbf{A}}_e, \tag{9.226}$$

we obtain the principal thermodynamic identity in the form $C_n^{\zeta}$:

$$\rho\, d\zeta + \rho\eta\, d\theta + \rho \overset{(n)}{\mathbf{A}}_e \cdot\cdot\, d(\overset{(n)}{\mathbf{S}}/\rho) - \overset{(n)}{\mathbf{S}} \cdot\cdot\, d\overset{(n)}{\mathbf{A}}_p - \overset{\circ}{\mathbf{S}} \cdot\cdot\, d\mathbf{O}^{\mathrm{T}} + w^*\, dt = 0. \quad (9.227)$$

Formulae (9.226) and (9.165) define different functions $\zeta$, for which nevertheless we use the same notation. In the model $C_n^{\zeta}$, the free energy $\zeta$ and the elastic deformation tensor $\overset{(n)}{\mathbf{A}}_e$ are considered to be functionals in the form similar to (9.1) but with co-rotational derivatives in place of the partial derivatives with respect to time:

$$\zeta = \overset{t}{\underset{\tau=0}{\zeta}} (\mathcal{R}(t),\ \mathcal{R}^h(t),\ \mathcal{R}^t(\tau),\ \mathcal{R}^{ht}(\tau)), \quad (9.228a)$$

$$\overset{(n)}{\mathbf{A}}_e = \overset{t}{\underset{\tau=0}{\mathbf{A}_e}} (\mathcal{R}(t),\ \mathcal{R}^h(t),\ \mathcal{R}^t(\tau),\ \mathcal{R}^{ht}(\tau)), \quad (9.228b)$$

$$\mathcal{R} = \{\overset{(n)}{\mathbf{S}}/\rho,\ \overset{(n)}{\mathbf{A}}_p,\ \mathbf{O},\ \theta\}, \quad h = \{V,\ S,\ J\}.$$

The tensors of equilibrium elastic deformations $\overset{(n)}{\mathbf{A}}{}_e^0$ and non-equilibrium elastic deformations $\overset{(n)}{\mathbf{A}}{}_e^1$ are introduced as follows:

$$\overset{(n)}{\mathbf{A}}{}_e^0 = \overset{t}{\underset{\tau=0}{\mathbf{A}_e}} (\mathcal{R}(t),\ 0,\ \mathcal{R}^t(\tau),\ \mathcal{R}^{ht}(\tau)), \quad \overset{(n)}{\mathbf{A}}{}_e^1 = \overset{(n)}{\mathbf{A}}_e - \overset{(n)}{\mathbf{A}}{}_e^0. \quad (9.229)$$

On substituting the functionals (9.228a), (9.228b), and (9.229) into the principal thermodynamic identity (9.227), we obtain the identity

$$\rho \left( \frac{\partial \zeta}{\partial \overset{(n)}{\mathbf{S}}/\rho} + \overset{(n)}{\mathbf{A}}{}_e^0 \right) \cdot\cdot\, d \left( \frac{\overset{(n)}{\mathbf{S}}}{\rho} \right) + \rho \left( \frac{\partial \zeta}{\partial \theta} + \eta \right) d\theta + \rho \frac{\partial \zeta}{\partial (\overset{(n)}{\mathbf{S}}/\rho)^h} \cdot\cdot\, d \left( \frac{\overset{(n)}{\mathbf{S}}}{\rho} \right)^h$$
$$+ \left( \rho \frac{\partial \zeta}{\partial \mathbf{O}} - \overset{\circ}{\mathbf{S}} \right) \cdot\cdot\, d\mathbf{O}^{\mathrm{T}} + \rho \frac{\partial \zeta}{\partial \mathbf{O}^h} \cdot\cdot\, d\mathbf{O}^{hT} + \rho \frac{\partial \zeta}{\partial \overset{(n)}{\mathbf{A}}{}_p^h} \cdot\cdot\, d\overset{(n)}{\mathbf{A}}{}_p^h$$
$$+ \rho \frac{\partial \zeta}{\partial \theta^h}\, d\theta^h + (w^* - \overset{(n)}{\mathbf{S}}_H \cdot\cdot \overset{(n)}{\mathbf{A}}{}_p^{\bullet} + \rho \overset{(n)}{\mathbf{A}}{}_e^1 \cdot\cdot (\overset{(n)}{\mathbf{S}}/\rho)^{\bullet} + \rho\delta\zeta)\, dt = 0, \quad (9.230)$$

where

$$\overset{(n)}{\mathbf{S}}_H = \overset{(n)}{\mathbf{S}} - \overset{(n)}{\mathbf{H}}_p, \quad \overset{(n)}{\mathbf{H}}_p = \rho(\partial \zeta / \partial \overset{(n)}{\mathbf{A}}_p) \quad (9.230a)$$

are the reduced quasienergetic stress tensors.

Since the differentials $d(\overset{(n)}{\mathbf{S}}/\rho)$, $d\theta$, $d\theta^h$, $d(\overset{(n)}{\mathbf{S}}/\rho)^h$, $d\mathbf{O}^{\mathrm{T}}$, $d\mathbf{O}^{hT}$, $d\overset{(n)}{\mathbf{A}}{}^h_p$, and $dt$ are independent, the identity (9.230) is equivalent to the following system of constitutive equations:

$$\overset{(n)}{\mathbf{A}}{}^0_e = -\partial\zeta/\partial(\overset{(n)}{\mathbf{S}}/\rho), \quad \overset{\circ}{\mathbf{S}} = \rho(\partial\zeta/\partial\mathbf{O}), \quad \eta = -\partial\zeta/\partial\theta,$$
$$\partial\zeta/\partial(\overset{(n)}{\mathbf{S}}/\rho)^h = 0, \quad \partial\zeta/\partial\theta^h = 0, \quad \partial\zeta/\partial\overset{(n)}{\mathbf{A}}{}^h_p = 0, \quad \partial\zeta/\partial\mathbf{O}^h = 0, \tag{9.231}$$
$$w^* = \overset{(n)}{\mathbf{S}}_H\cdot\cdot\overset{(n)}{\mathbf{A}}{}^{\bullet}_p - \rho\overset{(n)}{\mathbf{A}}{}^1_e\cdot\cdot(\overset{(n)}{\mathbf{S}}/\rho)^{\bullet} - \rho\delta\zeta.$$

A consequence of these relations is the fact that the potential $\zeta$ is independent of the co-rotational derivatives at the time $t$:

$$\zeta = \underset{\tau=0}{\overset{t}{\zeta}}\,(\mathcal{R}(t), \mathcal{R}^t(\tau), \mathcal{R}^{ht}(\tau)), \tag{9.232}$$

The model $C_n^\zeta$ of plastic continua is specified by three functionals: (9.228b), (9.232) and the functional relationship for the plastic deformation tensor $\overset{(n)}{\mathbf{A}}_p$.

## 9.3.2 *Constitutive Equations for Models $C_n^\zeta$ of Isotropic Plastic Continua*

**Theorem 9.5.** *For models $C_n^\zeta$ of isotropic plastic continua, the potential $\zeta$ and the tensor $\overset{(n)}{\mathbf{A}}{}^1_e$ are independent of the tensor $\mathbf{O}$, and the dissipation function takes the form*

$$w^* = \overset{(n)}{\mathbf{S}}_H\cdot\cdot\overset{(n)}{\mathbf{A}}{}^h_p - \rho\overset{(n)}{\mathbf{A}}{}^1_e\cdot\cdot(\overset{(n)}{\mathbf{S}}/\rho)^h - \rho\delta\zeta, \quad h = \{V, S, J\}. \tag{9.233}$$

▼ Since $\zeta(t)$ and $\overset{(n)}{\mathbf{A}}{}^1_e(t)$ are, in general, functions of $\mathbf{O}(t)$, the method used in Sect. 7.3.3. for continua of the differential type gives that for isotropic continua

$$\overset{\circ}{\mathbf{S}} = \rho(\partial\zeta/\partial\mathbf{O}) = \mathbf{0}, \quad \partial\overset{(n)}{\mathbf{A}}{}^1_e/\partial\mathbf{O} = \mathbf{0}; \tag{9.234}$$

i.e. $\zeta$ and $\overset{(n)}{\mathbf{A}}{}^1_e$ are actually independent of $\mathbf{O}$. Then, according to Theorem 4.9 (Sect. 4.2.18), the tensor $\overset{(n)}{\mathbf{S}}$ commutes with $\overset{(n)}{\mathbf{A}}$, i.e. these tensors are coaxial. According to Theorem 4.12, the principal thermodynamic identity (9.227) in this case can be represented in terms of the co-rotational derivatives:

$$\rho\zeta^h + \rho\eta\theta^h + \rho\overset{(n)}{\mathbf{A}}_e\cdot\cdot(\overset{(n)}{\mathbf{S}}/\rho)^h - \overset{(n)}{\mathbf{S}}\cdot\cdot\overset{(n)}{\mathbf{A}}{}^h_p + w^* = 0. \tag{9.235}$$

Since $\zeta^h = \dot{\zeta}$, so, choosing the functionals $\zeta$ and $\overset{(n)}{\mathbf{A}}{}_e^1$ in the form (9.228a), where reactive variables $\mathcal{R}$ no longer involve the tensor $\mathbf{O}$, and making the manipulations (9.230) once more, we obtain again the constitutive equations (9.231), but the dissipation function in them is expressed in terms of the co-rotational derivatives, i.e. formula (9.233) really holds. ▲

We apply now the Onsager principle to the models $C_n^\xi$ and form the specific internal entropy production (4.728) with taking (9.233) into account. As a result, we obtain the following constitutive equations being complementary to (9.231):

$$\begin{aligned} -\mathbf{q}/\theta &= L_{11}\cdot\nabla\theta + L_{12}\cdot\cdot\overset{(n)}{\mathbf{S}}_H + L_{13}\cdot\cdot(\overset{(n)}{\mathbf{S}}/\rho)^h,\\ \overset{(n)}{\mathbf{A}}{}_p^h &= L_{12}\cdot\nabla\theta + L_{22}\cdot\cdot\overset{(n)}{\mathbf{S}}_H + L_{23}\cdot\cdot(\overset{(n)}{\mathbf{S}}/\rho)^h,\\ \rho\overset{(n)}{\mathbf{A}}{}_e^1 &= L_{13}\cdot\nabla\theta + L_{23}\cdot\cdot\overset{(n)}{\mathbf{S}}_H + L_{33}\cdot\cdot(\overset{(n)}{\mathbf{S}}/\rho)^h, \end{aligned} \tag{9.236}$$

where $L_{\alpha\beta}$ are functionals in the form (9.228b), and $\mathcal{R} = \{\overset{(n)}{\mathbf{S}}/\rho,\ \overset{(n)}{\mathbf{A}}_p,\ \theta\}$.

The further derivation of a form of relations (9.236) is realized in the same way as it was done for models $A_n^\xi$ by formal replacing $\mathbf{T}\to\mathbf{S}$, $\mathbf{C}\to\mathbf{A}$ and $(\cdot)\to(h)$ (see Sects. 9.1.3–9.1.12). In particular, the *associated model* $C_n^\xi$ *of an isotropic plastic continuum* has the form similar to Eqs. (9.75)–(9.83):

$$\begin{aligned} &\zeta = \zeta(J_\gamma^{(I)}(\overset{(n)}{\mathbf{S}},\ \overset{(n)}{\mathbf{A}}_p),\ \theta,\ w_p),\quad f_\beta = f_\beta(J_1^{(I)},\ J_2^{(I)},\ J_4^{(I)},\ J_5^{(I)},\ J_7^{(I)},\ \theta,\ w_p),\\ &\overset{(n)}{\mathbf{A}}_e = \widetilde{\varphi}_1\mathbf{E} + \widetilde{\varphi}_2\overset{(n)}{\mathbf{S}} + \widetilde{\varphi}_3\overset{(n)}{\mathbf{S}}{}^2 + \widetilde{\varphi}_4\overset{(n)}{\mathbf{A}}_p + \varphi_5\overset{(n)}{\mathbf{A}}{}_p^2 + \varphi_6(\overset{(n)}{\mathbf{S}}\cdot\overset{(n)}{\mathbf{A}}_p + \overset{(n)}{\mathbf{A}}_p\cdot\overset{(n)}{\mathbf{S}}),\\ &\overset{(n)}{\mathbf{A}}{}_p^h = \widetilde{\psi}_1\mathbf{E} - \psi_2\overset{(n)}{\mathbf{S}} + \psi_7\overset{(n)}{\mathbf{A}}{}_p^h,\\ &\varphi_\gamma = -\rho(\partial\zeta/\partial J_\gamma^{(I)}),\quad \psi_\alpha = h\sum_{\beta=1}^{k}\dot{\varkappa}_\beta(\partial f_\beta/\partial J_\alpha^{(I)}),\quad w_p = \int_0^t \overset{(n)}{\mathbf{S}}\cdot\cdot\overset{(n)}{\mathbf{A}}_p\,d\tau. \end{aligned} \tag{9.237}$$

The functions $\widetilde{\varphi}_\alpha$ and $\widetilde{\psi}_\alpha$ are determined by (9.78) and (9.82), and the simultaneous invariants $J_\gamma^{(I)}(\overset{(n)}{\mathbf{S}},\ \overset{(n)}{\mathbf{A}}_p)$ – by formulae (9.75).

For the *Huber–Mises* model $C_n^\xi$, the relation for plastic yield results in

$$\overset{(n)}{\mathbf{P}}{}_p^h = 3\dot{\varkappa}hf_Y\mathbf{P}_H, \tag{9.238a}$$

where we have denoted the tensor deviators

$$\mathbf{P}_p = \overset{(n)}{\mathbf{A}}_p - \frac{1}{3}I_1(\overset{(n)}{\mathbf{A}}_p)\mathbf{E},\quad \mathbf{P}_H = \overset{(n)}{\mathbf{S}} - H\overset{(n)}{\mathbf{A}}_p - \frac{1}{3}I_1(\overset{(n)}{\mathbf{S}} - H\overset{(n)}{\mathbf{A}}_p)\mathbf{E}, \tag{9.238b}$$

and

$$f_Y = \frac{1}{2Y_H}\frac{\partial f}{\partial Y_H}, \quad f = f(Y_H,\ \theta,\ w_p), \quad Y_H^2 = \frac{3}{2}\mathbf{P}_H \cdot\cdot\, \mathbf{P}_H. \tag{9.238c}$$

*Remark 1.* In spite of the formal similarity of models $C_n^\xi$ and $A_n^\xi$ for isotropic plastic continua, we can see that constitutive equations of these models are not equivalent. □

*Remark 2.* The set of relations (9.234) is the additional requirement for constitutive equations (9.237). As shown in Sect. 4.2.18 (see Theorem 4.9), the relation $\overset{\circ}{\mathbf{S}} = 0$ is equivalent to the requirement that the tensors $\overset{(n)}{\mathbf{S}}$ and $\overset{(n)}{\mathbf{A}}$ be coaxial or commutative:

$$\overset{(n)}{\mathbf{S}} \cdot \overset{(n)}{\mathbf{A}} - \overset{(n)}{\mathbf{A}} \cdot \overset{(n)}{\mathbf{S}} = 0. \tag{9.239a}$$

According to the relation of additivity (9.224), this requirement should be replaced by the stronger one

$$\overset{(n)}{\mathbf{S}} \cdot \overset{(n)}{\mathbf{A}}_e - \overset{(n)}{\mathbf{A}}_e \cdot \overset{(n)}{\mathbf{S}} = 0, \quad \overset{(n)}{\mathbf{S}} \cdot \overset{(n)}{\mathbf{A}}_p - \overset{(n)}{\mathbf{A}}_p \cdot \overset{(n)}{\mathbf{S}} = 0. \tag{9.239b}$$

The relation (9.239a) always follows from (9.239b), but not conversely.

In order to verify whether relations (9.237) satisfy the conditions (9.239b), we express them in terms of the tensor $\overset{(n)}{\mathbf{S}}$:

$$\begin{gathered}\overset{(n)}{\mathbf{S}} = \bar{\psi}_1\mathbf{E} - \bar{\psi}_2\overset{(n)}{\mathbf{A}}{}^h_p + \bar{\psi}_3\overset{(n)}{\mathbf{A}}_p,\\ \bar{\psi}_1 = \widetilde{\psi}_1/\psi_2, \quad \bar{\psi}_2 = 1/\psi_2, \quad \bar{\psi}_3 = \psi_7/\psi_2.\end{gathered} \tag{9.240}$$

On substituting (9.240) into the second condition (9.239b), we obtain that this condition reduces to the requirement of commutativity for the tensors $\overset{(n)}{\mathbf{A}}_p$ and $\overset{(n)}{\mathbf{A}}{}^h_p$:

$$\overset{(n)}{\mathbf{A}}_p \cdot \overset{(n)}{\mathbf{A}}{}^h_p - \overset{(n)}{\mathbf{A}}{}^h_p \cdot \overset{(n)}{\mathbf{A}}_p = 0. \tag{9.241a}$$

Substituting relation (9.237) for $\overset{(n)}{\mathbf{A}}_e$ into the first condition (9.239b) and using (9.240), we find that two more conditions of commutativity must be satisfied:

$$\overset{(n)}{\mathbf{A}}{}^2_p \cdot \overset{(n)}{\mathbf{A}}{}^h_p - \overset{(n)}{\mathbf{A}}{}^h_p \cdot \overset{(n)}{\mathbf{A}}{}^2_p = 0, \quad \overset{(n)}{\mathbf{A}}_p \cdot (\overset{(n)}{\mathbf{A}}{}^h_p)^2 - (\overset{(n)}{\mathbf{A}}{}^h_p)^2 \cdot \overset{(n)}{\mathbf{A}}_p = 0. \tag{9.241b}$$

The conditions (9.241) are rather restrictive. Among all the co-rotational derivatives $h \in \{V, s, J\}$ considered in this section, only the co-rotational derivative $h = V$ in the basis of the left stretch tensor satisfies all these conditions, because from the definition (9.224a) of the tensors $\overset{(n)}{\mathbf{A}}_p$ it follows that only the derivative $\overset{(n)}{\mathbf{A}}{}^V_p$ has the eigenbasis coincident with $\mathbf{p}_\alpha$:

$$\overset{(n)}{\mathbf{A}}{}^V_p = \frac{1}{n - \mathrm{III}} \sum_{\alpha=1}^{3} (\overset{p}{\lambda}{}_\alpha^{n-\mathrm{III}} - \lambda_\alpha^{n-\mathrm{III}})^\bullet \mathbf{p}_\alpha \otimes \mathbf{p}_\alpha.$$

Therefore, for this derivative all relations of commutativity (9.241) are identically satisfied.

As follows from the above reasoning, among all the considered co-rotational derivatives, the application of the derivative in the eigenbasis of the left stretch tensor $h = V$ for models $C_n^\zeta$ of plastic continua is correct. □

### 9.3.3 General Representation of Constitutive Equations for Models $D_n^\zeta$ of Plastic Continua

Models $D_n^\zeta$ of plastic continua are multiplicative, because the relation of additivity of elastic and plastic deformations for them does not hold, and constitutive equations for these models are constructed in the same way as for the models $B_n^\zeta$.

Models $D_n^\zeta$ are based on the following representation of the stress power.

**Theorem 9.6.** *The stress power $w_{(i)}$ can always be represented in the following additive form:*

$$w_{(i)} = \overset{(n)}{\mathbf{S}} \cdot\cdot \overset{(n)}{\mathbf{g}}{}^\bullet + \overset{\circ}{\mathbf{S}} \cdot\cdot \dot{\mathbf{O}}^{\mathrm{T}} = \overset{(n)}{\mathbf{S}}_e \cdot\cdot \overset{(n)}{\mathbf{g}}{}^\bullet_e + \overset{\circ}{\mathbf{S}}_e \cdot\cdot \dot{\mathbf{O}}_e^{\mathrm{T}} + \mathbf{S}_v \cdot\cdot \dot{\mathbf{V}}_p + \mathbf{S}_{\mathrm{O}} \cdot\cdot \boldsymbol{\Omega}_p, \quad (9.242)$$

*where $\boldsymbol{\Omega}_p$ is the spin tensor of plastic rotation (9.158), $\overset{(n)}{\mathbf{S}}_e$ are the quasienergetic symmetric tensors of elastic stresses, $\overset{(n)}{\mathbf{g}}_e$ are the quasienergetic measures of elastic deformations determined similarly to the tensors $\overset{(n)}{\mathbf{S}}$ and $\overset{(n)}{\mathbf{g}}$ with the substitution $\mathbf{V} \to \mathbf{V}_e$ (see Table 9.2), $\overset{\circ}{\mathbf{S}}_e$ is the rotation tensor of elastic stresses, $\mathbf{S}_v$ is the symmetric quasienergetic tensor of plastic stretch stresses, $\mathbf{S}_{\mathrm{O}}$ is the skew-symmetric tensor of plastic rotation stresses, determined as follows:*

$$\overset{\circ}{\mathbf{S}}_e = \frac{1}{2}(\mathbf{V}_e \cdot \mathbf{T} \cdot \mathbf{V}_e^{-1} - \mathbf{V}_e^{-1} \cdot \mathbf{T} \cdot \mathbf{V}_e) \cdot \mathbf{O}_e,$$

$$\mathbf{S}_v = \frac{1}{2}(\mathbf{V}_p \cdot \mathbf{T}_e + \mathbf{T}_e^{\mathrm{T}} \cdot \mathbf{V}_p^{-1}),$$

$$\mathbf{S}_{\mathrm{O}} = \frac{1}{2}(\mathbf{V}_p^{-1} \cdot \mathbf{T}_e \cdot \mathbf{V}_p - \mathbf{V}_p \cdot \mathbf{T}_e^{\mathrm{T}} \cdot \mathbf{V}_p^{-1}). \quad (9.242a)$$

*Here $\mathbf{T}_e$ is determined by (9.157).*

**Table 9.2** Expressions for $\overset{(n)}{\mathbf{S}}_e$, $\overset{(n)}{\mathbf{g}}_e$ at $n = \mathrm{I}, \ldots, \mathrm{V}$

| $n$ | $\overset{(n)}{\mathbf{S}}_e$ | $\overset{(n)}{\mathbf{g}}_e$ |
|---|---|---|
| I | $\mathbf{V}_e \cdot \mathbf{T} \cdot \mathbf{V}_e$ | $-\frac{1}{2}\mathbf{V}_e^{-2}$ |
| II | $\frac{1}{2}(\mathbf{V}_e \cdot \mathbf{T} + \mathbf{T} \cdot \mathbf{V}_e)$ | $-\mathbf{V}_e^{-1}$ |
| III | $\mathbf{T}$ | $\mathbf{Y}_e$ |
| IV | $\frac{1}{2}(\mathbf{V}_e^{-1} \cdot \mathbf{T} + \mathbf{T} \cdot \mathbf{V}_e^{-1})$ | $\mathbf{V}_e$ |
| V | $\mathbf{V}_e^{-1} \cdot \mathbf{T} \cdot \mathbf{V}_e^{-1}$ | $\frac{1}{2}\mathbf{V}_e^2$ |

The tensor $\mathbf{Y}_e$ as well as $\mathbf{Y}$ is determined by its derivative and initial value $\mathbf{Y}_e^0$ (when there are no initial plastic deformations $\mathbf{Y}_e^0 = \mathbf{E}$):

$$\dot{\mathbf{Y}}_e = \frac{1}{2}(\dot{\mathbf{V}}_e \cdot \mathbf{V}_e^{-1} + \mathbf{V}_e^{-1} \cdot \dot{\mathbf{V}}_e), \quad \mathbf{Y}_e(0) = \mathbf{Y}_e^0. \tag{9.243}$$

▼ To prove the theorem we should use formula (9.242). The first summand on the right-hand side of the formula differs from the total stress power $\overset{(n)}{\mathbf{T}} \cdot\cdot \overset{(n)}{\mathbf{G}}{}^{\bullet}$ only by the appearance of subscript $e$. Then we can apply Theorem 4.5 to the summand $\overset{(n)}{\mathbf{T}}_e \cdot\cdot \overset{(n)}{\mathbf{G}}{}_e^{\bullet}$; this theorem ensures that there exist quasienergetic couples having in this case the subscript $e$:

$$\overset{(n)}{\mathbf{T}}_e \cdot\cdot \overset{(n)}{\mathbf{G}}{}_e^{\bullet} = \overset{(n)}{\mathbf{S}}_e \cdot\cdot \overset{(n)}{\mathbf{g}}{}_e^{\bullet} + \overset{\circ}{\mathbf{S}}_e \cdot\cdot \dot{\mathbf{O}}_e^{\mathrm{T}}. \tag{9.244}$$

The second summand in (9.163) can be modified by using the definition (9.156) of the tensor $\mathbf{T}_U$ and the relation

$$\mathbf{U}_p = \mathbf{O}_e^{\mathrm{T}} \cdot \mathbf{V}_p \cdot \mathbf{O}_p, \tag{9.245}$$

then

$$\begin{aligned}
\mathbf{T}_U \cdot\cdot \dot{\mathbf{U}}_p &= \frac{1}{2}(\mathbf{F}_p^{-1} \cdot \mathbf{T}_e \cdot \mathbf{O}_p + \mathbf{O}_p^{\mathrm{T}} \cdot \mathbf{T}_e \cdot \mathbf{F}_p^{-1\mathrm{T}}) \cdot\cdot (\mathbf{O}_p^{\mathrm{T}} \cdot \mathbf{V}_p \cdot \mathbf{O}_p)^{\bullet} \\
&= \frac{1}{2}(\mathbf{O}_p \cdot \mathbf{F}_e^{-1} \cdot \mathbf{T}_e + \mathbf{T}_e \cdot \mathbf{F}_p^{-1\mathrm{T}} \cdot \mathbf{O}_p^{\mathrm{T}}) \cdot\cdot \dot{\mathbf{V}}_p + \frac{1}{2}(\mathbf{O}_p \cdot \mathbf{F}_p^{-1} \cdot \mathbf{T}_e \cdot \mathbf{V}_p \\
&\quad - \mathbf{V}_p \cdot \mathbf{O}_p \cdot \mathbf{F}_p^{-1} \cdot \mathbf{T}_e + \mathbf{T}_e \cdot \mathbf{F}_p^{-1\mathrm{T}} \cdot \mathbf{O}_p^{\mathrm{T}} \cdot \mathbf{V}_p \\
&\quad - \mathbf{V}_p \cdot \mathbf{T}_e \cdot \mathbf{F}_p^{-1\mathrm{T}} \cdot \mathbf{O}_p^{\mathrm{T}}) \cdot\cdot (\dot{\mathbf{O}}_p \cdot \mathbf{O}_p^{\mathrm{T}}).
\end{aligned} \tag{9.246}$$

Here we have used the permutation rule for tensors in the triple scalar product and also the relationship $\mathbf{O}_p \cdot \dot{\mathbf{O}}_p^{\mathrm{T}} = -\dot{\mathbf{O}}_p \cdot \mathbf{O}_p^{\mathrm{T}}$.

Due to the relation $\mathbf{O}_p \cdot \mathbf{F}_p^{-1} = \mathbf{V}_p^{-1}$, we obtain

$$\mathbf{T}_U \cdot\cdot \dot{\mathbf{U}}_p = \mathbf{S}_v \cdot\cdot \dot{\mathbf{V}}_p + (\mathbf{S}_v \cdot \mathbf{V}_p - \mathbf{V}_p \cdot \mathbf{S}_v) \cdot\cdot \boldsymbol{\Omega}_p. \tag{9.247}$$

Substitution of formulae (9.244) and (9.247) into (9.163) gives

$$w_{(i)} = \overset{(\mathrm{n})}{\mathbf{S}}_e \cdot\cdot \overset{(\mathrm{n})}{\mathbf{g}}{}^{\bullet}_e + \overset{\circ}{\mathbf{S}}_e \cdot\cdot \dot{\mathbf{O}}_e^{\mathrm{T}} + \mathbf{S}_v \cdot\cdot \dot{\mathbf{V}}_p + (\mathbf{S}_v \cdot \mathbf{V}_p - \mathbf{V}_p \cdot \mathbf{S}_v + \mathbf{T}_o) \cdot\cdot \boldsymbol{\Omega}_p. \quad (9.248)$$

We can immediately verify that

$$\mathbf{S}_0 = \mathbf{S}_v \cdot \mathbf{V}_p - \mathbf{V}_p \cdot \mathbf{S}_v + \mathbf{T}_o, \quad (9.249)$$

where $\mathbf{S}_0$ and $\mathbf{S}_v$ are determined by formula (9.242a), and $\mathbf{T}_o$ – by (9.157). Then Eq. (9.248) actually yields the representation (9.242). ▲

On substituting the representation (9.242) for the stress power into PTI (4.124), we get

$$\rho\dot{\psi} + \rho\eta\dot{\theta} - \overset{(\mathrm{n})}{\mathbf{S}}_e \cdot\cdot \overset{(\mathrm{n})}{\mathbf{g}}{}^{\bullet}_e - \overset{\circ}{\mathbf{S}}_e \cdot\cdot \dot{\mathbf{O}}_e^{\mathrm{T}} - \mathbf{S}_v \cdot\cdot \dot{\mathbf{V}}_p - \mathbf{S}_0 \cdot\cdot \boldsymbol{\Omega}_p = 0. \quad (9.250)$$

Introducing the Gibbs free energy

$$\zeta = \psi - \frac{1}{\rho} \overset{(\mathrm{n})}{\mathbf{S}}_e \cdot\cdot \overset{(\mathrm{n})}{\mathbf{g}}_e, \quad (9.251)$$

we obtain the principal thermodynamic identity in the $D_n^{\zeta}$ form:

$$\rho\dot{\zeta} + \rho\eta\dot{\theta} + \rho \overset{(\mathrm{n})}{\mathbf{g}}_e \cdot\cdot \left( \frac{\overset{(\mathrm{n})}{\mathbf{S}}_e}{\rho} \right)^{\bullet} - \overset{\circ}{\mathbf{S}}_e \cdot\cdot \dot{\mathbf{O}}_e^{\mathrm{T}} - \mathbf{S}_v \cdot\cdot \dot{\mathbf{V}}_p - \mathbf{S}_0 \cdot\cdot \boldsymbol{\Omega}_p + w^* = 0. \quad (9.252)$$

Active and reactive variables may involve the following functions:

$$\Lambda = \{\zeta,\ \eta,\ \overset{(\mathrm{n})}{\mathbf{g}}_e,\ \overset{\circ}{\mathbf{S}}_e,\ \mathbf{S}_v,\ \mathbf{S}_0,\ w^*\}, \quad \mathcal{R} = \{\theta,\ \overset{(\mathrm{n})}{\mathbf{S}}_e/\rho,\ \mathbf{O}_e,\ \mathbf{V}_p,\ \mathbf{O}_p^{\mathrm{T}}\}. \quad (9.253)$$

In models $D_n^{\zeta}$ of plastic continua, active and reactive variables are connected by a functional relation in the form (9.167)

$$\Lambda = \underset{\tau=0}{\overset{t}{\Lambda}} (\mathcal{R}(t),\ \dot{\mathcal{R}}(t),\ \mathcal{R}^t(\tau),\ \dot{\mathcal{R}}^t(\tau)). \quad (9.254)$$

Splitting the measures $\overset{(\mathrm{n})}{\mathbf{g}}_e$ into the *measures of equilibrium elastic deformations* $\overset{(\mathrm{n})}{\mathbf{g}}{}^0_e$ and the measures of nonequilibrium elastic deformations $\overset{(\mathrm{n})}{\mathbf{g}}{}^1_e$:

$$\overset{(\mathrm{n})}{\mathbf{g}}{}^0_e = \underset{\tau=0}{\overset{t}{\mathbf{g}}} (\mathcal{R}(t),\ 0,\ \mathcal{R}^t(\tau),\ \dot{\mathcal{R}}^t(\tau)), \quad \overset{(\mathrm{n})}{\mathbf{g}}{}^1_e = \overset{(\mathrm{n})}{\mathbf{g}}_e - \overset{(\mathrm{n})}{\mathbf{g}}{}^0_e, \quad (9.255)$$

and substituting the functionals (9.254) and (9.255) into PTI (9.252), we reduce the identity to the form similar to (9.230), that gives the constitutive equations

$$\begin{cases} \overset{(n)}{\mathbf{g}}{}^{0}_{e} = -\partial\zeta/\partial(\overset{(n)}{\mathbf{S}}{}_{e}/\rho), \quad \overset{\circ}{\mathbf{S}}_{e} = \rho(\partial\zeta/\partial\mathbf{O}_{e}), \quad \eta = -\partial\zeta/\partial\theta, \\ \partial\zeta/\partial(\overset{(n)}{\mathbf{S}}{}_{e}/\rho)^{\bullet} = 0, \quad \partial\zeta/\partial\theta^{\bullet} = 0, \quad \partial\zeta/\partial\dot{\mathbf{V}}_{p} = 0, \\ \partial\zeta/\partial\dot{\mathbf{O}}_{p} = 0, \quad \partial\zeta/\partial\dot{\mathbf{O}}_{e} = 0, \\ w^{*} = \widetilde{\mathbf{S}}_{v}\cdot\cdot\dot{\mathbf{V}}_{p} + \widetilde{\mathbf{S}}_{\mathrm{O}}\cdot\cdot\boldsymbol{\Omega}_{p} - \rho\overset{(n)}{\mathbf{g}}{}^{1}_{e}\cdot\cdot(\overset{(n)}{\mathbf{S}}{}_{e}/\rho)^{\bullet} - \rho\delta\zeta. \end{cases} \tag{9.256}$$

Here we have introduced the notation

$$\widetilde{\mathbf{S}}_{v} = \mathbf{S}_{v} - \mathbf{N}_{p}, \quad \mathbf{N}_{p} = \rho(\partial\zeta/\partial\mathbf{V}_{p}),$$
$$\widetilde{\mathbf{S}}_{\mathrm{O}} = \mathbf{S}_{\mathrm{O}} - \mathbf{N}_{\mathrm{O}}, \quad \mathbf{N}_{\mathrm{O}} = \rho\mathbf{O}_{p}\cdot(\partial\zeta/\partial\mathbf{O}_{p}^{\mathrm{T}}). \tag{9.257}$$

From these equations it follows that, just as for models $C_n^{\zeta}$, the potential $\zeta$ is independent of the derivatives at the current time $t$, i.e.

$$\zeta = \overset{t}{\underset{\tau=0}{\zeta}}(\mathcal{R}(t),\ \mathcal{R}^{t}(\tau),\ \dot{\mathcal{R}}^{t}(\tau)). \tag{9.258}$$

Apply the Onsager principle to models $D_n^{\zeta}$ and form the specific internal entropy production (4.728) with the help of expression (9.256) for $w^{*}$. As a result, we obtain the following constitutive equations being complementary to (9.256)

$$\begin{cases} -\mathbf{q}/\theta = L_{11}\cdot\nabla\theta + L_{12}\cdot\cdot\widetilde{\mathbf{S}}_{v} + L_{13}\cdot\cdot\widetilde{\mathbf{S}}_{\mathrm{O}} + L_{14}\cdot\cdot(\overset{(n)}{\mathbf{S}}{}_{e}/\rho)^{\bullet}, \\ \dot{\mathbf{V}}_{p} = L_{12}\cdot\nabla\theta + L_{22}\cdot\cdot\widetilde{\mathbf{S}}_{v} + L_{23}\cdot\cdot\widetilde{\mathbf{S}}_{\mathrm{O}} + L_{24}\cdot\cdot(\overset{(n)}{\mathbf{S}}{}_{e}/\rho)^{\bullet}, \\ \boldsymbol{\Omega}_{p} = L_{13}\cdot\nabla\theta + L_{23}\cdot\cdot\widetilde{\mathbf{S}}_{v} + L_{33}\cdot\cdot\widetilde{\mathbf{S}}_{\mathrm{O}} + L_{34}\cdot\cdot(\overset{(n)}{\mathbf{S}}{}_{e}/\rho)^{\bullet}, \\ \rho\overset{(n)}{\mathbf{g}}{}^{1}_{e} = L_{14}\cdot\nabla\theta + L_{24}\cdot\cdot\widetilde{\mathbf{S}}_{v} + L_{34}\cdot\cdot\widetilde{\mathbf{S}}_{\mathrm{O}} + L_{44}\cdot\cdot(\overset{(n)}{\mathbf{S}}{}_{e}/\rho)^{\bullet}, \end{cases} \tag{9.259}$$

In *models $D_n^{\zeta}$ of plastic yield*, all the functionals in (9.259) are assumed to be only functions of their arguments and the cross–effects are neglected; as a result, relations (9.259) take a simpler form

$$\begin{cases} -\mathbf{q} = \boldsymbol{\lambda}\cdot\nabla\theta, \\ \dot{\mathbf{V}}_{p} = {}^{4}\mathbf{L}_{v}\cdot\cdot\widetilde{\mathbf{S}}_{v}, \\ \boldsymbol{\Omega}_{p} = {}^{4}\mathbf{L}_{\mathrm{O}}\cdot\cdot\widetilde{\mathbf{S}}_{\mathrm{O}}, \\ \overset{(n)}{\mathbf{g}}{}^{1} = 0. \end{cases} \tag{9.260}$$

Here $\zeta$ and $L_{22}, L_{23}, L_{33}$ are functions having the form

$$\zeta = \zeta(\mathcal{R}, w_\beta^p), \tag{9.261a}$$

$$^4\mathbf{L}_v = {}^4\mathbf{L}_v(\mathcal{R}, \dot{\mathcal{R}}, w_\beta^p), \quad {}^4\mathbf{L}_0 = {}^4\mathbf{L}_0(\mathcal{R}, \dot{\mathcal{R}}, w_\beta^p), \quad \mathcal{R} = \{\overset{(n)}{\mathbf{S}}_e/\rho, \mathbf{O}_e, \theta, \mathbf{V}_p, \mathbf{O}_p^{\mathrm{T}}\}. \tag{9.261b}$$

## 9.3.4 Constitutive Equations for Models $D_n^\zeta$ of Isotropic Plastic Continua

Consider an isotropic plastic continuum. Applying the method given in Sect. 9.3.2, we can show that the potential $\zeta$ in this case is independent of $\mathbf{O}_e$; i.e. $\overset{\circ}{\mathbf{S}}_e = 0$, and hence the tensors $\overset{(n)}{\mathbf{S}}_e$ and $\overset{(n)}{\mathbf{g}}_e$ must be coaxial. The constitutive relation (9.256) between $\overset{(n)}{\mathbf{g}}_e$ and $\overset{(n)}{\mathbf{S}}_e$, which satisfies this condition, has the usual form (see (9.205)):

$$\overset{(n)}{\mathbf{g}}_e = \widetilde{\varphi}_1 \mathbf{E} + \widetilde{\varphi}_2 \overset{(n)}{\mathbf{S}}_e + \varphi_3 \overset{(n)}{\mathbf{S}}{}_e^2, \tag{9.262}$$

$$\varphi_\gamma = -\rho(\partial\zeta/\partial I_\gamma(\overset{(n)}{\mathbf{S}}_e)), \quad \zeta = \zeta(I_\gamma(\overset{(n)}{\mathbf{S}}_e), J_\gamma^{(I)}(\mathbf{V}_p, \mathbf{O}_p^{\mathrm{T}}), \theta, w_p),$$

$$w_p = \int_0^t (\mathbf{S}_v \cdot\cdot \dot{\mathbf{V}}_p + \mathbf{S}_0 \cdot\cdot \boldsymbol{\Omega}_p)\, d\tau.$$

In the associated model $D_n^\zeta$ of plastic continua, relations (9.260) of plastic yield are connected to the yield surface by the gradient law:

$$\dot{\mathbf{V}}_p = h \sum_{\beta=1}^{k} \dot{\varkappa}_\beta (\partial f_\beta/\partial \widetilde{\mathbf{S}}_v), \qquad \boldsymbol{\Omega}_p = h \sum_{\beta=1}^{k} \dot{\varkappa}_\beta (\partial f_\beta/\partial \widetilde{\mathbf{S}}_0). \tag{9.263}$$

Here $f_\beta$ are the plastic potentials being functions (9.261a), which can be considered to depend on the following arguments:

$$f_\beta = f_\beta(\mathbf{S}_v, \mathbf{S}_0, \mathbf{V}_p, \mathbf{O}_p^{\mathrm{T}}, \theta, w_p). \tag{9.264}$$

For the model $B_n^\zeta$ of a plastic continuum with proper strengthening, $f_\beta$ do not depend explicitly on $\mathbf{V}_p$; they are functions of simultaneous invariants of the tensors $\mathbf{S}_v$ and $\mathbf{S}_0$:

$$f_\beta = f_\beta(J_\gamma^{(I)}(\mathbf{S}_v, \mathbf{S}_0), \theta, w_p). \tag{9.265}$$

Since the tensors $\mathbf{S}_v$ and $\mathbf{S}_0$ are combinations of the tensors $\mathbf{T}_e$ and $\mathbf{V}_e$, the model (9.265) describes plastic strengthening but in a special way.

Since the tensor $\mathbf{S}_0$ is skew-symmetric, we can connect the tensor to the vorticity vector by the relation

$$\boldsymbol{\omega}_0 = \frac{1}{2}\boldsymbol{\varepsilon} \cdot\cdot \mathbf{S}_0 \tag{9.266}$$

and consider the simultaneous invariants $J_\gamma^{(I)}$ as scalar functions of the symmetric tensor $\mathbf{S}_\nu$ and vector $\boldsymbol{\omega}_0$:

$$f_\beta = f_\beta(J_1^{(I)},\ J_2^{(I)},\ J_4^{(I)},\ \theta, w_p), \quad J_\gamma^{(I)} = J_\gamma^{(I)}(\mathbf{S},\ \boldsymbol{\omega}_0). \tag{9.267}$$

The simultaneous invariants $J_\gamma^{(I)}$ for an isotropic continuum are determined by formulae (9.202). Then, according to the expression (9.203), constitutive equations (9.263) take the forms

$$\dot{\mathbf{V}}_p = \widetilde{\psi}_1\mathbf{E} - \psi_2\mathbf{S}_\nu, \quad \boldsymbol{\Omega}_p = \psi_4\boldsymbol{\omega}_0 \cdot \boldsymbol{\varepsilon}, \tag{9.268}$$

$$\widetilde{\psi}_1 = \psi_1 + \psi_2 I_1(\mathbf{S}_\nu), \quad \psi_\gamma = h\sum_{\beta=1}^{k} \dot{\varkappa}_\beta \frac{\partial f_\beta}{\partial J_\gamma^{(I)}}.$$

The function $h$ is determined by (9.51), where the partial derivative with respect to time has the form

$$\frac{d' f_\beta}{dt} = \frac{\partial f_\beta}{\partial \mathbf{S}_\nu} \cdot\cdot \dot{\mathbf{S}}_\nu + \frac{\partial f_\beta}{\partial \mathbf{S}_0} \cdot\cdot \dot{\mathbf{S}}_0. \tag{9.269}$$

The system of $(9 + k)$ scalar equations (9.263) and

$$f_\beta = 0, \quad \beta = 1, \ldots, k, \tag{9.270}$$

is solved for $(9 + k)$ scalar unknowns: six independent components of the tensor $\mathbf{V}_p$, three components of the tensor $\mathbf{O}_p$ and $k$ scalar functions $\dot{\varkappa}_\beta$.

### 9.3.5 The Principles of Material Symmetry and Material Indifference for Models $C_n^\xi$ and $D_n^\xi$

1. The tensor $\overset{(n)}{\mathbf{A}}$ is $H$-invariant, hence the tensors $\overset{(n)}{\mathbf{A}}_e$ and $\overset{(n)}{\mathbf{A}}_p$ in models $C_n^\xi$ are $H$-invariant as well. Then the co-rotational derivatives $\overset{(n)}{\mathbf{A}}{}_p^h$, as shown in Sect. 4.7.5, will be $H$-invariant too. And since the tensors $\overset{(n)}{\mathbf{S}}$ are $H$-invariant, the constitutive equations (9.237) for models $C_n^\xi$ of isotropic plastic continua remain unchanged under the change of reference configuration $\overset{\circ}{\mathcal{K}} \to \overset{*}{\mathcal{K}}$; and hence the principle of material symmetry for the models $C_n^\xi$ holds true.

The tensors $\overset{(n)}{\mathbf{A}}_e$ and $\overset{(n)}{\mathbf{A}}_p$ as well as $\overset{(n)}{\mathbf{A}}$ are $R$-indifferent. Then, as shown in Sect. 4.7.5, all the co-rotational derivatives $\overset{(n)}{\mathbf{A}}{}_p^h$ $(h = \{V, S, J\})$ will be $R$-indifferent too. Hence, due to $R$-indifference of the tensors $\overset{(n)}{\mathbf{S}}$, constitutive equations (9.237) during the passage $\mathcal{K} \to \mathcal{K}'$ in rigid motion will transform consistently: i.e. relations between $\overset{(n)}{\mathbf{A}}_e{}'$, $\overset{(n)}{\mathbf{A}}_p{}'$ and $\overset{(n)}{\mathbf{S}}{}'$ in $\mathcal{K}'$ will be the same as the ones between tensors $\overset{(n)}{\mathbf{A}}_e$, $\overset{(n)}{\mathbf{A}}_p$ and $\overset{(n)}{\mathbf{S}}$ in $\mathcal{K}$. This means that the principle of material indifference for the models $C_n^\xi$ also holds.

2. Due to $H$-indifference of the tensors $\mathbf{T}_e$ and $\mathbf{V}_p$ (see Sect. 9.2.5), the tensors $\mathbf{S}_v$ and $\mathbf{S}_0$ (9.242) are also $H$-indifferent. The tensors $\overset{(n)}{\mathbf{S}}_e$, $\overset{\circ}{\mathbf{S}}_e$ and $\overset{(n)}{\mathbf{g}}_e$ are $H$-invariant, because, according to Theorem 9.4 (see Sect. 9.2.5), all their constituent tensors $\mathbf{T}$, $\mathbf{V}_e$ and $\mathbf{O}_e$ are $H$-invariant.

Then the constitutive equations (9.262) under $H$-transformations $\overset{\circ}{\mathcal{K}} \to \overset{*}{\mathcal{K}}$ remain unchanged, and relations (9.268) for plastic deformations change consistently: the relations between $\overset{*}{\mathbf{V}}{}_p^\bullet$, $\overset{*}{\boldsymbol{\Omega}}_p$ and $\overset{*}{\mathbf{S}}_V$, $\overset{*}{\mathbf{S}}_0$ in $\overset{*}{\mathcal{K}}$ remain the same as for $\mathbf{V}_p$, $\boldsymbol{\Omega}_p$, $\mathbf{S}$, $\mathbf{S}_0$ in $\overset{\circ}{\mathcal{K}}$, i.e. the principle of material symmetry for isotropic models $D_n^\xi$ holds.

In rigid motions $\mathcal{K} \to \mathcal{K}'$, the tensors $\overset{(n)}{\mathbf{S}}_e$ and $\overset{(n)}{\mathbf{g}}_e$ are $R$-indifferent due to $R$-indifference of their constituent tensors $\mathbf{T}$ and $\mathbf{V}_e$ (see the theorems from Sects. 4.10.4 and 4.10.5), and the tensors $\mathbf{S}_V$ and $\mathbf{S}_0$ are $R$-invariant due to $R$-invariance of their constituent tensors $\mathbf{T}_e$ and $\mathbf{V}_p$ (see Theorem 9.4 from Sect. 9.2.5). Then in rigid motions the relations (9.262) will change in the consistent way: relations between $\overset{(n)}{\mathbf{g}}{}_e'$ and $\overset{(n)}{\mathbf{S}}{}_e'$ in $\mathcal{K}'$ will be the same as the ones between $\overset{(n)}{\mathbf{g}}_e$ and $\overset{(n)}{\mathbf{S}}_e$ in $\mathcal{K}$. Moreover, since the tensors $\dot{\mathbf{V}}_p$ and $\boldsymbol{\Omega}_p$ are $R$-invariant, relations (9.268) are transformed in the consistent way as well; hence the principle of material indifference for isotropic models $D_n^\xi$ of plastic continua is satisfied.

Thus, all the models of plastic continua $A_n^\xi$, $B_n^\xi$, $C_n^\xi$ and $D_n^\xi$ given in this section are correct, because they satisfy all the fundamental principles of continuum mechanics.

## 9.4 Constitutive Equations of Plasticity Theory 'in Rates'

### *9.4.1 Representation of Models $A_n$ of Plastic Continua 'in Rates'*

To solve many problems of plasticity theory, especially dynamic problems, it is convenient to use the representation of constitutive equations 'in rates' (see Sect. 4.8.12) just as for elastic materials (see Sect. 6.1.3).

First, we consider the equations of models $A_n^\zeta$ of plastic continua, namely the associated models $A_n^\zeta$ of plasticity (9.74).

Differentiate the expression (9.74b) for the elastic deformation tensor with respect to $t$

$$\overset{(n)}{\mathbf{C}}{}^\bullet_e = \sum_{\gamma=1}^{r}\left(\sum_{\beta=1}^{r}\left(\frac{\partial\varphi_\gamma}{\partial I_\beta^{(s)}} I_{\beta\mathbf{T}}^{(s)}\cdot\cdot\left(\frac{\overset{(n)}{\mathbf{T}}}{\rho}\right)^{\bullet} + \frac{\partial\varphi_\gamma}{\partial w_\beta^p} J_\beta^{(s)}\right) + \frac{\partial\varphi_\gamma}{\partial\theta}\dot{\theta}\right) I_{\gamma\mathbf{T}}^{(s)}$$
$$+\sum_{\gamma=1}^{r}\varphi_\gamma\frac{\partial I_{\gamma\mathbf{T}}^{(s)}}{\partial\overset{(n)}{\mathbf{T}}/\rho}\cdot\cdot\left(\frac{\overset{(n)}{\mathbf{T}}}{\rho}\right)^{\bullet}. \qquad (9.271)$$

Introducing the tensors ${}^4\widetilde{\mathbf{N}}$ and $\overset{(n)}{\mathbf{C}}_\theta$ by formulae similar to (4.423):

$${}^4\widetilde{\mathbf{N}} = \sum_{\gamma,\beta=1}^{r}(\varphi_{\gamma\beta} I_{\gamma\mathbf{T}}^{(s)}\otimes I_{\beta\mathbf{T}}^{(s)} + \varphi_\gamma\delta_{\gamma\beta}{}^4\mathbf{I}_\gamma),$$

$$\overset{(n)}{\mathbf{C}}_\theta = \sum_{\gamma=1}^{r}\frac{\partial\varphi_\gamma}{\partial\theta} I_{\gamma\mathbf{T}}^{(s)}, \quad \overset{(n)}{\mathbf{C}}_{wp} = \sum_{\gamma,\beta=1}^{r}\frac{\partial\varphi_\gamma}{\partial w_\beta^p} J_\beta^{(s)} I_{\gamma\mathbf{T}}^{(s)}, \qquad (9.272)$$

where

$$\varphi_{\gamma\beta} = \frac{\partial\varphi_\gamma}{\partial I_\beta^{(s)}} = \frac{\partial^2\zeta}{\partial I_\gamma^{(s)}\partial I_\beta^{(s)}}, \quad I_{\gamma\mathbf{T}}^{(s)} = \frac{\partial I_\gamma^{(s)}(\overset{(n)}{\mathbf{T}}/\rho)}{\partial\overset{(n)}{\mathbf{T}}/\rho}, \quad {}^4\mathbf{I}_\gamma^{(s)} = \frac{\partial I_{\gamma\mathbf{T}}^{(s)}}{\partial(\overset{(n)}{\mathbf{T}}/\rho)}, \qquad (9.273)$$

and using the relation (4.103) between the rates of the energetic deformation tensors $\overset{(n)}{\mathbf{C}}{}^\bullet$ and the tensor of deformation rates $\mathbf{D}$:

$$\overset{(n)}{\mathbf{C}}{}^\bullet = {}^4\overset{(n)}{\mathbf{X}}\cdot\cdot\mathbf{D} = \overset{(n)}{\mathbf{C}}{}^\bullet_e + \overset{(n)}{\mathbf{C}}{}^\bullet_p, \qquad (9.274)$$

(the tensors ${}^4\overset{(n)}{\mathbf{X}}$ are determined by formulae (4.102)), we rewrite Eq. (9.271) in the form

$${}^4\widetilde{\mathbf{N}}\cdot\cdot(\overset{(n)}{\mathbf{T}}/\rho)^\bullet = {}^4\overset{(n)}{\mathbf{X}}\cdot\cdot\mathbf{D} - \overset{(n)}{\mathbf{C}}{}^\bullet_p - \overset{(n)}{\mathbf{C}}_\theta\dot{\theta} - \overset{(n)}{\mathbf{C}}_{wp}. \qquad (9.275)$$

Denote the tensor inverse of ${}^4\widetilde{\mathbf{N}}$ by ${}^4\mathbf{P}$:

$${}^4\mathbf{P}\cdot\cdot{}^4\widetilde{\mathbf{N}} = \mathbf{\Delta}, \qquad (9.276)$$

then Eq. (9.275) can be written in the form

$$\left(\frac{\overset{(n)}{\mathbf{T}}}{\rho}\right)^{\bullet} = {}^4\mathbf{P}\cdot\cdot\,({}^4\overset{(n)}{\mathbf{X}}\cdot\cdot\,\mathbf{D} - \overset{(n)}{\mathbf{C}}{}^{\bullet}_p - \overset{(n)}{\mathbf{C}}_{wp} - \overset{(n)}{\mathbf{C}}_{\theta}\dot{\theta}). \tag{9.277}$$

Taking in addition the relations (9.74c) and (9.74d) for the plastic deformation rates

$$\overset{(n)}{\mathbf{C}}{}^{\bullet}_p = \sum_{\alpha=1}^{z_1} \varphi_\alpha J^{(s)}_{\alpha\mathbf{T}}, \quad f_\beta = 0, \tag{9.278}$$

we obtain the desired constitutive equations 'in rates'.

Notice that, according to (9.272) and (9.74), the tensors occurring in (9.277) and (9.278) are functions of the same arguments as the relations (9.74):

$$\begin{aligned} &{}^4\mathbf{P} = {}^4\mathbf{P}\left(I^{(s)}_\gamma\left(\frac{\overset{(n)}{\mathbf{T}}}{\rho}\right),\ \theta,\ w^p_\beta\right), \quad f_\beta = f_\beta(J^{(s)}_\alpha(\overset{(n)}{\mathbf{T}},\ \overset{(n)}{\mathbf{C}}_p),\ \theta,\ w^p_\beta), \\ &\overset{(n)}{\mathbf{C}}_{wp} = \overset{(n)}{\mathbf{C}}_{wp}\left(I^{(s)}_\gamma\left(\frac{\overset{(n)}{\mathbf{T}}}{\rho}\right),\ \theta,\ w^p_\beta,\ J^{(s)}_\alpha(\overset{(n)}{\mathbf{T}},\ \overset{(n)}{\mathbf{C}}_p)\right). \end{aligned} \tag{9.279}$$

If we consider associated models $A^\zeta_n$ of an elastoplastic continuum with linear elasticity (see Sect. 9.1.5), then the tensor ${}^4\mathbf{P}$ has the form

$${}^4\mathbf{P} = \frac{1}{\rho}\,{}^4\mathbf{M} = \frac{1}{\overset{\circ}{\rho}}\,{}^4\overset{\circ}{\mathbf{M}}, \tag{9.280}$$

where ${}^4\overset{\circ}{\mathbf{M}}$ is the elastic moduli tensor (6.8a) being independent of $\overset{(n)}{\mathbf{T}}$, $\overset{(n)}{\mathbf{C}}_p$, $w^p_\beta$, and $\overset{(n)}{\mathbf{C}}_e$, and possibly dependent only on $\theta$ (see Exercise 9.4.1).

Notice also that with the help of the continuity equation (3.15) and formula (3.20) the left-hand sides of Eqs. (9.277) and (9.278) can be written in the forms

$$\frac{1}{\rho}\left(\frac{\partial\overset{(n)}{\mathbf{T}}}{\partial t} + \nabla\cdot(\mathbf{v}\otimes\overset{(n)}{\mathbf{T}})\right) = \left(\frac{\overset{(n)}{\mathbf{T}}}{\rho}\right)^{\bullet}, \tag{9.281}$$

$$\frac{\partial\rho\overset{(n)}{\mathbf{C}}_p}{\partial t} + \nabla\cdot(\rho\mathbf{v}\otimes\overset{(n)}{\mathbf{C}}_p) = \rho\overset{(n)}{\mathbf{C}}{}^{\bullet}_p. \tag{9.282}$$

Then the Eqs. (9.277) and (9.278) take the divergence form

$$\begin{cases} \dfrac{\partial\overset{(n)}{\mathbf{T}}}{\partial t} + \nabla\cdot(\mathbf{v}\otimes\overset{(n)}{\mathbf{T}}) = \rho\,{}^4\mathbf{P}\cdot\cdot\,({}^4\overset{(n)}{\mathbf{X}}\cdot\cdot\,\mathbf{D} - \sum_{\alpha=1}^{z_1}\varphi_\alpha J^{(s)}_{\alpha\mathbf{T}} - \overset{(n)}{\mathbf{C}}_{wp} - \mathbf{C}_\theta\dot{\theta}), \\ \dfrac{\partial\rho\overset{(n)}{\mathbf{C}}_p}{\partial t} + \nabla\cdot(\rho\mathbf{v}\otimes\overset{(n)}{\mathbf{C}}_p) = \sum_{\alpha=1}^{z_1}\rho\varphi_\alpha J^{(s)}_{\alpha\mathbf{T}}, \quad f_\beta = 0. \end{cases} \tag{9.283}$$

## Exercises for 9.4

**9.4.1.** Show that for the associated models $A_n^\xi$ of an elastoplastic continuum with linear elasticity (see Sect. 9.1.5), the tensor ${}^4\mathbf{P}$ can be represented in the form (9.280).

## 9.5 Statements of Problems in Plasticity Theory

### *9.5.1 Statements of Dynamic Problems for Models $A_n^\xi$ of Plasticity*

The general system of balance laws for plastic continua is the same as the one for other types of continua; and it can be written in the spatial description in the form (3.307). This system should be complemented with constitutive equations of plastic continua 'in rates' (for models $A_n^\xi$ of associated plasticity, these are relations (9.283)) and the differential equation of propagation of a boundary of domain $V(t)$ (5.124). As a result, we obtain the following system (when there are no phase transformations):

$$
\begin{aligned}
&\frac{\partial \rho}{\partial t} + \nabla \cdot \rho \mathbf{v} = 0,\\
&\frac{\partial \rho \mathbf{v}}{\partial t} + \nabla \cdot \rho \mathbf{v} \otimes \mathbf{v} = \nabla \cdot ({}^4\overset{(\mathrm{n})}{\mathbf{E}} \cdot\cdot \overset{(\mathrm{n})}{\mathbf{T}}) + \rho \mathbf{f},\\
&\frac{\partial \rho \eta}{\partial t} + \nabla \cdot \rho \mathbf{v} \eta = -(1/\theta)\nabla \cdot \mathbf{q} + (1/\theta)(\rho q_m + w^*),\\
&\frac{\partial \rho \mathbf{u}}{\partial t} + \nabla \cdot (\rho \mathbf{v} \otimes \mathbf{u}) = \rho \mathbf{v},\\
&\frac{\partial \rho \mathbf{F}^{\mathrm{T}}}{\partial t} + \nabla \cdot (\rho \mathbf{v} \otimes \mathbf{F}^{\mathrm{T}} - \rho \mathbf{F} \otimes \mathbf{v}) = 0,\\
&\frac{\partial \overset{(\mathrm{n})}{\mathbf{T}}}{\partial t} + \nabla \cdot (\mathbf{v} \otimes \overset{(\mathrm{n})}{\mathbf{T}}) = J\, {}^4\widetilde{\mathbf{M}} \cdot\cdot ({}^4\overset{(\mathrm{n})}{\mathbf{X}} \cdot\cdot \mathbf{D} - \sum_{\alpha=1}^{z_1} \varphi_\alpha J_{\alpha \mathrm{T}}^{(s)} - \overset{(\mathrm{n})}{\mathbf{C}}_{wp} - \mathbf{C}_\theta \theta),\\
&\frac{\partial \rho \overset{(\mathrm{n})}{\mathbf{C}}_p}{\partial t} + \nabla \cdot (\rho \mathbf{v} \otimes \overset{(\mathrm{n})}{\mathbf{C}}_p) = \rho \sum_{\alpha=1}^{z_1} \psi_\alpha J_{\alpha \mathrm{T}}^{(s)},\\
&f_\beta = 0, \quad \beta = 1, \ldots, k; \quad \frac{\partial f}{\partial t} + \mathbf{v} \cdot \nabla f = 0.
\end{aligned}
\tag{9.284}
$$

This system consists of $30 + k$ scalar equations for the unknown functions

$$
\rho,\ \theta,\ \mathbf{u},\ \mathbf{v},\ \mathbf{F},\ \overset{(\mathrm{n})}{\mathbf{T}},\ \overset{(\mathrm{n})}{\mathbf{C}}_p,\ \varkappa_\beta,\ f \ \parallel \ \mathbf{x},\ t,\ \ \beta = 1, \ldots, k,
$$

whose components are $30 + k$ scalar functions. In the system (9.284) as well as in the equation system (8.232) of viscoelasticity, the energy balance equation has been replaced by the entropy balance equation. In addition, to solve (9.284) we should use expressions (4.389) or (4.38) for the tensors of energetic equivalence ${}^4\overset{(n)}{\mathbf{E}}_A = {}^4\overset{(n)}{\mathbf{E}}$; expressions (9.28b) for $\eta$ and (9.40c) for $w^*$; expressions (4.102) for ${}^4\overset{(n)}{\mathbf{X}}$; expressions (9.65), (9.68), (9.272) for $\psi_\alpha$, $\overset{(n)}{\mathbf{C}}_{wp}$, $\mathbf{C}_\theta$, $f_\beta$, and also expressions for simultaneous invariants $J_\alpha^{(s)}$.

According to the classification introduced in Sect. 6.1.3, Eqs. (9.284) are called the T$\theta$RUVF-*system of the associated model* $A_n^\xi$ *of plasticity*.

Boundary conditions for the system (9.284) at the surface $\Sigma$ of a solid can be written by analogy with the corresponding ones for an elastic solid at large deformations (see Sect. 6.3.1); and when there are no phase transformations, the boundary conditions have the form

$$\begin{aligned} \Sigma_2: \quad & \mathbf{n}\cdot\mathbf{T} = \mathbf{t}_{ne}, \quad -\mathbf{n}\cdot\boldsymbol{\lambda}\cdot\nabla\theta = q_{ne}, \\ \Sigma_3: \quad & \mathbf{v} = \mathbf{v}_e, \quad \theta = \theta_e. \end{aligned} \tag{9.285}$$

Here $\mathbf{t}_{ne}$, $\mathbf{v}_e$, $q_{ne}$ and $\theta_e$ are given values.

At the symmetry planes $\Pi_\Sigma$, boundary conditions have the form (6.76); and at the contact surface, boundary conditions are written as follows (see (6.69)):

$$\Sigma_1: \quad \mathbf{n}\cdot[\mathbf{T}] = 0, \quad [\mathbf{v}] = 0, \quad \mathbf{n}\cdot[\boldsymbol{\lambda}\cdot\nabla\theta] = 0, \quad [\theta] = 0, \quad [\rho\mathbf{F}] = 0. \tag{9.286}$$

The system (9.284) should also be complemented with the initial conditions

$$\begin{aligned} t = 0: \quad & \rho = \overset{\circ}{\rho}, \quad \mathbf{v} = \mathbf{v}_0, \quad \mathbf{u} = \mathbf{0}, \quad \theta = \theta_0, \quad \mathbf{F} = \mathbf{E}, \\ & \overset{(n)}{\mathbf{C}}_p = \mathbf{C}_{p0}, \quad \mathbf{T} = \mathbf{T}_0, \quad \varkappa_\beta = \varkappa_{\beta 0}, \quad \beta = 1, \ldots, k, \end{aligned} \tag{9.287}$$

where $\overset{\circ}{\rho}$, $\mathbf{v}_0$, $\theta_0$, $\mathbf{C}_{p0}$, $\mathbf{T}_0$ and $\varkappa_{\beta 0}$ are given initial values of the functions.

The statement of the *dynamic* T$\theta$RUVF-*problem of thermoplasticity* $A_n^\xi$ *in the spatial description* consists of the equation system (9.284), boundary conditions (9.285), (9.286) and initial conditions (9.287).

The corresponding statement of the *dynamic* T$\theta$UVF-*problem of thermoplasticity* $A_n^\xi$ *in the material description* consists of T$\theta$UVF-system (the balance laws (3.310) and constitutive equations (9.277)–(9.279)):

$$\begin{aligned} \overset{\circ}{\rho}\frac{\partial\mathbf{v}}{\partial t} &= \overset{\circ}{\nabla}\cdot\mathbf{P} + \overset{\circ}{\rho}\mathbf{f}, \\ \overset{\circ}{\rho}\theta\frac{\partial\eta}{\partial t} &= \overset{\circ}{\nabla}\cdot\overset{\circ}{\boldsymbol{\lambda}}\cdot\overset{\circ}{\nabla}\theta + \overset{\circ}{\rho}q_m + \overset{\circ}{w}{}^*, \\ \frac{\partial\mathbf{u}}{\partial t} &= \mathbf{v}, \quad \frac{\partial\mathbf{F}}{\partial t} = \overset{\circ}{\nabla}\otimes\mathbf{v}^{\mathrm{T}}, \end{aligned} \tag{9.288}$$

$$\frac{\partial \overset{(n)}{\mathbf{T}}/\rho}{\partial t} = {}^4\mathbf{P} \cdot\cdot \left( {}^4\overset{(n)}{\mathbf{X}} \cdot\cdot \mathbf{F}^{-1\mathrm{T}} \cdot \overset{\circ}{\nabla} \otimes \mathbf{v} - \sum_{\alpha=1}^{z_1} \psi_\alpha J^{(s)}_{\alpha\mathbf{T}} - \overset{(n)}{\mathbf{C}}_{wp} - \mathbf{C}_\theta \frac{\partial \theta}{\partial t} \right),$$

$$\frac{\partial \overset{(n)}{\mathbf{C}}_p}{\partial t} = \sum_{\alpha=1}^{z_1} \psi_\alpha J^{(s)}_{\alpha\mathbf{T}}, \quad \mathbf{P} = \sqrt{g/\overset{\circ}{g}}\mathbf{F}^{-1} \cdot \overset{(n)}{\mathbf{E}}{}_A^{-1} \cdot \overset{(n)}{\mathbf{T}},$$

$$f_\beta = 0, \quad \beta = 1, \ldots, k,$$

the boundary conditions

$$\begin{aligned}
&\overset{\circ}{\Sigma}_1 : \quad \overset{\circ}{\mathbf{n}} \cdot [\mathbf{P}] = 0, \quad [\mathbf{v}] = 0, \quad \overset{\circ}{\mathbf{n}} \cdot [\overset{\circ}{\boldsymbol{\lambda}} \cdot \overset{\circ}{\nabla}\theta] = 0, \quad [\theta] = 0,\\
&\overset{\circ}{\Sigma}_2 : \quad \overset{\circ}{\mathbf{n}} \cdot \mathbf{P} = \overset{\circ}{\mathbf{t}}_{ne}, \quad -\overset{\circ}{\mathbf{n}} \cdot \overset{\circ}{\boldsymbol{\lambda}} \cdot \overset{\circ}{\nabla}\theta = \overset{\circ}{q}_{ne},\\
&\overset{\circ}{\Sigma}_3 : \quad \mathbf{v} = \mathbf{v}_e, \quad \theta = \theta_e,
\end{aligned} \tag{9.289}$$

and the initial conditions

$$t = 0: \ \mathbf{v} = \mathbf{v}_0, \ \mathbf{u} = 0, \ \theta = \theta_0, \ \mathbf{F} = \mathbf{E}, \ \overset{(n)}{\mathbf{C}}_p = \mathbf{C}_{p0}, \ \mathbf{P} = \mathbf{T}_0, \ \varkappa_\beta = \varkappa_{\beta 0}. \tag{9.290}$$

The following functions are unknown in this problem:

$$\mathbf{v}, \ \mathbf{u}, \ \theta, \ \mathbf{F}, \ \overset{(n)}{\mathbf{T}}, \ \overset{(n)}{\mathbf{C}}_p, \ \varkappa_\beta \ \| \ X^i, \ t, \quad X^i \in \overset{\circ}{V}, \quad t \in [0, \ t_{\max}]. \tag{9.291}$$

In this statement, just as in the viscoelasticity theory (see (8.245)), the entropy balance equation has been used in place of the energy balance equation.

For many problems, the influence of temperature on mechanical parameters of a problem (velocity, displacements, stresses) can be neglected; then we can exclude the entropy balance equation from the system (9.284)–(9.287) or (9.288)–(9.290) and temperature $\theta$ from the number of unknowns. As a result, we obtain *statements of dynamic problems of plasticity theory* $A_n^\xi$ for isothermal processes (in this case, a temperature is assumed to be constant $\theta = \theta_0$).

### 9.5.2 Statements of Quasistatic Problems for Models $A_n^\xi$ of Plasticity

For plastic continua as well as for ideal elastic media, one frequently uses the models of quasistatic processes (see Sect. 6.1.5) and statements of quasistatic problems.

The statement of the *quasistatic problem of plasticity theory* $A_n^\xi$ *in the spatial description* consists of the equilibrium equations (6.35), relations (4.37) between the

Cauchy stress tensor **T** and the energetic stress tensors $\overset{(n)}{\mathbf{T}}$, kinematic relations (2.77) and constitutive equations (9.74) of the associated model $A_n^\zeta$ of plastic continua with quasilinear elasticity (see Exercise 9.1.8):

$$
\begin{gathered}
\nabla \cdot \mathbf{T} + \rho \mathbf{f} = 0, \\
\mathbf{T} = {}^4\overset{(n)}{\mathbf{E}} \cdot \overset{(n)}{\mathbf{T}}, \quad \overset{(n)}{\mathbf{T}} = {}^4\mathbf{M} \cdot\cdot (\overset{(n)}{\mathbf{C}} - \overset{(n)}{\mathbf{C}}_p), \\
\overset{(n)}{\mathbf{C}} = \frac{1}{n - \mathrm{III}}((\mathbf{F}^{\mathrm{T}} \cdot \mathbf{F})^{(n-\mathrm{III})/2} - \mathbf{E}), \\
\mathbf{F}^{-1} = \mathbf{E} - \nabla \otimes \mathbf{u}^{\mathrm{T}}, \quad \overset{(n)}{\mathbf{C}}{}_p^{\bullet} = \sum_{\alpha=1}^{z_1} \psi_\alpha J_{\gamma \mathbf{T}}^{(s)}, \\
f_\beta = 0, \quad \beta = 1, \ldots, k,
\end{gathered}
\tag{9.292}
$$

and also the boundary conditions

$$
\Sigma_1 : \quad \mathbf{n} \cdot \mathbf{T} = \mathbf{t}_{ne}; \qquad \Sigma_2 : \quad \mathbf{u} = \mathbf{u}_e; \qquad \Sigma_3 : \quad \mathbf{n} \cdot [\mathbf{T}] = 0, \ \ [\mathbf{u}] = 0, \tag{9.293}
$$

where the vectors $\mathbf{t}_{ne}$ and $\mathbf{u}_e$ are given. The problem is solved for the following unknowns (their number determined by the number of independent components is $3 + 6 + k$):

$$
\mathbf{u}, \ \overset{(n)}{\mathbf{C}}_p, \ \varkappa_\beta \ \parallel \ \mathbf{x}, \ t, \quad \beta = 1, \ldots, k. \tag{9.294}
$$

Here expressions for ${}^4\overset{(n)}{\mathbf{E}}$ are determined by formulae (4.38), expressions for ${}^4\mathbf{M}$ are given in Exercise 9.1.8, and $\psi_\alpha$, $J_{\alpha \mathbf{T}}^{(s)}$ and $f_\beta$ are evaluated by the formulae of Sect. 9.1.6.

The system (9.292)–(9.294) should be complemented with relations (6.89), which allow us to find the unknown geometry of the domain $V$ in the actual configuration.

The corresponding statement of the quasistatic problem of thermoplasticity $A_n^\zeta$ in the material description has the form

$$
\begin{gathered}
\overset{\circ}{\nabla} \cdot \mathbf{P} + \overset{\circ}{\rho} \mathbf{f} = 0, \\
\mathbf{P} = (\overset{\circ}{\rho}/\rho) \mathbf{F}^{-1} \cdot {}^4\overset{(n)}{\mathbf{E}} \cdot \overset{(n)}{\mathbf{T}}, \quad \overset{(n)}{\mathbf{T}} = {}^4\mathbf{M} \cdot\cdot (\overset{(n)}{\mathbf{C}} - \overset{(n)}{\mathbf{C}}_p), \\
\overset{(n)}{\mathbf{C}} = \frac{1}{n - \mathrm{III}}((\mathbf{F}^{\mathrm{T}} \cdot \mathbf{F})^{(n-\mathrm{III})/2} - \mathbf{E}), \\
\mathbf{F} = \mathbf{E} + \overset{\circ}{\nabla} \otimes \mathbf{u}^{\mathrm{T}}, \quad \overset{(n)}{\mathbf{C}}{}_p^{\bullet} = \sum_{\alpha=1}^{z_1} \psi_\alpha J_{\gamma \mathbf{T}}^{(s)},
\end{gathered}
\tag{9.295}
$$

$$f_\beta = 0, \quad \beta = 1, \dots, k,$$

$$\mathring{\Sigma}_1 : \quad \mathring{\mathbf{n}} \cdot \mathbf{P} = \mathbf{t}_{ne}, \qquad \mathring{\Sigma}_2 : \quad \mathbf{u} = \mathbf{u}_e, \qquad \mathring{\Sigma}_3 : \quad \mathring{\mathbf{n}} \cdot [\mathbf{P}] = 0, \quad [\mathbf{u}] = 0,$$

for the following unknowns (i.e. for $3 + 6 + k$ independent components):

$$\mathbf{u}, \ \overset{(n)}{\mathbf{C}}_p, \ \varkappa_\beta \ \parallel \ X^i, \ t, \quad \beta = 1, \dots, k. \tag{9.296}$$

The domain $\mathring{V}$ of solving the problem (9.295)–(9.296) is known.

## 9.6 The Problem on All-Round Tension–Compression of a Plastic Continuum

### *9.6.1 Deformation in All-Round Tension–Compression*

To investigate peculiarities of the models of plastic continua we consider the problem on all-round tension–compression of a cube being the domain $V = \{x^\alpha : -h^\alpha/2 < x^\alpha < h^\alpha/2\}$ (Fig. 9.4).

The motion law for the cube is sought in the same form as the one for a beam (see (6.131)) but with the elongation ratios being equal to each other:

$$k_1 = k_2 = k_3 = k, \tag{9.297}$$

i.e.

$$x^\alpha = k(t)\, X^\alpha, \quad \alpha = 1, \ 2, \ 3. \tag{9.298}$$

The deformation gradient $\mathbf{F}$ and the energetic deformation tensors $\overset{(n)}{\mathbf{C}}$ have the same forms as the ones in the problem on tension of a beam (see Exercises 2.2.1 and 4.2.13); and, due to (9.297), we find that all of them are spherical tensors:

$$\mathbf{F} = k\mathbf{E}, \qquad \overset{(n)}{\mathbf{C}} = \frac{1}{n - \mathrm{III}}(k^{n-\mathrm{III}} - 1)\mathbf{E}. \tag{9.299}$$

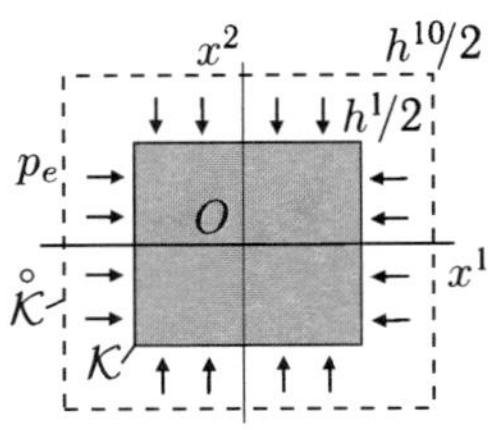


**Fig. 9.4** All-round compression of a cube

For this problem, the change in density is determined by

$$J = \rho/\overset{\circ}{\rho} = \det \mathbf{F}^{-1} = 1/k^3. \tag{9.300}$$

## 9.6.2 Stresses in All-Round Tension–Compression

Assume that a solid considered is described by the Huber–Mises model $A_n^\zeta$ of an isotropic elastoplastic continuum, which is defined by relations (9.84)–(9.87) and (9.92). Moreover, the model is assumed to have linear elasticity, and relations (9.79) for elastic deformations can be written in the form (see Exercise 9.1.9)

$$\overset{(n)}{\mathbf{T}} = J(l_1 I_1(\overset{(n)}{\mathbf{C}}_e)\mathbf{E} + 2l_2 \overset{(n)}{\mathbf{C}}_e). \tag{9.301}$$

From (9.301) we get the following relation between the first invariants:

$$I_1(\overset{(n)}{\mathbf{T}}) = J(3l_1 + 2l_2) I_1(\overset{(n)}{\mathbf{C}}_e). \tag{9.302}$$

Just as in the problem on tension of an elastic beam (see Sect. 6.4), the tensor $\mathbf{T}$ is connected to $\overset{(n)}{\mathbf{T}}$ by relations (6.137); but since the tensor $\mathbf{F}$ is spherical, these formulae in the problem considered take the form

$$\mathbf{T} = k^{n-\mathrm{III}} \overset{(n)}{\mathbf{T}}. \tag{9.303}$$

Expressions for the tensors $\overset{(n)}{\mathbf{C}}_e$ and $\overset{(n)}{\mathbf{C}}_p$ in this problem are sought to be independent of coordinates $\mathbf{x}$; then from (9.301) and (9.303) it follows that $\overset{(n)}{\mathbf{T}}$ and $\mathbf{T}$ are independent of $\mathbf{x}$ too, i.e. components $\bar{T}^{ij}$ of the tensor $\mathbf{T} = \bar{T}^{ij}\bar{\mathbf{e}}_i \otimes \bar{\mathbf{e}}_j$ are the same in the whole cube. Hence, the equilibrium equations in the system (9.292), when there are no mass forces ($\mathbf{f} = 0$), are identically satisfied.

From the boundary conditions at the cube surface ($\mathbf{n} \cdot \mathbf{T} = -p_e \mathbf{n}$) we obtain that

$$X^\alpha = h^\alpha/2: \quad \bar{T}^{\alpha i} = -p_e \delta^{\alpha i}, \quad \alpha = 1,\ 2,\ 3. \tag{9.304}$$

Since components $\bar{T}^{ij}$ are the same in the whole cube, the expression (9.304) still holds in the whole domain $V$ considered too. Thus, the tensor $\mathbf{T}$ is spherical; and hence the tensor $\overset{(n)}{\mathbf{T}}$ is also spherical:

$$\mathbf{T} = -p_e \mathbf{E}, \quad \overset{(n)}{\mathbf{T}} = -\widetilde{p}_e \mathbf{E}, \quad \widetilde{p}_e = p_e k^{\mathrm{III}-n}. \tag{9.305}$$

On substituting (9.305) into (9.301) and (9.302), we find the following expressions for elastic deformations:

$$I_1(\overset{(n)}{\mathbf{C}}_e) = -\frac{3p_e k^{\mathrm{III}-n}}{J(3l_1 + 2l_2)}, \tag{9.306}$$

$$\overset{(n)}{\mathbf{C}}_e = -\frac{p_e k^{\mathrm{III}-n}}{J(3l_1 + 2l_2)}\mathbf{E}, \tag{9.307}$$

i.e. the tensor $\overset{(n)}{\mathbf{C}}_e$ for this problem is spherical too.

### 9.6.3 The Case of a Plastically Incompressible Continuum

Let a considered medium be plastically incompressible, i.e. the rates of plastic deformations $\overset{(n)}{\mathbf{C}}{}^{\bullet}_p$ are determined by Eqs. (9.94). Since in this problem the tensors $\overset{(n)}{\mathbf{T}}$ are spherical, the deviators $\mathbf{P}_T$ of these tensors, by definition, will be zero: $\mathbf{P}_T \equiv 0$. Then the deviators $\mathbf{P}_H$ (9.85) coincide with the deviators $\mathbf{P}_p$ of the tensors of plastic deformation (9.87), and relations (9.94) for plastic deformation take the form

$$\dot{\mathbf{P}}_p = 3\dot{\varkappa} h f_Y \mathbf{P}_p. \tag{9.308}$$

These relations has the evident solution $\mathbf{P}_p \equiv 0$; and, due to plastic incompressibility of the continuum (9.93), we obtain that there are no plastic deformations in a plastically incompressible medium for the problem on all-round compression:

$$\overset{(n)}{\mathbf{C}}_p \equiv 0, \quad \overset{(n)}{\mathbf{C}} = \overset{(n)}{\mathbf{C}}_e. \tag{9.309}$$

Formulae (9.309) and (9.299) give

$$I_1(\overset{(n)}{\mathbf{C}}_e) = I_1(\overset{(n)}{\mathbf{C}}) = \frac{3}{n - \mathrm{III}}(k^{n-\mathrm{III}} - 1). \tag{9.310}$$

On substituting the expressions (9.310) and (9.300) into (9.306), we find the relations between $k$ and $p_e$:

$$p_e = \frac{3l_1 + 2l_2}{n - \mathrm{III}}(1 - k^{n-\mathrm{III}})k^{n-\mathrm{III}-3}. \tag{9.311}$$

In place of (9.311), one frequently uses the relation between $p_e$ and $J = \rho/\overset{\circ}{\rho}$:

$$p_e = \frac{3K}{n - \mathrm{III}}(1 - J^{(\mathrm{III}-n)/3})J^{1+\frac{\mathrm{III}-n}{3}}, \tag{9.312}$$

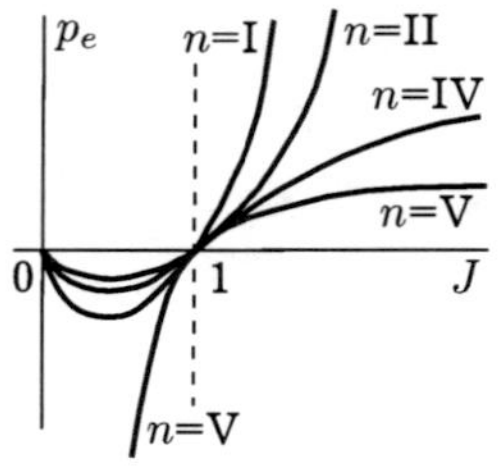


**Fig. 9.5** The functions $p_e(J)$ (9.312)

which follows from (9.300) and (9.311). Here

$$K = l_1 + \frac{2}{3}l_2,$$

is called the *modulus of volumetric compression.*

The functions (9.312) are shown in Fig. 9.5.

The values $J > 1$ correspond to all-round compression ($p_e > 0$), and $J < 1$ – to all-round tension ($p_e < 0$). When $J > 1$, for all $n$ the function $p_e(J)$ is monotonically increasing; for $n = $ I and II it is convex downwards, but for $n = $ IV and V it is convex upwards.

When $0 < J < 1$, the function $p_e(J)$ has an extremum for $n = $ I, II and IV and $p_e(0) = 0$; and for $n = $ V: $p_e \to -\infty$ as $J \to 0$.

Thus, different models $A_n$ give qualitatively different diagrams $p_e(J)$.

### 9.6.4 The Case of a Plastically Compressible Continuum

Let us consider now the Huber–Mises model $A_n^{\xi}$ for a plastically compressible medium (see Exercise 9.1.7). Since for this problem $\mathbf{P}_T = 0$ and $\mathbf{P}_H = \mathbf{P}_p$, the results of Exercise 9.1.7 give

$$\dot{\mathbf{P}}_p = 3\dot{\varkappa} h f_Y \mathbf{P}_p, \tag{9.313}$$

$$\dot{I}_1(\overset{(n)}{\mathbf{C}}_p) = \dot{\varkappa} h f_{1Y}, \quad f = 0. \tag{9.314}$$

The first relation has the zero solution: $\mathbf{P}_p = 0$, and the second one admits a non-trivial solution; hence the tensor $\overset{(n)}{\mathbf{C}}_p$ in this case is spherical:

$$\overset{(n)}{\mathbf{C}}_p = \frac{1}{3}Y_{1p}\mathbf{E}, \quad Y_{1p} \equiv I_1(\overset{(n)}{\mathbf{C}}_p). \tag{9.315}$$

The simultaneous invariant $Y_{1H}$ in this case, according to (9.305) and (9.315), becomes

$$Y_{1H} = (\overset{(n)}{\mathbf{T}} - H_1 \overset{(n)}{\mathbf{C}}_p) \cdot\cdot \mathbf{E} = -(3\widetilde{p}_e + H_1 Y_{1p}), \quad Y_H \equiv 0, \tag{9.316}$$

$$H_1 = H_1^0 Y_{1p}^{2n_1}, \quad Y_\pm = \frac{1}{2}(|Y_{1H}| \pm Y_{1H}). \tag{9.317}$$

Equation (9.314) serves for determination of $\dot{\varkappa}$, and the first invariant of the plastic deformation tensor $Y_{1p}$ can be found from the plasticity condition $f = 0$. If $f$ is assumed to have a quadratic form (see Exercise 9.1.7), then

$$f = \frac{Y_+^2}{\sigma_T'^2} + \frac{Y_-^2}{\sigma_C'^2} - 1 = 0. \tag{9.318}$$

On substituting the expressions (9.316) and (9.317) into (9.318), we find

$$H_1 Y_{1p} = \begin{cases} \sigma_C' - 3\widetilde{p}_e, & \text{if } Y_{1H} < 0, \\ -(\sigma_T' + 3\widetilde{p}_e), & \text{if } Y_{1H} > 0. \end{cases} \tag{9.319}$$

Here the yield strengths $\sigma_C'$ and $\sigma_T'$, by definition, are assumed to be positive: $\sigma_C' > 0$, $\sigma_T' > 0$, that leads to the proper choice of signs in (9.318) for $\sigma_C'$ and $\sigma_T'$: if $Y_{1p} = 0$, then in compression $\widetilde{p}_e = \sigma_C'/3 > 0$, and in tension $\widetilde{p}_e = -\sigma_T'/3 < 0$.

Remind that the equation $f = 0$ holds if and only if the change in plastic deformation is different from zero. If $f < 0$ (in the considered case this occurs when $Y_{1H} \geqslant -\sigma_C'$ or $Y_{1H} \leqslant \sigma_T'$), then $\overset{(n)}{\mathbf{C}}{}_p^{\bullet} = 0$, or that is the same: $\overset{(n)}{\mathbf{C}}_p(t) = \overset{(n)}{\mathbf{C}}{}_p^* =$ const, where $\overset{(n)}{\mathbf{C}}{}_p^* = \overset{(n)}{\mathbf{C}}_p(t^*)$ is the value reached in the preceding cycle of plastic loading up to time $t^*$ of the beginning of unloading.

Before initial loading, at $t = 0$ one usually assumes that there are no plastic deformations, therefore $\overset{(n)}{\mathbf{C}}{}_p^*(0) = 0$ (Fig. 9.6).

Hence, relations (9.319) hold true only under plastic loading. When $f < 0$, i.e. under unloading or under loading in an elastic domain these relations should be replaced by $Y_{1p}(t) = Y_{1p}^*$. Here $Y_{1p}^*$ is the value reached in the preceding cycle of plastic loading.

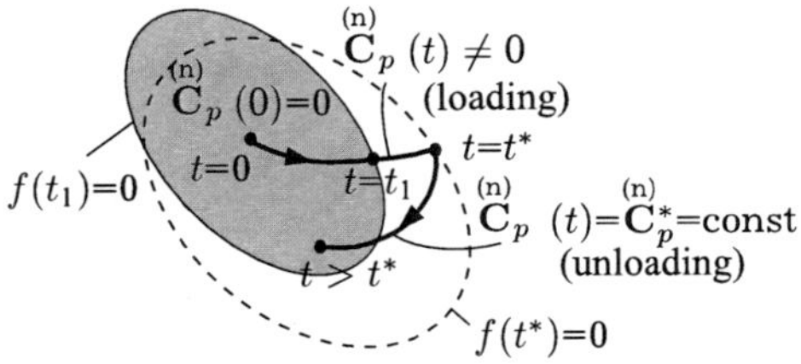


**Fig. 9.6** The cycle of plastic loading and subsequent unloading

On substituting the expression (9.316) for $H_1$ into (9.319), we find that

$$Y_{1p} = \begin{cases} \text{sign}\,(\sigma'_C - 3\widetilde{p}_e)\left(\dfrac{|\sigma'_C - 3\widetilde{p}_e|}{H_1^0}\right)^{\frac{1}{2n_1+1}}, & \text{if } Y_{1H} \leqslant -\sigma'_C, \\ -\text{sign}\,(\sigma'_T + 3\widetilde{p}_e)\left(\dfrac{|\sigma'_T + 3p_e|}{H_1^0}\right)^{\frac{1}{2n_1+1}}, & \text{if } Y_{1H} \geqslant \sigma'_T, \\ Y^*_{1p}, & \text{if } -\sigma'_C < Y_{1H} < \sigma'_T, \end{cases} \quad (9.320)$$

where

$$-Y_{1H} = 3\widetilde{p}_e + H_1^0|Y_{1p}|^{2n_1+1}\text{sign}\,Y_{1p}.$$

Since relations (9.306) and (9.307) for elastic deformations still hold for plastically compressible continuum, so, combining them with (9.315) and (9.320), we obtain

$$\overset{(n)}{\mathbf{C}} = \overset{(n)}{\mathbf{C}}_e + \overset{(n)}{\mathbf{C}}_p = -\frac{1}{3}\left(\frac{p_e k^{\text{III}-n+3}}{K} - Y_{1p}\right)\mathbf{E}. \quad (9.321)$$

Then, according to formula (9.299), we get

$$\frac{3}{n-\text{III}}(1 - k^{n-\text{III}}) = \frac{p_e}{K}k^{\text{III}-n+3} - Y_{1p},$$

or

$$p_e = K\left(\frac{3(1-k^{n-\text{III}})}{n-\text{III}} + Y_{1p}\right)k^{n-\text{III}-3}. \quad (9.322)$$

Combining (9.322) with expression (9.320), we find a nonlinear relation between $p_e$ and $k$. These relations depend on the path of loading of a plastic material. The following example illustrates this fact.

### *9.6.5 Cyclic Loading of a Plastically Compressible Continuum*

Let the function $\widetilde{p}_e \equiv p_e k^{\text{III}-n}$ be given in the form of nonmonotone dependence upon $t$ (Fig. 9.7), where $\widetilde{p}_e(0) = 0$ and $Y_{1p} = 0$, and $H_1 \geqslant 0$. Then from the expression (9.320) for different sections of the function $\widetilde{p}_e(t)$ we get

$$\text{OA}: \ 0 < \widetilde{p}_e \leqslant \sigma'_C/3, \ Y_{1H} = -3\widetilde{p}_e > -\sigma'_C, \ Y_{1p} = 0$$

– elastic compression,

$$\text{AB}: \ \widetilde{p}_e > \sigma'_C/3, \ Y_{1H} = -(3\widetilde{p}_e + Y_{1p}H_1) = -\sigma'_C, \ Y_{1p} = (\sigma'_C - 3\widetilde{p}_e)/H_1 < 0$$

– plastic compression,

$$\text{BC}: \ -\sigma^{*\prime}_T/3 \leqslant \widetilde{p}_e \leqslant \sigma^{*\prime}_C/3, \ Y_{1H} = -(3\widetilde{p}_e + Y^*_{1p}H^*_1) > -\sigma'_C, \ Y_{1p} = Y^*_{1p} < 0$$

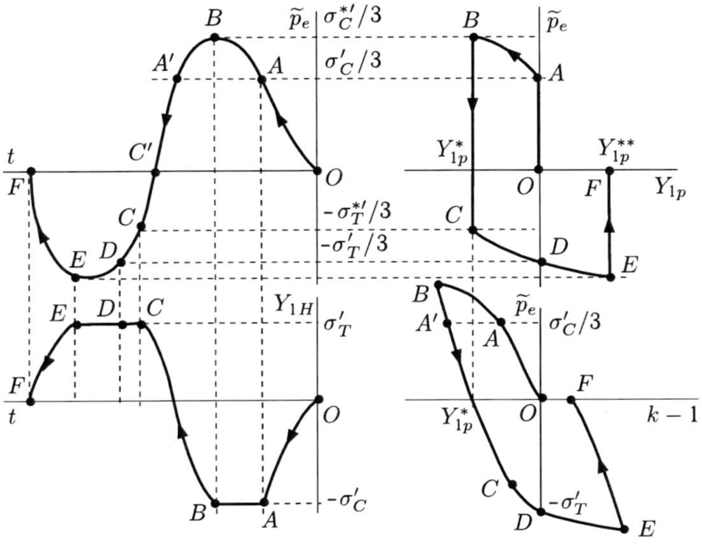


**Fig. 9.7** For the problem on all-round compression of a plastically compressible cube

– unloading–elastic loading,

$$\text{CD}: \quad -\sigma_T'/3 < \widetilde{p}_e < -\sigma_T^{*\prime}/3, \quad Y_{1H} = -(3\widetilde{p}_e + Y_{1p}H_1) = \sigma_T',$$
$$Y_{1p} = -(\sigma_T' + 3\widetilde{p}_e) < 0 \qquad (9.323)$$

– plastic tension,

$$\text{DE}: \quad \widetilde{p}_e \leqslant -\sigma_T'/3, \quad Y_{1H} = \sigma_T', \quad Y_{1p} = -(\sigma_T' + 3\widetilde{p}_e) \geqslant 0$$

– plastic tension,

$$\text{EF}: \quad \widetilde{p}_e > -\sigma_T'/3, \quad Y_{1H} = -(3\widetilde{p}_e + Y_{1p}^{**}H_1^{**}) < \sigma_T', \quad Y_{1p} = Y_{1p}^{**} > 0$$

– unloading.

As follows from these relations, after the appearance of plastic deformations at points, where values of $\widetilde{p}_e$ are the same, for example, at $A$ and $A'$, values of $Y_{1p}$ and $k$ are distinct. In other words, in the plasticity domain the diagrams $p_e(k)$ under loading and unloading are not coincident.

*Remark.* If after preceding plastic compression (sections $AB$ and $BC'$) (Fig. 9.7) there occurs a tension (the section $C'C$), then the yield strength in tension changes up to the value $\sigma_T^{*\prime}/3$ in comparison with the value of $\sigma_T'/3$ when there is no preceding plastic compression. This effect really occurs in many plastic materials and is called the *Bauschinger effect*. □

In the model considered, $\sigma_T^{*\prime}$ can be found from relation (9.323) at the point $C$:

$$Y_{1H} = -3\widetilde{p}_e\big|_C + Y_{1p}^* H_1^* = \sigma_T'. \tag{9.324}$$

Since $Y_{1p}^* H_1^*$ can be evaluated from the conditions (9.323) at the point $B$:

$$Y_{1p}^* H_1^* = \sigma_C' - 3\widetilde{p}_e\big|_B, \tag{9.325}$$

according to (9.324) and (9.325), we obtain

$$\sigma_T^{*\prime} \equiv -3\widetilde{p}_e\big|_C = \sigma_T' - \sigma_C' + 3\widetilde{p}_e\big|_B. \tag{9.326}$$

Thus, the yield strengths in tension $\sigma_T'$ and $\sigma_T^{*\prime}$ differ from one another by the value $(\sigma_C - 3\widetilde{p}_e)$ of excess of the load $3\widetilde{p}_e$ above the yield strength in compression. This result is a typical feature of the Huber–Mises model considered.

## 9.7 The Problem on Tension of a Plastic Beam

### *9.7.1 Deformation of a Beam in Uniaxial Tension*

Consider the above-mentioned classical problem on tension of a beam (see Example 2.1 and Exercises 2.1.1, 2.3.2, and 4.2.13), but for the case when the motion of a beam is described by the model $A_n^\xi$ (the Huber–Mises model) of isotropic elastoplastic continuum, i.e. by Eqs. (9.301) and (9.92).

The motion law for the problem is independent of type of a continuum and it is sought in the form (6.131):

$$x^\alpha = k_\alpha(t)\ X^\alpha, \quad \alpha = 1,\ 2,\ 3. \tag{9.327}$$

The deformation gradient $\mathbf{F}$ and the energetic deformation tensors $\overset{(n)}{\mathbf{C}}$ in the problem on tension of a beam (see Exercises 2.2.1 and 4.2.13) have the diagonal forms

$$\mathbf{F} = \sum_{\alpha=1}^{3} k_\alpha \bar{\mathbf{e}}_\alpha \otimes \bar{\mathbf{e}}_\alpha, \tag{9.328}$$

$$\overset{(n)}{\mathbf{C}} = \sum_{\alpha=1}^{3} \overset{(n)}{\bar{C}}_{\alpha\alpha} \bar{\mathbf{e}}_\alpha \otimes \bar{\mathbf{e}}_\alpha, \tag{9.329}$$

$$\overset{(n)}{\bar{C}}_{\alpha\alpha} = \frac{1}{n - \mathrm{III}}(k_\alpha^{n-\mathrm{III}} - 1). \tag{9.330}$$

We can find a change in the density for this problem as follows:

$$J = \rho/\overset{\circ}{\rho} = \det \mathbf{F}^{-1} = \frac{1}{k_1 k_2 k_3}. \qquad (9.331)$$

Tensors of elastic and plastic deformations $\overset{(n)}{\mathbf{C}}_e$ and $\overset{(n)}{\mathbf{C}}_p$ are also sought in the diagonal form:

$$\overset{(n)}{\mathbf{C}}_e = \sum_{\alpha=1}^{3} \overset{(n)}{C}{}^e_{\alpha\alpha} \bar{\mathbf{e}}_\alpha \otimes \bar{\mathbf{e}}_\alpha, \qquad (9.332)$$

$$\overset{(n)}{\mathbf{C}}_p = \sum_{\alpha=1}^{3} \overset{(n)}{C}{}^p_{\alpha\alpha} \bar{\mathbf{e}}_\alpha \otimes \bar{\mathbf{e}}_\alpha. \qquad (9.333)$$

Due to the additivity relation (9.3), we get

$$\overset{(n)}{C}_{\alpha\alpha} = \overset{(n)}{C}{}^e_{\alpha\alpha} + \overset{(n)}{C}{}^p_{\alpha\alpha}, \quad \alpha = 1,\ 2,\ 3. \qquad (9.334)$$

## 9.7.2 Stresses in a Plastic Beam

Equations (9.301) and (9.302) for the problem also hold true.

On passing to Cartesian components $\overset{(n)}{T}_{\alpha\beta}$ of the tensor $\overset{(n)}{\mathbf{T}}$, from (9.301), (9.302), and (9.332) we get that only diagonal components of this tensor are nonzero:

$$\overset{(n)}{T}_{\alpha\alpha} = J(l_1 I_1(\overset{(n)}{\mathbf{C}}_e) + 2l_2 \overset{(n)}{C}{}^e_{\alpha\alpha}), \quad \alpha = 1,\ 2,\ 3. \qquad (9.335)$$

$$\overset{(n)}{\mathbf{T}} = \sum_{\alpha=1}^{3} \overset{(n)}{T}_{\alpha\alpha} \bar{\mathbf{e}}_\alpha \otimes \bar{\mathbf{e}}_\alpha. \qquad (9.336)$$

A relation between $\mathbf{T}$ and $\overset{(n)}{\mathbf{T}}$ has the form (6.138)

$$\sigma_{\alpha\alpha} = k_\alpha^{n-\mathrm{III}} \overset{(n)}{T}_{\alpha\alpha}, \quad \alpha = 1,\ 2,\ 3; \quad \mathbf{T} = \sum_{\alpha=1}^{3} \sigma_{\alpha\alpha} \bar{\mathbf{e}}_\alpha \otimes \bar{\mathbf{e}}_\alpha. \qquad (9.337)$$

The equilibrium equations (9.292), when $\mathbf{f} = 0$, are satisfied identically.

Since the lateral surface $X^\alpha = \pm h^0_\alpha/2$ ($\alpha = 2, 3$) of the beam is assumed to be free of loads, from the boundary conditions at this surface ($\mathbf{n} \cdot \mathbf{T} = 0$), just as for an ideally elastic body, we obtain

$$\sigma_{\alpha\alpha} = 0, \quad \overset{(n)}{T}_{\alpha\alpha} = 0, \quad \alpha = 2,\ 3. \qquad (9.338)$$

Substituting the values (9.338) into (9.335) and summing these three relations (9.335), we find that

$$I_1(\overset{(n)}{\mathbf{C}}_e) = \frac{\overset{(n)}{T}_{11}}{J(3l_1 + 2l_2)}. \tag{9.339}$$

Substitution of the expression (9.339) for the first invariant into (9.335) at $\alpha = 1$ yields the relation between $\overset{(n)}{T}_{11}$ and $\overset{(n)}{C}{}^e_{11}$; and from formula (9.335) at $\alpha = 2, 3$ we find the relation between $\overset{(n)}{C}{}^e_{22}$, $\overset{(n)}{C}{}^e_{33}$, and $\overset{(n)}{C}{}^e_{11}$:

$$\overset{(n)}{T}_{11} = JE\overset{(n)}{C}{}^e_{11}, \tag{9.340}$$

$$\overset{(n)}{C}{}^e_{22} = \overset{(n)}{C}{}^e_{33} = -\nu\overset{(n)}{C}{}^e_{11}. \tag{9.341}$$

As before, here we have denoted the elastic modulus and Poisson's ratio at large deformations:

$$E = \frac{(3l_1 + 2l_2)l_2}{l_1 + l_2}, \qquad \nu = \frac{l_1}{2(l_1 + l_2)}. \tag{9.342}$$

### *9.7.3 Plastic Deformations of a Beam*

Let us consider now constitutive equations (9.92) and (9.93) for plastic deformations when an isotropic medium is assumed to be plastically incompressible. Due to the assumption (9.333) that the tensor $\overset{(n)}{\mathbf{C}}_p$ is diagonal, from (9.92) we get

$$\frac{d}{dt}\overset{(n)}{C}{}^p_{\alpha\alpha} = \frac{\varkappa h}{3\sigma_s^2}\left(\left(\overset{(n)}{T}_{\alpha\alpha} - \frac{1}{3}\overset{(n)}{T}_{11}\right) - H(\overset{(n)}{C}{}^p_{\alpha\alpha} - \frac{1}{3}Y_{1p})\right), \quad \alpha = 1, 2, 3. \tag{9.343}$$

Here the strengthening parameter $H$ and the first invariant $Y_{1p}$ have the forms

$$H = H_0 Y_p^{2n_0}, \qquad Y_{1p} = I_1(\overset{(n)}{\mathbf{C}}_p), \tag{9.344}$$

$$Y_p^2 = \frac{1}{2}((\overset{(n)}{C}{}^p_{11} - \overset{(n)}{C}{}^p_{22})^2 + (\overset{(n)}{C}{}^p_{22} - \overset{(n)}{C}{}^p_{33})^2 + (\overset{(n)}{C}{}^p_{11} - \overset{(n)}{C}{}^p_{33})^2).$$

Notice that Eqs. (9.343) and (9.344) are symmetric in indices $\alpha = 2, 3$; therefore, $\overset{(n)}{C}{}^p_{22} = \overset{(n)}{C}{}^p_{33}$, and from the condition of plastic incompressibility (9.93) ($Y_{1p} = 0$) we obtain

$$\overset{(n)}{C}{}^p_{33} = \overset{(n)}{C}{}^p_{22} = -\frac{1}{2}\overset{(n)}{C}{}^p_{11}. \tag{9.345}$$

Then the system (9.343) takes the form

$$\frac{d}{dt}\overset{(n)}{C}{}^{p}_{11} = \frac{\dot{\varkappa} h}{3\sigma_s^2}\left(\frac{2}{3}\overset{(n)}{T}{}_{11} - H\overset{(n)}{C}{}^{p}_{11}\right), \tag{9.346}$$

$$\frac{d}{dt}\overset{(n)}{C}{}^{p}_{\alpha\alpha} = \frac{\dot{\varkappa} h}{3\sigma_s^2}\left(-\frac{1}{3}\overset{(n)}{T}{}_{11} - H\overset{(n)}{C}{}^{p}_{\alpha\alpha}\right), \quad \alpha = 2,\ 3. \tag{9.347}$$

Equations (9.347) follow from (9.346) and (9.345), therefore, the system (9.346), (9.347) contains only one independent equation, namely (9.346). This equation allows us to evaluate the parameter of loading $\dot{\varkappa}$ if values of $\overset{(n)}{T}{}_{11}$ and $\overset{(n)}{C}{}^{p}_{11}$ are known.

To evaluate $\overset{(n)}{C}{}^{p}_{11}$ we use the yield surface equation (9.96), which for this problem becomes (here we have taken the expression for $Y_H$ from Exercise 9.1.1 into account):

$$\frac{1}{6\sigma_s^2}\big((\overset{(n)}{T}{}_{11} - H(\overset{(n)}{C}{}^{p}_{11} - \overset{(n)}{C}{}^{p}_{22}))^2 + (\overset{(n)}{T}{}_{11} - H(\overset{(n)}{C}{}^{p}_{11} - \overset{(n)}{C}{}^{p}_{33}))^2$$
$$+H^2(\overset{(n)}{C}{}^{p}_{22} - \overset{(n)}{C}{}^{p}_{33})^2\big) = 1. \tag{9.348}$$

Due to (9.345), this equation is simplified:

$$|\overset{(n)}{T}{}_{11} - \frac{3}{2}H\overset{(n)}{C}{}^{p}_{11}| = \sqrt{3}\sigma_s. \tag{9.349}$$

Since

$$H = H_0 Y_p^{2n_0} = H_0(\overset{(n)}{C}{}^{p}_{11} - \overset{(n)}{C}{}^{p}_{22})^{2n_0} = H_0\left(\frac{3}{2}\overset{(n)}{C}{}^{p}_{11}\right)^{2n_0}, \tag{9.350}$$

Eq. (9.349) takes the final form

$$|\overset{(n)}{T}{}_{11} - H_0\left(\frac{3}{2}\overset{(n)}{C}{}^{p}_{11}\right)^{2n_0+1}| = \sqrt{3}\sigma_s. \tag{9.351}$$

To find an expression for $\overset{(n)}{C}{}^{p}_{11}$ from (9.351), by analogy with formulae (9.318)–(9.320) we should consider separately the cases of plastic tension, plastic compression and leaving the yield surface. Then from (9.351) we get

$$\overset{(n)}{C}{}^{p}_{11}=\begin{cases}\dfrac{2}{3}\operatorname{sign}\,(\overset{(n)}{T}_{11}-\sqrt{3}\sigma_s)\left(\dfrac{|\overset{(n)}{T}_{11}-\sqrt{3}\sigma_s|}{H_0}\right)^{1/(2n_0+1)}, & \\ & \text{if } Y_{1H}\geqslant\sqrt{3}\sigma_s,\\ \dfrac{2}{3}\operatorname{sign}\,(\overset{(n)}{T}_{11}+\sqrt{3}\sigma_s)\left(\dfrac{|\overset{(n)}{T}_{11}+\sqrt{3}\sigma_s|}{H_0}\right)^{1/(2n_0+1)}, & \\ & \text{if } Y_{1H}\leqslant-\sqrt{3}\sigma_s,\\ \overset{(n)}{C}{}^{p*}_{11},\ \text{if } -\sqrt{3}\sigma_s<Y_{1H}<\sqrt{3}\sigma_s, & \end{cases}\tag{9.352}$$

where

$$Y_{1H}=\overset{(n)}{T}_{11}-H_0\left|\frac{3}{2}\overset{(n)}{C}{}^{p}_{11}\right|^{2n_0+1}\operatorname{sign}\,(\overset{(n)}{C}{}^{p}_{11}).$$

Here we have taken into account that $\sigma_s>0$, and $\overset{(n)}{C}_{11}{}^{p*}$ is the value of $\overset{(n)}{C}{}^{p}_{11}$ reached at the time $t^*$ of leaving the yield surface (at $t=0$: $\overset{(n)}{C}{}^{p*}_{11}=0$).

### 9.7.4 Change of the Density

According to Eqs. (9.334), (9.341) and (9.345), we can express $\overset{(n)}{C}_{22}$ in terms of $\overset{(n)}{C}_{11}$ and $\overset{(n)}{C}{}^{p}_{11}$:

$$\overset{(n)}{C}_{22}=\overset{(n)}{C}{}^{e}_{22}+\overset{(n)}{C}{}^{p}_{22}=-\nu\overset{(n)}{C}{}^{e}_{11}-\frac{1}{2}\overset{(n)}{C}{}^{p}_{11},$$

i.e.

$$\overset{(n)}{C}{}^{p}_{22}=-\left(\nu\overset{(n)}{C}_{11}+\left(\frac{1}{2}-\nu\right)\overset{(n)}{C}{}^{p}_{11}\right).\tag{9.353}$$

With the help of relations (9.330) we can represent $k_2$ in terms of $\overset{(n)}{C}_{22}$ and $\overset{(n)}{C}_{11}$ in terms of $k_1$:

$$\begin{aligned}k_2&=(1+(n-\mathrm{III})\overset{(n)}{C}_{22})^{1/(n-\mathrm{III})}=\left(1-(n-\mathrm{III})\left(\nu\overset{(n)}{C}_{11}+\left(\frac{1}{2}-\nu\right)\overset{(n)}{C}{}^{p}_{11}\right)\right)^{1/(n-\mathrm{III})}\\&=\left(1-\nu(k_1^{n-\mathrm{III}}-1)-(n-\mathrm{III})\left(\frac{1}{2}-\nu\right)\overset{(n)}{C}{}^{p}_{11}\right)^{1/(n-\mathrm{III})}.\end{aligned}\tag{9.354}$$

Thus, we have obtained the expression of $k_2$ in terms of $k_1$ and $\overset{(n)}{C}{}^{p}_{11}$.

Taking the equality $k_2=k_3$ into account and substituting (9.354) into (9.331), we get

$$J=\frac{\rho}{\overset{\circ}{\rho}}=\frac{1}{k_1}\left(1-\nu(k_1^{n-\mathrm{III}}-1)-(n-\mathrm{III})\left(\frac{1}{2}-\nu\right)\overset{(n)}{C}{}^{p}_{11}\right)^{-2/(n-\mathrm{III})}.\tag{9.355}$$

Notice that although the continuum considered is plastically incompressible, however, unlike the problem on all-round compression, in this case the density depends on plastic deformations.

### 9.7.5 Resolving Equation for the Problem

Equations (9.340) and (9.334) yield the relation between $\overset{(n)}{T}_{11}$, $\overset{(n)}{C}_{11}$, and $\overset{(n)}{C}{}^{p}_{11}$:

$$\overset{(n)}{C}_{11} = \overset{(n)}{C}{}^{p}_{11} + \frac{\overset{(n)}{T}_{11}}{EJ}. \tag{9.356}$$

1. Consider the case of initial plastic tension, when the conditions $\overset{(n)}{C}{}^{p*}_{11} = 0$ and $\overset{(n)}{T}_{11} \geq \sqrt{3}\sigma_s$ are satisfied. Then for the plastic deformation $\overset{(n)}{C}{}^{p}_{11}$, from the first line of formula (9.352) we find the expression

$$\overset{(n)}{C}{}^{p}_{11} = \left(\frac{\overset{(n)}{T}_{11} - \sqrt{3}\sigma_s}{\widetilde{H}_0}\right)^{1/(2n_0+1)}, \quad \text{where } \widetilde{H}_0 = H_0(3/2)^{2n_0+1}. \tag{9.357}$$

Substituting (9.357) into (9.356) and replacing $\overset{(n)}{C}_{11}$ by $k_1$ according to formula (9.330), and $\overset{(n)}{T}_{11}$ by $\sigma_{11}$ according to (9.337), we obtain

$$\frac{k_1^{n-\mathrm{III}} - 1}{n - \mathrm{III}} = \frac{k_1^{\mathrm{III}-n}\sigma_{11}}{EJ} + \frac{2}{3}\left(\frac{k_1^{\mathrm{III}-n}\sigma_{11} - \sqrt{3}\sigma_s}{\widetilde{H}_0}\right)^{1/(2n_0+1)}. \tag{9.358}$$

Combining this equation with expression (9.355) for $J$, into which formula (9.357) has been substituted:

$$J = \frac{1}{k_1}\left(1 - \nu(k_1^{n-\mathrm{III}} - 1) - (n - \mathrm{III})(\frac{1}{2} - \nu)\left(\frac{k_1^{\mathrm{III}-n}\sigma_{11} - \sqrt{3}\sigma_s}{\widetilde{H}_0}\right)^{\frac{1}{2n_0+1}}\right)^{-\frac{2}{n-\mathrm{III}}}, \tag{9.358a}$$

we find the main *resolving equation* for this problem in initial plastic tension. This equation has a form of the implicit relation $\Phi(\sigma_{11},\ k_1) = 0$ between $\sigma_{11}$ and $k_1$.

2. Under *initial loading in the elastic domain* there is no plastic deformation ($\overset{(n)}{C}{}^{p}_{11} = 0$), and formulae (9.358) and (9.358a) coincide with the resolving equation for models $A_n$ of elastic continua (6.148):

$$\sigma_{11} = E\frac{k_1^{n-\mathrm{III}-1}}{n - \mathrm{III}}(k_1^{n-\mathrm{III}} - 1)(1 - \nu(k_1^{n-\mathrm{III}} - 1))^{2/(\mathrm{III}-n)}. \tag{9.359}$$

3. If there is *initial plastic loading in the compression domain* when the conditions $\overset{(n)}{C}{}^{p*}_{11} = 0$ and $\overset{(n)}{T}_{11} \leq -\sqrt{3}\sigma_s$ are satisfied, then in formula (9.352) we choose the second condition. As a result, for plastic deformation we have

$$\overset{(n)}{C}{}_{11}{}^{p} = -\left(\frac{|k_1^{\mathrm{III}-n}\sigma_{11} + \sqrt{3}\sigma_s|}{\widetilde{H}_0}\right)^{1/(2n_0+1)}, \tag{9.360}$$

and instead of (9.358) and (9.358a), we get the resolving relation

$$\frac{k_1^{n-\mathrm{III}} - 1}{n - \mathrm{III}} = \frac{k_1^{\mathrm{III}-n}\sigma_{11}}{EJ} - \left(\frac{|k_1^{\mathrm{III}-n}\sigma_{11} + \sqrt{3}\sigma_s|}{\widetilde{H}_0}\right)^{1/(2n_0+1)}, \tag{9.361}$$

$$J = \frac{1}{k_1}\left(1 - \nu(k_1^{n-\mathrm{III}} - 1) - (n - \mathrm{III})(\frac{1}{2} - -\nu)\left(\frac{|k_1^{\mathrm{III}-n}\sigma_{11} + \sqrt{3}\sigma_s|}{\widetilde{H}_0}\right)^{\frac{1}{2n_0+1}}\right)^{\frac{2}{\mathrm{III}-n}}. \tag{9.361a}$$

4. At last, if there occurs *unloading* after plastic loading in the domain of tension or compression, then from Eqs. (9.355) and (9.356) we obtain the relation between $\sigma_{11}$ and $k_1$

$$\frac{k_1^{n-\mathrm{III}} - 1}{n - \mathrm{III}} = \frac{k_1^{\mathrm{III}-n}\sigma_{11}}{EJ} + \overset{(n)}{C}{}^{p*}_{11}, \tag{9.362}$$

$$J = \frac{1}{k_1}\left(1 - \nu(k_1^{n-\mathrm{III}} - 1) - (n - \mathrm{III})(\frac{1}{2} - \nu)\overset{(n)}{C}{}^{p*}_{11}\right)^{-1/(n-\mathrm{III})}. \tag{9.362a}$$

Here $\overset{(n)}{C}{}^{p*}_{11}$ is the maximum value of plastic deformations reached at the time $t_*$ of the beginning of unloading; this value is calculated by formula (9.357) for preceding plastic tension and by (9.360) – for compression.

### *9.7.6 Numerical Method for the Resolving Equation*

Transpose the first summand from the left-hand side of Eq. (9.358) onto the right-hand side, and then raise the obtained expression to power $1 + 2n_0$. As a result, we find the equation

$$A\left(\frac{k_1^{n-\mathrm{III}} - 1}{n - \mathrm{III}} - \frac{k_1^{\mathrm{III}-n}\sigma_{11}}{EJ}\right) = \frac{k_1^{\mathrm{III}-n}\sigma_{11} - \sqrt{3}\sigma_s}{\widetilde{H}_0}, \tag{9.363}$$

where we have introduced the notation

$$A(k_1, \sigma_{11}) \equiv \left(\frac{k_1^{n-\mathrm{III}} - 1}{n - \mathrm{III}} - \frac{k^{\mathrm{III}-n}}{J(k, \sigma_{11})}\sigma_{11}\right)^{2n_0}. \tag{9.364}$$

To solve numerically Eq. (9.363) we can apply the method of step-by-step approximations, where by given values of $k_1$ and values of $\sigma_{11}^{\{m-1\}}$ at the $(m-1)$th iteration we find a value of $\sigma_{11}^{\{m\}}$ at the $m$th iteration:

$$\widetilde{\sigma}^{\{m\}} = \frac{1}{1+\widetilde{H}_0 A^{\{m-1\}}/J^{\{m-1\}}}\left(\sqrt{3}\sigma_s + \frac{\widetilde{H}_0 A^{\{m-1\}} k^{n-\mathrm{III}}}{n-\mathrm{III}} k^{n-\mathrm{III}}\right), \quad (9.365)$$

where $J^{\{m-1\}} = J(k, \sigma_{11}^{\{m-1\}})$ and $A^{\{m-1\}} = A(k_1, \sigma_{11}^{\{m-1\}})$ are values of the functions at preceding $(m-1)$th iteration $(m = 1, 2, \ldots)$. As an initial value of $\sigma_{11}^{\{0\}}$ we can take the value of $\sigma_{11}^{\{m\}}$ computed at the preceding iteration cycle for the preceding value of $k_1$. The method is convergent for values of $n_0$ within the interval $-0.5 < n_0 < 1$.

We can solve Eq. (9.361) in the compression domain by a similar way. In the domains of elastic loading and unloading the stress $\sigma_{11}$ can be determined in the explicit way from Eqs. (9.359) and (9.362).

Figure 9.8 shows the functions $\sigma_{11}(k_1)$ obtained by the above-mentioned numerical method for different models $A_n$ and at different values of the parameter

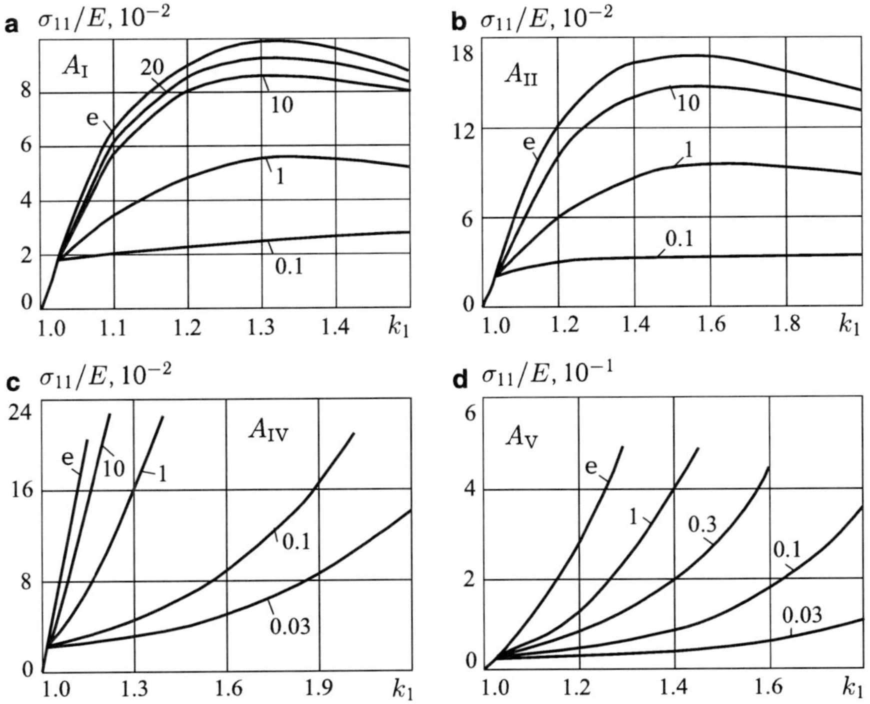


**Fig. 9.8** Diagrams $\sigma_{11}(k_1)$ for models $A_n$ of elastoplastic continuum in uniaxial tension (e – the elastic continuum model, numbers at curves are different values of parameters $\widetilde{H}_0/E, n_0 = 0.1$): **(a)** – model $A_{\mathrm{I}}$, **(b)** – model $A_{\mathrm{II}}$, **(c)** – model $A_{\mathrm{IV}}$, and **(d)** – model $A_{\mathrm{V}}$

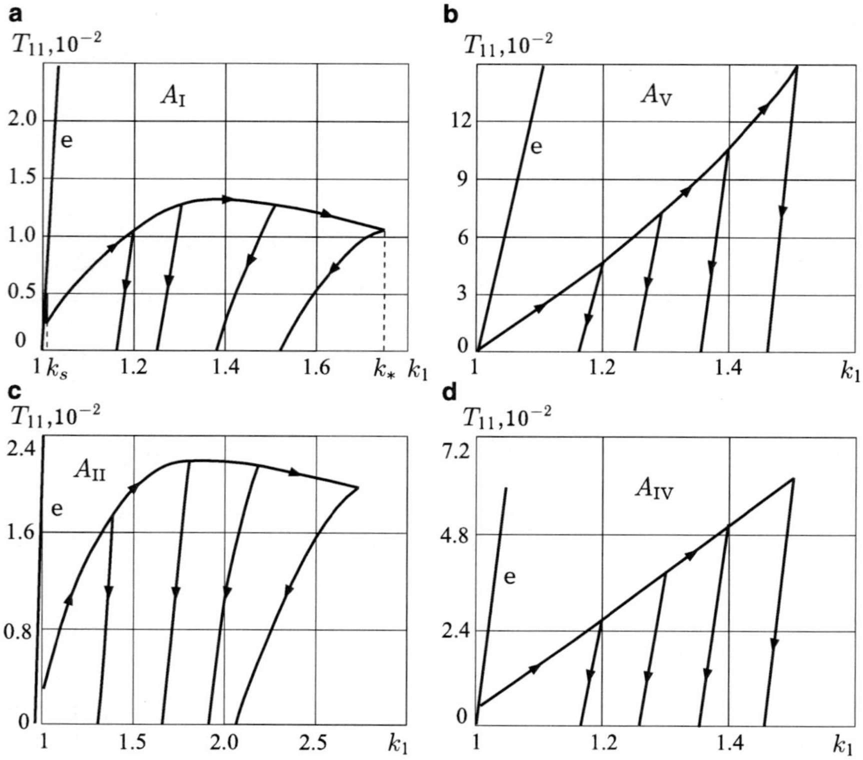


**Fig. 9.9** Diagrams of deforming $\sigma_{11}(k_1)$ for models $A_n$ of an elastoplastic continuum in uniaxial tension followed by unloading (e – the model of an elastic continuum, $\nu = 0.3$): (**a**) – model $A_{\mathrm{I}}$ ($\widetilde{H}_0/E = 0.2$, $n_0 = 0.1$, $\sigma_s/E = 0.001$), (**b**) – model $A_{\mathrm{V}}$ ($\widetilde{H}_0/E = 0.05$, $n_0 = 0.1$, $\sigma_s/E = 0.002$), (**c**) – model $A_{\mathrm{II}}$ ($\widetilde{H}_0/E = 0.15$, $n_0 = 0.1$, $\sigma_s/E = 0.002$), and (**d**) – model $A_{\mathrm{IV}}$ ($\widetilde{H}_0/E = 0.05$, $n_0 = 0.1$, $\sigma_s/E = 0.002$)

$\widetilde{H}_0$. For the models $A_{\mathrm{I}}$ and $A_{\mathrm{II}}$, the functions $\sigma_{11}(k_1)$ are convex upwards for a purely elastic continuum and for elastoplastic models; but for the models $A_{\mathrm{IV}}$ and $A_{\mathrm{V}}$ in the plastic domain (at $k \geqslant k_s$, where $k_s$ is the elongation at the yield point: $\overset{(n)}{T}_{11}(k_s) = \sqrt{3}\,\sigma_s$) the functions $\sigma_{11}(k_1)$ are convex downwards. For all the models $A_n$, the appearance of plastic deformations can cause considerable decreasing values of stresses $\sigma_{11}$ in comparison with an elastic continuum.

Figure 9.9 shows diagrams of deforming $\sigma_{11}(k_1)$ for different models $A_n$ of an elastoplastic continuum under plastic loading up to some limiting values $k_*$ and subsequent unloading, which have been computed by Eqs. (9.358) and (9.359). Let us note some important effects caused by large values of plastic deformations:

1. When $\sigma_s/E \leqslant 0.01$, diagrams of deforming in the elastic domain at $k < k_s$ and under unloading for values $k_*$ close to 1 ($k_* \lesssim 1.1$) are practically linear and

have the same slope; however, at higher values of $k_*$ ($k_* \gtrsim 1.2$) these diagrams are considerably distinct: diagrams of unloading become essentially nonlinear for all models $A_n$;
2. For the models $A_{\mathrm{I}}$ and $A_{\mathrm{II}}$, the slope of a tangent to the function $\sigma_{11}(k_1)$ under unloading decreases with growing $k_*$, and for the models $A_{\mathrm{IV}}$ and $A_{\mathrm{V}}$ it, on the contrary, increases.

### 9.7.7 Method for Determination of Constants $H_0$, $n_0$, and $\sigma_s$

The model $A_n^{\xi}$ of an isotropic plastically-incompressible continuum with the Mises potential (9.301), (9.97) contains five constants: $E$, $\nu$, $H_0$, $n_0$ and $\sigma_s$. If the experimental dependence $\sigma_{11} = \sigma_{11}^{(\mathrm{ex})}(\delta_1)$ (diagram of deforming) under active loading and corresponding curves of unloading are known, then the constant $\sigma_s$ can be determined by the formula $\sigma_s = k_1^{\mathrm{III}-n}\sigma_{11s}/\sqrt{3}$. Here $\sigma_{11s}$ is the value of stress $\sigma_{11}$ at the diagram of deforming, after unloading from which the residual elongation $\delta_1^p$ at $\sigma_{11} = 0$ takes on a value given a priori. For metals and alloys, and also for grounds and rocks, one usually assumes that $\delta_1^p = 0.2\,\%$, and the corresponding value $\sigma_{11s}$ is denoted by $\sigma_{0,2}$ (Fig. 9.10). This value is the *yield strength*. Since $\delta_1^p = 0.002 \ll 1$, the domain of elasticity for such materials corresponds to small deformations and $\sigma_s = \sigma_{11s}/\sqrt{3}$; the elastic modulus $E$ can be determined as a slope of the tangent to the initial section of the experimental diagram $\sigma_{11}^{(\mathrm{ex})}(\delta_1)$. Just as for elastic media, Poisson's ratio $\nu$ is usually calculated by formula (6.147) as the ratio of transverse elongation of a beam to its longitudinal elongation in the domain of small deformations.

The constants $\widetilde{H}_0$ and $n_0$ can be calculated by approximating the experimental diagram of deforming $\sigma_{11}^{(\mathrm{ex})}(\delta_1)$ under active plastic loading with the help of the relation $\sigma_{11}(\delta_1)$ (9.358). To do this, one should minimize the functional of mean-square distance between the experimental and theoretical curves at $N$ points:

$$\Delta = \left(\frac{1}{N}\sum_{i=1}^{N}\left|1 - \frac{\sigma_{11}(\delta_{1(i)})}{\sigma_{11}^{(\mathrm{ex})}(\delta_{1(i)})}\right|^2\right)^{1/2} \to \min. \qquad (9.366)$$

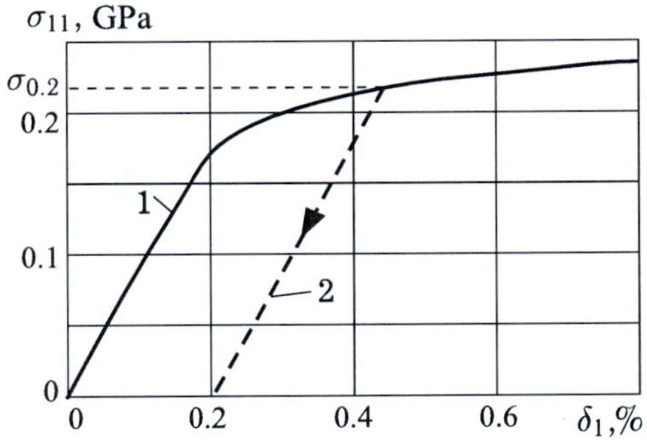


**Fig. 9.10** Diagram of deforming for aluminium alloy and method for determination of the yield strength $\sigma_{0,2}$: *1* – active loading, *2* – unloading

## 9.7.8 Comparison with Experimental Data for Alloys

Figure 9.11a shows experimental diagrams of deforming for steel alloy under temperatures 20°C and 800°C, and also their approximations by the method, mentioned in Sect. 9.7.6, with the help of different models $A_n^{\xi}$ according to formula (9.358). Table 9.3 gives values of the constants $E, \widetilde{H}_0, n_0$ and $\sigma_s$, calculated when $\nu = 0.35$.

An accuracy of approximation proves to be high for all the models $A_n^{\xi}$; the model $A_{\rm I}^{\xi}$ exhibits the least value of the deviation $\Delta$. Notice that the value of coefficient $n_0$ is negative, therefore the diagram of deforming is convex upwards in the vicinity of the yield point $\sigma_s$. Figures 9.11b and 9.12a show experimental diagrams of deforming for aluminium alloys D16 and AK4 in tension and their approximations by the models $A_n^{\xi}$ with the help of formula (9.358); and Table 9.3 gives values of the constants $E, \widetilde{H}_0, n_0$, and $\sigma_s$. An accuracy of approximation is also sufficiently high for all the models $A_n^{\xi}$.

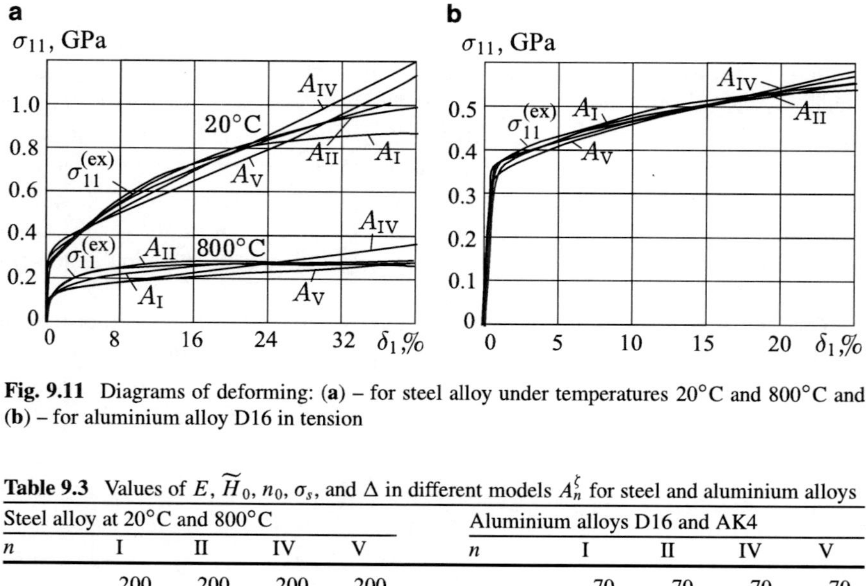


**Fig. 9.11** Diagrams of deforming: (**a**) – for steel alloy under temperatures 20°C and 800°C and (**b**) – for aluminium alloy D16 in tension

**Table 9.3** Values of $E, \widetilde{H}_0, n_0, \sigma_s$, and $\Delta$ in different models $A_n^{\xi}$ for steel and aluminium alloys

| Steel alloy at 20°C and 800°C | | | | |
|---|---|---|---|---|
| $n$ | I | II | IV | V |
| $E$, GPa | $\frac{200}{60}$ | $\frac{200}{60}$ | $\frac{200}{60}$ | $\frac{200}{60}$ |
| $\widetilde{H}_0$, GPa | $\frac{4.8}{0.72}$ | $\frac{3}{0.54}$ | $\frac{1.2}{0.18}$ | $\frac{0.6}{0.18}$ |
| $-n_0$ | $\frac{0.02}{0.26}$ | $\frac{0.08}{0.26}$ | $\frac{0.18}{0.38}$ | $\frac{0.3}{0.36}$ |
| $\sigma_s$, MPa | $\frac{145}{46.3}$ | $\frac{145}{46.3}$ | $\frac{145}{46.3}$ | $\frac{145}{46.3}$ |
| $\Delta$, % | $\frac{6.4}{3.7}$ | $\frac{2.5}{5.4}$ | $\frac{5.4}{14}$ | $\frac{5.3}{15.2}$ |

| Aluminium alloys D16 and AK4 | | | | |
|---|---|---|---|---|
| $n$ | I | II | IV | V |
| $E$, GPa | $\frac{70}{67.7}$ | $\frac{70}{67.7}$ | $\frac{70}{67.7}$ | $\frac{70}{67.7}$ |
| $\widetilde{H}_0$, GPa | $\frac{0.63}{1.42}$ | $\frac{0.63}{0.81}$ | $\frac{0.42}{0.4}$ | $\frac{0.42}{0.61}$ |
| $-n_0$ | $\frac{0.3}{0.08}$ | $\frac{0.28}{0.16}$ | $\frac{0.32}{0.24}$ | $\frac{0.3}{0.08}$ |
| $\sigma_s$, MPa | $\frac{130}{197}$ | $\frac{130}{197}$ | $\frac{130}{197}$ | $\frac{130}{197}$ |
| $\Delta$, % | $\frac{2.4}{1.7}$ | $\frac{3}{1.3}$ | $\frac{4}{1.9}$ | $\frac{6.5}{7.6}$ |

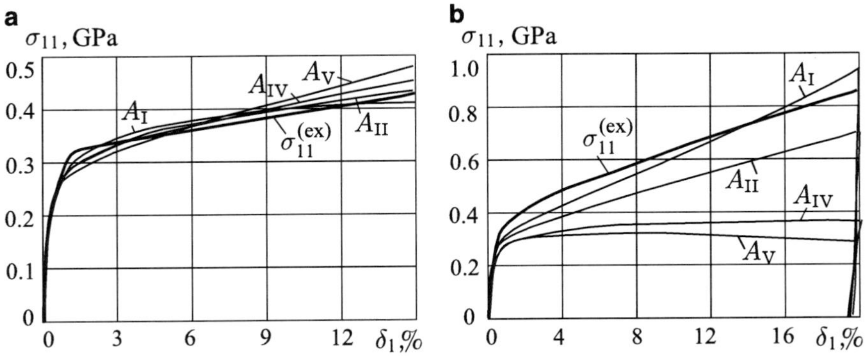


**Fig. 9.12** Diagrams of deforming for aluminium alloy AK4: in tension (**a**) and in compression (**b**)

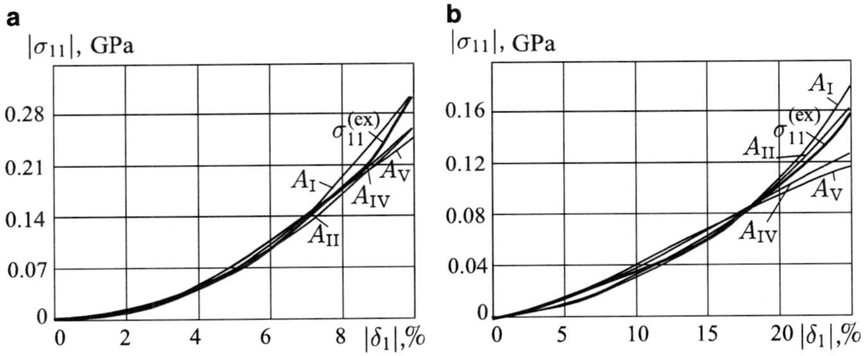


**Fig. 9.13** Diagrams of deforming in compression for sand grounds: wet sand (**a**) and dry sand (**b**)

Figure 9.12b shows experimental and computed diagrams of deforming for aluminium alloy AK4 in compression. Computations were performed by formula (9.360), and the constants $E$, $\widetilde{H}_0$, $n_0$, and $\sigma_s$ had been evaluated previously by the curve of deforming in tension. In this case the model $A_{\mathrm{I}}^{\xi}$ exhibits the best accuracy; the remaining models lead to a considerable error of approximation.

### 9.7.9 Comparison with Experimental Data for Grounds

Figures 9.13 and 9.14 show experimental diagrams of deforming in compression for sand grounds (for dry and wet sands), and also their approximations with the help of relationship (9.361). These diagrams differ from the corresponding compression

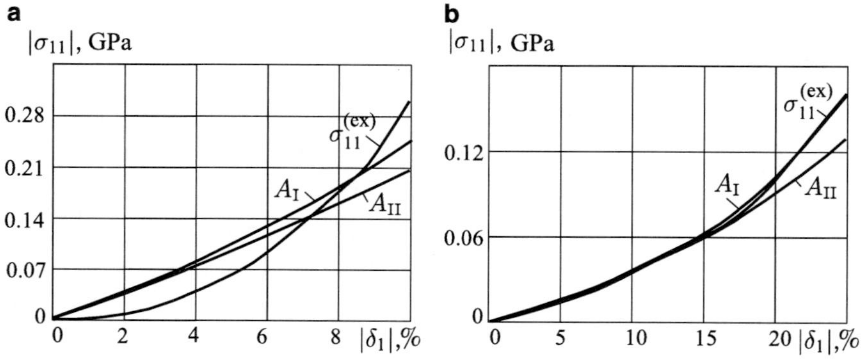


**Fig. 9.14** Approximation of diagrams of deforming for sand grounds in compression by the models with linear strengthening: wet sand (**a**) and dry sand (**b**)

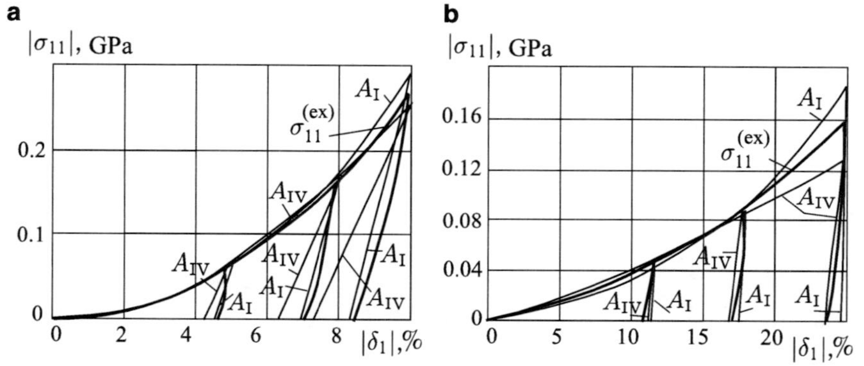


**Fig. 9.15** Diagrams of deforming under loading and subsequent unloading (experimental and computed by the models $A_{\mathrm{I}}$ and $A_{\mathrm{IV}}$ for sand grounds: wet sand (**a**) and dry sand (**b**)

diagrams for alloys (see Fig. 9.12b) by the presence of an intense convexity upwards (in absolute values of coordinates); and under unloading the deformation curve goes sharply downwards (Fig. 9.15).

For damp sands the accuracy of approximation with the help of all models $A_n^{\zeta}$ is sufficiently high, for dry sands the accuracy is lower. Models $A_{\mathrm{II}}^{\zeta}$ and $A_{\mathrm{IV}}^{\zeta}$ show the best approximation results. Unlike metallic alloys, for grounds the constant $n_0$ has positive values, and the ratio of the yield strength to the maximum magnitude of Cauchy stresses $\alpha_s = \sigma_s/\sigma_{max}$ is considerably smaller than for steels: $\alpha_s \approx 0.001$ and $\alpha_s \approx 0.1$, respectively.

A magnitude of $\sigma_s$ for grounds is usually such small that the initial elastic stage of deforming is not visible on the diagram of deforming (Figs. 9.13 and 9.15), although values of elastic modulus $E$ under loading and unloading in experiments prove to be close (in the considered models $A_n^{\zeta}$ they are coincident). Table 9.4 gives values of the constants $E$, $\widetilde{H}_0$, $n_0$ and $\sigma_s$ for the considered types of grounds.

**Table 9.4** Values of the constants $E$, $\widetilde{H}_0$, $n_0$, $\sigma_s$, and $\Delta$ in models $A_n^\xi$ for different grounds

| Dry sand | | | | | Wet sand | | | | |
|---|---|---|---|---|---|---|---|---|---|
| $n$ | I | II | IV | V | $n$ | I | II | IV | V |
| $E$, GPa | 10 | 10 | 10 | 10 | $E$, GPa | 10 | 10 | 10 | 10 |
| $\widetilde{H}_0$, GPa | 0.324 | 0.56 | 1.6 | 3.1 | $\widetilde{H}_0$, GPa | 42 | 45 | 260 | 600 |
| $n_0$ | 0.09 | 0.15 | 0.27 | 0.34 | $n_0$ | 0.6 | 0.6 | 0.8 | 0.9 |
| $\sigma_s$, MPa | 0.23 | 0.23 | 0.23 | 0.23 | $\sigma_s$, MPa | 0.29 | 0.29 | 0.29 | 0.29 |
| $\Delta$, % | 21.4 | 17.2 | 12.2 | 14 | $\Delta$, % | 55 | 56 | 55 | 56 |

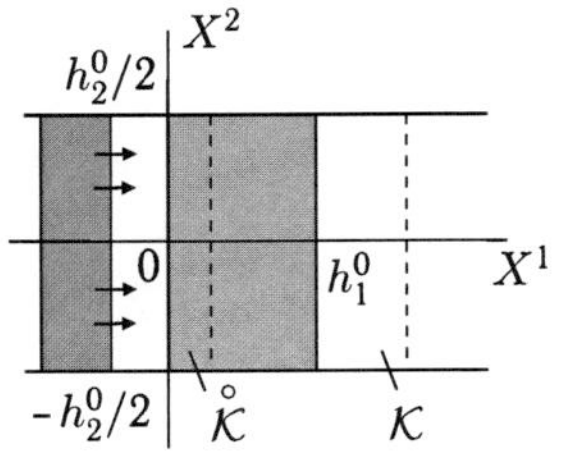


**Fig. 9.16** Propagation of a plane wave in the plate

Figure 9.14 shows the results of approximation of experimental diagrams of deforming for sand grounds with the help of the models of plastic continua with linear strengthening (see Sect. 9.1.8) when we assume a priori that $n_0 = 0$. The constant $H_0$ for these models can be determined by minimizing the mean-square deviation (9.405). By this method, we have get the following values: for wet sands, $H_0 = 2$ GPa for the models $A_{\mathrm{I}}^\xi$ and $A_{\mathrm{II}}^\xi$; and for dry sands, $H_0 = 0.24$ GPa for the model $A_{\mathrm{I}}^\xi$ and $H_0 = 0.3$ GPa for the model $A_{\mathrm{II}}^\xi$. The quality of approximation to the experimental diagrams of deforming by the models with linear strengthening is worse than that obtained by the models with power strengthening (9.86) especially for wet sands. However, in some cases this model proves to be more convenient for solving special problems (see Sect. 9.8).

## 9.8 Plane Waves in Plastic Continua

### *9.8.1 Formulation of the Problem*

Let us investigate now the dynamical problem of plasticity theory $A_n^\xi$ (9.288)–(9.290) in the material description and consider the problem on a *plane wave* in a plate, which is caused by high-speed (not quasistatic) loading on one of its surfaces $X^1 = 0$ (Fig. 9.16).

The end surface $X^1 = h_1^0$ of the plate is assumed to be free of loads; and on the lateral surfaces $X^\alpha = \pm h_\alpha^0/2$, $\alpha = 2, 3$, the condition of free slip (the symmetry condition) is given. Thus, boundary conditions for this problem have the form

$$
\begin{aligned}
X^1 = 0: &\quad P_{11} = -\frac{p_e(t)}{J}(F^{-1})_{11}, \quad P_{12} = P_{13} = 0,\\
X^1 = h_1^0: &\quad P_{\alpha 1} = 0, \quad \alpha = 1,\ 2,\ 3; \\
X^\alpha = \pm h_\alpha^0/2: &\quad x^\alpha = X^\alpha, \quad P_{\alpha 1} = 0, \quad \alpha = 2,\ 3.
\end{aligned} \tag{9.367}
$$

Here $P_{ij}$ are Cartesian components of the Piola–Kirchhoff tensor, and $(F^{-1})_{11}$ – of the inverse deformation gradient $\mathbf{F}^{-1} = (F^{-1})_{ij}\bar{\mathbf{e}}^i \otimes \bar{\mathbf{e}}^j$. Such boundary conditions approximately simulate the process of impact of a massive rigid slab on the investigated plate of plastic material slipping along a rectangular rigid pipe.

### 9.8.2 The Motion Law and Deformation of a Plate

A law of the motion of the plate is sought in the form

$$
\begin{cases}
x^1 = x^1(X^1, t),\\
x^2 = X^2,\\
x^3 = X^3,
\end{cases} \tag{9.368}
$$

where $x^1(X^1,\ 0) = X^1$.

The velocity $\mathbf{v}$ has only one non-zero component:

$$
\mathbf{v} = \frac{\partial x^i}{\partial t}\bar{\mathbf{e}}_i = v^1\bar{\mathbf{e}}_1, \qquad v^1 = \partial x^1/\partial t.
$$

The deformation gradient $\mathbf{F}$ has the form

$$
\mathbf{F} = \frac{\partial x^i}{\partial X^j}\bar{\mathbf{e}}_i \otimes \bar{\mathbf{e}}^j = \sum_{\alpha=1}^{3} k_\alpha \bar{\mathbf{e}}_\alpha^2, \tag{9.369}
$$

$$
k_1 = k_1(X^1,\ t) = \partial x^1/\partial X^1, \qquad k^\alpha = \partial x^\alpha/\partial X^\alpha = 1, \quad \alpha = 2,\ 3,
$$

and, as distinct from quasistatic tension of a beam (see (9.328)), here the elongation ratio $k_1(X^1, t)$ depends on $X^1$ linearly.

The energetic deformation tensors $\overset{(n)}{\mathbf{C}}$ also contain the only non-zero component:

$$
\overset{(n)}{\mathbf{C}} = \overset{(n)}{C}_{11}\bar{\mathbf{e}}_1^2, \quad \overset{(n)}{C}_{11} = \frac{1}{n - \mathrm{III}}(k_1^{n-\mathrm{III}} - 1), \quad \overset{(n)}{C}_{22} = \overset{(n)}{C}_{33} = 0, \tag{9.370}
$$

and changing density is determined by function $k_1$:

$$
J = \rho/\overset{\circ}{\rho} = \det \mathbf{F}^{-1} = 1/k_1. \tag{9.371}
$$

Tensors $\overset{(n)}{\mathbf{C}}_e$ and $\overset{(n)}{\mathbf{C}}_p$ are sought in the diagonal form

$$\overset{(n)}{\mathbf{C}}_e = \sum_{\alpha=1}^{3} \overset{(n)}{C}{}^e_{\alpha\alpha} \bar{\mathbf{e}}_\alpha^2, \quad \overset{(n)}{\mathbf{C}}_p = \sum_{\alpha=1}^{3} \overset{(n)}{C}{}^p_{\alpha\alpha} \bar{\mathbf{e}}_\alpha^2, \tag{9.372}$$

$$\overset{(n)}{C}{}^e_{11} + \overset{(n)}{C}{}^p_{11} = \overset{(n)}{C}_{11}, \quad \overset{(n)}{C}{}^e_{22} + \overset{(n)}{C}{}^p_{22} = 0, \quad \overset{(n)}{C}{}^e_{33} + \overset{(n)}{C}{}^p_{33} = 0,$$

where all the components $\overset{(n)}{C}{}^e_{\alpha\alpha}$ and $\overset{(n)}{C}{}^p_{\alpha\alpha}$ depend only on $X^1$ and $t$.

From the condition of plastic incompressibility we get

$$\overset{(n)}{C}{}^p_{22} = \overset{(n)}{C}{}^p_{33} = -\frac{1}{2}\overset{(n)}{C}{}^p_{11}, \tag{9.373}$$

$$\overset{(n)}{C}{}^e_{22} = \overset{(n)}{C}{}^e_{33} = \frac{1}{2}\overset{(n)}{C}{}^p_{11}, \tag{9.374}$$

i.e. transverse deformations $\overset{(n)}{C}{}^e_{22}$, $\overset{(n)}{C}{}^e_{33}$ are different from zero only in a plastic domain.

### 9.8.3 Stresses in the Plate

According to (9.335), the stress tensor $\overset{(n)}{\mathbf{T}}$ in this problem has the diagonal form

$$\overset{(n)}{\mathbf{T}} = \sum_{\alpha=1}^{3} \overset{(n)}{T}_{\alpha\alpha} \bar{\mathbf{e}}_\alpha^2, \tag{9.375}$$

where

$$\overset{(n)}{T}_{11} = J(l_1 + 2l_2)\overset{(n)}{C}_{11} - 2Jl_2\overset{(n)}{C}{}^p_{11}, \tag{9.376}$$

$$\overset{(n)}{T}_{22} = \overset{(n)}{T}_{33} = J(l_1\overset{(n)}{C}_{11} + l_2\overset{(n)}{C}{}^p_{11}); \tag{9.377}$$

all the values depend only on $X^1$ and $t$.

According to (9.375) and (9.369), the tensors $\mathbf{T}$ and $\mathbf{P}$ also have a diagonal form

$$\mathbf{T} = \sum_{\alpha=1}^{3} \sigma_{\alpha\alpha} \bar{\mathbf{e}}_\alpha^2, \quad \mathbf{P} = \sum_{\alpha=1}^{3} P_{\alpha\alpha} \bar{\mathbf{e}}_\alpha^2 = \frac{1}{J}\mathbf{F}^{-1} \cdot \mathbf{T}, \tag{9.378}$$

$$\sigma_{11} = k_1^{n-\mathrm{III}} \overset{(n)}{T}_{11}, \quad \sigma_{22} = \sigma_{33} = \overset{(n)}{T}_{22} = \overset{(n)}{T}_{33},$$

$$P_{11} = \sigma_{11}, \quad P_{22} = P_{33} = \frac{1}{k_1}\sigma_{22}.$$

### 9.8.4 The System of Dynamic Equations for the Plane Problem

Since

$$\overset{\circ}{\nabla}\cdot\mathbf{P}=\sum_{\alpha=1}^{3}\frac{\partial P_{\alpha\alpha}}{\partial X^{\alpha}}\bar{\mathbf{e}}_{\alpha}=\frac{\partial P_{11}}{\partial X^{1}}\bar{\mathbf{e}}_{1},$$

$$\overset{\circ}{\nabla}\otimes\mathbf{v}^{\mathrm{T}}=\frac{\partial v^{1}}{\partial X^{1}}\bar{\mathbf{e}}_{1}^{2},\quad \mathbf{F}^{-1\mathrm{T}}\cdot\overset{\circ}{\nabla}\otimes\mathbf{v}=\frac{1}{k_{1}}\frac{\partial v^{1}}{\partial X^{1}}\bar{\mathbf{e}}_{1}^{2}, \tag{9.379}$$

$${}^{4}\overset{(\mathrm{n})}{\mathbf{X}}\cdot\cdot\,\mathbf{F}^{-1\mathrm{T}}\cdot\overset{\circ}{\nabla}\otimes\mathbf{v}=k_{1}^{n-\mathrm{III}-1}\frac{\partial v^{1}}{\partial X^{1}}\bar{\mathbf{e}}_{1}^{2}$$

(here we have used the result of Exercise 4.2.16 for tensors $\overset{(\mathrm{n})}{\mathbf{X}}$), so the first, third, fourth and fifth equations of the system (9.288) in the considered problem take the forms

$$\overset{\circ}{\rho}\frac{\partial v^{1}}{\partial t}=\frac{\partial\sigma_{11}}{\partial X^{1}},\quad \frac{\partial x^{1}}{\partial t}=v^{1},\quad \frac{\partial k_{1}}{\partial t}=\frac{\partial v^{1}}{\partial X^{1}},$$

$$\frac{\partial}{\partial t}\left(\frac{\overset{(\mathrm{n})}{T}_{11}}{\rho}\right)=\frac{l_{1}+2l_{2}}{\overset{\circ}{\rho}}k_{1}^{n-\mathrm{III}-1}\frac{\partial v^{1}}{\partial X^{1}}-\frac{2l_{2}}{\overset{\circ}{\rho}}\frac{\partial\overset{(\mathrm{n})}{C}{}^{p}_{11}}{\partial t}. \tag{9.380}$$

Notice that the last equations of the system (9.380) can be obtained by immediate differentiation of relation (9.376).

It should also be noted that the fifth equation of the system (9.288) gives two more scalar expressions for $\partial\overset{(\mathrm{n})}{T}_{22}/\partial t$ and $\partial\overset{(\mathrm{n})}{T}_{33}/\partial t$; however, these are just expressions but not equations, and therefore they do not appear in the general system (9.380).

Taking the expression (9.371) for $\rho$ and expression (9.378) for $\sigma_{11}$ into account, we get

$$\frac{\partial}{\partial t}\left(\frac{\overset{(\mathrm{n})}{T}_{11}}{\rho}\right)=\frac{k_{1}^{\mathrm{III}-n+1}}{\overset{\circ}{\rho}}\frac{\partial\sigma_{11}}{\partial t}+\frac{\mathrm{III}-n+1}{\overset{\circ}{\rho}}\sigma_{11}k_{1}^{\mathrm{III}-n}\frac{\partial k_{1}}{\partial t}. \tag{9.381}$$

The last equation of the system (9.380) after substitution of the expression (9.381) takes the form

$$\frac{\partial\sigma_{11}}{\partial t}=\left((l_{1}+2l_{2})k_{1}^{2(n-\mathrm{III}-1)}-\frac{(\mathrm{III}-n+1)\sigma_{11}}{k_{1}}\right)\frac{\partial v^{1}}{\partial X^{1}}-2l_{2}k_{1}^{n-\mathrm{III}-1}\frac{\partial\overset{(\mathrm{n})}{C}{}^{p}_{11}}{\partial t}. \tag{9.382}$$

Just as in other one-dimensional problems considered above, to evaluate $\partial \overset{(n)}{C}{}^{p}_{11}/\partial t$ we use the equation of a yield surface $f = 0$ (see (9.96)), which, according to the results of Exercise 9.1.1, in this problem has the form

$$\frac{1}{3\sigma_s^2}(\overset{(n)}{T}_{11} - \overset{(n)}{T}_{22} - H(\overset{(n)}{C}{}^{p}_{11} - \overset{(n)}{C}{}^{p}_{22}))^2 = 1. \tag{9.383}$$

Substituting expressions (9.377) for $\overset{(n)}{T}_{22}$, (9.373) for $\overset{(n)}{C}{}^{p}_{22}$, (9.370) for $\overset{(n)}{C}_{11}$ and (9.350) for $H$ into (9.383), we obtain

$$|\overset{(n)}{T}_{11} - \overset{(n)}{\widetilde{T}}_{22} - \widetilde{H}\overset{(n)}{C}{}^{p}_{11}| = \sqrt{3}\,\sigma_s, \tag{9.384}$$

where

$$\widetilde{H} = \frac{3}{2}H + \frac{l_2}{k_1}, \quad \overset{(n)}{\widetilde{T}}_{22} = \frac{l_1}{k_1}\overset{(n)}{C}_{11} = \frac{l_1(k_1^{n-\mathrm{III}} - 1)}{(n - \mathrm{III})k_1}. \tag{9.385}$$

For the model of plasticity with linear strengthening when $n_0 = 0$ and $H = H_0$, we have $\widetilde{H} = \frac{3}{2}H_0 + \frac{l_2}{k_1}$, and Eq. (9.384) has the following analytical solution for $\overset{(n)}{C}{}^{p}_{11}$:

$$\overset{(n)}{C}{}^{p}_{11} = \frac{\overset{(n)}{T}_{11} - \overset{(n)}{\widetilde{T}}_{22}}{\widetilde{H}}\widetilde{h}_{+} - \frac{\sqrt{3}\,\sigma_s}{\widetilde{H}}\widetilde{h}_{-} + \overset{(n)}{C}{}^{p*}_{11}(1 - \widetilde{h}_{+}). \tag{9.386}$$

Here $h_{+}$ and $h_{-}$ are the Heaviside functions:

$$\widetilde{h}_{+} \equiv h_{+} + h_{-}, \quad \widetilde{h}_{-} = h_{+} - h_{-},$$

$$h_{+} = \begin{cases} 1, & Y_{1H} \geqslant \sqrt{3}\,\sigma_s, \\ 0, & Y_{1H} < \sqrt{3}\,\sigma_s, \end{cases} \quad h_{-} = \begin{cases} 1, & Y_{1H} < -\sqrt{3}\,\sigma_s, \\ 0, & Y_{1H} > -\sqrt{3}\,\sigma_s, \end{cases} \tag{9.387}$$

$$Y_{1H} = \overset{(n)}{T}_{11} - \widetilde{T}_{22} - \widetilde{H}\overset{(n)}{C}{}^{p}_{11},$$

and $\overset{(n)}{C}{}^{p*}_{11}$ is the value of plastic deformation reached at the time $t^*$ of leaving the yield surface.

Substituting the expressions (9.378) and (9.385) for $\overset{(n)}{T}_{11}$ and $\overset{(n)}{\widetilde{T}}_{22}$ into (9.386) and then differentiating (9.386) with respect to $t$ within differentiability sections, we obtain

$$\frac{\partial \overset{(n)}{C}{}^{p}_{11}}{\partial t} = b_0\widetilde{h}_{+}\frac{\partial \sigma_{11}}{\partial t} + ((b_1\sigma_{11} - b_2)\widetilde{h}_{+} - \sqrt{3}\,\sigma_s b_3\widetilde{h}_{-})\frac{\partial k_1}{\partial t}. \tag{9.388}$$

Here we have introduced the following functions of $k_1$:

$$b_0 = \frac{2k_1^{\mathrm{III}-n+1}}{3H_0k_1+2l_2}, \quad b_1 = \frac{2k_1^{\mathrm{III}-n}}{3H_0k_1+2l_2}\left(\mathrm{III}-n+\frac{2l_2}{3H_0k_1+2l_2}\right), \tag{9.389}$$
$$b_3 = \frac{4l_2}{(3H_0k_1+2l_2)^2}, \quad b_2 = \frac{2l_1}{(3H_0k_1+2l_2)k_1}\left(k_1^{n-\mathrm{III}} - \frac{3H_0k_1(k_1^{n-\mathrm{III}}-1)}{(3H_0k_1+2l_2)(n-\mathrm{III})}\right).$$

Substituting the expression (9.388) into (9.382) and grouping like terms, we find the equation

$$\frac{\partial \sigma_{11}}{\partial t} = (c_1 - \sigma_{11}c_2)\frac{\partial v^1}{\partial x^1}. \tag{9.390}$$

Here we have introduced the following notation for functions of $k_1$:

$$c_0 = 1 + \frac{4l_2\widetilde{h}_+}{3H_0k_1+2l_2},$$
$$c_1 = \frac{l_1+2l_2}{c_0}k_1^{2(n-\mathrm{III}-1)} + \frac{2l_2}{c_0}k_1^{n-\mathrm{III}-1}(b_2\widetilde{h}_+ + \sqrt{3}\,\sigma_s b_3\widetilde{h}_-),$$
$$c_2 = \frac{1}{c_0k_1}\left(1+\mathrm{III}-n+\frac{4l_2\widetilde{h}_+}{3H_0k_1+2l_2}\left(\mathrm{III}-n+\frac{2l_2}{3H_0k_1+2l_2}\right)\right). \tag{9.391}$$

Notice that for most plastic materials occurring in practice the following condition is satisfied:

$$\sigma_{11}c_2/c_1 \ll 1. \tag{9.392}$$

Therefore, the term $\sigma_{11}c_2$ in Eq. (9.390) can be neglected; and we may consider the simpler equation

$$\frac{\partial \sigma_{11}}{\partial t} = c_1^2\frac{\partial v^1}{\partial X^1}. \tag{9.393}$$

### 9.8.5 *The Statement of Problem on Plane Waves in Plastic Continua*

Substituting the relation (9.393) in place of the last equation in the set (9.380) and excluding the second equation from (9.380) (because coefficients in the equation set do not depend explicitly on $X^1$), we finally obtain the system

$$\begin{cases} \overset{\circ}{\rho}(\partial v/\partial t) = \partial T/\partial X, \\ \partial k/\partial t = \partial v/\partial X, \\ \partial T/\partial t = c_1(\partial v/\partial X), \end{cases} \tag{9.394}$$

which consists of three equations of the first order for the three functions $v, k, T \parallel X, t$, given in the domain $0 < X < h_1^0$, $0 < t < t_{\max}$. Here we have introduced the notation: $T \equiv \sigma_{11}$, $X \equiv X^1$, $v \equiv v^1$, $k \equiv k_1$.

This system should be complemented by boundary conditions (9.367), which, according to (9.369) and (9.378), have the form

$$X = 0: \quad T = -p_e(t), \tag{9.395a}$$

$$X = h_1^0: \quad T = 0, \tag{9.395b}$$

and also by the initial conditions

$$t = 0: \quad v = 0, \; k = 1, \; T = 0, \quad 0 < X < h_1^0. \tag{9.396}$$

As a result, we obtain a *statement of the dynamic problem on propagation of plane waves in plastic continua.*

### *9.8.6 Solving the Problem by the Characteristic Method*

To solve the problem stated we can apply the characteristic method, which is widely used for solving systems of partial differential equations of the hyperbolic type. Consider only the case of active loading when $dp_e/dt \geqslant 0$.

According to the characteristic method, consider differentials of the unknown functions

$$dv = \frac{\partial v}{\partial t}\, dt + \frac{\partial v}{\partial X}\, dX, \quad dk = \frac{\partial k}{\partial t}\, dt + \frac{\partial k}{\partial X}\, dX, \quad dT = \frac{\partial v}{\partial t}\, dt + \frac{\partial v}{\partial X}\, dX. \tag{9.397}$$

The system (9.394) and (9.397) consists of six equations being linear with respect to six unknown functions: $v_t = \partial v/\partial t$, $v_X = \partial v/\partial X$, $T_t = \partial T/\partial t$, $T_X = \partial T/\partial X$, $k_t = \partial k/\partial t$ and $k_X = \partial k/\partial X$. The system can be rewritten in the matrix form

$$\begin{pmatrix} \overset{\circ}{\rho} & 0 & 0 & -1 & 0 & 0 \\ 0 & -1 & 0 & 0 & 1 & 0 \\ 0 & -c_1 & 1 & 0 & 0 & 0 \\ dt & dX & 0 & 0 & 0 & 0 \\ 0 & 0 & 0 & 0 & dt & dX \\ 0 & 0 & dt & dX & 0 & 0 \end{pmatrix} \begin{pmatrix} v_t \\ v_X \\ T_t \\ T_X \\ k_t \\ k_X \end{pmatrix} = \begin{pmatrix} 0 \\ 0 \\ 0 \\ dv \\ dk \\ dT \end{pmatrix}. \tag{9.398}$$

There exists an unique solution of this system if and only if its determinant is different from zero. However, on the plane $(X, t)$ there are curves (called characteristics) where the solution uniqueness is violated. This case is realized when the determinant of system (9.398) vanishes:

$$\det(\,) = -\overset{\circ}{\rho}\, dX^3 + c_1\, dt^2\, dX = 0. \tag{9.399}$$

From (9.399) we find equations of two families of characteristics:

$$dX = \pm a\,dt, \tag{9.400}$$

where

$$a^2(k) \equiv c_1(k)/\overset{\circ}{\rho} \tag{9.401}$$

is the speed of sound in a plastic material considered.

Substitution of (9.400) into the first and second equations of system (9.394) yields conditions over the characteristics

$$\overset{\circ}{\rho}\,dv = dT\frac{dt}{dX} = \pm\frac{dT}{a}, \quad dk = dv\frac{dt}{dX} = \pm\frac{dv}{a},$$

or

$$\overset{\circ}{\rho}a\,dv \mp dT = 0, \quad a\,dk \mp dv = 0. \tag{9.402}$$

Introducing the functions

$$\xi(k) = \int_1^k c_1(k')\,dk', \quad \varphi(k) = \int_1^k a(k')\,dk', \tag{9.403}$$

for which $d\xi = c_1\,dk$, $d\varphi = a\,dk$, we can integrate Eq. (9.402):

$$\xi(k) - T = \text{const}, \qquad \varphi(k) \mp v = \text{const}. \tag{9.404}$$

Thus, there are two families of characteristics and two conditions over each of the characteristics:

$$\begin{cases} dX - a(k)\,dt = 0, \\ v - \varphi(k) = \varphi^0_+ = \text{const}, \\ T - \xi(k) = \xi^0_+ = \text{const}, \end{cases} \tag{9.405}$$

$$\begin{cases} dX + a(k)\,dt = 0, \\ v + \varphi(k) = \varphi^0_- = \text{const}, \\ T - \xi(k) = \xi^0_- = \text{const}, \end{cases} \tag{9.406}$$

Consider a point $\mathcal{M}$ belonging to the plane $(X, t)$. For this point there always exist two characteristics, to one of which a tangent is positive $(dt/dX = 1/a > 0)$ and to the other – negative $(dt/dX = -1/a < 0)$.

Choose such a point $\mathcal{M}$, whose both characteristics intersect the axis $OX$ (Fig. 9.17) at some points $P$ and $Q$. Since the points $P$ and $Q$ belong to the domain where initial conditions (9.396) are given, at these points we have $v = 0$, $k = 1$ and $T = 0$. Then the conditions (9.405) and (9.406) yield that along the

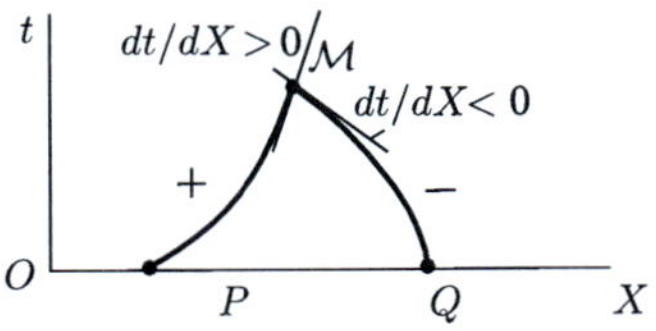

**Fig. 9.17** Characteristics at point $\mathcal{M}$

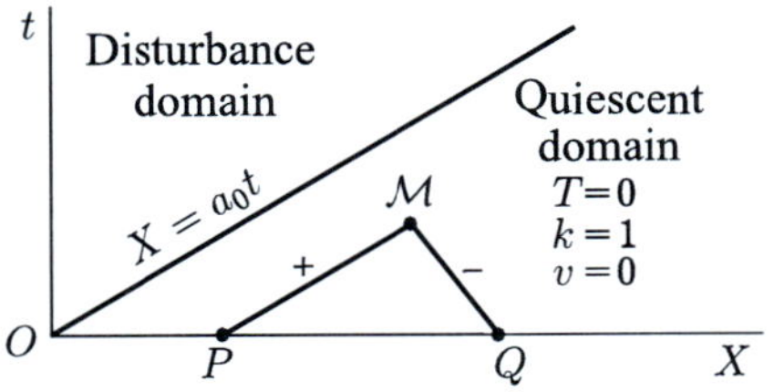

**Fig. 9.18** Characteristics at point $\mathcal{M}$ in the quiescent domain

characteristics $P\mathcal{M}$ and $\mathcal{M}Q$ the constants are zero: $\varphi^0_\pm = 0$ and $\xi^0_\pm = 0$ (as $\xi(1) = 0$ and $\varphi(1) = 0$). Hence, at the point $\mathcal{M}$, where the characteristics intersect, the following conditions are satisfied simultaneously:

$$v = \varphi(k) \quad \text{and} \quad v = -\varphi(k), \tag{9.407}$$

that is possible only if $k = 1$ and $v = 0$. Therefore, at the point $\mathcal{M}$: $T = \xi(1) = 0$.

Thus, at all points $\mathcal{M}$, whose both characteristics intersect the axis $OX$, there are no disturbances under any loading, i.e. this is a quiescent domain.

Moreover, since in a quiescent domain $k = 1$, we have $a(k) = a(1) = a_0 =$ const. Here $a_0$ is the speed of sound in a quiescent continuum considered, for which, according to (9.401) and (9.391), we have

$$a_0 = c_1(1)/\overset{\circ}{\rho} = (l_1 + 2l_2)/\overset{\circ}{\rho}, \tag{9.408}$$

because in the quiescent domain there are no plastic deformations, and $\widetilde{h}_+ = \widetilde{h}_- = 0$, $Y_{1H} = 0$.

Hence, all characteristics $\mathcal{M}P$ and $\mathcal{M}Q$ intersecting the axis $OX$ are straight lines determined by equations (Fig. 9.18)

$$P\mathcal{M}: \quad t - \frac{X}{a_0} = c_+, \qquad \mathcal{M}Q: \quad t + \frac{X}{a_0} = c_-, \tag{9.409}$$

where $c_\pm$ are constants being individual for each of the characteristics.

The characteristic containing the point $O$ being the origin of coordinates ($t = 0, X = 0$) separates the quiescent domain and the disturbance domain and is called the *front wave*. From (9.409) we obtain the equation of the front wave: $X = a_0 t$.

Let us choose now a point $\mathcal{M}$ belonging to the disturbance domain and consider its characteristics. According to the construction, the '+'-characteristic intersects the axis $Ot$ at point $P$, and the '−'-characteristic intersects the front wave at point

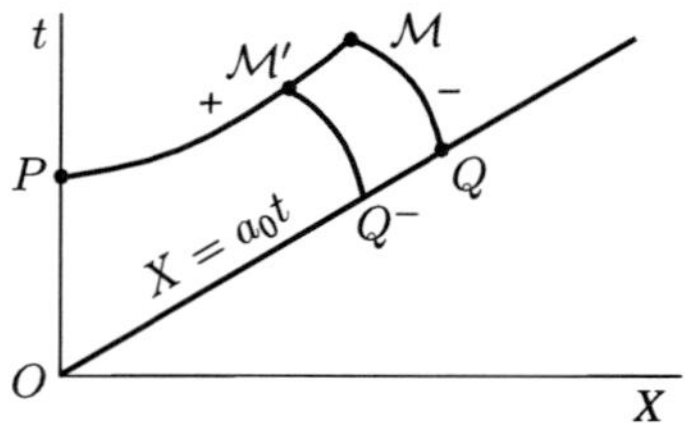


**Fig. 9.19** Characteristics at point $\mathcal{M}$ in the disturbance domain

$Q$ (Fig. 9.19). Since the point $Q$ occurs in the quiescent domain, at this point we have $v = 0$, $k = 1$, $T = 0$; and hence formulae (9.406) yield that over the '−'-characteristic $\varphi^0_- = 0$ and $\xi^0_- = 0$. Therefore, for the point $\mathcal{M}$ considered, from (9.405) and (9.406) we obtain the following relations along the characteristics $\mathcal{M}P$ and $\mathcal{M}Q$:

$$\mathcal{M}P: \quad v = \varphi(k) + \varphi^0_+, \quad T = \xi(k) + \xi^0_+, \tag{9.410a}$$

$$\mathcal{M}Q: \quad v = -\varphi(k), \quad T = \xi(k). \tag{9.410b}$$

From these equations we find relations at the point $\mathcal{M}$:

$$v = \varphi^0_+/2 = \text{const}, \qquad \xi^0_+ = 0,$$
$$k = k^0 = \text{const}, \qquad T = \xi(k^0) = T^0 = \text{const}. \tag{9.411}$$

If we choose an arbitrary point $\mathcal{M}'$ belonging to the '+'-characteristic $\mathcal{M}P$ and construct its '−'-characteristic $\mathcal{M}'Q'$, then for the point $\mathcal{M}'$ we obtain the same Eqs. (9.410) and (9.411) with the same constants $\varphi^0_+$ and $k^0$, because the value $\varphi^0_+$ is the same along the whole characteristic $P\mathcal{M}$. Hence, although values of the functions $v$, $k$ and $T$ are different from the corresponding quiescent values, they remain constant along the whole '+'-characteristic. This means that the speed of sound $a(k) = a(k^0) = \text{const}$ is also constant, and the characteristic $\mathcal{M}P$ is a straight line determined by the equation $t - (X/a(k^0)) = C$, $C = \text{const}$. Since $k^0$ and $T^0$ are values of the functions $k(X,t)$ and $T(X,t)$ along the characteristic $\mathcal{M}P$, for different values of $X$ and $t$ satisfying the equation $t - (X/a(k^0)) = C$ the following relations hold

$$\begin{aligned} &k^0 = k(C) = k\left(t - \frac{X}{a(k^0)}\right) = \text{const}, \\ \mathcal{M}P: \quad &T^0 = \xi(k^0) = \xi(k(C)) = \xi\left(k\left(t - \frac{X}{a(k^0)}\right)\right) = \text{const}. \end{aligned} \tag{9.412}$$

If these relations are applied to the point $P$, where $X = 0$ and where besides equations (9.411) the boundary condition (9.395a) must be satisfied, then from (9.412) and (9.395a) we get

$$P: \quad \xi(k(t)) = -p_e(t). \tag{9.413}$$

From this equation we find the function $k(t)$:

$$k(t) = \xi^{-1}(-p_e(t)), \tag{9.414}$$

assuming that there exists an inverse function $k = \xi^{-1}(T^0)$.

On substituting (9.414) into (9.412) and (9.410b), we find values of $k$ and $T$ at point $\mathcal{M}$ with coordinates $X$ and $t$:

$$\begin{aligned} k^0 &= k(X,t) = k(t - \frac{X}{a^0}) = \xi^{-1}\left(-p_e\left(t - \frac{X}{a^0}\right)\right), \\ T^0 &= T(X,t) = \xi\left(k(t - \frac{X}{a^0})\right) = -p_e\left(t - \frac{X}{a^0}\right), \\ v(X,t) &= -\varphi(k(X,t)) = -\varphi\left(\xi^{-1}\left(-p_e\left(t - \frac{X}{a^0}\right)\right)\right). \end{aligned} \tag{9.415}$$

Since the point $\mathcal{M}$ was arbitrary, the system (9.415) is a desired solution in the disturbance domain at those values of $t$ when the argument of functions (9.415) is nonnegative, i.e. when $t > X/a$. At times $t \leqslant X/a$ the disturbance does not reach the point $\mathcal{M}$ considered, and instead of (9.415) we have a solution corresponding to the quiescent state:

$$k = 1, \quad T = 0, \quad v = 0, \quad t < X/a. \tag{9.416}$$

The solution (9.415) and (9.416) holds only up to those values of $t$ until the front wave reaches the end surface $X = h_1^0$ of the plate, i.e. when $t < h_1^0/a_0$.

### 9.8.7 Comparative Analysis of the Solution for Different Models $A_n^\xi$

Let us analyze the solution (9.415) obtained. Under active loading when $dp_e/dt \geqslant 0$, we have $\overset{(n)}{C}{}^{p*}_{11} = 0$. Then for $t > X/a$ from (9.415) it follows that $d\sigma_{11}/dt = dT/dt \leqslant 0$; therefore, loading of the plate is its compression. Hence, relations (9.386), (9.387), and (9.391) in an elastic domain become

$$\begin{gathered} h_+ = 0, \quad h_- = 0, \quad \widetilde{h}_+ = \widetilde{h}_- = 0, \\ \overset{(n)}{C}{}^{p}_{11} = 0, \quad c_0 = 1, \quad c_1 = (l_1 + 2l_2)k_1^{2(n-\text{III}-1)}, \\ a = a_0 k^{n-\text{III}-1}, \quad a_0 = \sqrt{(l_1 + 2l_2)/\overset{\circ}{\rho}}, \\ \xi(k) = \frac{l_1 + 2l_2}{2(n - \text{III} - 1}(k^{2(n-\text{III})-1} - 1), \quad \varphi(k) = \frac{a_0}{n - \text{III}}(k^{n-\text{III}} - 1), \end{gathered} \tag{9.417}$$

if $\overset{(n)}{T}_{11} - \overset{(n)}{\widetilde{T}}_{22} > -\sqrt{3}\sigma_s$.

Substituting into formula (9.384) the expression (9.385) for $\overset{(n)}{\widetilde{T}}_{22}$ and in place of $\overset{(n)}{T}_{11}$ its expressions in terms of $k$ according to (9.378), (9.404), and (9.415): $\overset{(n)}{T}_{11} = k^{\mathrm{III}-n}T = k^{\mathrm{III}-n}\xi(k)$, we find the limiting value $k = k_s < 1$, from which plasticity starts in compression:

$$\frac{k_s^{\mathrm{III}-n}}{2(n-\mathrm{III})-1}(k_s^{2(n-\mathrm{III})-1}-1) - \frac{\nu(k_s^{n-\mathrm{III}}-1)}{(1-\nu)(n-\mathrm{III})k_s} = -\frac{\sqrt{3}\,\sigma_s}{l_1+2l_2}. \tag{9.418}$$

Here, as usual, $\nu = l_1/(2(l_1+l_2))$ is the Poisson ratio.

In the domain of plastic deformations (in compression) when $k \leqslant k_s < 1$, relations (9.386), (9.387), and (9.391) become

$$h_+ = 0, \quad h_- = 1, \quad \widetilde{h}_+ = 1, \quad \widetilde{h}_- = -1,$$

$$\overset{(n)}{C}{}^p_{11} = \frac{1}{\widetilde{H}}(\overset{(n)}{T}_{11} - \overset{(n)}{\widetilde{T}}_{22} + \sqrt{3}\,\sigma_s)$$

$$= \frac{l_1+2l_2}{\widetilde{H}}\left(k^{\mathrm{III}-n}\xi(k) - \frac{\nu(k^{n-\mathrm{III}}-1)}{(1-\nu)(n-\mathrm{III})k} + \frac{\sqrt{3}\sigma_s}{l_1+2l_2}\right),$$

$$c_0 = 1 + \frac{4l_2}{3H_0k+2l_2}, \quad c_1 = (l_1+2l_2)\frac{\widetilde{c}_1(k)}{c_0(k)}, \tag{9.419}$$

$$\widetilde{c}_1(k) = k^{2(n-\mathrm{III}-1)} + \left(\frac{1-2\nu}{1-\nu}\right)k^{n-\mathrm{III}-1}\left(b_2 - \frac{\sqrt{3}\,\sigma_s b_3}{l_1+2l_2}\right),$$

$$a = a_0\sqrt{\widetilde{c}_1(k)/c_0(k)}, \quad a_0 = \sqrt{(l_1+2l_2)/\overset{\circ}{\rho}},$$

$$\xi(k) = (l_1+2l_2)\int_1^k \frac{\widetilde{c}_1(k')\,dk'}{c_0(k')}, \quad \varphi(k) = a_0\int_1^k \sqrt{\frac{\widetilde{c}_1(k')}{c_0(k')}}\,dk'.$$

Typical dependences of the dimensionless speed of sound $a/a_0$ upon $k$, determined by functions (9.419), for different models $A_n^\xi$ are shown in Fig. 9.20 (here $\nu = 0.3$, $\sigma_s/m = 1.5\cdot 10^{-3}$, $H/m = 10^{-2}$ when $m = l_1+2l_2$). As follows from these graphs, there is a distinction in kind between the models $A_{\mathrm{I}}^\xi$, $A_{\mathrm{II}}^\xi$ and $A_{\mathrm{IV}}^\xi$, $A_{\mathrm{V}}^\xi$: for the models $A_{\mathrm{I}}^\xi$ and $A_{\mathrm{II}}^\xi$ the speed of sound increases with decreasing $k$ (in compression when $k < 1$), and for the models $A_{\mathrm{IV}}^\xi$ and $A_{\mathrm{V}}^\xi$ it does not grows (for $A_{\mathrm{IV}}^\xi$ it remains constant, and for $A_{\mathrm{V}}^\xi$ – decreases). For all models $A_n^\xi$, the function $a(k)$ in the case considered has a discontinuity at $k = k_s$ due to the assumption on linear strengthening of a continuum (see formula (9.386)).

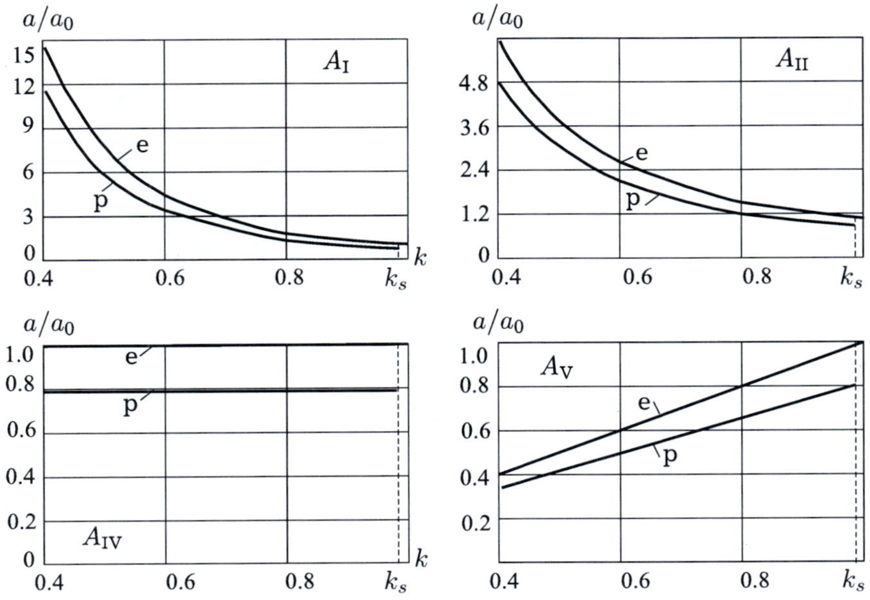


**Fig. 9.20** The speed of sound (dimensionless $a/a_0$) *versus* the ratio $k$ in compression for different models $A_n^\xi$ of elastic materials (e) and plastic materials (p)

## 9.8.8 *Plane Waves in Models* $A_{\mathrm{IV}}^\xi$ *and* $A_{\mathrm{V}}^\xi$

Since the function $\xi(k)$ is negative when $k < 1$ and monotonically diminishes within the interval of values of $k$ from 1 to 0 for all models $A_n^\xi$, the function $k(t) = \xi^{-1}(-p_e(t))$ is monotonically decreasing ($k(0) = \xi^{-1}(-p_e(0)) = 1$) (here we have taken into account that $p'(t) \geqslant 0$ by the assumption made). Thus, at the front surface $X = 0$ of the plate the function $k(t)$ monotonically decreases, and the function $a(k)$ grows for the models $A_{\mathrm{I}}^\xi$, $A_{\mathrm{II}}^\xi$ and diminishes for the models $A_{\mathrm{IV}}^\xi$, $A_{\mathrm{V}}^\xi$.

Then on the plane $(t, X)$ '+'-characteristics in the disturbance domain (as shown above, they are straight lines) have higher values of a slope from the axis $OX$ with growing $t$ for models $A_{\mathrm{IV}}^\xi$, $A_{\mathrm{V}}^\xi$ and smaller values – for models $A_{\mathrm{I}}^\xi$, $A_{\mathrm{II}}^\xi$ (Fig. 9.21). The decrease of a slope of characteristics $t = X/a(k)$ from the axis $OX$ means that characteristics with different values of $k$ may intersect at the front wave (Fig. 9.21). As a result, the solution becomes ambiguous, that is inadmissible under the assumptions made above on the absence of jumps of the functions $k$, $T$ and $v$ themselves (only jumps of their first derivatives are admissible). Thus, for the models $A_{\mathrm{I}}^\xi$ and $A_{\mathrm{II}}^\xi$, the solution obtained is inapplicable, and it will be constructed in another way (see Sect. 9.8.9.).

For the models $A_{\mathrm{IV}}^\xi$ and $A_{\mathrm{V}}^\xi$, characteristics do not intersect, and the solution (9.415) obtained actually holds. Figure 9.21 shows the graphic method of construction of the solution (9.415) with the help of given values of the function $p_e(t)$.

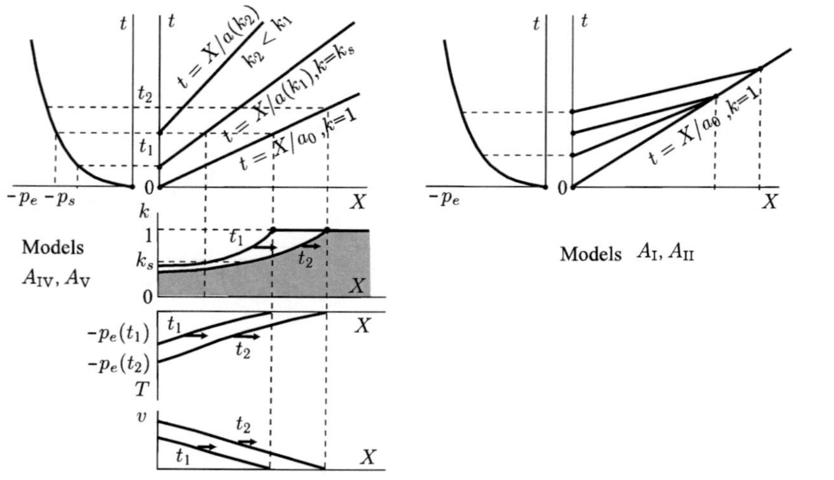


**Fig. 9.21** Graphic method of construction of the solution (9.415)

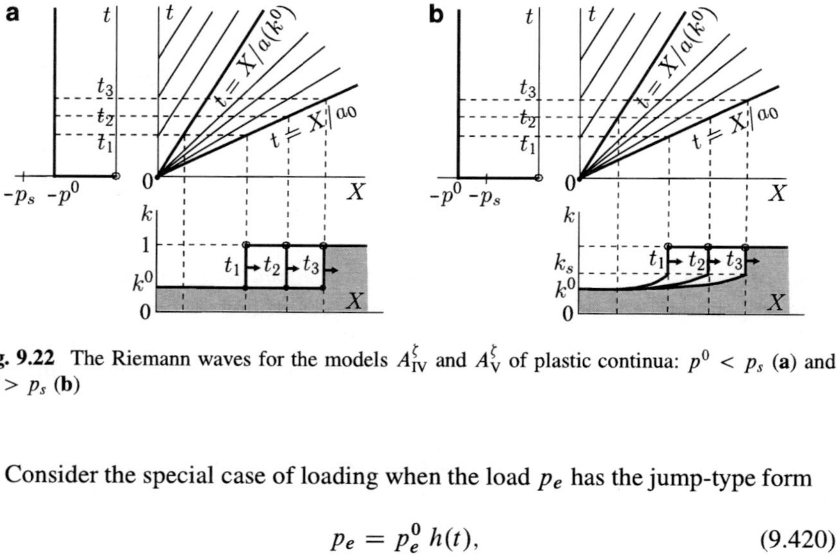


**Fig. 9.22** The Riemann waves for the models $A_{IV}^{\zeta}$ and $A_{V}^{\zeta}$ of plastic continua: $p^0 < p_s$ (**a**) and $p^0 > p_s$ (**b**)

Consider the special case of loading when the load $p_e$ has the jump-type form

$$p_e = p_e^0\, h(t), \tag{9.420}$$

where $h(t)$ is the Heaviside function. Then $T$ and $k$ at the face surface of the plate range from 0 and 1 to final values $T^0$ and $k^0 < 1$, respectively, and then remain constant for all $t > 0$. Thus, on the plane $(t, X)$ there appears an angle bounded by the characteristics $X = a_0 t$ and $X = a(k^0)t$ and filled with characteristics $X = a(k)t$ while $k^0 < k < 1$ (the fan of characteristics). Waves corresponding to these characteristics are called the *Riemann waves* by analogy with gas dynamics. These waves are characterized by the fact that they propagate without changes in amplitudes of $T$ and $v$ (Fig. 9.22).

If $p^0 < p_s$, where $p_s$ is the pressure of the beginning of plasticity in compression ($-p_s = \xi(k_s)$, here $k_s$ is determined by (9.418)), then the wave retains its shape (being a step) (Fig. 9.22). If $p^0 > p_s$, then the shape of the wave spreads: values of $k$ vary from $k^0$ up to $k_s$, but the maximum value $k_s$ remains constant.

### 9.8.9 Shock Waves in Models $A_{\mathrm{I}}^{\xi}$ and $A_{\mathrm{II}}^{\xi}$

Let us return now to the models $A_{\mathrm{I}}^{\xi}$ and $A_{\mathrm{II}}^{\xi}$ and consider only the case of stepwise loading of the plate (9.420). In this case the system (9.394) and (9.396) admits the trivial solution

$$T = -p = \text{const}, \quad k = \text{const}, \quad v = \text{const}. \tag{9.421a}$$

However, the boundary condition (9.395b) may be satisfied only if we assume that there is a jump discontinuity for the functions, followed by the trivial solution

$$T = 0, \quad k = 1, \quad v = 0. \tag{9.421b}$$

In other words, for the models $A_{\mathrm{I}}^{\xi}$ and $A_{\mathrm{II}}^{\xi}$ there is a solution in the form of a *shock wave*. The function $X = X_D(t)$ separating on the plane $(t, X)$ two solutions (9.421) is an equation of the front of the shock wave. To find this function and also to determine values of $k$ and $v$ (relationships over characteristics do not hold there), we should use relations (5.70) at a surface of a strong discontinuity in the material description. For the considered problem, these relations reduce to the following ones:

$$\begin{cases} \overset{\circ}{M} v - p = 0, & (9.422a) \\ \overset{\circ}{M}(k-1) + \overset{\circ}{\rho} v = 0, & (9.422b) \\ \overset{\circ}{M}(\frac{v^2}{2} + [e]) - pv = 0. & (9.422c) \end{cases}$$

Here we have taken into account that $[v] = v$, $[P] = P_{11} = \sigma_{11}$, $[F] = F_{11} - 1 = k - 1$, because on one side of the singular surface, according to (9.421b), the medium is quiescent. The mass rate $\overset{\circ}{M}$ is determined by (5.57) and (5.12):

$$\overset{\circ}{M} = \overset{\circ}{D}\overset{\circ}{\rho}, \quad \overset{\circ}{D} = \overset{\circ}{c} = \partial \overset{\circ}{x}_{\Sigma}/\partial t = dX_D/dt. \tag{9.423}$$

Assume that the temperature jump is zero across the shock wave: $[\theta] = 0$, then the result of Exercise 9.1.10 for the internal energy jump $[e]$ gives the expression

$$[e] = e - e_0 = \frac{l_1}{2\overset{\circ}{\rho}} I_1^2(\overset{(n)}{\mathbf{C}}_e) + \frac{l_2}{\overset{\circ}{\rho}} I_1(\overset{(n)}{\mathbf{C}}{}_e^2). \tag{9.424}$$

Here we have taken into account that, according to (9.421b), $e = e_0$ in a quiescent domain.

From (9.372)–(9.374) we obtain the following expressions for invariants $I_1(\overset{(n)}{\mathbf{C}}_e)$ and $I_1(\overset{(n)}{\mathbf{C}}{}_e^2)$:

$$I_1(\overset{(n)}{\mathbf{C}}_e) = \overset{(n)}{C}{}^e_{11} + 2\overset{(n)}{C}{}^e_{22} = \overset{(n)}{C}{}^e_{11} + \overset{(n)}{C}{}^p_{11} = \overset{(n)}{C}_{11},$$
$$I_1(\overset{(n)}{\mathbf{C}}{}_e^2) = (\overset{(n)}{C}{}^e_{11})^2 + 2(\overset{(n)}{C}{}^e_{22})^2 = (\overset{(n)}{C}_{11} - \overset{(n)}{C}{}^p_{11})^2 + \frac{1}{2}(\overset{(n)}{C}{}^p_{11})^2. \quad (9.425)$$

On substituting (9.425) into (9.424), we find the expression for the internal energy jump in terms of deformations $\overset{(n)}{C}_{11}$ and $\overset{(n)}{C}{}^p_{11}$:

$$[e] = \frac{1}{2\overset{\circ}{\rho}}(l_1\overset{(n)}{C}{}^2_{11} + 2l_2((\overset{(n)}{C}_{11} - \overset{(n)}{C}{}^p_{11})^2 + \frac{1}{2}(\overset{(n)}{C}{}^p_{11})^2))$$
$$= \frac{1}{2\overset{\circ}{\rho}}((l_1 + 2l_2)\overset{(n)}{C}{}^2_{11} - 4l_2\overset{(n)}{C}_{11}\overset{(n)}{C}{}^p_{11} + 3l_2(\overset{(n)}{C}{}^p_{11})^2). \quad (9.426)$$

For the case of initial plastic loading of compression when $\overset{(n)}{C}{}^{p*}_{11} = 0$, substitution of formula (9.376) for $\overset{(n)}{T}_{11}$ into (9.386) yields

$$\overset{(n)}{C}{}^p_{11} = \frac{1}{\widetilde{H}k}\left((l_1 + 2l_2)\overset{(n)}{C}_{11} - 2l_2\overset{(n)}{C}{}^p_{11} - l_1\overset{(n)}{C}_{11} + \sqrt{3}\,\sigma_s k\right). \quad (9.427)$$

From this equation we can easily express the plastic deformation $\overset{(n)}{C}{}^p_{11}$ as follows:

$$\overset{(n)}{C}{}^p_{11} = \frac{\sqrt{3}\,\sigma_s k + b\overset{(n)}{C}_{11}}{3(b + H_0 k/2)}, \quad b = \frac{1-2\nu}{2(1-\nu)}. \quad (9.428)$$

On substituting formula (9.428) for $\overset{(n)}{C}{}^p_{11}$ and formula (9.370) for $\overset{(n)}{C}_{11}$ into (9.426), we get a jump of internal energy $[e]$ as a function of $k$:

$$[e] = [e](k). \quad (9.429)$$

If loading occurs only in the elastic domain, where $\overset{(n)}{\mathbf{C}}_p = 0$, then

$$[e](k) = \frac{l_1 + 2l_2}{2\overset{\circ}{\rho}(n - \mathrm{III})^2}(k^{n-\mathrm{III}} - 1)^2. \quad (9.430)$$

With taking account of expression (9.429), three relationships (9.422) allow us to determine three unknown functions: $k$, $v$ and $\overset{\circ}{M}$ in terms of $p$. For this purpose, we express $v$ from Eq. (9.422b):

$$v = \overset{\circ}{M}(1-k)/\overset{\circ}{\rho}, \tag{9.431}$$

substitute (9.431) into (9.422a) and obtain the expression for $\overset{\circ}{M}$:

$$\overset{\circ}{M} = \sqrt{\overset{\circ}{\rho} p/(1-k)}. \tag{9.432}$$

We have chosen the positive sign of the square root according to the physical meaning of the solution: a shock wave must propagate in the positive direction of the axis $OX$.

Substituting of (9.432) into (9.431) yields

$$v = \sqrt{p(1-k)/\overset{\circ}{\rho}}. \tag{9.433}$$

On substituting (9.429), (9.432), and (9.433) into (9.422c), we get the equation for $k$:

$$[e](k) = \frac{p}{2\overset{\circ}{\rho}}(1-k). \tag{9.434}$$

The derived relationship (9.434) with the expression (9.426) for a jump of internal energy allows us to find $k$ as a function of $p$: $k = k(p)$.

Since $p$ is known from the problem condition, we can determine $k$ from the equation (9.434) and then find $\overset{\circ}{M}$ and $v$ with the help of formulae (9.431) and (9.432). As a result, from the first formula of (9.423) we obtain

$$\overset{\circ}{D} = \frac{\overset{\circ}{M}}{\overset{\circ}{\rho}} = \sqrt{\frac{p}{\overset{\circ}{\rho}(1-k)}}. \tag{9.435}$$

Thus, we have found a complete solution of the problem.

### 9.8.10 *Shock Adiabatic Curves for Models $A_{\mathrm{I}}^{\xi}$ and $A_{\mathrm{II}}^{\xi}$*

The function $k = k(p)$ (or $p = p(k)$) expressed by formula (9.434) is called a *shock adiabatic curve for an elastoplastic continuum*.

Since $k = \overset{\circ}{\rho}/\rho$ is the ratio of densities, we can introduce the specific volume $V = 1/\rho = k/\overset{\circ}{\rho}$ which is a function of $p$, i.e. we have a function $p = p(V)$

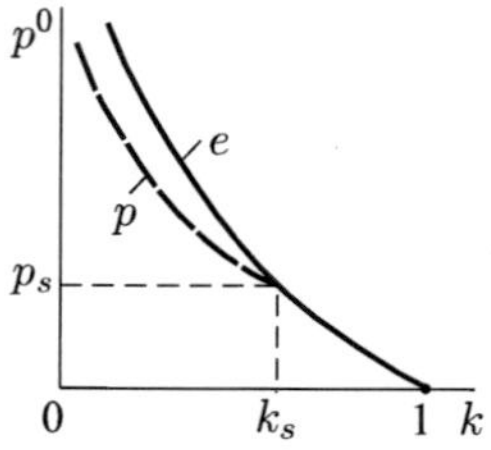


**Fig. 9.23** The shock adiabatic curve

(or $V = V(p)$). This function is well-known in gas dynamics as a *shock adiabatic curve* for similar problems.

If loading occurs only in an elastic domain, then, substituting (9.430) into (9.434), we derive the following equation for the shock adiabatic curve $p = p(k)$:

$$\frac{p}{l_1 + 2l_2} = \frac{1}{1-k}\left(\frac{k^{n-\text{III}} - 1}{n - \text{III}}\right)^2. \tag{9.436}$$

Figure 9.23 shows a graph of this function.

On substituting (9.426), (9.430), and (9.370) into (9.434), we obtain the equation of the shock adiabatic curve with taking account of plastic deformations:

$$\frac{p}{l_1 + 2l_2} = \frac{1}{1-k_1}(\overset{(n)}{C}{}^2_{11} - b\overset{(n)}{C}{}^p_{11}(4\overset{(n)}{C}_{11} - \overset{(n)}{C}{}^p_{11})), \tag{9.437}$$

where

$$\overset{(n)}{C}{}^p_{11} = \frac{\sqrt{3}\,\sigma_s k + b\overset{(n)}{C}_{11}}{3(b + H_0 k/2)},$$

$$\overset{(n)}{C}_{11} = \frac{1}{n - \text{III}}(k_1^{n-\text{III}} - 1). \tag{9.438}$$

Figure 9.23 shows the shock adiabatic curve $p(k)$.

Figure 9.24 exhibits shock adiabatic curves computed by formula (9.437) for dry and wet sand grounds. Constants $E$, $\nu$, $\sigma_s$, and $H$ for grounds have been taken from the experimental data for specimens of a beam form in uniaxial high-speed compression. The method of evaluation and values of these constants are given in Sect. 9.7.9; for approximation we have used the plasticity model with linear strengthening.

Let us ask the question: what is the speed $\overset{\circ}{D}$ of propagation of a shock wave in comparison with the sound speed $a_0$ in a quiescent continuum. To answer the question one should find a value of $k$ when the pressure values are small: $p/(l_1 + 2l_2) \ll 1$. Linearizing the left-hand side of Eq. (9.436) in a neighborhood of the value $k = 1$, we obtain

$$p = -(l_1 + 2l_2)\delta, \tag{9.439}$$

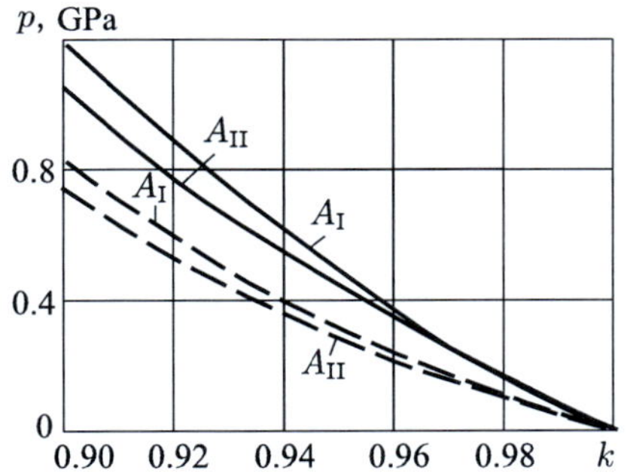


**Fig. 9.24** Shock adiabatic curves for sand grounds: solid lines correspond to wet sand, dashed lines – to dry sand

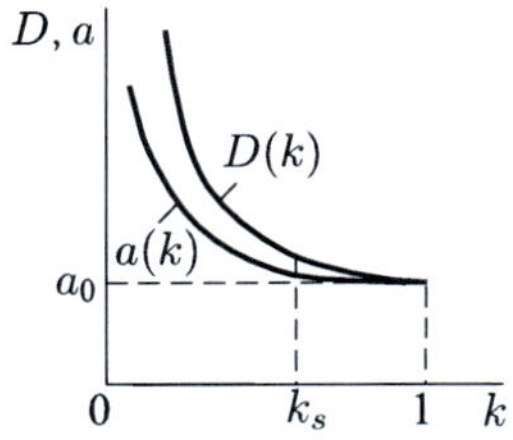


**Fig. 9.25** Dependence of the speed of a shock wave and the speed of sound upon the compression coefficient $k_1$ in a plastic continuum

where $k = 1+\delta$, $|\delta| \ll 1$, $\delta \leqslant 0$ for both the models $A_{\mathrm{I}}$ and $A_{\mathrm{II}}$. Then from (9.435) we find that

$$\overset{\circ}{D} = \sqrt{-\frac{p}{\overset{\circ}{\rho}\delta}} = \sqrt{\frac{\delta(l_1+2l_2)}{\overset{\circ}{\rho}\delta}} = \sqrt{\frac{l_1+2l_2}{\overset{\circ}{\rho}}} = a_0, \tag{9.440}$$

i.e. for the models $A_{\mathrm{I}}$ and $A_{\mathrm{II}}$ of elastoplastic continua, small disturbances propagate with the speed of sound.

As follows from Eq. (9.435), at finite values of $p/(l_1+2l_2)$ the speed $\overset{\circ}{D}$ of a shock wave proves to be supersonic (Fig. 9.25).

## *9.8.11 Shock Adiabatic Curves at a Given Rate of Impact*

The solution obtained in Sect. 9.8.10 remains valid for another case of boundary conditions: when on the surface $X^1 = 0$ of a plate one gives a constant rate $v$ of impact instead of the stress component $P_{11}$:

$$X^1 = 0: \quad v = v_0 = \text{const}, \quad P_{12} = P_{13} = 0; \tag{9.441}$$

the remaining conditions in (9.367) hold without changes. In this case, it is convenient to represent the shock adiabatic curve as a function $p = p(v)$. For this, we

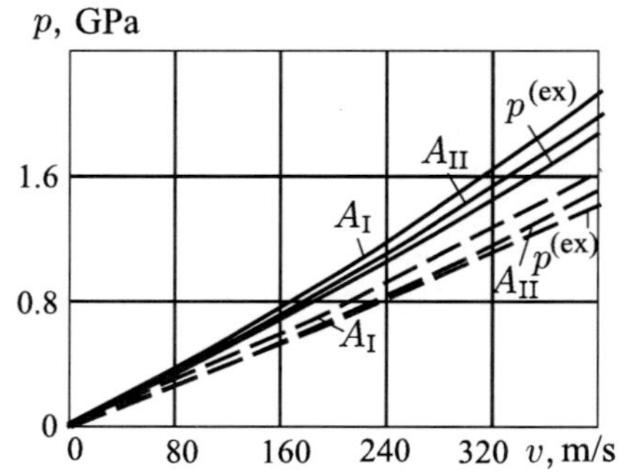


**Fig. 9.26** Computed and experimental shock adiabatic curves in coordinates $(p, v)$ for sand grounds: solid curves – wet sand, dashed lines – dry sand

should substitute the function $p(k)$ expressed by formulae (9.437) and (9.438) into Eq. (9.433). As a result, we get the equation

$$p(k) = \frac{v^2}{\overset{\circ}{\rho}(1-k)}, \tag{9.442}$$

which at a given value of $v$ can be resolved for $k$. Evaluating a root of the equation (when $0 < k \leqslant 1$, this root is unique), for example, by the method of bisecting an interval, we find the function $k = k(v)$.

On substituting this function into Eq. (9.437), we obtain the shock adiabatic curve $p = p(k) = p(k(v)) = p(v)$ in coordinates $(p, v)$ (the pressure versus the speed). Figure 9.26 shows graphs of the shock adiabatic curve $p = p(v)$ for sand grounds, they were computed for the models $A_{\mathrm{I}}^{\xi}$ and $A_{\mathrm{II}}^{\xi}$ of plastic continua with linear strengthening. The constants $E$, $\nu$, $\sigma_s$ and $H_0$ in the computations were taken also from the results of experiments in uniaxial tension (see Sect. 9.7.9). This figure also exhibits experimental shock adiabatic curves $p^{(\mathrm{ex})}(k)$ for dry and wet sand grounds. The computed curves determined by both the models $A_{\mathrm{I}}^{\xi}$ and $A_{\mathrm{II}}^{\xi}$ satisfactorily approximate the experimental data.

The shock adiabatic curve can be represented in another form, namely as a dependence of the speed $\overset{\circ}{D}$ on the rate of impact $v$. To obtain such a function $\overset{\circ}{D}(v)$, we should use formulae (9.422a) and (9.423) and substitute into them the dependence $p = p(k) = p(k(v)) = p(v)$ derived above; as a result, we obtain the desired form of a shock adiabatic curve:

$$\overset{\circ}{D} = \frac{p(v)}{\overset{\circ}{\rho} v}. \tag{9.443}$$

Figure 9.27 shows the shock adiabatic curves $\overset{\circ}{D}(v)$ computed for sand grounds by formula (9.443) and also experimental shock adiabatic curves $\overset{\circ}{D}{}^{(\mathrm{ex})}(v)$. The computed and experimental results are sufficiently close.

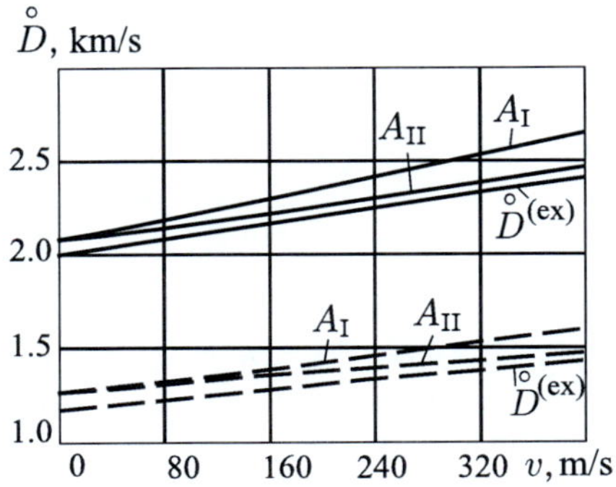


**Fig. 9.27** Shock adiabatic curves in coordinates $(D^o, v)$ for sand grounds: solid curves – wet sand, dashed lines – dry sand

## 9.9 Models of Viscoplastic Continua

### 9.9.1 *The Concept of a Viscoplastic Continuum*

In Chap. 7 we considered models of continua of the differential type, and in Sect. 7.4.4 it was shown that these models can be applied to describe the creep effect in some solids. The limitation of models of the differential type consists in the fact that they do not involve elastic deformations; therefore, with their help one cannot determine instantly elastic deformations under loading and subsequent unloading (see Figs. 7.2 and 7.3). In the cases when instantly elastic deformations cannot be neglected in comparison with creep deformations, one should use more complicated models, for example, models of viscoplastic continua, that will be considered in this section.

As a rule, a continuum is called *viscoplastic* if its plastic properties (the appearance of residual deformations) depend on time. The models of associated plastic continua (considered in Sects. 9.1.5–9.1.13) describe pure plastic properties, which are independent of time. Indeed, although the constitutive equations (9.52) of associated models formally involve time $t$, we can easily exclude the time if we will differentiate not with respect to time but with respect to the loading parameter (see Exercise 9.1.11).

From the experimental point of view, this fact means that if we perform experiments in uniaxial tension (see Sect. 9.7) with a constant rate of lengthening: $k_1(t) = bt$, then diagrams of deforming in elastic and plastic domains are independent of the lengthening rate $b$. For viscoplastic continua, this dependence takes place.

### 9.9.2 *Model $A_n$ of Viscoplastic Continua of the Differential Type*

Let us consider the simplest viscoplastic model, namely the *model $A_n$ of a viscoplastic continuum of the differential type*, which can be formally obtained from

Eqs. (9.52) if we assume that the expression for plastic deformation holds for all the times at both loading and unloading. In this case, the parameter $h$ should be assumed to be equal to 1:

$$\overset{(n)}{\mathbf{C}}{}_{\nu}^{\bullet} = \sum_{\alpha=1}^{k} \varkappa_{\alpha} (\partial f_{\nu\alpha} / \partial \overset{(n)}{\mathbf{T}}). \tag{9.444}$$

In this formula we have changed the notation of plastic deformation $\overset{(n)}{\mathbf{C}}_{p}$ by $\overset{(n)}{\mathbf{C}}_{\nu}$, which is called the *viscous deformation*, and the functions $\dot{\varkappa}_{\alpha}$ are replaced by $\varkappa_{\alpha}$. We can always make this substitution because functions $\dot{\varkappa}_{\alpha}$ in (9.48) have been introduced as ratio coefficients of plastic deformations to the yield surface gradient. For the associated models of plastic deformations, the functions $\dot{\varkappa}_{\alpha}$ can be found from Eqs. (9.52) with complementing by Eqs. (9.41) of the yield surface, but for a viscoplastic continuum of the differential type Eqs. (9.41) are absent and expressions for $\varkappa_{\alpha}$ and $f_{\nu\alpha}$ are given with the help of the additional relations

$$\varkappa_{\alpha} = \varkappa_{\alpha}(\overset{(n)}{\mathbf{T}}, \overset{(n)}{\mathbf{C}}_{p}, \theta), \quad f_{\nu\alpha} = f_{\nu\alpha}(\overset{(n)}{\mathbf{T}}, \overset{(n)}{\mathbf{C}}_{p}, \theta), \quad \alpha = 1, \ldots, k. \tag{9.445}$$

The functions $f_{\nu\alpha}$ will be called the *viscous potentials*.

*Remark 1.* For viscoelastic continua of the differential type the stress tensor $\overset{(n)}{\mathbf{T}}$ is a sum of equilibrium and viscous stresses (see (7.6)), but for viscoplastic materials of the differential type the deformation tensor $\overset{(n)}{\mathbf{C}}$ consists of elastic and viscous deformations, similarly to (9.3):

$$\overset{(n)}{\mathbf{C}} = \overset{(n)}{\mathbf{C}}_{e} + \overset{(n)}{\mathbf{C}}_{\nu}. \tag{9.446}$$

□

For the *unipotential model $A_n$ of a viscoplastic continuum of the differential type*, we assume that $k = 1$, $\varkappa_1 = 1$, and $f_{\nu 1} = f_{\nu}$; therefore,

$$\overset{(n)}{\mathbf{C}}{}_{\nu}^{\bullet} = \partial f_{\nu} / \partial \overset{(n)}{\mathbf{T}}, \quad f_{\nu} = f_{\nu}(\overset{(n)}{\mathbf{T}}, \overset{(n)}{\mathbf{C}}_{\nu}, \theta). \tag{9.447}$$

These relations should be complemented with equations for the elastic deformation tensor $\overset{(n)}{\mathbf{C}}_{e}$, for example, (9.59).

The equation system can be rewritten in terms of the invariants as follows:

$$\overset{(n)}{\mathbf{C}}{}_{\nu}^{\bullet} = \sum_{\alpha=1}^{z_1} \pi_{\alpha} J_{\alpha\mathbf{T}}^{(s)}, \quad \overset{(n)}{\mathbf{C}}_{e} = \sum_{\gamma=1}^{z} \varphi_{\gamma} I_{\gamma\mathbf{T}}^{(s)}, \tag{9.448}$$

where $\pi_{\alpha}$ and $\varphi_{\gamma}$ are scalar functions being the derivatives of the plastic and elastic potentials $f_{\nu}$ and $\zeta$ with respect to the invariants:

$$\pi_\alpha = \partial f_v/\partial J_\alpha^{(s)v}, \quad f_v = f_v(J_\alpha^{(s)}, \theta), \quad J_\gamma^{(s)} = J_\gamma^{(s)}(\overset{(\mathrm{n})}{\mathbf{T}}, \overset{(\mathrm{n})}{\mathbf{C}}_v), \quad J_{\alpha\mathbf{T}}^{(s)} = \partial J_\alpha^{(s)}/\partial \overset{(\mathrm{n})}{\mathbf{T}}, \tag{9.449}$$

$$\varphi_\gamma = -\partial\zeta/\partial I_\gamma^{(s)}, \quad \zeta = \zeta(I_\gamma^{(s)}, \theta), \quad I_\gamma^{(s)} = I_\gamma^{(s)}(\overset{(\mathrm{n})}{\mathbf{T}}/\rho), \quad I_{\gamma\mathbf{T}}^{(s)} = \partial I_\gamma^{(s)}/\partial(\overset{(\mathrm{n})}{\mathbf{T}}/\rho). \tag{9.450}$$

Due to the Onsager principle, the plastic potential $f_v$ is a quadratic function of the linear invariants $J_\gamma^{(s)}(\overset{(\mathrm{n})}{\mathbf{T}}, \overset{(\mathrm{n})}{\mathbf{C}}_v)$ and a linear function of the quadratic invariants $J_\gamma^{(s)}$.

### *9.9.3 Model of Isotropic Viscoplastic Continua of the Differential Type*

For isotropic viscoplastic continua of the differential type, the simultaneous invariants $J_\gamma^{(s)}(\overset{(\mathrm{n})}{\mathbf{T}}, \overset{(\mathrm{n})}{\mathbf{C}}_v)$ can be chosen in the form (9.75), then relations (9.448) take the forms

$$\overset{(\mathrm{n})}{\mathbf{C}}{}_v^{\bullet} = \widetilde{\pi}_1\mathbf{E} - \pi_2\overset{(\mathrm{n})}{\mathbf{T}} + \pi_7\overset{(\mathrm{n})}{\mathbf{C}}_v, \tag{9.451}$$

$$\overset{(\mathrm{n})}{\mathbf{C}}_e = \widetilde{\varphi}_1\mathbf{E} + (\widetilde{\varphi}_2/\rho)\overset{(\mathrm{n})}{\mathbf{T}} + (\varphi_3/\rho^2)\overset{(\mathrm{n})}{\mathbf{T}}{}^2. \tag{9.452}$$

Here $\widetilde{\pi}_1 = \pi_1 + \psi_2 I_1(\overset{(\mathrm{n})}{\mathbf{T}})$, and the potentials depend on the following invariants:

$$f_v = f_v(I_1(\overset{(\mathrm{n})}{\mathbf{T}}), I_2(\overset{(\mathrm{n})}{\mathbf{T}}), I_1(\overset{(\mathrm{n})}{\mathbf{C}}_p), I_2(\overset{(\mathrm{n})}{\mathbf{C}}_v), J_7^{(I)}(\overset{(\mathrm{n})}{\mathbf{T}}, \overset{(\mathrm{n})}{\mathbf{C}}_v), \theta), \tag{9.453}$$

$$\zeta = \zeta(I_1(\overset{(\mathrm{n})}{\mathbf{T}}/\rho), I_2(\overset{(\mathrm{n})}{\mathbf{T}}/\rho), I_3(\overset{(\mathrm{n})}{\mathbf{T}}/\rho), \theta), \tag{9.454}$$

If the plastic potential has been chosen in the Huber–Mises form (9.96)

$$f_v = \frac{Y_H^2}{3\mu}, \quad Y_H^2 = \frac{3}{2}\mathbf{P}_H \cdot\cdot \mathbf{P}_H,$$

$$\mathbf{P}_H = (\overset{(\mathrm{n})}{\mathbf{T}} - H_v\overset{(\mathrm{n})}{\mathbf{C}}_v) - \frac{1}{3}I_1(\overset{(\mathrm{n})}{\mathbf{T}} - H_v\overset{(\mathrm{n})}{\mathbf{C}}_v)\mathbf{E}, \quad H_v = H_{v0}Y_v^{2n_{v0}}, \tag{9.455}$$

$$Y_v^2 = \frac{3}{2}\mathbf{P}_v \cdot\cdot \mathbf{P}_v, \quad \mathbf{P}_v = \overset{(\mathrm{n})}{\mathbf{C}}_v - \frac{1}{3}I_1(\overset{(\mathrm{n})}{\mathbf{C}}_v)\mathbf{E},$$

where $H_{\nu 0}$, $n_{\nu 0}$ and $\mu$ are the constants, then, performing the manipulations of Sect. 9.1.8, we can rewrite relations (9.450) as follows:

$$\overset{(n)}{\mathbf{C}}{}^{\bullet}_{\nu} = \frac{1}{\mu}(\mathbf{P}_T - H_\nu \overset{(n)}{\mathbf{C}}_\nu), \quad I_1(\overset{(n)}{\mathbf{C}}_\nu) = 0. \tag{9.456}$$

For isotropic linear-elastic continua, Eqs. (9.452) with use of the additive relationship (9.3) take the form (see Exercise 9.1.9):

$$\overset{(n)}{\mathbf{T}} = J l_1 I_1(\overset{(n)}{\mathbf{C}})\mathbf{E} + 2 J l_2(\overset{(n)}{\mathbf{C}} - \overset{(n)}{\mathbf{C}}_\nu). \tag{9.457}$$

The equation system (9.456), (9.457) is the model $A_n$ of an isotropic viscoplastic continuum of the differential type with the Huber–Mises potential.

### 9.9.4 General Model $A_n$ of Viscoplastic Continua

One can say that this is the general *model $A_n$ of viscoplastic continua*, if the additive relation (9.3) is replaced by

$$\overset{(n)}{\mathbf{C}} = \overset{(n)}{\mathbf{C}}_e + \overset{(n)}{\mathbf{C}}_p + \overset{(n)}{\mathbf{C}}_\nu, \tag{9.458}$$

i.e. the deformation tensor $\overset{(n)}{\mathbf{C}}$ in this model is a sum of the three terms: elastic deformation $\overset{(n)}{\mathbf{C}}_e$, plastic deformation $\overset{(n)}{\mathbf{C}}_p$ and *viscous deformation* $\overset{(n)}{\mathbf{C}}_\nu$. For elastic and plastic deformations we assume the same relations as the ones for pure plastic continua (see Sect. 9.1). For example, for the *associated model $A_n$ of viscoplastic continua*, the relations (9.74) hold:

$$\overset{(n)}{\mathbf{C}}_e = \sum_{\gamma=1}^{r} \varphi_\gamma I^{(s)}_{\gamma\mathbf{T}}(\overset{(n)}{\mathbf{T}}/\rho), \tag{9.459a}$$

$$\overset{(n)}{\mathbf{C}}{}^{\bullet}_p = \sum_{\alpha=1}^{z_1} \psi_\alpha J^{(s)}_{\alpha\mathbf{T}}(\overset{(n)}{\mathbf{T}}, \overset{(n)}{\mathbf{C}}_p), \tag{9.459b}$$

$$\zeta = \zeta(I^{(s)}_\gamma(\overset{(n)}{\mathbf{T}}/\rho), \theta, w^p_\beta), \quad \gamma = 1, \ldots, r, \tag{9.459c}$$

$$f_\beta = f_\beta(J^{(s)}_\alpha(\overset{(n)}{\mathbf{T}}, \overset{(n)}{\mathbf{C}}_p), \theta, w^p_\beta), \tag{9.459d}$$

$$\varphi_\gamma = -\frac{\partial \zeta}{\partial I^{(s)}_\gamma}, \quad I^{(s)}_{\gamma\mathbf{T}} = \frac{\partial I^{(s)}_\gamma(\overset{(n)}{\mathbf{T}}/\rho)}{\partial \overset{(n)}{\mathbf{T}}/\rho}, \quad \psi_\alpha = h \sum_{\beta=1}^{k} \varkappa_\beta \frac{\partial f_\beta}{\partial J^{(s)}_\alpha}. \tag{9.459e}$$

And we assume that, just as for viscoplastic continua of the differential type, for viscous deformation Eqs. (9.448) and (9.449) still hold:

$$\overset{(n)}{\mathbf{C}}{}^{\bullet}_{v} = \sum_{\alpha=1}^{z_1} \pi_\alpha J^{(s)}_{\alpha\mathbf{T}}, \tag{9.460}$$

$$\pi_\alpha = \partial f_v/\partial J^{(s)}_\alpha, \quad f_v = f_v(J^{(s)}_\alpha, \theta), \quad J^{(s)}_\gamma = J^{(s)}_\gamma(\overset{(n)}{\mathbf{T}}, \overset{(n)}{\mathbf{C}}_v), \quad J^{(s)}_{\alpha\mathbf{T}} = \partial J^{(s)}_\alpha/\partial \overset{(n)}{\mathbf{T}}. \tag{9.461}$$

### 9.9.5 Model $A_n$ of Isotropic Viscoplastic Continua

Using the results of Sects. 9.1.7 and 9.9.4, from (9.460) and (9.461) we get that the equation system (9.81)–(9.83), (9.451)–(9.454) holds for isotropic viscoelastic continua. Choosing the Huber–Mises model (9.96) and (9.455) for plastic and viscous potentials and applying the linear elasticity model for elastic deformation, from (9.460) and (9.461) we get the equation system

$$\overset{(n)}{\mathbf{T}} = Jl_1 I_1(\overset{(n)}{\mathbf{C}})\mathbf{E} + 2Jl_2(\overset{(n)}{\mathbf{C}} - \overset{(n)}{\mathbf{C}}_p - \overset{(n)}{\mathbf{C}}_v), \tag{9.462}$$

$$\overset{(n)}{\mathbf{C}}{}^{\bullet}_{p} = \frac{\varkappa h}{\sigma_s^2}(\overset{(n)}{\mathbf{T}} - \frac{1}{3}I_1(\overset{(n)}{\mathbf{T}})\mathbf{E} - H\overset{(n)}{\mathbf{C}}_p), \quad I_1(\overset{(n)}{\mathbf{C}}_p) = 0, \tag{9.463}$$

$$f = \frac{1}{3}(Y_H/\sigma_s)^2 - 1 = 0, \quad H = H_0 Y_p^{2n_0},$$

$$\overset{(n)}{\mathbf{C}}{}^{\bullet}_{v} = \frac{1}{\mu}(\overset{(n)}{\mathbf{T}} - \frac{1}{3}I_1(\overset{(n)}{\mathbf{T}}) - H_v\overset{(n)}{\mathbf{C}}_v), \quad I_1(\overset{(n)}{\mathbf{C}}_v) = 0, \quad H_v = H_{v0} Y_v^{2n_{v0}}, \tag{9.464}$$

where the invariants $Y_H$, $Y_p$ and $Y_v$ are expressed by formulae (9.85), (9.87), and (9.455).

### 9.9.6 The Problem on Tension of a Beam of Viscoplastic Continuum of the Differential Type

As an example, consider the problem on uniaxial tension of a beam of viscoplastic material of the differential type, which is described by the motion law (9.327). The deformation gradient $\mathbf{F}$ and the deformation tensors $\overset{(n)}{\mathbf{C}}$ and $\overset{(n)}{\mathbf{C}}_e$ are determined by formulae (9.328)–(9.332), (9.341). For viscous deformations $\overset{(n)}{\mathbf{C}}_\nu$ we have formulae which are similar to Eqs. (9.334), and for plastic deformations – formulae (9.345):

$$\overset{(n)}{\mathbf{C}}_\nu = \sum_{\alpha=1}^{3} \overset{(n)}{C}{}^{\nu}_{\alpha\alpha} \bar{\mathbf{e}}_\alpha \otimes \bar{\mathbf{e}}_\alpha. \tag{9.465}$$

$$\overset{(n)}{C}_{\alpha\alpha} = \overset{(n)}{C}{}^{e}_{\alpha\alpha} + \overset{(n)}{C}{}^{\nu}_{\alpha\alpha}, \quad \alpha = 1,\ 2,\ 3. \tag{9.466}$$

$$\overset{(n)}{C}{}^{\nu}_{33} = \overset{(n)}{C}{}^{\nu}_{22} = -\frac{1}{2}\overset{(n)}{C}{}^{\nu}_{11}. \tag{9.467}$$

The stress tensors $\overset{(n)}{\mathbf{T}}$ are diagonal, for them relations (9.335)–(9.341) hold; in particular, the component $\sigma_{11}$ is expressed as follows:

$$\sigma_{11} = k_1^{n-\mathrm{III}} JE(\overset{(n)}{C}_{11} - \overset{(n)}{C}{}^{\nu}_{11}). \tag{9.468}$$

On substituting (9.465) into (9.456), we get the following equation for viscous deformation $\overset{(n)}{C}{}^{\nu}_{11}$:

$$\frac{d}{dt}\overset{(n)}{C}{}^{\nu}_{11} = \frac{1}{\mu}\left(\frac{2}{3}k_1^{\mathrm{III}-n}\sigma_{11} - H\overset{(n)}{C}{}^{\nu}_{11}\right), \tag{9.469}$$

$$H_\nu = H_{\nu 0} Y_p^{2n_{\nu 0}}, \quad Y_p^2 = (\frac{3}{2}\overset{(n)}{C}{}^{\nu}_{11})^2. \tag{9.470}$$

Changing density is determined by the same relationship (9.355)

$$J = \frac{\rho}{\overset{\circ}{\rho}} = \frac{1}{k_1}\left(1 - \nu(k_1^{n-\mathrm{III}} - 1) - (n - \mathrm{III})(\frac{1}{2} - \nu)\overset{(n)}{C}{}^{\nu}_{11}\right)^{-2/(n-\mathrm{III})}. \tag{9.471}$$

On substituting (9.468) into (9.469) and taking the expression (9.330) for $\overset{(n)}{C}_{11}$ into account, we derive the following final equation for viscous deformation $\overset{(n)}{C}{}^{\nu}_{11}$:

$$\frac{d}{dt}\overset{(n)}{C}{}^{\nu}_{11} = \frac{1}{\mu}\left(\frac{2}{3}JE(\frac{1}{n-\mathrm{III}}(k_1^{n-\mathrm{III}} - 1) - \overset{(n)}{C}{}^{\nu}_{11}) - H_{\nu 0}(\frac{3}{2}\overset{(n)}{C}{}^{\nu}_{11})^{2n_{\nu 0}}\overset{(n)}{C}{}^{\nu}_{11}\right). \tag{9.472}$$

If the elongation function $k_1(t)$ of the beam is given, then, on solving the equation, the stress $\sigma_{11}$ can be found from Eq. (9.468).

If, just as in the creep problem (see Sect. 7.4.4), the stress $\sigma_{11}(t)$ is given, then from Eq. (9.468) we can compute the dependence $k_1 = k_1(\sigma_{11},\ \overset{(n)}{C}{}^{\nu}_{11})$; and with the help of Eq. (9.469) we determine the viscous deformation $\overset{(n)}{C}{}^{\nu}_{11}$. A final expression for $k_1(t)$ is found by using the dependence $k_1(\sigma_{11}, \overset{(n)}{C}{}^{\nu}_{11})$ once again.

The considered model of an isotropic viscoplastic continuum of the differential type contains the five material constants: $E$, $\nu$, $\mu$, $H_{\nu 0}$, and $n_{\nu 0}$. The elastic modulus is determined by the initial section of the diagram $\sigma_{11}(k_1)$ at given lengthening $k_1(t)$, when the influence of viscous deformations can be neglected. Just as for elastic continua, the Poisson ratio is determined by relation (9.341) also for small times, when viscous deformations are small. The remaining three constants: $\mu$, $H_{\nu 0}$ and $n_{\nu 0}$ can be found with the help of experimental creep curves at the given stress $\sigma_{11}(t)$ varying as a step-function (7.146).

Figure 9.28 shows the experimental creep curve $|\delta_1^{(ex)}(t)|$ for Ni-alloy at temperature 1100°C and its approximation by Eqs. (9.468) and (9.469) for different models $A_n$. Constants $H_{\nu 0}$ and $n_{\nu 0}$ in the computations have been chosen to be zero, and the viscous coefficient $\mu$ has been determined by minimization of the mean-square

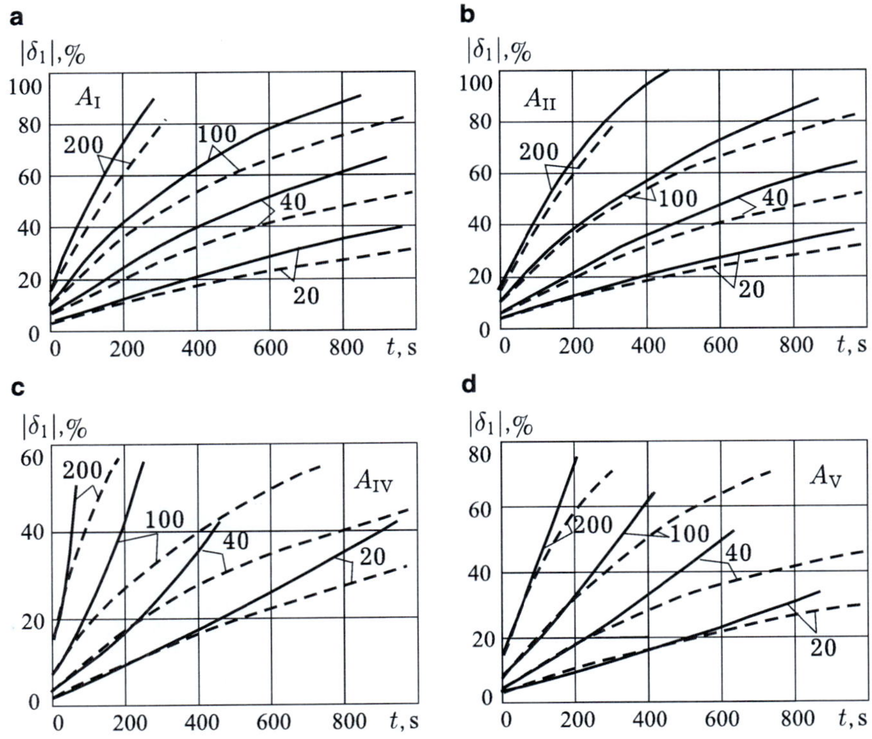


**Fig. 9.28** Creep curves for Ni-alloy at temperature 1,100°C and different values of compressing stress $\sigma^o$: dashed curves are experimental data, solid curves are computations by different models $A_n$ of viscoplastic continua of the differential type: **(a)** – model $A_{\mathrm{I}}$, **(b)** – model $A_{\mathrm{II}}$, **(c)** – model $A_{\mathrm{IV}}$, and **(d)** – model $A_{\mathrm{V}}$

distance between the computational $|\delta_1(t)| = |k_1(t)-1|$ and experimental $|\delta_1^{(ex)}(t)|$ creep curves at $\sigma^o = -20$ MPa for several times (see Sect. 7.4.3). We have obtained the following values of the constants:

$$\begin{aligned} E &= 2\text{GPa} \quad &&\text{for all models } A_n, \\ \mu &= 40 \text{ GPa}\cdot\text{s} \quad &&\text{for } n = \text{I}, \\ \mu &= 42 \text{ GPa}\cdot\text{s} \quad &&\text{for } n = \text{II}, \\ \mu &= 55 \text{ GPa}\cdot\text{s} \quad &&\text{for } n = \text{IV}, \\ \mu &= 60 \text{ GPa}\cdot\text{s} \quad &&\text{for } n = \text{V}. \end{aligned}$$

Figure 9.28 also exhibits computational and experimental creep curves at different values of stress $\sigma^o$. The model $A_{\text{II}}$ gives the best approximation quality (Fig. 9.28b).

## Exercises for 9.9

**9.9.1.** Show that the model of isotropic viscoplastic continuum (9.455), (9.456), when $H_0 = 0$, can be written as the differential equation

$$\overset{(n)}{\mathbf{T}}{}^{\bullet} = J(l_1 I_1(\overset{(n)}{\mathbf{C}}{}^{\bullet}) + \frac{2l_2}{3\mu} I_1(\overset{(n)}{\mathbf{T}}))\mathbf{E} + 2Jl_2\overset{(n)}{\mathbf{C}}{}^{\bullet} - \left(\frac{2Jl_2}{\mu} - \frac{\dot{J}}{J}\right)\overset{(n)}{\mathbf{T}}, \quad \frac{\dot{J}}{J} = -\nabla\cdot\mathbf{v}.$$

# References


1. Batra, R.C.: Elements of Continuum Mechanics. AIAA Education Series, Reston (2005)
2. Basar, Y., Weichert, D.: Nonlinear Continuum Mechanics of Solids. Springer, Berlin (2000)
3. Bonet, J., Wood, R.D.: Nonlinear Continuum Mechanics for Finite Element Analysis. Cambridge University Press, Cambridge (1997)
4. Borg, S.F.: Matrix Tensor Methods in Continuum Mechanics. World Scientific, London (1990)
5. Bowen, R.M.: Introduction to Continuum Mechanics for Engineers. Plenum Press, New York (1989)
6. Calcote, L.R.: Introduction to Continuum Mechanics. D. Van Nostrand, Princeton (1968)
7. Chadwick, P.: Continuum Mechanics: Concise Theory and Problems. Dover Publications (1999)
8. Chernykh, K.F.: An Introduction to Modern Anisotropic Elasticity, Moscow, Nauka, 1988 (in Russian). Begell Publishing House, New York (1998)
9. Coleman, B.D.: Thermodynamics of materials with memory. Arch. Ration. Mech. Anal. **17**(1), 1–46 (1964)
10. Cotter, B.A., Rivlin, R.S.: Tensors associated with time-dependent stress. Quart. Appl. Math. **13**(2), 177–182 (1955)
11. Dimitrienko, Yu.I.: Novel viscoelastic models for elastomers under finite strains. Eur. J. Mech. Solid. **21**(2), 133–150 (2002)
12. Dimitrienko, Yu.I.: Tensor Analysis and Nonlinear Tensor Functions. Kluwer Academic Publishers, Dordrecht-Boston-London (2002)
13. Ericksen, J.L., Rivlin, R.S.: Large elastic deformations of homogeneous anisotropic elastic materials. J. Ration. Mech. Anal. **3**(3), 281-301 (1954)
14. Eringen, A.C.: Nonlinear Theory of Continuous Media, McGraw–Hill, New York (1962)
15. Eringen, A.C.: Mechanics of Continuum. Wiley, New York (1967)
16. Ferrarese, G., Bini, D.: Introduction to Relativistic Continuum Mechanics (Lecture Notes in Physics). Springer, Berlin (2007)
17. Fung, Y.C.: A First Course in Continuum Mechanics. Prentice Hall, New Jersey (1977)
18. Gladwell, G.M.L.: Contact Problems in the Classical Theory of Elasticity, Sijthoff and Noordhoff, 1980. Chinese Edition published (1991)
19. Green, A.E., Zerna, W.: Theoretical Elasticity. Oxford University Press (1954)
20. Godunov, S.K., Romenskii, E.I.: Elements of Continuum Mechanics and Conservation Laws. Springer, Berlin (2003)
21. Goldstein, R.V., Entov, V.M.: Qualitative Methods in Continuum Mechanics. Longman, London (1994)
22. Gonzalez, O., Stuart, A.M.: A First Course in Continuum Mechanics (Cambridge Texts in Applied Mathematics). Cambridge University Press (2007)
23. Gurtin, M.E.: Introduction to Continuum Mechanics (Mathematics in Science and Engineering). Academic Press, New York (1981)
24. Gurtin, M.E.: Configurational Forces as Basic Concepts of Continuum Physics. Springer, Berlin (1999)

25. Heinbockel, J.H.: Introduction to Tensor Calculus and Continuum Mechanics. Trafford Publishing, Canada (2001)
26. Hill, R.: Constitutive inequalities for isotropic elastic solids under finite strain. Proc. Roy. Soc. Lond. A **314**, 1519 (1970)
27. Holzapfel, G.A.: Nonlinear Solid Mechanics: A Continuum Approach for Engineering. Wiley, New York (2000)
28. Jaumann, G.: Grundlagen der Bewegungslehre. Springer, Leipzig (1905)
29. Jaunzemis, W.: Continuum Mechanics. The Macmillan Company, New York (1967)
30. Jog, C.S.: Continuum Mechanics. Alpha Science (2007)
31. Kolmogorov, A.N., Fomin, S.V.: Elements of the Theory of Functions and Functional Analysis, Moscow, Nauka, 1976 (in Russian). Dover Publications, New York (1999)
32. Lai, W.M., Rubin, D., Krempl, E.: Introduction to Continuum Mechanics. Butterworth–Heinemann, Pergamon, New York (1993)
33. Lebedev, L.P., Vorovich, I.I., Gladwell, G.M.L.: Functional Analysis: Applications in Mechanics and Inverse Problems. Kluwer Academic Publishers, Dordrecht-Boston-London (1996)
34. Leigh, D.C.: Nonlinear Continuum Mechanics. McGrawHill, New York (1968)
35. Liu, I-Shih: Continuum Mechanics. Springer, Berlin (2002)
36. Lurie, A.I.: Nonlinear Theory of Elasticity. Amsterdam, North–Holland (1990)
37. Lurie, A.I.: Theory of Elasticity. Springer, Berlin (2005)
38. Malvern, L.E.: An Introduction to the Mechanics of a Continuous Media. Prentice–Hall, Englewood Cliffs (1970)
39. Mase, G.E.: Continuum Mechanics for Engineers. CRC Press, Boca Raton (1999)
40. McDonald, P.H.: Continuum Mechanics (PWS Series in Engineering). PWS Publishing, Boston (1995)
41. Murnaghan, F.D.: Finite Deformation of an Elastic Solid. Wiley, New York (1951)
42. Nemat–Nasser, S.: On nonequilibrium thermodynamics of continua. Mechanics Today **2**, 94–158 (1975)
43. Noll, W.: The Foundations of Mechanics and Thermodynamics, (Selected papers). Springer, Berlin (1974)
44. Oldroyd, J.G.: On the formulation of rheological equations of State. Proc. Roy. Soc. **200**(1063), 523–541 (1950)
45. Onsager, Lars.: Reciprocal relations in irreversible processes. Phys. Rev. **37**(2), 405–426 (1931)
46. Onsager, Lars.: Reciprocal relations in irreversible processes. Phys. Rev. Second series **38**(12), 2265–2279 (1931)
47. Reddy, J.N.: An Introduction to Continuum Mechanics. Cambridge University Press (2007)
48. Roberts, A.J.: A One-Dimensional Introduction to Continuum Mechanics. World Scientific, London (1994)
49. Roy, M.: Mecanique des millieux continus et deformables. Gauthier–Villars (1950)
50. Smith, D.R.: An Introduction to Continuum Mechanics (Solid Mechanics and Its Applications). Springer, Berlin (1999)
51. Spencer, A.J.M.: Continuum Mechanics. Dover Publications, New York (2004)
52. Talpaert, Y.R.: Tensor Analysis and Continuum Mechanics. Springer, Berlin (2003)
53. Temam, R., Miranville, A.: Mathematical Modeling in Continuum Mechanics. Cambridge University Press (2005)
54. Truesdell, C.A.: The Elements of Continuum Mechanics. Springer–Verlag, New York (1965)
55. Truesdell, C.A.: Rational Thermodynamics. McGrawHill, New York (1969)
56. Truesdell, C.A., Noll, W.: The Nonlinear–Field Theories of Mechanics. Springer, Berlin (2003)
57. Valanis, K.S.: Irreversible Thermodynamics of Continuous Media. Springer, New York (1971)
58. Vardoulakis, I., Eftaxiopoulos, D.: Engineering Continuum Mechanics: With Applications from Fluid Mechanics, Solid Mechanics, and Traffic Flow. Springer, Berlin (2007)
59. Wu Han-Chin: Continuum Mechanics and Plasticity (CRC Series–Modern Mechanics and Mathematics). Chapman & Hall/CRC, Boca Raton (2005)
60. Ziegler, H.: An Introduction to Thermodynamics. Amsterdam, North–Holland (1977)

# Basic Notation

| | |
|---|---|
| $\mathbf{A}$ | Left Almansi deformation tensor; |
| $\overset{(n)}{\mathbf{A}}$ | Quasienergetic deformation tensors; |
| $\mathbf{a}^{\mathrm{Ol}}$, $\mathbf{a}^{J}$, $\mathbf{a}^{\mathrm{CR}}$, $\mathbf{a}^{D}$, $\mathbf{a}^{d}$, $\mathbf{a}^{V}$, $\mathbf{a}^{U}$, and $\mathbf{a}^{S}$ | Co-rotational derivatives of Oldroyd, Jaumann, Cotter–Rivlin, left and right mixed, left and right in eigenbasis, and spin; |
| $\mathbf{B}$ | The third energetic deformation tensor; |
| $\mathbf{C}$ | Right Cauchy–Green deformation tensor; |
| $\overset{(n)}{\mathbf{C}}$ | Energetic deformation tensors; |
| $\overset{(n)}{\mathbf{C}}_G$ | Generalized energetic deformation tensors; |
| $\overset{(n)}{\mathbf{C}}_e$ and $\overset{(n)}{\mathbf{C}}_p$ | Tensors of elastic and plastic deformations; |
| $\widehat{\mathbf{c}}_i$ | Principal basis of anisotropy (orthonormal) of a solid in undistorted configuration $\widehat{\mathcal{K}}$; |
| $\overset{\circ}{\mathbf{c}}$ | The velocity of a singular surface in a reference configuration; |
| $\mathbf{D}$ | Deformation rates tensor; |
| $\overset{\circ}{D}$ and $D$ | The normal speed of propagation of a singular surface in configurations $\overset{\circ}{\mathcal{K}}$ and $\mathcal{K}$; |
| $\mathbf{E}$ | Unit (metric) tensor; |
| $E$ | Total energy of a body; |
| ${}^4\overset{(n)}{\mathbf{E}}$ | Tensors of energetic equivalence; |
| $e$ | Specific internal energy of a body; |
| $\bar{\mathbf{e}}_i$ | Basis of Cartesian coordinate system; |
| $\mathbf{F}$ | Deformation gradient; |
| $\mathbf{F}_e$ and $\mathbf{F}_p$ | Gradients of elastic and plastic deformations; |
| $\mathbf{f}$ | Specific mass force vector; |
| $\mathbf{G}$ | Right Cauchy–Green deformation measure; |
| $\overset{(n)}{\mathbf{G}}$ | Energetic deformation measures; |
| $\overset{(n)}{\mathbf{G}}_G$ | Generalized energetic deformation measures; |

| | |
|---|---|
| $\mathbf{g}$ | Left Almansi deformation measure; |
| $\overset{\circ}{g}_{ij}$ and $g_{ij}$ | Metric matrices in configurations $\overset{\circ}{\mathcal{K}}$ and $\mathcal{K}$; |
| $H$ | Entropy of a body; |
| $\tilde{\mathbf{H}}$ and $\overset{\circ}{\mathbf{H}}$ | Left and right Hencky logarithmic deformation tensors; |
| $\mathbf{H}$ | Tensor of $H$-transformation from one reference configuration $\overset{\circ}{\mathcal{K}}$ to another reference configuration $\overset{*}{\mathcal{K}}$; |
| $\mathbf{I}$ | Momentum vector of a body; |
| $I_1(\mathbf{C})$, $I_2(\mathbf{C})$ and $I_3(\mathbf{C})$ | Principal invariants of second-order tensor $\mathbf{C}$; |
| $I_1^{(s)}(\boldsymbol{\Omega})$ | Invariants of second-order tensor $\boldsymbol{\Omega}$ relative to an orthogonal group $\overset{\circ}{G}_s$; |
| $\overset{(n)}{i}_A, \overset{(n)}{i}_B, \overset{(n)}{i}_C, \overset{(n)}{i}_D, \overset{(n)}{i}_G, i$ | Different forms of enthalpy; |
| $\mathbf{J}$ | Left Cauchy–Green deformation tensor; |
| $J = \rho/\overset{\circ}{\rho}$ | Ratio of densities; |
| $K$ | Kinetic energy of a body; |
| ${}^4\mathbf{K}(t)$ | Tensor of relaxation cores; |
| $\mathcal{K}$ and $\overset{\circ}{\mathcal{K}}$ | Actual and reference configurations; |
| $\mathbf{L}$ | Velocity gradient; |
| ${}^4\mathbf{M}$ | Quasilinear tensor of elasticity; |
| $\mathbf{n}$ and $\overset{\circ}{\mathbf{n}}$ | Normal vectors in configurations $\mathcal{K}$ and $\overset{\circ}{\mathcal{K}}$; |
| $\mathbf{O}$ | Rotation tensor accompanying the deformation; |
| $\mathbf{P}$ | Piola–Kirchhoff stress tensor; |
| $\mathbf{P}_\alpha^{(\mathbf{C})}, \alpha = 1, \ldots, \bar{n}$, | Orthoprojectors of symmetric tensor $\mathbf{C}$; |
| $\mathbf{p}$ and $\overset{\circ}{\mathbf{p}}$ | Eigenvectors of stretch tensors $\mathbf{V}$ and $\mathbf{U}$; |
| $p$ | Pressure; |
| $Q$ | Rate of heating; |
| $\bar{Q}_m$ and $\bar{Q}_\Sigma$ | Entropy production by external mass and surface sources; |
| $\bar{Q}^*$ | Entropy production by internal sources; |
| $Q_\beta$ | Thermodynamic fluxes; |
| ${}^4\overset{(n)}{\mathbf{Q}}$ | Tensors of quasienergetic equivalence; |
| $\mathbf{q}$ | Heat flux vector; |
| $q_m$ and $q_\Sigma$ | Heat influxes due to mass and surface sources; |
| $q^*$ | Specific internal entropy production; |
| ${}^4\mathbf{R}(t)$ | Tensor of relaxation functions; |
| $\overset{\circ}{\mathbf{r}}_i$ and $\mathbf{r}_i$ | Vectors of local bases in configurations $\overset{\circ}{\mathcal{K}}$ and $\mathcal{K}$; |
| $\overset{(n)}{\mathbf{S}}$ | Quasienergetic stress tensors; |
| $\overset{\circ}{\mathbf{S}}$ | Rotation tensor of stresses; |
| $\mathbf{S}_G$ | Generalized rotation tensor of stresses; |
| $\mathbf{T}$ | Cauchy stress tensor; |

| | |
|---|---|
| $\mathbf{T}^h$ and $\mathbf{T}^H$ | General notation for co-rotational derivatives of a tensor in covariant and contravariant moving bases $\mathbf{h}_i$ and $\mathbf{h}^i$; |
| $\overset{(n)}{\mathbf{T}}$ | Energetic stress tensors; |
| $\overset{(n)}{\mathbf{T}}_G$ | Generalized energetic stress tensors; |
| $\mathbf{t}_n$ | Stress vector; |
| $\mathbf{U}$ | Right stretch tensor; |
| $U$ | Internal energy of a body; |
| $\mathbf{u}$ | Displacement vector; |
| $\mathbf{V}$ | Left stretch tensor; |
| $\mathbf{v}$ | Velocity; |
| $\mathbf{W}$ | Vorticity tensor; |
| $W_m$ and $W_\Sigma$ | Powers of external mass and surface forces; |
| $W_{(i)}$ | Power of internal surface forces; |
| $w^*$ | dissipation function; |
| $\mathring{\mathbf{x}}$ and $\mathbf{x}$ | Radius-vectors of a material point in configurations $\mathring{\mathcal{K}}$ and $\mathcal{K}$; |
| $X^i$ and $x^i$ | Lagrangian and Eulerian coordinates; |
| $X_\beta$ | Thermodynamic forces; |
| $Y_\alpha(\mathbf{T})$ | Spectral invariants of a symmetric second-order tensor; |
| $\Gamma_{ij}^m$ and $\mathring{\Gamma}_{ij}^m$ | Christoffel symbols in configurations $\mathcal{K}$ and $\mathring{\mathcal{K}}$; |
| $\delta_\alpha$ | Relative elongation; |
| $\varepsilon_{ij}$ | Covariant components of the deformation tensor; |
| $\overset{(n)}{\zeta}_A$, $\overset{(n)}{\zeta}_B$, $\overset{(n)}{\zeta}_C$, $\overset{(n)}{\zeta}_D$, $\overset{(n)}{\zeta}_G$, $\zeta$ | Different forms of the Gibbs free energy; |
| $\eta$ | Specific entropy; |
| $\theta$ | Temperature; |
| $\mathbf{\Lambda}$ | Right Almansi deformation tensor; |
| $\lambda_\alpha$ | Eigenvalues of stretch tensors $\mathbf{U}$ and $\mathbf{V}$; |
| $\boldsymbol{\lambda}$ and $\mathring{\boldsymbol{\lambda}}$ | Heat conductivity tensors in configurations $\mathcal{K}$ and $\mathring{\mathcal{K}}$; |
| $\pi$ | Thermodynamic potential; |
| $\rho$ and $\mathring{\rho}$ | Density in configurations $\mathcal{K}$ and $\mathring{\mathcal{K}}$; |
| $\boldsymbol{\tau}_\alpha$ | Unit tangent vectors to a surface $S$; |
| $\psi$ | Helmholtz free energy; |
| $\mathbf{\Omega}$ | Spin of rotation accompanying the deformation; |
| $\mathbf{\Omega}_U$ and $\mathbf{\Omega}_V$ | Spins of the right and left stretch tensors; |
| $\boldsymbol{\omega}$ | Vorticity vector; |
| $\nabla$ and $\mathring{\nabla}$ | Nabla-operators in configurations $\mathcal{K}$ and $\mathring{\mathcal{K}}$; |
| ▼ and ▲ | The beginning and the end of a proof of Theorems, respectively; |
| □ | The end of Examples and Remarks. |

# Index

**D**

## J



## L



## M

**U**

## V



## W



## Y